ASTRONOMICAL DATA

1 Light Year (L.Y.)	$= 9.461 \times 10^{15}$ m
1 Astronomical Unit (A.U.) (Earth-Sun distance)	$= 1.496 \times 10^{11}$ m
Earth-Moon distance	$= 3.844 \times 10^{8}$ m
Radius of Sun	$= 6.960 \times 10^{8}$ m
Radius of Earth	$= 6.378 \times 10^{6}$ m
Radius of Moon	$= 1.738 \times 10^{6}$ m
Mass of Sun	$= 1.989 \times 10^{30}$ kg
Mass of Earth	$= 5.974 \times 10^{24}$ kg
Mass of Moon	$= 7.35 \times 10^{22}$ kg

FREQUENTLY USED CONVERSION FACTORS

1 radian	$= 57.296°$
1 in.	$= 2.54$ cm (exactly)
1 yd	$= 0.9144$ m (exactly)
1 mi	$= 1.6093$ km
1 mi/h	$= 0.4470$ m/s
(mass weighing) 1 lb	$= 0.45359237$ kg (exactly)
1 lb (force)	$= 4.44822$ N
1 u	$= 1.6605 \times 10^{-27}$ kg
1 Cal	$= 4186$ J
1 hp	$= 745.7$ W

(See Appendix C for a more complete list.)

VECTOR RELATIONS

$$\mathbf{A} \cdot \mathbf{B} = |\mathbf{A}||\mathbf{B}| \cos \theta_{AB}$$

$$|\mathbf{A} \times \mathbf{B}| = |\mathbf{A}||\mathbf{B}| \sin \theta_{AB}$$

$$\mathbf{A} \times \mathbf{B} = \begin{vmatrix} \hat{\mathbf{i}} & \hat{\mathbf{j}} & \hat{\mathbf{k}} \\ A_x & A_y & A_z \\ B_x & B_y & B_z \end{vmatrix}$$

$$\mathbf{A} \times (\mathbf{B} \times \mathbf{C}) = (\mathbf{A} \cdot \mathbf{C})\mathbf{B} - (\mathbf{A} \cdot \mathbf{B})\mathbf{C}$$

$$(\mathbf{A} \times \mathbf{B}) \cdot \mathbf{C} = (\mathbf{C} \times \mathbf{A}) \cdot \mathbf{B} = (\mathbf{B} \times \mathbf{C}) \cdot \mathbf{A}$$

PRINCIPLES OF
Physics

 SAUNDERS GOLDEN SUNBURST SERIES

PRINCIPLES OF
Physics

Jerry B. Marion
and
William F. Hornyak

University of Maryland

SAUNDERS COLLEGE PUBLISHING

Philadelphia New York Chicago
San Francisco Montreal Toronto
London Sydney Tokyo Mexico City
Rio de Janeiro Madrid

Address orders to:
383 Madison Avenue
New York, NY 10017

Address editorial correspondence to:
West Washington Square
Philadelphia, PA 19105

Text Typeface: 10/12 Times Roman
Compositor: York Graphic Services, Inc.
Acquisitions Editor: John Vondeling
Developmental Editor: Jay Freedman
Project Editor: Lynne Gery
Copy Editor: Michael Boyette
Managing Editor & Art Director: Richard L. Moore
Art/Design Assistant: Virginia A. Bollard
Text Design: Phoenix Studios, Inc.
Cover Design: Lawrence R. Didona
Text Artwork: ANCO/Boston
Production Manager: Tim Frelick
Assistant Production Manager: Maureen Iannuzzi

Front cover credit: Photograph by E. Lessing/Magnum Photos, Inc.
Back cover credit: Photograph courtesy of National Aeronautics and Space Administration.

Library of Congress Cataloging in Publication Data

Marion, Jerry B.
    Principles of physics.

    Includes index.

    1. Physics. I. Hornyak, William F. (William
Frank)   . II. Title.
QC21.2.M366 1984   530   83-7709

ISBN 0-03-049481-8

PRINCIPLES OF PHYSICS   ISBN 0-03-049481-8

3456 032 987654321

CBS COLLEGE PUBLISHING
Saunders College Publishing
Holt, Rinehart and Winston
The Dryden Press

PREFACE

This is a text for an introductory physics course for students of science and engineering. It is intended for courses requiring calculus either as a prerequisite or as a corequisite. It is ideally suited to courses covering topics in classical physics in two semesters. Its more elaborate companion text, *Physics for Science and Engineering,* is directed to courses that allow ample time to explore many topics in great depth. This text, however, is no simple abridgment of the longer companion. To fit the shorter course schedules a careful selection of subject matter was made, restricting treatment to a smaller number of essential topics.

In this text, many sections appearing in the companion work are rewritten and rearranged. The following are examples of major changes. The chapters on rotation of rigid bodies and dynamics of rigid bodies are now combined into a single chapter restricted to only the simplest types of rotational motion, that about a fixed axis and rolling motion. The chapter on deformation of solids has been eliminated. The chapters on static fluids and fluid dynamics have been condensed into one chapter that appears earlier in the text; this allows the chapters on oscillatory motion, mechanical waves, and sound to be placed together. The chapters on electromagnetic waves and radiating systems have been combined into a shorter single chapter. The chapters on geometric optics and optical instruments have likewise been combined, shortened, and placed after the chapters on interference and diffraction.

Opting for a less encyclopedic treatment still allows the selected topics to be developed in adequate detail for solid understanding. We continue the style of the companion text, which is informal but at the same time is concise and crisp and has a no-nonsense character.

A hallmark of this text is again the realistic treatment of topics. For example, the sliding contact between smooth surfaces is *not* taken as synonymous with the absence of friction; quite the contrary is shown to be the case. The examples cited in most texts as applications of Bernoulli's equation are really invalid, since they seldom involve laminar flow; instead, as it should be, the energy equation for fluid flow is cited here. The rather accurate predictions based on the application of the ideal gas law to simple real gases (such as air at near normal conditions of temperature and pressure) is *not* due to the absence of short-range intermolecular phenomena, but is shown rather to be due to the fortuitous cancellation of opposing effects. As a general policy, the approximations necessary for a first order treatment of any phenomenon are clearly spelled out.

Mathematics. Advanced mathematical tools are introduced ''gently,'' and the need for their use is motivated in each case by the subject at hand. If calculus is begun as a corequisite, this approach also allows the mathematics course to catch up its use in physics. Even when the first course in calculus is a prerequisite, some students may benefit from this "need-to-know" approach.

Differentiation is introduced as a logical requirement for a more precise understanding of motion. The concept of work is used to introduce the dot product of vectors and the definite integral. The cross product of vectors is first met in the chapter on angular momentum. Integration along specific physical paths appears only in the most elementary context: in discussing work, thermodynamic processes, Ampère's law, and electromagnetic induction. Simple surface integration is introduced in connection with Gauss's law. Geometric symmetry is often used in the examples in order to reduce normally required multiple integration to one or more simple single integrations.

Optional Topics. This text contains three different types of optional material. There are "extended footnotes" in the body of the text (in smaller type and set off by triangle symbols). There are a few sections labelled *Optional* (generally at the end of a chapter). Any or all of these may be omitted without loss of continuity in later chapters. Finally, we have included three Enrichment Topics at the end of the text. These sections cover applications of general engineering importance (Practical Magnetic Materials) and topics of current interest (General Planetary Orbits and the Electric Dipole Antenna).

Examples, Questions, and Problems. We place strong emphasis on problem-solving ability. Indeed, we view the ability to solve problems as the best proof of a student's understanding of the text material. To achieve this aim, a large number of worked-out examples are presented. The examples span all levels of difficulty, and in each instance situations are selected that emphasize the text presentation. An effort has been made to relate the examples to one another in order to reveal different aspects of a given physical system.

We have sprinkled the text liberally with thought-provoking questions addressed to the student (within parentheses and marked with a small circle •). At the end of each chapter we have also included a set of more elaborate questions.

There are many problems at the end of each chapter, grouped together by section number. Moderately difficult problems are marked with a dot (•) and more difficult ones with two dots (••). Usually there is also a set called Additional Problems, which draws together concepts developed in several sections and may include a few problems of greater difficulty. The answers to approximately half of the numerical problems appear in the back of the book. We use SI units (sometimes referred to as metric or MKSA units) throughout the text, and conversions to other familiar units are mentioned when appropriate.

Acknowledgments. We have had the benefit of a great many reviews during the development and redrafting of the manuscript. Among those who have lent their expertise and encouragement are:

J. Clifton Albergotti, University of San Francisco
Henry E. Bass, University of Mississippi
Lawrence Evans, Duke University
Kenneth N. Geller, Drexel University
Philip J. Green, Texas A & M University
Charlie Harper, California State University, Hayward
Donald F. Holcomb, Cornell University
Carl A. Kocher, Oregon State University
David Markowitz, University of Connecticut
B. E. Powell, West Georgia College
Carl A. Rotter, West Virginia University
Stanley Shepherd, Pennsylvania State University

We would like especially to thank David Markowitz, University of Connecticut, for his close reading of the manuscript and the many very valuable suggestions that he made. We are also particularly indebted to Jay Freedman for his many valuable suggestions and editorial assistance. Our typists, Theresa Desch and Elinor Fisher, were indispensable in preparing the manuscript.

Jerry B. Marion
William F. Hornyak

IN MEMORIAM

It is with great sorrow that we note the untimely death of Jerry B. Marion. As his coauthor and friend, I join the physics community in giving tribute and praise for his many contributions to the successful teaching of physics. Sadly, he was unable to see this printed version of the manuscript he so diligently labored over.

W.F.H.

CONTENTS OVERVIEW

CONTENTS

CHAPTER 1

INTRODUCTION

The word *physics* is derived from the Greek word $\phi \bar{v} \sigma \iota \varsigma$, meaning ''nature.'' Indeed, physics is a branch of natural science that deals with the properties and interactions of matter and radiation. Because physics is concerned with such fundamental ideas as these, it is generally considered to be the most basic of all the sciences.

The key to progress in the understanding of Nature is to base conclusions on the results of *observations* and *experiments*, analyzed by *logic* and *reason*. This basic approach to science is called the *scientific method*. This ''method'' is not a prescription for learning scientific truth; instead, the scientific method represents a philosophy of discovery that emphasizes the importance of *measurement* when dealing with problems of the real physical world. Theories are valuable (indeed, indispensable) in organizing the facts that have been gathered about the behavior of Nature. But at its roots, physics is an experimental science. The ultimate answer to any question concerning natural phenomena must be the result of experimental measurement.

In order to describe the natural universe, we use *concepts, theories, models,* and *laws*. Generally, a *theory* attempts to explain why Nature behaves as it does. Paradoxically, to construct a theory, we introduce certain *unexplained* fundamental abstractions or *concepts*. Thus, we consider the concepts of *energy, time, space,* and *electric charge* as ''given,'' without offering an explanation for their existence. (Even so, we can still provide precise *definitions* for these concepts viewed as quantities.) The theory then asserts a connection between these concepts and some observed characteristics of interactions of matter. This connection is achieved by constructing a *model* to reflect the experimentally determined facts. Finally, the deductions from these models result in the *laws* of physics, which tell us *how* things behave in terms of the theory.

How are theories to be judged? If there are several contending theories relating to the same set of experimental facts, how do we decide which to adopt? There are three criteria for answering such questions—*predictive power, comprehensiveness,* and *simplicity*.

A theory, if it is to have merit, must be able to predict observable results of experiments that have not yet been performed. When the experiments *are* performed, the results must agree with the predictions of the theory within acceptable limits. Also, a greater degree of credence is associated with a theory that can relate to a wide variety of phenomena. Finally, we have a faith that Nature is inherently simple, so we are led to believe that a valid theory should be transparent, direct, and simple, with an economy of postulates and ad hoc assumptions. However, simplicity must not be equated with ease of comprehension. The theory of relativity is a model of simplicity and logical precision, based on only two fundamental postulates. But to comprehend fully the implications of this theory is a formidable task.

1

What is accomplished by constructing a theory? What does a theory really explain? The best answer is probably that our theories provide a point of view of Nature that permits us to assemble in a comprehensible form the essence of a variety of related facts and observations concerning physical phenomena. Theories provide a means of reducing an enormous number of experimental results to a manageable number of precise statements. Progress consists of refining these theories and discovering links among them as new observational information becomes available and as new insights are developed. No theory is perfect, and none is all-encompassing. But we are confident in the expectation that our theories will provide ever better descriptions of Nature through continual evolution.

1–1   UNITS AND STANDARDS OF MEASURE

To describe physical quantities and processes in a precise and orderly way, it is essential that a system of measure first be established. The *units* for quantities such as length, time, and mass must be defined, and standards for these units must be provided so that we can agree on the meaning of measurements. A measurement of some observable can then be quoted as a *numerical value* together with the appropriate *unit* for that quantity. We stress the necessity for developing the habit of *always* quoting physical quantities in terms of a numerical value and the *relevant unit of measure*. It makes no sense, for example, to quote a length as ''60,'' for there is a considerable difference whether the units are centimeters or miles!

Most scientific measurements and most of the world's commerce are carried out in *metric* units. Engineering practice in this country uses the metric system more and more, and there is even a slow conversion to metric usage in the public sector. Within a matter of years the United States may join the rest of the industrialized world in the exclusive use of the metric system.

The metric unit of length is the *meter* (m), originally defined as 10^{-7} of the distance from the Equator to the North Pole along a meridian passing through Paris. This ingenious but impractical standard was replaced in 1889 by the distance between two finely drawn,

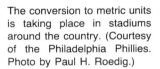

The conversion to metric units is taking place in stadiums around the country. (Courtesy of the Philadelphia Phillies. Photo by Paul H. Roedig.)

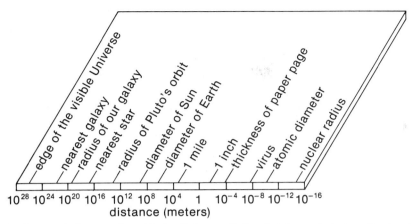

Fig. 1–1. Range of distances found in the Universe. Notice that the scale is *logarithmic*, with each scale division representing a factor of 10^4.

parallel scribe marks on a bar of platinum-iridium. Copies of this *standard meter bar* were distributed from the central depository near Paris to standards laboratories throughout the world. Even in the classical Greco-Roman world, it was necessary to disperse stone tablets throughout the empire to standardize the length of the *Roman foot*.

To provide a truly universal standard for the meter, in 1961 an international agreement was reached to base the definition of the meter on an atomic radiation. Because all atoms of a particular type are identical, such a definition allows every laboratory to prepare a standard for the meter in exactly the same way. Accordingly, the meter is now defined to be 1,650,763.73 wavelengths of a particular orange radiation from krypton atoms. The intercomparison of the practical laboratory meter with the standard optical wavelength requires the use of a special optical instrument called an interferometer, described in Chapter 35.

The range of distances encountered in the Universe is indicated in Fig. 1–1.

An atomic standard for *time* has also been established. Until recently, the *second* (s) was defined to be 1/86,400 of the mean solar day. The small but perceptible changes in the speed of the Earth's rotation make this definition of the second inadequate for precise experiments. In 1967, it was decided to define 1 s to be the time required for 9,192,631,770 complete vibrations of cesium atoms. Practical clocks may be calibrated using the cesium atomic standard by coupling the atomic vibrations to high-speed electronic counters that can determine the number of vibrations corresponding to any desired comparison time interval. (• What are some repetitive phenomena that are used in more common time-keeping devices?)

The range of time intervals found in the Universe is shown in Fig. 1–2.

The unit of mass in the metric system—the *kilogram* (kg)—was originally defined to be the mass of one liter $(1 \ \ell = 10^3 \ cm^3 = 10^{-3} \ m^3)$ of water. This definition has been refined by establishing a standard in the form of a platinum-iridium cylinder. The primary standard kilogram is maintained in the International Bureau of Weights and Measures, near Paris, and precision copies are available in most countries. It would, of course, be desirable to have an atomic standard for mass, just as we have for length and time. Although it is possible to compare the mass of one atomic species with that of another to extremely high precision, we do not yet have a sufficiently precise way to determine the mass of the standard kilogram in terms of the mass of an atom. (The precision is, at best, a few parts per million.) When technology has progressed to the point that we can make such comparisons with high precision, an atomic standard for mass will probably be adopted.

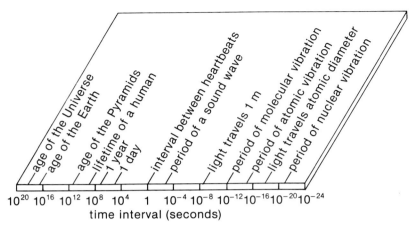

Fig. 1–2. Range of time intervals found in the Universe. The various periods indicated are representative only.

The range of mass values in the Universe is indicated in Fig. 1–3.

The units of all mechanical quantities can be expressed in terms of the basic units of length, time, and mass by using various products of these units. That is, the units of such quantities as energy, momentum, and force can all be expressed by combinations of meters, seconds, and kilograms. We say that these units are *derived* from the basic units. For example, speed (or velocity) is expressed in terms of a derived unit, namely, meters per second, m/s.

In some cases, for convenience, a special name is given to the derived unit for a quantity. The metric unit of force is the kg·m/s^2 (Section 5–2), and we call this unit the *newton* (N). But this does not alter the fact that only three basic units—those of length, time, and mass—are all that are required to describe any mechanical quantity.

Another type of derived unit is one that corresponds to some measured physical quantity that provides a useful and convenient unit. In descriptions of astronomical objects and events, the relevant distances, expressed in meters, are very large numbers indeed. For example, the distance to Alpha Centauri (a member of the star grouping nearest the Earth) is 4.07×10^{16} m. To deal with such enormous distances, it is convenient to define a new length unit equal to the distance that light will travel (at a speed of 2.998×10^8 m/s) during 1 year (3.156×10^7 s). This unit is called the *light-year*: 1 L.Y. = 9.46×10^{15} m. In terms of this unit, the distance to Alpha Centauri is 4.30 L.Y. A

Fig. 1–3. Range of mass values found in the Universe.

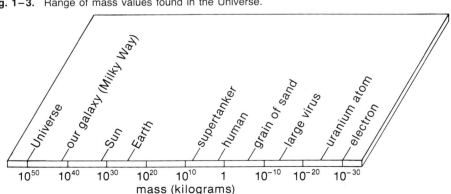

TABLE 1–1. Prefixes and Powers of 10

MULTIPLE	PREFIX	SYMBOL
10^{12}	tera-	T*
10^9	giga-	G*
10^6	mega-	M
10^3	kilo-	k
10^2	hecto-	h
10	deka-	da
10^{-1}	deci-	d
10^{-2}	centi-	c
10^{-3}	milli-	m
10^{-6}	micro-	μ
10^{-9}	nano-	n
10^{-12}	pico-	p
10^{-15}	femto-	f
10^{-18}	atto-	a

*In the United States, 1 *billion* means 10^9, but in Europe it means 10^{12}; the meaning of T and G are unambiguous, so the term *billion* should be avoided.

similar unit, useful for distance measurements in the solar system, is the *astronomical unit* (A.U.), defined to be the mean distance from the Earth to the Sun, 1.496×10^{11} m.

Systems of Measure.   The primary advantage of the metric system is that it is a *decimal* system. That is, the common multiples and subunits of the primary units are related by factors of 10.* Such units as the *centimeter* (10^{-2} m), the *kilometer* (10^3 m), and the *milligram* (10^{-3} g $= 10^{-6}$ kg) are often used. Sometimes a new term is employed instead of the standard prefix. For example, 10^3 kg is referred to as a *metric ton* instead of a megagram. The convention for use of prefixes and the corresponding powers of 10 are listed in Table 1–1.

In this book we use the particular metric units and symbols that have been prescribed by the commission on "Le Système International d'Unités." We refer to these as *SI units*. A summary of the SI units used in this book is given inside the rear cover.

A few parts of the world (most notably, the United States) have not yet converted to the metric system and still retain the British system of units. Sometimes it is necessary to convert from one of these systems to the other. In both systems, *time* is measured in seconds. The relationship connecting *length* can be expressed in several ways:

$$1 \text{ in.} = 2.54 \text{ cm (exactly)}$$
$$1 \text{ yd.} = 0.9144 \text{ m (exactly)}$$
$$1 \text{ mi.} = 1.609 \text{ km}$$

The situation with regard to the unit of *mass* is more confusing. To give a complete discussion requires us to distinguish between "mass" and "weight," a topic considered in Chapter 5. At this point we only note that an object with a weight of 1 pound (lb) has a mass of exactly 0.45359237 kg. Note also that the *metric ton* (or *tonne*, 1000 kg) is equal to 2204.6 lb (weight), which is approximately the same as the British long ton, or 2240 lb.

Dealing with Units.   We say that a physical quantity such as *distance* has dimen-

---

*The only exception is for time intervals longer than a second. Entrenched habit requires us to use minutes, hours, days, etc., for such multiples. However, subsecond units are always decimal—we use milliseconds, microseconds, nanoseconds, etc.

sions of *length* and is measured in *units* of *meters*; that is, if x represents distance,* $[x]$ = L. Every equation expresses an equality between quantities that have the same dimensions. Moreover, if any equation or formula involves a sum, each term in the sum must have the same dimensions. In addition, there are *dimensionless* quantities (numbers) such as π, $\sqrt{2}$, and sin 30° that also may appear in equations. Multiplying a quantity with dimensionality by a dimensionless number does not change the dimensionality.

Units and dimensions obey the familiar arithmetic of numbers. For example, suppose that we wish to write an equation that states, "speed multiplied by time is equal to distance." If distance is represented by x, speed by v, and time by t, we have $v \times t = x$. The dimensions of the quantities must be related by an equation of the same form, or

$$[v] \times [t] = [x]$$

Speed is measured in m/s, so the dimensionality of v is length per unit time, or LT^{-1}. Thus,

$$LT^{-1} \times T = L$$

Because $T^{-1} \times T = 1$, both sides of the equation have dimensions of length. When these quantities are expressed with units, we might have, for example,

$$\left(20 \; \frac{m}{s}\right) \times (4 \; s) = 80 \; \frac{m}{s} \times s = 80 \; m$$

The unit "s" appears both in the numerator and in the denominator of the result; cancellation leaves the final answer expressed in *meters,* as expected.

It is sometimes necessary to convert the value of a quantity expressed in one set of units to the equivalent value in a different set of units. To do this, we take any needed relationship of the form

$$1 \; mi = 5280 \; ft$$

and write it as a unity ratio:

$$\frac{5280 \; ft}{1 \; mi} = 1$$

Then, we can multiply or divide any term in the expression to be converted by unity factors of this type. By properly choosing these factors, a value can be converted from one set of units to another. Notice that this procedure does not change the dimensionality of the quantity.

For example, to convert 40 mi/h to m/s, we write

$$40 \; mi/h = \left(40 \; \frac{mi}{h}\right) \times \left(\frac{1.609 \; km}{1 \; mi}\right) \times \left(\frac{10^3 \; m}{1 \; km}\right) \times \left(\frac{1 \; h}{3600 \; s}\right)$$

$$= 17.9 \; m/s$$

Notice how miles, hours, and kilometers cancel to leave the answer expressed in units of m/s.

In this book, when writing equations, we usually supply the units for each quantity. Thus, we write

$$x = vt = (30 \; m/s)(5 \; s) = 150 \; m$$

---

*A square bracket is used to denote the dimensionality of a quantity; thus, $[x]$ = L is read as "the dimensionality of x is length L."

However, if the equation is lengthy, the explicit inclusion of the units may not be convenient. Then, we write

$$x = vt = (30)(5) = 150 \text{ m}$$

In these cases, it is implied that all values are expressed in the appropriate SI units. The units of the final answer are given explicitly.

In this text, we shall usually carry calculated numerical values to three significant figures as in the above examples. When a given quantity is quoted using fewer significant figures, assume a precision equivalent to three places. Thus, if given a time interval $t = 5$ s, interpret it as equivalent to $t = 5.00$ s.

1–2   DENSITY

One of the important physical properties that distinguishes one type of matter from another is *density*. Each sample of a particular substance, under identical physical conditions, has a mass that is directly proportional to its volume. Therefore, an important characteristic of a substance is the ratio of its mass to its volume—this is the density ρ:

$$\rho = \frac{M}{V} \tag{1–1}$$

The dimensionality of density is $[\rho] = ML^{-3}$. In the metric system, density is measured in units of kg/m^3. However, densities are often expressed in units of g/cm^3. Some useful densities are listed in Table 1–2.

TABLE 1–2. Densities of Some Common Substances*

SUBSTANCE	DENSITY kg/m^3	DENSITY g/cm^3
Gold (Au)	1.93×10^4	19.3
Mercury (Hg)	1.36×10^4	13.6
Lead (Pb)	1.13×10^4	11.3
Copper (Cu)	8.93×10^3	8.93
Iron (Fe)	7.86×10^3	7.86
Aluminum (Al)	2.70×10^3	2.70
Water (H_2O)	0.9998×10^3	0.9998
Ice (H_2O)	0.917×10^3	0.917
Alcohol (C_2H_5OH)	8.06×10^2	0.806
Air ($N_2 + O_2$) ⎫ normal	1.293	1.293×10^{-3}
Helium (He) ⎬ pressure	0.1786	1.786×10^{-4}
Hydrogen (H_2) ⎭	0.08994	8.994×10^{-5}

*Near 0°C.

QUESTIONS

1–1   Look up the definitions of *length* and *time* in a dictionary. Ignoring those definitions that do not deal with physical concepts, comment on the definitions that are relevant as we use the words in physics. Do these definitions give you a clear understanding of length and time? Try to devise better definitions.

1–2   How could you measure the length of a curved line?

1–3   Using a ruler with millimeter markings, devise a method for determining the thickness of a page of this book. Discuss the possible sources of error.

1–4   It is said that Galileo once used his own pulse rate to measure time intervals. Discuss the limitations and the accuracy of this technique.

1–5   Until the early sixteenth century, water clocks were the most accurate instruments available for measuring short time intervals. Such clocks usually determined the amount of water collected under controlled conditions during the time interval to be measured. Discuss the design of a possible water clock based on either the collected volume or mass of water, and estimate the accuracy with which an event lasting about 1 minute might be measured.

PROBLEMS

Section 1–1

1–1   In the discussion of unit conversion, a calculation was made to convert mi/h to m/s. Make this calculation again, using 1 in. = 2.54 cm (and other factors, as necessary) instead of 1 mi = 1.609 km.

1–2   When conversion to the metric system is complete in this country, what will the highway speed-limit signs now posted at 55 mi/h read?

1–3   Express 1 year in seconds.

1–4   What fraction of a mile is the "metric mile" of 1500 m?

1–5   Use the original definition of the meter and the fact that 1 mi = 1.609 km to find the diameter of the Earth in miles.

1–6   What is the factor that converts in.^{3} to cm^3?

1–7   An athlete runs at constant speed and completes the 100-yd dash in 9.52 s. What will be his time for the 100-m distance running at the same speed?

1–8   Machinists sometimes work in units of *mils* (10^{-3} in.) or *microinches* (10^{-6} in.). Express both of these units in meters.

1–9   Land that has an area of 1 mi^2 (called a *section* of land) contains 640 acres. In the metric system, land areas are measured in units of the *hectare*, equal to 10^4 m^2. How many acres equal 1 hectare?

1–10   Convert 3.65 yd^2 to cm^2.

1–11   In the *Gregorian calendar,* which we now use, every fourth year is a *leap year,* so the average length of a year is 365¼ days. The actual solar year in 1980 was 365 d, 5 h, 48 min, 45.6 s. Assume that the solar year is always this long (it varies slightly), and calculate how long it will be before our current calendar is out-of-step with the Sun by one day.

1–12   The distance x that an object moves during a time t is expressed as $x = vt + \frac{1}{2}at^2$. Find the dimensions of a.

1–13   If an object is dropped from a height h, the speed v with which it strikes the ground depends only on h and on g, the acceleration due to gravity, together with a numerical (i.e., dimensionless) factor. We know that $[h] = L$ and $[v] = LT^{-1}$, and we state that $[g] = LT^{-2}$. Use only this information to obtain an expression for the dependence of v on h and g. (Your expression will, of course, be missing the numerical factor that would appear in the complete equation.)

Section 1–2

1–14   A part for an aircraft engine is manufactured from steel (same density as iron) and has a mass of 4.86 kg. If, instead, a magnesium-aluminum alloy ($\rho = 2.55$ g/cm^3) is used, what will be the mass of the part?

1–15   A flat circular plate of copper has a diameter of 48.6 cm and a mass of 62 kg. What is the thickness of the plate?

1–16   What is the mass of air (at normal conditions) in a room that has dimensions 8 m × 6 m × 3.2 m?

1–17   An adult human inhales about 5 liters of air per minute. What is the mass of air inhaled by a person in 1 hour?

1–18   What is the average density of (a) the Earth and (b) the Sun. [*Hint:* Use the data in the table inside the front cover.]

1–19   First-century Imperial Roman coinage valued a gold aureus (typically 7.3 g) equivalent to 100 copper sestertii (typically 26.5 g). What was considered the relative value of gold to copper? If the average thickness of the sestertius was twice that of the aureus, what was the approximate ratio of the diameters?

CHAPTER 2

KINEMATICS OF LINEAR MOTION

We live in a Universe of continual motion. In every piece of matter, the atoms are in a state of unceasing agitation. We move around on the Earth's surface while the Earth moves in its orbit around the Sun. Although the stars in the sky seem to remain motionless as the Earth revolves beneath them, they too are in motion. Even the enormous collections of stars—the galaxies such as our own Milky Way—are moving through the vastness of space.

Because motion is such an important feature of the world around us, it is the logical subject with which to begin the study of physical phenomena. The ideas developed here are applied to all of the topics that follow—in describing planetary motion, in discussing fluids and electric current, and in studying the behavior of atoms and nuclei. Motion is an essential common feature in all physical processes.

In this chapter we begin the study of *mechanics,* a subject that is developed in much of the remainder of this text. It is appropriate that the discussion of physical topics start with mechanics because this was the first of the subdivisions of physics to be developed and has become one of the cornerstones of modern physical science.

Mechanics can be logically divided into two parts, *kinematics* and *dynamics.* Kinematics is concerned with the geometric *description* of motion; it does not address the question of the *cause* of motion. Dynamics, on the other hand, relates the motion of objects to their properties and to the forces that act on them. We begin by treating the kinematics of simple straight-line motion.

Abstractions and Idealizations.   In this chapter we begin dealing with the motion of objects. The real objects of everyday experience all have measurable size and recognizable physical structure. The actual motion of objects is in general quite complicated and involves rotations (changes in orientation) as well as translations (changes in spatial position). You can certainly appreciate this statement if you have ever witnessed the wobbly flight of a poorly thrown football!

The discussion of motion can be simplified by introducing the concept of a *particle,* which is the physical counterpart of the idea of the mathematical point. A particle has no physical size or structural features and is not capable of rotational motion. (However, we eventually endow these particles with certain properties, such as mass and electric charge.)

In our model for particle motion we imagine a trajectory or path that is followed by the particle. In time, the particle occupies *every point* on this path in *continuous succession,* just as we suppose real objects do. This motion occurs whether we observe it or not, and we imagine that it is possible to observe the particle by some method that in no way alters the motion. We therefore introduce the idealization of a "neutral observer." We take for granted the notion of a straight line and indeed all the propositions of Euclidean geometry.

Are these various abstractions realistic? Is the concept of a particle even plausible? First, we must appreciate that a *model* is developed to represent in some simple but approximate way an aspect of the real physical world. Models are used whenever we do not completely understand some physical process, or if the correct theory is so cumbersome that it does not permit calculations to be made easily. A model is considered successful if it describes the way Nature behaves to the level of accuracy we demand. An unsophisticated view of mechanics is sufficient to design a toy cart, for example, but a highly developed theory is necessary to account for the detailed motions of Earth satellites.

We should also note that apparently pointlike particles do exist in Nature. There are now theoretical speculations that *electrons*, particles found in all atoms, are no larger than about 10^{-18} m; in fact, with present experimental techniques, electrons have *no* measurable size. Thus, electrons can be considered to be ideal pointlike particles for all our purposes. So also for atoms and, usually, for grains of sand or pebbles. Indeed, the objects that we wish to represent as particles are only required to have a size that is small compared with the dimensions of the trajectory we are considering, or their sizes must be irrelevant to the motion studied. Thus, to a reasonable approximation, the Earth (and indeed any planet) may be considered to be a particle for purposes of discussing its orbital motion around the Sun. (The concept of the center of mass, to be introduced later, makes this approximation better than might be judged simply on the basis of the scale of sizes involved.) The Sun and stars, in turn, may be considered particles when galactic motions are involved.

It is extremely useful to model the behavior of Nature in terms of particles—if we keep in mind the limitations. The idea of particles is used in many of the discussions that follow.

2–1   POSITION AND DISPLACEMENT

Consider a straight-line scale along which the progress of a moving particle can be followed. On this *coordinate line*, locate an origin (which we usually label O) and mark off distance intervals in a convenient unit of length, such as meters. Label the intervals on one side of the origin with increasing positive numbers and label those on the other side with increasing negative numbers. In this way, we construct a *directed* coordinate line. The motion of a particle along such a coordinate line can be followed by recording the position s occupied by the particle at the corresponding time t. These time-position pairs are written as (t_1, s_1), (t_2, s_2), and so on, and are marked along the directed coordinate line as shown in Fig. 2–1.

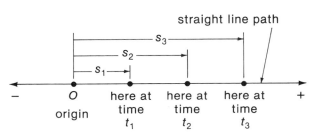

Fig. 2–1. The progress of a particle moving in a straight line is recorded in terms of the time-position pairs (t_1, s_1), (t_2, s_2), and so forth.

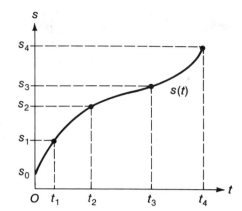

Fig. 2–2. Graph of a particle's position function, $s = s(t)$. Several time-position pairs are indicated on the curve.

It is evident from Fig. 2–1 that the position of the particle is determined by its distance from the origin O. The choice of some other point as the origin would have been equally acceptable. It seems clear that no essential feature of the description of a particle's motion should depend on how the origin is chosen.

It is usually convenient to pick arbitrarily a reference instant for time zero and to quote the times t_1, t_2, and so on, in terms of the *elapsed time* after the reference instant t_0. There is no requirement that the particle be at the coordinate origin O at time t_0 (although the location of the origin can be adjusted to make this correspondence).

The motion of a particle can be represented in graphic form as in Fig. 2–2, where the curve represents s as a continuous function of t, and we write

$$s = f(t) \qquad \text{or} \qquad s = s(t)$$

The notation in each of these equations means "the position s as a function of the time t."

To indicate that the position function $s(t)$ is to be evaluated at some particular time—for example, at $t = 10$ s—we can write with equivalent meaning

$$s(t = 10 \text{ s}), \qquad s(10 \text{ s}), \qquad s(10)$$

In the last instance we suppose it is clear from the context that the time value is given in seconds. This notation is used throughout the text.

To further facilitate the description of motion, we introduce a quantity called the *displacement*, the distance between two points actually occupied by the particle as it moves along the path. Thus, the (algebraic) distance, $s_2 - s_1$, is the displacement of the particle during the elapsed time interval $t_2 - t_1$.

Example 2–1

Suppose that the position of a particle traveling in a straight line is given by the expression

$$s = s(t) = bt + c$$

with $b = 5.0$ m/s and $c = -3.0$ m. What is the coordinate location of the particle at times $t = 0.5$ s and $t = 10$ s?

Solution:

For $t = 0.5$ s, we have

$$s(0.5 \text{ s}) = (5.0 \text{ m/s})(0.5 \text{ s}) - (3.0 \text{ m})$$
$$= (2.5 \text{ m}) - (3.0 \text{ m})$$
$$= -0.5 \text{ m}$$

The negative sign here indicates that the position of the particle at $t = 0.5$ s is to the *left* of the origin. For $t = 10$ s, we have

$$s(10 \text{ s}) = (5.0 \text{ m/s})(10 \text{ s}) - (3.0 \text{ m})$$

$$= 47.0 \text{ m}$$

Where was the particle at $t = -1$ s (that is, at a time 1 s *earlier* than the arbitrarily chosen reference instant)?

$$s(-1 \text{ s}) = (5.0 \text{ m/s})(-1 \text{ s}) - (3.0 \text{ m})$$

$$= (-5.0 \text{ m/s}) - (3.0 \text{ m})$$

$$= -8.0 \text{ m}$$

which is a point 8.0 m to the left of the origin.

2–2   VELOCITY

Suppose that the position function of a particle is $s = s(t)$. We now inquire about the *rate* of the motion described by this function. At a certain instant t_a, the particle is at the position P, for which $s = s(t_a)$; at some elapsed time Δt later, it is at the position Q, for which $s = s(t_a + \Delta t)$, as shown in Fig. 2–3.* During the time interval Δt, the particle has undergone a displacement, $\Delta s = s(t_a + \Delta t) - s(t_a)$. The average time rate of the motion is called the particle's *average velocity*† $\bar{v}$; it is defined to be the displacement divided by the elapsed time, or

$$\bar{v} = \frac{s(t_a + \Delta t) - s(t_a)}{(t_a + \Delta t) - t_a} = \frac{\Delta s}{\Delta t} \qquad \textbf{(2–1)}$$

According to Eq. 2–1, the dimensions of velocity are those of distance divided by time; that is, $[v] = LT^{-1}$. We usually measure velocity in units of *meters per second* (m/s or ms^{-1}), although we may sometimes use cm/s, km/s, or km/h.

The quantity $\bar{v}$ gives the *average* velocity of a particle that undergoes a total displace-

Fig. 2–3.   (a) The positions of a particle, P when $t = t_a$ and Q when $t = t_a + \Delta t$, along a coordinate line. (b) The graph of the position function $s = s(t)$, showing the relevant quantities.

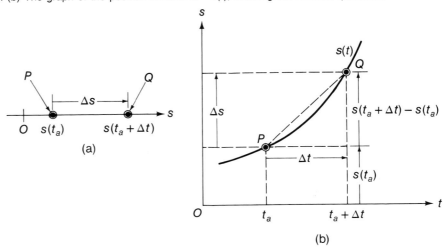

*The symbol Δ, upper-case Greek *delta*, stands for the difference or change in the quantity that follows it.
†An overbar on a symbol is used to represent an *average* value.

ment Δs during the time interval Δt. It should be noted that $\bar{v}$ depends on both t_a and Δt (see Fig. 2–3b). The information provided by the value of $\bar{v}$ concerning the progress of the motion is incomplete and may even be misleading. For example, Fig. 2–4 shows the motion of a particle that is at a position s_a at time t_a and moves to s_b at time t_b, finally returning to s_a at time t_a'. For the time interval, $\Delta t = t_a' - t_a$, the total displacement is $\Delta s = 0$. Thus, $\bar{v} = 0$ for this time interval even though the particle undergoes considerable motion between times t_a and t_a'. To give a means for describing this motion, we introduce the idea of *average speed*, defined as the *total path length* $\Delta \ell$ traveled by the particle divided by the corresponding time interval Δt:

$$\text{average speed} = \frac{\Delta \ell}{\Delta t}$$

For motion along a straight line, the average velocity and the average speed are equal unless the particle reverses its direction of motion during the time interval considered. Then, $\Delta \ell / \Delta t$ is always numerically greater than or equal to $\bar{v}$ (as in Fig. 2–4, where $\bar{v} = 0$).

We can give more precise information concerning the motion of a particle by defining the *instantaneous velocity*. Imagine a limiting process in which the time interval Δt is allowed to decrease to smaller and smaller values. Hold t_a fixed, and at each of the successively smaller values of Δt evaluate the average velocity $\bar{v}$. As Δt is made smaller and smaller, $\bar{v}$ changes less and less. In fact, $\bar{v}$ approaches a limiting value as $\Delta t \to 0$. This limiting value is called the *instantaneous velocity* at time t_a and is denoted by v (without an overbar).

Now that *average velocity* and *instantaneous velocity* have been defined, we really do not need the qualifier "instantaneous." The term *velocity* is used to mean the instantaneous value. When we require the velocity to be averaged over a particular time or space interval, the term *average velocity* is used.

For motion along a straight line, the velocity v may be either positive or negative; that is, the particle may be moving either to the right or to the left. Notice that the instantaneous values of the velocity and the speed are equal in magnitude. Speed is always positive (or zero), but velocity can be negative.

Fig. 2–4. (a) Motion of a particle, $s_a \to s_b \to s_a$, during the time interval $\Delta t = t_a' - t_a$. (b) A position-time graph of the motion for the same time interval, showing that the total displacement is $\Delta s = 0$, so the average velocity $\bar{v}$ is also zero. However, the average speed during the interval is *not* zero: $\Delta \ell / \Delta t = 2(s_b - s_a)/(t_a' - t_a)$.

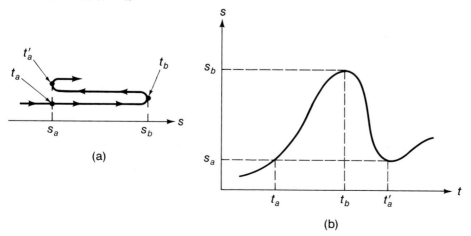

(a)

(b)

There is a simple mathematical method for determining the limiting value approached by $\bar{v}$ as $\Delta t \to 0$. It is based on the kind of graph shown in Fig. 2–3b. In Fig. 2–5a, the line connecting P and Q (called a secant or chord) makes an angle β with the horizontal. In view of the definition of the tangent of an angle (see Eq. A–18 in Appendix A), we find that

$$\bar{v} = \frac{\Delta s}{\Delta t} = \frac{s(t_a + \Delta t) - s(t_a)}{\Delta t} = \tan \beta$$

That is, $\bar{v}$ is equal to the *slope* of the line $\overline{PQ}$ that connects the end-points of the interval considered.

Figure 2–5b shows the limiting process for obtaining the velocity at time t_a. As $\Delta t \to 0$, the line $\overline{PQ}$ becomes the tangent line $\overline{TT'}$ to the curve $s(t)$ at the point P (that is, for time t_a). Thus,

$$v(t_a) = \lim_{\Delta t \to 0} \frac{\Delta s}{\Delta t} = \lim_{\Delta t \to 0} \frac{s(t_a + \Delta t) - s(t_a)}{\Delta t} = \tan \alpha \qquad \textbf{(2–2)}$$

This expression is just the definition of the *derivative* of $s(t)$ evaluated as t_a. For any function $s(t)$ for which the limit at $t = t_a$ exists (called a differentiable function), we have

$$v(t_a) = \lim_{\Delta t \to 0} \left. \frac{s(t + \Delta t) - s(t)}{\Delta t} \right|_{t=t_a}$$

In the usual calculus notation, we write

$$v(t) = \frac{ds}{dt} = \frac{d}{dt} s(t) \quad \text{and} \quad v(t_a) = \left. \frac{ds}{dt} \right|_{t=t_a} \qquad \textbf{(2–3)}$$

The velocity function $v(t)$ is the general time derivative of the position function. Note particularly, in contrast with the definition of $\bar{v}$, that the instantaneous velocity does not depend on Δt. The velocity v of an automobile, for example, is the quantity indicated by the speedometer reading. Figure 2–6 emphasizes the point that the velocity at a particular

Fig. 2–5. (a) The average velocity $\bar{v}$ between points P and Q is equal to the slope of the line $\overline{PQ}$. (b) The velocity v at time t_a is obtained by progressively decreasing the time interval Δt, thus moving the point at the end of the interval, given by $s(t_a + \Delta t)$, ever closer to the point P, which corresponds to $s(t_a)$. During the limiting process, we have $Q_1 \to Q_2 \to Q_3 \to \cdots \to P$, and $\beta \to \alpha$, where $\tan \alpha$ is the slope of the tangent line $\overline{TT'}$ at P.

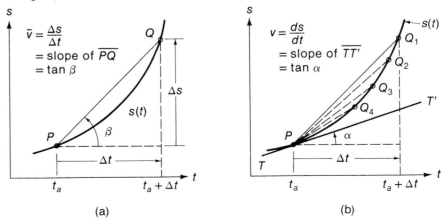

(a)   (b)

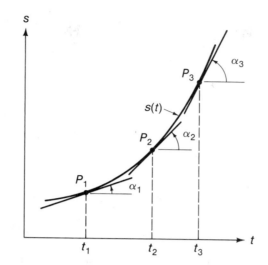

Fig. 2-6. The velocity of a particle at any time is equal to the slope of the distance-time curve at that time. Thus, $v(t_1) = \tan \alpha_1$.

time is equal to the slope of the distance-time curve at that time. Also, note that neither $\bar{v}$ nor v can depend on the choice for the location of the origin of the coordinate line because only the displacements Δs (that is, the differences in position coordinates) are involved in the calculations.

Example 2-2

Given the position function

$$s(t) = \tfrac{1}{4}gt^2 + 2bt$$

where $g = 1$ m/s² and $b = 1$ m/s, evaluate the average velocity $\bar{v}$ during elapsed time intervals, starting at $t_a = 2$ s, for which $\Delta t = 1$ s, 0.5 s, 0.1 s, 0.01 s, and 0.001 s.

Solution:

Evaluating $s(t_a) = s(2\ s)$, we find

$$s(2\ s) = \tfrac{1}{4}(1\ \text{m/s}^2)(2\ s)^2 + 2(1\ \text{m/s})(2\ s) = 5\ \text{m}$$

Then, using Eq. 2–1 for $\bar{v}$, we have

$$\bar{v} = \frac{s(2\ s + \Delta t) - (5\ \text{m})}{\Delta t}$$

For $\Delta t = 1$ s,

$$\bar{v} = \frac{s(3\ s) - (5\ \text{m})}{1\ s} = \frac{\tfrac{9}{4} + 6 - 5}{1}\text{m/s} = 3.250\ \text{m/s}$$

Substituting the remaining values of Δt, we obtain the following results for $\bar{v}$:

Δt (s)	1	0.5	0.1	0.01	0.001
$\bar{v}$ (m/s)	3.250	3.125	3.025	3.0025	3.00025

From these values for $\bar{v}$, it appears that in the limit $\Delta t \to 0$, we have $v = 3$ m/s (exactly).

We can see the dependence of $\bar{v}$ on Δt if we write

$$\Delta s = s(t_a + \Delta t) - s(t_a)$$

$$= [\tfrac{1}{4}g(t_a + \Delta t)^2 + 2b(t_a + \Delta t)] - [\tfrac{1}{4}gt_a^2 + 2bt_a]$$

$$= [\tfrac{1}{4}gt_a^2 + \tfrac{1}{2}gt_a\,\Delta t + \tfrac{1}{4}g(\Delta t)^2 + 2bt_a + 2b\,\Delta t]$$

$$- [\tfrac{1}{4}gt_a^2 + 2bt_a]$$

$$= (\tfrac{1}{2}gt_a + 2b)\,\Delta t + \tfrac{1}{4}g(\Delta t)^2$$

Then

$$\bar{v} = \frac{\Delta s}{\Delta t} = \tfrac{1}{2}gt_a + 2b + \tfrac{1}{4}g\,\Delta t$$

For $t_a = 2$ s, $g = 1$ m/s², and $b = 1$ m/s, we find

$$\bar{v} = (3 + \tfrac{1}{4}\,\Delta t)\ \text{m/s}$$

and the dependence of $\bar{v}$ on Δt for this case is exhibited explicitly. It is easy to verify that this expression leads to the values for $\bar{v}$ listed in the table above. It is also clear that in the limit $\Delta t \to 0$ the velocity at $t = 2$ s is indeed exactly 3 m/s.

In Example 2–2 the position function was

$$s(t) = \tfrac{1}{4}gt^2 + 2bt$$

and we obtained

$$\bar{v} = \frac{\Delta s}{\Delta t} = \tfrac{1}{2}gt + 2b + \tfrac{1}{4}g\,\Delta t$$

for the time interval, t to $t + \Delta t$. Then, the instantaneous velocity can be obtained by taking the limit,

$$v = \operatorname*{Lim}_{\Delta t \to 0}\left(\tfrac{1}{2}gt + 2b + \tfrac{1}{4}g\,\Delta t\right)$$

$$= \tfrac{1}{2}gt + 2b$$

The two terms in the expression for v can be obtained from the formula for the derivative of the product of a constant and a variable raised to a power n, namely,

$$\frac{d}{dt}At^n = nA\,t^{n-1} \tag{2–4}$$

Thus, we see here that

$$v = \frac{d}{dt}s(t) = \frac{d}{dt}(\tfrac{1}{4}gt^2 + 2bt)$$

$$= 2(\tfrac{1}{4}g)t^1 + (2b)t^0$$

$$= \tfrac{1}{2}gt + 2b$$

so that

$$v(2\text{ s}) = \tfrac{1}{2}(1\text{ m/s}^2)(2\text{ s}) + 2(1\text{ m/s})$$

$$= 3\text{ m/s}$$

just as we obtained in the limiting process.

(• Can you verify Eq. 2–4 for $s(t) = At^3$ by using the Δ process as in Example 2–2? You can expand $A(t + \Delta t)^3$ by using Eq. A–12 in Appendix A.)

Example 2–3

Use the Δ process (Example 2–2) to obtain the derivative of $s(t) = Ae^{\alpha t}$.

Solution:

We have

$$s(t + \Delta t) = Ae^{\alpha(t + \Delta t)} = Ae^{\alpha t}e^{\alpha\,\Delta t}$$

Then, using Eq. A–11 from Appendix A, we obtain

$$s(t + \Delta t) - s(t) = Ae^{\alpha t}(e^{\alpha\,\Delta t} - 1)$$

$$= Ae^{\alpha t}\left(1 + \frac{\alpha\,\Delta t}{1!} + \frac{\alpha^2(\Delta t)^2}{2!} + \cdots - 1\right)$$

$$s(t + \Delta t) - s(t) = Ae^{\alpha t}\left(\alpha\,\Delta t + \tfrac{1}{2}\alpha^2(\Delta t)^2 + \cdots\right)$$

and the derivative is

$$\frac{d}{dt}Ae^{\alpha t} = \operatorname*{Lim}_{\Delta t \to 0}\frac{Ae^{\alpha t}(\alpha\,\Delta t + \tfrac{1}{2}\alpha^2(\Delta t)^2 + \cdots)}{\Delta t}$$

$$= \operatorname*{Lim}_{t \to 0}Ae^{\alpha t}(\alpha + \tfrac{1}{2}\alpha^2\,\Delta t + \cdots)$$

$$= \alpha Ae^{\alpha t}$$

The derivative of any function can be obtained by following this general procedure. In particular, it has been used to derive the useful list of differentiation formulas inside the rear cover. Simple substitution of variables in these rules may be used to replace the more cumbersome Δ process.

Example 2–4

Suppose that the position function of a particle is

$$s(t) = \frac{v_0}{\alpha}(1 - e^{-\alpha t})$$

where $\alpha > 0$ and v_0 are constants. What are the dimensions of α and v_0? What is the velocity of the particle for $t = 0$, $t = 1/\alpha$, and $t \to \infty$?

Solution:

Because the exponent of e must be dimensionless (• Why?), we have $[\alpha] = \mathrm{T}^{-1}$. Also, the dimension of $s(t)$ is $[s] = \mathrm{L}$; hence, $[v_0/\alpha] = \mathrm{L}$, or $[v_0] = \mathrm{LT}^{-1}$. That is, the dimensions of v_0 are those of *velocity*.

Differentiating $s(t)$ with respect to the time, we obtain the velocity:

$$v(t) = \frac{d}{dt}s(t) = \frac{v_0}{\alpha}(-1)(-\alpha)e^{-\alpha t}$$

$$= v_0 e^{-\alpha t}$$

Notice that differentiating with respect to *time* leads to a derivative whose dimensions are those of the original function divided by time.

For the various values of t, we have

$$v(0) = v_0$$

$$v(1/\alpha) = v_0 e^{-1} \cong 0.368\, v_0$$

$$\mathrm{Lim}_{t \to \infty} = 0$$

2–3   ACCELERATION

Both the velocity and the position of a particle may be functions of time. A particle that moves in such a way that it "speeds up" or "slows down" undergoes accelerated motion, and the time rate of change of the velocity is called the *acceleration* of the particle.

The *average acceleration* $\bar{a}$ of a particle during a time interval Δt, starting at t, is defined to be

$$\bar{a} = \frac{v(t + \Delta t) - v(t)}{\Delta t} = \frac{\Delta v}{\Delta t} \tag{2–5}$$

The (instantaneous) *acceleration* a at the instant t is defined to be the limit of Eq. 2–5 as Δt approaches zero (see Fig. 2–7); thus,

$$a = \mathrm{Lim}_{\Delta t \to 0} \frac{v(t + \Delta t) - v(t)}{\Delta t}$$

and hence *the acceleration is the general time derivative of the velocity.*

Thus, in standard notation we have

$$a(t) = \frac{dv}{dt}$$

Now, the velocity is the time derivative of the position function, so the acceleration is the derivative of the derivative (or the *second* derivative) of the position function. We express this by writing

$$a(t) = \frac{dv}{dt} = \frac{d}{dt}\frac{ds}{dt} = \frac{d^2 s}{dt^2} \tag{2–6}$$

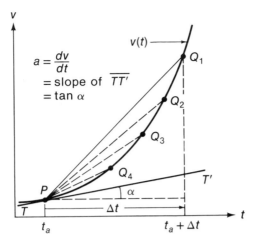

$$a = \frac{dv}{dt}$$
$$= \text{slope of } \overline{TT'}$$
$$= \tan \alpha$$

Fig. 2–7. The acceleration a at time t_a is obtained by progressively decreasing the time interval Δt. This moves the point Q at the end of the interval ever closer to P. In the limit, $\Delta v/\Delta t$ equals the slope of the tangent line $\overline{TT'}$ at the point P which corresponds to $v(t_a)$. Compare Fig. 2–5 for the case of velocity.

The velocity at a particular instant is equal to the slope of the position-time graph at that instant (Fig. 2–8a). In the same way, the acceleration at a particular instant is equal to the slope of the velocity-time graph at that instant (Fig. 2–8b). It is important to understand that $v = 0$ does *not* imply $a = 0$. (• Can you think of examples from everyday experience in which $v = 0$ but $a \neq 0$?)

Note that the dimensions of acceleration are $(LT^{-1})/T = LT^{-2}$. Acceleration is usually quoted in units of *meters per second per second* or *meters per second squared*, which is written as m/s^2.

Fig. 2–8. (a) Velocity is equal to the slope of the s-t graph. (b) Acceleration is equal to the slope of the v-t graph.

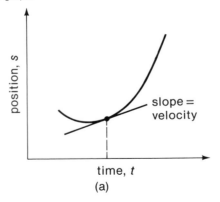

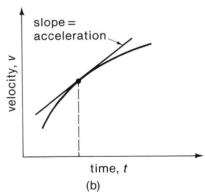

Example 2–5

As an example of a situation for which acceleration is not a constant, we consider a simple case of oscillatory motion. In Chapter 14, such motion is treated in detail.

When a particle is pulled back and forth by a freely vibrating spring, the position function is

$$s(t) = A \sin \omega t$$

where ω is a constant with units s^{-1}. To find the velocity, $v(t) = ds/dt$, we take the derivative of the sine function. In the table inside the rear cover we find

$$\frac{d}{dx} \sin x = \cos x$$

Substituting $x = \omega t$ and $dx = \omega \, dt$, we have

$$\frac{1}{\omega} \frac{d}{dt} \sin \omega t = \cos \omega t$$

or

$$\frac{d}{dt} \sin \omega t = \omega \cos \omega t$$

Thus,

$$v(t) = \frac{d}{dt} s(t) = \frac{d}{dt} A \sin \omega t$$

$$v(t) = A\omega \cos \omega t$$

In the same way, we find

$$a(t) = \frac{dv}{dt} = \frac{d^2 s}{dt^2} = -A\omega^2 \sin \omega t$$

(• Can you verify this result?) and the oscillating particle experiences a sinusoidal acceleration.

2–4 MOTION WITH CONSTANT ACCELERATION

Motion that takes place at constant velocity along a straight line is called *uniform linear motion*. Then, with $v = v_0 = $ constant, the acceleration is necessarily zero:

$$a = \frac{dv}{dt} = 0$$

for uniform motion.

When the velocity is constant, we have $\bar{v} = v_0$. Suppose that at time $t = 0$ the position of a particle is s_0. Then, after an elapsed time t its position is $s(t)$, and we have

$$\bar{v} = v_0 = \frac{s(t) - s_0}{t - 0}$$

or

$$s(t) = v_0 t + s_0 \tag{2-7}$$

Equation 2–7 is the position function of a particle that moves with constant velocity v_0 and passes through the position s_0 at $t = 0$. This equation includes the case $v_0 = 0$, that is, $s(t) = s_0$, in which the particle remains at a certain position and experiences no displacement with time.

Next, consider the motion that results when the acceleration is constant, so $\bar{a} = a$. If the velocity at $t = 0$ is v_0 and the velocity at time t is $v(t)$, we have

$$\bar{a} = a = \frac{v(t) - v_0}{t - 0}$$

or

$$v(t) = at + v_0 \tag{2-8}$$

We can determine the position function for this case by noting that when the velocity is a linear function of time, the average velocity during any time interval, as we see in Fig. 2–9, is just one half the sum of the initial and final velocities at the extremes of the interval.*

Now, the displacement during the time interval t is given by $\bar{v}t$. Therefore, we can write

$$s(t) - s_0 = \bar{v}t \tag{2-9}$$

Then, for $\bar{v}$ we use the expression (see Fig. 2–9)

$$\bar{v} = \tfrac{1}{2}[v(t) + v_0]$$

---

*Actually, it is not trivial to prove this proposition by algebraic means. In Section 2–6 we obtain the same results by integration, thereby avoiding any reference to average values.

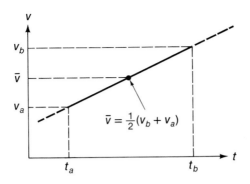

Fig. 2–9. The velocity of a particle varies linearly with time, with $v(t_a) = v_a$ and $v(t_b) = v_b$. The average velocity $\bar{v}$ for the interval $t_b - t_a$ is the mean between v_b and v_a; that is, $\bar{v} = \frac{1}{2}(v_b + v_a)$.

Next, we substitute for $v(t)$ from Eq. 2–8 and write

$$\bar{v} = \tfrac{1}{2}[(at + v_0) + v_0] = \tfrac{1}{2}at + v_0$$

Then, using the expression for $\bar{v}$ in Eq. 2–9, we find

$$s(t) = \tfrac{1}{2}at^2 + v_0 t + s_0 \qquad\qquad \textbf{(2–10)}$$

This equation gives the position of a particle that moves with constant acceleration a, with velocity v_0 and position s_0 when $t = 0$. Note that Eq. 2–10 remains correct when $a = 0$; then, the equation for $s(t)$ reduces to Eq. 2–7.

It should be noted that Eq. 2–10 is consistent with the previous definitions. If the expression for $s(t)$ is used in the defining equations for v and a, we find

$$v(t) = \frac{ds}{dt} = \frac{d}{dt}(\tfrac{1}{2}at^2 + v_0 t + s_0)$$

$$= at + v_0$$

and
$$a(t) = \frac{dv}{dt} = \frac{d}{dt}(at + v_0) = a = \text{constant}$$

Figure 2–10 shows the position function $s(t)$ for a particle moving with constant acceleration (Eq. 2–10); the curve $s(t)$ is a *parabola*. The position of the particle at $t = 0$

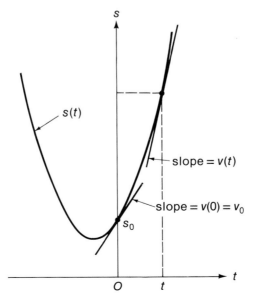

Fig. 2–10. The graph of the position function $s(t)$ for a particle moving with constant acceleration is a *parabola*. The position s_0 and the velocity v_0 at $t = 0$ are indicated.

is $s = s_0$. The velocity at $t = 0$ is v_0 and is equal to the slope of the curve at this point. At any other time t, the velocity of the particle is equal to the slope of the curve at the corresponding point.

It is sometimes useful to have an expression based on Eqs. 2–7 through 2–10 but which has no explicit reference to time. From Eq. 2–9 for $s(t)$, we have

$$s(t) - s_0 = \bar{v}t$$

Then, using $\bar{v} = \frac{1}{2}(v + v_0)$, we find

$$s - s_0 = \frac{1}{2}(v + v_0)t$$

From Eq. 2–8 for $v(t)$, we have $t = (v - v_0)/a$. Thus,

$$s - s_0 = \frac{1}{2}(v + v_0)\frac{(v - v_0)}{a}$$

and the variable t has been eliminated. Then,

$$s - s_0 = \frac{v^2 - v_0^2}{2a}$$

or

$$v^2 - v_0^2 = 2a(s - s_0) \qquad \textbf{(2–11)}$$

We summarize our results to this point:

For	$a = $ constant
we have	$v = at + v_0$
	$\bar{v} = \frac{1}{2}(v + v_0)$
	$s = \frac{1}{2}at^2 + v_0t + s_0$
also	$v^2 = v_0^2 + 2a(s - s_0)$
with	$v = v_0$ and $s = s_0$ at $t = 0$

$\textbf{(2–12)}$

Example 2–6

A particle is observed at time $t = 0$ to be at the coordinate position $x_0 = 5$ m and moving with a velocity $v_0 = 20$ m/s. The particle undergoes a constant deceleration (that is, an acceleration in the direction opposite to the velocity). If, 10 s after the initial observation, the particle has a velocity $v = 2$ m/s, what is the acceleration? What is the position function? How much time will elapse before the particle returns to $x = 5$ m?

Solution:

In this problem,

$$v_0 = 20 \text{ m/s}$$
$$v(10 \text{ s}) = 2 \text{ m/s}$$
$$x_0 = 5 \text{ m}$$

with a and $x(t)$ to be determined.

Using $v = at + v_0$ from Eq. 2–12, we have

$$a = \frac{v - v_0}{t} = \frac{(2 \text{ m/s}) - (20 \text{ m/s})}{10 \text{ s}} = -1.80 \text{ m/s}^2$$

The fact that a is negative indicates that the acceleration is indeed in the direction opposite to the velocity.

The position function, from Eq. 2–12, is

$$x(t) = (-0.9 \text{ m/s}^2)t^2 + (20 \text{ m/s})t + (5 \text{ m})$$

For the particle to return to $x = 5$ m, we must solve

$$5 \text{ m} = (-0.90 \text{ m/s}^2)t^2 + (20 \text{ m})t + (5 \text{ m})$$

or

$$t[(-0.90 \text{ m/s}^2)t + (20 \text{ m/s})] = 0$$

Hence, $t = 0$ and $t = \dfrac{20 \text{ m/s}}{0.90 \text{ m/s}^2} = 22.22$ s

The first solution, $t = 0$, is the initial condition, so the return to $x = 5$ m occurs at $t = 22.22$ s.

Example 2–7

In the American sport of drag racing, the performances are reported in British units instead of metric units. Thus, the length of the course is set at $\frac{1}{4}$ mile (mi) and the speeds are measured in miles per hour (mi/h or mph).

A drag racer in the Top Fuel Class sometimes achieves a speed of 240 mi/h at the end of the $\frac{1}{4}$-mi strip. What must be the acceleration (assumed constant) to reach this speed?

Solution:

We have

$$s - s_0 = \tfrac{1}{4} \text{ mi}$$

$$v_0 = 0$$

$$v = 240 \text{ mi/h}$$

with the acceleration a to be found.

Using Eq. 2–12 and solving for a, we have

$$a = \frac{v^2 - v_0^2}{2(s - s_0)} = \frac{(240 \text{ mi/h})^2 - 0}{2(\tfrac{1}{4} \text{ mi})}$$

$$= 115{,}200 \text{ mi/h}^2$$

Almost no one has a "feeling" for the magnitude of the acceleration expressed in these units. However, recall that acceleration is the change in the velocity per unit time. We can understand a velocity value expressed in miles per hour and a short time interval measured in seconds. Therefore, multiply the above acceleration value by $(1 \text{ h}/3600 \text{ s})$:

$$a = 115{,}200 \ \frac{\text{mi}}{\text{h}^2} \times \frac{1}{3600} \ \frac{\text{h}}{\text{s}}$$

$$= 32 \ (\text{mi/h})/\text{s}$$

which states that the velocity increases by 32 mi/h each second (or 51 km/h each second). Expressed in these mixed units, the value of the acceleration in this case is much easier to understand.

World Champion Shirley Muldowney reached a speed of 255.58 mi/h at the end of a quarter mile during the 1979 Winternationals. On three separate occasions she has driven that distance from a standing start in 5.77 seconds. (Courtesy of Atco Raceway and Chrysler Corporation.)

Example 2–8

A truck travels at a constant speed of 80 km/h and passes a more slowly moving car. At the instant the truck passes the car, the car begins to accelerate at a constant rate of 1.2 m/s² and passes the truck 0.5 km farther down the road. What was the speed of the car when the truck passed it?

Solution:

The elapsed travel time by both the truck and the car between the two passing situations is the same, and using the truck's speed, we have

$$t = \frac{0.5 \text{ km}}{80 \text{ km/h}} = 6.25 \times 10^{-3} \text{ h} = 22.5 \text{ s}$$

Thus, for the car, we have

$$s - s_0 = 0.5 \text{ km} = 500 \text{ m}$$

$$a = 1.2 \text{ m/s}^2$$

$$t = 22.5 \text{ s}$$

with v_0 the unknown to be found.
Using Eq. 2-12, we have

$$s - s_0 = \tfrac{1}{2}at^2 + v_0 t$$

or

$$(500 \text{ m}) = \tfrac{1}{2}(1.2 \text{ m/s}^2)(22.5 \text{ s})^2 + v_0(22.5 \text{ s})$$

Therefore,

$$v_0 = \frac{500 \text{ m} - 303.8 \text{ m}}{22.5 \text{ s}} = 8.72 \text{ m/s}$$

$$= (8.72 \text{ m/s})(3600 \text{ s/h})(10^{-3} \text{ km/m})$$

$$= 31.4 \text{ km/h}$$

2-5   FREE FALL

We now seek to apply the equations that have been developed to a familiar situation, namely, an object that falls freely near the surface of the Earth. The problem of falling bodies had been addressed in classical Greek times by Aristotle (384–322 B.C.), the most prominent of the early natural philosophers, whose teachings dominated scientific thought. Aristotle's approach to science was basically to set forth "philosophic truths" resulting from logical deductions, not "observational facts" obtained from a mathematical analysis of carefully conducted experiments. Some of Aristotle's ideas about physical phenomena therefore turned out to be terribly wrong. He relied on ordinary experience in a superficially qualitative way. For example, he claimed as perfectly natural that, under similar circumstances, a heavy object would fall to the Earth more rapidly than would a lighter object. Of the four "elements," Earth, Water, Air, and Fire, out of which all objects were supposed to be constituted, heavy objects contained more elemental Earth. Hence, the heavier objects were more strongly urged to seek their natural terrestrial level—the Earth's surface—and thus fall more rapidly.

It is true that a feather and a rock will fall at different rates, but this is due entirely to the effects of air resistance and does not represent the more fundamental effect of gravity. (• How would Aristotle compare the rates of fall of two rocks separately and when tied together?) Neither Aristotle nor his followers sought to perform the simple experiment that would have tested his basic premise (and proven it false). Nevertheless, so completely did Aristotelian doctrine influence succeeding generations that fundamental advances in physical science were stifled in Western civilization for nearly 2,000 years.

In the sixteenth century, a new attitude toward scientific thought was emerging. During this age of discovery, new ideas began to develop in science, and the Aristotelian philosophy was being overturned. Science began to be guided by conclusions based on experiment and observation, coupled with logic and reason—the scientific method.

By taking this new approach, Galileo Galilei (1564–1642), the great Italian physicist and astronomer, was able to establish mechanics as a mature science through his careful experiments and well-constructed logical arguments. Indeed, many of the concepts we have treated to this point had their origins in Galileo's monumental work, *Discourses and Mathematical Demonstrations Concerning Two New Sciences Pertaining to Mechanics and Local Motion*, published in 1638. In contradiction to Aristotelian lore, Galileo put forward the hypothesis that all objects falling near the Earth's surface accelerate at the same constant rate. One conclusion that can be drawn from this hypothesis is that all objects released from the same height above the Earth will fall to the ground in equal times. Galileo tested this prediction and found it to be true.*

---

*But not, as legend would have it, by dropping a cannon ball and a musket ball from the Tower of Pisa. The demonstration of the effect by dropping objects with different masses from a tower was carried out in 1586 by the Dutch mathematician, Simon Stevinus (1548–1620), not by Galileo. Galileo made ingenious use of metal balls rolling down inclined planes and water clocks to study the constant acceleration due to gravity.

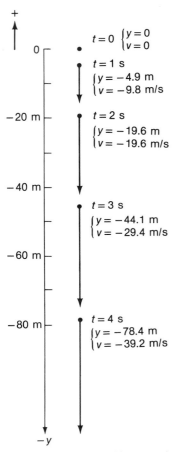

Fig. 2–11. The position s and the velocity v of an object released from rest.

Consider now the ideal case of an object that falls freely without any interfering effects due to wind or air resistance; this is the situation that would prevail in a vacuum. (For the case of retarded motion, see Section 6–3.) We postulate along with Galileo that all bodies (particles) that fall freely near the Earth's surface experience a constant downward acceleration caused by the pull of the Earth's gravity. The value of this acceleration is determined by experiment and is given the symbol g:

$$g = 9.80 \text{ m/s}^2 \tag{2–13}$$

In Section 5–3 we discuss the association of this quantity with the law of universal gravitation and the properties of the Earth.

The value of g at sea level varies slightly over the surface of the Earth. The main effect is an increase with latitude due to the Earth's rotation: at the Equator, $g = 9.78 \text{ m/s}^2$, whereas the polar value is 9.83 m/s^2. For middle latitudes, Eq. 2–13 is sufficiently accurate. For some engineering purposes, "standard gravity" means $g = 9.80665 \text{ m/s}^2$, a value appropriate for a latitude of approximately 45° at sea level. In British units the value of g that corresponds to 9.80 m/s^2 is 32.2 ft/s^2.

It is convenient to have a set of equations representing the application of Eqs. 2–12 to a particle moving under the influence of the Earth's gravity. The constant acceleration g is always directed downward. Therefore, if we choose the upward direction as positive, the sign of the particle's acceleration is negative, $a = -g$. Then,

$$
\begin{aligned}
\text{For} \qquad & a = -g \\
\text{we have} \qquad & v = -gt + v_0 \\
\text{and} \qquad & y = -\tfrac{1}{2}gt^2 + v_0 t + y_0 \\
\text{also} \qquad & v^2 = v_0^2 - 2g(y - y_0) \\
\text{with} \qquad & v = v_0 \quad \text{and} \quad y = y_0 \quad \text{at } t = 0
\end{aligned}
\tag{2–14}
$$

Notice also that we can use Eqs. 2–14 to obtain an expression for the velocity v of an object dropped (that is, $v_0 = 0$) through a height $h = y_0 - y$:

$$v = \sqrt{2gh} \tag{2–15}$$

With Eqs. 2–14 and 2–15 we can easily calculate the velocity and the position of an object that is released from rest and allowed to fall freely. Figure 2–11 shows the results for the first few seconds after release. (The upward direction is positive and $y_0 = 0$.)

Example 2–9

A stone is thrown vertically upward from the ground with an initial velocity of 25 m/s. How long is required for the stone to reach its maximum height, and how high does it rise? With what velocity does the stone strike the ground on its return? How long does the entire round trip take?

Solution:

In this problem, with the upward direction positive, we have

$$
\begin{aligned}
a &= -g = -9.80 \text{ m/s}^2 \\
v_0 &= +25 \text{ m/s} \\
y_0 &= 0
\end{aligned}
$$

The maximum height occurs at the time $t = T$ when the upward velocity becomes zero. (• Why?) Thus, from Eq. 2-14 for $v(t)$,

$$v(T) = 0 = -gT + v_0$$

or

$$T = \frac{v_0}{g} = \frac{25 \text{ m/s}}{9.80 \text{ m/s}^2} = 2.55 \text{ s}$$

The maximum height y_m can be calculated from Eq. 2-14, where $y_0 = 0$ and $y = y_m$ for $v = 0$:

$$0 = v_0^2 - 2g(y_m - 0)$$

or

$$y_m = \frac{v_0^2}{2g} = \frac{(25 \text{ m/s})^2}{2(9.80 \text{ m/s}^2)} = 31.9 \text{ m}$$

You should verify that the same result is obtained if $T = 2.55$ s is used in Eq. 2-14 for $y(t)$.

The velocity of the stone upon its return to ground level also can be determined from Eq. 2-14. Here, $y = y_0 = 0$, and we have

$$v^2 = v_0^2$$

so

$$v = \pm v_0$$

The desired value is $v = -v_0 = -25$ m/s because the negative sign indicates *downward* motion. (• What is the significance of the other solution, $v = +v_0$?)

To find the time of the return to ground level, we use Eq. 2-14 for $y(t)$ with $y = y_0 = 0$; then,

$$-\tfrac{1}{2}gt^2 + v_0t = 0$$

or

$$t(v_0 - \tfrac{1}{2}gt) = 0$$

This equation has the solutions

$$t = 0 \quad \text{and} \quad t = \frac{2v_0}{g} = 2T$$

Thus, $t = 2T = 5.10$ s is the duration of the round trip.

The diagrams show the motion graphically in terms of $y(t)$ and $v(t)$. In particular, note the symmetric behavior of the motion about $t = T$.

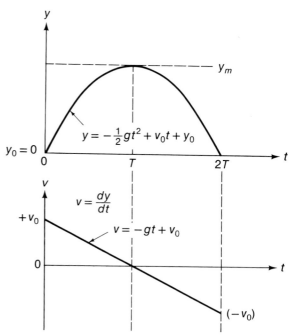

Example 2-10

A falling object requires 1.50 s to travel the last 30 m before hitting the ground. From what height above the ground did it fall?

Solution:

Let the velocity upon hitting the ground be v_2 and let the velocity be v_1 at the time $\Delta t = 1.50$ s earlier. Then, with the *downward* direction positive (so that $a = +g$), we have

$$v_2 = v_1 + g \Delta t = v_1 + (9.80 \text{ m/s}^2)(1.50 \text{ s})$$

$$= v_1 + 14.70 \text{ m/s} \qquad (1)$$

We also know that the distance Δy fallen during the time Δt can be related to the *average* velocity by

$$\Delta y = \tfrac{1}{2}(v_1 + v_2) \Delta t$$

so

$$v_1 + v_2 = \frac{2(30 \text{ m})}{(1.50 \text{ s})} = 40 \text{ m/s} \qquad (2)$$

Adding Eqs. (1) and (2), we find

$$2v_2 = (40 + 14.70) \text{ m/s}$$

or

$$v_2 = \tfrac{1}{2}(54.70 \text{ m/s}) = 27.35 \text{ m/s}$$

Finally, solving Eq. 2-15 for h, we have

$$h = \frac{v_2^2}{2g} = \frac{(27.35 \text{ m/s})^2}{2(9.80 \text{ m/s}^2)} = 38.16 \text{ m}$$

> The main body of the text contains both optional sections and extended parenthetical remarks set off with triangles (▶ ◀). Sections labeled (Optional) contain material that may be omitted or the treatment of which may be postponed. (Sections labeled Enrichment Topics, found in the back of this text, also contain some valuable supplemental material.)
>
> The integration process, although again referred to briefly in Chapter 4, is not used extensively until Chapter 7.

2–6 GENERAL EQUATIONS DESCRIBING MOTION (Optional)

The basic equations that we have been using (Eqs. 2–12 and 2–14) are valid *only* for cases in which the acceleration is *constant* (including the case $a = 0$). When the acceleration itself is a function of time, we must return to the basic definitions for the acceleration, velocity, and position functions.

We now ask, "If the acceleration $a(t)$ is given, how do we find $v(t)$ and finally $s(t)$?" We have

$$a(t) = \frac{d}{dt} v(t)$$

We require the function $v(t)$ such that its time derivative is $a(t)$. The necessary inverse operation involves the *antiderivative* or the *integration* operation, written symbolically as*

$$\int dv(t) = v(t) = \int a(t)\, dt \qquad \textbf{(2–16)}$$

The differential calculus provides a precise prescription for obtaining the derivative of any function, such as $s = s(t)$ or $v = v(t)$, by using the limiting process as in Example 2–3. There exists no corresponding prescription in integral calculus for performing the inverse operation. Each integration must be considered individually; the integration process is a tentative one, requiring verification that the result, if differentiated, does in fact reproduce the original integrand. The task of evaluating

integrals is made easier by the use of tables of standard integrals. To use such tables it is only necessary to reduce the integrand function of the desired integral to a form that is tabulated. You will find it worthwhile to become familiar with the use of integral tables.†

The integration process in Eq. 2–17 determines $v(t)$ only to within an arbitrary additive constant; that is, the result is $v(t) + C$, where C is any constant. This is immediately evident if we perform the test of differentiating the result of the integration to determine whether the initial function is recovered; that is,

$$\frac{d}{dt}[v(t) + C] = \frac{d}{dt}v(t) + \frac{dC}{dt} = \frac{d}{dt}v(t) + 0 = a(t)$$

Thus, if $v(t)$ is a solution to the integration of $a(t)$, then $v(t) + C$ is also a solution. The evaluation of the integration constant C requires that additional information be provided. This information is usually given in terms of the *initial conditions* of the problem.

Once we have determined $v(t)$ by integrating $a(t)$, we may proceed to find $s(t)$ by integrating $v(t)$; that is,

$$\int ds(t) = s(t) = \int v(t)\, dt \qquad \textbf{(2–17)}$$

This integration produces yet another (independent) constant of integration that requires another given initial condition for its evaluation. The final expression for the position function $s(t)$ is the equation describing the motion of the particle.

Example 2–11

Given that the acceleration of a particle is constant, $a(t) = a$, find the equation of motion if at $t = 0$ the velocity is v_0 and the position is s_0.

Solution:

Here,

$$v(t) = \int a\, dt = at + C$$

---

*The integral sign $\int$ has its historical origin as a distorted symbol for the letter S (implying a sum). We refer to $\int a(t)\, dt$ as the indefinite integral of the function $a(t)$, which is the *integrand*.

†A list of useful integrals appears inside the rear cover.

Now, $v(0) = v_0$, so that $C = v_0$. Thus,

$$v(t) = at + v_0$$

Then,

$$s(t) = \int (at + v_0) \, dt$$

or

$$s(t) = \tfrac{1}{2}at^2 + v_0t + C'$$

We know that $s(0) = s_0$; hence, $C' = s_0$. Finally,

$$s(t) = \tfrac{1}{2}at^2 + v_0t + s_0$$

This result is precisely Eq. 2–10 for $s(t)$. You can see that the method used here based on integral calculus is much more direct than the earlier derivation. Also, note that it was unnecessary to deal at all with average quantities.

QUESTIONS

2–1   Define or describe briefly the terms (a) position and position function, (b) displacement, (c) path length.

2–2   Give examples (if possible) for motion with (a) average speed less than average velocity, (b) average speed equal to average velocity, (c) average speed greater than average velocity.

2–3   Give an example of motion for which the average velocity is independent of the time interval used to define the average. Write the algebraic form of the position function $s(t)$.

2–4   Is it possible for a particle experiencing constant acceleration to have its velocity for all times directed opposite to the acceleration?

2–5   The position function $s(t)$ of a particle is shown in the graph. (a) What can be said about the velocities at times t_0, t_1, t_2, t_3, and t_4? (b) State whether the acceleration is positive, negative, or zero during the time intervals between t_0, t_1, t_2, t_3, t_4, and beyond t_4.

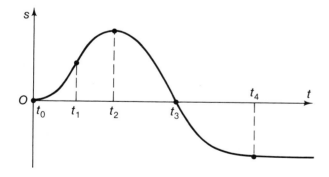

2–6   Sketch the graph of the velocity function $v(t)$ for a particle that initially has a constant velocity v_0, is suddenly decelerated at a constant rate for a length of time Δt, following which it is immediately subjected to an acceleration equally large in magnitude but acting for a time $\Delta t/2$, and arrives finally at a constant velocity $-v_0$.

2–7   Aristotle held that each of the four basic ''elements''—

Earth, Water, Air, and Fire—occupied a natural place or region in the world-system. The element Earth occupied the lowest position and Fire the highest, with Water and Air in between (see drawing). All objects contained various mixtures of these elements. Suggest possible admixtures of these elements for the objects involved in (a) the rising of bubbles in water, (b) the floating and sinking of objects in water, (c) the falling of rain and snow, (d) the falling of a stone and a feather.

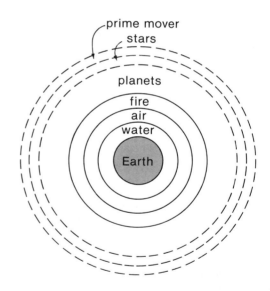

2–8   Galileo experimented with balls rolling down inclined planes in order to reduce the acceleration along the plane and thus slow the rate of descent of the balls. This provided for more accurate time measurements with his relatively crude water clocks. Considering in particular the extreme cases of planes A and B shown in the drawing on page 28, can you explain why you would expect the acceleration along the plane to decrease as the inclination angle θ decreases?

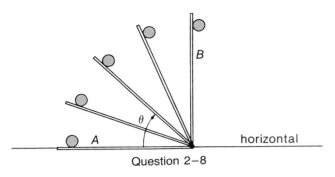

Question 2–8

the planes depended only on the difference in elevation h through which the ball descended, what trigonometric dependence on the angle of inclination θ would you deduce for the acceleration acting along the planes? [*Hint:* Consider applying Eq. 2–11 to the situations shown in the drawing.]

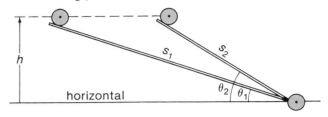

2–9   If, in using the various inclined planes of Galileo, you found that the velocity of a ball on reaching the bottom of

PROBLEMS

Section 2–1

2–1   Suppose that the straight-line motion of a particle is described by the position function $x(t) = \alpha t^3 + \beta t^2$, where $\alpha = \frac{1}{3}$ m/s^3 and $\beta = -1$ m/s^2. Make a graph of $s(t)$ by plotting the position of the particle at 0.5-s intervals for $|t| \leqslant 3$ s. Locate the particle at $t = \pm 0.7$ s and at $t = \pm 2.3$ s by using the graph.

2–2   A particle travels along a straight line according to the position function $s(t) = at^2 + b$, where $a = \frac{1}{2}$ m/s^2 and $b = 2$ m. What is the particle's displacement during the interval from $t = 0$ s to $t = 3$ s? From $t = 1$ s to $t = 3$ s? From $t = -3$ s to $t = 3$ s?

Section 2–2

2–3   The positions of a particle at 1-s intervals beginning at $t = 0$ are 0 m, 1.6 m, 5.6 m, 12.0 m, 20.8 m. Make a graph of $s(t)$. Use the graph to determine the speed at $t = 2$ s and at $t = 3$ s.

2–4   In the celebrated tortoise-and-hare race, the hare runs the first 100 m of a 300-m course in 40 s and then rests for 2 h. Upon awakening, the hare runs the remainder of the race in 1 min, only to have the tortoise just beat him at the finish line. (a) If the tortoise moved at constant speed throughout the race, what was this speed? (b) What were the average speeds of the hare during the first 40 s and during the last minute of the race?

2–5   A particle moves one half of the distance from A to B at a constant velocity of $v_1 = 10$ m/s and moves the remainder of the distance at a constant velocity of $v_2 = 40$ m/s. What was the average velocity $\bar{v}$ for the entire trip? Does $\bar{v}$ equal $\frac{1}{2}(v_1 + v_2)$? Explain.

2–6   An athlete can run a 1500-m race in 3 min and 50 s. (a) What is his average speed in meters per second? (b) If his actual speed near the finish line is 20 percent greater than his average speed, and if he wins over his nearest rival, who is running at almost the same speed, by an extra forward lean of 3.0 cm, what is the difference in time between first and second place? Does it make sense to measure the elapsed time of such races to an accuracy of a millisecond? Explain. [*Hint:* Near the finish line, small changes in position Δs are associated with small time intervals Δt such that $\Delta s / \Delta t \cong ds/dt = v$.]

2–7   The position function of a particle traveling a straight-line path is $s(t) = \frac{1}{2}at^2 + bt + c$, with $a = 1$ m/s^2, $b = -1$ m/s, and $c = 2$ m. (a) Make a sketch of $s(t)$. (b) With what average velocity does the particle travel from $s = \frac{3}{2}$ m to $s = \frac{7}{2}$ m? (c) With what velocity does it arrive at $s = \frac{7}{2}$ m? (d) What was the velocity at $s = \frac{3}{2}$ m? Does your answer have any particular significance? (e) Show on your sketch the lines whose slopes correspond to these velocities.

2–8   A particle moves according to the position function
• $s(t) = \alpha t^3 + \beta t^2 + \delta t$,   where $\alpha = 1$ m/s^3, $\beta = -9$ m/s^2, and $\delta = 24$ m/s. (a) Make a sketch of $s(t)$. (b) Determine $v(2$ s), $v(4$ s), and $v(4.5$ s). (c) Evaluate $\bar{v}$ and the average speed $\Delta \ell / \Delta t$ for the following time intervals: 0–2 s, 0–4 s, and 0–4.5 s.

2–9   A man whose height is 1.8 m walks directly away from a
••   street light that is 5.0 m above street level. If he walks with a constant velocity of 2.0 m/s, at what rate does his shadow lengthen? [*Hint:* Make appropriate use of the similar triangles present in the geometry of this problem.]

Section 2–3

2–10 A particle is moving with a velocity $v_0 = 60$ m/s at $t = 0$. Between $t = 0$ and $t = 15$ s the velocity decreases uniformly to zero. What was the average acceleration during this 15-s interval? What is the significance of the sign of your answer?

2–11 A particle travels with a velocity $v_0 = 10$ m/s during the interval from $t = 0$ to $t = 5$ s. Then, it travels for an additional 10 s at a uniformly increasing velocity, arriving at a final velocity of 40 m/s. What was the average acceleration for the total 15-s interval?

2–12 A tennis ball falling vertically strikes the ground with a velocity of 20 m/s and rebounds with a velocity of 14 m/s. If the total time of contact with the ground was 0.01 s, what was the average acceleration produced by the impact?

2–13 In Example 2–5, suppose that $A = 5$ cm and $\omega = 3$ rad/s. What are the values of the maximum speed and the maximum acceleration?

2–14 The position function of a particle is $s(t) = at^2 + bt + c$, with $a = 5$ m/s^2, $b = 2$ m/s, and $c = -1$ m. (a) What is the distance traveled by the particle during the time interval from $t = 1$ s to $t = 3$ s? (b) What is the average velocity $\bar{v}$ during this interval? (c) What is the average acceleration during this interval? Is $\bar{a} = \bar{v}/\Delta t$? Explain.

Section 2–4

2–15 A Cessna 150 aircraft has a lift-off speed of about 125 km/h. (a) What minimum constant acceleration does this require if the aircraft is to be airborne after a take-off run of 250 m? (b) What is the corresponding take-off time? (c) If the aircraft continues to accelerate at this rate, what speed will it reach 25 s after it begins to roll?

2–16 An automobile traveling initially at a speed of 60 m/s is accelerated uniformly to a speed of 85 m/s in 12 s. How far does the automobile travel during the 12-s interval?

2–17 A truck is moving with a speed of 60 m/s. The driver applies the brakes and slows the vehicle (uniformly) to a speed of 40 m/s while traveling a distance of 250 m. What was the acceleration of the truck?

2–18 The driver of a car traveling at 90 km/h observes a hazard on the road and applies the brakes, giving a constant deceleration of 2.5 m/s^2. (a) If the driver's reaction time is 0.2 s, how long does it take to stop after sighting the hazard? (b) What distance does the automobile travel before coming to rest?

2–19 An automobile is cruising at a speed of 100 km/h on a
• highway posted at 80 km/h. As the automobile passes a parked highway patrol car, the officer accelerates his car at a uniform rate, reaching 60 km/h in 10 s; he continues to accelerate at the same rate until he catches the speeding car. (a) How long did the chase last? (b) How far from the parked position of the patrol car was the speeder overtaken? (c) What was the speed of the patrol car as it overtook the speeder?

Section 2–5

2–20 A balloon is rising at a constant rate of 10 m/s when a stone is dropped from the gondola by simply releasing it to fall freely. (a) If the balloon is 80 m above the ground at that time, how long does it take for the stone to hit the ground? (b) With what velocity does the stone strike the ground? [*Hint:* Consider carefully the proper initial velocity to ascribe to the stone on its release.]

2–21 An irate physics student, on learning of his failing exam grade, wants to throw a rock through the glass skylight in the lecture-hall ceiling 10 m above the floor. If breaking the glass requires a projectile velocity of at least 7 m/s, with what minimum speed must the student hurl the rock if its flight starts from 2 m above the floor?

2–22 A body falling from rest travels one third of the total dis-
• tance of fall during the last second. Find (a) the height from which it was released and (b) the total time of fall.

2–23 A ball is thrown directly downward with an initial velocity of 8 m/s from a height of 30 m. When does the ball strike the ground?

2–24 A stone is dropped from a cliff 200 m high. If a second
• stone, thrown vertically downward from the cliff 1.50 s after the first stone is released, strikes the base of the cliff at the same instant as the first stone, with what velocity was the second stone thrown?

2–25 A startled student sees beer cans being thrown upward
•• past his dormitory window from a party below. The height of the window is 1.5 m from ledge to top. Using a stopwatch, the student determines that one of the cans is in view for a total of 0.6 s (upward plus downward flight). How high above the window ledge did the beer can rise before starting to fall? [*Hint:* What is the relationship between the time to rise through the distance of the window opening and the time to fall through the same distance?]

Section 2–6 (Optional)

2–26 Given that the acceleration of a particle is $a(t) = \beta t$, find the position function $x(t)$ if at $t = 0$ we have $x(0) = 0$, and if at $t = \tau$ we have $x(\tau) = \gamma$.

2–27 A particle moves with an acceleration $a(t) = 3\alpha t + 2\beta$,
• with $\alpha = 1$ m/s^3 and $\beta = 1$ m/s^2. Determine $s(t)$ such that $v(1$ s$) = 4$ m/s and $s(1$ s$) = 3$ m.

2–28 A computer-controlled engine drives a vehicle with an acceleration that increases proportionately with time and reaches a velocity of 20 km/h after 20 s, starting from rest. What distance does the vehicle travel in 1 min after starting?

2–29 Suppose that the velocity of a particle is given by
•
$$v(t) = \alpha(1 - e^{-\beta t})$$
Find (a) the acceleration and (b) the position function $x(t)$ if at $t = 0$, $x(0) = 0$. (c) What is the limit of $v(t)$ as $t \to \infty$ if $\beta > 0$? (β is real.) (d) What is the value of $v(0)$? (e) What is the limit of the acceleration as $t \to \infty$?

Additional Problems

2–30 Consider the position function
$$s(t) = \pm A \sqrt{\frac{t}{t - t_0}}$$
with $A = 1$ m and $t_0 = 1$ s. Make a plot of this function. For what values of t can a real particle follow this position function?

2–31 A car makes a 200-km trip at an average speed of 40 km/h. A second car starting 1 h later arrives at their common destination at the same time. What was the average speed of the second car?

2–32 A glider is approaching a landing, moving at a constant
• velocity along a path sloping 15° down from horizontal, with the Sun directly behind it. If the Sun is 45° above the horizon, how fast, in terms of its glide velocity, is the plane's shadow moving along the ground? [*Hint:* Use the law of sines.]

2–33 The position function of a particle is $s(t) = \alpha t^3 + \beta t$,
• with $\alpha = -\frac{1}{10}$ m/s^3 and $\beta = 6$ m/s. (a) What is the maximum positive excursion reached after starting ($t = 0$)? (b) With what velocity does the particle return to the origin? (c) What is the acceleration (in units of g) 10 s after starting?

2–34 In studies designed to investigate the physiologic effects of large accelerations on human beings, Lt. Col. John L. Stapp rode a rocket-propelled sled that was brought to rest from an achieved speed of 285 m/s within 1.5 s (New Mexico, 1954). What was the value of the deceleration (assumed constant) experienced by Col. Stapp? Express the result in terms of g. (He survived with minor blood-vessel damage.)

2–35 A motorist is traveling along an interstate highway at the illegal speed of 120 km/h. (The posted speed limit is 90 km/h.) At kilometer-post 38 his radar detector buzzes, having been activated by the beam from a highway patrol car at kilometer-post 40. If the reliable range of the patrol car's radar speed meter is 800 m, at what rate must the motorist decelerate to avoid a speeding ticket?

2–36 A boy runs as fast as he can to "hitch a ride" on the back
• of a truck. When he is still 20 m from the truck, the truck begins to move forward with a constant acceleration of 0.8 m/s^2. If the boy just catches the truck, how fast was he running? [*Hint:* Sketch the two position functions and inquire about the special significance of the fact that the boy *just* reached the truck.]

2–37 An ideally elastic ball bounces repeatedly from a hard surface, rising 10 cm vertically between each bounce. How many times per minute does the ball strike the surface?

2–38 An athlete can throw a certain object vertically to a height of 15 m in Austin, Texas, where $g = 9.7934$ m/s^2. How much does he fall short of this height in Stockholm, where $g = 9.8185$ m/s^2, if he throws the object with the same initial velocity in each case?

2–39 A ball is thrown vertically upward with an initial velocity
•• of 12 m/s. One second later a second ball is thrown upward directly in line with the first ball with a velocity of 16 m/s. (a) At what time and (b) at what height do the two balls collide? (c) Is the first ball rising or falling at the time of collision?

2–40 A toy rocket is equipped with an engine that provides a constant acceleration of 40 m/s^2. If the rocket is fired vertically and if the engine burns for 3 s, to what maximum height will the rocket rise? [*Hint:* After the fuel is exhausted, the rocket "coasts" upward.]

2–41 A ball is thrown directly upward from ground level with
• an initial velocity v_0. The ball rises to a height h and then lands on a roof at a height $\frac{1}{2}h$. If the entire motion required 3.5 s, find the velocity v_0 and the height h.

2–42 At a certain instant, a ball is thrown downward with a
•• velocity of 8 m/s from a height of 40 m. At the same instant, another ball is thrown upward from ground level directly in line with the first ball with a velocity of 12 m/s. (a) How long after release do the balls collide? (b) At what height does this collision occur? (c) In what direction is the second ball moving when the collision occurs?

2–43 Two trucks are approaching each other on a long, straight
•• stretch of highway in Arizona. Each proceeds at a constant speed of 50 km/h. When the trucks are 20 km apart,

a roadrunner (a strange bird that prefers to run on the ground at high speeds rather than fly) just in front of one truck runs at a speed of 70 km/h toward the other truck. When it arrives just in front of that truck, it turns and runs back toward the first truck. In cartoon fashion, it runs back and forth between the two trucks until the fateful final moment when it is sandwiched between the colliding trucks. Although the roadrunner makes an infinite number of trips before the ''end,'' it travels a finite distance in a finite time. What total distance does the roadrunner cover? Can you show that it makes an infinite number of trips in doing so? [*Hint:* Use Eq. A−3 in Appendix A for the sum of the geometric series.]

2−44 Captain Buck Spacewalker starts out from a deep-space
•• station in his ion-drive ship, which can provide a constant acceleration of 0.001 g throughout his journey. Exactly one week later, a chemical-drive messenger rocket is sent from the station to overtake him with a secret message. The rocket is capable of accelerating at 3.000 g for a maximum of 10 minutes and must coast at its final speed thereafter. (a) For how long should the messenger rocket fire its engine in order to match its velocity with that of Captain Spacewalker at the moment that it overtakes him? (b) How long after the captain's departure does the messenger rocket reach him? (c) What is the common velocity of the two ships when they meet?

CHAPTER 3

VECTORS

In the preceding chapter we discussed the motion of a particle confined to a straight-line path. To be more realistic we must allow motion in two and three dimensions. In this chapter we introduce the idea of *vectors* as an aid in describing these more complicated motions.

To locate a particle in space we use a *frame of reference* or *coordinate frame*. The familiar Cartesian* or rectangular coordinate frame is one possibility. In a Cartesian frame, such as that in Fig. 3–1, the position of a particle located at the point A would be given by quoting the three coordinates, x_A, y_A, z_A.

In physics, we are usually interested in *changes* that take place in the physical world. For example, we might be concerned with the time variation of some physical observable at point A of Fig. 3–1, or how this observable is related to the physical condition at some other point B. Or our interest might be the time-developed history of the motion (the *trajectory*) of a particle that passed through point A.

To help sort out various observables in space, we use an appropriate coordinate frame. We emphasize at the outset that such coordinate frames are products of our own invention, devised solely to facilitate the description of events that we study. Clearly, *any laws of Nature we discover cannot in any way depend on our special choice of a coordi-*

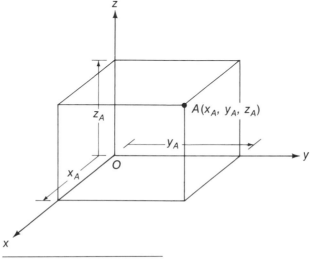

Fig. 3–1. The location of the point A is specified by the three coordinates, x_A, y_A, z_A.

---

*Named for the famous French mathematician, René Descartes (Latinized form: Renatus Cartesius) (1596–1650), who first made graphs of equations by plotting the curves on a rectangular coordinate system.

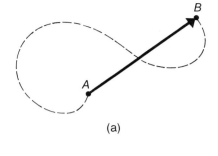

(a)

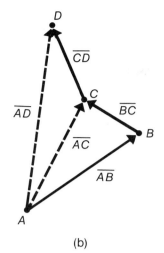

(b)

Fig. 3–2. (a) The directed-line representation of the displacement from A to B, $\overline{AB}$. (b) The representation of continued displacements, A to B to C to D, illustrating the notion of a sum of displacements.

nate frame (that is, on the choice of the spatial orientation of the frame or on the location of its origin). The laws of Nature must be *invariant* to the choice of the coordinate frame used to describe the situation.*

The description of the most general motion of an object clearly requires a three-dimensional coordinate frame. However, the great majority of the cases we discuss involve motion in a plane, either horizontal or vertical. Accordingly, we emphasize here the use of vectors in two dimensions.

3–1   DISPLACEMENT

When a particle moves from one position to another, we say that it undergoes a *displacement*. In Fig. 3–2a a particle moves from A to B, and we represent the displacement $\overline{AB}$ by the straight-line segment connecting A to B. An arrowhead is placed at B, indicating that the displacement proceeds *from A to B*. In so doing, we imply nothing concerning the actual path followed by the particle in moving from A to B; for example, the motion could have been along the curved path shown by the dashes in Fig. 3–2a. In this description we are concerned only with the *end-points A and B*.†

If the particle proceeds from B to a new position C, it undergoes a new displacement $\overline{BC}$, thereby producing a total displacement from A that is represented by the *single* displacement $\overline{AC}$, shown in Fig. 3–2b. Yet a further displacement from C to D, $\overline{CD}$, leads to a total displacement from A represented by the single displacement $\overline{AD}$. The total displacement $\overline{AD}$, although it consists of three separate parts, depends only on the location of point D relative to point A. In the same way, the displacement $\overline{AB}$ in Fig. 3–2a may be viewed as the sum of a large number of small displacements that follow the curved path from A to B.

With Fig. 3–2b, we intuitively arrive at the geometric concept of displacement (arrow) addition. The sum of the two displacements $\overline{AB}$ and $\overline{BC}$ is the resultant displacement $\overline{AC}$, obtained by drawing the arrow connecting point A to point C (with the arrow tip at C). This type of sum is clearly more complicated than the simple addition of two numbers.

---

*We restrict attention to nonaccelerating coordinate frames; see Section 5–2.

†The initial point A is the tail end (or *origin*) and the end-point B is the arrow tip (or *terminus*) of the displacement $\overline{AB}$.

These ideas lead us to the concept of *vectors*. Many quantities in Nature possess the same essential properties as displacement arrows, namely, *magnitude* and *direction*. In general, vectors are quantities that have magnitude and direction and obey the same rules of combination as displacements. Physical quantities in this class include velocity, acceleration, force, momentum, and many others.

Besides vectors, there are quantities in Nature that require only a magnitude (and an appropriate scale unit) for a complete specification. These quantities—called *scalars*—do not have any associated direction. Physical quantities in this class include mass, energy, temperature, electric charge, and many others.

3–2 BASIC VECTOR CONCEPTS

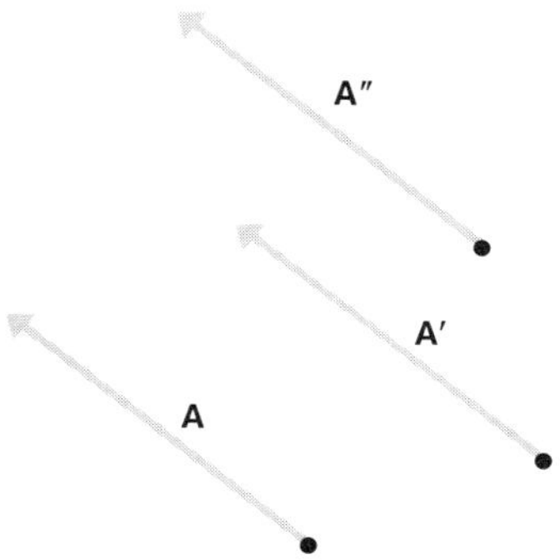

In printed matter, vectors are usually represented by boldface characters, such as **A**, **E**, **r**, and **ω**. In handwritten material the most convenient way to indicate vectors is by a small arrow above the symbol: $\vec{A}$, $\vec{E}$, $\vec{r}$, and $\vec{\omega}$. If we wish to refer only to the *magnitude* of the vector **A**, we use lightface type, A. When we want to call attention to the vector property of a quantity but need only its magnitude, we write $|\mathbf{A}|$; of course, we have, identically, $A \equiv |\mathbf{A}|$. Note that A, the magnitude of **A**, is always a nonnegative real (positive definite) quantity: $A \geq 0$.

We represent a vector graphically by means of a directed line segment (arrow). The length of the line segment (on some convenient scale) corresponds to the magnitude of the vector, and the arrow indicates the direction of the vector. The origin (tail) of the line segment may be *any* point. Thus, we can represent the *same* vector by different parallel line segments of equal length, as in Fig. 3–3. Vectors are characterized by magnitude and direction, nothing more.

Multiplication of a Vector by a Scalar. The product of a vector **A** and a dimensionless number $n > 0$ is defined to be a new vector $n\mathbf{A}$ having the *same* direction as **A** but altered in magnitude: $|n\mathbf{A}| = n|\mathbf{A}| = nA$. When $n < 0$, the product $n\mathbf{A}$ is defined to have a direction *opposite* to **A** and to have a magnitude $|n\mathbf{A}| = |n||\mathbf{A}| = |n|A$. When $n = -1$, the result is the *negative* of the original vector. In the event that n represents a physical quantity with dimensions, the product $n\mathbf{A}$ will correspond to a new physical quantity. Figure 3–4 illustrates several examples of vectors multiplied by scalars. The multiplication of a vector by a vector is discussed in Section 7–1 (the *scalar product*) and in Section 10–1 (the *vector product*).

Unit Vectors. We may associate with any vector **A** a *unit vector** $\hat{\mathbf{u}}_A$ that has the same direction as **A** but has a (dimensionless) length of 1, namely,

$$\hat{\mathbf{u}}_A \equiv \frac{\mathbf{A}}{|\mathbf{A}|} = \frac{\mathbf{A}}{A} \qquad (3\text{–}1)$$

Then, we can write $\mathbf{A} = A\hat{\mathbf{u}}_A$.

Unit vectors may also be selected to designate particular directions in space even though they are not associated with a specific vector. The utility of such choices will become clear as we proceed.

Addition and Subtraction of Vectors. The vector sum of two vectors, **A** and **B**, is defined by the graphic construction (geometric addition) shown in Fig. 3–5. To carry out this summation, first move the vector **B** parallel to itself (i.e., translate the vector **B**) until

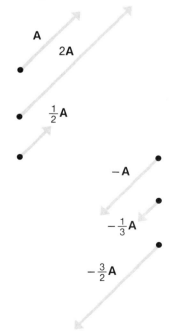

Fig. 3–3. The three directed line segments are all parallel and have the same length. Thus, **A**, **A′**, and **A″** are different representations of the *same* vector.

Fig. 3–4. The result of multiplying the vector **A** by positive and negative dimensionless numbers. The vector −**A** is the *negative* of the original vector **A**.

---

*Unit vectors in this book are always indicated by a caret (ˆ) over the symbol.

its tail end (origin) is at the arrow tip (terminus) of **A**, as shown in Fig. 3–5b. Then, the vector resultant **R** of the addition **R** = **A** + **B** is obtained by drawing a vector connecting the tail end of **A** to the arrow tip of **B**. Alternatively, the vector **A** can be translated parallel to itself until its tail end is at the arrow tip of **B**, as shown in Fig. 3–5c. Then, **R**′ = **B** + **A** is obtained by drawing a vector from the tail end of **B** to the arrow tip of **A**. Evidently, **R** = **R**′. This equality,

$$\mathbf{R} = \mathbf{A} + \mathbf{B} = \mathbf{B} + \mathbf{A} = \mathbf{R}' \qquad (3\text{--}2)$$

exhibits the *commutative* property of vector addition.

The so-called *parallelogram construction* for vector addition is illustrated in Fig. 3–5d; this method is *entirely equivalent* to the arrow method described above. The two vectors **A** and **B** are translated until their tail ends coincide. The resultant vector **R**, the included diagonal between **A** and **B**, is obtained by completing the construction of the parallelogram with the aid of the dashed lines which are parallel to **A** and **B**. The negative of a vector, −**B**, has already been defined, so we can now define the *subtraction* of vectors, **A** − **B**, to be the addition of **A** and −**B**, as indicated in Fig. 3–6a and b.

The parallelogram construction may also be used to depict vector subtraction. Again the two vectors are translated until their tail ends coincide. The resultant vector **R** is the diagonal connecting the arrow tip of **B** to the arrow tip of **A**, as shown in Fig. 3–6c. (• If **A** − **B** = **R** then **A** = **B** + **R**. Is this contention verified in Fig. 3–6c?)

Vector addition is *associative* as well as commutative. The associative property is expressed by writing

$$\mathbf{R} = \mathbf{A} + \mathbf{B} + \mathbf{C} = (\mathbf{A} + \mathbf{B}) + \mathbf{C} = \mathbf{A} + (\mathbf{B} + \mathbf{C}) \qquad (3\text{--}3)$$

The first equality, **R** = **A** + **B** + **C**, is an extension of the addition law to three vectors. Refer to Fig. 3–7. First, **B** is translated so that its tail end matches the arrow tip of **A**. Then, **C** is translated so that its tail end matches the arrow tip of **B**. Finally, the tail end of **A** is connected to the arrow tip of **C** to yield **R**. The use of parentheses in the expression (**A** + **B**) + **C** means that we first add **A** and **B**, obtaining an intermediate sum, to which we add **C**. The two algebraic laws expressed by Eqs. 3–2 and 3–3 tell us that when vectors are added, the result is the same for any order of the addition and for any sub-grouping of the added vectors.

Vectors in Physics. In a formal mathematical sense, the vector **C** is equivalent to the sum of the vectors **A** and **B** when **C** = **A** + **B**. It is an important fact concerning the way Nature behaves that this same statement applies to physical vectors. For example, when two forces, $\mathbf{F}_A$ and $\mathbf{F}_B$, are simultaneously applied to a particle, the particle responds precisely as if the single force $\mathbf{F}_C = \mathbf{F}_A + \mathbf{F}_B$ had been applied. This is an experimentally verified fact. This result is suggested, but not guaranteed, by the mathematics of vectors.

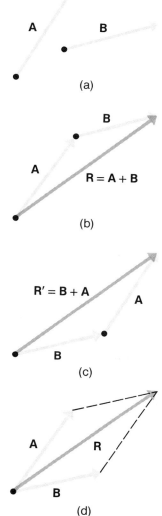

Fig. 3–5. The graphic or geometric addition of two vectors, **A** and **B**, also illustrating the commutative nature of vector addition.

Fig. 3–6. (a) The vectors **A** and **B**. (b) The graphic representation of **R** = **A** − **B** = **A** + (−**B**). (c) The parallelogram representation of **R** = **A** − **B**.

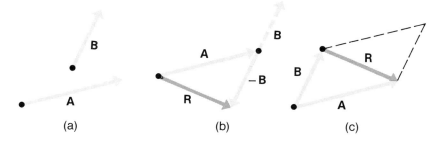

Fig. 3–7. (a) The successive addition of vectors to produce the sum, $\mathbf{R} = \mathbf{A} + \mathbf{B} + \mathbf{C}$. (b) The grouped addition $\mathbf{R} = (\mathbf{A} + \mathbf{B}) + \mathbf{C}$. (c) The grouped addition $\mathbf{A} + (\mathbf{B} + \mathbf{C})$.

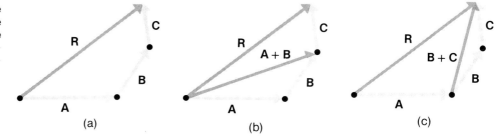

We have no way of knowing whether a particular mathematical result or technique has any applicability to the physical world until it has been established by experimental test. (Albert Einstein once remarked that he did not understand why mathematics, which is purely a product of the mind, is so well suited to describing the real physical world.)

Example 3–1

What is the resultant vector $\mathbf{R} = \mathbf{A} + \mathbf{B}$, if $A = 3$ and $B = 4$, and if $\mathbf{A}$ is perpendicular to $\mathbf{B}$? Give the direction of $\mathbf{R}$ relative to $\mathbf{A}$.

Solution:

First, translate $\mathbf{B}$ so that its tail end matches the arrow tip of $\mathbf{A}$. Then, draw the resultant vector $\mathbf{R}$, as shown in diagram (b). The Pythagorean theorem gives $R^2 = A^2 + B^2$ or $R = \sqrt{3^2 + 4^2} = 5$. Because $\mathbf{A}$ and $\mathbf{B}$ are at right angles, $\tan \theta = B/A = 4/3$, so that $\theta = 53.13°$.

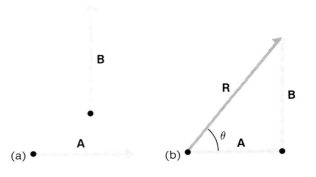

Example 3–2

If the vectors $\mathbf{A}$ and $\mathbf{B}$ $(A = 3, B = 4)$ have an included angle of $\alpha = 60°$ between them, what is the resultant, $\mathbf{R} = \mathbf{A} + \mathbf{B}$?

Solution:

The resultant vector $\mathbf{R}$ is constructed as in diagram (b). Using the law of cosines (see Eq. A–37 in Appendix A), we have

$$R^2 = A^2 + B^2 + 2AB \cos \alpha$$
$$= 3^2 + 4^2 + 24 \cos 60° = 37$$

Therefore, $R = \sqrt{37} = 6.08$

From the law of sines (see Eq. A–36 in Appendix A), we have

$$\frac{B}{\sin \theta} = \frac{R}{\sin (180° - \alpha)} = \frac{R}{\sin \alpha}$$

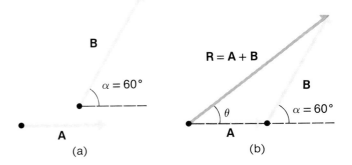

or $\sin \theta = \dfrac{B}{R} \sin \alpha = \dfrac{4}{\sqrt{37}} \sin 60° = 0.569$

from which $\theta = 34.7°$

3–3 VECTOR COMPONENTS

In some types of problems, there is a *preferred* direction in space—for example, the direction of motion or of a force or of an electric field. It is often necessary to determine the *component* of a vector **A** in the preferred direction (called the *projection* of **A** onto this axis). To do this, construct a unit vector **û** at the origin of the vector **A** and in the preferred direction. Then, the component of **A** in the direction of **û** is defined to be

$$A_u = A \cos \theta_u \tag{3-4}$$

where θ_u is the angle between **û** and **A** measured in the plane defined by these vectors. We also define the *vector component* of **A** in the direction of **û** to be

$$\mathbf{A}_u = (A \cos \theta_u)\mathbf{\hat{u}} \tag{3-5}$$

Note carefully that A_u has associated with it an algebraic sign; A_u is *not* simply the magnitude of $\mathbf{A}_u$. Figure 3–8 shows two cases illustrating this point. In Fig. 3–8a we see a vector **A** that has a vector component $\mathbf{A}_u$ in the direction of **û**; here, $A_u > 0$. In Fig. 3–8b we see a vector **B** that has a vector component $\mathbf{B}_u$ in the direction opposite to **û**; here $B_u < 0$. The proper sign of the component is always given by the term $\cos \theta_u$.

For a vector **A**, we have defined the following quantities:

magnitude of **A**: $A = |\mathbf{A}|$, a nonnegative real quantity

component of **A**: A_u, a (signed) quantity

vector component of **A**: $\mathbf{A}_u$, a vector quantity in the direction of **û**

A particular vector is *zero* only if its component in every direction vanishes.

Vectors in Two Dimensions. To describe general two-dimensional vector operations, it is convenient (although not essential) to introduce a set of coordinate axes. We define dimensionless unit vectors, **î** and **ĵ**, that are directed parallel to the positive x- and y-axes, respectively, as indicated in Fig. 3–9. Then, a vector **A** has vector components in the x- and y-directions given by

$$\mathbf{A}_x = A_x\mathbf{\hat{i}} = (A \cos \theta_x)\mathbf{\hat{i}}$$
$$\mathbf{A}_y = A_y\mathbf{\hat{j}} = (A \cos \theta_y)\mathbf{\hat{j}}$$

where θ_x and θ_y are the angles between **A** and **î** and between **A** and **ĵ**, respectively (Fig. 3–9a).

Because we have chosen x- and y-axes that are mutually perpendicular in two dimensions, we have, in Fig. 3–9a, $\theta_x + \theta_y = \pi/2$. Then, $\cos \theta_y = \cos (\pi/2 - \theta_x) = \sin \theta_x$. Thus, we need to specify only the angle θ_x, which we write without a subscript. That is, the vector components are

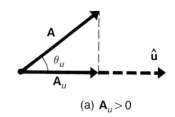

(a) $A_u > 0$

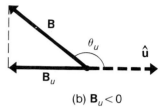

(b) $B_u < 0$

Fig. 3–8. The vector components in the direction of û, illustrated for two different vectors.

Fig. 3–9. Resolution of the vector **A** into rectangular vector components, $\mathbf{A}_x = A_x\mathbf{\hat{i}}$ and $\mathbf{A}_y = A_y\mathbf{\hat{j}}$. In (a), $A_x = A \cos \theta_x$ and $A_y = A \cos \theta_y$. In (b), $A_x = A \cos \theta$ and $A_y = A \sin \theta$. The angle θ is positive when measured in the counterclockwise sense, as shown. The unit vectors **î** and **ĵ** indicate only the directions of the x- and y-axes. These vectors can be freely translated to any point where the tail end of **A** might be located; this in no way influences the calculation of the vector components.

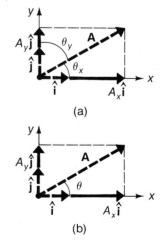

$$\mathbf{A}_x = A_x\hat{\mathbf{i}} = (A \cos \theta)\hat{\mathbf{i}}$$
$$\mathbf{A}_y = A_y\hat{\mathbf{j}} = (A \sin \theta)\hat{\mathbf{j}}$$

(3–6)

It is evident from Fig. 3–9 that

$$\mathbf{A} = \mathbf{A}_x + \mathbf{A}_y = A_x\hat{\mathbf{i}} + A_y\hat{\mathbf{j}}$$

(3–7)

Thus, Eqs. 3–6 and 3–7 are *inverse transformations,* one set giving the vector components of **A** and the other giving the vector **A** in terms of the components.

We also have, from Fig. 3–9,

$$A^2 = A_x^2 + A_y^2$$

(3–8)

and

$$\theta = \tan^{-1} \frac{A_y}{A_x}$$

(3–9)

The signs of A_x and A_y determine the quadrant in which θ lies. (Notice that Eq. 3–8 also follows from Eqs. 3–6 when we use the trigonometric identity, $\sin^2 \theta + \cos^2 \theta = 1$.)

By applying these ideas, we arrive at a simple method for computing analytically the sum of any number of given vectors. Consider Fig. 3–10, which shows the vector addition, $\mathbf{R} = \mathbf{A} + \mathbf{B}$. In terms of the x- and y-components of **R**, we have

$$\mathbf{R}_x = R_x\hat{\mathbf{i}} = (A_x + B_x)\hat{\mathbf{i}}$$
$$\mathbf{R}_y = R_y\hat{\mathbf{j}} = (A_y + B_y)\hat{\mathbf{j}}$$

(3–10)

so that

$$\mathbf{R} = R_x\hat{\mathbf{i}} + R_y\hat{\mathbf{j}} = (A_x + B_x)\hat{\mathbf{i}} + (A_y + B_y)\hat{\mathbf{j}}$$

(3–11)

Also,

$$R^2 = (A_x + B_x)^2 + (A_y + B_y)^2$$

(3–12)

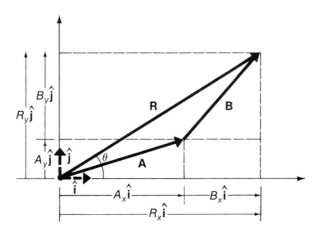

Fig. 3–10. A graphic representation of the vector addition $\mathbf{R} = \mathbf{A} + \mathbf{B}$, employing the components (A_x, A_y) and (B_x, B_y). We have $R_x = A_x + B_x$ and $R_y = A_y + B_y$.

and

$$\theta = \tan^{-1} \frac{A_y + B_y}{A_x + B_x}$$

(3–13)

Any number of coplanar vectors can be added by a straightforward extension of these equations.

Example 3–3

A ship travels a straight course (using land-based references) of 130 km directed 60° north of east. How far north and east of the starting point does the ship travel?

Solution:

Orient the x- and y-axes so that x corresponds to east and y to north. We then identify A_E with A_x and A_N with A_y. Thus,

$A_E = A \cos 60° = (130 \text{ km}) \cos 60° = 65.0 \text{ km}$

$A_N = A \sin 60° = (130 \text{ km}) \sin 60° = 112.6 \text{ km}$

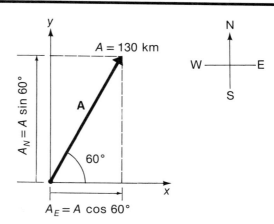

Example 3–4

In flat, open country a hiker travels due east for 5.12 km and then turns due south, proceeding for another 3.87 km before resting. Locate this rest position relative to the starting point.

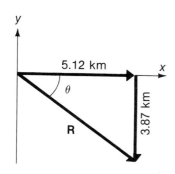

Solution:

We have

$$R^2 = (5.12 \text{ km})^2 + (-3.87 \text{ km})^2$$

or $R = \sqrt{(5.12)^2 + (-3.87)^2} \text{ km} = 6.42 \text{ km}$

$$\theta = \tan^{-1} \left(\frac{-3.87 \text{ km}}{5.12 \text{ km}} \right)$$

Because $A_x > 0$ and $A_y < 0$, the angle θ lies in the fourth quadrant. Thus,

$$\theta = -37.1°, \text{ south of east} \qquad (\text{or } +322.9°)$$

(• You might find it worthwhile to solve this problem graphically. Try a scale of 1 cm = 0.5 km. Try to estimate the accuracy of your results before comparing with the analytic solution.)

Example 3–5

Next, we return to Example 3–2 and obtain the solution in terms of the components of the two vectors. It is convenient to select a two-dimensional rectangular coordinate system with the x-axis in the direction of **A**. (Vectors **A** and **B** are both in the x-y plane.) In problems of this sort, with no additional information given, we are free to select the most convenient coordinate frame.

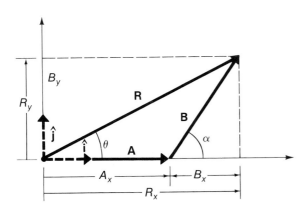

Solution:

According to the construction in the diagram, we have

$$R_x = A_x + B_x = A + B \cos \alpha = 3 + 4 \cos 60° = 5$$

$$R_y = A_y + B_y = 0 + B \sin \alpha = 4 \sin 60° = 3.46$$

Then, $\quad R = \sqrt{R_x^2 + R_y^2} = \sqrt{(5)^2 + (3.46)^2} = 6.08$

Finally, $\qquad \theta = \tan^{-1} \dfrac{R_y}{R_x} = \tan^{-1}\left(\dfrac{3.46}{5}\right)$

Because $A_x > 0$ and $A_y > 0$, the angle θ lies in the first quadrant. Thus,

$$\theta = 34.7°$$

The results in the two examples agree, as expected, but it is important to note that the introduction of the coordinate frame was *not* a necessity but *only* a convenience (particularly so if you did not remember the laws of cosines and sines!). This convenience would have become even more evident if we had the task of adding three or more vectors. (• What do you suppose would be the result of selecting the axes rotated by, say, 15° clockwise? See Problem 3–7.)

Vectors in Three Dimensions. The previous discussion of components and vector components is readily generalized to three dimensions. If we are required to add two vectors (no matter how they are oriented in space), we can translate one vector by parallel displacement so that its tail end matches the arrow tip of the other vector. These two vectors now define a plane, and a rectangular coordinate frame constructed in that plane permits us to use the methods we have just established for analyzing vectors in two dimensions. However, given three or more arbitrary vectors, the selection of a plane containing all the vectors is not always possible, and to add such vectors it is convenient to use a three-dimensional coordinate system.

We again make use of a rectangular or Cartesian coordinate frame and introduce the additional unit vector $\hat{\mathbf{k}}$ along the direction of the z-axis, as shown in Fig. 3–11. Then, the vector $\mathbf{A}$ is expressed as

$$\mathbf{A} = A_x\hat{\mathbf{i}} + A_y\hat{\mathbf{j}} + A_z\hat{\mathbf{k}} \qquad\qquad \textbf{(3–14)}$$

Fig. 3–11. The vector **A** is resolved into components $A_x\hat{\mathbf{i}}$, $A_y\hat{\mathbf{j}}$, and $A_z\hat{\mathbf{k}}$.

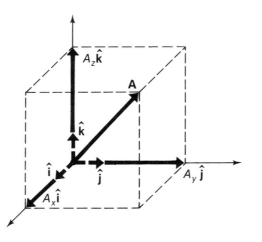

Two vectors are equal if and only if the three components of one are individually equal to the corresponding components of the other. That is, if $A_x = B_x$, $A_y = B_y$, and $A_z = B_z$, then $\mathbf{A} = \mathbf{B}$. Therefore, vector equations such as $\mathbf{A} = \mathbf{B}$, $\mathbf{A} = n\mathbf{B}$, or $\mathbf{A} = 0$ symbolically represent three separate equations involving the individual components.

3–4   RECTANGULAR AND POLAR COORDINATES

To this point we have used only rectangular coordinate systems in our discussions. Because of the symmetric nature of a particular problem—for example, the description of circular motion—it may prove more convenient to use a different type of coordinate system such as *plane polar coordinates*.

Figure 3–12 shows two equivalent ways to locate a point P with respect to a reference point (origin) O. In Fig. 3–12a the rectangular coordinates of the point are (x, y). Figure 3–12b shows the same point located in terms of the polar coordinates (ρ, ϕ), where the angle ϕ is measured counterclockwise from a reference line taken to coincide with the x-axis. The relationships connecting the various coordinates are

$$\begin{aligned} x &= \rho \cos \phi & \rho &= \sqrt{x^2 + y^2} \\ y &= \rho \sin \phi & \phi &= \tan^{-1}(y/x) \end{aligned} \tag{3–15}$$

In polar coordinates, the direction of a vector is specified in terms of the unit vectors, $\hat{\mathbf{u}}_\rho$ and $\hat{\mathbf{u}}_\phi$, as shown in Fig. 3–13. The unit vector $\hat{\mathbf{u}}_\rho$ is directed radially outward from the origin O. The unit vector $\hat{\mathbf{u}}_\phi$ is perpendicular to $\hat{\mathbf{u}}_\rho$ and is in the direction of increasing ϕ. An arbitrary vector $\mathbf{A}$ can always be written in terms of unit vectors. For rectangular and polar coordinates, we have

$$\begin{aligned} \mathbf{A} &= A_x \hat{\mathbf{i}} + A_y \hat{\mathbf{j}} \\ \mathbf{A} &= A_\rho \hat{\mathbf{u}}_\rho + A_\phi \hat{\mathbf{u}}_\phi \end{aligned} \tag{3–16}$$

The quantity A_ρ is the *radial component* of $\mathbf{A}$; A_ϕ is the *azimuthal component*. The squared magnitude of a vector is given by the Pythagorean rule; thus,

$$A^2 = A_\rho^2 + A_\phi^2 \tag{3–17}$$

The Position Vector. One of the most important physical vectors is the *position vector* $\mathbf{r}$. One major objective of mechanics is to describe the motions of particles under various influences. This may be accomplished by introducing a coordinate frame with an

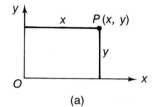
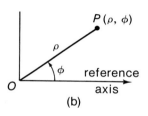

Fig. 3–12. (a) The point P is located with respect to the origin O by the coordinates (x, y). (b) The same point P is located in terms of the polar coordinates (ρ, ϕ).

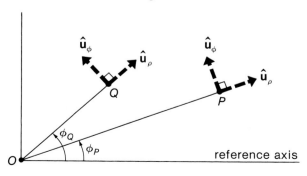

Fig. 3–13. Unit polar vectors at points P and Q. Notice that $\hat{\mathbf{u}}_\rho$ and $\hat{\mathbf{u}}_\phi$ have different directions for different points. However, $\hat{\mathbf{u}}_\rho$ and $\hat{\mathbf{u}}_\phi$ are always mutually perpendicular.

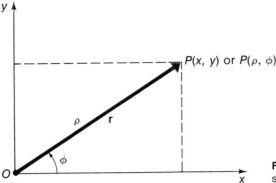

Fig. 3–14. The location P of a particle is specified by the position vector **r**.

origin and oriented axes. The vector that describes the location of a particle with respect to the origin O is the position vector **r**, obtained by placing the tail end at the origin and the arrow tip at the particle's coordinate $P(x, y)$ or $P(\rho, \phi)$, as shown in Fig. 3–14. The vector **r** is therefore fixed by these two points, and, in a sense, is only a different way of quoting $P(x, y)$ or $P(\rho, \phi)$ relative to the chosen coordinate frame.

Using either rectangular or polar components the position vector **r** would be written as

$$\mathbf{r} = x\hat{\mathbf{i}} + y\hat{\mathbf{j}}$$

or $\quad \mathbf{r} = \rho\hat{\boldsymbol{\mu}}_\rho$

(3–18)

QUESTIONS

3–1   Define or describe briefly the terms (a) scalar, (b) vector, (c) unit vector, and (d) components of a vector.

3–2   What can you conclude about two vectors if the resultant of their vector sum is zero?

3–3   The vector sum of three vectors gives a zero resultant. What can you conclude about the vectors?

3–4   Can a vector be zero if one of its components is nonzero?

3–5   What is the magnitude of the vector sum of three vectors $\mathbf{A} = a\hat{\mathbf{i}}$, $\mathbf{B} = a\hat{\mathbf{j}}$, and $\mathbf{C} = a\hat{\mathbf{k}}$?

3–6   The passage of time proceeds in a definite direction, from the past toward the future. Since a magnitude is also associated with time, do these properties permit us to classify time as a vector?

PROBLEMS

Section 3–2

3–1   A car is driven due west a distance of 30 km and then due south a distance of 50 km. Make an appropriate vector diagram, and use graphic means to determine the car's displacement from the starting point.

3–2   The vector **S** has a magnitude of 2.4 units and is directed due north. The vector **T** has a magnitude of 1.6 units and

is directed due west. Find the direction and magnitude of (a) the vector $\mathbf{S} + \mathbf{T}$ and (b) the vector $\mathbf{S} - 3\mathbf{T}$.

3–3   A vector of magnitude 3.6 units is added to a vector of magnitude 5.2 units, with an angle of 60° between the two vectors. (a) Find the magnitude and direction of the resultant vector by a graphic construction. (b) What is the magnitude and direction of a third vector which if added to the given two vectors would yield zero?

Section 3–3

3–4 Repeat Problem 3–1 using analytic means based on Eqs. 3–12 and 3–13. Compare the results.

3–5 The vector **C** has a magnitude $|\mathbf{C}| = 3$ units and a direction with respect to the unit vector $\hat{\mathbf{u}}$ specified by $\theta_u = 227°$, measured counterclockwise from $\hat{\mathbf{u}}$ to **C**. What is the component C_u?

3–6 Given two vectors, $\mathbf{a} = 3\hat{\mathbf{i}} + 2\hat{\mathbf{j}}$ and $\mathbf{b} = -\hat{\mathbf{i}} + 7\hat{\mathbf{j}}$, find a third vector **c** such that $\mathbf{a} + \mathbf{b} + \mathbf{c} = 0$.

3–7 Repeat Example 3–5 but take the coordinate frame orien-
• ted as shown in the diagram. Is either the magnitude of the resultant or the angle between the resultant and the vector **A** different? Comment.

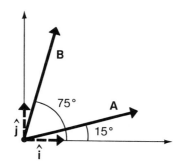

3–8 The vector **A** has a magnitude of 5 units and makes an angle of 20° with the x-axis. The vector **B** has a magnitude of 4 units and makes an angle of 50° with the x-axis. Find $\mathbf{C} = \mathbf{A} + \mathbf{B}$.

3–9 The addition of two vectors, **a** and **b**, yields a resultant **s** such that $\mathbf{s} = \mathbf{a} + \mathbf{b}$. What are the possible relationships of the original two vectors?

3–10 Given that $\mathbf{A} = 2\hat{\mathbf{i}} + 3\hat{\mathbf{j}}$ and $\mathbf{B} = 3\hat{\mathbf{i}} - 4\hat{\mathbf{j}}$, find the magnitudes and directions of (a) $\mathbf{C} = \mathbf{A} + \mathbf{B}$ and (b) $\mathbf{D} = 3\mathbf{A} - 2\mathbf{B}$.

3–11 Two vectors, **A** and **B**, lie in the x-y plane and are perpen-
•• dicular to each other. Show that their components are related by the expression, $A_x B_x = -A_y B_y$.

3–12 Find the angle between the two vectors $\mathbf{A} = 5\hat{\mathbf{i}} + \hat{\mathbf{j}}$ and $\mathbf{B} = -2\hat{\mathbf{i}} + 4\hat{\mathbf{j}}$.

3–13 What is the unit vector in the direction of the vector $\mathbf{A} = 4\hat{\mathbf{i}} + 3\hat{\mathbf{j}}$?

3–14 What is the component of the vector $\mathbf{A} = 2\hat{\mathbf{i}} + 3\hat{\mathbf{j}}$ in the direction of the unit vector $\hat{\mathbf{u}} = \dfrac{1}{\sqrt{5}}(2\hat{\mathbf{i}} + \hat{\mathbf{j}})$?

3–15 Given the vectors $\mathbf{A} = 3\hat{\mathbf{i}} - 2\hat{\mathbf{j}}$ and $\mathbf{B} = 4\hat{\mathbf{i}} + 3\hat{\mathbf{j}}$, determine the projection of $\mathbf{C} = \mathbf{A} + \mathbf{B}$ on the x-axis.

3–16 Given the vectors $\mathbf{A} = 2\hat{\mathbf{i}} + 3\hat{\mathbf{j}}$, $\mathbf{B} = 3\hat{\mathbf{i}} - 2\hat{\mathbf{j}}$, and $\mathbf{C} =$ $4\hat{\mathbf{i}} - 3\hat{\mathbf{j}}$, find the vector **D** such that $\mathbf{A} + \mathbf{B} + \mathbf{C} + \mathbf{D} = 0$.

3–17 Find the vector sum of the three vectors,

$$\mathbf{a} = -3\hat{\mathbf{i}} - 2\hat{\mathbf{j}} + 7\hat{\mathbf{k}}$$

$$\mathbf{b} = \hat{\mathbf{i}} + 3\hat{\mathbf{j}} + 3\hat{\mathbf{k}}$$

$$\mathbf{c} = -5\hat{\mathbf{k}}$$

Section 3–4

3–18 The point $P(x, y)$ is located in the coordinate frame S by
• $x = 6$ and $y = 3$. (a) What is the position vector **r** of this point? (b) A second coordinate frame, S', is constructed with its origin at $(x, y) = (3, 4)$ and with its x' and y' axes parallel to the x and y axes of S. What is the position vector, $\mathbf{r}'$, of the point P in S'? (c) A third coordinate frame, S'', is constructed with its origin at $(x, y) = (0, 0)$ and with its x'' axis rotated 30° counterclockwise from the x axis of S. What is the position vector, $\mathbf{r}''$, of the point P in S''?

3–19 The point $Q(x, y)$ is given in the coordinate frame S by
• $x = 9$ and $y = 7$. For each of the coordinate frames indicated in Problem 3–18, find the displacement vector $\overline{PQ} = \Delta\mathbf{r}$. Compare their magnitudes and directions.

3–20 Two sets of magnets are arranged along the sides of a pinball machine. The force exerted on the ball by one set of magnets is $\mathbf{A}(\mathbf{r}) = (a/x^2)\hat{\mathbf{i}} + b\hat{\mathbf{j}}$ and $\mathbf{B}(\mathbf{r}) = (a/y^2)\hat{\mathbf{i}} - b\hat{\mathbf{j}}$ by the other set. The constants a and b are suitable dimensional quantities. What is the total force $\mathbf{A} + \mathbf{B}$ exerted on the ball at any position $\mathbf{r} = x\hat{\mathbf{i}} + y\hat{\mathbf{j}}$? (A common way of indicating that quantities such as **A** and **B** depend on the position **r** is to write $\mathbf{A}(\mathbf{r})$ and $\mathbf{B}(\mathbf{r})$.)

3–21 The position function of a particle at time t is given by the expression $\mathbf{r} = \mathbf{a} + \mathbf{b}t$, where **a** and **b** are constant vectors. (a) Describe the motion of the particle. (b) What are the dimensions of the constants a and b?

3–22 A particle is located at the point $(x, y) = (3$ m, 5 m$)$. Locate the particle in plane polar coordinates.

3–23 A particle is located by the position vector $\mathbf{r} = (2$ m$)\hat{\mathbf{u}}_p$ and $\phi = 35°$. What are the rectangular coordinates of the particle?

3–24 Quito, Ecuador, and Kampala, Uganda, both lie approximately on the Equator. The longitude of Quito is 78°30′W and that of Kampala is 32°30′E. What is the distance from Quito to Kampala (a) along the shortest surface path and (b) along a direct (through-the-Earth) path? (The radius of the Earth is $6.38 × 10^6$ m.)

Additional Problems

3–25 Two particles are located at $\mathbf{r}_1 = 3\hat{\mathbf{i}} + 7\hat{\mathbf{j}}$ and $\mathbf{r}_2 = -2\hat{\mathbf{i}} + 3\hat{\mathbf{j}}$, respectively. Find the vector distance from the particle labeled 1 to the particle labeled 2. Give both the magnitude of the vector and its orientation with respect to the x-axis. [*Hint:* Find $\mathbf{r} = \mathbf{r}_1 - \mathbf{r}_2$.]

3–26 What is the relationship of three vectors, $\mathbf{A}$, $\mathbf{B}$, and $\mathbf{C}$, if $\mathbf{C} = \mathbf{A} + \mathbf{B}$ and $C^2 = A^2 + B^2$?

3–27 Vector $\mathbf{A}$ has a magnitude of 6 units and vector $\mathbf{B}$ has a magnitude of 5 units. What is the angle between $\mathbf{A}$ and $\mathbf{B}$ if the magnitude of $\mathbf{A} - \mathbf{B}$ is 4 units? [*Hint:* Let $\mathbf{A}$ or $\mathbf{B}$ be directed along the x-axis.]

3–28 The vector $\mathbf{A}$ has a magnitude of 4 units and makes an angle (measured counterclockwise) of $40°$ with respect to the x-axis. The vector $\mathbf{B}$ has a magnitude of 6 units and makes an angle of $125°$ with respect to the x-axis. Determine the magnitude and direction of $\mathbf{C} = \mathbf{A} + \mathbf{B}$.

3–29 Three vectors $\mathbf{A}$, $\mathbf{B}$, and $\mathbf{C}$, each have a magnitude of 10 units. The vectors make angles with respect to the x-axis (measured counterclockwise) of $\theta_A = 60°$, $\theta_B = 180°$, and $\theta_C = 300°$. Determine the magnitude and direction of (a) $\mathbf{A} + \mathbf{B} + \mathbf{C}$, (b) $\mathbf{A} - \mathbf{B} + \mathbf{C}$, and (c) $2\mathbf{A} + \mathbf{B} + 2\mathbf{C}$.

3–30 What is the geometric condition for three vectors with magnitudes of 7, 24, and 25 units to add to zero?

3–31 A regulation baseball diamond has the dimensions shown in the diagram, above right. During a game, the pitcher throws a fastball directly over home plate; the batter hits it directly to the second baseman (standing on second base), who throws the ball to the first baseman, completing a double play. Assume that the ball travels at all times in a horizontal plane. Calculate the total distance traveled, and the displacement from the origin at the pitcher's mound, for each point at which the ball changes direction. [*Hint:* The pitcher's mound is not at the center of the diamond.]

3–32 A rectangular room has the dimensions 4 m × 5 m × 6 m, as shown in the figure. A bug starting at point A in the corner crawls to point B along a straight-line path on the floor, then crawls up the wall from B to the corner

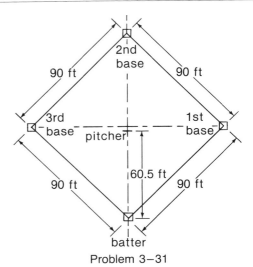

Problem 3–31

point C, also along a straight-line path. (a) What is the value of x such that the total path A to C is a minimum? [*Hint:* You may use either the calculus or geometric considerations based on a "developed plan" of the room with the walls folded onto the plane of the floor.] (b) A second insect flies directly from A to C. How much shorter is its trip than the minimum for the crawling bug?

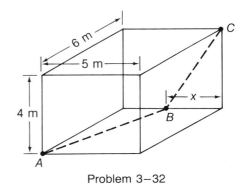

Problem 3–32

CHAPTER 4

MOTION IN A PLANE

In the discussion of kinematics in Chapter 2 we considered motion along a straight line. With the aid of vector methods we are now prepared to extend the discussion to motion in two and three dimensions. New and important features of motion are revealed when we proceed from one-dimensional motion to two-dimensional motion. However, when we proceed further to three-dimensional motion, we learn little new physics but must use more complicated mathematics to describe the motion. Accordingly, we concentrate now on the kinematics of two-dimensional motion.

It often happens that motion in space actually takes place in a plane. For example, a thrown ball moves in a vertical plane. In such cases we orient the axes of our coordinate frame so that one of the planes—*x-y*, *y-z*, or *x-z*—corresponds to the plane of motion. Then the problem is reduced to two dimensions.

We consider here the motion of (pointlike) particles, but the results are also applicable to rigid objects with dimensions—blocks, balls, and similar objects—if we restrict attention to *translational* motion and omit (for the moment) any consideration of *rotational* effects. Pure translation is illustrated in Fig. 4–1a. In this case, the orientation of the body remains fixed as the motion proceeds. That is, the two reference points, *P* and *R*, travel along parallel paths (the dashed and solid curves, respectively, in Fig. 4–1a). Knowledge of the motion of one point—for example, *R*—determines completely the motion of any other point, such as *P*, and thus determines the motion of the body as a whole. *In this sense,* the object moves as if it were a particle.

Fig. 4–1. Two-dimensional motion of a body (a) with only translational motion and (b) with both translational and rotational motion. The inset shows how the rotational motion produces changes in the relative orientation of two points, *R* and *P*, on the body.

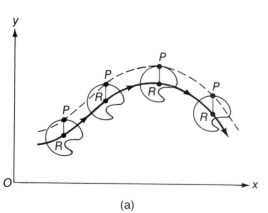

(a)

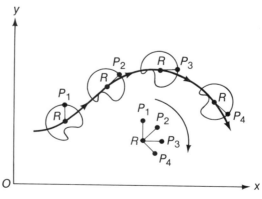

(b)

45

In Fig. 4–1b we show a case of motion with both translation and rotation. Here, the orientation of point P with respect to R changes continually because of the rotation. The inset shows this rotational aspect of the motion. The motion of point R alone no longer determines the motion of the body as a whole. This complication is treated in later chapters.

4–1   THE POSITION VECTOR

To describe the motion of a particle, we first adopt a convenient coordinate frame. At any particular instant the position of the particle is specified by the position vector $\mathbf{r}(t)$. As described in Section 3–4, we construct this vector with its tail end at the origin of the coordinate frame and its arrow tip at the instantaneous position of the particle. As the particle moves, the position vector (the tip) defines the path or *trajectory* of the particle with respect to the chosen coordinate origin. Figure 4–2 illustrates a particular trajectory in a Cartesian coordinate frame.

Quoting the position vector (or position function) $\mathbf{r}(t)$ is equivalent to specifying the x and y coordinates of the particle as functions of time, with

$$\mathbf{r}(t) = x(t)\hat{\mathbf{i}} + y(t)\hat{\mathbf{j}} \qquad (4-1)$$

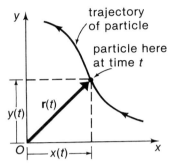

Fig. 4–2. The use of the position vector $\mathbf{r}(t)$ to describe the path followed by a particle.

We are interested in following the changes in $\mathbf{r}(t)$ as the particle proceeds along its trajectory. Consider a particle that moves in the x-y plane, as shown in Fig. 4–3. At a particular instant t_1, the particle is at the point P given by $\mathbf{r}(t_1)$; at a later time t_2, the particle is at point Q given by $\mathbf{r}(t_2)$. During the time interval $t_2 - t_1$, the particle undergoes a displacement corresponding to*

$$
\begin{aligned}
\mathbf{r}_{12} &= \mathbf{r}(t_2) - \mathbf{r}(t_1) \\
&= [x(t_2)\hat{\mathbf{i}} + y(t_2)\hat{\mathbf{j}}] - [x(t_1)\hat{\mathbf{i}} + y(t_1)\hat{\mathbf{j}}] \\
&= [x(t_2) - x(t_1)]\hat{\mathbf{i}} + [y(t_2) - y(t_1)]\hat{\mathbf{j}} \qquad (4-2)
\end{aligned}
$$

We can also imagine $t_2 - t_1$ to be a small incremental quantity Δt. The corresponding

---

*The double subscript on $\mathbf{r}$ is used to identify the initial and final points, respectively, or the displacement, either through the time parameter, as here, or through the spatial designations, P and Q. Thus, we could equally well write $\mathbf{r}_{12}$ or $\mathbf{r}_{PQ}$.

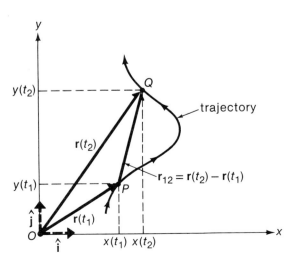

Fig. 4–3. The displacement $\mathbf{r}_{12}$ of a particle executing planar motion during the time interval $t_2 - t_1$.

incremental displacement is then $\Delta\mathbf{r}_{12}$ or, simply, $\Delta\mathbf{r}$.

$$\Delta\mathbf{r} = \Delta x\hat{\mathbf{i}} + \Delta y\hat{\mathbf{j}} \qquad (4-3)$$

This quantity is called the *incremental displacement vector* (in two dimensions).

Example 4-1

Suppose that the trajectory of a particle is $\mathbf{r}(t) = x(t)\hat{\mathbf{i}} + y(t)\hat{\mathbf{j}}$ with $x(t) = at^2 + bt$ and $y(t) = ct + d$, where a, b, c, and d are constants that have appropriate dimensions. What displacement does the particle undergo between $t = 1$ s and $t = 3$ s?

Solution:

Using Eq. 4-2, we have

$$\Delta\mathbf{r} = [x(3) - x(1)]\hat{\mathbf{i}} + [y(3) - y(1)]\hat{\mathbf{j}}$$

Assume for the sake of simplicity that the constant a is expressed in m/s^2, b and c in m/s, and d in m.

Then,
$$x(3) = 9a + 3b \qquad y(3) = 3c + d$$
$$x(1) = a + b \qquad y(1) = c + d$$

Hence,

$$\Delta\mathbf{r} = (8a + 2b)\hat{\mathbf{i}} + 2c\hat{\mathbf{j}} \text{ meters}$$

4-2   THE VELOCITY VECTOR

Suppose that the position vector of a particle is $\mathbf{r}(t_1)$ at time t_1 and is $\mathbf{r}(t_2)$ at time t_2. During the time interval $t_2 - t_1$ the particle undergoes a displacement, $\mathbf{r}_{12} = \mathbf{r}(t_2) - \mathbf{r}(t_1)$. In analogy with the definition of average velocity for straight-line motion (Eq. 2-1), we now define the *average vector velocity* during this time interval to be

$$\bar{\mathbf{v}} = \frac{\mathbf{r}(t_2) - \mathbf{r}(t_1)}{t_2 - t_1} = \frac{\mathbf{r}_{12}}{t_2 - t_1} \qquad (4-4)$$

Notice that $\bar{\mathbf{v}}$ is indeed a vector quantity because it involves the multiplication of a vector, $\mathbf{r}_{12}$, by a scalar, $1/(t_2 - t_1)$. The magnitude of $\bar{\mathbf{v}}$ is $|\mathbf{r}_{12}|/(t_2 - t_1)$ and the direction of $\bar{\mathbf{v}}$ is the same as that of $\mathbf{r}_{12}$. Of course, $\bar{\mathbf{v}}$ and $\mathbf{r}_{12}$ have different dimensions; nevertheless, we find it convenient on occasion to draw velocity vectors in the same figures that represent trajectories in coordinate space, as in Fig. 4-4.

We can rewrite Eq. 4-4 in the form

$$\bar{\mathbf{v}} = \frac{\mathbf{r}(t + \Delta t) - \mathbf{r}(t)}{\Delta t} = \frac{\Delta\mathbf{r}}{\Delta t} \qquad (4-5)$$

Aside from shifting emphasis, this equation has precisely the same meaning as Eq. 4-4 if we identify t with t_1, Δt with $t_2 - t_1$, and $\Delta\mathbf{r}$ with $\mathbf{r}_{12}$. The (instantaneous) *vector velocity* is now defined as the general time derivative,

$$\mathbf{v} = \lim_{\Delta t \to 0} \frac{\mathbf{r}(t + \Delta t) - \mathbf{r}(t)}{\Delta t} = \frac{d\mathbf{r}}{dt} = \frac{d}{dt}\mathbf{r}(t) \qquad (4-6)$$

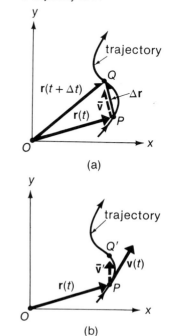

Fig. 4-4. (a) The finite displacement, $\overline{PQ} = \Delta\mathbf{r}$, and the average velocity $\bar{\mathbf{v}}$ (dashed line). (b) As the limit in Eq. 4-6 is developed, the point Q approaches P without limit. The velocity $\mathbf{v}(t)$ is tangent to the trajectory at P.

Figure 4–4 illustrates Eqs. 4–5 and 4–6. We see that **v** is always tangent to the trajectory at the point that locates the particle position. It is also evident from this figure that **v** is in general a function of time. Thus, at the time t we express the position vector of the particle as **r**(t) and the velocity as **v**(t). When we are interested only in the magnitude of **v**(t), we write $|\mathbf{v}(t)|$ or $v(t)$, which we call the *speed* of the particle. (Note that $|\mathbf{v}(t)| \geqslant 0$.)

We wish to emphasize that **r**(t) can vary with time *both* in magnitude and in direction. The velocity is a measure of both of these possible changes. Thus, for example, a particle that travels in a circular path with a radius R and centered at the origin has $|\mathbf{r}(t)| = R$, which is a constant, yet the particle has a nonzero velocity. We return to a discussion of this interesting type of motion later in this chapter.

The velocity vector can always be given in component form. Thus,

$$\mathbf{v}(t) = \frac{d}{dt}\mathbf{r}(t) = \frac{d}{dt}[x(t)\hat{\mathbf{i}} + y(t)\hat{\mathbf{j}}]$$

Now, for rectangular coordinates, the unit vectors $\hat{\mathbf{i}}$ and $\hat{\mathbf{j}}$ do not change with time; hence,

$$\mathbf{v}(t) = \frac{dx}{dt}\hat{\mathbf{i}} + \frac{dy}{dt}\hat{\mathbf{j}}$$

$$= v_x\hat{\mathbf{i}} + v_y\hat{\mathbf{j}} \tag{4–7}$$

where the velocity components are

$$v_x = \frac{dx}{dt} \quad \text{and} \quad v_y = \frac{dy}{dt}$$

The vector components of the velocity are

$$\mathbf{v}_x = \frac{dx}{dt}\hat{\mathbf{i}} \quad \text{and} \quad \mathbf{v}_y = \frac{dy}{dt}\hat{\mathbf{j}}$$

Also, for the speed v, we have

$$v = |\mathbf{v}| = \sqrt{v_x^2 + v_y^2} \tag{4–8}$$

Example 4–2

Suppose that the position vector function for a particle is given as $\mathbf{r}(t) = x(t)\hat{\mathbf{i}} + y(t)\hat{\mathbf{j}}$, with $x(t) = at + b$ and $y(t) = ct^2 + d$, where $a = 1$ m/s, $b = 1$ m, $c = \frac{1}{8}$ m/s², and $d = 1$ m.

(a) Calculate the average velocity during the time interval from $t = 2$ s to $t = 4$ s.

(b) Determine the velocity and the speed at $t = 2$ s.

Solution:

(a) For the average velocity, we have

$$\bar{\mathbf{v}} = \left(\frac{x(4) - x(2)}{4 \text{ s} - 2 \text{ s}}\right)\hat{\mathbf{i}} + \left(\frac{y(4) - y(2)}{4 \text{ s} - 2 \text{ s}}\right)\hat{\mathbf{j}}$$

$$\bar{\mathbf{v}} = \left(\frac{5 \text{ m} - 3 \text{ m}}{2 \text{ s}}\right)\hat{\mathbf{i}} + \left(\frac{3 \text{ m} - 1.5 \text{ m}}{2 \text{ s}}\right)\hat{\mathbf{j}}$$

$$= (\hat{\mathbf{i}} + 0.75\,\hat{\mathbf{j}}) \text{ m/s}$$

(b) For the velocity components, we have

$$v_x = \frac{dx}{dt} = a = 1 \text{ m/s}$$

$$v_y = \frac{dy}{dt} = 2ct = (\tfrac{1}{4} \text{ m/s}^2)t$$

Therefore,

$$\mathbf{v} = v_x\hat{\mathbf{i}} + v_y\hat{\mathbf{j}} = (1 \text{ m/s})\hat{\mathbf{i}} + (\tfrac{1}{4} \text{ m/s}^2)t\hat{\mathbf{j}}$$

$$\mathbf{v}(2) = (1 \text{ m/s})\hat{\mathbf{i}} + (0.5 \text{ m/s})\hat{\mathbf{j}}$$

and the speed is

$$|\mathbf{v}|_{t=2s} = \sqrt{(1\ \mathrm{m/s})^2 + (0.5\ \mathrm{m/s})^2} = 1.118\ \mathrm{m/s}$$

The position functions, $x(t)$ and $y(t)$, are shown in diagrams (a) and (b). We can eliminate the time parameter by substituting

$t = (x - b)/a$ into the expression for y. Then, we obtain

$$y = y(x) = \tfrac{1}{8}(x - 1)^2 + 1$$

with x and y both given in meters. This is the equation of a parabola and is illustrated in diagram (c).

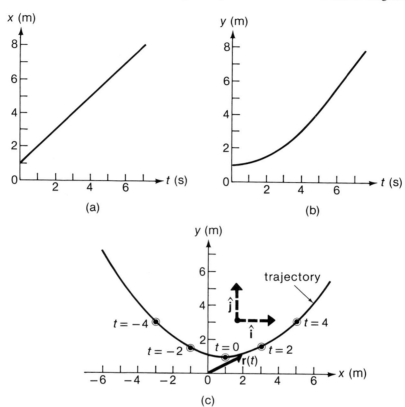

(a)

(b)

(c)

4-3   THE ACCELERATION VECTOR

Suppose that the velocity of a particle at time t is $\mathbf{v}(t)$ and that a time interval Δt later it is $\mathbf{v}(t + \Delta t)$, as shown in Fig. 4-5. We define the *average vector acceleration* during this time interval to be

$$\bar{\mathbf{a}} = \frac{\mathbf{v}(t + \Delta t) - \mathbf{v}(t)}{\Delta t} = \frac{\Delta \mathbf{v}}{\Delta t} \qquad (4\text{-}9)$$

The (instantaneous) *vector acceleration* is obtained by taking the limit of Eq. 4-9 as $\Delta t \rightarrow 0$; thus,

$$\mathbf{a} = \lim_{\Delta t \to 0} \frac{\mathbf{v}(t + \Delta t) - \mathbf{v}(t)}{\Delta t} = \frac{d\mathbf{v}}{dt} = \frac{d}{dt}\mathbf{v}(t) \qquad (4\text{-}10)$$

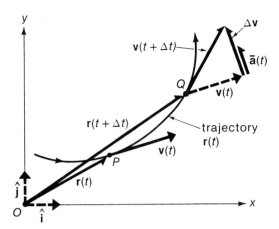

Fig. 4–5. The trajectory of a moving particle. The velocity increment is $\Delta\mathbf{v} = \mathbf{v}(t + \Delta t) - \mathbf{v}(t)$. The average acceleration $\bar{\mathbf{a}}$ is in the same direction as $\Delta\mathbf{v}$.

We again emphasize that $\mathbf{v}(t)$ can change with time *both* in magnitude and in direction. Each of these changes is reflected in the resulting vector $\mathbf{a}$. For example, a particle traveling in a circular path has a velocity vector that continually changes direction (always being tangent to the circular path). Thus, the particle undergoes an acceleration even if the speed remains constant.

Finally, it is useful to introduce the vector components for the acceleration $\mathbf{a}$. If we write

$$\mathbf{v}(t) = v_x(t)\hat{\mathbf{i}} + v_y(t)\hat{\mathbf{j}}$$

then

$$\mathbf{a} = a_x\hat{\mathbf{i}} + a_y\hat{\mathbf{j}}$$
$$= \frac{dv_x}{dt}\hat{\mathbf{i}} + \frac{dv_y}{dt}\hat{\mathbf{j}} \qquad (4\text{--}11)$$

where the acceleration components are

$$a_x = \frac{dv_x}{dt} = \frac{d^2x}{dt^2} \qquad \text{and} \qquad a_y = \frac{dv_y}{dt} = \frac{d^2y}{dt^2}$$

Example 4–3

A particle has the velocity function $\mathbf{v}(t) = at^2\hat{\mathbf{i}} + bt\hat{\mathbf{j}}$, with $a = 3 \text{ m/s}^3$ and $b = -4 \text{ m/s}^2$.

(a) What is the velocity at $t = 1$ s and at $t = 3$ s?

(b) What is the average acceleration during this time interval?

(c) What is the acceleration at $t = 1.5$ s?

Solution:

(a) We have

$$\mathbf{v}(1) = (3\hat{\mathbf{i}} - 4\hat{\mathbf{j}}) \text{ m/s}$$

$$\mathbf{v}(3) = (27\hat{\mathbf{i}} - 12\hat{\mathbf{j}}) \text{ m/s}$$

(b) The average acceleration can be calculated using Eq. 4–9 if we identify $t = 1$ s and $\Delta t = 3$ s $- 1$ s $= 2$ s; that is,

$$\bar{\mathbf{a}} = \frac{\mathbf{v}(3) - \mathbf{v}(1)}{2 \text{ s}} = \frac{(27 - 3)\hat{\mathbf{i}} \text{ m/s}}{2 \text{ s}} - \frac{(12 - 4)\hat{\mathbf{j}} \text{ m/s}}{2 \text{ s}}$$

$$= (12\hat{\mathbf{i}} - 4\hat{\mathbf{j}}) \text{ m/s}^2$$

(c) The acceleration is

$$\mathbf{a}(t) = \frac{d}{dt}\mathbf{v}(t) = 2at\hat{\mathbf{i}} + b\hat{\mathbf{j}}$$

Then,

$$\mathbf{a}(1.5) = (9\hat{\mathbf{i}} - 4\hat{\mathbf{j}}) \text{ m/s}^2$$

4–4   MOTION WITH CONSTANT LINEAR ACCELERATION

In many practical cases we consider a particle moving with a constant acceleration **a**. This means that in an arbitrarily chosen coordinate frame, the components a_x and a_y are both constants. By a straightforward extension of the discussions in Chapter 2, we can obtain the velocity functions:

$$\left.\begin{array}{l} v_x(t) = a_x t + v_{0x} \\ v_y(t) = a_y t + v_{0y} \end{array}\right\} \tag{4–12}$$

where v_{0x} and v_{0y} correspond to the components of the velocity at $t = 0$, namely, $\mathbf{v}_0$. We can also write Eqs. 4–12 in vector form as

$$\mathbf{v}(t) = \mathbf{a}t + \mathbf{v}_0 \tag{4–13}$$

Continuing the parallel development with Chapter 2 gives the position function components,

$$\left.\begin{array}{l} x(t) = \tfrac{1}{2}a_x t^2 + v_{0x} t + x_0 \\ y(t) = \tfrac{1}{2}a_y t^2 + v_{0y} t + y_0 \end{array}\right\} \tag{4–14}$$

or, in vector form,

$$\mathbf{r}(t) = \tfrac{1}{2}\mathbf{a}t^2 + \mathbf{v}_0 t + \mathbf{r}_0 \tag{4–15}$$

In these equations x_0 and y_0 correspond to the components of the particle's coordinate position $\mathbf{r}_0$ at $t = 0$.

The above equations also hold, of course, for $\mathbf{a} = 0$. In this case,

$$\mathbf{r}(t) = \mathbf{v}_0 t + \mathbf{r}_0$$

and the ensuing motion is simple uniform motion in the direction of the constant velocity $\mathbf{v}_0$.

Ballistic Trajectories.*   In any case of constant acceleration, the equations describing the motion will be simplified if we select one of the coordinate axes to be parallel to the direction of the acceleration vector. The motion of a projectile under the influence of gravitational acceleration is typical of this class of problems. Once launched, such objects as baseballs, golf balls, or bullets are subject only to the downward acceleration of gravity, if we neglect air-resistance effects.

Consider the flight path of a golf ball driven from a tee with an initial speed v_0. Orient a Cartesian coordinate frame so that the z-axis is taken to be positive in the upward direction. Thus, $\mathbf{a}_0 = -g\hat{\mathbf{k}}$, with $g = 9.80 \text{ m/s}^2$. The origin is most conveniently taken to be the tee, so that $\mathbf{r}_0 = 0$. Neglecting aerodynamic effects, the golf ball will travel in a vertical plane, which we take to be the y-z plane, as indicated in Fig. 4–6. In selecting the axes in this special way, the three-dimensional problem has been reduced to a two-

The rider and his motorcycle are in a ballistic trajectory. (Courtesy of *Cycle World*.)

---

*The path followed by an unpowered and unguided projectile is called a *ballistic trajectory*. A ballistic flight is one in which the projectile is given an initial boost and is then allowed to coast to its target.

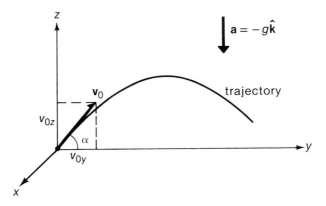

Fig. 4–6. The flight path of a struck golf ball. The initial position (the tee) is at the origin ($\mathbf{r}_0 = 0$); the initial velocity $\mathbf{v}_0$ and the ensuing motion are in the y-z plane. The angle between $\mathbf{v}_0$ and the horizontal plane is α.

dimensional one. The equations of motion become

$$
\left.
\begin{aligned}
y(t) &= v_{0y}t \\
z(t) &= -\tfrac{1}{2}gt^2 + v_{0z}t
\end{aligned}
\right\}
\tag{4–16}
$$

The angle α specifies the angle between the horizontal plane and the direction with which the ball was initially lofted. Therefore, $v_{0y} = v_0 \cos \alpha$ and $v_{0z} = v_0 \sin \alpha$. We can eliminate t from Eqs. 4–16 by substituting $t = y/(v_0 \cos \alpha)$ into the expression for $z(t)$. This yields

$$
z = -\left(\frac{g}{2\,v_0^2 \cos^2 \alpha} \right) y^2 + (\tan \alpha)y
\tag{4–17}
$$

which is the equation of a parabola. Figure 4–7 shows the position functions, $y(t)$ and $z(t)$, as well as the velocity vector at various points along the trajectory.

The time when the golf ball is at the level of the ground (assumed flat) is obtained by setting $z(t) = 0$ in Eq. 4–16. The two solutions are

$$
t = 0 \qquad \text{and} \qquad T = \frac{2v_{0z}}{g} = \frac{2v_0}{g} \sin \alpha
$$

The first solution is the time the ball was driven; the second solution corresponds to the end of the flight. The range R, defined as $R = y(T)$, is obtained either by substituting the value of T into $y(t)$ given by Eq. 4–16 or by setting $z = 0$ in Eq. 4–17. (• Why are these two methods equivalent?) The result is

$$
R = y(T) = \frac{2v_0^2}{g} \sin \alpha \cos \alpha = \frac{v_0^2}{g} \sin 2\alpha
\tag{4–18}
$$

where we have used $\sin 2\alpha = 2 \sin \alpha \cos \alpha$ (Eq. A–34 in Appendix A).

For a given initial velocity v, what angle α will produce the longest drive (that is, the largest R)? Evidently this occurs when $\sin 2\alpha = 1$, so that $2\alpha = \pi/2$. Thus, $\alpha = \pi/4$ (or 45°) gives the maximum range, namely,

$$
R_m = \frac{v_0^2}{g}
\tag{4–19}
$$

The correct description of the trajectories of projectiles (cannon balls) was first given by Galileo. He deduced that the maximum range would be achieved for $\alpha = 45°$ (neglecting air-resistance effects).

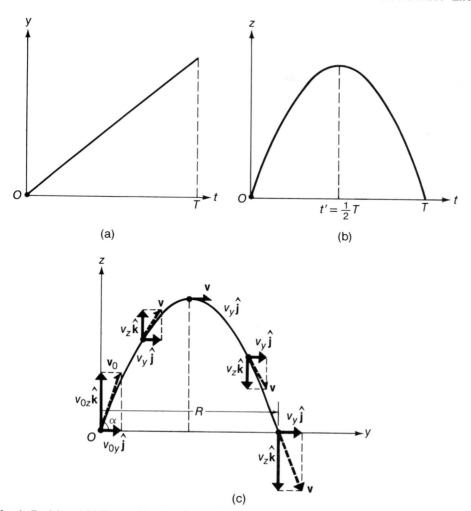

Fig. 4–7. (a) and (b) The position functions, $y(t)$ and $z(t)$, for the flight of a struck golf ball. The time T refers to the time at which the ball strikes the ground. (c) The trajectory z as a function of y, also showing the vector components of the velocity at various points. The distance R is the horizontal range.

For a given v_0 and angle α, at what time t' will the golf ball reach maximum height? To determine t' we set dz/dt equal to zero (• Can you see why?):

$$\frac{d}{dt} z(t) = -gt + v_{0z} = 0$$

or

$$t' = \frac{v_{0z}}{g} = \frac{v_0}{g} \sin \alpha \qquad (4\text{–}20)$$

Physically, this means that maximum height corresponds to $v_z = 0$. (• Why?) We see that

$$t' = \tfrac{1}{2} T$$

which emphasizes the symmetric nature of the quadratic function $z(t)$. The descending portion of $z(t)$ is simply a reversal of the motion that brought the ball to the top of its trajectory. The golf ball strikes the ground with exactly the same speed with which it was lofted, and it makes a descending angle with the horizontal exactly equal to α. (• Can you verify these assertions?)

The maximum height reached by the ball is readily determined by substituting the time t' (Eq. 4–20) into the equation for the vertical motion, $z(t)$; thus,

$$z_m = -\tfrac{1}{2}gt'^2 + v_{0z}t' = -\tfrac{1}{2}g\left(\frac{v_{0z}}{g}\right)^2 + v_{0z}\left(\frac{v_{0z}}{g}\right)$$

Hence,
$$z_m = \frac{v_{0z}^2}{2g} = \frac{v_0^2}{2g}\sin^2\alpha = \tfrac{1}{4}R\tan\alpha \qquad\qquad \textbf{(4–21)}$$

where the last equality follows from Eq. 4–18.

Example 4–4

A student stands at the edge of a cliff and throws a stone horizontally over the edge with a speed of 18 m/s. The cliff is 50 m above a flat horizontal beach, as shown in the diagram. How long after being released does the stone strike the beach below the cliff? With what speed and angle of impact does it land?

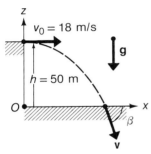

Solution:

We orient the coordinate axes as shown in the diagram so that the trajectory of the stone lies in the x-z plane. We have

$$x(t) = v_0 t, \qquad z(t) = h - \tfrac{1}{2}gt^2$$

When the stone lands on the beach, we have $z = 0$, so that the corresponding time t' is

$$t' = \sqrt{\frac{2h}{g}} = \sqrt{\frac{2(50\ \text{m})}{9.80\ \text{m/s}^2}} = 3.19\ \text{s}$$

On impact,

$$v_z = \frac{dz}{dt} = -gt' = -(9.80\ \text{m/s}^2)\times(3.19\ \text{s}) = -31.26\ \text{m/s}$$

$$v_x = \frac{dx}{dt} = v_0 = 18\ \text{m/s}$$

Therefore, the speed is

$$v = \sqrt{v_x^2 + v_z^2} = \sqrt{(31.26\ \text{m/s})^2 + (18\ \text{m/s})^2} = 36.1\ \text{m/s}$$

Also,
$$\tan\beta = \frac{v_z}{v_x} = \frac{-31.26\ \text{m/s}}{18\ \text{m/s}}$$

so that
$$\beta = -60.1°$$

Example 4–5

An often-quoted problem involves a jungle native with a blowpipe who is intent on hitting a falling tree-dwelling animal with his dart. He releases his dart at the precise instant that the animal lets go from a branch and starts to fall. The remarkable fact is that no matter what the height from which the animal falls and no matter what the speed of the released dart, the dart will find its mark only if the blowgun is aimed exactly at the animal at the start of its fall. Show that these statements are correct.

Solution:

We select the origin of a coordinate frame at the business end of the blow pipe and represent the acceleration of gravity by the vector $-g\hat{\mathbf{k}}$. Then, the initial position of the dart is $\mathbf{r}_{d0} = 0$, so the vector equation of motion for the dart is

$$\mathbf{r}_d(t) = -\tfrac{1}{2}gt^2\hat{\mathbf{k}} + \mathbf{v}_{d0}t$$

At first we imply nothing about either the magnitude v_{d0} or the direction of $\mathbf{v}_{d0}$. The falling animal is initially at the position $\mathbf{r}_{a0}$ and has an initial velocity of zero; that is, $\mathbf{v}_{a0} = 0$. Thus, the equation of motion for the animal is

$$\mathbf{r}_a(t) = -\tfrac{1}{2}gt^2\hat{\mathbf{k}} + \mathbf{r}_{a0}$$

If the dart is to strike the animal at some later time t', we must have

$$\mathbf{r}_d(t') = \mathbf{r}_a(t')$$

or
$$-\tfrac{1}{2}gt'^2\hat{\mathbf{k}} + \mathbf{v}_{d0}t' = -\tfrac{1}{2}gt'^2\hat{\mathbf{k}} + \mathbf{r}_{a0}$$

Therefore,
$$\mathbf{v}_{d0}t' = \mathbf{r}_{a0}$$

That is, the blow gun must be aimed *directly* at the animal as it begins to fall. The strike will occur at the time

$$t' = \frac{r_{a0}}{v_{d0}}$$

This time does, of course, depend on the initial distance and the

velocity imparted to the dart. Notice that the solution is obtained by using vector methods without the necessity of specifying a particular coordinate frame.

(• How would you work out the solution if components were used instead of vectors? Remember, at first you *do not know* that $\mathbf{v}_{d0}$ is parallel to $\mathbf{r}_{a0}$—this must be proven.)

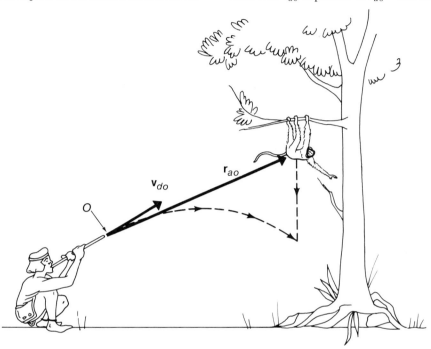

4–5 CIRCULAR MOTION

The motion of a particle that is constrained to move in a circular path is of sufficient general interest to warrant special attention. Although such motion can be analyzed quite rigorously by using Cartesian coordinates, a polar coordinate frame (Fig. 3–12) is a more natural choice. Consider motion for which $\rho = R$ is a constant. Therefore, we need only the single variable ϕ to describe the motion, and we write $\phi(t)$ to emphasize the time dependence. (In a Cartesian frame we would require both of the variables, x and y: $x = R \cos \phi$ and $y = R \sin \phi$.)

The instantaneous time rate of change of $\phi(t)$ is called the *angular frequency of rotation* (or the *angular speed*) and is designated by the symbol ω (Greek omega); thus,

$$\omega = \frac{d\phi}{dt} \qquad\qquad \textbf{(4–22)}$$

It is often convenient to measure ϕ in *radians** and ω in *radians per second* (rad/s). The

---

*See Section A–10 of Appendix A for a discussion of radian measure.

dimensionality of ω is $[\omega] = T^{-1}$, that is, *inverse time*. Unless otherwise specified, angular measure will always be understood to be in radians.

Motion at Constant Speed. The simplest type of circular motion is that in which the particle moves with a constant speed. Then, the angle ϕ increases by equal amounts in equal times, and, hence, ω is a constant. It follows that this type of motion—called *uniform circular motion*—is *periodic* or *cyclic*. The particle repeatedly arrives at any arbitrarily selected position at regular intervals of time. This repetition time is called the *period T*. The angle ϕ increases by 2π radians during each time interval T; therefore,

$$\omega = \frac{2\pi}{T} \qquad (4-23)$$

Another useful concept is the *frequency* of the motion, the number of cycles or revolutions executed per unit time. Frequency is usually measured in *cycles per second*, which we abbreviate as *hertz* (Hz); sometimes the unit *revolutions per minute* (rpm) is used. The frequency ν is related to the period T by

$$\nu = \frac{1}{T} = \frac{\omega}{2\pi} \qquad (4-24)$$

Fig. 4–8. Uniform circular motion. (a) The particle travels with a constant speed, so that $v(t) = v(t + \Delta t) = v$. (b) The velocity increment $\Delta \mathbf{v}$ is equal to $\mathbf{v}(t + \Delta t) - \mathbf{v}(t)$. The shaded triangles are similar because $-\mathbf{v}(t)$ is perpendicular to $\mathbf{r}(t)$ and $\mathbf{v}(t + \Delta t)$ is perpendicular to $\mathbf{r}(t + \Delta t)$.

A particle undergoing uniform circular motion travels an arc length $2\pi R$ (the circumference of the circular path) in a time equal to the period. Therefore, the speed v of the particle is

$$v = \frac{2\pi R}{T} = \omega R \qquad (4-25)$$

The velocity vector of a particle is always tangent to the particle's path. For the case of uniform circular motion, the velocity $\mathbf{v}$ is therefore always perpendicular to the particle's position vector $\mathbf{r}$, as indicated in Fig. 4–8. Although the velocity is constant in magnitude, its direction changes continually. Such motion implies the existence of an *acceleration*.

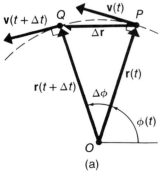

(a)

Figure 4–8 shows the finite changes in $\mathbf{r}$ and $\mathbf{v}$ that result from the motion during the time increment Δt. However, we are really interested in the limiting situation in which $\Delta t \to 0$ and, consequently, $\Delta \phi \to 0$. In this limit, the velocity vectors $\mathbf{v}(t + \Delta t)$ and $\mathbf{v}(t)$ become more nearly parallel, and $\Delta \mathbf{v}$ becomes directed more nearly toward the center of motion O. The acceleration of the particle is

$$\mathbf{a} = \frac{d\mathbf{v}}{dt} = \operatorname*{Lim}_{\Delta t \to 0} \frac{\Delta \mathbf{v}}{\Delta t} = \operatorname*{Lim}_{\Delta \phi \to 0} \frac{\Delta \mathbf{v}}{\Delta t}$$

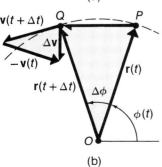

(b)

In this limit the acceleration is directed toward the center of the circle and is therefore called the *centripetal* (or *center-seeking*) acceleration. The magnitude of the acceleration is determined from the fact that the triangles involving $\mathbf{r}$ and $\mathbf{v}$ are similar isosceles triangles (see Fig. 4–8). (• Can you see that the included angles are equal?) Therefore, we can write

$$\frac{\Delta v}{v} = \frac{\Delta r}{R} \qquad \text{or} \qquad \frac{\Delta v}{\Delta r} = \frac{v}{R}$$

Then,*

$$a = \lim_{\Delta t \to 0} \frac{\Delta v}{\Delta t} = \lim_{\Delta t \to 0} \frac{\Delta v}{\Delta r} \cdot \frac{\Delta r}{\Delta t} = \frac{v}{R} \lim_{\Delta t \to 0} \frac{\Delta r}{\Delta t}$$

or

$$a = \frac{v^2}{R} = R\omega^2 \qquad (4-26)$$

where $a = |\mathbf{a}|$, with $\mathbf{a}$ directed *inward*.

Figure 4-9 shows the location of the particle at four different times, together with the corresponding vectors, $\mathbf{r}(t)$, $\mathbf{v}(t)$, and $\mathbf{a}(t)$. Notice that r, v, and a are all constant, whereas the vectors $\mathbf{r}$, $\mathbf{v}$, and $\mathbf{a}$ vary with time.

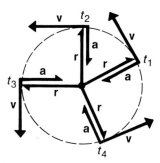

Fig. 4-9. The coordinate positions $\mathbf{r}(t)$ of a particle executing uniform circular motion at various selected times. Also shown superimposed are the vectors corresponding to the velocity $\mathbf{v}(t)$ and the acceleration $\mathbf{a}(t)$.

Example 4-6

A particle travels with uniform speed in a circular path with a radius $R = 2$ m and a period $T = 0.25$ s. Evaluate ω and ν. What is the speed? What is the centripetal acceleration?

Solution:

Using Eqs. 4-23 and 4-24, we have

$$\omega = \frac{2\pi}{T} = \frac{2\pi \text{ rad}}{0.25 \text{ s}} = 8\pi \text{ rad/s}$$

$$\nu = \frac{1}{T} = 4 \text{ cycles/s} = 4 \text{ Hz}$$

From Eq. 4-25, $v = \omega R = (8\pi \text{ rad/s})(2 \text{ m}) = 16\pi$ m/s and from Eq. 4-26, we find

$$a = R\omega^2 = (2 \text{ m})(8\pi \text{ rad/s})^2 = 128\pi^2 \text{ m/s}^2$$

Example 4-7

The Moon circles the Earth at a distance $R_M = 3.84 \times 10^8$ m, with a period $T_M = 27.3$ days. Compute the centripetal acceleration of the Moon in units of g. (Note that 1 day = 86,400 s.)

We can rewrite Eq. 4-26 as

$$a = \frac{v^2}{R} = \left(\frac{2\pi R_M}{T_M}\right)^2 \frac{1}{R_M} = \left(\frac{2\pi}{T_M}\right)^2 R_M$$

$$= \frac{4\pi^2 (3.84 \times 10^8 \text{ m})}{[(27.3 \text{ d})(8.64 \times 10^4 \text{ s/d})]^2} = 2.72 \times 10^{-3} \text{ m/s}^2$$

so that $\quad a/g = \dfrac{2.72 \times 10^{-3} \text{ m/s}^2}{9.80 \text{ m/s}^2} = 2.78 \times 10^{-4}$

4-6 CIRCULAR MOTION WITH CONSTANT ANGULAR ACCELERATION

The angular acceleration α is defined to be the rate of change of the angular speed ω; thus,

$$\alpha \equiv \frac{d\omega}{dt} = \frac{d^2\phi}{dt^2} \qquad (4-27)$$

---

*We make use of the algebra of incremental quantities and the fact that the limit of a product is the product of the limits (when the limits exist).

When the angular acceleration is constant, the equations of motion, $\omega = \omega(t)$ and $\phi = \phi(t)$, can be obtained by analogy with the case of linear motion with constant linear acceleration where we found general expressions for $v(t)$ and $s(t)$ (refer to Eqs. 2–12). Such a comparison yields for the angular frequency

$$\omega(t) = \alpha t + \omega_0 \tag{4–28}$$

where ω_0 is the angular frequency at the time $t = 0$. The corresponding equation for the angular position is

$$\phi(t) = \tfrac{1}{2}\alpha t^2 + \omega_0 t + \phi_0 \tag{4–29}$$

where ϕ_0 is the angular position at $t = 0$. Note that differentiating Eqs. 4–28 and 4–29 with respect to time yields the expected results contained in Eq. 4–27. Continuing the analogy, we may also write

$$\bar{\omega} = \tfrac{1}{2}(\omega + \omega_0) \tag{4–30}$$

and

$$\omega^2 = \omega_0^2 + 2\alpha(\phi - \phi_0) \tag{4–31}$$

While use is made of the formal similarity of the equations describing the two types of motion, it must be kept in mind that the dimensions in the two sets of equations are different.

▶ *Equations of Motion by Integration.* A more straightforward way of obtaining the equations of motion for constant angular acceleration during circular motion is to make use of the anti-differentiation or integration process. The first integration of Eq. 4–27 yields, with α a constant,

$$\omega = \int \alpha \, dt = \alpha t + C$$

If at $t = 0$ we have $\omega(0) = \omega_0$, then $C = \omega_0$, so that

$$\omega(t) = \alpha t + \omega_0$$

This result is Eq. 4–28. Integrating $\omega = d\phi/dt$, we have

$$\phi = \int \omega \, dt = \int (\alpha t + \omega_0) \, dt = \tfrac{1}{2}\alpha t^2 + \omega_0 t + C'$$

If at $t = 0$ we have $\phi(0) = \phi_0$, then $C' = \phi_0$, so that

$$\phi(t) = \tfrac{1}{2}\alpha t^2 + \omega_0 t + \phi_0$$

This result is Eq. 4–29.

The more general process of integration also allows treating the case in which α is not a constant but varies with time. In such a case, we have

$$\omega(t) = \int \alpha(t) \, dt \tag{4–32}$$

It is instructive to compare the present approach with that used in Section 2–6.   ◀

For motion in a circular path, the velocity vector (which is always tangent to the path) has only an azimuthal component; that is, $\mathbf{v} = v_\phi \hat{\mathbf{u}}_\phi$ (compare Eq. 3–16). The acceleration vector, on the other hand, has both a radial component (which is equal to the inward-directed centripetal acceleration) and an azimuthal component (which is due to the tangentially directed angular acceleration); thus,

$$\mathbf{a} = a_\rho \hat{\mathbf{u}}_\rho + a_\phi \hat{\mathbf{u}}_\phi \tag{4–33}$$

The radial acceleration is (compare Eq. 4-26)

$$a_\rho = -\frac{v^2}{R} = -R\omega^2 \qquad (4\text{-}34a)$$

where the negative sign indicates the inward direction of the centripetal acceleration.* The azimuthal acceleration is (using Eq. 4-25)

$$a_\phi = \frac{dv_\phi}{dt} = \frac{d}{dt}(R\omega) = R\frac{d\omega}{dt} = R\alpha \qquad (4\text{-}34b)$$

The radial and azimuthal vector components of the acceleration are mutually perpendicular, so the magnitude of the acceleration is (compare Eq. 3-17)

$$a = |\mathbf{a}| = \sqrt{a_\rho^2 + a_\phi^2} \qquad (4\text{-}35)$$

Note that if $a_\phi = 0$, then Eqs. 4-30 to 4-32 reduce to the case of uniform circular motion.

Example 4-8

At a particular instant, a particle traveling in a circular path with a radius of 0.5 m has a speed $v = 1.6$ m/s and an angular acceleration $\alpha = 16$ rad/s². What is the magnitude of the acceleration? What angle does the acceleration vector make with a tangent to the path?

Solution:

We have

$$a_\rho = -\frac{v^2}{R} = -\frac{(1.6 \text{ m/s})^2}{0.5 \text{ m}} = -5.12 \text{ m/s}^2$$

Also, $\quad a_\phi = R\alpha = (0.5 \text{ m})(16 \text{ rad/s}^2) = 8 \text{ m/s}^2$

Hence,

$$a = |\mathbf{a}| = \sqrt{a_\rho^2 + a_\phi^2} = \sqrt{(-5.12)^2 + (8)^2} = 9.50 \text{ m/s}^2$$

and $\qquad \tan \beta = \frac{a_\rho}{a_\phi} = \frac{-5.12}{8} = -0.64$

or $\qquad\qquad \beta = -32.6°$

Notice, in the diagram, the significance of the negative sign for β. If there were no angular acceleration ($\alpha = 0$), we would have $a_\phi = 0$ and $\beta = -90°$.

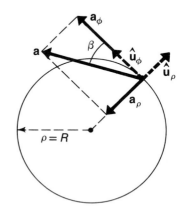

*Notice that in these equations we distinguish between *magnitudes* (such as $a = |\mathbf{a}|$) and *components* (such as a_ρ). The latter carry subscripts appropriate for the coordinate frame used and can have positive or negative values. The new coordinate frame and its variables ρ and ϕ (and unit vectors $\hat{\mathbf{u}}_\rho$ and $\hat{\mathbf{u}}_\phi$) are just features of the polar coordinates that you are familiar with from your earlier studies.

4–7 RELATIVE MOTION

We now consider how two observers who have a constant relative velocity describe the motion of the same particle in terms of their own coordinate frames. We assume that the relative orientation of the two frames does not change with time, thereby ruling out any possible rotational motion of the two observers. Figure 4–10 illustrates the geometry for two coordinate frames, S and S'. The relationship between the two position vectors $\mathbf{r}$ and $\mathbf{r}'$, reckoned in the frames S and S', respectively, can be expressed as

$$\mathbf{r} = \mathbf{R} + \mathbf{r}' \tag{4–36}$$

Here, $\mathbf{R}$ is the vector locating the origin O' (of the S' frame) in the S frame, and $d\mathbf{R}/dt = \mathbf{u}$, the constant velocity of S' with respect to S. We assume the existence of a "universal clock" to which both observers may refer simultaneously; formally, this means $t' = t$. Differentiating $\mathbf{r}$ with respect to this universal time t gives

$$\frac{d\mathbf{r}}{dt} = \frac{d\mathbf{R}}{dt} + \frac{d\mathbf{r}'}{dt}$$

or

$$\mathbf{v} = \mathbf{u} + \mathbf{v}' \tag{4–37}$$

where $\mathbf{v}$ and $\mathbf{v}'$ are the velocities of the particle in the frames S and S', respectively.

If the particle has nonzero acceleration, a further differentiation of the expression for $\mathbf{v}$ gives

$$\frac{d\mathbf{v}}{dt} = \frac{d\mathbf{u}}{dt} + \frac{d\mathbf{v}'}{dt}$$

Because $\mathbf{u}$ is assumed constant, $d\mathbf{u}/dt = 0$, so that

$$\mathbf{a} = \mathbf{a}' \tag{4–38}$$

These equations for $\mathbf{r}$, $\mathbf{v}$, and $\mathbf{a}$ (together with $t' = t$) constitute a set of transformation equations—the *Galilean transformation equations*—connecting the two frames, S and S'.

We can draw an important conclusion from the result contained in Eq. 4–38. If an

Fig. 4–10. Two frames, S and S', used to describe the motion of a particle at point P. The velocity of S' relative to S is $\mathbf{u}$, a constant.

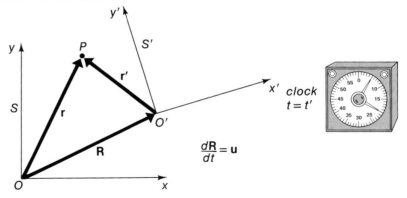

observer in a certain coordinate frame measures an acceleration **a** for a particle, any other observer whose coordinate frame moves *at constant velocity* with respect to the first frame will measure the same acceleration vector for the particle. Acceleration (unlike velocity or position) is therefore an *invariant* quantity.* Vectors such as **a** that are the same in all unaccelerated coordinate frames are called *free vectors*. We often use this important concept.

The derivation of the Galilean transformation equations suggests several questions. Does a "universal clock" actually exist? If each observer uses his own clock, how do they synchronize the clocks? How do the observers make "simultaneous" measurements of the particle's motion? These questions can be addressed properly only within the framework of the theory of relativity.

Example 4-9

A boat crosses a river 160 m wide in which the current flows with a uniform speed of 1.5 m/s. The steersman maintains a bearing (i.e., the forward heading of the boat) perpendicular to the river and a throttle setting to give a constant speed of 2 m/s with respect to the water. What is the velocity of the boat relative to a stationary shore observer? How far downstream from the initial position is the boat when it reaches the opposite shore?

Solution:

The key to solving problems of this type is to identify carefully the selection of the observers (and frames), S and S'. In this case it is convenient to take the shore observer to be S and to consider a *hypothetical observer* S' drifting with the river, as shown in the diagram. The "particle" being observed is the boat.

We are given that $u_x = 1.5$ m/s, $u_y = 0$, and $v'_y = 2$ m/s, $v'_x = 0$. Therefore, taking components of Eq. 4–37, we have

$$v_x = u_x + v'_x = 1.5 \text{ m/s}$$

$$v_y = u_y + v'_y = 2 \text{ m/s}$$

which give the velocity components as seen by the shore observer. The time to cross the river can be found using v_y:

$$t = \frac{w}{v_y} = \frac{160 \text{ m}}{2 \text{ m/s}} = 80 \text{ s}$$

Then we obtain for ℓ, the distance to the downstream landing point,

$$\ell = v_x t = (1.5 \text{ m/s})(80 \text{ s}) = 120 \text{ m}$$

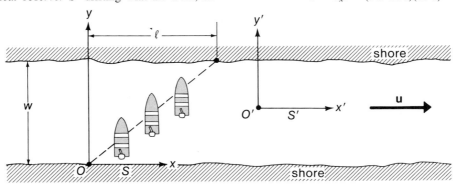

Example 4-10

A child in danger of drowning in a river is being carried downstream by a current that flows uniformly at a speed of 2.5 km/h. The child is 0.6 km from shore and 0.8 km upstream of a boat landing when a rescue boat sets out. If the boat proceeds at its maximum speed of 20 km/h with respect to the water, what heading relative to the shore should the boatman take? What angle does the boat velocity **v** make with the shore? How long will it take the boat to reach the child?

*This conclusion is completely general and applies also in three dimensions.

Solution:

Identify the frame S' (origin O') with the child and frame S (origin O) with a stationary observer at the boat landing. Then, the child sees the boat approaching along a straight line with a speed v', as shown in diagram (a). (• Can you see why this is the case?)

At $t = 0$, the separation of the origins is $R_0 = \sqrt{h_0^2 + \ell_0^2}$. Therefore, the time required to make the rescue is

$$t = \frac{R_0}{v'} = \frac{\sqrt{(0.6 \text{ km})^2 + (0.8 \text{ km})^2}}{20 \text{ km/h}} \times 60 \text{ min/h} = 3 \text{ min}$$

Note that this time is independent of the speed of the river current. (• Why is this so?)

The heading β' is determined from $\tan \beta' = h_0/\ell_0 = 0.75$ so that

$$\beta' = 36.9°$$

From the shore (frame S), the view is that shown in diagram (b). We have

$$v_x = v_x' - u = v' \cos \beta' - u = v' \frac{\ell_0}{R_0} - u$$

$$v_y = v_y' = v' \sin \beta' = v' \frac{h_0}{R_0}$$

Now, $R_0 = \sqrt{(0.6 \text{ km})^2 + (0.8 \text{ km})^2} = 1.00$ km, so we obtain

$$v_x = \frac{0.8 \text{ km}}{1.0 \text{ km}} (20 \text{ km/h}) - 2.5 \text{ km/h} = 13.5 \text{ km/h}$$

$$v_y = \frac{0.6 \text{ km}}{1.0 \text{ km}} (20 \text{ km/h}) = 12.0 \text{ km/h}$$

The direction of $\mathbf{v}$ with respect to the shoreline is given by

$$\tan \beta = \frac{v_y}{v_x} = \frac{12.0 \text{ km/h}}{13.5 \text{ km/h}} = 0.889$$

or

$$\beta = 41.6°$$

Steering any other course would necessarily take longer to rescue the drowning child. Note the similarity between this problem and Example 4–5.

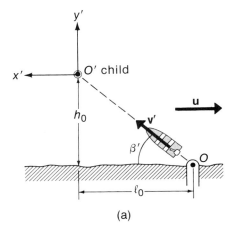

(a)

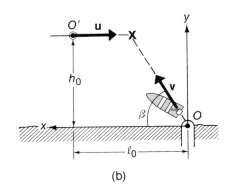

(b)

Example 4–11

A science student is riding on a flat car of a train traveling along a straight horizontal track at a constant speed of 10 m/s. The student throws a ball into the air along a path that he judges to make an initial angle of 60° with the horizontal and to be in line with the track. The student's professor, who is standing on the ground nearby, observes the ball to rise vertically. How high does the ball rise?

Solution:

Identify the student as the S' observer and the professor as the S observer. For the initial motion in S', we have

$$\frac{v_y'}{v_x'} = \tan 60° = \sqrt{3}$$

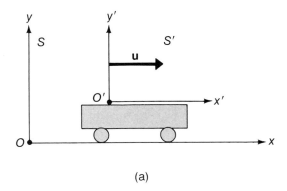

(a)

Then, because there is no x-motion in S, we can write

$$v_x = v_x' + u = 0$$

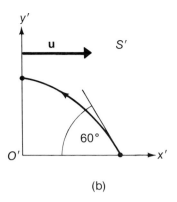

(b)

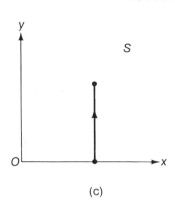

(c)

so that

$$v'_x = -u = -10 \text{ m/s}$$

Hence, the ball is thrown backward in S'. Then,

$$v_y = v'_y = \sqrt{3}|v'_x| = 10\sqrt{3} \text{ m/s}$$

Using

$$v_y^2 = 2gh$$

we find

$$h = \frac{(10\sqrt{3} \text{ m/s})^2}{2 \times (9.80 \text{ m/s}^2)} = 15.3 \text{ m}$$

The motion of the ball as seen by the student in S' is shown in diagram (b). The view of the professor in S is shown in diagram (c).

QUESTIONS

4–1   What is meant by the terms (a) position vector, (b) acceleration vector, and (c) ballistic trajectory?

4–2   At what point or points along the path of a fly ball hit into center field is the speed of the baseball a minimum? A maximum?

4–3   A successful hunter who is accustomed to aiming in front of a moving game target in order to get a "hit" questions the correctness of the conclusions reached in Example 4–5. Reconcile his doubts with the given analysis.

4–4   Describe how you would steer a car traveling at a constant speed (a) so that the acceleration is zero or (b) so that the magnitude of the acceleration is a constant.

4–5   Define or describe briefly the terms (a) angular frequency, (b) centripetal acceleration, and (c) angular velocity.

4–6   How would you modify Eqs. 4–28 through 4–31 if the angular measure were in degrees rather than in radians? How would you modify Eqs. 4–25 and 4–26?

4–7   An ice-skater is executing a perfect "figure eight," consisting of two equal, tangent circular paths. Throughout the first loop her speed is uniformly increased, and she coasts at a constant speed during the second loop. Make a sketch of the acceleration vector $\mathbf{a}$ at various times during the entire figure exercise.

4–8   A stunt pilot at an air show is executing a vertical (circular) loop-the-loop in a glider. Make a sketch showing the angular velocity vector of the glider during the descending arc of the flight and during the ascending arc. Also show the angular acceleration vector for the two cases. Be sure to consider the effects of the Earth's gravity.

4–9   Two observers (and frames) S and S' determine the same velocity for a moving object. What can you deduce about the relative motion of the two frames? What would your answer be if they reported different velocities for the object but the same attributed acceleration? Suppose even the acceleration of the object were observed to be different.

PROBLEMS

Section 4–1

4–1   The position vector of a particle is $\mathbf{r}(t) = \alpha t \hat{\mathbf{i}} + (\beta t^2 + \delta)\hat{\mathbf{j}}$, with $\alpha = 2$ m/s, $\beta = 8$ m/s^2, and $\delta = 1$ m.

The y coordinate is a quadratic function of x. Determine this function.

4–2   For the particle described in Problem 4–1, calculate

(a) the incremental displacement vector for the period from $t = 0.1$ s to $t = 0.2$ s and (b) the incremental displacement vector for the period from $t = 0.1$ s to $t = 0.3$ s. (c) What is the angle between the vectors found in (a) and (b)?

4–3 A particle moves in a circular path given by $x(t) = A \cos 2\omega t$, $y(t) = B \sin 2\omega t$, with $A = 6$ m, $B = 6$ m, and $\omega = 1$ s^{-1}. Determine the vector $\Delta \mathbf{r}$ that describes the displacement from point P (at $t = 0$) to point Q (at $t = \pi/4$ s).

Section 4–2

4–4 A particle, initially at the position $\mathbf{r}_1 = (6 \text{ m})\hat{\mathbf{i}} + (7 \text{ m})\hat{\mathbf{j}}$, is at the position $\mathbf{r}_2 = (4 \text{ m})\hat{\mathbf{i}} - (2 \text{ m})\hat{\mathbf{j}}$ one second later. Find the average velocity and the average speed of the particle during this time interval.

4–5 The position function of a particle is $\mathbf{r}(t) = \alpha t^3 \hat{\mathbf{i}} + \beta t^2 \hat{\mathbf{j}}$. (a) What is the velocity at $t = 1$ s? (b) At $t = 1$ s the particle is $\sqrt{5}$ m from the origin and has a speed of 5 m/s. How far from the origin is the particle at $t = 2$ s?

4–6 The position function of a golf ball is $\mathbf{r}(t) = x(t)\hat{\mathbf{i}} + y(t)\hat{\mathbf{j}} = ct\hat{\mathbf{i}} + (bt - at^2)\hat{\mathbf{j}}$, where a, b, and c are positive constants with appropriate dimensions. The x-axis is horizontal and the y-axis is vertical. At $t = 0$, the ball's trajectory makes a 45° angle with the horizontal. If the greatest height attained by the ball is h, prove that the horizontal distance reached at that time is $2h$. [*Hint:* The largest value of $y(t)$ occurs when $dy/dt = 0$.]

4–7 The position function of a particle is $\mathbf{r}(t) = 4t\hat{\mathbf{i}} + (2t^2 - 3t)\hat{\mathbf{j}}$, where r is in meters when t is given in seconds. (a) Where is the particle located at the instant when $v_y = 0$? (b) What is the particle's speed at that instant?

Section 4–3

4–8 A particle moves in accordance with the functions $x(t) = (3 \text{ m/s}^2)t^2 + (2 \text{ m/s})t$ and $y(t) = (2 \text{ m/s}^2)t^2 + (4 \text{ m/s})t + (3 \text{ m})$. Determine the velocity vector $\mathbf{v}(t)$ and the acceleration vector $\mathbf{a}(t)$. Find the magnitude and the direction of $\mathbf{a}$ at $t = 2$ s.

4–9 Measurements made on the multiple-flash photo, above right, show that the equation of motion of the ball is $\mathbf{r}(t) = at\hat{\mathbf{i}} + (h - bt^2)\hat{\mathbf{j}}$, where $a = 2.15$ cm/flash, $h = 41$ cm, and $b = 1.089$ cm/flash^2. The light source was flashing at a constant rate of 30 flashes/s during the exposure. Determine $\mathbf{v}(t)$ and $\mathbf{a}(t)$ for any t (in seconds).

4–10 A particle has the position function

$$\mathbf{r}(t) = (t^3 + 1)\hat{\mathbf{i}} + (t^3 - 6t^2 + 12t)\hat{\mathbf{j}}$$

where each term is expressed in meters when t is given in

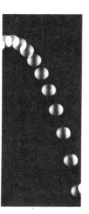

(Courtesy of Education Development Center, Newton, Mass.)

Problem 4–9

seconds. (a) Find the general expressions for the velocity and the acceleration. (b) At what time is the velocity in the y-direction a minimum? (c) Describe the behavior of $\mathbf{r}(t)$, $\mathbf{v}(t)$, and $\mathbf{a}(t)$ for times near the value found for part (b).

4–11 A particle has the cycloidal position function
•• $\mathbf{r}(t) = x(t)\hat{\mathbf{i}} + y(t)\hat{\mathbf{j}}$ with

$$x(t) = R(\omega t - \sin \omega t)$$

and $$y(t) = R(1 - \cos \omega t)$$

where R and ω are constants. Determine the velocity and the acceleration of the particle at any time t. For what values of t is the particle momentarily at rest? What are the coordinate locations of these momentary rest positions and what are the corresponding accelerations? Does the magnitude of the acceleration depend on time?

Section 4–4

4–12 A boy throws a ball into the air as hard as he can and then runs as fast as he can under the ball in order to catch it. If his maximum speed in throwing the ball is 20 m/s and his best time for a 20-m dash is 3 s, how high does the ball rise?

4–13 A baseball is released from the thrower's hand 2 m above
• the level of a flat playing field. The initial velocity is 20 m/s in a direction making an angle of 30° with the horizontal. (a) What is the maximum height reached by the ball? (b) How far from the thrower does the ball strike the ground? (c) Find the time and the velocity at impact.

4–14 A driven golf ball just clears the top of a tree that is 15 m
• high and 30 m from the tee, and then lands (with no roll or bounce) on the green, 180 m from the tee. What was the initial velocity imparted to the golf ball? [*Hint:* Use Eqs. 4–17 and 4–18 to find the angle α.]

4–15 A rifleman fires a bullet with a muzzle velocity of

500 m/s at a small stationary target 100 m away and at the same elevation as the rifle. At what angle of elevation must the rifle barrel be set? With this setting, how high above the target is the rifle aimed? (Neglect air-resistance effects.)

4–16 A ball on a floor is kicked, imparting to the ball an initial
• velocity v_0. The ball strikes a wall that is 6 m from the original position and then rebounds. The ball retraces its path and hits the floor at the exact spot from which it was kicked. What is the minimum speed v_0 that the ball must have to follow such a path? [*Hint:* At what angle must the ball strike the wall? What would have been the range of the ball if the wall had not been present?]

4–17 Two projectiles are launched with the same speed but at
•• angles $\dfrac{\pi}{4} \pm \beta \left(\beta < \dfrac{\pi}{4} \right)$ with the horizontal. Show that the projectiles have the same range. (This result was first deduced by Galileo.) Show also that the difference in flight times is

$$\Delta t_m = \frac{2\sqrt{2}v_0}{g} \sin \beta$$

Discuss the case $\beta = \pi/4$.

4–18 A projectile is to be catapulted over a fortress wall 20 m
• high by a siege *ballista* (a Roman catapult) capable of a launching speed of 22 m/s. The projectile is released at a height of 4 m above ground level. What is the farthest horizontal distance from the foot of the wall that the ballista may be set up? What elevation angle must be used at this distance? [*Hint:* Galileo would probably suggest that you first determine the elevation angle using Eq. 4–21.]

4–19 Water leaves the nozzle of a garden hose with a speed of
• 10 m/s. At what angle above the horizontal is the nozzle pointed if the water falls on a flower bed 10 m away horizontally and 1 m lower than the nozzle? Assume that the water stream behaves in the manner of projectiles. [*Hint:* Using $1/\cos^2 \alpha = 1 + \tan^2 \alpha$ in Eq. 4–17 reduces this equation to a quadratic in tan α.]

4–20 Photographs show that a competitive long jumper launches himself (i.e., his body center) at an angle of approximately 25° with respect to the horizontal. When a jumper lands in the pit, he doubles over to the extent that his body center is about 0.6 m lower than at take off. What must be a jumper's take-off velocity in order to match the record jump of 8.90 m by Robert Beamon at the 1968 Olympics in Mexico City? Is this a reasonable velocity? Compare with that of a sprinter. [*Note:* In Mexico City, $g = 9.786$ m/s^2.]

Section 4–5

4–21 Find the tangential velocity and the centripetal accelera-

tion of a particle glued to the tip of a fan blade that has a radius of 20 cm and rotates with an angular frequency of 1000 rpm.

4–22 A training simulator designed to expose astronauts to the accelerations encountered in rocket launches consists of a seat at the end of a 5-m boom that can rotate around a vertical axis. What rotational frequency (in rpm) is required to simulate a $7g$ acceleration? What is the tangential velocity at this acceleration?

4–23 The position function of a particle is $\mathbf{r}(t) = A \sin \omega t\,\hat{\mathbf{i}} + A \cos \omega t\,\hat{\mathbf{j}}$, where A and ω are constants. Describe the motion of the particle. (Show that $|\mathbf{r}|$ is independent of the time.) For the interval from $t = 0$ to $t = \pi/2\omega$, (a) determine the average speed of the particle and (b) determine the average velocity of the particle. (c) What is the average velocity for the interval from $t = 0$ to $t = \pi/\omega$?

Section 4–6

4–24 A disc with a radius of 50 cm "spins up" from rest to an
• angular speed such that the centripetal acceleration on the rim is equal to g. How many revolutions of the disc are required if the angular acceleration is 0.20 rad/s^2?

4–25 A disc with a radius $R = 0.1$ m starts from rest at $t = 0$ and is given a constant angular acceleration, $\alpha = \pi$ rad/s^2. A particle glued to a point on the circumference has an initial (counterclockwise) angular location of $\phi_0 = \pi/2$ rad, measured from a fixed reference line. At $t = 1$ s, find the following quantities for the particle: (a) the angular position, (b) the angular velocity, (c) the tangential velocity, (d) the tangential and radial accelerations, and (e) the resultant acceleration.

4–26 A particle glued to the rim of a wheel with a radius of 20 cm has an angular position (measured counterclockwise) with respect to a fixed reference line given by $\phi(t) = at^3 + bt^2 + \pi/2$ radians, with $a = 2$ s^{-3} and $b = -3$ s^{-2}. (a) Find the angular frequency and the acceleration of the particle. (b) At what time is the resultant acceleration entirely radial? What is its value at this instant? (c) At what time is the resultant acceleration entirely tangential? What is its value at this instant?

Section 4–7

4–27 A shopper in a department store can walk up a stationary (stalled) escalator in 30 s. If the normally functioning escalator can carry the standing shopper to the next floor in 20 s, how long would it take the shopper to walk up the moving escalator? Assume the same walking effort for the shopper whether the escalator is stalled or moving.

4–28 Two canoeists in identical canoes exert the same effort paddling in a river. One paddles directly upstream (and moves upstream), whereas the other paddles directly downstream. An observer on the riverbank reckons their speeds to be 1.2 m/s and 2.9 m/s. How fast is the river flowing?

4–29 A motorboat has two throttle positions on its engine: the high-speed position propels the boat at 10 km/h in still water, and the low position gives half the maximum speed. The boat travels downstream on a river at half speed and returns to its dock at full speed. It takes 15 percent longer to make the upstream trip than it does to make the downstream trip. How fast is the river flowing?

4–30 A train is moving east on a straight track with a speed of 12 m/s. Raindrops falling straight down with respect to

the ground make initial tracks on the train windows at an angle of 27° to the vertical. What is the (vertical) speed of the raindrops with respect to the ground? What is the speed of the raindrops with respect to the train?

4–31 What compass bearing should a pilot set if he wishes to fly due north when there is a wind blowing from the west at 100 km/h and the aircraft moves with an air speed of 350 km/h? What is the ground speed of the aircraft?

4–32 A boat requires 2 min to cross a river that is 150 m wide. The boat's speed relative to the water is 3 m/s, and the river current flows at a speed of 2 m/s. At what possible upstream or downstream points does the boat reach the opposite shore?

Additional Problems

4–33 The constant acceleration of a particle is $\mathbf{a} = (2 \text{ m/s}^2)\hat{\mathbf{i}}$. At $t = 1$ s the velocity of the particle is $\mathbf{v}(1) = (2\hat{\mathbf{i}} + 3\hat{\mathbf{j}})$ m/s, and the position of the particle at $t = 0$ is $\mathbf{r}_0 = (3 \text{ m})\hat{\mathbf{i}}$. Determine the position function for the particle at any time t.

4–34 Show that two projectiles launched simultaneously near
• the surface of the Earth will collide if and only if they would have collided without benefit of the acceleration due to the Earth's gravity. Give an expression for the time of impact. Are these results in accord with Example 4–5? [*Hint:* Use vector notation, as in Example 4–5.]

4–35 A football is kicked from a kicking tee with an initial velocity of 20 m/s at an angle of 45° to the horizontal. A receiver 60 m away in the direction of the kick starts to run toward the ball at the instant it is kicked. What is the slowest constant speed at which he can run in order to field the ball with a shoe-top catch?

4–36 A batter hits a pitched baseball 1 m above the ground, imparting to the ball a speed of 40 m/s. The resulting line drive is caught on the fly by the left fielder 60 m from home plate with his glove 1 m above the ground. If the shortstop, 45 m from home plate and in line with the drive, were to jump straight up to make the catch instead of allowing the left fielder to make the play, how high above the ground would his glove have to be?

4–37 A catapult hurls a stone with a velocity of 20 m/s at an
• angle of 40° to the horizontal from a wall that is 10 m above the level ground. What will be the increase in range of the projectile if the catapult is moved to a wall that is 20 m high? [*Hint:* Use Eqs. 4–16, but add a term z_0 to $z(t)$.]

4–38 A particle is traveling along a ballistic trajectory with ini-
• tial velocity v_0 and elevation angle α. It is asserted that at the top of its trajectory the particle behaves momentarily

as if it were traveling along a circular path with a speed equal to its actual speed. (a) What is the radius R of such a circular path? [*Hint:* What must be the centripetal acceleration at the top of the trajectory?] (b) A mathematician claims that this radius is simply the *radius of curvature* for the actual path at the highest point, that is,

$$R^{-1} = \left| \frac{d^2z}{dy^2} \right|_{\text{apex}}$$

Use Eq. 4–17 and verify this assertion.

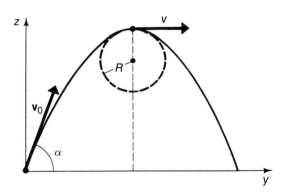

4–39 A particle originally at rest on top of a hemisphere with a
• radius R is given a horizontal velocity v_0. What is the *smallest* value of v_0 that will allow the particle to leave the surface immediately instead of starting to slide down the surface?

4–40 An athlete can throw a ball a maximum horizontal distance R. If air-resistance effects are negligible, show that the athlete can throw the same ball to a maximum height $\frac{1}{2}R$.

4–41 An LP record rotates at $33\frac{1}{3}$ rpm. What is the arc length along the record groove corresponding to one cycle of a tone with a frequency of 20,000 Hz (20 kHz) when the stylus is 10 cm from the record axis? (For comparison, the diameter of a human red blood cell is about 13×10^{-6} m.)

4–42 The value of the acceleration due to gravity varies with
• the distance above the Earth's surface. An artificial satellite moves uniformly in a circular orbit around the Earth. The period of the orbit is 12 h. What is the value of ''g'' at the position of the orbit?

4–43 An object is located on the Earth's surface at a latitude λ. Calculate the centripetal acceleration that the object experiences by virtue of the rotation of the Earth about its axis. Express your answer in terms of g.

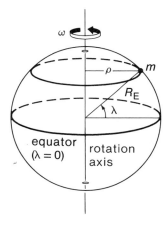

4–44 A particle moves in a circular path ($R = 0.80$ m) with a constant angular acceleration that carries the particle from rest at time zero to an angular speed of 2.0 rad/s at $t = 0.4$ s. Determine the following quantities at $t = 0.8$ s: the angular acceleration α, the radial acceleration a_ρ, the azimuthal acceleration a_ϕ, the acceleration a, and the angle β defined in the diagram of Example 4–8.

4–45 Some science students are canoeing on a river. While
•• heading upstream they accidentally drop overboard an empty beer can and continue paddling for 1 h, reaching a point 4 km further upstream. At this point they realize what has happened and turn around and head downstream. They catch up with and retrieve the floating beer can 10 km downstream from the turnaround point. (a) Assuming a constant paddling effort throughout, how fast is the river flowing? (b) What would be the canoe velocity in a still lake for the same paddling effort?

4–46 An elevator is moving upward at a constant velocity of
•• 2.0 m/s. A bolt in the elevator ceiling 3 m above the elevator floor works loose and falls. (a) How long does it take for the bolt to fall to the elevator floor? (b) What is the velocity of the bolt just as it hits the elevator floor, according to an observer in the elevator? (c) What would be the velocity according to an observer standing on one of the floor landings of the building? (d) What total distance do the two observers determine for the bolt's journey from ceiling to floor? [*Hint:* What is the initial velocity of the bolt?]

4–47 Two boats are moving along perpendicular paths on a still lake at night. One boat moves with a speed of 3 m/s; the other moves with a speed of 4 m/s. At $t = 0$, the boats are 300 m apart, and they manage to collide some time later. At what time did the collision occur? How far did each boat move between $t = 0$ and the time of collision?

4–48 An oceanliner on a calm sea is steaming due west at a speed of 20 km/h. Smoke from its funnel stretches out in a line 15° south of east. If the wind is blowing from the north, what is the speed of the wind? (Assume that the smoke particles take up the wind velocity as soon as they are emitted from the funnel.)

4–49 An airplane flies a straight-line course between two air-
•• ports at a constant air speed. Show that the time for a round trip is always increased by the presence of a wind blowing with a constant velocity during the entire flight, independent of the direction of the wind.

CHAPTER 5

NEWTON'S LAWS OF MOTION

We have by now examined in detail the methods for describing the motion of particles and objects that can be considered to behave as particles. In these discussions we have dealt with the time variations of the geometry of motion, the essential ingredient of the subject called *kinematics*. In this chapter we extend our view to include the question of *why* particles move as they do. The study of the way that other agencies or objects affect the motion of particles constitutes the subject called *dynamics*.

By the skillful combination of intuition and experimentation, Galileo began the modern development of the science of mechanics. Isaac Newton (who was born the same year that Galileo died, 1642) incorporated the findings of Galileo along with his own brilliant discoveries into a formulation of dynamics that has become known as *Newtonian mechanics*. Newton's theory is summarized in three laws that have wide validity and profound consequences. In the following pages we state and then discuss the implications of these laws.

Although Newton's laws have provided a remarkably simple and satisfying theory of mechanics, they nonetheless involve certain problems of logic. The particular line of development of the theory that we follow here is not the one taken by Newton. This discussion is neither more philosophically satisfying nor less subject to logical criticism. The difficulty that arises in any formulation of Newtonian theory is a result of the need to introduce simultaneously two concepts not considered in the subject of kinematics. These new concepts are *force* and *mass*. In stating his three laws of dynamics, Newton intermingled the definitions of these quantities with the rules of their behavior. No one has yet discovered an escape from this logical paradox. It is a tribute to Newton's insight that, in spite of this difficulty, he was able to formulate a set of laws that provides a correct description of the dynamics of bulk matter.

In the discussions here, an intuitive approach is taken in the introduction of the concept of force. By using an operational definition of this important quantity, we can avoid the force–mass dilemma. This procedure, although not completely free of criticism, offers a straightforward approach to Newtonian dynamics.

Newton's laws are not a perfect description of the way that Nature behaves in every domain. Within this century it has been discovered that Newtonian theory fails in the realm of atoms and nuclei, and it fails when extremely high velocities are involved. The investigation of these failures has led to the development of quantum theory and the theory of relativity. However, the durability of Newton's laws in satisfactorily accounting for observations in the everyday macroscopic world remains intact.

In this chapter we concentrate on developing and interpreting Newton's laws. In the next chapter, we direct our attention toward some interesting and important applications of Newtonian theory.

5–1 FORCE

The Basic Forces. We are all familiar with the everyday use of the word *force* to describe a push or a pull. Both animate and inanimate agencies can exert forces. Stretched springs and rubber bands or taut cables and ropes exert forces on objects attached to their ends. Liquids and gases exert forces on container walls and exert buoyant forces on immersed objects. Materials that are rubbed together exert frictional forces on each other. Machines can be used to exert forces that cut or press, and they can be used for locomotion by exerting forces against the ground. All of these examples involve *contact forces,* in which one object exerts a force on another object by coming into direct contact with it.

There is another type of interaction, exemplified by the *gravitational force,* in which a force acts between objects that need not be in physical contact with each other. Thus, the gravitational force is sometimes said to be an ''action-at-a-distance'' force. It is the gravitational force acting through the vacuum of space that maintains the Earth in its orbit around the Sun. It is also the force that keeps us ''pinned'' to the Earth's surface even as we move about, and it ''holds'' our atmosphere and oceans in captive attachment. The importance of the gravitational force in our lives is therefore self-evident, and it is the subject of a careful examination in this and in later chapters. Another important force that acts at a distance is the *electromagnetic force.* This is the basic force that acts between electrically charged objects. Much of modern life depends on devices that operate by utilizing the electromagnetic force. This too is a force that we examine in detail later.

Force—An Operational Definition. We all know from experience that pulling at both ends of a strand of an elastic material or a spring produces an elongation of the object. It is also easy to discover that the greater the pulling effort, the greater the elongation. This fact suggests a means for exerting a definite and reproducible force on an object.

To be specific, suppose that we have constructed several identical, standard spring scales such as the one shown schematically in Fig. 5–1a. The spring is imagined to have a natural or relaxed length ℓ_0. If we pull on the two (hook) ends, we can stretch the spring to

Fig. 5–1. (a) Schematic of a spring scale. The length ℓ of the spring is determined by the position of the pointer P. (b) Arrangement for determining the equivalence of two standard spring scales.

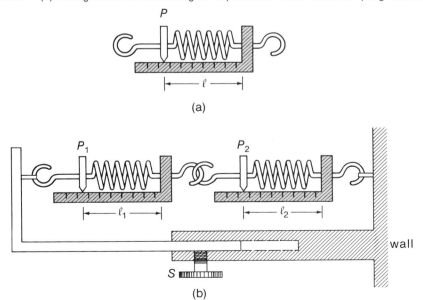

(a)

(b)

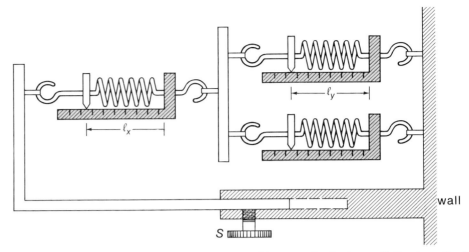

Fig. 5–2. Arrangement used to determine the relationship connecting the force with the elongation for standard springs.

a total length ℓ; that is, $\Delta\ell = \ell - \ell_0$ represents the elongation of the spring. The force F necessary to produce this elongation is expected to be a function of $\Delta\ell$. We now describe an empirical way to discover the relationship connecting F and $\Delta\ell$.

Quite apart from the identical construction of the springs, we can verify their genuine equivalence by a simple experiment. Refer to Fig. 5–1b. The two springs, which have relaxed lengths ℓ_{10} and ℓ_{20}, are connected as shown. By adjusting the slide S, we automatically stretch both springs. If for every setting of S we find that $\Delta\ell_1 = \ell_1 - \ell_{10}$ is equal to $\Delta\ell_2 = \ell_2 - \ell_{20}$, the two standard springs may be judged equivalent.* Imagine that several equivalent springs have been constructed and tested this way. We are now ready to determine the connection between F and $\Delta\ell$.

Refer next to Fig. 5–2 where one standard spring scale is shown balanced against two other standard springs. Suppose that in the earlier arrangement of Fig. 5–1b the slide S is set so that there results a specific elongation, $\Delta\ell_s = \ell - \ell_0$. (Because we have established that the two springs are equivalent, this elongation refers to either spring.) The elongation $\Delta\ell_s$ corresponds to some definite but unknown spring force F_s. Then, in the arrangement of Fig. 5–2, we set the slide S so that $\Delta\ell_y = \Delta\ell_s$ and measure $\Delta\ell_x = \ell_x - \ell_0$. The elongation $\Delta\ell_x$ now corresponds to a force $2F_s$ (exerted by the left-hand spring). Next, we reset S in Fig. 5–2 such that $\Delta\ell_x = \Delta\ell_s$ and determine a new value of $\Delta\ell_y$. This defines a force equal to $\frac{1}{2}F_s$ (exerted by each of the right-hand springs). (• Can you generalize this procedure to define forces $3F_s$ and $\frac{1}{3}F_s$, and so forth?)

From the spring measurements we can imagine obtaining data such as that shown in Fig. 5–3. This graph shows that F is directly proportional to $\Delta\ell$; that is,

Fig. 5–3. Results of the hypothetical experiment performed using the arrangement of springs shown in Fig. 5–2. The graph is linear, so the springs obey Hooke's law.

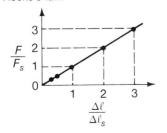

$$F = \kappa\Delta\ell = \kappa(\ell - \ell_0) \qquad (5-1)$$

The proportionality constant κ is the *force constant* for the particular set of standard springs. As long as they are not stretched too far, most springs obey Eq. 5–1.

The linear relationship connecting F with $\Delta\ell$ that is given by Eq. 5–1 is referred to

---

*Is this self-evident? Our intuition (based on experiences such as tug-of-war games) suggests that it is— and, indeed, that is correct. But there are sophisticated ideas involved in this simple statement. Think about this point again after studying Newton's third law.

as *Hooke's law.** By using a standard spring scale (being careful to confine the elongation to the range in which Hooke's law is valid), we have a simple way of applying any desired force. For example, if, in a second experiment, we wish to apply a force that is 2.46 times larger than that in a previous experiment, we simply set up conditions such that $\Delta\ell$ on the second occasion is 2.46 times $\Delta\ell$ in the first instance. We assert that our standard spring scale can be made to exert a force with a particular magnitude and with a direction along the spring axis (opposite to the direction of the spring extension).

In all of the discussion here, we have avoided making any statement as to what *force* actually is, except that it is a quantity with both magnitude and direction (a *vector* quantity). We have merely given an operational method for setting up forces that have any desired ratio, one to another. We now use these intuitive ideas concerning force in the discussion of Newton's laws of dynamics.

5-2   NEWTON'S LAWS OF MOTION

Frames of Reference.   Newton's laws of dynamics relate to the motion of objects. In order to specify positions, velocities, and accelerations, it is necessary first to establish a *frame of reference*. That is, we require a background coordinate system so that precise meaning can be given to all of the kinematic quantities that enter the theory.

What type of reference frame shall we choose? Newton understood the important and profound point that a frame suitable for expressing the laws of dynamics must be a *nonaccelerating* reference frame. It is easy to see why. Have you ever walked on a moving merry-go-round (which undergoes continual centripetal acceleration) and noticed the sensation of strange forces acting on your body? And how would you describe the motion of a person walking on the ground in terms of a coordinate frame attached to the merry-go-round?

Newton realized that a reference frame attached to the Earth is not an unaccelerated frame. The rotational motion of the Earth results in a centripetal acceleration at the Equator (where the value is maximum) of 3.4×10^{-2} m/s^2 or approximately 0.34 percent of g. Moreover, the Earth experiences a centripetal acceleration of 5.9×10^{-3} m/s^2 due to its orbital motion around the Sun. If these accelerations can be tolerated or ignored in a particular situation, an Earth-based reference frame will be adequate for describing the dynamics of an object or system. Thus, for many purposes, an Earth-based frame is a satisfactory "unaccelerated" reference frame.

Newton sought to avoid the difficulties associated with a rotating Earth-based frame by choosing as his "primary" reference frame one that is based on the positions of the stars. If the stars are in fixed positions in the sky, a reference frame based on these positions is well-defined and enduring. We now know that each star in the sky undergoes its own particular motion; however, this motion is difficult or impossible to detect for stars that lie at great distances from the Earth. Thus, we might base a primary reference frame on these distant stars. This procedure will be adequate for most purposes, but it cannot be justified as a fundamental part of a precise theory. (We return to this point later and give a definition of a proper unaccelerated reference frame.)

Once a primary unaccelerated reference frame has been established, it is possible to define any number of additional frames that, at most, move with constant velocity and

---

*Robert Hooke (1635–1703), an English physicist, stated the equivalent of this famous law in 1676 in the form of a Latin cryptogram, CEIIINOSSSTTUV, the solution of which is *ut tensio sic vis* ("as the extension, so the force").

without rotation with respect to the primary frame. All such frames are equally suitable for describing the dynamics of particles and systems.

Newton's First Law. According to Aristotelian doctrine, the "natural state" of an object is one of *rest,* and in order to sustain *any* motion, some outside agency must provide a propelling force. This must surely be true, it was argued, because all moving objects, of their own accord, eventually come to rest. In the seventeenth century, observations made by Galileo convinced him that this view of motion was incorrect. Instead, Galileo concluded that in the absence of any external applied forces, the state of motion of a body is one of uniform motion—that is, the motion takes place with constant velocity, with zero velocity included as a possibility. Galileo reasoned that an object sliding over a rough horizontal surface eventually comes to rest because a retarding surface—the force of friction—acts on the object. If the same object slides over a slippery surface, with the same initial velocity, it will move a greater distance before coming to rest. If this test is continued on smooth ice or an oiled surface, the stopping distance is increased. Further refinements are now possible. Standard modern laboratory equipment for performing nearly friction-free experiments includes the *air track* and the *air table.* These devices provide air cushions on which objects float, thereby permitting unimpeded horizontal motion. The remnant effect of friction in these experiments is several orders of magnitude smaller than if lubricants were used.

Using the results of his Earth-bound experiments, Galileo correctly arrived at a universal conclusion, namely, that in the absence of all friction, an object once set into motion will continue to move with constant velocity until acted upon by a force. Newton accepted Galileo's conclusion and incorporated it into his first axiom regarding the nature and causes of motion. We now refer to this statement as *Newton's first law of motion.* In Newton's words (translated from the original Latin version*):

I. Every body continues in its state of rest, or in uniform motion in a right line (straight line) *unless it is compelled to change that state by forces impressed upon it.*

In mathematical language Newton's first law becomes

$$I.\ \mathbf{F} = 0 \quad \textit{implies} \quad \mathbf{v} = \text{constant} \tag{5-2}$$

The first law does not really tell us anything about the nature of force. We have here only a statement concerning the behavior of objects in the *absence* of any force. But even so, the first law is not an empty statement. Consider a body that experiences no outside forces.† This situation can exist only in the complete absence of other matter, for contact or action-at-a-distance forces (such as the gravitational force) always originate with other material bodies. However, we can approximate the situation by considering an object isolated in deep space. With the zero-force condition of the first law satisfied owing to the isolation, we know that the object moves with constant velocity in an unaccelerated reference frame, which we call an *inertial reference frame.* Instead of relying on the

---

*The theory of dynamics developed by Isaac Newton (1642–1727) is contained in his monumental work, *Philosophiae Naturalis Principia Mathematica,* published in 1686. (We usually refer to this treatise simply as the *Principia.*) The translated passages (except for the editorial parenthetical inserts) are from Andrew Motte, 1729.

†Actually, it is not necessary to require that a body experience no outside force in order for the first law to be valid. It is required only that the *resultant* (or *net*) force be zero. But in following Newton's development of the theory there is, at this point, no clear definition of force. Therefore, one can safely consider only the case of no forces whatsoever.

distant stars to define an appropriate unaccelerated reference frame, we can take the attitude that *any* frame in which any particle subject to no force moves with **v** = *constant* is, in fact, an inertial frame. That is, the first law can be considered to *define* the concept of an inertial reference frame.

Notice how fragile the argument is here. First we recognize that a reference frame is necessary in order to describe motion. We choose a primary frame defined by the distant stars or any frame not undergoing acceleration with respect to the primary frame. We then use these frames to formulate the first law of motion. Finally, we use the first law to define a proper reference frame (an inertial frame) for describing dynamic effects. Then, because all inertial frames are equally valid, we can attach no meaning to the concept of a ''primary'' reference frame! Nevertheless, in spite of the circular logic, Newton's laws provide us with the correct relationship connecting forces and accelerations.

Because the first law is valid in any inertial reference frame, there can be no physical distinction between a state of rest and one of constant velocity for an object. If an object is subject to no force, an observer who chooses to view the object from an inertial reference frame attached to the object will, of course, see the object at rest. But another observer who views the same object from a second frame that moves with constant velocity with respect to the first frame will see the object in uniform motion. Neither interpretation of the situation is more basic or more correct than the other; both views are equally valid.

The state of motion of an isolated body will change only if it is acted upon by a force. This tendency to remain at rest or in uniform motion is due to a property of the body we call its *inertia*. Newton's first law is accordingly called the *law of inertia,* and we use the term *inertial reference frame* for any frame in which the law is valid.

The measure of the inertia of a body is its *mass*. In the second law we find how the motion of an object is determined by the force applied to it and by its mass.

Newton's Second Law. Common experience suggests how an object responds to the application of a force. We observe that the stronger the push or pull applied to an object, the greater is its altered motion or *acceleration* (remember, $\mathbf{a} = d\mathbf{v}/dt$). We also observe that the same magnitude of push or pull produces a different acceleration when applied to two different objects, such as a golf ball and a bowling ball.

In order to remove ourselves from various local extraneous influences, suppose that we conduct a series of experiments in deep space, completely isolated from all other objects. With a standard spring scale, we imagine applying a variety of controlled forces to some selected rigid test object, as indicated in Fig. 5-4. We observe the resulting acceleration of the object measured in an inertial frame as we apply a sustained force (by moving the disembodied hand shown in the figure). A definite strength of sustained force is set by maintaining a fixed value of $\Delta\ell = \ell - \ell_0$. We observe that the acceleration produced is constant in magnitude and is directed along the spring axis (that is, in line with the direction of the applied force). If we increase the force by doubling $\Delta\ell$, we observe an acceleration that is also twice the magnitude of the original acceleration. By repeating such measurements with our test object, we conclude that the acceleration produced is directly proportional to the applied force. If we use the vector symbol **F** to

Fig. 5-4. Experiment to determine the nature of the acceleration produced by an applied force, using a standard spring scale.

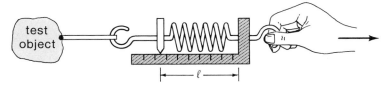

represent the applied force and **a** for the resulting acceleration, we have as our discovered empirical law $\mathbf{F} \propto \mathbf{a}$, which we can write as

$$II.\ \mathbf{F} = m\mathbf{a} \tag{5-3}$$

where m is the proportionality constant (not yet identified).

Next, we use several standard springs, each exerting a constant force and acting simultaneously on our selected test object. We discover that the resulting acceleration is in the direction of the *resultant* force and has a magnitude again given by Eq. 5–3 if **F** now stands for the resultant or net force on the object. That is, **F** is the *vector sum* of all individual external forces acting on the object: $\mathbf{F} = \Sigma_i \mathbf{F}_i$.

The conclusion expressed in Eq. 5–3 is known as *Newton's second law* of dynamics. In his own words:

II. The change in motion (acceleration*) *is proportional to the motive force impressed; and is made in the direction of the right line* (straight line) *in which that force is impressed.*

The scalar quantity m appearing in Eq. 5–3 was introduced simply as a constant of proportionality. Suppose that we alter the shape of the test object (without altering the amount of material present) and then maintain its rigidity in the new form. We observe that the relationship between **F** and **a** is unchanged; that is, the constant m is the same as before. Next, we consider a second test object, which may even consist of a material with a completely different chemical composition. When this object is tested by itself using various forces, the values obtained for the accelerations indicate the same value of m as for our first test object. Now, we physically attach the second object to the first, and we find that the appropriate value of m for the combined unit is twice that for either object alone. This means that, for the same force, only half as large an acceleration results. Thus, the quantity m has an inertial quality and we refer to m as the *inertial mass* of the object. Mass is an intrinsic property of matter, and, in a colloquial sense, we say that mass is a measure of the amount of matter in an object.

Because we have defined force in terms of the action of springs, Newton's second law permits a definition of mass in terms of the acceleration produced by a definite force. We can determine the mass of an object in terms of the mass of a standard (for example, the standard kilogram) by using a spring to apply the same force to each and then measuring the ratio of accelerations produced. If m_S represents the standard mass, the unknown mass m is

$$m = \frac{a_S}{a}m_S \tag{5-4}$$

This definition of mass is essential in the formulation of the theory of dynamics.

In SI units, mass is measured in kilograms (kg) and acceleration is measured in meters per second per second (m/s^2). Therefore, force is measured in units of $\text{kg}\cdot\text{m/s}^2$, to which we give the special name *newton* (N):

$$1\ \text{N} = 1\ \text{kg}\cdot\text{m/s}^2$$

That is, a force of 1 N applied to an object with a mass of 1 kg will produce an acceleration of 1 m/s^2.

---

*We examine more closely the meaning of Newton's words when we discuss momentum in Chapter 9.

Example 5–1

(a) What is the mass of an object if a net applied force of 7 N produces an acceleration of 3 m/s²?

Solution:

A direct application of Eq. 5–3 gives

$$m = \frac{F}{a} = \frac{7 \text{ N}}{3 \text{ m/s}^2} = \frac{7}{3} \text{ kg}$$

(b) If an additional force of 5 N is also applied to the same object, but at right angles to the 7-N force, what is the resultant acceleration?

Solution:

The resultant vector $\mathbf{F}_R$ has the magnitude

$$F_R = \sqrt{(7 \text{ N})^2 + (5 \text{ N})^2} = 8.60 \text{ N}$$

and makes an angle α with $\mathbf{F}_1$ given by

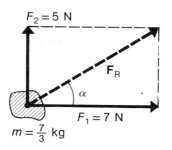

$$\tan \alpha = F_2/F_1 = 5/7$$

or

$$\alpha = 35.5°$$

The resulting acceleration is in the direction of $\mathbf{F}_R$, with a magnitude

$$a = \frac{F_R}{m} = \frac{8.60 \text{ N}}{(7/3) \text{ kg}} = 3.69 \text{ m/s}^2$$

Example 5–2

A stone with a mass of 100 g is tied to the end of a 2-m string and is whirled in a circular path at a constant speed of 3 m/s, as shown in the diagram. The other end of the string is fixed in an inertial frame. What is the force that the taut string exerts if this is the only force acting on the stone?

Solution:

The stone travels in a circular path at constant speed, so this is the special case, $\rho = R =$ constant (Section 4–5). Thus, the stone undergoes a constant centripetal acceleration (directed toward the center O) with magnitude (Eq. 4–26)

$$a_c = \frac{v^2}{R} = \frac{(3 \text{ m/s})^2}{2 \text{ m}} = 4.5 \text{ m/s}^2$$

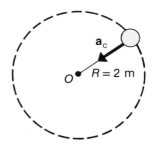

Therefore, the force exerted on the stone by the string is also directed toward the center O and has the constant magnitude

$$F = ma_c = (0.1 \text{ kg})(4.5 \text{ m/s}^2) = 0.45 \text{ N}$$

Example 5–3

A flat disc with a mass of 2 kg slides across the frozen surface of a lake with an initial speed of 5 m/s. If the frictional force on the disc has a constant value of 4 N, in what distance will the disc come to rest?

Solution:

Because the disc has no vertical motion, we can ignore all vertical forces acting on the disc. (That is, whatever vertical forces *are* acting, they must give a zero resultant force.)

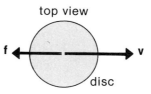

top view

disc

The given frictional force acts on the disc in a direction precisely opposite to its velocity, as indicated in the diagram. The acceleration produced by the frictional force is (taking the positive direction to be to the right)*

$$a = -\frac{f}{m} = -\frac{4 \text{ N}}{2 \text{ kg}} = -2 \text{ m/s}^2$$

We also know (Eq. 2–12) that $v^2 = v_0^2 + 2a(s - s_0)$; because

$v = 0$, we can write

$$(s - s_0) = -\frac{v_0^2}{2a} = -\frac{(5 \text{ m/s})^2}{2(-2 \text{ m/s}^2)} = 6.25 \text{ m}$$

Newton's Third Law. On the basis of his observations, Newton presumed that all forces in Nature act in *pairs*. Consider the simple case of two objects, *A* and *B*, that have a mutual interaction but are otherwise isolated. Newton reasoned that the interaction between these objects involves a pair of forces—the force exerted on *A* by *B* and that exerted on *B* by *A*. Let us call the first of these forces $\mathbf{F}_{AB}$ and the second $\mathbf{F}_{BA}$. Newton postulated that

$$III. \ \mathbf{F}_{AB} = -\mathbf{F}_{BA} \qquad\qquad (5-5)$$

This relationship constitutes *Newton's third law*, and in his words,

III. To every action (force) *there is always opposed an equal reaction* (force); *or, the mutual actions* (forces) *of two bodies upon each other are always directed to contrary parts* (in opposite directions).

This law states that the forces in an action-reaction pair have the same magnitude but opposite directions. The symmetry of the situation (attaching no more importance to object *A* than *B*) permits us to call either $\mathbf{F}_{AB}$ or $\mathbf{F}_{BA}$ the ''force'' to which the other is then the ''reaction force.''

▶ *Collisions.* Newton was partially led to the formulation of the third law by observing the collision of objects. Consider two objects essentially isolated from the surroundings, for example, two small flat discs *A* and *B* moving on a horizontal air table. Suppose that moving with velocities $\mathbf{v}_A$ and $\mathbf{v}_B$ their paths are such that they collide. During collision, a mutual force acts between them for a short duration, altering the individual motions. The two discs separate with new velocities $\mathbf{v}_A'$ and $\mathbf{v}_B'$. Newton's observations for similar situations led him to note, however, that the vector-sum quantity $m_A\mathbf{v}_A' + m_B\mathbf{v}_B'$ (called the *linear momentum*; see Chapter 9) after collision was always equal to the corresponding vector sum $m_A\mathbf{v}_A + m_B\mathbf{v}_B$ prior to collision. This is equivalent to stating that the quantity $m_A\mathbf{v}_A + m_B\mathbf{v}_B$ is a constant in time. Thus,

$$\frac{d}{dt}(m_A\mathbf{v}_A + m_B\mathbf{v}_B) = 0$$

or

$$m_A\frac{d\mathbf{v}_A}{dt} = -m_B\frac{d\mathbf{v}_B}{dt}$$

Noting that $d\mathbf{v}_A/dt = \mathbf{a}_A$ and $d\mathbf{v}_B/dt = \mathbf{a}_B$, and using the second law, we have

$$m_A\mathbf{a}_A = -m_B\mathbf{a}_B$$

or

$$\mathbf{F}_{AB} = -\mathbf{F}_{BA}$$

Thus, the third law is simply another way of stating that the linear momentum of the two interacting but otherwise isolated objects is conserved (i.e., independent of time). We return to this important conservation principle in Chapter 9. ◀

Consider a test object *A* and a standard spring scale *B* (as in Fig. 5–4). The force exerted on the test object by the spring is the force $\mathbf{F}_{AB}$, and the force exerted on the spring by the test object is the reaction force $\mathbf{F}_{BA}$. Figure 5–5 shows these two forces, which are related by Eq. 5–5: $\mathbf{F}_{AB} = -\mathbf{F}_{BA}$.

The third law tells us that forces always occur in pairs. But notice carefully that the

---

*We frequently deal with problems that involve vector quantities and we often write scalar equations for these quantities. When a particular coordinate frame or positive direction has been chosen, the equations contain *components of the vectors* so that each quantity has an associated algebraic sign. In the example here, the frictional force is directed toward the left, and we write $-f$ in the second-law equation.

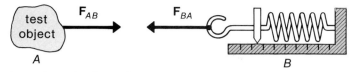

Fig. 5-5. Action and reaction forces for the situation illustrated in Fig. 5-4.

two forces in every action-reaction pair always act on *different* objects. In Fig. 5-5, the force $\mathbf{F}_{AB}$ acts on A, whereas the force $\mathbf{F}_{BA}$ acts on B. To determine the motion of an object, we need to know only the forces acting *on* that object; it does not matter that the object exerts forces on other bodies.

Example 5-4

In deep space, some agency (a disembodied hand) applies a constant force of 2 N to a block A that is in contact with a block B, as shown in diagram (a).

(a) Identify all mutual action-reaction force pairs.

(b) If the mass of block A is $m_A = 1$ kg and that of block B is $m_B = 2$ kg, what is the resulting acceleration of the two blocks? What force acts on block B?

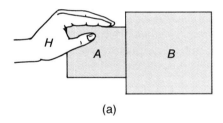

(a)

Solution:

Diagram (b) is an "exploded view" of the situation, isolating the three objects involved. (The hand and the two blocks remain in contact throughout the time during which we are considering the motion.)

(a) The force that the hand H exerts on block A is $\mathbf{F}_{AH}$; block A exerts the reaction force $\mathbf{F}_{HA}$ on the hand. Block A also exerts a force $\mathbf{F}_{BA}$ on block B, and block B exerts the reaction force $\mathbf{F}_{AB}$ on block A. We observe that there are two separate pairs of mutual forces: one pair is $\mathbf{F}_{AH} = -\mathbf{F}_{HA}$, and the other pair is $\mathbf{F}_{BA} = -\mathbf{F}_{AB}$.

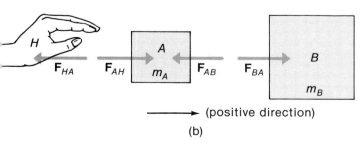

(positive direction)

(b)

(b) For the force on block B, using Eq. 5-3 and taking the direction to the right as positive, we have

$$F_{BA} = m_B a_B$$

Because the net force on block A is $F_{AH} - F_{AB}$, we again apply Eq. 5-3 with the result

$$F_{AH} - F_{AB} = m_A a_A$$

We add these two equations, and realizing that the blocks remain in contact so that $a_A = a_B = a$, we obtain

$$F_{AH} - F_{AB} + F_{BA} = (m_A + m_B)a$$

From the third law we know that F_{AB} and F_{BA} have the same magnitude; thus, we arrive at the result

$$F_{AH} = (m_A + m_B)a$$

Blocks A and B, because they are in contact, have the same motion. Consequently, we could consider them to be a single unit with a mass $m_A + m_B$ upon which there acts only the *external* force $\mathbf{F}_{AH}$, resulting in a common acceleration a, as indicated in diagram (c). Then, we would obtain exactly the same expression for F_{AH}. In this view, the forces $\mathbf{F}_{BA}$ and $\mathbf{F}_{AB}$ are simply canceling internal forces that behave in the same way as all the internal molecular forces present in the blocks.

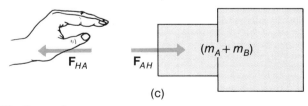

(c)

Finally, we have

$$a = \frac{F_{AH}}{(m_A + m_B)} = \frac{2\text{ N}}{(1\text{ kg} + 2\text{ kg})} = \frac{2}{3}\text{ m/s}^2$$

Also, from the above,

$$F_{BA} = m_B a = (2\text{ kg})\left(\frac{2}{3}\text{ m/s}^2\right) = \frac{4}{3}\text{ N}$$

We learn from this example that when several objects participate in a common motion, we may conveniently group them together into a single unit, provided that in using Eq. 5–3, we consider *only* the forces that are externally impressed on the combined unit. Naturally, if we wish to learn the force that one of these objects exerts on another (as in the present case), we must consider the objects separately.

5–3 MASS AND WEIGHT

In this section, we continue the discussion of the concept of mass and examine some of its important properties, particularly in the presence of the Earth's gravity.

The Gravitational Force.   Another of the important contributions of Isaac Newton was his formulation of the theory of gravitation. Newton postulated that every pair of particles exerts on one another a mutual attractive gravitational force that is directly proportional to the product of their masses and inversely proportional to the square of the distance between them. The magnitude of the gravitational force between two particles with masses m and M, separated by a distance r, can be expressed as

$$F_g = G\frac{mM}{r^2}$$ (5–6)

where G is the *universal gravitation constant* and has the value 6.673×10^{-11} N·m^2/kg^2. In Section 13–2 Newton's formulation of the theory of gravitation is summarized and the force law is given in full vector notation (see Eq. 13–1). It is also shown, following Newton, that for solid spherical objects,* the distance r is the distance between the *centers* of the objects.

▶ *The Determination of G.*   The value of the gravitation constant G can be determined by measuring the force between a pair of objects with known masses that are separated by a known distance. The first experiment from which a reasonably accurate value for G could be obtained was carried out by the English chemist Henry Cavendish (1731–1810), who reported his results to the Royal Society in 1798. For his measurement, Cavendish used an instrument (called a *torsion balance*) that was similar to the one invented independently by Charles Coulomb and used by him in studies of electric and magnetic forces. The type of torsion balance used by Cavendish (see Fig. 5–6) consists of a thin rod that carries two small balls with masses m_1 and m_2. The rod is attached to the lower end of a wire that is suspended from a fixed support. The apparatus is allowed to come to rest with the large balls, M_1 and M_2, some distance away. This equilibrium position is measured by directing a light beam from a stationary source onto a mirror that is attached to the suspension wire; the point at which the reflected beam is seen on the scale is recorded. Next, the large balls are moved close to the small balls; this corresponds to the situation shown in Fig. 5–6. Because M_1 attracts m_1 and M_2 attracts m_2, a torque† is exerted on the rod. The rod rotates through a small angle until the restoring torque due to the torsional force in the twisted wire just balances the gravitational torque. The angular displacement of the rod is determined by noting the new position of the light spot on the scale. The force necessary to twist the wire by a certain amount (the so-called torsional constant) is related to the period of oscillation of the system, which can be measured in a separate experiment. In this way Cavendish measured the gravitational force between known masses separated by a known distance. The value of G obtained from Cavendish's data is only 1 percent different from the value now accepted.   ◀

Weighing the Earth.   The gravitational force exerted by the Earth (mass M_E) on an object (mass m) located on or near the Earth's surface is the *attractive* force,

---

*Precisely, for spherical objects whose density depends at most on the radial distance from the center.
†The idea of torque is discussed in Chapter 10. A torque applied to an object tends to make the object rotate.

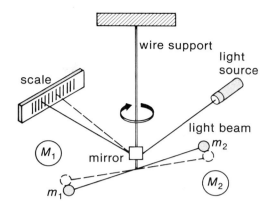

Fig. 5-6. Schematic diagram of the apparatus used by Henry Cavendish to determine the value of G.

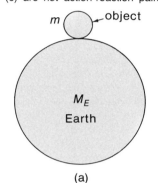

Fig. 5-7. (a) An object at rest on the Earth's surface. (b) The forces acting on the *object*. (c) The forces acting on the *Earth*. The forces shown together in (b) and (c) are *not* action-reaction pairs.

$$F_g = G\frac{mM_E}{R_E^2}$$

where R_E is the radius of the Earth. Now, in Section 2-5 we learned that near the Earth's surface any unsupported body falls downward with the constant acceleration g (assuming negligible air resistance). According to Newton's second law, accelerated motion requires the presence of a force; in this case, the force is the gravitational force,* and we write the second law as†

$$F_g = mg$$

If we equate these two expressions for the gravitational force on the object, we see that the mass m can be eliminated; then, solving for M_E, we find

$$M_E = \frac{gR_E^2}{G} \tag{5-7}$$

which expresses the mass M_E of the Earth in terms of the measurable quantities, g, R_E, and G. Substituting the values of these quantities, we find $M_E = 5.98 \times 10^{24}$ kg. Thus, the Cavendish experiment, in which the value of G was determined, is sometimes called "weighing the Earth."

Force Pairs. Consider an object with a mass m at rest on the Earth's surface. Figure 5-7 shows the situation, exaggerating the size of the object for illustrative purposes. In this condition there is no motion at all of the object relative to the Earth. As indicated in Fig. 5-7b, the object is acted upon by two forces—the Earth's attractive gravitational force $\mathbf{F}_g$ and the contact force $\mathbf{N}$ with the ground.‡ Because the object is at rest with respect to the Earth, $\mathbf{F}_g + \mathbf{N} = 0$, so that

$$\mathbf{N} = -\mathbf{F}_g \tag{5-8}$$

The Earth is also acted upon by two forces—the contact force $\mathbf{w}$ with the object and the attractive gravitational force $\mathbf{f}_g$ exerted by the object (Fig. 5-7c). Again, because there is

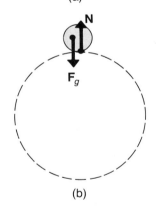

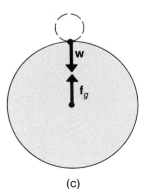

---

*The value of g varies slightly with geographic location and altitude. This variation is ascribed to the dependence of F_g on geometric features of the Earth and to kinematic effects due to the rotation of the Earth, *not* to any changes in m.

†In writing this expression, we make the assumption that the mass m that appears in the gravitational force equation (Eq. 5-6) is the same as the inertial mass m that appears in Newton's second law. This point is discussed further on page 80.

‡The contact force $\mathbf{N}$ is often referred to as the *normal* force because it is normal (i.e., perpendicular) to the surface of contact.

no relative motion, $\mathbf{w} + \mathbf{f}_g = 0$, so that

$$\mathbf{w} = -\mathbf{f}_g \tag{5-9}$$

Notice carefully that the forces appearing in Eq. 5–8 and those in Eq. 5–9 are *not* action-reaction pairs. The force pairs are

$$\mathbf{F}_g = -\mathbf{f}_g \qquad \text{and} \qquad \mathbf{N} = -\mathbf{w}$$

Comparing these equations with Eqs. 5–8 and 5–9, we conclude that all four forces— $\mathbf{F}_g$, $\mathbf{f}_g$, $\mathbf{N}$, and $\mathbf{w}$—have the same magnitude.

▶ *Inertial and Gravitational Mass.* In Section 5–2 we noted that the quantity m, the proportionality factor connecting the force applied to an object and its acceleration, is called the *inertial mass* of the object. The value of m determines the magnitude of the applied force necessary to produce a particular acceleration. An experiment designed to measure the inertial mass of an object could be carried out in deep space using a spring force. As shown in Fig. 5–8a, the spring force $\mathbf{F}_s$ applied to the object produces an acceleration $\mathbf{a}$. The gravitational force does not enter into such an experiment at all.

Next, imagine performing an experiment with the same spring scale and the same object at a point near the Earth's surface. The spring scale is attached to a post that is anchored to the Earth and the object is suspended from the lower end, as shown schematically in Fig. 5–8b. When the object hangs motionless, the downward gravitational force $\mathbf{F}_g$ is just balanced by the upward spring force $\mathbf{F}_s$. Let us suppose that we cleverly chose the conditions in the first case (Fig. 5–8a) so that the spring elongation is exactly the same in the two experiments. That is, the magnitude of the accelerating force in Fig. 5–8a is equal to the gravitational force in Fig. 5–8b.

Now, in the second experiment (Fig. 5–8b), there is no acceleration; hence, dynamic effects are not involved. The magnitude of the gravitational force depends on some property of the object and we call this property the *gravitational mass m′*. This is the mass that appears in Newton's universal law of gravitation (Eq. 5–6).

We now ask a profoundly important question: Is the gravitational mass of an object equal to its inertial mass? We can answer this question by performing a third experiment. In Fig. 5–8b, we cut the string that attaches the object to the spring scale and measure the acceleration of the object as it falls toward the Earth. During the fall, the force acting on the object (namely, $\mathbf{F}_g$) has exactly the same magnitude as the spring force $\mathbf{F}_s$ that acted in the first experiment. Therefore, if indeed we have $m = m'$, the acceleration due to gravity g must exactly equal the acceleration a measured in the first experiment. If the accelerations are not equal, we conclude that gravitational mass and inertial mass are not the same.

Clearly, the experiments we have been discussing would be very difficult to perform. However, there are equivalent experiments that *can* be performed with considerable precision to answer the same question. Newton was aware of this problem, and he conducted experiments with pendula to determine whether $m = m'$. He concluded that there was no difference between the two types of mass to within 1 part in 1000. In 1890 Eötvös* devised an ingenious method to test the equivalence of inertial and gravitational mass. Using two objects made from different materials, he compared the effect of the Earth's gravitational force with the inertial effect of the Earth's rotation. Eötvös was able to conclude that $m = m'$ to within about 5 parts in 10^9. In recent years, Robert Dicke of Princeton University and V. Braginskii at Moscow University have improved upon the Eötvös experiment and have established the equivalence of gravitational and inertial mass to within 1 part in 10^{12}.

There is nothing in Newtonian theory or in our experience that would lead us to expect gravitational and inertial mass to be the same. This is a remarkable result for which no one has a completely satisfactory explanation. However, the hypothesis of the *exact* equality of gravitational and inertial mass is an essential ingredient in the general theory of relativity. This important idea is called the *principle of equivalence.* ◀

Weight. Let us return to the discussion of the forces $\mathbf{w}$ and $\mathbf{N}$ that appear in Eqs. 5–8 and 5–9. The contact force $\mathbf{w}$ that an object exerts on whatever is supporting it is called the *weight* of the object. In the case illustrated in Fig. 5–7, the weight $\mathbf{w}$ of the object is the force it exerts on the Earth. Because there is no acceleration of the object in this instance, $\mathbf{w}$ is equal to the gravitational force $\mathbf{F}_g$ acting on the object. We stress, however, that $\mathbf{F}_g$ acts *on the object,* whereas $\mathbf{w}$ acts *on the surface of the Earth.* The magnitude of $\mathbf{w}$ is $w = mg$ (for this case of zero acceleration).

A different situation is shown in Fig. 5–9 where the mass m rests on the floor of an elevator that can be made to accelerate either up or down. Imagine that the elevator is accelerating upward with constant acceleration $\mathbf{a}$. Further, let us imagine that, compared

---

*Roland von Eötvös (1848–1919), a Hungarian baron, invented many scientific instruments, including a sensitive gravimeter used to determine the density of subsurface rock strata.

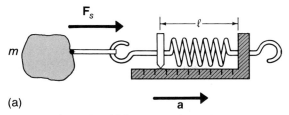

(a)

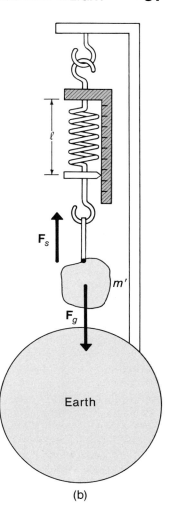

Fig. 5-8. A deep-space experiment in which a spring force $\mathbf{F}_s$ applied to an object produces an acceleration $\mathbf{a}$. (b) The same object and spring scale are suspended near the Earth's surface. If the spring elongation is the same as in (a), the gravitational force $\mathbf{F}_g$ has the same magnitude as the spring force $\mathbf{F}_s$ in (a). If the object in (b) is now released, it will have an acceleration g equal to a only if the gravitational mass m' of the object is equal to its inertial mass m.

with the gravitational force of the Earth, the gravitational attraction between the object and the elevator is negligible. Then, focusing attention on m, and taking the upward direction to be positive, we have

$$N - F_g = ma$$

or

$$N = F_g + ma = mg + ma = m(g + a)$$

Here, $\mathbf{N}$ is the force the elevator floor exerts on the object and $\mathbf{F}_g$ is the gravitational force the Earth exerts on the object. The reaction force to $\mathbf{N}$ is $\mathbf{w}$, the weight of the object, because this is the force the object exerts on the supporting elevator floor. These two forces have the same magnitude, so the weight of the object is

$$w = m(g + a) \qquad \textbf{(5-10a)}$$

The weight is therefore *increased* by the amount ma over the weight at rest. Note that the gravitational force on the object is still $F_g = mg$.

If the elevator accelerates downward with an acceleration a', we have

$$w = m(g - a') \qquad \textbf{(5-10b)}$$

The weight is therefore *reduced* by the amount ma' from the weight at rest. Again, the gravitational force on the object remains $F_g = mg$.

Fig. 5-9. (a) An object with a mass m is on the floor of an elevator that is accelerating upward. (b) Forces acting on the object. (c) Forces acting on the elevator.

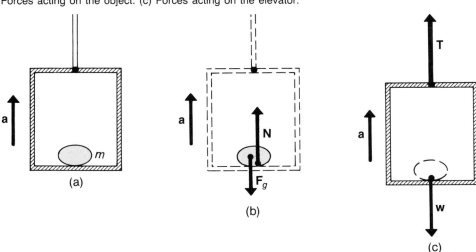

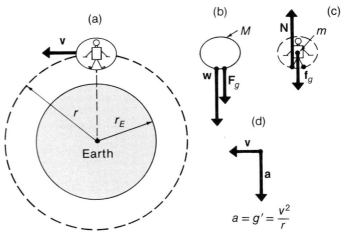

Fig. 5-10. An astronaut in an orbiting artificial satellite is *weightless*.

If the cable that supports the elevator were to break, the elevator would be in free fall and would accelerate downward with $a' = g$. According to Eq. 5-10b, we would have

$$w = 0 \quad \text{(for } a' = g\text{)}$$

In this situation, we say that the object is *weightless,* although the gravitational force on the object remains equal to mg. An object is weightless if it does not exert a force on any supporting platform or suspending rope.*

Consider an artificial satellite that moves around the Earth in a circular orbit with radius r (Fig. 5-10a). The satellite encounters no resistance, and even though the path is circular, the satellite undergoes free fall. At the position of the orbit, the acceleration due to gravity is g', which must be equal to the centripetal acceleration of the satellite, v^2/r, in order to maintain the circular orbital motion. Now, suppose that there is an astronaut in the satellite. The total force F_s acting on the satellite is the gravitational force F_g plus the weight w of the astronaut (Fig. 5-10b); this force is equal to the mass M of the satellite multiplied by its acceleration:

$$F_s = F_g + w = M\frac{v^2}{r} = Mg'$$

The total force f_a acting on the astronaut is the gravitational force f_g plus the (upward) contact force N (Fig. 5-10c); this force is equal to the mass m of the astronaut multiplied by his acceleration:

$$f_a = f_g - N = m\frac{v^2}{r} = mg'$$

Now, according to the usual definition, we have $F_g = Mg'$ and $f_g = mg'$. Substituting these expressions into the equations for F_s and f_a, we conclude that $w = 0$ and $N = 0$. Thus, the astronaut is weightless. This is the result we would expect: both the satellite and the astronaut are undergoing free fall, so the astronaut can exert no net force on the

---

*In most other texts you will find *weight* defined as the gravitational force acting on an object: $\mathbf{w} = m\mathbf{g}$. Then, in the case of acceleration, mg is called the *true weight* of the object and the force exerted by the object on any support is called the *apparent weight.* This definition seems unnecessarily complicated (and it is almost never used in a consistent manner); moreover, it is also contrary to the usual sense of weight as the force exerted *by* the object.

satellite. (The astronaut could push on the ''ceiling'' and the ''floor'' at the same time, but this would not result in the application of any net force and would not constitute *weight*.)

The definition of weight given here can be applied with a consistent interpretation in all situations. Even in complex circumstances involving accelerations **a** not in the direction of **g**—such as a block accelerating down an inclined plane or a person revolving in a fairground ride or an astronaut experiencing the ''artificial gravity'' of a rotating space station—the weight of the object or person is always equal to the (vector) force exerted *on* the supporting object or system. (See Problems 5–26 and 5–32.)

▶ *The Pound as a Unit of Force.* For many years engineers used almost exclusively the British system of units in which the *pound* is a unit of *force*. In this system, 1 pound (lb) is taken to be the weight of an unaccelerated object with a mass of exactly 0.45359237 kg at a location where the acceleration due to gravity has its ''standard'' value, $g = 9.80665$ m/s^2 = 32.1740 ft/s^2 (see Section 2–5). Thus, the relationship connecting the British and metric units of force is

$$w = mg$$

$$1 \text{ lb} = (0.45359237 \text{ kg})(9.80665 \text{ m/s}^2)$$

$$= 4.44822 \text{ N}$$

In the British system, the unit of mass, ingloriously called the *slug*, is defined as that mass which, when subjected to a net force of 1 lb, will undergo an acceleration of 1 ft/s^2.

A shopper in a grocery store must contend with an untidy and incorrect usage of the units of mass and weight regardless of the system that prevails. According to the *legal* definition, the pound is a unit of mass: 1 lb = 0.45359237 kg (exactly). However, in a store that operates under the British system, the shopper will find weight measured in *pounds*; if the metric system is used, weight will be measured in *kilograms*! Thus, an international shopper will have to reckon that a 1-pound steak bought in New York will be considered to ''weigh'' 0.454 kg in Paris markets.

We include mention of these points for reference when consulting older texts. In this book, we use only kilograms for mass and newtons for force.   ◀

Example 5–5

The spaceship shown in the diagram is traveling in deep space and is undergoing an acceleration, relative to an inertial reference frame, which happens to be 9.80 m/s^2. What is the weight of a 2-kg object that rests on the ''floor'' of the spaceship cabin?

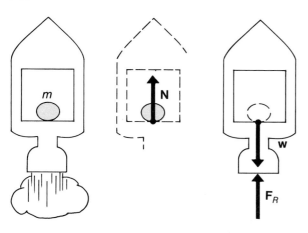

Solution:

According to an inertial-frame observer, the object ($m = 2$ kg) is accelerating (along with the spaceship) at a rate g. Hence, the contact force N with the floor (which is the *only* force acting on the object) must be

$$\mathbf{N} = m\mathbf{a}$$

Because $a = g$, we have $N = mg$. Also, the weight is the reaction force to N; therefore,

$$w = mg = (2 \text{ kg})(9.80 \text{ m/s}^2) = 19.6 \text{ N}$$

It would thus appear that, to an observer in the spaceship, the mass m rests on the floor with its natural (Earth-environment) weight. This result is consistent with the principle of equivalence. (• Suppose that an observer who is confined to the spaceship interior—with no windows—attempts to discover whether the ship is accelerating through space or is simply sitting on the Earth's surface. Can the observer, by performing experiments within the spaceship, distinguish between these two conditions? See Problem 5–16.)

5–4 SOME COMMENTS ON NEWTON'S LAWS

▶ In setting down his three laws of dynamics, Newton relied heavily on intuitive ideas concerning the terms used in the statements of the laws, terms otherwise not defined in any rigorous way. For example, the second law gives the relationship between force and mass, but it does not tell us what either force or mass really is. We might even paraphrase the statement of the second law to read: "An object will not accelerate until something makes it accelerate, and that something is defined as a force." Indeed, in one approach to Newton's laws, force is defined in terms of the acceleration it produces. Although this procedure has some merit, we have elected first to gain some intuitive appreciation for the concept of force by considering several simple experiments with springs. Additional experiments concerning the effect of force on the motion of objects then led us to the second law and the concept of mass. In this way we avoid the logical difficulties that confound us when Newton's laws are considered alone.

We have asserted that Newton's laws are valid in any inertial (nonaccelerating) reference frame. It is not difficult to see why this is so. In Section 4–6 we showed that two observers, S and S', in relative motion with a constant velocity will measure the same acceleration for a particle, $\mathbf{a} = \mathbf{a}'$. That is, acceleration is an invariant quantity, or a free vector, as quantities that are independent of the reference frame are called. If the frame of observer S is an inertial reference frame, he concludes that a force $\mathbf{F} = m\mathbf{a}$ is acting on the particle. Then, the frame of observer S' is also an inertial reference frame, and he concludes that a force $\mathbf{F}' = m\mathbf{a}'$, is acting on the particle. But $\mathbf{a} = \mathbf{a}'$, and we have

$$\mathbf{F} = m\mathbf{a} = m\mathbf{a}' = \mathbf{F}'$$

Thus, $\mathbf{F} = \mathbf{F}'$, and the second law has exactly the same form in each frame. We conclude that if $\mathbf{F} = m\mathbf{a}$ is valid in a frame S, it is also valid in any other frame S' that, at most, moves with a constant velocity and without rotation relative to S. Moreover, $\mathbf{F} = m\mathbf{a}$ is the invariant form of the second law in all inertial frames. *Force* $\mathbf{F}$ *and acceleration* $\mathbf{a}$ *are free vectors and have the same magnitude and direction regardless of the particular inertial reference frame used to describe them.* ◀

QUESTIONS

5–1 Webster's dictionary defines inertia as "disinclination to move." Explain how mass may be thought of as a measure of inertia.

5–2 How might a pre-Newtonian experimenter verify that a spring scale, such as that shown in Fig. 5–1a, obeys Hooke's law (and thereby gives an operational meaning to force ratios exerted by the spring) if the spring scale is suspended vertically from a beam and used to weigh many identically fabricated weights in various combinations? Contrast this procedure for defining force ratios with that given in Section 5–1, noting in particular that now the Earth's gravity is involved.

5–3 State Newton's three laws in your own words.

5–4 A pragmatic assertion concerning the Newtonian laws might be: "It doesn't matter that Newton did not lay out a perfect axiomatic system. What is important is the *program* Newton proposed: Find the general force laws operating in Nature and describe the ensuing motion of systems when subjected to them." Discuss the merits of this point of view.

5–5 A loaded wagon is set into motion by a tractor. According to Newton's third law, the tractor pulls on the wagon with a force that is equal and oppositely directed to the force with which the wagon pulls back on the tractor. Why does the wagon move? Why does the tractor move?

5–6 An object in deep space has two forces of equal magnitude acting on it. Describe the resulting acceleration as the angle between the two forces is varied slowly from 0° to 180°.

5–7 Describe the essential differences between the concepts of mass and weight.

5–8 Give some examples of an object in the state of weightlessness.

5–9 Two students exerting a steady pull on the two ends of a string are unable to break it no matter how hard they try. Yet, when they tie one end of the string to a sturdy post and together pull on the free end with the same maximum steady force, they succeed in breaking the string. Explain.

5–10 You are given a rope not quite strong enough to support a suspended heavy crate. Explain why it is possible to use the same rope to restrain the crate so that it slides slowly down even a very slippery inclined plane without breaking the rope. Would friction between the sliding surfaces help?

PROBLEMS

Section 5-2

5-1 A 2-kg object originally judged to be at rest in an inertial frame is subjected to a constant force of 3 N for a period of 10 s. What final speed does it reach? What displacement does it undergo during the first 10 s? During the first 20 s?

5-2 The velocity of a particle with a mass of 4 kg that moves along the x-axis is determined at various instants. The results are $v(t = 1\text{ s}) = 5$ m/s, $v(2) = 8$ m/s, $v(3) = 11$ m/s, and $v(4) = 14$ m/s. Write down the simplest possible expression for $v(t)$ and give the acceleration of the particle. What force acts on the particle?

5-3 A 0.50-kg particle has an acceleration $\mathbf{a} = \alpha\hat{\mathbf{i}} + \beta\hat{\mathbf{j}}$, with $\alpha = 4$ m/s^2 and $\beta = -2$ m/s^2. What is the magnitude of the force that acts on the particle?

5-4 A rifle bullet with a mass of 12 g, traveling with a speed of 400 m/s, strikes a large wooden block, which it penetrates to a depth of 15 cm. Determine the magnitude of the frictional force (assumed constant) that acts on the bullet.

5-5 An object with a mass of 500 g is held by a string in a circular orbit with a diameter of 1 m as it rotates on a horizontal air table. If the maximum tension the string can support before breaking is 100 N, with what linear speed will the object leave its orbit if it is spun ever faster until the string breaks?

5-6 What constant force is required to accelerate a 2000-kg automobile from rest to 90 km/h in 12.2 s?

5-7 What constant force must be applied to a 5000-kg space vehicle (in deep space) to accelerate it from rest to 10,000 km/h during a time interval of 1 y?

5-8 An automobile ($m = 2200$ kg) moves with a velocity ($v = 32$ m/s) on a level road. If a constant braking force of 6000 N is applied to the automobile wheels, what distance will the vehicle travel after the brakes are applied?

5-9 A particle with a mass $m = 4$ kg is known to be acted upon by the two forces F_1 and F_2, shown in the diagram (right). The particle moves with the constant velocity shown. What third force must be acting on the particle?

5-10 A freight train that consists of 100 equal-mass box cars is
• pulled along a straight stretch of horizontal track by a diesel engine delivering a constant force of 10^6 N to the coupling of the leading car. Neglecting any frictional drag due to the tracks or air resistance, what is the tension in the front coupling of the leading car? In the front coupling of the last car? In the front coupling of the nth car, where n runs from 1 at the front to 100 at the end?

5-11 A rope with a mass $m = 0.8$ kg is attached to a block with a mass $M = 4$ kg. The rope-block combination is pulled across a horizontal frictionless surface by a 12-N force that is applied to the free end of the rope. What is the acceleration of the system? What force does the rope exert on the block?

5-12 A particle with a mass of 0.80 kg moves in a circular path with a radius of 2 m, completing each revolution in a time of 4 s. What is the magnitude of the force acting on the particle? If, at some instant, the velocity of the particle is $\mathbf{v} = v\hat{\mathbf{i}}$, and the motion is in the x-y plane, what are the two possible vector expressions for the force? (Evaluate v.)

Section 5-3

5-13 An object with a mass of 2 kg is suspended from the ceiling of an elevator by a calibrated ideal (massless) spring scale, as shown in the diagram on page 86. (a) If the elevator is moving upward with a constant velocity v, what is the spring scale reading in newtons? (b) If the elevator is accelerating upward with a constant acceleration of 3 m/s^2, what is the spring scale reading? (c) If the elevator is accelerating downward with a constant acceleration of 2 m/s^2, what is the spring scale reading?

5-14 The Earth orbits the Sun at a mean distance of 1.496×10^8 km with a period of 365.25 days. (a) Use Eqs. 4-26, 5-3, and 5-6 to "weigh" the Sun. (b) The mean radius of the Sun is 6.96×10^5 km. Calculate the gravitational acceleration of a particle at that position. (For the Sun, a ball of gas, the concept of a "surface" doesn't have much meaning.)

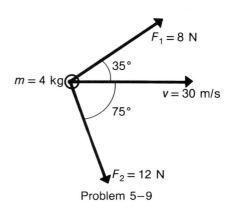

Problem 5-9

85

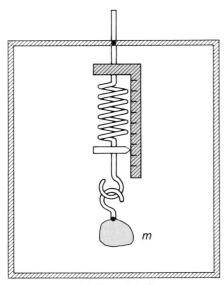

Problem 5–13

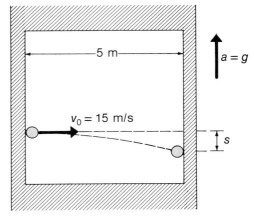

Problem 5–16

5–15 An experimental rocket with a mass of 10^4 kg is to be
• launched straight upward. At $t = 0$ the engine is started, and the force that it applies to the rocket increases from zero according to $F(t) = (3.92 \times 10^3 \text{ N/s}^2)t^2 + (8.82 \times 10^3 \text{ N/s})t$, where t is given in seconds. (a) Find the equation that gives the contact force exerted on the rocket by the ground. (b) At what time does the rocket begin to accelerate upward? (c) What is the rocket's acceleration at $t = 5$ s? (Neglect the change in the mass of the rocket as its fuel is ejected; this refinement will be considered in Section 9–5.)

5–16 Suppose the experimenter in the accelerating spaceship of
•• Example 5–5 throws a ball across the cabin from wall to wall (a distance of 5 m), giving the ball an initial velocity of 15 m/s parallel to the floor. If the spaceship is accelerating with $a = g$, how much closer to the floor is the impact point on the far wall than is the launching point? On the basis of your result, what answer would you give to the question asked at the end of Example 5–5? [*Hint:* Consider this problem from the point of view of an observer in an inertial frame. What motion does he ascribe to the ball? How long does it take for the ball to cross from wall to wall? How far did the laboratory itself move during this time?]

5–17 A vendor sets up shop in an elevator where he buys and
• sells radishes by the newton. (Formerly, he bought and sold by the pound.) The vendor buys a bundle of radishes when the elevator is accelerating downward at a rate of 3.5 m/s², and he sells the same bundle at the same price per newton when the elevator is accelerating upward at 2.5 m/s². What is the vendor's percentage profit?

Additional Problems

5–18 An object with a mass of 3 kg is pulled upward by a string that has a breaking strength of 50 N. What is the maximum acceleration the object can be given without breaking the string?

5–19 The largest-caliber antiaircraft gun operated by the Luftwaffe during World War II was the *12.8-cm Flak 40*. This weapon fired a 25.8-kg shell with a muzzle velocity of 880 m/s. What propulsive force was necessary to attain the muzzle velocity within the 6.0-m barrel? (Assume constant acceleration and neglect the Earth's gravitational effect.)

5–20 A slingshot accelerates a 12-g stone to a velocity of 35 m/s within a distance of 5 cm. To what (constant) force is the stone subjected during the acceleration?

5–21 A flat disc with a mass of 500 g is sent sliding along a frozen pond with an initial speed of 10 m/s. If the sliding frictional force exerted by the ice surface on the disc is a consant 0.5 N, in what distance will the disc come to rest?

5–22 A block with a mass of 1 kg slides along a horizontal
• surface while constrained to a circular path of 50-cm radius by a string connected at the other end to a stationary frictionless pivot. Assume that the block experiences a frictional force with a magnitude of 2 N and in a direction always exactly opposed to the instantaneous velocity. If the block is given an initial tangential velocity of 5 m/s, find the string tension as a function of time thereafter.

5–23 Two identical particle-like objects ($m = 15$ kg) are separated by a distance of 5 m in deep space. (a) What force does one object exert on the other? (b) What is the initial acceleration of either object?

5–24 A particle ($m = 2.5$ kg) moves along the x-axis with a
  • constant velocity ($v_x = 1.6$ m/s). In the region from
    $x = 5.0$ m to $x = 10.0$ m, the particle experiences a con-
    stant force $F_y = 8$ N that acts in the $+y$-direction. (a) If
    the particle was at $x = 0$ at $t = 0$, when will it escape
    the effects of the force? (b) What is the value of y when
    $x = 10.0$ m? (c) What is the velocity vector of the parti-
    cle after it escapes the effects of the force?

5–25 What is the acceleration due to gravity at a height above
    the Earth's surface equal to one Earth diameter?

5–26 Refer to Fig. 5–10. Deduce numerical values for all
  • quantities shown if the mass of the capsule is 2×10^3 kg
    and the mass of the astronaut is 90 kg. Assume that the
    satellite is orbiting in a circular path 10^3 km above the
    Earth's surface where $g' = 7.32$ m/s². (The Earth's ra-
    dius is 6.38×10^3 km.)

5–27 A small bob with a mass of 250 g is suspended by a string
    from the ceiling of a truck that is accelerating at a constant
    rate of 2 m/s² as it moves along a flat, straight road, as
    shown in the diagram. Assuming the bob to be suspended
    motionless with respect to the truck, what is the tension in
    the string? What angle does the string make with the verti-
    cal? (The force exerted on the bob by the string—the
    string tension—is directed along the string.)

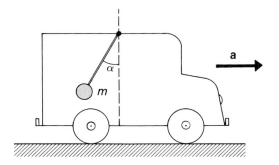

5–28 A subway car travels at a speed of 30 km/h around
    an unbanked (flat) curve. The freely suspended hand
    straps are observed to swing out and hang at an angle of
    10° from the vertical. What is the radius of the curve?

5–29 A grasshopper is observed to jump 2 m horizontally, ris-
  • ing to a maximum height of 1 m along the way. During
    takeoff, the entire body mass of the grasshopper, consid-
    ered to be concentrated at a single point, moves a distance
    of 1.5 cm while its hind feet are still in contact with the
    ground. What effective constant force, in terms of its
    weight, did the feet of the grasshopper exert on the
    ground? Could you do as well in terms of your body

weight? [*Hint:* Use your knowledge of ballistic trajectories to determine the grasshopper launch speed v_0. This
can be immediately related to the required (constant) acceleration imparted by its legs. The grasshopper's mass (a
few grams) is not needed to make the calculation.]

5–30 Consider a platform that is rotating at a constant angular
  •• speed $\omega = 2$ rad/s. The platform is located on the Earth's
    surface and its plane is perpendicular to **g**. A man
    ($m = 80$ kg) stands on and rotates with the platform.
    Make a diagram showing the forces that the man exerts on
    the platform when he stands a distance r from the axis of
    rotation. Identify the vector **w** that represents the man's
    weight. Calculate the weight w for $r = 1, 2, 4, 8, 12$ m.
    What direction does the man sense as "up"? Calculate
    the angle α that "up" makes with the vertical for the
    same values of r. Suppose that the platform is covered
    with soil and that seeds are planted uniformly throughout.
    If the platform continues to rotate while the plants grow,
    describe the appearance of the crop.

5–31 An object with a mass of 1 kg is at rest on the surface of
    the Earth at the Equator. What is the weight of the object?
    [*Hint:* Remember that the Earth rotates.]

5–32 A space station in the shape of a torus (a donut) and with
    an outer radius of 150 m revolves once every minute
    about its central axis. What is the weight of an 80-kg
    astronaut who is standing on the outer rim?

5–33 A pilot performing aerobatic maneuvers is in a vertical
    dive at a speed $v_0 = 600$ km/h. He pulls out of the dive
    along a circular arc maintaining constant speed, eventu-
    ally climbing vertically. (a) If the pilot is to experience an
    acceleration not to exceed $7g$, what is the minimum radius
    R of the circle? (b) How long after the start of the pullout
    at time t_0 will the aircraft be climbing vertically?
    (c) Construct a diagram of the forces acting on the air-
    craft when it has a heading 45° below the horizon.

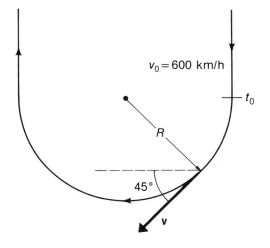

CHAPTER 6

APPLICATIONS OF NEWTON'S LAWS

In the preceding chapter we introduced and discussed Newton's laws of motion, and we presented a few examples to illustrate the principles involved. The range of application of Newtonian dynamics is surprisingly extensive. We now show the richness of these applications with further examples, some involving considerable detail. We emphasize that even complex problems can be solved by a systematic approach using Newton's laws. Of particular importance in this regard is the use of *free-body diagrams*.

6–1   FREE-BODY DIAGRAMS

In a number of instances in Chapter 5 we focused attention on a particular object among a collection of interacting objects. Then, we identified all of the forces acting on this singled-out object and, by applying Newton's laws, determined its motion. This procedure can be formalized by introducing the idea of the *free-body diagram*.

Consider the situation in Fig. 6–1a, in which a block with a mass m is being pulled along a horizontal surface by a rope. At least five objects play a role in this example—the block itself, the rope, the hand, the horizontal surface, and the Earth. In Fig. 6–1b we isolate the block and show a free-body diagram with *all* of the forces (and *only* the forces) acting *on* the block. The state of motion of an isolated object—in this instance, the block—is completely determined by the sum of all the forces acting on it. A common error in analyzing situations of this type is the failure to distinguish which of the many forces present act *directly* on the isolated object and which do not. In this case, there is, first of all, the force exerted on the block by the rope, which we identify as the tension $\mathbf{T}$ in the rope. The gravitational force on the block due to the Earth is $\mathbf{F}_g$, with magnitude mg. The unknown resultant force $\mathbf{R}$ exerted on the block by the surface is conveniently represented by its two vector components, $\mathbf{N}$ (the component perpendicular to the surface) and $\mathbf{f}$ (the component parallel to the surface). The effect of $\mathbf{N}$ and $\mathbf{f}$ taken together is completely equivalent to the effect of $\mathbf{R}$ alone. (Resist the temptation to show $\mathbf{R}$ as well as $\mathbf{N}$ and $\mathbf{f}$ in the free-body diagram!) The component $\mathbf{f}$ is the *frictional force* acting on the block, and $\mathbf{N}$ is the *normal* (that is, the perpendicular) *constraining force*.* (We consider these forces in detail later.) Note carefully that the hand does *not* exert a force directly on

---

*The force $\mathbf{N}$ is called a "constraining" force because it is responsible for confining (i.e., constraining) the motion within a horizontal plane.

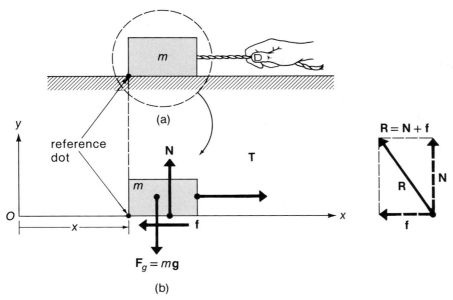

Fig. 6-1. Illustrating the concept of a *free-body diagram.* (a) The actual situation in which a block is pulled along a horizontal surface by a rope (and hand). (b) The block as the isolated object of interest, showing all the forces acting on the block *and* a convenient inertial coordinate frame of reference.

the block. The influence of the hand is transmitted through the rope; this is already accounted for in the tension **T**.

Finally, *internal forces,* such as those forces that act between different internal parts of an isolated object, are not to be included among the forces involved in the application of Newton's second law. They occur as action-reaction pairs and cancel each other. See Questions 6-1 and 6-2.

Next, we establish a convenient inertial coordinate frame of reference in order to describe the motion of the block. This motion is completely specified when we know the motion of any point on the block, such as the dot of paint shown in the lower left-hand corner of the block in Fig. 6-1. Thus, the motion of the block is described by giving at each instant the location of this dot with respect to the chosen coordinate frame.

Finally, we are ready to apply Newton's laws of motion to the isolated block. Recall that the second law, $\mathbf{F} = m\mathbf{a}$, is a *vector* equation. Usually, the easiest approach is to examine separately the various components of this equation.

Because no motion occurs in the vertical or y-direction, we have

$$N - mg = ma_y = 0$$

from which
$$N = mg$$

In the horizontal or x-direction, we have

$$T - f = ma_x$$

Therefore, the motion in the x-direction is described by

$$a_x = \frac{d^2x}{dt^2} = \frac{1}{m}(T - f)$$

If T and f are constant, then a_x is constant, and this equation is easily integrated. We find

(compare Eq. 2–11)

$$x(t) = x_0 + v_0 t + \frac{1}{2}\left[\frac{1}{m}(T - f)\right]t^2$$

Also, $y(t) = 0$

These equations completely describe the motion of the reference dot and, hence, the motion of the block.

Next, consider the more complicated case of two blocks with masses m_1 and m_2 that are rigidly coupled and are sliding across a horizontal plane, pulled by a force **F**, as shown in Fig. 6–2a. Figure 6–2b is the free-body diagram for the block m_1 in terms of an inertial coordinate frame that has its x-axis along the horizontal surface and its y-axis perpendicular to this surface. Figure 6–2c is the free-body diagram for the block m_2, shown in a similar way. Notice that the tensions exerted on the two blocks by the linkage have opposite directions. That is, $\mathbf{T}_1 = -\mathbf{T}_2$, and we write for the x-components, $T_1 = T = -T_2$.

For the block m_1, Newton's second law for the forces in the x- and y-directions are

Fig. 6–2. (a) Two rigidly coupled blocks slide across a horizontal plane, pulled by a force **F**. (b) The free-body diagram for block m_1 alone. (c) The free-body diagram for block m_2 alone. (d) The free-body diagram for both blocks considered as a single unit.

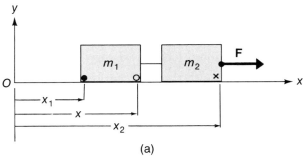

(a)

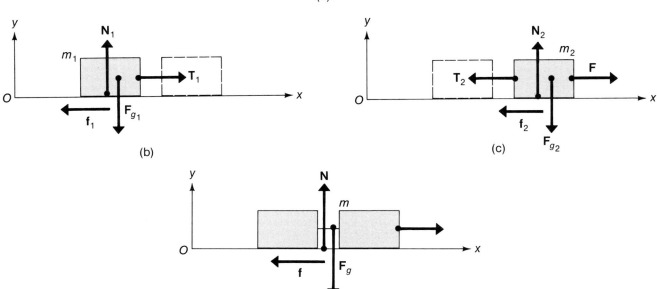

(b)   (c)

(d)

$$N_1 - m_1g = 0$$

and
$$T - f_1 = m_1a_x \qquad \qquad \textbf{(6-1)}$$

Similarly, for the block m_2, we have

$$N_2 - m_2g = 0$$

and
$$F - T - f_2 = m_2a_x \qquad \qquad \textbf{(6-2)}$$

When the two blocks together are considered as a unit (Fig. 6–2d), the tension T becomes an internal force and is *not* a force exerted on the isolated unit by any external agency. Thus, T does not appear in the free-body diagram for the coupled pair. Again taking x- and y-components, we find

$$N - mg = 0$$

and
$$F - f = ma_x \qquad \qquad \textbf{(6-3)}$$

where $N = N_1 + N_2$, $f = f_1 + f_2$, and $m = m_1 + m_2$.

A word is in order concerning the coordinate frames shown in Fig. 6–2b, c, and d. First, the frames are, in principle, independent of one another—they are required only to be inertial frames. At our option, we take them to be one and the same as shown in Fig. 6–2a; this is clearly a convenient choice in the present case. The coordinate x_1 refers to the position of some selected identifying mark (a reference dot) on block m_1; similarly, the coordinate x_2 refers to some selected mark on block m_2. The coordinate x (without subscript) can refer to a completely separate mark (on either block or on the coupling). These coordinates are related at all times by

$$x_1 = x + \ell_1$$
$$x_2 = x + \ell_2$$

where ℓ_1 and ℓ_2 are constants. It then follows that $v_{1x} = dx_1/dt$, $v_{2x} = dx_2/dt$, and $v = dx/dt$ are all equal. Likewise, the accelerations are all equal:

$$a_{1x} = a_{2x} = a_x$$

It was in anticipation of this result that we placed no identifying subscripts on the acceleration terms in Eqs. 6–1, 6–2, and 6–3.

If we add Eqs. 6–1 and 6–2, we see that we obtain precisely Eq. 6–3. If we divide Eq. 6–1 by m_1 and divide Eq. 6–2 by m_2, then subtract one result from the other, we obtain

$$T\left(\frac{1}{m_1} + \frac{1}{m_2}\right) = \frac{F}{m_2} + \frac{f_1}{m_1} - \frac{f_2}{m_2}$$

or
$$T = \frac{m_1F + m_2f_1 - m_1f_2}{m_1 + m_2}$$

What do we learn from this? If we are required only to find the acceleration a_x of the system, we should obviously consider the two blocks as a unit and construct directly the free-body diagram of Fig. 6–2d. In this case, there is no need to construct separate free-body diagrams for m_1 and m_2. It is not wrong to construct the separate diagrams and arrive at Eqs. 6–1 and 6–2. However, Fig. 6–2d provides a more direct means and eliminates the need for determining the unknown tension T; at the very least, this method simplifies the required algebra. (In this respect the situation here is very similar to that discussed in Example 5–4.)

On the other hand, if we want to find the tension in the coupling, there is no choice but to use the individual free-body diagrams. (• Can you see why it is permissible to use *any* two of the three diagrams?)

In summary, we give these general rules for solving problems in dynamics:

1. Draw carefully a diagram showing all the key features of the stated problem.
2. Depending on the questions raised, draw one or more *separate* free-body diagrams. In each instance show *all* the forces (and *only* the forces) acting *on* the isolated object through its interaction with external agencies. These, in general, will include contact forces and gravitational forces. (Do not show any *internal* forces.)
3. Select a set of inertial coordinate frames; you may select a different one for each free-body diagram. Of course, the constraints in the problem will relate these frames and the involved coordinates through various imposed conditions (such as those in the equations for x_1 and x_2 in the preceding example).
4. Apply Newton's laws of motion to *each* isolated object shown in its own free-body diagram, using component resolutions where appropriate.
5. An optional step, very useful in complicated situations, is to examine the behavior of the solutions for certain simplifying, extreme, or special conditions (for example, by allowing one of the masses or angles or velocities to become zero). This is particularly helpful for those cases in which you know or may be able to guess the result. At the very least, this step should provide additional insight into the significance of the solutions (and may reveal possible errors in the analysis).

Finally, a word of caution. Particularly when pressed for time, you may be tempted to forgo or combine some of these steps. Doing so usually turns out to be a false economy, because there is the great risk of omitting relevant forces, making sign errors, incorrectly identifying the geometry, and so forth. It pays to be neat and orderly.

6–2 PHYSICAL IDEALIZATIONS

Even simple physical systems that we select to study have complicating features that, however, do not materially alter the actual behavior of the system. Such minor complications usually involve strings, pegs, pulleys, and rods that connect separate components of the system in various ways. It is possible in many cases to ascribe idealized properties to such complicating items that considerably simplify the analysis.

Ideal Strings. In many problems we are confronted with various objects that are connected by light strings (or cords or ropes). An important property of a string (one that is correct to a high degree of accuracy in ordinary cases) is that it can sustain only a force of tension. That is, a string can exert a *pull* but not a *push*. A string has zero internal resistance to bending and will therefore align itself *along* the direction of the applied tension force vector. Any attempt to make a string exert a compressional force will simply cause the string to collapse into an irregular shape.

Although strings obviously can stretch under tension, these changes in length are usually small compared with the other lengths or displacements that occur in problems. For our present purposes, we assume that every string has a constant length.

Strings obviously also possess mass. The effect of a string's mass is most easily appreciated by considering an example. In Fig. 6–3 we show two blocks with masses m_1 and m_2 that are tied together by a string with a length L and a total mass m_s. Through a second string, a disembodied hand applies a force **F** to the first block and thereby pulls the two blocks in straight-line motion in deep space, far from any possible interacting body. The figure also shows separate free-body diagrams for each of the two blocks *and* the connecting string.

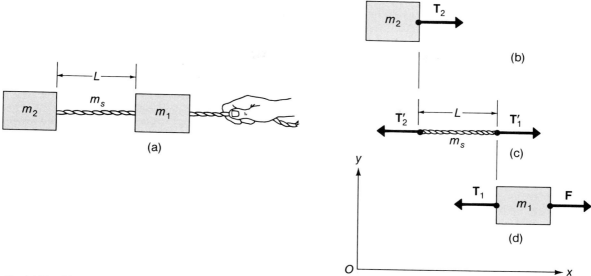

Fig. 6-3. (a) Two blocks that are tied together with a string are pulled along by a force **F**. The string has a length L and a mass m_s. (b), (c), and (d) Free-body diagrams for the block m_2, the connecting string m_s, and the block m_1, respectively.

We observe first that $\mathbf{T}_2$ and $\mathbf{T}'_2$ represent an action-reaction pair of forces, as do $\mathbf{T}_1$ and $\mathbf{T}'_1$. Then, taking components in the x-direction, we can write

$$F - T_1 = m_1 a_x$$

$$T_1 - T_2 = m_s a_x$$

$$T_2 = m_2 a_x$$

Adding all three equations gives

$$a_x = \frac{F}{m_1 + m_2 + m_s}$$

and using this result in the equation for $T_1 - T_2$ leads to

$$T_1 - T_2 = \frac{m_s}{m_1 + m_2 + m_s} F$$

For strings with small mass, $m_s \ll m_1$ and $m_s \ll m_2$, we note that the tensions T_1 and T_2 may be considered to be essentially equal; then, the expression for a_x becomes

$$a_x \cong \frac{F}{m_1 + m_2}$$

It is also instructive to divide the string shown in Fig. 6-3c into two unequal arbitrary parts with lengths ℓ and $L - \ell$. Figure 6-4 shows the separate free-body diagrams for the two parts (which are assumed to be homogeneous). Evidently,

$$T_\ell - T_2 = \frac{\ell}{L} m_s a_x$$

and

$$T_1 - T_\ell = \frac{(L - \ell)}{L} m_s a_x$$

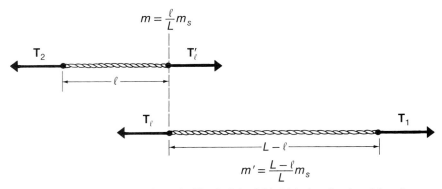

Fig. 6–4. The connecting string shown in Fig. 6–3 is divided into lengths ℓ and $L - \ell$.

Substituting $a_x = F(m_1 + m_2 + m_s)$ into the first of these equations, we obtain

$$T_\ell = T_2 + \frac{\ell}{L} \frac{m_s}{(m_1 + m_2 + m_s)} F \qquad (6-4)$$

We see that T_ℓ increases linearly with ℓ, from the value T_2 at $\ell = 0$ to the value T_1 at $\ell = L$. For the case, $m_s \ll m_1$ and $m_s \ll m_2$, all the tensions are essentially equal: $T_\ell \cong T_1 \cong T_2$.

This example demonstrates how free-body diagrams for the divided sections of an object may be used to learn about the *internal* stresses in an object. (Note that for each of the divided sections, T_ℓ is an *external* force!)

An *ideal string* is defined to be a string that has negligible mass and the same tension throughout its length; at each point the tension **T** is directed along the string.

Example 6–1

An elevator accelerates upward at a constant rate of 2 m/s². A uniform string with a length $L = 25$ cm and a mass $m = 2$ g supports a small block with a mass $M = 150$ g that hangs from the car ceiling. Find the tension along the string.

Solution:

The tension T_0 in the string at the ceiling support can be calculated by considering the free-body diagram for the entire string *plus* the block. If we take upward as the positive direction, we can write

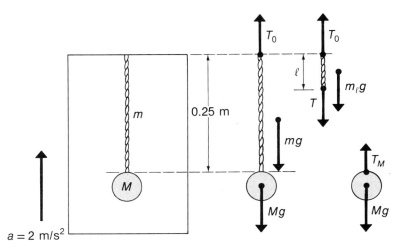

$$T_0 - (m + M)g = (m + M)a$$

or $$T_0 = (m + M)(g + a)$$

$$= (152 \times 10^{-3} \text{ kg})(9.80 \text{ m/s}^2 + 2.00 \text{ m/s}^2)$$

$$= 1.794 \text{ N}$$

The tension T at a point in the string a distance ℓ from the ceiling can be found by considering the free-body diagram for the upper portion of the string segment, which has a length of ℓ. The mass of this segment is $m_\ell = (\ell/L)m$, so

$$T_0 - T - m_\ell g = m_\ell a$$

or

$$T = T_0 - m(g + a)\frac{\ell}{L}$$

$$= 1.794 \text{ N} - (2 \times 10^{-3} \text{ kg})(9.80 \text{ m/s}^2 + 2.00 \text{ m/s}^2)\frac{\ell}{0.25 \text{ m}}$$

$$T = (1.794 - 0.094\ \ell) \text{ N}$$

with ℓ in meters.

Note that for $\ell = L$, that is, at the end attached to the block, we have

$$T = T_M = 1.770 \text{ N}$$

As a check, consider the free-body diagram for the block alone. Then,

$$T_M - Mg = Ma$$

or $$T_M = M(a + g) = (0.15 \text{ kg})(11.80 \text{ m/s}^2)$$

$$= 1.770 \text{ N}$$

in agreement with the result above. (• What would be the tension in the string if its mass m were zero?)

Ideal Pegs, Pulleys, and Rods. In order to simplify the treatment of certain classes of problems, we often make use of several ideal components in addition to ideal strings. For example, the function of an ideal peg or pulley is simply to change the direction of the force exerted by a string. If the string is also ideal (massless), the tension in the string on either side of the peg has the same magnitude. The situation is illustrated in Fig. 6–5. The tension in each part of the string is T, and the net contact force N exerted on the string by the peg is directed along the bisector AA'. The resultant of these three forces acting on the string is *zero*, so the entire effect of the peg is to change the direction of the tension. The magnitude of the contact force is $N = 2T \cos \alpha$. (• How would the analysis change if the mass of the string were not zero?)

Figure 6–6 shows an arrangement of blocks with the connecting string passing over an *ideal pulley*. An ideal pulley has zero mass and is supported by and rotates freely on frictionless bearings. As the string is pulled over the pulley, the string causes the pulley to rotate in such a way that there is no slippage between them. (That is, the dots of paint shown in Fig. 6–6 remain adjacent during the motion.) Slippage does not occur because the string exerts a tangential force on the pulley rim due to the friction that exists between them. There is no conceptual difficulty in allowing friction between an ideal string and an ideal pulley because there are no energy losses associated with this nonslipping type of friction. (See Section 6–2.) (There would be losses if friction were present and slippage did occur.)

Finally, as far as the relevant forces are concerned, the ideal pulley and string behave just as the ideal peg and string shown in Fig. 6–5. The string tensions at both free ends are equal, and the force exerted by the pulley is also $N = 2T \cos \alpha$ and is likewise along the bisecting line AA'. In the following problems ideal pegs and pulleys are interchangeable elements—the main effect of either is simply to redirect the action line of the string tension force.

Another ideal component we sometimes use is the *ideal rod*. Such rods are absolutely rigid structural members and are employed as supporting devices or to maintain a fixed separation between parts of a system. An ideal rod cannot be extended, compressed, bent, or twisted. The rod may have mass or it may be massless. Forces are transmitted undiminished through ideal massless rods.

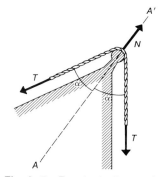

Fig. 6–5. Free-body diagram for an ideal (massless) string that passes over an ideal (frictionless) peg.

Fig. 6–6. The two blocks are connected by an ideal (massless) string that runs over an ideal pulley (massless and with a frictionless bearing). The line AA' bisects the angle between the ends of the string.

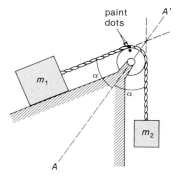

Example 6–2

Consider the arrangement shown in Fig. 6–6 in which a block with a mass $m_1 = 2$ kg slides on a frictionless, inclined plane with elevation angle 45°; the mass of the hanging block is $m_2 = 3$ kg. An ideal connecting string runs over an ideal pulley. What is the linear acceleration of the blocks, and what forces are exerted on the pulley and on the inclined plane?

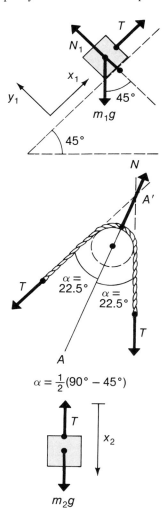

$$\alpha = \tfrac{1}{2}(90° - 45°)$$

Solution:

In the free-body diagrams for the blocks we take the coordinate x_1 to be along the inclined plane for m_1 and we take the coordinate x_2 to be vertically downward for m_2. Note that although the accelerations of the two blocks have different directions, they are numerically equal; we denote their common magnitude as a. Then, the equations for the force components in these directions are

$$T - m_1 g \sin 45° = m_1 a$$

$$m_2 g - T = m_2 a$$

Adding these equations gives

$$(m_2 - m_1 \sin 45°)g = (m_1 + m_2)a$$

from which

$$a = \frac{(m_2 - m_1 \sin 45°)}{(m_1 + m_2)} g = \frac{\left(3 - 2\dfrac{\sqrt{2}}{2}\right)}{(3 + 2)} (9.80 \text{ m/s}^2)$$

$$= 3.11 \text{ m/s}^2$$

From the free-body diagram for block m_1, the force components perpendicular to the plane obey the relationship

$$N_1 = m_1 g \cos 45° = (2 \text{ kg})(9.80 \text{ m/s}^2)\frac{\sqrt{2}}{2}$$

$$= 13.9 \text{ N}$$

The force exerted on the inclined plane by the block is the reaction force to N_1.

The string tension T can be obtained from the free-body diagram for block m_2:

$$T = m_2(g - a) = (3 \text{ kg})(9.80 \text{ m/s}^2 - 3.11 \text{ m/s}^2)$$

$$= 20.1 \text{ N}$$

The string tension is needed to calculate N, the force exerted on the string by the pulley.

From the free-body diagram for the string running over the pulley, we have

$$N = 2T \cos 22.5° = 2(20.1 \text{ N}) \cos 22.5°$$

$$= 37.1 \text{ N}$$

The force exerted on the pulley by the string is the reaction force to N.

Example 6–3

A block with a mass $m = 300$ kg is set into motion on a horizontal frictionless surface by using an ideal pulley-and-rope system, as shown in the diagram. What horizontal applied force F is required to produce an acceleration of 5 cm/s^2 for the block?

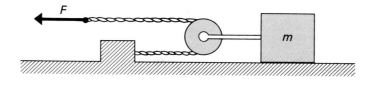

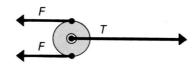

Solution:

The free-body diagram for the pulley is shown above. The force exerted on the pulley by the rope (this is equivalent to the reaction force to N in Fig. 6-5, with $\alpha = 0$) is simply the sum of the two forces F. Because the pulley is massless, we also have $T = 2F$. Moreover, $T = ma$, so that

$$F = \tfrac{1}{2}ma = \tfrac{1}{2}(300 \text{ kg})(0.05 \text{ m/s}^2)$$

$$F = 7.5 \text{ N}$$

Notice that with this *block-and-tackle* arrangement, the force required to produce a particular acceleration is only half that required if the force were applied directly to the block. This block-and-tackle system is said to have a *mechanical advantage* of 2 (that is, $T/F = 2$).

Example 6-4

A 0.50-kg stone is tied to the end of a string that has a length of 0.25 m and is rotated in a horizontal plane at a constant rate of 1.5 revolutions/s. The string makes an angle β with the vertical. Find the tension in the string and the angle β.

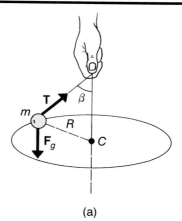

(a)

Solution:

The horizontal component of the tension, namely, $T \sin \beta$, provides the centripetal acceleration, $a_c = v^2/R = R\omega^2$, required for the circular motion. Therefore,

$$T \sin \beta = mR\omega^2 = mL\omega^2 \sin \beta$$

from which

$$T = mL\omega^2 = (0.50 \text{ kg})(0.25 \text{ m})(2\pi \times 1.5 \text{ s}^{-1})^2$$

$$= 11.10 \text{ N}$$

In the vertical direction there is no acceleration, so we have

$$T \cos \beta - F_g = 0$$

or $$\cos \beta = \frac{mg}{T} = \frac{(0.50 \text{ kg})(9.80 \text{ m/s}^2)}{11.10 \text{ N}} = 0.441$$

from which

$$\beta = 63.8°$$

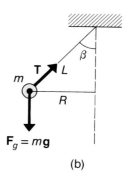

(b)

Example 6-5

A penny with an approximate mass of 2 g sits on a horizontal turntable at a distance of 25 cm from the spindle. The turn-table accelerates at a rate of $\alpha = \tfrac{1}{3}$ rad/s^2. What is the frictional force the table exerts on the penny 2 s after start-up?

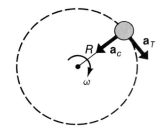

Solution:

The horizontal frictional force provides the total acceleration, which consists of both a radial (centripetal) component and a tangential component (see diagram). At 2 s after start-up, the angular speed is

$$\omega = \alpha t = (\tfrac{1}{3} \text{ rad/s})(2 \text{ s}) = \tfrac{2}{3} \text{ rad/s}$$

The corresponding centripetal acceleration is therefore

$$a_c = R\omega^2 = (\tfrac{1}{4} \text{ m})(\tfrac{2}{3} \text{ rad/s})^2 = \tfrac{1}{9} \text{ m/s}^2$$

The tangential acceleration is a constant equal to

$$a_T = R\alpha = (\tfrac{1}{4} \text{ m})(\tfrac{1}{3} \text{ rad/s}) = \tfrac{1}{12} \text{ m/s}^2$$

The total acceleration is the vector sum $\mathbf{a} = \mathbf{a}_c + \mathbf{a}_T$ with magnitude

$$a = \sqrt{a_c^2 + a_T^2} = \sqrt{(\tfrac{1}{9})^2 + (\tfrac{1}{12})^2} \text{ m/s}^2 = \tfrac{5}{36} \text{ m/s}^2$$

Therefore, the required frictional force has a magnitude

$$f = ma = (2 \times 10^{-3} \text{ kg})(\tfrac{5}{36} \text{ m/s}^2) = 2.78 \times 10^{-4} \text{ N}$$

6−3 FRICTION

When an effort is made to slide one object relative to another along the common surface of contact, the resistive force that acts along (tangent to) the interface is called *friction.** We have referred to friction and its effects on several occasions; now we give specific attention to this subject.

Friction can be a serious practical problem or it can be quite useful. Friction in machinery wastes energy and causes wear. However, walking would be impossible without friction and wheels would not roll across the ground. The details of the phenomenon of friction are complicated and not well understood in terms of any fundamental theory. Consequently, we give here only a qualitative overview of the subject.

Consider a book that is at rest on a flat tabletop. Suppose that you apply a horizontal force to the book by pushing on it with your finger in an effort to slide it along the tabletop. If the applied force is too small, the book does not move—it remains static. Evidently, there must be another horizontal force acting on the book, a force that just cancels your applied force. This force exerted on the book by the tabletop is referred to as the force of *static friction*. Note that this force is rather unusual in that it is exactly equal and opposite to the force you apply, and it vanishes when you stop pushing! If you continue to increase your pushing effort, at some point the book will finally ''break loose'' and start to slide. During the sliding motion it is clear that a frictional force is still acting because, if you stop pushing, the book comes to rest (perhaps after first traveling a short distance). If you are careful, you may be able to feel that the break-away force is slightly larger than the force you must exert to maintain the book sliding with a constant speed. The resisting horizontal force exerted on the sliding book by the tabletop is referred to as *sliding* (or *kinetic*) *friction*.

Although frictional processes are complex, simple empirical ''laws'' are found to apply in many situations. We should, however, not expect these laws to be of the same basic character as Newton's laws of dynamics or the law of gravitation. The laws of

*The first careful and systematic experimental study of friction was made by the French physicist Charles Augustin de Coulomb (1736−1806), who is more famous for his investigations of electric phenomena. Some of the earliest work on friction was by Leonardo da Vinci (1452−1519), whose clear understanding of the essentials of the subject is revealed by sketches and commentary in his famous *Notebooks*.

friction are only approximate descriptions of the way that matter behaves under some (but not all) conditions. When we lack a fundamental theory for a process or find the theory too complicated to use in practical situations, we are forced to rely on approximate or empirical descriptions. We have already used one such empirical law, namely, Hooke's law, and later we introduce another famous example, Ohm's law of electric conductivity. None of these empirical laws is valid in all circumstances, but each is sufficiently accurate for many engineering and practical applications.

When an object is in contact with a surface, the component of the externally applied force that acts tangentially to the surface of contact is exactly balanced by the force of static friction between the object and the surface, up to a certain maximum value (Fig. 6–7a, b). Experimental evidence shows that this maximum force is approximately proportional to the loading (or normal) force pressing the two objects together. For example, an ordinary wood block in contact with a flat (dry) wood board may generate frictional forces that will balance applied tangential forces up to about 0.4 of the normal force that is exerted on the block by the board. If the applied tangential force exceeds this value, the block will slide relative to the board surface. It has been found that this maximum (static) force is independent of the area of contact between the objects. That is, the maximum force of static friction between a wood block and a board will be the same when the block rests on its largest side as when it rests on its smallest side. Thus, in ordinary cases we can write a relationship of direct proportionality between the maximum force of static friction, $f_{s,\text{max}}$, and the normal force N:

$$f_{s,\text{max}} = \mu_s N \qquad (6\text{–}5)$$

where the proportionality constant μ_s is called the *coefficient of static friction*.

If the applied tangential force exceeds $f_{s,\text{max}}$, relative sliding occurs (Fig. 6–7c). Again, experimental evidence indicates that the force of sliding or kinetic friction is independent of the area of contact and is directly proportional to the normal force. Thus,

$$f_k = \mu_k N \qquad (6\text{–}6)$$

where μ_k is the *coefficient of kinetic friction*. The direction of the force $\mathbf{f}_k$ is always opposite to the direction of the relative velocity between the objects in contact.

Generally, $\mu_s > \mu_k$. (• Is this what you would expect?) Also, μ_k has some dependence on the relative velocity of the objects, the degree depending strongly on the specific nature of the objects, as we mention later. For some materials, the variation of μ_k with velocity is not pronounced, and it is often assumed (for ease of calculation!) that μ_k is constant.

The statements expressed by Eqs. 6–5 and 6–6, together with the observation that the frictional forces are independent of the area of contact, constitute the laws of friction. Notice carefully that the equation relating to static friction (Eq. 6–5) involves the maximum possible frictional force $\mu_s N$, whereas Eq. 6–6 expresses the force $\mu_k N$ that actually exists during sliding. It is important to distinguish clearly between these two situations.

An important effect associated with sliding friction is the conversion of some of the motional energy of the system into heat. For example, if you rub the palms of your hands together for a few seconds, the sensation of heat generation will be readily felt. We discuss the important topic of energy and its various forms in Chapter 7.

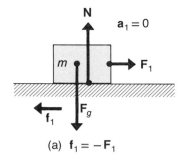

(a) $\mathbf{f}_1 = -\mathbf{F}_1$

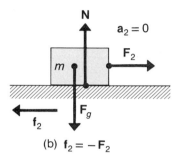

(b) $\mathbf{f}_2 = -\mathbf{F}_2$

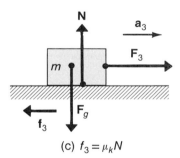

(c) $f_3 = \mu_k N$

Fig. 6–7. (a, b) For $f_s < f_{s,\text{max}}$, the frictional force exactly balances the applied force; then, there is no acceleration. (c) When a force sufficient to cause motion is applied, the frictional force is equal to $\mu_k N$ and the acceleration is $(F - \mu_k N)/m$.

▶ *Dragster Tires*. An automobile (or any driven vehicle) is propelled forward by virtue of the frictional force that exists between the tires and the road surface. (Some details are given later in this section.) In order to achieve maximum acceleration, it is clearly necessary for this frictional force to be as large as possible. We have shown that frictional forces for particular materials in contact depend only on the normal force. Why, then, do drag racers use tires that are much wider than normal tires? The extra width produces a greater area of contact between the tires and the road surface, but frictional forces do not depend on the contact area. When a drag racer accelerates his vehicle, starting from rest, the tires spin furiously. The best way to accelerate would be to maintain a drive power level that produces frictional forces just short of slipping (because $\mu_s > \mu_k$). But this is technically very difficult to achieve, so the drag racer simply applies maximum power and allows the tires to spin until the car's velocity increases to the point that slipping ceases. While the tires are slipping, considerable heat is generated, causing actual melting of the tire rubber (and possibly also of the road surface). The presence of liquid rubber (and asphalt) reduces the friction coefficient and the frictional force. With the wide dragster tires, however, the increased contact area distributes the heat load and keeps the temperature low enough so that excessive melting does not occur. A greater frictional force and therefore a greater acceleration are thus achieved by using wide tires instead of narrow ones. ◀

Example 6−6

A wooden crate with a mass of 100 kg rests on a flat, wooden floor. An effort is made to start the crate sliding by pulling on a rope that makes an angle of 30° with the horizontal, as shown in the diagram. If the coefficient of static friction between the crate and the floor is 0.4, what is the minimum rope tension that will start a sliding motion?

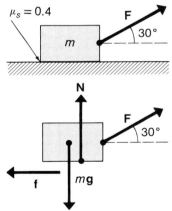

Solution:

It is important to recognize that, because F has an upward component, an increase in F will cause a decrease in N. From the free-body diagram, we can write for the vertical force components

$$N + F \sin 30° - mg = 0$$

so that

$$N = (100 \text{ kg})(9.80 \text{ m/s}^2) - F(0.5)$$

$$= (980 \text{ N}) - 0.5F$$

Just before sliding begins, we have $f = \mu_s N$. Then, for the horizontal force components, we have

$$F \cos 30° - f = 0$$

Thus,

$$F = \frac{\mu_s N}{\cos 30°} = \frac{0.4(980 - 0.5F)}{0.866}$$

Solving for F, we find

$$F = 368 \text{ N}$$

(• Calculate the magnitudes of N and f. How are they related to the weight of the block?)

Example 6−7

A block of wood rests on an inclined plane, also of wood, with an adjustable angle of inclination α as indicated in the diagram. The angle α is slowly increased from zero, and the block is observed to start sliding when $\alpha = 23°$. What is the coefficient of static friction μ_s between the block and the plane?

Solution:

We are looking for μ_s, which is the ratio f/N when the block just begins to move. In the direction normal to the plane (i.e., the y-direction), the force components are

$$N - mg \cos \alpha = 0$$

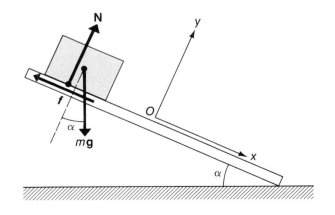

In the direction along the plane (i.e., the x-direction), we have

$$mg \sin \alpha - f = 0$$

Now, we form the ratio,

$$\frac{f}{N} = \frac{mg \sin \alpha}{mg \cos \alpha} = \tan \alpha$$

Hence, $\qquad (f/N)_{\max} = \mu_s = \tan 23° = 0.424$

This maximum angle is sometimes called the *angle of repose*. Note that it is independent of the mass of the block.

Example 6-8

The crate in Example 6-6 is pulled along at a constant speed v. To maintain this motion, the required force applied at 30° to the horizontal is $F = 330$ N. What is the value of the coefficient of sliding friction at the speed v?

Solution:

The same free-body diagram applies, and we have, in the horizontal and vertical directions,

$$F \cos 30° = \mu_k N$$

and $\qquad N = (980 \text{ N}) - 0.5F$

Thus, $\qquad \mu_k = \dfrac{(330 \text{ N})(0.866)}{(980 \text{ N}) - (0.5)(330 \text{ N})} = 0.351$

Example 6-9

Two blocks, $m_1 = 2$ kg and $m_2 = 4$ kg, are connected with a light string that runs over a frictionless peg to a hanging block with a mass M, as shown in diagram (a). The coefficient of sliding friction between block m_2 and the horizontal surface at the speeds involved is $\mu_k = 0.2$. The coefficient of *static friction* between the two blocks is $\mu_s = 0.4$. What is the maximum mass M for the hanging block if the block m_1 is *not* to slip on block m_2 while m_2 is sliding over the surface?

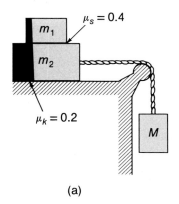

(a)

Solution:

This example involves both static and sliding friction. Assume that slipping of block m_1 is impending. The relevant free-body diagrams are shown in diagram (b). Using the diagram for the entire system, we can write

$$N - (m_1 + m_2)g = 0 \qquad \textbf{(1)}$$

$$T - f = (m_1 + m_2)a \qquad \textbf{(2)}$$

and taking the downward direction to be positive, we have for the hanging block,

$$Mg - T = Ma \qquad \textbf{(3)}$$

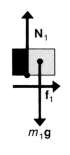
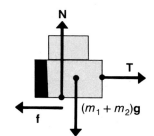

(b)

Adding Eqs. (2) and (3) gives

$$(M + m_1 + m_2)a = Mg - f$$

We also have, using Eq. (1),

$$f = \mu_k N = \mu_k(m_1 + m_2)g$$

Thus, $\qquad a = \dfrac{M - \mu_k(m_1 + m_2)}{M + m_1 + m_2}g$

From the free-body diagram for m_1, we have

$$N_1 - m_1 g = 0$$

$$f_1 = m_1 a$$

Notice that the only horizontal force acting on m_1 is the frictional force f_1; it is this force that accelerates m_1 to the right. Just before slipping occurs, we find

$$\frac{f_1}{N_1} = \mu_s$$

or

$$\mu_s = \frac{a}{g} = \frac{M - \mu_k(m_1 + m_2)}{M + m_1 + m_2}$$

Solving for M, we have

$$M = \frac{(\mu_s + \mu_k)(m_1 + m_2)}{1 - \mu_s}$$

Hence,

$$M = \frac{(0.4 + 0.2)(2 \text{ kg} + 4 \text{ kg})}{(1 - 0.4)} = 6 \text{ kg}$$

If M were to exceed this value, the resulting greater acceleration would require $f_1 = m_1 a$ to exceed the maximum sustainable value of the quantity, $\mu_s N_1 = \mu_s m_1 g$. Then, slipping between the blocks would result. (• What is the minimum value of M that will cause m_2 to slide?)

Example 6–10

A 5-kg block is initially at rest on an inclined plane, as shown in diagram (a). A force $F = 20$ N is applied to the block in a direction parallel to the plane. Determine the acceleration of the block if the coefficient of kinetic friction is $\mu_k = 0.42$.

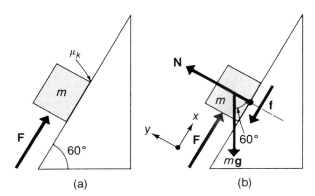

(a)   (b)

Solution:

Assume that the block will slide up the plane. Then, the free-body diagram is as shown in diagram (b), with the frictional force f directed down the plane. In the y-direction, we have

$$N = mg \cos 60° = (5 \text{ kg})(9.80 \text{ m/s}^2) \cos 60° = 24.5 \text{ N}$$

Hence, $f = \mu_k N = 0.42(24.5 \text{ N}) = 10.29 \text{ N}$

Then, in the x-direction, we have

$$F - f - mg \sin 60° = ma$$

$$(20 \text{ N}) - (10.29 \text{ N}) - (5 \text{ kg})(9.80 \text{ m/s}^2) \sin 60° = (5 \text{ kg})a$$

from which $a = -6.55 \text{ m/s}^2$

In this case it is *not* permissible to assume that the negative sign in the solution simply indicates that the acceleration is 6.55 m/s² *down* the inclined plane. The reason is that motion *down* the plane means that the frictional force **f** is directed *up* the plane, contrary to the free-body diagram that was used. Instead, the problem must be solved again, with **f** directed up the plane. Then, with the same coordinate frame, we find

$$F + f - mg \sin 60° = ma$$

$$(20 \text{ N}) + (10.29 \text{ N}) - (5 \text{ kg})(9.80 \text{ m/s}^2) \sin 60° = (5 \text{ kg})a$$

from which $a = -2.43 \text{ m/s}^2$

Now the negative sign does indeed correctly indicate that the block accelerates down the plane.

When solving problems that involve directions of motion with friction present, careful attention must be given to ensure that the frictional force is always directed opposite to the motion.

Example 6–11

An automobile with a mass M starts from rest and accelerates at the maximum rate possible without slipping on a road with a coefficient of static friction $\mu_s = 0.5$. If only the rear wheels are driven and half the weight of the automobile is supported on these wheels, how much time is required to reach a speed of 100 km/h?

Solution:

The normal load on both rear wheels is $N = \frac{1}{2}Mg$, and the maximum sustainable frictional force is

$$f = \mu_s N = \frac{1}{2}\mu_s Mg$$

Thus, the acceleration of the automobile is

$$a = \frac{f}{M} = \frac{1}{2}\mu_s g$$

Therefore,

$$t = \frac{v}{a} = \frac{2v}{\mu_s g} = \frac{2(100 \text{ km/h})(10^3 \text{ m/km})}{0.5(9.80 \text{ m/s}^2)(3600 \text{ s/h})}$$

$$= 11.3 \text{ s}$$

Notice that the mass of the automobile is not required to obtain the solution.

Example 6–12

A driver is attempting to negotiate a flat (i.e., unbanked) highway curve that has a radius of curvature $R = 100$ m. What is the maximum speed v_m that the driver may use if his vehicle is not to skid? ($\mu_s = 0.50$.)

Solution:

The force equations are

$$N - mg = 0$$

and

$$f = ma_c = m\frac{v^2}{R}$$

where we have used $a_c = v^2/R$ for the centripetal acceleration. It is evident that as the speed v is increased, the frictional force necessary to prevent skidding increases quadratically. The maximum value of f is $\mu_s N = \mu_s mg$, so we have

$$v_m = \sqrt{\mu_s Rg} = \sqrt{(0.50)(100 \text{ m})(9.80 \text{ m/s}^2)}$$

$$= 22.1 \text{ m/s} = 79.7 \text{ km/h}$$

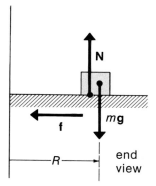

Example 6–13

If the highway curve in Example 6–12 is banked instead of flat, what angle of inclination with respect to the horizontal is required to reduce the frictional force to zero for a speed of 22.1 m/s (80 km/h)?

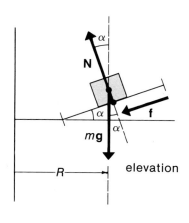

Solution:

The force equation in the radial direction is

$$f \cos \alpha + N \sin \alpha = m\frac{v^2}{R}$$

and in the vertical direction we have

$$N \cos \alpha - f \sin \alpha - mg = 0$$

In order to eliminate the unknown N, we multiply the first of these equations by $\cos \alpha$ and the second by $\sin \alpha$, then subtract the results. Using $\sin^2 \alpha + \cos^2 \alpha = 1$, we find

$$f = m\left(\frac{v^2}{R} \cos \alpha - g \sin \alpha\right)$$

This is the frictional force required to maintain the circular motion (no skidding) for the inclination angle α. By properly selecting the angle α, the force can be made zero. When f is set equal to zero, the bracketed term vanishes, and it follows that

$$\tan \alpha = \frac{v^2}{gR} = \frac{(22.1 \text{ m/s})^2}{(9.80 \text{ m/s}^2)(100 \text{ m})} = 0.498$$

from which $\alpha = 26.5°$

▶ *A Microscopic View of Friction.*   An important source of friction is the collection of adhesive molecular forces that act between the constituent molecules of the materials in contact.

The magnitude of these forces is influenced by the presence of oxide layers or accumulations of dirt and other foreign substances coating the surfaces (occasionally, as with lubricants,

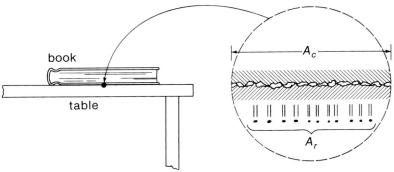

Fig. 6–8. A highly magnified view of the contacting surfaces of a book and a tabletop, showing that the actual or *real microscopic interaction* area, A_r, of the high points is only a small fraction of the gross, macroscopic, or *covering* contact area, A_c. The black dots in the right-hand diagram represent the individual microscopic areas of contact that sum to the area A_r. It is the area A_r that supports the normal load.

deliberately introduced). In fact, for ordinary sliding motion to be possible, the presence of such layers at the interface is essential. If two flat clean surfaces are brought together in a vacuum, they will tend to bond together, producing frictional forces ten or more times greater than would ordinarily act between the surfaces.

In the cases we consider in this book, it is assumed that the surfaces are reasonably smooth and that dirt, occluded air, and oxide layers are present. Under these conditions, simple sliding occurs readily.

At the atomic and molecular level, two surfaces in contact appear quite rough and have numerous gaps and voids, as illustrated in Fig. 6–8. Even metallic surfaces that are carefully polished will at best consist of a series of microscopic hills and valleys that range in cross section from about 100 to 1000 molecular diameters. Forces between molecules diminish rapidly with the distance of separation, essentially vanishing at distances greater than several molecular diameters. Consequently, the adhesion between the molecules at the touching, high points of the two surfaces provides the major contribution to the friction between the surfaces. During sliding, these bonds must be continually broken loose and then formed anew as different high points are brought into contact. Surface molecules in these interacting regions are set into rapid vibrational motion, generating heat (and occasionally sound). Also, pieces of one surface may be broken loose and then welded or bonded to the other surface. During this process, scratches and other forms of wear will appear.

In the previous discussion, the contact area we referred to is the gross macroscopic covering area A_c. The actual or real microscopic interaction area of contact, where the strong adhesive molecular forces are exerted, is the much smaller area A_r (Fig. 6–8). It is this area that supports the entire normal force load. Even under strong loading, $A_r \ll A_c$.

For a particular pair of surfaces, an increase in the normal force tends both to flatten the existing contact points by elastic and plastic deformations and to generate new contact points. Experiments show that in ordinary cases, the increase in the area

A_r is approximately proportional to the normal force, thereby leading to the observed relationship connecting the maximum sustainable static frictional force, $f_{s,\mathrm{max}}$, and the normal force N (Eq. 6–5).

It is often stated that the friction coefficient μ_k depends on the degree of smoothness of the contacting surfaces, becoming smaller if the surfaces are smoother. In fact, smooth surfaces are often said to be "frictionless." This is simply *not so*. To a surprising extent, μ_k is independent of the roughness of the sliding surfaces—at least, for finishes commonly employed in engineering practice, and even for surfaces showing moderate wear. For extremely smooth and flat (and clean) surfaces, μ_k is actually *larger* than for the same surfaces with rougher finish. For example, two pieces of glass have a larger value of μ_k if the surfaces are polished than if they are rough-ground! Atomically smooth surfaces of crystals, produced by cleavage, also have very large friction coefficients.

Finally, what about the dependence of the coefficient of kinetic friction on the relative velocity between the surfaces in contact? The values of μ_k for all material pairs exhibit some velocity dependence, often to a marked degree. At very slow sliding speeds, μ_k increases with speed. At some point, a maximum is reached, and thereafter μ_k decreases with increasing speed; quite low values of μ_k are found for metals at high speeds (greater than several hundred m/s). Figure 6–9 shows the dependence on speed of μ_k for two situations.

Wheels and Rolling Friction. Consider an isolated, perfectly circular wheel that rolls across a flat surface. It is our experience that any rolling wheel will eventually come to rest, and we understand that the reason is the friction between the wheel and the surface. This *rolling friction* is rather different from static or sliding friction.

We could construct the force diagram for the rolling wheel as in Fig. 6–10. Here we see the forces acting on the wheel— the gravitational force $\mathbf{F}_g$, the normal force $\mathbf{N}$, and the frictional force $\mathbf{f}$. In the ideal situation, the wheel remains perfectly round and the surface remains perfectly flat, so there is a single point of contact at which both $\mathbf{f}$ and $\mathbf{N}$ act.

Fig. 6–9. The dependence on speed of the coefficient of sliding friction μ_k for a steel slider on unlubricated indium and lead.

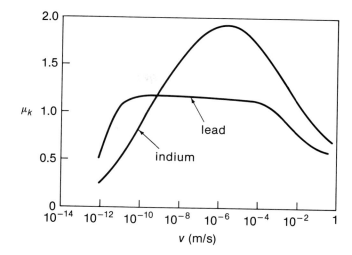

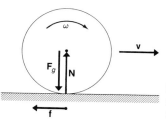

Fig. 6–10. The forces acting on a perfectly round wheel that rolls across a flat surface.

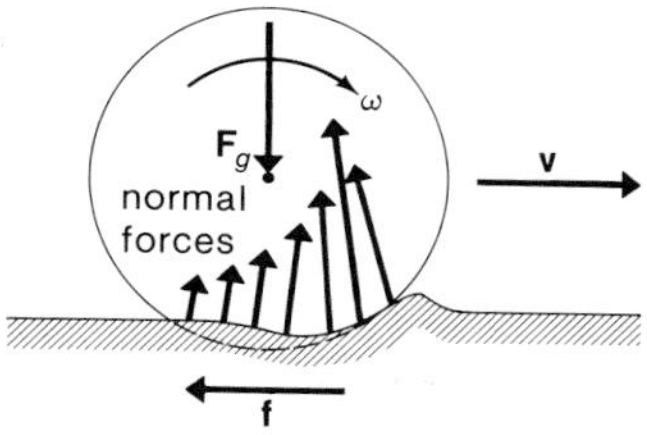

Fig. 6–11. Schematic behavior of a real wheel rolling across a surface. Both the wheel and the surface are deformed. (The deformations are exaggerated in the figure.)

Now look at Fig. 6–10 more closely. Although **f** is the rearward force that is required to slow the wheel, this force has a tendency* to make the wheel roll forward! That is, **f** tends to increase the rotational speed ω about the wheel axis. (Neither **N** nor $\mathbf{F}_g$ has this tendency.) Clearly, the surface cannot have such an effect on the wheel. We are forced to conclude that the frictional force is zero!

This is exactly correct. With the ideal conditions we have assumed, there is no instantaneous relative motion between the wheel and the surface at the point of contact (actually, a line), so there is no mechanism that can cause frictional losses. Thus, there is no dynamic friction, and the wheel will roll forward at constant velocity.

For a *real* case, the diagram shown in Fig. 6–10 is incorrect in several respects. First, both the wheel and the surface suffer deformations to an extent that depends on their particular elastic qualities. Figure 6–11 shows an exaggerated view of these deformations—the lower portion of the wheel is flattened and a shallow trench is formed in the surface. Thus, the line of contact between the wheel and the surface has expanded to a sizable area, and the normal force is now distributed over the contact area. As the wheel rolls forward, the leading portion of the wheel and the surface are compressed while the trailing portion becomes relaxed. Consequently, the deformations are greater for-

ward of the center of rotation. A large deformation is associated with a large force (Hooke's law); hence, the individual normal forces acting on various parts of the wheel are greater in the forward section, as indicated in Fig. 6–11.

The resultant of the rearward-acting frictional forces associated with the distributed normal force is shown as **f** in Fig. 6–11, and it is clear that this force tends to make the wheel roll forward, just as in Fig. 6–10. However, it is also clear that the net rotational effect of the normal force is in the opposite direction. The latter force counteracts the rotational effect of the frictional force and causes the observed rotational deceleration. Moreover, there is now a net horizontal force in the direction opposite to the motion, so the forward velocity of the wheel is also decreased.

The nature of the force of rolling friction is very complex. As the wheel and the surface deform during the rolling motion, some slippage does occur, so that sliding friction is present to a degree. The primary frictional effect over the area of contact, however, is one of static friction, which produces no energy losses.

In most cases, the important source of frictional losses due to rolling is in the deformation process. No real material is perfectly elastic, so there are always energy losses when a material is compressed and then relaxed. This continual compression-relaxation cycle generates heat in the wheel and in the surface. It

---

*This "tendency" is due to the *torque* exerted on the wheel by **f**. Although we do not discuss this quantity in detail until Chapter 10, the basic concept is easy to understand.

is for this reason that automobile tires become hot when run at high speeds.

What happens when the brakes are applied to a moving wheel? Figure 6–12 shows the situation in a schematic way. First, we must realize that normal drum-and-shoe (or disc) brakes cannot be effective on a single wheel or even on a pair of wheels connected by an axle; the braked wheel must be connected in some way to another wheel or other support in front or behind. (• Can you see why?) Therefore, we assume that Fig. 6–12 represents only one wheel of an automobile (or bicycle). When the brake shoes are pressed on the drum, the rotational motion of the wheel is retarded. This produces a forward force on the road, and the reaction force of the road on the tire is in the backward direction. This force greatly increases the force **f** shown in Fig. 6–12. The frictional force still attempts to increase the rotational motion of the wheel, but this is now opposed by both the normal force and the force exerted by the brakes.

The near-static conditions that exist at the tire-road interface can be upset if the braking action is severe. Then, skidding oc-

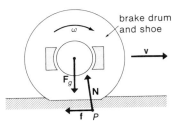

Fig. 6–12. The vector **N** here represents the effect of all individual normal forces indicated in Fig. 6–11. Also, **f** here represents the effect of all the individual frictional forces that were not shown in Fig. 6–11. When the brakes are applied, the frictional force **f** is increased significantly and the vehicle is slowed.

curs. Because $\mu_k < \mu_s$, skidding results in a decrease of the frictional force tending to slow the vehicle, and the vehicle moves with dangerously high speed without directional control.

If the wheel is *driven,* the additional force exerted by the tire on the road is in the backward direction and the reaction force of the road on the tire is in the forward direction. The result is a forward acceleration. ◄

6–4 VISCOUS FORCES

Galileo asserted that all objects falling freely near the surface of the Earth experience the same constant acceleration g. However, when an object moves through a fluid—even the air—it is observed that a resisting force or drag is present. This retarding force is the result of friction between the moving object and the molecules of the liquid or gaseous medium. Galileo was aware of these resistive effects and he knew that an object would fall "freely" only if they were negligible.

When a cannonball falls toward the Earth (the legendary experiment of Galileo at the Tower of Pisa), the effects of air resistance are small, and the acceleration of the cannonball is closely equal to g. However, there are many cases involving the motion of an object through a resisting (or *viscous*) medium in which the retarding force plays a crucial role. This is the case, for example, in the falling motion of a crumpled piece of paper or a raindrop.

Viscous drag is not a simple force; in fact, the magnitude of the force varies with the speed of the object in a complicated way. At relatively low velocities, the flow of a medium past a smooth object produces a viscous force f_v that is (approximately) directly proportional to the velocity. The force law that governs this case is called *Stokes' linear law of resistance**:

$$f_v = bv \tag{6-7}$$

When an object falls toward the Earth, the viscous force f_v is directed upward, so the equation of motion becomes

$$mg - bv = ma = m\frac{dv}{dt} \tag{6-8}$$

When the object is first released, we have $v = 0$, and hence the acceleration, $a = dv/dt$, is just equal to g. As a result of this acceleration during the first moments after release, the

*First formulated for the case of a sphere moving through a viscous medium by the British mathematician and physicist, Sir George Stokes (1819–1903), in 1845.

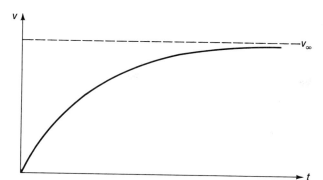

Fig. 6–13. The expected behavior of the solution $v(t)$ to Eq. 6–8.

velocity v increases from zero to some small value. The presence of this velocity reduces the left-hand side of Eq. 6–8, thereby reducing the acceleration dv/dt. Thus, in the subsequent moments, v increases at a slower rate. This process continues and, eventually, v becomes sufficiently large that the left-hand side of Eq. 6–8 is reduced essentially to zero. (Of course, bv can never exceed mg.) That is, in the limit $t \to \infty$, we have $mg - bv_\infty \to 0$, or $v_\infty = mg/b$. This limiting velocity v_∞ is called the *terminal velocity*. Thus, we expect the solution to Eq. 6–8 to have the behavior illustrated in Fig. 6–13. Indeed, when a crumpled piece of paper or a raindrop falls through air, it rapidly reaches terminal velocity and thereafter falls with a (relatively small) constant velocity. (If it were not for the retarding effect of the air, think of the disastrous result of a rain shower from a cloud at a height of a thousand meters or so!)

QUESTIONS

6–1   In considering the accelerated motion of an automobile, it is stated that, "Although the prime mover responsible for accelerating the automobile is its engine, the various forces operating in the engine and its power train are internal forces and do not enter in the application of Newton's second law applied to the automobile as the isolated object. It is the external forces which the roadway exerts on the tires, that is, the reaction forces to that which the engine-driven tires exert on the road, that propel the automobile forward." Examine this assertion in detail and either justify or refute it.

6–2   Suppose you increase your speed while jogging. Describe the forces causing you to accelerate when you are (a) running on horizontal ground, (b) running uphill, or (c) running downhill.

6–3   What makes a *Mexican jumping bean* jump?

6–4   Explain why it is easier to raise a sliding garage door if it is *counterweighted*.

6–5   Why will a string just strong enough to suspend a large rock snap if you attempt to jerk the rock suddenly upward with the string?

6–6   Suppose that when you jump from a certain height to the floor, it takes only one tenth as much time to come to rest

after impact as it took to fall. What is the average force exerted by the floor on your feet compared with your normal weight?

6–7   Explain why a plumb bob anywhere in the United States will not hang exactly in the direction of the Earth's gravitational force acting on it.

6–8   Since frictional forces are always supposed to oppose motion, explain why the frictional force the floor exerts on the soles of your shoes is in the same direction as your motion when you start to run.

6–9   Explain in your own words the distinction between the coefficient of static friction and the coefficient of kinetic friction.

6–10 The coefficient of kinetic friction is measured for two flat surfaces that slide over each other. The experiment is performed after each surface is progressively polished to a finer finish. At first, both surfaces are very rough. Describe and explain how you would expect μ_k to vary with the progress of the polishing.

6–11 Describe your braking procedure if you wanted to stop an automobile in the shortest distance on an icy road.

6–12 Why is it relatively safe to jump into deep water from a height of, say, 10 m yet very injurious to jump from the same height onto a cement floor?

6–13 Explain how a skydiver might change his terminal velocity while in free fall.

PROBLEMS

Sections 6–1 and 6–2

6–1 Two blocks with masses $m_1 = 3$ kg and $m_2 = 5$ kg are pushed up a frictionless inclined plane by a constant force $F = 100$ N which acts along the plane as shown in the diagram. (a) What is the acceleration of the blocks? (b) What force does the block m_1 exert on the other block?

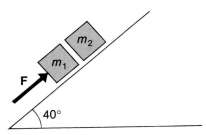

6–2 Two blocks with masses $m_1 = 2$ kg and $m_2 = 3$ kg are suspended vertically by a connecting ideal string that runs over an ideal pulley, as shown in the diagram. (a) Find the linear acceleration of the system. (b) What is the tension in the connecting string? (c) What is the tension in the string that supports the pulley?

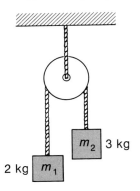

6–3 Two blocks, $m_1 = 5$ kg and $m_2 = 3$ kg, are connected by an ideal string that runs over an ideal peg. The blocks are free to slide on frictionless inclined planes, as shown in the diagram. (a) Find the linear acceleration of the sys-

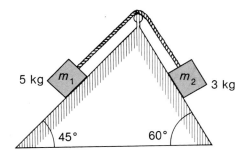

tem. (b) What is the string tension? (c) What is the force exerted on the peg?

6–4 Two blocks, $m_1 = 6$ kg and $m_2 = 4$ kg, are connected by a homogeneous rope that has a mass of 1 kg, as shown in the diagram. A constant vertical force, $F = 150$ N, is applied to the upper block m_1. (a) What is the acceleration of the system? (b) What is the tension in the rope at the top end? (c) At the bottom end? (d) At the midpoint of the rope?

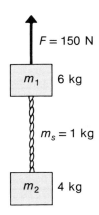

6–5 The compound block-and-tackle arrangement shown in the diagram on the opposite page is used to raise a block with a mass of 300 kg. Calculate the force F necessary at the free end of the rope to impart an upward acceleration of 0.2 m/s^2 to the block. What is the mechanical advantage of this arrangement? (Assume that the block-and-tackle system and the rope are ideal and massless, and that all forces act vertically.)

6–6 A ball with a mass $m = 1.5$ kg is attached to a string that is anchored in a rigid ceiling, as shown in the diagram. The length of the string is $\ell = 2.2$ m. A horizontal force F draws the ball aside until the string makes an angle $\alpha = 20°$ with the vertical. In this position the ball is at rest. Determine F. If the force is removed, what will be the initial acceleration of the ball?

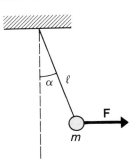

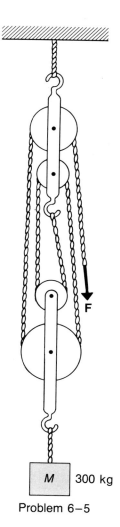

M | 300 kg

Problem 6–5

6–7 A rope has a length of 12 m and a mass of 16 kg. The rope hangs from a rigid support. A man whose mass is 80 kg slides down the rope at a constant speed of 0.8 m/s. What is the tension in the rope at a point 6 m from the top when the man has slid to this point?

6–8 Three blocks ($m_1 = 6$ kg, $m_2 = 4$ kg, and $m_3 = 8$ kg) are connected as in the diagram. The surfaces and the pulleys are frictionless. Determine the motion of the system and find the tension in each string.

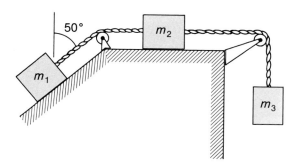

Section 6–3

6–9 A crate rests on the flat floor of a truck that is traveling up a hill at an angle of 15°. If the coefficient of static friction between the crate and the truck floor is $\mu_s = 0.4$, what is the maximum acceleration the truck may have if the crate is not to slip?

6–10 Imagine a long stretch of smooth ice (such as a frozen river) over which a hockey puck can slide. If the initial velocity of the puck is v_0 and if the coefficient of kinetic friction between the puck and the ice has the constant value μ_k, (a) derive an expression for the stopping time, (b) derive an expression for the distance the puck will travel before coming to rest, and (c) find the distance for the case in which $v_0 = 8$ m/s and $\mu_k = 0.12$.

6–11 Two blocks, $m_1 = 2$ kg and $m_2 = 5$ kg, are connected by an ideal string that runs over an ideal pulley. The block m_1 is free to slide on an inclined plane as shown. The angle of inclination of the plane is 30°, and $\mu_k = 0.30$. (a) Calculate the linear acceleration of the system starting from rest. (b) Calculate the tension in the connecting string.

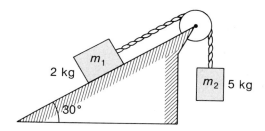

6–12 A block with a mass of 2 kg is pushed by a horizontal
  •   force, $F = 20$ N, as shown in the diagram. Assume that the coefficient of sliding friction is $\mu_k = 0.40$. Consider both possible directions of motion. (a) Find the acceleration of the block if $\alpha = 10°$. (b) What is the acceleration if $\alpha = 70°$?

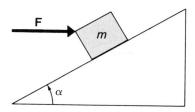

6–13 A crate is pulled along a horizontal floor with a rope that
  •   is maintained at a constant angle α above the horizontal. The coefficient of kinetic friction is μ_k. For a given constant force, show that the maximum acceleration along the floor results when $\tan \alpha = \mu_k$.

6–14 Refer to the diagram in Problem 6–12. Consider the situ-
  •   ation in which $\alpha = 30°$, $m = 2$ kg, and the coefficient of static friction is $\mu_s = 0.50$. Determine the maximum and minimum values of the force F for which the block will

remain stationary. What will happen if $F > F_{max}$ or $F < F_{min}$?

6-15 A block with a mass m will be pinned by static friction against the wall of a cylindric shell that rotates about a vertical axis if the angular speed of rotation exceeds a certain critical value ω_0. (a) Show that $\omega_0 = \sqrt{g/\mu_s R}$, where μ_s is the coefficient of static friction between the wall and the block and R is the radius of the cylinder. (b) A favorite application of this effect is a rather scary ride at some amusement parks in which the passengers are pinned against the wall of a large rotating cylinder. When the rotational speed exceeds ω_0, the floor on which the passengers were previously standing suddenly drops away! If $R = 5$ m, and if the coefficient of static friction between clothing and the wall is $\mu_s = 0.20$, what is the critical angular speed in revolutions per minute (rpm) for this ride?

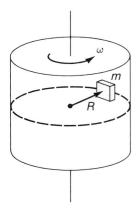

6-16 An automobile coasts in neutral and moves initially on a
• slick surface at a speed of 20 km/h while its wheels roll without slipping. The total mass of the car, $m = 1000$ kg, is equally distributed on its four wheels. The relevant co-efficients of friction between the tires and roadway are $\mu_s = 0.20$ and $\mu_k = 0.10$. (a) If a braking force is applied (rear wheels only) up to 80 percent of the maximum value supportable by static friction, what is the stopping distance and what is the stopping time? (b) In a panic application of the brakes, the rear wheels lock. Now what is the stopping distance and what is the stopping time? (c) If the force due to rolling friction for each of the four wheels is 2 percent of the force due to kinetic friction when a wheel is skidding, in what distance will the car roll to a stop, and how long will it take? (Assume all other forms of resistance to the motion are negligible.) (d) In view of the answers to part (c), is it permissible to neglect rolling friction when solving parts (a) and (b)? Explain.

6-17 In deep space, a particle ($m = 0.6$ kg) slides in a circular
• path on the inner surface of a cylinder, the radius of which is $R = 1.8$ m. The coefficient of kinetic friction for this case is $\mu_k = 0.20$. If the particle has a speed $v_0 = 2$ m/s at $t = 0$, find the speed at any later time t. Comment on the behavior of $v(t)$. [*Hint*: First show that $R\dfrac{dv}{v^2} = -\mu_k dt$. A table of integrals is given inside the rear cover.]

Section 6-4

6-18 An object is released from rest at $t = 0$ and falls through a
• viscous medium. Assume that the motion conforms to Stokes' law. Verify that the velocity

$$v(t) = v_\infty(1 - e^{-(b/m)t})$$

satisfies the equation of motion (Eq. 6-8), where $v_\infty = mg/b$. Verify also that the distance of fall is given by

$$y(t) = (v_\infty^2/g)(e^{-(b/m)t} - 1) + v_\infty t$$

[*Hint*: Show that $dy(t)/dt = v(t)$.]

Additional Problems

6-19 A painter whose mass is 80 kg is sitting in a light scaffold chair, as shown in the diagram at right. With what steady force must he pull on the rope to achieve an upward acceleration of 0.20 m/s²? Is there any limit to the acceleration he can achieve? (Neglect the mass of the chair, and take the rope and pulley to be ideal elements.)

6-20 Two blocks, $m_1 = 2$ kg and $m_2 = 5$ kg, are supported by
• two ideal pulleys and ideal connecting cords, as shown in the figure at right. (a) Determine the acceleration of each block. (b) What is the tension in each of the cords shown? [*Hint*: Remember that the cord running around the two pulleys has a fixed length. From this, deduce the relationship between the accelerations of the two blocks.]

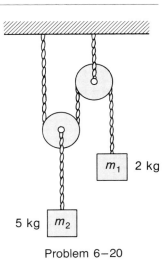

Problem 6-20

6–21 The three blocks shown in the diagram below have masses
•• $m_1 = 2$ kg, $m_2 = 3$ kg, and $m_3 = 5$ kg. The blocks are connected by ideal cords and pulleys, A and B. (a) Determine the acceleration of each block. (b) Determine the tension in each of the cords shown. [*Hint:* The cord that passes over pulley B has a fixed length; this means that the *upward* acceleration of m_1 *relative* to pulley B is equal to the *downward* acceleration of m_2 *relative* to pulley B. Newton's equations hold only in an inertial frame, so you must deduce the accelerations of m_1 and m_2 with respect to the ground in terms of the accelerations with respect to pulley B, using the acceleration of pulley B.]

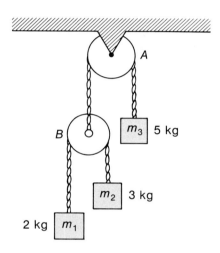

6–22 Two blocks, $m_1 = 2$ kg and $m_2 = 5$ kg, are initially
• resting on the floor. They are connected by an ideal cord running over an ideal pulley, as shown in the diagram. Find the acceleration of each block and the pulley if the upward force applied to the pulley is (a) 35 N, (b) 70 N, (c) 140 N. [*Hint:* Refer to the hints in the two previous problems.]

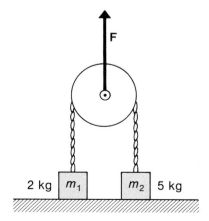

6–23 Two blocks are pulled by a 50-N force, as shown in the diagram. What is the minimum value of μ_s, the coefficient of static friction between the two blocks, that will prevent m_1 from slipping on m_2?

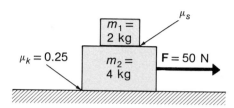

6–24 A block is free to slide down an inclined plane that is
•• equipped with wheels so that it can roll along a horizontal surface, as shown in the accompanying figure. If the angle of inclination of the plane is 45°, and if the coefficient of static friction between the block and the plane is $\mu_s = 2/5$, show that the cart must have an acceleration to the right of at least $(3/7)g$ if the block is not to slide down the plane.

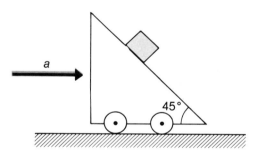

6–25 A daring skier slides freely down a 45° slope (without using ski poles or angling his skis). The coefficient of sliding friction between his skis and the snow is $\mu_k = 0.080$. If he starts from rest, calculate his speed 100 m down the slope. How long does it take him to travel this distance?

6–26 A block slides down an inclined plane with an initial speed $v_0 = 3$ m/s. The plane has an angle of 20° with respect to the horizontal, and the coefficient of kinetic friction is $\mu_k = 2/3$. In what distance does the block come to rest?

6–27 A small twig slides down the roof of a house pitched at an
 • angle of 30° with respect to the horizontal. The twig starts
 from rest at a point 1.5 m from the edge and strikes the
 ground 2 m from the house, which has its roof-line 4 m
 above the ground. What is the coefficient of kinetic friction between the twig and the roof? (Assume zero air resistance during the fall.)

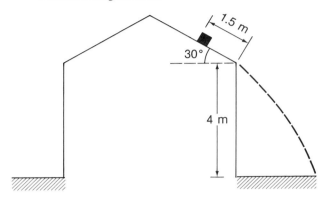

6–28 Two blocks with masses $m_1 = 5$ kg and $m_2 = 3$ kg are
 • tied together by an ideal string. The blocks slide down an
 inclined plane that makes an angle of 30° with the horizontal, as shown in the diagram. If the coefficient of kinetic friction between m_1 and the plane is 0.20, and that
 between m_2 and the plane is 0.30, find the acceleration of
 the system and the tension in the string. What would happen if the blocks were reversed in their positions, with m_2
 leading the way down?

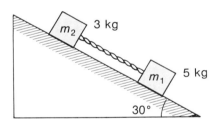

6–29 A block is pushed along a horizontal surface by a force F
 that makes an angle α with the horizontal, as shown in the
 diagram. Taking the coefficient of sliding friction to be
 μ_k, show that as α is increased, a critical angle will be
 reached at which forward motion is impossible, regardless
 of the magnitude of F. Find the expression for this critical
 angle.

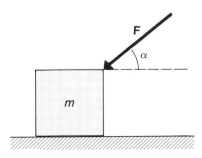

6–30 Two blocks, m_1 and m_2, are connected by an ideal string
 • that runs over an ideal pulley. The block m_1 is free to slide
 on an inclined plane with an adjustable inclination angle
 α, as shown in the diagram. It is observed that when α is
 less than 15°, the block m_1 slides up the incline, and when
 α is greater than 63°, it slides down the incline. (a) What
 is the coefficient of static friction between block m_1 and
 the inclined plane? (b) If $m_1 = 2$ kg, what is m_2?

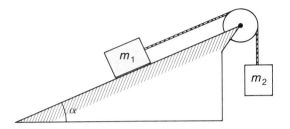

6–31 A penny is placed on a record turntable that revolves at
 $33\frac{1}{3}$ rpm. If the coefficient of static friction between the
 penny and the turntable is $\mu_s = 0.25$, how far from the
 center of rotation may the penny be placed if it is to remain stationary on the turntable?

6–32 A jet airliner, traveling with a ground speed of 700 km/h,
 banks into a turn that has a radius $R = 8$ km. (a) If passengers standing in the center aisle experience forces exerted on them by the floor that are entirely perpendicular
 to the floor, what is the banking angle α of the aircraft?
 (b) How is the weight of a passenger related to his mass?

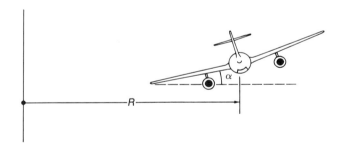

6–33 A lead ball and a piece of paper loosely crumpled into a
 • spherical shape are dropped simultaneously from a tower
 that is 20 m high. Assume the terminal velocity for the
 lead ball to be 600 m/s, and that for the paper ball to be
 60 m/s. How far above the ground is the paper ball when
 the lead ball strikes the ground? [*Hint:* Use the expression
 for $y(t)$ given in Problem 6–18, and use Eq. A–11 in
 Appendix A to expand the exponential to order t^3.]

6–34 Refer to the expression for $y(t)$ given in Problem 6–18 for
 an object falling in a viscous medium. Consider the motion during a short time interval after release (i.e.,
 $t \ll m/b$). Expand the exponential by using Eq. A–11 in
 Appendix A and show that

$$y(t \ll m/b) \cong \tfrac{1}{2}gt^2$$

Thus, the object begins to fall in the normal way during the short time before the viscous retarding force becomes important.

6–35 Equation 6–8 is actually a *differential equation*. Sometimes equations of this type cannot be solved in terms of simple functions. In these cases, the solutions must be obtained by numerical methods. Use this approach to solve Eq. 6–8. First, write the equation in terms of finite increments:

$$\Delta v = \left(g - \frac{b}{m} v \right) \Delta t = g \left(1 - \frac{v}{v_\infty} \right) \Delta t$$

Use the notation

$$v_{i+1} = v_i + \Delta v_i \quad \text{and} \quad t_{i+1} = t_i + \Delta t_i$$

in which

$$\Delta v_i = g \left(1 - \frac{v_i}{v_\infty} \right) \Delta t_i$$

Consider a case for which $v_\infty = 100$ m/s. Take all intervals to be $\Delta t_i = 2$ s and let $v_0 = 0$ at $t_0 = 0$ (that is, for $i = 0$). From $v_0 = 0$, we find v_1 to be

$$v_1 = v_0 + \Delta v_0 \quad ; \quad \Delta v_0 = 2g \left(1 - \frac{0}{v_\infty} \right)$$

Then, $v_2 = v_1 + \Delta v_1 \quad ; \quad \Delta v_1 = 2g \left(1 - \frac{v_1}{v_\infty} \right)$

and $v_3 = v_2 + \Delta v_2 \quad ; \quad \Delta v_2 = 2g \left(1 - \frac{v_2}{v_\infty} \right)$

and so forth. Calculate numerical values of v_i by this process up to at least $i = 10$. Compare your results with the exact values calculated from the actual solution:

$$v_i = v_\infty (1 - e^{-(b/m)t_i})$$

Suggest a way for improving the above numerical method, which is accurate only to about 10 percent.

6–36 A steel block with a mass of 100 g is set to sliding along a horizontal steel plane with an initial speed $v_0 = 300$ m/s. The coefficient of kinetic friction is

$$\mu_k(v) = 0.27 \frac{(1 + 0.0044 \ v)}{(1 + 0.064 \ v)}$$

where v is in m/s. (This expression is commonly used for high-speed, steel-on-steel friction.) (a) Calculate the stopping time by using a numerical method like that in Problem 6–35. [*Suggestion:* Use a linear extrapolation, $v_f = v_i - a_i \Delta t$, with a_i evaluated at v_i by using $\mu_k(v_i)$. Then, v_f becomes v_i for the next interval Δt, and so on. Start with $\Delta t = 100$ s, and switch to $\Delta t = 50$ s when $a_i \Delta t \cong \frac{1}{2} v_i$; switch again to $\Delta t = 25$ s when next this occurs; and continue in this manner. Halt the process when $v_f \cong 2$ m/s.] (b) Determine an "average" value for μ_k by using $\bar{v} = \frac{1}{2} v_0$ in the expression for $\mu_k(v)$. From this, calculate the corresponding deceleration and determine the stopping time. Compare this result with that in (a) and with the exact answer, $T = 670$ s. [The exact analytic solution for the stopping time, given the initial speed v_0, is $T = 5.498 \ v_0 - 1163.7 \ln (1 + 0.0044 \ v_0)$.] You should find that your result obtained by the numerical method is within 5 percent of the true answer, whereas the selected constant value of μ_k gives a result that is within 8 percent. You can therefore appreciate why there is such a strong temptation to assume that μ_k is independent of speed! If one selects some reasonable mean value for μ_k in the velocity range of interest, this approximation affords a useful first estimate of the solution. What would be the error in T if we used the low-speed limit for the friction coefficient, $\mu_k(v \rightarrow 0) = 0.27$?

CHAPTER 7

WORK AND KINETIC ENERGY

In previous chapters we applied Newton's laws of dynamics in a variety of situations involving the motion of particles. For the cases considered, solutions were obtained in which the particle's location, velocity, and acceleration were given as explicit functions of time. Thus, we were able to describe rather completely the motion of the particle through a set of equations for $\mathbf{r}(t)$, $\mathbf{v}(t)$, and $\mathbf{a}(t)$.

The concepts developed in this chapter and in the next permit us, in many cases, to express directly a relationship connecting position and velocity without obtaining a time-dependent solution—indeed, without any explicit reference to time whatsoever. For example, suppose that a block slides down a ramp. We wish to know the velocity of the block at the bottom of the ramp but we are not really interested in *when* the block arrives there. The techniques considered here allow us to obtain solutions to such problems in a simple and direct way. Although the information obtained is less than that contained in the complete time-dependent solution, nonetheless many interesting characteristics of a system and its motion can be understood more easily. Moreover, we can treat in a straightforward way cases in which the forces change as the particle changes position—cases that are often difficult to treat by applying Newton's laws directly.

The procedure that we follow introduces us to the profoundly important concept of *energy*. The importance of this idea lies in the fact that although energy can exist in many different forms, and processes may be devised that transform energy from one form to another, the total energy content of an isolated system remains constant. This is the well-known law of energy conservation.

We shall find in this chapter that there is a form of mechanical energy called *kinetic energy* that a body possesses by virtue of its motion. In the next chapter we introduce another form of mechanical energy, called *potential energy,* that a body possesses by virtue of its position or geometric configuration. There are many other forms of energy, such as heat energy, chemical energy, electrical energy, and the important relativistic concept of the energy equivalence of mass; these forms of energy are treated in later chapters. Another form of energy we meet in this chapter, called *work,* is energy in the process of being transferred by mechanical means from one form to another; "work is energy in transit."

The process of balancing energy accounts as demanded by the conservation of energy principle requires calculating certain mathematical quantities associated with the various forms of energy. We start to formulate such prescriptions by carefully refining our every-day concepts of mechanical energy and work.

7-1   WORK DONE BY CONSTANT FORCES

Work Involved in Infinitesimal Displacements.   We first consider in some detail the concept of *work*. Everyone is familiar with the use of this term in ordinary conversation. We usually think of work as involving physical activity and muscular effort or in terms of work done by machines. Indeed, work is done in all these situations, but the definition that we need turns out to be more restricted and more precise. If a force acts on an object and causes it to undergo a displacement, we say that work was done by the force. To avoid the necessity of treating rotational effects, we again confine attention to particles and consider only the work done by forces during translational displacements.

Figure 7-1 shows a block (considered as a particle in the sense of Chapter 4) that is acted upon by a force **F**. During the time that this force is acting, the block undergoes an infinitesimal displacement $\Delta \mathbf{s}$. We define the infinitesimal mechanical work ΔW done by the agency responsible for the force **F** by the equation

$$\Delta W = F \, \Delta s \cos \theta \qquad\qquad (7-1)$$

where θ is the angle between the directions of the vectors **F** and $\Delta \mathbf{s}$. (There may be other forces acting on the block, but we are interested here only in the work associated with the particular force **F**.) Although it is generated by two *vector* quantities—force and displacement—work is a *scalar* quantity.

Notice that $\Delta W = 0$ when $\cos \theta = 0$, that is, when $\theta = \pi/2$. When the force **F** and the displacement $\Delta \mathbf{s}$ are perpendicular to one another, no work is done. For example, when a block slides across a horizontal frictionless surface, the normal force **N** exerted by the surface does no work on the block because the force is perpendicular to the displacement.

If the displacement $\Delta \mathbf{s}$ is zero, the work done is necessarily also zero. For example, consider a student holding a ball in his outstretched hand. Because the ball is stationary, the student does no work on the ball; nevertheless, he feels tired—he is convinced that he has done work. Indeed, the student *has* done work. However, this work is *internal work,* done in the student's muscles. Work always involves (force) × (distance); in the case of internal muscular work, electrons and ions are moved by electric forces generated in the muscle fibers. Although work is done in the muscles, this work is not done on the ball and therefore does not cause any change in the position or physical characteristics of the ball.

The dimensionality of work is

$$[\Delta W] = [F][\Delta s] = (MLT^{-2})(L) = ML^2T^{-2}$$

The SI unit of work is the kg·m^2/s^2 or N·m. We introduce the special name *joule* (J) for the unit of work*:

$$1 \text{ J} = 1 \text{ N·m} = 1 \text{ kg·m}^2/\text{s}^2$$

The Scalar Product (or Dot Product) of Two Vectors.   There are many instances in which a *scalar* physical quantity is generated by a type of multiplication involving two *vectors.*† An important example is the scalar quantity *work,* which is defined by the vectors **F** and $\Delta \mathbf{s}$. The kind of multiplication that generates a scalar from two vectors is

---

*This unit is named in honor of the English physicist, James Prescott Joule (1818–1889), who contributed greatly to the understanding of the concepts of work and energy.

†In Section 10–1 we discuss another type of vector multiplication (the *cross product*) that produces a *vector* result.

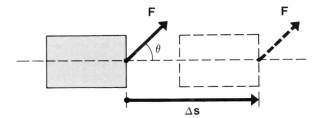

Fig. 7–1. The block (a particle) is acted upon by a force **F** during a displacement Δ**s**. The displacement is imagined to be infinitesimally small so that, throughout the motion, the force **F** remains constant.

called the *scalar product* or the *dot product* and is defined to be

$$\mathbf{A} \cdot \mathbf{B} = |\mathbf{A}||\mathbf{B}| \cos \theta_{AB} = AB \cos \theta_{AB} \qquad (7-2)$$

where θ_{AB} is the angle between the two vectors. This type of multiplication is symbolized by the dot between the vectors. (The multiplication cross is reserved for the product that yields a vector.)

It is often useful to interpret the dot product $\mathbf{A} \cdot \mathbf{B}$ as either $A(B \cos \theta_{AB})$ or $B(A \cos \theta_{AB})$. The first form suggests taking the component of **B** in the direction of **A** and multiplying by the magnitude of **A**. The second form suggests taking the projection of **A** onto the direction of **B** and multiplying by the magnitude of **B**. These two views are equally correct. A very useful identity is $\mathbf{A} \cdot \mathbf{A} = |\mathbf{A}||\mathbf{A}| \cos 0° = A^2$. Thus, $\mathbf{A} \cdot \mathbf{A}$ is just the magnitude of the vector **A** squared.

From the defining equation (Eq. 7–2), it is clear that the scalar product is commutative; that is,

$$\mathbf{A} \cdot \mathbf{B} = \mathbf{B} \cdot \mathbf{A} \qquad (7-3)$$

In addition, the scalar product obeys the distributive law of multiplication:

$$\mathbf{A} \cdot (\mathbf{B} + \mathbf{C}) = \mathbf{A} \cdot \mathbf{B} + \mathbf{A} \cdot \mathbf{C} \qquad (7-4)$$

The unit vectors $\hat{\mathbf{i}}$, $\hat{\mathbf{j}}$, and $\hat{\mathbf{k}}$ are mutually perpendicular. It therefore follows that they obey the relationships

and also

$$\left.\begin{aligned} \hat{\mathbf{i}} \cdot \hat{\mathbf{j}} = \hat{\mathbf{j}} \cdot \hat{\mathbf{i}} = \hat{\mathbf{i}} \cdot \hat{\mathbf{k}} = \hat{\mathbf{k}} \cdot \hat{\mathbf{i}} = \hat{\mathbf{j}} \cdot \hat{\mathbf{k}} = \hat{\mathbf{k}} \cdot \hat{\mathbf{j}} = 0 \\ \hat{\mathbf{i}} \cdot \hat{\mathbf{i}} = \hat{\mathbf{j}} \cdot \hat{\mathbf{j}} = \hat{\mathbf{k}} \cdot \hat{\mathbf{k}} = 1 \end{aligned}\right\} \qquad (7-5)$$

If we write **A** and **B** in Cartesian component form, we obtain

$$\begin{aligned} \mathbf{A} \cdot \mathbf{B} &= (A_x\hat{\mathbf{i}} + A_y\hat{\mathbf{j}} + A_z\hat{\mathbf{k}}) \cdot (B_x\hat{\mathbf{i}} + B_y\hat{\mathbf{j}} + B_z\hat{\mathbf{k}}) \\ &= A_xB_x + A_yB_y + A_zB_z \end{aligned} \qquad (7-6)$$

We see that if we know the components of two vectors, we have a useful alternate formulation to Eq. 7–2 to use in calculating the value of a scalar product.

Using the dot-product notation, Eq. 7–1 may now be written in compact form as

$$\Delta W = \mathbf{F} \cdot \Delta \mathbf{s}$$ (7–7)

Example 7–1

Calculate the scalar product of the two vectors $\mathbf{A} = 3\hat{\mathbf{i}} - 2\hat{\mathbf{j}}$ and $\mathbf{B} = 4\hat{\mathbf{i}} + 6\hat{\mathbf{j}} + 2\hat{\mathbf{k}}$.

Solution:

Using Eq. 7–6, we have

$$\mathbf{A} \cdot \mathbf{B} = (3\hat{\mathbf{i}} - 2\hat{\mathbf{j}}) \cdot (4\hat{\mathbf{i}} + 6\hat{\mathbf{j}} + 2\hat{\mathbf{k}})$$
$$= (3)(4) + (-2)(6) + (0)(2) = 0$$

(• What can you deduce about the relationship of the two vectors **A** and **B**?)

Example 7–2

A particle travels at constant speed, but with no restriction on its direction of motion. Show that for this situation the acceleration vector at any instant is always perpendicular to the velocity vector.

Solution:

Because $|\mathbf{v}(t)|$ is a constant, $v^2(t) = \mathbf{v}(t) \cdot \mathbf{v}(t)$ is also a constant. Thus,

$$\frac{d}{dt}(\mathbf{v} \cdot \mathbf{v}) = 0$$

But $\dfrac{d}{dt}(\mathbf{v} \cdot \mathbf{v}) = \mathbf{v} \cdot \left(\dfrac{d\mathbf{v}}{dt}\right) + \left(\dfrac{d\mathbf{v}}{dt}\right) \cdot \mathbf{v} = 2\mathbf{v} \cdot \left(\dfrac{d\mathbf{v}}{dt}\right)$

Therefore, $\mathbf{v} \cdot \dfrac{d\mathbf{v}}{dt} = \mathbf{v} \cdot \mathbf{a} = 0$

which (for $\mathbf{a} \neq 0$) ensures that we always have $\mathbf{v} \perp \mathbf{a}$.

To understand this result in simple terms, notice that any nonzero acceleration $\mathbf{a}$ that is not perpendicular to the velocity $\mathbf{v}$ must have a component in the direction of $\mathbf{v}$ or opposite to it. Such a component would have the effect of changing the speed. Thus, in order for $v(t)$ to be constant, we *must* have either $\mathbf{a} = 0$ or $\mathbf{v} \perp \mathbf{a}$.

Newton's second law states that any acceleration is in the same direction as the force that causes it. Also, the instantaneous displacement of an object is in the same direction as its instantaneous velocity. It follows from Example 7–2, then, that in the case of an accelerating object with constant speed, the total force at any instant must be perpendicular to the instantaneous displacement. Therefore, when $|\mathbf{v}(t)| = $ constant, we must have $\Delta W = \mathbf{F} \cdot \Delta \mathbf{s} = 0$ at every instant; that is, if there is no change in speed, no work is done.

Work Done by Several Forces Acting Simultaneously.   When several forces act on an object during a displacement, we can calculate the work done by each force individually. The total work done on the object is then the algebraic sum of the individual contributions:

$$\Delta W_{\text{total}} = \Delta W_a + \Delta W_b + \Delta W_c$$
$$= \mathbf{F}_a \cdot \Delta \mathbf{s} + \mathbf{F}_b \cdot \Delta \mathbf{s} + \mathbf{F}_c \cdot \Delta \mathbf{s}$$

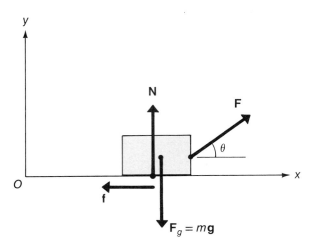

Fig. 7–2. Free-body diagram for a block that is pulled at constant velocity across a horizontal surface.

Alternatively, we can proceed by first finding the resultant **F** of the individual forces; thus,

$$\Delta W_{\text{total}} = (\mathbf{F}_a + \mathbf{F}_b + \mathbf{F}_c) \cdot \Delta \mathbf{s}$$
$$= \mathbf{F} \cdot \Delta \mathbf{s}$$

(7–8)

(That is, the dot product obeys the distributive law.)

Figure 7–2 shows the free-body diagram for a block being pulled at constant velocity across a horizontal surface by a force **F**. There are four forces acting on the block: **F**, **N**, $\mathbf{F}_g$, and the frictional force **f**. According to the definition, the amount of work done by the force **F** during a displacement $\Delta \mathbf{s}$ is $\Delta W_F = (F \cos \theta) \Delta s$. The normal force **N** and the gravitational force $\mathbf{F}_g$ act perpendicular to the surface, so $\theta = \pi/2$ for these forces, and $\Delta W = 0$ for each.

Negative Work. What is the work done by the frictional force **f** in Fig. 7–2? This force acts on the block in a direction *opposite* to that of the displacement; that is, $\theta = \pi$ and $\cos \theta = -1$ for **f**. Consequently, $\Delta W_f = -f \Delta s$.

The total work done *on the block* is the sum of the work done by all the forces. No work is done by either **N** or $\mathbf{F}_g$, so the total work is simply that done by **F** plus that done by **f**, namely,

$$\Delta W_{\text{total}} = \Delta W_F + \Delta W_f = F \Delta s \cos \theta - f \Delta s$$
$$= (F \cos \theta - f) \Delta s$$

Now, if the block moves with constant velocity, the horizontal forces that are acting must sum to zero; that is, $F \cos \theta - f = 0$. Thus, $\Delta W = 0$, and there is no net work done on the block.

How do we interpret the overall result that no net work is done? First, notice that the block's velocity is the same at the end of the displacement as at the beginning. Moreover, the block has not been deformed or altered in its composition. Thus, it seems reasonable that the net work done on the block is, in fact, zero. However, we know that the force **F** has done work on the block. Also, the force **f** has done work on the block, but this work is *negative*. This means that the *block* has actually done work. (We always interpret negative work in this way.) The block has done work in breaking the intermolecular bonds that extend between the block and the surface, bonds that are responsible for friction. In effect, then, the work done by the applied force **F** is done against friction. The result of this work is the generation of heat; this heat cannot be fully recovered and converted into mechanical work.

Work Along a Straight-Line Path. Later in this chapter, we discuss more complicated situations, but now consider the simple case in which a *particular* force $\mathbf{F}$ that acts
on an object is *constant* and the displacement s of the object is constrained to a straight
line. Other forces in addition to $\mathbf{F}$ may be present; for example, if $\mathbf{F}$ is not along $\mathbf{s}$, other
forces are required to maintain the straight-line motion. Then, using Eq. 7–8, the total
work done by the constant force $\mathbf{F}$ in a sequence of N straight-line displacements $\Delta\mathbf{s}_n$ is*

$$W = \sum_{n=1}^{N} (\Delta W)_n = \sum_n \mathbf{F} \cdot \Delta\mathbf{s}_n$$

$$= \mathbf{F} \cdot \sum_n \Delta\mathbf{s}_n$$

If the total displacement is $\mathbf{s} = \Sigma\, \Delta\mathbf{s}_n$, we have

$$W = \mathbf{F} \cdot \mathbf{s} \qquad (\mathbf{F} = \text{constant}) \qquad\qquad (7\text{–}9)$$

Example 7–3

Consider a block with a mass $m = 8$ kg that is being
pushed up a ramp, as shown in the diagram. The coefficient of
kinetic friction between the block and the ramp is $\mu_k = 0.40$.
What amount of work is done on the block by all the forces
acting as the block is pushed at constant velocity from point A to
point B, a distance of 0.65 m, if the angle of inclination of the
ramp is 32°?

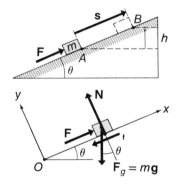

Solution:

From the free-body diagram, we have, for the y-direction,

$$N = mg \cos\theta = (8 \text{ kg})(9.80 \text{ m/s}^2) \cos 32°$$

$$= 66.5 \text{ N}$$

Then, the frictional force is

$$f = \mu_k N = (0.40)(66.5 \text{ N}) = 26.6 \text{ N}$$

For the x-direction, the acceleration is zero; thus, we have

$$F = f + mg \sin\theta$$

$$= 26.6 \text{ N} + (8 \text{ kg})(9.80 \text{ m/s}^2) \sin 32°$$

$$= 68.1 \text{ N}$$

The work done by the force F on the block is

$$W_F = Fs = (68.1 \text{ N})(0.65 \text{ m}) = 44.3 \text{ J}$$

The work done by the frictional force f on the block is

$$W_f = -fs = -(26.6 \text{ N})(0.65 \text{ m}) = -17.3 \text{ J}$$

The work done by the gravitational force F_g on the block is

$$W_g = -F_g s \sin\theta = -mgs \sin\theta$$

$$= -(8 \text{ kg})(9.80 \text{ m/s}^2)(0.65 \text{ m}) \sin 32°$$

$$= -27.0 \text{ J}$$

The work done by the normal force N on the block is zero:
$W_N = 0$. (• Why is the work done by this force zero?)
Notice that

$$W_F + W_f + W_g + W_N = 0$$

That is, the net work done on the block is zero, as it must be,
because the speed of the block is constant and hence no net force
acts on the block.

Only the externally applied force $\mathbf{F}$ does positive work on
the block. The frictional and gravitational forces both do negative work on the block. In effect, the work done by $\mathbf{F}$ is energy

---

*When a sum is understood to run over a particular range, we use a more compact notation in which we
abbreviate $\Sigma_{n=1}^{N} A_n$ to $\Sigma_n A_n$ or ΣA_n.

transferred to the agencies responsible for the forces **f** and **F**$_g$. The frictional force results in the dissipation of work to heat. But what happens to the work done against the gravitational force? This interesting question is addressed in Chapter 8.

7-2 WORK DONE BY VARIABLE FORCES DURING LINEAR MOTION

Next, we examine cases in which a particular force **F** acting on a particle depends on the position of the particle. We postpone discussion of the general case and restrict the inquiry in this section to situations in which the particle is constrained to move along a straight-line path, say, the x-direction. If the force **F** acts only in the x-direction but varies with x, we have

$$\Delta W = F(x)\,\Delta x$$

What is the total work done by $F(x)$ during the displacement of the particle from the point A ($x = x_A$) to point B ($x = x_B$)? Dividing the interval into N increments Δx_n, we can write

$$W = \sum_{n=1}^{N} F(x_n)\,\Delta x_n$$

Although $F(x)$ is not constant, we can give a graphic interpretation of this expression, as in Fig. 7-3, where $F(x)$ is shown as a function of x. A typical term in the sum, such as $F(x')\,\Delta x' = \Delta W'$, is seen to be equivalent to the small heavily shaded rectangular area. In the limit of an infinite number of vanishingly small infinitesimal steps, *keeping x_A and x_B fixed,* we have

$$W = \lim_{N \to \infty} \sum_{n=1}^{N} F(x_n)\,\Delta x_n = \int_{x_A}^{x_B} F(x)\,dx \qquad (7-10)$$

where the last expression is called the *definite integral* of $F(x)$ between $x = x_A$ and $x = x_B$. In Fig. 7-3, the small "error" areas (shown in black for $\Delta x'$) vanish in the limit,

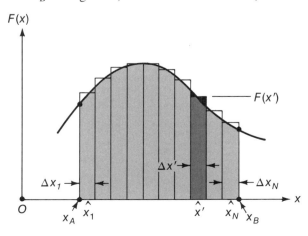

$F(x)$

Fig. 7-3. The graphic representation of $\Sigma_n\,F(x_n)\,\Delta x_n$ and $\int_{x_A}^{x_B} F(x)\,dx$.

$N \rightarrow \infty$, and the value of W is precisely the entire shaded area (or the *area under the curve*) between the limits x_A and x_B. The equivalence of the sum and the definite integral in Eq. 7-10 is called the "fundamental theorem of integral calculus."*

▶ If one is given the indefinite integral (recall Section 2-6)

$$\int f(x) \, dx = g(x) + C$$

the definite integral of $f(x)$ is defined to be

$$\int_a^b f(x) \, dx \equiv [g(x) + C]\Big|_a^b \equiv [g(b) + C] - [g(a) + C]$$

$$\equiv g(b) - g(a)$$

The quantities a and b are the *limits of the integral,* and the specified process involves first substituting b (the upper limit) into the indefinite integral, then subtracting the result of substituting a (the lower limit) into the same expression. The result is a definite value (not a function), which is the area under the curve $f(x)$ between $x = a$ and $x = b$. Note that interchanging the limits has the effect of changing the sign of the integral; thus,

$$\int_b^a f(x) \, dx = g(a) - g(b)$$

$$= -[g(b) - g(a)] = -\int_a^b f(x) \, dx$$

It is also possible to consider the definite integral to be a function of a variable upper limit. Thus, we have

$$\int_a^x f(x') \, dx' = g(x) - g(a)$$

which may be considered a function of the upper limit x. Since x' is a *dummy variable* (i.e., x' disappears in arriving at the final result), this same equation is sometimes written in a simpler form by dropping any separate label for the dummy variable appearing in the integrand, and we would have

$$\int_a^x f(x) \, dx = g(x) - g(a)$$

Be sure to distinguish clearly between the two separate meanings of x, one as the variable in the integrand and one as the variable upper limit.

The definite integral with a variable upper limit and a fixed lower limit may also be used instead of evaluating a constant of integration from an initial condition, as is required with the use of the indefinite integral. For example, we have for linear motion with a constant velocity v_0

$$dx/dt = v$$

and the corresponding integrals in the present case,

$$\int_{x_0}^x dx = \int_0^t v \, dt = v_0 \int_0^t dt$$

where the corresponding pair of upper and lower limits imply that $x = x_0$ when $t = 0$, and $x = x$ at $t = t$. Carrying out the integrals, we obtain

$$x - x_0 = v_0(t - 0)$$

or

$$x = x_0 + v_0 t$$ ◀

In the expression $\Delta W = \mathbf{F} \cdot \Delta \mathbf{s}$, the positive sense of $\Delta \mathbf{s}$ is in the direction of the physical displacement. Because of the algebraic sign implicitly associated with the order of the limits for the definite integral, the selection of the *lower limit* to correspond to the *initial position* of the object and the *upper limit* to correspond to the *final position* automatically accounts for the direction of the displacement. The algebraic sign of $F(x)$ depends on the coordinate frame selected and must be properly chosen in each case. Observe carefully in the examples that follow how these conventions apply.

In some portion of the interval $x_A < x < x_B$, the integrand $F(x)$ may be negative. Then, negative work is done in that segment of the path. When both positive and negative contributions exist, the total amount of work done is equal to the algebraic sum of the parts. The evaluation of the definite integral between the limits x_A and x_B will automatically give the correct sum. This point is illustrated in Fig. 7-4, where

$$W = \int_{x_A}^{x_B} F(x) \, dx = \int_{x_A}^{x_M} F(x) \, dx + \int_{x_M}^{x_B} F(x) \, dx$$

$$= W_1 + W_2 = A_1 - A_2$$

---

*This theorem was discovered independently and at about the same time by Isaac Newton and the German mathematician Gottfried Leibniz (1646-1716).

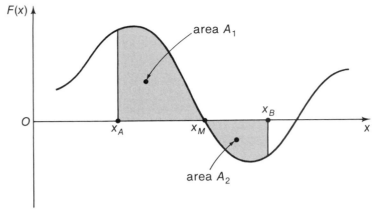

Fig. 7–4. The total amount of work done by the force $F(x)$ in the displacement $x_A \rightarrow x_B$ is $W = W_1 + W_2 = A_1 - A_2$.

Example 7–4

A coiled spring with one end fixed has a relaxed length ℓ_0 and a spring constant κ. What amount of work must be done to stretch the spring by an amount s?

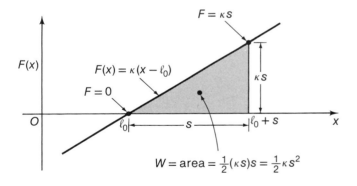

$$W = \text{area} = \tfrac{1}{2}(\kappa s)s = \tfrac{1}{2}\kappa s^2$$

Solution:

In order to stretch the free end of the spring to a point x, some agency must exert a force,

$$F(x) = \kappa(x - \ell_0)$$

The work done by the force in moving the free end of the spring from $x = \ell_0$ to $x = \ell_0 + s$ is

$$W = \int_{\ell_0}^{\ell_0 + s} F(x)\, dx = \int_{\ell_0}^{\ell_0 + s} \kappa(x - \ell_0)\, dx$$

$$= \tfrac{1}{2}\kappa(x - \ell_0)^2 \bigg|_{\ell_0}^{\ell_0 + s} = \tfrac{1}{2}\kappa[(\ell_0 + s - \ell_0)^2 - (\ell_0 - \ell_0)^2]$$

$$= \tfrac{1}{2}\kappa s^2$$

The diagram shows a graphic representation of this process. Note that $W = \tfrac{1}{2}(\kappa s)s$ is just the area of the triangle defined by $F(x)$ and the given limits.

Example 7–5

A spring with zero relaxed length and spring constant $\kappa = 50$ N/m moves a block by contracting from a stretched length of 30 cm to a length of 5 cm, as indicated in the figure. The block slides on a horizontal frictionless surface. What is the amount of work done on the block by the spring?

Solution:

The net horizontal force on the block is

$$F = F_S = -\kappa x$$

so the work done on the block by the spring is

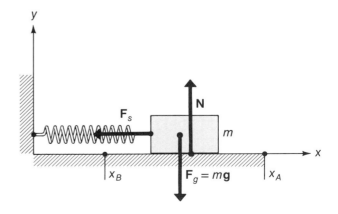

$$W = \int_{x_A}^{x_B} F(x)\,dx = \int_{x_A}^{x_B} (-\kappa x)\,dx = -\tfrac{1}{2}\kappa x^2 \Big|_{x_A}^{x_B} = \tfrac{1}{2}\kappa(x_A^2 - x_B^2)$$

$$= \tfrac{1}{2}(50\ \text{N/m})[(0.30\ \text{m})^2 - (0.05\ \text{m})^2] = 2.188\ \text{J}$$

Example 7–6

An object with a mass m is released from rest at a distance R_0 from the center of the Earth and falls freely toward the Earth. What amount of work has been done on the object by the gravitational force when the object strikes the Earth's surface? (Neglect the motion of the Earth and the effects of the Earth's atmosphere.)

Solution:

The only force on the object is the gravitational force, which is directed toward the center of the Earth. The initial velocity of the object is zero, so the path followed is a straight line, which we take to be the x-axis.

Now, the gravitational force is (Eq. 5–6)

$$F_g = -G\frac{mM_E}{x^2}$$

where the negative sign indicates that $\mathbf{F}_g$ is in the direction of negative x. According to Eq. 5–7, $g = GM_E/R_E^2$, so we can

rewrite F_g as

$$F_g = -mg\frac{R_E^2}{x^2}$$

Then, the work done by this force is (see the table of integrals inside the rear cover)

$$W = -mgR_E^2 \int_{R_0}^{R_E} \frac{dx}{x^2} = mgR_E^2 \cdot \frac{1}{x}\Big|_{R_0}^{R_E}$$

$$= mgR_E^2\left(\frac{1}{R_E} - \frac{1}{R_0}\right)$$

If the object is released from a point close to the Earth at a height $h = R_0 - R_E$, then we can write

$$W = mgR_E^2\frac{(R_0 - R_E)}{R_0 R_E} = mgh\frac{R_E}{R_0}$$

and since $R_E \cong R_0$, we obtain the final result,

$$W \cong mgh$$

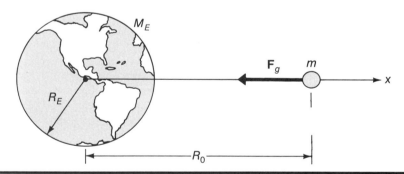

7–3 WORK DONE DURING TWO-DIMENSIONAL MOTION

Consider the work done by a force $\mathbf{F}$ on an object that moves in a plane. (Again, forces in addition to $\mathbf{F}$ may be acting on the object.) Suppose that the object moves along a specified path from an initial position A to a final position B, as indicated in Fig. 7–5. Divide the path into a large number N of small, straight-line displacements $\Delta \mathbf{s}_n$. Then, the work done by $\mathbf{F}$ during the movement $A \rightarrow B$ is (see Eq. 7–7)

$$W(A \rightarrow B) = \sum_{n=1}^{N} \mathbf{F}(\mathbf{r}_n)\bullet\Delta\mathbf{s}_n = \sum F(\mathbf{r}_n)\cos\theta_n\,\Delta s_n$$

$$= \sum_n F_s(\mathbf{r}_n)\,\Delta s_n$$

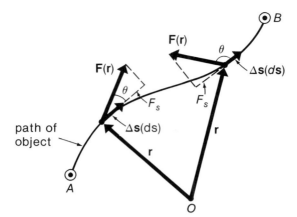

Fig. 7–5. Geometry of the displacement $A \rightarrow B$.

where $F_s(\mathbf{r}_n)$ is the component of $\mathbf{F}(\mathbf{r}_n)$ in the direction of $\Delta\mathbf{s}_n$. The notation implies that F_s is a function of $\mathbf{r}_n$. In the limit $N \rightarrow \infty$, we have (see Eq. 7–10)

$$
\begin{aligned}
W(A \rightarrow B) &= \int_A^B \mathbf{F}(\mathbf{r}) \cdot d\mathbf{s} \\
&= \int_A^B F(\mathbf{r}) \cos\theta\, ds \\
&= \int_A^B F_s(\mathbf{r})\, ds
\end{aligned}
\qquad\qquad (7\text{–}11)
$$

These alternate forms of Eq. 7–11 are all equivalent, and one may be more convenient than the others in a particular case.

Cartesian Coordinates. In many cases, Cartesian (rectangular) coordinates are best suited for describing the situation. The initial position is then specified by (x_A, y_A) and the final position by (x_B, y_B). The force vector is written as $\mathbf{F} = F_x\hat{\mathbf{i}} + F_y\hat{\mathbf{j}}$ and the *differential displacement* (see the following paragraph) as $d\mathbf{s} = dx\hat{\mathbf{i}} + dy\hat{\mathbf{j}}$. Then, the work done during the total displacement is

$$
W = \int_A^B (F_x\hat{\mathbf{i}} + F_y\hat{\mathbf{j}}) \cdot (dx\hat{\mathbf{i}} + dy\hat{\mathbf{j}})
$$

or

$$
W = \int_A^B F_x\, dx + \int_A^B F_y\, dy
\qquad\qquad (7\text{–}12)
$$

▶ *Differentials.* A useful concept in calculus is the differential. The basic idea is to treat a derivative as the quotient of very small incremental quantities. For a function $y = f(x)$ that describes the path of an object, we have

$$
\frac{dy}{dx} \cong \frac{\Delta y}{\Delta x} \quad \text{or} \quad \Delta y \cong \frac{dy}{dx}\Delta x
$$

as shown in Fig. 7–6. If the intention is eventually to proceed to the limit $\Delta x \rightarrow 0$, we may think in terms of either Δx or dx. The corresponding increment $\Delta y = [f(x + \Delta x) - f(x)]$ may also be considered to be either Δy or dy. As is convenient, we write infinitesimal quantities as either Δx or dx, and so forth.   ◀

The evaluation of the integrals in Eqs. 7–12 may be complicated because, in the general case, $\mathbf{F}$ is a function of both variables, x and y. Thus, the problem may require the

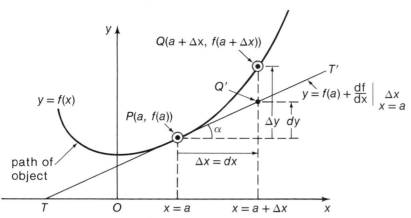

Fig. 7–6. The geometric representation of the increment Δy and the differential dy, evaluated at $x = a$, for the path function $y = f(x)$. Here, $\Delta y = f(a + \Delta x) - f(a)$ and $dy = df(x)/dx|_{x=a} \cdot \Delta x$. Also, $\tan \alpha = dy/dx$; TT' is the tangent to the path at P.

evaluation of integrals such as

$$\int_A^B F_x(x, y) \, dx \qquad \text{and} \qquad \int_A^B F_y(x, y) \, dy$$

Integrals with these forms may be given precise meaning, but advanced mathematical techniques are required. There are, however, a variety of common situations in which the integrals can be calculated readily.

Suppose, for example, that the force **F** is constant—that is, independent of the position of the object. Then, with F_x and F_y both constant, Eq. 7–12 becomes*

$$W = F_x \int_{x_A}^{x_B} dx + F_y \int_{y_A}^{y_B} dy$$

$$= F_x(x_B - x_A) + F_y(y_B - y_A)$$

$$= \mathbf{F} \cdot (\mathbf{r}_B - \mathbf{r}_A) \qquad\qquad (7\text{–}13)$$

The situation is illustrated in Fig. 7–7.

Fig. 7–7. Graphic representation of the work done by a constant force **F**, as a particle moves along the curved path from point A to point B in the x-y plane.

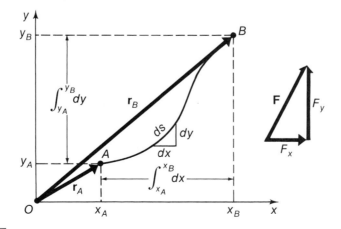

*The final result is also valid for general three-dimensional motion with a constant force.

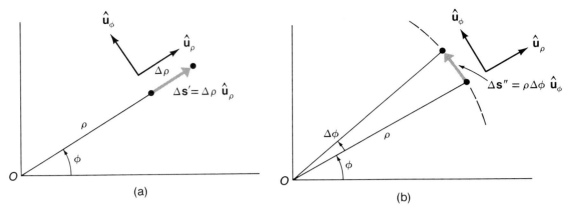

Fig. 7–8. Components of the displacement vector in polar coordinates. (a) The incremental displacement $\Delta\mathbf{s}'$ for ϕ = constant. (b) The incremental displacement $\Delta\mathbf{s}''$ for ρ = constant.

Equation 7–13 makes the important point that the work done by a constant force is *independent* of the path followed by the object during the movement $A \rightarrow B$. The work done depends only on the positions of the end-points A and B. We return to discuss this type of situation in the following chapter.

Polar Coordinates.   In polar coordinates (Section 3–4), the force vector is expressed in terms of a radial component F_ρ and an azimuthal component F_ϕ (see Eq. 3–16):

$$\mathbf{F} = F_\rho\hat{\mathbf{u}}_\rho + F_\phi\hat{\mathbf{u}}_\phi$$

where F_ρ and F_ϕ can be functions of ρ and ϕ.

Figure 7–8 shows how we obtain the expression for a differential displacement $\Delta\mathbf{s}$ in polar coordinates. For a displacement $\Delta\mathbf{s}'$ with the angle ϕ constant (Fig. 7–8a), the displacement vector is $\Delta\rho\hat{\mathbf{u}}_\rho$. If ρ is constant (Fig. 7–8b), the displacement vector $\Delta\mathbf{s}''$ is $\rho\,\Delta\phi\hat{\mathbf{u}}_\phi$. Thus, for an arbitrary displacement, we have $\Delta\mathbf{s} = \Delta\mathbf{s}' + \Delta\mathbf{s}''$, or

$$\Delta\mathbf{s} = \Delta\rho\hat{\mathbf{u}}_\rho + \rho\,\Delta\phi\hat{\mathbf{u}}_\phi$$

When the displacement $\Delta\mathbf{s}$ becomes vanishingly small, the increments are replaced by differentials:

$$d\mathbf{s} = d\rho\hat{\mathbf{u}}_\rho + \rho\,d\phi\hat{\mathbf{u}}_\phi \qquad \textbf{(7–14)}$$

Then, the work done along a path from A to B is

$$W = \int_A^B F_\rho\,d\rho + \int_A^B F_\phi\rho\,d\phi \qquad \textbf{(7–15)}$$

Example 7–7

A pendulum bob with mass m is displaced from its equilibrium position and is raised to a vertical height h, as shown in the figure. If the bob is released from this height at rest, what is the total amount of work done on the bob as it arrives at the bottom of its swing?

Solution:

Because $\mathbf{F}_g$ is constant, with $F_{gy} = -mg$ and $F_{gx} = 0$, we have, from Eq. 7–9, noting that $y_B - y_A = -h$,

$$W_g = F_{gx}(x_B - x_A) + F_{gy}(y_B - y_A)$$

$$= 0(x_B - x_A) + (-mg)(-h)$$

$$= mgh$$

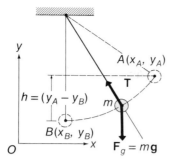

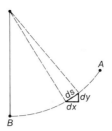

In this special case, we note that although **T** is neither constant in direction nor constant in magnitude, **T** does no work on the bob because Δ**s** and **T** are perpendicular at all times. Therefore, the total amount of work done on the bob is just mgh.

Example 7–8

A pendulum bob with a mass m is raised very slowly (with nearly zero acceleration) through a vertical height h from its equilibrium position A to the point B by the application of a force **F** that acts always tangent to the circular arc followed by the bob. Calculate the amount of work done by **F**.

Solution:

We use polar coordinates, ρ and ϕ, with the origin at the point of suspension. Here, the radial coordinate ρ is constant in magnitude and equal to the length ℓ of the pendulum. Therefore, the differential displacement is $ds = \rho\, d\phi\hat{\mathbf{u}}_\phi = \ell\, d\phi\hat{\mathbf{u}}_\phi$. The force **F** is in the direction $\hat{\mathbf{u}}_\phi$ or $\mathbf{F} = F(\phi)\hat{\mathbf{u}}_\phi$; then, since $\hat{\mathbf{u}}_\phi\cdot\hat{\mathbf{u}}_\phi = 1$, we have

$$W_F(A \to B) = \int_A^B \mathbf{F}\cdot d\mathbf{s} = \int_{\phi_1=0}^{\phi_2} F(\phi)\ell\, d\phi$$

Next, we need the expression for $F(\phi)$. For zero acceleration of the bob, we see from the diagram that

$$F = mg\,\sin\,\phi$$

Therefore,

$$W_F = mg\ell\int_0^{\phi_2}\sin\phi\, d\phi = -mg\ell\,\cos\,\phi\Big|_0^{\phi_2}$$
$$= mg\ell(1 - \cos\,\phi_2)$$

Now,

$$\cos\,\phi_2 = \frac{\ell - h}{\ell} = 1 - \frac{h}{\ell}$$

so that

$$W_F = mgh$$

which is the expected result.

Again, note that the tension **T** does no work. The gravitational force $\mathbf{F}_g$ does an amount of work that is just the negative of the work done by **F**; that is, $W_g = -W_F$. (• Can you see why?) The total amount of work done on the bob in this case is zero.

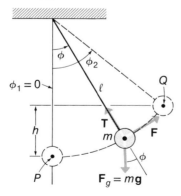

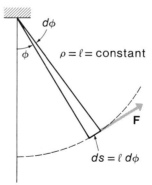

Example 7–9

In Example 7–6 we calculated the work done by gravity in moving a particle along a radial line in the vicinity of the Earth. We now consider a similar case, but one in which we allow displacement along an arbitrary path. (For convenience, we limit the displacement to two dimensions along a plane that contains the Earth's center, but the result is valid for an arbitrary three-dimensional displacement.)

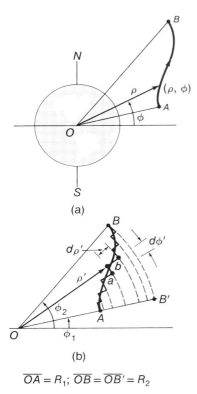

(a)

(b)

$$\overline{OA} = R_1; \quad \overline{OB} = \overline{OB'} = R_2$$

Solution:

Diagram (a) shows the two-dimensional path $A \rightarrow B$ that is followed by the particle. An arbitrary point on this path is identi-

fied by the angle ϕ (the latitude) and ρ (the radial distance from the center of the Earth). In diagram (b), this path has been subdivided into many small alternately azimuthal and radial steps which, in the limit of an infinite number of such steps, correspond exactly to the actual curved path $A \rightarrow B$. One such pair of steps, from point a to point b, is illustrated in diagram (b). The azimuthal step is $\rho' \, \Delta\phi'$ and the corresponding radial step is $\Delta\rho'$.

The gravitational force $\mathbf{F}_g$ is given by

$$\mathbf{F}_g = -G\frac{mM_E}{r^2}\,\hat{\mathbf{u}}_\rho$$

In proceeding from a to b, no work is done during the azimuthal part of the step because $\mathbf{F}_g$ is perpendicular to $\Delta\mathbf{s}$. During the radial part of the step, the work done is $F_g' \, \Delta\rho'$, where F_g' is the gravitational force at $r = \rho'$. The sum of all the azimuthal steps (along which no work is done) is equivalent to the arc $\overparen{B'B}$; the sum of all the radial steps is equivalent to the radial displacement $\overline{AB'}$. Thus, the total work done by $\mathbf{F}_g$ is the same for $A \rightarrow B$ as for $A \rightarrow B'$. Hence,

$$W(A \rightarrow B) = W(A \rightarrow B') = \int_{R_A}^{R_B} F_g \, d\rho = -GmM_E \int_{R_A}^{R_B} \frac{d\rho}{\rho^2}$$

$$= GmM_E\left(\frac{1}{R_B} - \frac{1}{R_A}\right)$$

Notice that $R_B > R_A$, so $W < 0$ and the work done *by* the field was negative; this means that work was done *on* the field during the displacement $A \rightarrow B$. Compare this expression with the result obtained in Example 7–6.

7–4 THE WORK-ENERGY THEOREM

Consider an object with mass m that is acted upon by a constant net force $\mathbf{F}$. This leads to a constant acceleration for the object, $\mathbf{a} = \mathbf{F}/m$. Choose a coordinate frame with the x-axis oriented along the direction of $\mathbf{F}$ (and $\mathbf{a}$). Initially, the object is at point A and has a velocity $\mathbf{v}_A = v_{Ax}\hat{\mathbf{i}} + v_{Ay}\hat{\mathbf{j}}$, as shown in Fig. 7–9. After an elapsed time t, the object is at point B, and we have

$$a = (v_{Bx} - v_{Ax})/t$$

$$s = \tfrac{1}{2}(v_{Bx} + v_{Ax})t$$

$$v_{Ay} = v_{By}$$

During this motion the work done by the force $\mathbf{F}$ is

$$W = Fs = mas$$

$$= \tfrac{1}{2}m\frac{(v_{Bx} - v_{Ax})}{t}(v_{Bx} + v_{Ax})t$$

$$= \tfrac{1}{2}m(v_{Bx}^2 - v_{Ax}^2)$$

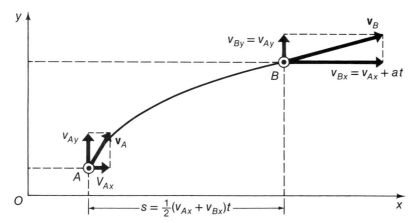

Fig. 7–9.   The path followed by an object that is acted upon by a constant net force in the x-direction.

Next, we add and subtract the quantity $\frac{1}{2}mv_{Ay}^2 = \frac{1}{2}mv_{By}^2$ (recall that $v_{Ay} = v_{By}$). The result is

$$W(A \rightarrow B) = \tfrac{1}{2}m[(v_{Bx}^2 + v_{By}^2) - (v_{Ax}^2 + v_{Ay}^2)]$$

Identifying the sums of the squares of the velocity components, we have

$$W(A \rightarrow B) = \tfrac{1}{2}m(v_B^2 - v_A^2) \qquad \text{(7–16)}$$

Thus, the total amount of work done by the *sum of all forces* acting on the object during a displacement from A to B is equal to the change in $\frac{1}{2}mv^2$. This result was obtained for the special case of a constant net force, but it is quite generally true. Note carefully that $\mathbf{v}_B$ need not be (and is generally not) parallel to $\mathbf{v}_A$. The work done depends only on the velocity magnitudes (the speeds) v_A and v_B.

The quantity $\frac{1}{2}mv^2$ is the *energy* associated with a particle by virtue of its motion and is called *kinetic energy* (K). From Eq. 7–12, we see that kinetic energy has the same dimensions as work and is therefore also measured in joules. The kinetic energy acquired by the particle is derived ultimately from the energy possessed by the agency responsible for the force **F**. Work is done in transferring the energy from this source to the particle where it appears as kinetic energy. Thus, work is a transitive form of energy; it does not persist. However, the kinetic energy of a particle *does* persist as long as no forces act on the particle to change its speed.

The relationship (Eq. 7–16) that connects the work done to the change in kinetic energy ΔK is called the *work-energy theorem:*

$$W(A \rightarrow B) = \tfrac{1}{2}mv_B^2 - \tfrac{1}{2}mv_A^2 = K_B - K_A \qquad \text{(7–17)}$$

The power of this theorem lies in the fact that it holds for variable forces as well as constant forces, even in the presence of forces of friction.

▶ The preceding treatment is adequate for practically all cases; however, it may be useful to give a more complete proof. A general proof of the work-energy theorem for motion in two dimensions can be constructed in the following way: The net force **F** is a general position-dependent force that, in a selected reference frame, can be expressed as $\mathbf{F} = F_x\hat{\mathbf{i}} + F_y\hat{\mathbf{j}}$. The work done by **F** on the body during the movement from A to B along a specified path is (Eq. 7–12)

$$W(A \rightarrow B) = \int_A^B F_x \, dx + \int_A^B F_y \, dy$$

Using Newton's law in the forms $F_x = m \, dv_x/dt$ and $F_y = m \, dv_y/dt$, the equation for the work done becomes

$$W(A \rightarrow B) = m \int_A^B \frac{dv_x}{dt}\, dx + m \int_A^B \frac{dv_y}{dt}\, dy$$

Now, we may write

$$\frac{dv_x}{dt}\, dx = \frac{dx}{dt}\, dv_x = v_x\, dv_x$$

$$\frac{dv_y}{dt}\, dy = \frac{dy}{dt}\, dv_y = v_y\, dv_y$$

The manipulation of these differentials may be more easily understood by thinking first in terms of the incremental quantities Δv_x, Δx, and Δt, where Δv_x and Δx are changes that occur during the interval Δt. Thus, $\Delta v_x = a_x\, \Delta t$ and $\Delta x = v_x\, \Delta t$; also

$$\frac{\Delta v_x}{\Delta t}\, \Delta x = \frac{\Delta x}{\Delta t}\, \Delta v_x = v_x\, \Delta v_x$$

The equation for the work done can now be written as

$$W(A \rightarrow B) = m \int_{v_{Ax}}^{v_{Bx}} v_x\, dv_x + m \int_{v_{Ay}}^{v_{By}} v_y\, dv_y$$

$$= \tfrac{1}{2}m(v_{Bx}^2 - v_{Ax}^2 + v_{By}^2 - v_{Ay}^2)$$

Again, identifying the sums of the squares of the velocity components, we have

$$W(A \rightarrow B) = \int_A^B F_x\, dx + \int_A^B F_y\, dy$$

$$= \tfrac{1}{2}mv_B^2 - \tfrac{1}{2}mv_A^2 = K_B - K_A$$

This result may be extended to the general case of motion in three dimensions. (• Can you see how to do this?) ◄

Example 7–10

Imagine the situation in Example 7–5 to be modified by adding a frictional force between the block and the surface; let the friction coefficient be $\mu_k = 0.2$ (constant for all velocities achieved during the motion). Calculate the total work done by all of the forces acting on the block during its motion from $x_A = 30$ cm to $x_B = 5$ cm. If the block is released from rest at x_A, what is its velocity at x_B? The block's mass is $m = 0.5$ kg.

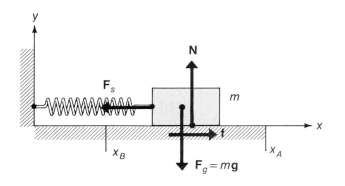

Solution:

The frictional force is

$$f = \mu_k N = \mu_k mg$$

and the work done by this force is

$$W_f = \mu_k mg \int_{x_A}^{x_B} dx = \mu_k mg(x_B - x_A)$$

$$= 0.20(0.5\ \text{kg})(9.80\ \text{m/s}^2)(0.05\ \text{m} - 0.30\ \text{m})$$

$$= -0.245\ \text{J}$$

The work done by the spring is the same as in Example 7–5:

$$W_S = \int_{x_A}^{x_B} (-\kappa x)\, dx = 2.188\ \text{J}$$

No work is done by the gravitational force or by the normal force. Therefore,

$$W = W_f + W_S = -0.245\ \text{J} + 2.188\ \text{J} = 1.943\ \text{J}$$

From Eq. 7–13 with $v_A = 0$, we have

$$v_B = \sqrt{\frac{2W}{m}} = \sqrt{\frac{2(1.943\ \text{J})}{0.5\ \text{kg}}}$$

$$= 2.788\ \text{m/s}$$

Example 7–11

Review the situation in Example 7–10 if the mass of the block is changed to 2.0 kg.

Solution:

Now, the work W_S is the same, but W_f increases to $2.0/0.5 = 4$ times the previous value.

$$W_S = 2.188\ \text{J} \quad \text{and} \quad W_f = -0.980\ \text{J}$$

so that

$$W = W_S + W_f = 1.208\ \text{J}$$

Then, $\quad v_B = \sqrt{\dfrac{2W}{m}} = \sqrt{\dfrac{2(1.208\ \text{J})}{2.0\ \text{kg}}} = 1.099\ \text{m/s}$

Notice that when the block arrives at $x_B = 5$ cm, the horizontal forces acting on the block are

$$F_S(x_B) = -\kappa x = -(50 \text{ N/m})(0.05 \text{ m})$$

$$= -2.50 \text{ N}$$

$$f = \mu_k mg = (0.20)(2.0 \text{ kg})(9.80 \text{ m/s}^2)$$

$$= 3.92 \text{ N}$$

That is, at x_B we have $|f| > |F_S|$ and the net force on the block at this point is directed toward the *right*. Thus, the block, which is

moving toward the *left* at x_B, is experiencing a net retarding force that acts to decrease its speed. The block reaches x_B with a velocity (1.099 m/s) that is smaller than the velocity found for $m = 0.5$ kg (2.788 m/s). (• At what point along the path $A \rightarrow B$ is the net horizontal force equal to zero?) The block will always arrive at x_B if $W > 0$ for the path $x_B \rightarrow x_A$. (• Can you see why? What is the maximum value of m for which the block will reach x_B?)

Example 7–12

(a) A particle with a mass m slides without friction down the complicated ramp shown in the diagram, never losing contact with the ramp. Let the heights (the values of y) at x_A, x_B, x_C, and x_D be, respectively, $h_A = 7$ m, $h_B = 4$ m, $h_C = 7.2$ m, and $h_D = -1$ m. The particle has an initial velocity, $v_A = 3$ m/s, directed downward and tangential to the ramp at x_A. What is the speed of the particle at x_B, x_C, and x_D? (b) How far up the ramp will the particle move after passing through x_D?

Solution:

(a) It is clear from the diagram that $\mathbf{N} \cdot d\mathbf{s}$ is always zero, so work is done only by the gravitational force. To find $W(A \rightarrow B)$, we must evaluate

$$W(A \rightarrow B) = \int_A^B \mathbf{F}_g \cdot d\mathbf{s} = \int_A^B (-mg)\, dy$$

$$= mg(h_A - h_B)$$

Then, using Eq. 7–17, we find

$$v_B^2 = v_A^2 + \frac{2W}{m} = v_A^2 + 2g(h_A - h_B)$$

$$v_B^2 = (3 \text{ m/s})^2 + 2(9.80 \text{ m/s}^2)(7 \text{ m} - 4 \text{ m})$$

$$= 67.8 \ (\text{m/s})^2$$

Thus, $\qquad\qquad v_B = 8.23 \text{ m/s}$

Also, $\quad v_C^2 = (3 \text{ m/s})^2 + 2(9.80 \text{ m/s}^2)(7 \text{ m} - 7.2 \text{ m})$

$$= 5.08 \ (\text{m/s})^2$$

so that $\qquad\qquad v_C = 2.25 \text{ m/s}$

Finally, $\quad v_D^2 = (3 \text{ m/s})^2 + 2(9.80 \text{ m/s}^2)(7 \text{ m} + 1 \text{ m})$

$$= 165.80 \ (\text{m/s})^2$$

with $\qquad\qquad v_D = 12.88 \text{ m/s}$

(b) The maximum height h_M will be reached at x_M where $v_M = 0$. That is,

$$0 = v_A^2 + 2g(h_A - h_M)$$

or $\qquad\qquad h_M = h_A + \dfrac{v_A^2}{2g}$

$$= 7 \text{ m} + \frac{(3 \text{ m/s})^2}{2(9.80 \text{ m/s}^2)}$$

$$= 7.46 \text{ m}$$

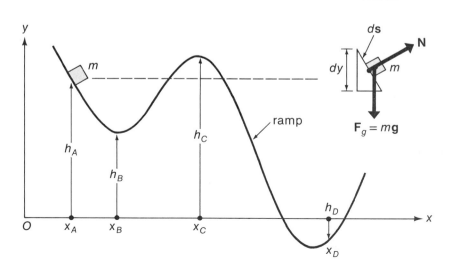

Example 7–13

A rock with a mass m is tied to the end of a string and whirled in a circular path that lies in a vertical plane, as shown in the diagram. What is the minimum (tangential) speed v_0 that the rock must have at the bottom of the path if the rock is to be able to pass the top position of the circle while maintaining the string taut? (Notice that the angular speed of the rock is *not* constant.)

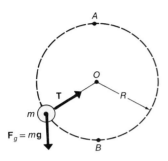

Solution:

This problem is one of a class that requires an important technique for obtaining the solution. Energy conservation alone is not sufficient; Newton's law must also be applied at some particular point. Other similar cases are examined in Problems 7–34 and 8–20.

At the top, the minimum speed occurs when the tension just vanishes. Then, the net downward force is $F_g = mg$. This force must give the centripetal acceleration required to maintain a circular path:

$$F_g = mg = m\frac{v_T^2}{R}$$

Using Eqs. 7–13 and 7–17 in which we identify the top position as A and the bottom position as B, we have for the work done by the gravitational force $\mathbf{F}_g$

$$W(A \rightarrow B) = mgh = K_B - K_A$$

so that
$$mg(2R) = \tfrac{1}{2}mv_0^2 - \tfrac{1}{2}mv_T^2$$

from which

$$v_0^2 = v_T^2 + 4gR = gR + 4gR = 5gR$$

Finally,
$$v_0 = \sqrt{5gR}$$

For $R = 1$ m, the rock's speed is 7.00 m/s at the bottom of the circle and 3.13 m/s at the top. Note that these results are independent of the mass of the rock. (• What does this imply for the case of a string that itself has mass but to which no rock is attached?)

7–5 POWER

Consider a force $\mathbf{F}(x, y)$ that is one of the forces acting on a particle. The work done by this force during a general two-dimensional displacement that requires a time Δt is

$$\Delta W = F_x\,\Delta x + F_y\,\Delta y$$

To obtain the rate at which work is done, divide this equation by Δt and take the limit, $\Delta t \rightarrow 0$; this gives

$$\frac{dW}{dt} = \lim_{\Delta t \to 0} \frac{\Delta W}{\Delta t} = \lim_{\Delta t \to 0}\left(F_x\frac{\Delta x}{\Delta t} + F_y\frac{\Delta y}{\Delta t}\right)$$

$$= F_x\frac{dx}{dt} + F_y\frac{dy}{dt}$$

$$= F_x v_x + F_y v_y$$

$$= \mathbf{F}\cdot\mathbf{v}$$

The rate at which work is done, dW/dt, is called the *power* P. Thus,

$$P = \frac{dW}{dt} = \mathbf{F}\cdot\mathbf{v} \qquad\qquad (7\text{–}18)$$

This result is also valid for general three-dimensional motion. (• Can you prove this assertion?)

The SI unit of power is the *joule per second* (J/s), to which we give the special name *watt* (W)*:

$$1 \text{ W} = 1 \text{ J/s}$$

The unit of power in the British system, namely, the *horsepower* (hp), is often used in various engineering applications:

$$1 \text{ hp} = 745.7 \text{ W} \cong \tfrac{3}{4} \text{ kW}$$

Example 7-14

The engine of an automobile develops 50 hp simply to overcome air and road resistance while maintaining a constant speed of 55 mi/h. What is the effective thrust (or force) developed by the engine (that is, the hypothetical force which, if pulling the automobile, would result in the same power expenditure)?

Solution:

First, convert the power and the speed to SI units:

$$P = 50 \text{ hp} = (50 \text{ hp})(746 \text{ W/hp}) = 37{,}300 \text{ W}$$

$$v = 55 \text{ mi/h} = (55 \text{ mi/h})(1609 \text{ m/mi})/(3600 \text{ s/h})$$

$$= 24.6 \text{ m/s}$$

Then, $F = \dfrac{P}{v} = \dfrac{37{,}300 \text{ W}}{24.6 \text{ m/s}} = 1516 \text{ N}$

Example 7-15

A block with a mass $m = 4$ kg slides, starting from rest, down an inclined plane ($\alpha = 35°$) from an initial height $h_0 = 6$ m. The friction coefficient is $\mu_k = 0.25$. Find the kinetic energy of the block and the power being expended against friction when the block has reached the height $h = 2$ m.

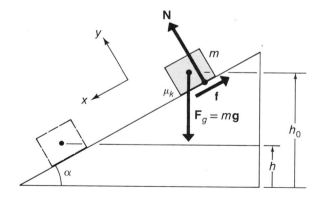

Solution:

The frictional force is

$$f = -\mu_k N = -\mu_k mg \cos \alpha$$

$$= -(0.25)(4 \text{ kg})(9.80 \text{ m/s}^2)(\cos 35°)$$

$$= -8.03 \text{ N}$$

The work done by this frictional force is

$$W_f = f(x - x_0)$$

where $x - x_0 = (h_0 - h) \sin \alpha$. The work done by the gravitational force F_g is

$$W_g = mg(h_0 - h)$$

If we use Eq. 7-17 and note that the work done by the normal force **N** is zero we have

$$K = W_f + W_g + W_N$$

$$= -\mu_k mg \frac{\cos \alpha}{\sin \alpha} (h_0 - h) + mg(h_0 - h) + 0$$

$$= mg(1 - \mu_k \operatorname{ctn} \alpha)(h_0 - h)$$

$$= (4 \text{ kg})(9.8 \text{ m/s}^2)[1 - (0.25)(\operatorname{ctn} 35°)][(6 \text{ m}) - (2 \text{ m})]$$

$$= 100.8 \text{ J}$$

---

*Named in honor of James Watt (1736–1819), the Scottish engineer who was responsible for the development of the first practical industrial steam engines.

The power expended against friction is

$$P = -fv = -f\sqrt{\frac{2K}{m}}$$

$$P = (8.03 \text{ N})\sqrt{\frac{2(100.8 \text{ J})}{4 \text{ kg}}}$$

$$= 57.0 \text{ W}$$

QUESTIONS

7-1 Explain briefly what is meant by the terms (a) work, (b) negative work, and (c) scalar product of two vectors.

7-2 When you take your rather large dog for a walk, the leash is kept taut by the dog leading the way. Is the work you do on the dog positive or negative? Explain.

7-3 A satellite moves in a circular orbit around the Earth. How much work is done by the Earth's gravitational force on the satellite? Explain.

7-4 Discs, balls, boxes, and all manner of objects sent sliding across a smoothly frozen surface of a lake eventually come to rest. What can you deduce about the direction of frictional forces?

7-5 A nervous backseat passenger in an automobile pushes on the back of the seat in front of him during the entire trip. At the end of the trip he announces, "Well, I might have been nervous but at least I helped the engine by the work I performed. After all, the force I exerted was applied over a large displacement." What is wrong with his reasoning?

7-6 Prove that the components of the vector **A** may be written $A_x = \mathbf{A} \cdot \hat{\mathbf{i}}$, $A_y = \mathbf{A} \cdot \hat{\mathbf{j}}$, and $A_z = \mathbf{A} \cdot \hat{\mathbf{k}}$.

7-7 Two springs with unequal spring constants are linked together at one end, and the combination is stretched by pulling on the two free ends. What is the ratio of the work expended in stretching each of the joined springs?

7-8 Define kinetic energy. Does the kinetic energy of a particle depend on the direction of its velocity? Can a particle ever have a negative kinetic energy?

7-9 Explain in your own words the physical content of the work-energy theorem.

7-10 A 50-kg girl can climb a 4-m-high flight of stairs in 4 s. What is her approximate rate of energy expenditure in watts (in horsepower)?

7-11 Explain why mountain hiking trails and motor roads have many winding switchbacks rather than going straight up or down.

7-12 Why are automobiles and trucks equipped with gearshifts?

PROBLEMS

Section 7-1

7-1 Determine the dot product of the two vectors $\mathbf{A} = 3\hat{\mathbf{i}} + 2\hat{\mathbf{j}}$ and $\mathbf{B} = 2\hat{\mathbf{i}} + 4\hat{\mathbf{j}}$. What is the angle θ_{AB} between **A** and **B**? [*Hint:* Write Eq. 7-2 as $\cos\theta_{AB} = \mathbf{A} \cdot \mathbf{B}/AB$.]

7-2 Show that the two vectors $\mathbf{B} = 6\hat{\mathbf{i}} + 2\hat{\mathbf{j}}$ and $\mathbf{C} = 3\hat{\mathbf{i}} - 9\hat{\mathbf{j}}$ are perpendicular to one another.

7-3 Suppose that the position function of a particle is

$$\mathbf{r}(t) = \tfrac{1}{2}\mathbf{a}t^2 + \mathbf{v}_0 t + \mathbf{r}_0$$

with **a**, $\mathbf{v}_0$, and $\mathbf{r}_0$ suitable constant vectors. Show that

$$\mathbf{v} \cdot \mathbf{v} = 2\mathbf{a} \cdot (\mathbf{r} - \mathbf{r}_0) + \mathbf{v}_0 \cdot \mathbf{v}_0$$

This expression is the vector generalization of Eq. 2-11.

7-4 A student pulls a block along a frictionless horizontal ice surface, as shown in the diagram. (The student manages to walk across the frictionless surface by wearing spiked shoes.) If the angle α is maintained at a constant value of 30° and if the tension in the ideal massless rope is main-

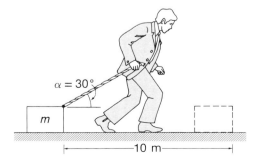

$\alpha = 30°$

m

$\vdash$————10 m————$\dashv$

tained constant at 3 N, how much work is done by the student in dragging the block a horizontal distance of 10 m? Assume that the block slides on the ice at all times. Calculate separately the work done on the block by the rope tension, the work done by the block on the rope, the work done on the student by the rope tension, and the work done by the student on the rope. Do your results give meaning to the idea that work is done on the block by the student, even though they are not in direct contact? (• What would be the effect of the rope having nonzero mass?)

7–5 Two balls with masses $m_1 = 10$ kg and $m_2 = 8$ kg hang from an ideal pulley, as shown in the diagram. What amount of work is done by the Earth's gravitational force on each ball separately during a downward vertical displacement of 50 cm by m_1? What is the *total* work done on each ball, including the work done by the string tension? Comment on any relationship you have discovered connecting these quantities.

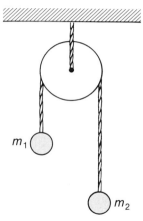

7–6 Two blocks with masses m_1 and m_2 are connected by a massless rope, as shown in the diagram. The ramp is frictionless. If m_1 rises in height by a distance h, how much work does gravity do on the two blocks during this motion? [*Hint:* Consider each block separately.]

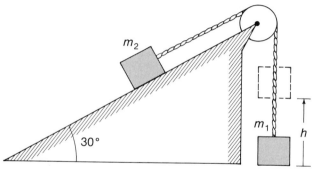

7–7 The excavation for a building is to be 40 m long, 15 m
• wide, and 10 m deep. The material to be removed has a

density of 4 g/cm³. Each bucket of dirt is hoisted to a height of 3 m above the *original ground level* and dumped into a truck. Compute the minimum amount of work required for the complete excavation.

7–8 A 25-kg crate is suspended from a rope and lowered from a height of 12 m to a height of 4 m, all the while experiencing a downward acceleration equal to 4.0 m/s². What amount of work is done on the crate by the rope?

Section 7–2

7–9 Evaluate the definite integrals (a) $\int_0^1 (x^2 + 2x + 1)\, dx$

and (b) $\int_1^2 e^{-x}\, dx$.

7–10 Show that (a) $\int_0^\pi \sin\theta\, d\theta = 2$ and (b) $\int_0^\theta \sin^2 x\, dx = \frac{1}{2}\theta - \frac{1}{4}\sin 2\theta$. [*Hint:* For (b), use Eq. A–32 in Appendix A.]

7–11 A technique that is frequently useful in solving integration
• problems is *trigonometric substitution*. Use the following steps to evaluate

$$\int_0^a \frac{a^2\, dx}{(x^2 + a^2)^{3/2}}$$

(a) From the diagram, you can see that $\cos\beta = a/(x^2 + a^2)^{1/2}$. Use this equation to express the integrand in terms of a and $\cos\beta$. (b) Again from the diagram, write $x = a\tan\beta$. Differentiate this to find $dx/d\beta$ (see the list of derivatives inside the rear cover). Treating the result as a quotient of differentials, find the expression for dx. (c) Use Eq. A–18 in Appendix A to write dx in terms of a and $\cos\beta$. Multiply this expression by the result of part (a) to obtain the complete integrand in terms of functions of β. (d) One more step is necessary before the definite integral can be evaluated: the limits of integration are still in terms of x and must be converted to β. Look at the diagram and write the values of β when $x = 0$ and when $x = a$. (e) Perform the integration and evaluate the definite integral.

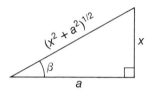

7–12 An object that is constrained to move along the x-axis is acted upon by a force $F(x) = ax + bx^2$, where $a = 5$ N/m and $b = -2$ N/m². The object is observed to proceed directly from $x = 1$ m to $x = 3$ m. How much work was done on the object by the force? Does the process of integration take into account the fact that the force $F(x)$ changes sign in this interval?

7–13 The block shown in the diagram has a mass of 2 kg and is acted upon by a horizontal force $F(x) = a - bx$, where $a = 20$ N and $b = 10$ N/m. The coefficient of sliding friction between the block and the flat surface is $\mu_k = 0.20$. The block is initially at $x = 0$. What distance must the block travel if the net work done on the block by $F(x)$ and friction combined is to be exactly zero? At what point along the way does the net force on the block reach zero?

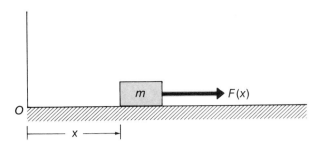

7–14 By adding a certain variable force, the block in Problem 7–13 is observed to proceed from $x = 0$ to $x = 7$ m and then to return to $x = 1$ m. What total amount of work was done in the combined motion by $F(x)$? How much work was done by the frictional force?

Section 7–3

7–15 A block with a mass m slides frictionlessly down a complicated ramp, as shown in the diagram. What amount of work is done by the gravitational force as the block slides down to the horizontal level, having started at a height h above this level? Are there any other forces acting on the block? If so, what amount of work is done by these forces?

7–16 A pendulum bob with a mass m is released from position 1 shown in the accompanying diagram. Some time later, after the support cord has intercepted the two horizontal fixed pegs A and B, the bob is observed to be at position 2. How much work was done on the bob by gravity?

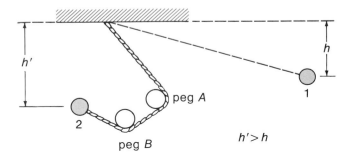

7–17 The pendulum bob in Example 7–8 is again raised very slowly through a height h but now by a force $\mathbf{F}$ that is horizontal at all times. Show that the work done by the force $\mathbf{F}$ is again equal to mgh.

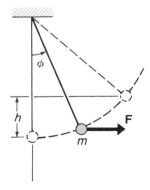

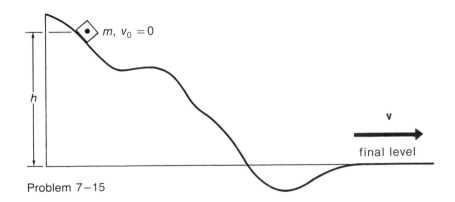

Problem 7–15

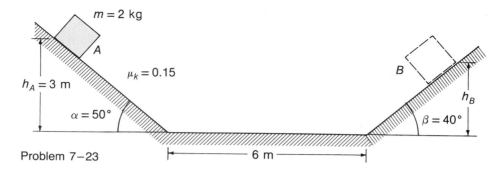

Problem 7–23

Section 7–4

7–18 A particle with a mass of 0.25 kg is dropped from rest and falls through a distance of 10 m. What amount of work is done on the particle by the gravitational force? What kinetic energy does the particle acquire? What is its velocity at the end of the 10-m fall? If the mass of the particle is changed to 2.5 kg, which of the values calculated will change and which will not change?

7–19 A particle with a mass m and an initial velocity v_0 begins to compress a spring. In what distance will the particle be brought to rest if the spring constant is κ?

7–20 A block with a mass m and an initial horizontal velocity v_0 slides across a horizontal surface with coefficient of friction μ_k. In what distance is the block brought to rest? (Use energy considerations here; compare Problem 6–10.)

7–21 A block with a mass $m = 10$ kg is lowered vertically by a rope, the other end of which is attached to a machine that exerts a tension $T = ky^2$, where $k = 2$ N/m^2 and where y is the distance by which the block has descended from its starting position. If the block starts from rest at $y = 0$, what velocity does it have after descending to $y = 10$ m?

7–22 In Problem 7–21, what is the farthest descent y of the
• block? At what position will the speed of the block be maximum?

Additional Problems

7–28 Given that $\mathbf{C} = \mathbf{A} + \mathbf{B}$, what is the geometric interpretation of $\mathbf{C} \cdot \mathbf{C}$? Compare with Eq. A–37 in Appendix A.

7–29 Evaluate $\displaystyle\int_0^\pi \sin^3 \theta \, d\theta$.

[*Hint:* Write $\sin^3 \theta = \sin^2 \theta \sin \theta = (1 - \cos^2 \theta) \sin \theta$ and let $x = \cos \theta$; make the corresponding change in the limits.]

7–23 If the block in the diagram above is released from rest at position A, to what height h_B will it rise before coming to rest momentarily at position B?

Section 7–5

7–24 A crane is to lift 2000 kg of material to a height of 150 m in 1 min at a uniform rate. What electric power is required to drive the crane motor if 35 percent of the electric power is converted to mechanical power? (• Is it reasonable to neglect changes in the kinetic energy during this process?)

7–25 A small automobile with a mass of 400 kg is powered by an engine delivering a maximum of 45 hp. At what maximum speed could it go up a 15° hill? (Neglect all frictional forces.)

7–26 A vehicle with a mass m is propelled in a straight horizon-
• tal line by an engine delivering a constant power P. If the vehicle starts from rest, show that the distance traveled in a time t is $s = \sqrt{8Pt^3/9m}$.

7–27 A block with a mass of 5 kg is pushed up a 20° inclined plane, starting from rest, by a constant *horizontal* force of 50 N. If the coefficient of sliding friction between the block and the ramp is $\mu_k = 0.20$, at what rate is heat generated by frictional losses when the block has been moved a distance of 0.5 m up the plane? At what rate is the block gaining kinetic energy at this point?

7–30 A block with a mass m slides down the rough surface of a
• plane that makes an angle α with the horizontal. If the block starts down the plane with a velocity v_0 and is observed to come to rest after traveling a distance ℓ (along the inclined plane), what is the coefficient of sliding friction?

7–31 A force F is applied to a block that was originally at rest,
• as in the diagram. After the block has moved a distance of
6 m to the right, the direction of the horizontal component
of the force F is reversed (but the magnitude remains the
same). With what velocity does the block arrive at its
starting point? [*Hint:* The block travels a total distance to
the right greater than 6 m. Divide the motion into appropriate parts and apply Eq. 7–12 successively.]

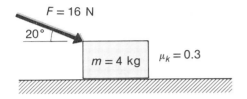

7–32 A particle with a mass of 0.5 kg is acted on by a force
• $\mathbf{F} = (ax + b)\hat{\mathbf{i}}$, where $a = 2$ N/m and $b = 4$ N. If the
initial velocity at $x = 0$ is $\mathbf{v}_0 = +(1.5 \text{ m/s})\hat{\mathbf{i}}$, what is the
velocity of the particle when it arrives at $x = 6$ m? Give a
general expression for $\mathbf{v}$ as a function of x.

7–33 A particle with a mass of 2 kg is constrained to move
•• along the x-axis. The particle is subjected to a force
$F(x) = (2 \text{ N/m})x - 4$ N. Consider the possibility of the
particle starting with an initial velocity of $+1.5$ m/s and
moving from $x = 0$ to $x = 6$ m. According to Eq. 7–12
the particle should arrive at $x = 6$ m with a velocity considerably greater than v_0. Yet the proposed trip is impossible. Why?

7–34 A small block with mass m slides down a frictionless
ramp at the bottom of which is a loop-the-loop track with
a radius R, as shown in the diagram. If the block is to stay
in contact with the track at all times during the loop, what
is the minimum height h from which the block can be
released from rest?

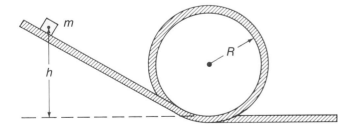

7–35 A soapbox derby cart has a length of 2 m and is required
• to have a total mass (including the driver) of 100 kg. This
particular cart and driver has a total mass of 70 kg, so it is
required to carry an additional load of 30 kg. Where
should the 30-kg block be placed to provide maximum
velocity along the horizontal section at the bottom of the
20° accelerating ramp? First, assume that by simply placing the block in the *middle* of the cart, the velocity at the
bottom of the ramp is 10 m/s. Then, determine the best
(and the worst) that one can do by placing the block at
other positions along the cart. (• If you were the driver of
this cart, where would you plan to sit during the race?)

7–36 The diagram below shows a ball with a mass m attached to
• a string with a length ℓ. The peg is located a distance d
directly below the support point. If the ball is to swing
completely around the peg, starting from the position
shown, prove that d must be greater than $3\ell/5$. [*Hint:*
Refer to Example 7–13.]

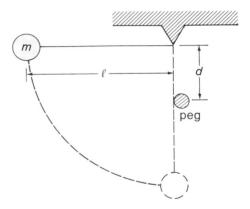

7–37 A thin rod with a mass m and a length ℓ is pivoted at one
• end by a frictionless bearing. The rod is released from
rest, standing erect on its pivoted end. What velocity does
the free end have at the bottom of the swing? What is the
tension in the bearing support at this instant? [*Hint:* Integrate over the length of the rod.]

CHAPTER 8

POTENTIAL ENERGY AND THE CONSERVATION OF ENERGY

It is often the case that the work done by a force acting on a body represents energy that can be recovered at a later time. For example, if you do work in raising a box from the floor to a shelf, this amount of work can be recovered in the form of kinetic energy by allowing the box to fall to the floor. Because of this feature of reversibility, the box, as a consequence of its elevated position, possesses stored energy called *potential energy*. (The box has the *potential* to do work by falling to the floor.)

By using the concept of potential energy we may define the *total mechanical energy* of an object or a system to be the sum of its kinetic and potential energies. In many important situations this sum is a constant even as the object under study moves from place to place. This condition frequently arises when friction may be neglected. Under such circumstances, the conservation of mechanical energy permits the easy solution of several classes of problems.

More generally, when all forms of energy are taken into account, there is a global principle of energy conservation that must be obeyed by all isolated systems.

8–1 POTENTIAL ENERGY IN ONE DIMENSION

The kinetic energy of an object is the energy associated with the *motion* of the object. We now introduce the idea that there is another form of energy—the potential energy—that can be associated with the *position* of the object.

Suppose that a student raises a box *very slowly* from the floor to a shelf at a height h (Fig. 8–1a). Raising the box "very slowly" means that at any instant the velocity of the box and hence its kinetic energy is essentially zero. The acceleration is then zero, so that the magnitude of the applied force $\mathbf{F}_a$ exerted on the box by the student is just equal to that of the gravitational force, $F_g = mg$. Then, because the displacement, $\mathbf{s}$, is in the same direction as the applied force $\mathbf{F}_a$, we have $\theta = 0$ for the angle between these vectors, and with $\cos \theta = 1$ the work W_a done *by* the student is

$$W_a = \mathbf{F}_a \cdot \mathbf{s} = F_a h = mgh$$

During this process, the work W_g done by the gravitational force is

$$W_g = \mathbf{F}_g \cdot \mathbf{s} = -F_g h = -mgh = -W_a$$

where the negative sign arises because $\mathbf{F}_g$ is directed opposite to the displacement $\mathbf{s}$.

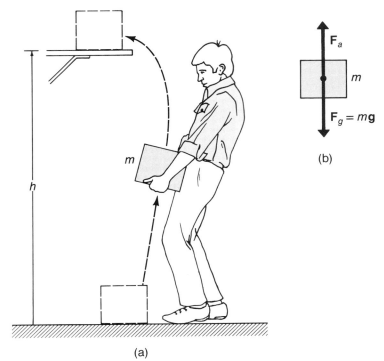

(a)

(b)

Fig. 8–1. (a) An object acquires a potential energy mgh by being raised to a height h. (b) Free-body diagram for the block as it is raised very slowly.

It is easy to see that the work done by the force $\mathbf{F}_a$ is the result of the expenditure of energy by the student. But what is the significance of the work W_g? We know from experience that if the shelf is suddenly removed and the box is left unsupported, it will fall to the floor. Just before impact the falling box has a kinetic energy acquired in the fall that is equal to W_g.

The energy possessed by the box on the shelf is a *positive* quantity relative to the level of the floor. We ascribe to the box a potential energy U_g equal to the negative of the work done by the gravitational force (which is itself a negative quantity); that is,

$$U_g = -W_g$$

This definition also permits us to associate the potential energy of the box exclusively with the gravitational force and to eliminate any further reference to the role of the student.

In many cases of interest an object is constrained to move along a straight-line path. The idea of potential energy can be applied to the case of motion restricted to one dimension, say, the x-axis. We imagine that a number of forces are acting. Because the motion takes place only along the x-direction, no work is done by the y- or z-components of any of these forces. Thus, for the work ΔW_n done on a body during the displacement Δx by the particular force $\mathbf{F}_n$ whose x-component is F_{nx}, we write

$$\Delta W_n = F_n \cos \theta_n \, \Delta x = F_{nx} \, \Delta x$$

As above, we introduce ΔU_n as the negative of ΔW_n:

$$\Delta U_n = -\Delta W_n$$

That is, ΔU_n is the negative of the incremental work done by the force $\mathbf{F}_n$ on the body.

First, consider forces $\mathbf{F}_n$ that depend *only* on the position of the body on which they act (that is, on x) and not, for example, on time or velocity (not even on the direction of velocity). Then,

$$\Delta U_n(x) = -F_{nx}(x)\,\Delta x \qquad (8-1)$$

Passing to the limit and integrating from point A to point B yields

$$U_n(B) - U_n(A) = -\int_{x_A}^{x_B} F_{nx}(x)\,dx \qquad (8-2)$$

If $\mathbf{F}_n$ satisfies the conditions stated above, then $U_n(B) - U_n(A)$ depends *only* on x_B and x_A. A further generalization is possible in which we define a function $U_n(x)$ such that

$$U_n(x) - U_n(x_0) = -\int_{x_0}^{x} F_{nx}(x)\,dx \qquad (8-3)$$

where x_0 is an arbitrary fixed point.* The function $U_n(x)$ is the value of potential energy at point x relative to the value at the reference point x_0. For example, in Fig. 8–1 we could take the floor to be the reference level; then, with x as the height of the box above the floor, we would have

$$U(x) - U(x_0) = mg(x - x_0)$$

Returning to Eq. 8–1, we may write

$$F_{nx}(x) = -\frac{\Delta U_n(x)}{\Delta x}$$

or, in the limit, $\Delta x \to 0$, we have the important relationship

$$F_{nx}(x) = -\frac{dU_n(x)}{dx} \qquad (8-4)$$

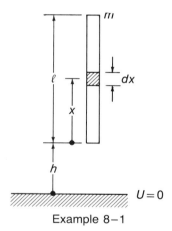

Thus, for one-dimensional motion with the type of forces we are discussing, it is equivalent to specify either the force acting $\mathbf{F}_n$ or the potential energy function $U_n(x)$. We refer to $U_n(x)$ as the *potential-energy function*.

Example 8–1

Example 8–1

Consider a uniform thin rod with a length ℓ and a mass m that is in a vertical position, as shown in the figure. What is the potential energy U of the rod with respect to the surface shown?

Solution:

The work dW done by gravity in raising the mass element $dm = (m/\ell)\,dx$ from the reference level to the position $h + x$ is $dW = -g(x + h)\,dm$; hence, $dU = -dW = (mg/\ell)(x + h)\,dx$.

Thus, the potential energy is

$$U = \int_0^\ell dU = \frac{mg}{\ell}\int_0^\ell (x + h)\,dx = \frac{mg}{\ell}(\tfrac{1}{2}x^2 + hx)\,\Big|_0^\ell$$

$$= \frac{mg}{\ell}(\tfrac{1}{2}\ell^2 + \ell h) = mg(h + \tfrac{1}{2}\ell)$$

The potential energy is the same as if we raised a total concentrated mass m from the reference level to a height $h + \tfrac{1}{2}\ell$.

---

*Notice that the integral is a function of its upper limit and generates a function of x, namely, $U_n(x)$.

8-2   CONSERVATION OF ENERGY

When the concept of potential energy is included in the statement of the work-energy theorem discussed in Chapter 7, a more revealing formulation of energy conservation is possible. However, not all forces in nature lead to definable potential-energy functions. The frictional force is one example of a force for which a potential energy cannot be specified. To make this distinction clear we begin at the outset by considering a case involving the presence of friction.

We can express the resultant F_x of the various forces acting on a body as

$$F_x = \sum_{n=1}^{N} F_{nx}(x) + f \tag{8-5}$$

where we have included the dissipative frictional force f (which, because the motion is in the x-direction, acts only in the x-direction*) as well as the applied forces $F_{nx}(x)$.

Using the work-energy theorem (Eq. 7–17) yields

$$W(A \rightarrow B) = \int_{x_A}^{x_B} F_x \, dx = \int_{x_A}^{x_B} \sum_{n} F_{nx}(x) \, dx + \int_{x_A}^{x_B} f \, dx = \tfrac{1}{2} m v_B^2 - \tfrac{1}{2} m v_A^2$$

Now, $$\int_{x_A}^{x_B} \sum_{n} F_{nx}(x) \, dx = \sum_{n} \int_{x_A}^{x_B} F_{nx}(x) \, dx = - \sum_{n} [U_n(x_B) - U_n(x_A)]$$

Therefore, $$\sum_{n} U_n(x_B) + \tfrac{1}{2} m v_B^2 = \sum_{n} U_n(x_A) + \tfrac{1}{2} m v_A^2 + \int_{x_A}^{x_B} f \, dx$$

Because the various $U_n(x)$ are simply functions of x, we can combine them all together as the single potential energy term

$$U(x) = \sum_{n} U_n(x) \tag{8-6}$$

Finally, $$U(x_B) + \tfrac{1}{2} m v_B^2 = U(x_A) + \tfrac{1}{2} m v_A^2 + \int_{x_A}^{x_B} f \, dx \tag{8-7}$$

Because the frictional force **f** is always directed opposite to the displacement, the integral in Eq. 8–7 is to be evaluated with the following rule:

$$f < 0 \quad \text{if} \quad x_B > x > x_A$$
$$f > 0 \quad \text{if} \quad x_B < x < x_A$$

(• Can you see why this rule is necessary?)

---

*The frictional force f is not simply a function of position, because it carries a sign that depends on the direction of motion at the point x. Technically, f is not a continuous function of x at a turnaround point, $x = x_T$, because $f(x_T)$ is not defined: $f(x \rightarrow x_T$, from the left$) = -f(x \rightarrow x_T$, from the right$)$.

Equation 8–7 expresses the law of *energy conservation* for a mechanical system. The quantity $E = U(x) + \frac{1}{2}mv^2$ (where v is the velocity at point x) is called the *mechanical energy* of the body. As before, we refer to $K = \frac{1}{2}mv^2$ as the *kinetic energy* (or motional energy). The total *potential energy* $U(x)$, or any of its individual parts $U_n(x)$, is simply the work that must be done to bring the body to the position x from some arbitrary reference point where all of the $U_n(x)$ are defined to be zero.* But as Eq. 8–7 shows, the basic law always involves the *difference*, $U(x_B) - U(x_A)$ [or the corresponding differences, $U_n(x_B) - U_n(x_A)$]; consequently, the position of this arbitrary reference level is immaterial.

In the case of zero friction ($f = 0$), Eq. 8–7 asserts that the mechanical energy of a body is a constant, independent of position and, hence, independent of the time (because $x = x(t)$ describes the path followed by the body). Such a quantity is called a *constant of the motion*. Thus, in the absence of frictional forces, the mechanical energy of a one-dimensional system is a constant of the motion.

There are many situations in which frictional forces can be neglected; in this approximation, we have

$$U(x) + \tfrac{1}{2}mv^2 = E = \text{constant} \qquad \text{(no friction)} \qquad (8\text{–}8)$$

where E is the constant total mechanical energy.†

For the case $f \neq 0$, the integral of $f\,dx$ in Eq. 8–7 is always negative. Therefore, the content of this equation can be stated as follows:

When a particle moves from a point x_A to a point x_B, the mechanical energy at x_B is equal to the mechanical energy at x_A less the dissipative work done by the particle against frictional forces.

Care must be exercised in evaluating the work done by friction. For example, consider the coordinates $x_A < x_B < x_C$; then, for motion that takes a particle along the path $x_A \to x_C \to x_B$, account must be taken of the fact that the sign of f along the path $x_C \to x_B$ is different from that along the path $x_A \to x_C$.

At what rate is energy expended against a frictional force? Return to Eq. 8–5 and write in differential form the work done in a displacement dx; thus,

$$F_x\,dx = \sum_n F_{nx}(x)\,dx + f\,dx$$

From Newton's second law, we have

$$F_x\,dx = m\,\frac{dv}{dt}\,dx$$

Also,

$$\sum_n F_{nx}(x)\,dx = -dU$$

---

*When we are interested in following the motion of a particular object, we usually refer to the potential energy (or the total energy) *of the body*. However, we should properly refer to the potential energy *of the system* because this energy is the result of the location of the body being studied relative to the body or bodies that produce the forces $\mathbf{F}_n$.

†Because $U(x)$ is defined only to within an additive constant, so is E.

so that the work done by friction is

$$f \, dx = dU + m \frac{dv}{dt} \, dx$$

Dividing by dt and using $dx/dt = v$, we find

$$fv = \frac{dU}{dt} + mv \frac{dv}{dt}$$

$$= \frac{dU}{dt} + \frac{d}{dt}[\tfrac{1}{2}mv^2]$$

$$= \frac{d}{dt}[U + \tfrac{1}{2}mv^2]$$

or, finally,

$$fv = \frac{dE}{dt} \qquad\qquad\qquad (8\text{--}9)$$

The frictional force $\mathbf{f}$ is always directed exactly opposite to the velocity $\mathbf{v}$; thus, $fv = \mathbf{f} \cdot \mathbf{v}$ is always negative, and we conclude that $dE/dt < 0$, so that E is a *decreasing* function of time. Hence, Eq. 8–9 states that the power dissipated in overcoming friction is equal to the rate at which mechanical energy is lost (to heat).

Equation 8–8 states that in the absence of friction, the total mechanical energy of a system remains constant. This in itself is a useful result because friction can be neglected in many situations. However, if friction *is* important, we must use the energy equation in which the work done against friction is explicitly included (Eqs. 8–7 and 8–9).

Does this mean that energy is no longer conserved because energy has been "lost" to friction? It does not—*if* we suitably redefine the system to include the places where the work done against friction has been absorbed.

It is a common experience that friction produces *heat*. In a later chapter we discuss in detail the fact that heat is a form of energy involving the internal kinetic energy associated with the motions of the atoms and molecules that comprise a sample. The work done against friction is the energy that must be expended to break the molecular bonds that form between the sliding surfaces (see Section 6–3). As a result, the molecules at these points become highly agitated. Moreover, the extra molecular energy is shared with neighboring molecules, so the regions of the samples in the vicinity of the sliding surfaces become heated. The idea of energy conservation holds in this situation if we are careful to include in the system energy all of the energy associated with heat in the sliding surfaces. In fact, it was the realization, in the mid-nineteenth century, that heat is a form of energy that led to the enunciation of the principle of energy conservation.*

As we examine a greater variety of situations involving different kinds of forces, we find that we must enlarge our view still further and allow for even more forms of energy. Indeed, we have come to adopt the attitude that if we do not find a balance of energy in a particular process, we invent a new form of energy that exactly makes up the difference!

---

*Credit for formulating the energy conservation principle is usually given to the German physiologist and physicist Hermann von Helmholtz (1821–1894) because of the detailed and forceful way in which he advanced the idea (1847). However, the concept had been announced earlier (1842) by the German physician and physicist Julius Robert Mayer (1814–1878), but he had had considerable difficulty in finding a journal that would publish his paper because the concept was too novel. When it finally did appear, Mayer's paper aroused no particular interest. Finally, the papers by the English physicist James Prescott Joule (1818–1889), which appeared a few years later, firmly established the connection between heat and other forms of energy, thereby setting the stage for Helmholtz's pronouncements. The credit for the energy conservation principle should probably be divided among these three scientists.

Following this way of thinking, we have invented, for example, chemical energy, electromagnetic energy, and nuclear energy. This is not really done to hide our ignorance about Nature, for once we have invented a new form of energy we must thereafter always include this energy in our calculations in the same way. If we have made a poor choice, we will rapidly come upon contradictions. Those forms that survive all tests become part of our concept of energy. Because of its enormous range of application, the energy conservation principle is one of the most important ideas in all of science.

Example 8-2

A particle with mass m moves across a horizontal surface under the influence of a potential, $U(x) = \frac{1}{2}\kappa x^2$. A constant frictional force acts on the particle, namely, $f(x) = \pm b$, where the sign depends on the direction of motion at x. The particle is released from rest at $x = x_0$. What is the speed of the particle as it passes $x = 0$ for the first time?

Solution:

In this case, we have $x_A = x_0$ and $x_B = 0$. Then, $U(x_A) = \frac{1}{2}\kappa x_0^2$ and $U(x_B) = 0$. Also, $v_A = 0$. Because $x_B < x_A$, we use $f(x) = +b$. Thus, Eq. 8-7 becomes

$$\frac{1}{2}mv_B^2 = \frac{1}{2}\kappa x_0^2 + \int_{x_0}^{0} b\, dx$$

$$= \frac{1}{2}\kappa x_0^2 - bx_0$$

so that
$$v_B = \sqrt{\frac{\kappa}{m}x_0^2 - \frac{2b}{m}x_0}$$

Notice that, by Eq. 8-4, the force $F(x)$ corresponding to the potential $U(x)$ in this example is $F(x) = -\kappa x$, which is Hooke's law for an ideal spring. Thus, the same answer could be obtained by the methods of Example 7-10; however, the potential-energy method is much simpler in this case. (See Problem 8-14.)

Example 8-3

Consider a block (a *particle*) with a mass m that slides down a frictionless inclined plane, as shown in the diagram. Describe the motion of the block by using the idea of energy conservation.

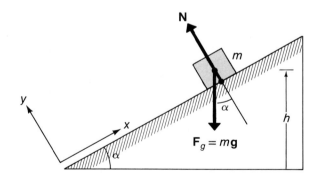

Solution:

The net force **F** acting on the block is directed down the plane:

$$F = F_x = -F_g \sin\alpha = -mg \sin\alpha = -\frac{dU(x)}{dx}$$

Then,
$$U(x) = mgx \sin\alpha + C$$

If we take $U(0) = 0$, we find $C = 0$. We know that

$$U(x) + \frac{1}{2}mv^2 = E = \text{constant}$$

so
$$mgx \sin\alpha + \frac{1}{2}mv^2 = E$$

Now, $x \sin\alpha = h$, so we can also write

$$mgh + \frac{1}{2}mv^2 = E$$

If the block slides, starting from rest at $x = x_0$ (with $x_0 \sin\alpha = h_0$), we have $E = mgh_0$ for the initial energy. Then,

$$mgh + \frac{1}{2}mv^2 = E = mgh_0$$

so that
$$\frac{1}{2}mv^2 = mg(h_0 - h)$$

The velocity at any height h is, therefore,

$$v = \sqrt{2g(h_0 - h)}$$

which is the same as the expression we obtained earlier for the case of free fall (see Eq. 2-15).

Notice that we can recover Newton's equation by differentiating the energy equation with respect to time. We have

$$\frac{d}{dt}(mgx \sin\alpha + \frac{1}{2}mv^2) = \frac{d}{dt}E$$

$$mg\frac{dx}{dt}\sin\alpha + mv\frac{dv}{dt} = 0$$

Canceling dx/dt with v and using $a = dv/dt$, we find

$$-mg \sin\alpha = ma$$

or
$$F = ma$$

as expected.

8−3 THE POTENTIAL ENERGY CURVE

It is instructive to examine a graphic representation of Eq. 8−8. Imagine that $U(x)$, for some particular physical situation, has the form shown in Fig. 8−2. The characteristics of the motion are determined by the value of the total mechanical energy of the system and by the shape of the potential energy function.

Solving Eq. 8−8 for v, we have

$$v = \pm \sqrt{\frac{2}{m}[E - U(x)]} \qquad \text{(8−10)}$$

It is clear that v has a physically meaningful value only when $E \geqslant U(x)$; otherwise, Eq. 8−10 would have an imaginary solution. This means that the object cannot have a mechanical energy less than E_0, as shown in Fig. 8−2. When $E = E_0$, the velocity v is identically zero (for all values of the time).

Suppose that the energy is slightly greater than E_0, say, $E_0 + \varepsilon$. Then, as the particle moves away from x_0 in either direction, a force acts to return the particle to the point $x = x_0$. Points such as x_0 are referred to as points of *stable equilibrium*.

If $E = E_1$, the particle may be found anywhere in the interval $x_1 \leqslant x \leqslant x_1'$, with the velocity at any point given by Eq. 8−10. In Fig. 8−2, v_1 (corresponding to $E = E_1$) is maximum for $x = x_0$ (that is, for $E_1 - U(x)$ a maximum). As $x \to x_1$, the velocity vanishes, but $F = -dU(x)/dx|_{x=x_1}$ is greater than zero; thus, at $x = x_1$, there is a force (and, hence, an acceleration) to the right. The particle is in effect "reflected" back to the right. A similar reflection occurs at $x = x_1'$, and the particle oscillates back and forth in the *potential well* between $x = x_1$ and $x = x_1'$. Points such as x_1 and x_1' are referred to as *turning points*.

The case $E = E_2$ is very interesting. According to Eq. 8−10, the particle is allowed to be anywhere in the regions $x_2 < x < x_2'$ and $x_2'' < x < x_2'''$; however, these are *disconnected regions*. If the system is initially prepared with the particle somewhere in the region $x_2 < x < x_2'$, it can never enter the region $x_2'' < x < x_2'''$, for it could do so only by

Fig. 8−2. The graphic representation of the one-dimensional potential energy function $U(x)$ for an object. Also shown are various values of the total mechanical energy.

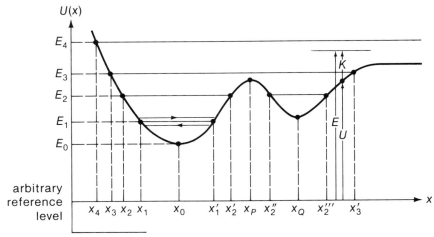

passing through the region $x_2' < x < x_2''$ where, from Eq. 8–10, v would have imaginary (i.e., unphysical) values. The region $x_2' < x < x_2''$ is referred to as a *potential barrier*. Barriers also exist for $x < x_2$ and $x > x_2'''$.

The case $E = E_3$ is similar to the case $E = E_1$, except that v now has a local minimum at $x = x_P$. (• Suppose that the particle is at $x = x_p$ (see Fig. 8–2) and has a total energy $E(x_p)$. What is the velocity of the particle? What is the character of the force that acts on the particle if it moves an incremental distance away from $x = x_P$? Compare with the character of the force in the vicinity of $x = x_0$. Can you see why points such as $x = x_P$ are called points of *unstable equilibrium*?) (• Suppose that the particle is at a point $x > x_3'$ (see Fig. 8–2) and is moving to the left with total energy E_4. Describe the motion of the particle.)

Example 8–4

Consider a particle with a mass $m = 2$ kg for which the potential energy function is $U(x) = 3x - 5 - (x - 3)^3$, where U is measured in joules when x is in meters. Describe the motion for two values of the total energy, $E_1 = 3$ J and $E_2 = 8$ J. For example, what is the velocity of the particle at $x = 3$ m?

Solution:

Most of the important features of the motion are evident in the energy diagram. For example, the minimum allowed value of the total energy in the vicinity of $x = 2$ m is 2.0 J. A particle, originally at $x = 2$ m, will undergo back-and-forth motion unless its energy is greater than 6.0 J; if $E > 6.0$ J, the particle will

escape from the potential well and will forever move off to the right.

At $x = 3$ m, there is no physically realizable motion for $E_1 = 3$ J. For $E_2 = 8$ J, we have

$$\tfrac{1}{2}mv^2 = E - U(x)$$

or $\tfrac{1}{2}mv_2^2 = 8 \text{ J} - U(3 \text{ m}) = 8 \text{ J} - 4 \text{ J} = 4 \text{ J}$

so that $v_2 = \sqrt{\dfrac{2(4 \text{ J})}{2 \text{ kg}}} = \pm 2$ m/s   at $x = 3$ m

(• What is the equation of the force $F(x)$ corresponding to the given $U(x)$? What is its behavior as $x \to \infty$? Is this physically realizable?)

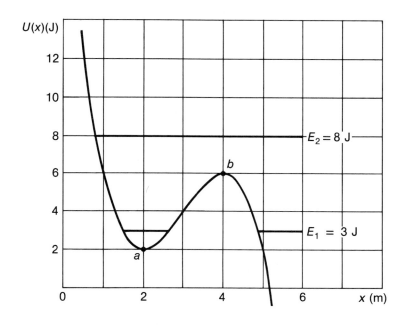

8–4 POTENTIAL ENERGY IN TWO AND THREE DIMENSIONS

We now extend the discussion and consider the general motion in two or three dimensions of an object subject to the influence of various forces. As the object moves about in accordance with Newton's laws, the acting forces perform work. However, it may not be possible to associate a potential energy with each of these forces because the work done may depend on the path followed by the object as well as on the locations of the initial and final points. In such cases, no generalization of Eq. 8–2 is possible. Examples of forces that *do* have well-defined potential energy functions are the gravitational force and the force exerted by an ideal spring.

Gravitational Potential Energy.   In Chapter 7 we discussed several examples in which objects followed various paths and made relatively small changes in elevation. The work done by the gravitational force of the Earth was found to be the same in each case, namely, $W_g = mgh$ (see Examples 7–7 and 7–8). This result is easy to understand by recalling the discussion that led to Eq. 7–13 for the work done by a constant force. Figure 8–3 illustrates the movement (by forces not shown) of a particle with mass m along a two-dimensional path from point A at (x_0, y_0) to point B at (x, y). For the gravitational force we have $dU_g = -dW_g = -\mathbf{F}_g \cdot d\mathbf{s}$, with $\mathbf{F}_g = -mg\hat{\mathbf{j}}$ and $d\mathbf{s} = dx\hat{\mathbf{i}} + dy\hat{\mathbf{j}}$. Then, $dU_g = mg\,dy$, from which

$$U_g(y) - U_g(y_0) = mg(y - y_0) = mgh \qquad (8\text{–}11)$$

Thus, the change in the gravitational potential energy is just mgh, independent of the path followed by the particle between the elevations y_0 and y.

Figure 8–4 shows a particle that is moved through a large elevation change in the vicinity of the Earth. In this case, $\mathbf{F}_g$ cannot be considered constant. Nevertheless, in view of the result obtained in Example 7–9, we expect to be able to define a gravitational potential energy function for this global situation.

Because the gravitational force does work only during the radial portions of any displacement of the particle m (see the diagram in Example 7–9), we have

$$dU_g = -dW_g = -F_g\,d\rho$$

Then,
$$U_g(B) - U_g(A) = GmM_E \int_{R_A}^{R_B} \frac{d\rho}{\rho^2}$$

Fig. 8–3. Between the points A and B the potential energy of the particle changes by mgh.

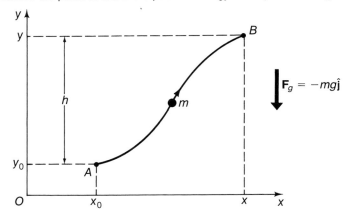

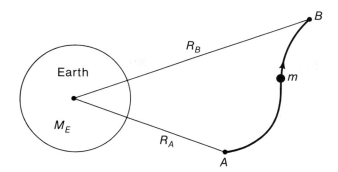

Fig. 8–4. A potential energy function can be defined even for a large-scale movement in the vicinity of the Earth.

or

$$U_g(B) - U_g(A) = GmM_E \left(\frac{1}{R_A} - \frac{1}{R_B} \right) \tag{8-12}$$

Thus, the change in the gravitational potential energy is independent of the path followed by the particle and depends only on the radial distances of the initial and final points measured from the center of the Earth.

It is convenient to define the potential energy for $R_A \rightarrow \infty$ to be zero. With this convention, we have for a general point in space ($R_B = r$) the potential energy function

$$U_g(r) = -G \frac{mM_E}{r} \tag{8-13}$$

For small elevation changes near the surface of the Earth, Eq. 8–13 yields the same result as Eq. 8–11. (• Can you show this equivalence? Refer to Example 7–6.)

Ideal Springs.   Stretching or compressing an ideal spring involves no internal friction. Consequently, the work done in stretching such a spring by an amount s from its relaxed length ℓ_0 is stored in the spring as elastic potential energy. Suppose that one end of an ideal spring is attached to a fixed universal pivot, as shown in Fig. 8–5. When the free end of the spring is moved along some path, the spring force is always radial and we have $dW = F_s(\rho)\, d\rho$. We assume that the spring obeys Hooke's law, so we have (see Example 7–4)

$$U(\ell_0 + s) - U(\ell_0) = \int_{\ell_0}^{\ell+s} \kappa(\rho - \ell_0)\, d\rho$$

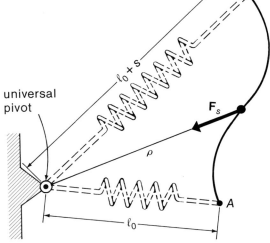

Fig. 8–5. The free end of an ideal spring is moved along a path from A to B while the other end swivels freely about the fixed universal pivot.

If we take $U(\ell_0) = 0$, we obtain

$$U(s) = \tfrac{1}{2}\kappa s^2 \qquad (8-14)$$

where we have written $U(s)$ as the abbreviation for $U(\ell_0 + s)$.

Concluding Remark. Suppose that an object moves along some path from point A to point B under the influence of forces F_p that have associated potential energy functions and forces F_q that have no such potential energy functions. Then, the work-energy theorem (Section 7–4) yields

$$\int_A^B \sum_p \mathbf{F}_p \cdot d\mathbf{s} + \int_A^B \sum_q \mathbf{F}_q \cdot d\mathbf{s} = \tfrac{1}{2}m(v_B^2 - v_A^2) \qquad (8-15)$$

The first term can be expressed in terms of a potential energy function; thus,

$$\int_A^B \sum_q \mathbf{F}_q \cdot d\mathbf{s} = \sum_p [U_p(B) - U_p(A)] + \tfrac{1}{2}m(v_B^2 - v_A^2) \qquad (8-16)$$

where the remaining integral depends on the particular path that is followed by the particle in moving from A to B and therefore influences the value of the difference in kinetic energies.

Example 8–5

A particle with mass m slides very slowly (that is, $\mathbf{a} \cong 0$) along a circular frictionless ramp from point P to point Q, as shown in the diagram. During the descent, the particle is restrained by a tangential force $\mathbf{F}$ that allows the motion to proceed slowly.

(a) Calculate explicitly the work done by this force.

(b) What is the work done if the coefficient of kinetic friction between the particle and the ramp is $\mu_k = 0.40$?

Solution:

(a) For zero acceleration we have

$$F = mg \cos \phi$$

The work done by $\mathbf{F}$ is

$$W_F = \int dW_F = -\int FR\,d\phi = -mgR \int_0^{\pi/2} \cos \phi\, d\phi$$

$$= -mgR \sin \phi \Big|_0^{\pi/2} = -mgR$$

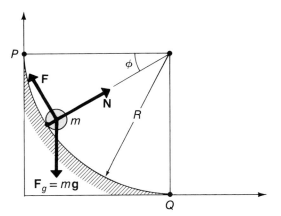

(b) The normal force is

$$N = mg \sin \phi$$

so the force of friction is

$$f = \mu_k N = \mu_k mg \sin \phi$$

For zero acceleration, we have

$$F + f = mg \cos \phi$$

Thus, $\qquad F = mg(\cos \phi - \mu_k \sin \phi)$

The work done by **F** is

$$W_F = -\int FR \, d\phi = -mgR \int_0^{\pi/2} (\cos \phi - \mu_k \sin \phi) \, d\phi$$

$$= -mgR(\sin \phi + \mu_k \cos \phi) \Big|_0^{\pi/2}$$

$$W_F = -mgR(1 - 0.40) = -0.60 \, mgR$$

The work done by the frictional force f is

$$W_f = -\mu_k mgR \int_0^{\pi/2} \sin \phi \, d\phi = -0.40 \, mgR$$

Hence, $\quad W_F + W_f = (-0.60 - 0.40)mgR = -mgR$

$$= U_g(Q) - U_g(P)$$

which agrees with Eq. 8–16 (in which $v_P = v_Q$).

Example 8–6

A small object with mass $m = 0.50$ kg is attached to the free end of an ideal massless spring with $\kappa = 10$ N/m. The other end of the spring is connected to a fixed frictionless universal pivot located at the origin O, as indicated in the diagram. The relaxed length of the spring is $\ell_0 = 0.10$ m. An unspecified force **F** carries the object, starting from rest, from the point A, which is 0.80 m along the x-axis, to the point B, which is 0.30 m along the y-axis. At the point B, the object has a speed $v_B = 4.5$ m/s. How much work was done by the force **F**?

Solution:

Using Eq. 8–16, we can write (with $v_A = 0$)

$$W_F = \int_A^B \mathbf{F} \cdot d\mathbf{s} = \tfrac{1}{2}mv_B^2 + [U_g(B) - U_g(A)] + [U_s(B) - U_s(A)]$$

We have

$$U_g(B) - U_g(A) = mgy_B = (0.50 \text{ kg})(9.80 \text{ m/s}^2)(0.30 \text{ m}) = 1.47 \text{ J}$$

$$U_s(B) - U_s(A) = \tfrac{1}{2}\kappa(y_B - \ell_0)^2 - \tfrac{1}{2}\kappa(x_A - \ell_0)^2$$

$$= \tfrac{1}{2}(10 \text{ N/m})[(0.20 \text{ m})^2 - (0.70 \text{ m})^2] = -2.25 \text{ J}$$

$$\tfrac{1}{2}mv_B^2 = \tfrac{1}{2}(0.50 \text{ kg})(4.5 \text{ m/s})^2 = 5.06 \text{ J}$$

Hence, $\qquad W_F = 5.06 \text{ J} + 1.47 \text{ J} - 2.25 \text{ J}$

$$= 4.28 \text{ J}$$

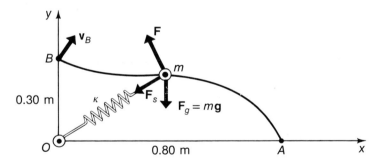

8–5 FORCE FIELDS

If some agency exerts a force on a particle, the particle will follow one or another of a large number of possible trajectories, depending on the initial conditions of position and velocity. As the particle passes through a specific point (x, y, z), the force on the particle has a magnitude and direction that we can express as $\mathbf{F}(x, y, z)$.* We imagine that we are allowed an unlimited number of trials so that we can test experimentally the force on the particle at every conceivable point. In this way we are led to the idea that there exists

---

*We assume that **F** is independent of time.

throughout space an agency that *would exert* a force $\mathbf{F}(x, y, z)$ on a particle *if* the particle were actually at the specified point. This is really what is meant by "the force $\mathbf{F}(x, y, z)$."

The idea that $\mathbf{F}(x, y, z)$ exists throughout space is called the *force-field* concept, and we refer to $\mathbf{F}(x, y, z)$ simply as the *force field*. For a particular particle, the force field may be viewed as a property of the space in which the particle moves. The actual path or trajectory that is followed by the particle depends, in addition to the details of the force field, on the initial conditions and on any constraints on the motion that might be imposed. The specification of the force field in which a particle moves is one of the important pieces of information that must be provided in order to solve problems in mechanics.

In the discussions in the preceding chapters we introduced various kinds of forces. Some of these were *contact forces*—forces resulting from actual contact between two objects. For example, we have considered the push exerted by a person's hand and the pull exerted by a spring or a string. Forces of another type are the so-called *action-at-a-distance forces*—forces that act between objects even though they are separated by empty space. The gravitational and the electromagnetic forces are action-at-a-distance forces. Actually, even contact forces, when examined at a microscopic level, can be seen to be the result of electromagnetic forces that act between the two surface layers of atoms of the objects in contact (i.e., in very close proximity to each other). Thus, to be somewhat more exact, contact forces could be called *action-at-a-small-distance forces*.

The gravitational and the electric force between pairs of particles at relative rest both have the form of an inverse-square law (that is, $F \propto 1/r^2$). The basic force law in each of these cases is stated in terms of the force between *particles*, but in some situations it is possible to use exactly the same force equation for objects with extent (finite size). An important case involves the gravitational force exerted on objects by the Earth. In Chapter 13, we discuss the fact that if two spherically symmetric objects interact via the gravitational force, they act in the same way as do particles. Then, the gravitational force that the Earth (mass M_E) exerts on a particle (mass m) located a distance r from the center of the Earth (with $r > R_E$, the radius of the Earth) is given by

$$\mathbf{F}_g = -G \frac{mM_E}{r^2} \, \hat{\mathbf{u}}_r \qquad (8\text{–}17)$$

where $\hat{\mathbf{u}}_r$ is the unit radial vector that is directed outward from the center of the Earth (Fig. 8–6). According to this equation, the particle m experiences, at any point in space, a force that has a direction toward the Earth's center and a magnitude inversely proportional to the square of the distance from the center. Figure 8–7 shows the *gravitational force field* that is experienced by a particle with a mass m attracted toward a spherical object with a mass M. Notice that the force vectors decrease in magnitude with increasing distance from the origin, as prescribed by Eq. 8–17.

The abstraction of a force field in empty space gains an added elegance when applied to the gravitational and electric forces. In the gravitational case we introduce a vector quantity called the *gravitational field strength* $\mathbf{g}$ associated with an object of mass M that

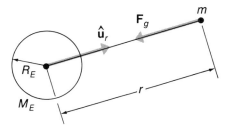

Fig. 8–6. The gravitational force $\mathbf{F}_g$ on m is directed toward the center of the Earth.

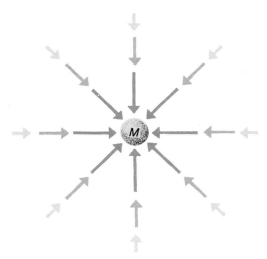

Fig. 8-7. Map of the gravitational force field surrounding a source object *M*. Each arrow represents the force on a particle at the position of the base of the arrow.

is the source of the field. Thus,

$$\mathbf{g} = \lim_{m \to 0} \frac{\mathbf{F}_g}{m} = -G\frac{M}{r^2}\hat{\mathbf{u}}_r \tag{8-18}$$

The field represented by $\mathbf{g}$ is a property only of the source object *M*. Placing a particle with a mass *m* at any point in this field results in a force on *m* given by $\mathbf{F}_g = m\mathbf{g}$, where $\mathbf{g}$ is the field strength at the point.

The essential feature of every force field is that associated with every point in space there exists a unique vector $\mathbf{F}(x, y, z)$ that specifies the force on a particle if placed at that point. This is the only aspect of force fields that concerns us at the moment. In particular, we now investigate some aspects of the work and energy associated with a particle that moves in various force fields. In later chapters we return to the discussion of the properties of action-at-a-distance forces.

8-6   CONSERVATIVE AND NONCONSERVATIVE FORCES

In this section we inquire about the requirement a force field must obey for the potential energy concept to be applied to it. We must keep in mind that potential energy *depends only on the position* of a particle in a force field. Recall that the difference in the potential energies of a particle at two positions in the field is defined as the negative of the work done by the force as the particle moves from one position to the other. It then follows that the work done by the force $\mathbf{F}$,

$$W(A \to B) = \int_A^B \mathbf{F} \cdot d\mathbf{s} \tag{8-19}$$

must be independent of the path followed by the particle from point *A* to point *B*. Only then may we associate a potential energy with the force field $\mathbf{F}$. Such forces are referred to as *conservative*. This name arises from the fact that a particle making a round trip $A \to B \to A$, following arbitrarily different paths from *A* to *B* and from *B* to *A*, will have the same potential energy at the end of the trip as it had at the beginning.

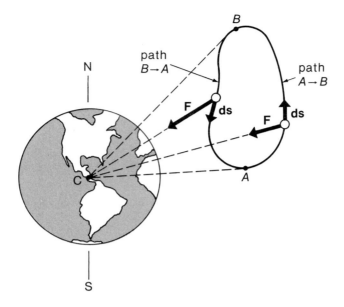

Fig. 8-8. The gravitational force is a conservative force. Hence, zero work is done in carrying a particle around the closed path *ABA*.

In Example 7-9 we showed explicitly that the work done by the gravitational force on a particle moving from point *A* to point *B* is independent of the path followed. Thus the gravitational field is a conservative force field. In Fig. 8-8 we show a particle following some arbitrary closed path *ABA* ($A \rightarrow B \rightarrow A$) in the Earth's gravitational field as an example. The net work done by the field is zero. The gravitational force field, and indeed any force field of the form

$$\mathbf{F}_1 = \frac{D}{r^2} \hat{\mathbf{u}}_\rho \qquad (8-20)$$

where D is a constant, is conservative, and has a corresponding potential energy function (referred to as $U(\infty) = 0$) given by

$$U_1(r) = \frac{D}{r} \qquad (8-21)$$

In contrast, consider a similar force field that acts in the direction of the unit vector $\hat{\mathbf{u}}_\phi$ instead of the unit vector $\hat{\mathbf{u}}_\rho$,

$$\mathbf{F}_2 = \frac{D}{r^2} \hat{\mathbf{u}}_\phi \qquad (8-22)$$

Such a force tends to induce a vortex, or swirling motion, to an object it acts on. To see that the force $\mathbf{F}_2$ is *nonconservative*, imagine carrying a particle around a closed loop path $r = R$ (a constant). Reference to Fig. 8-9 shows that everywhere along the path $\mathbf{F} \cdot \mathbf{ds} = F_2 R \, d\phi$ is positive; thus

$$W \text{ (closed circular loop, } r = R) = 2\pi R F_2$$

$$= \frac{2\pi D}{R} \neq 0$$

Thus the force $\mathbf{F}_2$ is nonconservative and there is no meaningful way to specify the potential energy of a particle in this field.

Do nonconservative force fields, such as $\mathbf{F}_2$ just discussed, exist in nature? We shall

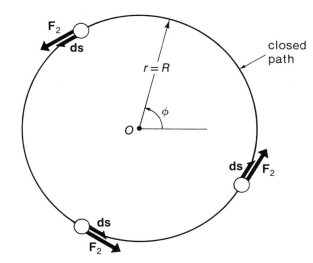

closed path

find in the chapter on electromagnetic induction that forces of just this type may be exerted on charged particles in the presence of time-varying magnetic fields. However, all the forces we study in the chapters on mechanics, excepting friction and velocity-dependent forces, are conservative forces.

QUESTIONS

8–1 Explain briefly what is meant by the terms (a) potential energy, (b) mechanical energy, and (c) force fields.

8–2 Explain why a bouncing ball can never rebound on the second bounce to a height greater than (or even equal to) the height reached on the first bounce, no matter how it started bouncing.

8–3 What happens to the potential energy a skier loses in coming down from the top of a hill when he stops at the end of the run?

8–4 Explain why potential energy is defined only up to an arbitrary constant. Illustrate your explanation for both the case of gravitational potential energy and the potential energy of a compressed or elongated ideal spring.

8–5 It is possible to launch a satellite into a stable orbit that is elliptic in shape. Explain why you would expect the speed of the satellite to be greatest when it is closest to the Earth's surface.

8–6 When the air in an inflated balloon is released, the balloon flies across the room. What was the source of its increased kinetic energy?

8–7 Describe in detail the energy of a pendulum bob as it swings back and forth. What role does the tension in the supporting string play? What role does gravity play? Why does the motion eventually stop?

8–8 Meteorites often burn up as they fall through the Earth's atmosphere. What is the source of the heat energy involved?

8–9 An electric heater in a house draws its power from a wall outlet. The electricity is provided by a nearby hydroelectric facility. Trace the flow of energy from whatever ultimate source you can identify to its final use.

8–10 A theoretically ideal automobile engine using all the chemical energy available in the combustion of its fossil fuel could achieve an "efficiency" of 150 miles per gallon (mpg). Practical values of mpg are substantially lower. Identify the energy losses present.

8–11 Describe briefly the meaning of the terms conservative and nonconservative forces. Why is the situation for general two- and three-dimensional motion more complicated than that for motion constrained to follow a fixed linear path? [*Hint:* In the latter case $\displaystyle\int_A^B F(x)\, dx = -\int_B^A F(x)\, dx$ always.]

8–12 Explain why it would be possible to describe a conservative force as one that does zero work on a particle when it is carried around any closed path.

8–13 Can velocity-dependent forces ever be conservative?

PROBLEMS

Section 8–1

8–1 A small block of mass m is pushed up an inclined plane with an inclination angle θ, as shown in the diagram. Calculate the work done by the Earth's gravitational force as the block moves along the inclined surface, and show that the gravitational potential energy acquired by the block is $U_g = mgh$ relative to the base of the inclined plane.

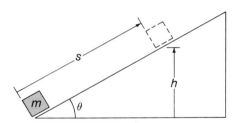

8–2 An ideal spring has a relaxed length $\ell_0 = 5$ cm and a spring constant $\kappa = 25$ N/m. (a) What amount of potential energy is stored in the spring if it is stretched by 1.5 cm? (b) What potential energy is stored if the spring is compressed by 1.5 cm? (c) Compare your two results and explain your findings.

8–3 Two uniform thin rods, each with length ℓ and mass $m/2$, are joined in a T shape as in the diagram. The assembly is held, first upright and then inverted, at a distance h above the reference surface. (a) What is its potential energy in each position? (b) Find the height to which a point with a mass m would have to be raised to gain the same potential energy in each case. Where is this point with respect to the joint between the two rods?

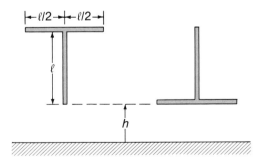

Section 8–2

8–4 A small 3-g mass is fired out of a toy gun aimed vertically upward. The mass leaves the gun having a kinetic energy of $1/10$ J. (a) What is the initial velocity of the mass? (b) How high above the gun will it rise into the air?

8–5 A small block with a mass of 10 g slides along a frictionless horizontal surface at a speed of 4 m/s. It encounters an ideal spring having a spring constant $\kappa = 25$ N/m, head-on as shown in the diagram. (a) What is the maximum compression of the spring? (b) What is the velocity of the block when the spring returns to one half of the maximum compression?

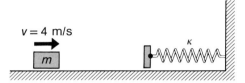

8–6 A small blob of putty having a mass of 10 g drops through
•   a height of 50 cm onto an ideal spring and sticks to its top, as shown in the diagram. If the spring constant is $\kappa = 12$ N/m, calculate the maximum compression of the spring.

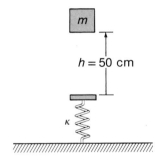

8–7 A pile driver is used to drive a stake into the ground (see diagram). The mass of the pile driver is $M = 3000$ kg, and it is dropped through a height of 10 m on each stroke. The resisting frictional force of the ground against the stake may be assumed to be a constant, equal to 5×10^6 N. How far into the ground is the stake driven on each stroke?

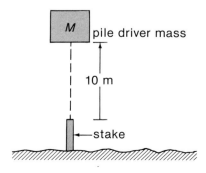

8–8 A block with a mass of 3 kg starts from rest at a height $h = 60$ cm on a plane having an inclination angle of 30°, as shown in the diagram. Upon reaching the bottom of the ramp, the block slides along a horizontal surface. If on both surfaces the friction coefficient is $\mu_k = 0.20$, how far will the block slide on the horizontal surface before coming to rest? [*Hint:* Divide the path into two straight-line parts.]

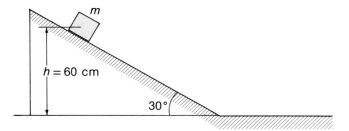

8–9 A 5-kg block is set into motion up an inclined plane, with an initial velocity of 8 m/s. The block comes to rest after traveling 3 m along the plane, as shown in the diagram. The plane is inclined at an angle of 30° to the horizontal. (a) Determine the change in kinetic energy. (b) Determine the change in potential energy. (c) Determine the frictional force on the block (assumed to be constant). (d) What is the coefficient of kinetic friction?

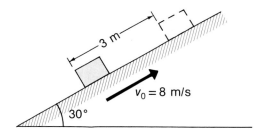

8–10 A limp rope with a mass of 0.2 kg and a length of 1 m is hung, initially at rest, on a frictionless peg that has a negligible radius, as shown in the diagram. What vertical velocity of the rope is reached just as the end slides off the peg? [*Hint:* Review Example 8–1.]

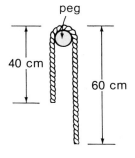

8–11 A block with a mass of 2 kg is lowered vertically, starting from rest, by a student who maintains the acceleration of the block at $\frac{1}{5}g$. How much work is done by the student in lowering the block 25 cm? What is the velocity of the block at this point?

8–12 A block starts from rest and slides along a plane with an inclination of 38°. After the block has traveled down the plane for a distance of 4 m, its velocity is 5 m/s. What is the coefficient of friction between the block and the plane?

8–13 A block of wood ($m = 1.6$ kg) is initially at the bottom of an inclined plane whose inclination angle is 25°. The coefficient of kinetic friction between the block and the plane is $\mu_k = 0.25$, and the coefficient of static friction is $\mu_s = 0.50$. The block is given an initial velocity of 2.5 m/s up the plane. How far along the plane will the block rise? Will the block then slide down the plane? Explain.

8–14 Solve the problem stated in Example 7–10 by using energy conservation (Eq. 8–7).

8–15 A uniform plank with a mass of 50 kg and a length of 3 m rests on a frictionless horizontal surface, as shown in the diagram. A student slowly raises the plank with the aid of the two ideal pulleys, so that it just dangles off the surface. What is the least amount of work the student must do?

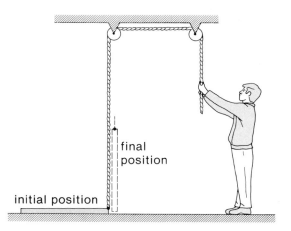

Section 8–3

8–16 A particle with a mass $m = 0.1$ kg moves under the influence of a force and has a potential energy of $U(x) = (\alpha + \beta x)x^2$, where $\alpha = 1$ J/m² and $\beta = -\frac{1}{2}$ J/m³. Where are the equilibrium points of the motion (i.e., where is $F(x) = 0$)? Are these points of stable or unstable equilibrium? What minimum total energy must the particle have in order to escape from the potential well and move to $x \to +\infty$?

8–17 A potential energy function often used to describe the interaction between two atoms in a diatomic molecule is the *Lennard-Jones potential*, $U(r) = U_0[(r_0/r)^{12} - 2(r_0/r)^6]$. (a) Sketch the function $U(r)$. (b) Show that $U(r)$ has a minimum value at $r = r_0$ and that this value is $-U_0$. (c) Find the values of r/r_0 for which $U(r) = -\frac{1}{4}U_0$.

Section 8–4

8–18 A particle with a mass m is tied to the end of a string and is whirled in a circular orbit in a vertical plane. If the *total energy* of the mass is a constant, what is the string tension when the particle is at its lowest point compared with that when the particle is at its highest point?

8–19 The world's records for throwing the shot put, the discus, and the javelin are approximately 22 m, 71 m, and 95 m, respectively. The corresponding masses of these missiles are 7.3, 2.0, and 0.80 kg. Assume that each throw is made at the optimum angle (at least, for the case of negligible air resistance) of 45°, and compute the initial kinetic energy in each case. (• Comment on the results and draw conclusions; for example, does the efficiency of the muscles in throwing depend on the mass of the object thrown?)

8–20 A particle with a mass m slides down a frictionless sphere
  • with a radius R, starting (essentially) from rest at the top. Through what vertical distance y does the particle move before it loses contact with the sphere?

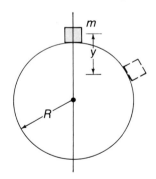

8–21 In the diagram at right, the two balls have masses
  •• $M = 2$ kg and $m = 0.5$ kg. Initially, the two balls are at positions A and B, and both are at rest. If the distance L is 60 cm, with what velocity does M hit the vertical wall?

Section 8–5

8–22 Sketch the force field for a springlike force, $\mathbf{F} = -\kappa(r - r_0)\hat{\mathbf{u}}_r$. Use an arrow diagram such as that in Fig. 8–7. Show the arrows on a plane that contains the origin.

Section 8–6

8–23 Given the two-dimensional force $\mathbf{F} = \dfrac{D}{r^2}\hat{\mathbf{u}}_\phi$, calculate the work done by this force along the closed path shown.

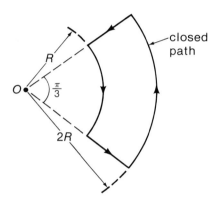

8–24 The particle shown in Fig. 8–9 is held to the circular path ($R = 50$ cm) by a taut string with an initial tension of 10 N. After one complete rotation the tension is found to increase to 30 N. What is the value of the constant D appearing in Eq. 8–22?

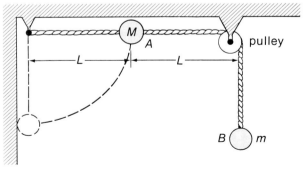

Problem 8–21

Additional Problems

8–25 A block with a mass of 2 kg slides across a horizontal surface. The coefficient of sliding friction is $\mu_k = 0.35$. The block collides with a horizontal spring, which it compresses by 50 cm before coming momentarily to rest. If the spring constant is 30 N/m, what was the velocity of the block at the instant of contact? How far back past this point does the recoiling block slide before coming to rest again?

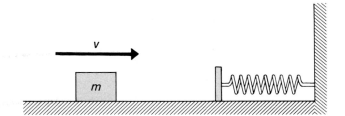

8–26 A student, using a lever of negligible mass with a length ℓ, very slowly raises a large boulder with a mass M to a height h by applying a constant vertical force F, as indicated in the diagram. If the pivot point (or fulcrum) is $\frac{1}{4}\ell$ from the boulder, how much work was done by the student? Considering the lever and the boulder together as a unit, show in a free-body diagram all the forces acting. Neglecting the kinetic energy of the lever and boulder, how much total work was done on this combined unit? Does the force at the pivot point do any work? Find the relationship between Mg and F.

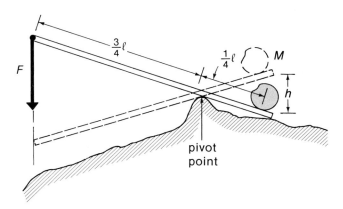

8–27 A slingshot consists of a Y-shaped fork with the tips
• 12 cm apart. The elastic band has a relaxed length of 40 cm. A draw that just doubles the length of the band requires a force of 50 N. What speed would this draw impart to a 100-g rock?

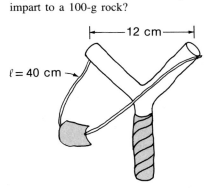

8–28 A student whose mass is 80 kg stands in the middle of a
• long, uniform plank that has a mass of 40 kg. He raises the center of the plank (and himself as well) 1 cm above the initial horizontal position by pulling on an ideal rope that runs over the two ideal pulleys, as shown in the diagram. The rope is attached to the plank at point B; assume that the contact point A remains fixed. Use the conservation of energy to calculate the tension in the rope. To what amount could the mass of the plank be increased before the student could no longer raise it off the ground?

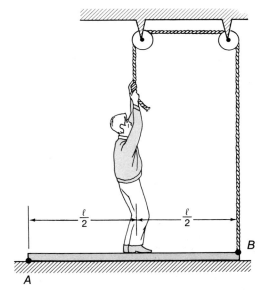

8–29 A limp chain with a length of 1 m and a total mass of 0.5 kg rests on a table with one third of its length hanging over the table edge. How much work is required to raise this part of the chain to the table top?

8–30 If the chain in Problem 8–29 is allowed to slide off the frictionless table, what is the velocity of the chain just as the trailing end leaves the table? (The height of the table top is >1 m.)

8–31 A student slowly raises a block by pulling on the end of a
 • massless rope as he walks to the right, as shown in the diagram. If the student moves a distance such that the angle α, initially $\alpha_0 = 30°$, is increased to $\alpha_f = 45°$, calculate how much work he does on the block by considering the motion of the block. Next, by direct calculation, find the amount of work done by the rope on the student. Assume an ideal pulley. Take $\ell = 10$ m and $m = 3$ kg. Compare the two answers.

8–32 Calculate the work done on a particle carried around the
 • closed path shown by each of the two forces $\mathbf{F}_1 = Kr\hat{\mathbf{u}}_\rho$ and $\mathbf{F}_2 = Kr\hat{\mathbf{u}}_\phi$. Is either force nonconservative?

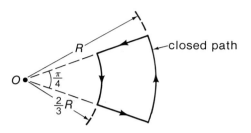

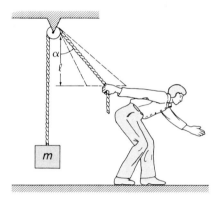

CHAPTER 9

LINEAR MOMENTUM AND SYSTEMS OF PARTICLES

The previous discussions have been devoted primarily to considerations of the motions of single particles. In this chapter we extend the treatment to particle pairs and then to systems of particles, allowing for more general relative motions. Of particular importance will be the consideration of simple two-particle systems that are isolated and therefore free of all external forces. In this situation we find that a powerful physical principle—the principle of momentum conservation—greatly simplifies the analysis. The importance of momentum conservation in treating dynamic situations lies in the fact that the principle is valid no matter how complicated the interaction forces between the particles.

9–1 THE LINEAR MOMENTUM OF A PARTICLE

In Section 5–2, some liberty was taken in translating Newton's words appearing in the *Principia*. His use of the Latin words *mutationem motus* or, literally, "change in the quantity of motion," was simply rendered as "acceleration" when we referred to the motion of a single particle. From the usage in the *Principia,* it is quite clear Newton had in mind that "quantity of motion" should be a kinetic property of mass and velocity taken together, namely, the product $m\mathbf{v}$. Therefore, Newton's second law should express the idea that force is equal to the rate of change of the quantity $m\mathbf{v}$; that is,

$$\mathbf{F} = \frac{d}{dt}(m\mathbf{v}) \qquad (9-1)$$

which is, in fact, the way that Newton expressed this law. Carrying out the time differentiation, we have

$$\mathbf{F} = \frac{d}{dt}(m\mathbf{v}) = m\frac{d\mathbf{v}}{dt} + \mathbf{v}\frac{dm}{dt}$$

For those cases in which the mass m of the object remains constant with time, we have $dm/dt = 0$, and the equation for the force reduces to

$$\mathbf{F} = m\frac{d\mathbf{v}}{dt} = m\mathbf{a} \qquad (9-2)$$

which is the way we previously expressed the second law. There are, however, situations

161

in which the mass *is* a function of time. We are then required to use Eq. 9–1 instead of Eq. 9–2; we discuss such a case in Section 9–4.

There exists today a vast quantity of empirical evidence that the concept embodied in Eq. 9–1, rather than that in Eq. 9–2, is the fundamental law of Nature, even in the subatomic, quantum, and relativistic realms of physics. We assert that Eq. 9–1 is correct because such empirical evidence demands it.

A word of caution is in order concerning the application of Eq. 9–1. In many problems involving so-called variable mass, only mass *exchange* between two parts of a system is actually involved. By viewing the system *as a whole,* the total mass is, in fact, conserved. We return to this point in Section 9–4.

It is useful to introduce a special term for the kinematic quantity $m\mathbf{v}$ that Newton called *quantity of motion*; we refer to $m\mathbf{v}$ as the *linear momentum* of a particle, and write

$$\mathbf{p} = m\mathbf{v} \tag{9-3}$$

In terms of the linear momentum, Newton's second law is expressed as

$$\mathbf{F} = \frac{d\mathbf{p}}{dt} \tag{9-4}$$

Impulse. There are situations in which the force applied to an object or particle lasts only for a time very short compared with the total time that we study the motion of the object. For example, when a golf ball is struck with a club or when a baseball is struck with a bat, the time during which the ball and the club or bat exert contact forces on each other is very short compared with the time the ball is in flight.

In most such cases the applied force is an unknown and complicated function of the time during the interval of contact or impact. The force is then best characterized by the total time integral of its effect. We rewrite Eq. 9–4, treating it as a relationship between differentials:

$$\mathbf{F}(t)\, dt = d\mathbf{p}$$

Then, if t_1 represents the onset of the interaction and t_2 its termination, we have

$$\mathbf{J} = \int_{t_1}^{t_2} \mathbf{F}(t)\, dt = \int_{\mathbf{p}_1}^{\mathbf{p}_2} d\mathbf{p} = \mathbf{p}_2 - \mathbf{p}_1 \tag{9-5}$$

The quantity $\mathbf{J}$ is the time integral of the force and is referred to as the *impulse*. Note carefully that Eq. 9–5 is a *vector* equation.

On occasion it is useful to consider the time average of the impulsive force. This is defined to be

$$\overline{\mathbf{F}} = \frac{1}{t_2 - t_1} \int_{t_1}^{t_2} \mathbf{F}(t)\, dt \tag{9-6}$$

where, as before, t_1 is the time at which the force is applied and t_2 is the time at which it is removed.

Finally, it is instructive to note the parallel structure of the two kinematic integrals we have so far considered, namely, the momentum change

$$\mathbf{p}_2 - \mathbf{p}_1 = m(\mathbf{v}_2 - \mathbf{v}_1) = \int_{t_1}^{t_2} \mathbf{F}(t)\, dt$$

and the change in kinetic energy (Eq. 7-17)

$$K_2 - K_1 = \tfrac{1}{2}m(v_2^2 - v_1^2) = \int_{\mathbf{r}_1}^{\mathbf{r}_2} \mathbf{F}(\mathbf{r}) \cdot d\mathbf{r}$$

The momentum change due to the action of the force $\mathbf{F}$ is obtained from an integral over *time*, whereas the change in kinetic energy due to the same force is obtained from an integral over *space*. These two relationships taken together completely specify the cumulative kinematic effects of the applied force $\mathbf{F}$.

Example 9-1

An impulsive force $F(t)$ is applied to an object with a mass $m = 200$ g that is initially at rest. The force increases linearly from zero to 500 N in a time of 0.02 and then decreases quadratically, becoming zero (with zero slope) 0.07 s after the initial application, as shown in the diagram. Determine (a) the impulse delivered, (b) the average force applied, and (c) the final velocity of the struck object.

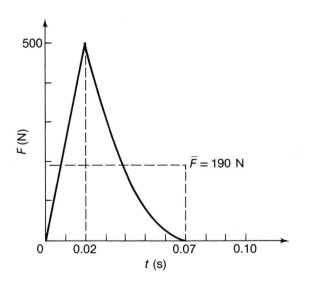

Solution:

The concepts of impulse and average force are particularly useful in cases such as the present one, where the force acts only during a very short time interval. The force $F(t)$ is described by

$$F(t) = \begin{cases} (2.5 \times 10^4 \text{ N/s})t, & 0 \le t \le 0.02 \text{ s} \\ (2.0 \times 10^5 \text{ N/s}^2)(t - 0.07 \text{ s})^2, & 0.02 \le t \le 0.07 \text{ s} \end{cases}$$

(a) The impulse is

$$J = \int_{t_1}^{t_2} F(t)\, dt$$

$$= (2.5 \times 10^4 \text{ N/s}) \int_0^{0.02} t\, dt$$

$$+ (2.0 \times 10^5 \text{ N/s}^2) \int_{0.02\,s}^{0.07\,s} (t - 0.07 \text{ s})^2 \, dt$$

$$= 2.5 \times 10^4 \cdot \tfrac{1}{2}t^2 \Big|_0^{0.02} + 2.0 \times 10^5 \cdot \tfrac{1}{3}(t - 0.07)^3 \Big|_{0.02}^{0.07}$$

$$= (5 + \tfrac{25}{3})\text{N} \cdot \text{s} = 13.33 \text{ kg} \cdot \text{m/s}$$

(b) The average force applied is

$$\bar{F} = \frac{J}{t_2 - t_1} = \frac{13.33 \text{ kg} \cdot \text{m/s}}{0.07 \text{ s}} = 190 \text{ N}$$

(c) The final velocity of the struck object is obtained from $J = m(v_2 - v_1)$:

$$v_2 = v_1 + \frac{J}{m} = 0 + \frac{13.33 \text{ kg} \cdot \text{m/s}}{0.2 \text{ kg}} = 66.6 \text{ m/s}$$

Note that any force $F(t)$ yielding the same value of J would result in the same final velocity v_2. In particular a constant force of 190 N (i.e., $\bar{F}$) acting for a duration of 0.07 s would also give $v_2 = 66.6$ m/s.

Finally, note that the calculated quantities J, $\bar{F}$, and v_2 are all vector quantities. In the present case they are all collinear and hence reference to their vector character was omitted. In the general case, $\mathbf{J}$, $\mathbf{v}_1$, and $\mathbf{v}_2$ need not be collinear.

Suppose that in Example 9–1 the applied impulse was vertically upward. Two important points should be noted. The first is that during the 0.07 s of the action of the impulse the average force of 190 N greatly exceeds the gravitational force of 1.96 N. The second point is that the flight time of the object in rising to its maximum height and returning back down is 13.6 s, or about 200 times the duration of the impulse. These two facts allow us to consider for all practical purposes the history of the object to consist of two phases. During the first 0.07 s the effects of the gravitational force may be neglected and the acquired velocity of the object may be calculated from the action of the impulsive force alone. During the second phase of the motion, lasting 13.6 s, the velocity changes relatively slowly under the influence of the weaker gravitational force. See also Problem 9–6.

9–2   THE TWO-PARTICLE INTERACTING SYSTEM

We first investigate the linear momentum of a completely isolated system consisting of two particles. Consider two particles with masses m_1 and m_2 that interact with each other. For example, the particles could be coupled together by an ideal inertia-less spring, but otherwise isolated in deep space. We consider the ideal inertia-less spring to be simply a schematic representation of the physical interaction between the particles. The particles and the spring are shown in Fig. 9–1.

The force exerted on m_1 by m_2 is $\mathbf{F}_{12}$ and that exerted on m_2 by m_1 is $\mathbf{F}_{21}$. Then, from Newton's third law, we can write

$$\mathbf{F}_{12} = -\mathbf{F}_{21}$$

Also, from the second law applied to each particle separately, we have

$$\mathbf{F}_{12} = \frac{d\mathbf{p}_1}{dt} \quad \text{and} \quad \mathbf{F}_{21} = \frac{d\mathbf{p}_2}{dt}$$

Adding these two expressions and using the third law, we find

$$\mathbf{F}_{12} + \mathbf{F}_{21} = \frac{d}{dt}(\mathbf{p}_1 + \mathbf{p}_2) = 0$$

or
$$\mathbf{p}_1 + \mathbf{p}_2 = \text{constant} \tag{9–7}$$

Thus, we see that although the individual momenta $\mathbf{p}_1$ and $\mathbf{p}_2$ might change with time because of the mutual interaction of the particles, the total momentum of the system,* $\mathbf{P} = \mathbf{p}_1 + \mathbf{p}_2$, is a constant of the motion; that is, the sum is independent of time. We refer to this result as the *conservation of linear momentum* of the isolated system.†

The nature of the force acting between the two particles did not enter into the considerations leading to Eq. 9–7. Therefore, it must be true that two otherwise isolated particles that happen to collide obeying some unknown and complicated force during impact will also obey the linear momentum conservation law.

Because force is a free vector, a particular force is represented by the same vector in

---

*We use $\mathbf{p}$ to indicate the momentum of a *particle* and $\mathbf{P}$ to indicate the momentum of a *system*.

†Notice that if we accept the validity of the third law, then linear momentum conservation is indeed a *result*. It is possible to begin by considering momentum conservation as an experimental fact and then to deduce the third law. This approach to dynamics is sometimes taken. A brief discussion of this line of reasoning was given in Section 5–2.

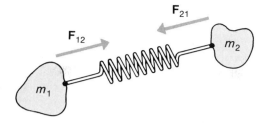

Fig. 9–1. An isolated system of two particles, m_1 and m_2, interacting with each other through a force, shown here schematically as due to an ideal inertia-less spring.

any inertial reference frame. This fact led us to the conclusion, in Section 5–4, that Newton's second law is valid in any inertial frame. The same reasoning applies to momentum conservation, expressed by Eq. 9–7. The specific statement is as follows:

In any inertial frame of reference, the linear momentum of an interacting isolated system of particles is conserved, i.e., is a constant of the motion.*

Collisions with the Earth.   We sometimes must deal with situations in which an object collides with the surface of the Earth or with another object that is attached to the Earth. For example, suppose that a ball is thrown downward and strikes the ground at an angle of 45°, then rebounds at the same angle. It is easy to see that the horizontal component of momentum is conserved in this process because the horizontal velocity is the same after the collision as before. However, the vertical component of the ball's momentum changes from $-mv_y$ to mv_y, so that there is a net momentum change amounting to $2mv_y$. To an observer standing on the ground, it would appear that momentum is not conserved. But to an observer stationed in space (and equipped with very precise measuring apparatus), the Earth would be seen to approach the ball and to rebound from it, thereby conserving momentum. Because the Earth is so massive compared with everyday objects, we can reasonably neglect the exceedingly small recoil of the Earth.

Elastic and Inelastic Collisions.   When one object collides with another object, deformations (generally, of both objects) occur during the time of contact. A striking example of deformation in a collision is shown in the high-speed photograph of a golf club colliding with a golf ball (at right). To produce a deformation of a solid object, work must be done against the elastic forces within the body. If the system is ideal, the deformed portions will recover completely and the stored energy will be returned to the colliding objects as kinetic energy. Then there is no change in the kinetic energy of the system caused by the collision and we say that the collision is *perfectly elastic*.

When we deal with real situations involving bulk matter, collisions are never perfectly elastic. A collision between two billiard balls or between two steel ball bearings is almost elastic, but there is always some loss of kinetic energy between the initial and final conditions. After all, we *hear* a click when two billiard balls collide, indicating that some energy has been converted into sound energy; additional energy is converted into nonacoustic forms. Any collision that is not elastic is called an *inelastic* collision. Any collision that results in a permanent deformation is necessarily inelastic.

In an inelastic collision, some of the initial kinetic energy is absorbed and converted into different forms of energy. If the amount of kinetic energy absorbed in a collision is the maximum that is allowed by momentum conservation, the collision is said to be *completely inelastic*. For example, if two railway cars collide and couple together, the collision is completely inelastic. (Any rebounding after collision will result in a larger value of the final kinetic energy; hence, coupling together provides for the maximum

(Courtesy of Dr. Harold E. Edgerton, MIT.)

---

*Although we have considered only a two-body system so far, we show later that the concept extends to any number of particles acting as an isolated system.

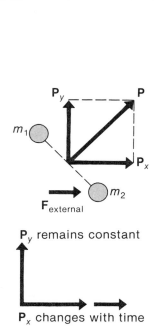

P_y remains constant

P_x changes with time

Fig. 9–3. If the external force acting on a system of particles has only an x-component, the y-component of the system's momentum will remain constant, but the x-component will change with time. (Of course, the total momentum **P** also changes with time.)

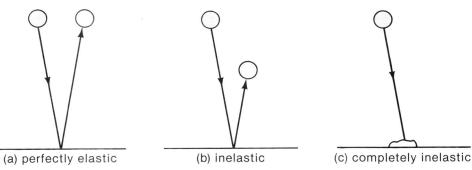

(a) perfectly elastic   (b) inelastic   (c) completely inelastic

Fig. 9–2. Three types of collision processes, illustrated by objects striking a hard floor.

absorption of kinetic energy. See Problem 9–43.) In all collisions involving isolated objects, momentum conservation is valid, even if the collision is inelastic.

Figure 9–2 shows a simple example of each of the three types of collision processes we have been discussing. The perfectly elastic case is an idealization.

Partially Isolated Systems. We should remember that Eq. 9–7 is a vector equation involving all three perpendicular components. We frequently encounter cases in which the system is acted upon by a force that is due to an agency outside the system. However, if this external force is constant *in direction,* Eq. 9–7 is still valid for components of the momentum in the plane perpendicular to the direction of this force (see Fig. 9–3). It may also happen in cases of collisions that, during the impulse, external forces are either absent or act in a particular constant direction. Again, total or partial use of Eq. 9–7 may be appropriate.

Of course, the use of Eq. 9–7 for such cases of partial isolation may not give a complete description of the motion. A more detailed study of the behavior of systems of particles in the presence of external forces is given in Section 9–4.

Example 9–2

A cannon is rigidly bolted to a railroad flatcar. They have a combined mass of 12,000 kg. The cannon is set to an elevation angle of 60° to the horizontal and fires a 50-kg shell with a muzzle velocity of 400 m/s.

(a) If the flatcar was originally at rest on a straight length of frictionless horizontal track, calculate the recoil speed V of the flatcar. (b) If it takes 2.5 ms to eject the shell, calculate the average impulsive force $\bar{F}$ the rails exert on the flatcar's wheels.

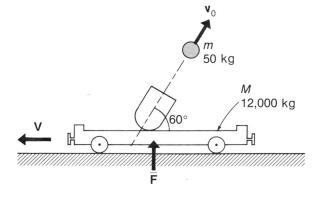

Solution:

(a) Since there are no external horizontal forces acting on the system consisting of the shell and the flatcar–cannon, this component of their combined linear momentum must be conserved. Thus,

$$0 = mv_0 \cos 60° - MV$$

or $$V = \frac{(50 \text{ kg}) \, 400 \text{ m/s} \cos 60°}{12{,}000 \text{ kg}} = 0.833 \text{ m/s}$$

(b) As the shell leaves the cannon, the system has an acquired vertical component of momentum provided by the impulsive vertical force exerted by the rail against the flatcar wheels. (This force is in addition to that required to support the weight of the flatcar.) We have

$$\bar{F} \, \Delta t = \Delta P = mv_0 \sin 60°$$

or $$\bar{F} = \frac{(50 \text{ kg}) \, 400 \text{ m/s} \sin 60°}{2.5 \times 10^{-3} \text{ s}} = 6.93 \times 10^6 \text{ N}$$

(• Compare the magnitude of $\bar{F}$ to that exerted by gravity on the shell. Is it permissible to neglect the gravitational force during the firing? Afterward?)

Example 9–3

A block with a mass m_1 slides frictionlessly across a horizontal surface with a velocity $\mathbf{v}_1$, directly toward a stationary block with a mass m_2. What are the velocities, $\mathbf{v}'_1$ and $\mathbf{v}'_2$, of the blocks after an elastic collision?

Solution:

From momentum conservation in the horizontal plane, we have

$$m_1 v_1 = m_1 v'_1 + m_2 v'_2 \qquad (1)$$

From energy conservation, we have (cancelling the common factor $\frac{1}{2}$)

$$m_1 v_1^2 = m_1 (v'_1)^2 + m_2 (v'_2)^2 \qquad (2)$$

Solve Eq. (1) for v'_2 and substitute into Eq. (2). After some algebraic manipulation, the result is

$$(v'_1)^2 - 2 \frac{m_1 v_1}{m_1 + m_2} v'_1 + \frac{m_1 - m_2}{m_1 + m_2} v_1^2 = 0$$

This equation can be factored; thus,

$$(v'_1 - v_1)\left(v'_1 - \frac{m_1 - m_2}{m_1 + m_2} v_1\right) = 0$$

The two solutions are

$$v'_1 = v_1$$

$$v'_1 = \frac{m_1 - m_2}{m_1 + m_2} v_1$$

The first of these solutions corresponds to the uninteresting case in which the block m_1 is in front of the block m_2, resulting in no collision! The second solution yields three interesting possibilities:

1. When $m_1 = m_2$, we have $v'_1 = 0$. That is, the collision causes m_1 to come to rest, while m_2 is set into motion with $v'_2 = v_1$. (• Verify the last conclusion.)
2. When $m_1 > m_2$, we have $v'_1 > 0$, so that m_1 (and also m_2) moves in the same direction as before the collision.
3. When $m_1 < m_2$, we have $v'_1 < 0$, so that the motion of m_1 is reversed while m_2 is set into motion with $v'_2 > 0$. (• Verify these conclusions.)

(• Verify that the relative velocity before collision, namely, $\mathbf{v}_1 - \mathbf{v}_2 = \mathbf{v}_1 - 0 = \mathbf{v}_1$, has the same magnitude as the relative velocity after collision, namely, $\mathbf{v}'_1 - \mathbf{v}'_2$. This is a general characteristic of elastic collisions.)

Example 9–4

Two equal-mass hockey pucks are free to slide on a horizontal frictionless ice surface. Initially, one puck has a kinetic energy K_0 with a velocity $\mathbf{v}_0$ directed at the second puck, which is at rest. After colliding, the first puck is observed to move in a direction making an angle α with the original direction of motion, as shown in the diagram. The struck puck recoils in a direction defined by the angle β.

(a) If the collision is perfectly elastic, what is the kinetic energy K_1 of the deflected puck?

(b) What is the relationship between the angles α and β?

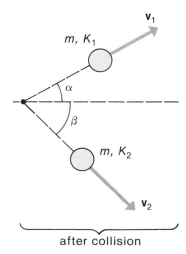

after collision

top view

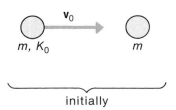

initially

Solution:

(a) From momentum conservation, we have

$$m\mathbf{v}_0 = m\mathbf{v}_1 + m\mathbf{v}_2 \qquad (1)$$

and for the components in the horizontal plane, we have

$$mv_0 = mv_1 \cos \alpha + mv_2 \cos \beta \tag{2}$$

$$0 = mv_1 \sin \alpha - mv_2 \sin \beta \tag{3}$$

From the conservation of energy, we have

$$K_0 = K_1 + K_2$$

$$\tfrac{1}{2}mv_0^2 = \tfrac{1}{2}mv_1^2 + \tfrac{1}{2}mv_2^2$$

or

$$v_0^2 = v_1^2 + v_2^2 \tag{4}$$

We may rearrange Eqs. (2) and (3) to give

$$v_2 \cos \beta = v_0 - v_1 \cos \alpha$$

$$v_2 \sin \beta = v_1 \sin \alpha$$

Squaring and adding yields

$$v_2^2 = v_0^2 - 2v_0v_1 \cos \alpha + v_1^2$$

Eliminating v_2 between this equation and the energy equation, Eq. (4), we find

$$v_1 = v_0 \cos \alpha$$

from which we obtain the desired expression for K_1, namely,

$$K_1 = K_0 \cos^2 \alpha \tag{5}$$

(b) Writing the momentum equation, Eq. (1), as $\mathbf{v}_0 = \mathbf{v}_1 + \mathbf{v}_2$ and squaring, we find

$$\mathbf{v}_0 \cdot \mathbf{v}_0 = v_0^2 = v_1^2 + v_2^2 + 2\mathbf{v}_1 \cdot \mathbf{v}_2$$

Comparing this result with Eq. (4), we conclude that

$$\mathbf{v}_1 \cdot \mathbf{v}_2 = 0$$

and hence the two vectors $\mathbf{v}_1$ and $\mathbf{v}_2$ are perpendicular to each other. In terms of the angles α and β

$$\cos (\alpha + \beta) = 0$$

or

$$\alpha + \beta = \pi/2$$

This condition that the directions of the final motions are at right angles is *true only* for elastic collision between two bodies with *equal masses*. In the present case the angle α describing the motion of the deflected puck can never exceed $\pi/2$. Only if the struck object has a mass greater than that of the incident object will a backward deflection (i.e., $\alpha > \pi/2$) be possible. (The case of unequal masses is treated in Problems 9–45 and 9–46.)

Finally, we note the two special cases corresponding to deflection angles for the incident puck of $\alpha = 0$ and $\alpha = \pi/2$. For $\alpha = 0$, Eq. (5) gives $K_1 = K_0$ and hence $K_2 = 0$. This situation with $\mathbf{v}_1 = \mathbf{v}_0$ and $v_2 = 0$ corresponds to no collision taking place. The other case with $\alpha = \pi/2$ yields $K_1 = 0$ using Eq. (5) and hence $K_2 = K_0$. This situation with $v_1 = 0$ and $\mathbf{v}_2 = \mathbf{v}_0$ corresponds to a head-on collision. (● Describe the behavior of the angle β for the struck object both when $\alpha \to 0$ and when $\alpha \to \pi/2$.)

Example 9–5

Two particles with masses $m_1 = 2.0$ kg and $m_2 = 5.0$ kg are free to slide on a straight horizontal frictionless guide wire. The particle with the smaller mass is moving with a speed of 17.0 m/s and overtakes the larger particle, which is moving in the same direction with a speed of 3.0 m/s. The larger particle has an ideal inertia-less spring ($\kappa = 4480$ N/m) attached to the side being approached by the smaller particle, as shown in the diagram.

(a) What is the maximum compression of the spring as the two particles collide?

(b) What are the final velocities of the particles?

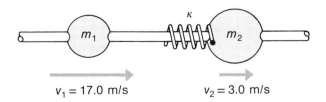

$v_1 = 17.0$ m/s $v_2 = 3.0$ m/s

Solution:

(a) When the spring is under maximum compression, the two particles have zero *relative velocity* and, hence, each may be assigned the same guide-wire velocity, say v_0. Because no external forces are acting in the horizontal direction, we may use the

conservation of linear momentum and write

$$m_1v_1 + m_2v_2 = (m_1 + m_2)v_0$$

Hence, $v_0 = \dfrac{(2.0 \text{ kg})(17.0 \text{ m/s}) + (5.0 \text{ kg})(3.0 \text{ m/s})}{2.0 \text{ kg} + 5.0 \text{ kg}}$

$$= 7.0 \text{ m/s}$$

Initially, with no interaction between the particles, the total kinetic energy is

$$K = \tfrac{1}{2}m_1v_1^2 + \tfrac{1}{2}m_2v_2^2 = \tfrac{1}{2}(2.0)(17.0)^2 + \tfrac{1}{2}(5.0)(3.0)^2$$

$$= 311.5 \text{ J}$$

At maximum compression of the spring, the total kinetic energy is

$$K_0 = \tfrac{1}{2}(m_1 + m_2)v_0^2 = \tfrac{1}{2}(2.0 + 5.0)(7.0)^2 = 171.5 \text{ J}$$

The difference in energy, $K - K_0 = 140.0$ J, must be stored in the spring. (● Why is this so?) Thus, the compression x of the spring is found from

$$\tfrac{1}{2}\kappa x^2 = K - K_0$$

or

$$x = \sqrt{\dfrac{2(140.0 \text{ J})}{4480 \text{ N/m}}} = 0.25 \text{ m}$$

(b) When the particles finally separate, both linear momentum and kinetic energy must be conserved (because the energy stored in the spring has been returned to the particles). Then,

$$m_1v_1' + m_2v_2' = m_1v_1 + m_2v_2 \tag{1}$$

and $$\tfrac{1}{2}m_1(v_1')^2 + \tfrac{1}{2}m_2(v_2')^2 = \tfrac{1}{2}m_1v_1^2 + \tfrac{1}{2}m_2v_2^2 \tag{2}$$

In the momentum equation, Eq. (1), we substitute the numerical values of the masses, obtaining

$$2v_1' + 5v_2' = (2.0)(17.0) + (5.0)(3.0) = 49 \text{ kg·m/s} \tag{3}$$

We also substitute for the masses in the kinetic energy equation, Eq. (2), with the result

$$2(v_1')^2 + 5(v_2')^2 = 2K = 623 \text{ J} \tag{4}$$

If we obtain v_2' in terms of v_1' from Eq. (3) and substitute this

expression into Eq. (4), after some simplification we find

$$(v_1')^2 - 14v_1' - 51 = 0$$

Factoring, we obtain

$$(v_1' - 17)(v_1' + 3) = 0$$

Thus, we find a *pair* of solutions,

$$v_1' = +17 \text{ m/s} \quad \text{and} \quad v_2' = +3 \text{ m/s}$$
$$v_1' = -3 \text{ m/s} \quad \text{and} \quad v_2' = +11 \text{ m/s}$$

Of these, only the second set is physically realizable. (• Explain why the first set is unphysical.) Verify that the kinetic energy K' after collision is equal to the kinetic energy K before collision.

Example 9–6

A device known as a *ballistic pendulum* affords another good example of the application of the momentum principle. A ballistic pendulum can be used to measure the velocity of such projectiles as bullets, arrows, and darts. The device consists of a rather massive block of wood that is suspended by parallel cords and is initially hanging at rest, as shown in the diagram. The test projectile (for example, a bullet) is fired horizontally into the block, which is thick enough to bring the bullet to rest, embedded inside it. (Notice that the collision of the bullet with the block is completely inelastic.) The block (with embedded bullet) swings up to a maximum elevation h. From the known masses and h, find the initial velocity of the bullet.

Solution:

Because the support cords are parallel, the pendulum swing of the block consists only of *translational motion*. Let m be the mass of the bullet and M be the mass of the block. During the very short time required to bring the bullet to rest within the block, the block moves only a negligible amount, and there are no external horizontal forces acting on the system of the block and bullet. Therefore, we may use the conservation of linear momentum in the horizontal direction during this interval. Just before the bullet enters the block, the momentum of the system is

$$P = p_{\text{bullet}} = mv$$

When the bullet has come to rest and the block begins to move

significantly, we have

$$P' = (m + M)V$$

where V is the horizontal velocity of the block and bullet combination. With $P = P'$, we find

$$V = \frac{m}{m + M}v$$

After the initial impact, the momentum of the system consisting of the block and embedded bullet is no longer conserved in the horizontal direction. (• What external horizontal force appears when $x > 0$? What happens to the momentum?)

The maximum height h of the block is reached when the velocity becomes zero. Using the conservation of energy, we can write

$$\tfrac{1}{2}(M + m)V^2 = (M + m)gh$$

Combining this with the previous result gives

$$v = \frac{m + M}{m}\sqrt{2gh}$$

If a 10-g bullet is fired into a 3-kg block and a height $h = 3.2$ cm is measured, the bullet velocity would be

$$v = \frac{3.010 \text{ kg}}{0.010 \text{ kg}}\sqrt{2(9.80 \text{ m/s}^2)(0.32 \text{ m})}$$

$$= 238 \text{ m/s}$$

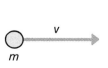

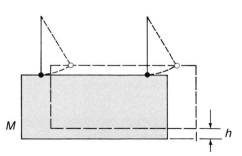

Momentum conservation allows us to obtain a result for the velocity of the bullet even though the force exerted on the bullet by the block during the stopping time is extremely complicated (even unknowable). To obtain the result from the time-developed equations of motion using Newton's second law would be exceedingly difficult (even impossible).

Note that the initial kinetic energy of the bullet, $\frac{1}{2}mv^2$, is in general much larger than the kinetic energy of the block (and bullet) at the instant the block begins to move. That is,

$$\tfrac{1}{2}mv^2/\tfrac{1}{2}(m + M)V^2 = \frac{m + M}{m} \gg 1$$

(• What happens to the lost kinetic energy?)

Note carefully that during the short stopping time of the bullet, momentum is conserved but kinetic energy is not (this is a characteristic of an inelastic collision), whereas later, as the pendulum begins to swing, energy is conserved but the momentum of the block changes because of the unbalanced forces that then begin to act. (• In what way would the preceding analysis be affected if the stopping time of the bullet were not negligibly short?)

9–3 CENTER OF MASS

When dealing with a system of particles or with a rigid body, we frequently inquire about the motion of the system "as a whole." In such situations it is useful to use the concept of the *center of mass*. The point that corresponds to the center of mass (C.M.) is the "effective mass center" of an object or system of particles. That is, if all the mass of the system or object were to be concentrated at this point, the resulting pointlike mass would have the same translational kinematic behavior as the extended mass system as a whole.

For a pair of equal-mass particles, we have the intuitive notion (which is quite correct) that the center of mass is located midway between the particles. Or, if one particle has a mass greater than that of the other, the C.M. is located closer to the more massive particle and on the line connecting the particles. The precise definition of the *center-of-mass vector* **R** of n particles is (see Fig. 9–4)

$$\mathbf{R} = \frac{m_1\mathbf{r}_1 + m_2\mathbf{r}_2 + \cdots + m_n\mathbf{r}_n}{m_1 + m_2 + \cdots + m_n} = \frac{1}{M}\sum_{i=1}^{n} m_i\mathbf{r}_i \qquad (9\text{–}8)$$

with

$$M = \sum_i m_i$$

It is important to realize that Eq. 9–8 is a *vector* equation and therefore represents three separate equations for the components of **R**, namely,

$$X = \frac{1}{M}\sum_i m_i x_i; \qquad Y = \frac{1}{M}\sum_i m_i y_i; \qquad Z = \frac{1}{M}\sum_i m_i z_i \qquad (9\text{–}9)$$

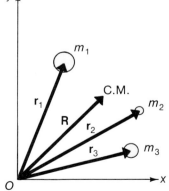

Fig. 9–4. The center-of-mass vector **R** for a collection of particles.

If the particles are in relative motion, these equations specify the C.M. coordinates at a particular instant.

Example 9-7

Locate the C.M. of the set of four particles shown in the diagram. The particles are on the corners of a 2-m square.

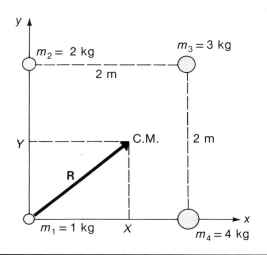

Solution:

This is a two-dimensional problem, so we are to find the coordinates (X, Y) of the C.M. We use Eqs. 9-9 with $M = 10$ kg; thus,

$$X = \frac{1}{10 \text{ kg}}[(1 \text{ kg})(0 \text{ m}) + (2 \text{ kg})(0 \text{ m})$$
$$+ (3 \text{ kg})(2 \text{ m}) + (4 \text{ kg})(2 \text{ m})]$$

$$= \frac{14 \text{ kg·m}}{10 \text{ kg}} = 1.4 \text{ m}$$

$$Y = \frac{1}{10}[(1)(0) + (2)(2) + (3)(2) + (4)(0)]$$

$$= \frac{10}{10} = 1.0 \text{ m}$$

These coordinates for the C.M. are shown in the diagram. Notice that the C.M. is *not* located at the center of the square. (• But the *y*-coordinate is equal to one half the side of the square. Can you see why?)

Center of Mass for a Solid Object. Equations 9-8 and 9-9 give the coordinates of the center of mass for a collection of particles. When the collection consists of a continuous distribution of matter, the sums are replaced by integrations. If *dm* represents a differential element of mass, we have

$$\mathbf{R} = \frac{1}{M} \int_M \mathbf{r} \; dm \qquad (9-10)$$

and
$$X = \frac{1}{M} \int_M x \; dm; \quad Y = \frac{1}{M} \int_M y \; dm; \quad Z = \frac{1}{M} \int_M z \; dm \qquad (9-11)$$

where the mass element is $dm = \rho \; dx \; dy \; dz$, with ρ equal to the mass density.

Example 9-8

Determine the location of the center of mass of a thin lamina in the shape of an isosceles triangle with a base length *a* and an altitude *b*.

Solution:

For convenience, orient the lamina in a coordinate frame as shown in the diagram. The *surface density* σ of the lamina of mass *M* is

$$\sigma = \frac{M}{\text{Area}} = \frac{2M}{ab}$$

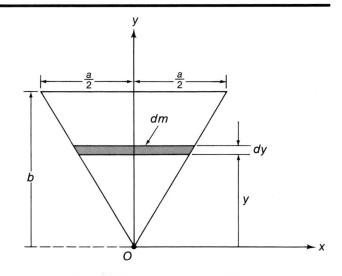

From symmetry we have $X = 0$. (• Prove this assertion.) Making use of Eq. 9–11 for Y, we have

$$Y = \frac{1}{M} \int y \, dm$$

and

$$dm = \sigma \frac{a}{b} y \, dy = \frac{2M}{b^2} y \, dy$$

Hence

$$Y = \frac{1}{M} \int_0^b \frac{2M}{b^2} y^2 \, dy = \frac{2}{b^2} \left(\frac{y^3}{3} \right) \Big|_0^b = \frac{2b}{3}$$

9–4 LINEAR MOMENTUM OF A SYSTEM OF PARTICLES

Suppose that the masses of the n particles in Eq. 9–8 remain constant as the history of the system unfolds. Then, differentiating with respect to time gives

$$M \frac{d\mathbf{R}}{dt} = \sum_i m_i \frac{d\mathbf{r}_i}{dt}$$

Now, $d\mathbf{R}/dt = \mathbf{V}$ is the velocity of the C.M. of the system, and $m_i(d\mathbf{r}_i/dt) = \mathbf{p}_i$ is the momentum of particle i. Thus, we can write

$$M\mathbf{V} = \sum_i \mathbf{p}_i = \mathbf{P} \qquad (9\text{–}12)$$

That is, the linear momentum $\mathbf{P}$ of a system of particles is equivalent to that of a hypothetical particle with the entire mass M of the system moving with the velocity of the C.M. Differentiating Eq. 9–12 with respect to time gives

$$\frac{d\mathbf{P}}{dt} = \sum_i \frac{d\mathbf{p}_i}{dt}$$

We suppose that there are two-body forces acting between each pair of the n particles; in addition there is an externally applied force $\mathbf{F}_i$ that acts on particle i. The total force on particle i is

$$\mathbf{F}_i = \frac{d\mathbf{p}_i}{dt} = \mathbf{F}_i^e + \sum \text{ (all two-body forces)}$$

The sum over the two-body forces involves the terms

$$\mathbf{F}_{12} + \mathbf{F}_{13} + \mathbf{F}_{14} + \cdots + \mathbf{F}_{21} + \mathbf{F}_{23} + \mathbf{F}_{24} + \cdots + \mathbf{F}_{31} + \mathbf{F}_{32} + \mathbf{F}_{34} + \cdots$$

where the unphysical "self-force" terms, $\mathbf{F}_{11}, \mathbf{F}_{22}, \mathbf{F}_{33}, \cdots$, have been eliminated. This series of terms can be grouped in the following way:

$$(\mathbf{F}_{12} + \mathbf{F}_{21}) + (\mathbf{F}_{13} + \mathbf{F}_{31}) + (\mathbf{F}_{14} + \mathbf{F}_{41}) + \cdots$$

According to Newton's third law, each of these terms in parentheses is zero; that is, $\mathbf{F}_{12} + \mathbf{F}_{21} = 0$, and so forth. Thus, the sum vanishes by the cancellation of the forces in pairs. Then, writing $\mathbf{F}^e = \Sigma \mathbf{F}_i^e$ for the total external force, we have

$$\mathbf{F}^e = \frac{d\mathbf{P}}{dt} = M \frac{d^2\mathbf{R}}{dt^2} = M \frac{d\mathbf{V}}{dt} \qquad (9\text{–}13)$$

Thus, the C.M. of the system moves in exactly the same way as would a particle with the same mass and subjected to the same force. In the event that there are no external forces acting, we find

$$\mathbf{P} = \text{constant} \qquad \text{(no external forces)} \qquad (9-14)$$

That is, the total linear momentum $\mathbf{P}$ of an isolated system of particles is a constant of the motion. This implies that the C.M. of the system simply moves with a constant velocity $\mathbf{V}$. (Stated otherwise: a system that interacts only internally cannot accelerate itself.) The only way in which the linear momentum of an individual particle in an isolated system can change is if there is an equal and opposite change in the momentum of the rest of the system.

Example 9–10

Two equal-mass cannon balls are tied together with a short chain (a favorite trick of medieval warriors) and are fired at an angle of 30° above the horizontal with a C.M. velocity of 120 m/s. Somewhere along the resulting trajectory, the chain breaks and one of the balls lands 9 s after the firing time, 1 km from the cannon and directly under its line of fire. If the chain has negligible mass compared with that of the cannon balls, where is the second ball at the instant the first ball lands?

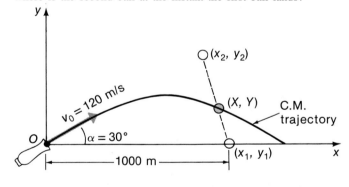

Solution:

The coordinates of the cannon ball that landed first are (x_1, y_1) with $x_1 = 1000$ m and $y_1 = 0$. The C.M. of the two balls follows the ballistic trajectory

$$X = V_0 t \cos \alpha$$

$$Y = V_0 t \sin \alpha - \tfrac{1}{2} g t^2$$

At $t = 9$ s,

$$X = (120 \text{ m/s})(9 \text{ s}) \cos 30° = 935.3 \text{ m}$$

$$Y = (120 \text{ m/s})(9 \text{ s}) \sin 30° - \tfrac{1}{2}(9.80 \text{ m/s}^2)(9 \text{ s})^2 = 143.1 \text{ m}$$

With balls of equal mass, the C.M. coordinates (Eq. 9–9) reduce to

$$X = \tfrac{1}{2}(x_1 + x_2) \qquad \text{and} \qquad Y = \tfrac{1}{2}(y_1 + y_2)$$

Solving for the coordinates of ball 2 at $t = 9$ s, we find

$$x_2 = 2X - x_1 = 2 \times 935.3 - 1000 = 870.6 \text{ m}$$

$$y_2 = 2Y - y_1 = 2 \times 143.1 - 0 = 286.2 \text{ m}$$

9–5   VARIABLE MASS

We now consider the behavior of an object that has variable mass. We must emphasize, however, that in every case there is a larger system that has constant mass. Thus, when we say that a certain object has ''variable mass,'' we mean that the object acquires mass from or loses mass to its surroundings. The object or system A that we study can be isolated from the global system of which it is a part by a bounding surface, as in Fig. 9–5. Then, mass (and momentum) can flow inward or outward through the boundary without affecting the mass of the global system. When mass flows inward, the time rate of change of the mass M of the system A is $dM/dt > 0$, and when mass flows outward, $dM/dt < 0$. A rocket is an important example of an object that has $dM/dt < 0$.

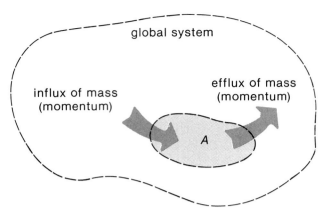

Fig. 9–5. A system A is isolated from the surrounding global system by a bounding surface through which mass (momentum) can flow. The mass of the global system, of which A is a part, remains constant.

The Rocket Equation. A rocket engine ejects a stream of rapidly moving gas particles through nozzles located at its rear. These hot gases are the result of burning a fuel in the combustion chamber of the rocket. The mass of the rocket continually decreases as the fuel and oxidizer are consumed and ejected.

Consider a rocket A that, at some time t, has a mass M and a velocity $\mathbf{v}$ in a certain inertial reference frame. An amount of gas with mass Δm is ejected from the nozzle during a time interval Δt, as indicated in Fig. 9–6. At the end of this interval, at time $t + \Delta t$, rocket A (now with mass $M - \Delta m$) has a velocity $\mathbf{v} + \Delta \mathbf{v}$, and the ejected gas has a velocity $\mathbf{u}$. Both of these velocities are measured in the inertial frame originally selected.

We suppose that an external force $\mathbf{F}^e$ acts on the rocket. This force might include gravity, air resistance, and so forth. Note carefully that $\mathbf{F}^e$ does *not* include any force of interaction between A and Δm; such an internal force would not contribute to the momentum change of the system as a whole.

Using Newton's second law (Eq. 9–4), we have

$$\mathbf{F}^e = \frac{d\mathbf{P}}{dt} \cong \frac{\Delta \mathbf{P}}{\Delta t} = \frac{\mathbf{P}_f - \mathbf{P}_i}{\Delta t}$$

where $\mathbf{P}_f$ and $\mathbf{P}_i$ stand for the final and initial momenta, respectively, of the entire system. Thus,

$$\mathbf{P}_f = (M - \Delta m)(\mathbf{v} + \Delta \mathbf{v}) + \mathbf{u}\,\Delta m$$

and
$$\mathbf{P}_i = M\mathbf{v}$$

Substituting these values into the equation for $\mathbf{F}^e$, neglecting the second-order infinitesimal $\Delta m\,\Delta \mathbf{v}$, and writing that $\dfrac{dm}{dt} = -\dfrac{dm}{dt}$, we have, in the limit $\Delta t \to 0$,

Fig. 9–6. A rocket A with mass M ejects a mass Δm of gas during a time interval Δt. The initial and final velocities are shown relative to the same inertial reference frame. The relative velocity of the exhaust gas with respect to the rocket, $\mathbf{v}_r = \mathbf{u} - \mathbf{v}$, is directed toward the *left*.

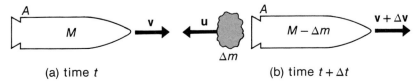

(a) time t (b) time $t + \Delta t$

$$M \frac{d\mathbf{v}}{dt} = \mathbf{F}^e + (\mathbf{u} - \mathbf{v}) \frac{dM}{dt}$$

$$= \mathbf{F}^e + \mathbf{v}_r \frac{dM}{dt}$$

(9–15)

where $\mathbf{v}_r = \mathbf{u} - \mathbf{v}$ is the relative velocity of the exhaust gas with respect to the rocket. Equation 9–15 is often called the *rocket equation*. The term $\mathbf{\Omega} = \mathbf{v}_r(dM/dt)$ is referred to as the *thrust* of the rocket engine. Note that $\mathbf{v}_r$, the velocity of the ejected gases relative to the rocket, is to the *left* in Fig. 9–6; because $dM/dt < 0$, the thrust gives a contribution to $M(d\mathbf{v}/dt)$ that is to the *right*, thereby producing a positive acceleration. To achieve large accelerations, design engineers must deal with ejection speeds v_r that are very high and with combustion chambers and nozzles that are capable of large throughput.

Example 9–8

A spacecraft is stationary in deep space when its rocket engine is ignited for a 100-s "burn." Hot gases are ejected at a constant rate of 150 kg/s with a velocity relative to the spacecraft of 3000 m/s. The initial mass of the spacecraft (plus fuel and oxidizer) is 25,000 kg.

(a) Determine the thrust of the rocket engine and the initial acceleration in units of g.

(b) What is the final velocity of the spacecraft?

Solution:

(a) We take the direction of motion of the spacecraft to be positive. The thrust is

$$\Omega = v_r \frac{dM}{dt} = (-3000 \text{ m/s})(-150 \text{ kg/s}) = 450,000 \text{ N}$$

From Eq. 9–15, the initial acceleration is

$$a = \frac{dv}{dt} = \frac{v_r}{M} \frac{dM}{dt} = \frac{\Omega}{M}$$

$$= \frac{450,000 \text{ N}}{25,000 \text{ kg}} \times \frac{g}{9.80 \text{ m/s}^2} = 1.84g$$

(b) We use Eq. 9–15 to write

$$M \frac{d\mathbf{v}}{dt} = \mathbf{v}_r \frac{dM}{dt}$$

or

$$\frac{dM}{M} = \frac{dv}{v_r}$$

Direct integration gives

$$\int_{M_i}^{M_f} \frac{dM}{M} = \frac{1}{v_r} \int_0^v dv$$

or

$$v = -v_r \ln (M_i/M_f)$$

Substituting numerical values for the 100-s burn, $M_i = 25,000$ kg, $M_f = M_i - (150 \text{ kg/s})(100 \text{ s}) = 10,000$ kg, and $v_r = -3000$ m/s, we find

$$v = -(-3000 \text{ m/s}) \ln \frac{25,000}{10,000} = 2749 \text{ m/s} \quad (6150 \text{ mi/h})$$

Notice that this ratio of initial to final mass, $M_i/M_f = 2.5$, produces a final velocity v that is nearly equal to the exhaust velocity. To exceed the exhaust velocity by any substantial factor requires that the ratio M_i/M_f be much greater; that is, the payload is a small fraction of the entire rocket mass.

QUESTIONS

9–1 Describe and explain briefly what is meant by the terms (a) linear momentum conservation, (b) impulse, (c) elastic and inelastic collisions, and (d) the center of mass.

9–2 Give several examples from everyday experience of impulsive forces. Make a rough estimate of the magnitude of the impulse involved in each case.

9–3 Explain the bone-shattering power of a karate blow in combat.

9–4 Explain how "impact-absorbing" safety devices such as a car bumper work.

9–5 In the egocentric view of the world considered to consist of only you and the Earth, explain the observed behavior of the system when you jump into the air. Discuss carefully the role of the system center of mass, the relative masses involved, and the conservation of linear momentum.

9–6 Explain how, in jumping down from a substantial height, you can reduce the average force acting on you during impact with the ground by bending your knees during the initial contact period. Your discussion will be aided by considering the motion of your center of mass.

9–7 Does the location of the center of mass of a rigid body have to be an interior point? Explain.

9–8 Discuss the extent to which conservation of linear momentum is applicable to (a) an object following a ballistic trajectory, (b) two masses joined together by a spring, in free fall, and (c) the firing of a railroad-mounted cannon supported on a frictionless track.

9–9 If you were stranded and unable to move on the completely frictionless surface of a frozen lake, what use could you make of your boots to reach shore?

9–10 In the discussion of the momentum conservation of a system of particles leading to Eq. 9–14, the effects of the two-body forces acting between the particles were required to cancel in pairs. Consider a system of three interacting particles and write out explicitly all the two-body forces acting. Demonstrate with the aid of Newton's third law the required exact pairwise cancellation. Derive the equation replacing Eq. 9–14 if the third law did not apply to one pair of interactions.

9–11 An isolated system of particles has its momentum conserved but not necessarily its mechanical energy. Explain by giving examples of how this might occur.

9–12 Explain how rocket propulsion works in deep space.

9–13 Contrast the operation of rocket propulsion with that of jet propulsion.

PROBLEMS

Section 9–1

9–1 Calculate the momentum in units of kg·m/s for (a) a 2,500-kg truck traveling at a speed of 100 km/hr; (b) a 100-kg fullback running at a speed of 50 km/hr; (c) a 12-g rifle bullet traveling at a speed of 400 m/s; (d) a 200-g stone with a kinetic energy of 40 J.

9–2 A particle with initial momentum $\mathbf{p}_0 = 30\hat{\mathbf{i}}$ kg·m/s is subjected to a constant force $\mathbf{F} = 20\hat{\mathbf{j}}$ N for a 2-s duration. What is the final momentum?

9–3 A large raindrop with a radius of 2 mm and a terminal speed of 30 m/s strikes the bald head of a professor. If it takes 2 ms for the drop to break into a low-velocity spray, what average force was exerted on the professor's head? What mass sitting on top of his head would exert the same force?

9–4 A garden hose is held in a manner producing a right-angle curve near the nozzle, as shown in the diagram at right. What force is necessary to hold the nozzle stationary if the discharge rate is 0.60 kg/s with a velocity of 25 m/s?

9–5 Sand is pouring onto a spring scale from a vertical height
 • of 1 m at a rate of 0.5 kg/s. The maximum pile of sand the scale pan can support is 1 kg, with excess sand spilling over the side of the pan with negligible velocity, as

Problem 9–4

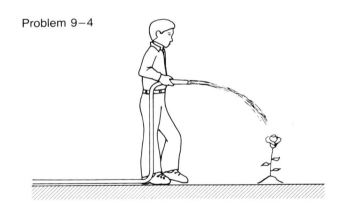

shown in the diagram at the top of the opposite page. When the equilibrium condition has been reached, what is the weight (force) reading of the scale?

9–6 The proper application of Eq. 9–5 requires that $\mathbf{F}(t)$ be the resultant of all the forces acting on the object. If the object in Example 9–1 is launched vertically upward by the stated applied force, determine the error in the acquired velocity at the end of 0.07 s resulting from ignoring the gravitational force.

Problem 9-5

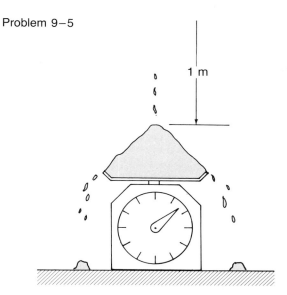

1 m

9-7 A ball with a mass of 0.15 kg is thrown directly against a wall with a velocity of 18 m/s, and it rebounds in the reverse direction with practically the same speed. If the ball was in contact with the wall for 1.5 ms, what average force did the wall exert on the ball?

9-8 A standard 2-ounce tennis ball (mass = 57 g) acquires a speed of 40 m/s during a volley. The racket is in contact with the ball for 5 ms during a horizontal stroke. (a) What impulse is imparted to the ball? (b) What is the average force between ball and racket?

9-9 A small ball is dropped from a height h onto a horizontal
• floor and is observed to rebound (on first bounce) to a maximum height ηh, with $0 \le \eta \le 1$. Show that the ratio of the total travel time from the instant of release back to the rebound height ηh, to the impulse time interval for an average impulsive force $\bar{F}$, is simply $\bar{F}/mg$.

Section 9-2

9-10 A block of wood with a mass of 0.6 kg is balanced on top of a vertical post 2 m high. A 10-g bullet is fired horizontally into the block. If the block and embedded bullet land at a point 4 m from the base of the post, find the initial velocity of the bullet.

9-11 Two blocks, with masses m_1 = 2 kg and m_2 = 3 kg, collide head-on while sliding on a frictionless horizontal surface. Initially their velocities are v_1 = 5 m/s and v_2 = 4 m/s, respectively. On collision they stick together and subsequently move as a single block. (a) What is the velocity of the combination after collision? (b) How much kinetic energy is lost in the collision? What happens to this energy?

9-12 A block with a mass of 2 kg rests on a horizontal frictionless surface and is attached to a wall with a horizontal ideal inertia-less spring under no initial compression. (The spring constant is κ = 250 N/m.) A blob of putty having a mass of 0.5 kg and a horizontally directed velocity **v** collides with the block and sticks to it, as shown in the diagram. Determine v if the spring undergoes a maximum compression of 10 cm.

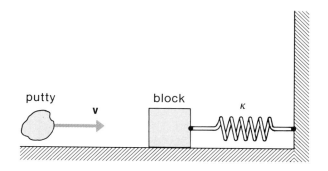

putty   block κ
          v

9-13 An automobile with a mass of 1500 kg is traveling due north at 90 km/h when it is struck by another automobile with a mass of 2000 kg and traveling due east at 60 km/h. If both cars lock together as a unit, find the resulting velocity (speed and direction) just after collision.

9-14 A ball with a mass m is attached to the end of a string with
• a length L = 50 cm and is released from a horizontal position, as shown in the diagram. At the bottom of its swing the ball strikes a block with a mass M = 2m resting on a frictionless table. Assume that the collision is perfectly elastic. (a) To what height does the ball rebound? (b) What is the velocity of the struck block after impact?

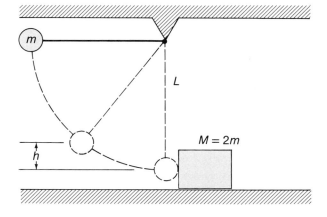

m

L

$M = 2m$

h

9-15 Two identical balls, each with mass m, undergo a colli-
•• sion. Initially, one ball is stationary and the other has a kinetic energy of 8 J. The collision is inelastic, and 2 J of

energy is converted to heat. What is the maximum deflection angle (α or β in the diagram) at which one of the balls may be observed?

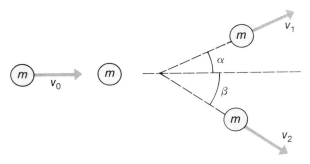

9–16 A 45-kg girl is standing on a plank that has a mass of
 • 150 kg. The plank, originally at rest, is free to slide on a frozen lake, which is a flat, frictionless supporting surface. The girl begins to walk at a constant velocity of 1.5 m/s *relative to the plank*. (a) What is her velocity relative to the ice surface? (b) What is the velocity of the plank relative to the ice surface?

9–17 In a collision process between two objects, the ratio of the relative velocity after collision to the relative velocity before collision is called the *coefficient of restitution* ϵ. A perfectly elastic collision has $\epsilon = 1$ and a completely inelastic process has $\epsilon = 0$; ordinary inelastic collisions have intermediate values of ϵ. If a ball is dropped from a height h onto a solid floor and rebounds to a height h', show that $h' = \epsilon^2 h$.

9–18 A particle with a mass m collides elastically with a larger object whose mass is M. The particle rebounds along its original line of motion. If the massive object is initially at rest, what fraction of the particle's original kinetic energy is transferred to the object? Discuss the result (a) for the case of a golf ball incident on a bowling ball ($M/m \cong 500$) and (b) for a 100-kg block dropped from a height of 10 m to the Earth's surface.

Section 9–3

9–19 Find the position of the C.M. of five equal-mass particles located at the five corners of a square-based right pyramid with sides of length ℓ and altitude h.

9–20 The mass of the Sun is 329,390 Earth masses, and the mean distance from the center of the Sun to the center of the Earth is 1.496×10^8 km. Treating the Earth and Sun as particles, with each mass concentrated at the respective geometric center, how far from the center of the Sun is the C.M. of the Earth-Sun system? Compare this distance with the mean radius of the Sun (6.960×10^5 km).

9–21 Determine the location of the center of mass of a thin lamina in the shape of an equilateral triangle with the length of its sides equal to a by (a) invoking the special geometric symmetry of the lamina and (b) using the result found in Example 9–8.

Section 9–4

9–22 A bomb is at rest with respect to an inertial reference frame in deep space. The bomb explodes, breaking into three pieces. Two pieces with equal mass fly off at equal speeds of 50 m/s in directions perpendicular to one another. What is the velocity of the third piece relative to the other two if it has three times the mass of either of the other pieces?

9–23 Consider two particles, $m_1 = 1$ kg and $m_2 = 2$ kg, the coordinates of which are known functions of time in a particular inertial frame: $\mathbf{r}_1 = 3t^2\hat{\mathbf{i}} + 2t\hat{\mathbf{j}}$ and $\mathbf{r}_2 = 3\hat{\mathbf{i}} + (2t^3 + 5t)\hat{\mathbf{j}} + 9\hat{\mathbf{k}}$, where the position vectors are given in meters for t in seconds. (a) What is the C.M. coordinate $\mathbf{R}$? (b) What is the C.M. velocity $d\mathbf{R}/dt$? (c) What is the C.M. acceleration $d^2\mathbf{R}/dt^2$?

9–24 A shell is fired with a uniform speed of 10 m/s at a stationary target 200 m away, in deep space. Midway to the target the shell explodes prematurely into two equal fragments. One fragment strikes the target 8 s after the explosion. (a) Where is the second fragment at the moment the first fragment strikes the target? (b) Will the second fragment also strike the target? If so, when?

9–25 A cannon fires an explosive shell for maximum range
 • with a muzzle velocity of 500 m/s. (See Section 4–4.) At the top of its trajectory the shell detonates prematurely and explodes into two fragments with equal masses. One fragment is observed to drop *from rest* to the ground directly below the explosion. (a) Make a sketch of the paths followed by the fragments and by the C.M. (b) Where does the second fragment land? (c) How long after the explosion does each fragment strike the ground (assumed to be level)?

9–26 A container is dropped from a balloon that drifts toward
 • the east with a uniform velocity of 50 m/s, holding a constant altitude of 2.5 km above the level ground. At some point in its descent, explosive bolts separate the container into two pieces, one with twice the mass of the other. Twenty seconds after the container is dropped, the lighter piece lands at a spot 200 m due south of a point directly under the balloon at the time when the piece impacts. Determine the position of the heavier piece when the lighter piece strikes the ground.

9–27 Two particles with masses $m_1 = 3$ kg and $m_2 = 2$ kg
 •• slide as a unit with a common velocity of 2 m/s on a level frictionless surface. Between them is a compressed massless spring with $\kappa = 50$ N/m. The spring, originally

compressed by 25 cm, is suddenly released, sending the two masses, which are not connected to the spring, flying apart from each other. The orientation of the spring with respect to the initial velocity is shown in the diagram. (a) What is the relative velocity of separation after the particles lose contact? (b) What is the velocity of the C.M. after separation? (c) What are the speeds of m_1 and m_2 with respect to the frictionless surface after separation?

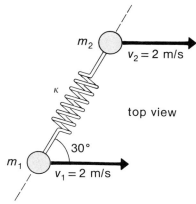

top view

Section 9–5

9–28 A rocket with a total mass of 160,000 kg sits vertically on a launch pad. At what rate must gases be ejected on ignition if the thrust is to just overcome the gravitational force on the rocket? Assume a velocity for the ejected gases of 3000 m/s. What rate is required on ignition if an initial acceleration of 2g is to be achieved? What thrust does the rocket engine deliver in these two cases?

9–29 A rocket with an initial total mass M_i is launched vertically from the Earth's surface. When the launch fuel has been completely burned, the rocket has reached an altitude that is small compared with the Earth's radius (so that the acceleration due to gravity may be considered

constant during the burn). Show that the final velocity is

$$v = -v_r \ln(M_i/M_f) - gt_b$$

where the time of burn is

$$t_b = (M_i - M_f)(dm/dt)^{-1}$$

(M_f is the final total mass of the rocket, v_r is the exhaust-gas velocity, and dm/dt is the constant rate of fuel consumption.)

9–30 A Titan II rocket has a total mass of 145 metric tons, and the first-stage rocket engines deliver a total thrust of 1.92×10^6 N. The fuel-oxidizer consists of kerosene and liquid oxygen and gives an exhaust velocity of 2400 m/s. If the first-stage burn time is 150 s in a vertical ascent, calculate (a) the initial acceleration, (b) the rate at which gas is ejected, for constant thrust, and (c) the velocity reached at the end of first-stage burn (use the results of Problem 9–29).

9–31 A rocket for use in deep space is to have the capability of boosting a payload (plus the rocket frame and engine) of 3.0 metric tons to an achieved speed of 10,000 m/s with an engine and fuel designed to produce an exhaust velocity of 2000 m/s. (a) How much fuel and oxidizer is required? (b) If a different fuel and engine design could give an exhaust velocity of 5000 m/s, what amount of fuel and oxidizer would be required for the same task? Comment.

9–32 An aircraft uses a rocket-assisted takeoff unit (RATO) that is capable of discharging 50 kg of gas in 10 s with an exhaust velocity of 1200 m/s. The 3000-kg aircraft, on ordinary takeoff without RATO, requires a distance of 400 m to reach a lift-off speed of 200 km/h with a constant propeller thrust. (a) What is the thrust developed by the rocket? (b) With both the RATO and the normal propeller thrust providing a constant acceleration, what is the distance required for lift-off?

Additional Problems

9–33 A horizontal stream of water 10 cm in diameter strikes a stationary vertical wall and breaks into a low-velocity spray. If the discharge rate against the wall is 10^3 liters/min, what is the force of the stream against the wall?

9–34 A golf ball ($m = 46$ g) is struck a blow that makes an angle of 45° with the horizontal. The drive lands 200 m away on a flat fairway. If the golf club and ball are in contact for a time of 7 ms, what is the average force of impact? (Neglect air-resistance effects.)

9–35 A hunter fires a shotgun that has a mass of 4 kg, discharging lead shot having a total mass of 35 g. The lightly held gun recoils with an initial speed of 2.9 m/s and is brought

to rest against the hunter's shoulder in 0.07 s, a long time compared with the time required for the shot to leave the gun barrel. (a) What is the average force against the hunter's shoulder? (b) What is the muzzle velocity of the shot? (c) If the length of the gun barrel is 75 cm and if the shot travels down the barrel with constant acceleration, how long does it take the shot to clear the barrel? Is this time short compared with 0.07 s, as claimed?

9–36 A small tin can has a mass of 250 g and is suspended from a string 2 m long. The can is observed to swing through a maximum angle of 10° when a bullet with a mass of 10 g is shot through the can. If the bullet enters the can horizontally with a velocity of 350 m/s, calculate its velocity upon emerging from the can.

9–37 An attack helicopter is equipped with a 20-mm cannon that fires 130-g shells in the forward direction with a muzzle velocity of 800 m/s. The fully loaded helicopter has a mass of 4000 kg. A burst of 160 shells is fired in a 4-s interval. What is the resulting average force on the helicopter and by what amount is its forward speed reduced? (Assume that the engine and rotors are not applying any force that would accelerate the helicopter horizontally.)

9–38 A loaded howitzer is rigidly bolted to a railroad flatcar. The mass of the combination is 20,000 kg. The flatcar is observed to be coasting along a level, straight (frictionless) track with a constant speed V. The howitzer is set to an elevation angle of 60° with the horizontal and then fired. After the howitzer is fired, the flatcar and attached howitzer are seen to recoil, moving in the opposite direction along the track with the same speed V. (a) If the muzzle velocity of the howitzer shell is 300 m/s and the shell has a mass of 50 kg, find V. (b) How high would this shell rise in the air?

9–39 A 75-kg man is standing in a 225-kg rowboat and is originally 10 m from a pier on a still lake. The man walks towards the pier a distance of 2 m relative to the boat and then stops. Assuming that the boat can move through the water without resistance, how far is the man now from the pier?

9–40 A horizontal force of 0.80 N is required to pull a 5-kg block across a tabletop at constant speed. With the block initially at rest, a 20-g bullet fired horizontally into the block causes the block to slide 1.5 m before coming to rest again. Determine the speed v of the bullet. Assume the bullet to be embedded in the block.

9–41 Two blocks rest on a frictionless surface with a compressed spring between them, as shown in the diagram. Block A has a mass of 1 kg and block B has a mass of 2 kg. The spring is initially compressed by 50 cm and has negligible mass. When both blocks are released simultaneously and the spring has dropped to the surface, block A is found to have a velocity of 2 m/s. (a) What is the velocity of block B? (b) How much energy was stored in the spring? (c) What fraction of the energy did each block receive? Why was it not divided evenly between the blocks? (d) What is the spring constant κ of the spring?

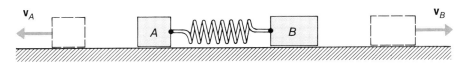

9–42 A puck with mass $M = 0.2$ kg is placed on a rough table at the midpoint between two identical springs with spring constant $\kappa = 200$ N/m. The springs are separated by a distance of 1 m. The coefficient of kinetic friction between the puck and the table is $\mu_k = 0.20$. The puck is set into motion by an impulse of 1 N·s in the direction of one of the springs. (a) Find the initial velocity and the initial kinetic energy of the puck. What frictional force acts on the puck? (b) What is the kinetic energy of the puck when it reaches the first spring? By how much is this spring compressed when the puck is momentarily at rest? (Ignore frictional forces over the distance of compression of the spring.) (c) With what kinetic energy does the puck arrive at the second spring? What will be the compression of this spring? (d) The puck runs back and forth between the two springs until it comes to rest as the result of the frictional forces. What total distance does the puck move before coming finally to rest?

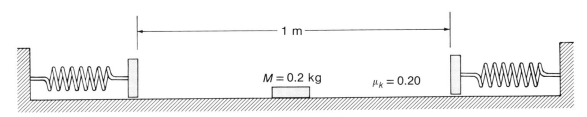

9-43 Two railway cars, m_1 and m_2, are moving along a track with velocities v_1 and v_2, respectively. The cars collide, and after the collision the velocities are v_1' and v_2'. Show that the change in kinetic energy, $K' - K$, will be maximum if the cars couple together. [*Hint:* Set

$$\frac{d(K' - K)}{dv_1'} = 0$$

and show that $v_1' = v_2'$.]

9-44 A particle with a mass m has an initial velocity v_0 and
• collides head-on with a block that has a mass M and is initially at rest. The collision is inelastic and an amount of energy E is transferred from the particle to the block. Derive an expression in terms of m, M, and E, for the minimum value of v_0 for which the process is possible. [*Hint:* Derive an expression for the velocity of the block; this velocity must be *real*.]

9-45 A particle with a mass m_1 is incident with a velocity $\mathbf{v}_1$ on
• a stationary particle with a mass $m_2 = 4m_1$. The incident particle is scattered at an angle $\theta_1 = 65°$, and the struck particle recoils at an angle $\theta_2 = 55°$. Determine the velocity ratios, v_1'/v_2' and v_1'/v_1. Compare the initial kinetic energy with the final kinetic energy and determine whether the collision is elastic.

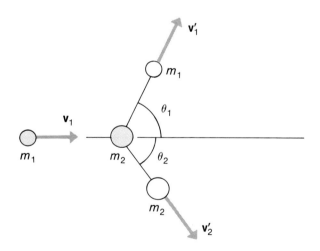

9-46 Refer to the diagram for Problem 9-45 and assume that
• $m_1 > m_2$. For an elastic collision, show that the maximum scattering angle θ_{1m} is obtained from $\cos^2 \theta_{1m} = 1 - m_2^2/m_1^2$.

9-47 A ball is dropped at $t = 0$ from a height $y_0 = 12$ m above
• a surface. At the instant of release a very massive platform is at a height $h = 4$ m above the same surface and is moving upward with a constant velocity $v = 3$ m/s. The ball rebounds elastically from the moving platform. (a) To what height does the ball rise? (b) At what time does the ball strike the platform for the second time?

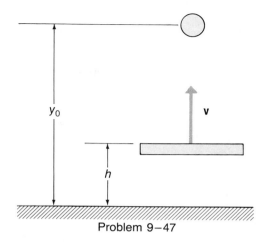

Problem 9-47

9-48 At $t = 0$, three particles are located and are moving as follows: $m_1 = 2$ kg: $(4, 2)$, $\mathbf{v}_1 = (3 \text{ m/s})\hat{\mathbf{j}}$; $m_2 = 3$ kg: $(-3, 2)$, $\mathbf{v}_2 = (-2 \text{ m/s})\hat{\mathbf{i}}$; and $m_3 = 5$ kg: $(2, -2)$, $\mathbf{v}_3 = (-1 \text{ m/s})\hat{\mathbf{j}}$. Determine the position of the C.M. of this system at $t = 0$, 1, 2, and 3 s. Make a graph of the motion of the C.M. What force acts on the system?

9-49 A shuffleboard puck strikes two identical pucks that are
• initially in contact and at rest on a frictionless horizontal surface. After impact, these two pucks travel with equal speeds that are each one third of the new speed of the incident puck, which moves in the same direction as before the collision. The struck pucks move in directions making equal angles of 60° with respect to the direction of motion of the incident puck. (a) Determine the speeds v_1', v_2', and v_3' in terms of v_1. (b) Calculate the kinetic energy of the system before impact and after impact. Is the collision elastic? (c) What is the velocity of the C.M.? (d) The diagram shows only the relative orientation of the velocity vectors. Make a drawing to some convenient scale that shows the positions of the three pucks and the C.M. with respect to the impact point at some time after the collision.

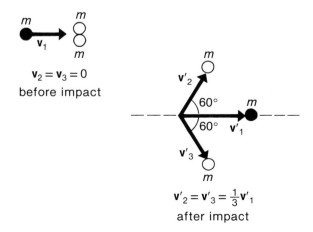

9–50 A two-stage rocket has the following specifications:

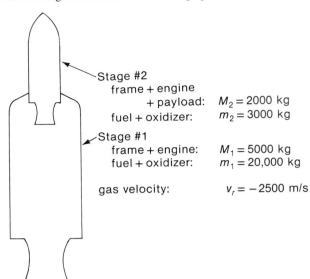

Stage #2
    frame + engine
       + payload: $M_2 = 2000$ kg
    fuel + oxidizer: $m_2 = 3000$ kg

Stage #1
    frame + engine: $M_1 = 5000$ kg
    fuel + oxidizer: $m_1 = 20{,}000$ kg

    gas velocity: $v_r = -2500$ m/s

The first stage of this rocket is fired while it is in deep space (initially at rest in some inertial frame). When this burn is completed and prior to igniting the second stage, the frame and engine of the first stage are jettisoned (at essentially zero separation velocity). (a) Show that at the end of the second-stage burn, a total achieved velocity v is

$$v = -v_r \, \ln \frac{(M_1 + M_2 + m_1 + m_2)(M_2 + m_2)}{(M_1 + M_2 + m_2)M_2}$$

(b) What is the numerical value of v? (c) What would be the achieved velocity of a rocket that has only a single stage with the same total mass, $M_1 + M_2 + m_1 + m_2$, and burns all the fuel, $m_1 + m_2$, in a single engine with the same gas velocity? (d) Recalculate (c) but subtract the 1000-kg mass of the redundant second-stage engine and accessories. Comment on the results of (b), (c), and (d).

CHAPTER 10

ROTATION AND ANGULAR MOMENTUM

In this chapter we introduce a new dynamic quantity called *angular momentum*, a quantity that obeys a conservation principle similar to that for linear momentum. The angular momentum concept is useful in the discussion of the dynamics of a single particle and systems of particles. In Chapter 11 we apply this idea to the motion of rigid bodies. The discussion of rotational dynamics is greatly simplified by introducing a new type of vector multiplication called the *cross product*.

10–1 THE VECTOR REPRESENTATION OF ROTATION

In the discussions of circular motion in earlier chapters we have considered the angular frequency ω to be a simple algebraic quantity. We may, in fact, define a *vector* counterpart $\boldsymbol{\omega}$ which is referred to as the *angular velocity*. The vector $\boldsymbol{\omega}$ is taken to have the magnitude of the angular frequency and a direction *perpendicular to the plane containing the circular path of a particle*. We use a right-hand rule to resolve the ambiguity of the two possible perpendicular directions. When the fingers of the right hand are curled in the direction of the (tangential) velocity, the thumb points in the direction of $\boldsymbol{\omega}$, as shown in Fig. 10–1. We usually draw the vector $\boldsymbol{\omega}$ along the axis of rotation.

The relationship connecting the quantities v, ω, and $|\mathbf{r}| = R$ is given by Eq. 4–25 (namely, $v = \omega R$). How do we write an equation that connects the *vector* quantities $\mathbf{v}$, $\boldsymbol{\omega}$, and $\mathbf{r}$? The angular velocity $\boldsymbol{\omega}$ is defined to have a direction perpendicular to the plane that contains the vector $\mathbf{v}$. Consequently, some new type of vector relationship is required to relate $\mathbf{v}$, $\boldsymbol{\omega}$, and $\mathbf{r}$. This new prescription involves a special kind of vector multiplication.

The Vector Product of Two Vectors. We now define a way to multiply two vectors so that a *vector* is produced. We call this type of multiplication the *vector product* or the *cross product*.* The vector product $\mathbf{C}$ of two vectors $\mathbf{A}$ and $\mathbf{B}$ is written as

$$\mathbf{C} = \mathbf{A} \times \mathbf{B}$$

The magnitude of $\mathbf{C}$ is defined to be

$$|\mathbf{C}| = |\mathbf{A}||\mathbf{B}| \sin \theta_{AB}$$

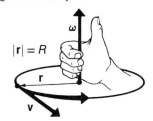

Fig. 10–1. Illustrating the right-hand-rule convention for determining the direction of the angular velocity vector $\boldsymbol{\omega}$ with respect to the position vector $\mathbf{r}$ and the tangential velocity $\mathbf{v}$.

---

*In Section 7–1 we defined a different type of vector multiplication that produces a *scalar* result called the *scalar* or the *dot product*.

183

or
$$C = AB \sin \theta_{AB} \qquad \textbf{(10–1)}$$

where θ_{AB} is the angle between the vectors **A** and **B** ($0° \leq \theta_{AB} \leq 180°$). The direction of **C** is defined to be *perpendicular* to a plane formed by **A** and **B**.* Because the direction of the perpendicular to a plane is ambiguous, we specify the direction by the *right-hand-rule* convention (as in Fig. 10–1). The perpendicular to the plane containing **A** and **B** is taken in the sense of advance of a right-handed screw rotated from the first vector **A** to the second vector **B** through the smaller angle between the vector directions (i.e., the angle that is less than 180°), as shown in Fig. 10–2a. Alternatively, if the right hand is held as in Fig. 10–2b, with the fingers curling in the direction of rotation carrying **A** into **B** through the smaller angle θ_{AB}, the thumb then points in the direction of **C**. (• Is it really necessary to specify that the rotation is through the *smaller* angle θ_{AB}? Will you obtain the correct direction for **C** if you rotate from **A** to **B** through the larger angle? Remember the significance of a vector with a negative sign.) Notice also that **A** × **A** = 0. (• Why is this so?)

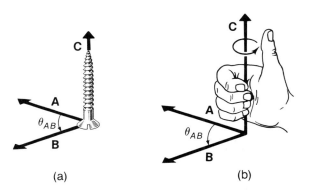

Fig. 10–2. The right-hand-rule convention for the vector product (cross product) **C** = **A** × **B**.

(a)   (b)

The cross product is not a commutative operation because

$$\textbf{A} \times \textbf{B} = -\textbf{B} \times \textbf{A} \qquad \textbf{(10–2)}$$

However, cross products obey the distributive law of multiplication:

$$\textbf{A} \times (\textbf{B} + \textbf{C}) = \textbf{A} \times \textbf{B} + \textbf{A} \times \textbf{C} \qquad \textbf{(10–3)}$$

A Cartesian coordinate system is said to be *right-handed* if the rotation of the *x*-axis toward the *y*-axis produces a right-hand-rule direction that corresponds to the direction of the *z*-axis. Then, in a right-handed system, the unit vectors obey the relationships†

$$\left. \begin{array}{l} \hat{\textbf{i}} \times \hat{\textbf{j}} = \hat{\textbf{k}}, \quad \hat{\textbf{j}} \times \hat{\textbf{k}} = \hat{\textbf{i}}, \quad \hat{\textbf{k}} \times \hat{\textbf{i}} = \hat{\textbf{j}} \\ \hat{\textbf{i}} \times \hat{\textbf{i}} = \hat{\textbf{j}} \times \hat{\textbf{j}} = \hat{\textbf{k}} \times \hat{\textbf{k}} = 0 \end{array} \right\} \qquad \textbf{(10–4)}$$

If we write **A** and **B** in Cartesian component form, we obtain

---

*If **A** and **B** are collinear, no plane is defined. But then $\theta_{AB} = 0$, so that $C = 0$ automatically.

†Much grief will be avoided if a right-handed coordinate frame is *always* used, for only then are Eqs. 10–4 valid.

$$\mathbf{A} \times \mathbf{B} = (A_x\hat{\mathbf{i}} + A_y\hat{\mathbf{j}} + A_z\hat{\mathbf{k}}) \times (B_x\hat{\mathbf{i}} + B_y\hat{\mathbf{j}} + B_z\hat{\mathbf{k}})$$
$$= (A_yB_z - A_zB_y)\hat{\mathbf{i}} + (A_zB_x - A_xB_z)\hat{\mathbf{j}} + (A_xB_y - A_yB_x)\hat{\mathbf{k}} \qquad \textbf{(10-5)}$$

▶ Perhaps the easiest way to reproduce Eq. 10–5 is by formal evaluation of the determinant

$$\mathbf{C} = \mathbf{A} \times \mathbf{B} = \begin{vmatrix} \hat{\mathbf{i}} & \hat{\mathbf{j}} & \hat{\mathbf{k}} \\ A_x & A_y & A_z \\ B_x & B_y & B_z \end{vmatrix} \qquad \textbf{(10-6)}$$

Developing this determinant by the first row* reproduces Eq. 10–5.

Example 10–1

Evaluate the cross product $\mathbf{A} \times \mathbf{B}$ with $\mathbf{A} = 3\hat{\mathbf{i}} + 7\hat{\mathbf{j}} - \hat{\mathbf{k}}$ and $\mathbf{B} = \hat{\mathbf{i}} - \hat{\mathbf{j}}$.

Solution:

We have

$$\mathbf{C} = \mathbf{A} \times \mathbf{B} = \begin{vmatrix} \hat{\mathbf{i}} & \hat{\mathbf{j}} & \hat{\mathbf{k}} \\ +3 & +7 & -1 \\ +1 & -1 & 0 \end{vmatrix}$$

or

$$\mathbf{C} = \hat{\mathbf{i}}[(+7)(0) - (-1)(-1)]$$
$$- \hat{\mathbf{j}}[(+3)(0) - (-1)(+1)]$$
$$+ \hat{\mathbf{k}}[(+3)(-1) - (+7)(+1)]$$
$$\mathbf{C} = -\hat{\mathbf{i}} - \hat{\mathbf{j}} - 10\hat{\mathbf{k}} \qquad ◀$$

The Vector Product for Circular Motion. We can use the definition of the vector product to write the velocity $\mathbf{v}$ as

$$\mathbf{v} = \boldsymbol{\omega} \times \mathbf{r} \qquad \textbf{(10-7)}$$

To see that this equation expresses the correct properties of $\mathbf{v}$ and $\boldsymbol{\omega}$, refer to Fig. 10–3. The description of the motion can be made with an arbitrarily selected origin. In Fig. 10–3a, the origin O is selected on the rotation axis and in the plane of rotation. In this case, $\boldsymbol{\omega}$ and $\mathbf{r}$ are perendicular; $|\mathbf{r}| = R$, so $v = R\omega$, and the direction of $\mathbf{v}$ is perpendicular to the plane containing $\boldsymbol{\omega}$ (the axis) and $\mathbf{r}$. Consequently, $\mathbf{v}$ is tangential to the path, as required.

In Fig. 10–3b, the origin is on the axis but at an arbitrary point O'. In this case, $\boldsymbol{\omega}$ and $\mathbf{r}$ have an included angle γ; thus, $v = r\omega \sin \gamma$. But $r \sin \gamma = \rho = R$; therefore, we

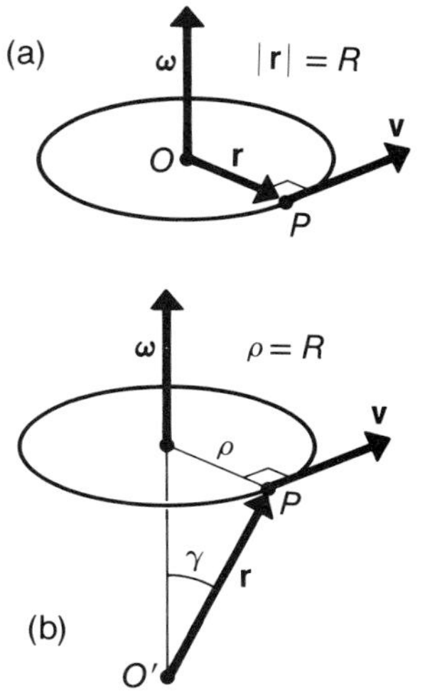

Fig. 10–3. The calculation of $\mathbf{v} = \boldsymbol{\omega} \times \mathbf{r}$ is the same, irrespective of the point on the rotation axis selected for the origin of the position vector $\mathbf{r}$. In (a), $v = R\omega$. In (b), $v = r\omega \sin \gamma = R\omega$.

(a)

(b)

---

*Developing a determinant by its first row means

$$\begin{vmatrix} a & b & c \\ d & e & f \\ g & h & i \end{vmatrix} = a\begin{vmatrix} e & f \\ h & i \end{vmatrix} - b\begin{vmatrix} d & f \\ g & i \end{vmatrix} + c\begin{vmatrix} d & e \\ g & h \end{vmatrix}$$

Then,

$$\begin{vmatrix} e & f \\ h & i \end{vmatrix} = ei - hf, \quad \text{and so forth.}$$

again have $v = R\omega$. The direction of **v** is seen to be the same as in Fig. 10–3a. (• Why is this so?)

For circular motion, we can also define an *angular acceleration vector* $\boldsymbol{\alpha}$, which is equal in magnitude to α; the direction of $\boldsymbol{\alpha}$ is parallel to $\boldsymbol{\omega}$ if α increases $|\boldsymbol{\omega}|$ and is antiparallel to $\boldsymbol{\omega}$ if α decreases $|\boldsymbol{\omega}|$.

10–2 ANGULAR MOMENTUM FOR A PARTICLE

In Chapter 9 we found that *linear momentum* is an extremely useful concept because of the conservation principle associated with it. We now introduce a quantity called *angular momentum* (or sometimes the *moment of momentum*). The importance and usefulness of angular momentum lie in the fact that it, too, has an associated conservation principle.

For a single particle, we define the angular momentum ℓ about the origin O of a reference frame to be

$$\boldsymbol{\ell} = \mathbf{r} \times \mathbf{p} = m\mathbf{r} \times \mathbf{v} \qquad (10-8)$$

where **r** is the position vector of the particle with respect to O and $\mathbf{p} = m\mathbf{v}$ is the linear momentum of the particle at the same instant. See Fig. 10–4.

The definition of the cross product, as prescribed in Section 4–6, requires the vector $\boldsymbol{\ell}$ to be perpendicular to the plane containing **r** and **p** and to have a direction given by the right-hand rule. Furthermore, the magnitude of $\boldsymbol{\ell}$ is*

$$|\boldsymbol{\ell}| = \ell = |\mathbf{r}||\mathbf{p}| \sin \theta = rp \sin \theta \qquad (10-9)$$

where θ is the smaller angle between the positive directions of **r** and **p**.

For many purposes, it is convenient to express the angular momentum of a particle in terms of its linear momentum components. In Fig. 10–5, **r** and **p** lie in the plane of the

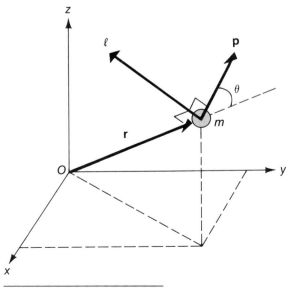

Fig. 10–4. Illustrating the definition of the angular momentum ℓ of a particle about the origin O. With $\boldsymbol{\ell} = \mathbf{r} \times \mathbf{p} = m\mathbf{r} \times \mathbf{v}$, we see that the vector $\boldsymbol{\ell}$ is perpendicular to the plane containing **r** and **p**. If θ is the angle between the positive directions of **r** and **p** (measured in the plane containing these vectors), we also have $|\boldsymbol{\ell}| = |\mathbf{r}||\mathbf{p}| \sin \theta$.

*Remember that, by definition, $0 \leqslant \theta \leqslant \pi$; therefore $\sin \theta \geqslant 0$.

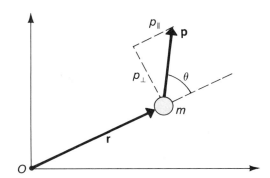

Fig. 10–5. Illustrating the components of the linear momentum **p**: $p_\parallel$ is parallel to **r** and $p_\perp$ is perpendicular to **r**. Both **r** and **p** lie in the plane of the page.

page, and we have

$$P_\perp = p \sin \theta$$

Using Eq. 10–9, we can now write

$$\ell = rp_\perp = mrv_\perp$$

Thus, the angular momentum of the particle has a magnitude equal to the product of its distance r from the origin and the transverse or perpendicular component of its linear momentum. The parallel component $p_\parallel$ makes no contribution to the angular momentum. Notice that by employing polar coordinates, we can also write

$$\ell = mrv_\phi$$

where $v_\phi = v_\perp$ is the azimuthal velocity component (see Eq. 3–17).

It is evident that the dimensions of angular momentum are

$$[\ell] = [r][m][v] = L \cdot M \cdot (L/T) = L^2 M T^{-1}$$

so the units can be expressed as kg·m²/s or as Joule-seconds (J·s).

In the important case of circular motion, with $|\mathbf{r}| = R = $ constant, the velocity **v** is perpendicular to the position vector **r**. Then, the tangential velocity, $v_t = v_\perp = v$, is equal to the azimuthal velocity v_ϕ, so that $\ell = mRv$. But we also have $v = R\omega$; thus,

$$\ell = mR^2\omega \qquad \text{circular motion} \qquad \textbf{(10–10)}$$

Example 10–2

A particle with a mass of 500 g rotates in a circular orbit with radius $R = 22.4$ cm at a constant rate of 1.5 revolutions per second. What is the angular momentum with respect to the center of the orbit?

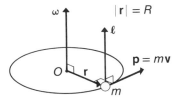

Solution:

As shown in the diagram, the angular momentum vector, $\ell = \mathbf{r} \times \mathbf{p}$, is parallel to $\boldsymbol{\omega}$ (which lies along the rotation axis). The constant magnitude of ℓ is $\ell = mR^2\omega$. With the motion repeating 1.5 times per second, we have

$$\omega = 2\pi(1.5 \text{ s}^{-1}) = 3\pi \text{ rad/s}$$

Thus, $\qquad \ell = (0.5 \text{ kg})(0.224 \text{ m})^2(3\pi \text{ s}^{-1})$

$$= 0.236 \text{ J·s}$$

Example 10–3

The conditions given in the previous example are exactly those of Example 6–4. In that example the stone was twirled at the end of a 25-cm-long string. To satisfy Newton's equations, the string was found to make an angle of 63.8° with the vertical. The radius of the circular orbit of the stone was, therefore, $R = L \sin \beta = (0.250 \text{ m}) \sin 63.8° = 0.224 \text{ m}$, the value that we used in Example 10–2.

To illustrate the importance of the selection of the origin in determining the angular momentum, calculate the angular momentum with respect to the stationary or hand-held end of the string, shown in diagram (a).

Solution:

We have $\mathbf{v} = \boldsymbol{\omega} \times \mathbf{r}$ (Eq. 10–7), so that $|\mathbf{v}| = v = \omega L \sin \beta = R\omega$, giving the same values as before. Now, however, for the angular momentum with respect to O (the hand), we have

$$\boldsymbol{\ell} = \mathbf{r} \times \mathbf{p} = m\mathbf{r} \times \mathbf{v}$$

which is *not* parallel to $\boldsymbol{\omega}$. In diagram (b), the vectors $\boldsymbol{\omega}$ (which corresponds to the axis of rotation) and $\mathbf{r}$ are instantaneously in the plane of the page. We see that $\boldsymbol{\ell}$ is inclined at an angle

$\beta = 63.8°$ to the horizontal. The angular momentum has a constant magnitude ℓ, but its vector direction rotates or *precesses* about a vertical axis with the angular frequency ω, as shown in diagram (c).

We have

$$|\boldsymbol{\ell}| = \ell = mRL\omega$$

$$= (0.5 \text{ kg})(0.224 \text{ m})(0.250 \text{ m})(3\pi \text{ s}^{-1})$$

$$= 0.264 \text{ J·s}$$

The *component* of $\boldsymbol{\ell}$ that is parallel to $\boldsymbol{\omega}$ (the axis of rotation) has a constant value given by

$$\ell_z = \ell \sin \beta = (0.264 \text{ J·s}) \sin 63.8°$$

$$= 0.236 \text{ J·s}$$

This is exactly the value found in the previous example. We commonly refer to this component of $\boldsymbol{\ell}$ as the angular momentum *about the axis of rotation*. For a particle that rotates about the z-axis, this axial component of the angular momentum is simply the tangential momentum mv_t multiplied by the perpendicular distance R to the y-axis. In the case here,

$$\ell_z = \ell \sin \beta = mR(L \sin \beta)\omega = mR^2\omega$$

$$= mv_t R$$

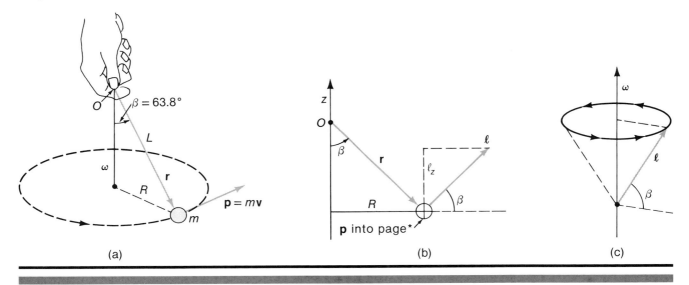

(a)   (b)   (c)

10–3   TORQUE

It is a common experience that an object can be set into rotation by a force that is applied "off center." For example, a tangential push on the rim of a wheel will cause rotation, but a similar push along a line through the axle will not. The tendency for a force

---

*When a vector in a figure is perpendicular to the plane of the page, it is convenient to represent the vector by the symbol ⊙ when it is pointing upward or out of the page, and with the symbol ⊕ when it is pointing downward or into the page. These symbols are intended to remind you of the front view of the tip of an arrow and the rear view of the feather-end of the arrow.

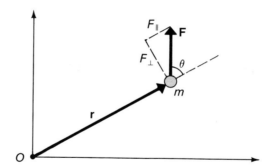

Fig. 10–6. Only the perpendicular component of **F** produces a torque about O. Both **r** and **F** lie in the plane of the page.

to produce rotation is proportional to the magnitude of the force and to the distance from the point of application to the center of rotation. The quantitative measure of this rotation tendency is in terms of the *torque* produced by the force.

Any vector can be expressed in terms of its component vectors. Therefore, a force **F** that is applied to a particle m, as in Fig. 10–6, can be expressed in terms of a component $\mathbf{F}_\perp$ perpendicular to the position vector **r** and a component $\mathbf{F}_\parallel$ parallel to **r**. Only the component $\mathbf{F}_\perp$ can produce a torque about the point O. The magnitude of the torque produced by **F** about O is defined to be

$$\tau = rF_\perp = rF \sin \theta \qquad (10-11)$$

where θ is the smaller angle between the positive directions of **r** and **F**. Using vector notation, we can write, in analogy with the definition of angular momentum,

$$\boldsymbol{\tau} = \mathbf{r} \times \mathbf{F} \qquad (10-12)$$

Torque is also referred to as *moment of force*. Note that the vector $\boldsymbol{\tau}$, which results from the cross product of two vectors, is perpendicular to the plane containing **r** and **F**, with a direction sense given by the right-hand rule (Fig. 10–7).

Another way to calculate the torque produced by a force is illustrated in Fig. 10–8. The vector direction of **F** is extended as the dashed ''line of action.'' The *lever arm* of **F** (namely, $r_\perp$) is the perpendicular to the line of action. Now, $r_\perp = r \sin \theta$, so Eq. 10–11 can also be written as $\tau = r_\perp F$.

The Connection Between Torque and Angular Momentum.   Newton's second law can be expressed by stating that the net force acting on a particle is equal to the time

Fig. 10–7. The torque about O produced by the force **F** acting on a particle at the point defined by **r** is $\boldsymbol{\tau} = \mathbf{r} \times \mathbf{F}$. The force **F** and the point of application lie in the *x-y* plane.

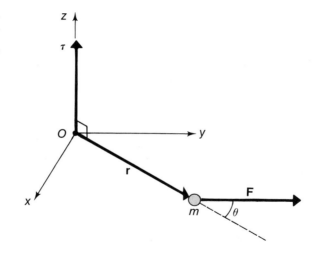

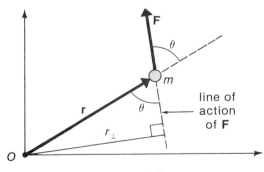

Fig. 10–8. The lever arm for the force **F** is $r_\perp$ and the torque about O is $\tau = r_\perp F = rF \sin\theta$. Both **r** and **F** lie in the plane of the page.

derivative of the particle's linear momentum:

$$\mathbf{F} = \frac{d\mathbf{p}}{dt}$$

If we take the cross product of this expression with **r** (from the left), we have

$$\mathbf{r} \times \mathbf{F} = \mathbf{r} \times \frac{d\mathbf{p}}{dt} \tag{10-13}$$

Next, we take the defining equation for angular momentum (Eq. 10–8) and differentiate with respect to time, obtaining

$$\frac{d\boldsymbol{\ell}}{dt} = \frac{d}{dt}(\mathbf{r} \times \mathbf{p}) = \frac{d\mathbf{r}}{dt} \times \mathbf{p} + \mathbf{r} \times \frac{d\mathbf{p}}{dt}$$

The first term on the right-hand side can be written as the cross product of a vector with itself, which therefore vanishes:

$$\frac{d\mathbf{r}}{dt} \times \mathbf{p} = m\frac{d\mathbf{r}}{dt} \times \mathbf{v} = m\mathbf{v} \times \mathbf{v} = 0$$

Then, using Eq. 10–13, we find

$$\frac{d\boldsymbol{\ell}}{dt} = \mathbf{r} \times \frac{d\mathbf{p}}{dt} = \mathbf{r} \times \mathbf{F}$$

But $\mathbf{r} \times \mathbf{F}$ is the torque, so we have

$$\boldsymbol{\tau} = \mathbf{r} \times \mathbf{F} = \frac{d\boldsymbol{\ell}}{dt} \tag{10-14}$$

That is, the torque is the time derivative of the angular momentum, just as force is the time derivative of linear momentum. Note carefully that *both* $\boldsymbol{\tau}$ and $\boldsymbol{\ell}$ are referred to the same origin that was selected to define the position vector **r**.

The Conservation of Angular Momentum. Equation 10–14 contains an important statement concerning the motion of a particle. The torque exerted on a particle is equal to the time rate of change of its angular momentum $\boldsymbol{\tau} = d\boldsymbol{\ell}/dt$. Consequently, if there is *no torque* acting, the angular momentum must remain constant:

$$\boldsymbol{\tau} = 0 \qquad \text{implies} \qquad \boldsymbol{\ell} = \text{constant} \tag{10-15}$$

Thus, in the absence of an applied torque, *angular momentum is conserved*.

An important case where an external force acting on a particle may produce significant changes in its motion yet leave the angular momentum constant is the case of a *central force,* that is, one that is always directed toward the origin used for determining the angular momentum.

It should be pointed out that the principle of angular momentum conservation is not a new law of Nature. The essential ingredient of the principle is already contained within (linear) Newtonian dynamics. All solvable problems in mechanics can be solved, at least in principle, by using Newton's equations. In many instances, however, it proves convenient to use angular momentum conservation to treat a particular situation, but it is not required, as is demonstrated by some of the examples we give.

Example 10–4

A small disc with mass m is guided by a light string to execute circular motion on a frictionless tabletop, as indicated in the diagram. The string passes through a small hole in the table, and the lower end is held by a hand. We describe the motion in plane polar coordinates. Initially, the disc is at a radius r_0 and twirls with an angular frequency ω_0. By pulling on the string *slowly,* the radius is reduced to r. What is the new value of ω? How much total work was done by pulling the string? What force F must be exerted to accomplish this change in radius?

Solution:

Consider the disc to be a particle executing circular motion about an origin O at the position of the hole. The gravitational force $m\mathbf{g}$ that acts on the disc is just balanced by the normal force $\mathbf{N}$ that the tabletop exerts on the disc; that is, $\mathbf{N} = mg$. These two forces have exactly cancelling torques about O. The other force that acts on the disc is the tension in the string which also generates no torque relative to O. Therefore, $\ell = mr^2\omega$ is a constant of the motion. Thus,

$$mr^2\omega = mr_0^2\omega_0$$

or

$$\omega = \left(\frac{r_0^2}{r^2}\right)\omega_0$$

The kinetic energy of the disc at any instant is the sum of azimuthal and radial terms (see Eq. 3–17):

$$K = \tfrac{1}{2}m(v_\phi^2 + v_\rho^2)$$

The string is pulled *slowly,* so that $v_\rho \ll v_\phi$. Thus, the term involving v_ρ may be neglected, and the azimuthal velocity is essentially the tangential velocity (nearly circular path). That is,

$$K = \tfrac{1}{2}mv_t^2 = \tfrac{1}{2}mr^2\omega^2$$

The angular frequency ω varies as $1/r^2$, so the kinetic energy K is *not* a constant but increases as r decreases. This increase in energy is supplied by the work done in pulling the string. Thus,

$$W = \tfrac{1}{2}mr^2\omega^2 - \tfrac{1}{2}mr_0^2\omega_0^2$$

or

$$W = \tfrac{1}{2}mr_0^2\omega_0^2\left(\frac{r_0^2}{r^2} - 1\right)$$

The force exerted is then

$$F = -\frac{dW}{dr} = \frac{mr_0^4\omega_0^2}{r^3} = mr\omega^2 = \frac{mv_t^2}{r}$$

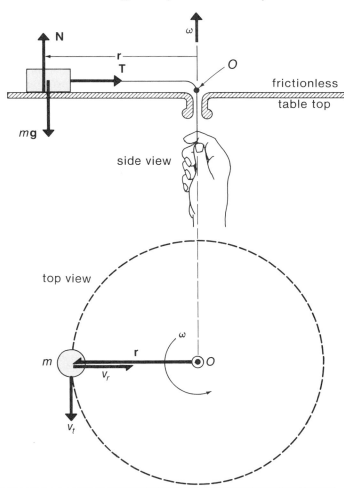

10-4 ANGULAR MOMENTUM OF A SYSTEM OF PARTICLES

To this point we have been discussing the angular momentum of a single particle. Now we wish to consider the case of a system of particles. The total angular momentum of the system is the vector sum of the individual angular momenta,*

$$L = \sum_i \ell_i \tag{10-16}$$

The total angular momentum as well as the individual momenta refer to a selected origin that is fixed in some inertial frame.

The time derivative of Eq. 10-16 gives

$$dL/dt = \sum_i d\ell_i/dt \tag{10-17}$$

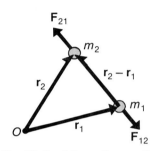

Fig. 10-9. If the action-reaction pair of forces F_{12} and F_{21} act along the line connecting the particles, they produce no net torque about any origin O.

We suppose that internal torques act as a result of the interaction forces between the particles of the system and that externally imposed torques are also present. The externally applied torques result from forces acting on the individual particles. Let us represent their sum by $\tau^e = \Sigma_i \tau_i^e$. If the internal forces that act between the particles are *central*, i.e., act along the line joining pairs of particles, then the sum of the internal torques vanishes, $\Sigma_i \tau_i$ (all internal two-body torques) $= 0$. To see why this is so, consider one pair of particles of the system (refer to Fig. 10-9). Together they produce a net torque about 0 equal to $\tau_1 + \tau_2 = r_1 \times F_{12} + r_2 \times F_{21}$. According to Newton's third law $F_{12} = -F_{21}$; therefore, $\tau_1 + \tau_2 = (r_2 - r_1) \times F_{21}$. If F_{21} is directed along a line connecting the two particles, it must be parallel or antiparallel to the vector $r_2 - r_1$ and hence $\tau_1 + \tau_2 = 0$. Summing over all internal torques in a like manner therefore gives a null result. Finally, in view of Eq. 10-14 we have

$$\tau^e = dL/dt \qquad \text{system of particles} \tag{10-18}$$

This important relationship expresses the way the total angular momentum of a system of particles varies with time under the action of applied external torques, $\tau^e = \Sigma_i \tau_i^e$.

Conservation of Angular Momentum. In the event that the system of particles is completely isolated so that $\tau^e = 0$, we find

$$L = \text{constant} \qquad \text{no external forces} \tag{10-19}$$

That is, the total angular momentum L of an isolated system of particles is a constant of the motion. Notice the analogous relationships that contrast $F^e = dP/dt$ (Eq. 9-13) with Eq. 10-18; and P = constant (Eq. 9-14) with Eq. 10-19 (when there are no external forces).

Attention is again called to the fact that Eqs. 10-18 and 10-19 are vector equations. Thus it may happen that the system is actually acted upon by an external force which, however, is constant *in direction*. In this case the component of the total angular momentum *in this direction* remains a constant of the motion. For example, if an external force

---

*We use ℓ to indicate the angular momentum of a *particle* and L to indicate the angular momentum of a system.

acts only in a direction parallel to the z-axis, $\mathbf{F}^e = F^e\hat{\mathbf{k}}$, then it necessarily produces zero torque about the z-axis. (• Prove this assertion by writing $\mathbf{r} = x\hat{\mathbf{i}} + y\hat{\mathbf{j}} + z\hat{\mathbf{k}}$ and determining $\boldsymbol{\tau}^e = \mathbf{r} \times \mathbf{F}^e$.) It then follows that $L_z = $ constant. This is an example of a partially isolated system of the type we discussed in Section 9–2.

In this chapter we shall treat only the simple case of two-particle systems. The most practical cases involving multiparticle systems occur for rigid bodies, which are treated in the next chapter.

Example 10–5

A small flat disc with a mass M of 2 kg slides on a frictionless horizontal surface. The disc is held in a circular orbit by a 1.5-cm string tied at one end to a pivot, as shown in the diagram. Initially, the disc has an angular velocity of $\omega = 3$ rad/s. A 1-kg mass m of putty is dropped onto the disc from directly above. If the putty sticks to the disc, what is the new period of rotation?

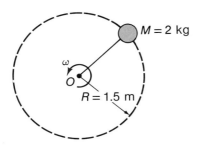

Solution:

Consider the two-particle system consisting of the rotating disc and the blob of putty. In this case the external forces that act consist of gravity, the tension in the string, and the vertical normal force on the disc. As the putty makes contact with the disc and deforms as it sticks, both the string tension and the normal force have a complicated impulsive behavior. However, none of the external forces produce torques about a vertical axis through the pivot point O. (• Prove this assertion.) Thus, the total angular momentum about this axis, taken to be the z-axis, remains constant; i.e., $L_z(\text{initial}) = L_z(\text{final})$, or

$$MR^2\omega_0 = (M + m)R^2\omega$$

The new period is $T = 2\pi/\omega$; hence

$$T = \frac{2\pi(M + m)}{\omega_0 M}$$

$$= \frac{2\pi(2 + 1)\text{kg}}{(3\text{s}^{-1})2 \text{ kg}} = \pi s$$

QUESTIONS

10–1 Describe and explain briefly what is meant by the terms (a) vector or cross product, (b) angular velocity vector, and (c) angular acceleration vector.

10–2 Under what circumstances can the cross product of two vectors be zero?

10–3 Describe and explain briefly what is meant by the terms: (a) angular momentum, (b) torque, (c) "line of action" of a force, and (d) lever arm of a force.

10–4 Confirm that the two sets of units kg·m²/s and J·s are both proper for angular momentum.

10–5 The diagram shows a wrench used to turn a nut, the hand supplying the force $\mathbf{F}$. For fixed values of ℓ and the magnitude of the force, what angle θ would result in the greatest turning effort? What angle would result in no

turning effort at all? For an arbitrary angle θ, resolve $\mathbf{F}$ into two mutually perpendicular components, one of which results in no turning effort and one of which gives a strong turning effort. What effect would changing the length ℓ have on the turning effort? Relate your answers to these questions to the definition of torque, Eq. 10–11.

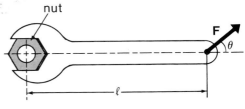

10–6 Consider the swinging of a pendulum consisting of a very small bob at the end of a massless string. (a) Is the angular momentum of the pendulum about the point of suspension a constant? (b) If not, at what point during the swing is it maximum? (c) When is it changing at the maximum time rate?

10-7   Write an expression for the angular momentum vector ℓ for a particle of mass m at rest on the surface of the Earth (at latitude λ) with respect to an origin at the center of the Earth. Be sure to give both the magnitude and the direction of ℓ.

10-8   Do you impart momentum to the Earth when you walk? Explain.

10-9   An Earth satellite in an elliptic orbit has the largest velocity when closest to the Earth's surface and the smallest velocity when furthest away. Explain.

10-10   Why does a helicopter require at least two sets of rotating blades?

10-11   Explain how a child on a swing can increase the amplitude of the swinging motion (a) by appropriate foot contact with the ground at the low point of the swing and (b) *without* such contact.

10-12   A professor sits in (splendid) isolation on a pivoted stool and holds a spinning top with its axis vertical; the stool is at rest. Risking life and limb, the professor grabs the top and stops the spinning. What is the result?

10-13   A shortstop leaps straight into the air to catch a hard-hit line drive baseball. Describe the motion of his body after the catch but before landing back on the ground.

PROBLEMS

Section 10-1

10-1   Find the cross product of the two vectors $\mathbf{C} = 2\hat{\mathbf{i}} - \hat{\mathbf{j}}$ and $\mathbf{D} = \hat{\mathbf{i}} + 2\hat{\mathbf{j}} - 3\hat{\mathbf{k}}$. What is the angle between these vectors?

10-2   Find the cross product of the two vectors $\mathbf{C} = 2\hat{\mathbf{i}} - 4\hat{\mathbf{j}} + 2\hat{\mathbf{k}}$ and $\mathbf{D} = -3\hat{\mathbf{i}} + 6\hat{\mathbf{j}} - 3\hat{\mathbf{k}}$. What special relationship exists between these two vectors?

10-3   A particle executing circular motion has a constant angular velocity $\boldsymbol{\omega} = \omega_0\hat{\mathbf{k}}$ with $\omega_0 = 5$ rad/s. (a) What is the particle's velocity $\mathbf{v}$ when its position vector measured from a point on the rotation axis is $\mathbf{r} = (3\hat{\mathbf{i}} + 4\hat{\mathbf{j}} + 5\hat{\mathbf{k}})$ meters? (b) Find the plane in which the particle moves and the radius of its circular orbit.

10-4   Define two vectors with unit length to be

$$\mathbf{a} = \cos\alpha\,\hat{\mathbf{i}} + \sin\alpha\,\hat{\mathbf{j}}$$

$$\mathbf{b} = \cos\beta\,\hat{\mathbf{i}} + \sin\beta\,\hat{\mathbf{j}}$$

By forming the dot product $\mathbf{a}\cdot\mathbf{b}$ and the cross product $\mathbf{a} \times \mathbf{b}$, derive the addition laws for the sine and cosine.

Section 10-2

10-5   Prove that the angular momentum of a particle with constant velocity about any arbitrary fixed point is a constant.

10-6   At time $t = 0$ a 3-kg particle is at the coordinate $\mathbf{r} = (2\hat{\mathbf{i}} + 3\hat{\mathbf{j}})$m in an inertial frame. The particle is moving with a constant velocity $\mathbf{v} = (2\hat{\mathbf{i}} - 3\hat{\mathbf{j}})$m/s. (a) Locate the particle at $t = 2$ s. (b) By direct calculation determine the angular momentum relative to the origin of the

coordinate frame at $t = 0$ and $t = 2$ s. (• Compare your results. Is angular momentum conserved?)

10-7   What is the orbital angular momentum of the Earth relative to the center of the Sun? [Refer to the table of astronomical data inside the front cover.]

10-8   Two particles move in opposite directions along parallel straight-line paths that are separated by a distance d. If the two linear momenta are equal in magnitude, show that the angular momentum of this two-particle system is the same for any choice of the origin.

Section 10-3

10-9   A particle with a mass of 4 kg moves along the x-axis with a velocity $v = 15t$ m/s, where $t = 0$ is the instant that the particle is at the origin. (a) At $t = 2$ s, what is the angular momentum of the particle about a point P located on the $+y$-axis, 6 m from the origin? (b) What torque about P acts on the particle?

10-10   An object with mass m falls freely, as shown in the diagram. Use Eq. 10-14 and calculate both $\boldsymbol{\tau}$ and ℓ with respect to the origin O. Thus, show that the linear acceleration of the object is $a_y = -g$.

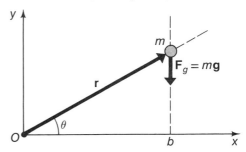

10–11 Consider a block with a mass m sliding down a frictionless plane set at angle α with the horizontal, as shown in the diagram. Find the acceleration of the block using Eq. 10–14. Is your result familiar? [*Hint:* Apply Eq. 10–14 twice, once with O and then with O' as the origin.]

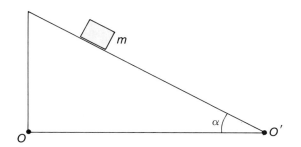

10–12 (All motion in this problem takes place in a vertical plane.) A ball with mass $m_1 = 4$ kg and an initial velocity $v_1 = 3$ m/s collides with another ball ($m_2 = 2$ kg), initially at rest, which is attached to a rigid massless rod of length 0.4 m that pivots freely about the opposite end. *Immediately after* the impact, the angular speed of m_2 about the pivot is ω, and m_1 moves with a speed $v_1' = 1.8$ m/s at an angle $\alpha = 35°$, as shown in the diagram.

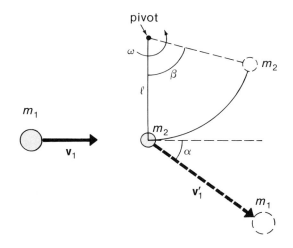

(a) Determine ω. (b) Find the maximum angle β that the rod will make with the vertical. (c) Is the collision elastic? If not, how much kinetic energy is transformed into heat? [*Hint:* During the collision, the linear momentum of m_2 is not conserved, whereas the angular momentum is conserved. • Can you see why?]

10–13 Two particles with equal masses m are attached to a folding crossarm on a shaft that rotates in frictionless bearings, as shown in the diagram. Originally, both ends of the crossarm are fully extended to radii $r_1 = 1$ m, and the rotational frequency is $\omega_1 = 10$ rev/s. When two springs in the crossarm are triggered, it folds so that both ends have radii $r_2 = 0.6$ m. What is the new angular velocity ω_2? Assume that the shaft and crossarm are massless.

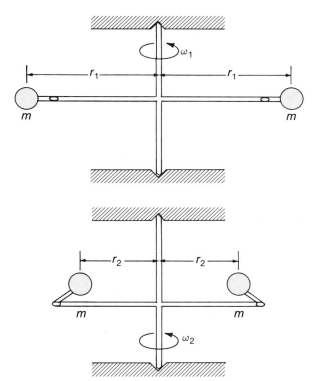

Additional Problems

10–14 Show that $\mathbf{A} \cdot (\mathbf{B} \times \mathbf{C}) = \begin{vmatrix} A_x & A_y & A_z \\ B_x & B_y & B_z \\ C_x & C_y & C_z \end{vmatrix}$

Show explicitly by vector methods that if $\mathbf{C} = \mathbf{A} \times \mathbf{B}$, then $\mathbf{C}$ is perpendicular to both $\mathbf{A}$ and $\mathbf{B}$.

10–15 A 300-g particle is confined to moving in a circular path of radius 25 cm. The application of a constant tangential force of 1.5 N causes the particle to decelerate from an initial speed of 8 m/s. (a) Make a sketch showing the angular acceleration and initial angular velocity vectors relative to the orbital plane of the particle motion. (b) Calculate the magnitude of α. (c) Determine the magnitude of ω 2 s after the continuous application of the force.

10–16 At time $t = 0$ a 5-kg particle is at the coordinate $\mathbf{r} = (2\hat{\mathbf{i}} + 3\hat{\mathbf{j}})$m in an inertial frame and has a velocity $\mathbf{v} = 8\hat{\mathbf{i}}$ m/s. A constant force $\mathbf{F} = -15\hat{\mathbf{j}}$ N acts on the particle. Refer to the origin of the frame in answering these questions: (a) What is the angular momentum of the particle at $t = 0$? (b) What torque acts on the particle? (c) What is the time rate of change of the angular momentum? (d) What is the angular momentum at time $t = 2$ s?

10–17 Two particles that have equal mass are connected by a rod with negligible mass and rotate about an axis through the center of the rod and perpendicular to it. The length of the rod is increased by 13 parts in 10^6 because of a temperature rise. By what fraction does the angular velocity change?

10–18 Show that the kinetic energy K of a particle of mass m rotating at a radius R about a fixed center is

$$K = \tfrac{1}{2}mR^2\omega^2 = \frac{\ell^2}{2mR^2}$$

where ω and ℓ are the instantaneous angular velocity and angular momentum, respectively.

10–19 A stone with a mass m is tied to the end of a string and
  •   twirled in a vertical plane. Show, using Eq. 10–14, that the angular momentum ℓ is related to the angle ϕ by the equation $\ell^2 = \ell_0^2 + 2m^2gR^3(1 - \cos\phi)$, where ℓ_0 is the angular momentum at the top of the circular path ($\phi = 0$). [*Hint:* Use O as the origin and note that $\dfrac{d\ell}{dt} = \dfrac{d\ell}{d\phi}\dfrac{d\phi}{dt} = \omega\dfrac{d\ell}{d\phi} = \dfrac{\ell}{mR^2}\dfrac{d\ell}{d\phi}.$] (• Can you also derive the desired result using conservation of energy? Refer to Problem 10–18.)

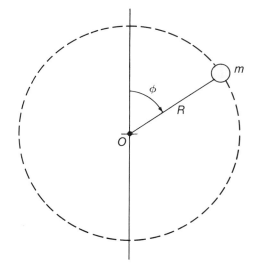

10–20 Two 2-kg masses are connected by a rod 1 m long hav-
  ••   ing negligible mass. The rod is free to rotate without friction about a perpendicular horizontal axis through its center, as in the diagram. The rod is originally horizontal and at rest, when a 1-kg mass of putty drops onto one of the masses with a speed of $v_0 = 5$ m/s and sticks to it. (a) What is the angular velocity of the system just after the putty sticks? (b) What is the loss in kinetic energy as a result of the impact? (Use the result of Problem 10–18.) (c) The system is unable to rotate completely around. How far around will it rotate before coming temporarily to rest?

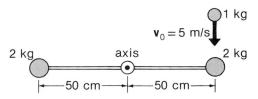

CHAPTER 11

THE MOTION OF RIGID BODIES

In discussing the dynamics of translational motion we recognize that a solid object—if it is not too large compared with the scale of its trajectory—will behave as a pointlike particle. In the previous discussions we have been careful to exclude from consideration any rotational motion of the objects about internal axes. In this way we reduced all situations involving solid objects to the case of particle motion.

Objects in the real world, however, are not pointlike particles. A real object has a mass distribution associated with its size and shape, and the motion of a real object usually involves both translation and rotation. In this chapter, we study the rotational motion of an object about some point or axis located within or associated with the object. We limit the discussion here to two simple cases. In the first case, a rigid body is considered to undergo pure rotation about an axis fixed in some inertial frame of reference. In the second case, the body may have simultaneous translational and rotational motions, with, however, the restriction that the rotation of the body relative to its own center of mass is about an axis that always moves parallel to itself. This type of combined translational and rotational motion occurs, for example, when a ball or a cylinder or a wheel rolls on a flat surface. Those cases in which torques act to *change* the orientation of the rotation axis are generally more complicated to treat than are the restricted cases we study here.

11–1 ANGULAR MOMENTUM

We first examine the dynamics of a rigid body that is constrained to rotate about an axis fixed in an inertial reference frame. A *rigid body* is one in which the relative coordinates connecting all of the constituent particles remain absolutely constant. That is, we allow the object to possess no elasticity, so that the constituent particles do not undergo any relative displacements when subjected to the accelerations of rotational or translational motions.

Consider first the rotation of a particle about a fixed axis identified with the z-axis, as in Fig. 11–1. The particle moves in a circular path, and its angular momentum *with respect to the origin O* is (Eq. 10–8)

$$\boldsymbol{\ell} = m\mathbf{r} \times \mathbf{v} \qquad (11\text{–}1)$$

The angular momentum vector $\boldsymbol{\ell}$ is perpendicular to both $\mathbf{r}$ and $\mathbf{v}$ and lies in the plane that contains the rotation axis and the position vector $\mathbf{r}$. As the particle rotates, $\boldsymbol{\ell}$ *precesses* (revolves) about the z-axis.

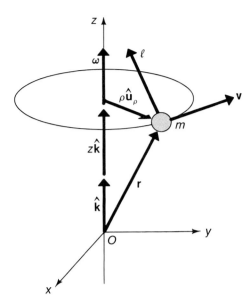

Fig. 11−1. The angular momentum of the particle is $\boldsymbol{\ell} = m\mathbf{r} \times \mathbf{v}$ and the z-component is $\ell_z = m\rho^2\omega$.

The vector $\mathbf{r}$ can be expressed as* (see Fig. 11−1)

$$\mathbf{r} = z\hat{\mathbf{k}} + \rho\hat{\mathbf{u}}_\rho$$

where ρ is the perpendicular distance from the z-axis to the particle and $\hat{\mathbf{u}}_\rho$ is a unit vector in the radial direction. Therefore, the angular momentum, $\boldsymbol{\ell} = m(z\hat{\mathbf{k}} + \rho\hat{\mathbf{u}}_\rho) \times \mathbf{v}$, consists of two components. One component, $mz\hat{\mathbf{k}} \times \mathbf{v}$, is directed radially inward from the particle toward the rotation axis. The other component, $m\rho\hat{\mathbf{u}}_\rho \times \mathbf{v} = m\rho v\hat{\mathbf{k}}$ is directed along the z-axis; that is,

$$\ell_z = m\rho v\hat{\mathbf{k}} \tag{11−2}$$

But $v = \rho\omega$, so we can write

$$\ell_z = m\rho^2\omega \tag{11−3}$$

which is equivalent to Eq. 10−10. Notice that ℓ_z does not depend on z; hence, for the purpose of calculating *this* particular component of the angular momentum, the location of the origin O along the rotation axis (the z-axis) is immaterial.

We now extend the discussion to the case of a rigid body whose motion is restricted to rotation about a fixed axis. It is again convenient to choose the z-axis to coincide with the rotation axis and to use cylindric coordinates. Figure 11−2 shows an irregularly shaped object rotating about an axis AA', which is the z-axis. The angular momentum of the differential mass element dm, just as the particle in Fig. 11−1, has a radial component and a z-component. Because we are dealing with a rigid body, only the z-component of the angular momentum influences the rotational motion of the object about the fixed axis. (The radial component is associated with the forces that the axis exerts on the support bearings.) Consequently, we single out for discussion the interesting z-component of the angular momentum.

For the differential mass element dm, we have in analogy with Eq. 11−3,

$$d\ell_z = \rho^2\omega \, dm$$

*To describe rigid-body motion, it is convenient to use *cylindric* coordinates, which are just plane polar coordinates (Section 3−4) with the addition of the perpendicular z-axis.

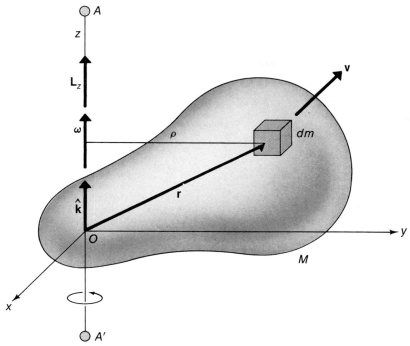

Fig. 11–2. The geometry of a rigid body undergoing rotation about a fixed axis AA' that coincides with the z-axis.

The z-component L_z of the angular momentum of the object is obtained by integrating the contributions of the individual mass elements. Thus,

$$L_z = \int d\ell_z = \omega \int_M \rho^2 \, dm \qquad (11–4)$$

where ω can be removed from the integral because it is the same for every mass element.

The quantity L_z is referred to as the *angular momentum about the axis of rotation*. Just as for ℓ_z, the exact position of the origin O along AA' is immaterial for the purpose of calculating L_z.

The integral in Eq. 11–4 depends only on the geometric properties of the object (and the position of the rotation axis) and is independent of any dynamic aspects of the motion. This integral is called the *rotational inertia* of the object and is denoted by the letter I:

$$I = \int_M \rho^2 \, dm \qquad (11–5)$$

To make clear that we have chosen the z-axis to correspond to the rotation axis we write I_z for the rotational inertia. Then, the angular momentum about the z-axis is

$$L_z = I_z \omega \qquad (11–6)$$

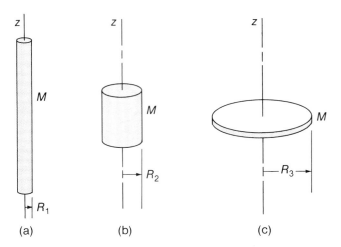

Fig. 11−3. Three different distributions about the z-axis of the same mass M of material. The rotational inertia values are $I_z(R_1) < I_z(R_2) < I_z(R_3)$.

The rotational inertia plays a role in rotational motion analogous to that of mass in translational motion. Thus, a large value of rotational inertia will require a large torque to give the object a rotational acceleration.

It is important to understand that the value of I_z for a particular body depends on the distribution of the body's mass about the z-axis. Figure 11−3a shows a long, slender rod with a radius R_1 and a mass M whose axis coincides with the z-axis. The value of I_z for this rod is relatively small because R_1 is small. If this rod is squeezed into a stubby cylinder, as in Fig. 11−3b, the rotational inertia will increase because more of the mass is now farther from the z-axis. (The total mass M remains the same.) Finally, if the material is compressed into the shape of a thin disc (Fig. 11−3c), the value of I_z will be increased still further.

11−2   ROTATIONAL INERTIA

In general, the determination of the rotational inertia for an object with a complicated shape involves a tedious calculation, or the value must be obtained by experiment. Fortunately, the calculation of the rotational inertia for homogeneous objects with simple geometric shapes is relatively straightforward. For example, a thin-walled cylinder or ring with mass M and radius R has all of its mass at a distance $\rho = R$ from the natural (or symmetry) axis of the cylinder. Consequently, the rotational inertia about this axis is (see Table 11−1e)

$$I = MR^2, \qquad \text{ring about symmetry axis} \qquad \textbf{(11−7)}$$

The following two examples illustrate the usual procedure for calculating the rotational inertia of objects possessing simple shapes.

Example 11−1

Find the rotational inertia of a thin uniform circular disc with radius R, mass M, and surface density $\sigma = M/\pi R^2$, about an axis through its center and perpendicular to the plane of the disc.

Solution:

Divide the disc into rings of differential width $d\rho$ and area $(2\pi\rho)\,d\rho$. Each ring thus has differential mass $dm = \sigma(2\pi\rho)\,d\rho$. Then, from Eq. 11−5,

$$I_z = \int_0^R \rho^2 \sigma(2\pi\rho)\, d\rho = 2\pi\sigma \int_0^R \rho^3 \, d\rho$$

$$= \tfrac{1}{2}\pi\sigma R^4 = \tfrac{1}{2}MR^2$$

(See Table 11–1i.)

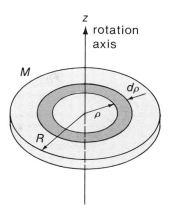

Example 11–2

Determine the rotational inertia of a uniform solid sphere with radius R, mass M, and density $\delta = M/(\tfrac{4}{3})\pi R^3$, about an axis through its center.

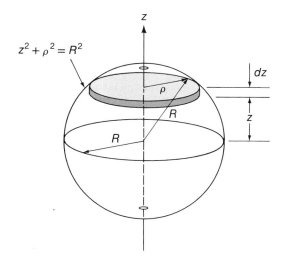

Solution:

By taking advantage of the cylindric symmetry, the rotational inertia can be evaluated with a single integration. Divide the solid sphere into discs with differential thickness dz and radius ρ. Then the volume of each disc is $\pi\rho^2\, dz$ and its differential mass is $dm = \pi\rho^2\delta\, dz$. According to the result of Example 11–1, the rotational inertia of a disc about its axis is $I_z = \tfrac{1}{2}MR^2$. Therefore, we can write

$$dI_z = \tfrac{1}{2}\rho^2\, dm = \tfrac{1}{2}\pi\rho^4\, \delta\, dz$$

Now, $z^2 + \rho^2 = R^2$, so we have $\rho^4 = (R^2 - z^2)^2$, which can be substituted in the preceding equation and integrated to find I_z.

$$I_z = \int dI_z = \tfrac{1}{2}\pi\delta \int_{-R}^{+R} (R^2 - z^2)^2 \, dz$$

$$= \tfrac{1}{2}\pi\delta[R^4 z - \tfrac{2}{3}R^2 z^3 + \tfrac{1}{5}z^5]_{-R}^{+R}$$

$$= \tfrac{8}{15}\pi\delta R^5 = \tfrac{2}{5}MR^2$$

(See Table 11–1g.)

Radius of Gyration.   Regardless of the shape of an object, we can always find a radial distance Γ from the axis of rotation at which the concentration of the entire mass M of the object would produce the same rotational inertia as that of the extended object itself. Then, the rotational inertia is equal to the value for a particle with a mass M at a distance Γ, namely, $I = M\Gamma^2$. We refer to this distance as the *radius of gyration* for the object about the particular axis; thus,

$$\Gamma^2 = \frac{I}{M} = \frac{1}{M}\int_M \rho^2 \, dm \qquad\qquad (11\text{–}8)$$

The Parallel-Axis Theorem. We now prove a useful theorem that allows the computation of the rotational inertia of an object about any axis if the rotational inertia is known for a *parallel axis that passes through the center of mass* of the object.

Refer to Fig. 11–4. We wish to express the rotational inertia I_A about the axis AA' in terms of the rotational inertia I_C about the parallel axis BB' through the center of mass C of the object. The perpendicular position vector from the axis AA' to the mass element dm is $\rho_A \hat{\mathbf{u}}_A$, where $\hat{\mathbf{u}}_A$ is the unit vector $\hat{\mathbf{u}}_\rho$ referred to the axis AA'. The corresponding vector for the axis BB' is $\rho_B \hat{\mathbf{u}}_B$. The vector $\mathbf{h}$ extends from AA' to BB' and is perpendicular to both axes. The various vectors are shown in Fig. 11–4b.

We can express I_A as

$$I_A = \int_M \rho_A^2 \, dm$$

Now, $\rho_A \hat{\mathbf{u}}_A = \mathbf{h} + \rho_B \hat{\mathbf{u}}_B$, so

$$\rho_A^2 = (\rho_A \hat{\mathbf{u}}_A) \cdot (\rho_A \hat{\mathbf{u}}_A) = (\mathbf{h} + \rho_B \hat{\mathbf{u}}_B) \cdot (\mathbf{h} + \rho_B \hat{\mathbf{u}}_B)$$
$$= h^2 + \rho_B^2 + 2\mathbf{h} \cdot (\rho_B \hat{\mathbf{u}}_B)$$

Therefore,

$$I_A = h^2 \int_M dm + \int_M \rho_B^2 \, dm + 2\mathbf{h} \cdot \int_M \rho_B \hat{\mathbf{u}}_B \, dm$$

The integral in the last term defines the position of the C.M. from an origin that is the C.M. itself (compare Eq. 9–11); consequently, this integral vanishes. Also,

$$\int_M dm = M \qquad \text{and} \qquad \int_M \rho_B^2 \, dm = I_C$$

Altogether, we have

$$I_A = I_C + Mh^2 \qquad\qquad (11-9)$$

Fig. 11–4. The axis BB' passes through the C.M. of the object and is parallel to the axis AA'. The vectors $\rho_A\hat{\mathbf{u}}_A$, $\rho_B\hat{\mathbf{u}}_B$, and $\mathbf{h}$ are all in a plane perpendicular to both axes, as shown in (b). The point C is the center of mass.

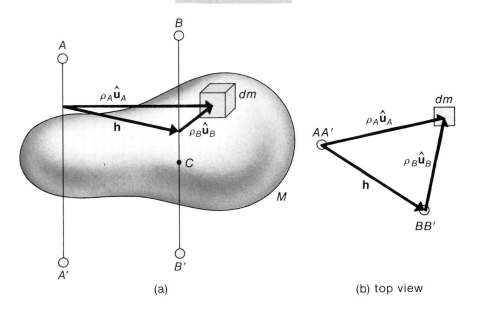

(a)   (b) top view

which is the statement of the parallel-axis theorem. The utility of this theorem lies in the fact that it is almost always easier to calculate the rotational inertia about an axis through the C.M. than to make a direct calculation for some other axis.

Example 11–3

Determine the rotational inertia of a thin uniform rod (a) about a perpendicular axis through its center and (b) about a perpendicular axis through one end.

Solution:

Diagram (a) shows the rod and the perpendicular axis BB' through the center. The rotational inertia is

$$I_C = \frac{M}{\ell} \int_{-\ell/2}^{+\ell/2} x^2 \, dx = \frac{M}{3\ell} x^3 \Big|_{-\ell/2}^{+\ell/2}$$

$$= \tfrac{1}{12} M\ell^2$$

(See Table 11–1a.)

The rotational inertia about a perpendicular axis AA' through one end of the rod can be calculated by using the geometry of diagram (b). We find

$$I_A = \frac{M}{\ell} \int_0^\ell x^2 \, dx = \tfrac{1}{3} M\ell^2$$

Alternatively, the same result may be obtained using the parallel-axis theorem and the result of (a). We obtain

$$I_A = I_C + M(\ell/2)^2 = \tfrac{1}{12} M\ell^2 + \tfrac{1}{4} M\ell^2 = \tfrac{1}{3} M\ell^2$$

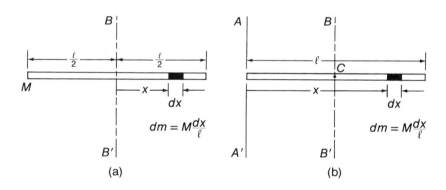

(a)   (b)

The results of Example 11–3, together with some other cases, are summarized in Table 11–1. Note the dependence of I on the location of the rotation axis. Entries (c) and (d) as well as (e) and (f) show how I depends on the choice of the rotation axis for the *same* object.

11–3   ROTATIONAL DYNAMICS

The rotational motion of a rigid body about a fixed axis depends on the nature of the external forces applied to the body. For each such force we can calculate the component of the torque that is in the direction of the rotation axis. This *axial torque* is the only torque component that can influence the rotational motion. (Other components of the applied torque are balanced by opposite torques provided by normal forces at the bearings of the rotation shaft.) We assume that the rotation axis is held in place by frictionless bearings.

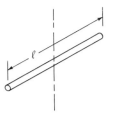

a) thin rod

$$I = \tfrac{1}{12} M\ell^2$$

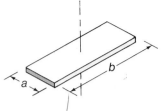

b) rectangular plate

$$I = \tfrac{1}{12} M(a^2 + b^2)$$

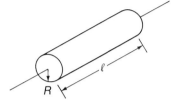

c) solid cylinder

$$I = \tfrac{1}{2} MR^2$$

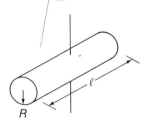

d) solid cylinder

$$I = \tfrac{1}{4} MR^2 + \tfrac{1}{12} M\ell^2$$

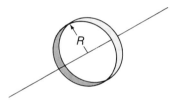

e) thin-walled cylinder or ring

$$I = MR^2$$

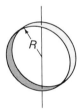

f) thin-walled cylinder or ring

$$I = \tfrac{1}{2} MR^2$$

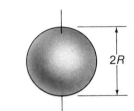

g) solid sphere

$$I = \tfrac{2}{5} MR^2$$

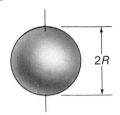

h) hollow spherical shell

$$I = \tfrac{2}{3} MR^2$$

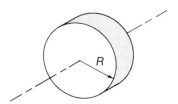

i) solid disc

$$I = \tfrac{1}{2} MR^2$$

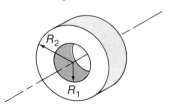

j) annular disc or cylinder

$$I = \tfrac{1}{2} M(R_1^2 + R_2^2)$$

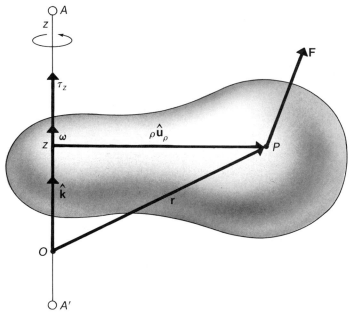

Fig. 11–5. The force **F** is applied to the body at a point P. The axial torque τ_z about O due to the force **F** does not depend on the location of O along the rotation axis AA'.

Such bearings can produce no axial torque and therefore cannot affect the rotational motion. Figure 11–5 shows a rigid body constrained to rotate about the axis AA'.

Following the same argument we made for ℓ_z in Section 11–1, we conclude that τ_z does not depend on z, and, hence, for the purpose of calculating the axial torque, the location along the axis of the origin O is immaterial. (• Carry out the analytic proof that τ_z is the same regardless of the position of the origin along the rotation axis.)

According to Eq. 11–6, $L_z = I_z \omega$. Differentiating with respect to time, we have (with I_z constant)

$$\frac{dL_z}{dt} = I_z \frac{d\omega}{dt} = I_z \alpha$$

where α is the angular acceleration. The time derivative of the angular momentum is equal to the applied torque (Eq. 10–14). Thus,

$$\tau_z = I_z \alpha \tag{11–10}$$

If several forces are acting, τ_z is the sum of all z-components of the torques (including those at the bearings if bearing friction is present).

In the event that the applied torque has no net axial component (that is, $\tau_z = 0$), the axial component of the angular momentum remains constant (that is, $L_z = $ constant). This is simply a consequence of angular momentum conservation. Notice that we consider here only the *axial* components of torque and angular momentum. However, for an unconstrained object, we may have $\tau_z = 0$ but one or both of the other components of **τ** may not be zero; then, $L_z = $ constant, but L_x and L_y can vary with time.

Example 11−4

A massless string is wrapped around a uniform solid cylinder that has a mass $M = 15$ kg and a radius $R = 6$ cm. The cylinder is free to rotate about its axis on frictionless bearings. One end of the string is attached to the cylinder and the free end is pulled tangentially by a force that maintains a constant tension of 2 N on the string.

(a) What is the angular acceleration α of the cylinder?

(b) What is the angular speed ω of the cylinder 2 s after the force is applied, if the cylinder was originally at rest?

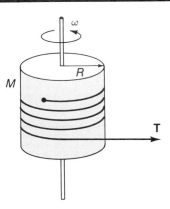

Solution:

(a) The torque applied to the cylinder is constant and has the magnitude

$$\tau = RT = (0.06 \text{ m})(2 \text{ N}) = 0.12 \text{ N·m}$$

The rotational inertia of the cylinder about its axis is

$$I = \tfrac{1}{2}MR^2 = \tfrac{1}{2}(15 \text{ kg})(0.06 \text{ m})^2$$

$$= 2.70 \times 10^{-2} \text{ kg·m}^2$$

Thus, using Eq. 11−10, we have

$$\alpha = \frac{\tau}{I} = \frac{0.12 \text{ N·m}}{2.70 \times 10^{-2} \text{ kg·m}^2} = 4.44 \text{ rad/s}^2$$

(b) The angular speed at $t = 2$ s is

$$\omega = \alpha t = (4.44 \text{ rad/s}^2)(2 \text{ s})$$

$$= 8.89 \text{ rad/s}$$

Example 11−5

(a) A grinding wheel with radius $R = 0.40$ m, mass $M = 200$ kg, and radius of gyration $\Gamma = 0.20$ m rotates about an axis through its center at an initial rate of 500 rpm (revolutions per minute). A tool to be sharpened is pressed against the rim and exerts a constant tangential frictional force of 25 N. In what time will the wheel be slowed to one half the initial rotation speed if no effort is made to maintain the speed? (Neglect bearing friction.)

(b) Suppose that the same grinding wheel, with the same initial rotation rate, requires 25 min to come to rest because of bearing friction. What is the (constant) magnitude of the axial torque produced by the bearing friction?

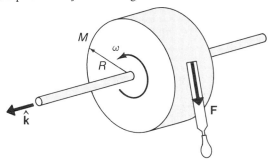

Solution:

(a) The axial torque produced by the tool is

$$\tau_z = -(0.40 \text{ m})(25 \text{ N}) = -10.0 \text{ N·m}$$

Now, $\quad I_z = M\Gamma^2 = (200 \text{ kg})(0.20 \text{ m})^2$

$$= 8.0 \text{ kg·m}^2$$

Thus, using Eq. 11−10, we have

$$\alpha = \frac{\tau_z}{I_z} = \frac{-10.0 \text{ N·m}}{8.0 \text{ kg·m}^2}$$

$$= -1.25 \text{ rad/s}^2$$

By integrating $d\omega/dt = \alpha$, we obtain

$$\omega = \omega_0 + \alpha t$$

Here, $\quad \omega_0 = 500 \text{ rpm} = (500 \text{ min}^{-1}) \times \dfrac{2\pi}{60 \text{ s/min}}$

$$= 52.36 \text{ rad/s}$$

Then, $\quad t = \dfrac{\omega - \omega_0}{\alpha} = \dfrac{-\tfrac{1}{2}(52.36 \text{ rad/s})}{-1.25 \text{ rad/s}^2}$

$$= 20.9 \text{ s}$$

(b) The final angular velocity is $\omega = 0$, so $\omega_0 + \alpha t = 0$. Thus, $\alpha = -\omega_0/t$, and

$$\tau_z = I_z\alpha = -\frac{I_z\omega_0}{t} = -\frac{(8.0 \text{ kg·m}^2)(52.36 \text{ rad/s})}{25(60 \text{ s})}$$

$$= -0.279 \text{ N·m}$$

Example 11—6

A massless string is wrapped around a pulley that has a radius $R = 10$ cm, a mass $M = 0.50$ kg, and a rotational inertia $I = 3.2 \times 10^{-3}$ kg about its axis. The free end of the string is attached to a hanging 2-kg block; the other end is fixed to a point on the rim of the pulley. Neglecting friction in the pulley bearing, find the linear acceleration of the hanging block and the force exerted by the bearing on the pulley.

Solution:

We assume that the string does not stretch, so the distance s that the block descends is equal to the arc length $R\phi$ through which the rim of the pulley moves; that is, $s = R\phi$. Then,

$$\frac{ds}{dt} = v = R\frac{d\phi}{dt} = R\omega \tag{1}$$

Also,

$$\frac{d^2s}{dt^2} = a = R\frac{d^2\phi}{dt^2} = R\alpha \tag{2}$$

Applying Newton's second law to the motion of the hanging block (with the positive direction taken downward), we have

$$mg - T = ma \tag{3}$$

For calculating the torque, we take the clockwise direction to be positive. From Eq. 11–10, $\tau = I\alpha$; then, using (2) we find

$$\tau = TR = I\alpha = \frac{Ia}{R} \tag{4}$$

Eliminating T between (3) and (4), we obtain

$$mR^2a + Ia = mgR^2$$

from which

$$a = \frac{mgR^2}{mR^2 + I} = \frac{g}{1 + I/mR^2} \tag{5}$$

$$= \frac{9.80 \text{ m/s}^2}{1 + (3.2 \times 10^{-3} \text{ kg·m}^2)/(2 \text{ kg})(0.10 \text{ m})^2}$$

$$= 8.45 \text{ m/s}^2$$

Then,

$$\alpha = \frac{a}{R} = \frac{8.45 \text{ m/s}^2}{0.10 \text{ m}} = 84.5 \text{ rad/s}^2$$

The force exerted by the bearing on the pulley is the normal force **N** shown in the free-body diagram for the pulley. The force equation for the vertical direction is

$$N - Mg - T = 0$$

We now know the acceleration a, so we can obtain the tension T from (3):

$$T = m(g - a)$$

Then,

$$N = T + Mg = m(g - a) + Mg$$

$$= (2 \text{ kg})(9.80 \text{ m/s}^2 - 8.45 \text{ m/s}^2) + (0.5 \text{ kg})(9.80 \text{ m/s}^2)$$

$$= 7.60 \text{ N}$$

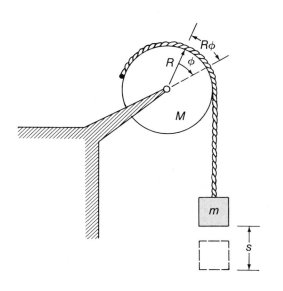

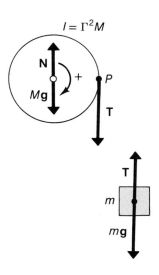

11-4   ROTATIONAL KINETIC ENERGY AND WORK

The kinetic energy of a rigid body rotating about a fixed axis can be given conveniently in terms of rotational quantities. For the situation illustrated in Fig. 11-6, the kinetic energy of the mass element dm is

$$dK = \tfrac{1}{2}v^2\,dm = \tfrac{1}{2}\rho^2\omega^2\,dm$$

and the kinetic energy for the entire body is, using Eq. 11-5,

$$K = \tfrac{1}{2}\omega^2 \int_M \rho^2\,dm = \tfrac{1}{2}I\omega^2 \tag{11-11}$$

We sometimes refer to $\tfrac{1}{2}I\omega^2$ as the *rotational kinetic energy* of the body.

If a rigid body undergoes rotational acceleration, there must be torques (Eq. 11-10) acting on the body. These torques are produced by forces, whose points of application are being displaced and which therefore do work on the body. The work W done by the applied torques results in a change of the body's rotational kinetic energy K. (We neglect bearing friction.) The rate at which this work is done is equal to the power expended, $P = dW/dt$, and is readily deduced from Eq. 11-11.

Differentiating $K = \tfrac{1}{2}I\omega^2$ with respect to time, we have

$$P = \frac{dW}{dt} = \frac{dK}{dt} = I\omega\,\frac{d\omega}{dt} = I\omega\alpha$$

Then, using $\tau = I\alpha$ (Eq. 11-10), there results

$$P = \tau\omega \tag{11-12}$$

This expression gives the instantaneous value of the power expended (τ and ω may both be time dependent). The total amount of work done is found by integration:

$$W = \int \tau\omega\,dt = \int \tau\,d\phi \tag{11-13}$$

Fig. 11-6. The kinetic energy of the mass element dm is $\tfrac{1}{2}\rho^2\omega^2\,dm$.

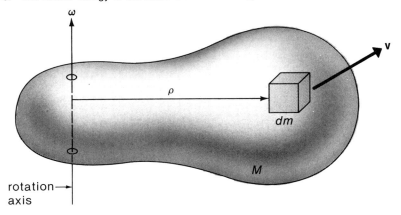

where $d\phi = \omega\, dt$ is the differential angle through which the body rotates during the interval dt. If the applied torque τ is constant, we have

$$W = \tau \int_{\phi_1}^{\phi_2} d\phi = \tau(\phi_2 - \phi_1) \qquad (11\text{--}14)$$

This equation is analogous to that for the work done by a constant force during a linear displacement $s_2 - s_1$, namely, $W = F(s_2 - s_1)$.

We have now obtained expressions for all of the dynamic quantities involved in rotational motion. In Table 11–2 these quantities are compared with the corresponding ones used to describe rectilinear motion. Notice that these quantities are *analogous* only—for example, the first six entries in the table have different dimensions in the two cases and therefore represent completely different physical quantities.

TABLE 11-2.   Comparison of Translational and Rotational Quantities

TRANSLATION (LINEAR MOTION)		ROTATION (ABOUT A FIXED AXIS)	
Displacement*	$s = s_0 + v_0 t + \frac{1}{2}at^2$	Angular displacement*	$\phi = \phi_0 + \omega_0 t + \frac{1}{2}\alpha t^2$
Speed	$v = ds/dt$	Angular speed	$\omega = d\phi/dt$
	$^* \qquad = v_0 + at$		$^* \qquad = \omega_0 + \alpha t$
Acceleration	$a = dv/dt$	Angular acceleration	$\alpha = d\omega/dt$
Mass	m	Rotational inertia	I
Force	F	Torque	τ
Momentum	$p = mv$	Angular momentum	$L = I\omega$
Dynamics	$F = dp/dt = ma$	Dynamics	$\tau = dL/dt = I\alpha$
Work	$W = \int F\, dx$	Work	$W = \int \tau\, d\phi$
Kinetic energy	$K = \frac{1}{2}Mv^2$	Kinetic energy	$K = \frac{1}{2}I\omega^2$
Power	$P = Fv$	Power	$P = \tau\omega$

*For constant acceleration (a or α).

Example 11-7

Refer again to the accelerated cylinder in Example 11–4.

(a) What is the rotational kinetic energy of the cylinder at $t = 2$ s?

(b) How much work was done by the applied force during the 2-s interval?

Solution:

(a) The rotational kinetic energy at $t = 2$ s is

$$K = \tfrac{1}{2}I\omega^2 = \tfrac{1}{2}(2.70 \times 10^{-2}\ \text{kg·m}^2)(8.89\ \text{rad/s})^2$$

$$= 1.07\ \text{J}$$

(b) From the work-energy theorem, the work done is equal to the increase in kinetic energy, namely, 1.07 J. However, it is instructive to calculate the work done by using Eq. 11–12 and $\omega = \alpha t$:

$$dW = P\, dt = \tau\omega\, dt = \tau\alpha t\, dt$$

Integrating,

$$\int_0^W dW = \tau\alpha \int_0^t t\, dt$$

so that

$$W = \tfrac{1}{2}\tau\alpha t^2 = \tfrac{1}{2}(0.12\ \text{N·m})(4.44\ \text{rad/s}^2)(2\ \text{s})^2$$

$$= 1.07\ \text{J}$$

as before.

Example 11—8

Refer to Example 11–6. Use energy conservation to obtain the acceleration of the block.

Solution:

Work is done on the system (block + pulley) corresponding to the decrease in the gravitational potential energy of the hanging block as it moves downward. Hence,

$$W = mgs$$

where s is the total distance of fall at time t.

The gain in kinetic energy of the system consists of two parts, namely, $\frac{1}{2}I\omega^2$ for the pulley and $\frac{1}{2}mv^2$ for the block:

$$K = \tfrac{1}{2}I\omega^2 + \tfrac{1}{2}mv^2$$

where v and ω are the values at time t. We know that $v = R\omega$, so equating K to the work done,

$$mgs = \tfrac{1}{2}\left(\frac{I}{R^2} + m\right)v^2$$

Differentiating this equation with respect to time and using $ds/dt = v$, we find

$$mgv = m\left(\frac{I}{mR^2} + 1\right)v\,\frac{dv}{dt}$$

Solving for $dv/dt = a$, we obtain

$$a = \frac{g}{1 + \dfrac{I}{mR^2}}$$

which is just Eq. (5) in Example 11–6.

11–5 ROTATION ACCOMPANIED BY TRANSLATION

The general motion of a system of particles is considerably simplified when the system constitutes a rigid body. Primarily, this is because the inertial properties of the rigid body can be expressed in terms of certain integrated mass distribution characteristics. The simple case of rotation about a fixed axis that we discussed earlier is one example of this feature. However, the problem is still relatively complicated for the completely general motion of a rigid body, such as the tumbling motion of a cylinder thrown into the air. In this case the inertial characteristics require that a nine-component mathematical quantity (or tensor) be defined in order for the required calculations to be made. We leave such problems to more advanced texts.

We shall, however, discuss a special case of the motion of a rigid body that simultaneously undergoes both translation and rotation. Figure 11–7 shows the situation. The

Fig. 11–7. The center of mass C of the rigid body moves with the velocity **V** in an inertial reference frame with origin O. The body rotates about an axis through C that is parallel to the z-axis; the angular velocity is $\boldsymbol{\omega} = \omega\hat{\mathbf{k}}$.

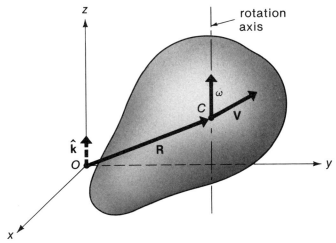

C.M. of the body executes general (possibly accelerated) motion. However, the rotation of the body is restricted to take place about an axis through the C.M. that moves in such a way that the axis remains parallel to its initial orientation. Examples of such motion might be a rolling cylinder or a flat object sliding on a horizontal surface and rotating about a vertical axis at the same time.

The relevant equations of motion for a rigid body under these conditions are relatively simple. The body as a whole moves in particle-like fashion with a mass-concentration center at the center of mass. In addition, it executes a rotation about its center of mass induced by external torques. A summary of these equations of motion is given in Eqs. 11–15a–f. They refer to an inertial frame with origin O in which $\mathbf{R} = \dfrac{1}{M}\displaystyle\int_M \mathbf{r}\,dm$ locates the center of mass. The body executes rotational motion, but only with angular velocity $\boldsymbol{\omega} = \omega\hat{\mathbf{k}}$ about the C.M. The coordinate frame, moving with the C.M., maintains its x, y, z-axes always parallel to the inertial-frame axes.

$$\mathbf{P} = M\,d\mathbf{R}/dt \qquad (11\text{–}15a)$$

$$\mathbf{L} = \mathbf{R} \times \mathbf{P} + I_C\omega\hat{\mathbf{k}} \qquad (11\text{–}15b)$$

$$d\mathbf{P}/dt = \mathbf{F}^e \qquad (11\text{–}15c)$$

$$d\mathbf{L}/dt = \mathbf{R} \times \mathbf{F}^e + I_C\alpha\hat{\mathbf{k}} \qquad (11\text{–}15d)$$

also $$\boldsymbol{\tau}_C = I_C\alpha\hat{\mathbf{k}} \qquad (11\text{–}15e)$$

and $$K = \tfrac{1}{2}MV^2 + \tfrac{1}{2}I_C\omega^2 \qquad (11\text{–}15f)$$

where $\mathbf{P} = M\,d\mathbf{R}/dt = M\mathbf{V}$ is the linear momentum of the C.M., $\mathbf{F}^e = \displaystyle\sum_n \mathbf{F}_n^e$ is the vector sum of the external forces, I_C is the rotational inertia about the C.M., and $\boldsymbol{\tau}_C$ is the sum of the external torques with respect to the C.M.

Equations 11–15 are true *whatever* the nature of the C.M. motion. If the C.M. of a body undergoes accelerated motion, any coordinate frame attached to the body is a noninertial frame; nevertheless, Eqs. 11–15 are still valid.

We have considered two different cases of rigid-body motion and have found them to be described by equations of the same form relating angular momentum and torque, namely, Eqs. 11–10 and 11–15e. In general, equations of this form do *not* apply for arbitrarily selected axes. In addition to the two special cases treated in this chapter, there is only a limited number of other situations in which similar equations are valid. However, the two cases discussed here are commonly met and are sufficient for our purposes.

Rather than give a general proof of Eqs. 11–15 we shall consider a simple but characteristic case of an object rolling on a flat surface. The main features contained in Eqs. 11–15 will be apparent in this treatment. However, we must first discuss rolling motion.

Rolling Motion.   A uniform cylinder or a sphere that rolls on a flat surface represents a practical situation that involves simultaneous translation and rotation of a rigid body. Figure 11–8 illustrates a rolling cylinder. (We consider only the case of rolling without slipping.) The translational velocity of the center of the cylinder is $\mathbf{V}$ and the rotational speed about the center is ω. During the time interval Δt, the center moves to a position C', a distance $V\,\Delta t$ from the original position C. During the same interval, the point P on the rim of the cylinder, which was originally in contact with the surface, rotates to the position P', such that $\Delta\phi = \omega\,\Delta t$. A different point Q' on the rim is now in contact

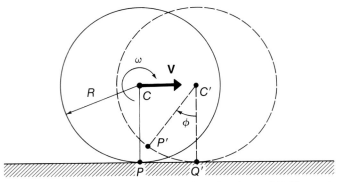

Fig. 11−8. Two positions of a rolling cylinder separated by a time interval Δt. The center of the cylinder moves with velocity **V**, and the angular speed is ω. Rolling without slipping requires **V** = Rω.

with the surface. We define rolling without slipping in the self-evident way, namely, that the length of the arc $\widehat{P'Q'}$ on the cylinder rim be equal to the distance of translational motion, $\overline{CC'}$. That is,

$$\widehat{P'Q'} = R \, \Delta\phi = R\omega \, \Delta t = \overline{CC'} = V \, \Delta t$$

so that
$$R\omega = V \qquad\qquad\qquad\qquad (11-16)$$

Rolling without slipping requires that Eq. 11−16 be satisfied *at all times*. Whenever $R\omega \neq V$, slipping occurs.

How may we describe the motion of a point on a cylinder when it rolls without slipping? To an observer moving with the center of the cylinder (which for a uniform cylinder is also its C.M.), each point a distance r_C from the center (C.M.) moves in a circle with speed $r_C\omega$. The velocity $\mathbf{v}_C$ relative to the C.M. for various points along the vertical line $\overline{PQ}$ passing through the instantaneous point of contact P is shown in Fig. 11−9b. The velocity of the C.M. relative to the point P is of course **V** (Fig. 11−9a).

Recalling the addition law for relative velocities, Eq. 4−37, we have

$$\mathbf{v}_P = \mathbf{V} + \mathbf{v}_C \qquad\qquad\qquad\qquad (11-17)$$

The instantaneous velocity of these same points along $\overline{PQ}$ relative to point P is shown in Fig. 11−9c. We see that the point on the cylinder opposite the point P is instantaneously at rest relative to point P, while the other points along $\overline{PQ}$ have horizontally directed velocities relative to point P that are equal in magnitude to $r_P\omega$. *These points appear to be*

Fig. 11−9. A cylinder rolling without slipping. (a) The velocity of the C.M. relative to the point P is **V**. The instantaneous velocity of points along the line $\overline{PQ}$, (b) relative to the moving C.M. located at point C, and (c) relative to the point of contact P.

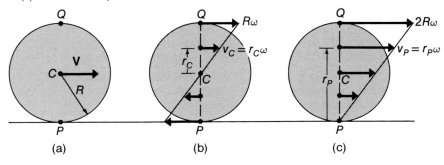

in instantaneous rotational motion about the point P with angular velocity ω. (A more general proof of this condition for points not on the line $\overline{PQ}$ is given in Problem 11–21.)

The kinetic energy of the rolling cylinder in the rest frame is therefore expected to be that appropriate for pure rotation about the momentarily stationary point of contact

$$K = \tfrac{1}{2}I_P\omega^2 \tag{11-18}$$

where I_P is the rotational inertia of the cylinder about a horizontal axis through point P and parallel to the cylinder axis. In view of the parallel-axis theorem (Eq. 11–9), the rotational inertia I_P may be written in terms of the rotational inertia about the cylinder axis I_C, namely

$$I_P = MR^2 + I_C \tag{11-19}$$

Thus
$$K = \tfrac{1}{2}(MR^2 + I_C)\omega^2 = \tfrac{1}{2}MV^2 + \tfrac{1}{2}I_C\omega^2 \tag{11-20}$$

Further, the angular momentum about the point P in the rest frame is expected to be

$$\mathbf{L}_P = I_P\omega\hat{\mathbf{k}} = (MR^2 + I_C)\omega\hat{\mathbf{k}} \tag{11-21a}$$

where the unit vector $\hat{\mathbf{k}}$ is parallel to the cylinder axis. The first term in Eq. 11–21a, $MR^2\omega\hat{\mathbf{k}} = MRV\hat{\mathbf{k}}$, may be interpreted to be the angular momentum of a particle-like mass M located at the C.M. and hence moving with velocity $\mathbf{V}$ while maintaining a constant distance R above the rolling plane. Thus, $MRV\hat{\mathbf{k}} = \mathbf{R}_P \times M\mathbf{V} = \mathbf{R}_P \times \mathbf{P}$* and hence

$$\mathbf{L}_P = \mathbf{R}_P \times \mathbf{P} + I_C\omega\hat{\mathbf{k}} \tag{11-21b}$$

We see that Eqs. 11–20 and 11–21b agree with the corresponding Eqs. 11–15f and 11–15b. The physical content of these equations is to allow the rigid-body motion to be considered to consist of a simple particle-like C.M. trajectory and a separate rigid-body rotation about a moving axis passing through the C.M.

An example to demonstrate the applicability of Eq. 11–15e, $\tau_C = I_C\alpha\hat{\mathbf{k}}$, requires consideration of an accelerating object. Consider a uniform cylinder rolling down an inclined plane without slipping. Figure 11–10 shows the free-body diagram for the cylinder. This example will also permit us to discuss the role of friction at the point of contact P; therefore, we have included the presence of a frictional force f.

The linear motion of the C.M. down the inclined plane is obtained from Eq. 9–13,

$$Mg \sin \beta - f = Ma \tag{11-22}$$

In view of the instantaneous rotation about point P we have $\tau_P = I_P\alpha$. Making use of the

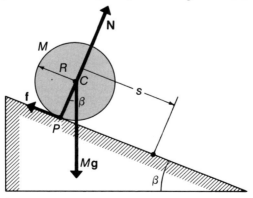

Fig. 11–10. A cylinder rolling down an inclined plane.

---

*Be sure to distinguish between R, the radius of the cylinder, and $\mathbf{R}_P$, the location of the C.M. with respect to the fixed point P. Note also that $\mathbf{V} = d\mathbf{R}_P/dt$.

parallel-axis theorem and the fact that the linear acceleration a of the cylinder is $a = R\alpha$, we have

$$MgR \sin \beta = I_C\alpha + MRa \qquad (11-23)$$

Multiplying Eq. 11–22 by R and subtracting the result from Eq. 11–23 yields

$$fR = I_C\alpha$$

or
$$\tau_C = I_C\alpha \qquad (11-24)$$

The result is a simple illustration of the general equation Eq. 11–15e.

This example is also very instructive in revealing the role of friction and the energy considerations that prevail. Using the result from above that $f = \dfrac{I_C}{R}\alpha = \dfrac{I_C}{R^2}a$ and the fact that $I_C = \frac{1}{2}MR^2$ for a uniform cylinder, we have

$$f = \tfrac{1}{2}Ma$$

Substituting this result into Eq. 11–22 gives the linear acceleration

$$a = \tfrac{2}{3}g \sin \beta$$

and for the frictional force we obtain

$$f = \tfrac{1}{3}Mg \sin \beta$$

The normal force N is

$$N = Mg \cos \beta$$

Then,

$$\frac{f}{N} = \tfrac{1}{3} \tan \beta$$

To prevent slipping, the coefficient of static friction must be equal to or larger than this value; thus,

$$\mu_s \geq \tfrac{1}{3} \tan \beta$$

(• Can you see why this is so? Refer to the discussion in Section 6–2.)

Next, let us examine the energy balance in this problem. We have, at any time,

$$K = \tfrac{1}{2}MV^2 + \tfrac{1}{2}I_C\omega^2$$
$$= \tfrac{1}{2}MV^2 + \tfrac{1}{4}MR^2\omega^2$$

Using $V = R\omega$, we find

$$K = \tfrac{3}{4}MV^2$$

Now, the increase in kinetic energy which we have just obtained is equal to the decrease in potential energy less the work done against friction W_f. If s represents the displacement along the plane, we can write

$$K = Mgs \sin \beta - W_f$$

or
$$\tfrac{3}{4}MV^2 = Mgs \sin \beta - W_f$$

Differentiating with respect to time, we obtain

$$\tfrac{3}{2}MVa = MgV \sin \beta - \frac{d}{dt}W_f$$

Using the previous result, $a = \frac{2}{3}g \sin \beta$, to substitute for a, we find

$$MgV \sin \beta = MgV \sin \beta - \frac{d}{dt} W_f$$

from which we see that $dW_f/dt = 0$. Thus, as the motion proceeds, W_f does not change, and we conclude that $W_f = 0$. This means that *no work is done against the static frictional force*. The reason is that in rolling without slipping there is no relative motion between the cylinder and the plane at the contact point, so no work is done by or against the static frictional force. (If the motion involved slipping, work would be done against the kinetic frictional force. Be certain that you understand the difference between these two situations.)

In Section 6–3 we argued that ideal rolling without slipping along a horizontal surface involves no friction between the rolling object and the surface. In this example, however, we have ideal rolling without slipping that *does* involve a frictional force. Notice how the presence of friction is *demanded* here in order to produce rolling. (• If the cylinder were rolling *up* the plane, what would be the direction of **f**?) The magnitude of the frictional force varies with the sine of the inclination angle and so vanishes for horizontal rolling, as we argued previously.

Example 11–9

Two massless ropes are wrapped around a uniform cylinder that has a mass M. The ropes are attached to the ceiling, as shown in the figure. Initially, the unwrapped portions of the ropes are vertical and the cylinder is horizontal. Find the linear acceleration of the falling cylinder and the tension in the ropes.

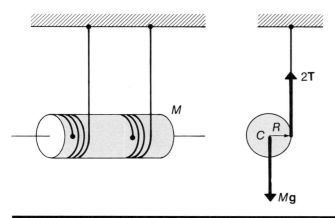

Solution:

Refer to the free-body diagram at the right of the figure. The tension in each rope is T, so the torque about the cylinder axis (which passes through the C.M.) is $2TR$. Also, $I_C = \frac{1}{2}MR^2$ (Table 11–1c); thus, $\tau_C = I_C\alpha$ becomes

$$2TR = \tfrac{1}{2}MR^2\alpha$$

If the downward direction is taken to be positive, the vertical acceleration a of the C.M. is obtained from

$$Mg - 2T = Ma$$

If the ropes do not slip on the cylinder, we have $a = R\alpha$. Then, eliminating T between the two preceding equations yields

$$a = \tfrac{2}{3}g$$

Finally, substituting this value for a into either equation containing T gives

$$T = \tfrac{1}{6}Mg$$

(• Why does the cylinder not swing to one side as it falls?)

Example 11–10

A uniform circular disc with a mass $M = 2$ kg and a radius $R = 10$ cm is laid flat on a horizontal frictionless surface. A constant force $F = 5$ N is applied to the end of a string that is wrapped around the disc, thereby causing the disc to rotate about a vertical axis and to translate in the horizontal direction.

(a) Find the linear acceleration a of the C.M. of the disc and the angular acceleration α about the C.M.

(b) Find the acceleration a_s of the free end of the string.

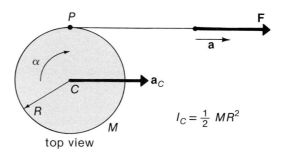

top view

(c) Show that the rate at which the applied force does work is equal to the rate at which the kinetic energy of the disc is increasing.

Solution:

(a) Since F is the only horizontal force acting on the disc, the linear acceleration of the C.M. is

$$a = \frac{F}{M} = \frac{5 \text{ N}}{2 \text{ kg}} = 2.5 \text{ m/s}^2 \qquad (1)$$

Equating the torque about a vertical axis through the C.M. to $I_C\alpha$, we find

$$\alpha = \frac{\tau_C}{I_C} = \frac{FR}{\frac{1}{2}MR^2} = \frac{2F}{MR}$$

$$= \frac{2(5 \text{ N})}{(2 \text{ kg})(0.10 \text{ m})} = 50 \text{ rad/s}^2 \qquad (2)$$

(b) To obtain the acceleration of the free end of the string with respect to the horizontal surface, we note that the velocity v_0 of the string with respect to the surface at, say, point P is equal to the velocity of the point P with respect to the C.M. (namely, $v_T = R\omega$) *plus* the velocity of the C.M. with respect to

the surface (namely, V):

$$v_0 = R\omega + V \qquad (3)$$

Differentiating this expression with respect to time gives

$$a_s = \frac{dv_0}{dt} = R\alpha + a$$

$$= (0.10 \text{ m})(50/\text{s}^2) + 2.5 \text{ m/s}^2$$

$$= 7.5 \text{ m/s}^2$$

(c) The rate of change of kinetic energy is

$$\frac{dK}{dt} = \frac{d}{dt}(\tfrac{1}{2}MV^2 + \tfrac{1}{2}I_C\omega^2) = MVa + I_C\omega\alpha$$

From Eq. (1) $Ma = F$, while from Eq. (2) $I_C\alpha = \tau_C = FR$, hence

$$\frac{dK}{dt} = F(V + R\omega) = Fv_0$$

where we have used (3). Thus, the rate of increase of kinetic energy of the disc is equal to the rate at which work is done by the force F on the string (and ultimately on the disc).

QUESTIONS

11-1 Describe and explain briefly what is meant by the terms (a) rotational inertia, (b) radius of gyration, and (c) parallel-axis theorem.

11-2 You are given two identical cans of food, both of the same weight (including contents). Although the labels have been lost, it is known that one contains a clear broth and the other a thick tomato paste. Devise an experiment to determine which can contains what food without opening the cans.

11-3 You are given two eggs, one raw and the other hard boiled. To test which is which, set each one spinning on a table top, momentarily stop it, and quickly release it. Contrast the expected behavior of the two eggs under this test.

11-4 A solid wooden cylinder and an iron cylindric shell or annulus have the same mass, outer radius, and axial length. Both are started down an inclined plane from rest and from the same height. Explain why one will roll down the plane faster than the other. Which cylinder will have the greater kinetic energy when arriving at the bottom of the plane?

11-5 The spool of thread shown in the diagram moves to the left when the thread is pulled in the direction ① and to

the right when pulled in the direction ②. Explain why. What happens when the thread is pulled in the direction ③, which is in line with the point of contact P?

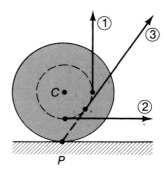

11-6 A mouse is sitting on the rim of a toy carousel (the rotation axis of which is vertical). The carousel is originally stationary. The mouse gets up and starts to walk around the rim. Describe the motion of the mouse and the carousel as seen by a child sitting on the floor nearby. The mouse finds that it has returned to its starting point after the carousel has turned through 180°. What can be said about the relationship between the rotational inertia of the carousel, the mass of the mouse, and the rim radius?

What happens if the mouse stops walking? What happens if the mouse then decides to jump off the carousel?

11–7 It is asserted that when a symmetric object such as a sphere or cylinder rolls on a horizontal surface with partial slipping, the time rate of loss of kinetic energy is given by the expression $\dfrac{dK}{dt} = f(V - R\omega)$, where $f = \mu_k N$ is the kinetic friction between the object and the surface, V is the C.M. velocity, ω is the angular speed of rotation about the C.M., and R is the radius. Explain or derive this expression.

PROBLEMS

Section 11–1

11–1 Show that the rotational inertia about the symmetry axis of a uniform annular cylinder with inner radius R_1, outer radius R_2, length ℓ, and mass M is $\tfrac{1}{2}M(R_1^2 + R_2^2)$. Thus, verify the value given in Table 11–1j.

11–2 A flat uniform circular ring with radius R is made from a thin wire that has a mass M. Prove that the rotational inertia about an axis along any diameter is $\tfrac{1}{2}MR^2$ (Table 11–1f). [*Hint:* Use the geometry shown in the diagram, where $dI_z = \rho^2\,dm$ and $dm = (M/2\pi)\,d\phi$.]

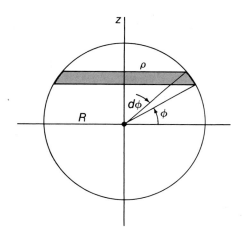

11–3 Determine the rotational inertia of a thin ring with a radius R and a mass M, about an axis that is tangent to the ring and lies in the plane of the ring by following the procedure of Example 11–3.

11–4 What is the rotational inertia of a cube with sides of length ℓ and a mass M, about an axis that coincides with one edge? [*Hint:* Start with the result for a rectangular plate—Table 11–1b.]

11–5 The armature of an electric motor is a uniform cylinder 15 cm in length and 6 cm in diameter; the density of the cylinder material is 8.5 g/cm³. When the rotation speed about the cylinder axis is 1800 rpm, what is the angular momentum about that axis?

Section 11–2

11–6 A solid uniform cylinder with a radius of 0.30 m and a mass of 12 kg is free to rotate about its axis, which is supported by frictionless bearings. A constant tangential force of 10 N is applied to the cylinder, originally at rest. (a) What is the torque applied to the cylinder? (b) What is the angular acceleration of the cylinder? (c) What is the angular velocity of the cylinder 3 s after the force is applied?

11–7 A phonograph turntable consists of a solid circular plate 30 cm in diameter that has a mass of 3.5 kg. The turntable is rotating at $33\tfrac{1}{3}$ rpm when the motor is switched off. A time of 105 s is required for the turntable to come to rest. (a) What is the (constant) angular acceleration? (b) How many revolutions does the turntable make after the motor is turned off? (c) What is the effective decelerating torque?

11–8 A massless string is wrapped around a solid disc that has
• a mass $M = 0.60$ kg and a radius $R = 8$ cm. The string slides (without friction) through a hook that supports a hanging block ($m = 2.4$ kg) and the end is attached to a fixed point Q, as shown in the diagram. Determine the linear acceleration of the hanging block. [*Hint:* First, find the relationship connecting the linear acceleration of the hanging block and the angular acceleration of the pulley.]

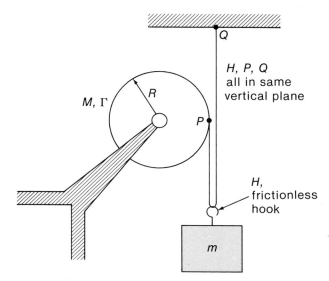

11–9 A block (m_1 = 5 kg) slides without friction on a plane inclined at an angle of 30°. This block is connected to a hanging block (m_2 = 4 kg) by a massless string that runs over a pulley (M = 0.50 kg) with a radius R = 3 cm and a radius of gyration Γ = 2 cm. The string does not slip on the pulley. (a) Find the linear acceleration of the system. (b) Find the tension in the string attached to each of the blocks. Why are the two tensions not equal? (c) Find the force exerted on the axle of the pulley by the bearings.

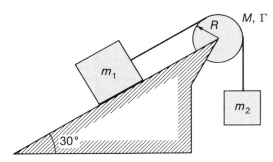

Section 11–3

11–10 An airplane engine delivers 3000 hp to a propeller that rotates at 2500 rpm. What is the torque exerted on the propeller shaft?

11–11 A solid uniform cylindric shaft has a mass of 4.0 kg and a diameter of 2.0 cm and is supported by frictionless bearings. A massless string with a length of 70 cm is wrapped around the shaft, which is set into rotation as the string is pulled with constant tension. The string unwinds, without slipping, in 2.0 s, and then pulls free. (a) What is the angular acceleration of the shaft? (b) What is the final angular speed? (c) Find the magnitude of the tension.

11–12 Assume that the Earth is a homogeneous sphere and calculate the kinetic energy of rotation about its axis. (Use the table of astronomic data inside the front cover.) For how long could this energy be tapped to supply the world's energy needs (assumed constant at 2×10^{20} J/y) if, in the process, the length of the day is not to be increased by more than 1 min?

11–13 A girl stands at the center of a rotating platform. Initially, her arms are extended horizontally and she holds a 3-kg dumbbell in each hand; the platform rotates once every 3 s. If the girl drops her hands to her sides, what will be the new angular speed? Assume that the rotational inertia of the girl is constant at 5 kg·m²; the rotational inertia of the platform is 2 kg·m². The dumbbells (assumed to be particles) are first held 80 cm from the rotation axis and, when the girl drops her hands to her sides, they are 20 cm from the axis. (Neglect friction.) Calculate the initial and final values of the rotational kinetic energy. Account for the difference.

11–14 Two blocks, with masses M_1 and M_2, are connected by a massless string that runs over a pulley without slipping. The pulley is a uniform solid disc with a mass m and a radius R that rotates on frictionless bearings. Determine the linear acceleration of the blocks by considering energy conservation. [*Hint:* Refer to Example 11–8.]

Section 11–4

11–15 A uniform cylinder with a radius R = 15 cm rolls without slipping on a horizontal surface with a linear speed of 2 m/s. To what maximum height h will the cylinder roll up the plane shown in the diagram?

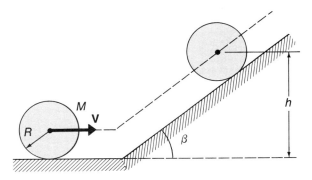

11–16 A spool has a mass of 2 kg, an inner radius R_1 = 3 cm, and an outer radius R_2 = 5 cm; the rotational inertia about the axis of the spool is I = 3.2×10^{-3} kg·m². A constant horizontal force of 5 N is applied to the free end of a massless thread that is wrapped around the inner cylinder of the spool. If the spool rolls without slipping, calculate the linear acceleration along a horizontal surface. What is the minimum coefficient of static friction required to prevent slipping?

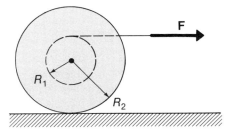

11–17 A marble with a radius b = 1 cm and a mass M = 15 g
•   is released from rest at the top of a loop-the-loop where h = 27 cm, as shown in the diagram at the top of page 219. The marble rolls without slipping completely around the track, which has an inner radius R = 10 cm. (a) Find the velocity of the marble at the top of the loop. (b) Find the reaction forces on the marble when at the top of the loop. (c) What is the minimum height h from which the marble can be released and still maintain contact with the track for the entire trip?

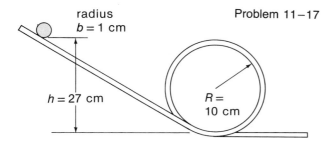

radius
b = 1 cm

Problem 11–17

h = 27 cm

R = 10 cm

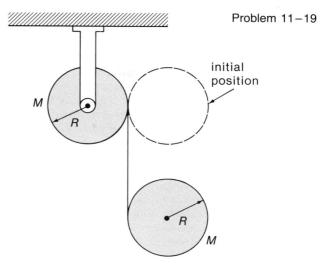

Problem 11–19

initial position

M

R

R

M

11–18 An object with cylindric symmetry has an outer radius R
• and a rotational inertia I about its axis. When the object is placed on its flat end on a plane with variable angle of inclination, sliding begins when the angle reaches 15°. When the object is allowed to roll down the plane, the maximum angle at which rolling without slipping takes place is 30°. What is the ratio MR^2/I? What is the coefficient of static friction? What possible type of simple object are we dealing with here?

11–19 Two identical uniform discs, each with radius $R =$
•• 8 cm and mass $M = 8.0$ kg, are connected by a massless string that is wrapped around the discs in opposite directions, as indicated in the diagram, above right. One disc is mounted on stationary frictionless bearings so that it can rotate freely about its axis. The other disc is initially held at the same height as the captive disc and then released to fall. During the fall, the string unwinds from both discs. Find the linear acceleration and the angular acceleration of the falling disc and the angular acceleration of the captive disc. Also, find the tension in the string.

11–20 A meter stick, initially at rest on a horizontal frictionless
• surface, is given a horizontal impulsive blow at a right angle to its long dimension. Just after the blow is delivered, one end of the stick (the end marked 0 cm) has zero velocity relative to the surface. At what point was

the blow delivered? [*Hint:* The impulsive blow imparts both translational and rotational motion to the stick, namely, $\Delta P = \bar{F} \Delta t$ and $\Delta L = \bar{F}d \Delta t$, where d is the distance from the C.M. to the point of the blow.]

11–21 Consider a cylinder that rolls without slipping. Prove
•• that the instantaneous velocity of an arbitrary point, such as Q in the diagram, corresponds to pure rotation about the contact point P with angular speed ω. [Hint: Note that $\mathbf{r} = \mathbf{R} + \mathbf{r}_C$ and form the vector product $\boldsymbol{\omega} \times \mathbf{r}$.]

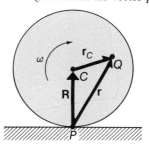

ω

$\mathbf{r}_C$

Q

C

R

$\mathbf{r}$

P

Additional Problems

11–22 Verify entry h in Table 11–1. [*Hint:* Use the geometry
• shown in Problem 11–2, adapted for the case of a sphere.]

11–23 It is proposed to power a passenger bus with a massive rotating flywheel that is periodically brought up to its maximum speed (3500 rpm) by means of an electric motor. The flywheel has a mass of 1200 kg, a diameter of 1.8 m, and is in the shape of a solid cylinder. (• This is not an efficient shape for a flywheel that is designed to power a vehicle. Can you see why?) (a) What is the maximum amount of kinetic energy that can be stored in the flywheel? (b) If the bus requires an average power of 30 hp, how long will the flywheel rotate?

11–24 The Sun rotates with a period of 24.7 days.* Suppose that the Sun were to collapse into a *neutron star* with a radius of 30 km. If the Sun is considered to be a rigid body in both conditions, what is the new rotation period? The radius of gyration of the Sun is approximately $0.3 R_S$. In the neutron-star state, assume that the density is uniform. Neglect the possibility of any mass ejection during the collapse. (This problem is intended only to establish an order-of-magnitude result, but neutron stars with solar mass have radii estimated to be as small as 10 km. One rotating neutron star—the *pulsar* in the Crab nebula—has a rotation period of 33 ms, the shortest period known.)

*The Sun is not a rigid body. This is the rotation period at the equator; the value increases with solar latitude.

11–25 A pair of rocket motors *AA* are used to move a space station into deep space. The space station is a hollow torus, as shown in the diagram, similar to an automobile inner tube, with a connecting section that lies along a diameter. (a) It is desired to rotate the station about the *BB'* axis so that the astronauts walk about feeling a simulated gravitational acceleration. If they are to experience a centripetal acceleration equal to *g* at the radius *R*, what must be the angular speed of the rotation? (b) The rocket motors *AA* are used to produce the rotation about the *BB'* axis. The rocket motors produce a constant tangential thrust. From a stationary start the station acquires the required angular speed in 1000 seconds. What is the thrust (force) exerted by each rocket motor? (c) When the required angular velocity ω has been achieved, the motors are retracted to the center of the station at *A'A'* for servicing and storage. The mass of each motor is $m = 255$ tonnes. What is the ratio of the original angular velocity ω to the final angular velocity ω'? (1 tonne = 10^3 kg.)

11–28 According to one model of the structure of the Sun, the
•• density varies with the radial distance from the center of the Sun in the way shown in the table below. The radius of the Sun is R_S. Assume that the Sun is a rigid body (this is not really a good assumption), and show that the rotational inertia about the solar rotation axis is approximately $0.09\, MR_s^2$. [*Hint:* Consider the Sun to be composed of a series of thin concentric shells and apply entry h of Table 11–1 to each shell. Your result will depend to some extent on the precise way you choose to treat the contribution of each shell.]

r/R_S	0	0.05	0.10	0.15	0.20
(g/cm^3)	148	125	86	56	36

r/R_S	0.30	0.40	0.60	0.80	1.00
(g/cm^3)	12	4	0.5	0.1	0

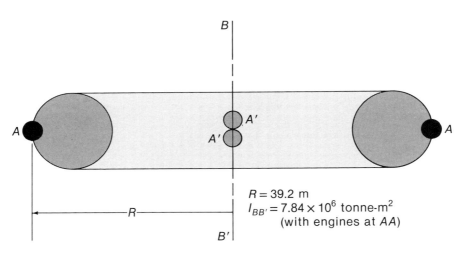

$R = 39.2$ m
$I_{BB'} = 7.84 \times 10^6$ tonne-m^2
(with engines at *AA*)

Problem 11–25

11–26 The lid of a box is a uniform rectangle with dimensions 0.5 m × 1.0 m; the hinge line is along one of the longer edges. The lid is allowed to fall, starting from a vertical position. What is the angular speed of the lid as it strikes the top of the box? [*Hint:* Use energy conservation.]

11–27 A string is wrapped around an unsupported solid cylinder that has a mass $M = 6$ kg and a radius $R = 5$ cm. A vertical force **F** is applied to the string such that the C.M. of the cylinder remains at the same height. (a) Show that the angular acceleration of the cylinder is $\alpha = 2g/R$. (b) What amount of work is done by the force during the 0.5-s interval after release? (c) Through what total distance does the free end of the string move during the same interval?

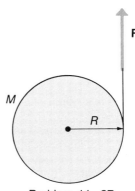

Problem 11–27

11–29 A light string is wrapped around a solid cylinder that has
• a radius $R = 5$ cm and a mass $M = 12$ kg. The cylinder
is on a plane with an inclination angle of 32°. The free
end of the string passes over an inertia-less and frictionless pulley to a block with mass $m = 2$ kg. Find the
linear acceleration of the cylinder and of the hanging
block. [*Hint:* What is the relationship between the linear
acceleration of the cylinder and that of the block?]

11–31 A uniform thin rod with a mass $M = 0.60$ kg and a
•• length of 0.30 m stands on the edge of a frictionless
table, as shown in the diagram. The rod is struck a horizontal impulsive blow, $J = 6$ N·s, at a point 0.20 m
above the table top, driving the rod directly off the table.
Determine the orientation of the rod and the position of
its C.M. 1 s after the blow is struck. [*Hint:* See Problem
11–20.]

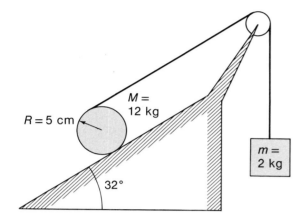

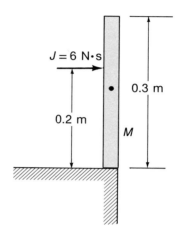

11–30 A cue stick strikes a cue ball and delivers a horizontal
• impulse J in such a way that the ball rolls without slipping as it starts to move. At what height above the ball's
center (in terms of the radius of the ball) was the blow
struck? [*Hint:* See Problem 11–20.]

CHAPTER 12

FORCES IN EQUILIBRIUM

It is possible for a rigid body to be subject to forces and yet not undergo any acceleration (either translational or rotational). A body in such a situation is said to be in *dynamic equilibrium*. Note that this broad definition *does not* require the body to be at rest—it may, in fact, be in translational motion with constant linear velocity, or in rotational motion with constant angular velocity, or both simultaneously. Of course, the conditions of dynamic equilibrium permit the object to be at rest, and this is an important case which has many applications.

12-1 CONDITIONS FOR DYNAMIC EQUILIBRIUM

A rigid body that undergoes neither translational acceleration of its C.M. nor rotational acceleration about its C.M. has $\mathbf{a}_C = 0$ and $\boldsymbol{\alpha}_C = 0$. Then, the equations of motion (Eqs. 11–15) reduce to

$$\frac{d\mathbf{P}}{dt} = 0 \qquad \text{and} \qquad \frac{d\mathbf{L}_C}{dt} = 0$$

where $\mathbf{P}$ is the linear momentum of the C.M. of the body and where $\mathbf{L}_C$ is the angular momentum relative to the C.M. These equations imply that the vector sums of all forces and all torques acting on the body must be zero:

$$\left. \begin{array}{c} \sum_n F_n^e = 0 \\[2em] \sum_n \boldsymbol{\tau}_{Cn} = 0 \end{array} \right\} \tag{12-1}$$

It is important to recognize that dynamic equilibrium requires that *both* of these conditions be satisfied simultaneously.

The Torque Theorem. We have stated the equilibrium conditions in terms of torques calculated with respect to the C.M. of the object. In fact, it is sufficient to calculate the torque with respect to any point, as we now prove.

Suppose that the sum of all forces $\mathbf{F}_n^e$ acting on a body is zero. Consider calculating the net torque produced by these same forces with respect to some arbitrary point P. The situation is illustrated in Fig. 12–1, which shows only one of the forces $\mathbf{F}_n^e$. The position vector $\mathbf{r}_n$ from the arbitrary point P to the point of application of the force $\mathbf{F}_n^e$ can be expressed in terms of $\mathbf{R}$, the vector from the C.M. to the point P, and $\mathbf{r}_{Cn}$, the vector that locates the point of application of the force with respect to the C.M. Evidently, we have

$$\mathbf{r}_{Cn} = \mathbf{R} + \mathbf{r}_n$$

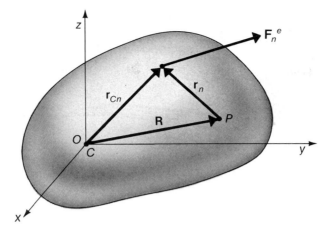

Fig. 12–1. The force $\mathbf{F}_n^e$ is *one* of the forces (whose vector sum is zero) acting on the object. The center of mass C is located at the origin, and P is any arbitrary point.

Then, the expression for the torque becomes

$$\sum_n \boldsymbol{\tau}_{Cn} = \sum_n \mathbf{r}_{Cn} \times \mathbf{F}_n^e$$

$$= \sum_n (\mathbf{R} + \mathbf{r}_n) \times \mathbf{F}_n^e$$

$$= \mathbf{R} \times \sum_n \mathbf{F}_n^e + \sum_n \mathbf{r}_n \times \mathbf{F}_n^e$$

The first term vanishes because the net applied force is zero. The second term is just the net torque calculated with respect to the point P. Therefore,

$$\sum_n \boldsymbol{\tau}_{Cn} = \sum_n \boldsymbol{\tau}_{Pn}$$

Thus, when the torque with respect to the C.M. vanishes as required by Eq. 12–1, the torque with respect to *any arbitrary point* also vanishes for the case in which the net applied force vanishes. The equilibrium conditions in their most general form become

$$\sum_n F_n^e = 0$$

general   **(12–2)**

$$\sum_n \boldsymbol{\tau}_n = 0$$

These equations specify the conditions of *dynamic equilibrium*. It is important to realize that these are *vector* equations and therefore represent *six* independent component equations. In many cases, the forces involved are confined to a plane. We can choose x- and y-axes in this plane, with the z-axis perpendicular to the plane. Then, the six equilibrium equations reduce to three:

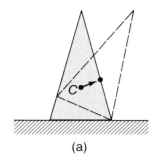

(a)

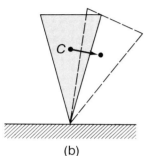

(b)

Fig. 12–2. (a) The cone is in a condition of stable equilibrium; a small tipping displacement *raises* the C.M. Upon release, the cone will return to its original position. (b) The cone is in a condition of unstable equilibrium; a small tipping displacement *lowers* the C.M. Upon release, the cone will fall.

$$\sum F_x^e = 0$$

$$\left. \sum F_y^e = 0 \right\} \quad \text{coplanar forces}$$

$$\sum \tau_z = 0$$

(12–3)

where the summation index n has been suppressed, and where the torques can be taken about any convenient point.

In many of the problems we consider, an object is acted upon by a gravitational force. In Section 13–3 we show that the gravitational forces acting on the individual constituents of an extended object with mass M can be represented by a single force $M\mathbf{g}$ acting at a point. If the acceleration due to gravity $\mathbf{g}$ is constant over the extent of the object (this is the case for ordinary objects near the Earth's surface), this point coincides with the center of mass of the object.

Stable and Unstable Equilibrium. In the discussion of the potential energy function in Section 8–1, it was pointed out that a particle is in a condition of *stable equilibrium* when the potential energy U has a (local) *minimum*. In such a case, any small displacement of the particle away from the equilibrium point produces a force that tends to restore the particle to the equilibrium position. The same is true for extended objects. Consider the cone in Fig. 12–2a, which rests on its base on a flat surface. If the cone is given a small displacement by tipping, the C.M. (point C) tends to rise—that is, work must be done to cause the displacement, and the potential energy increases. If the cone is released after a small displacement, it will return spontaneously to its original position. The resting condition in Fig. 12–2a is clearly one of stable equilibrium.

A particle—or an extended object—is in a condition of *unstable equilibrium* when the potential energy has a (local) *maximum*. Then, any small displacement, such as the tipping of the cone in Fig. 12–2b, will cause the C.M. to be lowered, and the potential energy will decrease. When released, the object will proceed spontaneously to seek a condition of still lower potential energy—it will fall. (• Consider a ballet dancer balanced on one toe. How does the dancer maintain this position?)

An object is said to be in a condition of *neutral equilibrium* when a small displacement causes no change in the potential energy. For example, a homogeneous sphere rolling on a horizontal surface is in a condition of neutral equilibrium.

12–2 APPLICATIONS OF THE EQUILIBRIUM CONDITIONS

We now give several examples of the applications of Eqs. 12–3 to situations involving dynamic equilibrium with coplanar forces. In these examples and the accompanying problems, we assume that the objects are all rigid bodies.

Example 12–1

A meter stick with a mass $m = 0.50$ kg is supported in a stationary horizontal position by vertical spring scales at each end. A block with a mass $M = 0.35$ kg is suspended from the meter stick by a massless string at the 25-cm mark. What are the force readings on the two scales?

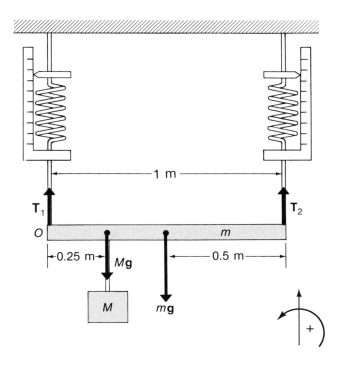

Solution:

There are no horizontal forces acting on the meter stick, so we have only one force equation:

$$\sum F_y^e = 0: \qquad T_1 + T_2 - Mg - mg = 0$$

from which $T_1 + T_2 = (0.35 \text{ kg} + 0.5 \text{ kg})(9.80 \text{ m/s}^2)$

$$= 8.33 \text{ N}$$

Thus, the force equation yields the *sum* of the force readings on the scales, but not the individual values. We find these by applying the torque condition for the torques about the zero-cm end of the meter stick. With counterclockwise torques taken to be positive, we have

$$\sum \tau_0 = 0:$$

$$(0 \text{ m})T_1 - (0.25 \text{ m})Mg - (0.50 \text{ m})mg + (1.00 \text{ m})T_2 = 0$$

$$0 - (0.25 \text{ m})(0.35 \text{ kg})(9.80 \text{ m/s}^2)$$
$$- (0.50 \text{ m})(0.5 \text{ kg})(9.80 \text{ m/s}^2) + (1.00 \text{ m})T_2 = 0$$

from which $T_2 = 0.86 \text{ N} + 2.45 \text{ N} = 3.31 \text{ N}$

Then, $T_1 = 8.33 \text{ N} - 3.31 \text{ N} = 5.02 \text{ N}$

By selecting the point O as the reference point for calculating the torques, we eliminated any contribution from T_1 and directly obtained the value of T_2. (• Calculate the torques with respect to some other point, such as the center of the meter stick or a point 1 m to the *left* of O. The resulting equation will involve both T_1 and T_2. Show that when this equation is combined with the force equation, the same values for T_1 and T_2 result.)

Notice that by writing *two* torque equations, one for each end of the meter stick, the values of T_1 and T_2 can be found without using the force equation at all. For any particular problem, a little thought should reveal the point or points about which the torques should be calculated to yield simple and useful relationships.

Example 12–2

A block with a mass $m = 20$ kg is suspended from one end of a 3-m plank. The plank is attached to one of the piers at point O and rests on the other pier. What forces, F_1 and F_2, do the piers exert on the plank?

Solution:

There is only one force equation, that for the vertical direction:

$$\sum F_y^e = 0: \qquad F_1 + F_2 - Mg - mg = 0$$

from which $F_1 + F_2 = (50 \text{ kg} + 20 \text{ kg})(9.80 \text{ m/s}^2)$

$$= 686 \text{ N}$$

We choose point O for calculating the torques and we let counterclockwise torques be positive. Then,

$$\sum \tau_0 = 0: \qquad (1 \text{ m})F_2 - (1.5 \text{ m})Mg - (3 \text{ m})mg = 0$$

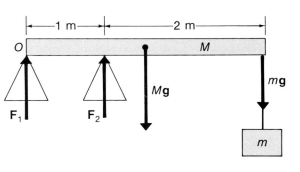

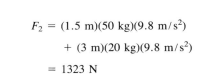

so that $F_2 = (1.5 \text{ m})(50 \text{ kg})(9.8 \text{ m/s}^2)$

$$+ (3 \text{ m})(20 \text{ kg})(9.8 \text{ m/s}^2)$$

$$= 1323 \text{ N}$$

Then,

$$F_1 = 686 \text{ N} - F_2 = 686 \text{ N} - 1323 \text{ N}$$

$$= -637 \text{ N}$$

We interpret the negative sign of F_1 in the following way. When we constructed the free-body diagram for the plank, we drew the vector $\mathbf{F}_1$ *upward*. The fact that the result for F_1 carries a negative sign means that the diagram is incorrect; $\mathbf{F}_1$ actually is directed *downward*. This is a general result: If the direction of a vector in a free-body diagram is chosen opposite to its actual direction, the result will contain a negative sign, thereby revealing the original incorrect choice.

Example 12–3

A long uniform plank rests against a frictionless vertical wall. The coefficient of static friction between the plank and the floor is $\mu_s = 0.35$. What is the smallest angle γ at which the plank can remain stationary?

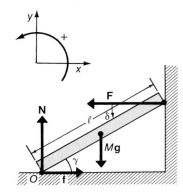

Solution:

We have

$$\sum F_x^e = 0: \qquad f - F = 0 \qquad (1)$$

$$\sum F_y^e = 0: \qquad N - Mg = 0 \qquad (2)$$

For the torques about point O, to which $\mathbf{N}$ and $\mathbf{f}$ make no contribution, we have

$$\sum \tau_O = 0: \qquad \ell(F \sin \gamma) - \tfrac{1}{2}\ell(Mg \cos \gamma) = 0 \qquad (3)$$

from which $\qquad F = \tfrac{1}{2}Mg \operatorname{ctn} \gamma$

Combining this result with (1) and (2) gives

$$\frac{f}{N} = \frac{1}{2} \operatorname{ctn} \gamma$$

As γ is decreased, f/N increases and eventually reaches the value $f/N = \mu_s = 0.35$, at which point the plank will begin to slip. Thus,

$$\operatorname{ctn} \gamma_{min} = 2 \mu_s = 0.70$$

so that $\qquad \gamma_{min} = 55.0°$

If friction exists between the top of the ladder and the wall, the system becomes indeterminate—the question cannot be answered from the given information. (• Can you see why?)

Example 12–4

A uniform sphere with a mass $M = 5$ kg is pulled (without rotating) at constant velocity up an inclined plane by a string that is attached to its surface. The string applies a force parallel to the plane, whose angle of inclination is 33°. If the coefficient of kinetic friction between the sphere and the plane is $\mu_k = 0.42$, find the string tension and the angle β that locates the point of attachment of the string, as shown in the diagram.

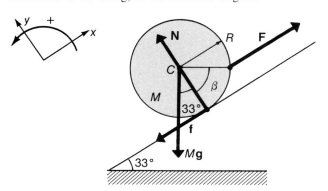

Solution:

Notice that although the sphere is in motion, there are no accelerations. Thus, the system is in dynamic equilibrium. The force equations are

$$\sum F_y^e = 0: \qquad N - Mg \cos 33° = 0$$

$$\sum F_x^e = 0: \qquad F - f - Mg \sin 33° = 0$$

Using $f = \mu_k N$, and solving for F, we have

$$F = Mg(\sin 33° + 0.42 \cos 33°)$$

$$= 43.95 \text{ N}$$

For the torques about the center C of the sphere,

$$\sum \tau_C = 0: \qquad RF \cos \beta - fR = 0$$

Substituting for f and using the above value for F, we find

$$\cos \beta = \frac{f}{F} = \frac{\mu_k Mg \cos 33°}{F}$$

$$\cos \beta = \frac{(0.42)(5 \text{ kg})(9.80 \text{ m/s}^2) \cos 33°}{43.95 \text{ N}} = 0.393$$

so that $\beta = 66.9°$

Example 12–5

A dresser with a mass $M = 60$ kg is pulled at constant velocity across a level floor by a force **F**, applied as in the diagram. If the coefficient of kinetic friction between the feet of the dresser and the floor is $\mu_k = 0.20$, find the force F and the normal forces N_F and N_R on the front and rear legs, respectively.

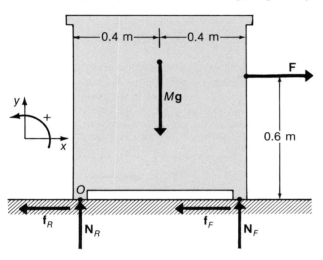

Solution:

Letting N_R and N_F represent the normal force on *both* rear

legs and *both* front legs, respectively, the force equations are

$$\sum F_y^e = 0: \quad N_R + N_F = Mg = (60 \text{ kg})(9.80 \text{ m/s}^2) = 588 \text{ N}$$

$$\sum F_x^e = 0: \quad -f_R - f_F + F = 0$$

Using $f = \mu_k N$, we have

$$F = \mu_k(N_R + N_F) = 0.20(588 \text{ N})$$

$$= 117.6 \text{ N}$$

The torques about the line connecting the rear legs (point O) give

$$\sum \tau_O = 0: \quad -(0.4 \text{ m})Mg - (0.6 \text{ m})F + (0.8 \text{ m})N_F = 0$$

from which

$$N_F = \frac{(0.4 \text{ m})(60 \text{ kg})(9.80 \text{ m/s}^2) + (0.6 \text{ m})(117.6 \text{ N})}{0.8 \text{ m}}$$

$$= 382.2 \text{ N}$$

Then, $N_R = 588 \text{ N} - N_F$

$$= 205.8 \text{ N}$$

Notice that $N_F > N_R$ because the force **F** tends to tip the dresser forward. (See also Problem 12–18.)

QUESTIONS

12–1 Describe and explain briefly what is meant by the terms (a) stable equilibrium, (b) unstable equilibrium, and (c) neutral equilibrium. Give several examples from everyday experience for each type of equilibrium.

12–2 A solid cylinder that has a height three times its diameter is resting on its base. Through what angle from the vertical may it be tipped before it will fall over?

12–3 Prove that for equilibrium with forces operating in a single plane, there can be at most three unknowns in the problem for a unique solution to be possible.

12–4 Three forces act on a rigid body. What conditions are necessary if the object is to be in dynamic equilibrium?

12–5 The forces acting on a rigid body are such that the torques about two distinct points P and Q vanish, i.e., $\tau_P = 0$ and $\tau_Q = 0$. The result is to provide six component equations. Does this guarantee that the body is in dynamic equilibrium? How does the situation change if the torques about a third distinct point R (not collinear with P and Q) also vanish, i.e., $\tau_R = 0$?

12–6 A three-legged stool rests on a level floor. Explain why in principle a unique solution for the load distribution on the legs can be given.

12–7 A four-legged chair rests on a level floor. Is the application of Eq. 12–2 sufficient to lead to a unique solution for the load distribution on the legs? Explain. It is a common observation that a four-legged chair does in fact have a unique load distribution on its legs if they are not too different in length. What additional factors operate in the real world to make this possible? [*Hint:* Are the component pieces of the chair completely rigid?]

12–8   A picture is hung on a wall using two eyebolts screwed into the frame, a connecting wire, and a hook attached to the wall, as in the diagram at right. It is found that unless sufficient slack exists in the length of the wire, the eyebolts may pull loose from the frame. Explain why.

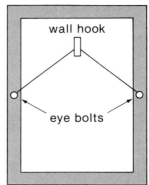

Question 12–8

PROBLEMS

12–1   A student manages to get his car stuck in a snow drift. Not at a loss, having studied physics, he attaches one end of a stout rope to the vehicle and the other end to the trunk of a nearby tree, allowing for a small amount of slack. The student then exerts a force **F** on the center of the rope in the direction perpendicular to the car-tree line, as shown in the diagram. If the rope doesn't stretch and if the magnitude of the applied force is 500 N, what is the force on the car? (Assume equilibrium conditions.)

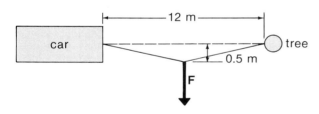

12–2   An automobile with a mass of 1500 kg has a wheel base (the distance between the axles) of 3.0 m. The C.M. of the automobile is on the center line at a point 1.2 m behind the front axle. Find the force exerted by the ground on each of the four wheels.

12–3   A block with a mass of 50 g is suspended from a meter stick at the 15-cm mark. The meter stick, which has a uniform density, will balance in a horizontal position if a pivot is placed at the 37-cm mark, as indicated in the diagram. What is the mass of the meter stick?

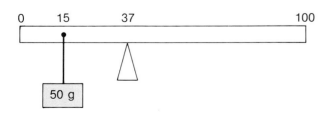

12–4   A trap door has a uniform density and a mass of 50 kg. The door is hinged along one side and is partially open,

making an angle of 30° with the horizontal. A rope that is attached to the side opposite the hinge and is parallel to the floor holds the door in this position, as shown in the diagram. (a) What is the tension in the rope? (b) What are the vertical and horizontal components of the force that the hinge exerts on the door?

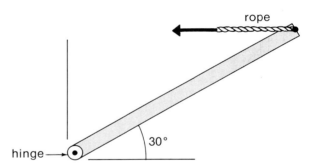

12–5   A homogeneous door has a height of 2.2 m, a width of 1.0 m, and a mass of 25 kg. The door is hinged at two points, one 0.30 m from the top and the other 0.30 m from the bottom. Assume that each hinge supports half of the door's weight. Find the horizontal and vertical components of the forces that the hinges exert on the door. (• Can you see why it is necessary to assume the equality of the vertical components of the hinge forces?)

12–6   To a reasonable approximation, the human arm may be considered to be a simple set of lever components, hinges, and tension-producing muscles. Suppose that a 10-kg block is held in the hand with the forearm horizontal and perpendicular to the upper arm, as shown in diagram (a), on opposite page. Use the dimensions given in diagram (b) and find (a) the tension T in the bicep, (b) both the compressional (vertical) and bending (horizontal) components of the force exerted on the humerus by the forearm at the elbow joint. Compare these results with the weight of the block. Can you see why it is tiring simply to hold a heavy block with the arm horizontal?

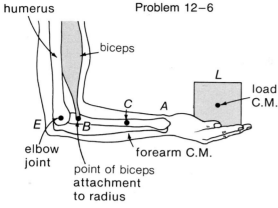

humerus Problem 12-6

biceps

L load
C.M.

C *A*

E *B*

elbow
joint

point of biceps
attachment
to radius

forearm C.M.

(a)

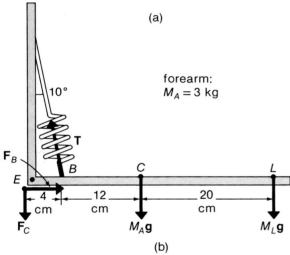

10°

forearm:
$M_A = 3$ kg

T

$\mathbf{F}_B$

B *C* *L*

E

4
cm

12
cm

20
cm

$\mathbf{F}_C$ $M_A\mathbf{g}$ $M_L\mathbf{g}$

(b)

12-7 A 100-kg lawn roller is to be raised over a curb that has a
height of 10 cm. The roller has a radius of 20 cm, and
essentially all of the mass is uniformly distributed within
the drum. The handle of the roller is attached to the ends
of the drum axis, as shown in the diagram. What is the
least force F that will succeed in pulling the roller over
the curb? What is the angle γ for this condition? [*Hint:*
Treat the problem as one of static equilibrium. Find the
force that just reduces to zero the normal force on the
drum at the lower ground level.]

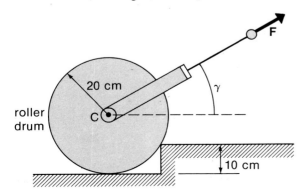

20 cm

F

γ

roller
drum

C

10 cm

12-8 A sphere with radius $R = 0.2$ m and mass $M = 12$ kg
rests between two smooth planks, as shown in the diagram. What force does each plank exert on the sphere?

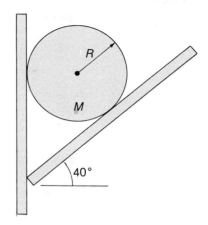

R

M

40°

12-9 A sphere is in static equilibrium as shown in the diagram. The elevation angle of the inclined plane is 35°.
What is the minimum coefficient of static friction necessary for the string to be vertical?

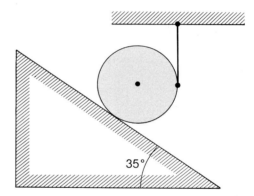

35°

12-10 A sphere rests on a plane that has an inclination angle of
35° while tied with a horizontal string, as shown in the
diagram. What is the minimum value of the coefficient
of static friction between the sphere and the plane that
will allow this equilibrium situation?

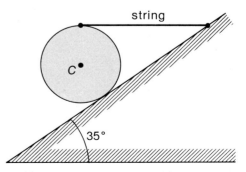

string

C

35°

12-11 Two planks, each with a length of 2.50 m, are joined
end-to-end by a hinge at point A, as shown in the dia-

gram. The planks stand on a frictionless horizontal surface, forming an isosceles triangle. A cord with a length of 1.5 m is attached to the midpoints of the two planks. Each plank has a mass of 10 kg. (a) Find the tension in the rope. (b) Find the hinge forces.

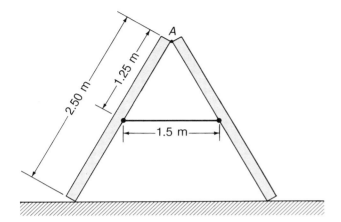

12–12 A 20-kg sign that has a uniform density hangs in front of an establishment. The sign is suspended by one hinge and a cable, as shown in the diagram. Find the tension in the cable and the components of the force that the hinge exerts on the sign.

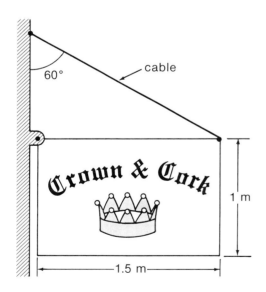

12–13 A ladder with a uniform density and a mass m rests against a frictionless vertical wall at an angle of 60°. The lower end rests on a flat surface where the coefficient of static friction is $\mu_s = 0.40$. A student with a mass $M = 2m$ attempts to climb the ladder. What fraction of the length L of the ladder will the student have reached when the ladder begins to slip? (See diagram at right.)

12–14 A post with a length L and a mass M rests with one end on the ground. The post is held in a vertical position by a thin wire that makes an angle γ with the horizontal and by a horizontal force **F** applied at a point that is a distance ℓ above the ground. The coefficient of static friction between the post and the ground is μ_s. Show that there is a minimum height ℓ for the point of application of the force **F** that allows a force of any magnitude to be applied without the bottom of the post slipping. Show that $\ell/L = 0.591$ for $\gamma = 60°$ and $\mu_s = 0.40$. This is an example of a *self-locking* system, a system that is particularly stable under the application of unbalancing forces. [*Hint:* Write the three equilibrium equations for the point O. Then, eliminate the tension T and solve for the ratio of f/N. Finally, take the limit, $F \rightarrow \infty$.]

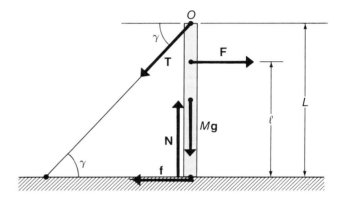

12–15 The boom of a crane has a length of 12 m and a mass that is small compared with all other masses in the structure. A 2000-kg load is suspended from the boom, as shown in diagram (a). The support cable is connected from the midpoint C of the boom to the winch at point Q, which is 3 m directly above the lower end of the boom at P. (a) What is the cable tension when the boom angle is 45°? (b) What are the components of the force that the pivot exerts on the boom at P? (c) At what point on the

Problem 12–13

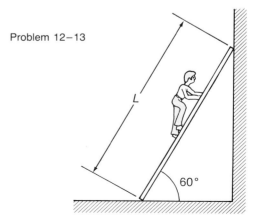

boom should the cable be attached in order to produce the smallest cable tension and still hold the load in equilibrium? [*Hint:* The geometry will be simplified by finding the missing entries—r, s, and α—in diagram (b).]

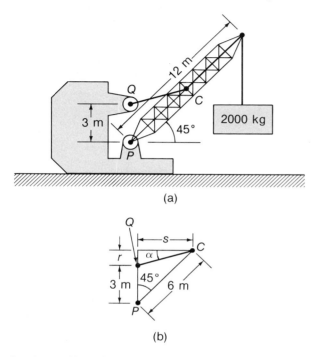

(a)

(b)

12–16 A uniform flexible cable has a mass of 100 kg and is suspended between two fixed points, A and B, at the same level, as shown in the diagram. At the support points the cable makes angles of 30°. (a) Find the components of the force exerted on each support by the cable. (b) Find the tension in the cable at the lowest point.

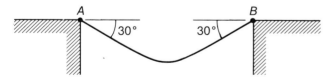

12–17 A rod that has a length ℓ and a negligible mass is placed against a vertical wall and is held in a horizontal position by a cord attached to the extended end. The cord makes an angle of 60° with the wall. A block with a mass m is suspended from the rod at a distance x from the wall. The coefficient of static friction between the end of the rod and the wall is $\mu_s = 0.20$. Show that the hanging block can have any mass whatsoever and the equilibrium will be maintained provided that $x \geq 5\ell/(5 + \sqrt{3})$.

12–18 Refer to Example 12–5. How large can the coefficient of kinetic friction μ_k become before the dresser will actually tip forward?

12–19 A massless string with a length ℓ is attached to a fixed
•• support at a point A and runs over a frictionless peg that is stationary at point B. The two points are at the same level and are separated by a distance $2b$. A block with a mass m is suspended from the free end of the string below the peg. Another identical block is attached to a pulley with negligible mass which rides at the center of the V formed by the string, as shown in the diagram. Use the energy principle and show that the system is in equilibrium when $\mathbf{y} = \ell - 4b/\sqrt{3}$.

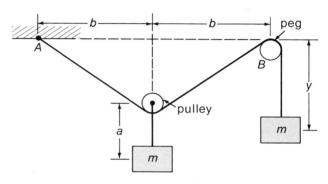

12–20 A hollow spherical shell is partially filled with an irregular mass of hardened cement and rests on a horizontal surface. Use the energy principle to show that the sphere will be in an equilibrium condition when the C.M. (point C) is directly below the center of the sphere (point O).

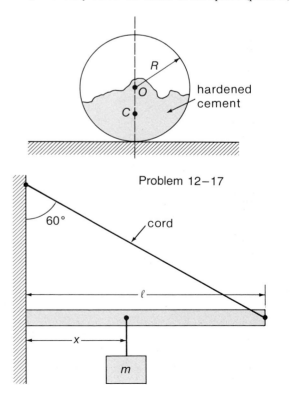

Problem 12–17

12–21 A solid homogeneous cube with sides of length b rests
•• balanced on the top of a cylindric surface that has a radius R, as shown in the diagram. Show that the system is
in a condition of stable equilibrium if the friction present
is sufficient to prevent sliding and if $R > \frac{1}{2}b$. [*Hint:*
Allow the cube to rotate without slipping through an
angle α. Find the new height of C. Take α to be small
and expand cos α and sin α.]

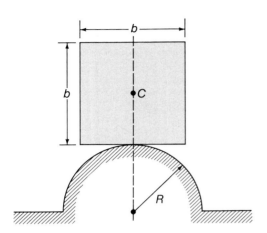

12–22 Determine the tension in the cable BC in the diagram.
Also, determine the horizontal and vertical components
of the force exerted on the massless strut AB at pin A.

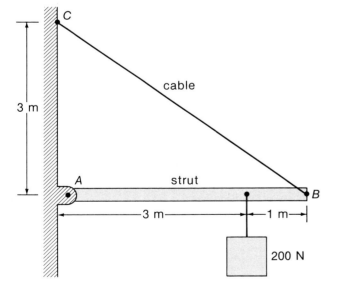

12–23 A uniform block with a height of 0.6 m and a width of
• 0.3 m is at rest on a plank, as shown in the diagram. The
coefficient of static friction between the block and the
plank is $\mu_s = 0.40$. (a) When one end of the plank is
slowly raised, will the block eventually slide or tip over?
At what angle θ will this occur? (b) Reconsider the situation for $\mu_s = 0.50$ and $\mu_s = 0.60$.

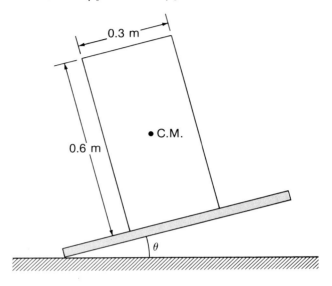

CHAPTER 13

GRAVITATION AND PLANETARY MOTION

Newton's theory of gravitation, first published in 1687, united the wealth of astronomical observations that had been accumulated over the centuries with the body of knowledge that is the science of physics. With Newton's theory, it suddenly became possible to explain in a simple way all of the observational data that previously had only been speculated upon. We begin the discussion of Newton's law of gravitation and its ramifications with a brief look at the historical background of astronomical observations and theories.

13–1 EARLY ASTRONOMY

Ancient astronomers believed in an Earth-centered solar system. Around 300 B.C., the Greek astronomers Heracleides and Aristarchus suggested that the Sun was the center of the solar system and that the Earth and the other planets revolved around the Sun in circular orbits. This suggestion was not supported by calculations and remained undeveloped. The idea was proposed again by the Polish astronomer Nicolaus Copernicus (1473–1543), who used the Sun-centered model to make detailed calculations of planetary positions. These calculations were still based on circular orbits, however.

The next major advance in the theory of planetary motion was made by the German astronomer Johannes Kepler (1571–1630). The data used by Kepler were obtained by the Danish astronomer Tycho Brahe (1546–1601), who built observatories in Denmark and later in Prague for the purpose of making precise measurements of planetary positions. Kepler tried a variety of planetary models in an effort to interpret the data that had been accumulated, particularly those concerning Mars. All of these attempts failed until Kepler finally abandoned the idea of circular orbits. Trying other geometric forms, he decided that planetary motions are best described in terms of *ellipses*.* Although he was mystically inclined toward the "perfect" circle, Kepler became the first to break with the ancient Greek idea of the uniform circular motion of planets and to propose a planetary model with elliptic orbits.

In 1609 Kepler published his famous work, *Astronomia Nova* ("New Astronomy"), which contained the first two of his three laws of planetary motion:

---

*The orbit of Mars has the largest deviation from a circle (eccentricity) among all the planets Kepler could have studied. If he had concentrated on any other planet, he might never have noticed that the orbit is elliptic, not circular.

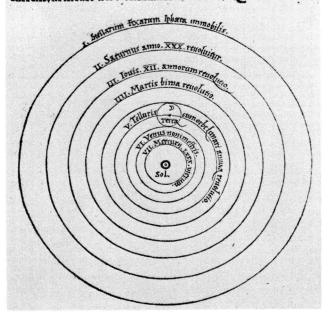

The Copernican heliocentric model of the solar system, as it appeared on a page in *De Revolutionibus.* (The epicycles on which the planets were presumed to move are not shown here.) (Courtesy of Chapin Library, Williams College.)

1. Every planet moves around the Sun in an orbit that is an ellipse, with the Sun located at one focus.

2. A straight line from the Sun to the planet sweeps out equal areas in equal time intervals.

Kepler's third law appeared in 1619 in *Harmonices Mundi* (''Harmony of the World''), buried amidst a mass of turgid mysticism that contained very little else of value:

3. The ratio of the square of the period of a planet to the cube of the length of the semimajor axis of its orbit is the same for all planets.

13–2   NEWTON'S GRAVITATION HYPOTHESIS

Elliptic planetary orbit about the Sun. The Sun is at the focus of the ellipse.

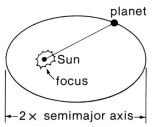

In his three laws of planetary motion, Kepler had given an accurate *description* of the way that planets orbit around the Sun. But by the time Newton turned his attention to the problem, nearly 50 years later, no one had provided a *fundamental theory* to support Kepler's laws. In 1665, Newton moved from Cambridge to his mother's farm in Woolsthorpe to escape the Black Plague that had begun to creep from London into the countryside. There, he studied quietly for two years, making important discoveries in optics, mathematics, and mechanics. During this period, Newton developed the theory of color, the differential calculus, and the first two laws of dynamics; in addition, he made a significant start on the theory of gravitation, which he finally completed and published in 1687.

Newton's law of gravitation is expressed by the equation*

$$F_{12} = -G\frac{m_1 m_2}{r_{21}^2}\hat{u}_r$$

(13–1)

where the vector notation is illustrated in Fig. 13–1. The vector F_{12} is the force exerted *on* m_1 by m_2. The location of m_1 relative to m_2 is specified by the vector $r_{21} = r_{21}\hat{u}_r$, where $\hat{u}_r$ is the unit vector in the direction from m_2 to m_1. Notice that $r_{21} = -r_{12}$, so that $F_{21} = -F_{12}$, in agreement with the requirement of Newton's third law.

The quantity G that appears in Eq. 13–1 is the *universal gravitation constant,* the best present value of which is

$$G = 6.673 \times 10^{-11} \text{ N·m}^2/\text{kg}^2$$

with an uncertainty of about 0.06 percent. (See Section 5–3 for a description of the Cavendish experiment, which leads to a value for G.)

Newton originally hypothesized that the effect of gravity due to a particular body should diminish with the distance from that body in the same way that the intensity of light diminishes with the distance from a source, namely, as the inverse square of the distance.† By this reasoning, Newton introduced the $1/r^2$ factor into the equation for the gravitational force. He went on to provide a basis for this assertion by using Kepler's third law of planetary motion. For a planet that revolves around the Sun with a period T in a circular or nearly circular orbit with a radius r, Kepler's third law can be expressed as $T^2 \propto r^3$. The period is $T = 2\pi r/v$ and the force F necessary to maintain the circular motion is the mass m of the planet multiplied by the centripetal acceleration, v^2/r. Combining these three relationships results in $F \propto m/r^2$. Newton understood this force to be the gravitational force exerted by the Sun on the planet. From Newton's third law, it is evident that the mass M of the Sun must enter the expression for the force in the same way that the planetary mass m enters. That is,

$$F = G\frac{mM}{r^2}$$

where G is the proportionality constant. This expression for F, deduced from Kepler's third law, is just the magnitude of the vector equation shown in Eq. 13–1.

Thus far we have made the tacit assumption that we are dealing with the gravitational interaction between *point* objects. Because such objects have no extent, there is no question about the meaning of the distance between objects. However, in the calculation of the gravitational force between the Earth and the Moon, Newton reasoned that, for extended

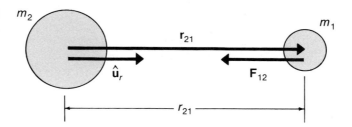

Fig. 13–1. The gravitational force on m_1 due to m_2 is F_{12}.

*The gravitational force is the only force we discuss in this chapter, so we dispense with the usual subscript g on F.

†The inverse-square law for light intensity was not formulated, even in rudimentary form, until 1604 when Kepler deduced the correct relationship.

spherical objects, the distance that enters the expression for the force law should be the distance between the *centers* of the objects. But Newton could not prove this assertion. With his brilliant theory incomplete, Newton chose not to make public the progress he had made. While devising the mathematical tools to solve the problem, Newton invented the integral calculus. Finally, he was able to make a proper calculation of the gravitational force between objects with finite size. The gravitational force law was then fully developed, and Newton included an exposition of the theory in the *Principia,* published in 1687.*

The importance of Newton's gravitational force law is not just that it correctly accounts for the motion of the planets around the Sun, but that it describes the gravitational force between *any* pair of objects. Newton's law of gravitation is truly a universal law of wide-ranging applicability.

13–3   THE GRAVITATIONAL FORCE ON AN EXTENDED OBJECT

Many types of problems involve the gravitational force on an extended object. Any such object (which we consider to be perfectly rigid) consists of a large number of particles (atoms). When we use the phrase "the force on the object," we really mean the vector sum of the forces $\mathbf{f}_i$ acting on all the individual particles in the object. The gravitational force due to the Earth on a particle with a mass m_i is $\mathbf{f}_i = m_i\mathbf{g}$, where the acceleration, $\mathbf{g} = -g\hat{\mathbf{k}}$, is directed downward (toward the center of the Earth), as in Fig. 13–2. For a relatively small everyday object, the vector $\mathbf{g}$ is uniform over the dimensions of the object. Then, the total gravitational force on the object is

Fig. 13–2. The individual particles m_i making up an object are acted upon by the gravitational forces $m_i\mathbf{g}$. The sum of all these individual forces is equivalent to a single force that acts at *C*, the C.M. of the object.

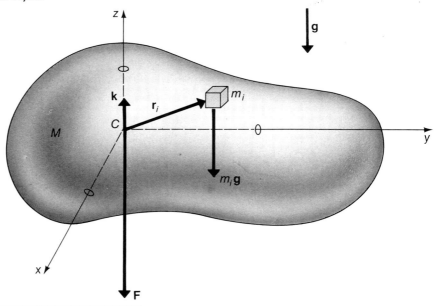

* During this period a more accurate value for the radius of the Earth also became available. This allowed for a better agreement between the gravitational acceleration at the surface of the Earth and at the Moon's orbit.

$$\mathbf{F} = \sum_i \mathbf{f}_i = \sum_i m_i\mathbf{g} = \left(\sum_i m_i\right)\mathbf{g} = M\mathbf{g}$$

where M is the total mass of the object.

Is there a single point on or in the object at which this effective force can be considered to act? Intuition suggests that there is such a point and that this point is the center of mass of the object. We can prove this suggestion to be correct in the following way. In Fig. 13–2, the point C is the C.M. of the object. If we calculate the total torque about C due to the gravitational force acting on each of the individual particles and sum the result, we find

$$\boldsymbol{\tau}_C = \sum_i \mathbf{r}_i \times \mathbf{f}_i = \sum_i \mathbf{r}_i \times m_i\mathbf{g} = \left(\sum_i m_i\mathbf{r}_i\right) \times \mathbf{g}$$

The term in parentheses (see Eq. 9–8) is proportional to the position vector of the C.M. in the center-of-mass coordinate frame and therefore vanishes; hence,

$$\boldsymbol{\tau}_C = 0$$

If a single force $\mathbf{F}$ is to represent the effect of all the individual forces $\mathbf{f}_i$, this force must also produce zero torque on the object. This will be the case only if the line of action of $\mathbf{F}$ passes through the C.M. of the object. Moreover, for any orientation of the object, $\mathbf{F}$ must still be directed through the center of mass. We conclude that $\mathbf{F}$ acts exactly *on* the C.M. of the object.

In the event that the object is so large that the gravitational acceleration vector $\mathbf{g}$ varies over the extent of the object, the point at which $\mathbf{F}$ acts no longer coincides with the C.M. of the object. This new point (called the *center of gravity*) is not unique because, in general, it depends on the orientation of the object with respect to $\mathbf{g}$. (• Can you see why this is so?) For the special case of a spherical object, such as the Earth, in which the density is either constant or a function of r (measured from the center) alone, the force $\mathbf{F}$ always acts at the center (the C.M.) even if $\mathbf{g}$ varies over the extent of the sphere. This is the crucial theorem that Newton proved with the aid of the newly invented integral calculus.

Example 13–1

The radius of the planet Mars is 3.40×10^6 m, and on the surface the acceleration due to gravity is 3.71 m/s². (Therefore, an astronaut on the Martian surface would have a weight of about one third his weight on the Earth's surface.) Use this information to calculate the mass M of Mars and its average density $\bar{\rho}$.

Solution:

In Section 5–3 we found that we can write two different equations for the gravitational force on an object with a mass m located near the Earth's surface, namely,

$$F = mg \quad \text{and} \quad F = G\frac{mM}{R^2}$$

We now interpret M and R as the values for Mars. Equating these

expressions for F and solving for M, we find

$$M = \frac{gR^2}{G}$$

Substituting the known values of g, R, and G, we find

$$M = \frac{gR^2}{G} = \frac{(3.71 \text{ m/s}^2)(3.40 \times 10^6 \text{ m})^2}{6.67 \times 10^{-11} \text{ N·m}^2/\text{kg}^2}$$

$$= 6.43 \times 10^{23} \text{ kg}$$

The volume of Mars is

$$V = \tfrac{4}{3}\pi R^3 = \tfrac{4}{3}\pi(3.40 \times 10^6 \text{ m})^3$$

$$= 1.646 \times 10^{20} \text{ m}^3$$

Then, the average density is

$$\bar{\rho} = \frac{M}{V} = \frac{6.43 \times 10^{23} \text{ kg}}{1.646 \times 10^{20} \text{ m}^3}$$

$$= 3.90 \times 10^3 \text{ kg/m}^3 = 3.90 \text{ g/cm}^3$$

which is slightly less than the average density of the Earth, namely, 5.50 g/cm³.

Example 13-2

Determine the gravitational force on a particle with mass m located a distance R from the center of a slender homogeneous rod with mass M and length ℓ, as shown in the diagram.

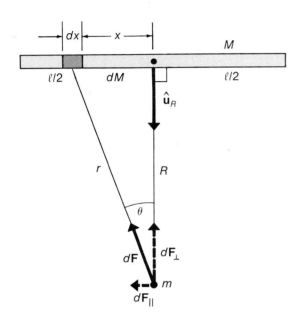

Solution:

The mass element $dM = (M/\ell)\, dx$ is a distance x from the center of the rod and a distance $r = \sqrt{x^2 + R^2}$ from the particle m. The force exerted on m by dM is $d\mathbf{F}$, the components of which are $d\mathbf{F}_\perp$ and $d\mathbf{F}_\parallel$. Another mass element at a distance x to the right of the midpoint of the rod also produces a perpendicular

force component $d\mathbf{F}_\perp$; however, the parallel component is directed opposite to $d\mathbf{F}_\parallel$. Because the particle m is directly opposite the midpoint of the rod, the parallel force components all cancel. The total force $\mathbf{F}$ can then be obtained by integrating $d\mathbf{F}_\perp$ from $x = 0$ to $x = \ell/2$ and doubling the result. We have

$$d\mathbf{F}_\perp = -dF \cos\theta\, \hat{\mathbf{u}}_R = -\frac{Gm(M/\ell)\, dx}{x^2 + R^2} \cdot \frac{R}{(x^2 + R^2)^{1/2}} \hat{\mathbf{u}}_R$$

$$= -\frac{GmMR}{\ell} \cdot \frac{dx}{(x^2 + R^2)^{3/2}} \hat{\mathbf{u}}_R$$

Then,

$$\mathbf{F} = 2\int d\mathbf{F}_\perp = -2\frac{GmMR}{\ell} \hat{\mathbf{u}}_R \int_0^{\ell/2} \frac{dx}{(x^2 + R^2)^{3/2}}$$

$$= -2\frac{GmMR}{\ell} \hat{\mathbf{u}}_R \cdot \frac{x}{R^2\sqrt{x^2 + R^2}}\Big|_0^{\ell/2}$$

$$= -2\frac{GmMR}{\ell} \hat{\mathbf{u}}_R \cdot \frac{\ell/2}{R^2\sqrt{(\ell/2)^2 + R^2}}$$

$$= -2\frac{GmM}{\ell R} \hat{\mathbf{u}}_R \cdot \frac{1}{\sqrt{1 + (2R/\ell)^2}}$$

Now, if the particle m is at a great distance from the rod, so that $R \gg \ell$, then in the denominator of the expression for $\mathbf{F}$, the factor unity can be neglected in comparison with $(2R/\ell)^2$. Thus, we have

$$\mathbf{F}(R \gg \ell) = -G\frac{mM}{R^2} \hat{\mathbf{u}}_R$$

which is the same as the result for a particle with mass M replacing the rod. This is to be expected because when $R \gg \ell$, the dimensions of the rod are no longer important and the rod is essentially pointlike.

13-4 GRAVITATIONAL POTENTIAL ENERGY

If a particle with a mass m is moved from a distance R_1 to a distance R_2 measured from another pointlike object with a mass M, the work done by the gravitational force is (see Example 7-9)

$$W = -GmM \int_{R_1}^{R_2} \frac{dr}{r^2} = GmM\left(\frac{1}{R_2} - \frac{1}{R_1}\right) \tag{13-2}$$

The associated change in potential energy is the negative of the work done (see Eq. 8–12); that is,

$$U_2 - U_1 = -GmM\left(\frac{1}{R_2} - \frac{1}{R_1}\right) \tag{13–3}$$

We define a convenient reference level for the potential energy by taking U to be zero for $R_1 \rightarrow \infty$. Then, for a separation $R_2 = r$, we have

$$U(r) = -G\frac{mM}{r} \tag{13–4}$$

Notice that we write $U(r)$ instead of $U(\mathbf{r})$ because the potential energy depends only on $r = |\mathbf{r}|$ when both objects are pointlike.

Suppose that the origin of a coordinate system is located at the position of the object M. The potential energy function for M and a particle m is defined for any point in space at which m might be located. At a particular point P, the force on m is related to the variation of U in the neighborhood of P. In general, the force will have x-, y-, and z-components that are related to the partial derivatives of U in the respective directions. However, in the event that M is a pointlike (or spherically symmetric) object, the situation is equivalent to a one-dimensional problem in that the potential energy function depends only on the distance r between m and M. Then, the force is equal to the negative of the derivative of $U(r)$ with respect to r.* Using the notation defined in Fig. 13–3, we have

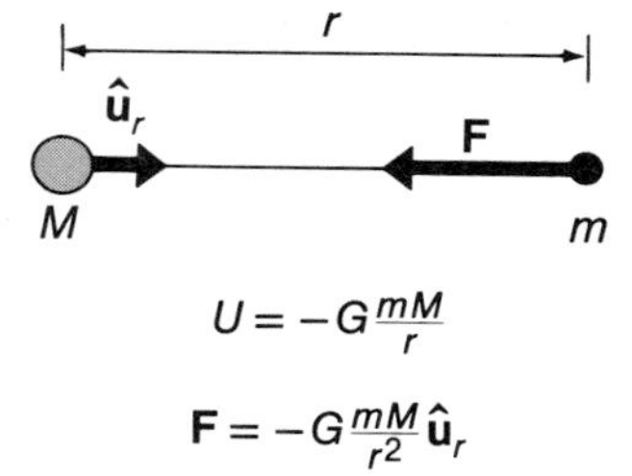

$$\mathbf{F}(\mathbf{r}) = -\frac{dU(r)}{dr}\hat{\mathbf{u}}_r = -G\frac{mM}{r^2}\hat{\mathbf{u}}_r \tag{13–5}$$

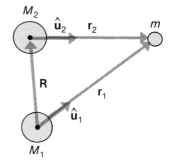

$$U = -G\frac{mM}{r}$$

$$\mathbf{F} = -G\frac{mM}{r^2}\hat{\mathbf{u}}_r$$

Fig. 13–3. Potential energy and force for a particle with a mass m in the vicinity of a pointlike object with a mass M.

which is the expected result and demonstrates the consistency of our definitions of work, potential energy, and force.

The Principle of Superposition. Throughout the discussions of force, we have assumed that the net effect of the application to an object of a number of forces, $\mathbf{F}_1$, $\mathbf{F}_2$, $\mathbf{F}_3$, $\cdots$, is simply the vector sum of the forces: $\mathbf{F}_{net} = \mathbf{F}_1 + \mathbf{F}_2 + \mathbf{F}_3 + \cdots$. This is actually one statement of a very important physical idea—the *principle of superposition*. It is a fact, established by experiment, that forces combine by ordinary (linear) vector addition. This result is not guaranteed by any theory; the introduction of a third particle into a two-particle system could conceivably alter the way in which the forces combine for two particles. However, according to experiment, when two or more particles exert gravitational (or other) forces on another particle, the net result is the vector sum of the individual forces, each calculated without regard for the presence of the other particle or particles. If the particles constitute an extended object, the summation procedure is replaced by an integration.

We assert that potential energy also combines by ordinary (linear) algebraic summation. For example, suppose that we assemble three gravitationally interacting particles into the configuration shown in Fig. 13–4. We imagine that the particles are originally dispersed infinitely far from one another. First, we bring M_1 into position; no work is

Fig. 13–4. The geometry of three gravitationally interacting pointlike objects.

---

*Since the potential energy $U(\mathbf{r})$ is only a function of $|\mathbf{r}| = r$, Eq. 8–4 relating the force to the derivative of the potential may be written $\mathbf{F} = -\dfrac{dU(r)}{dr}\hat{\mathbf{u}}_r$.

involved in this process, and the potential energy of M_1 by itself is zero:

$$U_1 = 0$$

Next, we bring M_2 into position; the potential energy of the M_1–M_2 combination is

$$U_2 = -G\frac{M_1 M_2}{R}$$

Finally, we bring the third particle m into position. The potential energy that results from this process, we assert, is

$$U_3 = -G\frac{M_1 m}{r_1} - G\frac{M_2 m}{r_2}$$

That is, U_3 is calculated by considering the potential energies of the M_1–m and M_2–m combinations *separately and independently*. Then, by the same reasoning, the potential energy of the three-particle system is

$$U = U_1 + U_2 + U_3$$

$$= -G\frac{M_1 M_2}{R} - G\frac{M_1 m}{r_1} - G\frac{M_2 m}{r_2}$$

If M_1 and M_2 are held fixed (that is, if $\mathbf{R}$ is maintained constant), the force on m at any point is obtained from*

$$\mathbf{F} = -\frac{dU}{dr_1}\hat{\mathbf{u}}_1 - \frac{dU}{dr_2}\hat{\mathbf{u}}_2$$

$$= -G\frac{M_1 m}{r_1^2}\hat{\mathbf{u}}_1 - G\frac{M_2 m}{r_2^2}\hat{\mathbf{u}}_2$$

Thus, the assumption of a linear algebra for the potential energy results in a linear force equation, as demanded by experiment. We conclude that potential energy also obeys the principle of superposition. (We can only infer this conclusion because forces are directly measurable whereas potential energies are not.)

Presumably, all of the basic forces in Nature, and their corresponding potential energies, obey the superposition principle—unless the forces involved are too large. It is difficult to give a precise meaning of "too large," but the strong force within nuclei and the gravitational force near extremely massive and dense objects (such as the curious astronomical objects called *black holes*), for example, seem to produce effects at variance with the principle of linear superposition or even the possibility of introducing a simple potential energy–force relationship. For all of the situations we consider, however, the principle is completely valid.

The Gravitational Field Strength.   Sometimes it is convenient to describe gravitational problems in terms of the *field* concept (see Section 8–5). An object with a mass M sets up in the space surrounding it a condition to which another object (a *test object*) will respond by experiencing a force directed toward M. This "condition" is the *gravitational field* of M, and we say that M is the *source* of the field. Assume that M is spherically symmetric so that we can treat it as a pointlike object. Then, if the mass of the test object is m, we formally define the gravitational field strength to be

---

*As the mass m moves about, both $\mathbf{r}_1$ and $\mathbf{r}_2$ change simultaneously. The relationship of the force to the potential requires taking both the derivative with respect to r_1 holding r_2 fixed and the derivative with respect to r_2 holding r_1 fixed.

$$g(\mathbf{r}) = \lim_{m \to 0} \frac{F(\mathbf{r})}{m} = -G\frac{M}{r^2}\hat{\mathbf{u}}_r$$

where r and $\hat{\mathbf{u}}_r$ refer to the center of M.

This form for the definition emphasizes the fact that the quantity $\mathbf{g}$, a specific measure of the gravitational field, depends only on the source mass M. The quantity $\mathbf{g}$ also represents the acceleration that is experienced by a test object placed in the field of M. The force on a test object m in the field $\mathbf{g}$ is

$$\mathbf{F} = m\mathbf{g}$$

which is the vector form of the gravitational force equation we have used in earlier chapters. When the source mass is the Earth and $r = R_E$, the magnitude of $\mathbf{g}$ is the familiar constant $g = 9.80$ m/s².

13–5   THE POTENTIAL ENERGY OF A SPHERICAL SHELL AND A PARTICLE

We now calculate the potential energy of a system that consists of a particle with a mass m at an arbitrary point P that is a distance r from the center of a thin spherical shell with mass M, thickness t, and radius R. We allow both $r > R$ and $r < R$. The geometry is shown in Fig. 13–5. We shall prove the theorem mentioned in Section 13–2, that the potential energy of m and the force acting on m due to the shell are the same as for a pointlike object with a mass M located at C when $r > R$. This interesting result also applies for the case of a solid sphere. (• Can you see how this extension can be made?)

We begin by considering a thin ring of material with density ρ (shown shaded in Fig. 13–5a), every part of which is a distance s from the point P located outside the shell. The width of this ring is $R \, d\theta$, its thickness is t, and its radius is $R \sin \theta$. Consequently, the mass dM of the ring is

$$dM = R \, d\theta \cdot 2\pi R \sin \theta \cdot t \cdot \rho$$
$$= 2\pi R^2 t\rho \sin \theta \, d\theta$$

Then, the potential energy of the ring and the particle m is (refer to Eq. 13–4)

$$dU = -Gm\frac{dM}{s} = -\frac{2\pi GmR^2 t\rho \sin \theta \, d\theta}{s}$$

where we again take $U = 0$ for infinite separation.

Fig. 13–5. Geometry for the calculation of the potential energy of a thin spherical shell and a particle. The thickness of the shell is t. (a) The particle location P is outside the shell. (b) The particle location P is inside the shell.

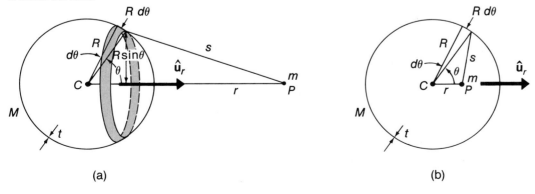

(a)   (b)

To obtain the potential energy due to the entire shell, we must integrate this expression for dU. It is easier to integrate over s instead of over the angle θ, so we make a change of variable. Referring to Fig. 13-5a, we see that the law of cosines can be used to write

$$s^2 = r^2 + R^2 - 2rR \cos \theta$$

Differentiating this expression with respect to θ for $R = $ constant and for a particular (fixed) value of r, we obtain

$$2s \frac{ds}{d\theta} = 2rR \sin \theta$$

so that

$$\frac{\sin \theta \, d\theta}{s} = \frac{ds}{rR}$$

Substituting this result into the equation for dU, we have

$$dU = -\frac{2\pi GmRt\rho}{r} ds \qquad \textbf{(13-6)}$$

We now integrate over s to obtain the potential energy for the entire spherical shell. For $r > R$, the variable s ranges from $s = r - R$ (corresponding to $\theta = 0$) to $s = r + R$ (corresponding to $\theta = \pi$). Thus, we find

$$U(r > R) = \int dU = -\frac{2\pi GmRt\rho}{r} \int_{r-R}^{r+R} ds$$

$$= -\frac{2\pi GmRt\rho}{r} [(r + R) - (r - R)]$$

$$= -\frac{Gm}{r} \cdot (4\pi R^2 t\rho)$$

Now, $4\pi R^2 t$ is just the volume V of the thin shell, and the mass is $M = V\rho$. Therefore, the expression for U becomes

$$U(r > R) = -G\frac{mM}{r} \qquad \text{outside shell} \qquad \textbf{(13-7)}$$

which is exactly the result we would obtain if the shell were a pointlike mass M located at the center (Eq. 13-4). Then, we obtain for the force on m

$$\mathbf{F(r)} = -\frac{dU}{dr}\hat{\mathbf{u}}_r = -G\frac{mM}{r^2}\hat{\mathbf{u}}_r \qquad (r > R) \qquad \text{outside shell} \qquad \textbf{(13-8)}$$

in agreement with Eq. 13-1.

Notice that a thick shell or even a solid sphere can be considered to be made up of a large number of thin shells. If each of these thin shells has uniform density, even though different shells may have different densities, the results for U and $\mathbf{F}$ still apply. That is, *the gravitational effect of a spherically symmetric distribution of matter on a particle outside the distribution is the same as that of a pointlike object with equal mass located at the center of the sphere.*

Next, consider the particle m to be located *inside* the shell ($r < R$); refer to Fig. 13-5b. We again must integrate Eq. 13-6, but now the variable s ranges from $s = R - r$ ($\theta = 0$) to $s = R + r$ ($\theta = \pi$). Thus,

$$U(r < R) = -\frac{2\pi GmRt\rho}{r} \int_{R-r}^{R+r} ds$$

$$= -\frac{2\pi GmRt\rho}{r} [(R + r) - (R - r)]$$

$$= -\frac{Gm}{R} \cdot (4\pi R^2 t\rho)$$

so that

$$U(r < R) = -G\frac{mM}{R} \qquad \text{inside shell} \qquad \text{(13-9)}$$

Because R is a constant, the force on m vanishes:

$$\mathbf{F(r)} = -\frac{dU}{dr}\hat{\mathbf{u}}_r = 0 \qquad (r < R) \qquad \text{inside shell} \qquad \text{(13-10)}$$

Thus, we have the remarkable result that for a uniform spherical shell the gravitational force on a particle inside the shell is *zero*. With the same argument that we used for $r > R$, we conclude that $\mathbf{F(r)} = 0$ within the hollow of any spherically symmetric shell of matter of whatever thickness.

What is the basic reason for the cancellation of the gravitational force inside the symmetric shell of matter? Referring to Fig. 13-5b, we see that most of the matter of the shell is located at distances $s > R$, resulting in a force contribution on the particle directed toward the center C. However, the matter in the shell located nearby with $s < R$, while containing less mass, is also closer to the particle. The force contribution due to this part of the shell is directed away from the center C, and because of the inverse-square-law nature of the universal law of gravitation, it just cancels the other contribution. No such cancellation occurs when the particle is outside the shell, since all the matter of the shell exerts a force directed toward the center C. Figure 13-6 illustrates the results we have obtained for the potential energy and the force. The graphs are shown for a thick shell with inner radius R_0 and outer radius R. (The case of a solid sphere corresponds to $R_0 = 0$.) For $R_0 < r < R$, the exact forms of the functions depend on the radial variation of the density. If the density is constant, the force $F(r)$ is given by the function

$$\mathbf{F}(r) = -\frac{GM}{R^3 - R_0^3}\left(r - \frac{R_0^3}{r^2}\right)\hat{\mathbf{u}}_r \qquad \text{(13-11)}$$

See Problem 13-13.

Notice that in these calculations there are two independent sets of variables. First, there are the coordinates of the mass element within the object M, over which the integration is carried out to obtain the potential energy. Second, there are the coordinates of the particle m which, although arbitrary, are nonetheless held fixed during the integration over M. Then, the force on m is obtained by differentiating the expression for the potential energy with respect to the coordinates of m (here given by r).

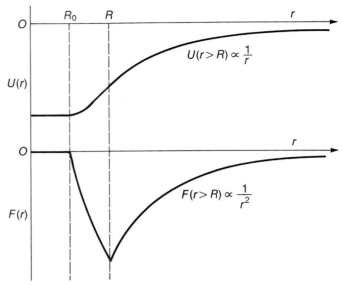

Fig. 13–6. The gravitational potential energy and the force on a particle due to a spherical shell of matter.

Example 13–3

A deep mine shaft is dug into the Earth with its lowest level at a distance r from the center of the Earth. What is the gravitational acceleration at this depth?

Solution:

Imagine dividing the matter of the Earth, M_E, into a solid sphere of radius r and mass M and a thick shell with inner radius r and outer radius R_E (the radius of the Earth) with mass $M_E - M$, as in the diagram. The terrestrial matter in the thick shell exerts no net force on a test particle located at the bottom of the shaft. The solid spherical portion of the Earth exerts a force directed toward the center of the Earth equivalent to that of a particle with the same mass, M, located at C. Thus the gravitational field strength at the bottom of the shaft is

$$\mathbf{g}(r) = -G\frac{M}{r^2}\hat{\mathbf{u}}_r$$

If we make the simplifying assumption of a constant terrestrial density, we then have $M = (r/R_E)^3 M_E$. (• Verify this expression for M.) Thus

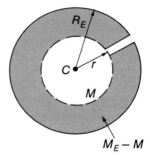

$$\mathbf{g}(r) = -G\frac{M_E}{R_E^2}(r/R_E)\hat{\mathbf{u}}_r = -g_E(r/R_E)\hat{\mathbf{u}}_r$$

where g_E is the acceleration of gravity at the Earth's surface. (• Show that Eq. 13–11 leads to the same result for $\mathbf{g}(r)$ if R_0 is set equal to zero.)

The deepest inhabitable shafts are only some 3 km down, so the decrease in g is only $\left(\dfrac{R_E - r}{R_E}\right) = 3/6370$ or approximately 0.05 percent.

Example 13–4

If an object is propelled upward from the surface of the Earth with a sufficient velocity, it will overcome its gravitational bond to the Earth and never fall back to its home planet. How-

ever, an object can escape from the Earth's gravity and still be bound within the solar system because of the enormous gravitational effect of the Sun. The minimum velocity required for an object to escape from the Earth (and yet remain within the solar system) is essentially the same as that required to remove the object to an infinite distance from the Earth, ignoring the effects of the Sun. (• Can you argue that this approximation is reasonable?) Subject to this approximation, what is the value of the *escape velocity* v_E for the Earth?

Solution:

If an object has an initial kinetic energy $\frac{1}{2}mv_E^2$ at the Earth's surface, this energy will be completely expended in raising the initial gravitational potential energy, $-GmM_E/R_E$, to zero as $r \to \infty$. That is, the total energy of the object, $K + U$, becomes exactly zero as $r \to \infty$, and so must also be exactly zero at the Earth's surface. That is,

$$K + U = \tfrac{1}{2}mv_E^2 - G\frac{mM_E}{R_E} = 0$$

from which

$$v_E = \sqrt{2GM_E/R_E}$$

$$= \sqrt{\frac{2(6.67 \times 10^{-11}\ \text{N·m}^2/\text{kg}^2)(5.98 \times 10^{24}\ \text{kg})}{6.38 \times 10^6\ \text{m}}}$$

$$= 1.12 \times 10^4\ \text{m/s}$$

$$= 11.2\ \text{km/s}$$

If an object is to be propelled to an infinite distance from the Earth, taking into account the effect of the Sun, the escape velocity increases to 43.5 km/s (see Problem 13–10). In fact, the gravitational potential energy of an object on the Earth's surface is about 93 percent *solar* and only about 7 percent *terrestrial*. (• Can you verify these numbers?)

13–6   PLANETARY MOTION

One of the important applications of Newtonian gravitation theory is in the study of planetary motion. The Sun is by far the most massive object in the solar system. Consequently, the motion of every planet is dominated by the gravitational effect of the Sun and is influenced very little by the presence of the other planets. Thus, planetary motion in the solar system is the result of a collection of essentially two-body, planet-plus-Sun interactions.*

The planet-plus-Sun system obeys Kepler's first law, namely, that the planet describes an elliptic orbit about the Sun. Figure 13–7 shows the appropriate geometry. The extent to which the elliptic orbit deviates from a circular path is measured by the parameter called the *eccentricity e*. Note that when $e = 0$, the *semimajor axis a* becomes equal to the *semiminor axis b* and the *focus F* moves to the center of the path C. The ellipse has become a circle with the Sun at the center.

The general elliptic orbital motion is discussed in the back of this book in the first of the Enrichment Topics. In the present section we discuss only the case of circular planetary orbits.

Circular Planetary Orbits.   The Earth's solar orbit (and those of Venus and the major planets Jupiter, Saturn, Uranus, and Neptune) are approximately circular.† A fairly accurate description of the motion of these planets may be obtained by considering the solar orbits to be circular with the Sun at the center of the circle, as shown in Fig. 13–8.

The first law of Kepler is satisfied by hypothesis since a circular orbit belongs to the class of ellipses (i.e., with zero eccentricity, $e = 0$).

The second law or areal law of Kepler follows from the fact that the angular momentum of the planet about the Sun is constant if we neglect the gravitational forces of the other planets. We have for the angular momentum

$$L = mrv_\phi = mr^2\omega \qquad (13\text{–}12)$$

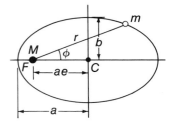

Fig. 13–7. The general elliptic planetary orbit. The Sun with mass M is located at the *focus* of the ellipse. The planet with mass m orbits about the Sun with coordinates r and ϕ. The ellipse is a symmetric figure with center C, semimajor axis a, semiminor axis b, and eccentricity $e = \dfrac{1}{a}\sqrt{a^2 - b^2}$. The focus F is a distance ae from C.

Fig. 13–8. Circular solar orbit of a planet with the Sun at the center. The orbital coordinates are r and ϕ, with r constant. The velocity of the planet relative to the Sun is $\mathbf{v} = v_\phi\hat{\mathbf{u}}_\phi$, since $v_r = 0$.

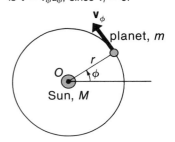

---

*That is not to say that planet-planet orbital perturbations are without interest. It was the perturbation in the orbital motion of Uranus that led to the prediction of the existence of Neptune and its later discovery in 1846. The planet Pluto also owes its discovery (1930) to predictions based on perturbation effects.

†The eccentricity for the Earth's orbit is only $e = 0.017$. The solar orbits of Mercury, Mars, and Pluto deviate notably from circular paths; for example, the eccentricity of Mercury is $e = 0.206$.

The area swept out by the line joining the planet and the Sun in a time Δt is $\Delta A = \frac{1}{2}r^2 \Delta\phi = \frac{1}{2}r^2\omega \Delta t$. Thus

$$dA/dt = \lim_{\Delta t \to 0} \Delta A/\Delta t = \frac{1}{2}r^2\omega = \frac{L}{2m} \qquad (13\text{--}13)$$

a constant.

Finally, the third law of Kepler is also straightforwardly demonstrated for the present simple case. Newton's second law requires that the centripetal acceleration should result from the planet-Sun gravitational force, or

$$\frac{mv_\phi^2}{r} = \frac{GmM}{r^2} \qquad (13\text{--}14)$$

With v_ϕ and r constants, the period T becomes

$$T = \frac{2\pi r}{v_\phi} \qquad (13\text{--}15)$$

Squaring Eq. 13–15 and eliminating v_ϕ appearing in Eq. 13–14 gives

$$T^2 = \frac{4\pi^2}{GM}r^3 \qquad (13\text{--}16)$$

For a circular orbit, radius r corresponds to the equivalent elliptic semimajor axis.

Binary Stars. A factor we have ignored so far is the effect on the motion of a gravitating pair of masses m_1 and m_2 if *neither* mass were much larger than the other. In such an event the two masses would orbit around their common center of mass, as shown in Fig. 13–9. In general, the shape of each orbit would be elliptic.

For the relevant masses the modification of the solar planetary orbits is very slight. For example, the C.M. of the Earth-Sun system is only 455 km from the center of the Sun. Since the solar radius is 6.96×10^5 km, this C.M. is well inside the Sun's interior. In comparison to these dimensions, the Earth's mean solar orbit radius is 1.496×10^8 km. However, this mass effect is quite significant for many astronomical systems.

One of the best-studied binary star systems is that of *Sirius*. One member of the pair, Sirius A, is a bright star, whereas Sirius B is so faint that it was not observed until its existence had been inferred by studying the motion of Sirius A. Figure 13–10 shows the intertwining tracks of the two stars as they moved across the sky during the period from 1900 to 1970. The tracks A and B represent the star motions; track C represents the motion of the C.M. (• Estimate the mass ratio M_A/M_B using Fig. 13–10.)

Fig. 13–9. (a) Two stars orbit around their common center of mass at O. (b) Elliptic motion of m_1 and m_2 around their C.M. at O. Note that O is the left-hand focal point of the orbit of m_2 and the right-hand focal point of the orbit of m_1.

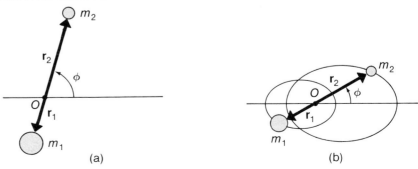

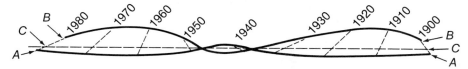

Fig. 13–10. The Sirius binary star system. The orbital plane is not perpendicular to our line of sight.

QUESTIONS

13–1 Define and describe briefly the terms (a) universal law of gravitation, (b) center of gravity, and (c) gravitational field strength.

13–2 The superior planets (those further from the Sun than the Earth) occasionally reverse their apparent motions against the star background as seen from the Earth, as shown in the diagram. Explain how such apparent motion is possible in the Copernican solar system. [*Hint:* The orbital angular speed of the planets decreases with increasing distance from the Sun.]

13–3 State Kepler's three laws in your own words. What is the major significance of the second law?

13–4 Explain how the planetary moons may be used to determine the mass of the parent planets. How could artificial satellites provide the same information?

13–5 Why must all stable satellite orbits be in planes that include the center of the Earth?

13–6 Explain how it might be possible for a satellite to be in a stable circular orbit about the Earth and yet remain always directly overhead at some point on the Earth. Why is this possible only for an equatorial point on the Earth?

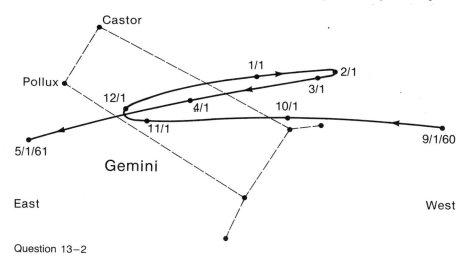

Question 13–2

PROBLEMS

Section 13–2

13–1 Two ocean liners, each with a mass of 40,000 metric tons, are moving on parallel courses 100 m apart. What is the magnitude of the acceleration of one of the liners toward the other due to the mutual gravitational attraction?

13–2 In the experiment Cavendish performed (see Section 5–3), he used lead spheres with radii of 1 in. and 4 in.

When the centers of two such spheres are separated by $r = 6$ in., the force of attraction is 2.76×10^{-8} N. Determine the value of G. (Refer to Table 1–2 for the density of lead.)

Section 13–3

13–3 A length of wire with a mass M is bent into an arc of a circle with radius R, as shown in the diagram on the next page. A particle with mass m is located at the center of the circular arc. What gravitational force does the wire exert on the particle?

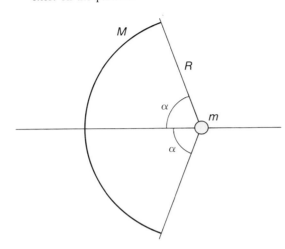

13–4   A straight thin rod has a mass M and a length L. Show that the gravitational force on a particle with a mass m located in line with the rod at a distance x from the near end is $F = GmM/x(x + L)$.

13–5   Calculate the fractional difference $\Delta g/g$ in the acceleration due to gravity at points on the Earth's surface nearest to and farthest from the Moon, taking into account the gravitational effect of the Moon. (This difference is responsible for the occurrence of the *lunar tides* on the Earth.)

13–6   What is the value of the acceleration due to gravity at the surface of (a) the Moon and (b) the Sun? Express the accelerations in units of g.

13–7   At what point along the line connecting the Earth and the Moon is there zero gravitational force on an object? (Ignore the presence of the Sun and the other planets.)

Section 13–4

13–8   What is the gravitational potential energy of a particle with a mass m located at the center of a thin homogeneous circular ring with a radius a and a mass M? What is the force on m? Explain physically the qualitative difference in the two results.

13–9   Show that the potential energy of a system consisting of four equal-mass particles M located at the corners of a square having sides of length d is $U = -(GM^2/d)(4 + \sqrt{2})$.

13–10 What minimum velocity must an object have at the surface of the Earth if it is to escape from the solar system (assumed to consist only of the Sun and the Earth) and

proceed infinitely far away? Is the *direction* of the velocity important?

13–11 A particle is in a circular orbit around the Earth at a height h above the Earth's surface. If the amount of work required to raise the particle from the surface to the orbit height is equal to the kinetic energy of the particle in its orbit, show that $h = \frac{1}{2}R_E$, where R_E is the radius of the Earth.

13–12 Suppose that a shaft is drilled through the Earth along a diameter. (a) If a particle is dropped into the shaft at the Earth's surface, with what speed will it pass through the center of the Earth? (b) If the particle is projected upward from the center of the Earth, what velocity is necessary to permit the particle to escape from the Earth?

Section 13–5

13–13 Derive the expression for the gravitational potential energy $U(r)$ of a particle of mass m and a homogeneous thick spherical shell of inner radius R_0, outer radius R, and mass M. Take $U(\infty) = 0$. Show that the force on the particle in the region $R_0 < r < R$ is given by Eq. 13–11.

13–14 Consider a thin disc with a radius R and a mass M. A
   •   particle with mass m is located on the disc axis at a distance r from the center of the disc. (a) What is the gravitational potential energy of the particle? (b) What is the force on the particle? (c) Allow r to become large compared with R and show that both U and $\mathbf{F}$ reduce to the expressions for particles.

13–15 Assume the Earth to be a perfect homogeneous sphere. Imagine that a narrow tunnel is drilled completely through the Earth along a diameter. (a) Show that a particle with a mass m located in the tunnel a distance x from the center of the Earth experiences a gravitational force that obeys Hooke's law, $F = -\kappa x$, where

$$\kappa = G\frac{mM_E}{R_E^3} = g\frac{m}{R_E}$$

(b) In Section 15–2 we show that a particle subject to a Hooke's-law force has a period of motion equal to $T = 2\pi\sqrt{m/\kappa}$ (Eq. 15–12). Verify that the time required for the mass m to travel from one side of the Earth to the other is 42.2 min.

Section 13–6

13–16 At what distance from the Earth's center will an artificial satellite in an equatorial circular orbit (that is, an orbit that lies entirely in the Earth's equatorial plane) have a period equal to 1 day? Such a *synchronous* (or *geostationary*) satellite moves in synchronism with the Earth, remaining in a fixed position above a point on the equator. Synchronous satellites are used extensively as fixed relay stations in the worldwide communications network.

13–17 An artificial Earth satellite is "parked" in an equatorial circular orbit at an altitude of 10^3 km. What is the minimum additional velocity that must be imparted to the satellite if it is to escape from the Earth's gravitational attraction? How does this compare with the minimum escape velocity for leaving from the Earth's surface?

13–18 Ganymede, the largest moon of Jupiter, revolves around the planet in a nearly circular orbit with a radius of 1.07×10^6 km and a period of 7.16 days. Using this information, determine the mass of Jupiter. Express the result in units of the Earth's mass.

13–19 A satellite is in a circular orbit just above the surface of the Moon. (The radius of the Moon is 1738 km.) (a) What is the acceleration of the satellite? (b) What is the speed of the satellite? (c) What is the period of the satellite orbit?

13–20 A satellite with a mass of 500 kg is in a circular orbit at an altitude of 500 km above the Earth's surface. Because of air friction, the satellite eventually is brought to the

Earth's surface, and it impacts with a velocity of 2 km/s. How much energy was absorbed by the atmosphere through friction?

13–21 Studies of the relationship of the Sun to the local galaxy—the Milky Way—have revealed that the Sun is located near the outer edge of the galactic disc, about 30,000 light-years (L.Y.) from the center. Furthermore, it has been found that the Sun has an orbital velocity of approximately 250 km/s around the galactic center. (a) What is the period of the Sun's galactic motion? (b) What is the approximate mass of the Milky Way galaxy? Using the fact that the Sun is a typical star, estimate the number of stars in our local galaxy.

13–22 The general solution to the problem of three or more
• mutually gravitating bodies cannot be expressed in terms of any known mathematical functions. Solutions are possible, however, for certain special situations. For example, consider three identical particle-like objects that have equal masses M and are located at the vertices of an equilateral triangle with sides of length h. Show that when these particles move in the plane of the triangle, their relative positions will be maintained if they travel in circular orbits about the common C.M. with angular frequency $\omega = \sqrt{3GM/h^3}$.

Additional Problems

13–23 Planet X has the same average density as the Earth but its mass is only one half that of the Earth. What is the value of g on the surface of planet X?

13–24 The acceleration due to gravity at a point varies inversely as the square of the distance of the point from the center of the Earth. (a) For heights h above the Earth's surface that are not too great, show that the fractional change in the acceleration is given approximately by $\Delta g/g \cong -2h/R_E$. (b) Use this expression to calculate the changes $\Delta g/g$ for $h = 10^3$ m and 10^6 m. Compare these values with those obtained from the exact equations.

13–25 Assume that the Earth's crust (approximately 30 km
•• deep) has a uniform density of 2.72 g/cm^3. (a) What is the fractional change in the acceleration due to gravity $\Delta g/g$ upon descending from the surface to a depth of 15 km? (b) A spherical deposit of pure iron (density 7.86 g/cm^3) has a diameter of 5 km and lies just beneath the Earth's surface. What is the fractional change $\Delta g/g$ measured on the surface immediately above the deposit? (c) A spherical cavity lies just under the Earth's surface and has a diameter of 5 km. What is the fractional

change $\Delta g/g$ measured on the surface immediately above the cavity?

13–26 Two identical objects with mass M are initially separated by a very large distance. If each object is released from rest and allowed to gravitate toward the other, show that when the separation of the particles is R their relative velocity is $2\sqrt{GM/R}$.

13–27 At the time of a lunar eclipse, a 1-kg object is moved from a point on the Earth's surface farthest from the Sun to a point on the Moon's surface nearest the Sun. What is the total change in gravitational potential energy in this process? What velocity at the surface of the Earth would be necessary to accomplish this movement? Explain the significance of the fact that this velocity differs by only a small amount from the escape velocity, $v_E = 11.2$ km/s.

13–28 Find the gravitational potential energy for a system of
• eight stars, each with solar mass, located at the corners of a cube whose sides are 1 L.Y. in length. [*Hint:* For n objects, there are $n(n-1)/2$ independent pairs; thus, there are 28 pairs in this star system, many of which are equivalent.]

CHAPTER 14

FLUIDS

Matter in the solid state generally offers considerable resistance to all changes in shape. Liquids and gases, in contrast, do not have rigid structure or form. These states of matter—which together we call *fluids*—are easily altered in shape. Liquids generally deform in shape without appreciable change in volume. Gases, on the other hand, readily change volume and expand to fill completely any container.

In this chapter we examine the properties of fluids that flow readily, such as water, other liquids with low viscosity, and gases under ordinary pressures. We exclude highly viscous substances, such as pitch, tar, wax, and glass, all of which flow so slowly that they usually do not deform significantly during the time intervals of interest in the discussions here.

14–1 FORCES IN A FLUID AT REST

An ideal fluid is a substance that offers substantially no resistance to slowly applied forces that change the shape of the substance while conserving its volume. Figure 14–1 shows a small parallelepiped of a fluid, which we imagine to be isolated as a free body within a larger volume of the fluid. The surrounding fluid exerts forces that act on the parallelepiped and that slowly deform it into the shape shown by the dashed line. The fluid offers no resistance to this volume-conserving deformation, so the forces do no work on the fluid. Therefore, the forces that act on the top and bottom surfaces of the parallelepiped (the shaded surfaces in Fig. 14–1) can have no tangential components. Thus, the only forces that can act on the shaded surfaces must be perpendicular to those surfaces. This conclusion is not altered by the orientation of the parallelepiped. It follows that *internal forces in a fluid exert only a normal force on any area located within the fluid*. The normal

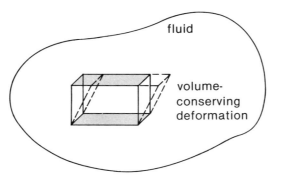

Fig. 14–1. The deformation of an isolated parallelepiped of fluid within a large volume of fluid.

force dF exerted per unit area dA is defined to be the *pressure p*:

$$p = \frac{dF}{dA} \qquad (14-1)$$

Units of Pressure.   Pressure is measured in N/m^2 or in the equivalent SI unit, the *pascal* (Pa):

$$1 \ N/m^2 = 1 \ Pa$$

Two other units are also acceptable in the SI system. *Standard atmospheric pressure* (1 atm) is defined to be

$$1 \ atm = 101{,}325 \ Pa$$

which is the average pressure due to the atmosphere at sea level. The other unit is the *bar* (for "barometric pressure"):

$$1 \ bar = 10^5 \ Pa$$

from which 1 atm = 1.013 bar.

Another (non-SI) pressure unit that is often used is the *torr* (named for Evangelista Torricelli, who invented the mercury barometer), which is the pressure exerted by a column of mercury 1 mm in height:

$$1 \ atm = 760 \ torr$$

In some engineering applications pressure is still measured in *pounds per square inch* ($lb/in.^2$ or psi),

$$1 \ atm = 14.70 \ lb/in.^2$$

and in inches of mercury,

$$1 \ atm = 29.92 \ in. \ Hg$$

Weather reports in the United States often give atmospheric pressures in units of inches of mercury.

The Hydrostatic Force on an Arbitrary Area.   Consider a differential element of area dA on a closed surface S, as indicated in Fig. 14–2. The unit normal vector $\hat{n}$ is perpendicular to the element of area, so the *directed* area element is represented by $\hat{n}dA$, where we take the positive sense of $\hat{n}$ to be *outward* from the surface. If the external pressure at the position of the differential area is p, the force exerted on the area dA is directed inward and is

$$d\mathbf{F} = -\hat{n}p \ dA \qquad (14-2)$$

The pressure in a fluid may vary from point to point. If an isolated volume of fluid (such as that shown in Fig. 14–2) is at rest, the integral of $d\mathbf{F}$ over the entire surface must be zero, in order to satisfy Newton's second law.

Figure 14–3 shows a segment of an arbitrary surface across which there exists a pressure difference $p_d = p_0 - p_1$. The force on the element of area ΔA is $\Delta\mathbf{F} = \hat{n}p_d \ \Delta A$, and the component of this force in the x-direction is

$$\Delta F_x = p_d \ \Delta A \cos \theta$$

An alternate way to interpret this equation is to recognize that the projection of ΔA onto a

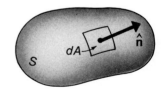

Fig. 14–2. A differential area of the surface S of an isolated fluid volume is represented by $\hat{n} \ dA$, where $\hat{n}$ is the outward-directed unit vector normal to the plane of dA.

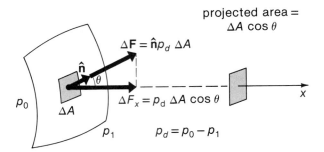

Fig. 14–3. Geometry for calculating the component of $\Delta\mathbf{F}$ in the x-direction.

plane perpendicular to the x-axis is $\Delta A \cos\theta$. Consequently, the component of the force in the x-direction, namely, ΔF_x, is equal to the pressure difference p_d multiplied by the projection of ΔA in the x-direction.

The result we have just obtained is also valid for curved surfaces. For example, consider a hemisphere that is immersed in a fluid with a uniform pressure p_0, as shown in Fig. 14–4. The projected area of the hemisphere is πR^2, so the x-component of the fluid force on the curved surface is $F_x = -\pi R^2 p_0$. This result can also be established by direct integration of Eq. 14–2 (see Problem 14–3). Because of the symmetry of the hemisphere about the x-axis as drawn in Fig. 14–4, the x-component of the force, $\mathbf{F}_x$, is the *total* force acting on the hemispheric surface. (• Can you see what happens to the other components of $\mathbf{F}$?)

Compressibility of Liquids. When a pressure is applied to a confined liquid, the volume will actually decrease by a small amount. The *compressibility* λ of a substance is defined to be the fractional change in volume per unit of applied pressure:

$$\lambda = -\frac{1}{V}\frac{dV}{dp} \tag{14–3}$$

Table 14–1 lists values of λ for various liquids. We leave for later chapters the discussion of the compressibility of gases.

Although liquids can be compressed, the fractional volume changes are relatively small even for pressures of several hundred atmospheres. Accordingly, when we discuss the properties of liquids, we usually consider them to be incompressible. Conversely, the compressible nature of gases is important in many situations.

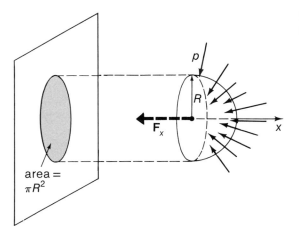

Fig. 14–4. The x-component of the fluid force on the hemispherical surface is $F_x = -\pi R^2 p$.

**TABLE 14–1.   Compressibility of Liquids
(at 20°C)**

| | $\lambda = -(1/V)\, dV/dp$ | |
LIQUID	$(10^{-11}\ \text{Pa}^{-1})$	$(10^{-6}\ \text{atm}^{-1})$
Acetone	125	127
Benzene	93	94
Ether	187	189
Glycerine	21	21
Mercury	3.7	3.8
Water	46	47

Example 14–1

What is the fractional change in the density of sea water between the surface (where the pressure is 1 atm) and a depth of 5.2 km (where the pressure is 500 atm)?

Solution:

The density is $\rho = M/V$, so $d\rho = -M\, dV/V^2$; dividing by $\rho = M/V$ gives

$$\frac{d\rho}{\rho} = -\frac{dV}{V}$$

and using Eq. 14–3 we have

$$\frac{\Delta\rho}{\rho} = \lambda\, \Delta p = [47 \times 10^{-6}\ (\text{atm})^{-1}] \times (500\ \text{atm} - 1\ \text{atm})$$

$$= 0.025\ \text{or}\ 2.5\ \text{percent}$$

Even at the tremendous pressure of 500 atm, the effect on the density is sufficiently small that it can be neglected for most purposes.

14–2   PRESSURE WITHIN A FLUID

We now consider the pressure at various points within a fluid column due to the gravitational force of the Earth acting on the fluid. Two cases are of importance, namely, incompressible liquids and compressible gases.

Hydrostatic Pressure.   Consider a tank that is filled with a liquid, as shown in Fig. 14–5. Select as a free body a thin horizontal slab of liquid at a depth z below the surface. The slab has a thickness dz and an area A. Let the pressure at depth z be p and the pressure at depth $z + dz$ be $p + dp$. The downward-exerted gravitational force on the slab is $g\, dm = g\rho A\, dz$. The slab is in equilibrium, so the force equation in the vertical direction is (with downward positive)

$$-(p + dp)A + \rho g A\, dz + pA = 0$$

Fig. 14–5.   A small horizontal slab of fluid within a tank of fluid is isolated as a free body.

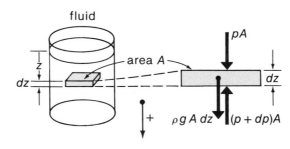

Simplifying, we obtain

$$dp = \rho g \, dz$$

We assume that the fluid is incompressible so that ρ is constant (see Example 14–1). Then, upon integrating, we find

$$p(z) = p_0 + \rho g z \qquad \text{incompressible fluid} \qquad \textbf{(14–4)}$$

where the integration constant is p_0, the pressure on the liquid surface ($z = 0$); if the surface is exposed to air, p_0 corresponds to atmospheric pressure. We see that the pressure within a liquid depends linearly on the depth.

If the tank is closed at the top by a piston that exerts a pressure p_a on the surface of the liquid, the constant p_0 in Eq. 14–4 is replaced by the applied pressure p_a. That is, the pressure at any depth z is $\rho g z$ plus the applied external pressure. This fact was first recognized by Blaise Pascal and is embodied in *Pascal's principle:*

A pressure that is applied to the surface of a confined (incompressible) liquid is transmitted undiminished to every point within the liquid.

Example 14–2

The rock crusher shown in the diagram is a device whose operation depends on Pascal's principle. Determine the force F' exerted on the rock when a force $F = 100$ N is applied to the piston in the smaller cylinder.

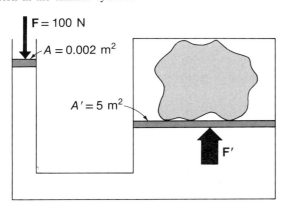

Solution:

The pressure exerted by the force F on the fluid in the smaller cylinder is

$$p = \frac{F}{A} = \frac{100 \text{ N}}{0.002 \text{ m}^2} = 5.0 \times 10^4 \text{ Pa}$$

This same pressure is transmitted undiminished to the larger cylinder, where the force exerted on the piston (and the force exerted by the piston on the rock) is given by

$$F' = P'A' = PA' = (5.0 \times 10^4 \text{ Pa})(5 \text{ m}^2)$$
$$= 2.5 \times 10^5 \text{ N}$$

Thus, a force of only 100 N applied to the smaller piston results in a rock-crushing force on the larger piston. (• Show that $F'/F = A'/A$.)

Atmospheric Pressure. Next, we obtain the expression for the decrease in pressure with height in the atmosphere. Because air is compressible, we must consider the density to be a function of the altitude y above sea level; that is, $\rho = \rho(y)$. Consider the thin slab of air in Fig. 14–6. The force equation is (with the upward direction positive)

$$pA - (p + dp)A - \rho(y)gA \, dy = 0 \qquad \textbf{(14–5)}$$

so that

$$dp = -\rho(y)g \, dy$$

To obtain an expression for $p(y)$, we need to know the way in which the density depends on y. Let us assume that the atmosphere is at a constant temperature so that the density varies in direct proportion to the pressure;* that is,

$$\rho(y) = \rho_0 \frac{p}{p_0} \tag{14-6}$$

where $\rho(0) = \rho_0$ and $p(0) = p_0$. Inserting this relationship into Eq. 14–5 and integrating, we have

$$\int_{p_0}^{p} \frac{dp}{p} = -\frac{g\rho_0}{p_0} \int_0^y dy$$

from which

$$\ln \frac{p}{p_0} = -\frac{g\rho_0}{p_0} y$$

or, finally,

$$p(y) = p_0 e^{-\alpha y} \qquad \text{isothermal atmosphere} \tag{14-7}$$

where $\alpha = g\rho_0/p_0$. Thus, in this approximation, the atmospheric pressure decreases exponentially with altitude. Equation 14–7 is called the *law of atmospheres*.

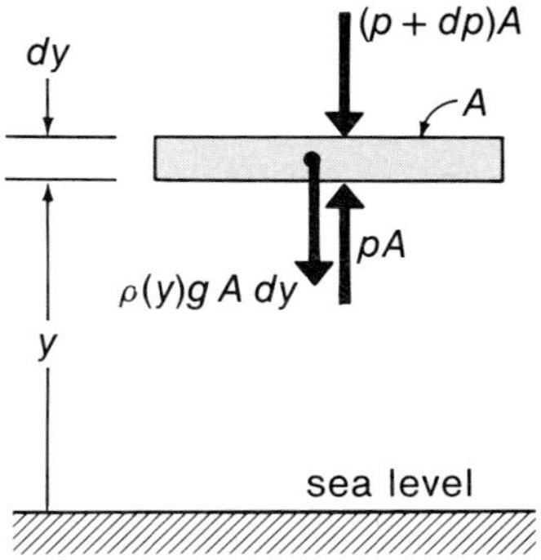

Fig. 14-6. A small horizontal slab of air within the atmosphere is isolated as a free body.

Example 14-3

(a) The greatest depth in the oceans is approximately 11.0 km (in the Mariana Trench, western Pacific). What is the pressure at this depth? The density of sea water is 1030 kg/m³.

(b) What is the pressure at an altitude of 11.0 km? (The density of air at sea level is 1.29 kg/m³ for a temperature of 0° C; see Table 1–2.)

Solution:

(a) Using Eq. 14–4, we have

$$p(z) = p_0 + \rho g z$$

Then,

$p(11.0 \text{ km}) =$

$$1 \text{ atm} + \frac{(1030 \text{ kg/m}^3)(9.80 \text{ m/s}^2)(11.0 \times 10^3 \text{ m})}{1.013 \times 10^5 \text{ Pa/atm}}$$

$$= 1097 \text{ atm}$$

The actual pressure measured at this depth is about 2 percent greater. The difference is due to the compressibility of water, which we have neglected here.)

(b) We use Eq. 14–7 in which the value of α is

$$\alpha = \frac{g\rho_0}{p_0} = \frac{(9.80 \text{ m/s}^2)(1.29 \text{ kg/m}^3)}{1.013 \times 10^5 \text{ Pa}}$$

$$= 1.25 \times 10^{-4} \text{ m}^{-1}$$

Then, $$p(y) = p_0 e^{-\alpha y}$$

so that $$p(11.0 \text{ km}) = (1 \text{ atm}) e^{-(1.25 \times 10^{-4} \text{ m}^{-1})(11.0 \times 10^3 \text{ m})}$$

$$= 0.253 \text{ atm}$$

(The "U.S. Standard Atmosphere" gives, for an altitude of 11.0 km, a pressure of 0.224 atm, about 10 percent smaller than the value we have obtained. The difference is due primarily to our neglect of the variation of temperature with altitude.)

---

*See Section 19–2. The air temperature does, of course, vary with altitude; consequently, the result we obtain here is only approximately correct.

14–3   THE MEASUREMENT OF PRESSURE

Manometers. A convenient device for measuring pressure, particularly gas pressure, is a *manometer*. This instrument consists of a static liquid column that is used to balance the pressures applied to the two ends of the column. Figure 14–7 shows two types of manometers: (a) an *open* manometer and (b) a *differential* manometer.

Fig. 14–7. (a) An open manometer. (b) A differential manometer.

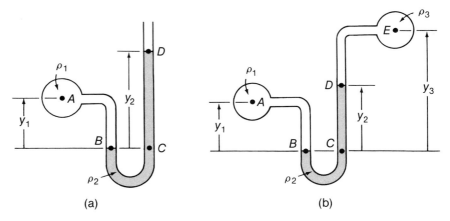

(a)   (b)

Fig. 14–7. (a) An open manometer. (b) A differential manometer.

In an open manometer, one end of the liquid column is open to the atmosphere. Thus, in Fig. 14–7a, the pressure at point D is atmospheric pressure p_0. At points B and C the pressure is the same (• Explain.) and is equal to $p_0 + \rho_2 g y_2$, where y_2 is the height of the manometer liquid that is supported by the pressure at C. The pressure at point A in the vessel is

$$p_A = p_0 + \rho_2 g y_2 - \rho_1 g y_1 \qquad (14–8)$$

If the fluid in the vessel is a gas, the density ρ_1 generally is sufficiently small that the last term in Eq. 14–8 can be neglected. Then,

$$p_A(\text{gas}) = p_0 + \rho_2 g y_2 \qquad (14–8a)$$

Notice that the pressure in the vessel can be less than atmospheric pressure (even a vacuum); then, point B will be above point D, and $y_2 < 0$.

Fig. 14–8. A mercury barometer (or Torricelli tube).

The pressure p_A in Eq. 14–8a is called the *absolute pressure*. The difference between p_A and atmospheric pressure, $p_A - p_0$, is called *gauge pressure*. Any type of gauge that relates the pressure in a vessel to atmospheric pressure shows the gauge pressure directly. Thus, in Fig. 14–7a, the gauge pressure in A is indicated directly by the column-height difference y_2.

In Fig. 14–7b, the difference in pressure between points A and E is

$$p_A - p_E = \rho_3 g(y_3 - y_2) + \rho_2 g y_2 - \rho_1 g y_1 \qquad (14–9)$$

Again, if the fluids in both vessels are gases, the terms involving ρ_1 and ρ_3 can be neglected, giving

$$p_A(\text{gas}) - p_E(\text{gas}) = \rho_2 g y_2 \qquad (14–9a)$$

The Barometer. Instruments that are designed to measure atmospheric pressure are called *barometers*. The common mercury barometer is essentially an open-tube manometer (Fig. 14–7a) in which the vessel is evacuated so that point B stands above point D. The usual construction of a mercury barometer, shown in Fig. 14–8, consists of a long

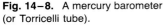

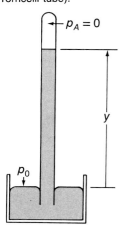

Fig. 14–8. A mercury barometer (or Torricelli tube).

tube closed at one end.* The tube is filled with mercury, then inverted and the open end placed in a reservoir of mercury. A vacuum space forms above the mercury column.† Then, Eq. 14–8a becomes (with $p_A = 0$ and $-y_2 = y$)

$$p_0 = \rho g y \qquad \text{barometric equation} \qquad \text{(14–10)}$$

where ρ is the density of mercury (13.5955×10^3 kg/m³ at 0° C). The height of the mercury column at standard atmospheric pressure (101,325 Pa) and standard gravity (9.80665 m/s²) is

$$y = \frac{p_0}{\rho g} = \frac{101{,}325 \text{ Pa}}{(13.5955 \times 10^3 \text{ kg/m}^3)(9.80665 \text{ m/s}^2)}$$
$$= 0.76000 \text{ m}$$

For this reason, 1 atm is sometimes written as 760 mm Hg. (See Section 14–1.)

Example 14–4

In the U-tube arrangement shown in the diagram, both chambers, A and B, are initially open to the atmosphere. The right-hand section of the tube contains a column of oil with an initial height y_B; the density of the oil is 0.835×10^3 kg/m³. The left-hand section contains water, which extends into the right-hand section, giving an effective height y_A for the water column. The density of water (at 20° C) is 0.998×10^3 kg/m³. Additional gas is now pumped into chamber B and the chamber is sealed. The new pressure in this chamber is p_B, and the inter-

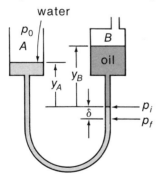

face between the oil and water has been lowered by an amount $P_f - P_i = \delta = 1$ cm. What is the pressure in chamber B? (The diameter of the U-tube is much smaller than that of the chambers; consequently, when the interface is depressed by 1 cm, the positions of the liquid surfaces change by negligible amounts.)

Solution:

If we use ρ_w for the density of water and ρ_{oil} for the density of oil, we can express the initial condition as

$$p_0 + \rho_w g y_A = p_0 + \rho_{oil} g y_B$$

After the pressure in chamber B has been increased to p_B, we can write

$$p_0 + \rho_w g (y_A + \delta) = p_B = \rho_{oil} g (y_B + \delta)$$

Subtracting the first equation from the second, we obtain

$$p_B - p_0 = (\rho_w - \rho_{oil}) g \delta$$
$$= (998 \text{ kg/m}^3 - 835 \text{ kg/m}^3)(9.80 \text{ m/s}^2)(0.01 \text{ m})$$

so that $\qquad p_B = 1 \text{ atm} + 15.97 \text{ Pa}$
$$= 1.000158 \text{ atm}$$

---

*The mercury barometer was invented in 1643 by Evangelista Torricelli (1608–1647), an Italian physicist and mathematician who was a student of Galileo and succeeded his mentor at the Florentine academy.

†Actually, this space contains a small amount of mercury vapor, amounting to a pressure of about 2×10^{-7} atm.

14-4 BUOYANCY AND ARCHIMEDES' PRINCIPLE

Consider an irregularly shaped object at rest within a fluid, as shown in Fig. 14–9a. Because the fluid pressure increases with depth, the force exerted by the fluid on the surface of the object is greater for those portions more deeply immersed. The net effect on the entire body is an upward or lifting force, which is called the *buoyant force*.

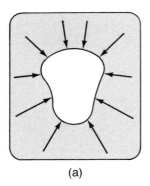
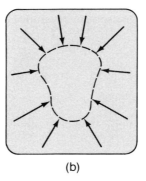

(a)   (b)

Fig. 14–9. (a) An irregularly shaped object immersed in a fluid. (b) The object replaced by an equal volume of fluid. The surface forces exerted by the surrounding fluid are the same in each case.

The magnitude and line of action of the buoyant force can be determined in the following way. First, note that the fluid pressure on the surface of the immersed object does not depend on the material of which the object is composed. For example, suppose that we remove the body and replace it with an exactly equal volume of the fluid, as in Fig. 14–9b. This replaced fluid is in static equilibrium under the combination of the same pressure-developed surface force that acted on the object (the buoyant force) and the gravitational force that acts vertically downward through the center of gravity of the fluid. Thus, the gravitational force precisely balances the pressure-developed or buoyant force. We conclude that the buoyant force F_b that is exerted by a fluid with a density ρ on an object with a volume V is

$$F_b = \rho V g \qquad \text{buoyancy} \qquad (14\text{--}11)$$

We can summarize the situation with the following statement:

A body that is entirely or partially submerged in a fluid is buoyed up by a force equal in magnitude to the weight of the displaced fluid and directed upward along a line that passes through the center of gravity of the displaced fluid.

This is the statement of *Archimedes' principle.*[*]

---

[*]The basic idea that is central to the principle of buoyancy was first conceived by Archimedes (c287 B.C.–212 B.C.), one of the greatest scientists of ancient Greek times. According to the legend (which is probably true), Archimedes made the discovery while in his bath. He had been asked by the king of Syracuse, Hieron II, to determine the purity of a new gold crown without destroying it. As he stepped into his bath, Archimedes noticed that the water level rose by an amount equivalent to the volume of the immersed part of his body and that he was buoyed up in the process. He saw that by comparing in this way the volume of the crown with the volume of an equal weight of known pure gold, he could determine the purity of the crown. He was so pleased with this discovery that he is supposed to have leaped from his bath and to have run naked to the palace shouting "Eureka!"—"I've got it!" Subsequently, measurements showed the crown to have been adulterated with silver, and the goldsmith was executed.

(• Can you see why the buoyancy principle is valid for a body partially immersed in a fluid as well as for a body that is completely immersed? • Can you see why the principle is valid for compressible as well as for incompressible fluids? • What is the buoyant force on an object immersed in a fluid that is undergoing free fall?)

Example 14–5

What fraction of an iceberg lies beneath the surface of the sea? (Assume the density of sea water to be $\rho_w = 1.028 \times 10^3$ kg/m^3 and that of sea ice to be $\rho_i = 0.917 \times 10^3$ kg/m^3. Both values actually depend on the salinity of the water.)

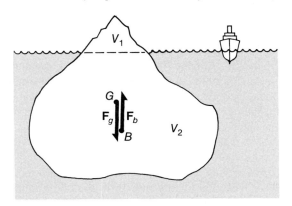

Solution:

For equilibrium conditions, the gravitational force F_g on the iceberg equals the buoyant force F_b.

$$F_g = \rho_i(V_1 + V_2)g \qquad \text{and} \qquad F_b = \rho_w V_2 g$$

so that

$$\frac{V_2}{V_1 + V_2} = \frac{\rho_i}{\rho_w} = \frac{0.917 \times 10^3 \text{ kg/m}^3}{1.028 \times 10^3 \text{ kg/m}^3} = 0.892$$

That is, only about 11 percent of the iceberg is visible above the water.

Example 14–6

Determine the density ρ of an irregular object by weight measurements, using a spring scale, when the object is in air and when the object is immersed in water with density ρ_w.

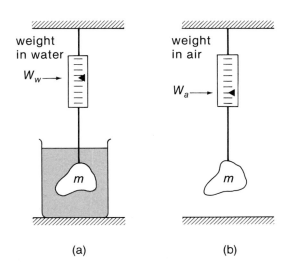

(a)   (b)

Solution:

When the object is immersed in water, as in diagram (a), there acts on the object a buoyant force equal to the weight of the displaced water, $F_b = \rho_w g V$, where V is the (unknown) volume of the object. When the object is in air, there is a similar buoyant force equal to the weight of the displaced air, but because $\rho_{\text{air}} \ll \rho_w$, we can neglect this buoyant force (see, however, Problem 14–47). Then, the two weight measurements give

$$w_w = mg - \rho_w g V \qquad \text{and} \qquad w_a = mg$$

Combining these expressions, we obtain

$$V = \frac{w_a - w_w}{\rho_w g}$$

The density of the object is

$$\rho = \frac{m}{V} = \frac{w_a/g}{(w_a - w_w)/\rho_w g}$$

$$= \frac{w_a}{w_a - w_w} \rho_w$$

14–5   GENERAL CONSIDERATIONS OF FLUID MOTION

We have seen that the description of a fluid at rest is quite simple. We devote the rest of this chapter to the more complicated subject of fluids in motion.

One way to describe the dynamics of fluids is by following the motion of a fluid "particle." Such a particle is defined to be a sample of the fluid with a size sufficiently small that the macroscopic properties of the fluid change in a continuous way from particle to particle. On the other hand, the size of a fluid particle must be sufficiently large that, on average, there are no net effects of the random thermal motions of the constituent molecules.

The trajectory of a fluid particle, as it moves in accordance with Newton's laws, is called a *path line*. We can imagine that a specific particle is identified by means of a colored speck or with a tiny droplet of dye. Then, a time-exposure photograph would reveal the path line of the particle (Fig. 14–10).

The study of fluids by following the motion of specific fluid particles was introduced by Lagrange.* An alternative point of view was developed by Euler, in which the description is made in terms of a *velocity field* $\mathbf{v}(\mathbf{r}, t)$ of the fluid. In the Eulerian scheme, we consider the velocity at a particular instant of all the particles in the fluid. In the condition referred to as *steady flow,* every fluid particle that passes through a particular point generates the same path line. The constant, gentle flow of water in a quiet stream is steady flow, as is the smooth, slow, and constant flow of a fluid through a pipe. But steady flow requires that $\mathbf{v}$ be independent of time, so that every particle passes through $P(\mathbf{r})$ with the same velocity $\mathbf{v}(\mathbf{r})$.

Nonsteady flow can involve *turbulent flow*. In this case the fluid particles that pass through a particular point do not have the same history but, instead, follow randomly related path lines. For example, the smoothly flowing water in a stream becomes turbulent when rocks are encountered and "white water" rapids occur.

The condition of a static fluid can be specified by giving the pressure $p(\mathbf{r})$ and the density $\rho(\mathbf{r})$ at every point within the fluid. In the event that the fluid is in motion, these quantities can have a time dependence, and then we write $p(\mathbf{r}, t)$ and $\rho(\mathbf{r}, t)$. When considering a liquid, we usually assume incompressibility, so that ρ = constant for all positions and times. We may also consider gases to have constant density if we exclude large-scale phenomena, such as meteorologic conditions, and if we restrict the velocities to less than about half the speed of sound in the gas (that is, $v \lesssim 150$ m/s for air under normal conditions).

An important force that influences the motion of fluids is the frictional or drag force called *viscosity*. Viscosity results from the forces exerted by one layer of a fluid on an adjacent layer as they slide past each other, which ultimately arise from the attractive forces that exist between the molecules in the two layers. Thus, viscosity is an inherent property of all real fluids. In some cases, however, when the effects of viscosity are relatively small, we neglect this complication in the interests of simplicity.

14–6   STREAMLINES AND CONTINUITY

Streamlines are used to describe the velocity field within a flowing fluid. At any point along a streamline, such as the points P, Q, and R in Fig. 14–11, the tangent to the line indicates the direction of the fluid velocity at that point. Notice that a *path line* refers to the trajectory of a single particle, whereas a *streamline* refers to the picture at a

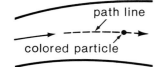

Fig. 14–10. A *path line* is the line traced out by a single particle as it moves with a flowing fluid.

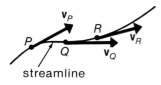

Fig. 14–11. The direction of the velocity of a fluid particle at a point, such as P, Q, or R, is the same as the direction of the tangent to the streamline through that point.

*Joseph Louis Lagrange (1736–1813) was a French-Italian mathematician, physicist, and astronomer.

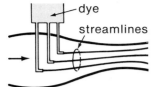

Fig. 14–12. (a) Streamlines in a flowing fluid are made visible by releasing dye into the liquid.

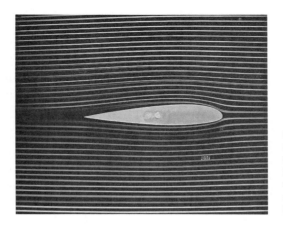

(b) The flow pattern around an airfoil is revealed by smoke particles that move along the streamlines. The flow is from right to left, corresponding to motion of the airfoil through the gas from left to right. (Courtesy of NASA.)

particular instant of the velocity directions of a number of particles. Along a streamline the velocity can vary (in magnitude and in direction). (• At what kinds of places will a streamline lead to a point where the velocity is *zero*?)

For any individual fluid particle, the tangent to the path line must have the same direction as the fluid velocity (the streamline) at every point. An essential feature of steady flow is that the streamlines are independent of time. Thus, the path line of a particle coincides with a streamline for the case of steady flow. This is not true for nonsteady flow.

A method for making streamlines visible is shown schematically in Fig. 14–12. A distinctive dye is released into the flowing liquid through tiny tubes located at various places within the containing pipe. The dye does not immediately mix with the water and is carried along on a streamline, which is thereby made visible. In the case of a flowing gas, as in a wind tunnel, the streamlines are often indicated by smoke streamers (see the photograph).

A concept related to the streamline is that of a *stream tube* or *tube of flow*. Imagine a tubelike surface generated by constructing all of the streamlines that pass through points on the perimeter C_1 of a small plane area ΔA_1, as shown in Fig. 14–13. This tube can be cut by another plane area ΔA_2 farther downstream, thereby creating a second perimeter C_2. The result is a volume defined by the tube wall and the two end areas, ΔA_1 and ΔA_2, whose perimeters are the curves C_1 and C_2. Fluid particles can never enter or leave a particular tube of flow by passing through the tube wall. (• Can you see why?)

In Fig. 14–13 we suppose that the areas ΔA_1 and ΔA_2 are sufficiently small that the flow velocity is essentially constant over each surface, being $\mathbf{v}_1$ at ΔA_1 and $\mathbf{v}_2$ at ΔA_2. Define two vector areas, $\hat{\mathbf{n}}_1 \, \Delta A_1$ and $\hat{\mathbf{n}}_2 \, \Delta A_2$, where $\hat{\mathbf{n}}_1$ and $\hat{\mathbf{n}}_2$ are unit normal vectors that are directed outward from the enclosed volume. The density of the fluid is ρ_1 at position 1 and is ρ_2 at position 2. Then, during a time Δt, a mass $\Delta m_2 = (\rho_2 \mathbf{v}_2 \cdot \hat{\mathbf{n}}_2 \, \Delta A_2) \, \Delta t$ passes through ΔA_2 (*leaving* the enclosed volume), and a mass $\Delta m_1 = -(\rho_1 \mathbf{v}_1 \cdot \hat{\mathbf{n}}_1 \, \Delta A_1) \, \Delta t$ passes through ΔA_1 (*entering* the enclosed volume). If there is no accumulation of mass

Fig. 14–13. A tube of flow defined by streamlines passing through the small closed flat curve C_1. The area defined by the perimeter of C_1 is ΔA_1, the unit normal to which is $\hat{\mathbf{n}}_1$, directed *outward* from the enclosed volume. The average fluid velocity through the surface ΔA_1 is $\mathbf{v}_1$. Similar considerations apply for C_2 and ΔA_2.

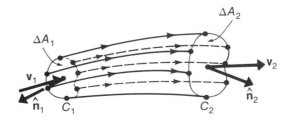

within the volume, as must be the case for an incompressible fluid, we have $\Delta m_1 = \Delta m_2$, or

$$\rho_1 \mathbf{v}_1 \cdot \hat{\mathbf{n}}_1 \, \Delta A_1 + \rho_2 \mathbf{v}_2 \cdot \hat{\mathbf{n}}_2 \, \Delta A_2 = 0$$

For incompressible fluids, we have $\rho_1 = \rho_2$, and hence,

$$\mathbf{v}_1 \cdot \hat{\mathbf{n}}_1 \, \Delta A_1 + \mathbf{v}_2 \cdot \hat{\mathbf{n}}_2 \, \Delta A_2 = 0 \qquad \text{incompressible fluids} \qquad (14\text{--}12)$$

Equation 14–12 is the *equation of mass continuity* and expresses the *continuity condition*.

In order to satisfy the continuity condition, the flow velocity must increase when the area is decreased and the streamlines bunch together. This is the case in the narrow throat in the pipe shown in Fig. 14–12.

14–7   THE ENERGY EQUATION

In the preceding section our discussion of (steady) laminar flow concerned the streamlines and tubes of flow present throughout the volume of the flowing fluid. Pipes, when present, served simply to confine the total flow. Laminar flow through a tube of flow and bulk flow through a pipe as a whole have similarities. For example, each offers an impenetrable confining surface channeling or guiding the flow. In a tube of flow there is no penetration of the tube surface because of the laminar nature of the flow. In a pipe the confinement is due to the physical presence of the pipe wall. Figure 14–14 shows a pipe having a cross-sectional area A_1 at location C_1 and a cross-sectional area A_2 at location C_2. If $\bar{v}_1$ is the average flow velocity entering the pipe at C_1 and $\bar{v}_2$ is the average flow velocity leaving the pipe at C_2, then for incompressible fluids a continuity condition similar to Eq. 14–12 results:

$$\bar{v}_1 A_1 = \bar{v}_2 A_2 \qquad\qquad (14\text{--}13)$$

Another condition that applies to all tubes of flow as well as to flow through pipes as a whole is the conservation of energy. Since the conservation of energy is more broadly applicable to flow through pipes and other devices conducting fluids, we begin by examining this more general case first. Later we impose the restrictions on the nature of the flow required for laminar flow and discuss the energy condition along individual tubes of flow. Consider the energy balance in any device or machine through which a fluid flows. The device could be an ordinary pipe or nozzle, or it could be a pump, compressor, turbine, or other complicated system. Figure 14–15 shows schematically a general device, the essential features of which are contained within the volume enclosed by the dashed line.

We make no restriction on the properties of the fluid or on the nature of the flow

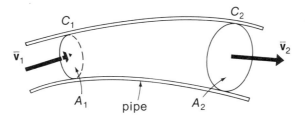

Fig. 14–14. Fluid flow through a pipe with average flow velocities $\bar{\mathbf{v}}_1$ and $\bar{\mathbf{v}}_2$ across cross-sectional areas A_1 and A_2, respectively.

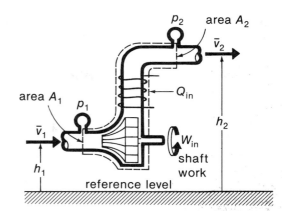

Fig. 14–15. Fluid flows through a generalized device (contained within the dashed line). The average entrance velocity of the fluid is $\bar{v}_1$ and the average exit velocity is $\bar{v}_2$. The device consists of a turbine/pump that may do work on the fluid or have work done on it, a heating element that can add heat energy to the fluid, and an increase in height that can exchange kinetic and potential energy within the fluid. The device could be as simple as a straight pipe, however. The bulbs represent manometers.

through the device. For example, the fluid could be viscous and the flow could be turbulent. Fluid with density ρ_1 and pressure p_1 enters the device with an average velocity $\bar{v}_1$ at a height h_1 above some convenient reference level. The corresponding quantities at exit are ρ_2, p_2, $\bar{v}_2$, and h_2. We do not need to include explicitly the variation in fluid pressure or gravitational potential energy across the diameter of the pipe. (See the remark following Eq. 14–14.)

Consider the flow through the system of a sample of fluid having a mass m. This fluid occupies a volume $V_1 = m/\rho_1$, so that its length at the entrance is $\ell_1 = m/\rho_1 A_1$, where A_1 is the input cross-sectional area. The work done *on* the fluid at the entrance to the device is equal to the product of the distance ℓ_1 and the force $p_1 A_1$ that moves the fluid, that is, $W_1 = mp_1/\rho_1$. The work done *by* the fluid at the exit is $W_2 = mp_2/\rho_2$. There might also be work done on the fluid by a mechanical rotor (W_{in}, per unit mass) or by heat supplied (Q_{in}, per unit mass), as indicated in Fig. 14–15. (Note that W_{in} and Q_{in} could be negative. • What do the negative signs tell us?) Then, the total energy input to the fluid from all these sources is

$$E_{in} = m\left(W_{in} + Q_{in} + \frac{p_1}{\rho_1} - \frac{p_2}{\rho_2}\right)$$

What net amount of energy is carried out of the system by the fluid? The net outflow of kinetic energy is $\frac{1}{2}m(\bar{v}_2^2 - \bar{v}_1^2)$. Also, there is a change in the potential energy of the fluid between entrance and exit amounting to $mg(h_2 - h_1)$. Finally, there may be a change in the internal energy of the fluid (the energy associated with thermal motions of the constituent molecules) due to heat gained or lost by the fluid. Let the internal energy per unit mass be u_1 at entrance and u_2 at exit. Then, the net amount of energy carried out of the system by the fluid is*

$$E_{out} = m(u_2 - u_1 + gh_2 - gh_1 + \tfrac{1}{2}\bar{v}_2^2 - \tfrac{1}{2}\bar{v}_1^2)$$

Conservation of energy requires $E_{in} = E_{out}$, so we have

$$W_{in} + Q_{in} + \frac{p_1}{\rho_1} - \frac{p_2}{\rho_2} = u_2 - u_1 + g(h_2 - h_1) + \tfrac{1}{2}(\bar{v}_2^2 - \bar{v}_1^2) \qquad \textbf{(14–14)}$$

Note that we have obtained this equation without assuming any special properties of the fluid and without any restriction on the nature of flow.

---

*We assume that no thermal energy leaves the fluid by the conduction of heat through the pipe walls, although we could explicitly include such a term in the energy equation.

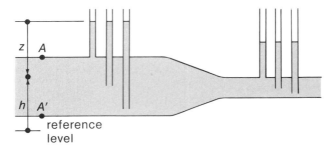

Fig. 14–16. Because $p + \rho gh$ is constant across any cross section of the pipe, the liquid rises to the same level regardless of the depth of the side pipe if the flow velocity is the same at each point.

In Section 14–2 we found that the fluid pressure at a depth z is ρgz (Eq. 14–14). This result means that the quantity $p/\rho + gh$ or $p + \rho gh$ appearing in Eq. 14–14 has the same value at any point on each cross section of the fluid, such as AA' in Fig. 14–16. The quantity $p + \rho gh$ is referred to as the *static pressure* and $\frac{1}{2}\rho v^2$ is called the *dynamic pressure*.

Bernoulli's Equation. Let us return to consider the case of steady, nonviscous, incompressible laminar flow discussed earlier in Section 14–6. We may apply the energy equation (Eq. 14–14) to the tube of flow between locations C_1 and C_2 shown in Fig. 14–13.

Moreover, we allow no mechanical or thermal energy input ($W_{in} = 0$ and $Q_{in} = 0$); then the internal energy of the fluid cannot change, so $u_2 - u_1 = 0$. (• Can you see why?) The result is

$$p_1 + \rho gh_1 + \tfrac{1}{2}\rho v_1^2 = p_2 + \rho gh_2 + \tfrac{1}{2}\rho v_2^2 \qquad (14\text{--}15)$$

This result is known as *Bernoulli's equation.**

Bernoulli's equation applies for very special flow along a streamline of an ideal fluid. The energy equation, on the other hand, is valid under the most general conditions, including the nonsteady, nonstreamline flow of real fluids. When the energy equation is specialized to the case of ideal streamline flow, no fluid can enter or leave a particular tube of flow (see the remarks concerning Fig. 14–13). Then there is no longer any distinction between the confining pipe and a tube of flow, so the energy equation becomes Bernoulli's equation.

In most of the examples and problems, we assume the ideal case of steady flow of a nonviscous, incompressible fluid, so we can use Bernoulli's equation.

Example 14–7

A tank is filled to a depth h_1 with a liquid that has a density ρ. The liquid escapes into air through a small hole in the bottom of the tank. What is the escape velocity v_2?

Solution:

Bernoulli's equation can be applied only to points that lie on the same streamline. However, in the situation here, we assume that the surface area A_1 is much larger than the orifice area A_2. Then the fluid particles on the surface have essentially zero velocity. Because the values of p and h are the same across the surface, all surface points, such as a, b, and c, are equivalent. Consequently, we can apply Bernoulli's equation between any point on the surface and a point in the orifice. With $v_1 = 0$, $h_2 = 0$, and $p_1 = p_2 = p_0$, Eq. 14–15 becomes

$$\rho gh_1 = \tfrac{1}{2}\rho v_2^2$$

---

*Usually attributed to the Swiss mathematician Daniel Bernoulli (1700–1782) in his *Hydrodynamica* (1738); however, Bernoulli's discussion of the relationship between pressure and velocity in this book was obscure. Equation 14–15 was actually first obtained by Leonhard Euler, who integrated the differential expression $dp + \rho g \, dh + \rho v \, dv = 0$ along a streamline from point 1 to point 2.

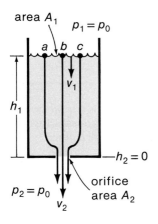

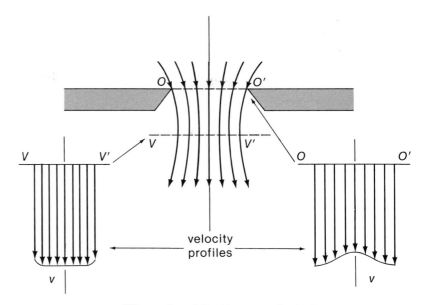

velocity profiles

from which

$$v_2 = \sqrt{2gh_1}$$

Thus, the velocity of the liquid emerging from the bottom of the tank is the same as that of an object dropped through a distance equal to the depth of the liquid. This result is known as *Torricelli's theorem*. Notice that the velocity of the emerging fluid is independent of direction; if the fluid were guided by a curved pipe, it would squirt to the original height h_1 (neglecting frictional effects).

When a jet of liquid escapes freely from an orifice, the streamlines, which converge toward the orifice, continue to converge for a short distance beyond the orifice. This effect is shown in the diagram above. The minimum cross section of the jet VV', where the streamlines are parallel, is called the *vena contracta*. The cross-sectional area at VV' is about 60 to 70 percent of that at the orifice OO'. (This contraction is not to be confused with the further contraction of the stream as it falls to lower elevations; see Problem 14–29.) The velocity profile across the *vena contracta* is much more uniform than in the plane of the orifice, as shown in the diagram.

Example 14–8

A *Venturi meter** is a device for measuring the flow rate of fluids in pipes. The construction is shown in the diagram. A manometer tube, containing a liquid with a density ρ', is placed between the main part of the pipe with area A_1 and a constricted part of the pipe with area A_2. In practice, A_2 is one third to one fifth of A_1. The fluid has a density ρ. Assume that the height difference h in the manometer is a true measure of the static pressure difference between the points of attachment to the pipe. Assume further that the fluid is incompressible and experiences no appreciable frictional losses in flowing from position 1 to position 2. Obtain the expression for the volume rate of flow of the fluid through the pipe in terms of h.

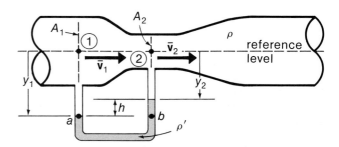

Next, equate the static pressures at points a and b in the manometer:

$$p_1 + \rho g y_1 = p_2 + \rho g y_2 + \rho' g h$$

Using the fact that $y_1 - y_2 = h$, we have

$$\frac{p_1 - p_2}{\rho} = gh\left(\frac{\rho'}{\rho} - 1\right)$$

For an incompressible fluid, $\bar{v}_1 = \bar{v}_2 A_2/A_1$. Combining these

Solution:

Apply Eq. 14–14 between positions 1 and 2 with the result

$$\tfrac{1}{2}\bar{v}_1^2 + \frac{p_1}{\rho} = \tfrac{1}{2}\bar{v}_2^2 + \frac{p_2}{\rho}$$

---

*After G. B. Venturi (1746–1822), an Italian physicist, whose studies inspired the invention of this device.

expressions, we find the volume rate of flow

$$q = \bar{v}_2 A_2 = A_2 \left[\frac{2gh(\rho'/\rho - 1)}{1 - (A_2/A_1)^2} \right]^{1/2}$$

Because this expression does not include frictional effects, the calculated value of q is always too large; nevertheless, the expression is accurate to better than 10 percent for a wide range of conditions.

Example 14–9

Water flows through a submerged orifice in a dam. The geometry is indicated in the diagram. Determine the amount of energy per unit mass of flow that is dissipated by friction in the fluid beyond the orifice.

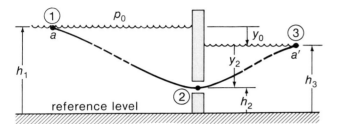

Solution:

We use the energy equation connecting the horizontal surface of the upstream basin (position 1), the orifice (position 2), and the horizontal surface of the downstream basin (position 3). The basins are sufficiently large that the particles on the surfaces have essentially zero velocity. Then, the portion of the energy equation (Eq. 14–14) that refers to position 1 is

$$(1) = p_0 + \rho g h_1 + \rho u_1$$

At position 2 the static pressure is $p_2 = p_0 + \rho g y_2$, and the average flow velocity is $\bar{v}_2$, so we can write

$$(2) = p_0 + \rho g y_2 + \rho g h_2 + \tfrac{1}{2}\rho \bar{v}_2^2 + \rho u_2$$

At position 3 we have $\bar{v}_3 = 0$; however, there is now an increase in the internal energy of the water due to the frictional dissipation of the kinetic energy of flow through the orifice. Thus,

$$(3) = p_0 + \rho g h_3 + \rho u_3$$

Now, equating (1) and (2) we have

$$u_2 - u_1 = g(h_1 - h_2 - y_2) - \tfrac{1}{2}\bar{v}_2^2$$
$$= g y_0 - \tfrac{1}{2}\bar{v}_2^2$$

We may reasonably assume that streamline flow exists between points 1 and 2; hence, $u_2 = u_1$, or (as in Example 14–7)

$$g y_0 = \tfrac{1}{2}\bar{v}_2^2$$

Equating (1) and (3), we have

$$u_3 - u_1 = g(h_1 - h_3) = g y_0$$

We conclude that the potential energy $g y_0$, which may be associated with the difference in surface-level heights on the two sides of the orifice, first appears as flow kinetic energy at the orifice position 2 and then later is converted by friction *entirely* into internal energy of the downstream water. Notice that this is always the case when water passes from a calm basin at one level to another calm basin at a lower level. (• Do you see why?)

This situation cannot be analyzed by assuming nonviscous flow along a hypothetical streamline aa', because such a streamline does not exist! (• Can you see why it is not possible to write a consistent set of streamline equations connecting points 1, 2, and 3?)

Example 14–10

One method of determining fluid velocities (particularly gas velocities) involves the use of a Pitot tube,* the geometry of one type of which is shown in the diagram. Obtain the expression for v_1 in terms of the manometer height difference h.

Solution:

First, notice that at the entrance to the tube that faces the fluid flow, the flow divides, producing a *stagnation point* where $v_2 = 0$.

Apply Eq. 14–14 between positions 1 and 2 with the result

$$p_1 + \tfrac{1}{2}\rho v_1^2 = p_2 \qquad (1)$$

Next, equate the static pressures at points a and b in the manometer:

$$p_a = p_1 + \rho g y_1 + \rho' g h = p_b = p_2 + \rho g(y_1 + h) \qquad (2)$$

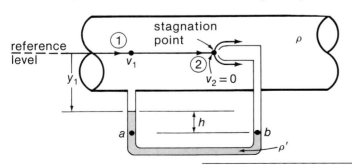

*Named for Henri Pitot (1695–1771), a French astronomer and mathematician, who first used such a device in 1732 to measure the flow velocity in the River Seine at Paris.

Solving Eq. 2 for $p_2 - p_1$ and substituting from Eq. 1, we find

$$p_2 - p_1 = gh(\rho' - \rho) = \tfrac{1}{2}\rho v_1^2$$

from which

$$v_1 = \sqrt{2gh\left(\frac{\rho'}{\rho} - 1\right)}$$

(• Why is the static pressure port, position 1, placed upstream and far from the central tube?)

Example 14-11

A pump delivers oil with a density of 850 kg/m^3 at a rate of 10 ℓ/s (10^{-2} m^3/s). The input to the pump is from a pipe with a diameter of 7.5 cm at a suction gauge pressure of -0.25 atm. The discharge of the pump is at a gauge pressure of 3.0 atm into a pipe with a diameter of 3.0 cm. There is negligible heat flow and the oil temperature remains constant. Determine the mechanical power output of the pump.

Solution:

Here we have $Q_{in} = 0$, $u_2 - u_1 = 0$, and $h_2 - h_1 = 0$. The average velocity of the fluid at the entrance to the pump is

$$\bar{v}_1 = \frac{10^{-2} \text{ m}^3/\text{s}}{\pi (3.75 \times 10^{-2} \text{ m})^2} = 2.26 \text{ m/s}$$

The average discharge velocity is

$$\bar{v}_2 = \frac{10^{-2} \text{ m}^3/\text{s}}{\pi (1.5 \times 10^{-2} \text{ m})^2} = 14.15 \text{ m/s}$$

The pressure difference between the output of the pump and the input is 3.00 atm $-$ (-0.25 atm) $=$ 3.25 atm $=$ 3.28 $\times$ 10^5 Pa. Then, using Eq. 14-14, we have (per kilogram of oil)

$$W_{in} = \tfrac{1}{2}[(14.15 \text{ m/s})^2 - (2.26 \text{ m/s})^2] + \frac{3.28 \times 10^5 \text{ Pa}}{850 \text{ kg/m}^3}$$

$$= 483.4 \text{ J/kg}$$

The power requirement is

$$P = W_{in} q \rho = (483.4 \text{ J/kg})(10^{-2} \text{ m}^3/\text{s})(850 \text{ kg/m}^3)$$

$$= 4109 \text{ W} \qquad (5.50 \text{ hp})$$

14-8   FRICTION IN FLUIDS

When one layer of a fluid slides over an adjacent layer, a drag force develops at the boundary between the layers. This fluid friction is called *viscosity*. As in the case of friction in the rubbing action between solids, viscous forces generate heat and result in the loss of mechanical energy.*

The frictional forces that are present at the interface between two solids sliding against one another result in very little transfer of matter from one surface to the other. In the case of sliding fluid layers, however, the viscous force exerted by one layer on the other does bring about the exchange of molecules between the layers. Some of the rapidly moving molecules from the higher-velocity layer are transferred (diffuse) to the lower-velocity layer, and vice versa. This amounts to a diffusion of momentum, which results in a decrease in the relative velocity between the layers and a conversion of transport kinetic energy into thermal energy.

In the case of steady flow, with viscosity present, the fluid layers slip past one another while maintaining their gross individual identity. This type of flow is called *laminar* (layered) *flow*. Although viscosity is the result of complicated molecular interactions, viscous effects in the laminar flow of all fluids (both gases and liquids) can be

---

*In the energy equation (14-14), notice that we distinguish between fluid energy u associated with internal thermal motion and transport kinetic energy $\tfrac{1}{2}v^2$, both per unit mass.

described by a simple expression that involves a single empirical coefficient for each substance.

The Viscosity Coefficient. When a solid moves through a fluid or when a fluid flows over a solid, the layer of fluid in direct contact with the solid surface experiences no motion with respect to the surface—*the relative velocity between the solid surface and the contacting layer of fluid is zero.*[*]

Consider a viscous fluid that fills the space between a pair of parallel plates. Suppose that the lower plate is held fixed while the upper plate is set into motion with a velocity v_0 that is parallel to the plate (Fig. 14–17). The velocity of the fluid in the layer in contact with the upper plate is v_0, whereas the fluid layer in contact with the lower plate remains at rest. At equilibrium, a linear velocity profile exists from P to Q, given by $v(y) = v_0(y/d)$, as shown in Fig. 14–17a. The block labeled A in Fig. 14–17b represents a small portion of the fluid at some instant. At subsequent times, this same quantity of fluid will have the shapes $B, C, D, \ldots$, and so forth. Thus, there exists a tangential displacement that increases uniformly with time at the rate v_0/d.

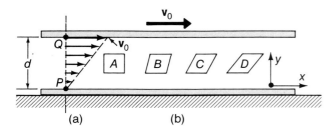

(a) (b)

Fig. 14–17. (a) Velocity profile for laminar flow of a viscous fluid. (b) A portion of fluid represented by the block A undergoes a tangential displacement that increases uniformly with time. The shapes B, C, and D represent the position and shape of block A at successive times.

In order to maintain the motion of the upper plate with velocity v_0, a tangential force $\mathbf{F}$ must be applied to the upper plate. (• What force must be applied to the lower plate to hold it stationary?) If the area of the upper plate is A, the tangential force per unit area applied to the fluid is F/A. The ratio of this force per unit area F/A to the displacement rate v_0/d is called the dynamic shear modulus or simply the *viscosity coefficient* η:

$$\eta = \frac{F/A}{v_0/d} \tag{14–16}$$

The quantity η is found to depend only on the molecular properties of the fluid (and on the temperature), and *not* on the geometry of the particular flow condition.

Equation 14–16 was obtained for the special case of a uniform velocity gradient, v_0/d. In general, the rate of change of the velocity with distance d is not uniform, so we substitute dv/dy for v_0/d. Thus, the tangential force required to maintain a certain value of dv/dy in the fluid is

$$F = \eta A \frac{dv}{dy} \tag{14–17}$$

It is evident from the expressions involving the viscosity coefficient η that the dimensions are those of pressure (F/A) multiplied by those of time (L/LT^{-1}). Thus, the SI units for the viscosity coefficient are Pa·s. Values of the viscosity coefficient are some-

---

[*] You can verify this proposition by attempting to blow away *all* of the dust on a smooth table top. Regardless of how hard you blow, there will remain a fine layer of dust. This indicates that the air layer immediately adjacent to the table surface remains at rest and is therefore unable to carry away the finest dust particles.

times stated in *poise* (P) or its submultiples:

$$1 \text{ Pa} \cdot \text{s} = 10 \text{ P} = 10^3 \text{ cP} = 10^7 \text{ } \mu\text{P}$$

The values of η for several substances are given in Table 14–2 and indicate the substantial variation of the viscosity coefficient with temperature. There is only a small dependence of η on pressure for liquids. Notice that η for liquids *decreases* with increasing temperature, whereas for gases η *increases* with temperature.

TABLE 14–2. Viscosity Coefficient η for Several Substances

T (°C)	GLYCERINE (Pa·s)	MERCURY (10^{-3} Pa·s)	WATER (10^{-3} Pa·s)	BENZENE (10^{-3} Pa·s)	AIR (10^{-5} Pa·s)	HYDROGEN (10^{-5} Pa·s)
0	12.11	1.68	1.787	0.912	1.71	0.84
20	1.49	1.55	1.002	0.652	1.81	0.87
40	0.35	1.45	0.653	0.503	1.90	0.91
60	0.12	1.37	0.466	0.392	2.00	0.95
80	—	1.30	0.355	0.329	2.09	0.98
100	—	1.24	0.282	—	2.18	1.02
200	—	1.05	—	—	2.58	1.21

Poiseuille's Law. One of the important applications of the equation for parallel flow (Eq. 14–17) is concerned with laminar flow in a straight cylindric pipe. Because of the cylindric symmetry of the pipe, such flow consists of a series of concentric, telescoping layers, with the central portion flowing most rapidly and with the outermost layer stationary at the pipe wall.

Consider a cylinder of fluid with a length L and a radius r centered on the pipe axis, as shown in Fig. 14–18. The pressure difference, $p_A - p_B$, drives the fluid; this motion is opposed by the drag force, $-\eta \cdot 2\pi rL \, dv/dr$, that acts on the surface of the cylinder of radius r. At equilibrium, which corresponds to a constant flow rate, we have

$$\pi r^2 (p_A - p_B) = -2\pi rL\eta \, dv/dr$$

Integrating from r to R, the pipe radius, we have*

$$\frac{p_A - p_B}{2\eta L} \int_r^R r \, dr = -\int_v^0 dv$$

so that

$$v(r) = \frac{p_A - p_B}{4\eta L} (R^2 - r^2) \qquad (14\text{–}18)$$

from which we see that the velocity profile is parabolic.

Fig. 14–18. A viscous fluid flow in a straight cylindric pipe. The drag forces shown act on the cylinder with surface area $2\pi rL$.

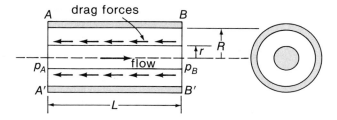

*Note the order of the limits of integration in the two integrals; v corresponds to r, and 0 corresponds to R.

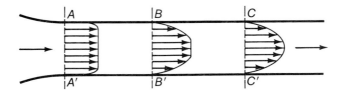

Fig. 14–19. Velocity profiles along a pipe for viscous laminar flow.

Figure 14–19 shows the fluid entering the pipe through a flared input. Then, at the input plane AA', the velocity profile is nearly uniform except for a thin layer adjacent to the pipe wall. Somewhat farther along the pipe, at plane BB', the flow has begun to develop a parabolic velocity profile but there is a core of residual uniform flow. Finally, at plane CC', parabolic flow, described by Eq. 14–18, is fully developed. The distance along the pipe required to develop parabolic flow depends on the pipe diameter and the flow velocity. A typical developmental length is 50 pipe diameters.

The volume flow rate through the pipe for fully developed parabolic laminar flow can be obtained by using Eq. 14–18 together with the continuity condition (Eq. 14–13a), which we express as $dq = v\,dA = v(r)\cdot 2\pi r\,dr$. Thus, the flow rate is

$$q = \frac{(p_A - p_B)\pi}{2\eta L} \int_0^R (R^2 - r^2)r\,dr$$

or

$$q = \frac{\pi R^4}{8\eta} \cdot \frac{p_A - p_B}{L} \qquad (14–19)$$

This result is known as *Poiseuille's equation.**

Example 14–12

A straight horizontal pipe with a diameter of 1 cm and a length of 50 m carries oil with a density of 930 kg/m^3 and a viscosity coefficient of 0.12 Pa·s. At the entrance to the pipe, the oil temperature is 20° C. The discharge rate is 0.80 kg/s at atmospheric pressure.

(a) Find the gauge pressure at the pipe input.

(b) Determine the maximum stream velocity of the oil.

(c) Show that the internal friction heating rate is

$$P = \frac{p_A - p_B}{\rho} \frac{dm}{dt}$$

where $dm/dt = q\rho$ is the rate of mass flow through the pipe and p_A and p_B are the pressures at the pipe entrance and exit, respectively. Determine the rate of heating in the 50-m length of pipe.

Solution:

(a) The flow rate q is

$$q = \frac{1}{\rho}\frac{dm}{dt} = \frac{0.80 \text{ kg/s}}{930 \text{ kg/m}^3} = 8.60 \times 10^{-4} \text{ m}^3/\text{s}$$

Then, using Eq. 14–19, the pressure difference is

$$p_A - p_B = \frac{8\eta Lq}{\pi R^4} = \frac{8(0.12 \text{ Pa·s})(50 \text{ m})(8.60 \times 10^{-4} \text{ m}^3/\text{s})}{\pi(5 \times 10^{-3} \text{ m})^4}$$

$$= 2.10 \times 10^7 \text{ Pa} = 207 \text{ atm} \qquad \text{(gauge)}$$

(b) The maximum stream velocity occurs at the center of the pipe. Using Eq. 14–18 with $r = 0$, we have

$$v_m = \frac{(p_A - p_B)R^2}{4\eta L} = \frac{(2.10 \times 10^7 \text{ Pa})(5 \times 10^{-3} \text{ m})^2}{4(0.12 \text{ Pa·s})(50 \text{ m})}$$

$$= 21.9 \text{ m/s}$$

(c) The difference between the input mechanical power and the output mechanical power represents the rate of heat generation by internal friction. According to Eq. 7–18, power is equal to the product of force and velocity. Here, the force is the pressure multiplied by the differential area $2\pi r\,dr$ of a cylindric lamina. Thus, the rate of heat generation is

$$P = 2\pi(p_A - p_B) \int_0^R v(r)r\,dr$$

---

* After Jean Louis Marie Poiseuille (1799–1869), a French physiologist noted for his study of blood flow. The result was derived and tested by Poiseuille in 1844. The poise was named in his honor.

Substituting for $v(r)$ from Eq. 14–18, we have

$$P = \frac{\pi(p_A - p_B)^2}{2\eta L} \int_0^R (R^2 - r^2) r \, dr$$

and using Eq. 14–19 to identify q, we obtain

$$P = q(p_A - p_B) = \frac{p_A - p_B}{\rho} \frac{dm}{dt}$$

(• Alternatively, this last equation could have been derived directly from Eq. 14–14. Work out this derivation.)

If the pipe is insulated against heat loss, all of the friction-generated heat appears as increased internal energy. This amounts to

$$P = \frac{2.10 \times 10^7 \text{ Pa}}{930 \text{ kg/m}^3} \times (0.80 \text{ kg/s})$$

$$= 18.1 \text{ kW}$$

QUESTIONS

14–1 Define and explain briefly the meaning of the terms (a) hydrostatic pressure, (b) Pascal's principle, (c) manometer, and (d) Archimedes' principle.

14–2 Why are dams built thicker at the bottom than the top?

14–3 Explain how a suction cup works.

14–4 An object is floating in a beaker of water. If taken from one place to another where the acceleration of gravity g is substantially different, what will happen to the depth of submersion of the floating object?

14–5 A log barely able to float in the waters of the upper Chesapeake Bay is carried by a current into the Atlantic Ocean. Will the log sink or float more easily? Explain.

14–6 Solid chunks of the metals listed in Table 1–2 are placed in a beaker of mercury. What will happen to each piece of metal?

14–7 Why does a rising bubble in a glass of beer get larger?

14–8 You are quietly floating in a swimming pool. Explain what motion you might undergo while breathing.

14–9 Define and explain briefly the meaning of the terms (a) streamline, (b) tube of flow, (c) continuity condition, (d) Bernoulli's equation, and (e) viscosity.

14–10 The pressures p_1 and p_2 appearing in Eqs. 14–14 and 14–15 are absolute pressures. Under what circumstances could they be replaced by gauge pressures?

14–11 In launching airplanes, should a carrier move with its flight deck heading into the wind or heading downwind? Explain.

14–12 Hold two pages of your open textbook with a finger in between, then blow between the pages. What happens? Explain.

14–13 Explain how a strong wind can cause a closed window to explode outward.

14–14 Explain how a pitched baseball can be made to "curve."

14–15 A beaker of water sitting on one arm of a balance is exactly counterbalanced by weights on the other arm of the balance. What happens if you stick your finger into the water without touching any part of the beaker?

PROBLEMS

Section 14–1

14–1 If a 1-megaton nuclear weapon is exploded at ground level, the peak overpressure (that is, the pressure increase above normal atmospheric pressure) will be 0.2 atm at a distance of 6 km. What force will be exerted on the side of a house with dimensions 4.5 m × 22 m as a result of such an explosion?

14–2 Use the fact that normal atmospheric pressure is 1.013×10^5 N/m² and that the radius of the Earth is 6.38 × 10⁶ m to calculate the mass of the Earth's atmosphere.

14–3 Refer to Fig. 14–4. Show by direct integration that the x-component of the fluid force on the hemispheric surface is $F_x = -\pi R^2 p_0$.

14–4 In 1654, Otto von Guericke of Magdeburg demonstrated the effect of air pressure by placing together two hemispheric steel shells with diameters of "nearly three-quarters of a Magdeburgian ell" (about 0.8 m) and pumping out the air from the enclosed volume. He then had two teams of eight horses pull in opposite directions on the hemispheres in an unsuccessful attempt to separate the shells. With what force would each team of horses have had to pull to break apart the shells? [*Hint:* Refer to Fig. 14–4.]

Section 14–2

14–5 (a) What is the difference in atmospheric pressure between the base and the top of the World Trade Center building (412.4 m)? [*Hint:* Write Eq. 14–5 in increment form and use $\rho(y) \cong \rho_0$.] (b) What is the atmospheric pressure on top of Mount Everest (8.85 km)?

14–6 What is the pressure on a diver at a depth of 40 m in sea water ($\rho = 1030$ kg/m^3)?

14–7 • The door to a room has dimensions 2.00 m × 0.70 m and is suspended on frictionless hinges. What force is required to open this door against an atmospheric pressure difference of 1 percent? Assume that the doorknob is at the open edge of the door. [*Hint:* This is a problem in rotational equilibrium.]

14–8 A swimming pool has dimensions 30 m × 10 m and a flat bottom. When the pool is filled to a depth of 2 m with fresh water, what is the total force due to the water on the bottom? On each end? On each side?

14–9 What must be the contact area between a suction cup (completely exhausted) and a ceiling in order to support the weight of an 80-kg student?

14–10 Describe the operation of a hydraulic jack, a schematic diagram of which is shown in the illustration. What are the functions of the various valves, V, V', and X? The load piston has a diameter of 30 cm and the pump piston has a diameter of 2 cm. What force f on the pump piston is required to raise a load of 5 tonnes (5000 kg)? If the hydraulic fluid has a compressibility $\lambda = 40 \times 10^{-11}$ (Pa)$^{-1}$, what is the fractional increase in density under this load?

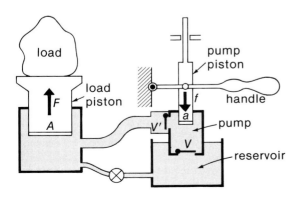

14–11 The hydraulic press shown in the diagram is used to raise the mass M through a height of 5.0 mm by performing 500 J of work at the small piston. The diameter of the large piston is 8.0 cm. a) What is the mass M? b) What is the pressure in the press?

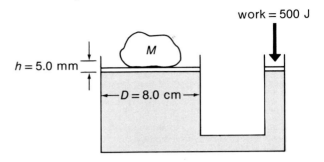

Section 14–3

14–12 Normal atmospheric pressure is 1.013×10^5 Pa. The approach of a storm causes the height of a mercury barometer to drop by 20 mm from the normal height. What is the atmospheric pressure? (The density of mercury is 13.59 g/cm^3.)

14–13 Blaise Pascal duplicated Torricelli's barometer using (as a Frenchman would) a red Bordeaux wine as the working liquid. The density of the wine was 0.984×10^3 kg/m^3. What was the height h of the wine column for normal atmospheric pressure? (Refer to Fig. 14–7 and use $g = 9.80$ m/s^2.) Would you expect the vacuum above the column to be as good as for mercury?

14–14 The mercury in a simple barometer (Fig. 14–8) stands at a height of 76.05 cm. When a small amount of air is allowed to enter the space above the mercury column, the height drops to 73.42 cm. What is the absolute pressure in the space above the column expressed in torr? Expressed in pascals? What is the gauge pressure in torr?

14–15 A simple U-tube that is open at both ends is partially filled with water. Kerosene ($\rho_k = 0.82 \times 10^3$ kg/m^3) is then poured into one arm of the tube, forming a column 6 cm in height, as shown in the diagram. What is the difference h in the heights of the two liquid surfaces?

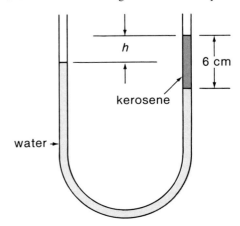

14−16 The open manometer shown in the diagram is used to determine the pressure in a vessel containing oil with a density of 0.85×10^3 kg/m^3. If the mercury stands at the heights indicated in the figure, find the gauge pressure at point A.

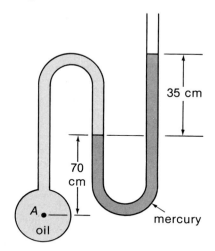

14−17 Determine the difference in pressure between points A and B shown in the diagram. (The density of kerosene is 0.82×10^3 kg/m^3.)

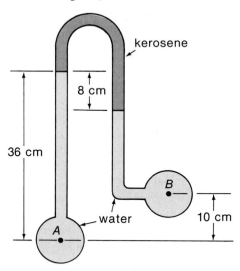

Section 14−4

14−18 A Styrofoam slab has a thickness of 10 cm and a density of 300 kg/m^3. What is the area of the slab if it floats just awash in fresh water when a 75-kg swimmer is aboard?

14−19 A crown that is supposed to be gold has a mass of 2.30 kg. When the crown is immersed in water and weighed, the apparent mass is 2.18 kg. Could the crown be of pure gold? (The density of gold is 19.3×10^3 kg/m^3.)

14−20 A frog in a hemispheric cockleshell just floats without sinking in a pea-green sea (density 1.35 g/cm^3). If the cockleshell has a radius of 6 cm and has negligible mass, what is the mass of the frog?

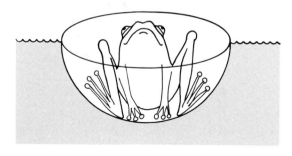

14−21 A hollow brass tube (diameter = 4.0 cm) is sealed at one end and filled with lead shot to give a total mass of 0.20 kg. When the tube is floated in pure water, what will be the depth z of the bottom of the tube?

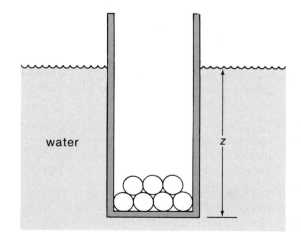

14−22 A beaker that is partly filled with water has a mass of 0.30 kg. The beaker rests on a platform scale. Next, a 250-g piece of copper ($\rho = 8.93 \times 10^3$ kg/m^3) is suspended so that it is completely immersed in the water but does not rest on the bottom of the beaker. What is the reading on the scale?

Section 14−6

14−23 Water flows at a rate of 2.0×10^{-4} m^3/s in a pipe with a diameter of 3 cm. A small crack (0.1 mm × 3 mm) develops, and the output flow rate decreases by 1 percent. (a) What is the average velocity of the water that squirts from the crack? (b) What is the average velocity of the water delivered by the pipe?

14−24 Rain falls vertically and accumulates at a rate of 10 cm/h. The rain is intercepted by a plane that has a surface area of 2.0 m^2 and is inclined to the horizontal by an angle of 20°. At what rate (in cm^3/s) does the rain water flow across the edge AA'?

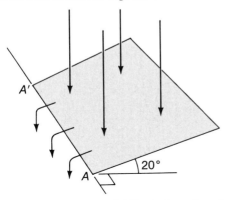

14–25 Air is forced at a rate of 150 g/s through a pipe with a diameter of 10 cm. What is the average flow velocity of the air? (The density of air is 1.293 kg/m^3.) Can the flow be considered incompressible in this case?

Section 14–7

14–26 A large water tank has a 1-cm^2 hole in its side at a point 2 m below the surface. What is the mass discharge rate (in kg/s) through the hole?

14–27 A stream of water (initially horizontal) flows from a small hole in the side of a tank. The stream is 2.5 m from the side of the tank after falling 0.7 m. How far below the surface of the water in the tank is the hole?

14–28 The surface of water in a tank stands 2.5 m above the bottom where a small hole allows water to drain out in a vertical stream. Ignore the dynamic contraction of the stream immediately below the hole (i.e., ignore the formation of the *vena contracta*). At what distance below the bottom of the tank will the stream have a cross-sectional area equal to one half the hole area?

14–29 A siphon is used to drain water from a tank, as indicated in the diagram at right. The siphon has a uniform diameter d. Assume steady flow. (a) Derive an expression for the volume discharge rate at the end of the siphon. (Select the reference level at point 3.) (b) What is the limitation on the height of the top of the siphon above the water surface?

14–30 A stream flows smoothly down a 2° slope. At one point the stream has a rectangular cross-section with a depth of 15 cm and a width of 60 cm. At another point, 100 m farther down the slope, the width is 100 cm. If the average flow velocity is 4 m/s at the upper point, what is the depth of the stream at the lower point? Assume steady flow.

14–31 The static pressure p in a tube filled with a flowing fluid can be measured with an open manometer, as shown in the diagram. Derive an expression for the static pressure at the tube wall in terms of h_1, h_2, ρ, ρ', and the atmospheric pressure p_0.

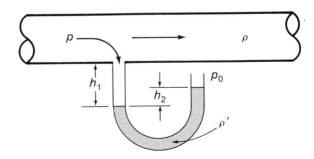

14–32 Water flows at a rate of 2 m^3/min through a pipe with a diameter of 10 cm. (a) What is the average flow velocity? (b) What is the average flow velocity in a region of the pipe that has a constriction with a diameter of 6 cm? (c) What is the static pressure difference (in atm) between the main pipe and the constriction?

14–33 A pump that delivers 10 hp of mechanical power is used to raise water through a height of 30 m at a mass flow rate of 20 kg/s. The water flows through 200 m of rough-surfaced pipe that has an inside diameter of 5.0 cm. If the suction and output pressures are both 1 atm, what is the rate of frictional heating of the water (in W/m)? What fraction of the pump power is dissipated through friction?

14–34 The *Prandtl tube*, shown in the diagram at the top of the next page, is a device similar to a Pitot tube (Example 14–10) for measuring fluid flow velocities. Show that the expression for the velocity v_1 is the same as that found for the Pitot tube. What advantage does the Prandtl tube have compared with the Pitot tube?

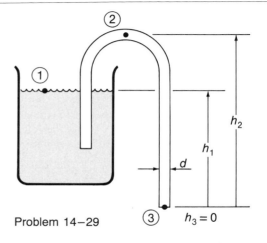

Problem 14–29

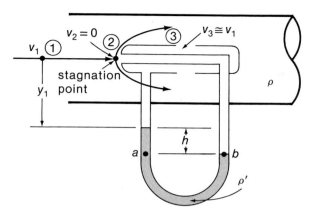

Problem 14–34

14–35 Two tubes that carry water have equal input pressure applied and both exhaust to air. Tube A has a diameter of 4 mm and a length of 10 m; tube B has a diameter of 6 mm and a length of 20 m. What is the ratio of the discharge rates q_B/q_A?

14–36 Water at 20° C is discharged into air at a rate of 2.0×10^{-2} m^3/s from a 20-m length of garden hose that has a diameter of 1.90 cm. What input pressure (in atm) is required?

14–37 A straight horizontal pipe with a diameter of 10 cm and a length of 20 m discharges water at a rate of 10 kg/s. The water temperature is 20° C. What is the pressure differential between the ends of the pipe?

Additional Problems

14–38 Show that for small changes in altitude y above sea level, the expression for the atmospheric pressure given by Eq. 14–7 reduces to $p(y) = p_0 - \rho_0 g y$. Explain why this result appears similar to Eq. 14–4.

14–39 What is the altitude below which lies 90 percent of the Earth's atmosphere (by mass)?

14–40 A cylindric diving bell (the bottom of which is open) has
• a height of 2 m. The bell is lowered into the sea until the water rises 0.80 m inside the bell. (a) At what depth is the top of the bell? (b) At this depth, what absolute air pressure inside the bell would prevent any water from entering the bell? [*Hint:* Assume that the pressure of a fixed quantity of gas at a constant temperature is inversely proportional to the confining volume. This assumption is the same as that expressed by Eq. 14–6.
• Can you see why?]

14–41 The upstream face of a dam makes an angle α with the
•• vertical, as shown in the diagram. The width of the dam is L, and the depth of water behind the dam is h. (a) Show that the water exerts on the dam a horizontal force, $F_x = \frac{1}{2}\rho g L h$ with a line of action $\frac{1}{3}h$ above the heel of the dam at H. (b) Show that the water exerts on the dam a downward force, $F_y = \frac{1}{2}\rho g L h^2 \tan \alpha$, with a line of action $\frac{1}{3}h$ to the right of the heel H.

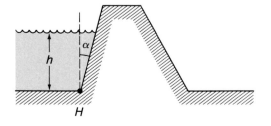

14–42 The mercury column in a simple barometer stands at a
•• height of 75.92 cm. An amount of pure water is introduced into the barometer column and forms a 1.20-cm layer on top of the mercury, as shown in the diagram. The mercury column now stands at a height of 74.08 cm. What is the pressure in the space above the water? (This is the *vapor pressure* of water at the particular temperature, namely, 20° C.)

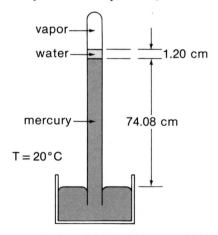

14–43 An evacuated steel ball with an outside diameter of
• 10.0 cm floats in fresh water almost completely submerged. The density of the steel is 7.86×10^3 kg/m^3. What is the wall thickness of the ball?

14–44 A *hydrometer*, shown schematically in the figure on the
• next page, is a device for measuring liquid density. The essential parts are a weighted volume V_0 and a uniform stem that floats partly out of the liquid. A particular hydrometer with a stem diameter $d = 8.5$ mm floats in fresh water at a depth measured by $x = 2.00$ cm. When

floating in methyl alcohol, $x = 11.00$ cm. The density of methyl alcohol is 794 kg/m³, and that of water at the same temperature is 998 kg/m³. Determine the volume V_0. What is the total mass of the hydrometer?

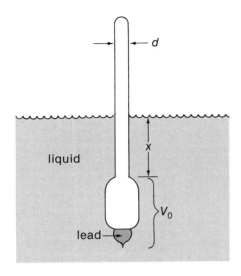

14–45 A beam balance is used to determine the mass of an
•   aluminum object. The balance weights are made of brass. What percentage error in the measurement would be introduced by neglecting air buoyancy effects? (The relevant densities are 2.70×10^3 kg/m³ for aluminum, 8.41×10^3 kg/m³ for brass, and 1.29 kg/m³ for air.)

14–46 A helium-filled balloon supports a total load (including
••   the mass of the balloon structure) of 1200 kg and is in equilibrium at an altitude where the atmospheric pressure is 0.58 atm. Assume an isothermal atmosphere at 0° C. (a) What is the altitude of the balloon? (b) What is the volume of the balloon? [*Hint:* The balloon material is not stretched, so the helium pressure is equal to atmospheric pressure. The gas densities are given in Table 1–2 for standard conditions. Recall that the gas densities vary in direct proportion to the pressure.] (c) If hydrogen is substituted for helium (equal volumes), what total load could the balloon support at the same altitude? [*Hint:* See Table 1–2.]

14–47 A cork is tied with a string to the bottom of a beaker and
••   floats totally submerged when water is poured into the beaker. The beaker is placed at the edge of a large platform that rotates with an angular frequency ω about a vertical central axis. Show that buoyancy effects cause the line of the string holding the cork to tip *inward*, to-

ward the center of rotation, by an angle $\alpha = \tan^{-1}(R\omega^2/g)$, where R is the equilibrium radius of the cork. (• What can you say about the shape of the water surface?)

14–48 The free surface of water in a tank is a distance h above ground level. At what distance below the free surface should a small hole be drilled into the tank wall so that the emerging water jet will strike the ground the greatest distance from the tank?

14–49 An enclosed cylindric tank with a radius R and a height h
••   is half filled with water. The air in the top part of the tank is maintained at an absolute pressure p. A hole with a diameter d at the bottom of the tank is then opened, as indicated in the diagram. (a) Derive an expression for the upward thrust exerted by the escaping water. (b) If $R = 30$ cm, $h = 2$ cm, and $d = 4$ cm, and if the tank mass is $m = 25$ kg, what is the minimum pressure p required to provide an upward acceleration of the tank? [*Hint:* Consult Section 9–5.]

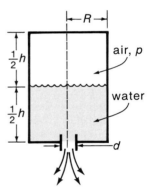

14–50 A suction pump located just above a water surface operates with an intake pressure of 1 atm and delivers 3.0 hp of mechanical power to rotating blades that raise water through a pipe with a uniform diameter. What is the maximum height above the surface to which 20 ℓ/s (2×10^{-2} m³/s) of water can be delivered? Neglect frictional effects.

14–51 The viscosity of a certain liquid is determined (at 40° C) by measuring the flow rate through a tube under a known pressure difference between the ends. The tube has a radius of 0.70 mm and a length of 1.50 m. When a pressure difference of 1/20 atm is applied, a volume of 292 cm³ is collected in 10 min. What is the viscosity of the liquid? What is the liquid?

CHAPTER 15

OSCILLATORY MOTION

There are many types of physical systems that undergo regular, repeating motions. Any motion that repeats itself at definite intervals, such as the motion of the Earth around the Sun, is said to be *periodic*. Any periodic motion that carries a system back and forth between alternate extremes is said to be *oscillatory*. Some familiar examples of oscillatory motion are illustrated in Fig. 15–1.

The common feature of oscillatory systems is that the motion is started by displacing an object from a position of *stable equilibrium*. When such a displacement is made, a *restoring force* acts to return the object to its equilibrium position (see Section 8–1). Upon release, the object moves toward the equilibrium position, but its inertia causes it to "overshoot" this point. The motion continues through the equilibrium position and beyond until the restoring force eventually stops the object and pulls it back toward the equilibrium position. The result is a continuing oscillatory motion back and forth through the position of equilibrium.

It is a common observation that any nondriven oscillating system (that is, one with no external energy source) eventually comes to rest. Such motion is said to be *damped;* the damping is due to energy-dissipating effects, such as friction. In order to maintain the oscillatory motion of a real system (such as the pendulum in a grandfather clock), energy must be continually supplied from a compressed spring or some other source. In this

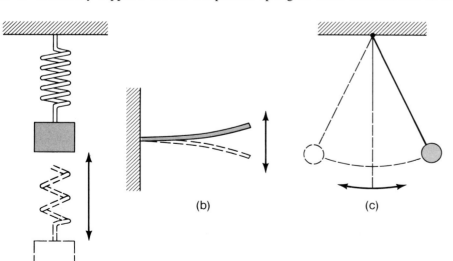

(a) (b) (c)

Fig. 15–1. Three examples of oscillatory motion: (a) a block attached to a spring, (b) a rod anchored at one end, and (c) a simple pendulum consisting of a bob attached to a string.

chapter we will first discuss oscillatory motion for the idealized case of no friction (that is, with no external energy source required to sustain the motion). Damped motion is discussed briefly at the end of the chapter.

15–1 RESTORING FORCES AND POTENTIAL ENERGY FUNCTIONS

Consider the simple one-dimensional motion of a particle subject to a conservative force (Section 8–6). Assume that there exists a point of stable equilibrium for the object. At the coordinate location ξ, the particle has a potential energy $U(\xi)$, as indicated in Fig. 15–2. According to the discussion in Section 8–1, the force on the particle at ξ can be expressed as

$$F(\xi) = -\frac{dU(\xi)}{d\xi} \tag{15-1}$$

Then, stable equilibrium occurs at the point (or points) for which $F(\xi) = 0$ or $dU(\xi)/d\xi = 0$, if, in addition, $d^2U(\xi)/d\xi^2 > 0$. (• Do you recall why this condition on the second derivative is necessary?) Evidently, the point ξ_0 in Fig. 15–2 is such a point of stable equilibrium.

If we confine attention to small excursions of the particle from the point $\xi = \xi_0$, it seems reasonable that the potential energy function $U(\xi)$ can be approximated by a constant term $U(\xi_0)$ plus a quadratic (parabolic) term that will duplicate the curvature at ξ_0. That is, we expect $U(\xi)$ to be of the form

$$U(\xi) = U(\xi_0) + \tfrac{1}{2}\kappa(\xi - \xi_0)^2 \tag{15-2}$$

where, for future convenience, we have used $\tfrac{1}{2}\kappa$ for the coefficient of the quadratic term. Equation 15–2 is the simplest expression that exhibits the essential features of the general potential energy function illustrated in Fig. 15–2. (• Notice that $U(\xi)$ does not contain a linear term in $(\xi - \xi_0)$. What would the existence of such a term imply about the behavior of the force near ξ_0?)

The force on the particle at the point ξ is found by differentiating $U(\xi)$; thus,

$$F(\xi) = -\frac{dU(\xi)}{d\xi} = -\kappa(\xi - \xi_0)$$

We use x to represent $\xi - \xi_0$, the (small) displacement from the equilibrium position. Then,

$$F(x) = -\kappa x \tag{15-3}$$

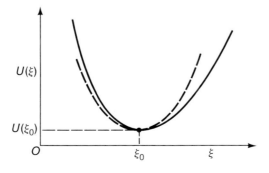

Fig. 15–2. The potential energy function $U(\xi)$ for a particle subject to a one-dimensional force. The dashed line represents the "best quadratic function" fitted to $U(\xi)$ at $\xi = \xi_0$.

which we recognize as the simple linear Hooke's-law relationship we introduced in Section 5–1. From this analysis we conclude

Any system that is displaced by a small amount from a position of stable equilibrium will experience (to first order) *a linear restoring force.*

This fact accounts for the remarkably similar behavior of quite different systems (such as those illustrated in Fig. 15–1) when they are displaced slightly from their equilibrium positions.

   If we apply Newton's second law to the motion of a particle with a mass m that is subject to the Hooke's-law force in Eq. 15–3, we have

$$F(x) = -\kappa x = ma = m \frac{dx^2}{dt^2}$$

so that the acceleration a is given by

$$\frac{d^2x}{dt^2} = -\frac{\kappa}{m}x \qquad (15-4)$$

This is the *differential equation* whose solution $x(t)$ describes the motion.

   It is important to note that the acceleration is directly proportional to the displacement x measured from the equilibrium position and is *oppositely* directed to the displacement (because $\kappa > 0$ and $m > 0$). Moreover, the acceleration is directly proportional to the strength of the restoring force, measured by κ, and is inversely proportional to the inertia of the system (which, for a particle, is the mass m). The description of *all* oscillatory motions involves these same two ingredients—the strength of the restoring force and an inertial factor.

15–2   FREE HARMONIC OSCILLATIONS

   We now turn our attention to a specific example of an oscillatory system. Consider the oscillatory motion of a block attached to a spring (Fig. 15–3). The other end of the spring is attached to a fixed point. The block has a mass m and slides without friction across a horizontal surface. The spring has a spring or force constant κ and a relaxed length ℓ. The figure shows the block displaced by an amount x measured from the equilibrium position O where the spring has its relaxed length. The oscillatory motion possible for such a spring-block system when no external force is present is referred to as *free oscillations*. If the spring is not stretched too far, the force exerted on the block by the spring can be represented approximately by the simple linear expression

$$F(x) = -\kappa x$$

   From the discussion in Section 8–1 we know that a spring stretched (or compressed) by an amount x possesses stored energy $U(x)$ equal to the work done in stretching the spring:

$$U(x) = \int_0^x F_a(x)\,dx = \int_0^x \kappa x\,dx = \tfrac{1}{2}\kappa x^2$$

(Note that F_a is the force *applied* to the spring and is equal to $+\kappa x$; in Eq. 15–3, F is

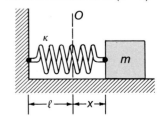

Fig. 15–3. A block with a mass m is attached to the free end of an ideal spring, the other end of which is fixed. The relaxed length of the spring is ℓ, corresponding to the block located at O ($x = 0$).

the force exerted on the particle *by* the spring and is equal to $-\kappa x$. F_a and F represent an action-reaction pair.)

At any instant the block has a kinetic energy equal to

$$K = \tfrac{1}{2}mv^2$$

Because friction is absent, the energy-conservation principle can easily be applied. The total energy E of the system consists of the potential energy of the spring and the kinetic energy of the block, $E = U(x) + K$, and this total energy remains constant:

$$\tfrac{1}{2}\kappa x^2 + \tfrac{1}{2}mv^2 = E = \text{constant} \qquad (15\text{-}5)$$

The energy equation is shown graphically in Fig. 15–4. The total energy is E, and at the arbitrary displacement x from the equilibrium position O, the components of E are the potential energy $U(x)$ and the kinetic energy K.

Notice in Fig. 15–4 that the kinetic energy K (and the velocity v) becomes zero when $x = \pm A$. Any further increase in the displacement would result in *negative* values for K. Because K cannot be negative (• Why?), we conclude that the maximum displacements from the equilibrium position occur for $x = \pm A$ for this particular value of E. The points on the energy curve labeled a and b are called the *turning points* of the motion (see Figs. 8–2 and EA–2 and the accompanying discussions). When, for example, the block moves toward the right in Fig. 15–4, it must reverse its motion at the point $x = +A$ and then move toward the left. (The block cannot stop and remain indefinitely at $x = +A$ because the spring exerts a force $F(A) = -\kappa A$ on the block at this point.) Thus, the displacement is limited to $-A \leqslant x \leqslant +A$, and the block moves back and forth within this interval.

Whenever the block is at $x = \pm A$ it has maximum potential energy U_m and zero kinetic energy. Thus,

$$U_m = U(\pm A) = \tfrac{1}{2}\kappa A^2 = E \qquad \text{at} \qquad x = \pm A \qquad (15\text{-}6)$$

Whenever the block passes through $x = 0$ it has maximum kinetic energy K_m and zero potential energy. Thus,

$$K_m = \tfrac{1}{2}mv_m^2 = E \qquad \text{at} \qquad x = 0 \qquad (15\text{-}7)$$

from which

$$v_m = \pm\sqrt{\frac{2E}{m}} \qquad \text{at} \qquad x = 0 \qquad (15\text{-}8)$$

(• What is the significance of the $\pm$ sign in this expression?)

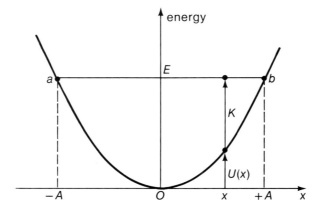

Fig. 15–4. Graph of the energy equation, Eq. 15–5. The total energy is E (a constant), and for any displacement x, we have $E = U(x) + K$. The maximum allowed excursion from the equilibrium position O is $\pm A$. The points a and b correspond to the turning points of the motion.

Simple Harmonic Motion.   The discussion thus far suggests that some type of oscillatory motion is possible for the spring-block system, but we have not yet obtained any detailed description of the motion (that is, an expression for $x(t)$). We now use the energy equation to find this description. If we substitute E from Eq. 15–6 into Eq. 15–5 and solve for v, we find

$$\tfrac{1}{2}mv^2 + \tfrac{1}{2}\kappa x^2 = \tfrac{1}{2}\kappa A^2$$

or
$$v = \frac{dx}{dt} = \pm \sqrt{\frac{\kappa}{m}(A^2 - x^2)} \qquad (15\text{–}9)$$

(• Discuss this result in terms of the general expression given by Eq. 8–10.) Integrating this equation over the time interval from 0 to t, for which the corresponding values of the displacement are x_0 and x, we have

$$\int_{x_0}^{x} \frac{dx}{(A^2 - x^2)^{1/2}} = \sqrt{\frac{\kappa}{m}} \int_{0}^{t} dt$$

where we select the positive value of the square root that appears in Eq. 15–9, because v is positive during this time interval. The integral over x leads to an arcsine solution (see the table of integrals inside the rear cover):

$$\left(\sin^{-1}\frac{x}{A}\right)\bigg|_{x_0}^{x} = \sqrt{\frac{\kappa}{m}}\,t$$

or
$$\sin^{-1}\frac{x}{A} = \sqrt{\frac{\kappa}{m}}\,t + \sin^{-1}\frac{x_0}{A}$$

It is now convenient to define the quantities

$$\phi_0 = \sin^{-1}\frac{x_0}{A} \qquad (0 \leq \phi_0 \leq 2\pi) \qquad (15\text{–}10a)$$

and
$$\omega_0 = \sqrt{\frac{\kappa}{m}} \qquad (15\text{–}10b)$$

so that the solution can be expressed as

$$x(t) = A \sin(\omega_0 t + \phi_0) \qquad (15\text{–}11)$$

The argument of the sine function, $\omega_0 t + \phi_0$, which we set equal to $\phi(t)$, is referred to as the *phase angle* of the motion. The constant ϕ_0 is called the *initial phase angle,* that is, the value of $\phi(t)$ when $t = 0$. The coefficient A represents the magnitude of the maximum displacement from the equilibrium position and is called the *amplitude* of the motion. If only x_0 and A are given, the value of ϕ_0 is not uniquely determined by Eq. 15–10a. Of the two possible solutions for ϕ_0, we select the one that is consistent with the sign of the initial value of $v = dx/dt$, a point we discuss further in the subsection "Initial Conditions."

Notice that for the special case $\phi_0 = \pi/2$, the solution becomes

$$x(t) = A \cos \omega_0 t \qquad (15\text{–}11a)$$

Both of the expressions for $x(t)$ in Eqs. 15–11 and 15–11a are called *sinusoidal functions.* By direct substitution, you can verify that $x(t)$, in either form, satisfies the differential equation, Eq. 15–4 (see Problem 15–3). Therefore, this solution describes the motion of *any* oscillating system that is subject to a Hookes'-law force.

The solution, Eq. 15–11 (or Eq. 15–11a), is indeed periodic because the value of $x(t)$ (including its sign!) is repeated whenever $\phi(t)$ is increased by 2π. The time interval required for this increase of 2π in $\phi(t)$ is called the *period T* of the motion. Thus,

$$\phi(t + T) = \phi(t) + 2\pi$$

or

$$\omega_0(t + T) + \phi_0 = \omega_0 t + \phi_0 + 2\pi$$

from which $\omega_0 T = 2\pi$, so that $\omega_0 = 2\pi/T$, as we found in Eq. 4–23. Then,

$$T = \frac{2\pi}{\omega_0} = 2\pi\sqrt{\frac{m}{\kappa}} \tag{15-12}$$

The motion of a system that takes place during a time interval T is referred to as one *cycle*. The number of such cycles that take place per unit of time is called the *frequency* of the motion, denoted by ν. Evidently,

$$\nu = \frac{1}{T} = \frac{1}{2\pi}\sqrt{\frac{\kappa}{m}} \tag{15-13}$$

The metric unit of frequency is the *hertz** (Hz), which stands for one cycle per second. For example, if an oscillatory motion exactly repeats itself after each time interval $T = 0.020$ s, the frequency is

$$\nu = \frac{1}{T} = \frac{1}{0.020 \text{ s}} = 50 \text{ Hz}$$

The frequency at which a system undergoes free oscillations is called the *natural* or *characteristic* frequency of the system.

We now see that the constant ω_0 may be interpreted as the *angular frequency* of the oscillatory motion, with units of radians per second:

$$\omega_0 = 2\pi\nu = \sqrt{\frac{\kappa}{m}} \tag{15-14}$$

Now that we have defined all of the quantities in Eq. 15–11, we resume the description of the motion of the system. The velocity of the block at any time t can be found by differentiating Eq. 15–11 with respect to time. We find

$$v(t) = \frac{dx}{dt} = A\omega_0 \cos(\omega_0 t + \phi_0) \tag{15-15}$$

The acceleration is obtained, in turn, by a further differentiation:

---

*Named in honor of Heinrich Hertz (1857–1894), the German physicist who made great contributions to the theory of electric oscillations.

$$a(t) = \frac{dv}{dt} = \frac{d^2x}{dt^2} = -A\omega_0^2 \sin(\omega_0 t + \phi_0) \tag{15-16}$$

The behavior of the solution, $x(t)$, conforms to the general description of oscillatory motion given earlier in this section. Equation 15-11 shows that the maximum displacements are $x = \pm A$. Equation 15-15 shows that these occur at times when $v = 0$. In addition, the velocity maxima occur when $x = 0$, as may be seen by comparing the expressions for $x(t)$ and $v(t)$ (Eqs. 15-11 and 15-15, respectively).

The time variations of the displacement, the velocity, and the acceleration are shown in Fig. 15-5 for the case $\phi_0 = 0$. Notice that the maxima (and the minima) of $x(t)$ and $v(t)$ are separated by intervals of $\pi/2$. (The sine and cosine functions are said to be *out of phase* by $\pi/2$.)

Motion described by sines and cosines of a single frequency is referred to as *simple harmonic motion* (SHM), and the associated system is called a *simple harmonic oscillator*.

Fig. 15-5. The variation with time (actually, $\omega_0 t$) of the displacement $x(t)$, the velocity $v(t)$, and the acceleration $a(t)$ for a simple harmonic oscillator. The curves are drawn for the special case $\phi_0 = 0$.

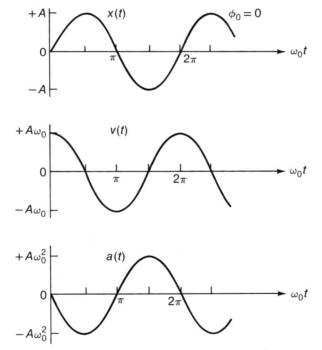

Initial Conditions. Equation 15-11 for the displacement $x(t)$ contains two adjustable parameters, namely, the amplitude A and the initial phase angle ϕ_0. To determine these parameters we must have additional information, usually given in the form of *initial conditions*. As this term suggests, x and v are given for $t = 0$. We designate these specific values x_0 and v_0.* By virtue of the definitions, we have

$$x_0 = A \sin \phi_0$$
$$v_0 = A\omega_0 \cos \phi_0 \tag{15-17a}$$

---

*Values of A and ϕ_0 may be determined by giving x and v for *any* specified time t, not just when $t = 0$. When this is done, the quoted conditions are referred to as the *boundary conditions*.

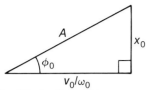

Fig. 15–6. Geometric representation of the parameters involved in SHM. Compare Eqs. 15–17.

Dividing the first of these equations by the second gives

$$\tan \phi_0 = \frac{\omega_0 x_0}{v_0} \qquad (15\text{–}17b)$$

Squaring the equations and adding gives

$$A^2 = x_0^2 + \frac{v_0^2}{\omega_0^2} \qquad (15\text{–}17c)$$

Fig. 15–7. Graph of the simple harmonic motion, $x(t) = C \sin (\omega_0 t + \pi/2)$. Values of $\phi = \omega_0 t + \pi/2$ are shown from $\omega_0 t = 0$ through a complete cycle in increments of $\pi/4$. Also, $v_m = \omega_0 C$ and $a_m = \omega_0^2 C$.

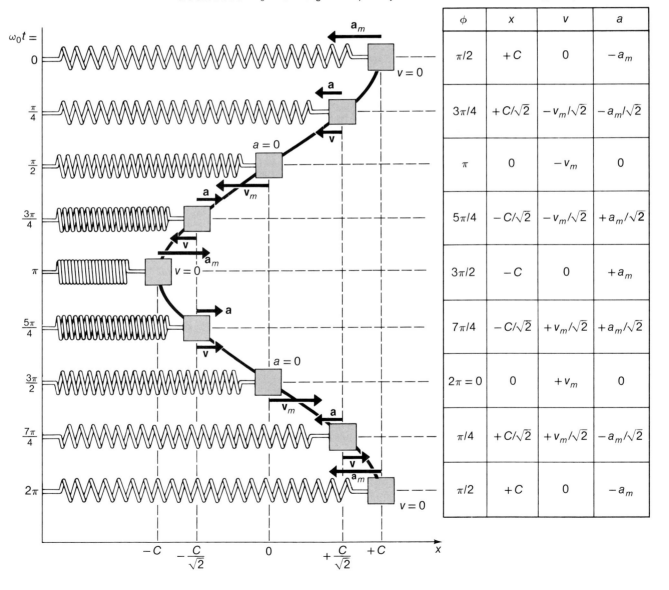

ϕ	x	v	a
$\pi/2$	$+C$	0	$-a_m$
$3\pi/4$	$+C/\sqrt{2}$	$-v_m/\sqrt{2}$	$-a_m/\sqrt{2}$
π	0	$-v_m$	0
$5\pi/4$	$-C/\sqrt{2}$	$-v_m/\sqrt{2}$	$+a_m/\sqrt{2}$
$3\pi/2$	$-C$	0	$+a_m$
$7\pi/4$	$-C/\sqrt{2}$	$+v_m/\sqrt{2}$	$+a_m/\sqrt{2}$
$2\pi = 0$	0	$+v_m$	0
$\pi/4$	$+C/\sqrt{2}$	$+v_m/\sqrt{2}$	$-a_m/\sqrt{2}$
$\pi/2$	$+C$	0	$-a_m$

Equations 15–17 uniquely determine both ϕ_0 and A in terms of x_0, v_0, and ω_0. These two relationships are conveniently represented geometrically by the right triangle shown in Fig. 15–6.

Suppose, for example, that a simple harmonic oscillator is displaced to $x = C$ ($C > 0$) and then released from rest at $t = 0$. Then, we identify $x_0 = C$ and $v_0 = 0$. From Eqs. 15–17 we find $\phi_0 = \pi/2$ and $A = C$. The phase angle becomes $\phi(t) = \omega_0 t + \pi/2$, and the motion proceeds as indicated in Fig. 15–7. This diagram shows one complete cycle of the motion, starting at $\omega_0 t = 0$ and with increments of $\pi/4$. The maximum velocity is $v_m = \omega_0 C$ and the maximum acceleration is $a_m = \omega_0^2 C$.

Example 15–1

A spring stretches by 3.0 cm from its relaxed length when a force of 7.5 N is applied. A particle with a mass of 0.50 kg is attached to the free end of the spring, which is then compressed horizontally by 5.0 cm from its relaxed length and released from rest at $t = 0$.

(a) What is the spring constant κ?

(b) Find the period, the frequency, and the angular frequency of the oscillations.

(c) Find the amplitude A and the initial phase angle ϕ_0 of the motion.

(d) Find the maximum velocity v_m and the maximum acceleration a_m.

(e) Find the total energy E of the system.

(f) Determine the displacement x, the velocity v, and the acceleration a when $\omega_0 t = \pi/8$.

Solution:

(a) The spring constant is

$$\kappa = -\frac{F}{x} = -\frac{(-7.5 \text{ N})}{0.03 \text{ m}} = 250 \text{ N/m}$$

(b) The angular frequency is

$$\omega_0 = \sqrt{\frac{\kappa}{m}} = \sqrt{\frac{250 \text{ N/m}}{0.50 \text{ kg}}} = 22.36 \text{ rad/s}$$

The frequency is

$$\nu = \frac{\omega_0}{2\pi} = \frac{22.36}{2\pi} = 3.559 \text{ Hz}$$

The period is

$$T = \frac{1}{\nu} = \frac{1}{3.559 \text{ s}^{-1}} = 0.281 \text{ s}$$

(c) The initial displacement is $x_0 = -5.0$ cm; also, $v_0 = 0$, so the amplitude is (Eq. 15–17c)

$$A = \sqrt{x_0^2 + (v_0^2/\omega_0^2)} = |x_0| = 0.050 \text{ m}$$

The initial phase angle is given by Eq. 15–17b, taking account of the fact that $x_0 < 0$,

$$\phi_0 = \tan^{-1}(\omega_0 x_0/v_0) = \tan^{-1}(-\infty) = 3\pi/2$$

(• Can you see why $x_0 < 0$ in Eq. 15–17a implies that $\pi \leq \phi_0 \leq 2\pi$?) Thus, we have

$$x(t) = (0.050 \text{ m}) \sin(22.36t + 3\pi/2)$$

(d) The maximum velocity is

$$v_m = A\omega_0 = (0.050 \text{ m})(22.36 \text{ s}^{-1}) = 1.118 \text{ m/s}$$

The maximum acceleration is

$$a_m = A\omega_0^2 = (0.050 \text{ m})(22.36 \text{ s}^{-1})^2 = 25.0 \text{ m/s}^2$$

(e) The total energy is (Eq. 15–6)

$$E = \tfrac{1}{2}\kappa A^2 = \tfrac{1}{2}(250 \text{ N/m})(0.050 \text{ m})^2 = 0.3125 \text{ J}$$

(f) When $\omega_0 t = \pi/8$ we have

$$\phi(t) = \omega_0 t + \phi_0 = \pi/8 + 3\pi/2 = 13\pi/8 = 292.5°$$

Then,

$$x = (0.050 \text{ m}) \sin 292.5° = -0.0462 \text{ m}$$

$$v = (1.118 \text{ m/s}) \cos 292.5° = +0.428 \text{ m/s}$$

$$a = -(25.0 \text{ m/s}^2) \sin 292.5° = +23.1 \text{ m/s}^2$$

Example 15–2

A horizontal spring-and-block system (Fig. 15–3) has an angular frequency $\omega_0 = 3\pi$ rad/s. Determine $x(t)$ if the block is displaced to $x_0 = 0.25$ m and given a velocity $v_0 = -1.5$ m/s at $t = 0$.

Solution:

The initial displacement is positive, so $0 \leq \phi_0 \leq \pi$ (Eq. 15–17a). Using Eq. 15–17b, we find

$$\phi_0 = \tan^{-1}\frac{\omega_0 x_0}{v_0} = \tan^{-1}\frac{(3\pi \text{ s}^{-1})(0.25 \text{ m})}{(-1.5 \text{ m/s})}$$

$$= \tan^{-1}(-1.571) = 2.138 \text{ rad} = 122.48°$$

From Eq. 15–17c, we obtain

$$A = \sqrt{(0.25 \text{ m})^2 + (-1.5 \text{ m/s})^2/(3\pi \text{ s}^{-1})^2} = 0.296 \text{ m}$$

Therefore, the displacement is given by

$$x(t) = (0.296 \text{ m}) \sin (3\pi t + 2.138)$$

Fig. 15–8. The representation of SHM as the projection of uniform circular motion onto a diameter corresponding to the y-axis: (a) the coordinate $y = \overline{OP}$ of the reference point Q at time t, (b) the y-component of the (tangential) velocity, $v = \omega A$, and (c) the y-component of the (centripetal) acceleration, $a = \omega^2 A$.

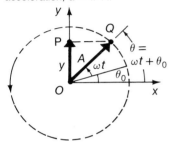

(a) $y = \overline{OP} = A \sin (\omega t + \theta_0)$

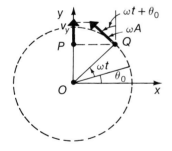

(b) $v_y = \omega A \cos (\omega t + \theta_0)$

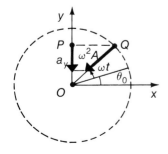

(c) $a_y = -\omega^2 A \sin (\omega t + \theta_0)$

15–3   RELATIONSHIP OF SHM TO UNIFORM CIRCULAR MOTION

The idea of angular frequency ω was originally introduced in connection with the discussion of circular motion (see Section 4–5). This quantity appeared as the time derivative of the angular position of a particle moving along a circular path. For uniform circular motion we found that ω is related to the period T by $\omega = 2\pi/T$. In the case of SHM we have a quantity ω_0 that also appears as the time derivative of an angle, namely, $\omega_0 = d\phi/dt$, where $\phi(t) = \omega_0 t + \phi_0$ is the phase angle of the motion.

There is, in fact, a high degree of similarity between simple harmonic motion and uniform circular motion. We can exploit this similarity to provide a simple means for visualizing the physical significance of the phase angle $\phi(t)$. Consider a particle that moves with a constant angular speed ω_0 around a circular path of radius A. Let the path lie in the x-y plane with the center at the origin O, as shown in Fig. 15–8. We refer to the instantaneous position Q of the particle as the *reference point* and to the circle in which the particle moves as the *reference circle*. The perpendicular projection of Q onto the y-axis is labeled P. Then, as the particle travels around the circle, the point P moves up and down along the y-axis. The line $\overline{OQ}$ makes an angle θ_0 with the x-axis at $t = 0$. At time t, this angle becomes $\omega t + \theta_0$, and the projection P of the reference point is described by the y-coordinate (Fig. 15–8a),

$$y(t) = A \sin (\omega t + \theta_0)$$

The (tangential) velocity of the particle is $v = \omega A$, and the y-component at time t is (Fig. 15–8b)

$$v_y(t) = \omega A \cos (\omega t + \theta_0)$$

The (centripetal) acceleration of the particle is $a = \omega^2 A$, and the y-component at time t is (Fig. 15–8c)

$$a_y(t) = -\omega^2 A \sin (\omega t + \theta_0)$$

If we identify ω with ω_0 and θ_0 with ϕ_0, these results for $y(t)$, $v_y(t)$, and $a_y(t)$ are identical with the corresponding equations (15–11, 15–15, and 15–16) for SHM. The analogy is therefore complete. *Simple harmonic motion can be described as the projection of uniform circular motion onto a fixed diameter of the circular path.* An observer who viewed the motion of the particle from the side, looking along the x-y plane from a distance $d \gg A$, would see the particle executing SHM.

15–4   EXAMPLES OF SIMPLE HARMONIC MOTION

We now give several additional examples of SHM. In each case we consider only motion in which the displacement from the equilibrium position has a small amplitude.

The Simple Pendulum. Consider a particle (the pendulum bob) that has a mass m and is suspended from a fixed support by a string of length ℓ. This system is in equilib-

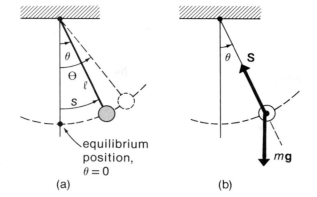

Fig. 15-9. (a) Geometry of a simple pendulum. The angular displacement θ is measured positive in a counterclockwise sense. (b) Free-body diagram for the pendulum bob.

rium when the bob hangs directly below the support point. A displacement away from this equilibrium position is conveniently measured in terms of the angle θ, as shown in Fig. 15-9a. The free-body diagram for the bob (Fig. 15-9b) shows the two forces that are acting. (We now use S for the tension, reserving T for the period.)

We now apply Newton's second law to the tangential component of the motion of the bob. The tangential acceleration is $a = d^2s/dt^2 = d^2(\ell\theta)/dt^2 = \ell(d^2\theta/dt^2)$, so we have

$$-mg \sin \theta = m\ell \frac{d^2\theta}{dt^2} \tag{15-18}$$

The general solution of an equation of this type is complicated. However, if the motion of the pendulum is restricted to angles sufficiently small that $\sin \theta$ may be replaced* by θ, the solution is readily obtained. In this event, we have

$$\frac{d^2\theta}{dt^2} = -\frac{g}{\ell} \theta \tag{15-19}$$

This equation has the same form as Eq. 15-4, so we have a case of SHM involving the angular coordinate θ. If Θ represents the amplitude of the angular motion, we can immediately write the solution as

$$\theta(t) = \Theta \sin (\omega_0 t + \phi_0) \tag{15-20}$$

where

$$\omega_0 = \sqrt{\frac{g}{\ell}} \tag{15-21}$$

The period of the motion is

$$T = \frac{2\pi}{\omega_0} = 2\pi\sqrt{\frac{\ell}{g}} \tag{15-22}$$

Notice that the period of the pendulum depends only on its length and not on the mass of the bob. (• Argue on physical grounds whether the approximate expression for the period, obtained by using $\sin \theta \cong \theta$, yields values for the period that are slightly larger or slightly smaller than the true value.)

Be certain that you understand the distinction between the angular coordinate $\theta(t)$,

---

*For example, when $\theta = 15° = 0.2618$ rad, θ (in radians) exceeds $\sin \theta$ by only 1.15 percent.

which describes the position of the bob and is confined to the relatively narrow range, $-\Theta \leqslant \theta \leqslant +\Theta$, and the phase angle, $\phi(t) = \omega_0 t + \phi_0$, which increases without limit as t increases.

The solution to the pendulum problem contained in Eqs. 15–20 through 15–22 is only approximate; the expressions are useful only when the amplitude Θ is sufficiently small. Methods for obtaining the general solution are beyond the scope of this text, so we give without proof an accurate expression for the period T in terms of a power series in $\sin(\Theta/2)$:

$$T = 2\pi\sqrt{\frac{\ell}{g}}\left[1 + \tfrac{1}{4}\sin^2\frac{\Theta}{2} + \tfrac{9}{64}\sin^4\frac{\Theta}{2} + \cdots\right] \qquad \textbf{(15–23)}$$

Notice that the first term in this series is just Eq. 15–22. In Example 15–3 we see that the error in the period due to the use of the approximate expression (Eq. 15–22) is less than 0.5 percent for $\Theta = 15°$.

▶ *Historical Note on the Pendulum.* The observation that the period of a simple pendulum is independent of the amplitude (or nearly so, for small amplitudes) was first made by Galileo in 1593. Galileo watched the oscillations of a lamp that was suspended from the ceiling of a cathedral in Pisa and timed the oscillations by using his own pulse as a clock. From this crude measurement, Galileo correctly deduced the laws of pendulum motion. (The lamp oscillations were damped by friction, but the period of a simple harmonic oscillator is insensitive to damping, if the damping is not too severe.)

About 1657, the Dutch physicist Christiaan Huygens (1629–1695) realized that a pendulum could be used to regulate the timekeeping of a clock. He also was the first to use a measurement of the period of a pendulum to determine the acceleration due to gravity g.

Notice that the mass m of the pendulum bob occurs on both sides of the Newtonian equation of motion (Eq. 15–18) and therefore cancels. This results in equations for $\theta(t)$ and the period T that are independent of the mass of the bob. However, Newton realized that the quantity m on the left-hand side of Eq. 15–18 arises from the gravitational force and is therefore *gravitational mass*, whereas m on the right-hand side represents the *inertial mass* associated with acceleration and the second law. There was nothing in the theory then (nor is there now) to specify the equality of the two types of mass. Newton devised an ingenious method for comparing these two masses (an experiment alluded to in Section 5–3). He constructed a hollow shell to use as a pendulum bob, and he placed in this shell different materials that he had determined by means of a balance to have the same weight (that is, the same gravitational mass). In this way he was able to alter the composition of the bob while maintaining identical external conditions (such as air resistance). Newton measured the period of the pendulum for several different materials in the shell, always finding the same result. He concluded that to the precision of his measurements (about 1 part in 10^3), gravitational and inertial mass are equivalent. Modern versions of Newton's experiment have improved the precision of the equivalence to 1 part in 10^{12} (see Section 5–3). ◀

Example 15–3

A simple pendulum with a length of 80 cm is displaced by 15° from its equilibrium position and then released from rest. The mass of the bob is 0.30 kg.

(a) Use Eqs. 15–21 and 15–22 to calculate the angular frequency ω_0 and the period T to four significant figures.

(b) Determine Θ and ϕ_0, and give the resulting equation of motion, $\theta(t)$.

(c) Determine the maximum value of the angular speed, $\eta = d\theta/dt$.

(d) Find the maximum value of the tension S in the string. At what angle θ does this occur?

(e) Use Eq. 15–23 to calculate an accurate value for the period. Compare with the result in (a).

Solution:

(a) In order to calculate the angular frequency and period to four-place accuracy, we use $g = 9.806$ m/s^2, the value appropriate for sea level at latitude 45°. Then,

$$\omega_0 = \sqrt{\frac{g}{\ell}} = \sqrt{\frac{9.806 \text{ m/s}^2}{0.80 \text{ m}}} = 3.501 \text{ rad/s}$$

$$T = 2\pi\sqrt{\frac{\ell}{g}} = 2\pi\sqrt{\frac{0.80 \text{ m}}{9.806 \text{ m/s}^2}} = 1.795 \text{ s}$$

(b) The angular amplitude of the motion is, clearly, the same as the initial displacement $\Theta = 15° = 0.262$ rad. Then, differentiating Eq. 15–20 and evaluating at $t = 0$, we have

$$\left.\frac{d\theta}{dt}\right|_{t=0} = \eta_0 = 0 = \omega_0\Theta \cos \phi_0$$

from which $\phi_0 = \pi/2$. (Compare the discussion of Fig. 15–6.) Therefore, the equation of motion is

$$\theta(t) = 0.262 \sin (3.50t + \pi/2)$$

(c) The angular speed, $\eta = d\theta/dt = \omega_0\Theta \cos \phi(t)$, is maximum when $\cos \phi(t) = \pm1$, that is, when $\phi(t) = \pi,\ 2\pi,\ 3\pi,\ \ldots,$ (which are the times when $\theta = 0$). Hence,

$$\eta_m = \omega_0\Theta = (3.50 \text{ s}^{-1})(0.262) = 0.917 \text{ rad/s}$$

(d) Refer to Fig. 15–9b. The radial component of the acceleration of the bob is the centripetal acceleration, $a = v^2/\ell = (\ell\, d\theta/dt)^2/\ell = \ell\eta^2$. Then, Newton's second law for the radial motion can be expressed as

$$S - mg \cos \theta = m\ell\eta^2$$

Both $\cos \theta$ and η are maximum when $\theta = 0$, so

$$S_m = m(g + \ell\eta_m^2)$$
$$= (0.30 \text{ kg})[9.806 \text{ m/s}^2 + (0.80 \text{ m})(0.917 \text{ s}^{-1})^2]$$
$$= 3.14 \text{ N}$$

(e) Using Eq. 15–23, with $\Theta = 15°$, we find

$$T = (1.795 \text{ s})[1 + \tfrac{1}{4} \sin^2 7.5° + \tfrac{9}{64} \sin^4 7.5° + \cdots]$$
$$= (1.795 \text{ s})[1 + 0.00426 + 0.00004 + \cdots]$$
$$= (1.795 \text{ s})[1.00430] = 1.803 \text{ s}$$

We see that the error in using the approximate expression, $T = 2\pi\sqrt{\ell/g}$, amounts to 0.43 percent for $\Theta = 15°$.

The Torsional Pendulum. If a long fiber or wire or thin rod is twisted about its cylindric axis, there is developed a restoring torque τ that follows a relationship similar to Hooke's law, namely,

$$\tau = -\Gamma\psi \qquad\qquad \textbf{(15–24)}$$

where ψ is the angle of twist and where Γ is the *torsional constant* for the system, a quantity that depends on the elastic properties of the twisted element.

Consider two spherical particles, each with mass m and held a distance ℓ apart by a rigid rod with negligible mass. This dumbbell-shaped object is suspended by a wire that is attached to the midpoint of the rod; the other end of the wire is clamped to a fixed support (Fig. 15–10). The rod is rotated by a small angle ψ away from its equilibrium position, thereby twisting the wire. When the rod is released, the system will execute simple harmonic torsional oscillations.

We consider the dumbbell to be a free body, so (Eq. 11–10) $\tau = I_0\alpha$, where I_0 is the rotational inertia of the system about the suspension point, and where α is the angular acceleration. Thus,

$$\tau = -\Gamma\psi = I_0\alpha = I_0 \frac{d^2\psi}{dt^2}$$

Fig. 15–10. A torsional pendulum. The rigid rod connecting the two spheres has negligible mass. The rod is shown displaced from its equilibrium position by a small angle ψ, and the support wire has been twisted by the same amount.

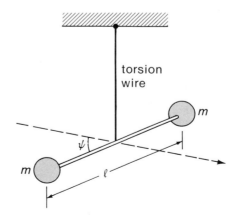

from which
$$\frac{d^2\psi}{dt^2} = -\frac{\Gamma}{I_0}\psi$$

This equation has exactly the same form as Eq. 15–4, so the angular motion of the system is SHM. The period is, therefore,

$$T = 2\pi\sqrt{\frac{I_0}{\Gamma}}$$
(15–25)

For the dumbbell system illustrated in Fig. 15–10, $I_0 = 2\,m(\ell/2)^2 = \frac{1}{2}m\ell^2$, so that $T = 2\pi\ell\sqrt{m/2\Gamma}$. A torsional system of this type was used by Henry Cavendish in his determination of the mass of the Earth (which also gave a value for the universal gravitation constant)—see Section 5–3.

($\bullet$ Suppose that you wish to determine the rotational inertia I_0 of an object with an axis of symmetry but with an otherwise complicated shape. Suppose also that you have a uniform cylinder with known dimensions and mass. How could you use torsional oscillations to determine I_0?)

The Physical Pendulum. We now examine the oscillatory motion of a system more realistic than the idealized simple pendulum we have been discussing. Consider the general rigid body shown in Fig. 15–11, which is suspended so that it can rotate on a horizontal axis, thereby becoming a *physical* (or *compound*) *pendulum*.

The torque about the point O is

$$\tau = -Mg\ell\sin\theta$$

$$= I_0\alpha = I_0\frac{d^2\theta}{dt^2}$$

where I_0 is the rotational inertia about the point O. Again, we consider only small angular motions, so that $\sin\theta$ can be replaced by θ. Then, we can write

$$\frac{d^2\theta}{dt^2} = -\frac{Mg\ell}{I_0}\theta$$
(15–26)

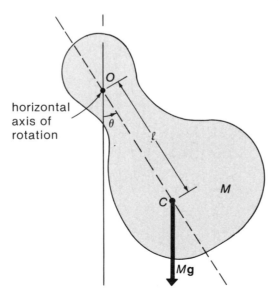

Fig. 15–11. A physical pendulum, consisting of a general rigid body that can rotate about a horizontal axis through the point O. Counterclockwise displacement angles are taken positive.

horizontal
axis of
rotation

This equation has the same form as Eq. 15–4 and therefore represents SHM with period

$$T = 2\pi \sqrt{\frac{I_0}{Mg\ell}} \qquad (15\text{-}27)$$

Notice that this general result for the period of a physical pendulum reduces to that for the simple pendulum when $I_0 = M\ell^2$, as is appropriate for a particle-like bob and a massless string.

Example 15–4

A slender uniform rod with a length of 2.0 m is suspended from one end. What is the period of the oscillatory motion that takes place in a vertical plane?

Solution:

The rotational inertia of a slender rod about one end is $I_0 = \frac{1}{3}M\ell^2$ (see Example 11–3). Thus, using Eq. 15–27,

we have

$$T = 2\pi \sqrt{\frac{I_0}{Mg\ell}} = 2\pi \sqrt{\frac{\frac{1}{3}M\ell^2}{Mg\ell}} = 2\pi \sqrt{\frac{\ell}{3g}}$$

$$= 2\pi \sqrt{\frac{2.0 \text{ m}}{3(9.80 \text{ m/s}^2)}} = 1.64 \text{ s}$$

Notice that the period of the rod is less than that of a simple pendulum of the same length by a factor of $\sqrt{3}$.

15–5   DAMPED HARMONIC MOTION
(Optional)

In the preceding sections we have considered only cases that are free from any effects that cause energy loss. To analyze a real case, however, we must take into account the presence of frictional or viscous drag forces. It is often permissible to approximate the effects of such forces by including in the equation of motion a force term that is proportional to the velocity of the particle. Then, the retarding force has the form

$$F_v = -bv = -b\frac{dx}{dt}$$

and the equation of motion becomes

$$ma = -\kappa x + F_v$$

or $\qquad m\dfrac{d^2x}{dt^2} + b\dfrac{dx}{dt} + \kappa x = 0 \qquad$ **(15–28)**

The form of the solution to Eq. 15–28 depends on the relative magnitudes of the quantities $b^2/4m^2$ and κ/m. We are mostly interested in cases that involve relatively small retarding forces. If the force coefficient b is sufficiently small that $b^2/4m^2 < \kappa/m$, the system undergoes *damped oscillatory motion* and is referred to as being underdamped. The solution for this case is

$$x(t) = Ae^{-\beta t} \sin(\omega^0 t + \phi_0) \qquad \text{underdamped} \qquad \textbf{(15–29)}$$

where $\qquad \beta = b/2m > 0 \qquad$ **(15–30a)**

and $\qquad \omega^0 = \sqrt{(\kappa/m) - (b/2m)^2} \qquad$ **(15–30b)**

Thus, the solution consists of an oscillatory (sinusoidal) term with a multiplicative coefficient (the exponential) that continually *decreases* with time (because $\beta > 0$). Figure 15–12 shows a typical form of Eq. 15–29 in which $\phi_0 = \pi/2$. Notice that the new angular frequency ω^0 is shifted away from the undamped value, $\omega_0 = \sqrt{\kappa/m}$, by an amount that depends on the force coefficient b. When $b = 0$, which corresponds to the frictionless case, we have $\beta = 0$ and $\omega^0 = \omega_0$; thus, Eq. 15–29 becomes identical to Eq. 15–11, as it must.

The constants A and ϕ_0 are specified by the initial conditions

$$x_0 = A \sin \phi_0 \qquad \textbf{(15–31a)}$$

$$v_0 = A(\omega^0 \cos \phi_0 - \beta \sin \phi_0) \qquad \textbf{(15–31b)}$$

Evidently, if ω^0 is to be *real*, the quantity within the radical sign of Eq. 15–30b must be positive. This leads to the stated restriction, $b^2/4m^2 < \kappa/m$. When $b^2/4m^2 \geq \kappa/m$, the solution does not represent oscillatory motion. The special condition when $\beta^2 = \dfrac{b^2}{4m^2} = \dfrac{\kappa}{m}$ leads to a type of nonoscillatory motion referred to as *critically damped* motion. The solution of Eq. 15–28 for $x(t)$ in this case is of the form

$$x(t) = (A + Bt)e^{-\beta t} \qquad \text{critically damped} \qquad \textbf{(15–32)}$$

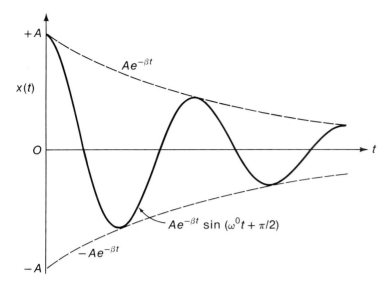

Fig. 15–12. The displacement $x(t)$ of an underdamped oscillator with initial phase angle $\phi_0 = \pi/2$. The dashed curves are the envelope functions, $\pm Ae^{-\beta t}$ with $\beta = b/2m$.

The two constants A and B are to be determined from the stated initial conditions.

Finally, the case when $\beta^2 = \dfrac{b^2}{4m^2} > \dfrac{\kappa}{m}$ also results in nonoscillatory motion, this time referred to as *overdamped*. It is convenient to introduce the quantity $\gamma = \sqrt{\left(\dfrac{b}{2m}\right)^2 - \left(\dfrac{\kappa}{m}\right)^2}$.

The solution to Eq. 15–28 for $x(t)$ is then

$$x(t) = e^{-\beta t}(Ce^{\gamma t} + De^{-\gamma t}) \qquad \text{overdamped} \qquad \textbf{(15–33)}$$

The two constants C and D are again to be determined from the initial conditions.

Example 15–5

Consider again the horizontal spring-block system illustrated in Fig. 15–3, with $m = 0.50$ kg and $\kappa = 250$ N/m, as in Example 15–1. Now, however, the surface exerts a resistive force, $-bv$, on the block. The system is started at $t = 0$ with $\phi_0 = 3\pi/2$ and $A = 5.0$ cm. (Notice that this implies $v_0 \neq 0$.) At $t = 2$ s, the envelope of the oscillatory function has decreased to ± 1.0 cm.

(a) Find the value of the force coefficient b.

(b) Determine the angular frequency ω^0 and compare with the corresponding frictionless value ω_0.

(c) Find the values of the initial velocity v_0 and the initial acceleration a_0.

Solution:

(a) The decrease in the envelope between $t = 0$ and $t = 2$ s tells us that $e^{-2\beta} = \frac{1}{5}$. Solving for β,

$$2\beta = \ln 5$$

or

$$\beta = 0.805 \text{ s}^{-1}$$

Using $\beta = b/2m$, we find

$$b = 2m\beta = 2(0.50 \text{ kg})(0.805 \text{ s}^{-1})$$

$$b = 0.805 \text{ kg/s} = 0.805 \text{ N·s/m}$$

(b) The angular frequency is

$$\omega^0 = \sqrt{\frac{\kappa}{m} - \left(\frac{b}{2m}\right)^2} = \sqrt{\frac{\kappa}{m}} \sqrt{1 - \frac{b^2}{4m\kappa}}$$

$$= \omega_0 \sqrt{1 - \frac{(0.805 \text{ s}^{-1})^2}{4(0.50 \text{ kg})(250 \text{ N/m})}} = \omega_0 \sqrt{1 - 0.00130}$$

$$\approx \omega_0[1 - 0.00065]$$

Now, $\omega_0 = \sqrt{\dfrac{\kappa}{m}} = \sqrt{\dfrac{250 \text{ N/m}}{0.50 \text{ kg}}} = 22.361$ rad/s

so $\omega^0 = (22.361 \text{ rad/s})(0.99935) = 22.346$ rad/s

Thus, ω^0 differs from ω_0 by less than 0.1 percent, even though the damping is sufficient to reduce the amplitude of the oscillations to 20 percent of the initial value after only seven full oscillations (2 s).

(c) Using Eq. 15–31b,

$$v_0 = A(\omega^0 \cos \phi_0 - \beta \sin \phi_0)$$

$$= (0.050)[(22.346) \cos (3\pi/2) - (0.805) \sin (3\pi/2)]$$

$$= +0.0403 \text{ m/s}$$

QUESTIONS

15–1 Define and explain briefly the meaning of the terms (a) restoring force, (b) free oscillation, (c) simple harmonic motion, (d) phase angle, and (e) natural frequency.

15–2 Give some examples from everyday occurrences of periodic motion. Are they examples of approximately simple harmonic motion?

15–3 At what points along the oscillatory motion of a spring-block simple harmonic oscillator is the transfer of kinetic energy to potential energy at the maximum rate? At what points is the energy transfer rate zero?

15–4 Define and explain briefly the meaning of the terms (a) reference circle, (b) simple pendulum, (c) torsional pendulum, and (d) damped harmonic motion.

15–5 A simple pendulum located on a valley floor is adjusted to have a period of exactly 1 s. How will the accuracy of its timekeeping be affected if it is moved to the top of a nearby mountain? Would the situation be any different if the timekeeping were to be done with a torsional pendulum?

15–6 Make a sketch of the potential energy of a simple pendulum as a function of the displacement angle θ.

15–7 As a simple pendulum oscillates back and forth, potential and kinetic energies are being exchanged. Is any work performed by the tension in the supporting string? Explain.

15–8 Two small bobs, one with twice the mass of the other, initially are at rest with a massless compressed spring between them, as shown in the diagram at right. The spring is suddenly released and the bobs fly apart with the spring dropping out of the way. (a) What is the ratio of the maximum angular displacements? (b) At what point will they collide as they return toward each other?

15–9 A *conical pendulum* consists of a hanging bob executing a simple horizontal circular orbit (see drawing). Contrast the motion of the bob when moving as a conical pendulum with its motion if it were moving as a simple pendulum swing in a vertical plane. Assume the same length of string ℓ in both cases and let $R/\ell \ll 1$.

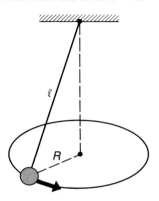

Question 15–9

15–10 Discuss the swinging motion of a bob supported by a string attached to the roof of a truck that is accelerating at a constant rate of $g/5$ on a horizontal road.

15–11 Give some examples from everyday situations where damping of any excited oscillatory motion is *desirable*. Give examples where it is *undesirable*.

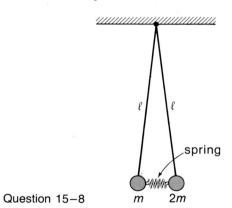

Question 15–8 m $2m$

PROBLEMS

Section 15–1

15–1 The potential energy function for a particle is $U(\xi) = a\xi^2 + b\xi + c$, with $a = 2$ J/m², $b = -4$ J/m, and $c = 1$ J. (a) Sketch the function $U(\xi)$. (b) Find any points of stable or unstable equilibrium. (c) Use the coordinate $x = \xi - \xi_0$ for the displacement from any point of stable equilibrium. Find the expression for the force

constant κ that appears in Eq. 15–3. [*Hint:* Complete the square in the expression for $U(\xi)$.]

15–2 Consider the gravitational potential energy of a pendulum bob (mass m) that is attached to a massless string with a length ℓ, as shown in the diagram. Use the angle θ as the coordinate variable, so that $\theta = 0$ represents the position of stable equilibrium. (a) Let the gravitational potential energy be zero for $\theta = 0$ and show that $U(\theta) = mg\ell(1 - \cos \theta)$. (b) The displacement of the bob is $s = \ell\theta$, measured along the circular path shown dashed in the diagram. Then, calculate $F(s) = -dU/ds$, and determine the value of κ for this case. What is the corresponding frequency? (Make the approximation of small angles.)

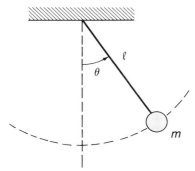

Section 15–2

15–3 Show by direct differentiation that the expression for $x(t)$ given by Eqs. 15–11 and 15–11a satisfies the differential equation, Eq. 15–4.

15–4 A particle executes SHM with an amplitude of 10 cm and a frequency $\nu = 3$ Hz. Find the following: (a) the values of the maximum velocity and the maximum acceleration; (b) the velocity and the acceleration when the displacement is 8 cm; and (c) the minimum time for the particle to move from the equilibrium position to a displacement of 6 cm.

15–5 A block with a mass of 0.50 kg is attached to the free end of an ideal spring that provides a restoring force of 40 N per meter of extension. The block is free to slide over a frictionless horizontal surface. The block is set into motion by giving it an initial potential energy of 2 J and an initial kinetic energy of 1.5 J. (a) Make a graph of the potential energy of the system for -0.5 m $\leq x \leq +0.5$ m. (b) Determine the amplitude of the oscillation—first from the graph, and then by algebraic calculation. (c) What is the speed of the block as it passes through the equilibrium position? (d) At what displacements is the kinetic energy equal to the potential energy? (e) Find the angular frequency ω_0 and the period T of the motion. (f) If the initial displacement was $x_0 > 0$ and if the initial velocity was $v < 0$, determine the initial phase angle ϕ_0. (g) Write the equation of motion $x(t)$.

15–6 Two blocks, with masses m_1 and m_2, are free to slide on a horizontal frictionless surface and are connected by an ideal spring with spring constant κ. Find the angular frequency of oscillatory motion for the system.

15–7 A particle that is attached to a vertical spring is pulled down a distance of 4.0 cm below its equilibrium position

and is released from rest. The initial upward acceleration of the particle is 0.30 m/s². (a) What is the period T of the ensuing oscillations? (b) With what velocity does the particle pass through its equilibrium position? (c) What is the equation of motion for the particle? (Choose the upward direction to be positive.)

15–8 A particle that hangs from an ideal spring has an angular frequency for oscillations, $\omega_0 = 2.0$ rad/s. The spring is suspended from the ceiling of an elevator car and hangs motionless (relative to the elevator car) as the car descends at a constant velocity of 1.5 m/s. The car then stops suddenly. (a) With what amplitude will the particle oscillate? (b) What is the equation of motion for the particle? (Choose the upward direction to be positive.)

15–9 A block with a mass m, when attached to a uniform ideal spring with spring constant κ and relaxed length ℓ, executes SHM with angular frequency, $\omega_0 = \sqrt{\kappa/m}$. The spring is then cut into two pieces, one with relaxed length $f\ell$ and the other with relaxed length $(1 - f)\ell$. The block is divided in the same fractions and the smaller part of the block is attached to the longer part of the spring; the remaining pieces are likewise joined. (a) What are the spring constants for the two parts of the spring? (b) What are the angular frequencies of oscillation for the two systems?

Section 15–3

15–10 A particle is executing uniform circular motion with radius $R = 30$ cm in a vertical plane. The period of the motion is 3 s. A distant overhead light source produces a shadow of the particle on a horizontal plane. The velocity of the shadow, which executes SHM, is $v = -v_0 \sin \omega t$. Determine v_0 and ω. What is the expression for x, the coordinate of the shadow measured from its central position?

15–11 A particle executes SHM along a vertical line according to $y(t) = (60$ m$) \sin (8t + \pi/8)$. (a) Determine the parameters for the corresponding particle that undergoes uniform circular motion in a vertical plane. Determine the position angle θ of the reference point when (b) the particle is at $y = +40$ m and the velocity is directed downward, and (c) the particle has a downward acceleration equal to g and a velocity directed upward. (d) What are the shortest time intervals after $t = 0$ required for the particle to reach the positions referred to in (b) and (c)?

Section 15–4

15–12 A thin rod has a mass M and a length $L = 1.6$ m. One end of the rod is attached by a pivot to a fixed support, and the hanging rod executes small oscillations about the pivot point. Find the frequency of these oscillations. If a particle with the same mass M is added to the lower end of the rod, by what factor will the period change?

15–13 A small bob with a mass of 0.20 kg hangs at rest from a massless string with a length of 1.40 m. At $t = 0$ the bob is given a sharp horizontal blow that delivers an impulse, $J = \int F \, dt = 0.15$ N·s. (a) Find the amplitude Θ of the ensuing oscillations. (b) Determine the equation of motion of the bob, $\theta(t)$.

15–14 A sensitive balance for determining the masses of small objects consists of a thin uniform crossbar with a length of 10 cm and a mass of 100 mg that is attached at its midpoint to a fused quartz torsional fiber. The quartz fiber is stretched between the arms of a frame, as shown in the accompanying figure. When the crossbar is set into torsional oscillations, the period (for small oscillations) is found to be 8.0 s. (a) What is the torsional constant of the fiber? (b) If a small object with a mass of 10 μg is attached to the end of the crossbar at point A, what is the resulting vertical deflection at equilibrium?

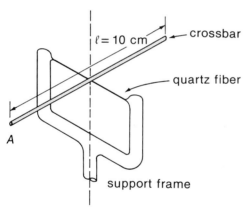

15–15 A straight uniform wire with a length of 25 cm is bent in
  • a right angle at its midpoint. The wire is hung on a horizontal knife edge that acts as an axis and is set in oscillation with a small angular amplitude, as indicated in the diagram. Determine the period of the oscillations.

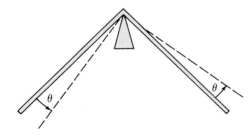

15–16 A thin uniform disc with a radius $R = 20$ cm swings
  • about an axis that consists of a thin pin driven through the disc, as shown in the diagram. At what distance r from the center of the disc must the pin be located in order that the period of oscillation be a minimum? What is the period for this location of the pin?

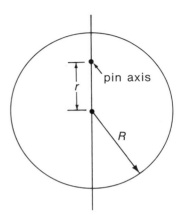

15–17 A physical pendulum consists of a bob with a radius $R = 5$ mm attached to a string whose mass is negligible compared with that of the bob. The distance from the point of support to the center of the bob is 25 cm. Find the period of this pendulum for oscillations with small amplitudes. Compare this period with that of the simple pendulum having the same length (but with a bob of negligible size).

Section 15–5

15–18 A horizontal spring-block system (as in Example 15–5) has $m = 0.50$ kg and $\kappa = 250$ N/m. The frictional force, $-bv$, between the block and the surface produces a damped oscillatory motion with a period that is 1.25 times the period of the corresponding undamped motion. (a) What is the value of the force coefficient b? (b) After what time interval will the amplitude envelope function have decreased to one-third the value at the beginning of the interval?

15–19 Derive an expression that gives the velocity as a function of time for a damped oscillator. Sketch $v(t)$ for the case $A = 1$ m, $\beta = 1$ s^{-1}, and $\omega^0 = 2$ rad/s.

Additional Problems

15–20 An interesting teeter-totter toy consists of a pointed peg
•• with two thin drooping arms having bobs of equal mass
at the ends, as shown in the diagram. A properly designed toy will rock back and forth about the pivot point
O without tipping over. For the sake of simplicity, consider the rocking motion confined to the plane containing
the arms and the pivot. (a) Show that the potential energy
of the toy, when the peg is displaced by an angle ϵ
from the vertical, is $U(\beta) = 2\ mg \cos \epsilon(L - \ell \cos \alpha)$.
Assume that the bobs are particles and choose the pivot
point O as the zero reference for the potential energy.
(b) Show that for stable equilibrium the bobs must hang
below the pivot point O when $\epsilon = 0$ (i.e., $\ell \cos \alpha > L$).
(c) For the case $\ell = 3L$ and $\alpha = 60°$, show that the period for small oscillations is $T = 2\pi\sqrt{14L/g}$.

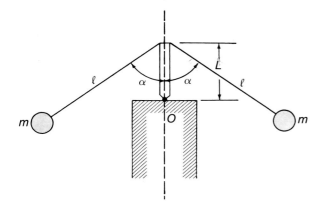

15–21 A hollow brass tube (diameter = 4.0 cm) is sealed at
one end and filled with lead shot to give a total mass of
0.20 kg (see drawing). The tube is depressed a small
amount from its position of equilibrium and is then released from rest. Ignoring frictional effects, show that
the tube undergoes vertical SHM. Determine the frequency of the motion.

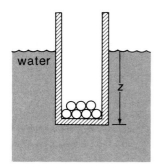

15–22 Show that Eq. 15–5 can be written in the form of an
equation for an ellipse in the coordinate variable x and
the momentum variable $p = mv$, namely,

$$\frac{p^2}{2Em} + \frac{x^2}{2E/\kappa} = 1$$

Sketch this ellipse in the x-p plane and determine the
semimajor and semiminor axes.

  Such representations of the history of oscillatory
motions are called *phase diagrams*. (• Show that the area
of the phase ellipse is $2\pi E/\omega_0$. As time advances, in
what sense (clockwise or counterclockwise) does the history of the oscillatory motion unfold in the x-p plane?)
(The area of an ellipse is πab, where a and b are the
semimajor and semiminor axes, respectively.)

15–23 A flat plate P executes horizontal SHM by sliding across
• a frictionless surface with a frequency $\nu = 1.5$ Hz. A
block B rests on the plate, as shown in the diagram, and
the coefficient of static friction between the block and
the plate is $\mu_s = 0.60$. What maximum amplitude of
oscillation can the plate-block system have if the block is
not to slip on the plate?

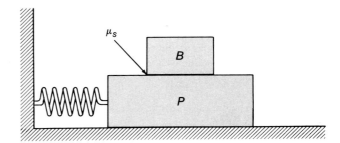

15–24 Consider a slender rod with mass $M = 4$ kg and length
$\ell = 1.2$ m that is pivoted on a frictionless horizontal
bearing at a point $\ell/4$ from one end, as shown in the
diagram. (a) Derive (from the definition) the expression
for the rotational inertia of the rod about the pivot.
(b) Obtain an equation that gives the angular acceleration
α of the rod as a function of θ. (c) Determine the period
of small-amplitude oscillations about the equilibrium
position.

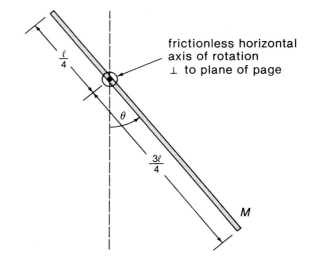

15–25 A block rests on a flat plate that executes vertical SHM with a period of 1.2 s. What is the maximum amplitude of the motion for which the block will not separate from the plate?

15–26 A block with a mass of 2 kg hangs without vibrating at the end of a spring ($\kappa = 500$ N/m) that is attached to the ceiling of an elevator car. The car is rising with an upward acceleration of $\frac{1}{3}g$ when the acceleration suddenly ceases (at $t = 0$). (a) What is the angular frequency of oscillation of the block after the acceleration ceases? (b) By what amount is the spring stretched during the time that the elevator car is accelerating? (c) What is the amplitude of the oscillation and the initial phase angle observed by a rider in the car? Take the upward direction to be positive.

15–27 A block with a mass $M = 0.50$ kg is suspended at rest from a spring with $\kappa = 200$ N/m. A blob of putty ($m = 0.30$ kg) is dropped onto the block from a height of 10 cm; the putty sticks to the block. (a) What is the period of the ensuing oscillations? (b) Determine the equation of motion for the block-and-putty combination. (Take the upward direction to be positive.) (c) What is the total energy of the oscillating system?

15–28 A *seconds pendulum* is one for which each back-and-forth vibration requires exactly 1 s (that is, $T = 2$ s). When such a pendulum is carried aloft in a balloon, it is found to "tick" only 59.914 times per minute. What is the height of the balloon above sea level? [*Hint:* Use differentials.]

15–29 A simple pendulum ($\ell = 2$ m) oscillates with a *horizontal* amplitude of 20 cm. A 100-mg ladybug clings to the bottom of the pendulum bob. What is the minimum holding force that the bug must exert during the motion to prevent being detached from the bob?

15–30 Find the period of a simple pendulum whose bob is near the Earth's surface and whose suspension string is infinitely long. Show that this period is equal to that of an artificial Earth satellite in a low-altitude orbit.

15–31 A narrow tunnel is cut through the Earth along a chord. Show that the period of a particle that oscillates in this tunnel is the same as that of a particle oscillating in a tunnel along a diameter (84.5 min). [*Hint:* Refer to Problems 13–12 and 15–36.]

15–32 Two springs, with spring constants κ_1 and κ_2, are connected to a block with mass m in the three different ways indicated in the diagram at right. Determine for each case the equivalent spring constant κ that would produce the same force on the block.

15–33 A thin rod with negligible mass supports at its lower end a particle-like object with a mass m. The upper end of the rod is free to rotate in the plane of the page about a

horizontal axis through the point O in the diagram. A horizontal spring is attached to a point on the rod, as shown. If the spring has its relaxed length when the rod is vertical, show that the period of small-amplitude oscillations is

$$T = 2\pi \left[\left(\frac{\kappa}{m} \right) \frac{\ell^2}{L^2} + \frac{g}{L} \right]^{-1/2}$$

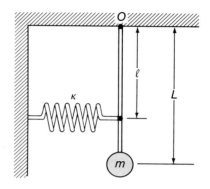

15–34 A horizontal spring-block system executes SHM with parameters A, $\phi_0 = 0$, and ω. At time $t = t_1$ (for which $\omega t_1 = \phi_1$), a blob of putty with mass m is dropped straight down on the block, sticking to it. (See diagram on page 298.) (a) Show that the new amplitude of oscillation is

$$A' = A \sqrt{\frac{M + m \sin^2 \phi_1}{M + m}}$$

Problem 15–32

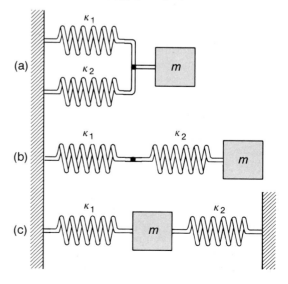

(b) Show that the new angular frequency is $\omega' = \omega\sqrt{M/(M + m)}$. [*Hint:* Use conservation of horizontal linear momentum and reevaluate the new total mechanical energy of the system.]

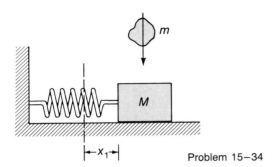

Problem 15–34

15–35 A wheel rotates uniformly at a rate of 300 rpm and drives a horizontal shaft by means of a Scotch yoke, as shown in the diagram. At $t = 0$ the drive pin is $\pi/4$ rad below the horizontal, as shown. (a) If the pin circle has a radius of 15 cm, write the equation for the displacement of the point P on the shaft, using the midpoint of its motion as origin. (b) If the mass of the driven shaft is 5 kg, what maximum force must the drive pin exert on the yoke? (Assume that the system is frictionless.)

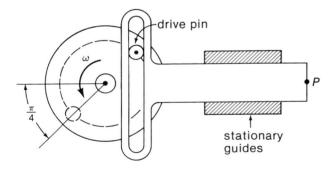

15–36 Show that a particle with a mass m moving in a narrow
•• tunnel through the Earth along a diameter has acting on it a gravitational force with a Hooke's-law form, namely, $F = -\kappa x$, where x is the distance from the Earth's center and where $\kappa = gm/R_E$. (a) Verify that such a freely moving particle has a period of 84.5 min. (b) At what constant speed would a low-altitude satellite have to travel so that the perpendicular projection of its motion onto the diameter of its circular orbit (made to correspond to the tunnel) matches the motion of the oscillating particle? (c) Is the satellite orbit referred to in (b) a stable orbit?

15–37 A thin uniform rod executes small-amplitude oscillations about a perpendicular horizontal axis through a point O that is a distance ℓ above the C.M. For what distance ℓ will the oscillation period be least?

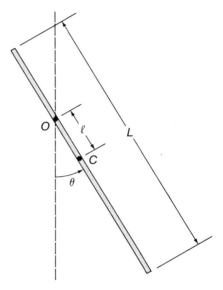

15–38 Show by direct substitution that $x(t)$ given by Eq. 15–29
• is in fact a solution of the equation of motion, Eq. 15–28. [*Hint:* The equation that results from the substitution must be valid at any time t.]

15–39 The oscillation of a simple pendulum is damped primarily by the effects of air resistance to the motion of the bob. It is stated that the damping will be small if for a spherically shaped bob with period T

$$T \ll \frac{2\pi}{g}v_\infty$$

where v_∞ is the terminal velocity of the bob in free fall. (a) Prove this assertion. (b) Estimate the magnitude of the term on the right-hand side of this inequality if $v_\infty \approx 200$ m/s.

CHAPTER 16

MECHANICAL WAVES

One of the important and interesting types of motion that occurs in nature is *wave motion*. In general, wave motion consists of the propagation of a physical condition possessing properties such as energy and momentum without the corresponding transmission of matter. In this chapter we consider in detail *mechanical wave motion* involving bulk matter. In this distinctive type of motion, the particles that constitute the bulk matter participate in the motion in a highly organized manner. Wave motion in a material medium consists of a local displacement from equilibrium that is transmitted successively from one part of the medium to the next by interactions among the particles. Mechanical waves can propagate through any material substance—solid, liquid, or gas.

A familiar example of mechanical wave motion is the occurrence of waves on the surface of a body of water. For example, when a stone is dropped into a calm lake or pool, a series of expanding ripples appears to radiate from the splash center. Evidently, traveling wave motion is involved. However, if you observe some bits of matter floating on the surface you will notice that they undergo an up-and-down bobbing motion as the waves move across the surface of the water. That is, the waves travel steadily in some direction without any corresponding transport of the water or floating matter. However, if a wave encounters a cork floating in still water, the cork will be set into motion as the result of receiving energy and momentum from the wave. Thus, the bulk matter through which a wave propagates acts as a medium for the transport of energy and momentum even though no part of the bulk matter participates in the long-range motion of the wave.

There are other important wave phenomena for which the wave-propagation mechanism is fundamentally different from that for mechanical waves. Examples are electromagnetic waves (such as light) and waves describing quantum phenomena. In these cases there is no need even for the existence of a medium to sustain the wave motion. However, they have in common with mechanical waves the essential feature of energy and momentum propagation.

16–1 GENERAL CHARACTERISTICS OF WAVES

Wave propagation can take place in systems that are one-dimensional, two-dimensional, or three-dimensional in character. A disturbance that propagates along a taut string or a spring constitutes a one-dimensional wave. Waves on the surface of a body of water are two-dimensional waves. A pressure disturbance (sound wave) that propagates through air is a three-dimensional wave.

Raindrops falling on a pond produce circular surface waves that propagate radially outward from the impact points. (Photo by Jay Freedman.)

The impacts of raindrops on a pond produce surface waves that travel outward in ever-widening rings. An imaginary circle that connects the points on a particular outward-moving wave crest constitutes a circular *wavefront*. Or, we could define different wavefronts by connecting other obviously related parts of the wave, such as the points in a particular ring of depression. These various wavefronts all propagate radially outward from the central impact point.

A sharp disturbance of the air, such as a hand clap or an explosion, will produce pressure waves that expand in three dimensions from the source. The wavefronts in such cases are concentric spheres that propagate radially outward.

Waves can also be classified according to the type of motion that the medium undergoes. If the displacement of the medium is at right angles to the direction of propagation of the wave, the wave is said to be *transverse*. For example, if the end of a taut horizontal string is flipped up and down, a transverse disturbance will propagate along the string (Fig. 16–1a). If the displacement of the medium is back and forth along the direction of propagation of the wave, the wave is called a *longitudinal* or *compressional* wave. For example, if the end of a coiled spring is pushed and then pulled, a longitudinal disturbance (or compressional wave pulse) will propagate along the spring (Fig. 16–1b).

Some waves are combinations of transverse and longitudinal displacements. Both types of displacement are involved, for example, in water waves and in *torsional* waves, which represent a propagating twisting motion of a rod or a spring.

One of the characteristics of a transverse wave is its *polarization*. If we select a coordinate frame with the x-axis along a taut string, the vectors representing the transverse displacements of the string from its equilibrium position will lie in the y-z plane. If all of these displacements, at every position along the string, are parallel, the wave is said to be

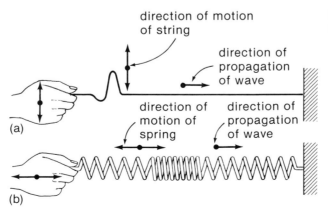

direction of motion
of string

direction of
propagation
of wave

direction of
motion of
spring

direction of
propagation
of wave

(a)

(b)

Fig. 16–1. (a) A transverse wave. (b) A longitudinal wave.

linearly polarized (Fig. 16–2). In the event that these parallel displacements are in the
y-direction, we describe the wave as a transverse wave propagating in the *x*-direction,
linearly polarized in the direction of the *y*-axis in the *y-z* plane. If the end of the string is
moved uniformly in a circular path, the resulting displacement vector will describe a circle
in the *y-z* plane; such a wave is said to be *circularly polarized*. The idea of polarization is
not applicable to longitudinal waves. (• Why?)

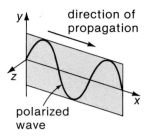

Fig. 16–2. A wave that is linearly
polarized in the direction of the
y-axis.

16–2   WAVE FUNCTIONS

The Phase.   A solitary wave pulse that propagates along a horizontal taut string can
be generated by jerking the end of the string up and down once (Fig. 16–1a). The
mathematical expression that describes the wave disturbance and its propagation as a
function of time is called the *wave function* of the pulse. Consider an infinitely long string
and select a coordinate frame O with the *x*-axis along the string, as in Fig. 16–3. Suppose
that a single linearly polarized pulse travels along the string in the direction of increasing
values of *x*. We let *y* represent the transverse displacement of the string. Then, $y = f(x, t)$
is the wave function for the pulse; that is, the displacement *y* is uniquely determined by the
two independent variables, *x* (propagation distance) and *t* (time).

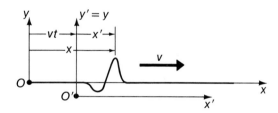

Fig. 16–3.   A single pulse travels along a
taut string with a velocity *v*.

In the ideal case, the pulse travels along the string with constant velocity while
maintaining its shape. Suppose that the pulse is viewed by an observer in a coordinate
frame O' that moves in the *x*-direction with the same velocity *v* at which the pulse moves
along the string (see Fig. 16–3). This observer would see a stationary pulse with a fixed
shape described by some function $y' = f(x')$. The connection between the coordinates in
the two frames is given by

$$x = x' + vt$$

$$y = y'$$

Consequently, the wave function in the frame O is

$$y = y' = f(x')$$

or   **(16–1a)**

$$y = f(x - vt)$$

If the pulse were traveling in the opposite direction and had a shape described in O' by
$y' = g(x')$, the wave function in O would be

$$y = g(x + vt) \qquad \textbf{(16–1b)}$$

Thus, the requirement that the pulse travel with a constant velocity while maintaining its
shape demands that the wave function depend on *x* and *t* combined in a specific way,
$u = x + vt$ or $u = x - vt$. The quantity $u = x \pm vt$ is called the *phase* of the wave.

The Phase Velocity.   Suppose that a wave pulse travels along a string in the positive
x-direction and has the wave function $y = f(x - vt)$. Fix attention on some feature of the

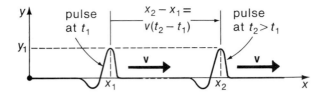

Fig. 16–4. A wave pulse travels to the right with a velocity v along a taut string. The location of the pulse is shown at times t_1 and t_2.

pulse, e.g., the peak, and follow its progress as the pulse moves. At time t_1 the pulse peak is at coordinate x_1, and at a later time t_2 the peak is at coordinate x_2 (Fig. 16–4). To give the same phase u at every instant, we must have $u_1 = u_2$, or

$$x_1 - vt_1 = x_2 - vt_2$$

so that

$$v = \frac{x_2 - x_1}{t_2 - t_1}$$

This velocity, which maintains a constant phase for the wave, is called the *phase velocity*. (• Can you obtain the same result by setting du/dt equal to zero?)

Superposition. In a medium that supports wave propagation there may be several sources of waves. For example, the photograph on page 300 shows many different wave sources and the resulting complicated pattern of wavefronts crossing one another. When two waves cross or merge, the instantaneous displacement of the medium is simply the linear sum of the individual displacements that would have occurred for each wave alone. When such a linear addition law is valid for combining waves, the waves are said to be *linear* and the addition obeys the *principle of superposition*.* We consider only systems that obey this principle.

If two separately generated waves with the same polarization are traveling in opposite directions on a string, the individual displacements can be expressed as $y_1 = f(x - vt)$ and $y_2 = g(x + vt)$. *At all times*—before, during, and after these waves overlap and combine—the total displacement of the string is the linear superposition of these waves; that is,

$$y = f(x - vt) + g(x + vt)$$

If the displacement y at the time of overlap is *less* than the displacement of either wave alone, the waves are said to exhibit *destructive interference* (Fig. 16–5a). If the waves combine to produce a displacement *greater* than that of either wave alone, the interference is *constructive* (Fig. 16–5b). To obtain *complete* destructive interference it is necessary that the two waves have exactly the same shape but opposite displacements. Even so, the complete cancellation exists only at the instant of overlap.

16–3 PROPAGATION OF TRANSVERSE WAVES ON A TAUT STRING

Consider a transverse wave pulse that travels along a taut string of mass μ per unit length. Figure 16–6 shows an isolated, symmetrically displaced segment of the string with length Δx and mass $\Delta m \cong \mu \, \Delta x$. The radius of curvature at the crest is R. (The displacement y is not the same as the radius R.)

---

*The superposition principle is valid for mechanical waves if the displacements are "sufficiently small." We give only this vague criterion because the way in which nonlinear effects enter is quite complicated. For electromagnetic waves in vacuum, however, linear superposition *always* holds.

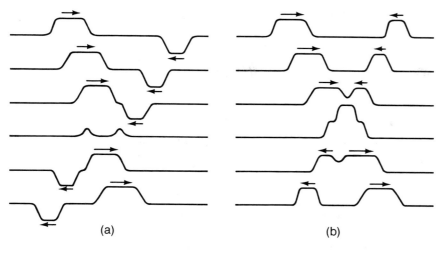

Fig. 16-5. The linear superposition of two waves traveling in opposite directions on a taut string. (a) Destructive interference. (b) Constructive interference.

In the frame of reference that moves with the pulse (that is, moves to the right with velocity $\mathbf{v}$), the string particles at the peak of the pulse move in a circular arc with tangential velocity $-\mathbf{v}$. Thus, the vertical force that provides the required centripetal acceleration of Δm is

$$2F \sin \alpha \approx 2F\alpha = 2F\left(\frac{\frac{1}{2}\Delta x}{R}\right) = F\frac{\Delta x}{R}$$

Setting this expression equal to $\Delta m \, a_c = \Delta m \, v^2/R = \mu \, \Delta x \, v^2/R$, we obtain

$$F\frac{\Delta x}{R} = \frac{\mu \, \Delta x \, v^2}{R}$$

from which $v^2 = F/\mu$, so that

$$v = \sqrt{\frac{F}{\mu}} \qquad\qquad (16\text{-}2)$$

This expression for the velocity exhibits a characteristic common to all mechanical waves. The velocity of a wave increases with increasing "strength" of the restoring force (here, the tension F), and the velocity decreases with increasing "inertia" of the system (here, the linear mass density μ).

Fig. 16-6. A symmetric pulse propagates to the right along a taut string. At its peak, the pulse has a radius of curvature R.

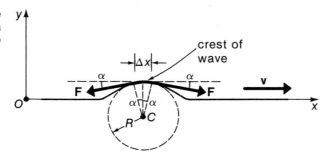

Example 16–1

A pulse is observed to propagate with a speed of 260 m/s along a taut string that has a linear mass density $\mu = 6.0$ g/m. What is the tension in the string?

Solution:

Using Eq. 16–3, we have

$$F = v^2\mu = (260 \text{ m/s})^2(6.0 \times 10^{-3} \text{ kg/m})$$
$$= 405.6 \text{ N}$$

Another important aspect of a traveling wave is the fact that it carries energy. Figure 16–7 shows a hypothetical cut in a displaced string at coordinate x. Assume that there is a wave pulse propagating to the right. At what rate does the string to the left of the cut do work on the part of the string to the right of the cut? This rate represents the *power* transmitted across the cut, and is equal to the product of the vertical component of the tension and the vertical component of the string velocity at x. (• Can you see why?) For small displacements, we have (see Fig. 16–7)

$$F_y = -F \sin \alpha \approx -F \tan \alpha = -F \frac{\partial y}{\partial x}$$

The power transmitted to the right is

$$P(x, t) = F_y v_y$$

Thus,

$$P(x, t) = -F \left(\frac{\partial y}{\partial x}\right)\left(\frac{\partial y}{\partial t}\right) \tag{16–3}$$

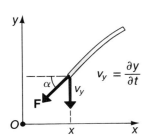

Fig. 16–7. The part of the string to the right of the hypothetical cut at x is driven by the force F which is due to the tension in the string to the left of the cut. At the cut, the velocity in the y-direction is $\partial y/\partial t$ and is downward.

For a cut at a point on the trailing edge of the pulse (as in Fig. 16–7), notice that $\partial y/\partial x > 0$ but $\partial y/\partial t < 0$, so that $P > 0$. If the cut is on the leading edge, we have $\partial y/\partial x < 0$ but $\partial y/\partial t > 0$, so again $P > 0$. To evaluate Eq. 16–3 it is necessary to know the wave function $y = f(x, t)$.

16–4 LONGITUDINAL WAVES IN A SOLID ROD

A second type of mechanical wave that we shall consider is a longitudinal or compressional wave transmitted along a slender solid rod. Suppose that a sharp axial blow is delivered to the end of a freely suspended rod. The material of the rod at the struck end undergoes a momentary axial compression. A compressional pulse then propagates through the rod as each cross section of the rod material moves back and forth parallel to the rod axis. Waves generated in this way are *longitudinal waves*. The velocity of propagation for such waves depends on the volume mass density ρ of the rod material and on its elastic properties expressed in terms of a quantity called *Young's modulus** Y. In fact, it

---

*Young's modulus is the proportionality factor that connects the force per unit area F/A applied to a material (a compressive or a tensile stress) and the fractional change in the dimension $\Delta \ell/\ell$ parallel to the applied force; that is, $F/A = Y(\Delta \ell/\ell)$. The units of Y are N/m^2 or Pa. Young's modulus is analogous to the force constant for a spring. (• Can you see why?)

is found that

$$v = \sqrt{\frac{Y}{\rho}}$$ (16-4)

This velocity, in which shear (sideways motion) and torsional (twisting motion) effects are neglected, is called the *extensional wave velocity*. Values of Y and v for several materials are given in Table 16–1.

TABLE 16–1. Young's Modulus and Extensional Wave Velocity for Thin Rods of Various Materials*

MATERIAL	$Y(10^{10}$ Pa) $\frac{N}{m^2}$	v(km/s)
Beryllium	30.8	12.87
Aluminum	7.0	5.00
Iron (cast)	15.2	4.48
Glass, pyrex	6.2	5.17
Copper	12.5	3.75
Platinum	16.7	2.80
Silver	7.5	2.68
Gold	8.12	2.03
Lucite	0.40	1.84
Lead	1.6	1.21
Polyethylene	0.076	0.92

*These are measured values, so small discrepancies may exist between the listed values and the prediction of the simple theory (Eq. 16–4).

16–5 HARMONIC WAVES

Suppose that one end of a long taut string is driven in the transverse direction by a simple harmonic force. (Recall from Chapter 15 that this means that the time dependence of the force is a sinusoidal function.) Then a sinusoidal wave (or simple harmonic wave, SHW) will propagate along the string. Strictly, the wave will be a *pure* sinusoidal wave only if the string is infinitely long and has been driven harmonically, since $t = -\infty$ so that a true steady-state condition exists. For a linearly polarized SHW propagating in the positive x-direction, the most general form of the wave function is

$$y(x, t) = A \cos [k(x - vt) + \phi_0]$$ (16-5)

In this expression, the argument of the cosine term, $\phi = k(x - vt) + \phi_0$, is called the *phase angle* of the wave and ϕ_0 is the *initial phase angle* (see Fig. 16–8); the quantity k is the wave number. The coefficient A is the *amplitude* of the wave, that is, the magnitude of the maximum transverse displacement; $-A \leq y \leq A$ for all t.

Another quantity that is useful for describing a SHW is the *wavelength* λ, the repetition length of the wave. The wavelength is defined to be the x-distance between any two successive corresponding points on the wave (that is, two points that have the same displacement and direction of motion). (• Why is it necessary to specify "and direction of motion"?) In Fig. 16–8, the displacement and motion at $y(a, t)$ is first repeated at $y(a', t)$, so that $\lambda = a' - a$. Because the point $x = a$ may be anywhere on the wave, we

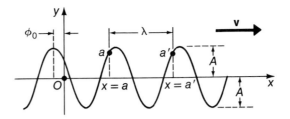

Fig. 16–8. A simple harmonic wave with amplitude A and wavelength λ shown at $t = 0$. The quantity ϕ_0 is the initial phase angle.

can write $y(x + \lambda, t) = y(x, t)$. Between $x = a$ and $x = a'$, the phase angle increases by 2π, so

$$k[(x + \lambda) - vt] + \phi_0 = k[x - vt] + \phi_0 + 2\pi$$

from which $k\lambda = 2\pi$, or

$$k = \frac{2\pi}{\lambda} \tag{16–6}$$

(• Now do you see the reason for the term "wave number"?)

The time factor kv can be simplified by introducing the *period T*, which is the repetition time of the wave. The period is defined to be the time interval between successive identical displacements and motions of the wave at a particular point; that is, we can write $y(x, t - T) = y(x, t)$. In terms of the phase angle, we have

$$k[x - v(t - T)] + \phi_0 = k[x - vt] + \phi_0 + 2\pi$$

from which $kvT = 2\pi$, or

$$kv = \frac{2\pi}{T}$$

The *frequency* of the wave is $\nu = 1/T$ (see Eq. 15–13). Then,

$$kv = \frac{2\pi}{\lambda} \cdot v = \frac{2\pi}{T} = 2\pi\nu$$

or

$$v = \lambda\nu \tag{16–7}$$

This equation expresses the fact that ν waves pass the (arbitrarily selected) origin each second, each wave having a length of λ meters. Thus, the feature that was at the origin at $t = 0$ is carried to a distance $\lambda\nu$ meters at $t = 1$ s. This corresponds to the same phase velocity that was defined in Section 16–2. Note also that Eq. 16–7 is of the proper dimensional form, namely, $[v] = LT^{-1} = [\lambda][\nu]$.

Equation 16–7 is valid for *all* types of simple harmonic waves. For each type of wave we must use the appropriate velocity v—for example, approximately 340 m/s for sound waves in air and 3×10^8 m/s for light waves in vacuum (or air).

The various quantities that describe the properties of waves can be used to express the wave function in several equivalent ways. If we include the idea of the *angular frequency*, $\omega = 2\pi\nu = kv$ (see Eq. 15–14), we have, for example,

$$v = \lambda\nu = \frac{2\pi}{k}\nu = \frac{\omega}{k} \tag{16–8}$$

Then,
$$y(x, t) = A \cos [k(x - vt) + \phi_0]$$
$$= A \cos [kx - \omega t + \phi_0]$$
$$= A \cos \left[2\pi \left(\frac{x}{\lambda} - \frac{t}{T} \right) + \phi_0 \right]$$

(16–9)

These expressions represent a SHW that propagates in the positive x-direction. (We assume that the end at $x = -\infty$ has been driven since $t = -\infty$.) (• What change must be made to describe a SHW that propagates in the negative x-direction?)

We can imagine that a harmonic wave is generated by driving the point $x = 0$ instead of one end. Then two identical waves are set up simultaneously, one traveling in the +x-direction and the other in the −x-direction. If we designate these waves y^+ and y^-, respectively, we have

$$y^+(x, t) = A \cos (kx - \omega t + \phi_0) \qquad x \geq 0 \qquad \textbf{(16–10a)}$$

$$y^-(x, t) = A \cos (-kx - \omega t + \phi_0)$$
$$= A \cos (kx + \omega t - \phi_0) \qquad x \leq 0 \qquad \textbf{(16–10b)}$$

The Wave Equation. Waves in a medium are described by a *wave equation* that gives the relationship between the space and time dependence of disturbances in that medium. A general derivation of this relationship is somewhat complicated; however, our harmonic waves may conveniently be used to indicate the essential steps. Consider the harmonic wave traveling in the positive x-direction:

$$y = A \cos [k(x - vt) + \phi_0]$$

We readily find that the second derivative with respect to x is

$$\partial^2 y / \partial x^2 = -Ak^2 \cos [k(x - vt) + \phi_0]$$

and that the second derivative with respect to t is

$$\partial^2 y / \partial t^2 = -Ak^2 v^2 \cos [k(x - vt) + \phi_0]$$

Comparing these two results shows that

$$\frac{\partial^2 y}{\partial x^2} = \frac{1}{v^2} \frac{\partial^2 y}{\partial t^2}$$

This result has the important feature that neither k nor in fact the shape of the wave function is involved. The only feature of the wave motion that explicitly appears is the phase velocity v. As Eq. 16–2 (or Eq. 16–4) shows, the phase velocity depends only on the physical properties of the medium and the internal forces that are acting.

The result we have found is generally true for *any* wave function (see also Problem 16–32). Thus, for waves that propagate in one dimension (corresponding to the x-direction), the wave equation is

$$\frac{\partial^2 \eta}{\partial x^2} = \frac{1}{v^2} \frac{\partial^2 \eta}{\partial t^2}$$

(16–11)

where $\eta(x, t)$ is a generalized physical variable and may represent any scalar quantity that propagates as a wave (for example, the displacement of a string, the pressure excursion in a gas, and so forth). Also, η can be any component of a vector that describes wave motion, such as the displacement vector **r** or the electric field vector **E**.

Because the physical variable η is necessarily a function of two independent variables, the wave equation is a *partial differential equation*. (• What would be the form of the wave equation for a three-dimensional wave?) The solution of this differential equation is the wave function $\eta(x, t)$.

Example 16–2

A SHW propagates in the positive x-direction along a taut string that has a linear mass density of 5 g/m and is under a tension of 100 N. The amplitude of the wave is 5 cm and the wavelength is 75 cm.
(a) What is the phase velocity of the wave?
(b) What is the frequency of the wave? What is the period?
(c) At $t = 0$, the wave has its maximum positive transverse velocity at $x = 0$. Evaluate the initial phase angle ϕ_0 and write the complete wave function.
(d) What is the phase angle for $x = 0$ at $t = 1$ ms? What is the corresponding displacement?
(e) What is the maximum magnitude of the transverse velocity?
(f) What is the average power transmitted along the string? (Average over one period.)
(g) What is the total energy per wavelength in the wave?

Solution:

(a) The phase velocity is (Eq. 16–2)

$$v = \sqrt{\frac{F}{\mu}} = \sqrt{\frac{100 \text{ N}}{5 \times 10^{-3} \text{ kg/m}}} = 141.4 \text{ m/s}$$

(b) Using Eq. 16–8, we have

$$v = \frac{v}{\lambda} = \frac{141.4 \text{ m/s}}{0.75 \text{ m}} = 188.5 \text{ Hz}$$

and $T = \dfrac{1}{v} = \dfrac{1}{188.5 \text{ Hz}} = 5.30 \times 10^{-3} \text{ s} = 5.30 \text{ ms}$

(c) Write the wave function in the form

$$y(x, t) = A \cos (kx - \omega t + \phi_0)$$

Then, $\quad v_y = \dfrac{\partial y}{\partial t} = A\omega \sin (kx - \omega t + \phi_0)$

If v_y is maximum at $x = 0$ for $t = 0$, the initial phase angle must be $\phi_0 = \pi/2$ (or an odd multiple thereof). Then,

$$y(x, t) = A \cos (kx - \omega t + \pi/2)$$

Notice that the magnitude of $y(x, t)$ is maximum, $|y(x, t)| = A$, when the velocity v_y is zero, and that the magnitude of v_y is maximum, $|v_y| = A\omega$, when $y(x, t)$ is zero.
(d) The phase angle for $x = 0$ and $t = 1$ ms is

$$\phi = kx - \omega t + \pi/2$$

$$= 2\pi\left(\frac{x}{\lambda} - \frac{t}{T}\right) + \frac{\pi}{2}$$

$$= 2\pi\left(0 - \frac{1.0 \times 10^{-3} \text{ s}}{5.30 \times 10^{-3} \text{ s}}\right) + \frac{\pi}{2}$$

$$= 0.385$$

Then, $\quad y = A \cos \phi = (5 \text{ cm}) \cos 0.385 = 4.63 \text{ cm}$

(e) The magnitude of the vertical velocity is

$$|v_y| = A\omega = 2\pi A v$$

$$= 2\pi(0.05 \text{ m})(188.5 \text{ s}^{-1})$$

$$= 59.2 \text{ m/s}$$

(f) The power transmitted through the position x at time t is given by Eq. 16–3:

$$P(x, t) = -F\left(\frac{\partial y}{\partial x}\right)\left(\frac{\partial y}{\partial t}\right)$$

With $\phi_0 = \pi/2$, we have $y(x, t) = -A \sin (kx - \omega t)$, so that

$$\frac{\partial y}{\partial x} = -Ak \cos (kx - \omega t)$$

$$\frac{\partial y}{\partial t} = A\omega \cos (kx - \omega t)$$

Then, $\quad P(x, t) = FA^2\omega k \cos^2 (kx - \omega t)$

Averaging over one period, we obtain

$$\overline{P} = \frac{1}{T} \int_t^{t+T} P(x, t) \, dt$$

$$= \frac{FA^2\omega k}{T} \int_t^{t+T} \cos^2 (kx - \omega t) \, dt$$

$$= \frac{FA^2\omega k}{T} \cdot \frac{T}{2} = \tfrac{1}{2}FA^2\omega k$$

(• Can you verify this integration by changing the variable to $u = kx - \omega t$? Would you obtain the same result by choosing $x = 0$ and integrating from 0 to T?)
Evaluating $\overline{P}$ using $k = 2\pi/\lambda$ and $\omega = 2\pi v$, we find

$$\overline{P} = \frac{2\pi^2 FA^2 v}{\lambda} = \frac{2\pi^2(100 \text{ N})(0.05 \text{ m})^2(188.5 \text{ s}^{-1})}{0.75 \text{ m}}$$

$$= 1.24 \text{ kW}$$

(g) The energy per wavelength is (• Justify the expression.)

$$E = \bar{P}T = (1.24 \times 10^3 \text{ W})(5.30 \times 10^{-3} \text{ s})$$

$$= 6.57 \text{ J}$$

Example 16–3

Two waves are propagating along a taut string that coincides with the x-axis; the first has the wave function $y_1 = A \cos [k(x - vt)]$ and the second has the wave function $y_2 = A \cos [k(x + vt) + \phi_0]$.

(a) What value of ϕ_0 will produce constructive interference at $x = 0$? Destructive interference?

(b) For each of the values of ϕ_0 found in (a), write the total wave function, $y = y_1 + y_2$.

(c) Find the points along the string that are always stationary.

Solution:

The linear superposition of the two waves gives

$$y = y_1 + y_2 = A\{\cos [k(x - vt)] + \cos [k(x + vt) + \phi_0]\}$$

Using Eq. A–28 in Appendix A, this expression simplifies to

$$y(x, t) = 2A \cos (kx + \tfrac{1}{2}\phi_0) \cos (kvt + \tfrac{1}{2}\phi_0)$$

(a) At $x = 0$, we have

$$y(0, t) = 2A \cos \tfrac{1}{2}\phi_0 \cos (kvt + \tfrac{1}{2}\phi_0)$$

For maximum displacement (either sign for y), $\cos \tfrac{1}{2}\phi_0$ must be ± 1. Thus, for constructive interference, we have

$$\phi_0(\text{constructive}) = 0, 2\pi, 4\pi, \ldots$$

$$= 2n\pi \qquad (n = 0, 1, 2, \ldots)$$

For destructive interference ($\cos \tfrac{1}{2}\phi_0 = 0$), we have

$$\phi_0(\text{destructive}) = \pi, 3\pi, 5\pi, \ldots$$

$$= (2n + 1)\pi \qquad (n = 0, 1, 2, \ldots)$$

(b) Substituting ϕ_0 for the case of constructive interference, we find

$$y_c(x, t) = 2A \cos kx \cos kvt$$

For the case of destructive interference, we have

$$y_d(x, t) = 2A \sin kx \sin kvt$$

(c) For both cases there are positions of zero displacement for all values of t. These occur at points for which $\cos kx = 0$ in the case of constructive interference and at points for which $\sin kx = 0$ in the case of destructive interference. That is,

$$\begin{rcases} kx = \tfrac{1}{2}(2n + 1)\pi & \text{(constructive)} \\ kx = n\pi & \text{(destructive)} \end{rcases} \quad n = 0, 1, 2, \ldots$$

Such points of permanent zero displacement are called *nodes*.

(• Suppose that one of the waves in this example is polarized in the z-direction and that the other remains polarized in the y-direction. How would this affect the analysis? The situation is not simple!)

16–6 THE REFLECTION AND TRANSMISSION OF WAVES AT A BOUNDARY

We have been considering waves propagating along strings (and rods) assumed to be infinite in length. We now study the effects that are introduced when a system is *bounded*. The boundaries may be in the form of terminations, with a fixed or a free end, or they may be the result of discontinuous changes in the density of the string. Consider first the latter possibility.

Suppose that two very long strings, with linear mass densities μ_1 and μ_2, are joined at $x = 0$, forming a single taut string system with tension F throughout. Imagine that a train of pulses generated at $x = -\infty$ is incident on the junction point from the left; one of these pulses is shown in Fig. 16–9a. After the initial pulse (the *incident wave*) has interacted with the junction, there are two additional possibilities for continuing disturbances of the string, namely, a wave that propagates from the junction toward the right (the

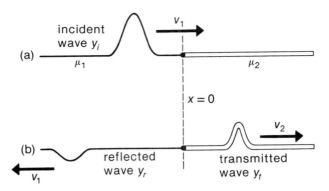

Fig. 16–9. Reflected and transmitted waves are generated when a wave encounters a discontinuity. The case illustrated corresponds to $\mu_2 = 4\mu_1$. The strings are under a uniform tension F.

transmitted wave) and a wave that propagates from the junction toward the left (the *reflected wave*), as indicated in Fig. 16–9b. Consequently, in the region $x < 0$, we must deal with the incident and reflected waves, whereas in the region $x > 0$, we have only the transmitted wave.

▶ *Boundary Conditions.* We can analyze the situation most readily (and with complete generality) by considering harmonic waves. For the incident wave, we write

$$y_i(x, t) = A_i \cos (k_1 x - \omega_1 t + \phi_0), \qquad x \leqslant 0 \quad \text{(16–12a)}$$

Because the junction point acts as a source for the transmitted and reflected waves, we use the wave functions given in Eq. 16–10; thus,

$$y_r(x, t) = A_r \cos (k_1 x + \omega_1 t - \phi_0), \qquad x \leqslant 0 \quad \text{(16–12b)}$$

$$y_t(x, t) = A_t \cos (k_2 x - \omega_2 t + \phi_0), \qquad x \geqslant 0 \quad \text{(16–12c)}$$

According to the superposition principle, we have

$$y(x, t) = \begin{cases} y_i(x, t) + y_r(x, t), & x \leqslant 0 \\ y_t(x, t), & x \geqslant 0 \end{cases} \quad \text{(16–13)}$$

Because the strings are connected, we have, at $x = 0$,

$$y_i(0, t) + y_r(0, t) = y_t(0, t) \quad \text{(16–14)}$$

This equation must hold for all values of t. Notice that $y_r(0, t) = A_r \cos (\omega_1 t - \phi_0) = A_r \cos (-\omega_1 t + \phi_0)$. Then, $y_r(0, t)$ will have the same phase as $y_i(0, t)$ and $y_t(0, t)$ only if $\omega_2 = \omega_1$, so this must be true. (Henceforth, we use ω for the common value of the angular frequency.) Equation 16–14 also imposes a condition on the amplitudes, namely,

$$A_i + A_r = A_t \quad \text{(16–15)}$$

The two string segments are assumed to be ideal strings, so they can sustain only tension. Then, Newton's third law requires identical slopes for the two segments at $x = 0$. (• Can you explain why?) That is,

$$\frac{\partial}{\partial x}(y_i + y_r)\bigg|_{x=0} = \frac{\partial}{\partial x}(y_t)\bigg|_{x=0}$$

This condition yields

$$A_i k_1 \sin (-\omega t + \phi_0) + A_r k_1 \sin (\omega t - \phi_0)$$
$$= A_t k_2 \sin (-\omega t + \phi_0)$$

from which $\qquad (A_i - A_r)k_1 = A_t k_2$

Combining this result with Eq. 16–15 gives

$$\frac{A_r}{A_i} = \frac{k_1 - k_2}{k_1 + k_2} \quad \text{and} \quad \frac{A_t}{A_i} = \frac{2k_1}{k_1 + k_2} \quad \text{(16–16)}$$

These expressions give the amplitudes of the reflected and transmitted waves in terms of the amplitude of the incident wave and the wave numbers for the two parts of the system. These ratios are referred to as the *amplitude reflection coefficient* and the *amplitude transmission coefficient*, respectively. ◀

Equations 16–16 were obtained without any reference to the physical properties of the string or to the tension. The results are, in fact, correct for *any* type of SHW that meets a discontinuity in a medium.

The angular frequency ω and the tension F are the same throughout the string system. Then, noting that $k = \omega/v = \omega\sqrt{\mu/F}$, we can express the amplitude reflection and transmission coefficients as

$$\frac{A_r}{A_i} = \frac{\sqrt{\mu_1} - \sqrt{\mu_2}}{\sqrt{\mu_1} + \sqrt{\mu_2}} \quad \text{and} \quad \frac{A_t}{A_i} = \frac{2\sqrt{\mu_1}}{\sqrt{\mu_1} + \sqrt{\mu_2}} \qquad (16\text{–}17)$$

Because of the requirement at $x = 0$ expressed by Eq. 16–14, the shapes of the incident, reflected, and transmitted pulses must all be the same in the time variable. However, the phase velocity is different for the two string segments, so the spatial shapes of the pulses scale directly with the velocities. That is, if $v_2 = \frac{1}{2}v_1$, the transmitted pulse will have a spatial width equal to one half that of the incident pulse. On the other hand, the reflected pulse will have the same spatial width as the incident pulse. Figure 16–9 illustrates just this case, namely, $\mu_2 = 4\mu_1$ and $v_2 = \frac{1}{2}v_1$. The amplitude ratios are $A_t/A_i = \frac{2}{3}$ and $A_r/A_i = -\frac{1}{3}$. Thus, the reflected pulse has a displacement opposite to that of the incident pulse, as shown in Fig. 16–9b.

Some special cases concerning Eqs. 16–17 are noteworthy. First, if $\mu_2 = \mu_1$, so that the discontinuity is removed, we find $A_r/A_i = 0$ and $A_t/A_i = 1$. That is, there is no reflected wave and the entire wave is transmitted unchanged, as we expect.

In the event that $\mu_2 > \mu_1$, the ratio A_r/A_i is negative and the displacement of the reflected pulse is opposite to that of the incident pulse. This means that the reflected pulse has a phase difference of π with respect to the incident pulse. (• Explain the origin of this phase difference.) When $\mu_2 < \mu_1$, the ratio A_r/A_i is always positive, and the displacement of the reflected pulse is the same as that of the incident pulse.

Next, suppose that the string is fastened to a fixed support at $x = 0$. This is equivalent to letting μ_2 become indefinitely large. Then, $A_r/A_i = -1$, and the entire wave is reflected with a phase change of π. This situation is illustrated in Fig. 16–10, where we represent the reflection by imagining that the incident wave disappears into the wall as the reflected wave emerges from the wall. The actual wave is the superposition of these two waves for $x < 0$.

Finally, we consider the reflection of a pulse from a free end, that is, $\mu_2 = 0$. We can provide a tension F at the free end ($x = 0$) and yet allow the string vertical movement by attaching the end to a ring that slides frictionlessly on a shaft that is perpendicular to

Fig. 16–10. Reflection of a pulse from a fixed support at $x = 0$. The reflected pulse (dashed line) is imagined to emerge from the wall as the incident pulse (dotted line) penetrates the wall. The superposition of these two pulses is the actual pulse (solid line).

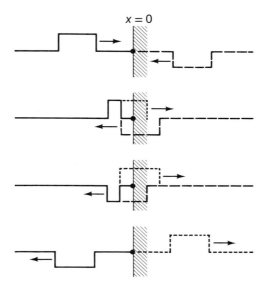

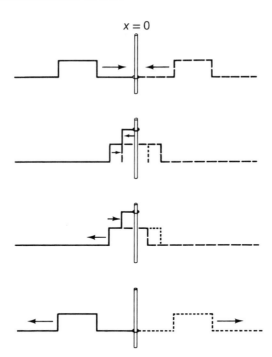

Fig. 16–11. Reflection of a pulse from a free end. Tension at the free end ($x = 0$) is provided by attaching the end to a ring that can slide frictionlessly on a fixed shaft.

the string. In this case, $A_r/A_i = 1$, and the entire wave is reflected with no phase change. This situation is shown in Fig. 16–11, using the same scheme as in Fig. 16–10.

▶ We have not mentioned the amplitude ratio A_t/A_i for either of the cases $\mu_2 \to \infty$ or $\mu_2 = 0$, because the situation must properly be analyzed by considering the flow of energy in the system. In this way, one finds that there is no transmitted wave in either case. ◀

Example 16–4

One end of a thin aluminum rod (diameter = 1 mm) is butt welded to the end of a copper rod with the same diameter. Both rods are very long and are under a uniform tension of 1000 N. The pertinent data are

$Y(\text{Cu}) = Y_1 = 12.5 \times 10^{10}$ Pa $Y(A\ell) = Y_2 = 7.0 \times 10^{10}$ Pa

$\rho(\text{Cu}) = \rho_1 = 8.93$ g/cm^3 $\rho(A\ell) = \rho_2 = 2.70$ g/cm^3

(a) Calculate the velocities for transverse waves on the two rods.

(b) Calculate the velocities for longitudinal waves on the two rods.

(c) What is the reflection factor A_r/A_i for a longitudinal pulse approaching the junction along the copper rod? For a transverse pulse?

Solution:

(a) For transverse wave velocities we have, with S equal to the cross-sectional area (Eq. 16–2)

$$v_1 = \sqrt{\frac{F}{\mu_1}} = \sqrt{\frac{F}{\rho_1 S}}$$

$$= \sqrt{\frac{1000 \text{ N}}{(8.93 \times 10^3 \text{ kg/m}^3)\pi(0.5 \times 10^{-3} \text{ m})^2}}$$

$$= 377.6 \text{ m/s}$$

$$v_2 = \sqrt{\frac{F}{\mu_2}} = \sqrt{\frac{F}{\rho_2 S}}$$

$$= \sqrt{\frac{1000 \text{ N}}{(2.70 \times 10^3 \text{ kg/m}^3)\pi(0.5 \times 10^{-3} \text{ m})^2}}$$

$$= 686.7 \text{ m/s}$$

(b) For longitudinal wave velocities we have (Eq. 16–4)

$$v_1 = \sqrt{\frac{Y_1}{\rho_1}} = \sqrt{\frac{12.5 \times 10^{10} \text{ N/m}^2}{8.93 \times 10^3 \text{ kg/m}^3}} = 3741 \text{ m/s}$$

$$v_2 = \sqrt{\frac{Y_2}{\rho_2}} = \sqrt{\frac{7.0 \times 10^{10} \text{ N/m}^2}{2.70 \times 10^3 \text{ kg/m}^3}} = 5092 \text{ m/s}$$

(c) Using $v = \omega/k$, we can express Eq. 16–16 as

$$\frac{A_r}{A_i} = \frac{k_1 - k_2}{k_1 + k_2} = \frac{\dfrac{1}{v_1} - \dfrac{1}{v_2}}{\dfrac{1}{v_1} + \dfrac{1}{v_2}} = \frac{v_2 - v_1}{v_2 + v_1}$$

Then, we have

Transverse: $\dfrac{A_r}{A_i} = \dfrac{v_2 - v_1}{v_2 + v_1} = \dfrac{686.7 - 377.6}{686.7 + 377.6} = 0.290$

Longitudinal: $\dfrac{A_r}{A_i} = \dfrac{5092 - 3741}{5092 + 3741} = 0.153$

16-7   STANDING WAVES

We now consider the interesting and important case of wave motion on a taut string with finite length, that is, a string with *both* ends terminated in some way. A familiar example of systems in this class is the family of stringed musical instruments.

Consider first the case of a string that is stretched between two fixed supports a distance L apart. We choose a coordinate frame in which the x-axis coincides with the string in its rest position. Imagine that the string is set into motion by delivering to it a transverse blow at some point along its length. This disturbance of the string initiates *two* waves that travel in opposite directions along the string. When these waves arrive at the end terminations, they will be reflected and then travel back along the string. If there are no energy losses, each reflection will be total and the waves will propagate back and forth along the string forever.

The wave propagation on the terminated string can be analyzed in a simple yet general way by assuming harmonic waves. Because waves travel in both directions along the string, we again use the wave functions of Eqs. 16–10, namely,

$$\left. \begin{array}{l} y^+(x, t) = A^+ \cos (kx - \omega t + \phi_0) \\ y^-(x, t) = A^- \cos (kx + \omega t - \phi_0) \end{array} \right\} \qquad \text{(16-18)}$$

We can expand these wave functions by using Eqs. A–22 and A–23 in Appendix A, obtaining

$$y^+(x, t) = A^+ \cos kx \cos (\omega t - \phi_0) + A^+ \sin kx \sin (\omega t - \phi_0)$$

$$y^-(x, t) = A^- \cos kx \cos (\omega t - \phi_0) - A^- \sin kx \sin (\omega t - \phi_0)$$

According to the superposition principle, the total displacement of the string is

$$y(x, t) = y^+(x, t) + y^-(x, t) \qquad \text{(16-19)}$$

$$= (A^+ + A^-) \cos kx \cos (\omega t - \phi_0)$$

$$+ (A^+ - A^-) \sin kx \sin (\omega t - \phi_0)$$

Take the origin $x = 0$ to be at the left-hand end support and take $x = L$ to be at the right-hand end support. Now, the displacement y must be zero at $x = 0$ for all values of t. This requires the first term in Eq. 16–19 to vanish for all t, and this is possible only if $A^+ = -A^- \equiv \frac{1}{2}A$. Then, the expression for $y(x, t)$ becomes

$$y(x, t) = A \sin kx \sin (\omega t - \phi_0) \qquad \text{(16-20)}$$

We also require y to be zero at $x = L$ for all t. Excluding the uninteresting case of $A = 0$, we must have

$$\sin k_n L = 0$$

that is, $k_n L = n\pi$, with $n = 1, 2, 3, \ldots$. If we recall that $k = 2\pi/\lambda$ (Eq. 16–6), then $2\pi L/\lambda_n = n\pi$; that is,

$$\lambda_n = 2L/n \qquad n = 1, 2, 3, \ldots \qquad \text{fixed end-points} \qquad \textbf{(16–21)}$$

For the string terminated by fixed supports at both ends, only those SHW solutions are possible for which the wavelength obeys Eq. 16–21. Thus, we can express the wave function as

$$y(x, t) = A \sin(n\pi x/L) \sin(\omega t - \phi_0) \qquad n = 1, 2, 3, \ldots \qquad \textbf{(16–22)}$$

Notice that at those positions x for which $\sin(n\pi x/L) = 0$, the displacement y is zero for all t. Such positions along the string are called *nodes*. The ends of the string, $x = 0$ and $x = L$, are always nodes. There are other positions, called *antinodes*, for which $\sin(n\pi x/L) = 1$. At these positions, the displacement has its maximum excursion, from $-A$ to $+A$, as time progresses.

These waves have features—the nodes and antinodes—that occur at fixed positions. Accordingly, they are called *standing waves*. When strings vibrate in this manner, sound waves are excited in the surrounding air, and we hear the tone generated by the string. (Thus, our assumption that there are no energy losses is not strictly correct!)

The standing wave that has $n = 1$ is called the *fundamental* (or *first harmonic*). The wave for $n = 2$ is the *second harmonic,* the wave for $n = 3$ is the *third harmonic,* and so forth. The wavelength λ_n is governed by the string length L; the frequency then depends on the phase velocity $v = \sqrt{F/\mu}$ (Eq. 16–2). Therefore, by alteration of the tension F, the string can be tuned to any particular frequency, $\nu_n = v/\lambda_n$. The first four harmonics that are possible for a string with a length L are shown in Fig. 16–12.

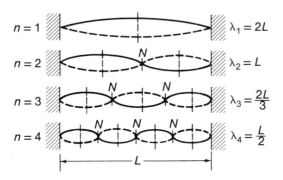

Fig. 16–12. Standing-wave patterns on a string with a length L. The fundamental has $n = 1$; the next three higher harmonics correspond to $n = 2, 3,$ and 4. The nodal points are designated by N. The curves represent the envelopes of the oscillations of the strings (i.e., the greatest excursion reached by each point of the string).

Energy in a Standing Wave. The mechanical energy of a wave on a string is the sum of the kinetic-energy and potential-energy contributions. As the string vibrates, it achieves its greatest excursion from equilibrium when the velocity is zero. (This condition corresponds to the turnaround point for a particle oscillating at the end of a spring.) Thus, the kinetic energy is zero, and the energy is entirely potential. Conversely, when the string passes through its equilibrium position, its potential energy is zero, so the energy is entirely kinetic. Thus, in a standing wave, the energy oscillates between being completely kinetic and completely potential; the time difference between these two states is one quarter period. Figure 16–13 illustrates this energy exchange over one complete cycle or period with $n = 1$ and $\phi_0 = -\pi/2$ in Eq. 16–22. At every instant the sum $U + K$ is constant (energy conservation).

$\omega t =$

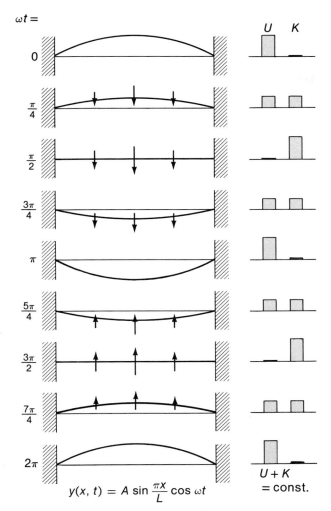

$y(x, t) = A \sin \dfrac{\pi x}{L} \cos \omega t$

$U + K$ = const.

Fig. 16-13. The exchange of energy between kinetic and potential forms during one period of the fundamental vibration of a stretched string. The case $n = 1$, $\phi_0 = -\pi/2$, is shown.

Example 16-5

Consider a string with a length L and a density μ that is under a tension F. One end of the string is fastened to a fixed support at $x = 0$, and the other end is attached to a ring that slides frictionlessly on a perpendicular shaft at $x = L$. Determine the possible standing-wave patterns.

Solution:

We start with the general wave functions, $y^+(x, t)$ and $y^-(x, t)$, of Eqs. 16-18. After applying the condition that a node exist at $x = 0$, we have (Eq. 16-20)

$$y(x, t) = A \sin kx \sin (\omega t - \phi_0)$$

At $x = L$, the ring slides frictionlessly, so the perpendicular component of the force is zero: $F_y = 0$. Now (see Fig. 16-7),

$F_y = -F \sin \alpha \cong -F \tan \alpha = -F(\partial y/\partial x)$. Thus, the boundary condition at $x = L$ is

$$\left.\frac{\partial y}{\partial x}\right|_{x=L} = 0$$

This gives $\cos k_n L = 0$, from which $k_n L = \frac{1}{2} n\pi$, with $n = 1, 3, 5, 7, \ldots$. Then, using $k = 2\pi/L$, the condition is

$$\lambda_n = 4L/n \qquad n = 1, 3, 5, 7, \ldots$$

Finally, the wave function for the standing wave is

$$y(x, t) = A \sin (n\pi x/2L) \sin (\omega t - \phi_0)$$
$$n = 1, 3, 5, 7, \ldots$$

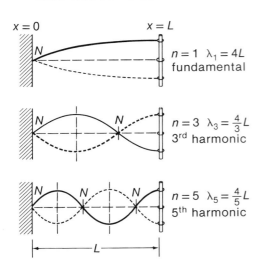

$x = 0$ $x = L$

N $n = 1$ $\lambda_1 = 4L$
fundamental

N N $n = 3$ $\lambda_3 = \frac{4}{3}L$
3rd harmonic

N N N $n = 5$ $\lambda_5 = \frac{4}{5}L$
5th harmonic

L

The diagram shows the fundamental standing-wave pattern and the next two lowest allowed harmonics. Notice the contrast with the standing waves shown in Fig. 16–12 for a string fixed at both ends. Only the *odd* harmonics are allowed in the present case.

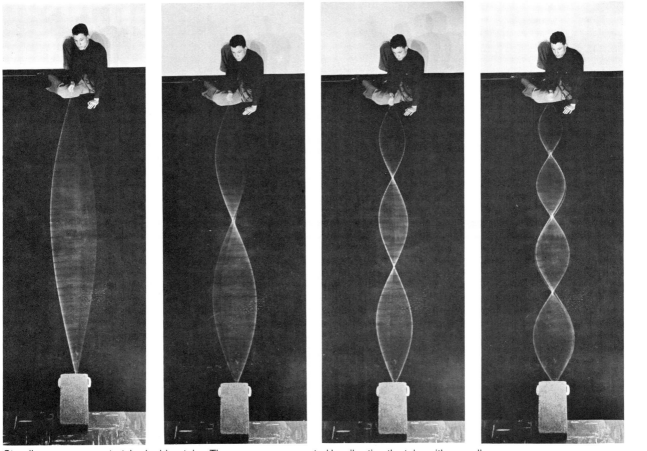

Standing waves on a stretched rubber tube. The waves are generated by vibrating the tube with a small amplitude at a point near a node. (From *PSSC Physics,* 2nd ed., 1965: D. C. Heath & Co., and Educational Development Center, Newton, Mass.)

16-8 COMPLEX WAVES (Optional)

Fourier Analysis. In 1807 the French mathematician and military engineer Joseph Fourier (1768–1830) pointed out that any function can be represented as an infinite series of sine and cosine terms. If a function $f(x)$ is periodic with a repetition length L, Fourier's theorem* states that

$$f(x) = \tfrac{1}{2}A_0 + \sum_{n=1}^{\infty} \left(A_n \cos \frac{2\pi n}{L}x + B_n \sin \frac{2\pi n}{L}x \right) \quad \textbf{(16–23)}$$

If the function $f(x)$ is relatively uncomplicated, the sum of only the first few terms in the series may give a satisfactory representation. For example, consider a rectangular step function with amplitude h and width $\tfrac{1}{2}L$, as shown in Fig. 16–14. This function is periodic, with repetition length L, and therefore can be represented by a Fourier series. Notice that in the interval $0 < x < L$ the function starts from zero, exhibits first a positive segment, then a negative segment, and finally ends at zero. This is the general behavior of a sine function, so we expect that the term $\sin (2\pi/L)x$ will be important in the series. Notice also that the rectangular step function is an *odd* function about the origin, that is, $f(-x) = -f(x)$. The sine function is also an *odd* function, $\sin(-x) = -\sin x$, whereas the cosine function is an *even* function, $\cos(-x) = +\cos x$. Therefore, we do not expect the cosine terms to appear in the Fourier series. Moreover, the rectangular step function is also an odd function about $x = \tfrac{1}{2}L$. Consequently, the even sine terms do not appear in the series. Then the series reduces to

$$f(x) = \frac{4h}{\pi} \sum_{n\,\text{odd}} \frac{1}{n} \sin \frac{2\pi n}{L}x \quad \textbf{(16–24)}$$

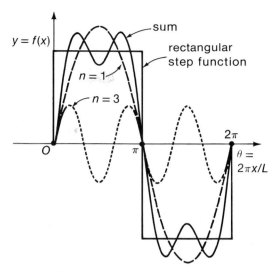

Fig. 16–15. The first two terms, and their sum, of the Fourier series representation of a rectangular step function.

Letting $\theta = 2\pi x/L$, it is found that

$$f(\theta)/h = 1.273 \sin \theta + 0.424 \sin 3\theta + 0.255 \sin 5\theta + \cdots$$

The first two terms of this series, their sum, and the rectangular step function are shown in Fig. 16–15. The two terms alone already give the general appearance of the step function.

The Fourier decomposition of a function is easy to visualize in an *amplitude spectrum,* a graphic display of the coefficients of the sinusoidal terms in the Fourier series. Figure 16–16 shows the amplitude spectrum for a rectangular step function. Only the odd harmonics are present and the amplitudes are proportional to $1/n$.

Fig. 16–14. A rectangular step function.

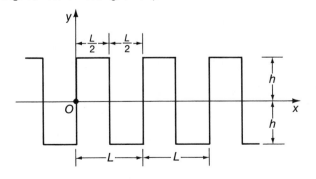

---

*Fourier's 1807 paper was not rigorous; P. G. L. Dirichlet (1805–1859) first gave a rigorous proof of this important theorem in 1829. If the function $f(x)$ is *not* periodic and is defined in the range of the entire x-axis, the series is replaced by an integral.

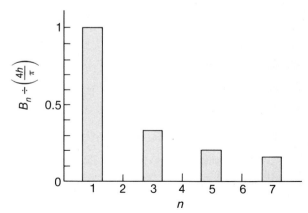

Fig. 16–16. Amplitude spectrum for the rectangular step function of Fig. 16–15.

We can also suppose that the rectangular step function is actually a propagating wave form. That is, Fig. 16–14 and Eq. 16–25 show the wave and its Fourier representation at time $t = 0$. To recover the propagating aspect of the wave it is only necessary to replace x by $x - vt$; thus,

$$y(x, t) = \frac{4h}{\pi} \sum_{n \text{ odd}} \frac{1}{n} \sin\left(\frac{2\pi n}{L}(x - vt)\right) \quad \textbf{(16–25)}$$

which is a wave train of rectangular pulses propagating to the right.

When a note is played on a musical instrument, the sound wave produced is usually rich in harmonic content. That is, the Fourier spectrum contains many terms with substantial amplitudes. The degree to which the various harmonics are present is different for different instruments. Thus, playing the note A (440 Hz) on a violin or on a piano will produce different intensities for the harmonics at 880 Hz, 1320 Hz, 1760 Hz, and so forth. Figure 16–17 shows the Fourier spectra for two different musical instruments.

Because the harmonics are produced with different intensities, a note played on a piano is easily distinguished from the same note played on a clarinet. Even the same type of instrument made in slightly different ways will have distinctive tonal qualities. Violins of the great makers—the Amati, Guarneri, and Stradivari families—are easily distinguished from ordinary fiddles.

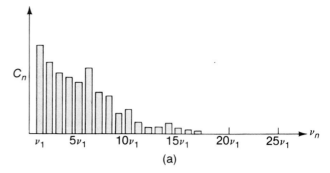

(a)

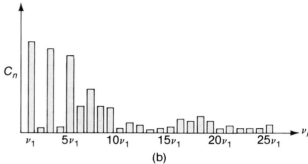

(b)

Fig. 16–17. (a) The harmonic spectrum for the note $B(\nu_1 = 493.88 \text{ Hz})$ played on a violin. (b) Spectrum for the note B flat $(\nu_1 = 233.08 \text{ Hz})$ played on a clarinet. In obtaining spectra such as these, it is customary to make the Fourier analysis in a series of terms of the form $C_n \cos(\omega_n t + \delta_n)$. Then, the coefficients C_n are related to the A_n and B_n in Eq. 16–23 according to $C_n = \sqrt{A_n^2 + B_n^2}$. The phase information contained in the δ_n is discarded in presenting the harmonic spectra. (Spectra courtesy of Richard E. Berg.)

QUESTIONS

16–1 Define and explain briefly the meaning of the terms (a) wavefront, (b) linear polarization, (c) wave function, (d) superposition, and (e) harmonic waves.

16–2 Explain the essential differences among transverse, longitudinal, and torsional waves. Give some examples of each type of wave motion.

16–3 Given a stopwatch and a loud cap pistol, devise a practical experiment for measuring the speed of sound in air.

16–4 Derive a formula for determining the distance to a lightning discharge based on the time delay between seeing the flash and hearing the thunder from the discharge.

16–5 Two transverse square wave pulses traveling in opposite directions, such as shown in Fig. 16–5a, are of the same shape. When they momentarily overlap exactly, there is no transverse displacement anywhere on the string (complete destructive interference). Explain what happened to the energy and momentum of the waves.

16–6 Is it possible to strike a taut string and send a single unidirectional wave pulse down the string?

16–7 Suppose a taut string has a mass per unit length that uniformly increases from one end to the other. Explain what would happen to the phase velocity, wavelength, and frequency of a transverse harmonic wave train traveling down the string.

16–8 Suggest an explanation for why distant ocean waves, when breaking on the beach, no longer constitute wave motion. [*Hints:* What is happening to the wavelength of the wave as it approaches the beach? Does the amplitude of the wave continue to remain small compared with the wavelength? Is there significant transport of water near the beach? What makes the sport of surfing possible?]

16–9 If energy loss due to frictional effects can be neglected for the expanding circular surface waves on water initiated by a sharp impact, how will the wave amplitude and wave energy vary with distance from the splash center?

16–10 Define and explain briefly the meaning of the terms (a) amplitude reflection and transmission coefficients, (b) standing waves, (c) nodes and antinodes, and (d) fundamental and harmonic frequencies.

16–11 A small lead bead is firmly attached to a taut string at some point. With the aid of a sketch such as Fig. 16–9,

explain what would happen to a transverse pulse encountering the bead.

16–12 A plucked guitar string eventually stops vibrating. What happens to the energy originally present in the standing wave?

16–13 Given a stroboscope (a flashing light source with adjustable flash rate), describe an experiment with a standing wave on a taut string that would enable you to determine the phase velocity of waves on the string.

PROBLEMS

Section 16–2

16–1 Suppose that you hear a thunder clap 16.2 s after seeing the associated lightning stroke. The speed of sound waves in air is 334 m/s, and the speed of light in air is 3.0×10^8 m/s. How far are you from the lightning stroke?

16–2 A stone is dropped into a deep canyon and is heard to strike the bottom 10.2 s after release. The speed of sound waves in air is 334 m/s. How deep is the canyon? What would be the percentage error in the depth if the time required for the sound to reach the canyon rim were ignored?

16–3 Two rectangular wave pulses travel along a string, each with a speed of 10 m/s. At $t = 0$ the pulses are approaching each other, as shown in the diagram. Sketch the shape of the string at times $t = 6 \times 10^{-3}$ s to $t = 18 \times 10^{-3}$ s in intervals of 2×10^{-3} s.

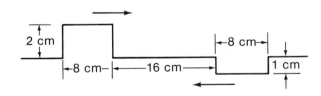

16–4 The wave function for a pulse is
•
$$\eta(x, t) = \frac{0.1\ C^3}{C^2 + (x - vt)^2}$$

with $C = 4$ cm. At some point x the displacement η is observed to decrease from its maximum value to one half the maximum value in 2×10^{-3} s. (a) What is the speed of the pulse? (b) What is the full width at half maximum displacement (FWHM) for the pulse? (c) Sketch the shape of the pulse at $t = 0$ by determining $\eta(x, 0)$ for $x = 0, \pm2$ cm, ±4 cm, ±6 cm, ±8 cm, and ±10 cm.

Section 16–3

16–5 A copper wire with a length of 5 m has a mass of 0.060 kg and is under a tension of 750 N. With what velocity will transverse waves propagate along the wire?

16–6 An earthquake on the ocean floor in the Gulf of Alaska induces a *tsunami* (sometimes called a tidal wave) that reaches Hilo, Hawaii, 4450 km distant, in a time of 9 h 30 min. Tsunamis have enormous wavelengths (100–200 km), and for such waves the propagation velocity is $v \approx \sqrt{g\bar{d}}$, where $\bar{d}$ is the average depth of the water. From the information given, find the average wave velocity and the average ocean depth between Alaska and Hawaii. (This method was used in 1856 to estimate the average depth of the Pacific Ocean long before soundings were made to give a direct determination.)

16–7 A string with a mass M and a length L hangs from a ceiling. Derive an expression for the velocity of transverse wave pulses on the string as a function of the distance x from the free end. [*Hint:* What is the tension in the string due to its own weight at a distance x from the bottom?]

16–8 In a gravity-free environment, a piece of string with its
•   ends joined can be made to assume a circular shape and to spin about its center with a tangential speed v_0. A sharp blow produces a small distortion in the string. Show that the distortion will travel around the string's path while maintaining a fixed position relative to the string. [*Hint:* First determine the tension in the undisturbed string.]

16–9 A triangular wave pulse on a taut string travels in the
••   positive x-direction with a velocity of 50 m/s. The linear mass density of the string is $\mu = 0.080$ kg/m. The shape of the pulse at $t = 0$ is shown in the diagram on the next page. (a) Using $u = x - vt$, verify that

$$y(x, t) = y(u) = (h/L)(L + u), \qquad -L < u < 0$$
$$= (h/L)(L - u), \qquad 0 < u < L$$
$$= 0, \qquad u > L \text{ or } u < -L$$

(b) Show that

$$P(x, t) = P(u) = Fv(h^2/L^2), \qquad -L < u < L$$
$$= 0 \qquad u > L \text{ or } u < -L$$

(c) Show that the total energy in the pulse is $E = 2Fh^2/L$. Find the value of E.

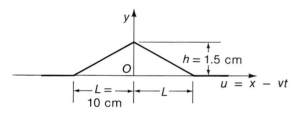

Section 16–4

16–10 Use the value of Young's modulus for platinum given in Table 16–1 and the density value of 21.4 g/cm³ to calculate the velocity of a pure compressional wave in a thin platinum rod. Compare your result with the value given in Table 16–1.

16–11 Young's modulus for copper can be obtained from the known density of copper (8.96 g/cm³) and a measurement of the propagation velocity of a compressional pulse generated by a sharp blow to one end of a thin copper rod. Suppose that the blow has a duration of $120\ \mu$s. The precision desired in the measurement of the modulus is 1 percent, and the timing accuracy is equal to the pulse duration. What is the minimum length of the rod used for the measurements? (You will need an approximate value of v for copper in order to design the experiment. Use the value given in Table 16–1.)

16–12 A coiled spring (such as a Slinky toy) has a mass
 • $M = 0.5$ kg, a length $\ell = 2$ m, and a spring constant $\kappa = 6$ N/m. Show that the extensional wave velocity is $v = \ell\sqrt{\kappa/M}$ and determine its value.

Section 16–5

16–13 The wave function for a linearly polarized wave on a taut string is (in SI units)

$$y(x, t) = (0.35 \text{ m}) \sin (10\pi t - 3\pi x + \pi/4)$$

(a) What is the velocity of the wave? (Give the speed and the direction.) (b) What is the displacement at $t = 0$, $x = 0.1$ m? (c) What is the wavelength and what is the frequency of the wave? (d) What is the maximum magnitude of the transverse velocity of the string?

16–14 A transverse SHW has a period $T = 25$ ms and travels in the negative x-direction with a velocity of 30 m/s. At $t = 0$, a particle on the string at $x = 0$ has a positive displacement of 2.0 cm and a velocity of $v_y = -2.0$ m/s. (a) What is the amplitude of the wave? (b) What is the initial phase angle? (c) What is the magnitude of the maximum transverse velocity? (d) Write the wave function for the wave.

16–15 A vibrating source located at $x = 0$ executes vertical SHM at a frequency of 100 Hz and excites transverse waves on an infinitely long taut string. At $t = 0$, the velocity of the source has one half of its maximum (positive) value of 30 m/s and is increasing. The string tension is 400 N and the linear mass density of the string is 50 g/m. (a) Write an equation for the SHM of the source that is compatible with Eqs. 16–10. What is the value of ϕ_0? (b) What is the phase velocity and what is the wavelength of the waves? (c) What is the amplitude of the waves? (d) Write the wave functions for the two waves that are generated (one for $x > 0$ and one for $x < 0$).

16–16 What is the average rate at which power is transmitted along the string in Problem 16–13 if $\mu = 75$ g/m? What is the total energy per wavelength in the wave?

16–17 A SHW on a taut string with $\mu = 0.10$ kg/m has a total energy of 4 J per wavelength. The maximum transverse velocity of the string is 15.0 m/s, which is one half of the propagation velocity of the wave. (a) What is the wavelength of the wave? [*Hint:* First, show that $E = \frac{1}{2}\mu\lambda v_{y,\text{max}}^2$.] (b) What is the tension in the string? (c) What is the frequency of the wave? (d) What is the amplitude of the wave?

Section 16–6

16–18 Two strings, one with twice the linear mass density of the other, are joined to make a single taut line. A SHW with an amplitude of 10 cm travels along the lighter string and approaches the junction. (a) What is the amplitude and phase (with respect to the incident wave) of the transmitted wave? (b) What are the same quantities for the reflected wave? (c) If the incident wave carries an average power of 100 W, what is the reflected average power? [*Hint:* Refer to Example 16–2 for an expression for $\bar{P}$.]

16–19 A long string with a linear mass density μ has a section
 • of string with density 2μ inserted into its length, as shown in the diagram at the top of page 321. A single pulse travels along the string toward the heavier section. What is the magnitude and the shape of the first pulse to be transmitted past the heavy section? (• Why is it necessary to specify the "first" pulse?)

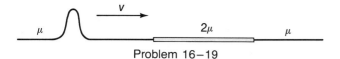

Problem 16–19

Section 16–7

16–20 Two strings with equal lengths and densities are held fixed at their ends. What is the ratio of the tensions in the strings that is required to give the same vibration frequency when one string vibrates in a mode with two segments (that is, with one node between the ends) and the other string vibrates in three sections (that is, with two nodes between the ends)?

16–21 A violin string has a length of 35 cm and a mass of 0.41 g; the string tension is 90 N. What are the lowest three harmonic frequencies for the open string (i.e., with no fingering)?

16–22 A string on a cello has a length of 0.70 m, and the tension is adjusted to sound concert A ($\nu_A = 440$ Hz) when played without fingering (this is the fundamental). (a) By what amount must the tension be increased if the fundamental tone is to be an octave higher ($\nu'_A = 2\nu_A$)? (b) By what amount must the length be shortened by fingering, with the original tension, to produce a fundamental tone that is an octave higher than middle C ($\nu_C = 261.6$ Hz)?

16–23 A thin rod, clamped at its midpoint, has a fundamental longitudinal vibration frequency of 1000 Hz. At what points could the clamp be placed to produce a fundamental frequency of 4000 Hz? [*Hint*: The position of the clamp is necessarily a node, and each end is an antinode.]

16–24 Two strings with different linear mass densities are
• joined together and stretched between two fixed supports. The string system is set into vibration with the standing-wave pattern shown in the diagram. Determine the length ratio L_2/L_1 if $\mu_2/\mu_1 = \frac{5}{4}$.

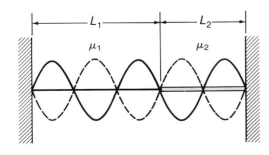

16–25 A 50-kg ball is attached to one end of a 1-m cord that has
•• a mass of 0.10 kg. The other end of the cord is attached to a ring that can slide frictionlessly on a horizontal

shaft, as shown in the diagram. A horizontal blow is delivered to the cord and excites the fundamental vibration with a maximum transverse velocity of 15 m/s. Assume that the ball remains essentially stationary as the cord vibrates. (a) What is the frequency of the fundamental vibration? (b) What is the amplitude of the motion? (c) If the blow is delivered as an impulse to a stationary cord, show that the wave function for the cord is $y(x, t) = A \sin (\pi x/2L) \sin \omega t$, where the ball is located at $x = 0$. (d) Determine the period of the pendulum motion of the hanging ball and compare with the period of the vibrating string. Is it reasonable to assume that the hanging ball remains stationary?

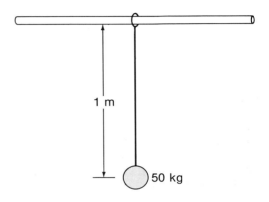

1 m

50 kg

Section 16–8

16–26 A taut string with a length L is displaced by raising the
• center point of the string a distance h so that the string has the shape of an equilateral triangle with base L and altitude h, as shown in the diagram. The Fourier expansion for the string is

$$A_n = 0, \quad B_n = \frac{8h}{n^2 \pi^2} \sin \frac{n\pi}{2}$$

Compare the given triangular shape with accumulated Fourier series sum as additional terms are added. (The sum should be carried to at least $n = 5$. A calculator with memory storage would be useful.)

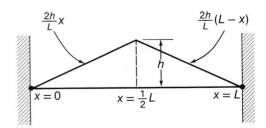

$\frac{2h}{L} x$ $\frac{2h}{L}(L-x)$

h

$x = 0$ $x = \frac{1}{2}L$ $x = L$

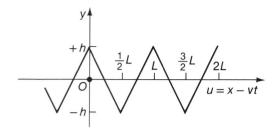

Additional Problems

16-28   The point $x = 0$ on an infinitely long taut string is displaced in the manner shown in the diagram. The wave velocity on the string is 12 m/s. Sketch the shape of the propagating waves at $t = 3 \times 10^{-3}$ s. (One wave propagates to the right and one to the left. • Do you see why?)

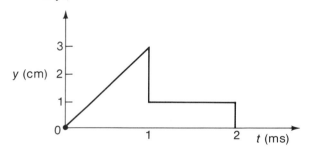

16-29   A transverse wave pulse, $y(x, t) = Ae^{-\sigma(x-vt)^2}$, travels
••   along a taut string, with $A = 3$ cm, $\sigma = 10$ m^{-2}, and $v = 10$ m/s. (a) Sketch the shape of the pulse at $t = 0$. (This is a *Gaussian function*.) (b) Determine the radius of curvature of the wave at $x = 0$ for $t = 0$. [*Hint:* When $\partial y / \partial x \ll 1$, then $R \approx (\partial^2 y / \partial x^2)^{-1}$. Is this condition met at $x = 0$, $t = 0$?] (c) What is the centripetal acceleration (in units of g) of a small portion of the string at $x = 0$, $t = 0$? (d) What vertical force acts on a 10-g piece of the string at $x = 0$, $t = 0$? (e) The linear density of the string is $\mu = 0.5$ kg/m. Express the result of (d) in terms of the string tension.

16-30   Transverse pulses propagate with a velocity of 200 m/s along a taut copper wire that has a diameter of 1.5 mm. What is the tension in the wire? (The density of copper is given in Table 1-2.)

16-31   The wave function of a wave is (in SI units) $y(x, t) = 0.20 \sin (80\pi t - 4\pi x)$. Rewrite this wave function in the three forms given in Eq. 16-9. Determine the values of v, λ, ω, ν, k, T, and ϕ_0.

16-32   In the discussion leading to the wave equation (Eq. 16-
•   11), a single harmonic wave traveling in the positive x-direction was used as an example. Repeat these steps for the simultaneous presence of two harmonic waves

traveling in opposite directions; for example, take $y = y_1 + y_2$ used in Example 16-3. Explain why the wave equation *must* involve the second derivatives and not the first derivatives!

16-33   In Example 16-12, we found that the power in a wave,
•   averaged over one period, is $\bar{P} = \frac{1}{2}FA^2k\omega$, where A is the wave amplitude. Apply this result to Eqs. 16-16 and show that, averaged over one period, the power in the transmitted wave plus that in the reflected wave is equal to the power in the incident wave.

16-34   A steel wire in a piano has a length of 0.70 m and a mass of 4.3×10^{-3} kg. To what tension must this wire be stretched in order that the fundamental vibration correspond to concert middle C, $\nu_C = 261.6$ Hz?

16-35   A violin string has a length of 0.35 m and is tuned to concert G, $\nu_G = 392$ Hz. Where must the violinist place his finger to play concert A, $\nu_A = 440$ Hz? If this position is to remain correct to one half the width of a finger (i.e., to within 0.6 cm), by what fraction may the string tension be allowed to slip?

16-36   The Fourier representation of the asymmetric "saw-
••   tooth" wave shown in the diagram is

$$\eta(x, t) = \frac{2h}{\pi} \sum_{n=1}^{\infty} \frac{(-1)^{n+1}}{n} \sin \left[\frac{2\pi}{L} (x - vt) \right]$$

Sketch the sum of the first three terms and compare with the original wave. Take $t = 0$ and x measured in units of L.

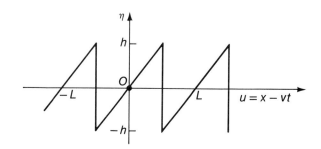

16-27   Consider the propagating triangular (or "sawtooth")
••   wave shown in the diagram. The Fourier representation of this wave is

$$y(x, t) = \frac{8h}{\pi^2} \sum_{m=1}^{\infty} \frac{1}{m^2} \cos \frac{2\pi m}{L} (x - vt), \qquad m \text{ odd}$$

Compare the given wave shape with the Fourier series sum up to $m = 5$ as a function of time. Take $x = \dfrac{L}{4}$, $L = 25$ cm, and $v = 10$ cm/s.

CHAPTER 17

SOUND

The propagation of pressure variations through a medium such as air is particularly interesting and important because pressure changes can be detected by the auditory apparatus of the ear. If the frequency of the pressure variations is in the range of about 15 Hz to about 20,000 Hz, the human ear produces the physiologic sensation of *hearing*. Waves that produce such a response are called *sound waves*. Many species of animals and insects can detect pressure waves with frequencies much higher than those that can be heard by humans. Bats and dolphins, for example, can hear sounds with frequencies above 100,000 Hz. However, we restrict attention here to the human audible range from 15 Hz to 20,000 Hz.*

Sound waves with frequencies above 20,000 Hz are referred to as *ultrasonic* waves. By using special laboratory apparatus it is possible to produce sound waves with frequencies of up to about 1 GHz $= 10^9$ Hz. Such waves have wavelengths in air as small as 300 nm, comparable with the wavelength of ultraviolet light. If the frequency of a sound wave is below 15 Hz, it is called an *infrasonic* wave. The motions of large-scale matter, such as those produced by earthquakes and volcanic activity, can generate infrasonic waves.

17-1   THE SPEED OF SOUND WAVES

Compressional Waves.   We consider first the propagation of sound waves for the one-dimensional case of a narrow tube that contains a gas and is closed at one end by a freely sliding piston (Fig. 17–1a). The result we obtain for the speed of sound in this special case is also valid for propagation in an unlimited (unbounded) gas.

Suppose that the gas in the tube is at equilibrium with the ambient (outside) pressure p_0 and with the ambient density ρ_0. Consider what happens when the piston moves through a simple cycle. Starting from rest, the piston first moves to the right, stops instantaneously at the point of maximum excursion, and then returns to the original position. While the piston moves to the right, the gas in front of the piston is compressed, so the pressure and the density are raised above the equilibrium values. During the other half of the cycle, while the piston moves to the left, the pressure and density are lowered below

---

*These limits vary widely from person to person, depending on age and on the history of exposure to high-intensity sounds. Usually, a person's ability to hear high-frequency sounds deteriorates with age; by middle age, most individuals have a high-frequency limit of 12,000 to 14,000 Hz. Persons who have been exposed for long periods to high-intensity sounds (such as performers in rock bands and their fans, or persons who work near aircraft) often have severely impaired high-frequency hearing.

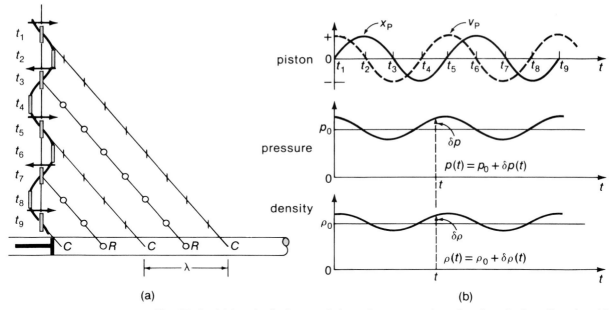

Fig. 17–1. (a) Longitudinal wave of alternating compressions C and rarefactions R produced by the oscillatory motion of a piston that executes SHM. The displacement of the piston relative to the wavelength λ is exaggerated. (b) The piston displacement x_P and velocity v_P and the gas pressure and density are shown as functions of time. The quantities $\delta p(t)$ and $\delta\rho(t)$ represent the deviations from the equilibrium values for the pressure and density, respectively. The deviations, δp and $\delta\rho$, compared with p_0 and ρ_0 are exaggerated. The quantity δp is called the *acoustic pressure*.

the equilibrium values. The result of the piston movement is a pressure variation that propagates along the tube in the form of an overpressure pulse or *compression,* followed by an underpressure pulse or *rarefaction.* If the piston continually repeats this motion as SHM, a simple harmonic wave consisting of alternating compressions and rarefactions in the gas will propagate steadily along the tube.

For a piston speed that is less than the wave speed, the pressure variation is in phase with the piston velocity. If the ratio of the wavelength λ of the wave to the tube diameter D is $\lambda/D \gtrsim 2\pi$, the wave propagates along the tube as a plane wave; that is, the amplitude and phase depend only on the longitudinal position x and on the time t, and *not* on position within the tube on any cross section. Figure 17–1 represents this case.

The Wave Equation. We wish to find the wave equation that describes the longitudinal pressure variations in a gas-filled tube. By examining this case in detail we can relate the wave velocity v that appears in the wave equation to the density and the elastic properties of the gas.

Consider a section Δx of the gas in a tube, as shown in Fig. 17–2. Let $y(x)$ represent the displacement of the gas matter from equilibrium at the coordinate x for a particular instant. The excursion of the pressure from the equilibrium pressure p_0 is the *acoustic pressure* δp, and the total pressure at x is $p = p_0 + \delta p$. The equilibrium value of the volume of the gas in the region considered is $V = S\,\Delta x$, and the change in volume of the original quantity of gas due to the acoustic pressure is $\Delta V = S(\partial y/\partial x)\,\Delta x$.* Thus, the fractional change in the volume is

$$\frac{\Delta V}{V} = \frac{\partial y}{\partial x}$$

---

*The displacement y is also a function of time, so we specify the partial derivative with respect to x.

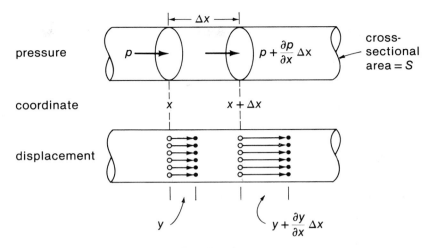

Fig. 17–2. Details of the pressure p and the matter displacement y in a section Δx of gas in a tube.

Now, the ratio of the applied pressure δp to the fractional volume change is called the *bulk modulus* β of the substance; that is

$$\beta = -\frac{\delta p}{\Delta V / V} \qquad (17\text{–}1)$$

(When $\delta p > 0$, the volume decreases, $\Delta V < 0$, so that the bulk modulus is a positive quantity.) Then, we can write

$$\frac{\partial y}{\partial x} = -\frac{\delta p}{\beta} \qquad (17\text{–}2)$$

The net force acting on the quantity of gas is the difference between the total pressures at the ends of the segment, multiplied by the cross-sectional area:

$$F = S\left[p - \left(p + \frac{\partial p}{\partial x}\, \Delta x\right)\right]$$

$$= -S\,\frac{\partial p}{\partial x}\, \Delta x$$

(The negative sign means that when the pressure increases to the *right,* the force on the gas element is to the *left.*)

The equilibrium density of the gas is ρ_0, so the mass in the section Δx is $\rho_0 S\,\Delta x$. The acceleration of the gas is the second derivative of displacement with respect to *time*. Thus, Newton's second law, $F = ma$, becomes

$$-S\,\frac{\partial p}{\partial x}\, \Delta x = \rho_0 S\,\Delta x\,\frac{\partial^2 y}{\partial t^2}$$

Now,

$$\frac{\partial p}{\partial x} = \frac{\partial}{\partial x}\,(p_0 + \delta_p) = \frac{\partial(\delta p)}{\partial x} = -\beta\,\frac{\partial^2 y}{\partial x^2}$$

where we have used Eq. 17–2 for δp. Then, Newton's equation becomes

$$\frac{\partial^2 y}{\partial x^2} = \frac{\rho_0}{\beta}\,\frac{\partial^2 y}{\partial t^2} \qquad (17\text{–}3)$$

which is the wave equation that describes the displacement $y(x, t)$. Note that it has exactly the form of Eq. 16–11 for generalized wave motion in one dimension.

The Velocity of Compressional Waves. In Section 16–5 we learned that a wave equation of the form of Eq. 17–3 is satisfied by any function $y(x, t) = y(u) = y(x \pm vt)$, where the wave velocity v is obtained from the coefficient of the second time derivative.

$$v = \sqrt{\frac{\beta}{\rho_0}} \qquad (17-4)$$

The compression of a gas produced by an applied pressure has the effect of raising the temperature of the gas unless the heat generated can be conducted away (see Section 19–2). For the situations we consider here, the heat conductivity of any gas is insufficient to transport the heat generated in the regions of compression to the cooler regions of rarefaction, at least for frequencies below about 10^5 Hz. Thus, we neglect heat flow between the various regions of a gas that supports an acoustic wave. That is, we assume conditions described as *adiabatic* (i.e., with no heat transfer), and we need to use in our equations the adiabatic bulk modulus β_a, which can be shown to be

$$\beta_a = \gamma p_0 \qquad (17-5)$$

where γ is a number very close to 1.40 for all gases that consist of diatomic molecules, such as H_2, O_2, N_2, and air.* Thus, the phase or wave velocity for sound in diatomic gases is

$$v = \sqrt{\frac{\beta_a}{\rho_0}} = \sqrt{\frac{1.40 \, p_0}{\rho_0}} \qquad \text{for diatomic gases} \qquad (17-6)$$

For air at $0°$ C and normal atmospheric pressure, we have $p_0 = 1.013 \times 10^5$ Pa and $\rho_0 = 1.293$ kg/m^3. Then the value calculated from Eq. 17–6 is $v_{calc} = 331.2$ m/s, which is to be compared with the measured value, $v_{meas} = 331.5$ m/s. At a temperature T (in degrees C), the speed of sound $v(T)$ in an ideal gas† is

$$v(T) = v(0) \sqrt{1 + \frac{T}{273}} \qquad \text{ideal gas} \qquad (17-7)$$

where $v(0)$ is the speed at $T = 0°$ C and where $T + 273$ is the absolute temperature of the gas.

Table 17–1 lists the speed of sound in various unbounded media at $0°$ C. For the solid media, the speeds given are for longitudinal waves.

---

*In Section 21–3 we show that the adiabatic condition for such molecules is given by the expression $pV^\gamma = $ constant. Differentiating this expression gives $p\gamma V^{\gamma-1} \, dV + V^\gamma \, dp = 0$, or

$$\beta_a = -\frac{dp}{dV/V} = \gamma p$$

† See Section 19–2 for the definition of an ideal gas.

TABLE 17-1. Speed of Sound (at 0° C)

MEDIUM	v (m/s)
Gases (at normal atmospheric pressure)	
Air	331.5
Argon	319
Carbon dioxide	259
Chlorine	206
Helium	965
Hydrogen	1284
Liquids	
Fresh water	1437
Sea water	1471
Mercury	1484
Solids*	
Aluminum	6420
Gold	3240
Lead	1960
Glass, Pyrex	5640
Lucite	2680
Polyethylene	1950

*Notice that these values for longitudinal waves in unbounded media are different from the values listed in Table 16-1 for extensional waves in thin rods.

Example 17-1

The wave function for a longitudinal sound wave in an air-filled tube is

$$y(x, t) = A \cos (kx - \omega t + \phi_0)$$

The amplitude of the motion of the driving piston is 2 mm, and the frequency is 1000 Hz. The temperature of the air is 20° C and the equilibrium density is $\rho_0 = 1.20$ kg/m³.

(a) Obtain expressions for the velocity of the gas matter, the acoustic pressure δp, and the density excursion $\delta \rho$.

(b) Obtain an expression for the power (averaged over a period) that is transported along the tube.

(c) Determine the initial phase angle ϕ_0 for the case in which the piston, whose average position is $x = 0$, undergoes SHM described by $x(t) = B \sin \omega t$.

(d) Show that the average power delivered to the piston is equal to the average power transported along the tube (b).

(e) Determine the speed of the wave and the wavelength.

(f) Obtain the amplitudes of the gas-matter velocity, the acoustic pressure, and the density excursion.

(g) Find the *intensity* of the wave, $I = \bar{P}/S$, which is the average power per unit cross-sectional area.

Solution:

(a) The velocity v_g of the gas matter is

$$v_g = \frac{\partial y}{\partial t} = A\omega \sin (kx - \omega t + \phi_0)$$

The acoustic pressure δp is (using Eq. 17-2)

$$\delta p = -\beta_a \frac{\partial y}{\partial x} = Ak\beta_a \sin (kx - \omega t + \phi_0)$$

where β_a is the adiabatic bulk modulus.

To obtain an expression for the density excursion, we note that $\rho = M/V$, so $\Delta \rho / \rho = -\Delta V / V$. Then, using $\Delta V / V = -\delta p / \beta_a$ (Eq. 17-1), we can write

$$\delta \rho = \frac{\rho_0}{\beta_a} \delta p = Ak\rho_0 \sin (kx - \omega t + \phi_0)$$

From these results we see that δp and $\delta \rho$ are in phase with v_g and that all three quantities differ in phase by $\pi/2$ compared with y.

(b) The power transported past a cross section of the tube at a point x at time t is (compare Eq. 16-3)

$$P(x, t) = \mathbf{F} \cdot \mathbf{v}_g = S \, \delta p v_g = -S\beta_a \left(\frac{\partial y}{\partial x}\right)\left(\frac{\partial y}{\partial t}\right)$$

$$= S\beta_a A^2 \omega k \sin^2 (kx - \omega t + \phi_0)$$

Now, the average of $\sin^2 u$ over a period is $\frac{1}{2}$ (see Example 16–2). Thus,

$$\overline{P} = \tfrac{1}{2}S\beta_a A^2 \omega k$$

(c) The motion of the piston, $Y(t)$, is transferred directly to the adjacent air molecules; thus, $B = A$ and $dY/dt = v_g$. With $Y = A \sin \omega t$, we have

$$\frac{dY}{dt} = A\omega \cos \omega t = v_g(0, t) = A\omega \sin (-\omega t + \phi_0)$$

from which we conclude that $\phi_0 = \pi/2$ because $\sin (\pi/2 - \omega t) = \cos \omega t$.

(d) The instantaneous power delivered to the piston is (using $\phi_0 = \pi/2$)

$$P(t) = S \, \delta p \, \frac{dY}{dt} = SA^2 k\beta_a \omega \cos^2 \omega t$$

The average power delivered per cycle is

$$\overline{P} = \tfrac{1}{2}SA^2 k\beta_a \omega$$

which is the same as the average power transported along the tube (b), as it must be.

(e) For the speed of the sound wave, we use Eq. 17–7 and the value of $v(0)$ for air in Table 17–1:

$$v(20°\ C) = (331.5\ \text{m/s})\ \sqrt{1 + \tfrac{20}{273}} = 343.4\ \text{m/s}$$

The wavelength is

$$\lambda = \frac{v}{\nu} = \frac{343.4\ \text{m/s}}{1000\ \text{s}^{-1}} = 0.3434\ \text{m}$$

(f) For the various amplitudes we have*

$$v_{gm} = A\omega = (2 \times 10^{-3}\ \text{m})(2\pi \times 10^3\ \text{s}^{-1}) = 12.57\ \text{m/s}$$

$$\delta p_m = Ak\beta_a = A\left(\frac{2\pi}{\lambda}\right)(\gamma p_0) = (2 \times 10^{-3}\ \text{m})\left(\frac{2\pi}{0.3434\ \text{m}}\right) \cdot$$

$$(1.40 \times 1.013 \times 10^5\ \text{Pa})$$

$$= 5.190 \times 10^3\ \text{Pa}$$

$$\delta\rho_m = Ak\rho_0 = A\left(\frac{2\pi}{\lambda}\right)\rho_0 = (2 \times 10^{-3}\ \text{m})\left(\frac{2\pi}{0.3434\ \text{m}}\right) \cdot$$

$$(1.20\ \text{kg/m}^3)$$

$$= 0.04391\ \text{kg/m}^3$$

Notice that the maximum acoustic pressure δp_m and the maximum density excursion $\delta\rho_m$ are small compared with the equilibrium values:

$$\frac{\delta p_m}{p_0} = \frac{5.190 \times 10^3\ \text{Pa}}{1.013 \times 10^5\ \text{Pa}} = 5.1\ \text{percent}$$

$$\frac{\delta\rho_m}{\rho_0} = \frac{0.04391\ \text{kg/m}^3}{1.20\ \text{kg/m}^3} = 3.7\ \text{percent}$$

(g) The intensity of the wave is $I = \overline{P}/S$, so that

$$I = \tfrac{1}{2}A^2 k\beta_a \omega \qquad\qquad \textbf{(17–8)}$$

For the present case, we have

$$I = \tfrac{1}{2}A^2\left(\frac{2\pi}{\lambda}\right)(\gamma p_0)\omega$$

$$= \tfrac{1}{2}(2 \times 10^{-3}\ \text{m})^2\left(\frac{2\pi}{0.3434\ \text{m}}\right)(1.40 \times 1.013 \times 10^5\ \text{Pa}) \cdot$$

$$(2\pi \times 10^3\ \text{s}^{-1})$$

$$= 3.26 \times 10^4\ \text{W/m}^2$$

This is an extremely high sound intensity; it might be encountered in special industrial applications but not in normal listening (see Table 17–2).

Example 17–2

A *Helmholtz resonator* is an acoustic device that consists of a rigid enclosure with a volume V coupled to the atmosphere through an open pipe with radius r and length ℓ, as indicated in the diagram. Derive an expression for the natural frequency ν_0 of oscillation of the system.

Solution:

The diagram shows a hypothetical displacement of the air mass in the pipe intruding a distance x into the main volume of the resonator. This intrusion compresses the air in the resonator

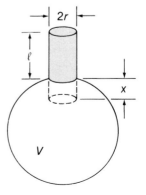

---

*As usual, a subscript m indicates *maximum value*, which, in these cases, is the amplitude.

by an amount $\Delta V = \pi r^2 x$. This compression, in turn, produces a pressure on the intruding mass given by

$$\delta p = -\beta_a \frac{\Delta V}{V} = -\beta_a \frac{\pi r^2 x}{V} = -\frac{v^2 \rho_0 \pi r^2}{V} x$$

The restoring force is

$$\delta F = \pi r^2 \, \delta p = -\frac{v^2 \rho_0 \pi^2 r^4}{V} x = -\kappa x$$

where the force constant κ for the volume of gas is

$$\kappa = v^2 \rho_0 \pi^2 r^4 / V$$

The mass of air oscillating in the pipe is $m = \pi r^2 \ell \rho_0$. The natural frequency is therefore (see Eq. 15–13)

$$\nu_0 = \frac{1}{2\pi} \sqrt{\frac{\kappa}{m}} = \frac{v}{2\pi} \sqrt{\frac{\pi r^2}{V\ell}}$$

(To obtain this expression we have made the implicit assumption that the wavelength, $\lambda_0 = v/\nu_0$, is large compared with all linear dimensions of the system. • Can you see why this assumption is necessary?)

When the system is excited, there is actually a small mass of air just beyond the open pipe that is also set into motion. Taking this into account increases the effective length of the pipe and gives a more accurate expression for the resonant frequency:

$$\nu_0 = \frac{v}{2\pi} \sqrt{\frac{\pi r^2}{V(\ell + 8r/3\pi)}}$$

Notice that even if $\ell = 0$, there is an effective mass, $8r^3 \rho_0 / 3$, associated with the vent opening.

17–2   STANDING WAVES IN PIPES

We now consider the propagation of longitudinal waves in gas-filled pipes with finite length and investigate the effects of the pipe termination (whether open or closed) on these waves. We first examine the case of a pipe closed at $x = 0$, as shown in Fig. 17–3a. The wave function for the wave incident on the closed end from the left is $y_i(x, t)$; the reflected wave is described by $y_r(x, t)$. The gas matter immediately adjacent to the closed end has zero displacement. Thus, the required boundary condition on the waves, using superposition, is

$$y_i(0, t) + y_r(0, t) = 0 \quad \text{(closed end)} \qquad \textbf{(17–9)}$$

This condition is analogous to that discussed in Section 16–6 for the case of transverse waves on a taut string terminated by a rigid support.

Next, consider a pipe terminated at $x = 0$ with an open end, as shown in Fig.

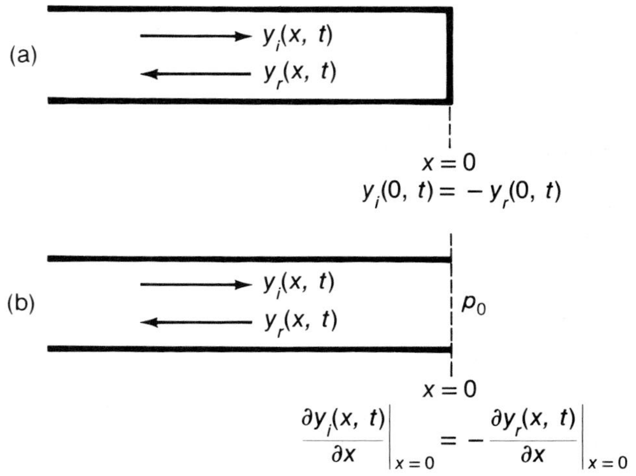

Fig. 17–3. Reflection of incident longitudinal waves $y_i(x, t)$ from the closed end of a pipe (a) and from the open end of a pipe (b).

(a)

$y_i(x, t)$

$y_r(x, t)$

$x = 0$

$y_i(0, t) = -y_r(0, t)$

(b)

$y_i(x, t)$

$y_r(x, t)$

p_0

$x = 0$

$$\left. \frac{\partial y_i(x, t)}{\partial x} \right|_{x=0} = -\left. \frac{\partial y_r(x, t)}{\partial x} \right|_{x=0}$$

17–3b. Now, we assume that the pressure is described by a continuous function of both x and t and that at $x = 0$ it has a constant value equal to the ambient pressure p_0. (This assumption is valid only if the pipe diameter is small compared with the wavelength.) This requires the acoustic pressure δp to be zero at $x = 0$. Because $\delta p = -B_a\,\partial y/\partial x$, we can use superposition to write the boundary condition as

$$\left.\frac{\partial y_i(x,\,t)}{\partial x}\right|_{x=0} + \left.\frac{\partial y_r(x,\,t)}{\partial x}\right|_{x=0} = 0 \qquad \text{(open end)} \qquad \textbf{(17–10)}$$

This condition is analogous to that discussed in Section 16–6 for the case of a transverse wave on a taut string terminated by a frictionless ring that slides on a fixed shaft.

The restriction that the pipe diameter be small compared with the wavelength is equivalent to specifying that there be no substantial sound radiation from the open end. Any such radiation would remove energy from the wave in the pipe, and consequently the reflected wave would have an amplitude smaller than that of the incident wave. Requiring the total absence of radiation from the end of the pipe is equivalent to requiring the total absence of friction for the case of the string terminated with a sliding ring. Moreover, the assumption that the sliding ring has zero mass is equivalent to ignoring the vibration of the mass of air just outside the open end of the pipe. These radiation and mass effects also introduce a phase difference between the incident and reflected waves.

Waves that propagate in terminated gas-filled pipes must obey the condition of Eq. 17–9 at every closed end and the condition of Eq. 17–10 at every open end. For particular wavelengths, standing waves are set up in pipes, just as they are on terminated strings. An interesting example is the case of organ pipes. Figure 17–4 shows the lowest-frequency standing wave, the fundamental mode, of an organ pipe that is open at the nondriven end. Figure 17–4a indicates schematically the flow pattern that produces the turbulent excitation of the air in the pipe as air is blown against the lip edge and escapes through the slot at O. Although the opening in the pipe at O is not as large as the completely open end opposite, the pressure at the driven end is essentially constant and equal to the ambient pressure p_0. Also shown in Fig. 17–4 are the variations of y, v_g, δp, and $\delta\rho$. (• Can you explain the reason for each shape?) Figure 17–5 illustrates the same quantities for an organ pipe that is closed at the nondriven end. (• Can you explain why only the *odd* harmonics are allowed in this case?)

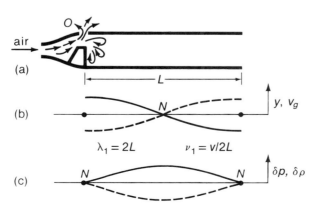

Fig. 17–4. (a) An organ pipe open at the nondriven end. (b, c) The graphs show the variation of y, v_g, δp, and $\delta\rho$ for the fundamental mode of vibration. The nodes are indicated by N. The wavelengths of the harmonics are given by $\lambda_n = 2L/n$, $n = 1, 2, 3, 4, \ldots$. (Compare Eq. 16–21.)

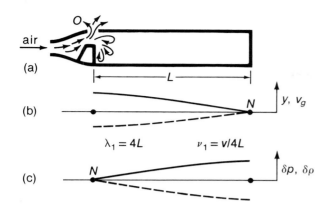

(a)

(b)

$\lambda_1 = 4L$ $\nu_1 = v/4L$

(c)

Fig. 17-5. (a) An organ pipe closed at the nondriven end. (b, c) The graphs show the variation of y, v_g, δp, and $\delta\rho$ for the fundamental mode of vibration. The nodes are indicated by N. The wavelengths of the harmonics are given by $\lambda_n = 4L/n$, $n = 1, 3, 5, 7, \ldots$. (Compare Example 16-5.) (Musicians refer to the next highest possible frequency after the fundamental as the *first overtone* (in the present case, the third harmonic, $n = 3$), the next highest possible frequency as the second overtone (here $n = 5$), and so on.)

One of the largest pipe organs in the world is played daily in the John Wanamaker department store in Philadelphia. Designed by George Audsley for the Louisiana Purchase Exposition in St. Louis, it was rebuilt in Wanamaker's store in 1911. Various additions brought the total to 30,067 pipes, the largest of which is a wooden gravissima 64 ft (19.5 m) in length. (Courtesy of John Wanamaker's, Philadelphia.)

Example 17-3

A third harmonic standing wave ($n = 3$) is set up in a 3-m organ pipe that is open at the nondriven end. The amplitude of the acoustic pressure is 0.5 percent of atmospheric pressure.

(a) Write expressions for $y(x, t)$, $v_g(x, t)$, $\delta p(x, t)$, and $\delta\rho(x, t)$.

(b) Determine the values of the amplitudes of the quantities in (a). (The temperature is $0°$ C.)

(c) What is the frequency of the standing wave?

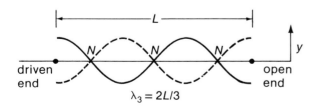

driven end

open end

$\lambda_3 = 2L/3$

Solution:

The diagram shows the displacement curve for the $n = 3$ standing wave. Here, we have $\lambda_3 = 2L/3$.

(a) The displacement can be represented as

$$y(x, t) = A \cos \frac{3\pi}{L} x \sin \frac{3\pi v}{L} t$$

Contrast this expression with Eq. 16-22. (● Do you see why the cosine function is required in the present case?) Then, with $v_g(x, t) = \partial y/\partial t$, we find

$$v_g(x, t) = \frac{3\pi v}{L} A \cos \frac{3\pi}{L} x \cos \frac{3\pi v}{L} t$$

For the acoustic pressure we have

$$\delta p(x, t) = -\beta_a \frac{\partial y}{\partial x} = -v^2 \rho_0 \frac{\partial y}{\partial x}$$

$$= \frac{3\pi v^2}{L} \rho_0 A \sin \frac{3\pi}{L} x \sin \frac{3\pi v}{L} t$$

Finally, the density excursion is (see Example 17-1)

$$\delta\rho(x, t) = \frac{\rho_0}{\beta_a} \delta p = \frac{1}{v^2} \delta p$$

$$= \frac{3\pi}{L} \rho_0 A \sin \frac{3\pi}{L} x \sin \frac{3\pi v}{L} t$$

(● Where are the nodes and antinodes of δp and $\delta\rho$?)

(b) Using $v = 331.5$ m/s, $p_0 = 1.013 \times 10^5$ Pa, and $\rho_0 = 1.293$ kg/m^3, the various amplitudes are

$$\delta p_m = (0.5 \times 10^{-2})(1.013 \times 10^5 \text{ Pa})$$

$$= 506.5 \text{ Pa}$$

$$A = \delta p_m \frac{L}{3\pi v^2 \rho_0} = (506.5 \text{ Pa})\frac{3 \text{ m}}{3\pi(331.5 \text{ m/s})^2(1.293 \text{ kg/m})^3}$$

$$= 1.135 \times 10^{-3} \text{ m}$$

$$v_{gm} = \frac{3\pi v}{L}A = \frac{3\pi(331.5 \text{ m/s})}{3 \text{ m}}(1.135 \times 10^{-3} \text{ m})$$

$$v_{gm} = 1.182 \text{ m/s}$$

$$\delta\rho_m = \frac{3\pi}{L}\rho_0 A = \frac{3\pi}{3 \text{ m}}(1.293 \text{ kg/m}^3)(1.135 \times 10^{-3} \text{ m})$$

$$= 4.610 \times 10^{-3} \text{ kg/m}^3$$

(c) The wavelength for the third harmonic wave is $\lambda_3 = 2L/3$, so the frequency is

$$\nu_3 = \frac{v}{\lambda_3} = \frac{3v}{2L} = \frac{3(331.5 \text{ m/s})}{2 (3 \text{ m})} = 165.8 \text{ Hz}$$

17–3 WAVES IN THREE-DIMENSIONAL MEDIA

We now investigate the propagation of sound waves in an unbounded gaseous medium that is isotropic (i.e., having the same physical properties in all directions) and homogeneous. Sound traveling in open air is one example. Consider the wave generated by the pulsations of a sphere that executes SHM about some mean radius. This wave is a harmonic *spherical wave* that propagates outward with the phase velocity v in all directions from the center of the pulsating sphere located at the origin.

Because of the spherical symmetry, the intensity of the wave (that is, the average power transmitted per unit area) is uniform over any spherical surface centered on the origin. The wave traveling through a small volume surrounding some point Q on one of these spherical surfaces is approximately a *plane wave*, as indicated in Fig. 17–6.

We can obtain an expression for the wave function of a spherical wave in the following manner. First, because of the symmetry, the wave propagates radially outward in the same way in all directions. Thus, the wave function is given by $y(r, t) = A \cos (kr - \omega t + \phi_0)$, where the amplitude A and its possible dependence on r are still to be determined.

According to Eq. 17–8, the intensity of a wave (the average power transported per unit area) is $I = \frac{1}{2}A^2 k\beta_a\omega$. Using $\beta_a = \rho_0 v^2$ and $k = \omega/v$, the expression for I becomes

$$I = \frac{\overline{P}}{S} = \tfrac{1}{2}A^2 v\rho_0\omega^2 \qquad (17\text{–}11)$$

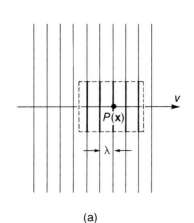

(a)

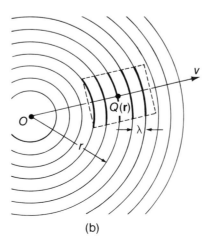

(b)

Fig. 17–6. Three-dimensional waves. (a) A plane wave, in which the wavefronts are parallel planes. (b) A spherical wave, in which the wavefronts are concentric spheres. In each case the wavefronts are separated by the wavelength λ of the wave. In a small volume surrounding point $Q(\mathbf{r})$ in (b), the wavefronts are nearly parallel, as in the vicinity of point $P(\mathbf{x})$ in (a). As the distance from O to Q increases, the wave more closely resembles a plane wave.

The total acoustic power per period that is transported by the wave through a spherical surface with radius R and area $S = 4\pi R^2$ centered on the source at the origin is

$$\overline{P} = SI = 4\pi R^2 I$$

If there are no losses of energy due to absorption, the total average power $\overline{P}$ must be independent of the arbitrarily selected radius R. This condition requires that the amplitude A be inversely proportional to r. Then, introducing a new constant B to measure a radius independent amplitude with $A = B/r$, we can express the wave function as

$$y(r, t) = \frac{B}{r} \cos (kr - \omega t + \phi_0) \qquad \text{spherical wave} \qquad \textbf{(17-12a)}$$

where Eq. 17-11 has been used to write

$$B = \sqrt{P/2\pi v \rho_0 \omega^2} \qquad \textbf{(17-12b)}$$

The purely radial oscillation of a sphere is called the *breathing mode* of the sphere's vibrations and it sets up a spherical outgoing wave. Other distinctive vibration modes and associated waves are possible; for example, one hemisphere could have a decreasing radius while the other hemisphere has an increasing radius, and vice versa—this is a *dipole* oscillation and produces dipole radiation.

17-4 INTERFERENCE EFFECTS*

Beats. An interesting case of interference comes about when two waves with slightly different frequencies combine to produce *beats*. Figure 17-7 shows two such waves and their superposition. Because the amplitudes of the waves are equal, completely destructive interference occurs at regular intervals and the total wave amplitude reaches zero.

Let the two displacement functions be

$$y_1 = A \cos (k_1 x - \omega_1 t)$$

$$y_2 = A \cos (k_2 x - \omega_2 t)$$

Fig. 17-7. (a, b) Two waves with the same amplitude but with slightly different frequencies. (c) The waves of (a) and (b) shown on the same axis. (d) The superposition of the two waves. The wavelength of the rapid oscillation is about the same as those of the original waves, but the amplitude is modulated on a longer time scale, as indicated by the dashed line.

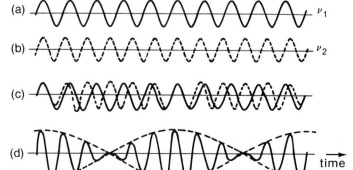

*The general subject of interference is discussed in detail in connection with light phenomena in Chapter 35.

The superposition at $x = 0$ yields

$$y = y_1 + y_2 = A(\cos \omega_1 t + \cos \omega_2 t)$$

because $\cos (-\alpha) = \cos \alpha$. Now we use the identity (see Eq. A–28 in Appendix A), $\cos \alpha + \cos \beta = 2 \cos \frac{1}{2}(\alpha + \beta) \cos \frac{1}{2}(\alpha - \beta)$, to write

$$y = 2A \cos \frac{\omega_1 + \omega_2}{2} t \cos \frac{\omega_1 - \omega_2}{2} t$$

The first cosine term involves the average frequency, $\bar{\omega} = \frac{1}{2}(\omega_1 + \omega_2)$, and the second term involves the difference frequency, $\Delta\omega = \omega_1 - \omega_2$. Thus,

$$y = (2A \cos \tfrac{1}{2} \Delta\omega t)\cos \bar{\omega} t \tag{17-13}$$

Because $\Delta\omega \ll \bar{\omega}$, the term $\cos \bar{\omega} t$ varies rapidly compared with $\cos \frac{1}{2} \Delta\omega t$. We can therefore consider the term in parentheses in Eq. 17–13 to be a slowly varying amplitude, $A'(t) = 2A \cos \frac{1}{2} \Delta\omega t$, that *modulates* the *carrier* term $\cos \bar{\omega} t$.

When the quantity $|A'(t)|$ is large, the sound intensity is large, and when $A'(t) = 0$, the intensity is zero. Each successive recurrence of the maximum of $|A'(t)|$ is called a *beat*. Two such maxima occur during each cycle, so the beat frequency is equal to the frequency difference, $\nu_1 - \nu_2$. For example, if two nearly identical tuning forks have natural frequencies of 440 Hz and 441.8 Hz, the combination of these tones will produce a beat frequency of 1.8 Hz, or one beat every 0.556 s. Beat frequencies as large as about 7 Hz are discernible by the ear. For higher beat frequencies, the ear does not respond to the modulations and no tone variation is heard.

The occurrence of beats is used in the tuning of stringed musical instruments, such as pianos and violins. (• Can you explain how this is done?) Notice also that two strings with fundamental tones that differ by nearly one octave (that is, a frequency ratio of 2 to 1) will produce beats between the string with the higher fundamental and the second harmonic of the other string. This effect can also be used in tuning instruments.

Interference of Two Point Sources. Consider two small identical pulsating spherical sources, S_1 and S_2, that are separated by a distance d. The sources are driven in SHM exactly in phase and with the same amplitude. Harmonic spherical waves (described by Eq. 17–12a) are radiated by each source. These spreading waves overlap and combine, producing a pattern of constructive and destructive interference effects, as indicated in Fig. 17–8. Points in space, such as the one represented by r_1 and r_2 in the diagram, that have $\Delta r = |r_1 - r_2|$ equal to an integer number of wavelengths, constitute a family of

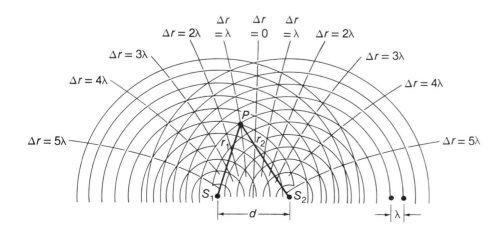

Fig. 17–8. The interference pattern produced by two small spheres pulsating in phase. Each circular arc represents a wavefront crest with a particular algebraic sign and is separated from the next arc by one wavelength λ. Whenever two such wavefronts intersect, there is constructive interference. Such a point is P, where $|r_1 - r_2| = \lambda$.

surfaces on which constructive interference occurs. Thus, the condition for constructive interference is

$$\Delta r = |r_1 - r_2| = n\lambda \qquad n = 0, 1, 2, 3, \ldots$$

In Fig. 17–8 the curves indicate the lines along which these surfaces intersect with a plane that passes through the sources, S_1 and S_2. Interlaced between these curves (actually, surfaces) are those along which destructive interference takes place. The condition for destructive interference is

$$\Delta r = |r_1 - r_2| = (n + \tfrac{1}{2})\lambda \qquad n = 0, 1, 2, 3, \ldots$$

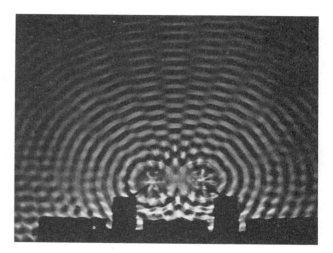

Interference of water waves radiating from two in-phase sources. Note the similarity with Fig. 17–8. (Along the "lines" that diverge from the sources, the interference is destructive.) (From *PSSC Physics*, 2nd ed., 1965: D. C. Heath & Co., and Educational Development Center, Newton, Mass.)

At large distances from the sources, the interference curves asymptotically approach straight lines that radiate from a point midway between S_1 and S_2. (• What is the nature of the surface that corresponds to one of these straight interference lines?) The geometry of this situation is shown in Fig. 17–9 where the origin O is the midpoint between S_1 and S_2.

We construct a line QQ' that is perpendicular to $\mathbf{r}$ and approximately perpendicular to r_1 and r_2. Thus, $\Delta r = |r_1 - r_2| \cong d \sin \theta$. If the sources pulsate in phase with equal amplitudes, the interference conditions at point $P(r, \theta)$ are

$$\left. \begin{array}{l} \text{Constructive: } d \sin \theta = n\lambda \\[4pt] \text{Destructive: } \; d \sin \theta = (n + \tfrac{1}{2})\lambda \end{array} \right\} \quad n = 0, 1, 2, 3, \ldots \qquad \textbf{(17–14)}$$

Fig. 17–9. Geometry for calculating interference effects at large distances from the sources: $r_1 \gg d$, $r_2 \gg d$. In this limit, $|r_1 - r_2| = d \sin \theta$.

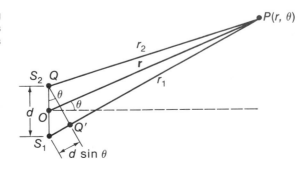

Example 17-4

An infinitely long flat rigid ribbon piston of width w is mounted in an infinite wall and is driven harmonically. Determine the smallest angle θ for complete destructive interference, resulting in zero radiation intensity for that direction.

Solution:

Consider two narrow strips each of width Δx, one—labeled a—at the top edge of the ribbon piston, and the other—labeled a'—just below the midpoint O. The radiation intensity at $P(r, \theta)$ due to these two sources will be zero if $\frac{1}{2}w \sin \theta = \frac{1}{2}\lambda$. This is the minimum angle for complete destructive interference. (• Can you see why?) Assuming that $r \gg w$, a second pair of strips—b and b'—just below the first pair will obey the same condition. The same will be true for successive pairs of strips, extending completely over the ribbon. (We have, in effect, performed an integration over the width of the ribbon.) The result is that there will be zero intensity at the angle θ if $w \sin \theta = \lambda$.

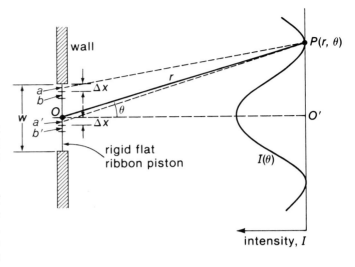

17-5 THE DOPPLER EFFECT

It is a common experience that when an ambulance or other vehicle with a siren passes you on the street, you hear a dramatic decrease in the frequency of the siren at the moment the vehicle passes—the familiar *whee-oo* sound. This frequency change is a result of the *Doppler effect*, named for the Austrian physicist Christian Johann Doppler (1803–1853), who first made extensive studies of this phenomenon and in 1842 developed the theory we now discuss.

Suppose that a sound source at rest emits waves with a frequency ν_0 that travel through still air with a velocity v to a stationary observer some distance away, as indicated in Fig. 17–10a. The observer hears a tone with the frequency $\nu_0 = v/\lambda_0$. Now, suppose that the source moves toward the observer with a velocity v_s, as in Fig. 17–10b. The distance moved by the source during one period $T_0 = 1/\nu_0$ is $v_s T = v_s/\nu_0$. Therefore, when two successive wave crests are emitted, the distance between the crests will be less

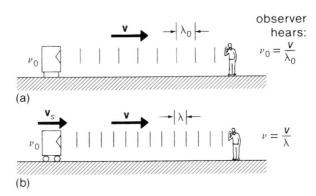

(a)

(b)

Fig. 17–10. (a) The stationary observer hears a tone with the same frequency ν_0 as that emitted by the stationary source. (b) The same observer hears a tone with a higher frequency when the source is approaching.

than λ_0 by the amount of movement v_s/ν_0. Therefore, the wavelength perceived by the observer is

$$\lambda = \frac{v}{\nu} = \lambda_0 - \frac{v_s}{\nu_0} = \frac{v}{\nu_0} - \frac{v_s}{\nu_0}$$

Thus, the tone heard by the observer has the frequency

$$\nu = \nu_0 \frac{v}{v - v_s} \qquad \text{source moving} \qquad \textbf{(17–15)}$$

When $v_s > 0$—that is, the source is approaching the observer—Eq. 26–16 shows that $\nu > \nu_0$, so that the observer hears a tone with a frequency higher than the natural frequency ν_0 he would hear if the source were at rest. On the other hand, if the source recedes from the observer, we have $v_s < 0$ and $\nu < \nu_0$. Consequently, when a source moves past an observer, the frequency of the tone drops suddenly.

Equation 17–15 gives the frequency of the tone heard by an observer when the source moves directly toward or away from the stationary observer. In the event that the *source* is stationary and the *observer* moves directly toward or away with a velocity u relative to the source, the perceived frequency is (see Problem 17–21)

$$\nu = \nu_0 \frac{v + u}{v} \qquad \text{observer moving} \qquad \textbf{(17–16)}$$

where $u > 0$ for the observer approaching the source and $u < 0$ for the receding observer.

▶ *Shock Waves and Sonic Booms.* In the discussion of the Doppler effect we considered only subsonic velocities for the source and the observer—$v_s < v$ and $u < v$. However, some objects *can* move through the air with supersonic velocities—for example, jet aircraft, rifle bullets, and rockets.

Consider an object S that moves through air with a constant velocity v_s that exceeds v. The source object runs ahead of the outgoing waves, and the pileup of waves produces a cone-shaped *shock front*, as illustrated in Fig. 17–11. The faster the object moves, the greater is the degree to which the front is swept back

Fig. 17–11. When an object moves with a velocity $v_s > v$, a swept-back cone-shaped shock front is generated. An object at P at time t moves to S at time $t + \Delta t$; during this interval the disturbance at P has propagated outward as the spherical wavefront A.

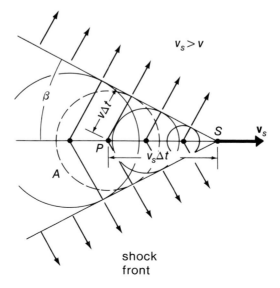

and the smaller is the half-angle β of the cone. This angle β is given by

$$\sin \beta = \frac{v}{v_s} \qquad (17-17)$$

An observer on the ground who is under the flight path of a supersonic jet aircraft will experience a sharp pressure increase when the shock front arrives. This is followed by a decrease in the pressure to a value below the ambient pressure. Finally, when the shock front associated with the rear of the aircraft passes, there is another sharp increase in pressure back to the ambient value. Typically, the time difference between the arrival of the two shock waves is $\frac{1}{50}$ s or so. Consequently, the observer hears a single loud *crack* or *sonic boom*. The pressure difference at the shock front is sufficiently large that nearby buildings may experience some minor structural damage (usually limited to broken windows). For this reason supersonic aircraft (SSTs and military jets) are permitted supersonic flight only over uninhabited regions.

The ratio of the speed of an object to the speed of sound for the same conditions is called the *Mach number*, $N_M = v_s/v$. Thus, a moving object produces a shock front whenever $N_M \geqslant 1$. ◀

Example 17-5

An automobile horn has a frequency $v_0 = 400$ Hz. The horn is sounded continuously as the automobile is driven in a straight line with a velocity of 90 km/h (25 m/s) past a stationary observer; the minimum distance between the automobile and the observer is 100 m. Determine the frequency of the tone heard by the observer as a function of the position of the automobile along its path.

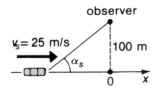

Solution:

The observer is stationary, but the source does not move directly toward him. Consequently, in Eq. 17-15, v_s must be replaced by $v_s \cos \alpha_s$, the component of v_s that is directed toward the observer. Also, the distance x of the source from the point O of closest approach is $x = (-100 \text{ m}) \operatorname{ctn} \alpha_s$. For exam-

ple, take $v = 343.4$ m/s (see Example 17-1e) and choose $\alpha_s = 60°$:

$$v = v_0 \frac{v}{v - v_s \cos \alpha_s}$$

$$= (400 \text{ Hz}) \frac{343.4 \text{ m/s}}{(343.4 \text{ m/s}) - (25 \text{ m/s}) \cos 60°} = 415.1 \text{ Hz}$$

and $\qquad x = (-100 \text{ m}) \operatorname{ctn} 60° = -57.7 \text{ m}$

The diagram below shows the variation in v calculated in this way.

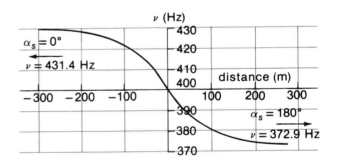

Example 17-6

An observer rides with a moving sound source directly toward a large vertical wall. The vehicle has a constant speed of 30 km/h (8.33 m/s), and the source has a frequency of 100 Hz. What beat frequency is heard by the observer as a result of the combination of the direct and reflected sounds?

Solution:

First, consider an observer at the wall. For him, the velocity of the source is $v_s = u$, and his own velocity is zero. Therefore, Eq. 17-15 gives

$$v_w = v_0 \frac{v}{v - u}$$

for the frequency heard by the observer at the wall. The reflected sound now comes from a stationary source and is heard by the moving observer. Then, Eq. 17-16 gives

$$v = v_w \frac{v + u}{v}$$

Substituting for v_w, we find

$$v = v_0 \frac{v + u}{v - u}$$

The beat frequency is

$$\nu - \nu_0 = \nu_0 \left(\frac{v + u}{v - u} - 1 \right) = \nu_0 \frac{2u}{v - u} = (100 \text{ Hz}) \frac{2(8.33 \text{ m/s})}{(343.4 \text{ m/s} - 8.33 \text{ m/s})} = 4.97 \text{ Hz}$$

17-6 SOUND INTENSITY

In order to describe the enormous range of sound intensities that occur in Nature, it is customary to introduce a logarithmic scale. Two sounds that differ by a factor of 10 in intensity (average acoustic power per unit area) are said to differ by *1 bel* (1 B).* A more commonly used unit is the *decibel* (dB), equal to 0.1 B. Given a sound with an intensity I (W/m^2) and a reference intensity I_0, the *sound intensity level* (in dB) is expressed as†

$$\beta = 10 \log (I/I_0) \qquad \text{in dB} \tag{17-18}$$

In acoustic work the reference intensity is usually taken to be

$$I_0 = 10^{-12} \text{ W/m}^2$$

which corresponds to the threshold of hearing in most individuals (see Table 17–2). At the other extreme, a rock band can produce a sound intensity level of about 115 dB, just below (?) the threshold of pain.

TABLE 17-2. Sound Intensity Levels

SOURCE	β (dB)	I (W/m^2)
(Physical damage)	140	100
(Painful)	120	1
Very loud thunder	110	0.1
Subway train	100	10^{-2}
Noisy factory	90	10^{-3}
Heavy street traffic	80	10^{-4}
Average factory	70	10^{-5}
Conversation	60	10^{-6}
Quiet home	40	10^{-8}
Quiet whisper	20	10^{-10}
Rustling leaves	10	10^{-11}
Threshold of hearing	0	10^{-12}

▶ According to Eq. 17–8, the intensity of a wave can be expressed as $I = \frac{1}{2}A^2 k \beta_a \omega$. By using the relations $\delta p_m = A k \beta_a$, $v^2 = \beta_a/\rho_0$, and $\omega = kv$, we find that

$$I = \frac{1}{2} \delta p_m^2 / v \rho_0 \tag{17-19}$$

That is, the intensity of a sound wave expressed in terms of the acoustic pressure amplitude δp_m is *independent* of the frequency (or wavelength). Thus, we can compare the intensities of sound waves by examining the pressure ratio, even if the frequencies are different.

*This unit is named in honor of Alexander Graham Bell (1847–1922), Scottish-American inventor of the telephone.

†Notice that log (I_1/I_0) gives the sound intensity level in *bels*. The logarithm here is to base 10.

We define a reference-level root-mean-square (rms) acoustic pressure amplitude to be*

$$\delta p_{rms} = \frac{1}{\sqrt{2}}\,\delta p_{m0} = \sqrt{I_0 v \rho_0}$$

$$= \sqrt{(10^{-12}\ \text{W/m}^2)(331.5\ \text{m/s})(1.293\ \text{kg/m}^3)}$$

$$\delta p_{rms} = 2.07 \times 10^{-5}\ \text{Pa}$$

This standard (often taken to be simply 2×10^{-5} Pa) corresponds to the intensity level 0 dBA (0 dB *absolute*). Then,

$$\beta = 20 \log \frac{\delta p_m}{\delta p_{rms}} \qquad \text{(in dBA)} \qquad \textbf{(17–20)} \blacktriangleleft$$

QUESTIONS

17–1 Define and explain briefly the meaning of the terms (a) compression and rarefaction, (b) sound intensity, (c) acoustic pressure, and (d) standing wave (in sound).

17–2 When a fast jet airplane flies overhead, the sound you hear appears to come from a considerable distance behind the visual position of the plane. Explain this effect, which is called the *aberration of sound*.

17–3 Although sound waves may travel long distances in air, they eventually die out. Give as many examples as you can of the processes by which the sound energy is absorbed as the sound wave travels.

17–4 Discuss the relationship of the sizes of the violin, viola, cello, and bass viol to their characteristic tones.

17–5 Tuning a violin (or other stringed instrument) consists of changing the string tension. Although this leaves the wavelength of the fundamental tone the same, the pitch of the sound is altered. Explain.

17–6 The auditory canal of the adult human ear is about 2.7 cm long. This canal is open to air at the auricle or external ear end and closed at the other end by the eardrum. Explain why the sensitivity of the ear is greatest at a frequency near 3200 Hz.

17–7 Why would a small room with reflecting walls be unsuitable for listening to music that is particularly rich in bass frequencies?

17–8 Define and explain briefly the meaning of the terms (a) beats, (b) interference, (c) Doppler effect, (d) sonic boom, and (e) decibel.

17–9 Two students, one leaning out the fourth-floor dormitory window and the other in the paved courtyard directly below, are carrying on a conversation. Why is it easier for the student leaning out the window to hear the student below than the other way around?

17–10 Six speakers are mounted in a baffle in a linear array (see drawing below). If they are driven in phase and are to be used to reinforce sound from the stage in a wide auditorium, how should the baffle be oriented? [*Hint:* Consult Example 17–4.]

17–11 In Example 17–6 the reflecting wall was assumed to act as a new stationary source emitting sound having the same frequency as that of the incident sound. Explain.

17–12 A vibrating tuning fork is tied to the end of a string and spun in a circular overhead arc by a student holding the other end of the string. Describe what a second student standing nearby hears. Does this correspond to what the student twirling the tuning fork hears?

17–13 A source, when stationary, emits sound at a frequency ν_0. Would the frequency of the sound you hear be higher, lower, or equal to ν_0 if (a) both you and the source were moving in the same direction at the same speed? (b) Both you and the source were stationary but a steady wind was blowing in a direction from the source toward you? Blowing in exactly the other direction?

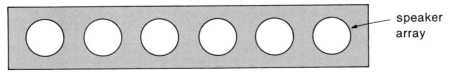

speaker array

Question 17–10

---

* The root-mean-square value of a variable is a type of average, defined as the square root of the mean of the square of the variable's values. For a sinusoidal function $y = A \sin x$, the mean value of y^2 over one complete cycle (or any integral number of cycles) is

$$\frac{A^2}{2\pi} \int_0^{2\pi} \sin^2 x\ dx = \frac{A^2}{2}$$

so the rms value is $A/\sqrt{2}$. Root-mean-square values can be obtained for nonsinusoidal functions as well; the multiplying factor will vary for different functions.

PROBLEMS

Section 17–1

327

17–1 What is the ratio of the speed of sound in air on a hot summer day $(T = 38°\ C)$ to that on a very cold winter day $(T = -28°\ C)$?

17–2 Calculate the speed of sound in helium at $0°\ C$ and normal atmospheric pressure. The density of helium for these conditions is $0.1786\ \text{kg/m}^3$; the value of γ for helium (a monatomic gas) is 1.67. Compare your result with that in Table 17–1.

17–3 A long thin copper rod is given a sharp compressional blow at one end. The sound of the blow, traveling through air at $0°\ C$, reaches the opposite end of the rod 6.4 ms later than the sound transmitted through the rod. What is the length of the rod? (Refer to Table 16–1.)

17–4 A sound wave in air at $0°\ C$ has an intensity of $10^{-6}\ \text{W/cm}^2$ (a *loud* sound). (a) What is the amplitude of the acoustic pressure? (b) If the wavelength of the wave is 0.80 m, what is the displacement amplitude?

17–5 An empty 2-liter jug is used as a Helmholtz resonator. The neck of the jug has a diameter of 5.0 cm and a length of 4.0 cm. (a) What is the resonant frequency ν_0? (Use the expression that includes the length correction.) (b) What amount of liquid must be left in the jug if the resonant frequency is to be middle C $(\nu_C = 261.6\ \text{Hz})$?

Section 17–2

17–6 The longest pipe on an organ that has pedal stops is often 16 ft (4.88 m). What is the fundamental frequency (at $0°\ C$) if the nondriven end of the pipe is (a) closed? (b) If the nondriven end is open? (c) What will be the frequencies at $20°\ C$?

17–7 The fundamental of an open organ pipe corresponds to middle C (261.6 Hz). The third harmonic of a closed organ pipe has the same frequency. What are the lengths of the two pipes?

17–8 A tuning fork whose natural frequency is 440 Hz is placed immediately above a tube that contains water. The water is slowly drained from the tube while the vibrations of the tuning fork excite the air in the tube above the water level. It is found that the sound is enhanced when the air column has a length of 58.0 cm and next when it has a length of 96.6 cm. Use this information to obtain a value for the speed of sound in air. What is the temperature of the air?

17–9 A clever engineering student sets out to design an organ pipe whose pitch (frequency) will be independent of temperature. Having read Section 18–3, he knows that most materials expand approximately linearly with increasing temperature so that the length L at a temperature T (in degrees C) is related to the length L_0 at $0°\ C$ according to $L = L_0(1 + \alpha T)$, where α is a constant. The student therefore decides to use a material for the pipe that will increase in length with temperature to compensate for the increase in the speed of sound with temperature. Determine the value of α that will permit this scheme to work. Most materials have values of α in the range from about 10^{-4} to about 10^{-6} per degree C. Is the scheme really practical?

17–10 A standing wave in a pipe with a length of 0.80 m is described by

$$y(x,\ t) = A \sin \frac{2\pi}{L} x \sin \left(\frac{2\pi v}{L} t + \frac{\pi}{4} \right)$$

(a) Locate the positions of all the nodes N and antinodes $\bar{N}$ in y, v_g, δp, and $\delta \rho$. (b) How is the pipe terminated? (c) What is the wavelength of the wave? What is the frequency? Is this the fundamental mode of vibration? If not, what is the fundamental frequency? (d) At $t = 0$ the acoustic pressure at $x = \frac{1}{2}L$ is 0.2 percent of atmospheric pressure. What is the value of the displacement amplitude A?

Section 17–3

17–11 What is the average acoustic power radiated into air at $20°\ C$ by a sphere that pulsates with SHM between radii of 14.8 cm and 15.2 cm with a frequency of 1200 Hz? (The density of air at $20°\ C$ is $1.20\ \text{kg/m}^3$.)

17–12 What must be the displacement amplitude (i.e., the radial excursion) of a sphere with a radius of 18 cm if the radiated acoustic power is to be twice that of a sphere with a radius of 10 cm and a displacement amplitude of 1.5 mm, both spheres having the same pulsation frequency?

Section 17–4

17–13 A tuning fork with unknown frequency beats against a standard tuning fork with frequency 440 Hz, producing a beat pulsation every 200 ms. What are the possible frequencies for the first tuning fork? A small amount of wax is placed on a prong of the first tuning fork, and the beat period increases. How do you choose between the possible values?

17–14 Two identical piano strings are intended to have middle C (261.6 Hz) as the fundamental. When played together the strings produce a beat frequency of 4.1 Hz. By what fractional amount must the tension in one string be changed so that the strings will vibrate with the same frequency?

17–15 Two small spheres that pulsate harmonically are separated by a distance of 0.5 m and are driven in phase with a frequency of 3000 Hz. (a) Determine the locus of points on a sphere (5.0 m radius with center midway between the vibrating spheres) that correspond to the first two interference maxima on either side of the equator. (The equator corresponds to $\theta = 0°$ in Fig. 17–9.) (b) Do the same for the first two minima. (c) If each source radiates an average power of 0.5 W, what is the sound intensity (W/m^2) on the equator of the sphere? [Hint: Should you add amplitudes or intensities?]

17–16 Two probes that touch the surface of a pool of water
• vibrate in phase and set up expanding circular wavefronts. The waves produce an interference pattern, as shown in Fig. 17–8. (a) If the two probes are separated by a distance $d = 4\lambda$, write an equation for the branch of the interference maxima that has $\Delta r = \lambda$. Identify the type of curve. (b) Does the asymptotic limit of this curve for large distances agree with Eq. 17–14?

17–17 A loudspeaker has a rectangular vibrating surface, 20 cm × 150 cm, and is mounted in the center of a stage facing an audience seated on a single horizontal floor. (Refer to Example 17–4.) (a) For the most efficient radiation of sound to the audience, should the long dimension of the loudspeaker be oriented vertically or horizontally? (b) For a tone with a frequency of 1000 Hz, determine the angle of the first intensity minimum in a vertical plane. (Assume the orientation found in (a).)

Section 17–5

17–18 A trumpet player traveling in an open convertible plays concert A (440 Hz) on his horn. A stationary observer

directly in the path of the automobile hears the note as $A^{\#}$ (466.16 Hz). What is the speed of the automobile?*

17–19 A sound source with $\nu_0 = 200$ Hz moves relative to still air with a speed of 15 m/s. Determine the wavelengths that would be measured by a stationary observer in the directions $\alpha_s = 0°$, 45°, 90°, 135°, and 180°. What are the corresponding frequencies? (Take $v = 331.5$ m/s.)

17–20 The engine noise from a jet aircraft seen directly overhead appears to originate at a point 15° behind the aircraft. What is the speed of the aircraft?

17–21 Derive Eq. 17–16. Show also that when both the source and the observer are moving (with respect to the still air) along the line connecting them, the perceived frequency is

$$\nu = \nu_0 \frac{v + u}{v - v_s}$$

17–22 A bullet fired from a rifle travels at Mach 1.38 (that is, $N_M = 1.38$). What angle does the shock front make with the path of the bullet?

17–23 By altering the ingredients in the propellant charge used to fire a 12-g bullet, the shock front angle is decreased from 26° to 21°. By how much was the kinetic energy of the bullet increased?

Section 17–6

17–24 An explosive charge is detonated at a height of several kilometers in the atmosphere. At a distance of 400 m from the explosion, the acoustic pressure reaches a maximum of 10 W/m^2. Assuming that the atmosphere is homogeneous over the distances considered, what will be the sound intensity level (in dB) at 4 km from the explosion? (Sound waves in air are absorbed at a rate of approximately 7 dB/km.)

17–25 Determine the maximum excursion of air molecules (at 0° C) in a 1-kHz sound wave that has an intensity (a) of 0 dB and (b) of 120 dB.

Additional Problems

17–26 By what percentage does the speed of sound change when a cold front passes and the air temperature drops from 34° C to 11° C?

17–27 Suppose that the volume V of the resonator in Example
• 17–2 is a continuation of the cylinder with radius r and has an additional length ℓ'. Consider the case in which

$\ell = \ell' = \frac{1}{2}L$, where L is the total length of the cylinder. Ignore the length correction, $\Delta\ell = 8r/3\pi$, and determine the oscillation frequency ν_0. Compare this result

---

*When Doppler first tested his theory (1842), he used musicians playing on a moving railway flatcar, with trained observers on the ground identifying the notes played.

with the fundamental frequency for a pipe of length L with one end open. Could the small discrepancy be due to the fact that λ_0 is not sufficiently large compared with L?

17–28 At low frequencies, a loudspeaker that is mounted on a
• very large wall (an *infinite baffle*) acts as a simple mechanical oscillator with a natural frequency ν_0. Consider a speaker with a diameter of 20 cm, an effective moving mass (speaker piston plus the voice coil) of 10 g, and a spring constant (provided by the piston mounting) of 2000 N/m. (a) What is the natural frequency ν_0, neglecting any damping resistance? (b) The speaker is next converted into an *acoustic-suspension* speaker by mounting it on one wall of a completely enclosed cabinet that has dimensions 0.5 m × 0.5 m × 0.2 m. What is the spring constant of the air enclosed in the cabinet for small displacements of the speaker piston? (c) The total effective spring constant for the acoustic-suspension speaker is the sum of the two spring constants. (• Can you explain why it is appropriate to add the spring constants?) What is the natural frequency of the system?

17–29 A large-diameter loudspeaker (called a *woofer*) is to be
• mounted in a *bass reflex* enclosure in order to give optimum damping for the woofer's free-air resonance at $\nu_0 = 45$ Hz. When the enclosure is a Helmholtz resonator with the same natural frequency ν_0, the energy absorbed from the oscillating woofer by the air in enclosure is maximum, which results in the required optimal damping. Find the dimensions of the enclosure cabinet if height:width:depth = 5:3:2 and if the port opening is a circular hole with a diameter of 16 cm. (Neglect the conical volume of the woofer.)

17–30 An organ pipe sounds concert A (440 Hz). If the pipe were filled with carbon dioxide, what would be the new frequency of the tone?

17–31 An organ pipe is tuned to middle C (261.63 Hz) at 0° C. At what temperature would the pitch rise to $C^{\#}$ (277.18 Hz)? (Neglect any change in length of the pipe.)

17–32 The two pulsating spheres shown in Fig. 17–9 are driven with a phase difference of π (exactly out of phase). Determine the conditions for interference maxima and minima at the point $P(r, \theta)$.

17–33 A stationary tuning fork that produces a pure tone with $\nu_0 = 250$ Hz is approached directly by an observer traveling with a speed of 10 m/s. What is the frequency of the tone heard by the observer?

17–34 A supersonic aircraft is flying parallel to the ground. When the aircraft is directly overhead, an observer sees a rocket fired from the aircraft. Ten seconds later the observer hears the sonic boom, followed 2.8 s later by the sound of the rocket engine. What is the Mach number of the aircraft?

17–35 A small sphere pulsates with SHM and radiates an average acoustic power of 10 W. (a) What is the intensity in dB (relative to 10^{-12} W/m²) at a distance of 20 m from the sphere? (b) What is the acoustic pressure amplitude at 20 m? (c) At what distance would the intensiy be 3 dB lower than at 20 m?

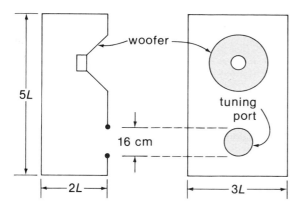

CHAPTER 18

TEMPERATURE AND HEAT

This chapter begins a series of five chapters in which we discuss a number of important topics concerning the form of energy* called *heat*. This general subject, referred to as *thermodynamics,* deals with the thermal properties and the related behavior of macroscopic (bulk) matter as well as the underlying microscopic (molecular) basis for this behavior. In this introductory chapter we give a quantitative description of temperature and heat for macroscopic systems. A later chapter is devoted to the molecular view contained in the *kinetic theory of matter.*

The discussion of macroscopic thermodynamics begins with the discovery that the three fundamental dimensions—length, time, and mass—are not sufficient for our purposes. It is necessary to introduce a new and important variable called *temperature.* Qualitatively, the temperature of an object indicates how ''hot'' or ''cold'' the object is.† Our first task, therefore, is to give a reliable, precise, and reproducible definition of temperature, one that permits the numerical specification of the degree of ''hotness'' or ''coldness.'' In short, we must devise a useful *thermometer* and define a *temperature scale* by which it can be calibrated.

18–1 TEMPERATURE AND THERMAL EQUILIBRIUM

When a substance is heated and becomes ''hotter''—that is, when the temperature of the substance is increased—several of its physical properties undergo measurable change. These property changes include the expansion of solids, liquids, and gases maintained at constant pressure, the increase in pressure of gases at constant volume, the increase in electric resistance of wires, the generation of thermoelectric currents in thermocouple circuits, and the change in color of glowing objects. These *thermometric properties* have all been used in the construction of thermometers. One selects as the *thermometric substance* some material that conveniently exhibits one of these properties. The addition of a scale or other numerical readout converts the material into a working thermometer.

We could, for example, use the thermal volume expansion of alcohol (suitably colored with dye to make it easily visible) for the operation of a thermometer. A small amount of the alcohol is sealed in a glass tube that has a narrow bore and a reservoir bulb at one end (Fig. 18–1). The level at which the alcohol stands in the tube is a measure of the total volume of alcohol. When the temperature changes, so does the level of the alcohol. If the height of the alcohol is calibrated above some reference zero in terms of the

---

*Review Chapter 7 for the important ideas relating to energy and energy conservation.

†In Chapter 20 we give a *microscopic* definition of temperature and relate it to the energy of the molecules in the sample.

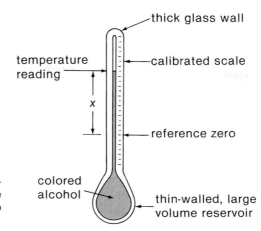

Fig. 18–1. A common, colored-alcohol thermometer. The alcohol in the capillary tube stands at a height x above some reference zero and provides a measure of the temperature.

response to known temperatures, the device becomes a thermometer. In Section 18–2 we explore how this calibration is accomplished. A thermometer of similar design commonly found in student laboratories uses mercury as the thermometric substance.

Thermal Equilibrium.   It is useful in thermodynamics to introduce the notion of a *system*. A system may be as simple as a small lump of matter or as complex as a locomotive engine. A system is in *thermal equilibrium* with its surroundings when it neither gains heat energy from nor loses heat energy to those surroundings. Equilibrium is a thermally quiescent state characterized by no discernible change with time of any thermometric property of the system. Two systems that are in thermal contact* are in thermal equilibrium with each other when neither system gains heat energy from the other.

We have the freedom to specify the boundaries of a system in any convenient way. For example, we could define as a system a certain volume of ice water that consists of solid ice and liquid water in intimate contact and in a state of thermal equilibrium. The important point here is not that the thermometric properties of the ice and the water are the same (they are not), but that neither set of properties changes with time. On the other hand, we could equally well consider the ice as one system and the water as a second system, with the two in thermal contact and in equilibrium. From this we can infer that for any system (simple or complex) to be in a state of equilibrium requires that there be no heat transfer between any of the various parts of the system. (• Can you see why the two views of the ice-water system justify this inference?)

There is a further generalization: Two systems that are in thermal contact and in equilibrium are considered to be still in equilibrium even if they are later separated and thermally isolated from each other (and from their surroundings).

Suppose that system C is in a state of equilibrium, as determined by monitoring its thermometric properties. Another system A, also in equilibrium and monitored, is brought into thermal contact with system C. If none of the thermometric properties of either system change as a result of the contact, we know that the two systems are in thermal equilibrium with each other. A similar test is subsequently performed with system C and another system B, indicating that C and B are also in equilibrium with each other. What can be said about the relative thermal states of systems A and B? Experiments have demonstrated that the thermal equilibrium relationship is a *transitive property*, so that A is also in equilibrium with B. That is, for the case of thermal equilibrium, $C \rightleftarrows A$ and

---

Thermal contact means that there is the possibility of heat flow, by some means, between the systems due simply to their proximity to each other. The systems need not be in physical contact, although this is the usual situation.

$C \rightleftarrows B$ implies $A \rightleftarrows B$. This conclusion is referred to as the *zeroth law of thermodynamics**:

If two systems are each in thermal equilibrium with a third system, the two systems are also in thermal equilibrium with each other.

Suppose that the system C we have been discussing is a thermometer such as the one shown in Fig. 18–1. The height of the alcohol would be the same whether the thermometer were placed in thermal contact with system A or with system B. Thus, all three systems—A, B, and C (the thermometer)—would be at the same temperature. That is, the zeroth law also implies that systems in equilibrium with each other are at the same temperature. (If this were not the case, temperature would not be a useful concept!)

18–2   THE MEASUREMENT OF TEMPERATURE

We now return to the discussion of how a thermometer can be calibrated. Imagine that we use an alcohol thermometer such as that illustrated in Fig. 18–1 (or a similar thermometer filled with mercury). We select two easily reproducible and convenient physical systems and *arbitrarily* associate with each a numerical value for the temperature. Two such choices are the *ice point* and the *steam point* for water. At the ice point, ice and air-saturated pure water are in equilibrium at a pressure of 1.00 atm. The temperature of the ice point is assigned the value 0.00 degrees on the Celsius scale† (0.00° C). At the steam point (assigned a temperature of 100.00° C), steam and boiling water are in equilibrium at a pressure of 1.00 atm.

To be calibrated, the thermometer is first inserted into a mixture of ice and water. When the thermometer has reached equilibrium, as indicated by a steady height of the liquid column, this height is marked and labeled 0.00° C. Next, the thermometer is inserted into boiling water, and when the new equilibrium column height is reached, this height is marked and labeled 100.00° C.

The bore of the thermometer tube is assumed to be uniform and to have a volume that is negligible compared with that of the reservoir bulb. We also make the *assumption* that there exists a linear relationship between the liquid column height and the temperature.‡ Then, the distance between the marks that correspond to 0° C and 100° C may be divided into 100 equal intervals, each representing a temperature interval of one degree Celsius (1 deg C). Thus, if the distance between the 0° C and 100° C marks is ℓ, a column height x, referred to the 0° C mark, corresponds to a temperature

$$t_C = 100\frac{x}{\ell} \ °C \tag{18–1}$$

By extending the column-height scale to negative values of x (as suggested in Fig. 18–1), we can measure temperatures below 0° C.

---

*So called because of its importance in establishing the fundamental thermodynamic concept of temperature on a firm foundation. The other numbered laws of thermodynamics are discussed in later chapters.

†Originally devised in 1742 by the Swedish astronomer and inventor Anders Celsius (1701–1744). The name of this temperature scale was changed, by international agreement in 1948, from centigrade to Celsius.

‡We make the assumption of linearity provisionally. When we eventually introduce a theoretically superior definition of temperature—namely, Lord Kelvin's definition based on the Carnot cycle (see Section 22–2)—linearity will be postulated for that Kelvin scale. Then, in principle, a judgment of the linearity of all other scales (for example, our scale based on the length of the alcohol column) can be based on this theoretical standard.

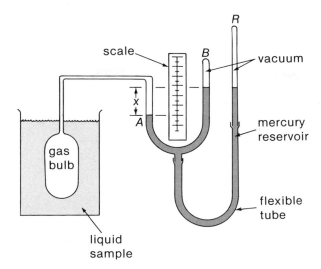

Fig. 18-2. A constant-volume gas thermometer. The difference in the heights of the mercury columns in *A* and *B* is a direct measure of the pressure of the gas.

The Constant-Volume Gas Thermometer. It is found experimentally that when a fixed amount (mass) of gas is confined in a container with constant volume, the pressure of the gas increases as the temperature is raised. This thermometric property of a gas may be used as the basis of an accurate thermometer that possesses, in addition, considerable intrinsic thermodynamic interest.

The essential details of a constant-volume gas thermometer are illustrated schematically in Fig. 18-2. The thermometer bulb, shown immersed in a sample liquid at some temperature, is connected by a small-bore tube to a mercury manometer, which is used to measure the gas pressure. For each measurement, the reservoir column *R* is raised or lowered to place the mercury level in column *A* opposite the reference zero on the scale. In this way, the volume of the gas is maintained at a constant value. The difference in mercury levels in columns *A* and *B* is a direct measure of the gas pressure. Each millimeter of height difference corresponds to a pressure of 1 torr (where 760 torr = 1 atm). Notice that the manometer registers *absolute pressure* and is therefore independent of atmospheric pressure. (• What feature of its construction assures that it measures absolute pressure?)

A precise and readily reproducible temperature has been selected by international agreement to serve as the primary temperature standard. This is the temperature at which water, ice, and water vapor exist together in thermal equilibrium—the so-called *triple point of water*. (A triple-point cell of the type used in practical thermometry is shown in Fig. 18-3.) The three phases of water exist together *only* at a pressure of 4.58 torr. The

Fig. 18-3. A triple-point cell. The cell contains pure water and is sealed after all air has been removed. The temperature of the thermal bath is adjusted until it is visually observed that water, ice, and water vapor exist together in the cell.

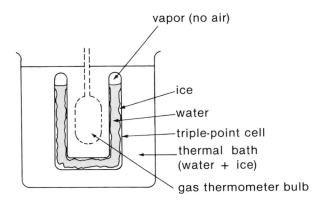

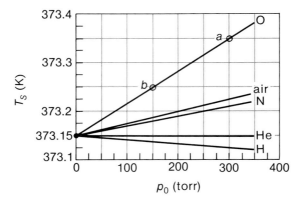

Fig. 18–4. Results of the determination of the temperature T_s of the steam point using different gases in a constant-volume gas thermometer. The pressure p_0 is the pressure of the gas at the triple point. For very dilute gases ($p_0 \to 0$), all results tend toward a value of 373.15 K.

corresponding temperature, which is also unique, is arbitrarily defined to be 273.16 degrees on a new temperature scale called the *Kelvin scale* (or *absolute scale*). The size of the degree on this scale (the kelvin, K) is the same as the Celsius degree. On the Kelvin scale, the temperature of the triple point of water is designated 273.16 K (read: 273.16 *kelvins*; formerly, 273.16 *degrees Kelvin*).

Consider performing the following experiments. The bulb of a constant-volume gas thermometer is filled with oxygen. The bulb is placed inside a triple-point cell, and the pressure p_0 in the bulb determined by the manometer reading is, say, 300 torr. Next, the gas thermometer is used to determine the steam-point temperature of water; this is done by observing the new manometer pressure p and using the assumed linear relationship*

$$T = (273.16 \text{ K}) \cdot \frac{p}{p_0} \qquad (18\text{--}2)$$

The result for the steam point is $T_s = 373.35$ K. This temperature is indicated by point a in Fig. 18–4, where the temperature T_s of the steam point determined by using Eq. 18–2 is plotted as a function of the gas pressure p_0 at the triple point. If the initial pressure p_0 is changed to 150 torr by removing some of the oxygen from the bulb, the steam-point temperature decreases to 373.25 K (point b in Fig. 18–4). The results of many such measurements for different pressures p_0 at the triple point are represented by the straight line for oxygen (O) in Fig. 18–4. The temperature intercept ($p_0 = 0$) is $T_s = 373.15$ K. Also shown in the figure are the results obtained using air, nitrogen (N), helium (He), and hydrogen (H). The important result is that for any gas used in the thermometer, the temperature intercept is the same, namely, 373.15 K. That is, all dilute gases exhibit the same pressure-temperature behavior at constant volume. (This is one of the properties we ascribe to an imaginary substance called an *ideal gas*. We make extensive use of the concept of an ideal gas in later discussions.) A modification of Eq. 18–2 now gives a precise definition for temperature in terms of pressure measurements with a constant-volume gas thermometer with any gas as the thermometric substance:

$$T = (273.16 \text{ K}) \cdot \lim_{p_0 \to 0} \frac{p}{p_0} \qquad (18\text{--}3)$$

Temperature Scales. By using a gas thermometer in the manner just described, the ice-point temperature is determined to be $T_i = 273.15$ K. Thus, the triple point for water

---

*We use T to represent temperature on the Kelvin scale.

is 0.01 K higher than the ice point. We also observe that $T_s - T_i = 100$ K, the same difference as for the Celsius scale. In fact, the coefficient 273.16 K in Eq. 18–3 was chosen to give exactly this result. The relationship between the two scales is

$$T = t_C + 273.15 \tag{18-4}$$

where T is measured in kelvins (K) and t_C is measured in °C.

Another temperature scale still used in some English-speaking countries is the *Fahrenheit scale** in which the temperature t_F is measured in *degrees Fahrenheit* (°F). The connection between the Fahrenheit and Celsius scales is given by

$$t_F = \tfrac{9}{5}t_C + 32.00° \tag{18-5}$$

On the Fahrenheit scale, the ice-point temperature is 32.00° F and the steam-point temperature is 212.00° F.

In engineering practice, the *Rankine temperature scale* is sometimes used. The size of the Rankine degree is the same as that of the Fahrenheit degree, but the zero point of the Rankine scale is the same as that of the Kelvin scale. Then, on the Rankine scale, the triple-point temperature is $\tfrac{9}{5} \cdot 273.16 = 491.69°$ R. In general, we have

$$t_R = \tfrac{9}{5}T = t_F + 459.68° \tag{18-6}$$

Practical Thermometry. Any particular type of thermometer has a limited range of usefulness. For example, there is a practical limit on the lowest temperature that can be measured with a gas thermometer because any gas will liquefy at a sufficiently low temperature. The gas best suited for low-temperature measurements is helium, which has the lowest liquefication temperature of any gas (4.2 K at normal atmospheric pressure).

To make measurements over a wide temperature range, it is necessary to use various thermometric devices, even some that lack a strict linear relationship connecting the thermometric property and the temperature. The use of such devices requires that a number of temperature reference points be established for purposes of calibration. Some of the reference temperatures on the International Practical Temperature Scale (IPTS) are given in Table 18–1.

The electric resistance of a metallic conductor varies with temperature, so this thermometric property can be used to construct a thermometer. Platinum has excellent chemical and physical properties and is widely used in *resistance thermometers* that are used in the temperature range from about −260° C to about 600° C. Very precise and sensitive methods have been developed for measuring resistance values, and it is possible to detect temperature changes as small as 10^{-4} deg C by using this technique.

It is found that a circuit that consists of two wires of different metals and connected with two junctions will develop a voltage within the circuit if the junctions are at different temperatures. This is the operating principle of the *thermocouple*. Usually, the cold junction of the thermocouple is placed in a bath that is maintained at the ice point, whereas the hot junction is placed in contact with the sample whose temperature is to be determined (see Fig. 18–5). Thermocouples constructed from wires of platinum and (90 percent−10

---

*Named for Gabriel Fahrenheit (1686–1736), German-Dutch physicist who devised the first precision thermometer, a mercury-filled instrument.

TABLE 18–1. Some Reference Temperatures on the International Practical Temperature Scale (IPTS)

EQUILIBRIUM STATE	T (K)	t_C (°C)
Triple point of water[a]	273.16	0.01
Triple point of hydrogen	13.81	−259.34
Boiling point of hydrogen[b]	20.28	−252.87
Boiling point of neon[b]	27.402	−246.048
Triple point of oxygen	54.361	−218.789
Boiling point of oxygen[b]	90.188	−182.962
Boiling point of water[b]	373.15	100.00
Freezing point of zinc[b]	692.73	419.58
Freezing point of antimony[b]	903.89	630.74
Freezing point of silver[b]	1235.08	961.93
Freezing point of gold[b]	1337.58	1064.43

[a] Primary standard.
[b] At normal atmospheric pressure.

percent) platinum-rhodium are often used for measurements in the temperature range from about 600° C to about 1000° C.

Very high temperatures—even above the melting points of most metals—can be determined by optical means that do not require contact with the sample. Devices for making such measurements are called *optical* (or *radiation*) *pyrometers*. An electrically heated filament in a pyrometer is viewed through a suitable filter simultaneously with the sample, which might be, for example, the interior of a furnace. The current through the comparison filament is adjusted so that its image vanishes when superimposed on the image of the sample. The "color temperature" of the filament and sample are then the same, and the sample temperature is determined from the current calibration of the filament. Optical pyrometers are commonly used to measure temperatures above about 1000° C; they are useful for temperatures up to about 3600° C (the melting point of tungsten).

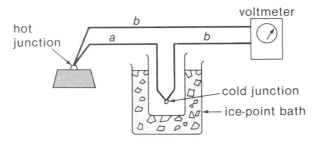

Fig. 18–5. A thermocouple circuit. One junction (the cold junction) of two wires, *a* and *b*, made from different metals is placed in an ice-point bath; the other junction (the hot junction) is in thermal contact with the sample.

Absolute Zero. The form of Eq. 18–3 clearly suggests that there is a lower limit to the temperature scale, namely, 0 K. However, this temperature cannot be reached in practice and can only be judged to be the theoretical limit of the extrapolation of experimental data. This limit is called the *absolute zero* of temperature. The impossibility of achieving absolute zero is known as the *Nernst heat theorem*, which can be stated in the following way:

It is not possible by any procedure, no matter how idealized, to cool a system to an equilibrium state of absolute zero in a finite number of operations.

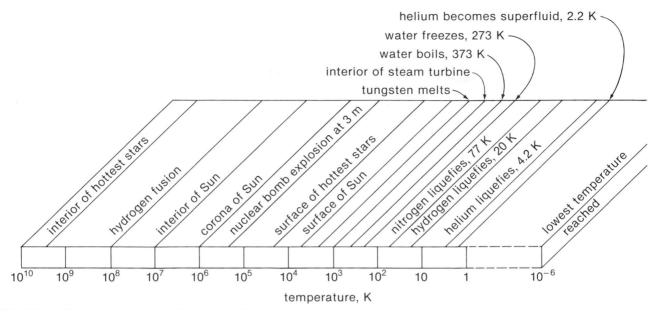

Fig. 18-6. Range of temperatures found in the Universe.

Various experimental techniques can be used to reach temperatures very close to absolute zero. For example, pumping away the vapor above a sample of liquid helium can reduce the temperature of the helium below 1 K. Still lower temperatures can be reached, for very small samples, by cyclic magnetization procedures. By using the ionic magnetization of crystals, temperatures near 10^{-3} K can be reached, and nuclear magnetization methods permit temperatures as low as 10^{-6} K to be achieved!

Example 18-1

In a laboratory experiment, a student uses a simple constant-volume gas thermometer to determine the Celsius temperature of absolute zero. The student finds the gas pressure at the ice point to be 182 torr, and at the steam point she finds $p = 246$ torr. What value does the student calculate for the Celsius temperature of absolute zero?

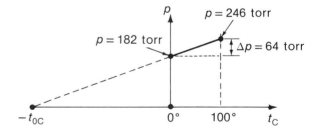

Solution:

The two measurements are shown on the diagram. From the two similar triangles, we can write

$$\frac{|t_{0C}|}{182 \text{ torr}} = \frac{100 \text{ deg}}{(246 - 182) \text{ torr}}$$

or $|t_{0C}| = (100 \text{ deg}) \cdot \dfrac{182}{64} = 284 \text{ deg}$

Hence, the student finds $t_{0C} = -284°$ C, compared with the actual value, $t_{0C} = -273.15°$ C. This difference is typical of the results obtained with the apparatus usually available in an elementary physics laboratory. Notice that in this experiment, the careful extrapolation to zero pressure (Fig. 18-4) was ignored.

18-3 THERMAL EXPANSION

Most substances expand when heated at constant pressure. (A notable exception, as discussed later, is water in the temperature range from $0°$ C to $4°$ C.) We restrict attention here to liquids and solids; the behavior of gases requires special attention (see Chapter 19). We also omit discussion of cases that involve a change of phase of the substance — for example, cases in which melting or boiling occurs.

Consider a thin rod with a length ℓ_0 that is heated, thereby causing its equilibrium temperature to change from T_0 to T. If the length at the temperature T is ℓ, a length change, $\Delta\ell = \ell - \ell_0$, has occurred because of the temperature change, $\Delta T = T - T_0$. Then, the *average coefficient of linear expansion* $\bar{\alpha}$ is defined to be

$$\bar{\alpha} = \frac{1}{\ell_0}\frac{\Delta\ell}{\Delta T} \tag{18-7}$$

(We could equally well express $\bar{\alpha}$ in terms of Δt_C.) Because the linear expansion coefficient varies somewhat with temperature, it is more precise to define α for a particular temperature by writing the *coefficient of linear expansion* as

$$\alpha = \alpha(T) = \frac{1}{\ell}\frac{d\ell}{dT} \tag{18-7a}$$

where it is understood that the derivative is evaluated at the temperature T. For example, the value of α for stainless steel is 0.81×10^{-5} deg^{-1} at $-100°$ C and is 1.43×10^{-5} deg^{-1} at $700°$ C. The value of α for most substances depends only weakly on pressure.

We can also define expansion coefficients for area and for volume. For example, the coefficient of volume expansion is defined to be

$$\beta(T) = \frac{1}{V}\frac{dV}{dT} \tag{18-8}$$

For a particular substance, the volume expansion coefficient is just three times the linear expansion coefficient: $\beta = 3\alpha$. To see this, consider a parallelepiped with sides ℓ_1, ℓ_2, and ℓ_3, so that the volume is $V = \ell_1\ell_2\ell_3$. Then,

$$\beta = \frac{1}{V}\frac{dV}{dT} = \frac{1}{\ell_1\ell_2\ell_3}\left[\ell_2\ell_3\frac{d\ell_1}{dT} + \ell_1\ell_3\frac{d\ell_2}{dT} + \ell_1\ell_2\frac{d\ell_3}{dT}\right]$$

$$= \frac{1}{\ell_1}\frac{d\ell_1}{dT} + \frac{1}{\ell_2}\frac{d\ell_2}{dT} + \frac{1}{\ell_3}\frac{d\ell_3}{dT}$$

$$= \alpha + \alpha + \alpha = 3\alpha \tag{18-9}$$

This is actually a general result and does not depend on the geometric shape of the object (see Problem 18-9).

Some typical values of α for solids and β for liquids are given in Table 18-2.

TABLE 18-2. Thermal Expansion Coefficients for Some Substances at 20° C[a]

SUBSTANCE	α (deg^{-1})	SUBSTANCE	β (deg^{-1})
Aluminum	2.30×10^{-5}	Alcohol (ethyl)	1.12×10^{-3}
Brass, yellow	1.90×10^{-5}	Alcohol (methyl)	1.20×10^{-3}
Concrete	$1.2 \ \times 10^{-5}$	Benzene	1.24×10^{-3}
Copper	1.67×10^{-5}	Carbon tetrachloride	1.24×10^{-3}
Glass, Pyrex	$3.2 \ \times 10^{-6}$	Gasoline	$9.5 \ \times 10^{-4}$
Gold	1.42×10^{-5}	Glycerine	$5.1 \ \times 10^{-4}$
Invar	$1.2 \ \times 10^{-6}$	Mercury	1.82×10^{-4}
Iron, wrought	1.12×10^{-5}	Olive oil	$7.2 \ \times 10^{-4}$
Lead	2.87×10^{-5}	Turpentine	$9.7 \ \times 10^{-4}$
Quartz, fused	$4.2 \ \times 10^{-7}$	Water	2.07×10^{-4}
Silver	1.90×10^{-5}		
Steel, stainless	1.05×10^{-5}		
Tungsten	$4.5 \ \times 10^{-6}$		

[a] Notice that the units of α are deg^{-1}; we do not need to specify whether this means *per degree Celsius* or *per absolute degree* (kelvin) because they are the same.

In this discussion, we have assumed that the object being heated (or cooled) is unconfined and can freely change its dimensions. We also made the implicit assumption that all temperature changes take place slowly. When these conditions are not met, stresses with considerable magnitudes may develop in an object. In the construction of bridges, railroad tracks, concrete walls and roads, and other large structures that are exposed to wide temperature variations, thermal expansion must be allowed for to prevent severe damage from buckling.

In some applications, materials are required that have very small expansion coefficients. For example, chronometer parts, pendulum rods, balance arms of scales, and other precision measuring devices must have dimensions that are stable when the temperature changes. For these applications, fused quartz or special alloys such as Invar (36 percent nickel steel) are often used because they have expansion coefficients that are $\frac{1}{10}$ or less of the coefficients for ordinary materials.

The Expansion of Water. When a quantity of water is cooled from the steam point, its volume decreases, as is the case for normal liquids, until a temperature of 4° C is reached. Further cooling, from 4° C to 0° C, causes the volume to *increase*. It is convenient to show this anomalous behavior of water in terms of the variation with temperature of the *specific volume* or volume per unit mass (the inverse of *density*). Figure 18-7 shows the specific volume (in units of cm^3/g) for water in the temperature range from 0° C to 100° C.

In winter, an open body of water, such as a pond or lake, cools primarily because of exposure to colder air. The cold surface water, with higher density, sinks to the bottom. But when the surface temperature falls below 4° C, the density of the surface water is *less* than that of the underlying layers, so the colder water no longer sinks. The water with temperature nearest to 0° C remains on the surface. Consequently, it is the *surface* of a body of water that freezes first. This permits the water near the bottom to remain liquid, thereby protecting aquatic life. If water behaved as a normal liquid, the coldest water would always be found at the bottom. Then, a pond or a lake would freeze from the bottom up, with an increased likelihood that the entire volume would freeze solid.

Not only does water expand as it cools from 4° C to 0° C, but upon freezing it expands still further. Again, this is contrary to the behavior of most substances. The

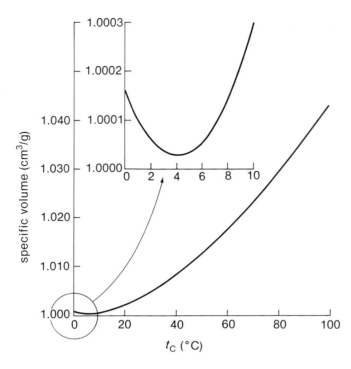

Fig. 18–7. The temperature variation of the *specific volume* (the inverse of *density*) for water. The behavior between 0° C and 4° C is anomalous. Specific volume is commonly expressed in cm³/g although these are not proper SI units. Sometimes the specific volume is expressed in mℓ/g; then, the value at 4° C is 1.0000 because 1 mℓ = 1.000027 cm³.

expansion upon freezing is substantial, amounting to about 8.3 percent. The freezing of water trapped in cracks of rock or roadway materials can often exert sufficient stress to enlarge the existing fissures. In this way, the material suffers progressive deterioration with the freeze-thaw cycle.

Example 18–2

The measurement of the coefficient of volume expansion for liquids is complicated by the fact that the container also changes size with temperature. The diagram shows a simple means for overcoming this difficulty. With this apparatus, $\bar{\beta}$ for a liquid is determined by measurements of the column heights, h_0 and h_t, of the liquid columns in the U-tube. One arm of the tube is maintained at 0° C in a water-ice bath, and the other arm is maintained at the temperature t_C in a thermal bath. The connecting tube is horizontal. Derive the expression for $\bar{\beta}$ in terms of h_0 and h_t.

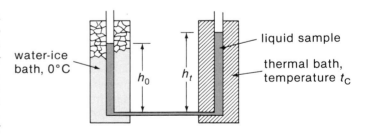

Solution:

The pressure at the free surface of each liquid column is the same (atmospheric pressure). The system is static (no liquid flow), so the pressure at the bottom of each column is also the same. Hence, $\rho_t g h_t = \rho_0 g h_0$, or $\rho_0/\rho_t = h_t/h_0$. Now the volume V_0 of a unit mass of liquid at 0° C is $V_0 = 1/\rho_0$, and at t° C it is

$V_t = 1/\rho_t$. Then, we have for a unit mass,

$$\frac{\rho_0}{\rho_t} = \frac{V_t}{V_0} = 1 + \bar{\beta} t_C = \frac{h_t}{h_0}$$

Solving for $\bar{\beta}$, we find the desired relationship,

$$\bar{\beta} = \frac{h_t - h_0}{h_0 t_C}$$

18–4   HEAT

Not until the nineteenth century did scientists accept the view that *heat* is simply one form of energy. Indeed, this realization is crucial in the formulation of the energy conservation principle (see Section 8–2). In earlier times, however, the prevailing theory viewed heat as a massless and invisible substance called *caloric*. According to this theory, every body contains an amount of caloric that depends on the temperature of the body—a hot object contains more caloric than does a cold object. Also, caloric was assumed to repel itself. Therefore, when a hot object is placed in contact with a cold object, caloric should flow from the hot object to the cold object until a common temperature is reached; experiment shows that this is indeed the behavior of objects at different temperatures that are placed in contact.

The caloric theory could also explain a number of other observations concerning heat and temperature effects. For example, the theory held that when a body is heated, caloric flows in and surrounds the atoms. Because caloric repels itself, as more caloric is added, the atoms are pushed farther apart and the body expands; again this prediction of the theory agrees with experiment. (• How could you use the caloric theory to account for the conversion of water to steam?)

The caloric theory appeared to be a comprehensive and useful description of heat phenomena. But the observations made in 1798 by Count Rumford (Benjamin Thompson, American-British physicist, 1753–1814) led eventually to the downfall of the theory. While overseeing the boring of cannon barrels in a Munich arsenal, Rumford was impressed by the amount of heat that was generated in the process. Every object was supposed to contain a definite amount of caloric, but Rumford observed that heat continued to be generated as long as the boring tool scraped away at the barrel. In fact, if the cannon had actually contained that much caloric at the beginning of the boring process, it should have melted! Rumford concluded that heat could not be a material substance. He believed that the scraping of the boring tool inside the barrel set into vibratory motion the particles comprising the cannon; this vibratory motion was then responsible for the sensation of heat. Thus, the generation of heat was due to the mechanical work done on the boring tool and continued as long as this work was being done.

Let us now consider carefully the concept of heat on the macroscopic level. We must first make a clear distinction between *heat* (energy) and *internal energy*. It was precisely the failure to appreciate this difference that was the basic fault of the caloric theory.

Heat is energy that is transferred from one body or system to another solely as the result ——— *of the temperature difference between them.*

Thus, a "quantity of heat" only has meaning in the sense that an energy flow exists between bodies at different temperatures. The direction of this heat (energy) flow is always from the body at the higher temperature to the body at the lower temperature.

A statement that a body or system *contains* a certain amount of heat is, in a strict sense, meaningless.

The energy content of a system in equilibrium is its internal energy.   ———

Heat energy can be extracted from a system in a manner that reduces its internal energy, and, conversely, the internal energy can be increased by the addition of heat energy. However, the internal energy of a system can be changed by many means in addition to the flow of heat energy into or out of the system. For example, mechanical, acoustic, electromagnetic, gravitational, nuclear, or other forms of energy exchange can be used to increase or decrease the internal energy of a system. A very important feature of the internal energy of a system is that it depends *only on the equilibrium temperature** and *not at all* on the means of energy exchange by which that temperature was achieved.

It is possible for a system continually to supply heat energy to other systems at lower temperatures while the supplying system remains at constant temperature, provided there is energy input to the system by some means, not necessarily thermal in character. This was a feature lacking in the caloric theory, and precisely the point realized by Rumford in his observation of heat generation in the boring of cannon barrels.

Specific Heat and Heat Capacity.   The specific heat c of a substance is defined as the amount of heat per unit mass that must be added to the substance to raise its temperature by one Celsius degree (1 deg C). To give a complete definition of specific heat, it is also necessary to indicate the experimental conditions under which the heating process is carried out. The two most common methods involve heating at constant volume or at constant pressure. For liquids and solids it is usually much easier to make the necessary measurements at constant pressure (usually atmospheric pressure). It is possible to relate the results of measurements made in this way to the results of measurements made at constant volume by a theoretical calculation. Usually, the differences in the two types of measurement are slight, owing to the smallness of the volume-expansion coefficients for liquids and solids. For gases, however, the two methods of measurement yield substantial differences that cannot be ignored. Therefore, we will consider at first only liquids and solids; we will come to the discussion of the specific heat of gases in Chapter 20.

The thermodynamic reference standard for quantity of heat is the heat required to raise the temperature of 1 kg of pure water from 14.5° C to 15.5° C at a pressure of 1 atm; this amount of heat is designated *one Calorie* (1 Cal). (A smaller unit, based on the heat required to raise the temperature of one *gram* of water by one degree, is called the *calorie,* with a lower-case c; evidently, 10^3 cal = 1 Cal.) Thus, the specific heat of water at 15° is equal to 1.0000 Cal/kg·deg.

The specific heat of water actually varies slightly with temperature—but by less than 1 percent between 0° C and 100° C. For example, to raise the temperature of 1 kg of water from 79.5° C to 80.5° C requires 1.00229 Cal. The variation with temperature of the specific heat of water is shown in Fig. 18–8. The fact that the specific heat of water is nearly constant is very useful in the determination of specific heat values for other substances in the temperature range from 0° C to 100° C, as we demonstrate later in this section.

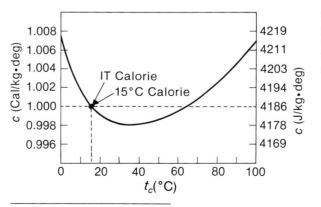

Fig. 18–8. The variation with temperature of the specific heat of water. On the scale of this diagram, the IT Calorie and the 15° C Calorie are indistinguishable.

---

*The dependence of the internal energy exclusively on temperature is exactly true only for an ideal gas but is approximately true for real substances. (See Section 20–2.)

Because heat is a form of energy, quantity of heat should properly be expressed in SI energy units, namely *joules*. On page 362 we describe an experiment for determining the relationship between the Calorie and the joule. The Calorie we have just defined—called the 15° C Calorie—is found to equal 4186 J:

$$1 \ (15° \ C) \ Cal = 4186 \ J$$

There is also defined, by international agreement, a secondary or practical standard called the International Table Calorie (IT Cal):

$$1 \ (IT) \ Cal = \frac{1 \ kWh}{860} = \frac{3.60 \times 10^6 \ J}{860} = 4186 \ J$$

To the number of significant figures given, these two definitions of the Calorie are equivalent. There is yet another definition, the *thermochemical Calorie,* arbitrarily set exactly equal to 4184 J:

$$1 \ (th) \ Cal = 4184 \ J$$

Some engineering calculations require a conversion from *British thermal units* (BTU) to calories:

$$1 \ Cal = 3.968 \ BTU$$

In this book we use SI units for most purposes, and we usually express thermal energies in joules. However, heat measurements have traditionally been made in units of calories. We sometimes use the *kilogram calorie* or Calorie as an alternate unit to the joule, because this unit involves the specification of mass in kilograms, the SI unit of mass. Table 18–3 lists specific heat values for various substances both in J/kg·deg and in Cal/kg·deg. (Again, it does not matter whether "deg" stands for "deg C" or "kelvin.")

TABLE 18–3.   Specific Heat Values for Various Substances (at a pressure of 1 atm)

SUBSTANCE	TEMPERATURE RANGE (°C)	SPECIFIC HEAT c (OR $\bar{c}$)	
		Cal/kg·deg	J/kg·deg
Aluminum	0–100	0.216	904
Brass	20–100	0.0917	384
Carbon (amorphous)	25–80	0.168	703
Copper	15–100	0.0931	390
Ethyl alcohol	20	0.572	2394
Glass (crown)	10–50	0.161	674
Glass (flint)	10–50	0.117	490
Gold	0–100	0.0316	132
Ice	−10	0.530	2219
Iron	0–300	0.122	511
Lead	0–300	0.0326	136
Mercury	20–100	0.0330	138
Methyl alcohol	20	0.600	2512
Silver	20–100	0.0560	234
Tin	0–100	0.0556	233
Wood (typical)	—	0.42	1760

The specific heat values in Table 18–3 are average values $\bar{c}$ for the quoted temperature ranges. That is,

$$\bar{c} = \frac{1}{m} \frac{\Delta Q}{\Delta T} \tag{18–10}$$

where ΔQ is the heat required to raise the temperature of the substance by ΔT. The specific heat at a particular temperature T is defined to be*

$$c = c(T) = \frac{1}{m} \frac{dQ}{dT} \tag{18–10a}$$

The specific heat c is defined for a chemically pure or homogeneous substance. In the event that a system consists of a mixture of substances, we refer to a quantity H, called the *heat capacity* of the system (an unfortunate term!), where

$$H = \sum_i m_i \bar{c}_i \tag{18–11}$$

The quantity of heat Q that must be supplied to such a mixture in order to raise the temperature by an amount ΔT is

$$Q = \left(\sum_i m_i \bar{c}_i \right) \Delta T = H \, \Delta T \tag{18–12}$$

Latent Heat. The familiar transitions of liquid water to steam and of ice to liquid water are called *phase transitions*. These phase changes for water are typical of the (gas) $\rightleftharpoons$ (liquid) $\rightleftharpoons$ (solid) transitions that occur suddenly at well-defined temperatures. These transitions are called *first-order phase transitions* and involve a striking change in the long-range order of the atoms and molecules in the substance. This type of phase transition is always accompanied by a change in internal energy. In all cases of phase transitions, we imagine that heat is added or removed slowly so that the substance is always at a uniform temperature.

There are many forms of phase transitions in addition to the obvious first-order changes. For example, at some particular temperature a substance might undergo an alteration in its crystal structure, or a sudden change in the electric conductivity might occur, transforming an ordinary conductor into a superconductor. A further consideration is the fact that some substances—plastics, glasses, tars, and so forth—undergo (solid) $\rightleftharpoons$ (liquid) transitions not at a well-defined temperature but over a broad range of temperatures. Although these cases are very interesting, we confine the discussions here to simple first-order phase transitions.

---

*Usually, the derivative (and related differentials) refer to changes in various finite (noninfinitesimal) quantities. Thus, we would ordinarily write

$$\Delta Q = Q(T + \Delta T) - Q(T) \qquad \text{and} \qquad \Delta Q = \frac{dQ}{dt} \Delta T$$

and we would interpret $Q(T + \Delta T)$ to be the heat content at a temperature $T + \Delta T$ and $Q(T)$ to be the heat content at a temperature T. However, we have emphasized that it is incorrect to refer to the heat content of a body; instead, we use the concept of internal energy. Although dQ is indeed a differential quantity, it is so only in the sense that it takes the system from one thermodynamic state to another. In some texts this distinction is emphasized by using the symbol $đQ$ (meaning the "inexact differential" of Q). However, we use ΔQ and dQ, with the understanding that this special interpretation holds.

Suppose that we have an isolated sample of ice at some temperature below the freezing point, say, $-25°$ C. The sample is maintained at atmospheric pressure, and heat is supplied at a slow constant rate. The recorded temperature as a function of time is shown in Fig. 18–9. From the initial condition at a, the ice warms up to the melting point ($0°$ C) at b. The temperature remains constant from b to c as the ice melts. At any particular time between b and c there is a definite fraction of the sample in the form of ice—100 percent at b and 0 percent at c. At the intermediate point b', 80 percent of the sample is in the form of ice. If the heat input were arrested at this point, an equilibrium state would persist (because the system is thermally isolated), with the fraction of ice constant at 80 percent. The system would then be in a condition called *phase equilibrium*—that is, the coexistence of two different phases in equilibrium.

From c to d, the temperature of the sample, now all water, increases at about half the rate that occurred during the interval a to b. The temperature continues to increase until the boiling temperature ($100°$ C) is reached at d. From d to e, heat is added to boil the water while the temperature remains constant. Boiling takes place as the system proceeds through an infinite number of states that are in phase equilibrium between the liquid and gaseous (steam) phases. At e all of the water has been converted into steam, and any further heating causes a continual increase in the temperature of the steam.

If the path from a to f is reversed by cooling the system, the temperature will first decrease until the condensation point of steam is reached and liquid water begins to appear. This temperature is the same as the boiling temperature, namely, $100°$ C. To be precise, we should refer not to the boiling or condensation temperature, but to the temperature at which phase equilibrium exists between the gaseous and liquid phases. Further cooling of the system carries it to c, where freezing begins. Again, the freezing and melting temperatures are the same ($0°$ C), and, properly, this temperature should be referred to as the temperature for phase equilibrium between the solid and liquid phases.

The conversion of ice to water involves changes in the long-range order of the water molecules. These changes are associated with a change in the internal energy of the system. The amount of energy that is required to convert 1 kg of ice to water at $0°$ C is called the *latent heat of fusion, L_f*; thus,

$$L_f \text{ (water)} = 79.71 \text{ Cal/kg}$$

$$= 3.337 \times 10^5 \text{ J/kg}$$

This is the energy that must be *supplied* to melt ice completely at $0°$ C, and it is the energy that is *released* when water freezes completely at $0°$ C.

The energy involved in the complete phase change at $100°$ C is called the *latent heat of vaporization, L_v*:

$$L_v \text{ (water)} = 539.12 \text{ Cal/kg}$$

$$= 2.2567 \times 10^6 \text{ J/kg}$$

Fig. 18–9. A temperature-time curve for water heated at a constant rate at atmospheric pressure. The sample is thermally isolated from its surroundings (except for the heat source). The lines *ab, cd,* and *ef* have different slopes, reflecting the different specific heat values for ice, water, and steam.

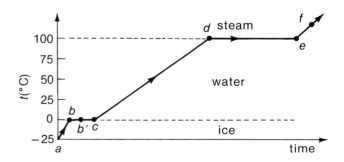

This is the energy that must be *supplied* to convert water completely to steam at 100° C, and it is the energy that is *released* when steam condenses to water at 100° C.

Table 18–4 gives the values of the *latent heats of transition*, L_f and L_v, for several substances at atmospheric pressure.

TABLE 18–4. Latent Heats of Transition (at 1 atm)

| | FUSION | | | VAPORIZATION | | |
| | | L_f | | | L_v | |
SUBSTANCE	t_C (°C)	Cal/kg	J/kg	t_C (°C)	Cal/kg	J/kg
Nitrogen[a]	−210.00	6.13	2.567×10^4	−195.80	47.64	1.994×10^5
Ethyl alcohol	−114.4	24.9	1.042×10^5	78.3	204	8.54×10^5
Mercury	−38.86	2.730	1.143×10^4	356.58	70.62	2.956×10^5
Water	0.00	79.71	3.337×10^5	100.00	539.12	2.2567×10^6
Sulfur[b]	115.18	12.74	5.33×10^4	444.60	68.5	2.869×10^5
Lead	327.3	5.533	2.316×10^4	1750	205.0	8.581×10^5
Zinc	419.58	26.98	1.1295×10^5	911	422.2	1.7675×10^6
Antimony	630.74	38.99	1.632×10^5	1440	134.0	5.61×10^5
Silver	961.93	21.07	8.82×10^4	2163	555.1	2.323×10^6
Gold	1064.43	15.8	6.61×10^4	2808	406.34	1.7009×10^6
Copper	1083.1	49.2	2.06×10^5	2566	1129	4.726×10^6

[a] At a pressure of 93.9 torr for fusion.
[b] For the fusion transition from monoclinic crystal to liquid.

Example 18–3

A block of ice with a mass $m = 10$ kg is moved back and forth over the flat horizontal top surface of a large block of ice. Both blocks are at 0° C, and the force that produces the back-and-forth motion acts only horizontally. The coefficient of kinetic friction (wet ice on ice) is $\mu_k = 0.060$. What is the total distance traveled by the upper block relative to the supporting block if 15.2 g of water is produced?

Solution:

The 15.2 g of water arises from the melting of an equivalent amount of ice (Δm) due to the friction between the blocks. The amount of heat required is $Q = L_f \Delta m$ and the amount of work done is $W = f \Delta s$, where f is the frictional force. Since $\Delta m \ll m$, we may consider $N = mg$ throughout the process; hence, $f = \mu_k N = \mu_k mg$. Setting Q equal to W and solving for Δs, we find

$$\Delta s = \frac{L_f \Delta m}{\mu_k mg} = \frac{(3.337 \times 10^5 \text{ J/kg})(15.2 \times 10^{-3} \text{ kg})}{0.060(10 \text{ kg})(9.80 \text{ m/s}^2)}$$

$$= 863 \text{ m}$$

Calorimetry. Figure 18–10 is a schematic drawing of a multipurpose device called a *calorimeter*. The inner vessel C is the calorimeter proper, and the outer jacket is a heat shield that reduces heat losses from the inner vessel. The water in the jacket is usually maintained at the average temperature prevailing in the calorimeter during an experiment. The calorimeter proper is filled with a measured amount of water, and the heat capacity for the container, stirrer, and thermometer combined is obtained from separate experiments. The device is used for the determination of the heat energy delivered to the calorimeter by any object immersed in the water by measuring the change in temperature of the water. Then, the value of an unknown quantity in Eq. 18–12 can be obtained in the manner described in the following example.

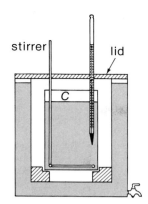

stirrer lid

C

Fig. 18–10. The essential features of a calorimeter, in which the temperature change between initial and final equilibrium states can be measured.

Example 18–4

A calorimeter of the type shown in Fig. 18–10 contains 0.80 kg of water at a temperature of 15.0° C. The heat capacity of the container, stirrer, and thermometer combined is $H = 0.0102$ Cal/deg. A 0.40-kg sample of molten lead is poured into the calorimeter. The final (equilibrium) temperature of the system is 25.0° C. What was the initial temperature of the lead? (The specific heat of molten lead in the temperature range here is $\bar{c}' = 0.0375$ Cal/kg·deg.)

Solution:

In the temperature interval, 15° C to 25° C, the average specific heat of water is $\bar{c}_w = 0.999$ Cal/kg·deg (see Fig. 18–8). Therefore, the amount of heat Q_c absorbed by the calorimeter is

$$Q_c = (\bar{c}_w m_w + H)\,\Delta T_{cal}$$

$$= [(0.999 \text{ Cal/kg·deg})(0.80 \text{ kg})$$

$$+ (0.0102 \text{ Cal/deg})](10.0 \text{ deg})$$

$$= 8.094 \text{ Cal}$$

The molten lead cools from the unknown initial temperature t_C to its freezing temperature of 327° C (see Table 18–4), thereby releasing an amount of heat Q_1:

$$Q_1 = \bar{c}'m_{Pb}\,(t_C - 327° \text{ C})$$

$$Q_1 = (0.0375 \text{ Cal/kg·deg})(0.40 \text{ kg})(t_C - 327° \text{ C})$$

$$= (0.015 \text{ Cal})(t_C - 327° \text{ C})$$

The complete freezing of the lead releases an amount of energy Q_2:

$$Q_2 = L_f m_{Pb} = (5.533 \text{ Cal/kg})(0.40 \text{ kg})$$

$$= 2.213 \text{ Cal}$$

where we have used the value of L_f given in Table 18–4. Cooling the solid lead from 327° C to 25° C releases an amount of energy Q_3. For this calculation we use the value of $\bar{c}$ for lead given in Table 18–3. Thus,

$$Q_3 = \bar{c}m_{Pb}\,(327° \text{ C} - 25° \text{ C})$$

$$= (0.0326 \text{ Cal/kg·deg})(0.40 \text{ kg})(302 \text{ deg})$$

$$= 3.938 \text{ Cal}$$

Then,

$$Q_c = Q_1 + Q_2 + Q_3$$

$$8.094 \text{ Cal} = (0.015 \text{ Cal})(t_C - 327° \text{ C})$$

$$+ 2.213 \text{ Cal} + 3.938 \text{ Cal}$$

from which

$$t_C = 456.5° \text{ C}$$

The Mechanical Equivalent of Heat. One of the important steps in the development of the concept of heat as a form of energy and the energy conservation principle was the determination by Joule* in 1849 of the ratio of heat-energy units to mechanical-energy

---

*James Prescott Joule (1818–1889), an English physicist. Joule was the son of a wealthy Manchester brewer and had the advantage of a home laboratory supplied by his father. Here he conducted a long series of experiments that demonstrated conclusively the equivalence of heat energy and mechanical energy. Joule's 1849 value was 1 Cal = 4154 J, within 1 percent of the present accepted value.

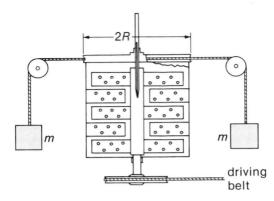

Fig. 18–11. A calorimeter used to measure the mechanical equivalent of heat.

units. Figure 18–11 shows schematically a modern version of the type of apparatus used by Joule (and others) for the determination of the mechanical equivalent of heat. The water-filled calorimeter contains a rotating set of paddles that are driven at constant speed by an external motor. The calorimeter is fitted with a set of interlacing stationary paddles to enhance the churning of the water. The input mechanical energy is transformed into heat through the effects of the viscosity of the water. The calorimeter container itself is restrained from rotating about the drive shaft by the torque provided by two weights, each with mass m, that hang from two cords connected tangentially to the rim of the calorimeter. The amount of mechanical energy input after n rotations of the movable paddles is (see Section 11–4)

$$E = \tau \, \Delta\phi = (2mgR)(2\pi n) = 4\pi nmgR$$

where R is the radius of the calorimeter rim. After the value of E is corrected for the amount of heat absorbed by the walls of the calorimeter, the measured increase in temperature of the water provides a means of determining the mechanical equivalent of heat.

18–5   HEAT TRANSFER

There are three basic and distinguishable mechanisms by which heat can be transferred from a hotter to a colder body or from a hotter to a colder portion of a body. These heat transfer mechanisms are *convection, radiation,* and *conduction.* We discuss briefly the first two of these processes and then devote most of our attention to the case of conduction.

Convection.   Convection takes place primarily in fluids (gases and liquids). In these media, heat is transferred by warmed matter that moves as the result of either forced or naturally induced currents. Most substances expand when heated so that their densities decrease. Thus, the heated portion of a fluid is buoyed upward and rises, delivering heat to the cooler fluid above. If the shape of the container allows it, this buoyant action sets up a natural circulation called *free* (or *natural*) *convection.* Convection involves the actual macroscopic flow of matter, which can be readily detected. For example, convective effects are responsible, in part, for the movement of water in the oceans and air in the atmosphere. Heated matter can also be forced to move by using pumps or fans; this type of circulation is called *forced convection.* Figure 18–12 illustrates several convective circumstances in a house heated by a hot-water system.

The mathematical analysis of convective processes in very complicated engineering designs of convective systems relies primarily on empirical information.

Radiation.   In the radiative transfer of heat, energy is carried by *electromagnetic*

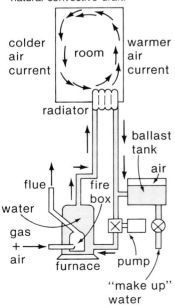

Fig. 18–12. Convective circulation in a house heated by a hot-water system. The water heated by the furnace is circulated through the radiator by forced convection; in the case here, this circulation is assisted by natural convection. The room air is heated primarily by natural convection. The products of combustion that result from the burning of the fuel in the furnace leave through a flue driven by a slight pressure difference produced by the injected gas, but primarily by a natural convective draft.

radiation. Every body, at whatever temperature above absolute zero, radiates electromagnetic waves from its surface. At each temperature this radiation is distributed over a characteristic wavelength spectrum. For temperatures below about $10^{4°}$ C, the radiation consists primarily of visible light and radiation with longer wavelengths.* Between temperatures of about 300° C and about 3000° C, the major portion of radiation has wavelengths longer than visible light and is called infrared (IR) radiation. At the temperature of the Sun's surface — about 6000° C — the radiation is primarily in the visible portion of the spectrum. For temperatures above about 10^4 °C, the dominant spectral component in the emitted radiation has a wavelength shorter than that of visible light; such radiation is called *ultraviolet* (UV) radiation.

Not only does the wavelength spectrum of emitted thermal radiation change with the temperature of the radiating body, but so does the amount of radiated energy. In fact, the radiated power increases with the fourth power of the absolute temperature.

Suppose that two objects at different temperatures are placed, without touching, inside an evacuated container that has perfectly reflecting inner walls. The two objects will exchange energy by radiation (and only by this process). The colder object will emit less energy than it receives from the hotter object, and vice versa. The process of energy transfer by radiation will continue until an equilibrium state is reached in which each object emits as much energy as it absorbs. Then the objects will be at the same temperature.

Conduction. Heat transfer by conduction is usually significant only for substances in the solid phase. (Some conduction does take place in fluids by a process very similar to *diffusion*. We discuss diffusion later in connection with the kinetic theory of matter.) In the conduction process, heat is transferred between two objects or between two parts of an object through the intervening matter without any macroscopic (visible) motion of the object or its parts. The random thermal atomic motion that constitutes the internal energy of the substance is transferred from one region to another by atom-atom collisions. No long-range displacements of the individual atoms occur during this process.

Isotherms are the familiar curved lines on weather maps that connect ground stations reporting the same temperature. This concept may be extended to the three-dimensional case of solid matter. We define an *isothermal surface* to be a surface within a solid that connects points all at the same temperature. Heat will flow through the solid in the direction in which the rate of change of the temperature is greatest. These lines, which are perpendicular to the isothermal surfaces, are called the *temperature gradient lines*. They may be viewed as heat-transfer flow lines or thermal current lines.

Consider the flow of heat along a homogeneous rectangular bar (Fig. 18–13) that connects two surfaces at temperatures T_1 and T_2. Assume that the surfaces have infinite heat capacity and that they therefore remain at constant temperature throughout the flow process. Assume further that the bar has been in place long enough to establish an equilibrium temperature distribution in the bar, and that there is negligible heat loss through the sides of the bar. In this case, the isothermal surfaces are parallel to the end faces of the bar. With $T_2 > T_1$, the heat flow is directed from the end at T_2 toward the end at T_1 along lines that are parallel to the length of the bar; that is, the flow is parallel to $\hat{n}$. Then the rate of heat flow dQ/dt must be constant along the length of the bar. (• Can you see why? Make an analogy to fluid flow in a uniform pipe.)

The rate of heat transfer between the two isothermal surfaces labeled ① and ② in Fig. 18–13 is shown by experiment to be proportional to the cross-sectional area A and to the negative of the temperature gradient, $-dT/dx$, evaluated at the position x of the

---

*The visible spectrum of light extends from about 700 nm (7000 Å) at the red end to about 400 nm (4000 Å) at the violet end.

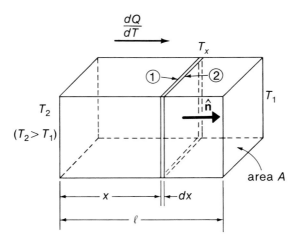

Fig. 18–13. Conduction of heat through a rectangular bar that connects two surfaces at constant temperatures, T_2 and T_1. Note that $dT/dx < 0$.

surfaces. (Notice that the flow direction $\hat{n}$ is positive in the direction opposite that of the temperature gradient.) The constant of proportionality κ is called the *thermal conductivity* of the substance. Altogether, we have

$$\frac{dQ}{dt} = -\kappa A \frac{dT}{dx} \qquad (18\text{–}13)$$

The conductivity κ has units of Calories (or joules) per second per meter per degree: Cal/s·m·deg.

Table 18–5 lists some thermal conductivity values for various materials. The thermal conductivity for a substance usually varies considerably with temperature. The values given are average values for the temperature range from 0° C to 100° C. A "perfect thermal conductor" would have $\kappa = \infty$, whereas a "perfect thermal insulator" would have $\kappa = 0$.

In building construction, a major concern is the prevention of heat flow through walls and ceilings. For these purposes it is more convenient to measure the effectiveness of heat insulators in terms of a quantity that expresses their resistance to heat flow. Since heat-insulating materials such as fiber glass and Styrofoam are commercially available in the form of sheets or slabs of a particular thickness ℓ, their resistance is rated in terms of their *R-value* ($\mathcal{R}$), where

$$\mathcal{R}(\text{R-value}) = \ell/\kappa \qquad (18\text{–}14)$$

Appropriate to the present construction trade, R-value is expressed in units of ft²·°F·h/BTU. (• Can you derive the conversion factor from SI units to these trade units?)

The heat flow through an insulator of area A and a temperature difference of $T_2 - T_1$ (see Fig. 18–13) is

$$\frac{dQ}{dt} = \frac{A}{\mathcal{R}}(T_2 - T_1) \qquad T_2 > T_1 \qquad (18\text{–}15)$$

Thus, the R-value measures the heat flow in BTU/h through a 1-ft² area of the particular insulator slab or sheet per °F of temperature difference. For buildings in the northern United States, insulation with an R-value of 30 ft²·°F·h/BTU (expressed as R30) is recommended in outside walls and attic floors.

Heat insulation often consists of several layers of different materials. In such cases the net R-value is the sum of the individual R-values; that is,

$$\mathcal{R} = \sum_i \mathcal{R}_i \qquad (18\text{--}16)$$

may be used in Eq. 18–15 (see Problem 18–21).

TABLE 18–5. Thermal Conductivities for Various Materials[a]

SUBSTANCE Metals	κ (Cal/s·m·deg)	SUBSTANCE Gases[b,e]	κ (Cal/s·m·deg)
Aluminum	0.050	Air	6.2×10^{-6}
Brass	0.026	Hydrogen	44.2×10^{-6}
Copper	0.092	Oxygen	6.4×10^{-6}
Gold	0.070	Nitrogen	6.2×10^{-6}
Lead	0.0083		
Silver	0.1000	**Composite Materials**	
Tin	0.0150	Asbestos (fiber)	0.19×10^{-4}
Liquids[b]		Brick (common red)	1.5×10^{-4}
		Concrete	0.81×10^{-4}
Alcohol (ethyl)[c]	0.400×10^{-4}	Cork	0.72×10^{-4}
Alcohol (methyl)[c]	0.483×10^{-4}	Fiber glass	0.11×10^{-4}
Transformer oil[d]	0.424×10^{-4}	Glass (window)	2.5×10^{-4}
Water 0° C	1.35×10^{-4}	Ice	4.0×10^{-4}
20° C	1.43×10^{-4}	Paper	0.70×10^{-4}
40° C	1.50×10^{-4}	Porcelain	2.5×10^{-4}
60° C	1.56×10^{-4}	Rock wool	0.10×10^{-4}
100° C	1.60×10^{-4}	Sawdust	0.12×10^{-4}
		Snow (compact)	0.51×10^{-4}
		Styrofoam	0.024×10^{-4}
		Wood (fir) ∥ to grain	0.30×10^{-4}
		⊥ to grain	0.09×10^{-4}

[a] Average values for 0° C ⩽ t_C ⩽ 100° C.
[b] Convection suppressed.
[c] 20° C.
[d] 70° C–100° C.
[e] 27° C; essentially independent of temperature and pressure (⩽ 10 atm).

Example 18–5

A steam pipe with an outside diameter of 5.00 cm has a uniform temperature of 100° C. The pipe, which was installed many years ago, is insulated with a 6.00-cm layer of asbestos fiber. (Because of the environmental problems associated with asbestos, different materials are now being used.) If the room temperature is 20° C, what is the heat loss to the room per hour per meter of pipe length?

Solution:

The system has cylindric symmetry, so we conclude that the isothermal surfaces are concentric cylinders and that heat flows radially outward along lines parallel to the normal vectors $\hat{n}$, as shown in the diagram on page 366. Let the temperature at radius r be T_r and set $dx = dr$ and $A = 2\pi rL$. When equilibrium conditions have been reached, the total outward heat flow must be independent of r. (• Can you see why?) Using

$$\frac{dQ}{dt} = -\kappa \frac{dT}{dr} \cdot 2\pi rL$$

we can write $\quad r\dfrac{dT}{dr} = -\dfrac{1}{2\pi L\kappa}\dfrac{dQ}{dt} = a$ (constant) **(1)**

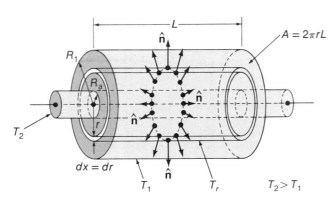

Next, we write
$$dT = a\,\frac{dr}{r}$$

Integrating, we find $T_r = a \ln r + b$

At $r = R_2$, we have $T_r = T_2$, and at $r = R_1$, we have $T_r = T_1$. Thus,

$$T_2 = a \ln R_2 + b$$

$$T_1 = a \ln R_1 + b$$

Subtracting one of these equations from the other gives an expression for a, which can be substituted into Eq. (1) with the result

$$\frac{dQ}{dt} = \frac{2\pi L\kappa(T_2 - T_1)}{\ln (R_1/R_2)} \tag{2}$$

We have the additional result that

$$T_r = T_2 - (T_2 - T_1)\,\frac{\ln (r/R_2)}{\ln (R_1/R_2)}$$

That is, the temperatures of the isothermal surfaces vary logarithmically with the radial distance r.

Evaluating the heat flow, using the appropriate value of κ from Table 18–5, we find

$$\frac{dQ}{dt} = \frac{2\pi(1\text{ m})(0.19 \times 10^{-4}\text{ Cal/s·m·deg})(80\text{ deg})}{\ln (8.50\text{ cm}/2.50\text{ cm})}$$

$$= 7.80 \times 10^{-3}\text{ Cal/s} = 28.1\text{ Cal/h}$$

for a 1-m length of the pipe. (• What is the temperature in the middle of the layer of insulation?)

QUESTIONS

18–1   Define and explain briefly the meaning of the terms (a) thermometric property, (b) zeroth law of thermodynamics, (c) triple point of water, (d) Kelvin temperature, and (e) absolute zero of temperature.

18–2   Explain how thermal contact may exist between two or more objects without there being any physical contact between them.

18–3   Discuss the practicality of Celsius' choice of the ice point and steam point of water to establish his temperature scale. Can you suggest any improvements on his choice?

18–4   Why should the bore of a thermometer tube be small?

18–5   Why are the metal hoops that are used to hold wooden beer casks together heated before being placed around the slats?

18–6   Why must concrete city sidewalks be interrupted by a series of gaps every few meters?

18–7   Concrete structural members often have reinforcing iron rods embedded in them for increased strength. Using Table 18–2, explain why this procedure is reasonable even in situations where there are substantial temperature changes. Could aluminum rods be substituted for the iron rods?

18–8   Why is it important to prevent the water in a car radiator from freezing in the winter?

18–9   Define and explain briefly the meaning of the terms (a) heat, (b) internal energy, (c) specific heat, (d) calorimetry, and (e) R-value.

18–10   Explain the seeming contradiction in the statement that a body cannot be viewed to contain any quantity of heat yet can transfer heat to another body with which it is in thermal contact.

18–11   Explain why water is commonly used for heat storage in passive-solar home-heating systems. Refer to Table 18–3 and Fig. 18–8.

18–12   Explain how sweating helps to cool us on a hot day.

18–13   Explain why blowing on your hands on a cold day can warm them, yet blowing on a hot cup of tea can cool it.

18–14   Explain how the condensation of steam warms the walls of a shower.

18–15   Why are the handles of metal pots and pans made of wood or plastic?

18–16   Explain why the ground temperature drops to low values on cloudless nights. (This accounts for the chilly nights in desert areas that normally have cloudless skies.)

PROBLEMS

Section 18-2

18-1   During a vacation in Budapest you become ill and the doctor reports that your temperature is 39.6° C. Should you ask to be rushed to the hospital? (Normal body temperature is 98.6° F.)

18-2   The highest recorded temperature on Earth was 136° F, at Azizia, Libya, in 1922. The lowest recorded temperature was −127° F, at Vostok Station, Antarctica, in 1960. Express these temperature extremes in degrees Celsius.

18-3   A helium gas thermometer indicates a pressure of 86.3 torr at the triple point of water and a pressure of 4.4 torr at the temperature of some sample of interest. What is the sample temperature in degrees Celsius? (Assume that the pressures involved are sufficiently low that Eq. 18-3 applies.)

Section 18-3

(When necessary, assume that α and β are independent of temperature, so that $\alpha = \bar{\alpha}$ and $\beta = \bar{\beta}$.)

18-4   A copper telephone wire is strung, with little sag, between two poles that are 35 m apart. How much longer is the wire on a summer day with $t_C = 35°$ C than on a winter day with $t_C = -20°$ C?

18-5   The New River Gorge bridge in West Virginia is a steel arch bridge 518 m in length. What total expansion must be allowed for between temperature extremes of −20° C and 35° C?

18-6   A simple pendulum clock made entirely of brass is adjusted to keep perfect time at a room temperature of 18° C. During a power failure the air conditioner is out of service and the temperature rises to 27° C. By how much per day does the clock run slowly? [*Hint:* Refer to Section 15-4.]

18-7   A hole with a diameter of 2.00 cm is drilled through a stainless-steel plate at room temperature (20° C). By what fraction will the radius of the hole be larger when the plate is heated to 150° C?

18-8   A cylindric steel plug is inserted into a circular hole with a diameter of 2.60 cm in a brass plate. When the plug and the plate are at a temperature of 20° C, the diameter of the plug is 0.010 mm smaller than the diameter of the hole. At what temperature will the plug be squeezed tight?

18-9   Equation 18-9 was derived for a parallelepiped. Show that the same result is obtained for a sphere.

18-10   The mercury column in a barometer stands at a height of 76.02 cm (and therefore indicates a pressure of 760.2 torr) when the temperature is 0° C. What will be the indicated pressure (column height) on a warm day when the temperature is 30° C and the actual atmospheric pressure is the same as on the colder day?

Section 18-4

18-11   A metal block is made from a mixture of aluminum (2.4 kg), brass (1.6 kg), and copper (0.8 kg). What amount of heat (in Cal) is required to raise the temperature of this block from 20° C to 80° C?

18-12   A sheet of mill iron with an area of 1 m^2 and a thickness of 5 mm is cooled from 600° C to 250° C by sprayed water delivered at 25° C. Assume that the water is converted to steam at 100° C upon contact with the iron. How much water is required to cool the iron? (Take $\bar{c}$ for iron in this temperature range to be 0.145 Cal/kg·deg.)

18-13   A friction brake is used to stop a flywheel that has a mass of 10 kg and a radius of gyration of 0.40 m. The initial rotational speed of the flywheel is 2000 rpm. (a) How much heat (in Cal) is generated in the slowing process? (b) If the brake unit, with a mass of 6.0 kg and an initial temperature of 60° C, absorbs all of the heat generated, what temperature will it reach? (Take $\bar{c}$ for the brake to be 0.15 Cal/kg·deg.)

18-14   A 0.450-kg sample of copper at a temperature of 87.2° C is immersed in 0.350 kg of water at 10.00° C contained in an aluminum calorimeter vessel with a mass of 30.2 g. The final equilibrium temperature of the system is 18.12° C. What is $\bar{c}$ for copper in this temperature range? (Neglect the heat capacity of the stirrer and the thermometer.)

18-15   A 300-g piece of tin is placed on a large block of ice, and this system is thermally isolated from its surroundings. As the tin cools to the temperature of the block (0° C), 18.2 g of ice melts. What was the original temperature of the piece of tin?

18-16   A heated iron rod ($m = 0.10$ kg) is immersed in a calorimeter that contains 0.20 kg of water at a temperature of 20.0° C. The heat capacity of the calorimeter (excluding the water) is 0.015 Cal/deg. If the final temperature of the system is 45.0° C, what was the original temperature of the iron rod?

18-17   Water is heated from 10° C to 82° C in a residential hot-water heater at a rate of 2.5 liters per minute (ℓ/min). Natural gas, with a density of 1.12 kg/m^3, is used in the heater, which has a transfer efficiency of 32 percent. What is the gas-consumption rate in m^3/h? (When burned, natural gas produces 8400 Cal/kg.)

18–18 An automobile engine delivers 60 hp of power while using 5.2 gallons of gasoline per hour. What is the overall efficiency of the engine? (1 U.S. gal = 3.79 ℓ. Gasoline has a density of 740 kg/m^3 and produces 11,500 Cal/kg upon combustion.)

Section 18–5

18–19 (a) The wall of a factory, with dimensions 6.0 m × 15.0 m, is made from common brick and has a thickness of 0.25 m. If the outside temperature is a steady −7.0° C while the inside temperature is constant at 24° C, what amount of heat is lost by conduction through the wall per hour? (b) How many gallons (1 U.S. gal = 3.79 ℓ) of crude oil, burned at 30 percent efficiency per 8-h working day, does this heat loss represent? (Crude oil has a density of 919 kg/m^3 and releases 11,500 Cal/kg when burned.)

18–20 A closed cubical box with sides of length 0.50 m is constructed from wood that has a thickness of 2.0 cm. When the outside temperature is 20.0° C, an electric heater inside the box must supply 110 W of power to maintain a uniform air temperature of 85.0° C within the box. What is the average thermal conductivity of the wood used?

18–21 A compound slab with a cross-sectional area A is composed of a thickness ℓ_1 of a material with a thermal conductivity κ_1, and a thickness ℓ_2 of a material with κ_2. The temperatures of the surfaces are maintained at T_1 and T_2, as indicated in the diagram. (a) Show that the rate of heat flow through the compound slab can be expressed as $dQ/dt = A(T_2 - T_1)/\mathcal{R}$, where $\mathcal{R} = \mathcal{R}_1 + \mathcal{R}_2 =$

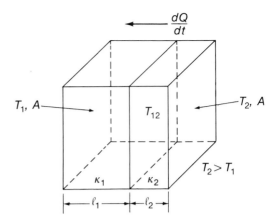

$(\ell_1/\kappa_1) + (\ell_2/\kappa_2)$. (b) Show that the temperature T_{12} of the interface is $T_{12} = T_1 + (T_2 - T_1)\mathcal{R}_1/(\mathcal{R}_1 + \mathcal{R}_2)$.

18–22 (a) Suppose that the wall in Problem 18–19 is insulated with 3.0 cm of rock wool. For the same temperature conditions, what is the ratio of heat-loss rate with the insulation in place to the original heat-loss rate? (b) What is the saving in crude oil per 20-day working month? (c) What is the R-value of the rock-wool insulation?

18–23 A completely enclosed calorimeter made of copper with a wall thickness of 1.0 mm has a surface area of 335 cm^2. Mechanical work is delivered to the water at a rate of $\frac{1}{4}$ hp by means of stirring paddles. When equilibrium is achieved, the water in the calorimeter has a uniform temperature. What is the difference between the water temperature and the outside temperature?

18–24 One end of an aluminum rod that has a length of 35.0 cm and a diameter of 1.0 cm is immersed in boiling water at atmospheric pressure. The other end is in thermal contact with a large block of ice at 0° C. After the system has reached equilibrium (with no losses through the sides of the rod), what is the melting rate of the ice (in kg/h)?

Additional Problems

18–25 At what absolute temperature are the Celsius and Fahrenheit temperatures numerically equal? At what absolute temperature is the Fahrenheit temperature twice the Celsius temperature?

18–26 Express the temperature of the triple point of water in °F and in °R.

18–27 Imagine a steel belt that encircles the Earth at the Equator with a snug fit when the temperature of the belt is a uniform 20° C. What would be the uniform space between the belt and the surface of the Earth if the temperature of the belt were increased to 21° C?

18–28 (a) Show that the length ℓ of a rod can be expressed as

$$\ell = \ell_0\left(1 + \int_{t_{0C}}^{t_C} \alpha(t_C)\, dt_C\right)$$

(b) In the temperature range from 0° C to 100° C, the thermal expansion coefficient for aluminum is given approximately by $\alpha(t_C) = (2.25 \times 10^{-5}\ \deg^{-1}) + (2.3 \times 10^{-8}\ \deg^{-2})t_C$. Find the change in length, $\Delta\ell = \ell - \ell_0$, when a 1-m aluminum wire is heated from 0° C to 100° C. (c) Determine $\alpha(t_C = 20°\ C)$ and also $\bar{\alpha}$ for the temperature range from 0° C to 100° C. Compare with the entry for aluminum in Table 18–2.

18–29 A very thin spherical shell of silver has a radius of 5.00 cm at a temperature of 10° C. If the shell is heated until it has a uniform temperature of 30° C, find (a) the change in radius, (b) the change in surface area, and (c) the change in volume.

18–30 A glass container with a volume V is filled to the brim with a liquid. The temperature of both the container and the liquid is T. The system is then heated to a uniform temperature $T + \Delta T$. (a) Show that the volume ΔV of liquid that overflows is $\Delta V = V(\bar{\beta}_\ell - \bar{\beta}_g) \Delta T$, where $\bar{\beta}_\ell$ and $\bar{\beta}_g$ are the average volume-expansion coefficients of the liquid and the glass, respectively. (b) A 1-ℓ carafe is filled to the brim with white wine at the uniform (refrigerated) temperature of 5° C. The carafe is made of ordinary glass ($\bar{\beta}_g = 3.2 \times 10^{-5}$ deg^{-1}), and 6.8 cm^3 of wine overflows when the carafe is warmed to room temperature (25° C). What is the average coefficient of volume expansion $\bar{\beta}_\ell$ for the wine?

18–31 A mercury thermometer, similar in design to the one
 • shown in Fig. 18–1, has a reservoir bulb that is 5 mm in diameter and 2 cm in length and a capillary tube that has a bore diameter of 0.20 mm. The thermometer is made of Pyrex glass, and only the bulb is subjected to temperature changes when in use. What is the total distance of movement of the upper surface of the mercury column in the tube when the bulb temperature changes from 0° C to 100° C?

18–32 An approximate expression for the specific heat of water
 •• in the temperature range $0°$ C $\leq t_C \leq 100°$ C is $c = 1.00680 - (5.509 \times 10^{-4})t_C + (9.277 \times 10^{-6})t_C^2 - (3.768 \times 10^{-8})t_C^3$. The coefficients carry appropriate units so that when the temperature is expressed in °C, the specific heat is given in Cal/kg·deg. (a) Calculate the amount of heat required to raise the temperature of 1 kg of water from 0° C to 100° C. (b) What is the average value of c in this temperature interval? (c) Determine the temperature in this interval at which the specific heat is minimum. Compare your result with the graph shown in Fig. 18–8.

18–33 A calorimeter contains 80 g of ice and 300 g of water in equilibrium. The heat capacity of the remainder of the calorimeter is 0.0193 Cal/deg. If a 550-g block of aluminum at an initial temperature of 100° C is dropped into the calorimeter, what will be the final temperature of the system?

18–34 A 4-ounce (118.3-ml) whiskey drink (half ethyl alcohol and half water by mass), originally at a temperature of 25.0° C, is cooled by adding 20 g of ice at 0° C. If all the ice melts, what is the final temperature of the drink? (Assume that there are no heat losses and that the specific heat can be calculated as for a simple mixture. The density of the whiskey is 0.919 g/cm^3.)

18–35 The Trümmelbach waterfall in Switzerland has a height of approximately 400 m. Assume that all of the available gravitational potential energy of the water at the top of

the falls is converted to heat energy in the water at the bottom of the falls. What should be the temperature increase of the water due to the 400-m drop? (Joule actually made such a measurement while on his honeymoon in 1847.)

18–36 A 45-g lead bullet is fired with a velocity of 250 m/s at a steel plate. The bullet is stopped completely by the plate, with $\frac{9}{10}$ of the initial kinetic energy appearing as heat in the bullet. What is the temperature increase of the bullet?

18–37 A simplified version of the calorimeter shown in Fig. 18–11 is used in an elementary physics laboratory for the determination of the mechanical equivalent of heat. Each counterbalancing weight has a mass of 2.50 kg and the calorimeter rim has a radius of 8.00 cm. The calorimeter proper contains 1.200 kg of water, and the heat capacity of the remainder of the calorimeter is 0.072 Cal/deg. A total of 470 rotations of the paddle shaft produces a temperature increase of the water from 14.00° C to 16.21° C. Use this information to determine the mechanical equivalent of heat. (• Why might this simple experiment yield too small a value?)

18–38 A long tungsten wire with a diameter of 0.0115 mm is encased in a porcelain sheath that has a diameter of 3.00 mm. The wire is heated electrically, and at equilibrium the temperature of the wire is 1200° C and the temperature of the outer surface of the porcelain sheath is 150° C. What power (in W/m) is being supplied to the tungsten wire by the electric current that flows through it?

18–39 Each of two panes of glass has a thickness of 3 mm.
 • Compare the rate of heat flow through a double-pane window constructed by leaving a 3-mm air gap between the panes with the rate of heat flow through a single 3-mm pane for the same temperature difference. Are double-pane windows good thermal insulators? (In actual practice, double-pane windows are not as effective as this calculation would suggest because of the substantial heat transfer in the trapped air by convection.)

18–40 Suppose that the steam pipe in Example 18–5 is fitted
 •• with a 2.00-cm-thick layer of Styrofoam insulation surrounding the asbestos insulation. (a) What is the new rate of heat loss (in Cal/h) per meter of length? (b) What is the temperature at the interface between the asbestos and the Styrofoam insulation?

18–41 The cold air outside a cabin has a temperature T_0. This
 • air blows through a crack at an average speed of 2.5 m/s and enters the cabin while still at the temperature T_0. A strip of paper with a thickness of 0.18 mm is pasted over the crack. What is the ratio of heat loss by the cabin air without the patch and with the patch? (The specific heat of air is 0.496 Cal/kg·deg.)

18–42   Consider a simplified model of a house heating problem. The outside surface area of the house is 300 m², and the wall thickness is 10 cm with an effective heat conductivity $\bar{\kappa} = 5.0 \times 10^{-6}$ Cal/s·m·deg. The interior contents of the house (including the occupants) have a combined mass of 2000 kg and an effective specific heat $\bar{c} = 0.20$ Cal/kg·deg. Take the outside nighttime temperature to be a constant 0° C. Assume that during the day, the house and its contents have been heated to a uniform interior temperature of 20° C. (a) If the furnace is turned off for the night, what interior temperature would the occupants find upon waking up 6 h later? (Assume that the house cools uniformly. Neglect the heat capacity of the walls.) (b) How much heat must the furnace supply in the morning to heat the interior to 20° C again? (Assume that this heat is supplied in such a short time interval that none escapes to the outside.) (c) If the furnace had been left on for the 6-h period to maintain a constant interior temperature of 20° C, how much heat would the furnace have provided? (d) How much energy was saved by shutting off the furnace for the night?

18–43   On a cold winter day, the uppermost layer of water in a lake is at a temperature of 0° C. Colder air blows in and maintains the surface at −15° C, thereby causing the lake to freeze. Consider the ice formation to proceed by heat loss due to conduction through the ice already formed. Show that the rate of formation of the ice layer of thickness x is $dx/dt = (\kappa_{ice}/x)(\Delta T/L_f)$. What is the thickness of the ice 2 h after freezing begins?

18–44   A radioactive source generates heat at a rate of 296 W in a spherical capsule with a diameter of 2.0 cm. The capsule is embedded in the center of a brass sphere that has a diameter of 20.0 cm. Assume that all of the nuclear radiation is confined within the capsule. What is the temperature of the capsule if the surface of the brass sphere has a constant equilibrium temperature of 20° C?

18–45   Thermal insulation is provided by two layers of insulators of equal thickness. One material is asbestos; the other is fiber glass. What common thickness is required for R30 equivalent heat insulation of the two layers?

CHAPTER 19

THE BEHAVIOR OF GASES

In this chapter, we are concerned primarily with the description of the *macroscopic* behavior of gases in various states of thermal equilibrium. We consider first the properties of an *ideal gas* and then the behavior of real substances. Although we concentrate on the gaseous phase of matter, we are also interested in the transitions to other phases.

It is our goal in the following chapter to give a description of the *microscopic* or molecular behavior of gases. Some of these ideas are introduced in this chapter, so we begin with a brief presentation of some of the important features of the atomic theory of matter.

19–1 THE ATOMIC THEORY OF MATTER

According to the atomic theory, any sample of bulk matter consists of an enormous number of tiny bits of matter called *atoms*. Because there is obviously a great variety in the types of bulk matter, there must be at least some variety in the types of atoms. It is now known that atoms do indeed exist in a variety of types called *elements*, of which there are more than 100 chemically differentiable species. An atom is defined as *the smallest unit of matter that can be identified as a particular chemical element*.

Atomic and Molecular Sizes. A number of different types of measurements have demonstrated that atoms have sizes ("diameters") that range from about 0.15 nm (1.5 Å) to about 0.35 nm (3.5 Å), with the more massive atoms tending to have the larger diameters. An atomic "size" is a rough measure of the extent of the region around the atomic center (or *nucleus*) that is occupied by the atom's orbiting electrons. Most of the mass of an atom is concentrated, in the form of *protons* and *neutrons*, in the atomic nucleus, even though the size of the nucleus is much less than that of the atom. Nuclear diameters range from about 1.5 fm to about 15 fm. (Note that 0.15 nm = 1.5×10^{-10} m, whereas 1.5 fm = 1.5×10^{-15} m.) To better appreciate the relative sizes of atoms and nuclei, imagine a baseball to be magnified to the size of the Earth, with the constituent atoms undergoing the same magnification. Then an atom of medium size would become as large as a baseball. If this atom were further magnified to the size of the largest stadium (diameter about 250 m), the nucleus at this new magnification would be about as large as a small pea.

A combination of two or more atoms bound together is called a *molecule*. If the atoms are of different chemical elements, the molecule represents a chemical *compound*. The uniform properties of pure chemical substances, such as water or ordinary table salt, suggest a model for molecules in which a definite number of atoms of a particular type is associated with each molecular species. Thus, a molecule of water consists of two hydrogen atoms (symbol H) and one oxygen atom (symbol O), and we use the chemical notation H_2O to represent the water molecule (Fig. 19–1). Also, the basic unit of salt contains

Fig. 19–1. Schematic representation of the chemical reaction that converts hydrogen and oxygen into water. A chemical reaction is simply the rearrangement of the constituent atoms into a new molecular form.

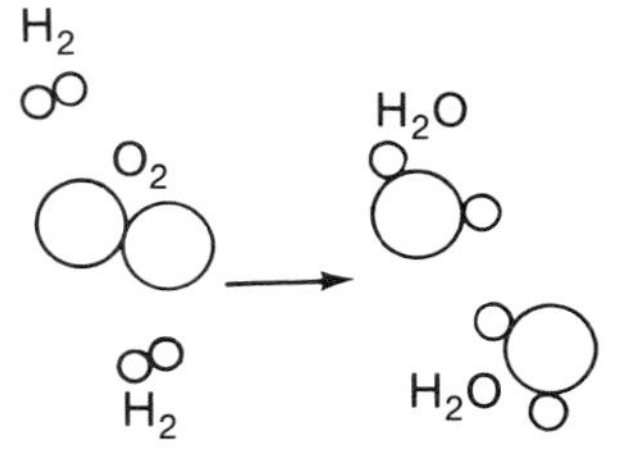

371

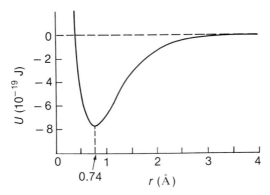

Fig. 19-2. The potential energy U of the H_2 molecule as a function of the distance between the atomic centers.

one sodium atom (symbol Na) and one chlorine atom (symbol Cl), and we write NaCl for sodium chloride. Sodium can be prepared in elemental (atomic) form, Na, but hydrogen, oxygen, and chlorine all occur as diatomic molecules, H_2, O_2, and Cl_2.

The force that acts between two atoms is a complicated function of the distance between the centers of the interacting atoms. The most meaningful way to discuss such interactions is in terms of the corresponding potential energy function. Figure 19-2 shows the potential energy of the diatomic hydrogen molecule H_2 as a function of the separation of the atomic centers. The force between the two atoms is related to the potential energy by the relationship $F = -dU/dr$. In Fig. 19-2 we see that U is minimum at $r = 0.74$ Å and therefore that $F = 0$ at this value of r. Thus, $r = 0.74$ Å is the equilibrium separation distance. For separations smaller than this value, the interatomic force is increasingly *repulsive*. For larger separations the force is *attractive* and begins to decrease rapidly with distance for $r \gtrsim 1$ Å. The force is effectively zero for separations greater than about 4 Å.

The equilibrium separation of the atoms in the hydrogen molecule is somewhat less than the "diameter" of the hydrogen atom (1.02 Å); that is, the atoms in a molecule slightly overlap at the equilibrium separation. In solids and liquids, the molecules also "touch" (see Fig. 19-3). In contrast, the gas molecules in a dilute gas are, on the average, many molecular diameters apart.

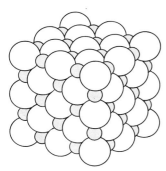

Fig. 19-3. This model of "touching" atoms was constructed by William Barlow in 1898 to represent the structure of the sodium chloride crystal. The small spheres are sodium atoms (actually, *ions*) and the large spheres are chlorine atoms (ions).

Atomic and Molecular Masses. The chemical properties of a particular element are determined by the number of electrons in the atoms of that element. The number of electrons in an electrically neutral atom is exactly equal to the number Z of protons in the nucleus; Z is called the *atomic number* of the element.

The nuclei of a particular element may contain different numbers N of neutrons. Such nuclei, all with the same number of protons but different numbers of neutrons, are called *isotopes* of that chemical element.

The total number of particles in a nucleus is called the *mass number A* of the particular isotope: $A = Z + N$. To distinguish one isotope of an element from another, it is customary to write the mass number as a superscript to the chemical symbol. Thus, the carbon ($Z = 6$) that is found in Nature consists of the isotopes ^{12}C (with $N = 6$), ^{13}C (with $N = 7$), and ^{14}C (with $N = 8$). The isotope ^{12}C (the "normal" isotope) has a natural abundance of 98.89 percent (that is, ^{12}C constitutes 98.89 percent of the carbon atoms in a natural sample).

Most hydrogen (99.985 percent) occurs as ^1H; the nucleus of ^1H is a single proton (Fig. 19-4a). The remainder of natural hydrogen (0.015 percent) is "heavy hydrogen," ^2H, or *deuterium*. The deuterium nucleus contains one neutron in addition to the proton (Fig. 19-4b). A third isotope of hydrogen, ^3H, has two nuclear neutrons (Fig. 19-4c). This isotope, called *tritium*, is radioactive and only traces of it are found in Nature.

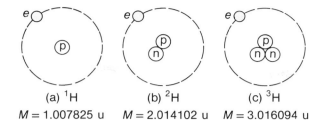

Fig. 19–4. Schematic diagrams of the three isotopes of hydrogen: (a) ^1H, (b) ^2H, and (c) ^3H.

(a) ^1H (b) ^2H (c) ^3H
$M = 1.007825$ u $M = 2.014102$ u $M = 3.016094$ u

Using modern techniques, it is possible to measure the mass of one isotope relative to another with high precision. For example, the mass of a hydrogen atom (^1H) relative to the mass of a carbon atom (^{12}C) is known to an accuracy of one part in 10^8. Atomic masses are usually expressed in terms of a special unit called the *unified atomic mass unit* (symbol, u), where 1 u is defined to be exactly $\frac{1}{12}$ of the mass of the isotope ^{12}C. In terms of u, the mass of every atom is approximately (but not exactly) equal to its mass number

When the chemical elements are arranged in order of increasing atomic number (that is, the number of protons in the nucleus), many of their chemical and physical properties display a recurring or periodic character. Beginning with the independent work of Dimitri Mendeleev and Lothar Meyer in the 1870's, successive refinements and discoveries of unknown elements have brought the periodic table to the form shown here. The table shows the symbol, atomic number, and atomic weight of each element. The atomic weight is an average of the number of protons and neutrons in the atoms of a natural sample of the element.

PERIODIC TABLE OF THE ELEMENTS

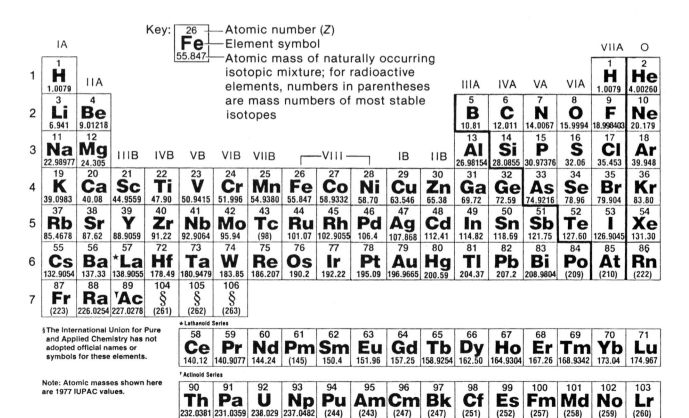

A. For example,

$$M(^1H) = 1.007825 \text{ u}$$

$$M(^{59}Co) = 58.9332 \text{ u}$$

$$M(^{197}Au) = 196.9666 \text{ u}$$

In terms of the kilogram, we have

$$1 \text{ u} = 1.66057 \times 10^{-27} \text{ kg}$$

with an uncertainty of about one part per million. Atomic masses can usually be compared with a precision at least 10 times greater than they can be given in kilograms.

The elements that occur in Nature consist of a mixture of isotopes. (Several elements—for example, fluorine, aluminum, cobalt, and gold—have only a single stable isotope.) Consequently, the quoted atomic mass of a particular element is actually a weighted sum of the masses of the various isotopes. The atomic masses listed in the table on page 373 correspond to the naturally occurring isotopic mixtures.

When two or more atoms combine to form a molecule of a chemical compound, the *molecular mass* of the substance is simply the sum of the atomic masses of the constituents.

Example 19—1

There are two stable isotopes of copper with the isotopic masses and natural abundances given below. What is the atomic mass of natural copper?

^{63}Cu: 62.9296 u   (69.09%)

^{65}Cu: 64.9278 u   (30.91%)

Solution:

The weighted sum is

$$M(Cu) = (62.9296 \text{ u})(0.6909) + (64.9278 \text{ u})(0.3091)$$

$$= 63.5472 \text{ u}$$

This is the atomic mass of copper as found in Nature.

Example 19—2

Using the atomic masses in the table on page 373, determine the molecular masses of (a) water, H_2O, (b) methyl alcohol, CH_3OH, and (c) uranium oxide, U_3O_8.

Solution:

We have

(a) $M(H_2O) = [2 \times (1.00797 \text{ u})] + [1 \times (15.9994 \text{ u})]$

$$= 18.0153 \text{ u} \cong 18 \text{ u}$$

(b) $M(CH_3OH) = [4 \times (1.00797 \text{ u})]$

$$+ [1 \times (12.01115 \text{ u})]$$

$$+ [1 \times (15.9994 \text{ u})]$$

$$= 32.0424 \text{ u} \cong 32 \text{ u}$$

(c) $M(U_3O_8) = [3 \times (238.03 \text{ u})] + [8 \times (15.9994 \text{ u})]$

$$= 842.085 \text{ u} \cong 842 \text{ u}$$

Avogadro's Number and the Mole.   In molecular theory, particularly in the study of gases, it proves convenient to measure the quantity of a substance in terms of the number of molecules instead of the mass. For this purpose we employ a unit called the *mole* (abbreviation, *mol*). We use the term *molecule* to mean the basic unit of a substance, whether an element or a compound, that occurs naturally. Thus, a molecule of hydrogen is H_2, but a molecule of helium is He (an atom). With this meaning, we define the mass in grams of 1 mol of any substance to be equal to the molecular mass of that substance expressed in u.* Thus, the mass of 1 mol of the isotope ^{12}C (molecular mass $M = 12$ u) is just 12 g, exactly. As further examples, 1 mol of H_2O ($M \cong 18$ u) contains approximately 18 g and 1 mol of CH_3OH ($M \cong 32$ u) contains approximately 32 g (refer to Example 19-2).

Since the individual molecular masses expressed in the units of u relate to their actual relative masses, it follows that the number of molecules in a mole must be the same for all substances. This number is called *Avogadro's number,* in honor of the Italian physicist Amedeo Avogadro, Count of Quaregna (1776–1856).† The presently accepted value of Avogadro's number is

$$N_0 = 6.02205 \times 10^{23} \text{ molecules/mol}$$

with an uncertainty of about 3 units in the last decimal place. Notice particularly that Avogadro's hypothesis is not valid if stated in terms of *atoms* instead of *molecules.* For example, it is *not* true that there are N_0 atoms in 1 mol of water.

19-2   THE EQUATION OF STATE (IDEAL GASES)

Consider a quantity of a homogeneous, chemically pure substance in a state of thermal equilibrium. We assert that the variables that are needed to define completely the *thermodynamic state* of the system are the temperature T, the pressure p, the volume V, and the mass m. The functional relationship that connects these variables,

$$F(T, p, V, m) = 0$$

is called the *equation of state* of the system. We may, of course, write this relationship in other forms, such as

$$T = f(p, V, m)$$

The function F (or f) must be determined by experiment.

One of the earliest relevant studies was carried out in 1662 by Robert Boyle (1627–1691), an Irish chemist, who investigated the pressure dependence of the volume of a confined gas (air) at constant temperature. Boyle discovered that, for a large range of values for p and V separately, the product pV remained constant (for constant temperature‡). Thus, for a fixed quantity of gas, any two states (1) and (2) with $T_1 = T_2$ have the pressure-volume relationship

$$p_1 V_1 = p_2 V_2$$

---

*Chemists refer to the mole as the *gram molecular weight.* (The definition actually refers to *mass,* not *weight,* but the distinction is not important here.)

†Avogadro, in 1811, put forward the hypothesis that equal volumes of all gases (at the same temperature and pressure) contain equal numbers of particles. (This brilliant idea was not generally accepted until 50 years later.) Avogadro was careful to state that the particles to which his hypothesis refers are *molecules* (as we have defined the term). One of the reasons that Avogadro's hypothesis was not quickly accepted was the vigorous opposition expressed by the prominent English chemist John Dalton (1766–1844), whose views were the result of his misunderstanding of the distinction between atoms and molecules. Indeed, Avogadro coined the term *molecule* and gave it the modern interpretation.

‡Boyle was rather lucky in his experiments because he took no precautions to maintain the temperature constant; however, his data indicate that there was little temperature variation of his apparatus. In fact, Boyle initiated the practice of carefully describing experiments and reporting data so that others could check the results.

This equality expresses *Boyle's law*. Accordingly, we can combine the variables p and V and write the equation of state as $T = f(p, V, m) = f(pV, m)$; solving for pV, this relation can be expressed as

$$pV = g(T, m)$$

Avogadro's hypothesis states that equal volumes of all gases, under the same conditions of pressure and temperature, contain the same number of molecules. Using this idea, pV can be expressed as a function of the temperature as

$$pV = n\phi(T)$$

where n is the number of moles in the sample.

Two French physicists, Jacques Charles (1746–1823) and Joseph Louis Gay-Lussac (1778–1850), discovered independently that a fixed amount of any confined gas at a constant pressure has a volume that depends linearly on the temperature. Thus, we can revise further the expression for the equation of state and write

$$pV = n(a + bT)$$

where the constants a and b depend on the scale used to measure the temperature. This relationship is known as the *law of Charles and Gay-Lussac*. In Section 18–2, we found that when the Kelvin temperature scale is used, the pressure of a gas becomes (extrapolates to) zero at zero temperature; thus, if we use the Kelvin scale, the constant a is zero. Also, the constant b is traditionally written as R, so that the expression for the equation of state becomes

$$pV = nRT \tag{19–1}$$

This equation expresses the *ideal gas law*.

The quantity R—called the *universal gas constant*—must be determined by experiment; the result is

$$R = 8.3144 \text{ J/mol·K}$$

with an uncertainty of about 3 in the last decimal place.

It is important to remember when applying the ideal gas law equation that p refers to the *absolute pressure* (not the gauge pressure[*]) and that T refers to the *kelvin* or *absolute temperature*.

How well do gases obey the ideal gas law? In Section 18–2, it was found that the temperature of the gas in a constant-volume gas thermometer is proportional to the pressure in the limit of low pressures (that is, for dilute gases). (See Fig. 18–4 on page 348.) Thus, real gases can be expected to obey the ideal gas law only in this limit. A hypothetical gas for which Eq. 19–1 is exactly correct under all conditions is called an *ideal gas*.

Dalton's Law of Partial Pressures. When a gas sample consists of a mixture of gases—n_1 moles of type 1, n_2 moles of type 2, and so forth—the ideal gas law equation becomes

$$pV = (n_1 + n_2 + n_3 + \ldots)RT$$

We may define the so-called *partial pressures* by the relations

$$p_1 V = n_1 RT, \qquad p_2 V = n_2 RT, \qquad \text{etc.}$$

_____

[*]Recall that $p(\text{absolute}) = p(\text{gauge}) + p(\text{atmospheric})$.

Then, we have

$$p = p_1 + p_2 + p_3 + \dots$$

This equation expresses *Dalton's law of partial pressures:* the total pressure of a mixture of gases is equal to the sum of the individual partial pressures, each obtained as if it were the only gas in the volume V.

Example 19-3

Determine the volume of 1 mol of an ideal gas at a temperature of $0°$ C and a pressure of 1 atm (standard temperature and pressure, commonly abbreviated as STP).

Solution:

The temperature that corresponds to $t_C = 0°$ C is $T = 273.15$ K; also, 1 atm $= 1.01325 \times 10^5$ Pa. Hence,

$$V = \frac{nRT}{p} = \frac{(1 \text{ mol})(8.3144 \text{ J/mol·K})(273.15 \text{ K})}{1.01325 \times 10^5 \text{ Pa}}$$

$$= 2.2414 \times 10^{-2} \text{ m}^3$$

$$= 22.414 \; \ell$$

This is a useful number to remember: 1 mol (at STP) $\rightarrow$ 22.4 ℓ.

Example 19-4

The ideal gas law equation (19-1) is valid for a real gas if the pressure is sufficiently low. Look at the upper line in Fig. 18-4, which describes the results of experiments performed with oxygen. If the initial (triple-point) pressure is 150 torr (point b), the temperature determined for the steam point is 373.25 K, only 0.10 K in error. This implies that oxygen at pressures near 150 torr and at temperatures between $0°$ C and $100°$ C behaves in a manner closely resembling an ideal gas. Use the ideal gas law equation to determine the density of oxygen at a pressure of 150 torr and a temperature of 273.16 K.

Solution:

From the table of atomic masses on page 373 we find $M(O_2) = 32.00$ u $= 32.00$ g/mol. Consider 1 mol of O_2 under the prescribed conditions. Then, using the ideal gas law equation, we find

$$\rho = \frac{m}{V} = \frac{m}{nRT/p}$$

$$= \frac{(32.0 \times 10^{-3} \text{ kg})(150/760)(1.01325 \times 10^5 \text{ Pa})}{(1 \text{ mol})(8.3144 \text{ J/mol·K})(273.16 \text{ K})}$$

$$= 0.282 \text{ kg/m}^3$$

Example 19-5

A *McLeod gauge* is a simple device for measuring low pressures. As shown in part (a) of the diagram, the gauge consists of a large bulb V and a capillary tube C that has a narrow bore. The relatively large volume V of the bulb (down to point A) is connected directly to the system whose pressure p is to be determined. The mercury reservoir R is now raised and the volume V is sealed off at point A. As the reservoir is raised further, the gas trapped in V is compressed into the smaller volume V' of the capillary tube, as indicated in part (b) of the diagram. Show that the pressure p is given by

$$p \cong \frac{V'}{V} \rho g h$$

where h is the difference in height of the mercury column (as shown in the diagram), and where ρ is the density of mercury. If $h = 1.2$ cm and $V/V' = 4 \times 10^4$, determine the pressure p.

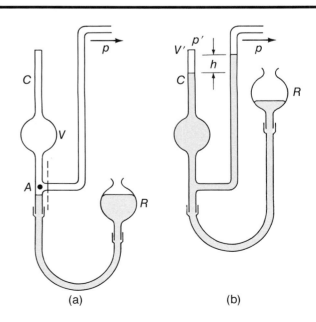

(a)   (b)

Solution:

The difference in pressure between p' in the volume V' and the pressure p in the system is

$$p' = p + \rho g h$$

We assume that the temperature is uniform throughout the system, so we can use Boyle's law to write

$$p'V' = pV$$

Combining these two equations yields

$$p = \frac{V'}{V - V'} \cdot \rho g h \cong \frac{V'}{V} \rho g h$$

where we make the approximation, $V \gg V'$. (• Can you see that this is equivalent to $p \ll \rho g h$?) Using the values given, we find

$$p = (4 \times 10^4)^{-1}(13.6 \times 10^3 \text{ kg/m}^3)(9.80 \text{ m/s}^2)(0.012 \text{ m})$$

$$= 0.0400 \text{ Pa} = 3.95 \times 10^{-7} \text{ atm} = 3.00 \times 10^{-4} \text{ torr}$$

19–3 THE EQUATION OF STATE (REAL GASES)

The equation of state that connects the thermodynamic variables p, V, and T for a fixed mass m of a substance defines a three-dimensional surface. We can imagine a Cartesian representation of such a surface by identifying V with the x-axis, T with the y-axis, and p with the z-axis. Before considering the complicated equation-of-state surfaces for real substances, we first examine the simpler situation for an ideal gas.

The Ideal Gas. The equation-of-state surface that represents the ideal gas law (Eq. 19–1), $pV = nRT$, is shown in Fig. 19–5.* Contour lines are indicated for equally spaced values of constant temperature (*isothermal lines*), such as ab, for equally spaced values of constant volume (*isochoric lines*), such as cd, and for equally spaced values of constant pressure (*isobaric lines*), such as ef. Also shown in the figure are the projections of the isothermal lines onto the p-V plane and the projections of the isochoric lines onto the p-T plane. The constant-volume lines in the p-T plane are straight lines that radiate from the point that represents absolute zero. The constant-temperature lines in the p-V plane are concentric rectangular hyperbolas. This latter representation of the isothermal lines is quite useful and is shown again in more detail in Fig. 19–6. The temperatures here—T_1, T_2, T_3, and T_4—are equally spaced, with $T_2 = 2T_1$, $T_3 = 3T_1$, and $T_4 = 4T_1$. According to the ideal gas law for a gas with a fixed volume V_1, the pressures that

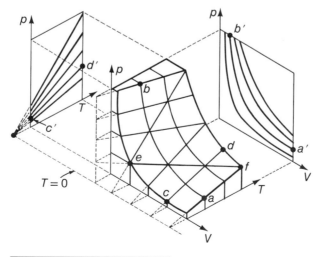

Fig. 19–5. The equation-of-state (or p-V-T) surface for an ideal gas. Also shown are the projections of constant-temperature lines onto the p-V plane and of constant-volume lines onto the p-T plane.

*A point that defines an equilibrium state of the gas lies *on* the surface.

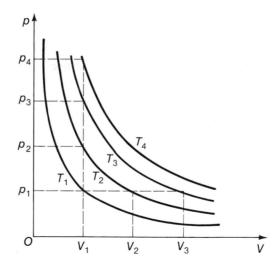

Fig. 19–6. The p-V diagram for an ideal gas, showing several isothermal lines.

correspond to (V_1, T_1), (V_1, T_2), (V_1, T_3), and (V_1, T_4) are also equally spaced, so that $p_2 = 2p_1$, $p_3 = 3p_1$, and $p_4 = 4p_1$, as shown in Fig. 19–6. Similarly, the points (p_1, T_1), (p_1, T_2), and (p_1, T_3) correspond to the equally spaced volumes V_1, $V_2 = 2V_1$, and $V_3 = 3V_1$.

Real Substances.   The p-V-T surface for a real, homogeneous substance is much more complicated than the simple surface that describes an ideal gas. For mixtures, solutions, and alloys, the situation is even more complex. Therefore, we restrict attention here to a typical chemically pure substance. Figure 19–7 illustrates the p-V-T surface for such a case. This diagram shows a number of contour lines such as gh, which represents a constant-temperature line (*isotherm*), and ij, which represents a constant-pressure line (*isobar*).

In Fig. 19–7 the region $ABCDE$ specifies the pure gaseous phase of the substance. We use this terminology because if the temperature is greater than the *critical temperature* T_c (corresponding to the isotherm BCD), no liquefaction of the gas is possible, regardless of how much the pressure is increased. (That is, no liquid droplets would ever form to indicate the presence of a liquid-gas interface.) For temperatures substantially above T_c—such as the isotherm AE—the relationship of p, V, and T closely approximates that specified by the ideal gas law. The projection of this isotherm onto the p-V plane results in a line $A'E'$ that is nearly the same as the rectangular hyperbola, $pV = nRT = $ constant.

For low pressures and below the critical temperature—at a point such as k—the substance is also in the gaseous phase. But as the pressure is increased (and simultaneously heat is removed to maintain constant temperature), a point e is reached at which liquefaction begins (liquid droplets appear) and a liquid-gas interface is formed. If heat removal is continued, even without a further increase in pressure, the fraction of liquid present continues to increase and the volume decreases due to the condensation. Finally, at point d, all of the substance has been converted to the liquid phase. If, at some point along the line ed, the removal of heat is stopped and the system is thermally isolated, the substance will remain in this state of two-phase equilibrium. At the point e, there is zero liquid and the substance is said to be a *saturated vapor*. At the point d, the substance is all liquid (and is called a *saturated liquid*).

The behavior of the substance in the gaseous region CFD is no different from that in the region $ABCDE$. However, because the substance can be liquefied at the temperatures that exist in region CFD, this region is called the *vapor phase* of the substance. The highest pressure p_c at which the liquid and vapor phases coexist (at the temperature T_c) is

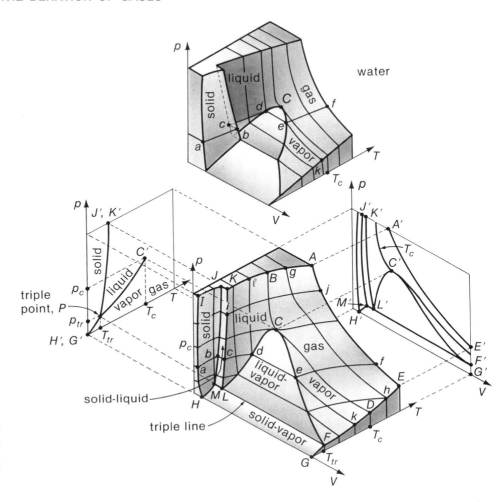

Fig. 19–7. The p-V-T surface for a typical real substance. The diagram for water (at the top) illustrates the case of a substance that expands upon freezing.

called the *critical pressure*. The point (p_c, T_c) is called the *critical point* (point C in Fig. 19–7). The critical point corresponds to the condition in which the densities of the liquid and vapor phases have become equal. Then there is no longer any distinction between the two phases; the substance behaves essentially as a gas but has a density closer to that of a liquid. (The density of water at the critical point is 0.325 g/cm³.) (• What is the value of the latent heat of vaporization at the critical point?)

If the pressure on a substance in the liquid phase is increased (at constant temperature), the state of the system will be described by a line such as $d\ell$. Eventually, a point will be reached (above ℓ and not shown in Fig. 19–7) at which solidification (or crystallization) begins. At first a two-phase, solid-liquid system is formed; finally, when the heat of fusion has been removed, the substance is entirely solid.

Figure 18–9 (page 359) shows a constant-pressure heating curve for water. The points that correspond to the labeled points on the heating curve are indicated in the p-V-T diagram at the top of Fig. 19–7, namely, *abcdef*. The break in the temperature rise *bc* in Fig. 18–9 is identified with the solid-liquid phase line *bc* in Fig. 19–7. Similarly, the break *de* corresponds to the liquid-vapor line *de* in Fig. 19–7. Water does not behave like most substances at the solid-liquid phase transition, as can be seen by comparing this region on the two p-V-T surfaces in Fig. 19–7. (• Does this mean that the heating curve for a normal substance will have an appearance different from that of Fig. 18–9?) The

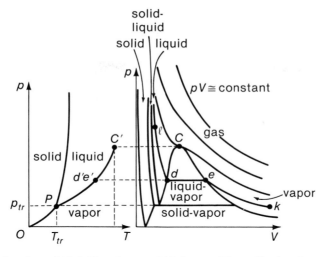

Fig. 19-8. Projections of the p-V-T surface for a typical pure substance onto the p-T and p-V planes. Compare Fig. 19-7.

critical point for water occurs for $T_c = 647.2$ K and $p_c = 218.3$ atm. Thus, if a heating curve were constructed for a very high pressure ($>p_c$), the break de would disappear.

    If one starts with a substance in the gaseous phase at sufficiently low temperature, it is possible to convert the substance directly to the solid phase, bypassing liquefaction, by increasing the pressure and removing heat. This process (and its inverse) is called *sublimation*. In Fig. 19-7, *LMHGF* corresponds to the dual-phase, solid-vapor region. Notice that sublimation can occur only for temperatures below the triple-point temperature T_{tr}. The *triple line* on the p-V-T surface is the isotherm for the triple-point temperature along which the three phases of the substance can exist together. (This line corresponds to the single point P—the *triple point*—in the projection onto the p-T plane.)

    In the projection of the p-V-T surface onto the p-T plane, the liquid-vapor phase boundary generates a line that appears to terminate abruptly at the critical point C'. However, it is clear in the view of the complete surface that there is no physical discontinuity at this point. Notice that each dual-phase boundary—solid-vapor, liquid-vapor, and solid-liquid—projects into a line in the p-T plane.

    Information concerning the p-V-T surface of a substance is often presented in the form of projections onto the p-T and p-V planes. These important projections are shown again in more detail in Fig. 19-8.

    Table 19-1 lists the critical-point temperature T_c and pressure p_c as well as the triple-point temperature T_{tr} and pressure p_{tr} for several pure substances.

    van der Waals' Equation.   The ideal gas law is remarkably accurate for very dilute gases.* However, the law becomes progressively less accurate as the pressure is increased

TABLE 19-1.   Critical-Point and Triple-Point Data

SUBSTANCE	T_c (K)	p_c (atm)	T_{tr} (K)	p_{tr} (torr)
Hydrogen	33.2	12.8	13.84	52.8
Nitrogen	126.	33.5	63.18	94.
Oxygen	154.6	50.1	54.35	1.14
Ammonia	405.6	112.5	195.40	45.57
Carbon dioxide	304.	72.9	216.55	3880.
Sulfur dioxide	430.9	77.7	197.68	1.256
Water	647.2	218.3	273.16	4.58

---

*The theoretical basis for the ideal gas law is discussed in the next chapter.

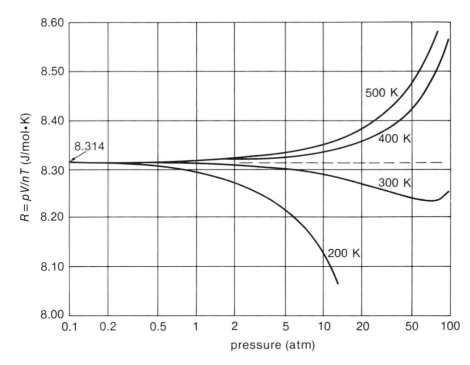

Fig. 19-9. Pressure-temperature characteristics of air. A value of $pV/nT = 8.314$ J/mol·K is expected for an ideal gas. At low pressures, air closely resembles an ideal gas. Note the logarithmic pressure scale.

and the mean spacing between the molecules is reduced. This failure is due primarily to the increasing importance of intermolecular forces at the smaller average distances and to the finite size of the molecules themselves.

Figure 19-9 illustrates the behavior of air under a variety of temperature and pressure conditions. The curves show the quantity pV/nT which, for an ideal gas, should equal $R = 8.314$ J/mol·K. This value is attained for all of the temperatures shown for pressures less than about 0.3 atm. Thus, air closely resembles an ideal gas at these pressures.

In 1873, the Dutch physicist Johannes van der Waals (1837–1923) showed that real substances can be more accurately described if the ideal gas law equation is modified to the form

$$\left(p + \frac{a}{\mathcal{V}^2}\right)(\mathcal{V} - b) = RT \qquad (19-2)$$

where the parameters a and b are only slightly dependent on temperature and must be determined empirically for each substance, and where $\mathcal{V} = V/n$ is the volume per mole or *molar volume* of the substance. The term $a/\mathcal{V}^2$ that augments the externally imposed pressure p arises from the attractive nature of the intermolecular force. The quantity b that modifies the molar volume $\mathcal{V}$ is approximately equal to N_0 multiplied by the volume of a single molecule. Thus, $\mathcal{V} - b$ represents the free volume available to each mole of the substance.

Figure 19-10 shows various isotherms on a p-$\mathcal{V}$ diagram according to the van der Waals' equation of state. Notice that Eq. 19-2 is a cubic equation in $\mathcal{V}$, so for any particular value of p, we expect three roots for $\mathcal{V}$. At the critical temperature T_c, all three roots are equal for the critical pressure p_c. For $T < T_c$ there are three distinct real roots for $\mathcal{V}$, whereas for $T > T_c$ there is only one real root (the other two are complex and are therefore physically unrealizable).

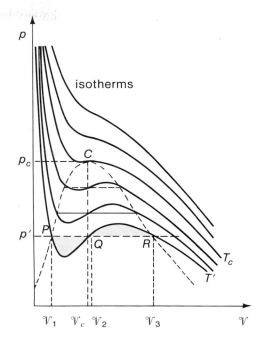

Fig. 19–10. Isotherms of a typical substance based on van der Waals' equation. The point C is the critical point. The three roots ($\mathcal{V}_1$, $\mathcal{V}_2$, and $\mathcal{V}_3$) of the cubic equation are shown for an isotherm, $T' < T_c$. In the region of the liquid-vapor phase (inside the dashed line), van der Waals' equation produces curved cubic lines instead of the straight isotherms that actually occur in this region (see Fig. 19–7). Maxwell showed that each straight isotherm, such as PQR, divides the curved cubic lines into two regions that have equal areas (the shaded regions in the diagram). This determines the positions of the roots.

Equation 19–2 may be expanded in powers of p with the result (see Problem 19–15)

$$\frac{pV}{nT} = R + \left(b - \frac{a}{RT}\right)\frac{p}{T} + \text{(terms of order } p^2 \text{ and higher)} \qquad \textbf{(19–3)}$$

For air the quantities a and b are such that the term linear in p vanishes for a temperature of 348 K (called the *Boyle temperature, T_B*). Figure 19–9 shows that for T_B and for nearby temperatures not too different from room temperature, the ideal gas law is indeed valid even for rather high pressures. *That is, the unusual accuracy of the ideal gas law for air at temperatures near the Boyle temperature is* not *due to the absence of the effects of short-range forces and finite molecular size; instead, it is due to the near cancellation of these two important effects!*

19–4 PHASE TRANSITION BEHAVIOR

Vapor Pressure. Suppose that an amount of water is introduced into a sealed container from which the air has been completely evacuated. If the amount of water is small, it will completely vaporize, producing a certain pressure of the vapor within the container. However, if a larger amount of water is introduced into the container, the vaporization will not be complete—a quantity of liquid water will remain on the bottom of the container. The space above the water surface can contain no additional vapor at that temperature. This is the condition of *saturation,* and the pressure of the vapor is called the *saturated vapor pressure* (or often, simply the *vapor pressure*) for that particular temperature. Saturation exists along a line such as *de* in the *p-V-T* diagram shown in the upper part of Fig. 19–7 (or the same line in Fig. 19–8 for a normal substance). The temperature is constant along this line, and the liquid and vapor phases coexist in equilibrium. Notice that, if the volume of the container is increased (with the temperature held constant by the addition of heat), a greater fraction of the water will be vaporized until finally, at *e*, no liquid water remains. Any further increase in the volume will result in the lowering of the pressure below the saturated value along the curve *ek*.

TABLE 19-2.   Saturated Vapor Pressure and Vapor Density for Water at Various Temperatures

t_C (°C)	p_v (torr)	ρ (kg/m³)	t_C (°C)	p_v (torr)	ρ (kg/m³)
0	4.58	0.00485	70	233.53	0.1984
10	9.21	0.00941	80	355.1	0.2938
15	12.79	0.0128	90	525.8	0.4241
20	17.51	0.0173	100	760.0	0.598
25	23.76	0.0230	110	1074.5	0.827
30	31.71	0.03035	120	1488.9	1.122
40	55.13	0.0511	130	2025.6	1.498
50	92.30	0.0832	140	2709.5	1.968
60	149.19	0.1305	150	3568.7	2.550

As a function of temperature, the saturated vapor pressure follows the line PC' shown in the p-T diagrams in Figs. 19–7 and 19–8. For most substances, the saturated vapor pressure p_v is a rapidly increasing function of T. Some values of p_v for water are given in Table 19–2. Notice that p_v exceeds atmospheric pressure (760 torr) when the temperature is greater than 100° C.

Although we have phrased the discussion of vapor pressure in terms of the vaporization of a liquid into an evacuated space, this is actually an unnecessary restriction. If two or more different gases that do not interact chemically (for example, water vapor and air) are present in a container, their partial pressures contribute independently to the total pressure (Dalton's law of partial pressures). When sufficient water vaporizes to reach saturation in a vessel that contains air, it is often said that "the air is saturated with water vapor." However, the saturation condition does not depend at all on the presence of the air. Saturation is reached when the rate at which liquid molecules vaporize from the liquid surface is equal to the rate at which molecules strike the surface and are absorbed. This equilibrium (saturation) condition is independent of the existence of other gases in the vessel.

Boiling.   At the *boiling point* of a liquid, the vapor pressure of the liquid is equal to the external pressure. Water has a vapor pressure equal to normal atmospheric pressure (760 torr) at a temperature of 100° C (Table 19–2), so this is the normal boiling-point temperature for water. However, if the ambient pressure is reduced (for example, on a mountain top), the boiling-point temperature is correspondingly lowered. At a height of 6 km (20,000 ft), normal atmospheric pressure is approximately 355 torr, and the boiling-point temperature of water is about 80° C. (Cooking times for foods in boiling water are therefore somewhat longer at mountain altitudes than at sea level.)

If the temperature of a liquid is increased to the boiling-point temperature by the addition of heat, the continued heating of the liquid will not result in any further increase in temperature. Instead, the liquid will vaporize to completion at the boiling-point temperature. During boiling, vapor bubbles are produced throughout the liquid volume and cause the familiar agitation of the liquid that we recognize as active boiling.

The boiling-point temperature of a substance is affected not only by pressure but by the addition of a nonvolatile solute to the solvent liquid. Such an addition lowers the vapor pressure by a small amount and causes the boiling-point temperature to increase slightly. The addition of 1 mol of any (nonelectrolyte) substance, such as glucose, sucrose, or ethylene glycol, to 1 kg of water (thereby producing a *1-molal* solution) will elevate the boiling-point temperature by 0.512 deg when the pressure is 1 atm. For example, ethylene glycol ($C_2H_6O_2$) has a molecular mass of 62.0 u = 62.0 g/mol. If 100 g of glycol is added to 1 kg of water, the solution will be 100/62.0 = 1.6 *molal*. Then, the new boiling-point temperature will be $100.00°$ C $+ (1.6)(0.512$ deg$) = 100.82°$ C. The molal boiling elevation constant depends on the composition of the solvent but not on the solute.

Melting. An increase in pressure always increases the boiling-point temperature of a substance. On the other hand, a pressure increase will lower the melting-point temperature of some substances and will raise it for others. If a substance expands in volume on melting, a pressure increase will cause the melting-point temperature to be raised. However, if a substance (such as water) contracts on melting, a pressure increase will cause the melting-point temperature to be lowered. This difference is due to the difference in slope, $\partial p/\partial T$, of the solid-liquid dual-phase lines for the two types of substances (refer to the two p-V-T surface diagrams in Fig. 19–7).

The change in melting-point temperature with pressure is a small effect for most substances. For water, the average rate of decrease of the melting-point temperature with pressure (for $p \lesssim 1000$ atm) is approximately 9.1×10^{-3} deg/atm. Because ice has a lower melting-point temperature when under great pressure, an ice surface near $0°$ C becomes very slippery for an ice skater. The relatively high pressure beneath the skate blades causes the ice to melt and provides a thin layer of water on which the blades travel with little friction. After the skater passes, the pressure returns to atmospheric pressure and the water refreezes.

Just as the boiling-point temperature is raised by the addition of a nonvolatile solute to the liquid, so is the melting-point temperature lowered in the same way. Both effects are caused by a decrease in the vapor pressure due to the addition of the solute. (• Can you see why this change in the vapor pressure results in the observed effects?) A 1-molal solution of nonelectrolyte substance in water will depress the melting (freezing) point to a temperature of $-1.86°$ C. Thus, a 5-molal solution of glycol in water will result in a freezing-point temperature of $-9.3°$ C. (Ethylene glycol is the primary ingredient of *antifreeze*.) Again, the molal freezing depression constant is a function of the solvent only.

For substances that are electrolytes (that is, ionizing substances)—such as NaCl and acids—the situation is somewhat more complicated. In general, these substances affect both the boiling-point and freezing-point temperatures to a greater extent than do nonelectrolyte substances (such as glycol and glucose). A practical application of the depression of the melting-point temperature by the addition of solutes is the spreading of calcium chloride $(CaCl_2)$ on icy road surfaces; this lowers the freezing-point temperature and promotes the melting of the ice at temperatures below $0°$ C.

Humidity. The partial pressure of water vapor in air at ordinary temperatures is small. For example, the maximum partial pressure (that is, the saturated vapor pressure) at $20°$ C is only about 2 percent of atmospheric pressure (see Table 19–2). Nevertheless, water vapor in air has a great effect on climatic conditions and on personal comfort.

The ratio of the actual partial pressure of water vapor in air to the saturated vapor pressure at the same temperature is called the *relative humidity*. When the relative humidity is low, the air is *dry*. Then, water evaporates quickly because the rate at which water molecules are vaporized is much greater than the rate at which they return to the water surface and are absorbed. In low-humidity conditions, body perspiration readily evaporates and we feel cooled by this process. (• What produces the cooling sensation?) However, if the relative humidity is high, the evaporation rate is low and one has a feeling of oppressive mugginess. If the relative humidity reaches 100 percent, there can be no net evaporation—wet clothing or damp skin will never dry under such a condition.

Example 19–6

When the air temperature is 30° C, a piece of metal is cooled until, at 20° C, water droplets appear on the surface. What is the relative humidity?

Solution:

The air in the immediate vicinity of the metal is cooled by conduction to the temperature of the metal. The appearance of water droplets means that the cooled air (at 20° C) is saturated at this temperature. Thus, the partial pressure at 30° C is equal to the saturated vapor pressure at 20° C, namely, 17.5 torr (see Table 19–2). Hence,

$$\text{relative humidity} = \frac{\text{actual partial pressure (30° C)}}{\text{saturated vapor pressure (30° C)}}$$

$$= \frac{17.5 \text{ torr}}{31.7 \text{ torr}} = 0.55 = 55\%$$

The temperature at which condensation first appears is called the *dew point,* and this technique for determining the relative humidity is called the *dew-point method.*

QUESTIONS

19–1   Define and explain briefly the meaning of the terms (a) unified atomic mass unit, (b) mole, (c) ideal gas, (d) equation of state, and (e) Dalton's law of partial pressures.

19–2   Describe simple experimental arrangements that you might use to verify Boyle's law and the law of Charles and Gay-Lussac. Discuss possible sources of error.

19–3   Converting water into steam requires supplying energy in the form of heat even though the temperature remains constant. Explain in physical terms what is happening to the added energy.

19–4   Define and explain briefly the meaning of the terms (a) isothermal line, (b) critical point, (c) triple line, (d) sublimation, and (e) humidity.

19–5   What is the distinction between the two regions of the p-V-T surface labeled gas and vapor?

19–6   What special physical significance is associated with the triple line and the triple point of a substance?

19–7   In describing the critical point for the van der Waals' equation of state (see Fig. 19–10), it was stated that at the critical pressure p_c and critical temperature T_c, the cubic equation (Eq. 19–2) in $\mathcal{V}$ has three equal roots $\mathcal{V} = \mathcal{V}_c$. Prove that, equivalently, the critical point corresponds to the conditions $\partial p/\partial \mathcal{V} = 0$ and $\partial^2 p/\partial \mathcal{V}^2 = 0$.

19–8   Explain why gases such as air obey the ideal gas law as well as they do at normal temperatures and pressures, even though under these circumstances such gases are far from being dilute.

19–9   Explain in physical terms why you expect the boiling point of a liquid to increase with increased pressure.

19–10   Explain how the use of a ''pressure cooker'' reduces the time needed to cook foods.

19–11   Explain why, on a cold winter day, frost can form on the *inside* of a windowpane even though the room is warm.

19–12   A dish with a small amount of water is placed in a bell jar, and the air inside the jar is pumped away. At first bubbles appear in the water, and then the water freezes. Explain.

19–13   What is the difference between evaporation and boiling?

19–14   Why is it difficult to dry clothes in the open air on a humid day?

PROBLEMS

Section 19–1

19–1   What is the mass (in kg) of the light isotope of hydrogen, ^{1}H?

19–2   Natural lithium consists of two isotopes, ^{6}Li (7.42 percent abundance) and ^{7}Li (92.58 percent abundance). The atomic mass of the isotope ^{7}Li is 7.016004 u. What is the atomic mass of ^{6}Li? [*Hint:* Use the information given in the Table on page 373.]

19–3   The element chlorine has two stable isotopes, ^{35}Cl (34.96885 u) and ^{37}Cl (36.96590 u). What are the natural abundances of these isotopes? [*Hint:* Use the information given in the Table on page 373.]

19-4   How many grams of hydrogen are there in 1 kg of ethyl alcohol (C_2H_5OH)?

19-5   The density of butyl alcohol (C_4H_9OH) is 0.80567 g/cm^3. How many molecules of butyl alcohol are there in 1 cm^3?

19-6   Lord Rayleigh (1842-1919), an English physicist, was able to estimate the size of an oil molecule by allowing a droplet of the oil ($m = 8 \times 10^{-4}$ g, $\rho = 0.90$ g/cm^3) to spread out over a water surface. He found that the maximum area covered was 0.55 m^2. When he attempted to spread the oil over a greater area, the film tore apart. Rayleigh concluded that the area of 0.55 m^2 represented a monomolecular layer, so that its thickness corresponded to the size of the oil molecule. What did Rayleigh find for the size of the molecule?

19-7   Estimate the "size" of a water molecule by assuming that each molecule in a sample occupies a cubical box that is in contact with all of the neighboring boxes.

Section 19-2

19-8   Gas is confined in a tank at a pressure of 10.0 atm and a temperature of 15° C. If one half of the gas is withdrawn and the temperature is raised to 65° C, what is the new pressure in the tank?

19-9   An automobile tire is inflated using air originally at 10° C and normal atmospheric pressure. During the process, the air is compressed to 28 percent of its original volume and the temperature is increased to 40° C. What is the tire pressure? After being driven at high speed, the tire air temperature rises to 85° C and the interior volume of the tire increases by 2 percent. What is the new tire pressure? Express each answer in Pa (absolute) and in lb/in.^{2} (gauge). [1 atm = 14.70 lb/in.^{2}]

19-10   A diving bell in the shape of a cylinder with a height of 2.50 m is closed at the upper end and open at the lower end. The bell is lowered from air into sea water ($\rho = 1.035$ g/cm^3). The air in the bell is initially at a temperature of 20° C. The bell is lowered to a depth (measured to the bottom of the bell) of 45.0 fathoms or 82.3 m. At this depth the water temperature is 4° C, and the bell is in thermal equilibrium with the water. (a) How high will sea water rise in the bell? (b) To what minimum pressure must the air in the bell be raised to expel the water that entered?

19-11   A 2.0-ℓ flask is filled with 1.50 g of helium gas (He) and sealed. The temperature of the flask is then reduced to $-150°$ C. What is the pressure of the helium gas?

19-12   A mixture of oxygen (O_2) and helium (He) at a temperature of 15° C and at a pressure of 1 atm has a density of 0.980 kg/m^3. (a) What is the partial pressure of each gas? (b) What fraction (by mass) of the mixture is oxygen?

19-13   Show that when the three roots of van der Waals' equation are all equal (at the critical point), we have $V_c = 3b$, $p_c = a/27b^2$, and $T_c = 8a/27Rb$. [Hint: Expand $(V - V_c)^3 = 0$; write Eq. 19-2 in powers of V and compare like terms.] Then, we have

$$\frac{p_c V_c}{RT_c} = \frac{3}{8}$$

This relationship connecting the critical-point parameters should be valid for *any* substance that follows the van der Waals' model. Actual values of $p_c V_c/RT_c$ for common substances vary from 0.23 to 0.32; for example, nitrogen has a value of 0.292. It is not surprising that the prediction of this simple theory does not agree precisely with experiment.

19-14   Assume that the molecules of diatomic gases such as nitrogen and oxygen consist of touching, hard-sphere atoms. Use the fact that the critical molar volumes for nitrogen and oxygen are, respectively, 90.0 cm^3/mol and 78.0 cm^3/mol, and determine the radii of nitrogen and oxygen atoms.

19-15   The smooth behavior of the curves in Fig. 19-9 suggests that we can make a power-series expansion of the form,

$$\frac{pV}{n} = pV = RT + B_1(T)p + B_2(T)p^2 + \ldots$$

Show that for the van der Waals' equation of state, Eq. 19-2, we have $B_1(T) = b - a/RT$. [Hint: Expand Eq. 19-2 and in the term involving $1/V$ substitute the approximate value $1/V \cong p/RT$.]

Section 19-4

19-16   A steel tank is partially filled with water and is sealed when the temperature is 20° C and atmospheric pressure is 1 atm. What is the gauge pressure in the tank when the temperature is raised to 120°? (Neglect any change in volume of the tank or of the water.)

19-17   At what temperature will water boil on the top of Mt. Whitney, California ($h = 4.42$ km)? [Hint: Use the law of atmospheres, $p = p_0 e^{-\alpha y}$, with $\alpha = 0.125$ km^{-1}; see Example 14-3b.]

19-18   A 250-g sample of glucose ($C_6H_{12}O_6$) is dissolved in 3.0 kg of water. What is the boiling point of the solution?

19-19   What is the relative humidity when the air temperature is 25° C and the dew point is 15° C?

19-20   A warehouse has a volume of 1800 m^3. At a temperature of 30° C, the relative humidity is 70 percent. How much water must be removed from the air to reduce the relative humidity to 40 percent?

Additional Problems

19–21 A cylinder with a tight-fitting movable piston originally contains gas at atmospheric pressure and a temperature of 20° C. The piston is moved so that the gas is compressed to $\frac{1}{15}$ of its original volume, at which point the pressure is 22 atm (absolute). What is the final temperature of the gas?

19–22 A bubble of marsh gas rises from the bottom of a freshwater lake, at a depth of 4.2 m and a temperature of 5° C, to the surface where the water temperature is 12° C. What is the ratio of the bubble diameters at the two locations? (Assume that the bubble gas is in thermal equilibrium with the water at each location.)

19–23 A classroom with dimensions 8 m × 6 m × 3.5 m contains air at a temperature of 18° C. What is the mass of air in the room? (The effective molecular mass of air is 28.96 u.)

19–24 What is the average distance between molecules in air at a temperature of 25° C and a pressure of 1 atm?

19–25 A 2.0-ℓ flask at a temperature of 15° C contains a mixture of argon (monatomic, Ar) and carbon dioxide (CO₂). The partial pressure of the argon is 350 torr and that of the carbon dioxide is 850 torr. (a) What is the

mass of each gas in the flask? (b) What is the density of the mixture?

19–26 A simple mercury barometer has an inside diameter of 6.5 mm, with a height of 8.0 cm above the mercury surface. The vapor pressure of mercury at 15° C is 7.8×10^{-4} torr. (a) What error in the determination of the atmospheric pressure is introduced by ignoring the presence of the mercury vapor? (b) What is the mass of the mercury vapor in the space above the mercury column?

19–27 A hospital sterilizer contains saturated steam at a gauge pressure of 2.50 atm. What is the temperature inside the unit? [*Hint:* Interpolate in Table 19–2.]

19–28 A 0.50-kg sample of sucrose (C₁₂H₂₂O₁₁) is dissolved in 1.00 kg of water. (a) What is the expected boiling-point temperature of the sugar solution? (b) What is the expected freezing-point temperature of the solution?

19–29 A high-pressure cell with a volume of 10 cm³ contains
• 1.90×10^{-2} mol of CO₂ gas at a temperature of 90° C. (a) If the gas obeyed the ideal gas law, what would be the cell pressure? (b) What is the actual pressure, using the empirical isotherms for CO₂ shown in the accompanying graph? The next three parts refer to maintaining a

Problems 19–29, 19–30, 19–31

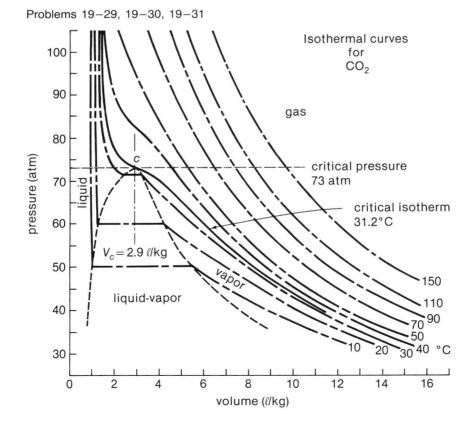

constant pressure equal to the actual pressure given by your answer to part (b). (c) By what amount must the volume be reduced to begin liquefication? (d) At what volume will the CO_2 be all liquid? (e) Describe the equilibrium state when the volume has been reduced to 2.5 cm^3.

19-30 (a) Refer to the graph of the empirical isotherms for CO_2
• on page 388 and determine the critical point values of p_c(Pa), T_c(K), and V_c(m^3/mol). (b) Using these values, show that $p_c V_c / RT_c$ is approximately $\frac{3}{8}$. (c) Evaluate the van der Waals' constants a (m^{6}Pa/mol^2) and b (m^3/mol). (d) The van der Waals' equation may be written in a power series

$$\frac{pV}{RT} = R + \left(Rb - \frac{a}{T}\right)\frac{1}{V} + \frac{Rb^2}{V^2} + \frac{Rb^3}{V^3} + \ldots$$

See Problem 19-15. Determine the Boyle temperature for which the term in $1/V$ vanishes.

19-31 (a) Using values of $a = 0.363$ m^{6}Pa/mol^2 and $b =$
•• 4.25×10^{-5} m^3/mol for CO_2 in the van der Waals equation, calculate the pressure in atmospheres for the isotherm corresponding to 40° C with the specific volume expressed in ℓ/kg as the variable. (b) Determine the corresponding isotherm using the ideal gas law. (c) Compare your results with the observed values shown in the graph or the following values:

$V(\ell/kg)$	2	3	4	5	6	8	10	12	14
p(atm)	91	82	77	70	64	53	46	39	35

(d) Repeat these calculations at 150° C.

19-32 On a winter day, a room with dimensions 6 m × 5 m × 3.5 m is heated to a temperature of 20° C with an accompanying reduction in relative humidity to 20 percent. How many liters of water must be vaporized to raise the relative humidity to 40 percent?

19-33 What is the mass of water vapor in a room with dimensions 8 m × 6 m × 2.8 m when the temperature is 20° C and the relative humidity is 45 percent?

19-34 A 2.5-m^3 sample of air at 40° C is sealed in a container. When the temperature is lowered to 10° C, 35 cm^3 of water collects in the bottom of the container. What was the original relative humidity of the air sample?

19-35 A room with a volume of 200 m^3 is at a temperature of 25° C and relative humidity of 60 percent. A bucket of water is placed in the room, which is then sealed. How many liters of water will eventually evaporate from the bucket?

19-36 A tropical air mass at a temperature of 30.0° C passes
•• over the Gulf of Mexico and becomes saturated with water vapor. The pressure is then 755.5 torr. The air mass moves northward and eventually reaches the Appalachians, where it abruptly rises; as a result, the temperature drops to 10° C and the pressure decreases to 710.0 torr. If the air remains saturated, how much water vapor (in g/m^3) was released by the air mass as rain?

CHAPTER 20

KINETIC THEORY

In the preceding two chapters the behavior of gases was discussed from a *macroscopic* point of view. We now examine the underlying *microscopic* description of gases and establish the relationship between these two ways of viewing gas dyamics. When certain quantum effects are included, it is revealed that these two descriptions are completely equivalent—a remarkable triumph of modern physical theory. A complete discussion of the details of a microscopic model is, however, beyond the scope of this book, and we limit consideration here to the essential features and a few applications.

From the previous discussions we know that dilute gases are well described by the ideal gas law, at least for temperatures above the triple-point temperature. This suggests the possibility that real gases can be described in terms of an uncomplicated microscopic model. Accordingly, we begin by making some simplifying assumptions.

First, we accept the atomic hypothesis that a gas consists of a very large number of small particles (molecules) and that for a particular pure stable substance these molecules have a definite atomic structure and are identical in all respects. Molecules have sizes that are small compared with the mean separation of the molecules in a dilute gas. Consequently, we ignore (at first) the molecular size and treat molecules as ideal pointlike particles. Likewise, the shortness of the range of intermolecular forces (also small compared with the mean separation of the molecules in a dilute gas) permits us to ignore the effects of these forces on the average.

In view of these simplifying assumptions, we can consider a molecule in a gas to move in a straight-line path until it collides with another molecule (or with a wall of the confining vessel). During such encounters, both elastic and inelastic collisions may occur. Inelastic collisions involve the conversion of some kinetic energy into internal atomic or molecular excitations. In other collisions the inverse process may occur, with atomic or molecular excitation energy converted into kinetic energy. These processes involving internal excitations can be important even at modest temperatures and for dilute gases, as we will see later. However, at first we assume that only elastic collisions take place. These molecular encounters, even though elastic, can result in significant changes in the speeds and directions of motion of the molecules. Consequently, any particular molecule follows a zigzag path, with abrupt changes in its motion at short and random time intervals. We assume that Newton's laws of motion are obeyed in all of these processes.

If a gas is in a state of thermal equilibrium, the total kinetic energy of the gas is necessarily constant. Because the number of molecules present in any macroscopic gas sample is very large, the *fraction* of the molecules with any particular speed and direction of motion will remain constant. Thus, there is present an *average velocity distribution*.

Only when phenomena involving a very small sample of molecules are considered do fluctuations from this average distribution become important (as, for example, in the case of Brownian motion, the jiggling about of dust or similar visible particles due to molecular impacts).

20–1   KINETIC THEORY OF DILUTE GASES

We first establish one of the important links between the macroscopic and microscopic views of gas behavior. The pressure exerted by a gas on the walls of a confining vessel may be imagined to result from many collisions that the gas molecules make with the walls. We assume that these collisions are elastic, so only the component of the molecular velocity perpendicular to the wall is changed (for example, from v_x to $-v_x$). The resulting change in linear momentum of the molecule is associated with the perpendicular force exerted on the molecule by the wall during the collision. The reaction to this force—the force exerted on the wall by the molecule—contributes to the total gas pressure on the wall.

Consider a small area A of a confining wall that we take to coincide with the y-z plane (Fig. 20–1). Let us examine the collisions between the wall and those molecules that have velocities with x-components between v_x and $v_x + \Delta v_x$, and with *any* velocity components in the y- and z-directions. For brevity, we refer to "the molecules with velocity components v_x."*

Construct a rectangular box with end areas A and sides with length $v_x \, \Delta t$ parallel to the x-axis. Let $\Delta \mathcal{N}(v_x)$ be the number of molecules per unit volume with velocity components v_x and any components in the y- and z-directions. The number of such molecules that strike the wall area A during the time Δt is just the number that occupy the volume $v_x \, \Delta t A$ shown in Fig. 20–1. This number is $v_x \, \Delta t A \, \Delta \mathcal{N}(v_x)$. For molecules that move exactly along the x-direction, this expression is clearly valid. However, molecules within the volume $v_x \, \Delta t A$ might have velocities with y- or z-components that will cause the molecules to leave the volume through the sides, top, or bottom before striking the area A. Even though we consider an incremental volume, it is still necessary to deal with a very large number of molecules. For every molecule that leaves the volume, there is another with the same velocity components, originally outside the volume, that will enter and strike A. The same reasoning applies to molecules within the volume and moving toward A that collide with other molecules and are scattered out of the volume without striking A. Just as many molecules are scattered *into* the volume and strike A as are scattered *out of* the volume.

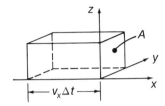

Fig. 20–1. Geometry for considering the collisions of gas molecules with a wall.

When a molecule collides elastically with the wall, the x-component of the velocity changes from v_x to $-v_x$, and the momentum changes by an amount $2\mu v_x$, where μ is the mass of the molecule. The force exerted on the area A during the time interval Δt is ΔF; then, using Newton's second law in the form that equates $\Delta F \, \Delta t$ to the change in momentum, we have

$$\Delta F \, \Delta t = [2\mu v_x][v_x \, \Delta t A \, \Delta \mathcal{N}(v_x)]$$

Thus, the pressure on A due to these collisions is

$$\Delta p = \frac{\Delta F}{A} = 2\mu v_x^2 \, \Delta \mathcal{N}(v_x) \qquad \textbf{(20–1)}$$

Because of the complete lack of long-range molecular order,† a dilute gas has a velocity distribution that is *isotropic;* that is, the velocity distributions for the x-, y-, and z-components are exactly the same. Also, the distributions for the components $+v_x$ are the

---

*There is no molecule that has a velocity *precisely* equal to some exactly specified value of v_x (for example, 7.68310765 . . . m/s); consequently, the only meaningful way to refer to the molecular velocities is to count the number of molecules whose velocities lie within a specified range, say, from v_x to $v_x + \Delta v_x$.

†We assume that the confining vessel has ordinary laboratory dimensions, so we rule out (for now) any gravitational effects.

same as for the components $-v_x$, and similarly for the y- and z-directions. Moreover, if for some reason the initial distribution were not isotropic, the random collisions between molecules would rapidly bring about an equilibrium condition with an isotropic velocity distribution. Thus, we have

$$\Delta \mathcal{N}(v_x) = \Delta \mathcal{N}(-v_x) = \tfrac{1}{2} \Delta \mathcal{N}(|v_x|)$$

so that Eq. 20–1 becomes

$$\Delta p = \mu v_x^2 \, \Delta \mathcal{N}(|v_x|)$$

and the total pressure due to all velocity components v_x is

$$p = \mu \sum_{|v_x|=0}^{\infty} v_x^2 \, \Delta \mathcal{N}(|v_x|) \tag{20–2}$$

If we consider v_x^2 and calculate the average over the distribution of velocity components v_x, we have*

$$\langle v_x^2 \rangle = \frac{\Sigma v_x^2 \, \Delta \mathcal{N}(|v_x|)}{\Sigma \, \Delta \mathcal{N}(|v_x|)} \tag{20–3}$$

where the denominator is equal to the total number of molecules per unit volume:

$$\sum_{|v_x|} \Delta \mathcal{N}(|v_x|) = \mathcal{N} \tag{20–4}$$

Combining Eqs. 20–2, 20–3, and 20–4, there results

$$p = \mu \mathcal{N} \langle v_x^2 \rangle \tag{20–5}$$

The isotropy condition is expressed as

$$\langle v_x^2 \rangle = \langle v_y^2 \rangle = \langle v_z^2 \rangle$$

Moreover, the square of the molecular speed is

$$v^2 = \mathbf{v} \cdot \mathbf{v} = v_x^2 + v_y^2 + v_z^2$$

Thus, we can write

$$\langle v_x^2 \rangle = \langle v_y^2 \rangle = \langle v_z^2 \rangle = \tfrac{1}{3} \langle v^2 \rangle$$

so that in terms of the total velocity distribution,

$$p = \tfrac{1}{3} \mu \mathcal{N} \langle v^2 \rangle \tag{20–6}$$

or, using $\mu \mathcal{N} = \rho$ for the mass density, we have†

$$p = \tfrac{1}{3} \rho \langle v^2 \rangle \tag{20–7}$$

---

*We use angled brackets to indicate an average over a *distribution;* an overbar continues to represent an average over time or space.

†Note carefully the differences between such quantities as $\langle x \rangle^2$, $\langle |x| \rangle^2$, and $\langle x^2 \rangle$. For example, suppose that the values of x are 1, -1, 3, and -3; then we have

$$\langle x \rangle^2 = [\tfrac{1}{4}\{1 + (-1) + 3 + (-3)\}]^2 = 0$$
$$\langle |x| \rangle^2 = [\tfrac{1}{4}\{1 + 1 + 3 + 3\}]^2 = 4$$
$$\langle x^2 \rangle = [\tfrac{1}{4}\{(1)^2 + (-1)^2 + (3)^2 + (-3)^2\}] = 5$$

This equation expresses a remarkably simple relationship between a microscopic property, namely, the molecular velocity distribution (properly, the speed distribution), and a macroscopic concept, namely, the pressure. Note also that, ignoring gravitational effects, the pressure on the area A is the same as the pressure on any part of the interior surface of the confining vessel, regardless of the shape. (• Can you see why the pressure should behave this way?)

It is important to realize that the velocities (speeds) that appear in the derivation of Eq. 20–7 are the speeds of the molecules *between* collisions. If a label could somehow be attached to an individual molecule, we would see it progress through the gas along a zigzag path, as mentioned earlier. The actual *transport speed* measuring the progress of the molecule through the gas would be only a small fraction of its average (or *root-mean-square*) speed, $v_{rms} = \langle v^2 \rangle^{1/2}$. Thus, if you open a bottle of perfume in one corner of a room by popping the cork, a considerably longer time will be required for a person across the room to detect the perfume by smell* than to hear the sound of the bottle being opened. (At atmospheric pressure and near room temperature, v_{rms} for most common gases is in the range from about 400 m/s to about 1800 m/s; these speeds are about the same as the speed of sound in the same gases.) The process by which individual molecules move long distances through a gas (or liquid or solid) is called *diffusion*.

Example 20–1

Dry air at 15° C and 1 atm consists of nitrogen N_2 with a partial pressure of 0.791×10^5 Pa and a mass fraction of 75.5 percent, oxygen O_2 with a partial pressure of 0.212×10^5 Pa and a mass fraction of 23.2 percent, plus a number of other small molecular fractions. Calculate the root-mean-square speeds of the nitrogen and oxygen molecules. Take the density of dry air at the stated conditions to be $\rho_0 = 1.225$ kg/m³.

Solution:

The density of nitrogen is $0.755\rho_0 = (0.755)(1.225$ kg/m³$) = 0.925$ kg/m³. Therefore, solving Eq. 20–7 for

$v_{rms} = \langle v^2 \rangle^{1/2}$, we find

$$v_{rms}(N_2) = \left[\frac{3p}{\rho} \right]^{1/2} = \left[\frac{3(0.791 \times 10^5 \text{ Pa})}{0.925 \text{ kg/m}^3} \right]^{1/2} = 506 \text{ m/s}$$

Similarly, for oxygen, the density is $(0.232)(1.225$ kg/m³$) = 0.284$ kg/m³. Therefore,

$$v_{rms}(O_2) = \left[\frac{3(0.212 \times 10^5 \text{ Pa})}{0.284 \text{ kg/m}^3} \right]^{1/2} = 473 \text{ m/s}$$

Notice that the oxygen molecules, which are more massive than the nitrogen molecules (32 u per molecule of O_2 versus 28 u per molecule of N_2), have a lower rms speed.

Kinetic Interpretation of Temperature. It is useful to continue the analysis beyond the result expressed in Eq. 20–6. The kinetic energy of a single molecule is $K = \frac{1}{2}\mu v^2$, so we recognize $\frac{1}{2}\mu\langle v^2 \rangle = \langle K \rangle$ as the *average translational kinetic energy* per molecule.† If the volume of the gas in question is V and there are n moles of gas present, we have $\mathcal{N} = nN_0/V$, where N_0 is Avogadro's number. Thus, using Eq. 20–6 we have

$$\frac{1}{2}\mu\langle v^2 \rangle = \frac{3}{2}\frac{p}{\mathcal{N}} = \frac{3}{2}\frac{pV}{nN_0}$$

Now, for a sufficiently dilute gas, the ideal gas law equation is always found to be valid,

---

*The human olfactory nerve is quite sensitive for some substances. As small a concentration as one molecule of methyl mercaptan (CH_3SH) in 10^{12} air molecules can be detected by the nostrils.

†Remember that μ is measured in kg, *not* in atomic mass units.

so we substitute $pV = nRT$, and obtain

$$\tfrac{1}{2}\mu\langle v^2\rangle = \tfrac{3}{2}\,\frac{R}{N_0}\,T$$

We define a new constant, the *Boltzmann constant*[*] k, to be

$$k \equiv \frac{R}{N_0} = \frac{8.3144 \text{ J/mol·K}}{6.02205 \times 10^{23} \text{ mol}^{-1}} = 1.38066 \times 10^{-23} \text{ J/K}$$

That is, the Boltzmann constant is the universal gas constant *per molecule*. Thus, we arrive at the important relationship,

$$\langle K\rangle = \tfrac{1}{2}\mu\langle v^2\rangle = \tfrac{3}{2}kT \qquad\qquad (20\text{–}8)$$

Using Eq. 20–8, we can express the rms speed as[†]

$$v_{\text{rms}} = \langle v^2\rangle^{1/2} = \sqrt{\frac{3kT}{\mu}} \qquad\qquad (20\text{–}9)$$

In the derivation of Eq. 20–8, we made the implicit assumption that both the containing vessel and the center of mass of the gas are at rest in some inertial reference frame. Thus, any kinetic energy associated with the C.M. of the gas does not influence its thermodynamic behavior. (For example, placing the vessel on a platform that moves with constant velocity does not alter the temperature of the gas.)

The total translational kinetic energy of n moles of gas is $nN_0\langle K\rangle$, which we identify as the *internal energy U* of the gas:

$$U = nN_0\langle K\rangle = \tfrac{3}{2}nRT \qquad\qquad (20\text{–}10)$$

This expression is valid for an ideal gas that possesses only translational kinetic energy. If rotational and vibrational forms of internal energy are involved, we write a more general expression for the internal energy that includes the contribution of these modes (see Eq. 20–14).

Equation 20–10 represents the important fact that the internal energy of an ideal gas *depends only on the temperature*. Thus, to describe an equilibrium state of such a gas, it is necessary to know *only* its temperature; it is *not* necessary to know the history of how the state was formed, whether by adding or removing heat or by adding or removing energy in any other form. The internal energy is therefore called a *state function*.[‡]

Quantum effects cannot be ignored completely in interpreting Eq. 20–10. For example, if the temperature of a substance approaches absolute zero, the internal energy does not reduce to zero, as this equation implies. According to quantum theory, there is a residual *zero-point energy* that remains even in the limit $T \to 0$ K.

---

[*] Named in honor of Ludwig Boltzmann (1844–1906), Austrian physicist and pioneer developer of modern statistical thermodynamics.

[†] Equation 20–9 explains the point raised at the end of Example 20–1.

[‡] The internal energy of a real substance is also a state function. However, in addition to the dependence on temperature, the internal energy of a real substance may have some dependence on the other state variables (pressure and volume); see Problem 21–31.

20-2   DEGREES OF FREEDOM AND THE EQUIPARTITION OF ENERGY

We have thus far considered only the translational motion of gas molecules, which involves the energy terms $\frac{1}{2}\mu v_x^2$, $\frac{1}{2}\mu v_y^2$, and $\frac{1}{2}\mu v_z^2$. But molecules may undergo motions in addition to translation. For example, a molecule might rotate about its C.M., thereby introducing energy terms such as $\frac{1}{2}I_x\omega_x^2$, $\frac{1}{2}I_y\omega_y^2$, and $\frac{1}{2}I_z\omega_z^2$. Or the component atoms of the molecule might execute vibrational motion in the molecular center-of-mass frame, giving energy terms of the form $E_{\text{vib}} = \frac{1}{2}\mu'v_{\text{vib}}^2 + \frac{1}{2}\kappa r^2$, where the second term models the bond energy as a Hooke's-law spring. (• What is the interpretation of μ' in this expression?) Each such independent mode of possible motion is referred to as a *degree of freedom* of the molecule.

It is a characteristic of the various modes of motion that each involves an energy term that depends on the square of the relevant dynamic quantity. In fact, we can express the dynamic energy of a molecule as

$$E = \sum_{i=1}^{\nu} \tfrac{1}{2}\lambda_i\xi_i^2 \tag{20-11}$$

where λ_i is some (usually constant) molecular property, such as an effective mass, a rotational inertia, or an effective spring constant; and where ξ_i is either a velocity variable or a displacement variable of an excited mode. Each term in the sum is called an *energy partition*. Translational motion of the molecule as a whole contributes three energy partitions (corresponding to the directions of the three coordinate axes) to Eq. 20-11. Each of the three rotational degrees of freedom, when excited, contributes an additional energy partition. However, each vibrational degree of freedom contributes *two* energy partitions—one involving the vibrational kinetic energy, $\frac{1}{2}\mu'v_{\text{vib}}^2$, and the other involving the Hooke's-law potential energy, $\frac{1}{2}\kappa r^2$. The quantity ν in Eq. 20-11 represents the total number of energy partitions for the system.

A system that consists of molecules with only the translational degrees of freedom excited has three energy partitions ($\nu = 3$), and Eq. 20-11 becomes

$$E = \sum_{i=1}^{3} \tfrac{1}{2}\mu v_i^2 = \tfrac{1}{2}\mu v_x^2 + \tfrac{1}{2}\mu v_y^2 + \tfrac{1}{2}\mu v_z^2$$

$$= \tfrac{1}{2}\mu v^2$$

Averaging over the molecules in the system and using Eq. 20-8, we can write

$$\langle E \rangle = \tfrac{1}{2}\mu\langle v_x^2 \rangle + \tfrac{1}{2}\mu\langle v_y^2 \rangle + \tfrac{1}{2}\mu\langle v_z^2 \rangle$$

$$= \tfrac{1}{2}\mu\langle v^2 \rangle = \tfrac{3}{2}kT$$

We have argued that in an isotropic system containing many molecules in thermal equilibrium, the averaged energy terms are all equal; that is, $\frac{1}{2}\mu v_x^2 = \frac{1}{2}\mu v_y^2 = \frac{1}{2}\mu v_z^2$. It then follows that an average energy of $\frac{1}{2}kT$ is associated with each translational degree of freedom; that is,

$$\tfrac{1}{2}\mu\langle v_x^2 \rangle = \tfrac{1}{2}\mu\langle v_y^2 \rangle = \tfrac{1}{2}\mu\langle v_z^2 \rangle = \tfrac{1}{2}kT$$

This important conclusion regarding the average energy associated with translational motion can be extended, by using statistical arguments, to other forms of molecular motion as well. That is, in equilibrium, the average energy associated with *each* partition that is excited has the same value, namely, $\frac{1}{2}kT$. Thus, we can write

$$\langle E \rangle = \frac{1}{2}\nu kT \qquad (20-12)$$

This general conclusion is referred to as the *equipartition of energy theorem.**

The determination of the number of degrees of freedom that a molecule possesses requires some quantum considerations. The simplest molecules are those of the noble gases—helium, neon, argon, krypton, xenon, and radon. These are monatomic elements, so the *molecules* are actually *atoms*. The electrons of these elements are arranged in closed shells of great stability.† As a result, the atoms of the noble gases have featureless structures—that is, they are *spherically symmetric*. One might imagine that such a spherical atom could undergo rotations about a central axis, as could a symmetric beach ball. However, for a quantity to be physically meaningful, there must be some way to *measure* it; this is a basic postulate of quantum theory. If an atom is spherically symmetric, there is no reference mark to use for a measurement of its rotation. We thus must conclude that such an atom *cannot rotate.*‡

A diatomic molecule possesses cylindric symmetry about an axis that passes through both atomic centers. Rotation about the symmetry axis is not measurable, but rotations about the two mutually perpendicular axes are measurable. Consequently, a diatomic molecule has two rotational degrees of freedom and two corresponding energy partitions. If the molecule is rigid, so that vibration is not possible, the total number of partitions is $\nu = 3 + 2 = 5$. If a diatomic molecule can undergo vibration, with the atoms moving back and forth along the line that connects their centers, this corresponds to an additional degree of freedom and contributes two additional partitions—one associated with the vibrational kinetic energy and one associated with the potential energy. Altogether, there are three translational partitions, two rotational partitions, and two vibrational partitions. Thus, $\nu = 7$ and $\langle E \rangle = \frac{7}{2}kT$.

For a triatomic molecule such as H_2O, in which the three atoms are not collinear, there exists no axis of spatial symmetry, so there are three rotational degrees of freedom. In addition, three independent modes of vibration are possible (see Fig. 20–2), and there are two energy partitions associated with each mode. Therefore, $\nu = 3 + 3 + (2 \times 3) = 12$. (If the molecule is rigid, $\nu = 3 + 3 = 6$.)

Characteristic Excitations.   Does a collection of molecules exhibit all of its possible motions at any temperature? To answer this question we must again invoke a quantum feature that occurs in the atomic domain. The possible energies of the various modes of molecular motion can occur only in discrete units or *quanta*. The smallest excitation of a particular mode (corresponding to one quantum of energy) depends on the specific molecule. Some typical values of the minimum excitation energies for different modes are given in Table 20–1. Also given are the orders of magnitude of the corresponding temperatures, $T_\theta = E_{\min}/k$. The first line of the table indicates the quantization of energy imposed on the translational motion by the finite size of the containing vessel. This limita-

Fig. 20–2. The three independent modes of vibration of a triatomic molecule. In each mode, the C.M. of the molecule remains fixed (or moves with constant velocity).

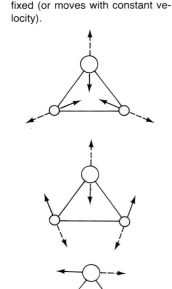

*First deduced by the great Scottish mathematical physicist James Clerk Maxwell (1831–1879), who was also responsible for developing electromagnetic theory. Important contributions to the statistical interpretation of energy were made by Boltzmann and by J. Willard Gibbs (1839–1903), the noted American physicist.

†It is for this reason that the noble gases are "noble"—namely, there are no valence electrons to form chemical bonds with other atoms, so they normally remain aloof and do not participate in chemical reactions, nor do they form diatomic molecules after the fashion of H_2, N_2, I_2, and so forth.

‡*Q.* "But suppose that the atom really *does* rotate, even though we cannot detect the rotation. What then?"

A. "If there is no conceivable way to detect the rotation or its effects, it is meaningless to claim that rotation exists."

TABLE 20–1. Typical Molecular Energies Imposed by Quantum Conditions

MODE	APPROXIMATE MINIMUM ENERGY (J/molecule)	APPROXIMATE CORRESPONDING TEMPERATURE T_θ (K)
Translation (in a box)	10^{-38}	10^{-15}
Rotation	10^{-21}	10^2
Vibration	10^{-20}	10^3
Internal (electronic)	10^{-19}	10^4

tion, for vessels of laboratory size, is negligible, as the values indicate. The last line in Table 20–1 is included to show the energy of a typical internal atomic excitation involving the structural rearrangement of the valence electrons of the atom. Most molecules dissociate into their atomic constituents before such excitations take place; it was for this reason that the partitions associated with these excitations were not included in the previous discussion.

Suppose that a sample of n moles of a diatomic gas is heated, beginning at a temperature below about 100 K. At such a low temperature the energy is not sufficient to excite either the rotational or vibrational modes (see Table 20–1). Thus, the sample consists almost exclusively of molecules executing only translational motion, with an internal energy $U = \frac{3}{2}nRT$. As the temperature is raised (but still below 100 K), rotational motion begins to occur because some inelastic collisions involving the more rapidly moving molecules have sufficient C.M. energy to excite a rotational mode in one (or perhaps both) of the colliding molecules. The small population of such states is continually added to and depleted as a result of the collisions. At each temperature the molecules that possess rotational energy represent some fixed percentage of the total sample. At a temperature somewhat above 100 K, most of the molecules execute rotational motion; then the full value of the partition associated with the rotational degree of freedom is developed. As a temperature of 1000 K is approached, the collisions become more energetic and more frequent, and the population of vibrating molecules begins increasing. At a temperature somewhat above 1000 K, most of the molecules participate in vibrational motion. Then the full value of the partition for a diatomic molecule, $\nu = 7$, is obtained.

20–3 SPECIFIC HEATS OF GASES

The equipartition principle has an important influence on the specific heats of gases. To facilitate comparisons, consider the *molar specific heat C,* which is the specific heat per mole of the gas sample. In particular, consider the specific heat for the case in which the gas is confined to a fixed volume. To distinguish this specific heat from that obtained under constant pressure conditions, we use a subscript v and write C_v for the molar specific heat at constant volume. With the volume constant, any flow of heat into or out of the gas directly affects the internal energy U (see Section 21–2). Then we define C_v as (compare Eq. 18–10a)

$$C_v(T) = \frac{1}{n}\frac{dU}{dT}\bigg|_V \tag{20–13}$$

where the notation emphasizes that the changes take place at constant volume.

Now, it is convenient to express the total internal energy of n moles of gas as (compare Eq. 20–10)

$$U(T) = \tfrac{1}{2}\nu(T; T_\theta)nRT \tag{20–14}$$

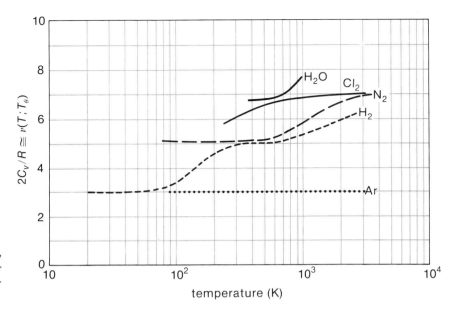

Fig. 20–3. Values of $2C_v/R = \nu$ as a function of temperature for several gases. Note the logarithmic temperature scale.

where $\nu(T; T_\theta)$ is a slowly varying function of T that exhibits change primarily near the temperatures T_θ (see Table 20–1). Thus, $\partial\nu/\partial T \cong 0$ between different values of T_θ. Then, Eq. 20–13 becomes

$$C_v(T) \cong \tfrac{1}{2}\nu(T; T_\theta)R \qquad (T \text{ not near } T_\theta) \qquad \textbf{(20–15)}$$

Figure 20–3 shows $2C_v/R = \nu$ as a function of temperature for several gases. As expected, for the monatomic gas argon (Ar), we find $\nu = 3$. The curve for hydrogen (H_2) clearly exhibits the transition from $\nu = 3$ to $\nu = 5$. Similarly, the curves for nitrogen (N_2) and chlorine (Cl_2) show the transition from $\nu = 5$ to $\nu = 7$, and that for steam (H_2O) shows the beginning of the departure from $\nu = 6$. The empirical value for the characteristic temperature of the vibrational mode for chlorine is 820 K, whereas for nitrogen it is 3400 K and for hydrogen it is 6300 K. Thus, at a temperature of 1000 K, the curve for chlorine has almost reached its full equipartition value of $\nu = 7$ due to the addition of the full contribution from the vibrational mode, whereas the nitrogen curve is about halfway to its full value and the hydrogen curve has just begun to indicate the presence of a small population of molecules in the vibrational state.

Example 20–2

An amount of heat $Q = 1.365$ Cal is required to raise the temperature of a 0.500-kg sample of iodine vapor (I_2) from 300° C to 400° C. It is also determined that C_v for iodine vapor is constant (to within 1 part in 10^3) from the boiling-point temperature (185.2° C) to about 1500° C. From this information, determine the molecular mass of I_2.

Solution:

Because the specific heat is constant over such a large temperature range, we can reasonably suppose that the full equipartition value, $\nu = 7$, has been developed. Thus,

$$C_v = \tfrac{1}{2}\nu R = \tfrac{1}{2}(7)(8.314 \text{ J/mol·K})$$

$$= 29.10 \text{ J/mol·K}$$

(The experimental value is 29.01 J/mol·K.)

The heat required to raise the temperature by ΔT for a sample with mass m and molecular mass M is (with $n = m/M$)

$$Q = \frac{m}{M} C_v \, \Delta T$$

Thus,

$$M = \frac{mC_v \, \Delta T}{Q} = \frac{(0.500 \text{ kg})(29.10 \text{ J/mol·K})(100 \text{ K})}{(1.365 \text{ Cal})(4186 \text{ J/Cal})}$$

$$M = 254.6 \times 10^{-3} \text{ kg/mol}$$

$$= 254.6 \text{ u}$$

According to the Table on page 373, the *atomic* mass of iodine is 126.9045 u. Hence, the molecular mass is $M = 253.81$, which is quite close to the value just calculated.

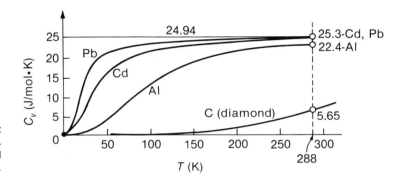

Fig. 20–4. The molar specific heats C_v for four solids that have different values of T_D (see Table 20–2). Values of C_v for solids are usually obtained by making small theoretical corrections to the experimentally determined values of C_p.

▶ *Specific Heats of Solids.* The theory of specific heats for solids is simple only for crystalline substances and for temperature ranges that do not involve phase transitions. The atoms in a crystalline solid are held in nearly fixed positions in a lattice whose structure varies from one substance to another. Thus, translational and rotational motions are suppressed; only vibrational thermal motion is allowed. Because vibrations can take place independently in three mutually perpendicular directions, the fully developed energy partition number is $\nu = 2 \times 3 = 6$.

The quantum nature of the vibrations of atoms in solids can also be expressed by a characteristic temperature called the *Debye temperature* T_D. The theory of solids developed by Debye* allows for the strong interaction of the vibrations of nearby atoms in a crystal. This coupling of the vibrating atoms plays the same role as the molecular collisions in gases in producing an equilibrium thermodynamic state on a macroscopic scale. The full theory requires incorporating the quantum feature of discrete energy excitations.

In the Debye theory, the specific heat C_v is a function of temperature T and the Debye temperature T_D, with the additional feature that $C_v \to 0$ as $T \to 0$. Because $\nu = 6$, the expected value of C_v when $T \gg T_D$ is $C_v = \frac{1}{2}\nu R = 3R \doteq 24.94$ J/mol·K. For many substances, $T_D < 250$ K, so that C_v should be close to 25 J/mol·K at room temperature. The observation

that many molar specific heats are approximately equal to this value was first made by Dulong and Petit in 1819.

If the variation of the specific heat of a crystalline solid can be characterized by a single temperature—namely, T_D—it should be possible to construct a single curve to represent the temperature dependence of the specific heat of all such solids by plotting C_v versus T/T_D. Figure 20–4 shows the temperature variation of C_v for four substances with rather different values of T_D (see Table 20–2). At a temperature of 288 K (15° C), there is a wide spread in the values of C_v. However, when the data for

TABLE 20–2.   The Debye Temperature T_D for Several Solids

SUBSTANCE	T_D (K)
Lead (Pb)	88
Cadmium (Cd)	168
Silver (Ag)	215
Aluminum (Al)	398
Nickel (Ni)	413
Calcium fluoride (CaF_2)	474
Carbon (C, diamond)	1860

---

*Peter Debye (1884–1966), Dutch-American physicist and chemist, winner of the 1936 Nobel Prize in chemistry.

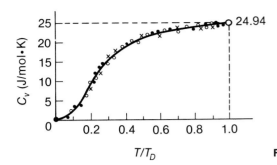

different substances are plotted as a function of T/T_D (Fig. 20–5), there is a remarkable agreement.

Although C_v approaches zero as $T \to 0$, it should not be inferred that the internal energy also tends to zero at the absolute zero of temperature. As in the case of gases, the zero-point energy remains and the atoms of a solid continue to vibrate about the stable lattice points even at $T = 0$.

Fig. 20–5. Values of C_v for different substances plotted as a function of T/T_D. ◄

20–4 MOLECULAR SPEEDS IN GASES

In the preceding discussions, we have made use of the fact that the molecules of a gas are in a state of agitated random motion with the molecular speeds described by some distribution whose rms value depends only on the temperature. The expression for the molecular speed distribution was first obtained by James Clerk Maxwell in 1860. Maxwell's derivation results in a function $dN(v)$, which gives the number of molecules in a gas sample with speeds between v and $v + dv$. The original 1860 derivation was made rigorous by Boltzmann, and it is now known as the *Maxwell-Boltzmann distribution*. We state the result without proof:

$$dN(v) = 4\pi N \left[\frac{\mu}{2\pi kT} \right]^{3/2} v^2 e^{-\frac{1}{2}\mu v^2/kT} \, dv \qquad (20-16)$$

where $N = \int dN(v)$ is the total number of molecules in the sample, μ is the mass of a single molecule, and T is the absolute temperature of the gas. Notice that for a particular molecular species, the distribution is a function only of the temperature.

The Maxwell-Boltzmann distribution has a number of interesting properties. The expression for $v_{\rm rms}$ (Eq. 20–9) is made use of in deriving Eq. 20–16, so the *root-mean-square speed* is known to be

$$v_{\rm rms} = \langle v^2 \rangle^{1/2} = \sqrt{\frac{3kT}{\mu}} \qquad (20-17a)$$

We can also show (see Problem 20–12) that the *most probable speed* (that is, the speed for which $dN(v)/dv$ is maximum) is

$$v_m = \sqrt{\frac{2kT}{\mu}} \qquad (20-17b)$$

Finally, the *average speed* can be obtained by writing

$$\langle v \rangle = \frac{\int v \, dN(v)}{\int dN(v)}$$

$$= 4\pi \left[\frac{\mu}{2\pi kT} \right]^{3/2} \int_0^\infty v^3 e^{-\frac{1}{2}\mu v^2/kT} \, dv$$

This is a standard integral that can be found in integral tables:

$$\int_0^\infty x^3 e^{-\beta x^2} \, dx = \frac{1}{2\beta^2}$$

Using this result, we find

$$\langle v \rangle = \sqrt{\frac{8kT}{\pi\mu}} \qquad (20\text{–}17c)$$

These speeds are in the ratios

$$v_m : \langle v \rangle : v_{\text{rms}} = \sqrt{\frac{2}{3}} : \sqrt{\frac{8}{3\pi}} : 1$$

$$= 0.8165 : 0.9213 : 1$$

Figure 20–6 shows $dN(v)/dv$ as a function of v for nitrogen molecules at three different temperatures. As the temperature is increased, the peak of the distribution shifts toward higher speeds and the curve becomes more spread out.

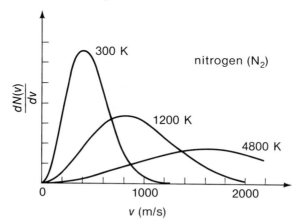

Fig. 20–6. The Maxwell-Boltzmann distribution for nitrogen molecules at three different temperatures.

A universal function to express the Maxwell-Boltzmann distribution for all cases can be obtained by writing the speed coordinate in relative units, $\xi = v/v_{\text{rms}}$, and by multiplying $dN(v)/dv$ by v_{rms}/N. Thus,

$$\frac{v_{\text{rms}}}{N} \cdot \frac{dN(v)}{dv} = 6\sqrt{\frac{3}{2\pi}} \, \xi^2 e^{-\frac{3}{2}\xi^2} \qquad (20\text{–}18)$$

This function is shown in Fig. 20–7. Notice the locations of v_m, $\langle v \rangle$, and v_{rms}.

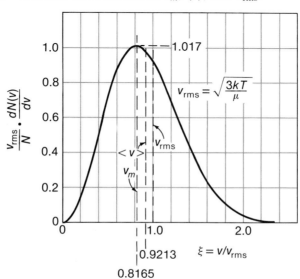

Fig. 20–7. The universal Maxwell-Boltzmann distribution function, Eq. 20–18.

The Boltzmann Statistical Factor. The analytical methods of statistical thermodynamics that yield the equipartition of energy principle also lead to another general principle of great importance. This principle states that for a collection of molecules in equilibrium, the number of molecules N_i with *total energy* E_i is proportional to the *Boltzmann statistical factor;* that is

$$N_i = Ae^{-E_i/kT} \qquad\qquad (20-19)$$

Thus, the ratio of the number of molecules with energy E_i to the number with energy E_j is

$$\frac{N_i}{N_j} = \frac{e^{-E_i/kT}}{e^{-E_j/kT}} = e^{-(E_i - E_j)/kT}$$

If we consider only the translational motion of gas molecules in equilibrium at a temperature T, the number of molecules $dN(v)$ within the speed interval dv is proportional to the product of two statistical factors. The first is the volume of *velocity space* that corresponds to the speed interval dv, namely, $dv_x\, dv_y\, dv_z$ in rectangular coordinates or $4\pi v^2\, dv$ in spherical coordinates. The second is the Boltzmann factor, which now becomes $e^{-\frac{1}{2}\mu v^2/kT}$. Thus, we have

$$dN(v) = 4\pi A v^2 e^{-\frac{1}{2}\mu v^2/kT}\, dv$$

When the constant A is evaluated by setting $\int dN(v) = N$, we obtain the Maxwell-Boltzmann distribution, Eq. 20–16! That is, the statistical approach of Boltzmann yields exactly the same result as Maxwell's original derivation.

The Boltzmann factor has broad applicability. A many-component system in thermal equilibrium has a certain number of degrees of freedom that contribute to the possible total energies E_i. The energy distribution for the components of the system is described by a function that contains the Boltzmann factor $e^{-E_i/kT}$ multiplied by the volume elements associated with the degrees of freedom. When only the translational degrees of freedom are considered, the appropriate volume element is $dv_x\, dv_y\, dv_z$, as cited above.

Example 20–3

Consider a gas sample in a vessel of fixed volume and in equilibrium at a temperature T. Let the sample be in a gravitational field with a uniform downward acceleration $\mathbf{g}$. Determine the pressure in the vessel as a function of height.

Solution:

Let the confining vessel have dimension $a \times b \times c$, as shown in the diagram on page 403. Each molecule in the gas sample has a kinetic energy $\frac{1}{2}\mu v^2 = \frac{1}{2}\mu v_x^2 + \frac{1}{2}\mu v_y^2 + \frac{1}{2}\mu v_z^2$, and a potential energy, μgy, referred to the plane $y = 0$, indicated in the diagram. According to Eq. 20–19, the number of molecules within a spatial volume, $dx\, dy\, dz$, and simultaneously in a velocity volume, $dv_x\, dv_y\, dv_z$, is

$$dN(\mathbf{v}, \mathbf{r}) = Ae^{-[\frac{1}{2}\mu(v_x^2 + v_y^2 + v_z^2) + \mu gy]/kT}\, dx\, dy\, dz\, dv_x\, dv_y\, dv_z$$
$$= A\{e^{-\frac{1}{2}\mu(v_x^2 + v_y^2 + v_z^2)/kT}\, dv_x\, dv_y\, dv_z\}\cdot\{e^{-\mu gy/kT}\, dx\, dy\, dz\}$$

where the second equality emphasizes the fact that the six relevant coordinates are all *independent*. If we ask first for the speed distribution of the molecules in *any* part of the vessel, we must perform the appropriate integration over x, y, and z. This integration leaves unaltered the terms in the first bracket (the velocity terms). Thus, the speed distribution does not depend on the space coordinates and is exactly the same as if $\mathbf{g}$ were zero. This is to be expected because we assumed the gas to be in equilibrium at a particular temperature (that is, the temperature is constant throughout the vessel).

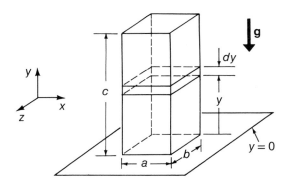

If we ask next for the number of molecules within a volume $(ab)\,dy$ at a height y and with any velocity whatsoever, we must integrate over v_x, v_y, and v_z (in the range $-\infty$ to $+\infty$) and over x and z (in the ranges, $0 \leqslant x \leqslant a$ and $0 \leqslant z \leqslant b$). The result in terms of a suitable constant $\mathcal{N}_0$ is

$$dN(y) = \mathcal{N}_0 ab e^{-\mu gy/kT}\, dy$$

Then, the number of molecules per unit volume is

$$\mathcal{N}(y) = \frac{1}{ab} \cdot \frac{dN(y)}{dy} = \mathcal{N}_0 e^{-\mu gy/kT}$$

so that $\mathcal{N}_0$ is seen to be the value of $\mathcal{N}(y)$ at $y = 0$.

The analysis we have made could be applied to a container of gas on an accelerating truck (• Can you see why?) or to the isothermal law of atmospheres discussed in Section 15–2. To convert the equation for $\mathcal{N}(y)$ into the form of Eq. 15–7, we write the ideal gas law equation as

$$p = \frac{n}{V}RT = \frac{N}{N_0 V}RT = \mathcal{N}kT$$

Using this equation for $\mathcal{N}(y)$, we obtain

$$p(y) = p_0 e^{-\mu gy/kT}$$

Now, the density of the gas sample is $\rho = \mathcal{N}\mu$. Thus, the exponent becomes

$$\frac{\mu gy}{kT} = \frac{\rho_0}{\mathcal{N}_0}\frac{gy}{kT} = \frac{\rho_0}{p_0}gy = \alpha y$$

Therefore, we have

$$p(y) = p_0 e^{-\alpha y} \qquad (\alpha = \rho_0 g / p_0)$$

which is just Eq. 14–7.

QUESTIONS

20–1 Define and explain briefly the meaning of the terms (a) root-mean-square speed, (b) translational internal energy, (c) degrees of freedom, (d) equipartition of energy, and (e) Debye theory of specific heat of solids.

20–2 Use the results expressed in Eqs. 20–7 and 20–8 to explain Boyle's law and the law of Charles and Gay-Lussac.

20–3 Justify Dalton's law of partial pressures on the basis of the kinetic theory.

20–4 A container at rest is filled with gas consisting of a single molecular species in thermal equilibrium. Then the vector sum of all the molecular velocities must vanish. Explain why this must be so and how it is then possible for the average molecular speed to be nonzero.

20–5 A high-pressure cylinder of air develops a small leak through which air slowly escapes. If the air inside the cylinder does not exchange heat with the cylinder walls, the temperature of the confined gas is observed to decrease. Give an explanation based on the kinetic theory. [*Hint:* Which molecules are most likely to find their way to the hole and leak out?]

20–6 Define and explain briefly the meaning of the terms (a) Maxwell-Boltzmann speed distribution, (b) Boltzmann statistical factor, (c) mean free path, and (d) transport phenomena.

20–7 Air consists mostly of a mixture of nitrogen and oxygen with small amounts of carbon dioxide, hydrogen, and argon. How would you expect the composition of air in terms of these components to vary with altitude?

20–8 In the derivation of the pressure an ideal gas exerts on the walls of its container, elastic collision between a molecule and a perfectly rigid (continuously solid) wall was assumed. The walls, of course, also consist of molecules that are executing thermal motions. Give a qualitative explanation of why, if the walls of the container are at the same temperature as the gas, the two models of the wall give the same result. [*Hint:* Reconsider the momentum interchange at the wall leading to Eq. 20–1 and the Boltzmann equipartition of energy theorem.]

20–9 In the derivation of the pressure in a gas using kinetic theory, the action of gravity was neglected, whereas in Example 20–3 it was specifically allowed for. Give a criterion for when gravitational effects may be neglected and when they must be included.

20–10 A fixed amount of helium is taken from an equilibrium state to another with twice the volume and at twice the pressure. What is the change in the rms speed of the helium atoms? What if only the pressure changed to twice its original value but the volume remained the same?

20–11 A piece of chalk is thrown into the air. Its kinetic energy, and hence that of the molecules it contains, varies throughout the flight. Does this mean that its temperature also varies accordingly? Why does its temperature change when it hits the ground?

PROBLEMS

Section 20–1

20–1 The speed distribution of 25 molecules is

$v(10^2$ m/s)	1	2	3	4	5	6	7	8	9	10	>10
$\Delta \mathcal{N}(v)$	1	2	4	5	4	3	3	2	0	1	0

(a) Plot the distribution function $\Delta \mathcal{N}(v)$ as a function of v. (b) Find the average speed, $\langle v \rangle$. (c) Find the root-mean-square speed, v_{rms}. (d) Determine the ratio $v_{rms}/\langle v \rangle$, and compare with $\sqrt{3\pi/8}$, which is the value appropriate for an ideal gas.

20–2 A vessel contains 5.0 mg of nitrogen (N_2) at a pressure of 0.10 atm. An experiment shows that the rms speed is $v_{rms} = 300$ m/s. (a) Determine the volume of the container. (b) What is the average spacing between molecular centers? Compare your value with the diameter of a nitrogen atom (1.06×10^{-10} m).

20–3 At a temperature of 15.0° C and a pressure of 1 atm, 5.551 g of xenon (Xe, a monatomic noble gas) occupies 1.000 ℓ. (a) What is the atomic mass of xenon? (b) What is the rms speed of the xenon atoms in the container? (c) Assume that the confining vessel is a cube, 10 cm on a side. What is the difference in the gravitational potential energy of an atom at the top of the cube and one at the bottom of the cube? Compare this result with the average kinetic energy of an atom ($\frac{1}{2}\mu\langle v^2\rangle$). Comment.

20–4 Determine the rms speed for (a) hydrogen (H_2), helium (He), and carbon dioxide (CO_2), at 0° C, and (b) oxygen (O_2), argon (Ar), and ammonia (NH_3), at 100° C.

Section 20–2

20–5 A vessel contains 10 mg of helium gas at a temperature of 20° C. (a) What is the average kinetic energy of a helium atom? (b) What is the total internal energy of the gas?

20–6 A sample of hydrogen gas is at a temperature of 500 K. Assume a full partition value of $\nu = 5$ for rotational motion. (a) What is the average translational kinetic energy of an H_2 molecule? (b) What is the average rotational kinetic energy of a molecule? (c) If the mean separation between the two hydrogen atoms in the molecule is 0.74 Å, what is the average angular momentum of a molecule?

20–7 A vessel contains 3 moles of nitrogen gas at a temperature of 3000 K. Assume that the full partition energy for the diatomic nitrogen molecule is reached ($\nu = 7$). (a) What is the total internal energy of the gas? (b) What is the average translational kinetic energy of an N_2 molecule? (c) What is the average vibrational potential energy of an N_2 molecule?

20–8 Very small particulate matter dispersed in a gas will share in the thermal motion of the gas molecules. As a result such particles largely behave as if they were simply "heavy molecules." The observable chaotic motion of such particles is referred to as *Brownian motion*. Consider fine smoke particles with a mass of order 10^{-13} g suspended in air at room temperature. (a) What is the average translational kinetic energy of a smoke particle? (b) What is its average speed? (c) Compare its average thermal kinetic energy with a change in its gravitational potential energy corresponding to an altitude change of 10 m. On the basis of this comparison, is it reasonable to expect such smoke particles to remain indefinitely suspended in the air? See also Problem 20–14.

Section 20–3

20–9 Calculate the specific heat at constant volume c_v (in Cal/kg·deg) for hydrogen sulfide (H_2S) at 15° C. Base your calculation on Eq. 20–15 and assume that no substantial vibrational motion is present. (The experimental value is 0.178 Cal/kg·deg.)

20–10 Calculate the specific heat at constant volume for air at 15° C using Eq. 20–15. Assume air to consist of 76 percent (by mass) nitrogen N_2 and 24 percent oxygen O_2. Assume also that neither component molecule undergoes any significant vibrational motion. Compare your result with the observed value of 0.171 Cal/kg·deg.

20–11 According to the Debye theory, the molar specific heat of a crystalline solid is one half the Dulong-Petit value (that is, $C_v = 12.47$ J/mol·K) for $T/T_D = 0.250$. At what temperature will diamond, aluminum, cadmium, and lead have this half value? Compare your results with Fig. 20–4.

Section 20–4

20–12 Show that the most probable speed for the Maxwell-Boltzmann distribution is $v_m = \sqrt{2kT/\mu}$.

20–13 In Example 13–4 the "escape velocity" for particles to completely leave the Earth's gravitational field was discussed. (a) Calculate the temperature at which the average speed of a hydrogen molecule would be equal to its escape velocity from the Earth's surface. (b) Repeat the calculation for an oxygen molecule. (c) Repeat these calculations for the surface of the Moon and the planet Mars. (d) On the basis of these calculations, explain why there is so little hydrogen compared to oxygen in the Earth's atmosphere and why neither the Moon nor Mars has any substantial atmosphere.

20–14 When fine particles suspended in air reach *sedimentation*

equilibrium, they obey a particle concentration distribution having the same form as the law of atmospheres

$$\mathcal{N}(y) = \mathcal{N}_0 e^{-\alpha y}$$

where $\alpha = \mu g/kT$ and μ is the particle mass; see Example 20–3. Determine the value of α for the smoke particles referred to in Problem 20–8 and readdress the question raised in part (c) of that problem.

20–15 A vessel contains 1000 molecules of helium at a temper-
•   ature of 250 K. Find the number of molecules expected to have speeds in the ranges 0–250 m/s, 250–500 m/s, 500–750 m/s, . . . , up to 2250–2500 m/s. [*Hint:* First determine v_{rms}, then transform the speed at the midpoint of each interval into the equivalent value of $\xi = v/v_{rms}$. Use Fig. 20–7 and find $\Delta N(v) = N\,\Delta v/v_{rms}$. Are your results consistent with $\Sigma\Delta N(v) = N$?]

20–16 An elevator with a height of 2 m is at rest. The air in the
••   elevator is at normal atmospheric pressure and has a density of 1.29 kg/m^3. The elevator is now sealed (airtight) and accelerates upward at $3g$. (a) Show that the position of the center of mass of the air in the elevator is

$$Y = \frac{1}{\alpha} + h(1 - e^{\alpha h})^{-1}$$

where $\alpha = 4g\rho_0/p_0$ and where h is the height of the elevator. (The required integral can be found in the table inside the rear cover.) (b) Show that $\Delta Y = \frac{1}{2}h - Y \cong \frac{1}{12}\alpha h^2$, when $\alpha h \ll 1$. [*Hint:* Expand e^x and $(1 + x)^{-1}$.] (c) Evaluate ΔY.

CHAPTER 21

THE FIRST LAW OF THERMODYNAMICS

In this chapter we are concerned primarily with a comprehensive version of the energy conservation principle known as the *first law of thermodynamics*. At first, attention is focused on a simple system that consists of a confined gas sample. Heat energy is allowed to enter or leave the sample. Also, the confining volume is allowed to change so that mechanical work can be done on or by the gas. Any temperature changes that accompany these processes indicate changes in the internal energy of the sample. We seek to understand the relationship connecting these variables as the system undergoes change.

In these discussions we often refer to processes by which a system changes from one thermodynamic state to another. That such a change can occur means that the system is not in an equilibrium state. However, if the change is made sufficiently slowly, the system remains arbitrarily close to equilibrium at all stages in the process. A process of this type is said to be *quasi-static*.

If a system in equilibrium is disturbed, it will achieve a new equilibrium state in some characteristic time (called the *relaxation time*) after the disturbance is removed. For a process to be quasi-static, it must take place during a time that is long compared with the relaxation time. During such a slow process the system can be considered to pass through a large (strictly, infinite) number of equilibrium states. This slow evolution allows a unique temperature to be defined for every point in the process. (To slow a process from its natural rate of progress requires deliberate intervention or control on the part of an external agency.)

A quasi-static process is an idealization of a real process that is valid (or nearly so) only for differential changes in equilibrium states. A finite change can never be truly quasi-static. We are forced to accept this lack of precision because it is impossible to maintain a system indefinitely in an equilibrium thermodynamic state. (This would require a perfect thermal insulator, which does not exist.)

21-1 MECHANICAL WORK DONE BY A GAS SYSTEM

Consider a system that consists of a gas confined in a cylinder closed with a frictionless, tight-fitting piston. The cylinder is immersed in a large constant-temperature heat bath or reservoir that constitutes the environment of the system (Fig. 21-1). Heat flows readily through the cylinder walls so that the gas is maintained always at the temperature

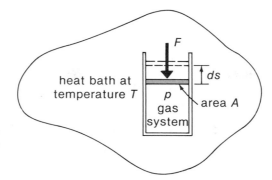

Fig. 21–1. A gas system is confined in a cylinder with a movable piston. The gas is maintained at the temperature T by the surrounding heat bath.

of the surrounding heat bath.* The gas (sufficiently dilute) obeys the ideal gas law.

Suppose that the gas is in thermal equilibrium at the temperature T and is in mechanical equilibrium by virtue of an external force F applied to the piston. The gas pressure is then $p = F/A$, where A is the surface area of the piston. Let the external force be slowly reduced so that the gas expands slightly and the piston moves upward by a differential displacement ds. During this process an amount of work $F\,ds = (pA)\,ds$ is done by the gas on the agency responsible for the force F. Now, $A\,ds = dV$ is the amount of volume expansion, so the work done by the gas is

$$dW = p\,dV \qquad (21-1)$$

Initially, the gas is in a thermal equilibrium state (p_a, V_a, T) and is in mechanical equilibrium with the force $F_a = p_a A$. By a large number of small, slow reductions in the external force, the system is allowed to proceed through a corresponding large number of quasi-static states, eventually arriving at the final equilibrium condition, (p_b, V_b, T) and $F_b = p_b A$. At each step in the process, heat is exchanged between the gas and the heat bath by whatever amount necessary to maintain the gas at the temperature T. (In this case, heat was actually supplied to the gas by the heat bath.) This constant-temperature, or *isothermal*, process is shown in Fig. 21–2. The quantity of gas in the cylinder is fixed, so throughout the process we have $pV = nRT =$ constant. Thus, the total amount of work *done by the gas* in expanding from volume V_a to V_b is

$$W_{ab} = \int_{V_a}^{V_b} p\,dV = nRT \int_{V_a}^{V_b} \frac{dV}{V}$$

from which

$$W_{ab} = nRT \ln \frac{V_b}{V_a} \qquad \text{isothermal process} \qquad (21-2)$$

That is, the work done, W_{ab}, is equal to the definite integral that corresponds to the area under the p-V curve between V_a and V_b; this is the shaded area in Fig. 21–2.

Because Eq. 21–1 is always valid, we can integrate this equation and write a general

---

*Ideal walls that are perfect heat conductors are called *diathermic walls*.

Fig. 21–2. The slow isothermal expansion of a gas from an initial state (p_a, V_a, T). The work done in the differential expansion dV at the pressure p is $dW = p\,dV$. The total work done is the sum of all such contributions and corresponds to the shaded area under the p-V curve. The gas obeys the ideal gas law; therefore, once the initial state (p_a, V_a, T) is specified, only p_b or V_b (but not both) is required to define the final state reached after the isothermal expansion.

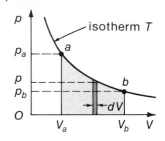

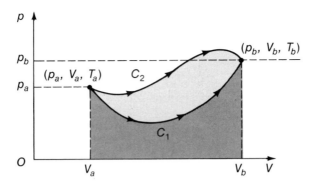

Fig. 21–3. The work done by an expanding gas depends on the path $p(V, T)$ that the system follows.

expression for the work done by a gas during an arbitrary expansion that carries the system from a state (p_a, V_a, T_a) to a state (p_b, V_b, T_b) along a thermodynamic path specified by $p = p(V, T)$:

$$W_{ab} = \int_{V_a}^{V_b} p(V, T)\, dV \tag{21-3}$$

Figure 21–3 illustrates two possibilities for the work done by an expanding gas between the same two states. In each case the work done, W_{ab}, is equal to the area under the particular curve $p = p(V, T)$. It is clear that $W_{ab}(C_2) > W_{ab}(C_1)$ and that, in general, the work done depends on the thermodynamic process involved (that is, on the path followed between the states).

When a thermodynamic path consists of a smooth progression of small changes specified by an infinite number of quasi-static intervening states, a reversal of the external conditions will return the system (and the environment) to its original state. Such a process is said to be *reversible*. In Fig. 21–3, if the imposed conditions carry the system from the state (p_a, V_a, T_a) to the state (p_b, V_b, T_b) along the path C_1, a reversal of the conditions will cause a compression that will carry the system along C_1 back to the state (p_a, V_a, T_a). When this process is completed, both the system and the environment will be in exactly the original condition. *A perfectly reversible process is an idealization that cannot be achieved by a real system.*

Let the ideal gas we are considering be confined in a cylinder with a tight-fitting movable piston. The initial position of the piston defines the volume V_a. The piston is suddenly pulled outward to a position that defines the volume V_b. In the natural (that is, highly probable) course of events, the gas will fill the newly created volume, $V_b - V_a$. Such an *uncontrolled* expansion is *irreversible*. Next, consider the unnatural (that is, highly improbable, though energy-conserving) process in which the gas occupying the volume $V_b - V_a$ suddenly and spontaneously collapses into the volume V_a, whereupon the piston is drawn inward to its original position defining the volume V_a. This hypothetical process is also irreversible. The *reversible (controlled)* path C_1 discussed above is any one of an infinite number of possible series of small steps proceeding through quasi-static states that connect the same initial and final equilibrium states.

Suppose that the gas system in Fig. 21–3 is carried from the state (p_a, V_a, T_a) to the state (p_b, V_b, T_b) along the reversible path C_2 and then is returned (by a compression) along the reversible path C_1 (in a direction opposite to that shown in the figure). Such a combined process that returns a system to its original state is called a *thermodynamic cycle*. For the case illustrated in Fig. 21–3, the gas system does an amount of work represented by the shaded area enclosed by the two curves. (• Can you see why this area represents work done *by* the gas? How would you interpret the diagram if the system followed first path C_1 and then the reverse of C_2?) In the event that the cycle path encloses a nonzero area (as in the case here), the environment does not return to its original condition when the cycle is completed. (• How has the environment changed?) We consider in detail several interesting and practical thermodynamic cycles later in this chapter.

Example 21–1

A system that consists of n moles of an ideal gas is in an initial state (p_a, V_a, T_a). The system is carried along the straight-line path C_{ab} shown in the diagram from the initial state to the state (p_b, V_b, T_b) with $p_b = 2p_a$ and $V_b = 2V_a$. The pressure is then reduced (at constant volume) so that the system follows the path C_{bc} to the state (p_c, V_c, T_c). Finally, the system undergoes an isobaric compression along path C_{ca} back to the state (p_a, V_a, T_a). All of these processes are carried out in a reversible manner.

(a) How much work was done by the gas along each of the paths C_{ab}, C_{bc}, and C_{ca}?

(b) What was the net amount of work done by the gas during the complete cycle?

(c) What are the temperatures T_b and T_c?

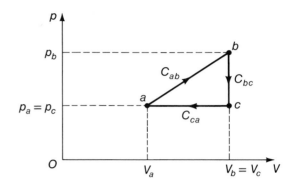

Solution:

(a) The equation for the straight-line path C_{ab} is $p = p_a(V/V_a)$. Thus, Eq. 21–3 becomes

$$W_{ab} = \int_{V_a}^{V_b} p\, dV = \frac{p_a}{V_a}\int_{V_a}^{2V_a} V\, dV = \frac{1}{2}\frac{p_a}{V_a}V^2\Big|_{V_a}^{2V_a} = \frac{3}{2}p_aV_a$$

$$W_{bc} = 0 \qquad \text{(• Can you see why this result follows?)}$$

$$W_{ca} = \int_{V_c}^{V_a} p\, dV = p_a\int_{2V_a}^{V_a} dV = p_aV\Big|_{2V_a}^{V_a} = -p_aV_a$$

The fact that $W_{ca} < 0$ means that work was done *on* the gas during the compression. (Notice that the required negative sign appears automatically if the lower limit of the integral is taken to correspond to the initial state of the process and if the upper limit is taken to correspond to the final state. *This practice should always be followed.*)

(b) The net amount of work done by the gas was

$$W = W_{ab} + W_{bc} + W_{ca} = (\tfrac{3}{2} + 0 - 1)p_aV_a = \tfrac{1}{2}p_aV_a$$

This is just the area of the triangle enclosed by the paths C_{ab}, C_{bc}, and C_{ca}.

(c) Using the ideal gas law equation, we have

$$nR = \frac{p_aV_a}{T_a} = \frac{p_bV_b}{T_b} = \frac{(2p_a)(2V_a)}{T_b}$$

so that

$$T_b = 4T_a$$

Also,

$$\frac{p_aV_a}{T_a} = \frac{p_cV_c}{T_c} = \frac{p_a(2V_a)}{T_c}$$

so that

$$T_c = 2T_a$$

Notice that, in order to evaluate the integral in Eq. 21–3, you must be able to substitute an analytical expression for p in terms of V. There is no such expression for irreversible processes, because in that case there is no well-defined function describing the intermediate states.

21–2   THE FIRST LAW OF THERMODYNAMICS

Simple Gas Systems. In the discussion of gas systems we have considered three forms of energy. *Heat* is the energy transferred between systems because of an existing temperature difference. *Internal energy* is the energy associated with the random thermal motions of the molecules in the system. *Mechanical work* is done as the result of the volume expansions or compressions that the system undergoes. For a differential change in the thermodynamic state of a gas system during a quasi-static process we have the following quantities and sign conventions:

dQ: the heat *added to* the system

$dW = p \, dV$: the mechanical work *done by* the system

dU: the *increase* in the internal energy of the system

The energy conservation principle can now be written as

$$dQ = dW + dU = p \, dV + dU \tag{21-4}$$

This expression is the differential form of the *first law of thermodynamics* for a gas system. There is no new physical idea contained in this law—it is simply the energy conservation principle applied to a thermodynamic system. For instance, if energy enters the system in the form of heat, some or all of it may be removed as work; the rest goes to increase the internal energy of the system. Integrating Eq. 21–4, we have

$$Q_{ab} = \oint_{V_a}^{V_b} p(V, T) \, dV + (U_b - U_a) \tag{21-5}$$

where the system changes from an initial state (p_a, V_a, T_a) to the final state (p_b, V_b, T_b) via the path C defined by $p = p(V, T)$. We call attention to the fact that the work done depends on the path followed by writing the integral sign as $\oint$. Whenever a process involves a volume expansion, the work done by the system is $W_{ab} > 0$, whereas $W_{ab} < 0$ for a volume contraction or compression.

In the previous discussions we have emphasized the fact that the internal energy U of an ideal system depends only on its equilibrium temperature. Thus, the internal energy is a function of the thermodynamic state; U is a *state function*, as are p, V, and T. Consequently, the change in internal energy during a process does *not* depend on the process path, but *only* on the difference in temperature between the initial and final states. This feature of the internal energy is indicated in Eq. 21–5, where we write $U_b - U_a$ for the integrated change in U.

By using Eq. 20–14 we can express the change in internal energy for a gas system as

$$U_b - U_a = \tfrac{1}{2}\nu n R (T_b - T_a) \tag{21-6}$$

where ν is the energy partition number, which is $\nu = 3$ for monatomic gases (He, Ne, Ar, and so forth) and $\nu = 5$ for diatomic gases (H_2, N_2, O_2, and so forth) near room temperature (see Section 20–2). A temperature increase always means an increase in internal energy; a temperature decrease always means a decrease in internal energy.

Let us examine the idea of internal energy in a different way. First, rewrite Eq. 21–5 in the form

$$U_{ab} = Q_{ab} - W_{ab} \tag{21-7}$$

Imagine that experiments are performed in which externally imposed conditions carry a system from an initial state (p_a, V_a, T_a) to the state (p_b, V_b, T_b) by various processes (Fig. 21–4). The paths C_1 and C_2 represent possible quasi-static processes, whereas I represents an *irreversible* process. (An irreversible process cannot properly be indicated on a p-V diagram by a specifiable single curve. The process does not proceed through

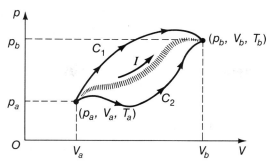

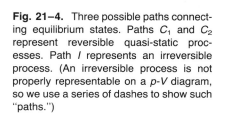

Fig. 21–4. Three possible paths connecting equilibrium states. Paths C_1 and C_2 represent reversible quasi-static processes. Path I represents an irreversible process. (An irreversible process is not properly representable on a p-V diagram, so we use a series of dashes to show such "paths.")

uniquely defined equilibrium states of the system; i.e., it does not follow quasi-static steps.) The quantities Q_{ab} and W_{ab} are measured, and it is discovered that $Q_{ab} - W_{ab}$ has the same value for each process! This energy difference—which is independent of the path and even whether or not the process is reversible—therefore depends only on the initial and final states. Thus, on the basis of experimental evidence, we can ascribe a quantity U to each state such that $U_b - U_a = U_{ab} = Q_{ab} - W_{ab}$. (Notice the similarity of this thermodynamic case to that of a conservative mechanical system for which a potential energy can be defined; see Chapter 8.) Because the work done W_{ab} depends on the path followed between the initial and final states, whereas U_{ab} is independent of the path, we must conclude that Q_{ab} also depends on the path followed.

The General Behavior of Systems.   Because of the universal nature of the energy conservation principle, much of the discussion of simple gas systems applies also to general systems that undergo any type of process. The system isolated for consideration may consist of any combination of gaseous, liquid, or solid matter. In addition, many different forms of energy exchange between the system and its environment may be present. Other types of work may be done in addition to the mechanical work $p\,dV$ associated with volume changes. For example, electric or magnetic work could be done, or a chemical reaction or phase change could involve energy changes. In spite of the variety of energy forms that can be involved, we can still express the first law for the system in the form

$$Q_{ab} = W_{ab} + U_{ab} \qquad\qquad \textbf{(21–8)}$$

where W_{ab} includes all of the possible ways that work can be done by the system. Depending on the circumstances, the quantities that appear in Eq. 21–8, including the separate terms that contribute to W_{ab}, can be either positive or negative. It is therefore advantageous to view Eq. 21–8 as an energy balance record:

The energy that enters a system (in the form of heat) is equal to the energy that leaves the system (in the form of work done by the system on the environment) plus the increase in the internal energy of the system.

Notice that the processes that connect two states of the system (such as a and b in Fig. 21–4) are not restricted to quasi-static ones. It is required only that the initial and final states be equilibrium states so that each has a well-defined temperature. Only for states in equilibrium is internal energy a well-defined concept.

The first law of thermodynamics has an important limitation. The law states whether energy considerations permit a particular process to carry a system from one equilibrium state to another. But the law does *not* state whether this process will actually occur. That is, a certain process might be entirely consistent with the energy conservation principle and still not take place. Questions of this sort are addressed by the *second law of thermodynamics,* which we discuss in the next chapter.

Example 21−2

One mole of an ideal monatomic gas is carried by a quasi-static isothermal process (at 400 K) to twice its original volume.
  (a) How much work W_{ab} was done by the gas?
  (b) How much heat Q_{ab} was supplied to the gas?
  (c) What is the pressure ratio, p_b/p_a?
  (d) Suppose that a constant-volume process is used to reduce the original pressure p_a to the same final pressure p_b. Determine the new values for W'_{ab}, Q'_{ab}, and U'_{ab}.

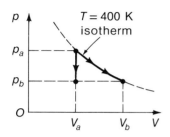

Solution:

(a) We use Eq. 21−2 to find the work done in an isothermal expansion:

$$W_{ab} = nRT \ln \frac{V_b}{V_a}$$
$$= (1 \text{ mol})(8.314 \text{ J/mol·K})(400 \text{ K}) \ln 2 = 2305 \text{ J}$$

This is the work done by the gas during the expansion.
  (b) For an isothermal process, $U_{ab} = 0$; therefore,

$$Q_{ab} = W_{ab} = 2305 \text{ J}$$

That is, energy enters the system in the form of heat, and exactly the same amount of energy is expended as work; no energy is added to or subtracted from the system's internal energy because the temperature does not change.
  (c) Using the ideal gas law equation with $T_a = T_b$, we have

$$\frac{p_b}{p_a} = \frac{V_a}{V_b} = \tfrac{1}{2}$$

(d) For the constant-volume process, we have $W'_{ab} = 0$. The new final temperature is (using $V'_b = V_a$ and $p'_b = p_b = \tfrac{1}{2}p_a$)

$$T'_b = T_a \frac{p'_b}{p_a} = (400 \text{ K})(\tfrac{1}{2}) = 200 \text{ K}$$

Then, using Eq. 21−6, with $\nu = 3$, we have

$$U'_{ab} = \tfrac{1}{2}\nu nR(T'_b - T_a)$$
$$= \tfrac{1}{2}(3)(1 \text{ mol})(8.314 \text{ J/mol·K})(200 \text{ K} - 400 \text{ K})$$
$$= -2494 \text{ J}$$

This represents a *decrease* in the internal energy of the gas. Finally, using the first law, we find

$$Q'_{ab} = W'_{ab} + U'_{ab} = -2494 \text{ J}$$

That is, during this process an amount of heat is rejected by the system to the environment and the internal energy decreases by the same amount.

21−3 APPLICATIONS OF THE FIRST LAW OF THERMODYNAMICS

Specific Heats of an Ideal Gas. In Section 20−3 we discussed the specific heat of a gas for constant-volume processes. We can now understand the importance of this restriction. Because the volume remains constant, the mechanical work done is zero. Then, according to the first law, $U_{ab} = Q_{ab}$; that is, the heat added to the gas appears entirely in the form of internal energy. Thus, the heat dQ_v added at constant volume in order to raise the temperature from T to $T + dT$ is

$$dQ_v = nC_v \, dT = dU \tag{21−9}$$

where C_v is the molar specific heat at constant volume (compare Eq. 20−13). The process that corresponds to Eq. 21−9 is shown in Fig. 21−5 as path C_1.
  The constant-pressure path C_2 in Fig. 21−5 also connects the initial state with a state at the temperature $T + dT$ and therefore gives the same increase in the internal energy. But work is done along the path C_2, so we have

$$dQ_p = p \, dV + dU \tag{21−10}$$

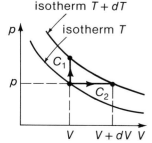

Fig. 21−5. Constant-volume (C_1) and constant-pressure (C_2) paths used to determine the relationship between the specific heats, C_v and C_p.

The molar specific heat at constant pressure is defined to be

$$C_p = \frac{1}{n} \frac{dQ_p}{dT} \tag{21-11}$$

Combining Eqs. 21–9, 21–10, and 21–11, we find

$$nC_p \, dT = nC_v \, dT + p \, dV$$

Moreover, along path C_2 the ideal gas law can be expressed as

$$p \, dV = nR \, dT$$

Together, these last two equations (cancelling $n \, dT$ throughout) provide the important connection between the specific heats C_v and C_p, namely,

$$C_p - C_v = R \tag{21-12}$$

Table 21–1 gives experimental values of the molar specific heats for various gases. It is apparent that Eq. 21–12—derived for the case of an ideal gas—is a correct description of the specific heats of real gases, to an accuracy of a few percent.

According to the discussion in Section 20–2, the value of ν that appears in Eq. 21–6 is expected to be $\nu = 3$ for a monatomic gas. For a diatomic gas the fully developed equipartition values are expected to be $\nu = 5$ (rigid) or $\nu = 7$ (nonrigid). For a triatomic gas, if the atoms in the molecule are not in line, we expect $\nu = 6$ (rigid) or $\nu = 12$ (nonrigid). Combining Eqs. 21–6 and 21–9, we see that (compare Eq. 20–15)

$$C_v = \tfrac{1}{2}\nu R \tag{21-13}$$

Thus, for monatomic gases, we expect $C_v = \tfrac{3}{2}R = 12.47$ J/mol·K and also

$$\gamma = \frac{C_p}{C_v} = \frac{\tfrac{3}{2}R + R}{\tfrac{3}{2}R} = \tfrac{5}{3} = 1.67$$

TABLE 21–1. Specific Heats for Various Gases (at 1 atm and 15° C)

	GAS	C_p (J/mol·K)	C_v (J/mol·K)	$C_p - C_v$ (J/mol·K)	$\gamma = C_p/C_v$
Monatomic gases	Ar	20.80	12.47	8.33	1.67
	He	20.80	12.47	8.33	1.67
	Ne	20.76	12.64	8.12	1.64
Diatomic gases	Air	29.01	20.68	8.33	1.403
	Cl_2	34.12	25.16	8.96	1.36
	H_2	28.76	20.43	8.33	1.41
	HCl	29.93	21.39	8.54	1.40
	N_2	29.05	20.72	8.33	1.40
	NO	29.26	20.93	8.33	1.40
	O_2	29.47	21.10	8.37	1.40
Polyatomic gases	CO_2	36.63	28.09	8.54	1.30
	H_2S	34.12	25.45	8.67	1.34
	NH_3	37.30	28.46	8.84	1.31
	CH_4	35.50	27.08	8.42	1.31
	C_2H_4	42.19	33.61	8.58	1.26

TABLE 21–2.
Summary of Nomenclature

$dp = 0$	isobaric
$dV = 0$	isochoric
$\left.\begin{array}{l} dT = 0 \\ dU = 0 \end{array}\right\}$	isothermal
$dQ = 0$	adiabatic

These predicted values for C_v and γ are in good agreement with the experimental results. (• Make similar comparisons for the diatomic and triatomic gases in Table 21–1.)

Adiabatic Expansion. In Section 21–1 we discussed the expansion of a simple gas system during an isothermal quasi-static process. Another important type of quasi-static expansion process is one that occurs with no exchange of heat. Such a process is called an *adiabatic* expansion.

Again, we consider a gas that is confined in a cylinder with a frictionless, tight-fitting piston (Fig. 21–6a). The cylinder walls and piston are imagined to be perfect heat insulators, so that there is no heat exchange between the gas system and the environment. The appropriate path equation for the adiabatic expansion involves $dQ = 0$, $dW = p\,dV$, and $dU = nC_v\,dT$. Thus, the first law becomes

$$0 = p\,dV + nC_v\,dT \qquad (21–14)$$

Expressing the ideal gas law equation, $pV = nRT$, in differential form, we have

$$p\,dV + V\,dp = nR\,dT \qquad (21–15)$$

Substituting $dT = -(p/nC_v)\,dV$ from Eq. 21–14 into Eq. 21–15, we obtain

$$C_v p\,dV + C_v V\,dp = -pR\,dV \qquad (21–16)$$

Using $C_p = C_v + R$ (Eq. 21–12) and $\gamma = C_p/C_v$, Eq. 21–16 reduces to

$$\gamma\,\frac{dV}{V} = -\frac{dp}{p}$$

Integrating, we find

$$\gamma \ln V = -\ln p + \text{constant}$$

from which

$$pV^\gamma = \text{constant} \qquad (21–17)$$

This is the adiabatic path equation that we require (Fig. 21–6b). The value of the *constant* in Eq. 21–17 depends on the initial state (p_a, V_a, T_a) of the system.

Figure 21–7 shows several isothermal and several adiabatic paths constructed for a sample consisting of 1 mol of an ideal diatomic gas ($\gamma = 1.40$). These curves provide an

Fig. 21–6. (a) The insulated cylinder allows no heat exchange between the gas system and the environment. (b) As the gas expands adiabatically, it follows a path, $p = p(V, T)$, that cuts across the family of isotherms.

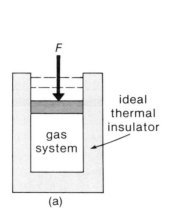

(a)

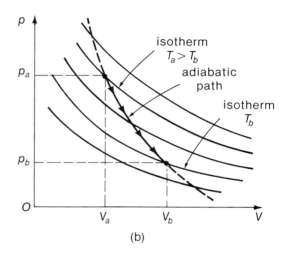

(b)

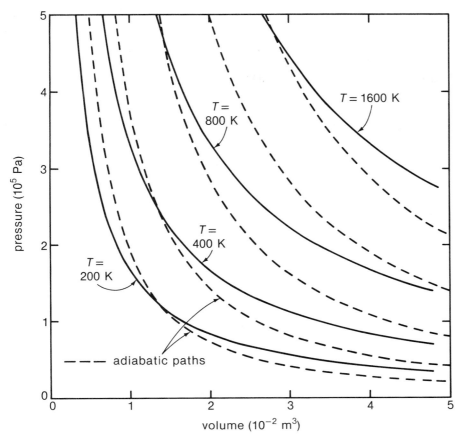

Fig. 21-7. The solid and dashed lines show, respectively, isothermal and adiabatic paths for a 1-mol sample of an ideal diatomic gas (for which $\gamma = 1.40$).

accurate picture of the behavior of such a sample. In other illustrations, however, we give only schematic representations of isothermal and adiabatic paths in order that the diagrams may be easily interpreted.

According to the ideal gas law, we can write $p_aV_a/T_a = p_bV_b/T_b$; and from Eq. 21–17, we have $p_aV_a^\gamma = p_bV_b^\gamma$. Therefore, once the initial state has been specified, we require one (and only one) of the quantities p_b, V_b, or T_b to identify uniquely the final state of an adiabatic process. Because $\gamma > 1$, the adiabatic paths in a p-V diagram decrease with increasing V more rapidly than do the isotherms; every adiabatic path therefore cuts across the family of isotherms (Fig. 21–7).

The work done during an adiabatic expansion can be expressed in terms of the temperature change, $T_b - T_a$. We have $Q_{ab} = 0$ for the adiabatic process, so the first law can be written as

$$W_{ab} = -U_{ab} = -nC_v(T_b - T_a) \tag{21-18}$$

Because $T_b < T_a$ (see Fig. 21–6b), we see that $W_{ab} > 0$, as expected for an expansion process.

The work done can also be expressed in terms of p and V. From the ideal gas law equation, we can write $nT_a = p_aV_a/R$ and $nT_b = p_bV_b/R$. Substituting these expressions into Eq. 21–18, the result is

$$W_{ab} = \frac{C_v}{R}(p_aV_a - p_bV_b) \tag{21-19}$$

Using $R = C_p - C_v$ (Eq. 21–12) and $\gamma = C_p/C_v$, Eq. 21–19 can be expressed as

$$W_{ab} = \frac{p_aV_a - p_bV_b}{\gamma - 1} \qquad \text{adiabatic process} \qquad \textbf{(21–20)}$$

This result can also be obtained by the direct integration of $p\,dV$, using $p = (\text{constant})/V^\gamma$ (see Problem 21–9).

Example 21–3

How much work is required to compress 5 mol of air at 20° C and 1 atm to $\frac{1}{10}$ of the original volume by (a) an isothermal process and (b) an adiabatic process? (c) What are the final pressures for the two cases?

Solution:

(a) The work done on the gas is $-W_{ab}$. For the isothermal process, we use Eq. 21–2:

$$-W_{ab} = -nRT \ln \frac{V'_b}{V_a}$$

$$= -(5 \text{ mol})(8.314 \text{ J/mol·K})(293 \text{ K}) \ln \tfrac{1}{10}$$

$$= 2.80 \times 10^4 \text{ J}$$

(b) For the adiabatic process, let us use Eq. 21–18. We must first determine the final temperature T_b. From Eq. 21–17, we have

$$p_bV_b^\gamma = p_aV_a^\gamma$$

and from the ideal gas law equation, we have

$$\frac{p_bV_b}{T_b} = \frac{p_aV_a}{T_a}$$

Dividing the first of these two equations by the second gives

$$T_b = T_a \left(\frac{V_a}{V_b} \right)^{\gamma-1} = (293 \text{ K})(10)^{0.403} = 741 \text{ K}$$

where we have used $\gamma(\text{air}) = 1.403$ (see Table 21–1). Thus, the

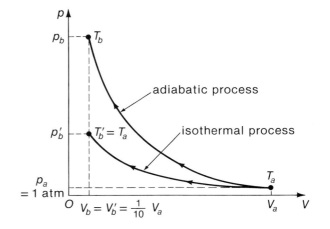

work done on the gas is (Eq. 21–18)

$$-W_{ab} = nC_v(T_b - T_a)$$

$$= (5 \text{ mol})(20.68 \text{ J/mol·K})(741 \text{ K} - 293 \text{ K})$$

$$= 4.63 \times 10^4 \text{ J}$$

where we have used the value of C_v for air given in Table 21–1.

(c) For the isothermal process, we have $p'_bV'_b = p_aV_a$; thus,

$$p'_b = p_a \left(\frac{V_a}{V'_b} \right) = (1 \text{ atm})(10) = 10 \text{ atm}$$

For the adiabatic process, we have $p_bV_b^\gamma = p_aV_a^\gamma$; thus,

$$p_b = p_a \left(\frac{V_a}{V_b} \right)^\gamma = (1 \text{ atm})(10)^{1.403} = 25.3 \text{ atm}$$

A Two-Phase System. A quantity of water is placed in a cylinder that is closed with a frictionless, tight-fitting piston. The cylinder has diathermic (perfectly conducting) walls. The entire assembly is immersed in a heat bath that has an adjustable temperature T. The external pressure on the piston is maintained at 1 atm. In the initial condition, the piston is in contact with the water. That is, the system under consideration consists entirely of the original mass of water.

The temperature of the heat bath is increased to 100° C sufficiently slowly that the

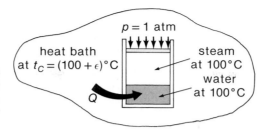

Fig. 21-8. Heat flows through the diathermic walls of the cylinder from the heat bath into the water and causes the vaporization of the water. This evaporation produces a two-phase, water-plus-steam system.

temperature of the water is always the same as that of the heat bath. During this process heat flows from the heat bath into the water. Next, the heat bath temperature is raised by an infinitesimal amount to $(100 + \epsilon)°$ C and maintained at this value. Heat flows into the water and causes vaporization. The system now consists of liquid water plus steam. This situation is illustrated in Fig. 21-8.

The amount of heat Q required to vaporize a mass m of the water is

$$Q = mL_v$$

where L_v is the latent heat of vaporization for water (see Table 18-4). Due to the production of steam, the volume of the system increases and the piston is raised (see Fig. 21-8). At 1 atm and 100° C, the density of steam is ρ_s and the density of water is ρ_w. Then, the volume expansion of the system when a mass m of water is converted into steam at 100° C is

$$\Delta V = m\left(\frac{1}{\rho_s} - \frac{1}{\rho_w}\right)$$

The work done by the system at the constant pressure p_0 (1 atm) is

$$W = p_0\,\Delta V = mp_0\left(\frac{1}{\rho_s} - \frac{1}{\rho_w}\right)$$

Using the first law, the change in internal energy ΔU for the water can be expressed as

$$\Delta U = Q - W = mL_v - mp_0\left(\frac{1}{\rho_s} - \frac{1}{\rho_w}\right)$$

Then, using $\rho_s = 0.598$ kg/m^3 and $\rho_w = 1040$ kg/m^3, we find

$$\frac{\Delta U}{m} = (2.257 \times 10^6 \text{ J/kg}) - (1.013 \times 10^5 \text{ Pa})\left(\frac{1}{0.598 \text{ kg/m}^3} - \frac{1}{1040 \text{ kg/m}^3}\right)$$

$$= 2.088 \times 10^6 \text{ J/kg}$$

Although an amount of heat equal to 2.257×10^6 J is required to convert 1 kg of water to steam at 1 atm and 100° C, only 2.088×10^6 J actually appears in the form of thermal motion of the molecules in the steam. The remainder of the energy, 0.169×10^6 J, is the work done by the steam in expanding at constant pressure. We may view the first amount of energy (2.088×10^6 J/kg) as the energy required to perform the microscopic work of separating the water molecules against the attractive intermolecular forces. The second amount of energy (0.169×10^6 J/kg) is the energy required to perform the macroscopic work of expansion.

Free Expansion of a Gas.   In various discussions it has been implicit that the macroscopic definition of internal energy contained in the first law (Eq. 21-7) and the microscopic definition from kinetic theory (Eq. 21-6) refer to exactly the same quantity.

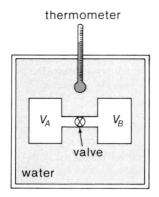

thermometer

Fig. 21–9. Schematic representation of the free-expansion apparatus of Joule.

An interesting test of this assumption was first devised and carried out by Gay-Lussac; the experiment was later repeated by Joule and subsequently refined still further by others.

The test consists basically of verifying that the internal energy of a gas defined by the first law does not depend on its volume. The experimental arrangement is shown schematically in Fig. 21−9. The gas is originally confined in the volume V_A, which is separated from the evacuated volume V_B by a closed valve. The entire system—defined to include *both* V_A and V_B—is immersed in a water bath. The valve is suddenly opened and the gas expands freely into the volume V_B. This expansion process (referred to as a *free expansion*) is uncontrollably rapid and does not proceed through a sequence of quasi-static states. A free expansion is an irreversible process.

The temperature of the water bath is measured carefully before and after the free expansion. The early measurements lacked precision and were inconclusive. However, refined versions of the experiment have demonstrated that the expansion produces essentially no change in the temperature of the water bath. From this result it follows that $\Delta Q = 0$ for the entire system $V_A + V_B$. Because there was no change in the total volume, no work was done; that is, $W = 0$. Using the first law, we must conclude that $\Delta U = 0$. Thus, the internal energy of the gas does not depend on the volume in which it is confined or on the pressure.*

We may examine further the implications of the result of the free-expansion experiment by noting that according to the microscopic view, $\Delta U = 0$ means that the average molecular speed $\langle v^2 \rangle$ also remains constant. Suppose that $V_A = V_B$; then, after expansion, the gas density is smaller by a factor of two, but the pressure is also smaller by the same factor (because $p = \frac{1}{3}\rho \langle v^2 \rangle$, according to kinetic theory—Eq. 20−7). Because the gas now occupies twice the volume, we have $p_f = \frac{1}{2}p_i$ and $V_f = 2V_i$. Thus, the relationship connecting the initial and final states is $p_iV_i = p_fV_f$. This, of course, is just Boyle's law!

21–4 CYCLIC THERMODYNAMIC PROCESSES

A cyclic thermodynamic process is one in which a system acts through a series of steps, eventually returning to its original equilibrium state. The steps that constitute the cycle may be reversible or irreversible. If the system consists of a single homogeneous substance (the *working substance*) and if all of the steps are reversible, the cycle can be represented by a closed curve in a p-V diagram (Fig. 21−10).

The Carnot Cycle. The cycle that is most important to the theory of thermodynamics is the idealized reversible cycle of Carnot.† Figure 21−11 gives three different views of a *Carnot cycle*. Figure 21−11a is a schematic diagram of a Carnot engine; Fig. 21−11b shows the p-V plot for a Carnot cycle, and Fig. 21−11c is the corresponding heat-flow diagram.

Choose the beginning of the cycle at the equilibrium state a, characterized by (p_a, V_a, T_1), with the cylinder that contains the working substance in contact with a hot thermal reservoir at a temperature T_1 (Fig. 21−11a). Imagine that the base of the cylinder is perfectly conducting (*diathermic*) and that the walls and the piston are ideally nonconducting. The working substance undergoes a slow quasi-static isothermal expansion to state b, characterized by (p_b, V_b, T_1). This is accomplished by allowing heat, as needed, to flow from the hot reservoir through the diathermic base of the cylinder into the working substance. Let the total amount of heat delivered to the working substance during the

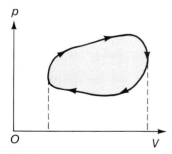

Fig. 21–10. The p-V diagram for a general reversible thermodynamic cycle of a homogeneous system. The shaded area represents the net work done by the system per cycle.

---

*At sufficiently high pressures and low temperatures, the internal energy of a real gas exhibits some dependence on pressure and volume.

†Nicolas Léonard Sadi Carnot (1796–1832), French military engineer and physicist.

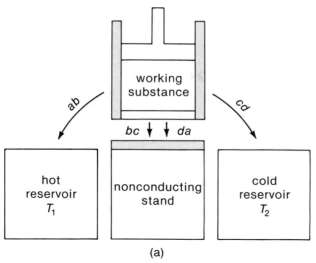

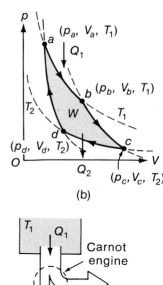

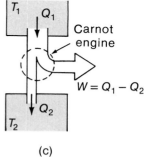

Fig. 21-11. The Carnot cycle. (a) Schematic diagram of the engine. The two heat reservoirs and the nonconducting stand are successively placed in contact with the diathermic base of the cylinder. (b) The p-V plot of the cycle. The paths ab and cd are along isotherms; bc and da are adiabatic paths. (c) The corresponding heat-flow diagram. The difference, $Q_1 - Q_2$, is equal to the work done by the system per cycle.

expansion ab be Q_1, as indicated in Fig. 21-11b. During this part of the cycle the temperature and therefore also the internal energy of the working substance are constant. Consequently, the heat-energy input is converted directly to the work done by the moving piston on whatever mechanical system is attached.

With the system at b, the cylinder is removed from contact with the hot reservoir and placed on the nonconducting stand. The system now undergoes a further expansion along the adiabatic path bc to the point c, characterized by (p_c, V_c, T_2). During this part of the cycle there is no heat exchange with the working substance; however, work is done at the expense of reducing the internal energy, and the temperature is thereby decreased to T_2.

Next, the cylinder is placed in contact with the cold thermal reservoir at temperature T_2. An external agency now slowly compresses the working substance isothermally to a predetermined volume V_d at point d, characterized by (p_d, V_d, T_2). During this part of the cycle the temperature and therefore also the internal energy of the working substance are constant. The work done on the working substance results in the rejection of an amount of heat Q_2 to the cold reservoir, as indicated in Fig. 21-11b.

Finally, the cylinder is removed from contact with the cold reservoir and again placed on the nonconducting stand. The volume V_d to which the system was carried by the isothermal compression cd was selected to allow an adiabatic compression to connect state d with the initial state a. This additional compression is also the result of work done on the working substance by an external agency and increases the system temperature from T_2 to T_1. This step completes the Carnot cycle and returns the system to its initial condition.

According to the first law, the amount of work W done by a system during a thermodynamic process is equal to the heat Q supplied to the system, less the change in internal energy ΔU of the system (see Eq. 21-4). If a system is carried through a complete thermodynamic cycle, the initial and final temperatures are equal, so $\Delta U = 0$. For a system that undergoes a Carnot cycle, no heat is supplied to or rejected by the system during the adiabatic paths, bc and da (Fig. 21-11b). An amount of heat $Q_1 = Q_{ab}$ ($Q_1 > 0$) is supplied to the system during the isothermal expansion ab, and an amount $Q_2 = -Q_{cd}$ ($Q_2 > 0$) is rejected during the isothermal compression da. (The heat *supplied* to the system during this process is *negative*; that is, $Q_{cd} < 0$.) Thus, the first law can be written as

$$W = Q_1 - Q_2 \quad \text{all heat engines} \tag{21-21}$$

Notice that Eq. 21–21 is applicable for a complete cycle of *any* heat engine (any device that operates in a cycle and converts heat energy into mechanical work) if we interpret Q_1 as the net amount of heat supplied to the engine (during whatever part of the cycle) and Q_2 as the net amount of heat rejected by the engine to some thermal reservoir.

The energy conservation statement contained in Eq. 21–21 is illustrated in the heat-flow diagram of Fig. 21–11c. The dashed circle is imagined to contain the Carnot engine. An amount of heat Q_1 is supplied to the engine by the hot reservoir at temperature T_1, and an amount of heat Q_2 is rejected to the cold reservoir at temperature T_2. The energy difference, $Q_1 - Q_2$, appears as work done by the system on an external mechanical agency.

Only a portion of the heat energy supplied to any engine is converted into mechanical work. The *thermal efficiency e* of an engine is defined to be the ratio of the work done (the *output*) to the heat supplied (the *input*). Thus,

$$e = \frac{W}{Q_1} = \frac{Q_1 - Q_2}{Q_1} = 1 - \frac{Q_2}{Q_1} \quad \text{all heat engines} \tag{21-22}$$

This equation is applicable for any heat engine that operates in a cycle.

For a Carnot engine, the heat exchanges take place during the isothermal processes. Using Eq. 21–22, we have*

$$Q_1 = nRT_1 \ln (V_b/V_a)$$

$$Q_2 = nRT_2 \ln (V_c/V_d)$$

from which
$$\frac{Q_2}{Q_1} = \frac{T_2}{T_1} \frac{\ln (V_c/V_d)}{\ln (V_b/V_a)} \tag{21-23}$$

Along the isothermal paths we can write

$$p_a V_a = p_b V_b \quad \text{and} \quad p_c V_c = p_d V_d$$

and along the adiabatic paths we have

$$p_b V_b^\gamma = p_c V_c^\gamma \quad \text{and} \quad p_d V_d^\gamma = p_a V_a^\gamma$$

If we multiply together these four equations and cancel the common term $p_a p_b p_c p_d$, we find

$$V_a V_b^\gamma V_c V_d^\gamma = V_a^\gamma V_b V_c^\gamma V_d$$

or
$$(V_a V_c)^{1-\gamma} = (V_b V_d)^{1-\gamma}$$

so that
$$\frac{V_c}{V_d} = \frac{V_b}{V_a} \tag{21-24}$$

---

*Note that $Q_2 > 0$; we ensure this by writing the logarithmic term as $\ln (V_c/V_d)$.

Using this result in Eq. 21–23, we obtain

$$\frac{Q_2}{Q_1} = \frac{T_2}{T_1} \qquad \text{Carnot engine} \qquad \textbf{(21–25)}$$

Consequently, the efficiency of a Carnot engine can be expressed as

$$e = 1 - \frac{T_2}{T_1} = \frac{T_1 - T_2}{T_1} \qquad \text{Carnot engine} \qquad \textbf{(21–26)}$$

This is the efficiency of an idealized Carnot engine (no heat losses through the walls, no losses due to friction). In Section 22–2 it is demonstrated that no heat engine of any type that operates between the temperatures T_1 and T_2 can have an efficiency greater than that specified by Eq. 21–26.

It is evident from Eq. 21–26 that the greater the temperature difference, $T_1 - T_2$, the greater will be the efficiency of a Carnot engine. Indeed, this is true of all thermodynamic cycles that convert heat into mechanical work (heat engines).

A heat engine can be imagined to run backward, thereby operating as a *refrigerator*. Figure 21–12 shows the heat-flow diagram for a refrigerator. The input of mechanical work W is used by the working substance to transfer heat from a cold reservoir at temperature T_2 to a hot reservoir at temperature T_1.* The p-V plot for the Carnot refrigerator cycle is the same as that for the Carnot heat-engine cycle (Fig. 21–11b), except that the paths are traversed in the opposite directions. (Each step in a Carnot cycle is *reversible,* so the entire cycle is reversible.) It is customary to express the first law for a refrigeration cycle in the form

$$Q_1 = Q_2 + W \qquad \text{refrigerator} \qquad \textbf{(21–27)}$$

where Q_1 is the amount of heat delivered to the hot reservoir and where Q_2 is the amount of heat removed from the cold reservoir by the input of the work W. Equation 21–27 appears to be a simple algebraic rearrangement of the heat-engine equation (21–21), but note that each of the quantities in Eq. 21–27 is the negative of the corresponding quantity in Eq. 21–21. We discuss some practical thermodynamic cycles in Section 21–5.

Fig. 21–12. Heat-flow diagram for a refrigerator—a heat engine operating backward.

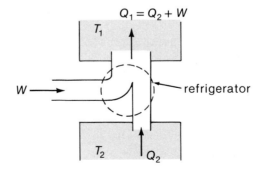

------

*It is convenient to use T_1 for the temperature of the *hot* thermal reservoir whether the device is a heat engine or a refrigerator.

Example 21–4

A Carnot engine uses as a working substance 0.15 mol of air (considered to be an ideal gas with $\gamma = 1.40$) and operates between the temperatures $T_1 = 550$ K and $T_2 = 300$ K. The maximum pressure in the cycle is $p_a = 20$ atm and the isothermal volume expansion ratio is $V_b/V_a = 5$ (see Fig. 21–11b).

(a) Determine the volume and the pressure at each of the points a, b, c, and d in the cycle (Fig. 21–11b).

(b) Calculate the amounts of heat Q_1 and Q_2 exchanged per cycle.

(c) Find the net amount of work W done per cycle.

(d) Determine the thermal efficiency e of the engine.

Solution:

(a) The volume V_a can be obtained from the ideal gas law equation in which $T_a = T_1$:

$$V_a = \frac{nRT_1}{p_a} = \frac{(0.15 \text{ mol})(8.314 \text{ J/mol·K})(550 \text{ K})}{20 \times (1.013 \times 10^5 \text{ Pa})}$$

$$= 3.386 \times 10^{-4} \text{ m}^3 = 0.3386 \text{ } \ell$$

Then, $V_b = 5V_a = 5(0.3386 \text{ } \ell) = 1.693 \text{ } \ell$

For the adiabatic path bc, we have $p_b V_b^\gamma = p_c V_c^\gamma$, which can be rewritten in the form $p_b V_b V_b^{\gamma-1} = p_c V_c V_c^{\gamma-1}$. Then, using the ideal gas law, this becomes

$$T_1 V_b^{\gamma-1} = T_2 V_c^{\gamma-1}$$

from which $V_c = V_b \left(\frac{T_1}{T_2}\right)^{1/\gamma-1}$

$$= (1.693 \text{ } \ell)\left(\frac{550}{300}\right)^{1/0.40} = 7.705 \text{ } \ell$$

Solving Eq. 21–24 for V_d, we have

$$V_d = V_c \frac{V_a}{V_b} = (7.705 \text{ } \ell)(\tfrac{1}{5}) = 1.541 \text{ } \ell$$

Also,

$$p_b = \frac{p_a V_a}{V_b} = \tfrac{1}{5} p_a = \tfrac{1}{5}(20 \text{ atm}) = 4.00 \text{ atm}$$

$$p_c = p_b \left(\frac{V_b}{V_c}\right)^\gamma = (4.00 \text{ atm})\left(\frac{1.693 \text{ } \ell}{7.705 \text{ } \ell}\right)^{1.40} = 0.479 \text{ atm}$$

$$p_d = \frac{p_c V_c}{V_d} = p_c \frac{V_b}{V_a} = 5p_c$$

$$= 5(0.479 \text{ atm}) = 2.395 \text{ atm}$$

(b) Using Eq. 21–2, we have

$$Q_1 = nRT_1 \ln \frac{V_b}{V_a}$$

$$= (0.15 \text{ mol})(8.314 \text{ J/mol·K})(550 \text{ K}) \ln 5$$

$$= 1104 \text{ J} \quad \text{(per cycle)}$$

From Eq. 21–25, we find

$$Q_2 = Q_1 \frac{T_2}{T_1} = (1104 \text{ J})\left(\frac{300 \text{ K}}{550 \text{ K}}\right) = 602 \text{ J} \quad \text{(per cycle)}$$

(c) Using the results of (b), we obtain

$$W = Q_1 - Q_2 = (1104 \text{ J}) - (602 \text{ J}) = 502 \text{ J} \quad \text{(per cycle)}$$

(d) The efficiency of the engine is

$$e = \frac{T_1 - T_2}{T_1} = \frac{(550 \text{ K}) - (300 \text{ K})}{550 \text{ K}} = 0.455 = 45.5 \text{ percent}$$

or

$$e = \frac{W}{Q_1} = \frac{502 \text{ J}}{1104 \text{ J}} = 0.455$$

Example 21–5

Determine the thermal efficiency of an engine that operates on the reversible cycle shown in the diagram. The cycle consists of two isothermal volume changes alternating with two constant-volume heat exchanges.

Solution:

The amount of heat added during the constant-volume step da is $Q_{21} = Q_{da} = nC_v(T_1 - T_2)$, and the amount rejected during the constant-volume step bc is $Q_{12} = -Q_{bc} = -nC_v(T_2 - T_1)$; thus, $Q_{21} = Q_{12}$. We take the attitude that the heat added Q_{21} is simply stored in the system until it is required to be returned as Q_{12}. Then, the only heat input is Q_1 and the net amount

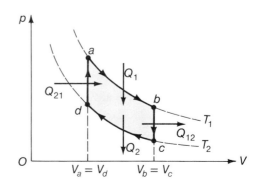

of heat added per cycle is $Q_1 - Q_2$. The first law therefore becomes (Eq. 21–21) $W = Q_1 - Q_2$. The thermal efficiency is

$$e = \frac{W}{Q_1} = \frac{Q_1 - Q_2}{Q_1} = 1 - \frac{Q_2}{Q_1}$$

For the isothermal step ab we have $Q_1 = nRT_1 \ln (V_b/V_a)$, and for cd we have $Q_2 = nRT_2 \ln (V_c/V_d)$. Also, $V_b = V_c$ and $V_d = V_a$, so that $Q_2/Q_1 = T_2/T_1$, and the efficiency becomes

$$e = 1 - \frac{T_2}{T_1}$$

which is the same as the efficiency of a Carnot engine operating between the same two temperatures (Eq. 21–26).

Example 21–6

A heat engine operates in a cycle that consists of an adiabatic expansion, an isothermal compression, and a constant-volume heating step, as indicated in the diagram. As in Example 21–4, the working substance is air (considered to be an ideal gas) and the operating temperatures are $T_1 = 550$ K and $T_2 = 300$ K. Determine the thermal efficiency of the engine.

Solution:

The efficiency e is given by Eq. 21–22

$$e = 1 - \frac{Q_2}{Q_1}$$

so we need to calculate the amount of input heat Q_1 and the amount of rejected heat Q_2. The input occurs at constant volume, so $Q_1 = nC_v(T_1 - T_2)$; the rejection occurs during an isothermal process, so $Q_2 = nRT_2 \ln (V_b/V_a)$. According to the calculation of V_c in Example 21–4, we can write

$$V_b = V_a \left(\frac{T_1}{T_2} \right)^{1/(\gamma-1)}$$

Thus,

$$Q_2 = nRT_2 \ln \left(\frac{T_1}{T_2} \right)^{1/(\gamma-1)} = \frac{nRT_2}{\gamma - 1} \ln (T_1/T_2)$$

Therefore,

$$e = 1 - \frac{RT_2 \ln (T_1/T_2)}{C_v(\gamma - 1)(T_1 - T_2)}$$

Using $\gamma = C_p/C_v$, we have $C_v(\gamma - 1) = C_p - C_v = R$. Then,

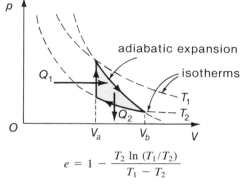

$$e = 1 - \frac{T_2 \ln (T_1/T_2)}{T_1 - T_2}$$

(As in the case of the Carnot engine, Eq. 21–26, the efficiency here also increases as $T_1 - T_2$ is increased; refer to Problem 21–19.)

Substituting for T_1 and T_2, we find

$$e = 1 - \frac{(300 \text{ K}) \ln (550/300)}{(550 \text{ K}) - (300 \text{ K})} = 0.273 = 27.3 \text{ percent}$$

The compression ratio in this cycle is

$$\frac{V_b}{V_a} = \left(\frac{T_1}{T_2} \right)^{1/(\gamma-1)} = \left(\frac{550 \text{ K}}{300 \text{ K}} \right)^{1/0.4} = 4.55$$

which is close to the value of 5 for the case in Example 21–4. Also, the values of T_1 and T_2 are the same. Nevertheless, the efficiency of the engine considered here is only *half* that of the Carnot engine. The reason is that the engine does not operate between two fixed temperatures (because the heat input Q_1 takes place over the range of temperatures from T_2 to T_1), and the efficiency is correspondingly smaller than the Carnot efficiency.

21–5   PRACTICAL THERMODYNAMIC CYCLES

The Gasoline Engine (Otto Cycle).   The gasoline internal-combustion engine operates in a cycle that consists of six parts, only four of which, called *strokes*, involve motion of the piston. Although portions of the cycle are actually irreversible, and friction and heat losses are present, we may simulate the entire engine cycle by the idealized quasi-static cycle shown in Fig. 21–13. This cycle is known as the *air-standard Otto cycle*. Throughout this idealized cycle the working substance is treated as if it were air alone, ignoring the fuel vapor and the combustion products (see Problem 21–34).

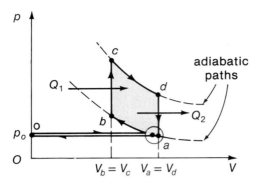

Fig. 21–13. The air-standard Otto cycle for internal-combustion gasoline engines. This cycle describes the usual "four-cycle" (or four-stroke) automobile engine. (The description of the "two-cycle" engine used in motorboats and lawn mowers is slightly different.) The shaded area represents the work done per cycle: $W = Q_1 - Q_2$.

The six steps in the cycle are as follows:

1. *Isobaric intake stroke oa.*   The gasoline-air mixture is drawn into the cylinder at atmospheric pressure p_0; the cylinder inlet valve remains open throughout this stroke. The volume increases from zero to V_a.

2. *Adiabatic compression stroke ab.*   The inlet valve is closed and the gasoline-air mixture is compressed adiabatically from volume V_a to V_b.

3. *Constant-volume ignition bc.*   The gasoline-air mixture (heated during the compression stroke) is ignited by an electric spark, thereby further increasing the temperature. The combustion process takes place in such a short time that there is essentially no motion of the piston; thus, $V_c = V_b$. This irreversible process is indicated in the diagram as an idealized constant-volume step; it does not constitute one of the strokes.

4. *Adiabatic expansion or power stroke cd.*   The heated air and combustion products expand against and drive the piston; the volume increases from V_c to V_d. This is the stroke that delivers power to the crankshaft.

5. *Constant-volume exhaust da.*   At the end of the power stroke, the exhaust valve is opened; this causes a sudden partial escape of the cylinder gases. There is essentially no piston motion during this part of the cycle, so we indicate da in the diagram as an idealized constant-volume step; thus, $V_a = V_d$.

6. *Isobaric exhaust stroke ao.*   The exhaust valve remains open and the moving piston drives out the residual cylinder gases as the volume decreases from V_a to zero.

The heat input to the actual (changing) working substance of a real engine is the energy released by the combustion of the gasoline vapor in the cylinder during step bc. In a true engineering sense this is the available energy, but it is only partially utilized in doing work. Some of the available energy is lost during the valve exhaust step da. For the idealized air-standard Otto cycle using equivalent moles of air, we have

$$Q_1 = nC_v(T_c - T_b)$$
$$Q_2 = nC_v(T_d - T_a)$$
and
$$W = Q_1 - Q_2$$

The thermal efficiency e of the engine is

$$e = \frac{W}{Q_1} = \frac{Q_1 - Q_2}{Q_1} = 1 - \frac{T_d - T_a}{T_c - T_b}$$

It is easy to show that $(T_d - T_a)/(T_c - T_b) = (V_b/V_a)^{\gamma-1}$ (see Problem 21–23). The ratio $r = V_a/V_b$ is called the *compression ratio* of the engine. Thus,

$$e = 1 - \frac{1}{r^{\gamma-1}} \qquad \text{air-standard Otto cycle} \qquad (21-28)$$

In practice, the compression ratio must be held below about 7 (additives permit a higher ratio), because if $r > 7$, the temperature T_b that is reached at the end of the compression stroke ab would be sufficiently high to ignite the fuel mixture before the occurrence of the spark. Such *preignition* causes "knocking" and loss of power. Using $r = 5$ and $\gamma(\text{air}) = 1.40$, we find

$$e = 1 - \frac{1}{(5)^{0.40}} = 0.475 \text{ or } 47.5 \text{ percent}$$

The actual realized efficiency of a properly tuned gasoline engine is usually no more than about 25 or 30 percent because of heat losses, friction, finite acceleration times for engine parts, and the fact that the efficiency of an irreversible cycle falls short of that of the idealized reversible cycle that we have assumed.

The Refrigerator (Two-Phase Cycle). The most common type of refrigerator operates in a two-phase cycle using the liquid and vapor phases of the working substance. We follow a given mass of the working substance (or *refrigerant*) through a complete cycle: compressor → condenser → liquid storage → throttle → evaporator → compressor. The usual refrigerant in ordinary household refrigerators is Freon-12 (CCl_2F_2); sulfur dioxide (SO_2) is often used in large commercial units. A schematic diagram of a two-phase refrigerator is shown in Fig. 21–14. In an ordinary household refrigerator only the evaporator is located inside the cold compartment. During each cycle the compressor does an amount of work W on the refrigerant using energy from an external source.

The idealized cycle for a two-phase refrigerator is shown on a p-V diagram in Fig. 21–15. This cycle closely resembles the cycle for a Carnot refrigerator that operates between the temperatures T_1 and T_2. Starting at point a in the diagram, the saturated liquid in the storage tank at (p_1, T_1) is allowed to undergo an irreversible throttled expansion ab into the evaporator coils (see Fig. 21–14). During this process the pressure and temperature drop to (p_2, T_2), both of which are much lower than (p_1, T_1). The expansion causes

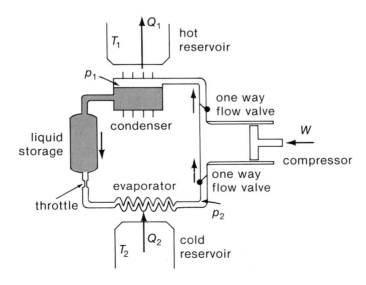

Fig. 21–14. Essential features of a common two-phase refrigerator.

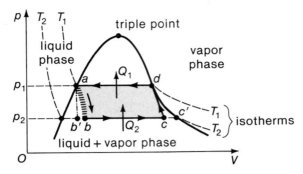

Fig. 21–15. A two-phase refrigerator cycle. The shaded area represents the work done on the working substance by the compressor. The irreversible throttling process *ab* could be replaced by an adiabatic, piston-controlled expansion *ab'*. However, this results in only a slight improvement in refrigerator performance and is therefore not economically worthwhile.

vaporization of the liquid, which continues until point c is reached, close to vapor saturation. This evaporation process bc takes place at essentially constant pressure and constant temperature, and it removes an amount of heat Q_2 from the cold reservoir (the cold compartment of the refrigerator). During the step cd, the refrigerant, which is a nearly saturated vapor at c, is adiabatically compressed to saturated vapor at d. It is then delivered at the higher pressure p_1 and higher temperature T_1 to the condenser coils. In passing through the coils (following the path da), the refrigerant gives up an amount of heat Q_1 to the hot reservoir (room air) at the temperature T_1. When this process is completed, the refrigerant is a saturated liquid at point a.

Most refrigerators are designed to maintain a temperature of a few degrees below 0° C in the cold compartment when the hot reservoir is at room temperature. It is therefore necessary to choose for the refrigerant a substance that will vaporize at about −5° C at low pressure and will condense at about 20° C at high pressure. Many substances have been used as refrigerants, the most common of which is Freon-12.

What is the "efficiency" of a refrigerator? Because a refrigerator performs a function opposite that of an engine, its efficiency cannot be defined in the same way. Instead, the performance of a refrigerator is specified by a figure of merit K_r, called the *coefficient of performance:*

$$K_r = \frac{Q_2}{W} = \frac{Q_2}{Q_1 - Q_2}$$

The quantity Q_2 is the amount of heat absorbed from the cold compartment at the temperature T_2. The quantity Q_1 is the amount of heat rejected into the room air; Q_1 is equal to the heat absorbed Q_2 plus the work W done on the refrigerant by the compressor. To the extent that the actual cycle may be represented by a reverse Carnot cycle, we have

$$K_r = \frac{T_2}{T_1 - T_2} \qquad \text{Carnot cycle} \qquad \textbf{(21–29)}$$

For a household refrigerator, typical temperatures are $T_1 = 300$ K (80° F) and $T_2 = 260$ K (10° F); then the theoretical coefficient of performance is K_r(Carnot) = $260/(300 - 260) = 6.5$. An actual refrigerator operating between the same two temperatures might have a coefficient of performance of 4 or 5.

Heat Pumps. Because a refrigerator rejects more heat to the room air than it absorbs from the cold reservoir ($Q_1 > Q_2$), a refrigerator can be used to pump heat from the *outside* air (now used as the cold reservoir) to the *inside* air. Thus, a device operating on a refrigeration cycle can be used as an air conditioner during summer (pumping heat from inside to outside) and as a *heat pump* during winter (pumping heat from outside to inside).

When operating as a heat pump, the figure of merit for the device is

$$K_h = \frac{Q_1}{W} = \frac{T_1}{T_1 - T_2} \qquad \text{Carnot cycle} \qquad \textbf{(21-30)}$$

for $T_1 = 300$ K (80° F) and $T_2 = 260$ K (10° F), we find K (Carnot) = 7.5. In fact, $K_h = K_r + 1$. (• Can you prove this relationship?) An actual heat pump might have $K_h = 5$ or 6 for these temperatures.

Example 21-7

A house loses heat through the exterior walls and roof at a rate of 5000 J/s = 5 kW when the interior temperature is 22° C and the outside temperature is −5° C. Calculate the electric power required to maintain the interior temperature at 22° C for the following two cases:

(a) The electric power is used in electric resistance heaters (which convert all of the electricity supplied into heat).

(b) The electric power is used to drive an electric motor that operates the compressor of a heat pump (which has a coefficient of performance equal to 60 percent of the Carnot cycle value).

Solution:

(a) The amount of power required is just equal to the heat loss rate, namely, 5 kW.

(b) We have

$$K_h = 0.60 \times K_h(\text{Carnot}) = 0.60 \times \frac{295}{295 - 268} = 6.6$$

Thus, the power required is

$$P = \frac{W}{t} = \frac{Q_1/K_h}{t} = \frac{5 \text{ kW}}{6.6} = 0.758 \text{ kW}$$

That is, the heat pump supplies 6.6 times as much heat to the house as do the electric heaters for the same expenditure of electric power. (This ignores the fact that the compressor motor is not 100 percent efficient in converting electricity to mechanical work; with a motor efficiency of 80 percent, however, the heat pump still requires less than 1 kW to heat the house.)

QUESTIONS

21-1 Define and explain briefly the meaning of the terms (a) quasi-static process, (b) reversible process, (c) internal energy, (d) isothermal process, (e) adiabatic process, and (f) free expansion of a gas.

21-2 Give several examples of irreversible processes from everyday experience.

21-3 The diagram (right) shows a gas system with an initial volume V that undergoes an arbitrary expansion by an amount dV. Noting that the change in volume for the illustrated portion is $d^2V = (dA\hat{n})\cdot d\mathbf{s} = (dA)(ds) \cos \theta$, and the corresponding work done during this volume change is $d^2W = (dF)(ds) \cos \theta = p(dA)(ds) \cos \theta$, show that quite generally $dW = p\, dV$.

21-4 Many practical thermodynamic processes are nearly adiabatic. Although such processes may be irreversible, the

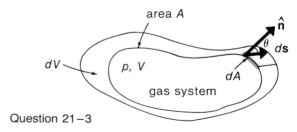

Question 21-3

application of the first law permits a precise statement concerning the work done in proceeding from one equilibrium state to another. Explain in detail why such a prediction is possible, and explain the necessary role of the adiabatic condition.

21-5 Explain, in physical terms, why the process must be specified (such as a constant-pressure or a constant-volume process) when defining the specific heat for gases, whereas the need for such exact specification is not too important for most solids and liquids. When exact specification of the process for adding heat to a solid or liquid is required, why is a constant-pressure process the practical choice?

21-6 Invent a thermodynamic process that requires *less* added heat to raise the temperature of a gas from T_1 to T_2 than a constant-volume process between the same two temperatures. Could the required heat be zero? How would you define the specific heat for these processes? Could the specific heat ever be negative?

21-7 What would be the physical consequence of having possible p-V isotherms with positive slopes?

21-8 An adiabatic expansion is always accompanied by a drop in temperature. Explain.

21-9 Define and explain briefly the meaning of the terms (a) Carnot cycle, (b) diathermic wall, (c) thermal efficiency, (d) Otto cycle, and (e) coefficient of performance.

21-10 The Carnot-cycle efficiency may be improved by either raising the upper temperature T_1 or lowering the lower temperature T_2. An equal amount of change in these two temperatures does not, however, give the same benefit. Explain.

21-11 In the idealized air-standard Otto cycle, work must be done on the air-fuel mixture during the adiabatic pressure stroke *ab*. What part of a real engine supplies this work? Which stroke of the Otto cycle is manifestly irreversible?

21-12 Heat pumps can deliver more heat energy to a house than the energy required to run the pump. Explain why this is consistent with the conservation of energy.

PROBLEMS

Section 21-1

21-1 Show that the work done by a gas during an isobaric process is $W_{ab} = p(V_b - V_a)$.

21-2 A sample of ideal gas is expanded to twice its original volume of 1 m³ in a quasi-static process for which $p = \alpha V^2$, with $\alpha = 5.0$ atm/m⁶, as shown in the diagram. How much work was done by the expanding gas?

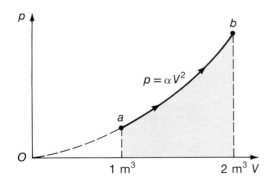

21-3 Argon gas ($m = 5.0$ g) undergoes an isothermal expansion at 20° C to twice its original volume. (a) How much work was done by the gas? (b) If the original pressure was 1 atm, what is the final pressure?

21-4 A 1-mol sample of an ideal gas is carried around the thermodynamic cycle shown in the diagram. The cycle consists of three parts—an isothermal expansion ($a \rightarrow b$), an isobaric compression ($b \rightarrow c$), and a constant-

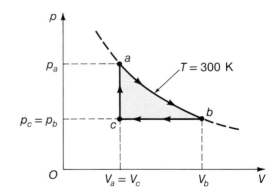

volume increase in pressure ($c \rightarrow a$). If $T = 300$ K, $p_a = 5$ atm, and $p_b = p_c = 1$ atm, determine the work done by the gas per cycle.

21-5 The *compressibility* of a gas is defined to be $-(1/V)(dV/dp)$. Show that the isothermal compressibility of an ideal gas is $1/p$.

Section 21-2

21-6 Many of the thermodynamic ideas we have discussed
  • apply to liquids and solids as well as to gases. A beaker contains 1.2 ℓ of mercury at 20° C. The beaker is open to the atmosphere. Heat is slowly added to the mercury and its temperature is increased to 100° C. During this process the pressure is constant at 1 atm. Neglect any thermal expansion of the beaker. Use the tabulated val-

ues for $\bar{\beta}$ (Table 18–2) and $\bar{c}$ (Table 18–3) for mercury. (Note that $\bar{c}$ is the specific heat at constant pressure.) (a) Calculate the work done by the expanding mercury. (b) Determine the amount of heat added to the mercury. (c) What is the increase in the internal energy of the mercury? (d) In view of these results, is there any substantial difference between the specific heat at constant pressure and the specific heat at constant volume for mercury near atmospheric pressure?

21–7 A quantity of an ideal gas ($n = 4$ mol) is confined within a cylinder that has a movable piston and diathermic walls. The cylinder is immersed in a heat bath that is maintained at a temperature of 300 K. An external force F_0 is required to hold the piston in mechanical equilibrium. This force is suddenly increased to 10 F_0 and remains at this value until the volume has decreased to one half of the original value. (The force is then adjusted to maintain this volume.) After this irreversible process has been completed, a new equilibrium is established at the smaller volume. (a) How much work was done by the external force? (b) What happened to this energy?

21–8 One mole of argon is confined in a cylinder with a movable piston at a pressure of 1 atm and at a temperature of 300 K. The gas is heated slowly and isobarically to a temperature of 400 K. The measured value of the molar specific heat at constant pressure for argon in this temperature range is $C_p = 2.5043$ R, and the measured value of pV/nT is 0.99967 R. Calculate in units of R, carried to two decimals, the following quantities: (a) the work done by the expanding gas; (b) the amount of heat added to the gas; (c) the increase in the internal energy of the gas (using the first law); and (d) the increase in the internal energy of the gas (using kinetic theory—Eq. 21–6). Compare the results of (c) and (d).

Section 21–3

21–9 Show by direct integration of $p\,dV$, using $p = $ (constant)/V^γ, that the work done during an adiabatic process is given by Eq. 21–20.

21–10 The following measured values are determined for argon at 1 atm and 15° C:

$$pV/T = 208.0 \text{ J/deg (for a 1-kg sample)}$$

$$c_p = 0.1253 \text{ Cal/kg·deg}$$

$$\gamma = 1.668$$

Using these data alone (together with the assumption that argon is an ideal gas), determine the mechanical equivalent of heat and compare it with the accepted value, 4186 J/Cal. (This is essentially the method used by J. R. Mayer in 1842.)

21–11 The *compressibility* of a gas is defined to be $-(1/V)(dV/dp)$. Show that the adiabatic compressibility of an ideal gas is $1/\gamma p$. (Compare Problem 21–5.) [*Hint*: Differentiate the adiabatic path equation, Eq. 21–17.]

21–12 The speed of sound in a gas is $v = \sqrt{\gamma p/\rho}$ (see Eq. 17–6). Assume that air is an ideal diatomic gas and calculate the speed of sound in air at 15° C. (The effective molecular mass of air is 28.96 u.)

Section 21–4

21–13 An average adult consumes about 90 Cal while running 1 mi. A cheeseburger from a typical fast-food service has an energy content of about 350 Cal. Is it more cost-effective to run a mile (powered by a cheeseburger) or to drive a mile in an automobile (powered by gasoline) that travels 12 miles per gallon? (Check current prices.)

21–14 A Carnot engine operates between the temperatures $T_1 = 100°$ C and $T_2 = 30°$ C. By what factor is the theoretical efficiency of the engine increased if the temperature of the hot reservoir is increased to 580° C (a boiler temperature common in power plants)?

21–15 A Carnot engine operates with a thermal efficiency of 40 percent, taking in 5 Cal of heat per cycle from a hot reservoir at a temperature of 250° C. (a) What is the temperature of the cold reservoir? (b) What amount of heat is rejected per cycle to the cold reservoir? (c) How much work is done by the engine per cycle? (d) If the working substance is 0.8 mol of an ideal monatomic gas, what is the expansion ratio during the isothermal expansion? (e) What is the expansion ratio during the adiabatic expansion?

21–16 An engine operates in a reversible cycle that consists of an isobaric expansion, an adiabatic expansion, and an isothermal compression, as shown in the diagram. Show that the efficiency of this engine is the same as that found in Example 21–6.

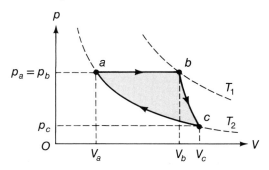

21–17 An engine operates in a reversible cycle that consists of an adiabatic expansion from a temperature $T_1 = 500°$ C to a temperature $T_2 = 200°$ C, followed by an isobaric compression and a constant-volume heating to close the cycle, as shown in the diagram. The working substance consists of 2.5 mol of a diatomic ideal gas. (a) Show that the efficiency of the engine is

$$e = 1 - \gamma \frac{T_2 - T_3}{T_1 - T_3}$$

(b) Derive an expression relating the three temperatures and determine the value of T_3. (c) Evaluate the efficiency e. (d) How much work is done by the engine per cycle?

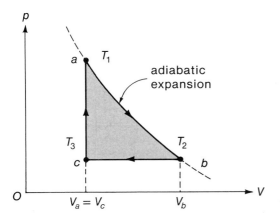

21–18 A Carnot refrigerator is used to freeze 50 kg of water at 0° C by discharging heat into a room that acts as a hot reservoir with a temperature of 27° C. (a) What is the total amount of heat discharged into the room during the process? (b) How much energy must be supplied to operate the refrigerator? For how long must the 1-kW refrigerator motor run in order to accomplish the freezing of the water?

21–19 In Example 21–6, an equation was derived for the effi-
  •   ciency of a particular heat engine:

$$e = 1 - \frac{T_2 \ln (T_1/T_2)}{T_1 - T_2}$$

It was claimed that e increases as $T_1 - T_2$ increases. Prove the claim. [*Hint:* Let $T_1/T_2 = x$ and show that $\dfrac{de}{dx} > 0$ (when $x > 1$). You will need to know that $\ln x > \dfrac{x - 1}{x}$ for all values of $x > 1$.]

Section 21–5

21–20 A 50-hp crane is driven by an engine that operates in a cycle between a boiler temperature of 200° C and an exhaust temperature of 100° C. The cycle rate of the engine is 500 rpm. The realized efficiency of the engine

is 40 percent of that of a reversible engine operating between the same temperatures. What is the rate of heat input required to operate the engine?

21–21 An electric-generating plant produces 1250 MW of
  ••   power by burning oil to drive steam turbines. Cooling water is supplied by a large river whose flow rate is 1800 m³/s. (Refer to diagram.) The electric generators have a combined efficiency of 95 percent; the thermal efficiency of the turbines is 40 percent; and the efficiency of transferring the fuel heat to the boiler is 80 percent. Assume that $\frac{1}{10}$ of the river's flow is used for cooling, and assume that the heat of combustion of the fuel oil is 4.40×10^7 J/kg. (a) What is the total output power of the turbines? (b) What are the input and exhaust heat flow rates for the turbines? (c) If the steam enters the turbines at 580° C and leaves at 110° C, what is the efficiency of the turbines if they operate in a reversible cycle? Compare this with the actual overall efficiency. (d) How many barrels (bbl) of fuel oil must be burned per hour to produce the output? (The density of the fuel oil is 920 kg/m³ and the volume of 1 bbl is 0.159 m³.) (e) What is the temperature rise of the cooling water that flows through the turbine condenser? If half of the heat delivered to the cooling water is dissipated into the air, what is the temperature rise of the river water when the cooling water has been thoroughly mixed?

21–22 A gasoline-powered automobile engine with a compres-
  •   sion ratio $r = 6.8$ delivers 30 hp. The thermal efficiency of the engine is 45 percent of the theoretical value for an engine operating on the air-standard Otto cycle, and all of the chemical energy of the gasoline is realized as heat. What is the rate of gasoline consumption in ℓ/h? (The heat of combustion of the gasoline is 4.81×10^7 J/kg and the density is 740 kg/m³.) If the automobile travels at an average speed of 95 km/h, what "mileage" does the automobile achieve in km/ℓ?

21–23 Show that for the air-standard Otto cycle, we have
  • $1/r^{\gamma-1} = (T_d - T_a)/(T_c - T_b)$, with the notation of Fig. 21–13. The quantity r is the compression ratio, V_a/V_b.

21–24 The small diagram on page 431 shows the *air-standard Sargent cycle*, which consists of two adiabatic steps ab and cd, a constant-volume heat input step bc, and an isobaric compression da. Show that the efficiency of an engine operating in this cycle is

$$e = 1 - \gamma \frac{T_2 - T_3}{T_1 - T_4}$$

21–25 The heat discharged by an electrically driven refrigerator amounts to 250 Cal/h. The cold compartment is maintained at a temperature of $-10°$ C and the room air has a constant temperature of 27° C. The coefficient of per-

Problem 21–21

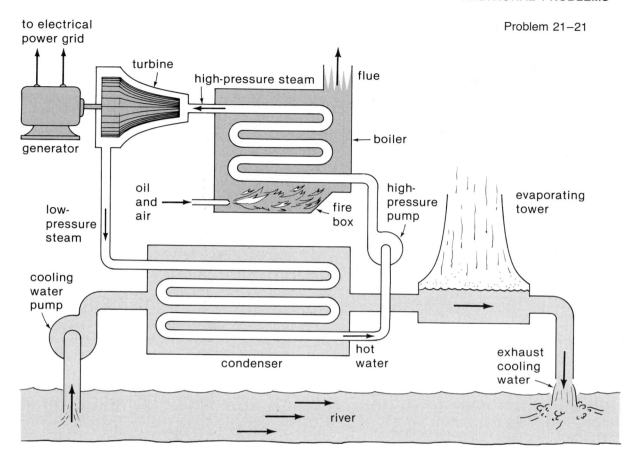

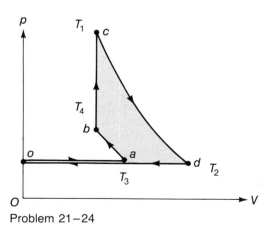

Problem 21–24

formance for the refrigerator is 55 percent of that for a Carnot refrigerator. What is the average power consumption of the motor that drives the compressor?

21–26 A small office building requires a heating rate of 10^5 Cal/h to maintain an inside temperature of $27°$ C during a winter day when the outside temperature is $5°$ C. The heat is provided by a heat pump that has a coefficient of performance K_h equal to 55 percent of that for an ideal Carnot system. What input power is required to drive the heat pump?

Additional Problems

21–27 Nitrogen gas ($m = 1.00$ kg) is confined in a cylinder with a movable piston exposed to normal atmospheric pressure. A quantity of heat ($Q = 25.0$ Cal) is added to the gas in an isobaric process. (a) How much work was done by the gas? (b) What is the increase in the internal energy of the gas? (c) What was the change in volume? (d) What was the change in temperature? [*Hint:* First find the ratio of the work done to the increase in internal energy.]

21–28 An ideal gas is carried around the cycle shown in the diagram, which consists of alternating constant-volume and constant-pressure paths. (a) How much work is done by the gas per cycle? (b) What amount of heat is supplied to the gas per cycle?

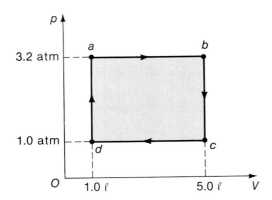

21–29 Argon gas ($n = 3$ mol), initially at a temperature of $20°$ C, occupies a volume of 10 ℓ. The gas undergoes a slow expansion at constant pressure to a volume of 25 ℓ; then, the gas expands adiabatically until it returns to its initial temperature. (a) Show the process on a p-V diagram. (b) What quantity of heat was supplied to the gas during the entire process? (c) What was the total change in internal energy of the gas? (d) What was the total amount of work done during the process? (e) What is the final volume of the gas?

21–30 A gas obeys the van der Waals equation of state (Eq. 19–2),

$$\left(p + \frac{a}{\mathcal{V}^2} \right)(\mathcal{V} - b) = RT$$

where $\mathcal{V} = V/n$ is the molar volume. Derive an expression for the work done per mole of gas in an isothermal expansion.

21–31 For a gas that obeys the van der Waals equation of state (see Problem 21–30 or Eq. 19–2), it can be shown that

$$C_p - C_v \cong R\left(1 + \frac{9}{4} \frac{\mathcal{V}_c T_c}{\mathcal{V} T} \right)$$

where $\mathcal{V}$ and T are the molar volume $\mathcal{V} = V/n$ and the temperature, respectively, and where $\mathcal{V}_c$ and T_c are the corresponding values at the critical point. (Notice that the expression for $C_p - C_v$ implies a slight dependence of the internal energy on volume.) For argon, we have $T_c = 150.7$ K and $\mathcal{V}_c = 75.3$ cm³/mol. Calculate the correction term (the deviation from the ideal-gas value) in the expression for $C_p - C_v$ for argon at $0°$ C. [*Hint:* Use the fact that 1 mol of any gas at $0°$ C and 1 atm occupies 22.41 ℓ.]

21–32 An ideal monatomic gas is confined in a cylinder with a cross-sectional area $A = 75$ cm² by means of a spring-loaded piston, as shown in the diagram. The relaxed length of the spring is ℓ_0 (no pressure difference across the piston); the force constant of the spring is $\kappa = 7500$ N/m. With the gas at its initial pressure, $p_a = 10^5$ N/m², the spring is compressed by an amount x_a, and the length of the gas-filled part of the cylinder is $\ell_a = 30$ cm. Heat is slowly added to the gas until it has expanded and compressed the spring by an additional 10 cm; that is, $\ell_b = 40$ cm. At any point during the process, we have $\ell = \ell_a + (x - x_a)$. (a) Show that $p = p_a + (\kappa/A^2)(V - V_a)$. Plot this path from V_a to V_b in a p-V diagram. (b) Show that the work done by the expanding gas is

$$W_{ab} = \oint_{V_a}^{V_b} p \, dV = \tfrac{1}{2}(V_b - V_a)(p_b + p_a)$$

(c) Show that the result obtained in (b) is just equal to the work done on the spring, namely, $\tfrac{1}{2}\kappa(x_b^2 - x_a^2)$. (d) Evaluate W_{ab} by using the expressions in both (b) and (c). (e) Calculate the increase in internal energy U_{ab} of the gas. (f) Determine the amount of heat Q_{ab} supplied to the gas during the process.

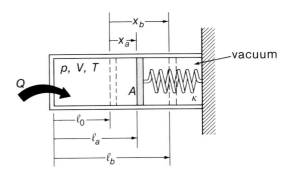

21–33 A cylinder with a length of 1 m is divided into two compartments by a thin, tight-fitting, diathermic piston that is clamped in position, as shown in the diagram. The left-hand compartment has a length of 0.4 m and contains 4 mol of neon at a pressure of 5 atm. The right-hand compartment contains helium at a pressure of

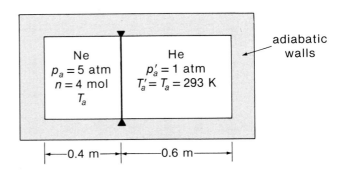

1 atm. The entire cylinder is surrounded by an adiabatic insulator. The initial temperature of each gas is 20° C. Now the piston is released and a new equilibrium condition is eventually reached. Locate the new position of the piston, which remains tight-fitting throughout.

21–34 A cylinder with adiabatic walls contains 0.40 ℓ of air at 20° C and 30 atm confined by a movable piston. A small quantity of gasoline (approx. 0.1 g) is introduced into the cylinder and the mixture is ignited. The constant-volume ignition yields a chemical energy of 1.2 Cal. After equilibrium is reached, the gas is expanded adiabatically to a volume of 2.0 ℓ. (a) How many moles of air did the cylinder originally contain? (b) Consider the gas in the cylinder to be air both before and after the ignition. What equilibrium temperature and pressure is reached after the ignition (but before the expansion)? (c) What is the temperature of the air after the adiabatic expansion? (d) Determine the amount of work done by the gas. (e) What was the efficiency of the conversion of chemical energy into mechanical work?

21–35 The top of a vertical cylindric tank is closed by a tight-fitting piston with negligible mass. The column of ideal gas below the piston has a height h. A fluid with a density ρ is slowly poured onto the piston, and the piston descends a distance y before fluid spills over the rim of the cylinder. (a) Suppose that the system is perfectly insulated. Show that

$$h = \frac{y}{1 - \left(1 + \dfrac{\rho g}{p} y\right)^{-1/\gamma}}$$

(b) Suppose that the process is isothermal. Show that

$$h = y + p/\rho g$$

(c) Suppose again that the process is isothermal. How much work is done in compressing the gas?

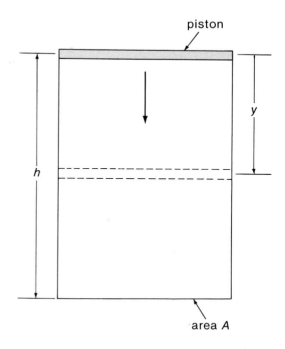
piston

h

y

area A

CHAPTER 22

THE SECOND LAW OF THERMODYNAMICS

In the preceding chapter, the principle of energy conservation was applied to thermodynamic systems by using a generalized form of the principle called the *first law of thermodynamics*. This law permits us to determine whether or not a particular thermodynamic process—reversible or irreversible—can possibly take place. However, the first law alone is not sufficient to determine whether or not the process will actually occur; the fact that a process is consistent with energy conservation does not imply that the processes will take place.

Natural processes—those that occur spontaneously, without any deliberate external intervention or control—are almost always irreversible. When such uncontrolled processes involve changes of state, they do not proceed through quasi-static equilibrium states (see Section 21–1). These uncontrolled processes have the important common feature that they take place in only one direction; the reverse process—even if it is energetically allowed—does not occur. For example, the spontaneous flow of heat is always from a hot object or region to a cooler object or region; heat does not flow spontaneously in the other direction. The free expansion of a gas will fill an evacuated volume; the gathering together of all the molecules into the original space is never observed. If two containers with different gases are connected, the gases will mix together; the separation of a mixture of gases into its molecular components will not occur spontaneously. All of these processes can be made to conform to the first law of thermodynamics, and yet they proceed only in one direction. The *second law of thermodynamics*—the principal subject of this chapter—is a precise statement of a broad generalization derived from observations of unidirectional processes. In the course of this study we shall also discover the intimate connection between the second law of thermodyamics and thermodynamically reversible processes.

22–1 THE SECOND LAW OF THERMODYNAMICS

It would clearly be advantageous to have a perfect heat engine, an engine capable of transforming heat into mechanical work with an efficiency of 100 percent. (Imagine never wasting *any* energy—your car might have a mileage of more than 150 miles per gallon of gasoline!) Figure 22–1 contrasts the heat-flow diagrams for an ordinary heat engine and a perfect heat engine. In the latter case, no heat is rejected to the cold reservoir ($Q_2 = 0$), so $W = Q_1$; consequently, $e = 1$.

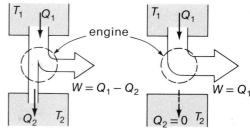

Fig. 22–1. Heat-flow diagrams for an ordinary heat engine and a perfect heat engine. In the latter case, $Q_2 = 0$ and $W = Q_1$ so that $e = 1$.

(a) ordinary heat engine

(b) perfect heat engine

The counterpart to the perfect heat engine—the perfect refrigerator—would require no mechanical work to remove heat from a cold reservoir and deliver it to a hot reservoir. Figure 22–2 shows the heat-flow diagrams for an ordinary refrigerator and a perfect refrigerator. In the latter case, $W = 0$ and $Q_1 = Q_2$.

Experience indicates that neither of these ideally perfect heat devices is possible. The statement that all such perfect devices are physically impossible is contained within the *second law of thermodynamics*. There are several equivalent statements of this important principle. The first formulation of the second law that we give was proposed by Kelvin (and by Planck):

It is impossible to remove heat from a system at a single temperature and have as the sole effect the conversion of this heat into mechanical work.

The Kelvin-Planck statement of the second law directly rules out the possibility of a perfect heat engine, as illustrated in Fig. 22–1b.

A second form of the law was proposed by Clausius*:

It is impossible to transfer heat continuously from a colder system to a warmer system as the sole result of any process.

The Clausius statement of the second law directly rules out the possibility of a perfect refrigerator, as illustrated in Fig. 22–2b.

It is interesting to note that the Kelvin and Clausius statements of the second law are equivalent in the sense that if either were false, the other would also be false. This proposition can be demonstrated in the following way. Suppose that the Clausius state-

Fig. 22–2. Heat-flow diagrams for an ordinary refrigerator and a perfect refrigerator. In the latter case, $W = 0$ and $Q_1 = Q_2$.

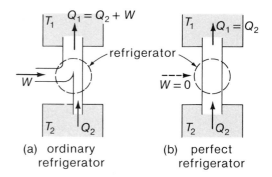

(a) ordinary refrigerator

(b) perfect refrigerator

---

*Rudolf Clausius (1822–1888), German physicist and one of the founders of thermodynamics. His statement of the second law was given in 1850.

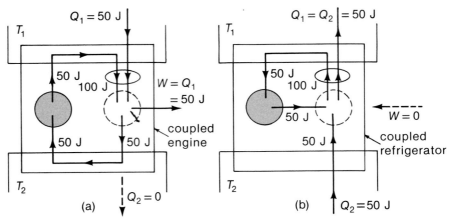

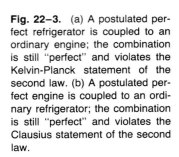

Fig. 22–3. (a) A postulated perfect refrigerator is coupled to an ordinary engine; the combination is still "perfect" and violates the Kelvin-Planck statement of the second law. (b) A postulated perfect engine is coupled to an ordinary refrigerator; the combination is still "perfect" and violates the Clausius statement of the second law.

ment is false. Imagine that an ordinary heat engine performs work with the rejection of heat to a cold reservoir (Fig. 22–3a). A perfect refrigerator, now possible, could transfer the rejected heat to the hot reservoir without the expenditure of any work. This heat would then be available to the engine to do more work. Using such a coupled device (Fig. 22–3a), all of the heat Q_1 removed from the hot reservoir would eventually appear as work; that is, $W = Q_1$ and $Q_2 = 0$. This would violate the Kelvin-Planck statement of the second law.

Now, suppose that the Kelvin-Planck formulation of the second law is false and that a perfect heat engine exists. Imagine that this engine is coupled to an ordinary refrigerator, which removes heat from a cold reservoir. The refrigerator delivers to a hot reservoir an amount of heat equal to the sum of the heat removed from the cold reservoir plus the work done by the engine (Fig. 22–3b). This excess heat is exactly the amount required to operate the perfect engine that drives the refrigerator. The net effect of the coupled engine and refrigerator is to transfer heat continuously from a cold reservoir to a hot reservoir without the need for any work provided by an external agency; that is, $Q_2 = Q_1$ and $W = 0$. This would violate the Clausius statement of the second law.

▶ *Perpetual-Motion Machines.* One often hears of claims concerning perpetual-motion machines. A perpetual-motion machine of the *first kind* is a device that would violate the first law by producing more work than the energy supplied to it. Claims for such machines are easy to refute by using an energy-conservation argument. A device that would extract heat from the environment and do work without benefit of any temperature difference could be consistent with the first law. But such a machine—a perpetual-motion machine of the *second kind*—would violate the second law and, hence, is unrealizable. ◀

22–2   APPLICATIONS OF THE SECOND LAW OF THERMODYNAMICS

Carnot's Theorem.   With remarkable insight for a time when the caloric theory was still in vogue, Sadi Carnot in 1824 stated a powerful thermodynamic theorem:

The efficiency of all reversible engines that absorb and reject heat at the same two temperatures is the same, and no irreversible engine working between the same two temperatures can have an efficiency as great.

Carnot's original proof of this theorem was based on caloric arguments and was therefore faulty. We now give a valid proof formulated by Clausius and Kelvin about 30 years later.

Consider two reversible engines, E and E', that operate between thermal reservoirs with temperatures T_1 and T_2 (with $T_1 > T_2$). Let the engine E' run backward, thereby functioning as a refrigerator. Also, let the two engines be coupled so that the cycles have equal duration. (We can imagine that this is accomplished even though the working substances and pressures of the two engines may be different.) The devices are adjusted so that the work done by the engine is exactly that required to operate the refrigerator. This situation is illustrated schematically in Fig. 22-4.

Suppose, at first, that the efficiency e of the engine E is greater than the efficiency e' of the engine E', which running backward constitutes the coupled refrigerator; that is, $e > e'$. Thus, with the notation of heat-flow quantities shown in Fig. 22-4, we can write (Eq. 21-22)

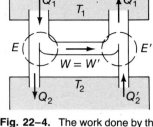

Fig. 22-4. The work done by the engine E is exactly that required to operate the coupled refrigerator E'.

$$e = \frac{Q_1 - Q_2}{Q_1} > \frac{Q_1' - Q_2'}{Q_1'} = e'$$

Now, the work W done by E is equal to the work W' required by E'. Thus,

$$W = Q_1 - Q_2 = Q_1' - Q_2' = W' \qquad \textbf{(22-1)}$$

Using this result, the efficiency equation yields

$$\frac{1}{Q_1} > \frac{1}{Q_1'} \qquad \text{or} \qquad Q_1' > Q_1$$

Equation 22-1 can be expressed as $Q_1' - Q_1 = Q_2' - Q_2$; then, it follows that $Q_2' > Q_2$. The net result of the coupled process is that an amount of heat $\Delta Q = Q_2' - Q_2 = Q_1' - Q_1$ is transferred from the cold reservoir (T_2) to the hot reservoir (T_1) without any work having been done by an external agency. But this violates the second law, so we must conclude that *e cannot be greater than e'*.

Suppose next that the operation of each (reversible) engine is reversed so that E' drives E, which now becomes a refrigerator. The same argument leads to the conclusion that e' cannot be greater than e. These two results imply that $e = e'$, so that the first part of Carnot's theorem is proved.

Finally, suppose that E' is a reversible engine whereas E is an irreversible engine. We can again show that e cannot be greater than e' (reversible), but the converse *cannot* be shown because E cannot be reversed. This leads to the second part of Carnot's theorem, namely, that for two engines operating between the same two temperatures,

$$e(\text{irreversible}) < e(\text{reversible})$$

In the following section we discuss the idealized, reversible analogs of various types of practical heat engines, all of which are actually irreversible to some extent. Therefore, we realize from the outset that the calculated efficiencies, based on reversible cycles, represent only upper limits to the true efficiencies.

▶ *The Thermodynamic Temperature Scale.* It is possible to define a thermodynamic temperature scale (independent of the properties of any real substance) in terms of a Carnot engine operating in a cycle between the temperatures T_1 and T_2. Lord Kelvin proposed the definition of thermodynamic temperature, inspired by Eq. 21-25:

Any two temperatures, T_1 and T_2, are in the same ratio as the amounts of heat, Q_1 and Q_2, that are absorbed and rejected, respectively, by an engine executing any Carnot cycle between these two temperatures; that is, $Q_1/Q_2 = T_1/T_2$.

Thus, the ratio of the thermodynamic temperatures of two heat reservoirs can be specified in terms of the amounts of heat exchanged in a Carnot cycle. If one of these reservoirs is at the temperature of the triple point of water, arbitrarily taken to be 273.16 K, the Kelvin temperature is defined to be

$$T = (273.16 \text{ K}) \frac{Q}{Q_{tr}} \qquad \textbf{(22-2)}$$

where T is the temperature of the reservoir that supplies the amount of heat Q to the engine, and where Q_{tr} is the amount of heat rejected to the triple-point reservoir (see Fig. 22-5).

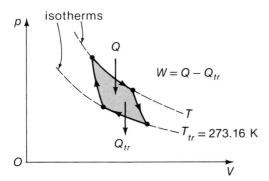

Fig. 22–5. A Carnot cycle used in the definition of the Kelvin temperature. The temperature T may be either greater than or less than T_{tr}.

Only for an ideal gas is the temperature defined by Eq. 22–2 identical to the temperature T that appears in the ideal gas law equation. For a real gas we use the asymptotic definition of temperature (see Eq. 18–3),

$$T = (273.16 \text{ K}) \lim_{p_0 \to 0} \frac{p}{p_0} \qquad \textbf{(22-3)}$$

The determination of temperature by using Eq. 22–3 is not always feasible, particularly for temperatures below 1 K or so. However, there are many techniques for measuring amounts of heat exchanged, so that the use of Eq. 22–2 is possible. In these discussions we make no further distinction between the temperatures defined by Eq. 22–2 and Eq. 22–3; we continue to use T to represent the Kelvin (absolute) temperature.   ◄

22–3   ENTROPY

We now return to the discussion of the Carnot cycle, making one important change in notation. In Fig. 21–11 and similar diagrams, we considered the heat rejected to the cold reservoir to be a positive quantity, $Q_2 > 0$; we recognized that this heat *leaves* the system by using a negative sign when expressing the work done as $W = Q_1 - Q_2$. (See the discussion preceding Eq. 21–21.) We now adopt the convention that any Q represents heat *added* to the system; thus, rejected heat carries a negative sign, and we write the work done as $W = Q_1 + Q_2$. With this convention, Eq. 21–25 can be expressed as

$$\frac{Q_1}{T_1} + \frac{Q_2}{T_2} = 0$$

Figure 22–6 shows a complicated thermodynamic cycle, *abcdefgha*, that consists of four isothermal steps and four adiabatic steps. By constructing the continuation of the adiabatic steps along bg' and $c'f$, the shaded region is divided into three Carnot cycles. Because there are no heat exchanges during the adiabatic steps, the three cycles combined yield

$$\frac{Q_1}{T_1} + \frac{Q_2}{T_2} + \frac{Q_3}{T_3} + \frac{Q_4}{T_4} = 0$$

Fig. 22–6. A multipart cycle that consists of isothermal and adiabatic steps.

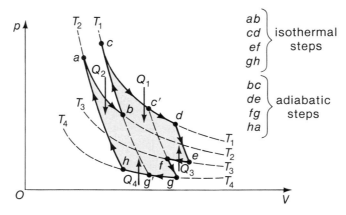

where $Q_1 = Q_{cc'} + Q_{c'd}$ is the amount of heat added at the temperature T_1 during the isothermal step $cc' + c'd$; similarly, $Q_4 = Q_{gg'} + Q_{g'h}$. Altogether, we can write

$$\sum_{i=1}^{4} \frac{Q_i}{T_i} = 0 \qquad (22-4)$$

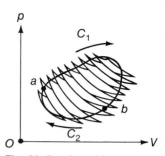

Fig. 22-7. An arbitrary reversible cycle is separated into a large number of adjacent Carnot cycles.

We can extend this idea of connected Carnot cycles to *any reversible cycle,* such as the one shown in Fig. 22-7. Imagine that this cycle is separated into a very large number of adjacent Carnot cycles, as indicated in the diagram. In the limit of an infinite number of cycles, the sum in Eq. 22-4 becomes an integral, and we have

$$\oint \frac{dQ}{T} = 0 \qquad \text{reversible cycle} \qquad (22-5)$$

where the symbol $\oint$ emphasizes that the integration is to be carried out around the entire path. Select any two states of the system, such as those represented by the points a and b. Then, Eq. 22-5 can be expanded to

$$\oint \frac{dQ}{T} = \oint_1 \int_a^b \frac{dQ}{T} + \oint_2 \int_b^a \frac{dQ}{T} = 0 \qquad (22-6)$$

The symbol $\oint$, introduced in Section 21-2, is again used to specify the particular path taken in passing between states a and b. The process is reversible, so we can write

$$\oint_2 \int_b^a \frac{dQ}{T} = -\oint_2 \int_a^b \frac{dQ}{T}$$

Using this result in Eq. 22-6 we find

$$\oint_1 \int_a^b \frac{dQ}{T} = \oint_2 \int_a^b \frac{dQ}{T}$$

An infinite number of cycles may be drawn through the points that represent the definite states a and b, so we conclude that

$$\oint \int_a^b \frac{dQ}{T} = \text{constant} \ (a, b) \qquad (22-7)$$

This result is valid for *any reversible path* that connects the equilibrium states (p_a, V_a, T_a) and (p_b, V_b, T_b). The value of the constant depends only on the properties of these states, a point we emphasize by writing the right-hand side of Eq. 22-7 as "constant (a, b)."

Example 22-1

Calculate the constant appearing in Eq. 22-7 for the two reversible paths shown in the diagram. State a is for n moles of an ideal monatomic gas specified by (p_a, V_a, T_a), while state b is specified by (p_b, V_b, T_b). Path Ⓐ consists of a constant-pres-

sure process followed by a constant-volume process. The intermediate state c is specified by (p_c, V_c, T_c) with $p_c = p_a$ and $V_c = V_b$. Path Ⓑ consists of an isothermal compression followed by a constant-pressure process passing through state d, specified by (p_d, V_d, T_d) with $p_d = p_b$ and $T_d = T_a$.

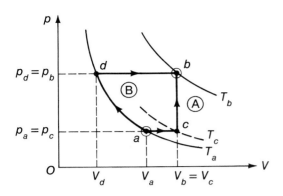

Solution:

For path $\textcircled{A}$ we have $\oint_A \dfrac{dQ}{T} = \oint_a^c \dfrac{dQ}{T} + \oint_c^b \dfrac{dQ}{T}$. Now along path ac, $dQ = nC_p\, dT$, while along path cb, $dQ = nC_v\, dT$. Also we have $C_v = \tfrac{3}{2}R$ and $C_p = \tfrac{5}{2}R$; hence,

$$\oint_A \frac{dQ}{T} = \tfrac{5}{2}nR \int_{T_a}^{T_c} \frac{dT}{T} + \tfrac{3}{2}nR \int_{T_c}^{T_b} \frac{dT}{T}$$

Evaluating the integrals gives

$$\oint_A \frac{dQ}{T} = nR\left[\ln\left(\frac{T_c}{T_a}\right)^{5/2} + \ln\left(\frac{T_b}{T_c}\right)^{3/2} \right]$$

Using the ideal gas law, $\dfrac{T_c}{T_a} = \dfrac{V_b}{V_a}$ and $\dfrac{T_b}{T_c} = \dfrac{T_b V_a}{T_a V_b}$; hence,

$$\oint_A \frac{dQ}{T} = nR\, \ln\left(\frac{V_b}{V_a}\right)\left(\frac{T_b}{T_a}\right)^{3/2}$$

For path $\textcircled{B}$ we have $\oint_B \dfrac{dQ}{T} = \oint_a^d \dfrac{dQ}{T} + \oint_d^b \dfrac{dQ}{T}$. We note that $dQ = dW + dU = p\, dV$ along isothermal path ad, since $dU = 0$ and $dW = p\, dV$. Using the ideal gas law $\dfrac{dQ}{T} = \dfrac{dW}{T} = \dfrac{p\, dV}{T} = nR\dfrac{dV}{V}$. Along path db we have $dQ = C_p n\, dT = \tfrac{5}{2}nR\, dT$. Thus,

$$\oint_B = nR \int_{V_a}^{V_d} \frac{dV}{V} + \tfrac{5}{2}nR \int_{T_a}^{T_b} \frac{dT}{T}$$

or

$$\oint_B = nR\left[\ln \frac{V_d}{V_a} + \ln\left(\frac{T_b}{T_a}\right)^{5/2} \right]$$

Now $\dfrac{V_d}{V_a} = \dfrac{T_a}{T_b}\dfrac{V_b}{V_a}$; hence,

$$\oint_B = nR\, \ln\left(\frac{V_b}{V_a}\right)\left(\frac{T_b}{T_a}\right)^{3/2}$$

We thus see that both paths $\textcircled{A}$ and $\textcircled{B}$ result in the same value for the constant in Eq. 22–7, namely

$$\text{constant } (a, b) = nR\, \ln\left(\frac{V_b}{V_a}\right)\left(\frac{T_b}{T_a}\right)^{3/2}$$

In a later discussion we shall discover that *any* path between states a and b gives exactly the same result.

We now assert that there exists a useful thermodynamic quantity, called the *entropy* S, which is a *state function* and therefore depends *only* on the nature of the equilibrium state with which it is associated. The *difference* in entropy between two states a and b is identified with the value of the integral of dQ/T between these states. Thus, we write Eq. 22–7 as

$$\oint_a^b \frac{dQ}{T} = S_b - S_a \qquad \text{reversible path} \qquad \textbf{(22–8)}$$

For a differential process, we have

$$\frac{dQ}{T} = dS \qquad \textbf{(22–9)}$$

Notice that the units of entropy are J/K (or Cal deg^{-1}).

It is important to realize that the path C along which the integral in Eq. 22–8 is carried out can represent *any reversible process* that connects the two states. Thus, we may choose a path that makes easier any calculation that is required. Even if the actual path that carries the system from state a to state b is irreversible, the entropy difference, $S_b - S_a$, can still be evaluated by choosing a reversible path from a to b. This is possible because the entropy is a state function; the value of $S_b - S_a$ depends only on the properties of the states a and b, not on the path that connects them. In this regard, entropy is similar to internal energy (see Section 21–2).

Entropy and the Second Law. We wish to develop an important relationship between the entropy change in an isolated system that is undergoing a thermodynamic change and the second law of thermodynamics. Suppose a system originally in equilibrium at point a in the p-V diagram of Fig. 22–8 undergoes an *irreversible adiabatic* process to the equilibrium state b. What can be said about the entropy change $S_b - S_a$? We can bring the system back to state a by a series of *reversible paths:* first an adiabatic compression to some state c at temperature T, then placing the system in thermal contact with a heat reservoir at temperature T and allowing an isothermal expansion cd to take place, carrying the system to a state d from which a final adiabatic expansion da can return the system to state a. In this cyclic process $abcda$, the only heat to enter the system is Q during the isothermal expansion. Since there has been no net change in the internal energy of the system in completing the cycle, the first law of thermodynamics requires that the work done by the system W must equal Q; i.e.,

$$W = Q \qquad (22-10)$$

We also note that because steps bc and da are reversible adiabatic steps, $S_c = S_b$ and $S_d = S_a$; thus,

$$S_a - S_b = S_d - S_c = Q/T \qquad (22-11)$$

This last equality follows because step cd is a reversible isothermal process.

The content of Eq. 22–10 if $Q > 0$ is that an amount of work W was done during a cycle by converting an equal amount of heat energy Q taken from the reservoir at temperature T, without any other change having taken place (the system returned to its initial state a). But this is precisely a violation of Kelvin's statement of the second law of thermodynamics. It follows therefore that $Q \le 0$. [Note that if $Q = W < 0$ the cycle has resulted in the work done on the system having been rejected as heat to the reservoir. There is, of course, no prohibition against this by the second law.] With $Q \le 0$, Eq. 22–11 becomes

$$S_b - S_a \ge 0 \qquad \text{general adiabatic process} \qquad (22-12)$$

Fig. 22–8. A cycle consisting of an *irreversible* adiabatic step *ab* followed by the *reversible* steps: an adiabatic compression *bc*, an isothermal expansion *cd*, and a return to state *a* by an adiabatic expansion *da*.

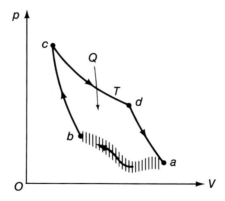

The inequality in Eq. 22–12 refers to an irreversible adiabatic process *ab*, whereas the null result pertains if the step is reversible.

The same type of reasoning may be applied to the more general case where the system consists, initially, of many parts that are *not in thermal equilibrium* with each other. If such a system (of many parts) is isolated from the remaining environment and undergoes any internal thermodynamic process, the system as a whole proceeds by an adiabatic step, since isolation implies no heat exchange with the environment. If the internal parts of the system interact reversibly, then the entropy of the system as a whole remains the same. If the internal parts interact irreversibly, then the entropy of the system as a whole increases.

In summary, the second law of thermodynamics imposes an important restriction on processes that may occur inside an *isolated* (or *closed*) system; i.e., one having no thermal contact with its surroundings. If the internal processes occurring between the component parts are all reversible, then the entropy of the system as a whole remains the same. If any of the steps are irreversible, the entropy of the system as a whole must increase. Thus, we have

$$S_b \geq S_a \qquad \text{isolated system} \qquad (22\text{--}13)$$

No finite process in a real system is strictly reversible, so the inequality applies for all practical purposes. Therefore, no isolated system can undergo a *spontaneous* transition from a state of higher entropy to one of lower entropy.

Suppose that a system is prepared in some arbitrary initial state and then allowed to undergo whatever spontaneous processes are available to it. As these processes occur, the system approaches ever more closely a truly stable equilibrium state among its component parts and the entropy of the system increases monotonically with time. Thus, the second law of thermodynamics leads to a correlation of natural or spontaneous processes with the direction of the evolution of time (''time's arrow''). In the examples and problems that follow, we find that the natural processes we expect to occur all involve increases in entropy. The entire Universe may be considered to be an isolated system, so it follows from the second law that the entropy of the Universe increases continually with time. Thus, the Universe is proceeding toward a state of uniform temperature in which heat engines can no longer do work because no temperature differences exist.

Free Expansion of a Gas.   Consider again (as in Section 21–3) the free expansion of an ideal gas originally confined to a volume V_A and separated from an evacuated second volume V_B by a closed valve (Fig. 22–9a). Before the valve is opened, the system (which is defined to consist of both volumes) is in an equilibrium state *a*. Opening the valve initiates a spontaneous irreversible process over which we have no further control. The system is isolated, so that $Q = 0$ and $W = 0$; it therefore follows from the first law that $\Delta U = 0$ and that the free expansion results in no temperature change. When both volumes are filled and the flow of gas has ceased, the system is in a truly equilibrium final state *b*.

We cannot calculate the entropy change between states *a* and *b* by considering the free expansion process itself. (There is no well-defined path along which dQ/T can be integrated.) However, because the entropy change depends only on the initial and final states, we can make the calculation by choosing *any reversible path* that connects the same two states. For this purpose we choose an isothermal process in which the slow movement of a piston increases the volume of the gas from V_A to $V_A + V_B$ (Fig. 22–9b). We use the differential form of the first law, namely,

$$dQ = dU + p \, dV$$

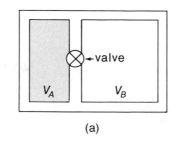

(a)

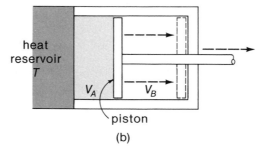

Fig. 22–9. (a) Container for a free expansion. (b) The slow movement of the piston achieves the same final state as the free expansion.

(b)

But the temperature T is constant, so $dU = 0$; then, dividing by T and using $dQ/T = dS$, we have

$$dS = \frac{dQ}{T} = \frac{p}{T}\, dV = nR\, \frac{dV}{V}$$

where we have also made use of the ideal gas law. Thus,

$$S_b - S_a = \int_a^b \frac{dQ}{T} = nR \int_{V_A}^{V_A + V_B} \frac{dV}{V} = nR\, \ln \frac{V_A + V_B}{V_A} > 0 \qquad \textbf{(22–14)}$$

so that $S_b > S_a$. Thus, the irreversible spontaneous free expansion process results in an increase in entropy for the isolated system.

Both the free expansion and the reversible isothermal process carry the system from state a to state b with the same entropy change. What, then, distinguishes these two quite different processes? The answer is found by examining the entropy change in the *surroundings* of the system. In the free expansion, there is no heat exchange between the system and the surroundings, so the surroundings undergo no entropy change:

$$\text{Irreversible free expansion:} \quad \begin{cases} \Delta S_{\text{system}} = +nR\, \ln \dfrac{V_A + V_B}{V_A} \\[2mm] \Delta S_{\text{surroundings}} = 0 \end{cases}$$

However, during the isothermal expansion, the surroundings of the system (that is, the heat reservoir at temperature T) delivers to the system an amount of heat Q equal to the work W done by the gas to accomplish the expansion, namely (see Eq. 21–2), $Q = W = nRT\, \ln\,[(V_A + V_B)/V_A]$. Because an amount of heat Q was *removed* from the surroundings, the entropy of the surroundings changes (decreases) by an amount $-Q/T = -W/T = -nR\, \ln\,[(V_A + V_B)/V_A]$. Thus,

$$\text{Reversible isothermal expansion:} \quad \begin{cases} \Delta S_{\text{system}} = +nR\, \ln \dfrac{V_A + V_B}{V_A} \\[2mm] \Delta S_{\text{surroundings}} = -nR\, \ln \dfrac{V_A + V_B}{V_A} \end{cases}$$

We may consider the *Universe* to consist of the *system* we are discussing and its *surroundings*. Then, the entropy change of the Universe is*

$$\Delta S_{\text{Universe}} = \Delta S_{\text{system}} + \Delta S_{\text{surroundings}} \qquad \text{(22–15)}$$

For the irreversible free expansion, we find $\Delta S_{\text{Universe}} = +nR \ln [(V_A + V_B)/V_A]$; however, for the reversible isothermal expansion, we have $\Delta S_{\text{Universe}} = 0$. These are general results. For any irreversible process the entropy of the Universe increases, but for any reversible process the entropy of the Universe does not change. In a complex irreversible process, some part of the system may actually experience a decrease in entropy, but the net result of the process must be an increase in entropy for the system and for the Universe.

Temperature Equalization.   Consider two objects, with object 1 at a higher initial temperature than object 2. Both are thermally isolated from each other and from their surroundings. The objects are brought into contact but are still isolated from the surroundings. Heat is transferred from the hotter to the colder object and they eventually reach thermal equilibrium at some intermediate temperature. At all times during the equalization process, we have $T_1 > T_2$. Let an amount of heat ΔQ be transferred between the objects when one temperature is T_1 and the other is T_2. Then, the change in entropy ΔS of the system of two objects is

$$\Delta S = \Delta S_1 + \Delta S_2$$

where
$$\Delta S_1 = -\Delta Q/T_1 \qquad \text{and} \qquad \Delta S_2 = \Delta Q/T_2$$

(The sign difference arises because heat is transferred from object 1 to object 2.) Thus,

$$\Delta S = \frac{T_1 - T_2}{T_1 T_2} \Delta Q > 0$$

This result is valid for each incremental transfer of heat, so we conclude that the entropy of the system increases continually throughout the process. Equilibrium is reached when the entropy change has achieved its maximum value. (See Problem 22–15.)

Entropy of an Ideal Gas.   The differential form of the first law is

$$dQ = dU + p \, dV$$

Dividing by T and using $dS = dQ/T$ and the ideal gas law, we obtain

$$\frac{dQ}{T} = \frac{dU}{T} + nR \frac{dV}{V} = dS$$

Assume that the system is an ideal monatomic gas. Then, we have from Eq. 20–14 that $U = \frac{3}{2}nRT$ and $dU = \frac{3}{2}nR \, dT$. Thus,

$$dS = nR\left(\frac{3}{2}\frac{dT}{T} + \frac{dV}{V}\right)$$

---

*We simply assert this provable proposition, namely, that the entropy change of a system (here, the Universe) is equal to the sum of the entropy changes of its constituent parts.

This expression for dS is an exact differential, which can be immediately integrated to give

$$S(T, V) = nR(\tfrac{3}{2} \ln T + \ln V) + S_0 \qquad (22-16)$$

where we indicate explicitly that $S = S(T, V)$ is a function of T and V. The constant of integration S_0 is called the *chemical constant* and depends only on the amount and the nature of the gas. According to the *Nernst heat theorem*, the absolute value of the entropy $S(T, V)$ of the system approaches zero as $T \to 0$. However, we deal here only with entropy *differences*.

Consider the free expansion of a fixed amount of gas from the volume V_A (state a) to the volume $V_A + V_B$ (state b) at a constant temperature. According to Eq. 22–16, we have

$$S_a = nR(\tfrac{3}{2} \ln T + \ln V_A) + S_0$$

$$S_b = nR[\tfrac{3}{2} \ln T + \ln (V_A + V_B)] + S_0$$

Then, $\qquad \Delta S = S_b - S_a = nR \ln [(V_A + V_B)/V_A]$

which is just the result we found in Eq. 22–14.

Entropy and the Availability of Work.   How much energy can be extracted as work from a *finite* heat source? Consider a thermal reservoir (the *cold* reservoir) at a temperature T_r and an object at a higher temperature T which acts as a *finite* hot reservoir. Imagine that a reversible engine operating on an incremental Carnot cycle is connected to the hot object and to the cold reservoir. As the engine does work, the temperature of the object decreases continually, eventually reaching the reservoir temperature T_r. At this point no temperature difference exists and no further work can be done by the engine.

Let the total amount of heat delivered to the engine from the object be Q; let the total amount of work done by the engine be W; and let the amount of heat rejected by the engine to the cold reservoir be $Q_r(Q_r < 0)$. During each incremental cycle we have

$$\Delta W = \Delta Q + \Delta Q_r$$

and also (from Eq. 21–25)

$$\frac{\Delta Q}{\Delta Q_r} = -\frac{T}{T_r}$$

where the negative sign is necessary because $\Delta Q_r < 0$. Thus, we find that

$$\Delta W = \Delta Q - \frac{\Delta Q}{T} T_r$$

As the object is drained of its available internal energy, the quantity $-\Delta Q/T = \Delta S$ represents a continual entropy decrease of the object per cycle. Thus, with $\Delta S < 0$, we have

$$\Delta W = \Delta Q + T_r \, \Delta S$$

For the complete process, we have $\Sigma \, \Delta W = W$, and $\Sigma \, \Delta Q = Q$; also, $\Sigma \, \Delta S = S_f - S_i$, which is the change in the entropy of the object between the final state and the initial state, where $S_f < S_i$. Thus, with $S_i - S_f > 0$, we have

$$W = Q - (S_i - S_f)T_r \qquad (22-17)$$

Fig. 22–10. (a) *T-S* diagram for an arbitrary reversible process. The area under the curve represents the heat added. (b) *T-S* diagram for a Carnot cycle.

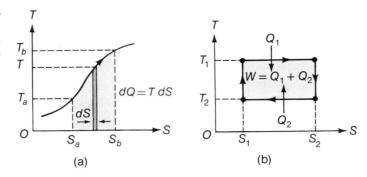

(a) (b)

That is, the amount of heat Q that is extracted from the object is not entirely converted into work by the engine. The amount of work actually done is reduced by the entropy change of the (finite) hot reservoir. An even smaller amount of work would be realized if the engine were not perfectly reversible.

Temperature-Entropy Diagrams. The area under a reversible thermodynamic path, $p = p(V, T)$, that is plotted in a p-V diagram represents the work done by the system during the corresponding process, namely, $W_{ab} = \int_a^b p \, dV$. If the entropy S is determined for every quasi-static state through which the system passes, the process can also be illustrated in a T-S diagram (Fig. 22–10a). The area under such a curve is equal to the heat added to the system during the process; that is, $Q_{ab} = \int_a^b T \, dS$. The T-S diagram for a Carnot cycle is rectangular (Fig. 22–10b). The horizontal parts of the cycle are the isothermal paths, and the vertical parts are the adiabatic paths ($\Delta S = 0$).

Example 22–2

Consider an isolated system that consists initially of 0.50 kg of water in a beaker at a temperature of 3° C and a 10-g ice cube at 0° C (not in contact with the water). The ice cube is placed in the water and eventually the system reaches equilibrium with all of the ice melted. Calculate the change in entropy of the system and demonstrate that $\Delta S > 0$ for this process. (Neglect the heat capacity of the beaker.)

Solution:

The latent heat of fusion for water is $L_f = 79.7$ Cal/kg and the specific heat of water near 0° C is $c_W = 1.01$ Cal/kg·deg. The equilibrium temperature t_C is determined by equating the energy given up by the water to the energy gained by the ice. Thus,

$$m_i L_f + m_i c_W (t_C - 0° \text{ C}) = m_W c_W (3° \text{ C} - t_C)$$

$$(0.010 \text{ kg})(79.7 \text{ Cal/kg}) + (0.010 \text{ kg})(1.01 \text{ Cal/kg·deg})t_C$$

$$= (0.50 \text{ kg})(1.01 \text{ Cal/kg·deg})(3° \text{ C} - t_C)$$

from which $\quad t_C = \dfrac{718}{515.1} = 1.394° \text{ C}$

The melting process is isothermal, so the change in entropy of the ice as it melts at temperature T_0 and then warms to t_C is

$$\Delta S_i = \left(\frac{\Delta Q}{T_0}\right)_{\text{melting}} + \left(\int \frac{dQ}{T}\right)_{\text{heating}}$$

where $dQ = m_i c_W \, dT$. Then, with $T_0 = 0° \text{ C} = 273.15$ K, we have

$$\Delta S_i = \frac{m_i L_f}{T_0} + m_i c_W \int_{T_0}^{T_0 + t_C} \frac{dT}{T}$$

$$= \frac{(0.010 \text{ kg})(79.7 \text{ Cal/kg})}{273.15 \text{ K}}$$

$$+ (0.010 \text{ kg})(1.01 \text{ Cal/kg·deg}) \ln \frac{273.15 + 1.394}{273.15}$$

$$= 2.969 \times 10^{-3} \text{ Cal deg}^{-1}$$

The change in entropy of the original amount of water is

$$\Delta S_W = m_W c_W \int_{T_0 + 3°C}^{T_0 + t_C} \frac{dT}{T}$$

$$= (0.50 \text{ kg})(1.01 \text{ Cal/kg·deg}) \ln \frac{273.15 + 1.394}{273.15 + 3.00}$$

$$\Delta S_W = -2.945 \times 10^{-3} \text{ Cal deg}^{-1}$$

Hence, the total entropy change for the system is

$$\Delta S = \Delta S_i + \Delta S_W = 2.4 \times 10^{-5} \text{ Cal deg}^{-1} \quad \text{or} \quad 0.100 \text{ J/K}$$

Thus, there is a net increase in the entropy of the system.

The system is isolated, so there is also a net increase in the entropy of the Universe, as we would expect for an irreversible process. Energy conservation (the first law) would not be violated if the process were to run backward, thereby reconstituting the ice cube; however, such a process would involve an entropy *decrease* and would therefore violate the second law.

QUESTIONS

22–1   Define and explain briefly the meaning of the terms (a) Carnot's theorem, (b) natural processes, (c) entropy, and (d) temperature-entropy diagram.

22–2   Explain the distinction between perpetual-motion machines of the first and second kinds.

22–3   Give several examples of thermodynamic processes that do not violate the principle of conservation of energy yet are barred from happening by the second law of thermodynamics.

22–4   Explain why it is possible to convert mechanical work entirely into heat but impossible to convert heat entirely into mechanical work.

22–5   According to Eq. 21–26, a Carnot engine that operates with a cold reservoir at $T_2 = 0$ K would have an efficiency $e = 1$. Yet the second law forbids an efficiency of unity. What does this imply about the possibility of constructing such a reservoir?

22–6   Is it possible to run power plants using heat engines and completely eliminate heat pollution of the environment?

22–7   Apply Eq. 22–16 to Example 22–1 and show directly that

$$\oint_a^b \frac{dQ}{T} = nR \ln \left(\frac{V_b}{V_a} \right) \left(\frac{T_b}{T_a} \right)^{3/2}$$

22–8   In Example 22–1 we found that

$$\oint_a^b \frac{dQ}{T} = \text{constant } (a, b)$$

for *any* process carrying the ideal gas from an equilibrium state *a* to an equilibrium state *b*. Yet for any specific process, the analytic form of the path had to be known in order to evaluate the (line) integral explicitly. Recall that in Section 8–4 we found that the work done on a particle by a conservative force field depended only on the initial and final positions of the particle. Yet here, too, if one wished to evaluate the line integral of the work explicitly, an analytic form prescribing the path had to be given. Discuss both the similarities and the differences in these two line-integral calculations.

22–9   Explain the relationship between the change in entropy of the Universe and the reversibility of a thermodynamic cycle.

22–10   In situations involving living cells, the replicating process consists of *decreasing* the cells' entropy. This results from the fact that a more orderly complex molecular arrangement is produced from constituent simple amino-acid molecules that are ingested into the cell than would result from a chance rearrangement of these ingested molecules. In view of this, discuss the change in entropy of a closed system consisting of a vessel of nutrient broth and a growing bacterial population.

PROBLEMS

Section 22–1

22–1   Prove that two reversible adiabatic paths cannot intersect. [*Hint:* Assume that the paths do intersect and use them to construct a thermodynamic cycle that violates the second law.]

Section 22–2

22–2   An engine operates in a cycle between the temperatures $T_1 = 180°$ C and $T_2 = 100°$ C and exhausts 4.8 Cal of heat per cycle while doing 1500 J of work per cycle. Compare the actual efficiency of the engine with that of a reversible engine operating between the same temperatures.

22–3   What is the minimum amount of work that must be done to extract 0.10 Cal of heat from a massive object at 0° C while rejecting heat to a hot reservoir at a temperature of 20° C?

22–4   In the tropics, the temperature of the ocean water at the surface averages about 30° C, whereas at a depth of 1 km the temperature is about 10° C. What is the maximum theoretical efficiency of an *ocean thermal gradient engine* that operates between these temperatures?

22–5   (a) A solar power plant uses heat from a solar collector to make steam that drives turbine generators. The steam comes from the collector at 500° C and leaves the turbines at 150° C. What is the maximum theoretical efficiency of the turbines? (b) The collector has an area of 6×10^6 m², and the average solar energy (*insolation*) reaching the collecting panels is 0.68 kW/m². The electrical output of the plant is 1250 MW. Calculate the overall efficiency of the plant and compare it with the theoretical efficiency of the turbines alone.

22–6   A special low-temperature refrigerator that operates on a Carnot cycle is found to deliver 0.50 Cal of heat to a hot reservoir at a Kelvin temperature of $\theta_1 = 5.21$ K while doing 1750 J of work. Determine the temperature θ_2 of the cold reservoir.

Section 22–3

22–7   Determine the increase in entropy of 1 kg of water that is heated from a temperature of 0° C to 100° C.

22–8   A 3-mol sample of helium is carried by some unspecified process from an equilibrium state with a pressure of 1 atm and a temperature of 0° C to a state with a pressure of 10 atm and a temperature of 100° C. What is the change in entropy of the helium?

22–9   A 0.30-kg piece of aluminum at a temperature of 85° C is placed in 0.50 ℓ of water at a temperature of 25° C. The system is thermally isolated. Determine the increase in entropy of the system when it has reached equilibrium.

22–10 As in Problem 21–18, a Carnot refrigerator is used to freeze 50 kg of water at 0° C, discharging heat into a room whose temperature remains constant at 27° C. (a) What is ΔS for the water? (b) What is ΔS for the surroundings? (c) What is ΔS for the Universe? (d) The process is repeated with a real (irreversible) refrigerator that has an efficiency 60 percent as great as that of the Carnot refrigerator. Calculate the same three entropy changes for this process.

Additional Problems

22–11 A 2-mol sample of air (considered to be an ideal gas) is used in a Carnot engine that operates between temperatures of 450 K and 300 K. During the isothermal expansion step, the volume increases from 0.8 ℓ to 3.5 ℓ. (a) What amount of heat is absorbed from the hot reservoir per cycle? (b) What amount of heat is rejected to the cold reservoir per cycle? (c) What amount of work is done per cycle?

22–12 A house is heated by burning heating oil and thereby
   •   directly delivering an amount of heat Q_1 to the house. Kelvin pointed out that by using a heat engine and a heat pump in combination (as shown in the diagram) the amount of heat delivered to the house could be increased to

$$Q = Q_1 \cdot \frac{T_2}{T_1}\left(\frac{T_1 - T_3}{T_2 - T_3}\right)$$

(a) Derive this equation. (b) Suppose that $T_1 = 200°$ C, $T_2 = 20°$ C, and $T_3 = 0°$ C. Show that $Q = 6.2\,Q_1$ so that, using Kelvin's scheme, 1 gal of heating oil gives the same heating effect as the direct heat from burning 6.2 gal of oil.

22–13 Determine the increase in entropy of the Universe when 1 kg of water slowly evaporates from an open beaker at a temperature of 20° C. (The latent heat of vaporization of water at 20° C is 585 Cal/kg.)

22–14 A Carnot engine that operates between temperatures of 250° C and 30° C delivers 2000 J of work per cycle. What is the change in entropy during one of the isothermal steps in the cycle? Sketch the cycle on a *T-S* diagram.

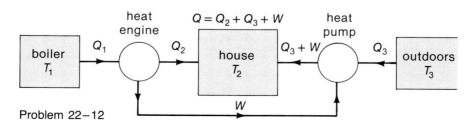

Problem 22–12

22–15 Consider two objects, one with mass m_1, specific heat
•• c_1, and temperature T_1, and the other with mass m_2, specific heat c_2, and temperature T_2, with $T_2 < T_1$. The objects are brought into thermal contact inside an adiabatic container. The temperature of object 1 is decreased to T and the temperature of object 2 is increased to T' (consistent with energy conservation). (a) Show that the entropy of the system increases by an amount

$$\Delta S = m_1 c_1 \ln \frac{T}{T_1} + m_2 c_2 \ln \frac{T'}{T_2}$$

where $\quad m_1 c_1 (T_1 - T) = m_2 c_2 (T' - T_2)$.

(b) Show that $d(\Delta S)/dT = 0$ requires $T = T'$ and therefore that this equilibrium condition is reached when ΔS is maximum. [*Hint:* Use

$$d(\Delta S)/dT = d(\Delta S)/dT' \cdot (dT'/dT)$$

in the second term in the expression for ΔS; obtain dT'/dT from the energy condition.]

22–16 An Otto-cycle engine (Section 21–5) uses 0.25 mol of
• helium as the working substance. The intake stroke occurs at $p_0 = 1$ atm, ending at a volume of $V_a = 6.0 \ \ell$ and a temperature of 20° C. The adiabatic compression ab (see Fig. 21–13) reduces the volume to $V_b = V_a/5$, and the constant-volume step bc raises the pressure to $p_c = 25$ atm. (a) Calculate the heats of Q_1 and Q_2 and the amount of work done per cycle. (b) Calculate the entropy change for each step of the cycle except the intake and exhaust strokes.

CHAPTER 23

ELECTRIC CHARGE

Experimentation concerning *electromagnetic* phenomena dates from the time of ancient Greece. The accumulated wealth of information about electric and magnetic effects was finally summarized in a definitive theory of electromagnetism formulated during the 1860's by the great Scottish mathematical physicist James Clerk Maxwell. These developments have led directly to the giant electric and electronic industries of today, which furnish us with an ever-growing list of electromagnetic devices. Indeed, the application of electromagnetism to human requirements is the foundation of our modern technologically oriented society.

Perhaps an even more compelling reason for the study of electromagnetism is the hidden role it plays even in our everyday experiences with force. When examined on an atomic level, all contact forces between laboratory-sized objects are basically electromagnetic in origin. Thus, except for gravity, all forces we experience in everyday activities are electromagnetic. In the next several chapters we investigate the nature of electromagnetic phenomena.

23–1   EARLY HISTORY

Electrification.   The first recorded observation of an electric effect was made in about 600 B.C. by the Greek mathematician and astronomer Thales of Miletus (c640–c546 B.C.). He discovered that a piece of amber, having been rubbed, would attract small bits of straw and feathers. The word *electric* is derived from the Greek word for amber (*elektron*). The next significant advance in the study of electric effects did not occur until the sixteenth century, when William Gilbert (1544–1603), physician to Queen Elizabeth I, investigated a variety of substances that exhibit the same electrification properties as amber. These substances he called *electrics,* materials that we now recognize to be *insulators*. Other materials, which are difficult to electrify by rubbing, he called *nonelectrics*; these materials are *conductors*.

In the early eighteenth century, Stephen Gray (1696–1736), an English scientist, discovered that the state of electrification can be transferred from one object to another when the objects are connected by a piece of metal. Gray thereby demonstrated that the agency responsible for electrification can be caused to flow. Thus, electrification was shown to be a transitory, not an intrinsic, property of matter.

The discovery that two different types of electrification are possible was made by Charles François Du Fay (1698–1739), a French chemist and physicist. Substances such as glass and mica were classified as *vitreous* materials, whereas amber, sulfur, hard rubber, and resins were classified as *resinous* materials. When rubbed with silk or cotton, the vitreous and resinous materials become oppositely electrified. Bodies with opposite electrification attract each other, but bodies with the same electrification repel each other. Figure 23–1 illustrates Du Fay's experiments. An electrified amber rod is affected oppositely by an electrified resinous material (amber) and an electrified vitreous material (glass).

The most able early American scientist was Benjamin Franklin,* whose numerous experiments and keen insight advanced the understanding of electric phenomena. The status of electrical science at the time (1748) of the publication of Franklin's *Experiment and Observations on Electricity* can be summarized as follows:

(a) All objects contain a substance called by Franklin an *electric fire* or *electricity*. The total amount of this substance in the world is constant.

(b) Rubbing together two objects, such as glass and silk, results in the transfer of electricity from one object to the other; one object gains a certain amount of electricity and the other loses an equal amount. Thus, glass and silk become oppositely electrified. In Franklin's convention, the glass becomes electrified in the positive (+) sense by *gaining* electricity, and the silk becomes equally electrified in the negative (−) sense by *losing* the electricity gained by the glass. However, when amber is rubbed with silk, the amber loses electricity and becomes electrified negatively, whereas the silk gains electricity and becomes electrified positively. According to Franklin, *positive* electricity is always transferred in such situations.

(c) Two objects electrified in the same sense, (+) and (+) or (−) and (−), exert repulsive forces on each other. Two objects electrified in the opposite sense, (+) and (−), exert attractive forces on each other.

(d) The mutual forces due to electrification are increased if the amount of electrification is increased or if the objects are brought closer to each other.

(e) Some materials (conductors), such as metals, readily conduct electricity, whereas others (nonconductors or insulators), such as wax and dry wood, do not. Most materials can be classified in one or the other of these categories. Even a metal can be electrified to some extent by rubbing with silk or cotton, provided it is held apart from other objects by a nonconductor. A metal object electrified in this way will lose its electrification if any point of the object is touched by a conductor that is connected to the ground.

Fig. 23–1. (a) There is a repulsion between rods that have the same electrification. (b) There is attraction between rods that have different electrification.

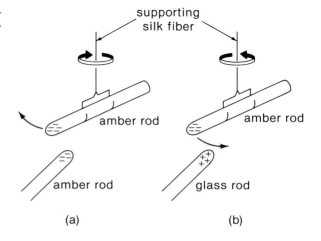

(a)

(b)

---

*Benjamin Franklin (1706–1790), American scientist, patriot, and statesman, is best known in the field of science for his experiments with lightning and for the invention of the lightning rod.

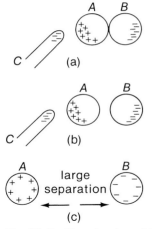

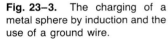

Fig. 23-2. The charging of two metal spheres by induction.

Franklin did not distinguish between the terms *electrified* and *charged*. We adopt the modern usage and henceforth refer to objects being *charged,* either positively or negatively.

Charging by Induction and Conduction.   The process of charging an object by induction* is illustrated in Fig. 23–2. Two conducting metal spheres, A and B, are placed in contact and a negatively charged rod C is brought near (but not touching) sphere A (Fig. 23–2a). According to Franklin's view, positive charge is attracted to the side of A nearer C, thereby leaving an equal negative charge on the opposite side of B. With C still present, A and B are separated. This leaves A with an excess positive charge and B with an equal negative charge (Fig. 23–2b). The charged rod C is now removed, and the charge states of the spheres remain the same (Fig. 23–2c).

The induction method illustrated in Fig. 23–2 produces charges on the spheres that are equal in magnitude and opposite in sign. How do we arrange for the two spheres to carry the *same* charge? Suppose that we have a positively charged rod; however, instead of *inducing* a charge on the sphere, suppose that the rod C touches sphere A and transfers charge to it by *conduction.* While spheres A and B (assumed to be identical) are still in contact, the rod is removed. The charge placed originally on A becomes equally distributed between the identical spheres. (• Why should the distribution be equal?) When the spheres are separated, they carry equal charges of the same sign (+).

Induction can also be used to charge a single conducting object. A positively charged glass rod is brought near an uncharged metal sphere without touching it. The Franklin positive charges on the sphere are repelled by the rod, leaving the side nearer the rod with an equal number of negative charges (Fig. 23–3a). A *ground wire*† is now brought into contact with the side of the sphere opposite the rod, thereby leading the positive charges to ground (Fig. 23–3b). With the charged rod still in place, the ground wire is disconnected (Fig. 23–3c). Finally, when the rod is removed, the sphere remains with a net negative charge, which becomes uniformly distributed over the sphere (Fig. 23–3d).

Fig. 23-3. The charging of a metal sphere by induction and the use of a ground wire.

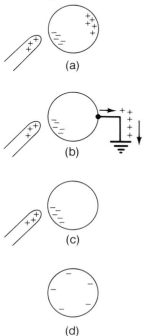

23-2   MATTER AND CHARGE

According to the atomic theory, all ordinary matter is made up of atoms, and every atom, in turn, consists of a cloud of electrons that surrounds a tiny nucleus of protons and neutrons. Every atom, in its normal condition, contains exactly as many electrons as protons. Because the proton charge and the electron charge are equal in magnitude‡ but opposite in sign, and because the neutron carries no charge,§ all ordinary matter is therefore electrically neutral. The sign convention for electric charge (derived from Franklin's convention) is that a proton carries a *positive* charge $(+e)$, whereas an electron carries a *negative* charge $(-e)$.

The distribution of electric charge is usually accomplished by the movement of *electrons;* the more massive positively charged nuclei remain essentially stationary in electric processes involving solids.‖ Thus, an object is given a negative charge by the addition of excess electrons or a positive charge by the removal of electrons; the number of *atoms* is not changed in either case.

---

*Induction charging was discovered in 1753 by an English physicist, John Canton (1718–1772).

†A ground wire is a conducting metal wire, one end of which is in contact with the Earth, usually through a metal stake or pipe driven into the ground. The wire is then used to connect other objects electrically with the Earth, which acts as a virtually infinite reservoir to supply or absorb charge as needed.

‡Experiments have established this equality to an accuracy of better than 1 part in 10^{18}.

§The limit established for any neutron charge (of either sign) is 2×10^{-15} of the electron charge.

‖ Positive charge often moves in electric processes in liquids and gases.

In certain types of materials (notably metals) a fraction of the atomic electrons are not bound to any particular atom but are free to move about in the material. When these free electrons move through macroscopic distances, we say that they conduct electricity, and these materials are called *conductors*. Even a weak external electric force applied to a conductor readily produces an electron flow (or *current*). (• Can you now see why the charges are distributed uniformly over the surfaces of the metal spheres in Figs. 23–2c and 23–3d?)

In a nonconducting material (an *insulator*) essentially all of the electrons are attached to atoms—that is, there are almost no free (conduction) electrons in insulators. (• Can you explain why the induction methods used to charge *conducting* objects, as shown in Figs. 23–2 and 23–3, cannot be used with *nonconducting* objects?)

When two different nonconducting materials, such as glass and silk, are rubbed together,* electrons are transferred from the surface of one material to the surface of the other. Thus, the material that gains electrons becomes negatively charged, and the material that loses electrons becomes positively charged by the same amount. The charged sites on the two insulators remain essentially fixed until the charge is *neutralized* (i.e., excess electrons are removed by some process, or "missing" electrons are added) or slowly leaks away (because no material is a perfect insulator). Phenomena involving fixed (static) charges are described as *electrostatic,* whereas those involving moving charges are called *electrodynamic.*

If you compare this qualitative summary of some of the features of modern electrical science with that for Franklin's time, made in the preceding section, the similarity is clear. The primary exception is in regard to the sign of the charge that moves. Franklin chose as *positive* the charge that resides on glass when rubbed with silk, and he chose to consider charging effects to be the result of the movement of *positive* charge. It is now known that the sign of the charge on glass is the same as that of the proton charge. It is also known that charging effects are due to the movement of the negative charges or electrons rather than the positive charges or protons. Nevertheless, common practice adheres to Franklin's choice for charge signs and for the direction of charge movement. Consequently, we consider electric current to flow in the direction in which positive charge *would* move, even though current is actually due to the motion of electrons in the opposite direction!

Charge Quantization. In 1897, J. J. Thomson (1856–1940), an English physicist, demonstrated that there are particles (called *electrons*) that are common to all atoms and have a fixed and unique ratio of electric charge to inertial mass. In 1909, R. A. Millikan† showed that all electrons, from whatever source, carry the same charge (which we designate $-e$). This is one example of the remarkable property of discreteness (or *quantization*) that exists in Nature. Thus, when an object becomes charged as the result of the addition or removal of some number of electrons, the object must carry a charge that is an integer multiple of e. In most situations, the number of electrons transferred is so large that a change of one electron, more or less, is undetectable. For this reason we often consider the variation of charge to be continuous.

It is now known that there exists in Nature, in addition to the electron and the proton, a host of additional particles. More than 100 of these *elementary particles* are known. Without exception, the charge observed for each of these particles is $+e$, $-e$, or 0. Thus,

---

*Actually, the transfer of charge between two different nonconducting materials comes about because of the strong interaction between the surface atoms that are brought into close contact. Thus, charging occurs without any rubbing; the rubbing serves only to cause surface contact over a larger area. This effect is called *triboelectrification.*

†Robert A. Millikan (1868–1953), American physicist, was awarded the 1923 Nobel Prize in physics for his experiments with electrons. J. J. Thomson had received the 1906 Prize for his pioneering work.

charge appears to obey a universal quantization rule, with only charges $+e$, $-e$, and 0 occurring in Nature.

Charge Conservation.   An object can be given an electric charge by the addition or removal of electrons. The charge that is *gained* by one object is *lost* by another. It is one of the fundamental conservation principles of nature that *the total amount of charge in an isolated system remains constant.*

Under certain conditions and subject to certain restrictions, it is possible to *create* a pair of electric charges, one of which is negative ($-e$) and one of which is positive ($+e$). If the negative charge is carried by an electron, then the positive charge is carried by a *positron* (a particle, not found in ordinary matter, that is identical to the electron except that it carries a positive charge*). Such creation processes are possible, in part, because they do not violate the law of charge conservation, since the total charge of the created pair is *zero*. Thus, the net charge of the system that contains the newly created pair is not altered.

No positive or negative charge has ever been observed to be created or destroyed *by itself*. These processes *always* involve equal positive and negative charges so that the charge conservation principle is never violated.

23–3   COULOMB'S LAW

In 1785, the French physicist Charles Augustin de Coulomb (1736–1806) made the first direct measurements of the forces between charged objects. Coulomb performed these experiments with a torsion balance of his own design. The apparatus (Fig. 23–4) was similar to that used by Cavendish 12 years later in his determination of the universal gravitation constant (see Section 5–3). Coulomb's balance did not need to be as sophisticated as the delicate instrument used by Cavendish because the forces measured were much larger than those in the gravitational experiments.

As shown in Fig. 23–4, a and b are small metal spheres that can be charged, and the mutual electrostatic force of attraction or repulsion can be measured for various separations and amounts of charge. Sphere a is connected to a counterweight by an insulating rod that is supported by a long fiber from an adjustable suspension head. The rod can pivot about the fiber axis, and the torsional twist of the fiber produces a restoring torque that opposes the torque due to the electrostatic force between a and b. Thus, an equilibrium condition is reached with the rod at some angle θ with respect to its neutral position (see the inset in Fig. 23–4). The torsional constant of the fiber is measured in a separate experiment involving the oscillation frequency of the system, as in the Cavendish experiment. Then, the electrostatic force can be computed in terms of the deflection angle θ (see Section 15–4).

The separation between the charged spheres can be changed by turning the suspen-

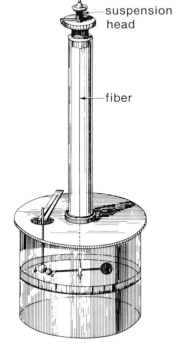

suspension head

fiber

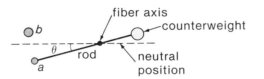

Fig. 23–4. Diagram of Coulomb's torsion balance as it appeared in his 1785 paper (New York Public Library). The inset shows the top view of the rod that is supported by the fiber.

---

*Electrons and positrons have opposite electric *and* magnetic properties. In the reverse of the creation process, an electron and positron that collide under certain conditions will *annihilate* one another, leaving zero charge and releasing energy.

sion head or by moving the support rod of sphere *b*. Some changes in the amounts of charge on *a* and *b* can be made readily. For example, suppose that a measurement is made of the force between *a* and *b* for some arbitrary charge states of the spheres. Sphere *b* is now removed from the apparatus and brought into contact with an identical, originally uncharged metal sphere. The two spheres share the charge equally, so when *b* is returned to the apparatus it carries exactly half the original charge. The effect on the force of reducing the charge on *b* by one half can now be determined.

On the basis of measurements of this type, Coulomb was able to formulate the law of electrostatic force, now called *Coulomb's law:*

The mutual electrostatic force between two point charges is directed along the line between the charges; the force is directly proportional to the product of the charges and is inversely proportional to the square of the distance between them. The force is repulsive if the charges have the same sign and is attractive if the charges have opposite signs.

In equation form, Coulomb's law is

$$\mathbf{F}_{12} = K\frac{q_1 q_2}{r_{21}^2}\hat{\mathbf{u}}_r \qquad (23-1)$$

where $\mathbf{F}_{12}$ is the force exerted *on* charge q_1 *by* charge q_2, and where $\hat{\mathbf{u}}_r$ is a unit vector directed from q_2 toward q_1, as shown in Fig. 23-5. The proportionality factor K is the *electrostatic force constant*. The force exerted on q_2 by q_1 is directed opposite to $\mathbf{F}_{12}$; thus,

$$\mathbf{F}_{21} = -\mathbf{F}_{12}$$

which is consistent with Newton's third law.

The SI unit of charge is the *coulomb* (C), which is the amount of charge carried by 6.2414×10^{18} electrons.* That is, the electron charge has the magnitude

$$e = 1.6022 \times 10^{-19}\ C$$

With the unit of charge defined, the electrostatic force constant K can be determined by experiment. When the charges are in vacuum (or air, for all practical purposes), we have

$$K = 8.98755 \times 10^9\ N \cdot m^2/C^2$$

Fig. 23-5. The geometry for Coulomb's law. Both charges are positive, and $\hat{\mathbf{u}}_r = \mathbf{r}_{21}/|\mathbf{r}_{21}|$.

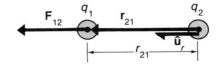

*The definition of the coulomb is best made in terms of electric current; this definition is given in Section 29-4.

A related constant is ϵ_0, called the *permittivity of free space:*

$$\epsilon_0 = \frac{1}{4\pi K} = 8.85419 \times 10^{-12} \ C^2/N\cdot m^2$$

In terms of ϵ_0, Coulomb's law is written as

$$\mathbf{F}_{12} = \frac{1}{4\pi\epsilon_0} \frac{q_1 q_2}{r_{21}^2} \hat{\mathbf{u}}_r \qquad (23-2)$$

Apart from the substitution $K = 1/4\pi\epsilon_0$, the other quantities are the same as in Eq. 23–1.

It is verified by experiment that electrostatic forces have the same linear character as the forces we have already studied; that is, electrostatic forces obey the principle of linear superposition (see Section 13–3). Thus, when more than two point charges are present, the net force on any one charge is the linear vector sum of the electrostatic forces individually exerted by each of the other charges. If three charges are present, the force on q_1 is

$$\mathbf{F}_1 = \mathbf{F}_{12} + \mathbf{F}_{13} = K\frac{q_1 q_2}{r_{21}^2}\hat{\mathbf{u}}_2 + K\frac{q_1 q_3}{r_{31}^2}\hat{\mathbf{u}}_3$$

where $\hat{\mathbf{u}}_2 = \mathbf{r}_{21}/|\mathbf{r}_{21}|$ and $\hat{\mathbf{u}}_3 = \mathbf{r}_{31}/|\mathbf{r}_{31}|$.

Experiments have demonstrated that the electrostatic force on an object combines with forces of a nonelectric nature by simple vectorial addition.

Example 23–1

(a) Determine the magnitude of the electrostatic force F_e between two electrons separated by a distance of 10^{-10} m (1 Å), a typical atomic dimension.

(b) Determine the ratio, F_e/F_g, of the electrostatic and gravitational forces that act between two stationary electrons. (The electron mass is 9.11×10^{-31} kg.)

Solution:

(a) Using Eq. 23–1, we have for the magnitude of the force,

$$F_e = K\frac{e^2}{r^2} = \frac{(8.99 \times 10^9 \ N\cdot m^2/C^2)(1.602 \times 10^{-19} \ C)^2}{(10^{-10} \ m)^2}$$

$$= 2.31 \times 10^{-8} \ N$$

(b) The two forces are

$$F_e = K\frac{e^2}{r^2} \qquad \text{and} \qquad F_g = G\frac{m_e^2}{r^2}$$

Therefore,

$$\frac{F_e}{F_g} = \frac{K}{G}\left(\frac{e}{m_e}\right)^2$$

$$= \frac{(8.99 \times 10^9 \ N\cdot m^2/C^2)}{(6.67 \times 10^{-11} \ N\cdot m^2/kg^2)}\left(\frac{1.602 \times 10^{-19} \ C}{9.11 \times 10^{-31} \ kg}\right)^2$$

$$= 4.17 \times 10^{42}$$

Thus, the electrostatic force between electrons is enormously greater than the gravitational force and, consequently, the latter can be neglected in all calculations involving electrostatic forces between atomic constituents.

Example 23–2

Two identical small Styrofoam balls ($m = 5$ g) are each charged uniformly to $+1.0 \ \mu C$. The balls are suspended from a fixed point by nonconducting threads, each with a length $\ell = 1.0$ m and with negligible mass. At equilibrium, what is the angle 2θ between the threads?

Solution:

From the free-body diagram for either charge, we have

$$F_e = T \sin\theta \qquad \text{and} \qquad mg = T \cos\theta$$

Hence,

$$F_e = mg \tan\theta = K\frac{q^2}{(2\ell \sin\theta)^2}$$

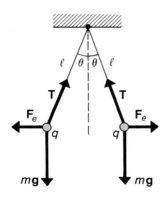

from which

$$\tan \theta \sin^2 \theta = \frac{Kq^2}{4mg \ \ell^2}$$

$$= \frac{(8.99 \times 10^9 \ \text{N·m}^2/\text{C}^2)(1.0 \times 10^{-6} \ \text{C})^2}{4(5 \times 10^{-3} \ \text{kg})(9.80 \ \text{m/s}^2)(1.0 \ \text{m})^2}$$

$$= 0.0459$$

This equation for θ can be solved graphically or by a successive-approximation procedure (see Problem 23–13). The result is $\theta = 0.358 = 20.5°$. Thus, the angle between the threads is $2\theta = 41.0°$.

Example 23–3

Three point charges, each carrying a charge of $+1.5 \ \mu\text{C}$, are located at the vertices of an equilateral triangle with sides $\ell = 30$ cm. Calculate the force exerted on one of the charges (for example, number 1) by the other two.

Solution:

Using the superposition principle, we can write for the force on charge 1

$$\mathbf{F}_1 = \mathbf{F}_{12} + \mathbf{F}_{13} = (F_{12x} + F_{13x})\hat{\mathbf{i}} + (F_{12y} + F_{13y})\hat{\mathbf{j}}$$

From the geometry it is evident that

$$F_{12x} = +F_{13x} \quad \text{and} \quad F_{12y} = -F_{13y}$$

Hence, $\mathbf{F}_1 = 2F_{12} \cos 30°\hat{\mathbf{i}} = \sqrt{3}F_{12}\hat{\mathbf{i}}$

Now,

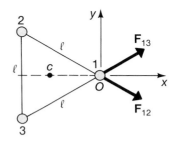

$$F_{12} = K\frac{q^2}{\ell^2} = \frac{(8.99 \times 10^9 \ \text{N·m}^2/\text{C}^2)(1.5 \times 10^{-6} \ \text{C})^2}{(0.30 \ \text{m})^2}$$

$$= 0.225 \ \text{N}$$

Thus, we have $\mathbf{F}_1 = 0.390\hat{\mathbf{i}} \ \text{N}$

That is, the force on any charge is 0.390 N, directed away from the center of the triangle at C.

QUESTIONS

23–1 Define and explain briefly the meaning of the terms (a) electrification, (b) conductor, (c) insulator, (d) charge quantization, and (e) charge conservation.

23–2 Explain how you would go about charging a metal sphere by induction. How would you charge the metal sphere by conduction?

23–3 Explain how an electrified plastic comb is able to attract small bits of paper even though the bits of paper have no net charge. Explain why this simple experiment in electrostatics is difficult if not impossible to perform on a humid summer day and rather easy to perform on a dry, crisp winter day.

23–4 Given electrified insulating rods of various sorts, how would you produce an attractive force between two con- ducting uncharged spheres suspended near each other by silk threads? You are not allowed to touch the rods to the spheres or the spheres to each other. How would you produce a repulsive force between the spheres?

23–5 The drawing on page 458 shows the essential parts of the *gold-leaf electroscope*. It consists of a metal ball connected to a metal rod, which in turn supports a wire structure from which hangs a thin gold-leaf sheet folded over in the middle. To avoid air currents and accidental discharge of the metal structure, the frame and gold leaf are inside a glass flask. When an electrified rod of either charge is brought near the ball, the gold-leaf halves sepa- rate (as shown). If the metallic structure of the electro- scope is uncharged and no electrified object is nearby, the gold-leaf halves collapse and hang vertically. If an

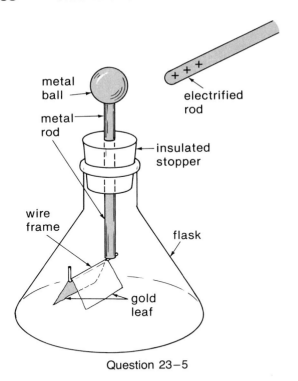

metal ball

metal rod

electrified rod

insulated stopper

wire frame

flask

gold leaf

Question 23–5

electrified rod *touches* the ball of the electroscope, the gold-leaf halves remain separated even after the rod is withdrawn. If the electroscope is charged in this manner on a humid day, the gold-leaf halves slowly collapse. Explain these observations.

23–6   How do we know that ordinary bulk matter contains equal amounts of positive and negative charge? [*Hint:* Refer to Example 23–1b.]

23–7   What effects are observed in the Coulomb balance (Fig. 23–4) when (a) the charge on sphere b is doubled, (b) the charge on sphere b is halved, (c) the distance between a and b is doubled, and (d) the distance between a and b is halved?

23–8   Given two pointlike charges that can be placed at any distance from each other, what does the graph of F_{12} versus r_{21} look like? Is it possible to make $F_{12} = 0$ exactly? What is the minimum number of charges in an isolated system such that each charge experiences zero net force? Make a sketch of your proposed arrangement.

23–9   Compare Coulomb's law with Newton's law of universal gravitation. Be sure to include both similarities and differences.

PROBLEMS

23–1   Two pointlike objects carry charges of $+2.5\ \mu C$ and $-1.5\ \mu C$ and are separated by a distance of 20 cm. Determine the magnitude and the direction of the mutual electrostatic force between the objects. Suppose that each object is a conducting sphere with a diameter of 5 cm. What qualitative effect on the force does this change make if the separation distance (now measured between centers) remains the same? Place an upper limit on the new value of the force.

23–2   Repeat Problem 23–1 with the separation distance increased to 2.0 m. Comment on the importance of the size of the charged objects.

23–3   Two pointlike objects carry charges of $q_1 = +4.0\ \mu C$ and $q_2 = -3.0\ \mu C$ and are separated by a fixed distance of 50 cm. (a) Determine the point (other than infinity) at which a pointlike positive charge will experience zero net electrostatic force as a result of the two fixed charges. (b) Repeat (a) for a pointlike negative charge. (c) Comment on these results.

23–4   Repeat Problem 23–3 if q_2 is changed to $+3.0\ \mu C$.

23–5.   Four pointlike charges, $q_1 = +8.0\ \mu C$, $q_2 = -4.0\ \mu C$, $q_3 = +4.0\ \mu C$, and $q_4 = -2.0\ \mu C$, are located in sequence at the corners of a square with sides 25 cm in length. Find the electrostatic force on q_2.

23–6   Four arbitrary pointlike charges are located at the corners of a rigid square plate. Prove that the electrostatic forces produce no net torque about the center of the plate.

23–7   Three identical small Styrofoam balls ($m = 2$ g) are suspended from a fixed point by three nonconducting threads, each with a length of 50 cm and negligible mass. At equilibrium the three balls form an equilateral triangle with sides of 30 cm. What is the common charge q that is carried by each ball?

23–8   In a simplified classical model of the hydrogen atom (the *Bohr model*) the electron orbits around the nuclear proton in a stable circular path with a radius of 0.53 Å (5.3×10^{-11} m). Consider the mass of the proton to be much greater than the mass of the electron. Calculate the orbital period of the electron and express the orbital speed as a fraction of the speed of light ($c = 3 \times 10^8$ m/s).

23–9   Suppose that 1 g of hydrogen is separated into electrons and protons. Suppose also that the protons are placed at the Earth's North Pole and that the electrons are placed at the South Pole. What is the resulting compressional force on the Earth?

23–10   A particle with mass m and carrying a charge $+q$ is lo-
  •   cated midway between two fixed particles that carry equal charges $+q$ and are separated by a distance 2ℓ.

Determine the frequency of oscillation of the middle charge when it is slightly displaced along the line joining the fixed charges. (The oscillating charge is restricted to move along the line joining the fixed charges.)

23–11 Two equal point charges separated by a distance 2ℓ are placed symmetrically on opposite sides of a large uniformly charged nonconducting plane, as shown in the diagram. Determine the locations on the plane where the net electrostatic force on a charge Q due to the two charges is maximum.

charged plane

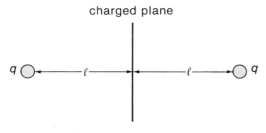

23–12 A total charge q is divided equally among $2n$ particles that are arranged with equal spacing around the perimeter of a circle with radius R. A point charge q' is placed a distance d from the plane of the circle along a perpendicular axis through the center, as shown in the diagram. We arrive at the idealization of a uniformly charged ring by allowing $n \rightarrow \infty$ while maintaining q fixed. Show that the force on the charge q' is

$$F = K\frac{qq'}{(R^2 + d^2)^{3/2}}$$

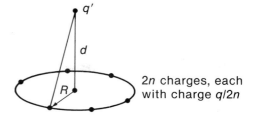

2n charges, each with charge $q/2n$

regardless of whether the charges are discrete or constitute a uniformly charged ring.

23–13 Refer to Example 23–2. Obtain the value for θ by using an approximation procedure to solve the equation $\tan \theta \sin^2 \theta = 0.0459$. Find a starting value θ_1 by assuming $\tan \theta \cong \sin \theta$, so that $\sin^3 \theta_1 = 0.0459$. Then write

$$\sin \theta_{n+1} = (0.0459/\tan \theta_n)^{1/2}$$

Use the value of θ_1 for θ_n and find θ_2; use the value of θ_2 to find θ_3; and so forth. Continue listing each value of θ_n until a value is found that yields the same value for θ_{n+1}. You must be careful when setting up a successive-approximation procedure. Try to solve the equation by writing $\tan \theta_{n+1} = 0.0459/\sin^2 \theta_n$. List the values of θ_n obtained using this procedure. What do you conclude? Finally, plot $\tan \theta$ and $0.0459/\sin^2 \theta$ versus θ. Show that the intersection corresponds to the value of θ obtained by numerical calculation.

CHAPTER 24

THE ELECTRIC FIELD

In Chapters 8 and 13 we discussed the field concept associated with the gravitational force. We now extend these considerations to the electromagnetic force and begin by examining the electrostatic case.

When an electric charge is placed at some point in space, the charge establishes everywhere around it a state of electric stress called an *electric field*. Another charge that is placed at any other point will interact with the field that has been set up at that point by the first charge. This interaction is identified with the electrostatic force that the first charge exerts on the second charge.

We restrict considerations of the electric field produced by a charge to a description in a coordinate frame at rest with respect to the charge. We ignore questions relating to the fact that some time is required for the field to propagate to all points in space; indeed, this is the essential meaning of the electro*static* field.

24–1 ELECTRIC FIELD STRENGTH

The electric field produced by a stationary charge* Q can be investigated by measuring the force on a *test charge* q_0 placed at various points in space. (We imagine that the spatial extent of q_0 is sufficiently small that the field is sampled by the test charge essentially at a point.) The *electric field intensity* or the *electric field strength* $\mathbf{E} = \mathbf{E}(\mathbf{r})$ at the point $\mathbf{r}$ is defined to be†

$$\mathbf{E} = \underset{q_0 \to 0}{\text{Lim}} \frac{\mathbf{F}_E}{q_0} \qquad (24\text{–}1)$$

where $\mathbf{F}_E = \mathbf{F}_E(\mathbf{r})$ is the force on the charge q_0 at the point $\mathbf{r}$. For convenience we construct a coordinate system whose origin coincides with the position of the charge Q. According to Coulomb's law (Eq. 23–2), the force on q_0 due to the charge Q is

$$\mathbf{F}_E = \frac{1}{4\pi\epsilon_0} \frac{Qq_0}{r^2} \hat{\mathbf{u}}_r \qquad (24\text{–}2)$$

---

*We have by now exhausted the supply of symbols to identify uniquely the various quantities we discuss. Consequently, in this and in the following chapters new meaning is given to some symbols already used. For example, the symbol Q (which represented *heat* in Chapters 18 through 22) is now used for *charge*. Whenever there is a chance for confusion, an explicit note is made.

†We take the limit $q_0 \to 0$ so that the test charge q_0 will not influence in any way the charges that are responsible for the field $\mathbf{E}(\mathbf{r})$.

where $\hat{\mathbf{u}}_r = \mathbf{r}/r$ (see Fig. 23–5). Thus, for the electric field strength at a point $\mathbf{r}$ due to a charge Q at the origin, we have, combining Eqs. 24–1 and 24–2,

$$\mathbf{E} = \frac{1}{4\pi\epsilon_0} \frac{Q}{r^2} \hat{\mathbf{u}}_r \qquad (24\text{–}3)$$

Note that if Q is positive, $\mathbf{E}$ is directed radially outward, whereas if Q is negative, $\mathbf{E}$ is directed radially inward. The direction of $\mathbf{E}$ at a particular point is always the same as the direction of the force on a *positive* test charge placed at that point. Because we have defined the field in the limit $q_0 \to 0$, our space consists only of the charge Q at the origin and the electric field that is everywhere defined by Eq. 24–3. If a charge q is now placed at the point $\mathbf{r}$, the force on q is

$$\mathbf{F}(\mathbf{r}) = q\mathbf{E}(\mathbf{r}) \qquad (24\text{–}4)$$

The direction of $\mathbf{F}(\mathbf{r})$ with respect to $\mathbf{E}(\mathbf{r})$ is determined by the sign of q.

It is evident from Eq. 24–4 (or Eq. 24–1) that the units of electric field strength are *newtons per coulomb*, N/C. In Section 25–1 we introduce equivalent units for $\mathbf{E}$ (namely, *volts per meter*, V/m).

Because the superposition principle is valid for electrostatic forces (see Section 23–3), the electric field strength, $\mathbf{E} = \mathbf{E}(\mathbf{r})$, at the point $P(\mathbf{r})$ produced by a set of charges Q_i that are simultaneously present is (see Fig. 24–1)

$$\mathbf{E} = \mathbf{E}_1 + \mathbf{E}_2 + \cdots + \mathbf{E}_i + \cdots$$

or

$$\mathbf{E} = \sum_i \mathbf{E}_i = \frac{1}{4\pi\epsilon_0} \sum_i \frac{Q_i}{r_i^2} \hat{\mathbf{u}}_i \qquad (24\text{–}5)$$

where r_i is the distance from Q_i to the point $P(\mathbf{r})$ and where $\hat{\mathbf{u}}_i$ is the unit vector $\mathbf{r}_i/r_i$.

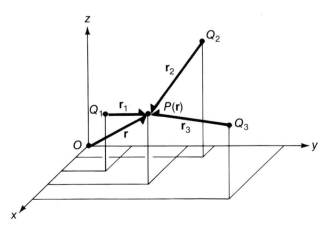

Fig. 24–1. The electric field at $P(\mathbf{r})$ is the superposition of the fields due to the individual charges Q_1, Q_2, and Q_3.

Example 24–1

Two positive point charges, $Q_A = Q$ and $Q_B = 3Q$, are separated by a distance $\overline{AB} = 2\ell$. Determine the electric field strength $\mathbf{E} = \mathbf{E}(Q_A) + \mathbf{E}(Q_B)$ at an arbitrary point on the bisector of the line $\overline{AB}$.

Solution:

Locate the coordinate axes as shown in the diagram so that the y-axis is the bisector of the line $\overline{AB}$. Then, the x-component of the field strength at the point $P(0, y)$ due to the charge Q_A is

$$E_x(Q_A) = \frac{1}{4\pi\epsilon_0} \frac{Q}{(\ell^2 + y^2)} \sin\theta = \frac{1}{4\pi\epsilon_0} \frac{Q\ell}{(\ell^2 + y^2)^{3/2}}$$

Similarly, $\qquad E_x(Q_B) = -\frac{3}{4\pi\epsilon_0} \frac{Q\ell}{(\ell^2 + y^2)^{3/2}}$

Thus, the x-component of the total field strength is

$$E_x = E_x(Q_A) + E_x(Q_B) = -\frac{2}{4\pi\epsilon_0} \frac{Q\ell}{(\ell^2 + y^2)^{3/2}}$$

For the y-component, we find

$$E_y = E_y(Q_A) + E_y(Q_B)$$

$$= \frac{1}{4\pi\epsilon_0} \frac{Q}{(\ell^2 + y^2)} \cos\theta + \frac{3}{4\pi\epsilon_0} \frac{Q}{(\ell^2 + y^2)} \cos\theta$$

$$= \frac{4}{4\pi\epsilon_0} \frac{Qy}{(\ell^2 + y^2)^{3/2}}$$

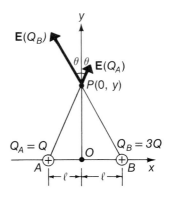

Now, $\mathbf{E}(0, y) = E_x\hat{\mathbf{i}} + E_y\hat{\mathbf{j}}$, so that

$$\mathbf{E}(0, y) = \frac{1}{4\pi\epsilon_0} \frac{Q}{(\ell^2 + y^2)^{3/2}}(-2\ell\hat{\mathbf{i}} + 4y\hat{\mathbf{j}})$$

What is the behavior of $\mathbf{E}(0, y)$ when $y^2 \gg \ell^2$? Then, the x-component becomes negligible in comparison with the y-component; also, $(\ell^2 + y^2)^{3/2} \to (y^2)^{3/2} = y^3$.

Thus, $\qquad \mathbf{E}(0, y) = \frac{1}{4\pi\epsilon_0} \frac{4Q}{y^2}\hat{\mathbf{j}}, \qquad y^2 \gg \ell^2$

This is just the result we would obtain for a *single* charge, $Q_A + Q_B = 4Q$, located at the origin.

In Example 24–1 we found that the field strength at a large distance from the charge system reduces to that for a single charge equal to the sum of the individual charges. This result is, in fact, an example of a broader generalization:

If a number of charges are located within a region of space, then at distances r that are large compared with the dimensions of the region, the electric field strength $\mathbf{E}(\mathbf{r})$ *is approximately equal to the distant solution*

$$\frac{1}{4\pi\epsilon_0} \frac{\hat{\mathbf{u}}_r}{r^2} \sum_i Q_i$$

where $\mathbf{r}$ *is measured from any convenient origin within the region, and where the sum is an* algebraic *sum over all of the charges in the region.*

A similar result was obtained for the gravitational case in Section 13–2. Because gravitational forces are always attractive, the corresponding sum over the masses is never zero; however, there are important electric cases for which $\Sigma Q = 0$. When $\Sigma Q \neq 0$, the preceding expression for $\mathbf{E}(\mathbf{r})$ is referred to as the *monopole* (or *singlet*) approximation, and ΣQ is called the *monopole strength* of the charge distribution.

If a particular charge distribution has $\Sigma Q = 0$, its monopole strength vanishes, but the distant field strength is not necessarily zero. At large distances, the field strength can have no term with a $1/r^2$ dependence because such terms are always associated with ΣQ; however, there may be terms in the expression for the field strength that depend on $1/r^3$, $1/r^4$, and so forth.

24–2 LINES OF FORCE

A convenient way to visualize the electric field produced by a charge or a charge distribution is to construct a map of the *field lines* or *lines of force* of the field. We consider here only the electrostatic case (charges at rest). Lines of force are continuous curves, the tangent to which at any point gives the direction of $\mathbf{E}(\mathbf{r})$ at that point. When the field lines are drawn to a regular scale, the density of the lines in a particular region is proportional to the magnitude of $\mathbf{E}(\mathbf{r})$ in that region. Where the lines are closely spaced the field is strong; where the lines are sparse the field is weak.

Figure 24–2 shows the lines of force associated with a positive point charge. The lines radiate from the charge and spread out uniformly in three dimensions. Field lines originate on positive charges and terminate on negative charges. In the electrostatic case, a field line never closes on itself; every field line connects a positive charge with a negative charge. (Sometimes we imagine one of these charges to be at infinity.) Field lines never intersect, and they behave as if there were a tendency for the lines to repel one another.

Figure 24–3 uses lines of force to illustrate the field produced by a positive charge $+3Q$ in combination with a negative charge $-Q$. The field is symmetric about the line connecting the charges, so the three-dimensional aspect of the field can be imagined by considering similar field lines to exist in planes rotated about this symmetry axis. The individual field strengths $\mathbf{E}_+$ and $\mathbf{E}_-$ due to the two charges are shown at the point P. The vector sum $\mathbf{E} = \mathbf{E}_+ + \mathbf{E}_-$ is tangent to the field line that passes through P. The number of field lines that originate on the charge $+3Q$ is three times the number that terminate on

Fig. 24–2. The lines of force for a positive point charge. The lines radiate uniformly outward from the charge.

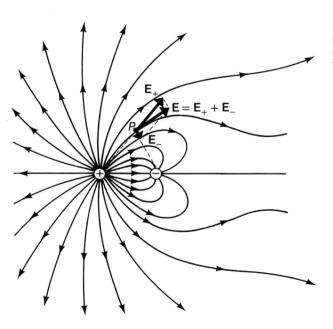

Fig. 24–3. Field lines for a charge distribution consisting of a positive charge $+3Q$ and a negative charge $-Q$.

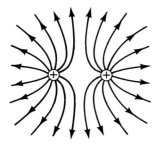

Fig. 24–4. Field lines for two equal positive point charges.

the charge $-Q$. Thus, two thirds of the lines originating on the charge $+3Q$ terminate at infinity where we imagine there exists an amount of charge $-2Q$ spread out uniformly; in this way we maintain an electrically neutral Universe. When viewed from a distance that is large compared with the separation of the charges, the outward-spreading lines become uniformly divergent; this tendency is evident in Fig. 24–3. (• Is this distant behavior consistent with the fact that the two charges have a monopole strength of $+2Q$?)

Figure 24–4 shows the lines of force associated with two equal positive point charges. In this case the tendency for the field lines to repel one another is quite evident. Note also that all of the lines diverge toward infinity and, at large distances, head directly away from a point located midway between the charges.

An important case of a two-charge distribution is the *electric dipole,* which consists of two equal and opposite charges, $+Q$ and $-Q$, separated by a fixed distance ℓ (Fig. 24–5). It is convenient to describe the field of an electric dipole in terms of a vector **p**, called the *dipole moment.* The *dipole strength* is $p = |\mathbf{p}| = \ell Q$, and the direction of **p** is from $-Q$ to $+Q$, as indicated in the figure.

At a point P specified by the position vector **r** ($r \gg \ell$) measured from the center of the dipole, the electric field $\mathbf{E}(\mathbf{r})$ due to a dipole with moment **p** is given by*

$$\mathbf{E}(\mathbf{r}) = \frac{1}{4\pi\epsilon_0}\left(-\frac{\mathbf{p}}{r^3} + \frac{3(\mathbf{p}\cdot\mathbf{r})\mathbf{r}}{r^5}\right), \qquad r \gg \ell \tag{24–6}$$

This expression is valid far from the dipole (the *distant-field* region, $r \gg \ell$); in the near-field region the expression for the field is more complicated. Notice that Eq. 24–6 con-

Fig. 24–5. An electric dipole **p** and the associated field vectors **E** at three points.

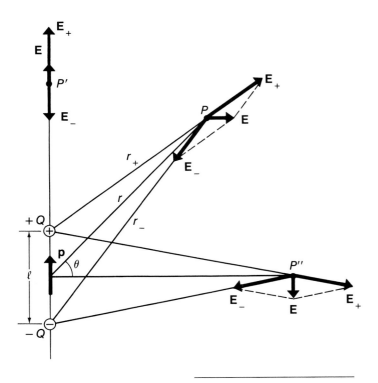

---

*The electric field due to a dipole is derived in Section 25–2.

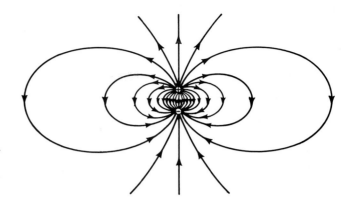

Fig. 24–6. The field lines for an electric dipole. For distances $r \gg \ell$, the pattern of field lines corresponds to Eq. 24–6. (See Fig. 25–5 for the field of a *point* dipole; in such a case, Eq. 24–6 is valid for all r.) The lines that leave the top of the diagram eventually curve back and join the appropriate lines that enter the bottom of the diagram.

tains no $1/r^2$ term (as expected for a distribution that has $\Sigma Q = 0$); both terms in the expression vary as $1/r^3$.

The field lines for an electric dipole are illustrated in Fig. 24–6. Notice that all of the field lines originate on the positive charge and terminate on the negative charge. No lines proceed indefinitely far away without returning (this would imply a nonzero monopole strength).

Example 24–2

Calculate the electric field intensity E at a point, such as P' in Fig. 24–5, that is in line with the dipole moment vector $\mathbf{p}$ (i.e., $\theta = 90°$) and is at a distance $r \gg \ell$.

Solution:

Referring to Fig. 24–5, we have

$$E = \frac{Q}{4\pi\epsilon_0}\left[\frac{1}{r_+^2} - \frac{1}{r_-^2}\right]$$

$$= \frac{Q}{4\pi\epsilon_0}\left[\frac{1}{(r - \ell/2)^2} - \frac{1}{(r + \ell/2)^2}\right]$$

$$= \frac{Q}{4\pi\epsilon_0}\cdot\frac{1}{r^2}\left[\frac{1}{(1 - \ell/2r)^2} - \frac{1}{(1 + \ell/2r)^2}\right]$$

Now, $r \gg \ell$, so we can expand the terms in the brackets according to

$$E = \frac{Q}{4\pi\epsilon_0}\cdot\frac{1}{r^2}\left[\left(1 + \frac{\ell}{r} - \cdots\right) - \left(1 - \frac{\ell}{r} + \cdots\right)\right]$$

$$= \frac{Q}{4\pi\epsilon_0}\cdot\frac{1}{r^2}\cdot\left(2\frac{\ell}{r}\right) = \frac{Q\ell}{2\pi\epsilon_0}\cdot\frac{1}{r^3}$$

Finally, using $p = Q\ell$, we have

$$E = \frac{1}{2\pi\epsilon_0}\cdot\frac{p}{r^3}, \qquad r \gg \ell$$

Notice that the same result is obtained from Eq. 24–6 by substituting $\theta = 90°$ and taking the magnitude of the resulting expression.

It is possible to construct charge distributions for which both the monopole strength and the dipole strength are zero. Figure 24–7 shows two particularly simple distributions that have this property. These are examples of what are called *quadrupole distributions*. For either of these quadrupoles, $\mathbf{E}(\mathbf{r})$ varies as $1/r^4$ when $r \gg \ell$; see Problem 24–6.

In another important field configuration, called the *uniform field*, the electric field strength is constant in magnitude and direction; for example, $\mathbf{E}(\mathbf{r}) = E_0\hat{\mathbf{i}}$, in which case the field lines are uniformly distributed parallel straight lines, all in the $+x$-direction. Such a field would exist between two imaginary parallel uniformly charged infinite planes, one located at $x = -\infty$ and positively charged, the other located at $x = +\infty$ and negatively charged. We often have occasion to discuss uniform fields and their effects.

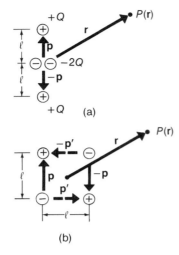

Fig. 24–7. Two different electric quadrupole configurations. In each case both the monopole strength ΣQ and the net dipole moment, $\mathbf{p} + (-\mathbf{p})$, vanish. In (b), the net dipole moment can be considered to be $\mathbf{p} + (-\mathbf{p})$ or $\mathbf{p}' + (-\mathbf{p}')$; in either case, the sum is zero. A charge distribution may have a quadrupole strength even though the monopole strength and the net dipole moment do not vanish.

24–3 ELECTRIC FLUX

The concept of flux is naturally appropriate for discussing the matter flow rate associated with the velocity vector field for the laminar flow of incompressible fluids. Where the density of streamlines is high, there is a large transport rate or flux of fluid matter through a given cross-sectional area of flow; where the density of streamlines is low, there is a relatively lower mass transport rate. Quite generally, the mass transport flux across any differential area $\hat{\mathbf{n}} \Delta A$ is given by the expression

$$\Delta(\text{flux of mass}) = \rho \mathbf{v} \cdot \hat{\mathbf{n}} \, dA$$

Fig. 24–8. The electric flux through an area ΔA. The angle between $\hat{\mathbf{n}}$ and $\mathbf{E}$ through ΔA is θ. (a) When $\theta < \pi/2$, the flux of $\mathbf{E}$ through ΔA is positive. (b) When $\theta = \pi/2$, the flux is zero. (c) When $\theta > \pi/2$, the flux is negative.

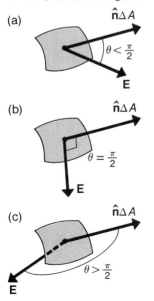

where ρ is the mass density of the fluid. (Refer to the discussion of the continuity equation in Chapter 14.) The concept of flux may be adapted to describing any vector field, even though there may be no actual matter-transport involved. For example, in many respects the streamline patterns for fluid flow resemble corresponding electric-field line patterns if we imagine positive charges to play the role of pointlike sources and negative charges the pointlike sinks of *electric flux*.

The magnitudes of the charges simply measure the strength of the sources or sinks. Figure 24–6 for equal positive and negative charges may be thought of as involving an amount of flux "injected" into the field at the positive charge that is exactly equal to the flux "drained" from the field at the negative charge. (• In the analogous case of fluid flow, could any streamline leave the source $+Q$ and fail to end on the sink $-Q$?) In general, all electric field lines that originate on the positive charges must also end on the negative charges. In Fig. 24–3 we imagine the flux originating at the source $+3Q$ to be three times larger than that entering the sink $-Q$. Thus two thirds of the flux originating at the positive charge would leave the field via sinks distributed uniformly at infinity.

For vector fields such as the electric or magnetic fields there is no transport or flow of matter; however, the analogy with fluid flow is very useful and leads to an important general law of electrostatics.

Consider an incremental surface with area ΔA located in the field $\mathbf{E}$ (Fig. 24–8). We define an incremental scalar quantity, $\Delta\varphi$, by the equation

$$\begin{aligned}\Delta\varphi &= \mathbf{E} \cdot \hat{\mathbf{n}} \, \Delta A \\ &= E \, \Delta A \cos \theta\end{aligned} \tag{24–7}$$

$\Delta\varphi$ is called the *flux* of $\mathbf{E}$ through the surface area ΔA (Fig. 24–8). This element of area is represented by the vector $\hat{n}\,\Delta A$, where $\hat{n}$ is the unit vector perpendicular to the tangent plane that contains the area ΔA. When ΔA is part of a *closed* surface, $\hat{n}$ is chosen to be directed *outward*.

Suppose that a charge Q is located at the origin of some coordinate frame. Imagine that the charge is surrounded by a spherical surface with radius r and with its center at the origin. Everywhere on this surface we have

$$\mathbf{E} = \frac{Q}{4\pi\epsilon_0 r^2}\,\hat{u}_r \qquad (24\text{-}8)$$

The unit normal vector $\hat{n}$ is directed outward from the surface and, in fact, for the spherical surface it is identical to the unit radial vector $\hat{u}_r$. Therefore, with ΔA located on the surface, as indicated in Fig. 24–9, we can write

$$\hat{n}\,\Delta A = \hat{u}_r\,\Delta A$$

Then, the *total* flux of $\mathbf{E}$ through the entire spherical surface, $S_0 = 4\pi r^2$, is*

$$\varphi = \lim_{\Delta A_i \to 0} \sum_i \mathbf{E}\cdot\hat{n}_i\,\Delta A_i = \oint_{S_0} \mathbf{E}\cdot\hat{n}\,dA \qquad (24\text{-}9)$$

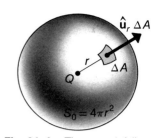

Fig. 24–9. The area ΔA lies on the spherical surface that surrounds the charge Q.

Note that Eq. 24–9 is applicable to any field originating within any closed surface. Using the expression for the field of a point charge, namely,

$$\mathbf{E}\cdot\hat{n} = \left(\frac{Q}{4\pi\epsilon_0 r^2}\,\hat{u}_r\right)\cdot\hat{n} = \frac{Q}{4\pi\epsilon_0 r^2}$$

we have, for the spherical surface,

$$\varphi = \frac{Q}{4\pi\epsilon_0 r^2}\oint_{S_0} dA = \frac{Q}{4\pi\epsilon_0 r^2}\cdot 4\pi r^2 = \frac{Q}{\epsilon_0} \qquad (24\text{-}10)$$

Suppose that a point charge Q is surrounded by both a spherical surface S_0 centered on Q as before and an irregular surface S, as shown in Fig. 24–10a. Evidently, every one of the force lines that radiate from Q passes through S as well as through S_0. It follows that the total flux φ is given by

$$\varphi = \oint_S \mathbf{E}\cdot\hat{n}\,dA = \frac{Q}{\epsilon_0} \qquad (24\text{-}11)$$

This result is *independent* of the shape of S; it is required only that S enclose Q.

Next, suppose that the closed surface does *not* surround Q, as indicated in Fig. 24–10b. Then, every line of force that enters S' must also leave the surface. *Entering* the surface implies that the angle between $\mathbf{E}$ and $\hat{n}$ is $\theta > \pi/2$ (as at P) so that $\mathbf{E}\cdot\hat{n}\,dA < 0$. *Leaving* the surface implies that $\theta < \pi/2$ (as at P') so that $\mathbf{E}\cdot\hat{n}\,dA > 0$. Thus, the *net flux* that passes through S' is *zero*. The generalization of these results leads to *Gauss's law*, which is discussed in Section 24–4.

Note carefully that the simple geometric results obtained here are a direct consequence of the fact that the electric field strength of a point charge depends on the inverse

---

*We use the integral symbol $\oint_S$ to indicate a surface integration over a *closed* surface S. Examples of the explicit evaluation of surface integrals are found in Examples 24–3 and 24–4.

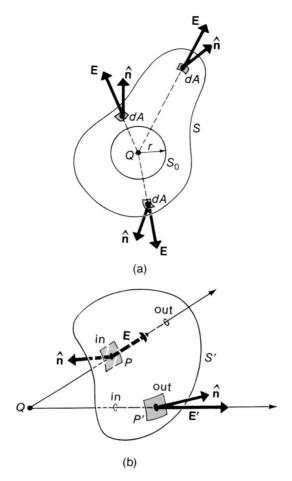

(a)

(b)

Fig. 24–10. (a) Every line of force that passes through S_0 also passes through S. The net flux through either surface is Q/ϵ_0. (b) Every line of force that enters S' also leaves the surface. The net flux through S' is zero.

square of the distance from the charge, so that the flux has no dependence on distance (Eq. 24–10). This, of course, is ultimately related to the $1/r^2$ dependence of the electric force between two point charges, as contained in Coulomb's law.

24–4 GAUSS'S LAW

When Eq. 24–11 is generalized to include the idea of superposition of electric fields from more than one charge, the result is *Gauss's law,** which gives an expression for the electric flux through a closed surface S that surrounds a distribution of charge:

$$\varphi_{\text{net}} = \oint_S \mathbf{E} \cdot \hat{\mathbf{n}} \, dA = \begin{cases} \dfrac{1}{\epsilon_0} \sum Q_{\text{in}} \\[2mm] \dfrac{1}{\epsilon_0} \displaystyle\int_V \rho \, dv \end{cases} \tag{24–12}$$

*Karl Friedrich Gauss (1777–1855), German mathematician, astronomer, and physicist.

where ΣQ_{in} represents the algebraic sum of all the discrete charges *inside* the surface S. In the event that there is a continuous distribution of charge with a volume density ρ, the sum must be replaced by the integral of $\rho\,dv$ over the volume V occupied by the charge. (The volume V is surrounded by the surface S.) Any closed surface S that surrounds a charge distribution is called a *Gaussian surface*. Gauss's law is a powerful tool for analyzing the field strengths due to various charge distributions.

Example 24–3

It is known that an (unspecified) array of charges produces a uniform electric field, $\mathbf{E} = E_x\hat{\mathbf{i}} + E_y\hat{\mathbf{j}}$, in a certain region of space. Show that there is no net flux through a cubical Gaussian surface that is located in a charge-free portion of this region, thereby verifying the validity of Gauss's law for this case.

Choose the coordinate axes to lie along the edges of the cube, as shown in the diagram.

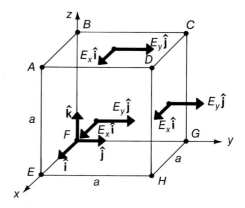

Solution:

The integral that must be evaluated, namely, $\oint_S \mathbf{E}\cdot\hat{\mathbf{n}}\,dA$, consists of the sum of contributions from each of the six faces of the cube. First, for the face *DCGH*, we have (refer to the diagram)

$$\int_{DCGH} \mathbf{E}\cdot\hat{\mathbf{n}}\,dA = (E_x\hat{\mathbf{i}} + E_y\hat{\mathbf{j}})\cdot\hat{\mathbf{j}}\int_{DCGH} dA = +E_y a^2$$

For the face *ADHE*, we have

$$\int_{ADHE} \mathbf{E}\cdot\hat{\mathbf{n}}\,dA = (E_x\hat{\mathbf{i}} + E_y\hat{\mathbf{j}})\cdot\hat{\mathbf{i}}\int_{ADHE} dA = +E_x a^2$$

For the face *ABCD*, we have

$$\int_{ABCD} \mathbf{E}\cdot\hat{\mathbf{n}}\,dA = (E_x\hat{\mathbf{i}} + E_y\hat{\mathbf{j}})\cdot\hat{\mathbf{k}}\int_{ABCD} dA = 0$$

For each corresponding parallel face the integral changes sign; thus,

$$\int_{ABFE} \mathbf{E}\cdot\hat{\mathbf{n}}\,dA = -E_y a^2$$

$$\int_{BCGF} \mathbf{E}\cdot\hat{\mathbf{n}}\,dA = -E_x a^2$$

$$\int_{EFGH} \mathbf{E}\cdot\hat{\mathbf{n}}\,dA = 0$$

The net flux is the sum of these six integrals, which vanishes:

$$\oint_S \mathbf{E}\cdot\hat{\mathbf{n}}\,dA = 0$$

Thus, Gauss's law is verified.

Example 24–4

A charge Q is located a perpendicular distance a from the center of a square surface with dimensions $2a$ by $2a$, as shown in diagram (a). Calculate the flux through this square surface.

Solution:

The field strength due to the charge Q is

$$\mathbf{E} = \frac{Q}{4\pi\epsilon_0}\frac{\hat{\mathbf{u}}_r}{r^2}$$

Now $\hat{\mathbf{u}}_r = \mathbf{r}/r$, so we can write

$$\mathbf{E} = \frac{Q}{4\pi\epsilon_0}\frac{\mathbf{r}}{r^3} = \frac{Q}{4\pi\epsilon_0 r^3}(x\hat{\mathbf{i}} + y\hat{\mathbf{j}} + z\hat{\mathbf{k}})$$

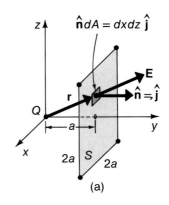

(a)

and $r^2 = x^2 + y^2 + z^2$.

The element of area on the surface S is $\hat{\mathbf{n}}\, dA = dx\, dz\, \hat{\mathbf{j}}$. Also, on S, we have $y = a$. Now, $\hat{\mathbf{i}} \cdot \hat{\mathbf{j}} = \hat{\mathbf{k}} \cdot \hat{\mathbf{j}} = 0$ and $\hat{\mathbf{j}} \cdot \hat{\mathbf{j}} = 1$; hence,

$$\mathbf{E} \cdot \hat{\mathbf{n}}\, dA = \frac{Q}{4\pi\epsilon_0} \frac{a}{r^3}\, dx\, dz$$

so that the flux through the open surface S is

$$\varphi = \int_S \mathbf{E} \cdot \hat{\mathbf{n}}\, dA$$

$$= \frac{Qa}{4\pi\epsilon_0} \int_{-a}^{a} \int_{-a}^{a} \frac{dx\, dz}{(x^2 + z^2 + a^2)^{3/2}}$$

$$= \frac{Qa}{\pi\epsilon_0} \int_0^a \int_0^a \frac{dx\, dz}{(x^2 + z^2 + a^2)^{3/2}}$$

This double integral can be evaluated by referring to the list of integrals inside the rear cover. The result is (see Problem 24–9)

$$\varphi = \frac{Qa}{\pi\epsilon_0} \cdot \frac{1}{a} \tan^{-1} \frac{z}{\sqrt{z^2 + 2a^2}} \Big|_0^a = \frac{Q}{\pi\epsilon_0} \tan^{-1} \frac{1}{\sqrt{3}}$$

$$= \frac{Q}{6\epsilon_0}$$

Because of the particular geometry of this problem, the result for φ can be obtained much more easily by making use of

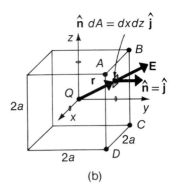

(b)

Gauss's law. Let the surface S be one face ($ABCD$) of a cube that surrounds the charge Q, as shown in diagram (b). The charge is at the *center* of the cube and is therefore located symmetrically with respect to each face. The net flux through the cubical surface is just Q/ϵ_0, so the flux through any one of the identical faces is one sixth as much. Thus,

$$\varphi = \frac{Q}{6\epsilon_0}$$

which is the result we obtained by the more laborious direct calculation.

24–5   ELECTRIC FIELDS FOR CONTINUOUS DISTRIBUTIONS OF STATIC CHARGE

We now extend the discussion and consider the fields produced by continuous distributions of charge.* We examine various cases in which the charges are held fixed by unspecified means. Later, we discuss fields arising from more realistic cases of charges deposited on various arrangements of insulators and conductors.

The Field of a Uniformly Charged Infinite Filament.   Consider an infinitely long filament that holds a uniform line distribution of charge (Fig. 24–11). Let the positive charge per unit length of the filament be λ. Then the differential piece of filament, dz, contains the charge $dQ = \lambda\, dz$. At a point P that is a perpendicular distance R from the filament, the electric field strength due to the charge on the segment dz at z is

$$d\mathbf{E}' = \frac{dQ}{4\pi\epsilon_0 r^2} \hat{\mathbf{u}}_r$$

where $r^2 = R^2 + z^2$. Evidently, there is a second piece of the filament, also with length dz, located at $-z$, which produces a field strength $d\mathbf{E}''$ at P. The z-component of $d\mathbf{E}''$, by symmetry, just cancels the z-component of $d\mathbf{E}'$; the two fields give equal contributions in

---

*As we noted in Section 23–2, charge is in fact quantized into discrete units e, so that treating a charge distribution as continuous must be an approximation. However, the value of e is so small that a charge of only 1 picocoulomb consists of 6.24×10^6 such units. Thus; we expect our approximation to be very good indeed.

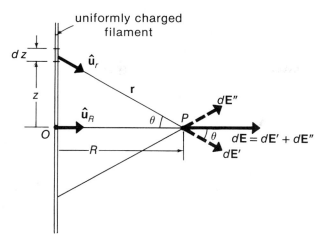

Fig. 24-11. An infinitely long filament carries a uniform charge λ per unit length. The amount of charge in the length dz is $dQ = \lambda\, dz$.

the radial ($\hat{\mathbf{u}}_R$) direction. Thus, at P, we have

$$d\mathbf{E} = d\mathbf{E}' + d\mathbf{E}'' = \frac{2\lambda\, dz \cos\theta}{4\pi\epsilon_0 r^2}\hat{\mathbf{u}}_R$$

or

$$d\mathbf{E} = \frac{\lambda R\, dz}{2\pi\epsilon_0(z^2 + R^2)^{3/2}}\hat{\mathbf{u}}_R$$

Integrating from $z = 0$ to $z = \infty$ we find (refer to the list of integrals inside the rear cover)

$$\mathbf{E}(R) = \frac{\lambda R\hat{\mathbf{u}}_R}{2\pi\epsilon_0}\int_0^\infty \frac{dz}{(z^2 + R^2)^{3/2}}$$

or

$$\mathbf{E}(R) = \frac{\lambda}{2\pi\epsilon_0}\cdot\frac{\hat{\mathbf{u}}_R}{R} \qquad \text{charged filament} \qquad (24\text{--}13)$$

The filament is infinitely long, so the field strength $\mathbf{E}$ has only a radial component and no z-component. Note that $\mathbf{E}$ depends inversely on the *first power* of the distance R.

This same result can be obtained by a straightforward application of Gauss's law. We deduce from the symmetry of the situation that $\mathbf{E}$ must be everywhere radially away from the filament. (• Can you see why $\mathbf{E}$ behaves this way?) Then, the dependence of $\mathbf{E}$ on R is easily found by using Gauss's law. As shown in Fig. 24-12, we surround a length ℓ of the filament with a cylindric closed Gaussian surface whose axis coincides with the filament and which has circular end caps. Let the radius of the cylinder be R. The charge contained within the Gaussian surface is $q = \lambda\ell$. Because $\mathbf{E}$ is everywhere in the direction of $\hat{\mathbf{u}}_R$, there is no flux through either end cap. The outward flux through the cylinder wall is just $2\pi R\ell E(R)$; thus,

$$2\pi R\ell E(R) = \frac{q}{\epsilon_0} = \frac{\lambda\ell}{\epsilon_0}$$

Fig. 24-12. A length ℓ of an infinite line charge is surrounded by a cylindric Gaussian surface.

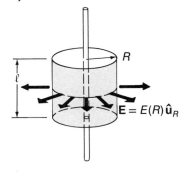

or

$$E(R) = \frac{\lambda}{2\pi\epsilon_0}\cdot\frac{1}{R}$$

and vectorially, we have

$$\mathbf{E}(R) = \frac{\lambda}{2\pi\epsilon_0}\cdot\frac{\hat{\mathbf{u}}_R}{R}$$

which is the same as the result obtained by explicit integration (Eq. 24-13).

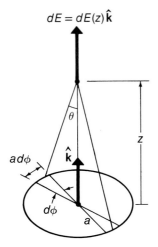

$$dE = dE(z)\hat{k}$$

Fig. 24–13. The geometry for calculating the electric field strength along the axis of a uniformly charged ring.

Notice that Gauss's law can be applied to obtain the result easily in this case because the geometry is particularly simple, so that the pattern of field lines can be readily deduced and a Gaussian surface constructed that permits a straightforward integration. Gauss's law is generally quite useful in situations that, like this one, possess a high degree of symmetry.

The Field of a Uniformly Charged Circular Ring. Next determine the electric field strength along the axis of a uniformly charged ring. The geometry is shown in Fig. 24–13. If the ring carries a total positive charge Q, the linear charge density along the ring is $\lambda = Q/2\pi a$, where a is the radius of the ring. In a small segment of the ring with arc length $a \, d\phi$ the amount of charge is $dQ = \lambda a \, d\phi$. This charge, combined with the charge on a similar segment of the ring at the other end of a common diameter (see Fig. 24–13), produces a contribution $d\mathbf{E}$ to the electric field that is directed along the ring axis (the direction of $\hat{k}$) because the transverse components cancel, as in the case of the infinite filament. Then, for the pair of ring segments, we have

$$d\mathbf{E} = \frac{2\lambda a \, d\phi \cos \theta}{4\pi\epsilon_0(z^2 + a^2)}\hat{k}$$

and

$$\mathbf{E} = \frac{\lambda a z}{2\pi\epsilon_0(z^2 + a^2)^{3/2}}\hat{k} \int_0^\pi d\phi$$

from which, substituting $\lambda = Q/2\pi a$,

$$\mathbf{E}(z) = \frac{Qz}{4\pi\epsilon_0(z^2 + a^2)^{3/2}}\hat{k} \qquad \text{charged ring} \qquad (24\text{–}14)$$

The value of $E(z)$ as a function of z is shown in Fig. 24–14. The field strength reaches its maximum value at $z = a/\sqrt{2}$ where $E_m = Q/(6\sqrt{3}\pi\epsilon_0 a^2)$ (see Problem 24–12). In this case Gauss's law cannot be used because the situation does not possess a sufficiently high degree of symmetry to permit the geometry of the field lines to be readily deduced.

Note that when $z^2 \gg a^2$, we have $(z^2 + a^2)^{3/2} \cong z^3$, so that

$$\mathbf{E}(z) \cong \frac{Q}{4\pi\epsilon_0 z^2}\hat{k}, \qquad z^2 \gg a^2$$

which is just the distant-field (monopole) expression.

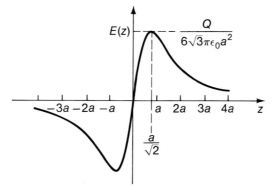

Fig. 24–14. The field strength E as a function of z along the axis of a uniformly charged ring.

The Field of a Uniformly Charged Infinite Flat Sheet. Consider an infinite flat sheet that carries a uniform positive surface charge density σ (Fig. 24–15), so that the

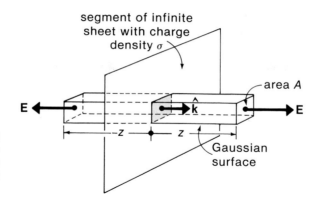

Fig. 24-15. A Gaussian surface in the shape of a rectangular box is constructed perpendicular to an infinite flat sheet that carries a uniform charge density σ.

charge on any finite area A is $Q = \sigma A$. From the symmetry it is clear that **E** must be everywhere perpendicular to the sheet. That is, with z measured perpendicular to the sheet, we have $\mathbf{E} = E(z)\hat{\mathbf{k}}$ on the side $z > 0$ and $\mathbf{E} = -E(z)\hat{\mathbf{k}}$ on the side $z < 0$; however, the function $E(z)$ is not yet known.

Construct a Gaussian surface in the shape of a closed rectangular box that is perpendicular to the sheet and extends a distance z on either side. The ends each have area A. Because there is no flux through the sides of the Gaussian surface, Gauss's law is expressed as

$$2A\, E(z) = \frac{\sigma A}{\epsilon_0}$$

or

$$E(z) = \frac{\sigma}{2\epsilon_0}$$

which shows that the field strength is independent of z. In vector form,

$$\mathbf{E} = \frac{\sigma}{2\epsilon_0}\hat{\mathbf{k}} \qquad \text{charged sheet} \qquad (24\text{-}15)$$

For the practical case of a uniformly charged flat sheet with large (but not infinite) dimensions, the field is uniform throughout the central region.

Fields of Uniform Spherical Charge Distributions. Consider a thin spherical shell with a radius R that carries a total positive charge Q distributed uniformly over the surface (Fig. 24-16). Because the geometry and the charge distribution both have spheri-

Fig. 24-16. Application of Gauss's law to a spherical charge distribution.

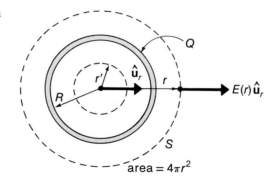

cal symmetry, we expect the electric field strength to be everywhere radially outward, that is, $\mathbf{E} = E(r)\hat{\mathbf{u}}_r$. To determine the function $E(r)$, construct a spherical Gaussian surface with a radius r that is concentric with the charge distribution. Consider the possibilities $r > R$ and $r < R$.

For the case $r > R$, Gauss's law yields

$$4\pi r^2 E(r) = \frac{Q}{\epsilon_0}$$

so that

$$\mathbf{E}(r) = \frac{Q}{4\pi\epsilon_0}\frac{\hat{\mathbf{u}}_r}{r^2}, \qquad r > R \qquad \text{(24–16a)}$$

Thus, the electric field strength in the region *outside* the charged sphere (that is, $r > R$) is the same as that produced by a point charge Q located at the center of the spherical shell.

For the case $r < R$, the Gaussian surface encloses no charge (see Fig. 24–16); hence,

$$\mathbf{E} = 0, \qquad r < R \qquad \text{(24–16b)}$$

The results expressed by Eqs. 24–16 are analogous to those obtained for the gravitational case (see Section 13–5). (• Why should the two cases give similar results?) Notice that the use of Gauss's law here gives the result immediately, whereas the evaluation of a complicated integral was required in the method used in Section 13–5.

Example 24–5

Consider a spherical distribution of charge with a volume charge density $\rho(r)$, measured in C/m^3, that depends at most on r. The charge is confined within a region with a radius R. Determine the electric field strength everywhere in space.

Solution:

The diagram shows the charge distribution and two Gaussian surfaces, one with radius $r < R$ and one with radius $r > R$. Again, from the spherical symmetry we argue that $\mathbf{E}$ can depend only on r and not any tangential angles; $\mathbf{E}$ has a component only in the direction $\hat{\mathbf{u}}_r$. Thus, for $r < R$, we have

$$4\pi r^2 E(r) = \frac{Q'}{\epsilon_0}$$

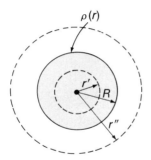

Then, the electric field strength at r is

$$\mathbf{E}(r) = \frac{Q'}{4\pi\epsilon_0 r^2}\hat{\mathbf{u}}_r$$

where Q' is the total charge within a sphere with radius r; that is,

$$Q' = \int_0^r 4\pi r^2 \rho(r)\, dr$$

For $r > R$, the Gaussian surface contains the entire charge Q; thus,

$$\mathbf{E}(r) = \frac{Q}{4\pi\epsilon_0 r^2}\hat{\mathbf{u}}_r, \qquad r > R$$

The field strength $E(r)$ can be evaluated only if the density function $\rho(r)$ is known. For the special case in which $\rho(r)$ is uniform, $\rho = Q/(4\pi R^3/3)$, we readily find that

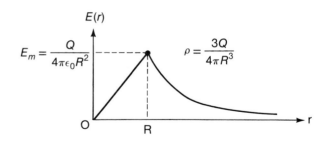

$$Q' = \left(\frac{r}{R}\right)^3 Q$$

Then, $\mathbf{E}(r) = \dfrac{rQ}{4\pi\epsilon_0 R^3}\hat{\mathbf{u}}_r,   \quad r < R$

The magnitude of the field strength for this case is shown in the diagram. Again, it is instructive to compare with the gravitational case (Section 13–5).

QUESTIONS

24–1 Define and explain briefly the meaning of the terms (a) test charge, (b) electric field strength, (c) lines of force, and (d) dipole moment.

24–2 Suppose you knew that the electric field at some point between two point charges vanished. What could you deduce about the sign of the charges? What if the electric field vanished at some point beyond the region between the charges? Can the electric field due to two charges ever vanish except along a line connecting the two charges?

24–3 Is it possible for electric lines of force to intersect?

24–4 Is it possible for an electric line of force to form a closed loop?

24–5 Let the dipole **p** shown in Fig. 24–5 be centered at the origin of a Cartesian coordinate frame, with its direction along the y-axis. Thus, with $\mathbf{p} = Q\ell\hat{\mathbf{j}}$ and $\mathbf{r} = x\hat{\mathbf{i}} + y\hat{\mathbf{j}} + z\hat{\mathbf{k}}$, show that

$$E_x = \frac{Q\ell}{4\pi\epsilon_0}\cdot\frac{3xy}{r^5}$$

$$E_y = \frac{Q\ell}{4\pi\epsilon_0}\cdot\left(-\frac{1}{r^3} + \frac{3y^2}{r^5}\right)$$

and $E_z = \dfrac{Q\ell}{4\pi\epsilon_0}\cdot\dfrac{3yz}{r^5}$

(See also Problem 24–3.)

24–6 Define and explain briefly the meaning of the terms (a) electric flux, (b) Gauss's law, and (c) Gaussian surface.

24–7 Suppose that Q in Figs. 24–9 and 24–10 represents a point source of incompressible fluid flow. The flow leaves symmetrically and is drained off uniformly at infinity. Discuss the volume flow rate across the various hypothetical surface boundaries. Defend the analogous

theoretical connection between the flow-velocity field $\mathbf{v}(\mathbf{r})$ and the electric field $\mathbf{E}(\mathbf{r})$. Does the magnitude of $|\mathbf{v}|$ decrease with the inverse square of the radius? [*Hint:* Make a suitable connection with the continuity equation for fluids.]

24–8 In discussing the electric field inside the uniformly charged spherical shell shown in Fig. 24–17, a Gaussian surface inside the shell was used to deduce that $\mathbf{E} = 0$ for $r < R$. However, a uniform field inside the shell would also result in no net flux through any interior closed Gaussian surface. How can you rule out the presence of such an interior uniform field? Discuss in general terms how the charge distribution on the thin shell should be distributed so that a uniform interior field may result. Is Gauss's law still applicable in this case for interior surfaces?

24–9 Three equal positive charges are located at the corners of an equilateral triangle. Does Gauss's law hold for this configuration? If so, could it be used to deduce the electric field? Sketch the electric field pattern of the lines of force. Discuss the geometric requirements for charge distributions if the electric field is to be determined by application of Gauss's law alone.

24–10 Discuss the applicability of Gauss's law to the gravitational field. Are there sinks as well as sources for the gravitational field?

24–11 A complicating detail not shown in Fig. 24–3 concerns the behavior of the field lines near the point where $\mathbf{E} = 0$. The actual behavior of the field lines associated with this point is shown in the figure on page 476. Discuss the features shown. What can you say about the electric flux leaving the charge $+3Q$ within a cone of half angle θ? How could you use this fact to determine θ?

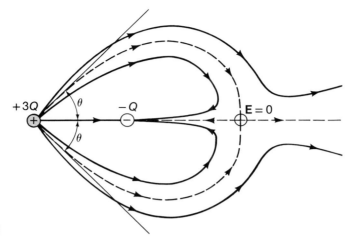

Question 24-11

PROBLEMS

Sections 24-1 and 24-2

24-1 A point charge $Q_1 = +3\ \mu C$ is placed at the origin, and a point charge $Q_2 = -8\ \mu C$ is placed at $x = +0.20$ m. (a) Determine the electric field strength, $\mathbf{E}(x) = E(x)\hat{\mathbf{i}}$, at all points along the x-axis. (b) At what points, if any, other than $|x| = \infty$, does $E(x) = 0$?

24-2 Three equal positive point charges Q are located in the x-y plane at coordinates (x, y) equal to $(a, 0)$, $(-a, 0)$, and $(0, a)$. (a) Determine the electric field strength along the y-axis. (b) At what points, if any, other than $|y| = \infty$, does the electric field vanish? (c) For large values of y (i.e., $|y| \gg a$), what effective charge located at the origin would give the same value of the electric field?

24-3 Show that Eq. 24-6 for the electric field strength of a dipole can be expressed in rectangular coordinates

as $\qquad E_x = \dfrac{3p}{4\pi\epsilon_0 r^3}\ \sin\theta\cos\theta$

and $\qquad E_y = \dfrac{p}{4\pi\epsilon_0 r^3}(3\cos^2\theta - 1)$

Also, show that

$$|\mathbf{E}| = (E_x^2 + E_y^2)^{1/2} = (p/4\pi\epsilon_0 r^3)(3\cos^2\theta + 1)^{1/2}$$

24-4 Refer to Fig. 24-5. Calculate the electric field intensity E at a point, such as P'', that is at an angle $\theta = 0°$ and a distance $r \gg \ell$. Compare your result with Eq. 24-6.

24-5 Consider a dipole in which each charge has $|Q| = 3\ \mu C$ and for which $2\ell = 2$ cm. (a) Determine the electric field strength at distances of 1 m and 10 m from the dipole along the dipole axis. (b) What would be the field strength at these distances for the positive charge alone?

(c) What is the ratio of the monopole field strength to the dipole field strength at the two distances?

24-6 Calculate the electric field strength at points along the
• x-axis ($|x| \gg \ell$) for the quadrupole charge distribution shown in the diagram.

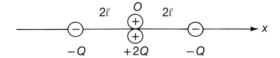

Section 24-3

24-7 A cube with sides of 0.20 m has its center at the origin of a rectangular coordinate frame and its faces perpendicular to the coordinate axes. The electric field strength is $\mathbf{E} = Kx\hat{\mathbf{i}}$, with $K = 100$ N/C·m. Determine the amount of charge within the cube.

24-8 Positive charge with a uniform linear density λ is carried
• on four parallel infinite filaments that pass through the corners of a square, as shown in the diagram. (a) Sketch the lines of force in a plane $ABFG$ that is perpendicular to the filaments. (b) Determine the amount of electric flux that passes through the area $ABCD$.

Section 24-4

24-9 Evaluate the double integral in Example 24-4 and obtain $\varphi = Q/6\epsilon_0$. For the integration over x you will need the integral of $dx/(x^2 + b^2)^{3/2}$, where $b^2 = z^2 + a^2$. This integral and the one for the resulting integration over z are contained in the list inside the rear cover.

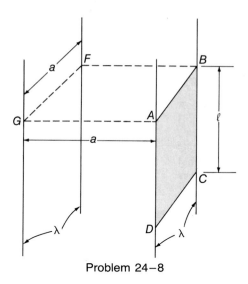

Problem 24–8

24–10 Consider two infinite parallel sheets that are separated by a distance ℓ. Each sheet carries a uniform surface density σ of positive charge. Use (a) Gauss's law and (b) superposition to determine the electric field strength both between and outside the sheets.

24–11 Rework Problem 24–10 if the sheets carry opposite charge densities, $+\sigma$ and $-\sigma$. Use superposition.

Additional Problems

24–15 Air will break down (lose its insulating quality) and sparking will result if the field strength is increased to about 3×10^6 N/C. (This field strength is also expressed as 3×10^6 *volts per meter*.) What acceleration will an electron experience in such a field? If the electron starts from rest, in what distance will it acquire a speed equal to 10 percent of the speed of light?

24–16 The electric field strength of a dipole at a distance of 10 m along the dipole axis is observed to be twice the value anticipated for a pair of charges, $+Q$ and $-Q$, separated by a distance of 2 cm. It is suspected that the reason for this discrepancy is that the positive charge is actually somewhat larger in magnitude than the negative charge. What fractional increase $\delta Q/Q$ of the positive charge would account for the observed field strength?

24–17 Construct a spherical Gaussian surface S with radius R
  • around a point electric dipole. Show by explicit calculation that the total flux $\oint_S \mathbf{E} \cdot \hat{\mathbf{n}} \, dA = 0$, where $\mathbf{E}$ represents the distant field (Eq. 24–6). [*Hint:* Proceed by first showing that the radial component of $\mathbf{E}$ is

Section 24–5

24–12 Show that the maximum field strength E_m along the axis of a uniformly charged ring occurs at $z = a/\sqrt{2}$ (see Fig. 24–14) and has the value $Q/(6\sqrt{3}\pi\epsilon_0 a^2)$.

24–13 Two infinite parallel filaments carry opposite linear
  •• charge densities, $+\lambda$ and $-\lambda$, and are separated by a distance ℓ. Determine the electric field strength at any distant point $P(x, y)$, that is, any point for which $x^2 + y^2 \gg \ell^2$. Express the result vectorially in terms of the dipole moment per unit length, $p_\ell = \ell\lambda\hat{\mathbf{j}}$.

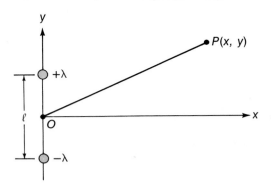

24–14 An infinitely long cylinder has a radius R and a uniform volume charge density ρ. Determine the electric field strength for $r < R$ and for $r > R$.

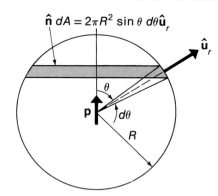

$\hat{\mathbf{n}} \, dA = 2\pi R^2 \sin\theta \, d\theta \hat{\mathbf{u}}_r$

$E_r = \mathbf{E} \cdot \hat{\mathbf{u}}_r = \dfrac{p}{2\pi\epsilon_0} \cdot \dfrac{\cos\theta}{R^3}$. Then select the area $\hat{\mathbf{n}} \, dA$ to be that corresponding to the ring with circumference $2\pi R \sin\theta$ and width $R \, d\theta$, as in the diagram. Finally, evaluate $\oint_S \mathbf{E} \cdot \hat{\mathbf{n}} \, dA = \dfrac{p}{\epsilon_0} \int_0^\pi \cos\theta \sin\theta \, d\theta$.]

24–18 If a molecule AB is completely ionic in character, it consists of ions A^+ and B^- separated by a certain internuclear distance ℓ. For singly charged ions (charges of $+e$ and $-e$) the dipole moment would be ℓe. (It is convenient to measure molecular dipole moments in units of $e\text{Å}$, that is, the product of the electronic charge, 1.602×10^{-19} C, and the angstrom unit, 10^{-10} m.) Intermolecular separations can be determined by a variety of methods. Then, any deviation of the measured dipole moment below the value ℓe indicates the degree to which the molecular bond is less than completely ionic. That is, $100 \ (p_{\text{meas}}/\ell e)$ = percentage ionic character. (Most bonds are partly ionic and partly covalent.) The lithium fluoride (LiF) gas molecule has an internuclear separation of 1.52 Å and a dipole moment of 1.39 $e\text{Å}$. The hydrogen iodide (HI) molecule has an internuclear separation of 1.62 Å but a dipole moment of only 0.080 $e\text{Å}$. What is the degree of ionic bonding in these two molecules?

24–19 The water molecule has the structure shown in the diagram with O—H separations of 0.97 Å. The observed dipole moment of H_2O is 0.387 $e\text{Å}$. What fraction of the O—H bond is ionic in character? (Refer to Problem 24–18.)

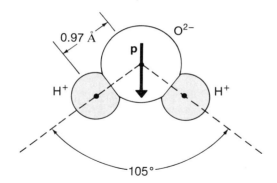

24–20 A wire is bent into the shape of a square with sides $2a$ and carries a uniform linear charge density λ. Determine the electric field strength at a point that is a perpendicular distance a from the center of the square. [*Hint:* Begin by determining the field strength along the perpendicular bisector of a length $2a$ of uniformly charged wire.]

24–21 A negative pointlike charge $-q$ with a mass m is constrained to move along the axis of a ring with a radius R that carries a uniformly distributed positive charge $+Q$. Locate the equilibrium position for the negative charge and show that the frequency for small oscillations about this position is $\omega = \sqrt{qQ/4\pi\epsilon_0 mR^3}$.

CHAPTER 25

ELECTRIC POTENTIAL

In this chapter we study the energy associated with the interactions of various arrangements of charge. We emphasize the case in which the charges are at rest in the coordinate frame used for the description. When a test charge or a particle is allowed unrestrained motion, it is always assumed that the electric field to which the moving charged particle responds is produced by source charges that remain at rest in some coordinate frame. Moreover, this is the frame that we select for describing the motion of the particle. We make this choice instead of selecting a frame that moves with the particle because the latter would involve a moving (time-varying) electric field. The description of such fields involves complications that we will not introduce until Chapter 29. Thus, for the present we restrict attention to the electrostatic case.

25–1 ELECTROSTATIC POTENTIAL ENERGY

We have seen that an electric charge exerts a force on any other nearby charge, and that the magnitude and direction of the force vector can be described in terms of the field concept. We will now see that this electric force, like the gravitational force, is *conservative*. Further, this fact implies that a potential-energy function can be defined for the electric field. We will treat several special cases of increasing complexity.

Single Source Charge. Consider a point charge Q that is at rest at the origin of some coordinate frame. A second point charge q is observed to move along a curved path from point A to point B (Fig. 25–1). The actual path C that is followed by the charged particle is determined by the field in which it moves, that is, by the Coulomb force exerted on it by Q, as well as by other unspecified forces.

We now calculate the work done on q by the Coulomb force field of Q, ignoring all other energy exchanges that might take place. The fundamental definition of work, in

Fig. 25–1. The path C followed by a point charge q in the electrostatic field of the point charge Q is divided into differential radial and circular arc segments.

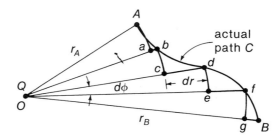

differential form, is

$$dW = \mathbf{F} \cdot d\mathbf{r} \qquad (25\text{-}1)$$

where $\mathbf{F}$ here is the Coulomb force,

$$\mathbf{F} = \frac{qQ}{4\pi\epsilon_0 r^2}\hat{\mathbf{u}}_r \qquad (25\text{-}2)$$

The total work done on q by this force as the particle moves from A to B is*

$$W(A \rightarrow B) \equiv W_{AB} = \frac{qQ}{4\pi\epsilon_0}\oint_A^B \frac{\hat{\mathbf{u}}_r \cdot d\mathbf{r}}{r^2} \qquad (25\text{-}3)$$

Although it might appear that the value of the integral in Eq. 25–3 depends on the path C that connects the points A and B, we now show that the integral is, in fact, completely independent of the path. We follow a procedure employed earlier in Example 7–9 for the work done by the gravitational force. Imagine that the actual path $C = \widehat{AB}$ is divided into a large number of alternating differential radial and circular arc segments. In the limit of an infinite number of segments, the broken path becomes identical to the actual path. For clarity, Fig. 25–1 shows only a few such segments; thus, $\widehat{AB} = \widehat{Aa} + \widehat{ab} + \widehat{bc} + \ldots + \widehat{gB}$. Along the circular arc segments, such as $\widehat{Aa}$ and $\widehat{bc}$, we have $d\mathbf{r} = r\,d\phi\,\hat{\mathbf{u}}_\phi$. Because $\hat{\mathbf{u}}_r \cdot \hat{\mathbf{u}}_\phi = 0$, these segments make no contribution to the line integral; that is, the force is perpendicular to the displacement, so the work done is zero. Along the radial segments, such as $\overline{ab}$ and $\overline{cd}$, we have $d\mathbf{r} = dr\hat{\mathbf{u}}_r$. Because $\hat{\mathbf{u}}_r \cdot \hat{\mathbf{u}}_r = 1$, the expression for W_{AB} reduces to

$$W_{AB} = \frac{qQ}{4\pi\epsilon_0}\int_{r_A}^{r_B} \frac{dr}{r^2} = -\frac{qQ}{4\pi\epsilon_0} \cdot \frac{1}{r}\Big|_{r_A}^{r_B}$$

or

$$W_{AB} = \frac{qQ}{4\pi\epsilon_0}\left(\frac{1}{r_A} - \frac{1}{r_B}\right) \qquad (25\text{-}4)$$

Thus, the work done, W_{AB}, depends only on r_A and r_B, the radial distances from Q to the points A and B, respectively; W_{AB} does not depend on the shape of the path that connects these points.

Suppose that the path C begins at A, follows some curve, and ends again at A; then the points A and B coincide so that the line integral vanishes. That is,†

$$W = \oint \mathbf{F} \cdot d\mathbf{r} = \frac{qQ}{4\pi\epsilon_0}\oint \frac{\hat{\mathbf{u}}_r \cdot d\mathbf{r}}{r^2} = 0 \qquad (25\text{-}5)$$

In Section 8–6 it was pointed out that any force for which the closed-path line integral vanishes, as in Eq. 25–5, is a *conservative force*. For such forces the idea of *potential energy* is useful. Accordingly, we define the electrostatic potential energy $U(\mathbf{r})$

---

*We take this attitude throughout; when work is done *by* the field, $W > 0$, and when work is done *on* the field, $W < 0$. The symbol $\oint$ is used to specify the path followed. (In Sections 21–2 and 22–3 such integrals specified thermodynamic paths.)

†Again, after its introduction for thermodynamic paths, we use the symbol $\oint$ to indicate an integration to be carried out around an entire closed path.

for the q-Q charge system as

$$U(B) - U(A) = -W_{AB}$$

or
$$U(B) - U(A) = \frac{qQ}{4\pi\epsilon_0}\left(\frac{1}{r_B} - \frac{1}{r_A}\right) \tag{25-6}$$

Notice that W_{AB} is equal to the *negative* of the change in potential energy. (Compare the development in Section 8–2.) As in the case of mechanical systems, only *differences* in electric potential energy are physically meaningful because the potential energy of a system at a particular point is defined only to within an arbitrary additive constant. In the SI system, potential energy is measured in joules (J).

We may consider r_A to define the base reference level from which potential energy is measured. It is often convenient to let point A be at an infinite distance, $r_A = \infty$, so that $U(A) = U(\infty) = 0$. Then, with $r_B = r$ representing a general point, we have

$$U(r) = \frac{qQ}{4\pi\epsilon_0}\cdot\frac{1}{r} \tag{25-7}$$

$U(r)$ is the electrostatic potential energy of the charges q and Q when they are separated by a distance r.

Suppose that the point charge q is initially at a point defined by the position vector $\mathbf{r}_1$. If the particle is released from rest, it will move directly toward or away from Q (depending on the relative signs of the two charges) and will be accelerated by the Coulomb force due to Q. Upon arriving at the point $\mathbf{r}_2$, the particle will have acquired a kinetic energy K given by

$$K = \tfrac{1}{2}mv^2 = W_{12} = U(r_1) - U(r_2) = \frac{qQ}{4\pi\epsilon_0}\left(\frac{1}{r_1} - \frac{1}{r_2}\right)$$

That is, some of the original potential energy has been converted into kinetic energy. Notice that $U(r_2) < U(r_1)$ whether the particle moves toward or away from Q. (• Verify this assertion by considering the sign of qQ and the corresponding quantity $(1/r_1) - (1/r_2)$.) For the mechanical counterpart of this situation, refer to Example 7–6.

Several Source Charges. Next, consider the electrostatic potential energy associated with a group of point charges. First, examine the simple case of three charges, Q_1, Q_2, and q, shown in Fig. 25–2. To calculate the potential energy of this configuration, imagine that the charges are initially separated by infinite distances from each other and from the final position of the group. In accordance with the convention that $U(\infty) = 0$, we assign zero potential energy to the initial arrangement of the charges.

The electrostatic force between each pair of charges obeys the superposition principle, so there is a potential energy associated with each pair in the final arrangement. The potential energy of the assembled charges is equal to the negative of the work done by the field in bringing the charges together into the final configuration (Fig. 25–2). First, bring the charge Q_1 to its position defined by the vector $\mathbf{r}_{01}$; no work is done during this process because the only field in the region is the field of Q_1 itself. Next, charge Q_2 is brought to the position $\mathbf{r}_{02}$; the negative work done by the field during this process, namely, $-W = Q_1Q_2/4\pi\epsilon_0R$, is equal to the potential energy contributed to the system. Finally, when q is brought to the position $\mathbf{r}$, there are two additional contributions to the total potential energy, namely, $qQ_1/4\pi\epsilon_0r_1$ and $qQ_2/4\pi\epsilon_0r_2$. Thus, the total potential energy of the

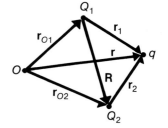

Fig. 25–2. A static configuration of the charges Q_1, Q_2, and q.

system is

$$U(R, r_1, r_2) = \frac{q}{4\pi\epsilon_0}\left(\frac{Q_1}{r_1} + \frac{Q_2}{r_2}\right) + \frac{Q_1 Q_2}{4\pi\epsilon_0 R}$$

Suppose that Q_1 and Q_2 are held in fixed positions and q is allowed to move around in the electrostatic field generated by Q_1 and Q_2. Then, the Q_1-Q_2 contribution to the potential energy remains constant and the reference level can be redefined so that

$$U(\mathbf{r}) = \frac{q}{4\pi\epsilon_0}\left(\frac{Q_1}{r_1} + \frac{Q_2}{r_2}\right) \tag{25-8}$$

This expression is for the case in which only q is free to move and represents the relevant potential energy when q is at the point defined by $\mathbf{r}$ (see Fig. 25–2).

If many point charges, $Q_1, Q_2, Q_3, \ldots$, are present in a fixed array in which only the point charge q is free to move, Eq. 25–8 can be generalized to

$$U(\mathbf{r}) = \frac{q}{4\pi\epsilon_0}\sum_i \frac{Q_i}{r_i} \tag{25-9a}$$

where we again ignore the constant potential energy of the fixed array of charges Q_i. If the array is a continuous distribution with a charge density ρ, the sum in Eq. 25–9a is replaced by an integral:

$$U(\mathbf{r}) = \frac{q}{4\pi\epsilon_0}\int_Q \frac{dQ}{r_c} = \frac{q}{4\pi\epsilon_0}\int_V \frac{\rho\, dv}{r_c} \tag{25-9b}$$

where r_c is the distance to q from the differential charge element $dQ = \rho\, dv$, and where the integration is over the entire distribution. Notice that in Eqs. 25–8 and 25–9 we have written $U(\mathbf{r})$, where $\mathbf{r}$ is the position vector of the charge q. In each case, the sum or integration involves the coordinates $r_i = |\mathbf{r} - \mathbf{r}_{0i}|$ or $r_c = |\mathbf{r} - \mathbf{r}_{0c}|$. Because the coordinates $\mathbf{r}_{0i}$ and $\mathbf{r}_{0c}$ are *fixed*, the result is a function of $\mathbf{r}$ alone. The examples at the end of this section clarify this point.

Because the superposition principle is valid, the electric force on a charge that is free to move in the electrostatic field of an array of fixed charges is necessarily a conservative force.

The Electrostatic Potential.   In Section 24–1 the electric field strength $\mathbf{E}$ was defined to be the limit of $\mathbf{F}_E/q_0$ as $q_0 \to 0$. We now introduce a related quantity, the *electrostatic potential* Φ, using a similar definition, namely,

$$\Phi(\mathbf{r}) \equiv \lim_{q \to 0} \frac{U(\mathbf{r})}{q} \tag{25-10}$$

Combining this definition with Eqs. 25–9 gives

$$\Phi(\mathbf{r}) = \frac{1}{4\pi\epsilon_0}\sum_i \frac{Q_i}{r_i} \tag{25-11a}$$

and

$$\Phi(\mathbf{r}) = \frac{1}{4\pi\epsilon_0} \int_Q \frac{dQ}{r_c} = \frac{1}{4\pi\epsilon_0} \int_V \frac{\rho \, dv}{r_c}$$ (25–11b)

We frequently refer to the *potential difference* between two points, A and B, in which case we write $\Phi(B) - \Phi(A)$ or $\Phi_B - \Phi_A$. As in the case of the electric potential energy, we can choose the zero reference level for $\Phi(\mathbf{r})$ to be at infinity so that $\Phi(\infty) = 0$.

Potential and potential difference are measured in *joules per coulomb* (J/C). To this unit we give the special name *volt*[*] (V); thus, $1 \text{ V} = 1 \text{ J/C}$. Because $1 \text{ J} = 1 \text{ N} \cdot \text{m}$, it follows that electric field strength, previously specified in N/C, can be expressed in the equivalent units of *volts per meter* (V/m).

When we wish to emphasize that the potential difference between two points is just the *voltage* between those points, we often use an equivalent notation, writing $V_{BA} \equiv \Phi_{BA} \equiv \Phi_B - \Phi_A$. If no confusion can arise, we simplify still further and write V for V_{BA}.

Example 25–1

Determine the electric potential due to two point charges, Q_1 and Q_2, along a perpendicular bisector of the line joining the charges.

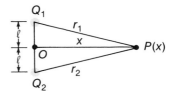

Solution:

The distances from Q_1 and Q_2 to $P(x)$ are $r_1 = r_2 = (x^2 + \ell^2)^{1/2}$, so that using Eq. 25–11a we have

$$\Phi(x) = \frac{1}{4\pi\epsilon_0} \left(\frac{Q_1}{(x^2 + \ell^2)^{1/2}} + \frac{Q_2}{(x^2 + \ell^2)^{1/2}} \right)$$

$$= \frac{Q_1 + Q_2}{4\pi\epsilon_0} \cdot \frac{1}{(x^2 + \ell^2)^{1/2}}$$

Because ℓ is constant, the potential is a function only of x. When $x^2 \gg \ell^2$, the expression for the potential reduces to

$$\Phi(x) \cong \frac{Q}{4\pi\epsilon_0} \cdot \frac{1}{x}, \qquad x^2 \gg \ell^2$$

where $Q = Q_1 + Q_2$. Thus, $\Phi(x^2 \gg \ell^2)$ is the same as the potential of a single charge Q located at the origin O.

Example 25–2

Derive an expression for the electric potential at large distances from an electric dipole.

Solution:

We first write an *exact* expression for the potential at a point P, as shown in the diagram on page 484. Thus,

$$\Phi(\mathbf{r}) = \frac{Q}{4\pi\epsilon_0} \left(\frac{1}{r_+} - \frac{1}{r_-} \right) = \frac{Q}{4\pi\epsilon_0} \left(\frac{r_- - r_+}{r_+ r_-} \right)$$

Because the potential is required only at large distances (that is, $r \gg \ell$), we can use the approximate relationships (refer to the diagram),

$$r_- - r_+ \cong \ell \cos \theta \qquad \text{and} \qquad r_+ r_- \cong r^2$$

Then the potential becomes

$$\Phi(\mathbf{r}) = \frac{Q}{4\pi\epsilon_0} \frac{\ell \cos \theta}{r^2}$$

Recall (Section 24–2) that the electric dipole moment is

---

[*]Named in honor of Count Alessandro Volta (1745–1827), Italian physicist, who constructed the first chemical battery or *voltaic cell*.

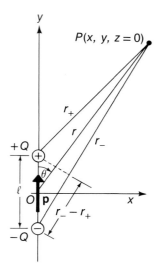

defined as $\mathbf{p} = Q\ell\hat{\mathbf{j}}$, so that $Q\ell\cos\theta = \mathbf{p}\cdot\mathbf{r}/r$. Thus, we have

$$\Phi(\mathbf{r}) = \frac{\mathbf{p}\cdot\mathbf{r}}{4\pi\epsilon_0 r^3}$$

The geometry of the dipole has cylindric symmetry about the direction of $\mathbf{p}$; hence, the vector $\mathbf{r}$ in the expression for $\Phi(\mathbf{r})$ may be considered to be a general three-dimensional position vector.

Example 25–3

Calculate the electric potential along the axis of a uniformly charged ring with radius a.

Solution:

The distance from the ring segment to the point P is $r_c = (z^2 + a^2)^{1/2}$. The charge on this segment is $dQ = \lambda a\,d\phi$, where $\lambda = Q/2\pi a$ is the linear charge density on the ring. Using Eq. 25–11b, we have

$$\Phi(z) = \frac{1}{4\pi\epsilon_0}\int_Q \frac{dQ}{r_c} = \frac{a\lambda}{4\pi\epsilon_0(z^2+a^2)^{1/2}}\int_0^{2\pi} d\phi$$

$$= \frac{2\pi a\lambda}{4\pi\epsilon_0(z^2+a^2)^{1/2}} = \frac{Q}{4\pi\epsilon_0}\cdot\frac{1}{(z^2+a^2)^{1/2}}$$

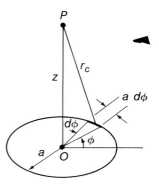

Again, notice that when $z^2 \gg a^2$, the potential $\Phi(z)$ is the same as that for a point charge Q located at the origin O.

Example 25–4

Determine the electric potential along the axis of a uniformly charged disc with radius R.

Solution:

The surface charge density of the disc is $\sigma = Q/\pi R^2$, so the charge on the small annular ring shown in the diagram is $dQ = 2\pi r\sigma\,dr$. Using the result of Example 25–3 together with $r_c = (z^2 + r^2)^{1/2}$, we have

$$\Phi(z) = \frac{\sigma}{2\epsilon_0}\int_0^R \frac{r\,dr}{(z^2+r^2)^{1/2}}$$

$$\Phi(z) = \frac{\sigma}{2\epsilon_0}(z^2+r^2)^{1/2}\Big|_0^R = \frac{\sigma}{2\epsilon_0}[(z^2+R^2)^{1/2} - z]$$

$$= \frac{Q}{2\pi\epsilon_0 R^2}[(z^2+R^2)^{1/2} - z]$$

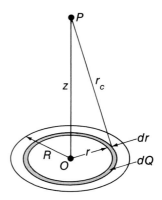

To investigate the behavior of the potential when $z^2 \gg R^2$, we expand the radical in the expression,

$$\Phi(z) = \frac{Q}{2\pi\epsilon_0 R^2}[z(1 + R^2/z^2)^{1/2} - z]$$

Thus, we obtain

$$\Phi(z) = \frac{Q}{2\pi\epsilon_0 R^2}[z(1 + \tfrac{1}{2}\frac{R^2}{z^2} + \ldots) - z]$$

from which $\Phi(z) \cong \dfrac{Q}{4\pi\epsilon_0 z}, \quad z^2 \gg R^2$

which is the expected result.

25–2 RELATIONSHIPS BETWEEN POTENTIAL AND FIELD STRENGTH

In the preceding section we obtained expressions for the direct determination of the electric potential for any static distribution of charge. The existence of an electric potential was assured in each case by the conservative nature of the force field associated with the static charges.

Suppose now that the electric field strength $\mathbf{E}(\mathbf{r})$ is given. If it is *also known* that $\mathbf{E}$ is produced solely by a static charge distribution, thereby guaranteeing that the field is conservative, we may write*

$$U(B) - U(A) = -W_{AB} = -\int_A^B \mathbf{F} \cdot d\mathbf{r}$$

or

$$U(B) = -q\int_A^B \mathbf{E} \cdot d\mathbf{r} + U(A) \qquad \textbf{(25–12)}$$

where we have used $\mathbf{F} = q\mathbf{E}$. Now, the electric potential is defined by $\Phi(\mathbf{r}) = \underset{q\to 0}{\text{Lim}}\ U(\mathbf{r})/q$ (Eq. 25–10). Then, Eq. 25–12 becomes

$$\Phi(B) = -\int_A^B \mathbf{E} \cdot d\mathbf{r} + \Phi(A) \qquad \textbf{(25–13)}$$

Finally, let point B be an arbitrary point in the field defined by the position vector $\mathbf{r}$, whereas the point A serves to establish the reference level for the potential. Then we can write

$$\Phi(\mathbf{r}) = -\int_A^\mathbf{r} \mathbf{E} \cdot d\mathbf{r} + \Phi(A) \qquad \textbf{(25–13a)}$$

Notice that the path C connecting point A and $\mathbf{r}$ can be *any* path, even a path selected solely for convenience. This is the case because the field $\mathbf{E}$ is produced by static charges only and is therefore a conservative field. The principal advantage of describing the situation in terms of the *potential* instead of the *potential energy* is that we are then able to deal with a quantity—the potential—that depends everywhere in space only on the source charges and not on the test charge q.

One common case is exceptionally simple—that of the *uniform* electric field with constant magnitude and direction at all points.† For such a field in a coordinate frame oriented with its y-axis parallel to the field lines, we have $\mathbf{E} = -E_0\hat{\mathbf{j}}$ with E_0 a constant.

---

*We no longer employ the special symbol $\oint$ for the integral since its value is independent of the path followed between points A and B.

†The external charge distribution that produces such a field is shown in Fig. 25–11.

Then, for two points A and B with $\mathbf{r}_A(x_A, y_A, z_A)$ and $\mathbf{r}_B(x_B, y_B, z_B)$, we have

$$U_B = -q \int_A^B (-E_0\hat{\mathbf{j}}) \cdot d\mathbf{r} + U(A) = + qE_0(y_B - y_A) + U(A)$$

and

$$\Phi_B = -\int_A^B (-E_0\hat{\mathbf{j}}) \cdot d\mathbf{r} + \Phi(A) = +E_0(y_B - y_A) + \Phi(A)$$

where $y_B - y_A$ is simply the y-elevation difference between the points A and B. (• Compare this expression for the potential energy to that for the potential energy of an object in a gravitational field. Which quantities in the two situations are analogous?)

If the line integral of $\mathbf{E} \cdot d\mathbf{r}$ is calculated along a closed path (so that points A and B coincide), the integral vanishes; thus,

$$\oint \mathbf{E} \cdot d\mathbf{r} = 0 \qquad (25\text{–}14)$$

This is a general result for a static distribution of charge (a conservative field).

Example 25–5

Determine the electric potential at all points in space for a uniform spherical charge distribution that has a radius R.

Solution:

Let us calculate $\Phi(\mathbf{r})$ from $\mathbf{E}(\mathbf{r})$, which is readily obtained by using Gauss's law. From the results of Example 24–5 (page 474), we have

$$\mathbf{E}(\mathbf{r}) = \begin{cases} \dfrac{Q}{4\pi\epsilon_0 r^2}\hat{\mathbf{u}}_r, & r > R \\[2ex] \dfrac{Qr}{4\pi\epsilon_0 R^3}\hat{\mathbf{u}}_r, & r < R \end{cases}$$

We use Eq. 25–13a,

$$\Phi(\mathbf{r}) = -\int_A^{\mathbf{r}} \mathbf{E} \cdot d\mathbf{r} + \Phi(A)$$

and select a reference level for Φ such that $\Phi(A) = 0$ at $r_A = \infty$. Because the electric field is due to static charges and is therefore conservative, we may integrate $\mathbf{E} \cdot d\mathbf{r}$ along any convenient path; we choose a radial path with $\hat{\mathbf{u}}_r \cdot d\mathbf{r} = dr$. Then, for $r \geq R$, we have

$$\Phi(r) = -\frac{Q}{4\pi\epsilon_0} \int_\infty^r \frac{dr}{r^2} = \frac{Q}{4\pi\epsilon_0} \cdot \frac{1}{r}\bigg|_\infty^r$$

$$= \frac{Q}{4\pi\epsilon_0} \cdot \frac{1}{r}, \qquad r \geq R \qquad (1)$$

For $r < R$, we have

$$\Phi(r) = -\frac{Q}{4\pi\epsilon_0 R^3} \int_R^r r\, dr + \Phi(R)$$

where $\Phi(R)$ is obtained from Eq. (1). Thus,

$$\Phi(r) = -\frac{Q}{4\pi\epsilon_0 R^3} \cdot \frac{r^2}{2}\bigg|_R^r + \frac{Q}{4\pi\epsilon_0 R}$$

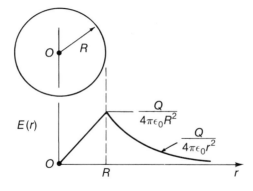

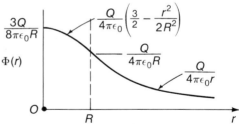

$$\Phi = \frac{Q}{4\pi\epsilon_0 R}\left(\frac{3}{2} - \frac{r^2}{2R}\right), \quad r < R \qquad (2)$$

These results for $E(r)$ and $\Phi(r)$ are summarized in the accompanying diagram.

The Gradient. Suppose that $\mathbf{E(r)}$ is to be determined for a given potential $\Phi(\mathbf{r})$. Consider first the effect of a small displacement in the x-direction from a point $P(x, y, z)$ to a point $P(x + \Delta x, y, z)$, as shown in Fig. 25–3. From Eq. 25–13 we have

$$\Phi(x + \Delta x, y, z) - \Phi(x, y, z) = -\int_x^{x+\Delta x} E_x \, dx = -\bar{E}_x \, \Delta x$$

$$\text{where } \bar{E}_x = -\frac{\Phi(x + \Delta x, y, z) - \Phi(x, y, z)}{\Delta x}$$

In the limit, $\Delta x \to 0$, $\bar{E}_x$ becomes $E_x(x, y, z)$ and the quantity on the right-hand side becomes the partial derivative* of Φ with respect to x; that is,

$$E_x(x, y, z) = -\frac{\partial\Phi(x, y, z)}{\partial x} \qquad (25\text{–}15\text{a})$$

In a similar way we can show that

$$E_y(x, y, z) = -\frac{\partial\Phi(x, y, z)}{\partial y} \qquad (25\text{–}15\text{b})$$

$$E_z(x, y, z) = -\frac{\partial\Phi(x, y, z)}{\partial z} \qquad (25\text{–}15\text{c})$$

The result we are seeking is therefore

$$\mathbf{E}(x, y, z) = -\left[\frac{\partial\Phi}{\partial x}\hat{\mathbf{i}} + \frac{\partial\Phi}{\partial y}\hat{\mathbf{j}} + \frac{\partial\Phi}{\partial z}\hat{\mathbf{k}}\right] \qquad (25\text{–}16)$$

Fig. 25–3. A change in the position of the point P results in an incremental change in the potential.

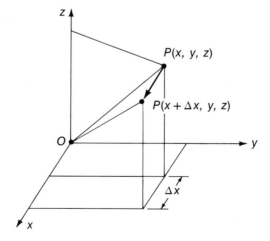

---

*The partial derivative $\partial\Phi/\partial x$ is equal to the derivative of Φ with respect to x when the independent variables y and z are held constant.

The (vector) quantity within the brackets on the right-hand side of Eq. 25–16 is called the *gradient* of Φ. Hence, we say that the field strength **E** is the negative of the gradient of the potential Φ. Thus, to determine $\mathbf{E} = \mathbf{E}(x, y, z)$ at the point $P(x, y, z)$ we must know $\Phi(x, y, z)$ everywhere in the neighborhood of P so that the required differentiation can be carried out.

▶ It is convenient to introduce a notation called the *del* or *gradient operator*, which is defined as

$$\nabla \equiv \hat{\mathbf{i}}\frac{\partial}{\partial x} + \hat{\mathbf{j}}\frac{\partial}{\partial y} + \hat{\mathbf{k}}\frac{\partial}{\partial z} \qquad (25\text{–}17)$$

The application of this operator to a scalar quantity $f(x, y, z)$ that is differentiable in x, y, z means

$$\nabla f(x, y, z) = \frac{\partial f}{\partial x}\hat{\mathbf{i}} + \frac{\partial f}{\partial y}\hat{\mathbf{j}} + \frac{\partial f}{\partial z}\hat{\mathbf{k}} \qquad (25\text{–}18)$$

Notice that when ∇ operates on a *scalar* quantity, the result is a *vector* quantity.

Equation 25–16 can now be expressed in compact form as

$$\mathbf{E} = -\nabla\Phi \qquad (25\text{–}19) \ \blacktriangleleft$$

Equipotentials and the Gradient.   Consider a simple two-dimensional case in which the electric potential is $\Phi(x, y)$. The curves described by $\Phi(x, y) = $ constant are called *equipotential curves* (Fig. 25–4a). In three dimensions, $\Phi(x, y, z) = $ constant describes *equipotential surfaces*.

The gradient of a potential function at a point is always normal to the equipotential curve or surface that passes through that point (see Fig. 25–4b). The direction of the normal, and hence the direction of the gradient of Φ, is the same as the direction of the maximum rate of change of Φ with displacement. Moreover, because **E** is the negative gradient of Φ, the direction of **E** is always perpendicular to the equipotential at the same point.

Figure 25–5 is a field diagram for a *point dipole* with strength $|\mathbf{p}| = 10^{-10}$ C·m. (A *point dipole* is produced by allowing $\ell \to 0$ in such a way that ℓq, and hence p, remains constant; the distant-field condition, $r \gg \ell$, is therefore met at all points in space.) Several of the field lines are shown (compare Fig. 24–6); notice the line with the indicated strength of 6 V/m. Also shown are the equipotential surfaces (corresponding to potentials of 3 V, 9 V, and 15 V). These equipotentials (and all others) are everywhere perpendicular to the field lines.

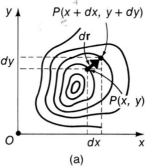

(a)

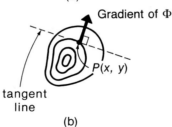

Gradient of Φ

tangent line

(b)

Fig. 25–4.   (a) The displacement d**r** connecting points on two different equipotentials. (b) The gradient of Φ at the point $P(x, y)$ is perpendicular to the equipotential curve that passes through $P(x, y)$.

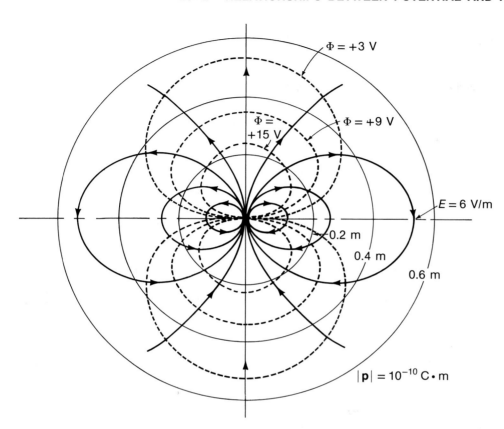

Fig. 25–5. Field lines and equipotentials for a point dipole with a strength of 10^{-10} C·m. The dipole axis is directed vertically upward in the diagram. The full three-dimensional field is obtained by rotating the pattern of field lines about the dipole axis. Notice that the field lines and equipotentials are everywhere perpendicular.

$\Phi = +3$ V

$\Phi = +15$ V

$\Phi = +9$ V

$E = 6$ V/m

0.2 m

0.4 m

0.6 m

$|\mathbf{p}| = 10^{-10}$ C·m

Example 25–6

The electric potential at large distances from a dipole $\mathbf{p}$ is (see Example 25–2)

$$\Phi(\mathbf{r}) = \frac{\mathbf{p} \cdot \mathbf{r}}{4\pi\epsilon_0 r^3}$$

Show that the gradient of this potential leads to Eq. 24–6, namely,

$$\mathbf{E}(\mathbf{r}) = \frac{1}{4\pi\epsilon_0} \left(-\frac{\mathbf{p}}{r^3} + \frac{3(\mathbf{p} \cdot \mathbf{r})\mathbf{r}}{r^5} \right)$$

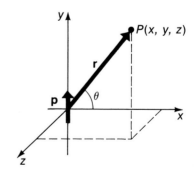

Solution:

With $\mathbf{p} = \ell Q\hat{\mathbf{j}}$, and using the geometry shown in the diagram, we have

$$\Phi(x, y, z) = \frac{\ell Q}{4\pi\epsilon_0 r^3}\hat{\mathbf{j}} \cdot \mathbf{r} = \frac{\ell Q y}{4\pi\epsilon_0 (x^2 + y^2 + z^2)^{3/2}}$$

The required partial derivatives are

$$-\frac{\partial \Phi}{\partial x} = \frac{\ell Q}{4\pi\epsilon_0} \cdot \frac{3xy}{(x^2 + y^2 + z^2)^{5/2}} = \frac{\ell Q}{4\pi\epsilon_0} \cdot \frac{3xy}{r^5}$$

$$-\frac{\partial \Phi}{\partial z} = \frac{\ell Q}{4\pi\epsilon_0} \cdot \frac{3zy}{r^5}$$

and

$$-\frac{\partial \Phi}{\partial y} = \frac{\ell Q}{4\pi\epsilon_0} \cdot \frac{3y^2(x^2 + y^2 + z^2)^{1/2} - (x^2 + y^2 + z^2)^{3/2}}{(x^2 + y^2 + z^2)^3}$$

$$= \frac{\ell Q}{4\pi\epsilon_0} \left(\frac{3y^2}{r^5} - \frac{1}{r^3} \right)$$

Therefore,

$$E = -\left[\frac{\partial \Phi}{\partial x}\hat{i} + \frac{\partial \Phi}{\partial y}\hat{j} + \frac{\partial \Phi}{\partial z}\hat{k}\right]$$

$$= \frac{\ell Q}{4\pi\epsilon_0}\left[-\frac{\hat{j}}{r^3} + \frac{3y}{r^5}(x\hat{i} + y\hat{j} + z\hat{k})\right]$$

$$E = \frac{1}{4\pi\epsilon_0}\left[-\frac{\mathbf{p}}{r^3} + \frac{3(\mathbf{p \cdot r})\mathbf{r}}{r^5}\right]$$

which is Eq. 24–6.

Example 25–7

Use the gradient of the potential to determine the electric field strength on the axis of a uniformly charged ring.

Solution:

The expression for the potential along the ring axis that was obtained in Example 25–3 does not apply for a general point $P(x, y, z)$. Nevertheless, this expression can be used to find $\mathbf{E}$ on the axis. The reason is that the axial symmetry ensures that $\mathbf{E}$ has vanishing components in directions perpendicular to the axis. If the ring axis corresponds to the z-axis, only E_z is nonvanishing; thus, we require only the z dependence of Φ to obtain the necessary partial derivative, $\partial\Phi/\partial z$. From the result of Example

25–3, we have

$$\Phi(z) = \frac{Q}{4\pi\epsilon_0(z^2 + a^2)^{1/2}}$$

Then,

$$-\frac{\partial \Phi}{\partial z} = -\frac{Q}{4\pi\epsilon_0}\left[\frac{1}{2}\frac{(-2z)}{(z^2 + a^2)^{3/2}}\right]$$

from which

$$\mathbf{E} = -\frac{\partial \Phi}{\partial z}\hat{k} = \frac{Qz}{4\pi\epsilon_0(z^2 + a^2)^{3/2}}\hat{k}$$

This is the same expression we obtained in Eq. 24–14.

25–3 CONDUCTORS

It has been pointed out that conducting materials (as distinct from nonconducting materials or insulators) contain large numbers of electrons—usually one or two per atom—that are free to move about within the material.* These free (conduction) electrons move in response to applied electric forces, thereby accounting for the conductive property of such materials. To charge a conductor negatively, one adds electrons; to charge a conductor positively, one removes some of the conduction electrons.

We now adopt the concept of an ideal homogeneous, isotropic, perfectly conducting substance, which we will simply call a *conductor*. We restrict attention to isolated conductors, either charged or neutral, possibly in the presence of externally generated electric fields. We further assume that only electric forces are important. In every case, we describe the situation only after the charges—placed on the conductors by some external agency—have achieved an equilibrium distribution. The electric field is then everywhere the result of the superposition of externally generated fields, if any, and the field produced by the conductor charge distributions themselves.

In no case involving an equilibrium distribution of charge can there be a net electric field within the volume of a conductor.† If such a net interior field existed it would exert a force on the conduction electrons that would produce a current of moving charges in violation of the assumption of an equilibrium (static) charge distribution. Any static

---

*Not all materials can be classified as either conductors or insulators; an important exception is the group of *semiconducting* materials. Electrical properties of conductors, semiconductors, and insulators are contrasted in Section 27–2.

†There *are* nonzero electric fields on an atomic scale of distance in all forms of matter. However, we refer here only to the macroscopic field obtained by averaging over a region of the conductor large enough that the effects of atomic structure have been removed.

charge that is carried by a conductor must therefore reside entirely on the surface of the conductor. Moreover, the distribution of this charge must be such that the field it produces, when superposed with any external field, results in zero field within the conductor. When no external field is present, the surface charges are distributed in such a way that there is no internal electric field.

Figure 25–6 shows the result of placing an uncharged solid conducting sphere in an originally uniform electric field E_0 (Fig. 25–6a). In the presence of the field, equal amounts of negative and positive charge are drawn toward the polar regions of the sphere ($\theta \cong 0$ and $\theta \cong \pi$). The equilibrium charge distribution that is induced on the surface of the sphere produces a uniform electric field E_C within the sphere (Fig. 25–6b). The field E_C exactly cancels the external field E_0 everywhere inside the sphere; thus, **E** (inside) = $E_0 + E_C = 0$ (Fig. 25–6c).

Obtaining the solution (that is, the expression for E_C) for even a simple problem such as that of a conducting sphere in a uniform field is complicated because the induced charge distribution and the final field configuration are both unknown. The method of solution must deal simultaneously in a self-consistent way with both of these related unknown quantities. Advanced techniques beyond the scope of this text are required to develop the solution for the general case of a conductor in an external field.

The basic characteristic we have discovered concerning the electrostatic case is that the electric field strength must be exactly zero everywhere within an isolated conductor. A number of important implications follow from this fact. For example, suppose that we construct a closed Gaussian surface that lies entirely within a conductor, as in Fig. 25–7. Then, because **E** = 0 everywhere within the surface, we have

$$\oint_S \mathbf{E} \cdot \hat{n} \, dA = 0 = \frac{Q}{\epsilon_0}$$

and the volume defined by the surface can contain no net charge; indeed, there can be no net charge anywhere inside the conductor.*

Next, consider the calculation of the potential, which is the integral of $\mathbf{E} \cdot d\mathbf{r}$, around a closed path such as the path *abcda* shown in Fig. 25–8. The segment $\widehat{ab}$ of the path lies just outside and parallel to the surface of a conductor. The segment $\widehat{cd}$ lies just inside and parallel to the surface. The two end segments, with lengths Δx and $\Delta x'$, complete the closed path. The electric field strength just outside the surface can be resolved into a tangential component E_t and a normal component E_n, as indicated in Fig. 25–8. Orient the integration path so that $\widehat{ab}$ is parallel to E_t. The closed-path line integral is (see Eq. 25–14)

$$\oint_{abcda} \mathbf{E} \cdot d\mathbf{r} = \widehat{ab} E_t - E_n \Delta x' + \widehat{cd} \cdot 0 + E_n \Delta x = 0$$

where we have explicitly indicated that the field strength inside the conductor (segment $\widehat{cd}$) is zero. In the limit $\Delta x \to 0$ and $\Delta x' \to 0$, and with $\widehat{ab}$ small but finite, we have

$$E_t = 0 \quad \text{on the surface of a conductor} \qquad \textbf{(25–20)}$$

That is, the resultant field E at the surface must be everywhere perpendicular to the surface (as noted in Fig. 25–6).

Finally, let us find an expression for E_n, the normal component of E. Consider a

---

*We exclude the possibility that a charge exists in a hollow cavity in the conductor.

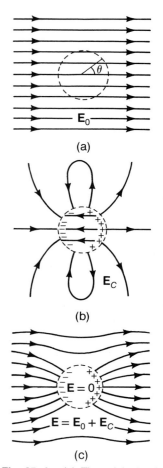

(a)

(b)

(c)

Fig. 25–6. (a) The original uniform field E_0. (b) The (dipole) field E_C produced by the charge distribution induced on the surface of the sphere. (c) The resultant field, $E = E_0 + E_C$, with E(inside) = 0. Notice that on the surface of the sphere neither the field E_0 (a) nor the field E_C (b) is normal to the surface; only the *true resultant field* $E = E_0 + E_C$ (c) must satisfy this requirement.

Fig. 25–7. The electric field is zero inside the surface S, so no net charge is contained within the surface.

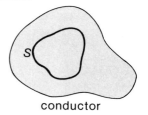

conductor

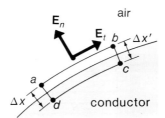

Fig. 25–8. Integration path to show that $E_t = 0$ at the surface of a conductor.

small cylindric Gaussian surface S that encloses a segment of a conducting surface, as shown in Fig. 25–9. The top cap of the cylinder, with area ΔA, lies just outside and parallel to the surface of the conductor. The bottom cap lies just below the surface, and the cylinder side wall of height Δx is perpendicular to the surface. Let the density of charge on the conductor surface be σ. Then, Gauss's law can be expressed as

$$\oint_S \mathbf{E}\cdot\hat{\mathbf{n}}\, dA = \frac{Q}{\epsilon_0}$$

$$E_n\, \Delta A = \frac{\sigma\, \Delta A}{\epsilon_0}$$

where the only contribution to the integral arises from the top cap because the field is zero within the conductor and there is no tangential component to pierce the cylinder wall. Thus, we conclude that

$$E_n = \frac{\sigma}{\epsilon_0} \qquad \text{on the surface of a conductor} \qquad \text{(25–21)}$$

The fact that the electric field vector on the surface of a conductor is everywhere normal to the surface means that *conductor surfaces are equipotentials*.

The basic properties of ideal isolated conductors under *electrosatic* conditions can be summarized by the following statements:

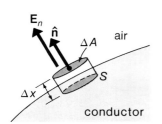

Fig. 25–9. The Gaussian surface used to evaluate E_n.

1. *The field strength within a conductor is everywhere zero.*
2. *The surfaces of conductors are equipotentials.*
3. *On the surface of a conductor the field strength vector is everywhere normal to the surface:* $\mathbf{E} = \mathbf{E}_n = \dfrac{\sigma}{\epsilon_0}\hat{\mathbf{n}}$.

Charged Conducting Spherical Shell. Consider a thick conducting spherical shell with inner radius R_1 and outer radius R_2 that carries a net charge Q (see Fig. 25–10). The interior of the shell is empty and there are no additional fields due to external sources.

First, construct a spherical Gaussian surface with radius $r' < R_1$. Because there is no charge in the hollow region of the shell, it follows that

$$\mathbf{E} = 0, \qquad r < R_1$$

From this fact and the result stated in Eq. 25–21, we conclude that there can be no charge on the inner surface of the shell, $r = R_1$. The region of the shell itself is the interior of a conductor, so the field there is also zero:

$$\mathbf{E} = 0, \qquad R_1 \leqslant r < R_2$$

Fig. 25–10. A conducting spherical shell that carries a net charge Q.

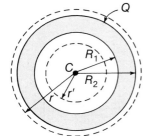

Finally, construct a spherical Gaussian surface with radius $r > R_2$. The net charge within this surface is equal to the charge Q on the shell. Thus,

$$4\pi r^2 E = \frac{Q}{\epsilon_0}$$

or

$$\mathbf{E} = \frac{Q}{4\pi\epsilon_0}\cdot\frac{\hat{\mathbf{u}}_r}{r^2}, \qquad r \geqslant R_2$$

When $r = R_2$, we have $E = E_n = \sigma/\epsilon_0$, where the surface charge density is $\sigma = Q/4\pi R_2^2$.

It is often convenient to choose arbitrarily the zero reference level for electric potential and electric potential energy to be at infinity. When dealing with conductors, however, the Earth itself is sometimes chosen as the zero reference for potential. Although the Earth is not an ideal conductor, it represents a readily available and essentially inexhaustible reservoir for supplying (or accepting) electrons. An object can be connected by a wire conductor to a metal stake or pipe that is driven into the Earth. Such an object is said to be *grounded*. The object, the wire, the stake, and the Earth are then all at the same potential, which we choose to be $\Phi = 0$. A grounded object does not necessarily have zero surface charge; induction effects may cause electrons to be drawn from or discharged to the Earth reservoir.

Parallel Charged Conducting Planes. Next, consider the important case of two infinite parallel conducting planes that contain opposite charges and are separated by a distance ℓ (Fig. 25-11). When ℓ is sufficiently small, the results also apply for finite planes; this is the case for a *parallel-plate capacitor* (see Section 26-1).

The surface charge densities on the planes are uniform with values $+\sigma_0$ and $-\sigma_0$. The lower plane is grounded, so $\Phi(0) = 0$. From the geometry it is clear that **E** must everywhere be normal to the planes. Construct a Gaussian surface as shown in Fig. 25-11 with cap areas ΔA and cylinder height Δy. The upper cap is within the upper conductor, where $\mathbf{E} = 0$. There is no flux through the cylinder wall, and the entire flux is through the lower cap. Thus, $E\,\Delta A = (\sigma_0/\epsilon_0)\,\Delta A$, so that

$$\mathbf{E} = -\frac{\sigma_0}{\epsilon_0}\hat{\mathbf{j}} \qquad (25-22)$$

The field strength is independent of y. Applying Eq. 25-13a, we have

$$\Phi(y) = -\int_0^y E\,dy + \Phi(0)$$

or

$$\Phi(y) = \frac{\sigma_0}{\epsilon_0}y \qquad (25-23)$$

Thus, the equipotential surfaces are seen to be parallel to the conducting planes, with $\Phi(0) = 0$ at the grounded plane and $\Phi(\ell) = (\sigma_0/\epsilon_0)\ell$ at the upper plane. (This result is expected because the equipotential surfaces must everywhere be normal to the lines of

Fig. 25-11. Equipotentials and lines of force for a pair of oppositely charged infinite parallel conducting planes.

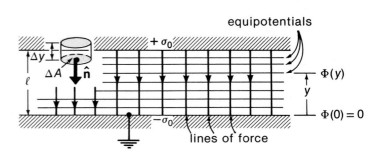

force.) Let V represent the *potential difference* between the planes (measured in *volts*):

$$V = \Phi(\ell) - \Phi(0) = \frac{\sigma_0}{\epsilon_0}\ell$$

or

$$V = E\ell \tag{25-24}$$

This simple result is useful in the discussion of capacitors (Section 26–1).

25–4   THE MOTION OF CHARGES IN ELECTROSTATIC FIELDS

Consider now the general motion in an electrostatic field of an unconstrained particle with charge q and mass m. The field is produced by an array of fixed source charges, and the motion is described using a coordinate frame at rest with respect to these charges. If conductors or insulators are present, we assume that induced charge effects produced by the moving particle under study are negligible (see Problem 25–16).

If the electrostatic field strength is $\mathbf{E}(x, y, z)$, the force on the charged particle is $\mathbf{F} = q\mathbf{E}$; hence, Newton's equation, $m\mathbf{a} = q\mathbf{E}$, yields

$$\mathbf{a}(x, y, z) = \frac{q}{m}\mathbf{E}(x, y, z) \tag{25-25}$$

The problem of finding the equation of motion for a particle in a given electrostatic field reduces to the mechanics problem in which $\mathbf{a}(\mathbf{r})$ and $\mathbf{v}_0$ are given and it is required to determine $\mathbf{r}(t)$. Notice that only the charge-to-mass ratio q/m of the particle enters in Eq. 25–25. Consequently, no study of the motion of the particle in an electrostatic field will reveal the value of either q or m separately.

The Electron Volt.   Suppose that an electron is released from rest in a uniform electrostatic field (see Fig. 25–11). Let the field be in the direction of the negative x-axis, so that $\mathbf{E} = -E_0\hat{\mathbf{i}}$. The electron ($q = -e$) will be accelerated in the positive x-direction, with $a_x = (e/m)E_0$. The velocity of the electron after an elapsed time t will be $\mathbf{v}(t) = (e/m)E_0t\hat{\mathbf{i}}$ and the position will be $x(t) = \frac{1}{2}(e/m)E_0t^2 + x_a$, where x_a is the position at $t = 0$. In moving from the initial position x_a to a point x_b, the electron will have acquired a kinetic energy equal to the work done by the field, namely,

$$\tfrac{1}{2}mv^2 = eE_0(x_b - x_a)$$

But $E_0(x_b - x_a)$ is just the potential difference V_{ba} between the two points; hence,

$$\tfrac{1}{2}mv^2 = e \times V_{ba}$$

If an electron (or any other particle with a charge $\pm e$) is accelerated through a potential difference of 1 V, it will acquire an increase in kinetic energy equal to

$$e \times V = (1.602 \times 10^{-19}\ \text{C}) \times (1\ \text{V})$$

$$= 1.602 \times 10^{-19}\ \text{J}$$

This unit of work or energy is given the special name *electron volt*, abbreviated *eV*:

$$1\ \text{eV} = 1.602 \times 10^{-19}\ \text{J}$$

Energies associated with atomic phenomena are typically a few eV. For example, an energy of 5.1 eV is required to remove the least bound electron from an atom of sodium. X rays can have energies of a few thousand eV (keV). Nuclear reactions and transitions frequently involve energies of a few million eV (MeV), and elementary particle phenomena often involve energies of billions of eV (GeV). The electron volt is a general energy unit and can be used to specify the energy of any object, whether charged or not.

Electrostatic Deflection. The operation of devices such as television picture tubes depends on the deflection of an electron beam in an electric field. Consider a pencil-like electron beam that enters a uniform electrostatic field in a direction perpendicular to the field vector **E** (Fig. 25–12). The field is confined to the region $-\frac{1}{2}\ell \leq y \leq \frac{1}{2}\ell$ and $0 \leq x \leq d$; the extent in the z-direction is arbitrary. The field is produced by establishing two equipotentials, $\Phi = 0$ along $y = -\frac{1}{2}\ell$ and $\Phi = V_d > 0$ along $y = +\frac{1}{2}\ell$. As indicated in Fig. 25–12, the beam enters the field with each electron having a velocity $\mathbf{v}_0$ directed along the positive x-axis.

The velocity components are

$$\left.\begin{aligned} v_x &= v_0 = \text{constant} \\ v_y &= a_y t = (e/m)Et, \quad t = x/v_0 \end{aligned}\right\} \tag{25–26}$$

At the point $P(d, y)$ where the beam emerges from the field, we have

$$v_y = \frac{eE}{m}\frac{d}{v_0}$$

$$y = \tfrac{1}{2}a_y t^2 = \frac{eE}{2m}\frac{d^2}{v_0^2} \tag{25–27}$$

The angle of deflection α is given by

$$\tan \alpha = \frac{v_y}{v_x} = \frac{eE}{m}\frac{d}{v_0^2} \tag{25–28}$$

The point O' from which the deflected beam appears to emanate can be found by noting that $\tan \alpha = y/\xi$. Then, using Eqs. 25–27 and 25–28, we obtain $\xi = \frac{1}{2}d$. Thus, the apparent origin of the deflected beam is midway through the field in the x-direction.

It is usually the case in *electron guns* that electrons emitted by a heated cathode are accelerated to a high velocity v_0 by causing the electrons to pass through a potential difference V_a. Thus, $\frac{1}{2}mv_0^2 = eV_a$. These high-velocity electrons are then aimed into a deflecting field. The strength E of the deflecting field is related to the potential difference V_d by $E = V_d/\ell$ (Eq. 25–24). Substituting for v_0^2 and E in Eq. 25–28, we find

$$\tan \alpha = \frac{d}{2\ell}\frac{V_d}{V_a} \tag{25–29}$$

Fig. 25–12. An electron beam is deflected through an angle α by a uniform electrostatic field.

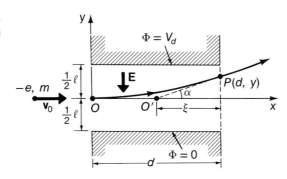

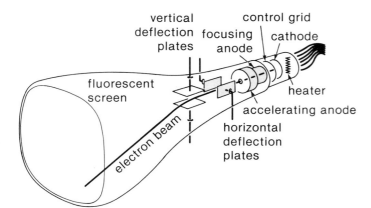

Fig. 25–13. Essential features of a cathode-ray tube. Plates are included for both horizontal and vertical deflection so that time-varying signals $V_d(t)$ can be displayed.

Figure 25–13 is a schematic diagram of a *cathode-ray tube* (CRT) similar to those found in oscilloscopes and television receivers. The accelerating potential difference V_a in the electron gun is usually in the range from a few thousand volts to about 30,000 V in color-television picture tubes.

Example 25–8

Determine the deflection angle α of an electron beam and the transit time through a uniform electrostatic deflecting field for the case in which $V_a = 2000$ V, $d = 4$ cm, $\ell = 1.5$ cm, and $V_d = 50$ V.

Solution:

Using Eq. 25–29, we have

$$\tan \alpha = \frac{d}{2\ell} \frac{V_d}{V_a} = \frac{(0.04 \text{ m})(50 \text{ V})}{2 (0.015 \text{ m})(2000 \text{ V})} = \frac{1}{30}$$

so that $\alpha = 1.91°$

To obtain the transit time from $T = d/v_0$ we need v_0:

$$v_0 = \sqrt{2\frac{e}{m}V_a}$$

For electrons, $e/m = 1.76 \times 10^{11}$ C/kg; thus,

$$v_0 = \sqrt{2 \ (1.76 \times 10^{11} \text{ C/kg})(2000 \text{ V})}$$
$$= 2.65 \times 10^7 \text{ m/s}$$

Then, $\quad T = \dfrac{d}{v_0} = \dfrac{0.04 \text{ m}}{2.65 \times 10^7 \text{ m/s}} = 1.51$ ns

Electric Dipole in a Uniform Electrostatic Field. Consider an electric dipole that consists of charges $+q$ and $-q$ separated by a distance ℓ, where $\frac{1}{2}\ell$ is the distance of each charge from the center of mass (C.M.) of the dipole, as indicated in Fig. 25–14. The dipole is placed in a uniform electrostatic field $\mathbf{E}$ with the dipole moment $\mathbf{p}$ at an angle θ with respect to $\mathbf{E}$. The force on each charge has the magnitude $F = qE$, with directions along and opposite to $\mathbf{E}$ as shown in Fig. 25–14. The net force on the dipole is zero, so there is no motion of the C.M.; however, there is a torque about the C.M. with magnitude

$$\tau_0 = \ell qE \sin \theta$$

which acts in the clockwise direction. This result can be expressed in compact vector notation as

$$\tau_0 = \mathbf{p} \times \mathbf{E} \qquad (25\text{–}30)$$

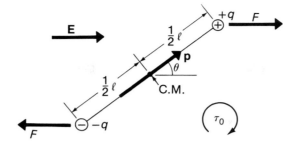

Fig. 25–14. An electric dipole with moment $p = \ell q$ in a uniform electrostatic field.

where $p = \ell q$. The dipole is in a state of stable equilibrium when $\theta = 0$. (• What is the situation when $\theta = \pi$?)

We now derive an expression for the potential energy of a dipole in a uniform electric field. The differential change in potential energy, dU, is the negative of the differential work done by the field during the change in position (or orientation), or $dU = -dW$. The torque produced by the field is clockwise about the C.M., while θ is measured as positive in the counterclockwise direction (refer to Fig. 25–14); hence,

$$dU = -dW = -(-\tau_0\,d\theta) = \tau_0\,d\theta$$

Thus,

$$U(\theta) - U(\theta_0) = \int_{\theta_0}^{\theta} \tau_0\,d\theta = \ell qE \int_{\theta_0}^{\theta} \sin\theta\,d\theta$$

$$= \ell qE(\cos\theta_0 - \cos\theta)$$

We let the potential energy of the dipole U_0 correspond to $\theta_0 = \pi/2$. Thus, U_0 corresponds to the work required to assemble the dipole in the absence of the field. For this choice of U_0 we have

$$U - U_0 = -\ell qE \cos\theta = -pE \cos\theta$$

or, in vector notation,

$$U = -\mathbf{p} \cdot \mathbf{E} + U_0 \qquad\qquad \textbf{(25–31)}$$

QUESTIONS

25–1 Define and explain briefly the meaning of the terms (a) electrostatic potential energy, (b) electrostatic potential, (c) the volt, and (d) gradient.

25–2 A positive charge is carried in a direction of increasing electric potential. Is the work done on the charge by the field positive or negative? What would be the case if the charge were negative?

25–3 Is it possible for the electric potential to be zero where the electric field is nonzero? Is it possible for the electric field to be zero where the potential is nonzero?

25–4 Describe the electric field in a region of space where the electric potential is a constant. Describe the potential in a region of space where the electric field vector is a constant.

25–5 Equipotentials for a point charge are closed spherical surfaces. Explain why you would expect equipotential surfaces for any arbitrary stationary charge distribution to be closed surfaces as well.

25–6 Can equipotential surfaces intersect?

25–7 Describe the equipotential surfaces for a finite electric dipole and compare them with those for a point dipole, shown in Fig. 25–5. Taking the potential to be zero at infinity, what is the potential at the point midway between the charges of the finite dipole? Describe a way to bring a test charge from infinity to this central point without requiring any work along the path followed.

25–8 What is the essential physical property that distinguishes a conductor from an insulator?

25–9 Define and explain briefly the meaning of the terms (a) parallel-plate capacitor, (b) electron volt, (c) electrostatic deflection, and (d) cathode-ray tube.

25–10 A hollow conductor is initially negatively charged to a total charge $-Q$. Explain what would happen if a positive charge $+Q'$ were deposited on the interior surface of the conductor. Consider the three cases $Q' > Q$, $Q' = Q$, and $Q' < Q$.

25–11 Two conducting spheres of equal radius are brought into contact and charged to a positive potential relative to infinity. In view of the Coulomb forces acting between the free charges on the conducting surfaces, how would you expect the charge density to be distributed? Sketch the expected field lines and the equipotential surfaces.

25–12 The accompanying drawing shows the charge distributions for (a) a single charge layer, (b) a double charge layer, and (c) an oppositely charged double layer. Verify by explicit application of Gauss's law that the electric field strengths shown are correct.

25–13 In our discussion of the motion of a charged particle in a stationary electric field (Section 25–4), induction effects on any conductors or insulators that might be present were neglected. Describe in physical terms what the expected results from such polarization effects would be if they were present and taken into account.

25–14 Explain why the motion of a charge released in an electric field does not necessarily follow a field line. Under what circumstances would the charge exactly follow a field line?

25–15 During a storm a "live" high-voltage power line falls on your car, both charging and immobilizing the car. Are

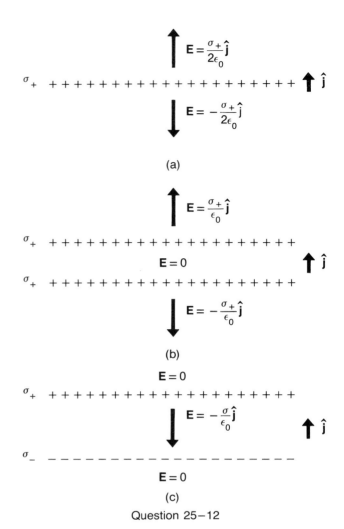

Question 25–12

you safe while inside? What would be the consequence of trying to get out of the car?

25–16 Why aren't birds electrocuted when they land on a high-voltage power line? Would the wing span of birds have to be taken into account when determining the minimum distance between high-voltage power lines?

PROBLEMS*

Section 25–1

25–1 A small spherical object carries a charge of 8 nC. At what distance from the center of the object is the potential equal to 100 V? 50 V? 25 V? Is the spacing of these equipotentials proportional to the change in Φ?

25–2 Two point charges, $Q_1 = +5$ nC and $Q_2 = -3$ nC, are separated by 35 cm. (a) What is the potential energy of the pair? What is the significance of the algebraic sign of your answer? (b) What is the electric potential at a point midway between the charges?

---

*Assume the zero reference level for potential to be at infinity unless the statement of the problem requires otherwise.

25–3 Four equal charges, $Q = 2\ \mu C$, are located at the corners of a square that is 0.50 m on a side. (a) What is the potential energy of the array? (b) What is the electric potential at the center of the square? (c) What is the electric field strength at the center of the square?

25–4 Two point charges, $Q_1 = +5\ \mu C$ and $Q_2 = -2\ \mu C$, are separated by 50 cm. (a) Where, along a line that passes through both charges, is the potential zero (in addition to $R = \infty$)? (b) Determine the electric field strength at the point found in (a). If $\Phi = 0$ at this point, why is E not also equal to zero?

25–5 Determine the electric potential along a perpendicular bisector of a straight line charge with length L and uniform linear charge density λ. Show that when the distance becomes large compared with L, your result reduces to that for a point charge $Q = \lambda L$ located at the center of the line charge.

Section 25–2

25–6 The electric field on the axis of a uniformly charged disc of radius R and total charge Q is

$$\mathbf{E} = \frac{Q}{2\pi R^2 \epsilon_0} \left[1 - \frac{z}{(z^2 + R^2)^{1/2}} \right] \hat{\mathbf{k}}$$

where z is the distance measured from the center of the disc. Using this expression for $\mathbf{E}$, find the potential $\Phi(z)$ when $\Phi(\infty)$ is taken to be zero. Compare your result with that found in Example 25–4.

25–7 The potential for a certain electric field is (in cylindric coordinates) $\Phi(\rho, \phi, z) = A(\rho^2 \sin \phi \cos \phi + z^2 \tan \phi)$, where A has suitable dimensions. Obtain an expression for the field strength $\mathbf{E}$ in rectangular coordinates. [*Hint:* Use $x = \rho \cos \phi$, $y = \rho \sin \phi$, $z = z$.]

25–8 At a certain distance from a point charge the field intensity is 500 V/m and the potential is -3000 V. (a) What is the distance to the charge? (b) What is the magnitude of the charge?

25–9 Use the expression obtained in Example 25–3 for the potential on the axis of a uniformly charged ring and determine the field strength along the axis. (• Why is it possible to do this?) Compare your answer with Eq. 24–14.

25–10 The potential for a certain two-dimensional electric field is $\Phi(x, y) = Axy$, where A has suitable dimensions. (a) Determine the field strength $\mathbf{E}$ everywhere in the x-y plane. (b) Determine the slope of the field line at a point (x, y) by calculating $\tan \beta = E_y/E_x$. (c) Determine the slope of the equipotential at a point (x, y) by calculating dy/dx for $\Phi = $ constant. (d) Are the field lines and the equipotentials mutually perpendicular? (e) Sketch some of the field lines and equipotentials.

25–11 Given the electric potential $\Phi(x, y) = \kappa(x^2 + y^2)$, where κ is a constant with suitable dimensions, show explicitly that the electric field strength $\mathbf{E}$ is perpendicular to the equipotential curves at any arbitrary point (x, y). Sketch the equipotentials and show $\mathbf{E}$ at some point.

25–12 For the uniform electric field $\mathbf{E} = \dfrac{E_0}{\sqrt{2}}(\hat{\mathbf{i}} + \hat{\mathbf{j}})$, prove by explicit calculation that $\oint \mathbf{E} \cdot d\mathbf{r} = 0$ for the rectangular path shown in the drawing.

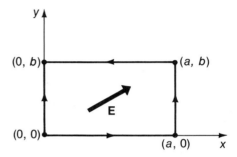

Section 25–3

25–13 What charge Q must be placed on a conducting sphere with a radius of 75 cm to produce a potential on the surface of 2×10^6 V?

25–14 A point charge Q is placed at the center of a conducting spherical shell with inner radius R_1 and outer radius R_2 and which carries no net charge. Determine the field strength $\mathbf{E}$ everywhere. Sketch some of the lines of force.

25–15 The potential difference between two oppositely charged large conducting plane parallel plates is 300 volts, and the separation between the plates is 2 mm. (a) What is the electric field between the plates? (b) What is the charge density, C/m^2, on a plate?

Section 25–4

25–16 A particle with a charge $q = 12$ nC and a mass $m = 2$ mg is released from rest at a point 5 cm from a charge $Q = 50$ nC that is stationary. (a) What is the initial acceleration of the particle in units of g? (b) What is the initial potential energy of the pair of charges? (c) What is the total amount of work done by the field in moving the particle q to infinity? (d) What is the speed of the particle q when it is a great distance from Q?

25–17 In Example 25–8, what is the magnitude of the transverse acceleration of the electrons in the beam as it passes between the deflecting plates? (Express the acceleration in units of g.)

25–18 Two parallel conducting plates, each 4 cm × 4 cm, are separated by 1.5 cm. (a) What magnitude of charge must be placed on each plate to produce a potential difference of 50 V? (b) An electron beam passes between the plates at a rate of 10^{-7} C/s. If each electron has a kinetic energy of 2.0 keV, calculate the amount of electronic charge that is between the plates at any instant. (c) Compare the two charges found in (a) and (b). Are induced charge effects in the conducting plates likely to be of any significance in this case? [Note that induced charges are never larger in magnitude than the charges producing the induction effect.]

25–19 An electric dipole **p** is placed in a uniform electric field, $\mathbf{E} = E\hat{\mathbf{i}}$, that is confined between the planes $x = 0$ and $x = L$, as shown in the diagram. Calculate the work done by the field as the dipole is assembled at the indi-

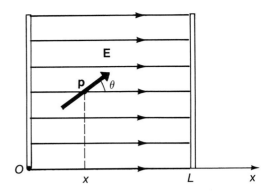

cated location, and show that the potential energy is $-\mathbf{p} \cdot \mathbf{E}$. (Bring the charges $\pm q$ separately from $x = \infty$.)

Additional Problems

25–20 All of the corners, except one, of a 1-m cube are occupied by charges of $+1$ μC. What is the electric potential at the empty corner?

25–21 Equal charges ($q = +2$ μC) are placed at 30° intervals around the equator of a sphere with a radius of 1.2 m. What is the electric potential (a) at the center of the sphere and (b) at the north pole of the sphere?

25–22 Three point charges are located as follows: $-Q$ at $x = -\ell$, $+2Q$ at $x = 0$, and $-Q$ at $x = +\ell$. Calculate the potential along the x-axis in the distant region ($|x| \gg \ell$).

25–23 (a) Determine the potential along the axis to the
•• right of the uniform hemispheric surface charge distribution shown in the diagram. There is no charge on the flat face of the hemisphere. [*Hint:* Use $dQ = 2\pi\sigma R^2 \sin \theta \, d\theta$.] (b) Show that $\Phi(x) \cong Q/4\pi\epsilon_0 x$, for $x \gg R$, where $Q = 2\pi R^2 \sigma$. (c) Calculate the field strength E at any point along the x-axis to the right of O.

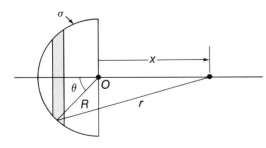

25–24 A metal sphere is to be charged to a potential of 10^6 V. The field strength is not to exceed 3×10^6 V/m, the point at which the insulating quality of air breaks down

and corona discharge or sparking takes place. (a) What is the minimum radius of the sphere? (b) For the minimum radius, what is the charge on the sphere?

25–25 Consider two infinitely long concentric cylinders. The
• inner cylinder has a radius R_1 and carries a linear charge density $+\lambda$. The outer cylinder has an inner radius R_2 and carries a linear charge density $-\lambda$. (a) Where on the outer cylinder does the charge reside? (b) Determine the electric field strength everywhere. (c) What is the potential difference, $V = \Phi(R_1) - \Phi(R_2)$? (d) Show that

$$\mathbf{E} = \frac{V}{\ln (R_1/R_2)} \cdot \frac{1}{\rho} \hat{\mathbf{u}}_\rho$$

where $\hat{\mathbf{u}}_\rho$ is the unit vector in the radial direction.

25–26 A dipole consists of two particles, one with charge Q and
•• mass m, and the other with charge $-Q$ and mass $2m$, separated by a distance L, as shown in the diagram. Show that the angular frequency of small oscillations of the dipole about its equilibrium position is $\omega = \sqrt{3QE/2mL}$ when placed in a uniform field E.

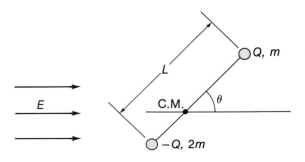

CAPACITANCE AND DIELECTRICS

In this chapter we continue the investigation of the electrostatic fields produced by charged conductors. We include the effects of induced charges and add the possibility of insulators being present in the systems studied. When an insulating material is introduced into an electric field (of ordinary strength), the electrons in the material remain bound within their respective atomic or molecular domains. However, the field can alter the atomic-charge distributions and produce *polarization* effects that cause substantial changes in the electric field at the macroscopic level. Insulators that exhibit polarization are referred to as *dielectrics*.

It is possible to determine the electrostatic field configuration for a system of insulators and charged conductors if the polarizability characteristics of the dielectric materials are known and if some geometric constants of the system can be determined. These geometric constants are associated with the idea of *capacitance*. Accordingly, we begin by establishing some of the features of capacitance in simple cases.

26–1 THE PARALLEL-PLATE CAPACITOR

A simple yet conceptually important device for storing electric charge is the *parallel-plate capacitor*. This device consists of two parallel conducting planes (the *capacitor plates*) that usually have the same shape and are spaced apart by a fixed distance s. One plate carries a charge $+Q$ and the other plate carries a charge $-Q$, as indicated in Fig. 26–1.

The charging of a capacitor can be accomplished by connecting the plates with a wire and then inducing a charge on one of the plates with a charged rod. An equal charge of the opposite sign will then reside on the other plate. When the wire is disconnected and the charged rod is removed, one plate will carry a charge $+Q$ and the other plate will carry a charge $-Q$.

If the dimensions of the plates are large compared with the spacing (that is, if $b \gg s$ in Fig. 26–1), the charges on the plates will be distributed over the inside surfaces in an essentially uniform manner. Thus, each plate, with area A, will have a surface charge density $\sigma = Q/A$. The resulting electric field between the plates will be practically the same as that for a pair of infinite parallel planes that carry the same charge density σ (Fig. 25–11). If we ignore the edge effects associated with finite plates, this arrangement constitutes an *ideal parallel-plate capacitor*.

In practice, the usual way of charging a capacitor is by means of a battery. The composition and operation of a battery are discussed in Section 27–4; for our present

Fig. 26–1. A parallel-plate capacitor with an arbitrary shape. The distance b represents a typical dimension of the capacitor.

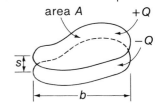

purposes we need only to know that a battery produces a constant potential difference between its terminals. The standard symbol for a battery is

$$B \bullet \!\!\! \begin{array}{c} + \\ \dashv\!| \; |\!\vdash \end{array} \!\!\! \bullet A$$

$$V_{BA}$$

where terminal B (the *positive* terminal) is at a higher electric potential than terminal A (the *negative* terminal); that is, $\Phi_B - \Phi_A = V_{BA} > 0$. When no confusion may arise, we simply write V for the potential difference V_{BA}. Figure 26–2 shows a parallel-plate capacitor connected to a battery so that a potential difference $\Phi_B - \Phi_A = V$ is established between the plates. Throughout this discussion the system that consists of the capacitor and the battery is assumed to possess no net charge and to be far removed from other conductors and charges. (Such a system is said to be "floating.")

If the spacing s between the plates is small compared with the dimensions of the plates, the electric field strength is essentially uniform and we can write (Eq. 25–22)

$$\mathbf{E} = \frac{\sigma}{\epsilon_0}\hat{\mathbf{n}} = \frac{Q}{\epsilon_0 A}\hat{\mathbf{n}} \tag{26-1}$$

where $\hat{\mathbf{n}}$ is a unit vector normal to the plates and in the direction from the positively charged plate toward the negatively charged plate. The potential difference between the plates is (see Eq. 25–24)

$$\Phi_B - \Phi_A = V = -\oint_A^B \mathbf{E} \cdot d\mathbf{r} = Es \tag{26-2}$$

Combining Eqs. 26–1 and 26–2, we have

$$Q = \frac{\epsilon_0 A}{s} V \tag{26-3}$$

Thus, the charge Q stored on either plate is related to the potential difference V between the plates by a geometry-dependent factor, $\epsilon_0 A / s$. In general, the proportionality constant that connects Q and V is called the *capacitance* C of the device or system; that is,

$$Q = CV \tag{26-4}$$

For the case of the parallel-plate capacitor, we have

$$C = \frac{\epsilon_0 A}{s} \qquad \text{parallel-plate capacitor} \tag{26-5}$$

From Eq. 26–4 we see that the units of capacitance are *coulombs per volt*, C/V, to

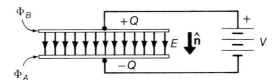

Fig. 26–2. A battery is connected to a parallel-plate capacitor to establish a potential difference $\Phi_B - \Phi_A = V$ between the plates.

Fig. 26–3. (a) The fringing field near the edge of a parallel-plate capacitor. The charge density $\pm\sigma'$ on the outside of the plates is highly nonuniform. (b) The idealized field of a parallel-plate capacitor. All of the charge is on the inside surfaces.

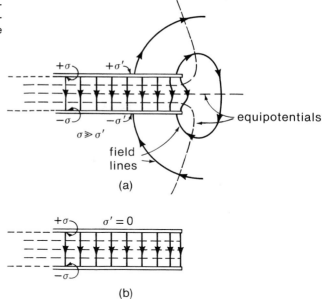

which is given the special name *farad** (F); thus,

$$1 \text{ F} = 1 \text{ C/V}$$

The farad is a rather large unit, so we usually find capacitive circuit elements with capacitance values measured in μF (10^{-6} F) or pF (10^{-12} F).

▶ *Fringing Fields.* Because a real parallel-plate capacitor has a finite size, the electric field lines at the edges are not exactly straight and uniformly spaced as in the ideal case, nor is the field contained entirely between the plates. Figure 26–3a shows the field lines near the edge of a parallel-plate capacitor; this nonuniform portion of the field is referred to as the *fringing field*. Notice also the behavior of the equipotentials in this region. Figure 26–3b shows the idealized situation: the field terminates abruptly at the edge of the capacitor.

Figure 26–3a suggests that the actual charge Q on a parallel-plate capacitor is slightly greater than the charge that results when the value of C given by Eq. 26–5 is used in Eq. 26–4. The charge Q on each plate includes *all* of the charge on *all* of the surfaces of each plate, most of which usually resides on the inside surface. In the "floating" condition that we have assumed, the total charge on each plate is the same in magnitude and opposite in sign. When a fringing field is present, as in Fig. 26–3a, we can write

$$C = \frac{\epsilon_0 A}{s} f \qquad \textbf{(26–5a)}$$

where $f \geq 1$ is a correction factor that can be determined for a specific plate geometry. For a circular capacitor with radius R, $f = 1.167$ when $s/R = 0.10$ and $f = 1.023$ when $s/R = 0.01$; of course, in the limit, $s/R \to 0$, we have $f = 1$. ◀

Example 26–1

A circular parallel-plate capacitor with a spacing $s = 3$ mm is charged to produce an electric field strength of 3×10^6 V/m. What plate radius R is required if the stored charge Q is 1 μC?

Solution:

The potential difference between the plates is

$$V = Es = (3 \times 10^6 \text{ V/m})(3 \times 10^{-3} \text{ m}) = 9 \times 10^3 \text{ V}$$

---

*Named in honor of Michael Faraday (1791–1867), the great English physicist about whom we learn more in later chapters.

The capacitance value is

$$C = \frac{Q}{V} = \frac{1 \times 10^{-6} \text{ C}}{9 \times 10^3 \text{ V}} = 1.11 \times 10^{-10} \text{ F} = 111 \text{ pF}$$

From Eq. 26–5 we have $A = \pi R^2 = Cs/\epsilon_0$, so that*

$$R = \sqrt{\frac{Cs}{\pi \epsilon_0}} = \sqrt{\frac{(1.11 \times 10^{-10} \text{ F})(3 \times 10^{-3} \text{ m})}{\pi(8.85 \times 10^{-12} \text{ F/m})}}$$

$$R = 0.109 \text{ m} = 10.9 \text{ cm}$$

With $s/R = (3 \times 10^{-3} \text{ m})/(0.109 \text{ m}) = 2.75 \times 10^{-2}$, the fringing field correction can be calculated to be $f = 1.056$. Thus, the required area is 5.6 percent smaller and the required radius is 2.8 percent smaller than calculated with Eq. 26–5. The actual required radius is, therefore, $R' = 10.6$ cm.

26–2 THE SPHERICAL CAPACITOR

The fringing-field effects associated with a parallel-plate capacitor are not present in a *spherical* capacitor because the field is totally confined within the region between the plates. Figure 26–4a shows a spherical capacitor that consists of an inner conducting sphere with a radius R_1 surrounded by a concentric conducting spherical shell with an inner radius R_2 and an outer radius R_3. We again assume an isolated floating system with no net charge.

Figure 26–4a indicates how the charging battery is connected to the spherical capacitor; Fig. 26–4b shows the potentials and the charges on the three surfaces. Because the outer shell of the capacitor is a solid conductor, we have $\Phi_2 = \Phi_3$ and $E = 0$ within the material of the shell. Applying Gauss's law using a spherical surface with radius between R_2 and R_3, we conclude that $Q_2 = -Q_1$. The system has no net charge, so we have $Q_3 = 0$. Next, applying Gauss's law using a spherical surface with radius greater than R_3, we conclude that $E = 0$ in the region outside the shell. If we take the potential at infinity to be $\Phi_\infty = 0$, it follows that $\Phi_2 = \Phi_3 = 0$ and $\Phi_1 = V$.

Next, surround the inner sphere with a spherical Gaussian surface of radius r ($R_1 \leqslant r \leqslant R_2$); then, Gauss's law gives

$$\mathbf{E} = \frac{Q_1}{4\pi\epsilon_0} \cdot \frac{1}{r^2} \hat{\mathbf{u}}_r \qquad (26-6)$$

Using Eq. 25–13a, the potential is

$$\Phi(r) = \Phi(R_1) - \frac{Q_1}{4\pi\epsilon_0} \int_{R_1}^{r} \frac{dr}{r^2} = V - \frac{Q_1}{4\pi\epsilon_0}\left(\frac{1}{R_1} - \frac{1}{r}\right)$$

Fig. 26–4. (a) An isolated spherical capacitor. (b) Electric field lines and potential values for the spherical capacitor in (a).

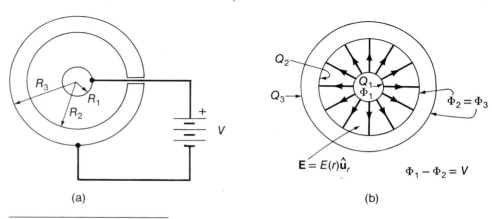

(a)

(b)

---

* Verify that the units of ϵ_0 (C²/N·m²) can be expressed as F/m.

When $r = R_2$, we have $\Phi(R_2) = 0$; hence,

$$Q_1 = 4\pi\epsilon_0 V \frac{R_1 R_2}{R_2 - R_1} \qquad (26\text{–}7)$$

For the mode of charging illustrated in Fig. 26–4a, in which $Q_2 = -Q_1$ and $Q_3 = 0$, we can again define a stored charge Q with $Q = CV$. Notice that the inner conductor has only one surface (at $r = R_1$) and carries a charge Q_1. The outer conductor has two surfaces (at $r = R_2$ and $r = R_3$) and carries a total charge $Q_2 + Q_3 = Q_2 = -Q_1$ (because $Q_3 = 0$). Thus, the coefficient of V in Eq. 26–7 is identified as the capacitance of the spherical capacitor.*

$$C = 4\pi\epsilon_0 \frac{R_1 R_2}{R_2 - R_1} \qquad \text{spherical capacitor} \qquad (26\text{–}8)$$

In Fig. 26–4a the potential is zero on both surfaces of the shell of the spherical capacitor (that is, $\Phi_2 = \Phi_3 = 0$). Thus, we can define $\Phi(\infty) = 0$ in a practical way by imagining that R_2 and R_3 become indefinitely large. If we surround a piece of laboratory equipment with a large conducting cage that is grounded, we thereby create an equipotential that effectively corresponds to $\Phi(\infty) = 0$. In the case of the spherical capacitor, allowing $R_2 \to \infty$ yields Lim $(R_2 \to \infty)$ $C = 4\pi\epsilon_0 R_1$. Thus, in general, the capacitance of a sphere with radius R can be expressed as

$$C = 4\pi\epsilon_0 R \qquad \text{sphere} \qquad (26\text{–}9)$$

▶ The manner in which a capacitor is grounded or brought near other conductors or charges can alter the charge distribution on the capacitor and can even influence the magnitude of the charge. In practical laboratory situations the presence of the Earth must be taken into account. The Earth may be assumed to be a conductor at zero potential relative to infinity. Thus, the outer conductor of the spherical capacitor in Fig. 26–4 may be brought to zero potential (relative to infinity) by connecting the outer shell to a ground wire. The solution for this situation will be exactly the same as that for the isolated state discussed above. That is, we will have $\Phi_2 = \Phi_3 = 0$ and $\Phi_1 = V$. Thus, we have

$Q_3 = 0$ and $Q_2 = -Q_1$ as before. (• Can you see why?) However, if the central conductor is brought to zero potential (relative to infinity) by using a ground connection, $\Phi_2 = \Phi_3 \neq 0$ and the outer surface at radius R_3 will acquire a charge, so that $Q_3 \neq 0$. Moreover, the total charge on the outer shell, $Q_2 + Q_3$, will *not* equal $-Q_1$. Thus, in cases such as this one, where the total charge on two conductors forming a capacitor is not zero, the simple concept of capacitance as a single geometric parameter does not apply. A more advanced treatment involving several so-called capacity coefficients is required.   ◀

Example 26–2

Determine the capacitance of the cylindric capacitor shown in the diagram. (The symbol below the battery indicates a grounding connection.)

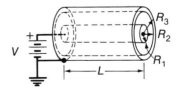

Solution:

We assume that the spacing between the cylinders is small; that is, $(R_2 - R_1)/R_2 \ll 1$ and $R_2/L \ll 1$. Then, the field is essentially radial and uniform in the region $R_1 \leq r \leq R_2$; thus, fringing effects can be ignored.

---

* About a hundred years ago, the American physicist Henry Rowland (1848–1901) constructed a precision spherical capacitor and found agreement between the measured and the calculated capacitance to within two parts in 10^5.

The electric field strength in the gap is easily determined by constructing a concentric cylindric Gaussian surface with radius $R_1 < r < R_2$ and length L. We find

$$\mathbf{E} = \frac{Q_1}{2\pi L \epsilon_0} \cdot \frac{1}{r} \hat{\mathbf{u}}_r$$

where Q_1 is the total charge on the inner conductor. For $R_1 \leqslant r \leqslant R_2$, we have (using Eq. 25–13a)

$$\Phi(r) = \Phi(R_1) - \frac{Q_1}{2\pi L \epsilon_0} \int_{R_2}^{r} \frac{dr}{r} = V - \frac{Q_1}{2\pi L \epsilon_0} \ln \frac{r}{R_1}$$

Now, $\Phi_2 = \Phi_3 = 0$, as in the case of the spherical capacitor. Then, $\Phi(r = R_2) = 0$, and we obtain

$$Q_1 = \frac{2\pi L \epsilon_0}{\ln (R_2/R_1)} V$$

Also, as in the case of the spherical capacitor, we have $Q_2 = -Q_1$ and $Q_3 = 0$. The capacitance of the cylindric capacitor is again defined through the relationship $Q = CV$; thus,

$$C = \frac{2\pi L \epsilon_0}{\ln (R_2/R_1)}$$

Example 26–3

Two spherical conductors with radii R_1 and R_2 are separated by a sufficiently large distance that induction effects are negligible. The spheres are connected by a thin conducting wire and are brought to the same potential V relative to $\Phi(\infty) = 0$.

(a) Determine the capacitance C of the system, where $C = (Q_1 + Q_2)/V$.

(b) What is the charge ratio Q_1/Q_2?

Solution:

(a) If induction effects are negligible, the charges on the two spheres are distributed uniformly, although with different charge densities. According to Eq. 26–9, the common potential V of the spheres can be expressed as

$$V = \frac{Q_1}{4\pi \epsilon_0 R_1} = \frac{Q_2}{4\pi \epsilon_0 R_2}$$

Thus, with $Q = Q_1 + Q_2$, we have

$$Q = 4\pi \epsilon_0 (R_1 + R_2)V$$

and, hence,

$$C = \frac{Q_1 + Q_2}{V} = 4\pi \epsilon_0 (R_1 + R_2)$$

(b) From the equation expressing the equality of the potentials, the charge ratio is

$$\frac{Q_1}{Q_2} = \frac{R_1}{R_2}$$

Notice that the charge *density* is in the inverse ratio to the charge itself; that is,

$$\frac{\sigma_1}{\sigma_2} = \frac{Q_1/4\pi R_1^2}{Q_2/4\pi R_2^2} = \frac{R_2}{R_1}$$

From the result of Example 26–3 it is easy to understand that the charge density at any point on a conductor with arbitrary shape will be inversely proportional to the radius of curvature of the surface at that point. The charged conducting surface shown in Fig. 26–5 consists of a cone that has spherical end caps with quite different radii. The charge density and, hence, the electric field is more intense at the smaller end. If the radius of the tip is made very small (needlelike), the field strength for even a modest charge may be sufficiently high that corona discharge will take place. This, in fact, is the way that lightning rods work. Lightning rods are well-grounded, pointed conductors placed high above surrounding objects they are to protect. During electrical storms the large cloud-to-ground potential differences induce a steady quiet corona discharge. This reduces the electric field in the immediate neighborhood and helps prevent a direct lightning stroke.

Fig. 26–5. The charge density is larger and, hence, the electric field is more intense at the end of the object that has the smaller radius of curvature.

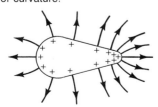

26–3   ELECTROSTATIC FIELD ENERGY

Consider an uncharged two-conductor capacitor with a capacitance C. Imagine charging such a capacitor by moving small amounts of charge dq from one conductor to the other. Let q be the total amount of charge transferred before a particular instant, and let $V(q)$ be the potential difference between the conductors at that instant. The work that must be done to transport a charge dq from one conductor to the other when the potential difference is $V(q)$ is

$$dW = V(q)\, dq = \frac{q}{C}\, dq$$

The total amount of work required to move a total charge Q is therefore

$$W = \int dW = \int_0^Q \frac{q}{C}\, dq$$

so that

$$W = \frac{1}{2}\frac{Q^2}{C} \tag{26-10}$$

We take the attitude that this work appears in the form of potential energy U stored in the electric field of the capacitor system. Thus,*

$$U = \frac{1}{2}\frac{Q^2}{C} = \frac{1}{2}QV = \frac{1}{2}CV^2 \tag{26-11}$$

If the capacitor is discharged, for example, by connecting an electric motor between the conductors, current will flow through the motor, the shaft will turn, and mechanical work will be done. When the capacitor is completely discharged, the electric field will have disappeared and a total amount of mechanical work $\frac{1}{2}Q^2/C$ will have been done.

We have by now twice encountered the idea of potential energy associated with electrostatic fields. It is important to distinguish carefully between these two situations because they represent quite different ways of generating potential energy. Earlier, it was pointed out that the electric potential $\Phi(\mathbf{r})$ produced by a fixed array of charges determines the difference in the potential energy, $U(B) - U(A) = q[\Phi(\mathbf{r}_B) - \Phi(\mathbf{r}_A)]$, of a point charge q between the positions specified by $\mathbf{r}_A$ and $\mathbf{r}_B$. The potential energy U in Eq. 26–11, however, is the energy required to create the field of the charged array itself. (We ignore any energy involved in the creation of the elementary charged particles that constitute the charges in the array. We consider only the energy required to move these charges in order to establish the desired field.)

Energy Density. Consider an ideal parallel-plate capacitor with capacitance $C = \epsilon_0 A/s$ that is charged to a potential difference V. According to Eq. 25–24, we have $V = Es$, where E is the magnitude of the field strength between the plates. Substituting these expressions for C and V into the last form of Eq. 26–11, we obtain

$$U = \tfrac{1}{2}CV^2 = \tfrac{1}{2}\epsilon_0 E^2 sA$$

The electric field between the plates is uniform, so we can define the *electrostatic energy density* to be $u = U/(\text{volume}) = U/sA$, or

$$u = \tfrac{1}{2}\epsilon_0 E^2 = \tfrac{1}{2}\epsilon_0 \mathbf{E} \cdot \mathbf{E} \tag{26-12}$$

---

*Notice that W in Eq. 26–10 represents the work done *on* the field by an outside agency. The work done *by* the field is $W' = -W$, and the potential energy stored in the field is $U = -W' = W$.

This result was obtained for the special case of a parallel-plate capacitor. However, the expression for u contains none of the parameters of the capacitor, so we expect the expression to be generally true. Indeed, Eq. 26–12 gives the correct electrostatic energy density for any electric field $\mathbf{E}$ in vacuum. (Compare Problem 26–24.)

The result contained in Eq. 26–12 provides us with the concept that *electric fields possess energy*. This view of associating energy with fields plays an important role for both electric and magnetic fields and particularly for the phenomenon of electromagnetic waves. We shall also discover that such waves carry not only energy but momentum as well.

Example 26–4

Use Eq. 26–12 to make an explicit calculation of the energy stored in the field of a simple spherical capacitor. Show that $U = \frac{1}{2}Q^2/C$.

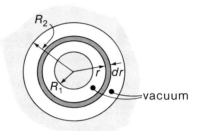

Solution:

According to Eq. 26–6, the electric field strength within the capacitor is

$$E(r) = \frac{Q}{4\pi\epsilon_0} \cdot \frac{1}{r^2}$$

Then, the energy stored in a spherical shell between r and $r + dr$ is

$$dU = u\, d(\text{volume}) = u(4\pi r^2\, dr)$$

Using Eq. 26–12 and the expression for $E(r)$, we have

$$dU = \frac{Q^2}{8\pi\epsilon_0} \cdot \frac{dr}{r^2}$$

from which we obtain

$$U = \frac{Q^2}{8\pi\epsilon_0} \int_{R_1}^{R_2} \frac{dr}{r^2} = \frac{Q^2}{8\pi\epsilon_0} \cdot \frac{R_2 - R_1}{R_1 R_2}$$

Now, $C = 4\pi\epsilon_0 R_1 R_2 / (R_2 - R_1)$ for a spherical capacitor (Eq. 26–8); therefore,

$$U = \frac{1}{2} \frac{Q^2}{C}$$

which is the desired result.

26–4   CAPACITOR NETWORKS

In problems that involve several capacitors, it is often convenient to introduce the idea of an *effective* or *equivalent capacitor*. That is, we seek a single capacitor that, when it replaces some network of capacitors, causes no change in the behavior of the remainder of the circuit to which it is attached. In diagrams we use the following descriptive symbol for a capacitor:

Series Connection.   Consider two capacitors, C_1 and C_2, connected in *series* between the terminals A and D, as shown in Fig. 26–6a. When the switch S is closed, the capacitors, initially uncharged, become charged as the battery supplies an amount of charge Q. This charge accumulates on plate d so that $Q_d = +Q$. When equilibrium is reached, an induced charge, $Q_c = -Q_d = -Q$, appears on plate c. Because plates c and b (and the connecting wire) are isolated, we must have $Q_b = -Q_c = +Q$. Finally, the charge induced on plate a is $Q_a = -Q_b = -Q$. Thus, there are equal charges on the two capacitors, $Q_1 = Q_2 = Q$.

The middle pair of plates, b and c, have the same potential, $\Phi_b = \Phi_c$, so that the potential difference across the capacitor system can be expressed as $\Phi_D - \Phi_A = \Phi_d - \Phi_a = (\Phi_d - \Phi_c) + (\Phi_b - \Phi_a) = V$. Now, we have

$$\Phi_d - \Phi_c = \frac{Q_1}{C_1} \quad \text{and} \quad \Phi_b - \Phi_a = \frac{Q_2}{C_2}$$

Hence,

$$V = \frac{Q_1}{C_1} + \frac{Q_2}{C_2} = Q\left(\frac{1}{C_1} + \frac{1}{C_2}\right)$$

If the equivalent capacitor C shown in Fig. 26–6b is to duplicate exactly the effect in the circuit of the C_1-C_2 combination, the same charge Q must appear on the plates when the switch is closed and the capacitor is charged to the voltage V. Thus,

$$V = \frac{Q}{C}$$

Comparing these two expressions for V, we see that

$$\frac{1}{C} = \frac{1}{C_1} + \frac{1}{C_2}$$

We can extend this argument to the case of N capacitors connected in series, with the result

$$\frac{1}{C} = \frac{1}{C_1} + \frac{1}{C_2} + \frac{1}{C_3} + \cdots + \frac{1}{C_N} \qquad \text{series connection} \qquad \textbf{(26–13)}$$

Another way to view the series connection of two capacitors is to imagine the wire between them shrinking to zero length. The two common plates, which carry equal charges of opposite sign, combine into a single plate with zero effective thickness. Then, the separation between the two outer plates is $s_1 + s_2$, and

$$\frac{1}{C} = \frac{s_1 + s_2}{\epsilon_0 A} = \frac{s_1}{\epsilon_0 A} + \frac{s_2}{\epsilon_0 A} = \frac{1}{C_1} + \frac{1}{C_2}$$

Parallel Connection. Consider two capacitors, C_1 and C_2, connected in parallel between the terminals A and D, as shown in Fig. 26–7a. When the switch S is closed, the capacitors, initially uncharged, become charged as the battery supplies an amount of charge Q. This charge is shared by the two capacitors, so that $Q_1 + Q_2 = Q$. The potential differences, $\Phi_b - \Phi_a$ and $\Phi_d - \Phi_c$, are both equal to V; thus, we have

$$Q = Q_1 + Q_2 = C_1 V + C_2 V = (C_1 + C_2)V$$

The battery must supply the same charge Q to the equivalent capacitor C. That is,

$$Q = CV$$

Comparing these two expressions for Q, we see that

$$C = C_1 + C_2$$

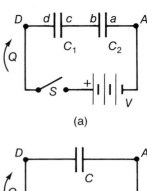

(a)

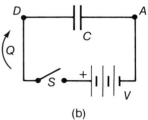

(b)

Fig. 26–6. (a) Two capacitors connected in series. (b) The capacitor C is equivalent to the series connection of C_1 and C_2.

Fig. 26–7. (a) Two capacitors connected in parallel. (b) The capacitor C is equivalent to the parallel connection of C_1 and C_2.

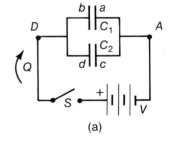

(a)

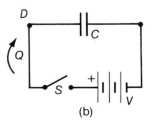

(b)

We can extend this argument to the case of N capacitors connected in parallel, with the result

$$C = C_1 + C_2 + C_3 + \cdots + C_N \qquad \text{parallel connection} \qquad \textbf{(26-14)}$$

Again, there is an alternative way of viewing the parallel connection of two or more capacitors. Imagine moving the plates together until they are joined at the edges. Then, the area of each combined plate is $A_1 + A_2$, and

$$C = \frac{\epsilon_0(A_1 + A_2)}{s} = \frac{\epsilon_0 A_1}{s} + \frac{\epsilon_0 A_2}{s} = C_1 + C_2$$

Example 26-5

Show that the potential differences across the two capacitors in Fig. 26-6a are in inverse proportion to their capacitance values.

Solution:

In the discussion of the series-connected capacitors C_1 and C_2, we concluded that the charge appearing on each capacitor

has the same magnitude; thus

$$Q_1 = Q_2$$

or

$$C_1 V_1 = C_2 V_2$$

It follows that

$$\frac{V_1}{V_2} = \frac{C_2}{C_1}$$

Example 26-6

(a) The diagram shows a network of capacitors between the terminals A and B. Reduce this network to a single equivalent capacitor.

(b) Determine the charge on the 4-μF and 8-μF capacitors when the capacitors are fully charged by a 12-V battery connected to the terminals.

(c) Determine the potential difference across each capacitor.

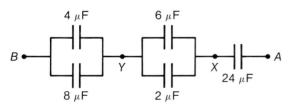

Solution:

(a) First, reduce the two parallel capacitors between B and Y to a single equivalent capacitor C_1:

$$C_1 = 4 \ \mu\text{F} + 8 \ \mu\text{F} = 12 \ \mu\text{F}$$

Next, reduce the two parallel capacitors between Y and X to a single equivalent capacitor C_2:

$$C_2 = 6 \ \mu\text{F} + 2 \ \mu\text{F} = 8 \ \mu\text{F}$$

The result of these two reductions is the equivalent network shown in the diagram below.

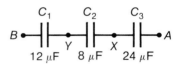

Finally, the three capacitors C_1, C_2, and C_3 can be reduced to the single equivalent capacitor C, where

$$\frac{1}{C} = \frac{1}{12 \ \mu\text{F}} + \frac{1}{8 \ \mu\text{F}} + \frac{1}{24 \ \mu\text{F}} = \frac{1}{4 \ \mu\text{F}}$$

Thus, $$C = 4 \ \mu\text{F}$$

(b) The total charge supplied by the battery to the capacitor network is $Q = CV = (4 \ \mu\text{F})(12 \ \text{V}) = 48 \ \mu\text{C}$. The equivalent capacitors C_1, C_2, and C_3 all have this same stored charge. The real capacitors that comprise C_1—the 4-μF and the 8-μF capacitors—share 48 μC of charge in direct proportion to their capacitance values. Therefore, the charge on the 4-μF capacitor is

$$Q_4 = \tfrac{4}{12} \times (48 \ \mu\text{C}) = 16 \ \mu\text{C}$$

and the charge on the 8-μF capacitor is

$$Q_8 = \tfrac{8}{12} \times (48 \ \mu\text{C}) = 32 \ \mu\text{C}$$

The charges on the 2-μF and 6-μF capacitors are shared in the same way. The 24-μF capacitor carries the full 48 μC of charge.

(c) The potential difference between points B and A is 12 V. That is,

$$\Phi_B - \Phi_A = (\Phi_B - \Phi_Y) + (\Phi_Y - \Phi_X) + (\Phi_X - \Phi_A)$$

$$= V_1 + V_2 + V_3 = V = 12 \text{ V}$$

Each of the capacitors C_1, C_2, and C_3 carries the same charge Q. Thus, we have

$$V_1 = \frac{Q_1}{C_1} = \frac{Q}{C_1} = \frac{48 \ \mu C}{12 \ \mu F} = 4 \text{ V}$$

$$V_2 = \frac{Q_2}{C_2} = \frac{Q}{C_2} = \frac{48 \ \mu C}{8 \ \mu F} = 6 \text{ V}$$

$$V_3 = \frac{Q_3}{C_3} = \frac{Q}{C_3} = \frac{48 \ \mu C}{24 \ \mu F} = 2 \text{ V}$$

and $V_1 + V_2 + V_3 = 12$ V, as expected.

Although the reduction of the network in this example was straightforward, it should be noted that some types of two-terminal networks cannot be treated in this simple manner using only the series and parallel addition rules. For example, the network shown in the diagram below cannot be reduced using these techniques. (However, see Problem 26–30 for a readily solvable special case.)

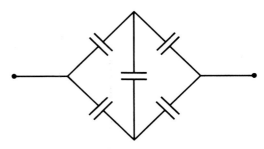

26–5 DIELECTRICS AND POLARIZATION

At about the time of Franklin's experiments with electricity it was discovered that the amount of charge that could be accumulated on a capacitor depends both on the geometry and on the insulating material between the conducting surfaces.* This effect can be demonstrated by inserting a slab of insulating material (a *dielectric*) between the plates of a parallel-plate capacitor. When a glass slab completely fills the space between the plates, it is found that a charge Q is accumulated on the plates for an applied potential difference only about one quarter of that required without the glass slab (Fig. 26–8). This implies that the capacitance value of the plates is increased by about a factor of 4 when the glass slab is used as a dielectric. The ratio of the capacitance C of a capacitor with a dielectric inserted to the capacitance C_0 with vacuum (or air) between the plates is called the *dielectric constant* κ of the material. That is,

$$\kappa = \frac{C}{C_0} \qquad \qquad \textbf{(26–15)}$$

Experiments shown that $\kappa > 1$ for all dielectric materials. (Values of κ for some materials are given in Table 26–1 on page 515.)

Because a capacitor with a dielectric requires a potential difference of only $1/\kappa$ of that required to accumulate the same charge without the dielectric present, it follows that the electric field strength $\mathbf{E}$ in the dielectric is smaller by the same factor. That is,†

$$\mathbf{E} = \frac{1}{\kappa} \mathbf{E}_0 \qquad \qquad \textbf{(26–16)}$$

We now inquire how this reduction in field strength comes about.

---

*In Franklin's day electric charge was stored in a primitive type of capacitor called a *Leyden jar*. This device consists of a glass jar with inner and outer conducting surfaces of metal foil. A highly charged Leyden jar can be dangerous and must be handled with great care.

†We consider only homogeneous isotropic dielectric materials and ignore the complications that may arise for crystalline substances. In the latter case, the dielectric ''constant'' is actually a tensor.

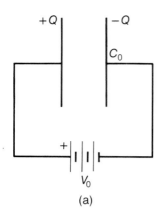

$+Q$ $-Q$

C_0

$+$

V_0

(a)

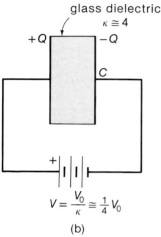

glass dielectric

$\kappa \cong 4$

$+Q$ $-Q$

C

$+$

$V = \dfrac{V_0}{\kappa} \cong \tfrac{1}{4} V_0$

(b)

Fig. 26–8. (a) A parallel-plate capacitor with vacuum (or air) between the plates is charged to a potential difference V_0. (b) When a glass dielectric is inserted between the plates, a potential difference of only about $\tfrac{1}{4}V_0$ is required to place the same charge on the plates.

Fig. 26–9. (a) In the absence of an external electric field the positive charge center ($\oplus$) and the negative charge center ($\ominus$) coincide for a nonpolar molecule. (b) When a field is applied, the charge centers are displaced by an amount ℓ.

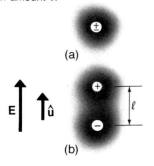

$\pm$

(a)

$+$

$-$

ℓ

E $\hat{u}$

(b)

Consider the effect on the molecules of a dielectric material when an external macroscopic electric field is applied. We distinguish two types of behavior, depending on whether the dielectric consists of a *polar* or a *nonpolar* material. In a polar material the electrons are distributed with respect to the positive nuclear charges in such a way that the individual molecules possess permanent dipole moments. In a nonpolar material, on the other hand, the charge distribution does not have this characteristic and the molecules do not possess permanent dipole moments. In the presence of an electric field, however, the positive and negative molecular charges in a nonpolar material experience oppositely directed forces and become displaced from each other by some average distance $\boldsymbol{\ell} = \ell\hat{u}$, where $\hat{u}$ is a unit vector in the direction of the applied field (Fig. 26–9b). The direction of $\boldsymbol{\ell}$ is taken to be from the negative charge center of the molecule to the positive charge center. Let the total individual molecular charge (of either sign) be q_0. Then the quantity $q_0\ell$ is defined to be the induced molecular *dipole moment* $\mathbf{p}$ (compare Fig. 24–5):

$$\mathbf{p} = q_0\boldsymbol{\ell} \qquad (26\text{–}17)$$

Further, the dipole moment density or *polarization* $\mathbf{P}$ for a dielectric material is defined to be

$$\mathbf{P} = \mathcal{N}\mathbf{p} \qquad \text{nonpolar dielectric} \qquad (26\text{–}18)$$

where $\mathcal{N}$ is the number density of molecules (molecules per cubic meter). Experiments show that under ordinary circumstances the polarization $\mathbf{P}$ is proportional to the strength of the applied electric field.

In a polar dielectric material each molecule possesses a permanent dipole moment $\mathbf{p}_0$ that results from its internal charge distribution. In the absence of an external electric field, however, the individual moments are randomly oriented so that the material as a whole has no dipole moment (Fig. 26–10a). When an external field is applied, the molecular dipoles experience a torque that tends to align them with the field direction (see Section 25–4). Total alignment is not achieved because of the disorienting effects of thermal agitation (Fig. 26–10b). The resulting average alignment $\langle\mathbf{p}_0\rangle$* is in the direction $\hat{u}$ of the field and, under ordinary circumstances, is proportional to the strength of the applied field. For a polar dielectric material the polarization $\mathbf{P}$ is expressed as

$$\mathbf{P} = \mathcal{N}\langle\mathbf{p}_0\rangle \qquad \text{polar dielectric} \qquad (26\text{–}19)$$

The formal similarity of the expressions for $\mathbf{P}$ (Eqs. 26–18 and 26–19) does not mean that the behavior of polar and nonpolar molecules is the same. In fact, the dielectric constants for polar materials are about an order of magnitude larger than those for nonpolar materials.

Imagine that the space occupied by a dielectric (either polar or nonpolar) is divided into a large number of identical cells (Fig. 26–11) with length L in the direction of the polarization vector $\mathbf{P}$ and with cross-sectional area d^2. Within each cell the actual molecular dipoles are replaced by fictitious charges $\pm q_P$ distributed on the front and rear faces of the cell, as shown in Fig. 26–11. These are *bound* charges that are not free to migrate through the dielectric. We consider only cases in which the polarization $\mathbf{P}$ in the medium is constant throughout. Then, for two adjacent cells, the common wall carries a charge $+q_P$ from one cell and $-q_P$ from the other. Thus, in the interior of the dielectric material

---

*The slanted brackets around $\mathbf{p}_0$ imply an average over the dipole moment distribution. (See the footnote on page 392.)

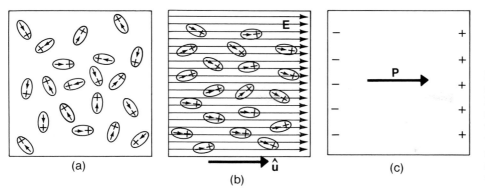

Fig. 26–10. (a) In the absence of an external electric field, the molecular dipoles are randomly oriented. (b) When a field is applied, the dipoles are partially aligned in the field direction û. (c) The net result is the appearance of bound surface charges on the ends of the material.

these charges cancel everywhere. However, bound charges remain on the macroscopic surface of the dielectric, as indicated in Fig. 26–11.

The polarization P is the dipole moment density. Therefore, the total dipole moment of the cell is equal to $P \cdot (Ld^2)$. Alternatively, the total dipole moment of the cell can be expressed as $q_P L$. From the two expressions for the total dipole moment we see that $P = |q_P|/d^2$. (The sign of the polarization charge $q_P \rightarrow \sigma_P$ is, of course, different at the two ends of the cell.)

In Fig. 26–11 the polarization vector **P** is perpendicular to the cell face with dimensions d by d. The actual (slanted) surface area S makes an angle θ with the cell face such that $S \cos \theta = d^2$. The surface charge density that appears on the macroscopic surface of the dielectric with normal unit vector **n̂** is, therefore,

$$\sigma_P = \frac{q_P}{S} = \frac{q_P}{d^2} \cos \theta = P \cos \theta$$

or

$$\sigma_P = \mathbf{P} \cdot \mathbf{\hat{n}} \qquad (26\text{–}20)$$

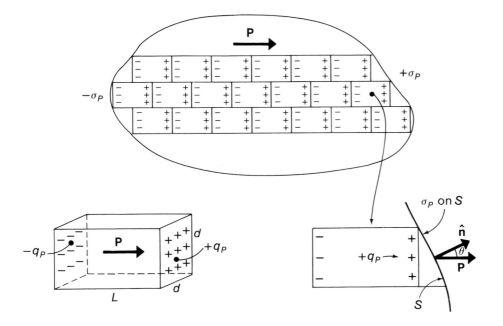

Fig. 26–11. A dielectric substance is divided into a large number of identical cells, each with a dipole moment density **P**. In the bulk material, the bound charges on the cell walls cancel throughout the interior, leaving a surface polarization charge density $\pm \sigma_P$.

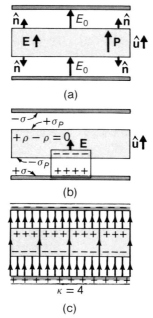

Fig. 26–12. (a) A dielectric slab is placed between the plates of a parallel-plate capacitor. The field strength in the gap is $E_0 = \sigma/\epsilon_0$. (b) A Gaussian surface is constructed to show that the field strength in the dielectric is $E = (\sigma - \sigma_P)/\epsilon_0 = \sigma/\epsilon_0\kappa$. (c) Field lines for the case $\kappa = 4$.

This effective polarization charge is still *bound* to neutral molecules in the dielectric. It is important to distinguish carefully between this bound charge and any *free* charge that might be deposited on the surface.

The induced surface charge on the dielectric is the key to understanding the reduction in the electric field strength of a capacitor that occurs when the dielectric is inserted. The induced charges are opposite in polarity to the charges on the adjacent capacitor plates, and they establish an electric field opposed to the field E_0 of the capacitor. The superposition of the two fields is less than that of the capacitor alone. We will now look at the quantitative details for the case of the parallel-plate capacitor.

Figure 26–12 illustrates a case of uniform polarization in which a slab of a dielectric material is inserted between the plates of an ideal parallel-plate capacitor. Let the surface density of free charge on the plates be $\pm\sigma$ with a corresponding electric field $\mathbf{E}_0 = E_0\hat{\mathbf{u}}$ in the gaps between the plates and the dielectric slab (Fig. 26–12a). As before, we have $E_0 = \sigma/\epsilon_0$. The polarization $\mathbf{P}$ produces a polarization surface charge density, $\sigma_P = \mathbf{P}\cdot\hat{\mathbf{n}}$, on the two parallel faces of the dielectric. Figure 26–12b shows the construction of a Gaussian surface with faces that are parallel and perpendicular to $\hat{\mathbf{u}}$. The face within the lower conducting plate is in a region where the field is zero. The electric field strength inside the dielectric slab is $\mathbf{E} = E\hat{\mathbf{u}}$. Therefore, there is a net flux EA through the face of the Gaussian surface that lies within the dielectric. The total charge inside the surface is $(\sigma - \sigma_P)A$, so that $EA = (\sigma - \sigma_P)A/\epsilon_0$. Thus,

$$E = (\sigma - \sigma_P)/\epsilon_0 \qquad (26\text{–}21)$$

According to Eq. 26–16, we have $\mathbf{E} = \mathbf{E}_0/\kappa$ and, hence, $E = \sigma/\epsilon_0\kappa$. Combining this last expression with Eq. 26–21 gives

$$\sigma_P = \sigma\left(1 - \frac{1}{\kappa}\right) \qquad (26\text{–}22)$$

Because $\kappa > 1$, we have $\sigma_P < \sigma$; however, for moderately large values of $\kappa(\kappa \geqslant 10)$, σ_P is nearly equal to σ. The presence of a large polarization charge greatly diminishes the field E within the dielectric, with the result $E \ll E_0$. Figure 26–12c shows the field lines for a glass dielectric with $\kappa = 4$. Notice that the field lines all originate on positive charges (free and polarization) and terminate on negative charges.

We can summarize these results with the expressions

$$\frac{1}{\epsilon_0}\mathbf{P} = \mathbf{E}_0 - \mathbf{E}$$

$$= \left(1 - \frac{1}{\kappa}\right)\mathbf{E}_0 \qquad (26\text{–}23)$$

$$= (\kappa - 1)\mathbf{E}$$

$$= \chi\mathbf{E}$$

where, in the last equality, the *electric susceptibility* χ of a dielectric material is defined to be

$$\chi = \kappa - 1 \qquad (26\text{–}24)$$

Values of the dielectric constant κ for various substances are listed in Table 26–1.

TABLE 26–1. Dielectric Properties of Various Substances

GASES*	χ	SOLIDS‡	κ
Helium, He	65×10^6	Paraffin	2.20
Hydrogen, H_2	254×10^6	Polyethylene	2.3
Air (dry, CO_2 free)	536×10^6	Polystyrene	2.56
Carbon dioxide, CO_2	922×10^6	Sulfur	3.44
LIQUIDS†	κ	"Ruby" mica	≈5.4
		Pyrex glass	≈5.6
Transformer oil	2.24	Glass (ordinary)	4 to 6
Benzene, C_6H_6	2.28	Sodium chloride	5.62
Chlorobenzene, C_6H_5Cl	5.71	Neoprene rubber	≈6.7
Ethyl alcohol, C_2H_5OH	24.3	High alumina porcelain	≈8.8
Methyl alcohol, CH_3OH	33.6	Titanium dioxide	173
Water, H_2O	80.4	Barium titanate	≈1500

*For 20° C, 1 atm.
†For 20° C.
‡Near room temperature.

Because κ is nearly unity for gases, it is more convenient to indicate the dielectric properties of gases in terms of the electric susceptibility χ.

For any dielectric material, there is a maximum electric field strength that the material can withstand without breaking down and losing its insulating quality. This maximum tolerable field strength is called the *dielectric strength* of the material. Values of the dielectric strength for some materials are given in Table 26–2.

TABLE 26–2. Dielectric Strengths for Various Substances

SUBSTANCE	DIELECTRIC STRENGTH (10^6 V/m)
Air	3
Epoxy cast resin	16
Glass (ordinary)	9
Lucite	18–20
Paraffin	10
Polyethylene	18
Transformer oil	12

Example 26–7

A wafer of titanium dioxide has an area of 1 cm^2 and a thickness of 0.10 mm. Aluminum is evaporated on the parallel faces to form a parallel-plate capacitor.

(a) Calculate the capacitance.

(b) When the capacitor is charged with a 12-V battery, what is the magnitude of charge delivered to each plate?

(c) For the situation in (b), what are the free and polarization surface-charge densities?

(d) What is the electric field strength E?

Solution:

(a) Combining Eqs. 26–5 and 26–15, we have

$$C = \kappa C_0 = \frac{\kappa \epsilon_0 A}{s} = \frac{(173)(8.85 \times 10^{-12} \text{ F/m})(10^{-4} \text{ m}^2)}{0.10 \times 10^{-3} \text{ m}}$$

$$= 1.53 \times 10^{-9} \text{ F}$$

(b) The battery delivers the *free* charge Q:

$$Q = CV = (1.53 \times 10^{-9} \text{ F})(12 \text{ V}) = 18.4 \text{ nC}$$

(c) The surface density of free charge is

$$\sigma = \frac{Q}{A} = \frac{18.4 \times 10^{-9} \text{ C}}{10^{-4} \text{ m}^2} = 1.84 \times 10^{-4} \text{ C/m}^2$$

The surface density of polarization charge is, using Eq. 26–22,

$$\sigma_P = \sigma\left(1 - \frac{1}{\kappa}\right) = \sigma\left(1 - \frac{1}{173}\right) = 0.9942 \ \sigma$$

(d) We have $E = E_0/\kappa$ and $E_0 = V/s$; hence,

$$E = \frac{V}{\kappa s} = \frac{12 \text{ V}}{(173)(10^{-4} \text{ m})} = 694 \text{ V/m}$$

26–6   THE DISPLACEMENT VECTOR

In the general case, the polarization **P** within a dielectric can be nonuniform. Then, within a volume V in the dielectric there can be a net polarization charge Q_P. For the case of free charge, we have $\mathbf{E}\cdot\hat{\mathbf{n}} = \sigma/\epsilon_0$ and a corresponding expression for Gauss's law, namely, $\oint_S \mathbf{E}\cdot\hat{\mathbf{n}} \ dA = Q/\epsilon_0$. Thus, the polarization expression $\mathbf{P}\cdot\hat{\mathbf{n}} = \sigma_P$, suggests that there exists a Gauss's-law relationship involving the polarization **P**. We assert that this relationship is

$$\oint_S \mathbf{P}\cdot\hat{\mathbf{n}} \ dA = -Q_P \tag{26–25}$$

where the sign is opposite to that in the expression for Gauss's law for **E** because the polarization vector **P** is directed from negative to positive charges, whereas the electric field vector **E** is directed from positive to negative charges. Also, we have used the standard convention that $\hat{\mathbf{n}}$ is directed outward from the volume V. Notice that when **P** is uniform, the surface integral vanishes for any completely interior surface. (• Can you see why?) In this case we have $Q_P = 0$, in agreement with the discussion in Section 26–5.

Figure 26–13 shows a Gaussian surface S at the boundary of a dielectric material. If the surface density of polarization charge is σ_P, Gauss's law can be expressed as

$$\mathbf{P}\cdot\hat{\mathbf{n}}' \ \Delta A = -\sigma_P \ \Delta A$$

because the polarization is zero on the part of the Gaussian surface that lies outside the dielectric. Thus, we have

$$\mathbf{P}\cdot\hat{\mathbf{n}}' = -\sigma_P \tag{26–26}$$

Notice that this result is the same as that expressed by Eq. 26–21 when we realize that $\hat{\mathbf{n}}$ in Eq. 26–20 is an outward normal at the dielectric surface, whereas $\hat{\mathbf{n}}'$ in Eq. 26–26 is directed inward (see Fig. 26–13).

When there is present an effective polarization charge Q_P as well as a free charge Q, we have

$$\oint_S \mathbf{E}\cdot\hat{\mathbf{n}} \ dA = \frac{1}{\epsilon_0} (Q + Q_P)$$

Shifting the term Q_P/ϵ_0 to the left-hand side of the equation and using Eq. 26–25, we obtain

$$\oint_S (\epsilon_0 \mathbf{E} + \mathbf{P})\cdot\hat{\mathbf{n}} \ dA = Q \tag{26–27}$$

This expression is often useful because the surface integral is expressed in terms of only the free charge Q, which can be specified easily.

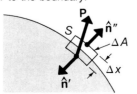

Fig. 26–13. A Gaussian surface S at the boundary of a dielectric material. The sides of the volume, with thickness Δx, are perpendicular to the boundary.

The sum $\epsilon_0 \mathbf{E} + \mathbf{P}$ defines a new vector, called the *electric displacement* **D**:

$$\mathbf{D} = \epsilon_0\mathbf{E} + \mathbf{P} \qquad \text{(26-28)}$$

Using this definition in Eq. 26-27, we have a new form of Gauss's law, namely,

$$\oint_S \mathbf{D}\cdot\hat{\mathbf{n}}\, dA = Q = \int_V \rho\, dv \qquad \text{(26-29)}$$

The displacement **D** represents a vector field that depends only on the *free charge;* even in the presence of matter, Q does not involve polarization charges.

According to Eq. 26-23, we have $\mathbf{P} = \epsilon_0(\kappa - 1)\mathbf{E}$; hence, we can also express **D** as

$$\mathbf{D} = \epsilon_0\kappa\mathbf{E} = \epsilon\mathbf{E} \qquad \text{(26-30)}$$

where $\epsilon = \epsilon_0\kappa$ is called the *dielectric permittivity.*

These equations are valid for dielectric materials that are homogeneous, isotropic, and linear (that is, $\mathbf{P} \propto \mathbf{E}$); such materials are called *Class A dielectrics.*

QUESTIONS

26-1 Define and explain briefly the meaning of the terms (a) capacitance, (b) fringing field, (c) spherical capacitor, (d) dielectric constant, (e) polarization, and (f) dielectric strength.

26-2 The drawing shows the lines of force between two conductors with equal and opposite charges $\pm Q$. The potential difference between the conductors is V. What would happen to the surface charge distributions if the two charges were doubled in magnitude? In what way would the pattern of the lines of force change? What would happen to the potential difference?

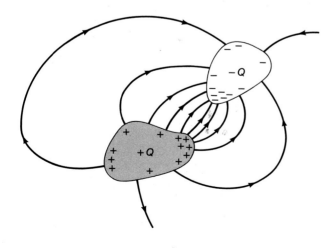

26-3 Assume that the drawing for the previous question represents two insulators with surface charges as shown. If the magnitude of the total charges were doubled by adding more charges distributed on the two surfaces in an arbitrary way, what would be the effect on the pattern of the lines of force? Would the concept of capacitance be relevant for the two insulators?

26-4 Why is it important to avoid sharp edges on metallic components in a high-voltage circuit such as that found in a television receiver?

26-5 You are given three capacitors, $C_1 = 1\ \mu\text{F}$, $C_2 = 2\ \mu\text{F}$, and $C_3 = 2\ \mu\text{F}$. Using these capacitors either singly or in combinations of two or three together, how many different effective capacitors can be formed?

26-6 Explain in physical terms the difference between free charges and polarization charges.

26-7 Describe what happens to the potential, the charge, the electric field, and the stored energy for a parallel-plate capacitor when (a) the charging battery is disconnected and the spacing between the plates is then doubled, (b) the charging battery is disconnected and the space between the plates is filled with oil ($\kappa = 2.5$), (c) the charging battery is left connected and the spacing between the plates is doubled, (d) the charging battery is left connected and the space between the plates is filled with oil.

26–8   Explain in physical terms why an external force is necessary to hold the plates of a parallel-plate capacitor apart.

26–9   A clever student claims that force between the plates of a parallel-plate capacitor may be derived as follows: The electric field at the position of the lower plate in the drawing due to *only* the charge on the upper plate is $E = \frac{Q}{2\epsilon_0 A}\hat{n}$. The force on the lower plate is therefore

its charge, Q, multiplied by this field or $\mathbf{F} = \frac{Q^2}{2\epsilon_0 A}\hat{n}$.

Comment on this derivation.

26–10   Define and explain briefly the meaning of the terms (a) displacement vector and (b) dielectric permittivity.

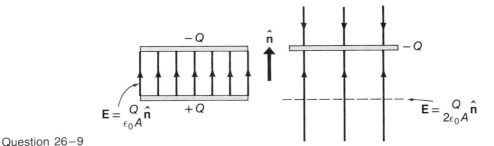

Question 26–9

PROBLEMS

Section 26–1

26–1   When a potential difference of 150 V is applied to the plates of a parallel-plate capacitor, the plates carry a surface-charge density of 30 nC/cm². What is the spacing between the plates?

26–2   A small object with a mass of 350 mg carries a charge of 30 nC and is suspended by a thread between the vertical plates of a parallel-plate capacitor. The plates are separated by 4 cm. The thread is observed to make an angle of 15° with the vertical. What is the potential difference between the plates?

26–3   One plate of a parallel-plate capacitor (with area $A = 100$ cm²) is grounded. The ungrounded plate carries a fixed charge Q. When the ungrounded plate is moved 0.5 cm farther away from the grounded plate, the potential difference between the plates increases by 200 V. Determine the magnitude of the charge Q.

26–4   The ratio of the plate spacing to the radius for a circular parallel-plate capacitor is 1/10. Include fringing effects and determine the capacitance for a radius $R = 10$ cm.

Section 26–2

26–5   A spherical capacitor consists of a conducting ball with a diameter of 10 cm that is centered inside a grounded conducting spherical shell with an inner diameter of 12 cm. What capacitor charge is required to achieve a potential of 1000 V on the ball?

26–6   Show that a spherical capacitor for which the surface radii are nearly equal (that is, $R_2 - R_1 \ll R_1$) is equivalent to a parallel-plate capacitor with a spacing $s = R_2 - R_1$ and a plate area $A = 4\pi R_1^2$. If this is so, the electric field strength in the gaps of the two equivalent capacitors should be the same for the same capacitor charge. Is this actually the case?

26–7   (a) A small spherical droplet of water has a radius of 2.5 mm. What is the charge on the droplet if its potential relative to $\Phi(\infty) = 0$ is 470 V? (b) Two such charged droplets unite to form a single larger spherical droplet without the loss of any charge. What is the potential on the larger droplet? (In each case, assume that the charge resides on the droplet surface.)

26–8   Compare the capacitance of a spherical capacitor having $R_1 = 7$ cm and $R_2 = 8$ cm with that of a cylindric capacitor having the same radii and an axial length of 15 cm. (In each case the outer conductor is grounded.) Why are the capacitance values nearly equal?

26–9   Show that the expression found in Example 26–2 for the capacitance of a cylindric capacitor reduces to that for a parallel-plate capacitor with dimensions $L \times 2\pi R_1$ when $s = R_2 - R_1$ and $s/R_1 \ll 1$. Argue that this result is reasonable.

26–10   Two conducting spheres with diameters of 0.40 m and 1.0 m are separated by a distance that is large compared with the diameters. The spheres are connected by a thin wire and are charged to 7 μC. (a) How is this total charge shared between the spheres? (Neglect any charge on the wire.) (b) What is the potential of the system of spheres relative to $\Phi(\infty) = 0$?

Section 26-3

26-11 Under standard conditions of temperature and pressure, dry air will break down when a field strength of approximately 3×10^6 V/m is reached. (a) What is the field energy density u under these conditions? (b) If this field strength were achieved in a parallel-plate capacitor with a plate area of 10 cm^2, what would be the magnitude of the force tending to pull the plates together? [*Hint:* See Question 26-9.]

26-12 Use $u = \frac{1}{2}\epsilon_0 E^2$ to calculate explicitly the electrostatic field energy stored in a cylindric capacitor that carries charges $\pm Q$. Show that your expression leads to $U = Q^2/2C$.

26-13 A 4-μF capacitor is charged to a potential difference of 150 V. (a) What amount of energy is stored in the capacitor? (b) How much work would have to be done by an external agency to double the amount of charge on the capacitor?

Section 26-4

26-14 (a) Determine the equivalent capacitance for the capacitor network shown in the diagram. (b) If the network is connected to a 12-V battery, calculate the potential difference across each capacitor and the charge on each capacitor.

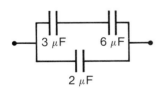

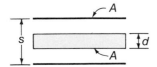

26-15 How should four 2-μF capacitors be connected to have a total capacitance of (a) 8 μF, (b) 2 μF, (c) 1.5 μF, and (d) 0.5 μF?

26-16 If in each of the networks of Problem 26-15, a 12-V battery is connected to the outside terminals of the combinations, what are the potential differences and charges on each capacitor?

26-17 A conducting slab with a thickness d and area A is inserted into the space between the plates of a parallel-plate capacitor with spacing s and area A, as shown in the

diagram. What is the value of the capacitance of the system?

Section 26-5

26-18 A coaxial cable consists of a central conductor with a diameter of 1 mm inside a thin outer conductor with a diameter of 7 mm. The dielectric material between the conductors is polystyrene. Calculate the capacitance per meter of length of the cable.

26-19 A rectangular parallel-plate capacitor has a dielectric
• slab that partially fills the space between the plates, as shown in the diagram. Show that the capacitance is

$$C = (\epsilon_0 b/s)[\kappa a - (\kappa - 1)x]$$

Does this expression reduce to the expected values when $x = 0$ and when $x = a$?

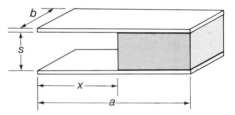

26-20 A parallel-plate capacitor has a spacing $s = 1.6$ mm. The capacitor, with air between the plates, is charged to a potential difference of 800 V and then isolated. Next, a slab of polystyrene is inserted, filling the space between the plates. (a) What is the electric field strength in the dielectric? (b) Determine the polarization P in the dielectric. (c) Calculate the surface densities of free charge (σ) and polarization charge (σ_P).

26-21 A parallel-plate capacitor can be charged to 3 C before breakdown occurs when air is between the plates. What maximum charge can the plates carry if the space between the plates is filled with (a) glass ($\kappa = 5$) or (b) polyethylene?

Section 26-6

26-22 Repeat the discussion leading to Eq. 26-12 for the case
• in which a dielectric material with dielectric constant κ completely fills the space between the plates of a parallel-plate capacitor. Show that $u = \frac{1}{2}\mathbf{E}\cdot\mathbf{D}$. (This result is valid only for *linear* dielectrics.)

26-23 Show that the capacitance of a parallel-plate capacitor in
•• which the region between the plates is partially filled with a dielectric slab, as shown in the diagram on page 520, is

$$C = \frac{A\epsilon_0}{t + d/\kappa}$$

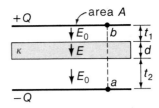

Problem 26–23

$2\kappa_1 Q/(\kappa_1 + \kappa_2)A$ and $\sigma_2 = 2\kappa_2 Q/(\kappa_1 + \kappa_2)A$, where $A = ab$. (b) Show that $V = \sigma_1 s/\epsilon_0 \kappa_1 = \sigma_2 s/\epsilon_0 \kappa_2$. (c) Show that $C = (\kappa_1 + \kappa_2)\epsilon_0 A/2s$.

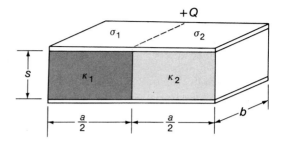

26–24 Calculate explicitly the field energy stored in the capacitor of Problem 26–23 by using the result of Problem 26–22, namely, $u = \frac{1}{2}\mathbf{E} \cdot \mathbf{D}$. Does the stored energy equal $Q^2/2C$, with C equal to the value given in Problem 26–23?

26–25 One half of the space between the plates of a rectangular
•• parallel-plate capacitor is filled with a dielectric material κ_1 and the other half is filled with a material κ_2, as indicated in the diagram. (a) Show that $\sigma_1 =$

26–26 Derive the capacitance value given in Problem 26–25c by considering the system to consist of two capacitors connected in parallel.

Additional Problems

26–27 A small object has a mass of 12 mg and carries a charge of 8 nC. When this object is released between the horizontal plates of a parallel-plate capacitor, it experiences a downward acceleration of $\frac{1}{3}g$. If the plate spacing of the capacitor is $s = 2.5$ cm, what is the potential difference between the plates?

26–28 Consider the Earth to be an isolated spherical conductor. (a) Calculate the capacitance of the Earth. (b) What excess charge on the Earth would produce a potential of 10^6 V relative to $\Phi(\infty) = 0$?

26–29 The field energy of a uniform *volume* distribution of
• charge Q with radius R is greater than that for a uniform *surface* distribution of the same charge with the same radius. (• Can you see qualitatively why this is so?) Show that the field energy of the volume distribution is $6/5$ times that of the surface distribution. [*Hint:* In each case, first calculate $E(r)$, then evaluate $\int 4\pi r^2 u(r)\, dr$.]

26–30 Five capacitors, each with $C = 2\ \mu$F, are connected to
• form the network shown in the diagram. Determine the equivalent capacitance between terminals A and B. [*Hint:* From the symmetry, deduce the potential difference $\Phi_a - \Phi_b$ when the network is connected to a battery.]

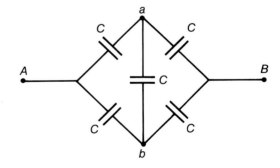

CHAPTER 27

CURRENT, RESISTANCE, AND DC CIRCUITS

Under equilibrium conditions—that is, when the charge distribution is static—the surface of a conductor is an equipotential and all of the field lines are perpendicular to the surface. Within the conductor the electric field is zero and no net charge is present. If this equilibrium condition is disturbed—for example, by placing a potential difference across the conductor—the electric field within the conductor is no longer zero. The existence of an internal field means that charges will move within the conductor; that is, an electric *current* will flow. Moreover, the conductor surface is no longer an equipotential.

In this chapter we study the special case in which there is a steady current within a conductor. This is known as *direct current* (DC). After developing the concept of resistance, we consider single-loop circuits involving resistors and batteries. Such circuits are easy to analyze because the same current flows through each of the circuit elements. But most circuits met in practice are more complicated and may involve several interconnecting loops. (A multiloop system within a circuit is called a *network*.) We begin by examining methods for reducing complicated networks to simpler ones. Then, general techniques for solving circuit problems are discussed. Attention here is restricted to direct-current (DC) situations in which all currents are constant in time. In Chapter 32 we discuss cases involving nonsteady currents, with emphasis on alternating-current (AC) circuits.

27–1 ELECTRIC CURRENT

The instantaneous rate at which positive charge passes through a surface is called the *electric current* through the surface. Figure 27–1 shows a surface S that intersects a conductor through which charge is flowing. The current I through the surface S is equal to the rate dq/dt at which positive charge passes through the surface. That is,

$$I = \frac{dq}{dt} \qquad (27-1)$$

In this chapter we are concerned with cases for which I is constant.

From Eq. 27–1 we see that the dimensions of current are charge per unit time. The units of current are *coulombs per second* (C/s), to which is given the special name

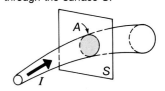

Fig. 27–1. A current $I = dq/dt$ flows along the conductor and through the surface S.

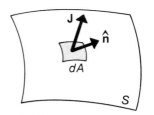

Fig. 27–2. The current dI through the differential element $\hat{n}$ dA of the surface S is $dI = \mathbf{J} \cdot \hat{n}\, dA$.

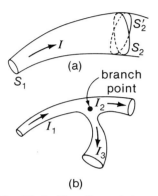

Fig. 27–3. The effects of charge conservation for a steady current. (a) The same current I flows through S_1 as flows through S_2 or S_2'. (b) The current I_1 divides at the branch point into two parts such that $I_1 = I_2 + I_3$.

ampere*:

$$1 \text{ ampere (A)} = 1 \text{ C/s}$$

Current Density. In many situations it is convenient to refer to the *current density* $\mathbf{J}$ in a conductor. For the case illustrated in Fig. 27–1, the *average* current density in the conductor as it passes through the intersecting surface S is $J = I/A$ and the direction of $\mathbf{J}$ is in the direction of the current flow. In general, the current density is defined in terms of the flow through a differential element, $\hat{n}\, dA$, of a surface S, as indicated in Fig. 27–2. Here, we have

$$dI = \mathbf{J} \cdot \hat{n}\, dA \qquad (27\text{–}2)$$

and the total current through any surface S is

$$I = \int_S \mathbf{J} \cdot \hat{n}\, dA \qquad (27\text{–}3)$$

Usually, we are concerned with *open* surfaces. Current density is measured in A/m^2.

Figure 27–3 illustrates the behavior imposed on the flow of current by the law of charge conservation. If I represents a steady current in Fig. 27–3a, no charge can accumulate in the volume defined by the surfaces S_1 and S_2 and the conductor wall. It follows that the same total current I flows through S_1 as through S_2. Moreover, the surface S_2 can be replaced by any surface, such as S_2', whose perimeter is everywhere on the conductor wall, without affecting the value of the current.

In Fig. 27–3b the current I_1 divides into two parts, I_2 and I_3, that flow through separate strands of the conductor. If the current is steady, charge conservation requires $I_1 = I_2 + I_3$.

It should be apparent that electric current bears a strong similarity to fluid flow (see Chapter 14). The conditions we have discussed here—called the *continuity conditions*—are the result of steady current flow and charge conservation; the corresponding conditions for fluid flow are the result of incompressibility and matter conservation.

27–2 RESISTIVITY

On the microscopic level, electric current is due to the movement of the conduction electrons (charge, $-e$) or of ions (charge, one or more times $\pm e$). The movements of these individual charge carriers are governed by statistical thermodynamic factors, so we imagine that all macroscopic equations, such as Eq. 27–1, are obtained by appropriate averages over microscopic thermal effects.

In a conducting wire, such as that illustrated in Fig. 27–4, the atoms remain in fixed positions, so the only charge carriers are the free electrons. Let n be the number of free electrons per unit volume. Suppose that an electric field is imposed on the wire so that a conventional positive current flows to the right. In addition to their random thermal motions, the electrons experience a steady drift to the left in response to the electric field. We introduce the average free electron velocity† $\langle \mathbf{v} \rangle$ and identify this with the electron

---

*Named in honor of the French physicist André-Marie Ampère (1775–1836), one of the founders of the subject of electrodynamics. In Section 29–4 we give a precise definition of the ampere.

†The slanted brackets around $\mathbf{v}$ imply an average taken over the velocity distribution. See the footnote on page 392.

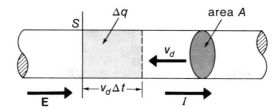

Fig. 27–4. The current *I* that flows to the right in the wire is the result of electrons drifting to the left with the average velocity $\langle \mathbf{v} \rangle = \mathbf{v}_d$.

drift velocity, $\mathbf{v}_d = \langle \mathbf{v} \rangle$. Then, in a time interval Δt, the amount of charge that passes through the surface S is the charge Δq contained within the volume $v_d A \, \Delta t$ (see Fig. 27–4); thus,

$$\Delta q = e n v_d A \, \Delta t \qquad (27\text{–}4)$$

The current I that flows in the wire is defined to be the net amount of positive charge that passes through the surface S per unit time. The *negative* electrons move toward the *left,* so the conventional current is directed toward the *right.* We have

$$I = \frac{\Delta q}{\Delta t} = e n v_d A \qquad (27\text{–}5)$$

Conduction in Metals.   To continue the discussion of charge movement in matter we adopt a simple model for metallic conductors.* In this model it is supposed that some of the electrons are detached from their parent atoms and become free to move about within the material in much the same way as an electron gas. When an electric field is applied, these free electrons drift with an average velocity $\mathbf{v}_d$ and give rise to a current expressed by Eq. 27–5. The current density in the material is

$$\mathbf{J} = -e n \mathbf{v}_d \qquad (27\text{–}6)$$

where the direction of $\mathbf{J}$ is *opposite* to the direction of the electron drift velocity $\mathbf{v}_d$.

The electrons are imagined to move about within the material and to collide frequently with the lattice structure of the material. Between collisions, the electrons move in accordance with the force produced by the imposed electric field. Consider an electron that has a velocity $\mathbf{v}_0$ as the result of a particular collision. If the local field within the conductor is $\mathbf{E}$, the electron velocity at time t after collision will be

$$\mathbf{v} = \mathbf{v}_0 - (e/m_e)\mathbf{E}t \qquad (27\text{–}7)$$

where m_e is the electron mass. The drift velocity $\mathbf{v}_d$ that appears in Eq. 27–6 is just the average of Eq. 27–7 between collisions. The progress of an electron undergoing random collisions in the presence of an electric field is shown schematically in Fig. 27–5. Be-

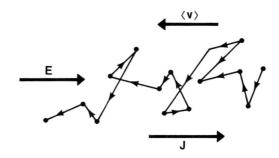

Fig. 27–5. Schematic representation of the path followed by a free electron in a conductor in which there is an electric field. Notice that the direction of the drift velocity $\mathbf{v}_d = \langle \mathbf{v} \rangle$ is opposite to the directions of $\mathbf{E}$ and $\mathbf{J}$.

---

*The consequences of this model were first worked out in 1900 by the German physicist Paul Drude (1863–1906), using the ideas of classical physics.

cause $\mathbf{v}_0$ is distributed randomly in space, it follows that $\langle \mathbf{v}_0 \rangle = 0$. Denoting by τ the average time between collisions (the *mean free time*), the average of $\mathbf{v}$ in Eq. 27–7 becomes

$$\langle \mathbf{v} \rangle = \mathbf{v}_d = -\frac{e\tau}{2m_e}\mathbf{E} \qquad (27\text{–}8)$$

(• Do you see how the factor $\frac{1}{2}$ arises when calculating $\langle \mathbf{v} \rangle$?) Substituting this expression for $\mathbf{v}_d$ into Eq. 27–6 yields

$$\mathbf{J} = \frac{e^2 n\tau}{2m_e}\mathbf{E}$$

which we can write as

$$\mathbf{J} = \frac{1}{\rho}\mathbf{E} = \sigma\mathbf{E} \qquad (27\text{–}9)$$

with

$$\rho = \frac{1}{\sigma} = \frac{2m_e}{e^2 n\tau} \qquad (27\text{–}10)$$

The quantity ρ is called the *resistivity* of the material and the reciprocal quantity σ is called the *conductivity*.* Many metals obey the relationship expressed in Eq. 27–9, which is predicted by the simple Drude model. These materials are referred to as *linear conductors* to emphasize the linear relationship between $\mathbf{J}$ and $\mathbf{E}$. In such materials, a large resistivity (or low conductivity) means that the current density will be small for a given imposed field; conversely, a material with a low resistivity (large conductivity) will exhibit a larger current density for the same imposed field. Because e and m_e are constant for all materials, the determining factors in the magnitude of ρ are the number of free electrons per unit volume, n, and the mean free time, τ.

Resistivity ρ is measured in units of *ohm-meters* (Ω·m), where $1\ \Omega = 1$ V/A (see Section 27–3). Conductivity σ is measured in the reciprocal units, namely, $(\Omega \cdot \mathrm{m})^{-1}$, for which the SI term is *siemens per meter* (S/m).

Table 27–1 gives values of n (the free-electron number density) for several metals. Also listed are the mean collision times τ calculated from Eq. 27–10 using experimental values for ρ. The quantity ξ represents the number of free electrons per atom that was assumed to obtain the value of n.

▶ *Modern Atomic View of Electrical Conductivity.* The classical theory of electrical conductivity attempted to treat the free conduction electrons as a noninteracting Maxwell-Boltzmann gas. Many detailed predictions based on this model were in serious error. According to modern theory the conduction electrons obey quite a different velocity-distribution law, one that is based on quantum restrictions imposed on the behavior of all electron systems. These restrictions force the average velocities for electrons at ordinary temperatures to be usually an order of magnitude larger than that predicted for an equivalent classical gas.

In addition, the conduction electrons are imagined to travel through a perfectly ordered crystalline lattice as free unhindered waves. In such a perfect crystal the conductivity would be infinite. Scattering of the electron waves, and thus the introduction of a finite conductivity, occurs only because of the presence of *imperfections* in the crystal lattice.

This is quite unlike the classical theory, in which the main source of electron scattering was assumed to be the collisions with the ionic lattice centers of the crystal. (In the quantum treatment the effects of the ionic lattice centers themselves are already taken into account in determining the propagation of the unhindered wave.) The scattering imperfections in quantum theory may be chemical impurities, dislocations in the lattice, polycrystalline boundaries, or thermal vibrations of the ionic centers

---

*Although ρ and σ have previously been used as symbols for different quantities, the context should readily identify which usage is meant.

TABLE 27-1. Electron Number Densities n and Mean Collision Times τ for Various Metals

	METAL	ξ^a	n (10^{28} m^{-3})	τ (10^{-14} s) $T = 77$ K	$T = 273$ K	$T = 373$ K
Alkali Metals	Li	1	4.70	14.6	1.76	1.2
	Na	1	2.65	34	6.4	Molten
	K	1	1.40	36	8.2	
	Rb	1	1.15	28	5.6	
	Cs	1	0.91	17.2	4.2	
Noble Metals	Cu	1	8.47	42	5.4	3.8
	Ag	1	5.86	40	8.0	5.6
	Au	1	5.90	24	6.0	4.2
Group 2A Metals	Be	2	24.7	—	1.0	0.54
	Mg	2	8.61	13.4	2.2	1.5
	Ca	2	4.61	—	4.4	3.0
	Sr	2	3.55	2.8	0.88	—
	Ba	2	3.15	1.32	0.38	—
Other Metals	Al	3	18.1	13.0	1.6	1.1
	Pb	4	13.2	1.14	0.28	0.20
	Bi	5	14.1	0.144	0.046	0.032

[a]Assumed number of free electrons per atom.

about their equilibrium positions. The resulting mean distance traveled between collisions (called the mean free path) for the conduction electrons in quantum theory is usually two orders of magnitude longer than predicted for the classical counterpart. The higher average velocity combined with less frequent collisions gives predicted mean collision times, τ, about one order of magnitude longer than in classical theory. These predicted values are in very good agreement with experimental values. ◄

Temperature Coefficients. Experiments show that the resistivity of a particular metal shows a nearly linear variation with temperature. Figure 27-6 shows this linear dependence for the case of copper. Because the number density n of free electrons in metals varies only slightly with temperature, Eq. 27-10 indicates that the mean collision time τ should be inversely proportional to the temperature. (The entries for τ in Table 27-1 reflect this variation.) This classical theory of metals (the Drude theory) is unable to predict correctly this relationship between τ and T. Calculations based on modern quantum theory, however, can account quite satisfactorily for the temperature dependence of τ (and, hence, also of ρ).

Fig. 27-6. Measured values of the resistivity for copper show the linear dependence on the temperature (except near absolute zero).

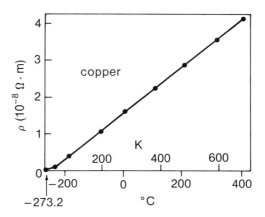

The rate at which the resistivity of a substance varies with temperature at a particular temperature can be expressed in terms of a *resistivity temperature coefficient* α:

$$\alpha = \frac{1}{\rho_0} \frac{d\rho}{dT}\bigg|_{T_0}$$

(27–11)

where $\rho_0 = \rho(T_0)$. If $\rho(T)$ were exactly linear, α would be a constant, and we would have

$$\rho(T) = \rho_0[1 + \alpha(T - T_0)]$$

(27–12)

In general, however, α is not strictly constant, so we define an *average* temperature coefficient $\bar{\alpha}$ for a temperature range near T_A and write:

$$\rho(T) = \rho_A[1 + \bar{\alpha}(T - T_A)]$$

(27–12a)

where $\rho_A = \rho(T_A)$. Values of the resistivity and the temperature coefficient are given for some selected substances in Table 27–2. Notice the extremely wide variation in ρ for different types of materials:

pure metals	$\sim 10^{-8}$ $\Omega \cdot$m
alloys	$\sim 10^{-6}$ $\Omega \cdot$m
semiconductors	$\sim 10^{-5} - 10^3$ $\Omega \cdot$m
insulators	$\sim 10^8 - 10^{17}$ $\Omega \cdot$m

Notice also that several of the values of α in Table 27–2 are *negative*, which means that the resistivity decreases with increasing temperature.

TABLE 27–2. Values of the Resistivity and the Temperature Coefficient for Various Substances

SUBSTANCE		ρ ($\Omega \cdot$m)	α (deg^{-1})
Conductors:		ρ (at 0° C)	$\bar{\alpha}$ (0° C–100° C)
Metals	Ag	1.47×10^{-8}	4.1×10^{-3}
	Cu	1.56×10^{-8}	4.3×10^{-3}
	Al	2.63×10^{-8}	4.4×10^{-3}
	Ni	6.38×10^{-8}	6.6×10^{-3}
	Li	8.55×10^{-8}	4.7×10^{-3}
	Fe	8.85×10^{-8}	6.5×10^{-3}
	Pt	9.83×10^{-8}	3.9×10^{-3}
Alloys	Manganin	4.8×10^{-7}	2×10^{-5}
	Constantan	4.9×10^{-7}	-1×10^{-5}
	Nichrome	1.1×10^{-6}	4×10^{-4}
Semiconductors:		ρ (at 20° C)	α (at 20° C)
Pure	C	3.5×10^{-5}	-5×10^{-3}
	Ge	0.45	-4.8×10^{-2}
	Si	640	-7.5×10^{-2}
Insulators:		ρ (at 20° C)	
	Amber	5×10^{14}	
	Glass	$10^{10} - 10^{14}$	
	Lucite	$>10^{13}$	
	Quartz	7.5×10^{17}	
	Sulfur	10^{15}	
	Wood	$10^8 - 10^{11}$	

Example 27-1

A length of No. 8 copper wire, which has a diameter of 3.26 mm, carries a steady current of 20 A. (The maximum current in No. 8 copper wire with rubber insulation that is permitted by the National Electrical Code is 35 A.) (a) What is the average current density J in the wire? (b) Determine the average drift velocity v_d of electrons in the wire. (c) Calculate the electric field strength E (assumed constant) along the wire.

Solution:

(a) The average current density is

$$J = \frac{I}{A} = \frac{I}{\pi R^2} = \frac{20 \text{ A}}{\pi (1.63 \times 10^{-3} \text{ m})^2} = 2.40 \times 10^6 \text{ A/m}^2$$

(b) In Table 27-1 we find $n(\text{Cu}) = 8.47 \times 10^{28} \text{ m}^{-3}$; then, using Eq. 27-6, we have

$$v_d = \frac{J}{en} = \frac{2.40 \times 10^6 \text{ A/m}^2}{(1.60 \times 10^{-19} \text{ C})(8.47 \times 10^{28} \text{ m}^{-3})}$$

$$= 1.77 \times 10^{-4} \text{ m/s}$$

or less than $\frac{1}{5}$ of a millimeter per second. This very small drift velocity is superimposed on the much larger random thermal speeds of the electrons.

(c) From Table 27-2, $\rho(\text{Cu}) = 1.56 \times 10^{-8} \text{ }\Omega\cdot\text{m}$; then, using Eq. 27-9, we obtain

$$E = \rho J = (1.56 \times 10^{-8} \text{ }\Omega\cdot\text{m})(2.40 \times 10^6 \text{ A/m}^2)$$

$$= 0.037 \text{ V/m}$$

27-3 OHM'S LAW

When a potential difference V is applied across a conducting object and causes a current I to flow, the ratio V/I is called the *resistance R* of the object:

$$R = \frac{V}{I} \qquad (27\text{-}13)$$

The unit of resistance is the ohm (Ω), where $1 \text{ }\Omega = 1 \text{ V/A}$.

The value of R for a particular conducting object (a *resistor*) depends on the material of which it is composed and on the geometry of its construction. However, for a large class of materials it has been found that R does *not* depend on the applied voltage (unless the voltage is very high). That is, for a resistor made from such a material, the ratio V/I is constant for all (ordinary) applied voltages V. Materials that follow this prescription are said to obey *Ohm's law*.*

We hasten to add that there are numerous materials and devices that do *not* obey Ohm's law. Some of these—for example, diodes, vacuum tubes, thermistors, and transistors—are specifically designed to have a nonlinear *V*-versus-*I* behavior. Indeed, for a sufficiently high voltage, any material can be made to deviate from Ohm's law. Thus, Ohm's law is by no means a universal law of Nature; instead, it is an approximate relationship that is obeyed by many materials under restricted conditions of applied voltage and temperature.

Resistivity and Resistance. Consider a homogeneous rectangular metal bar with a length ℓ and cross-sectional area A. As shown in Fig. 27-7, let the left-hand end of the bar be at a constant potential Φ_a and the right-hand end at a constant potential $\Phi_b < \Phi_a$. Then the field vector $\mathbf{E}$ is directed along the bar, and the current density $\mathbf{J}$ is also in the direction of the unit normal vector $\hat{\mathbf{n}}$. Thus,

$$\mathbf{J} = \frac{\mathbf{E}}{\rho} = -\frac{\hat{\mathbf{n}}}{\rho} \frac{d\Phi}{dx}$$

---

*After Georg Simon Ohm (1787–1854), the German physicist who discovered this effect in 1826.

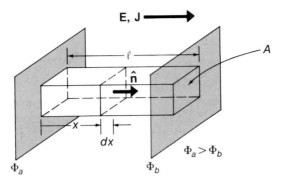

Fig. 27–7. The current density in the bar is $\mathbf{J} = -(\hat{n}/\rho)\, d\Phi/dx$.

The negative sign means that the current is directed opposite to the potential gradient $d\Phi/dx$.

The current that flows through any plane of area A that is perpendicular to the long dimension of the bar is

$$I = \mathbf{J} \cdot \hat{n} A = JA$$

Thus,

$$I = -\frac{A}{\rho}\frac{d\Phi}{dx} = -\sigma A \frac{d\Phi}{dx} \qquad (27\text{–}14)$$

Notice the similarity between Eq. 27–14 and the expression for heat conduction (Eq. 18–13). (• Identify the corresponding quantities in the two equations, and comment on how they are analogous to one another.)

When Eq. 27–14 is integrated, there results

$$\Phi(x) = \Phi_a - \frac{\rho I}{A} x$$

Thus, the equipotential surfaces, as expected, are uniformly spaced between Φ_a and Φ_b. When $x = \ell$, we have $\Phi(\ell) = \Phi_b$. Therefore, solving for I gives

$$I = \frac{A}{\rho \ell}(\Phi_a - \Phi_b) \qquad (27\text{–}15)$$

Equation 27–15 can be simplified by using the notation

$$V_{ab} = \Phi_a - \Phi_b \qquad (27\text{–}16)$$

This notation means that $V_{ab} > 0$ when $\Phi_a > \Phi_b$; that is, V_{ab} represents the *potential drop* in going from point a to point b. (In Section 26–1 we used $V_{ba} = \Phi_b - \Phi_a$ to represent the *potential increase* in going from point a to point b; thus, $V_{ab} = -V_{ba}$.) We now write Eq. 27–15 as

$$I_{ab} = \frac{A}{\rho \ell} V_{ab} \qquad (27\text{–}17)$$

where I_{ab} represents the (positive) current that flows from a to b. When no confusion can result, we drop the subscripts and write

$$I = \frac{A}{\rho \ell} V \qquad (27\text{–}17a)$$

Comparing this expression with Eq. 27–13 shows that we have

$$R = \frac{\rho \ell}{A} \quad \text{(straight bar)} \qquad\qquad \textbf{(27–18)}$$

If the resistivity ρ does not depend on the voltage, the resistance R is constant and the resistor (the conducting bar) obeys Ohm's law.

Example 27–2

(a) Calculate the resistance per meter of length for No. 20 copper wire at 20° C.

(b) Calculate the resistance at 0° C of a 2-m length of insulated manganin wire with a diameter of $\frac{1}{10}$ mm that is wrapped in a tight coil to form a *wire-wound resistor*.

(c) Determine the change in the resistance of the manganin resistor in (b) if the temperature is increased to 100° C.

Solution:

(a) No. 20 copper wire (a common size of wire, often used as electronic hookup wire) has a diameter of 0.812 mm. Using the value in Table 27–2 for the resistivity ρ_0 of copper at 0° C, we find

$$\frac{R_0}{\ell} = \frac{\rho_0}{A} = \frac{1.56 \times 10^{-8}\ \Omega\cdot\text{m}}{\pi(0.406 \times 10^{-3}\ \text{m})^2} = 0.0301\ \Omega/\text{m}$$

and using Eq. 27–12a we can write

$$\frac{R_{20}}{\ell} = \frac{R_0}{\ell}(1 + \bar{\alpha}\,\Delta T)$$

$$\frac{R_{20}}{\ell} = (0.0301\ \Omega/\text{m})[1 + (4.3 \times 10^{-3}\ \text{deg}^{-1})(20\ \text{deg})]$$
$$= 0.0327\ \Omega/\text{m}$$

(b) For the manganin coil, again using the resistivity value in Table 27–2, we find

$$R_0 = \frac{\rho_0 \ell}{A} = \frac{(4.8 \times 10^{-7}\ \Omega\cdot\text{m})(2\ \text{m})}{\pi(0.5 \times 10^{-4}\ \text{m})^2} = 122.2\ \Omega$$

(c) At a temperature of 100° C we have

$$R_{100} = R_0 (1 + \bar{\alpha}\,\Delta T)$$

so that

$$\Delta R = R_{100} - R_0 = R_0 \bar{\alpha}\,\Delta T$$
$$= (122.2\ \Omega)(2 \times 10^{-5}\ \text{deg}^{-1})(100\ \text{deg}) = 0.24\ \Omega$$

Manganin (84 percent Cu, 12 percent Mn, and 4 percent Ni) is a useful alloy for the construction of resistors when the application requires the resistance values to be stable under temperature changes.

Example 27–3

The annular gap between a coaxial pair of copper tubes is filled with pure silicon. The outer conductor is grounded ($\Phi_b = 0$), and the inner conductor is at a potential $\Phi_a = +2.0$ V. Calculate the total current for the dimensions shown in the diagram.

Solution:

Because the resistivity of copper is small compared with that of silicon, the copper surfaces may be considered to be equipotentials. Then, the electric field and the current density are everywhere radial; that is, $\mathbf{E} = E\hat{\mathbf{u}}_r$ and $\mathbf{J} = J\hat{\mathbf{u}}_r$. (• What are the shapes of the equipotential surfaces in the silicon?)

We use the differential form of Eq. 27–18, namely, $dR = \rho\, d\ell / A$, with $d\ell = dr$ and $A = 2\pi rL$. (• Why is it possible to use Eq. 27–18, which was obtained for the case of a

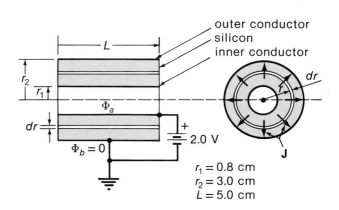

$r_1 = 0.8$ cm
$r_2 = 3.0$ cm
$L = 5.0$ cm

straight bar?) Thus,

$$dR = \frac{\rho\, dr}{2\pi rL}$$

The current flows radially outward through successive cylindric layers of material. Therefore, the total resistance is obtained by integrating dR from r_1 to r_2. We find

$$R = \int_{r_1}^{r_2} dR = \frac{\rho}{2\pi L} \int_{r_1}^{r_2} \frac{dr}{r} = \frac{\rho}{2\pi L} \ln(r_2/r_1)$$

Using $\rho(\text{Si})$ from Table 27–2 and the dimensions given in the diagram, we have

$$R = \frac{640 \ \Omega\cdot m}{2\pi(0.050 \ m)} \ln(3.0/0.8) = 2693 \ \Omega$$

Then, the current is

$$I = \frac{V}{R} = \frac{2.0 \ V}{2693 \ \Omega} = 7.43 \times 10^{-4} \ A$$

27–4   EMF

We now consider methods of providing a constant potential difference between two points so that a steady current (DC) can flow between the points. One device capable of producing a constant potential difference is a *chemical cell,* or *battery.* The essential common feature of all batteries is the immersion of two dissimilar metals in an electrolyte (ionic) solution or the sandwiching of an electrolytic paste between two plates of dissimilar metals. The first battery (Volta's "pile") consisted of a stack of zinc and copper discs separated by moistened pasteboard. The usual automobile battery consists of six lead–sulfuric-acid cells connected in series.

A chemical cell, when operating with a completed external circuit, converts chemical energy into electric energy. If the external circuit is disconnected, the chemical reactions within the cell will proceed until a certain potential difference (or *terminal voltage*) is developed between the electrodes. (• Can you see why the chemical reactions must eventually stop?) As long as there is no external connection, this potential difference will remain constant, with the *anode* (positive terminal) at a potential higher than that of the cathode (negative terminal). This potential difference is called the *EMF* of the cell.* (For a typical chemical cell the EMF is a volt or so.)

When an external circuit is connected to the cell, current flows through the cell; at the same time, the voltage between the electrodes drops below the EMF of the cell because the cell has an effective internal resistance. (For most cells, this internal resistance is usually between 0.75 Ω and 1.0 Ω.) Because of this effect, standard cells (that is, cells used as standards in the measurement of the EMF of other cells) are always used in circuits in such a way that no current flows through the cell; then, the voltage between the terminals is known to be the EMF of the cell.

Any device that converts nonelectric energy into electric energy is a source of EMF. Thus, chemical cells and batteries are sources of EMF, as are electric generators that convert mechanical energy into electric energy. Also, thermoelectric devices convert thermal energy into electric energy and solar cells convert radiant energy into electric energy.

27–5   SINGLE-LOOP CIRCUITS

We now examine the potential-current behavior of a simple electric circuit that consists of a battery and an external resistive load. We represent the resistive load (a *resistor*)

---

*Historically, this potential difference was given the unfortunate name *electromotive force*. But this quantity is not a force and does not carry the dimensions of force. We use here only the initials EMF as a name for this quantity and forgo completely the use of the earlier, misleading term.

by the sawtooth symbol,

We imagine that all connections are made with hookup wire that has negligible resistance. (Example 27–2a showed that a meter of No. 20 copper wire has a resistance of only 0.03 Ω, so this assumption is generally valid.) The battery has a distributed internal resistance, due primarily to the resistivity of the electrolyte and the accumulation of polar ions on the electrodes; however, it is conventional to represent this distributed resistance as a single equivalent resistance r, as indicated in Fig. 27–8. For a fully charged low-voltage battery the internal resistance is usually less than 1 Ω, and the EMF (symbol, ε) varies between 0.75 V and 2.75 V per cell.

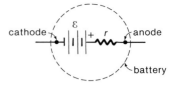

Fig. 27–8. Schematic representation of a battery. The EMF of the battery is ε and the internal resistance (actually a distributed resistance) is r.

Consider the circuit shown in Fig. 27–9, where the internal resistance of the battery is indicated explicitly. The battery terminals are labeled a and b. The potential Φ_a at point a is, for convenience, set equal to zero by grounding this point. The continuity condition on the current means that the same current I flows through each part of the circuit.

Figure 27–10 shows the variation of the electric potential Φ around the circuit. Because the current is constant (steady-state condition), we have $\oint \mathbf{E} \cdot d\mathbf{r} = 0$ (Eq. 25–14). Evaluating this integral in steps, we have

$$\oint \mathbf{E} \cdot d\mathbf{r} = \left(\int_a^{a'} + \int_{a'}^b + \int_b^c + \int_c^d + \int_d^a \right) E \, dr$$

$$(\Phi_{a'} - \Phi_a) + (\Phi_b - \Phi_{a'}) + (\Phi_c - \Phi_b) + (\Phi_d - \Phi_c) + (\Phi_a - \Phi_d) = 0$$

$$\varepsilon \qquad -V_{a'b} \qquad +0 \qquad -V_{cd} \qquad +0 \quad = 0$$

Using Eq. 27–13, the voltage drops can be expressed as $V_{a'b} = Ir$ and $V_{cd} = IR$. Thus, we have

$$\varepsilon - Ir = IR \qquad\qquad \text{(27–19)}$$

By writing the result in this form we emphasize the fact that the terminal voltage of the battery is $\varepsilon - Ir$ when a current I flows through the battery. Alternatively, we can write

$$\varepsilon = Ir + IR \qquad\qquad \text{(27–19a)}$$

This form of the result emphasizes the following rule:

In any circuit loop the sum of the rises in potential around the circuit equals the sum of the drops in potential.

(This is actually Kirchhoff's second circuit rule; see Section 27–7.)

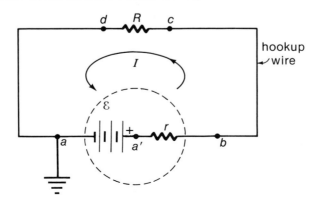

Fig. 27–9. A simple electric circuit that consists of a battery and an external load resistor R.

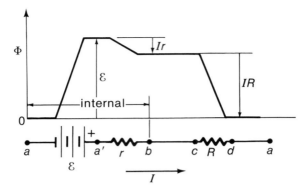

Fig. 27–10. The lower part of the diagram shows the circuit of Fig. 27–9 laid out in a straight line. The upper part of the diagram shows the variation of the potential Φ matched to the circuit. The net change in potential around the circuit (point a to point a) is zero.

Energy Relationships. When a charged particle that carries a charge $+q$ moves in a vacuum from a point a to a point b, thereby falling through a potential difference $V_{ab} - \Phi_a - \Phi_b > 0$, the particle acquires a kinetic energy equal to qV_{ab}. If, instead, this amount of charge passes through a resistor across which there is a potential difference $V_{ab} > 0$, the energy that would have appeared as kinetic energy is now dissipated in collisions within the resistor and appears as heat. Thus, the differential amount of ohmic loss or resistive (or Joule) heating within the resistor due to passage of an amount of charge dq is

$$dW_R = V_{ab} \, dq$$

The rate at which heat is generated is the power, $P_R = dW_R / dt$; hence, $P_R = V_{ab} \, dq / dt$. Since $dq / dt = I$, by writing V_{ab} simply as V we have $P_R = VI$. If, in addition, the resistance obeys Ohm's law, $V = IR$, the power dissipated in the resistor is

$$P_R = VI = I^2R \tag{27–20}$$

Similarly, the power dissipated in the internal resistance r is

$$P_r = I^2r \tag{27–21}$$

Inside a battery, the charge dq is driven through a potential increase $\mathcal{E}$ by the chemical reactions in the cells. In this process the amount of chemical energy converted into electric energy is

$$dW_C = \mathcal{E} \, dq$$

Thus, the power delivered by the battery is $P_C = dW_C / dt = \mathcal{E} \, dq / dt$, or

$$P_C = \mathcal{E}I \tag{27–22}$$

The power dissipated in a resistor R when a current I flows is I^2R (Eq. 27–20). Using the Ohm's-law relationship (Eq. 27–13), $V = IR$, we can express the power dissipation variously as

$$P = VI = I^2R = \frac{V^2}{R} \tag{27–23}$$

where V is the potential decrease across the resistor when the current I flows. Power is measured in *watts* (W), and from Eq. 27–23 we see that $1 \text{ W} = 1 \text{ A}^2 \cdot \Omega = 1 \text{ V} \cdot \text{A} = 1 \text{ V}^2/\Omega$.

Conservation of Energy. If we multiply Eq. 27–19a by I, we have $\mathcal{E}I = I^2 r + I^2 R$. Then, substituting for each of these terms from Eqs. 27–20, 27–21, and 27–22, we find

$$P_C = P_r + P_R \qquad\qquad \textbf{(27–24)}$$

This equation is the statement of energy conservation for the system, namely, the rate P_C at which chemical energy is converted to electric energy is equal to the rate, $P_r + P_R$, at which Joule heating losses occur.

Example 27–4

Two batteries and two external resistors are connected as shown in the diagram, using hookup wire with negligible resistance.

(a) Determine the current I.

(b) Calculate the power dissipated or generated in each of the circuit elements.

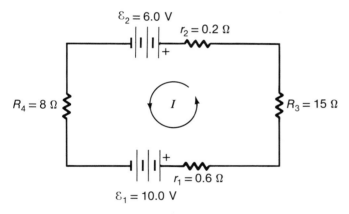

Solution:

Note that the batteries are connected so that they attempt to drive current in opposite directions around the circuit. Because $\mathcal{E}_1 > \mathcal{E}_2$, the current I flows counterclockwise, as indicated in the diagram. Battery 1 *provides* power to the circuit, whereas battery 2 *absorbs* power. The excess power, $(\mathcal{E}_1 - \mathcal{E}_2)I$, appears as Joule heating.

(a) We have

$$\mathcal{E}_1 - \mathcal{E}_2 = I(r_1 + r_2 + R_3 + R_4)$$

so that

$$I = \frac{10.0 \text{ V} - 6.0 \text{ V}}{0.6 \ \Omega + 0.2 \ \Omega + 15 \ \Omega + 8 \ \Omega} = 0.1681 \text{ A}$$

(If we had guessed that the current flow is in the opposite direction, we would have found $I < 0$, thereby indicating an incorrect choice for the direction. [• Can you see why this is the case?])

(b) Battery 1 is converting chemical energy into electric energy at a rate

$$P_{C1} = \mathcal{E}_1 I = (10.0 \text{ V})(0.1681 \text{ A}) = 1.681 \text{ W}$$

Battery 2 is converting electric energy into chemical energy at a rate

$$P_{C2} = \mathcal{E}_2 I = (6.0 \text{ V})(0.1681 \text{ A}) = 1.008 \text{ W}$$

Battery 1 is *discharging*, whereas battery 2 is being *charged*. The power dissipated in the various resistive elements is

$$P_{r1} = I^2 r_1 = (0.1681 \text{ A})^2 (0.60 \ \Omega) = 0.017 \text{ W}$$

$$P_{r2} = I^2 r_2 = (0.1681 \text{ A})^2 (0.20 \ \Omega) = 0.006 \text{ W}$$

$$P_{R3} = I^2 R_3 = (0.1681 \text{ A})^2 (15 \ \Omega) = 0.424 \text{ W}$$

$$P_{R4} = I^2 R_4 = (0.1681 \text{ A})^2 (8 \ \Omega) = 0.226 \text{ W}$$

The energy balance equation is

$$\mathcal{E}_1 I = \mathcal{E}_2 I + I^2 r_1 + I^2 r_2 + I^2 R_3 + I^2 R_4$$

$$1.681 \text{ W} = 1.008 \text{ W} + 0.017 \text{ W} + 0.006 \text{ W} + 0.424 \text{ W}$$
$$+ 0.226 \text{ W}$$

which does indeed balance.

Example 27–5

A power supply with an EMF $\mathcal{E}$ and an internal resistance r provides current to a load resistor R. What must be the value of R for maximum power to be delivered to the load by the supply?

Solution:

The current in the circuit is

$$I = \frac{\mathcal{E}}{r + R}$$

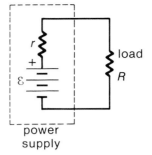

power
supply

Then, the power delivered to the resistor R is

$$P = I^2 R = \frac{\mathcal{E}^2 R}{(r + R)^2}$$

Differentiating, we obtain

$$\frac{dP}{dR} = \mathcal{E}^2 \left(\frac{(r + R)^2 - 2R(r + R)}{(r + R)^4} \right)$$

Setting dP/dR equal to zero, we find that the power is maximum when $R = r$. That is, maximum power is delivered when the load is matched to the internal resistance. This is a particularly important result in AC circuits. (See, for instance, Example 32–8.)

27–6   RESISTOR NETWORKS

We seek a method whereby a network of resistors can be replaced by a single effective or equivalent resistor in such a way that the remainder of the circuit is undisturbed. That is, if the network has terminals* a and b, the current between a and b and the potential difference V_{ab} must be the same with the equivalent resistor in place as for the original network.

First, consider a network of resistors connected in parallel. Figure 27–11 shows three resistors, R_1, R_2, and R_3, connected in parallel between terminals a and b. The potential difference (the voltage drop) V_{ab} is common to all three resistors. Moreover, the current I that enters the network is divided among the three resistors such that $I_1 + I_2 + I_3 = I$ (continuity of current). Then, using $I_1 = V_{ab}/R_1$ and the same for the other two resistors, we have

$$I = V_{ab} \left(\frac{1}{R_1} + \frac{1}{R_2} + \frac{1}{R_3} \right) \qquad (27-25)$$

When the equivalent resistor is connected between the terminals a and b, the same current I must result for the same voltage drop V_{ab}. Thus,

$$I = \frac{V_{ab}}{R} \qquad (27-26)$$

Comparing Eqs. 27–25 and 27–26, we see that

Fig. 27–11. The parallel connection of three resistors (a) can be replaced by the single equivalent resistor R (b).

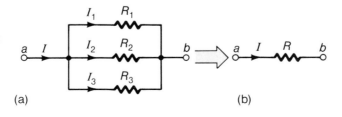

(a)   (b)

*We can imagine complicated networks that are connected to circuits through any number of terminals. In this section we consider only two-terminal networks.

$$\frac{1}{R} = \frac{1}{R_1} + \frac{1}{R_2} + \frac{1}{R_3} \qquad \text{parallel connection} \qquad (27\text{–}27)$$

Note also that the power dissipated in the equivalent resistor, $P = V_{ab}^2/R$, must equal the sum of the amounts of power dissipated in the individual resistors of the original network, namely, $V_{ab}^2/R_1 + V_{ab}^2/R_2 + V_{ab}^2/R_3$. Equating these two expressions for the power P, we have

$$\frac{V_{ab}^2}{R} = \frac{V_{ab}^2}{R_1} + \frac{V_{ab}^2}{R_2} + \frac{V_{ab}^2}{R_3}$$

which leads to the same result for R (Eq. 27–27). Clearly, the expression for the equivalent resistance R can be generalized to any number of parallel resistors.

Next, consider three resistors connected in series, as in Fig. 27–12. Now, the same current I flows through each resistor. We write the voltage drop V_{ab} between the terminals a and b as the sum of the voltage drops across each resistor, namely, $V_{aA} = IR_1$ and similarly for the other two resistors. Thus,

$$V_{ab} = V_{aA} + V_{AB} + V_{Bb}$$
$$= I(R_1 + R_2 + R_3) \qquad (27\text{–}28)$$

For the equivalent resistor, we have

$$V_{ab} = IR \qquad (27\text{–}29)$$

Comparing Eqs. 27–28 and 27–29, we see that

$$R = R_1 + R_2 + R_3 \qquad \text{series connection} \qquad (27\text{–}30)$$

(• Can you obtain the same result by considering the power dissipation in the network and in the equivalent resistor?) Notice carefully that resistors and capacitors do *not* follow the same rules for series and parallel combinations; compare Eqs. 26–14 and 26–13 with Eqs. 27–27 and 27–30.

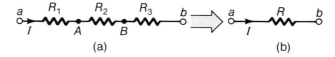

Fig. 27–12. The series connection of three resistors (a) can be replaced by the single equivalent resistor R (b).

Example 27–6

Determine the currents I_1, I_2, and I_3 in the circuit illustrated in diagram (a).

Solution:

First, replace the two parallel resistors, R_2 and R_3, with the equivalent resistor R:

$$\frac{1}{R} = \frac{1}{5\ \Omega} + \frac{1}{10\ \Omega} = \frac{3}{10\ \Omega} \qquad \text{or} \qquad R = \tfrac{10}{3}\ \Omega$$

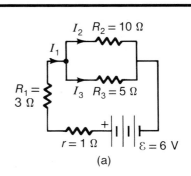

This reduction is shown in diagram (b).

Now, the three resistors, R_1, R, and r, are in series, so the overall equivalent resistance is

$$R' = R_1 + R + r$$

$$= 3 \ \Omega + \tfrac{10}{3} \ \Omega + 1 \ \Omega = \tfrac{22}{3} \ \Omega$$

This reduction is shown in diagram (c).

The current I_1 can now be calculated:

$$I_1 = \frac{\mathcal{E}}{R'} = \frac{6 \ \text{V}}{\tfrac{22}{3} \ \Omega} = \tfrac{9}{11} \ \text{A}$$

The currents through R_2 and R_3 are obtained by noting that the voltage drop across each of these resistors is $V = I_1 R$. Thus,

$$I_2 = \frac{V}{R_2} = \frac{I_1 R}{R_2} = \frac{(\tfrac{9}{11}\text{A})(\tfrac{10}{3} \ \Omega)}{10 \ \Omega} = \tfrac{3}{11} \ \text{A}$$

$$I_3 = \frac{V}{R_3} = \frac{I_1 R}{R_3} = \frac{(\tfrac{9}{11}\text{A})(\tfrac{10}{3} \ \Omega)}{5 \ \Omega} = \tfrac{6}{11} \ \text{A}$$

Also, $I_2 + I_3 = I_1$, as required.

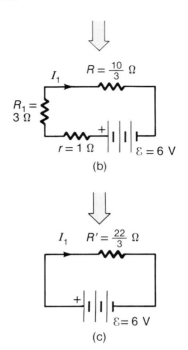

(b)

(c)

27–7 KIRCHHOFF'S RULES

When an electric circuit consists of a number of interconnected loops with several sources of EMF, a simple reduction of the resistances is not sufficient to obtain a solution. For example, consider the circuit shown in Fig. 27–13. Suppose that values are given for both EMFs and for all the resistances. It is necessary to determine the currents, I_1, I_2, and I_3. *Kirchhoff's rules** provide a systematic procedure for obtaining a sufficient number of linear equations to solve such problems.

Rule 1: The sum of the currents flowing away from any point in a circuit is equal to the sum of the currents flowing toward that point.

Rule 2: In traversing any closed path in a circuit, the algebraic sum of the EMFs is equal to the algebraic sum of the voltage (IR) drops.

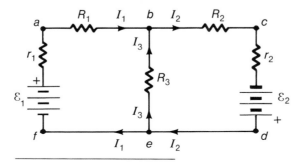

Fig. 27–13. What are the currents I_1, I_2, and I_3?

*After the German physicist, Gustav Kirchhoff (1824–1887).

The first rule is simply the current continuity condition; in Fig. 27–13 this rule applied at point b (or point e) means $I_1 + I_3 = I_2$.

The second rule is just the statement of $\oint \mathbf{E} \cdot d\mathbf{r} = 0$. Because the direction of the actual current flow may not be evident at the outset, how do we determine the proper algebraic signs to use in applying the second rule? (Notice that a selected path may be traversed in either of two directions.) To avoid the use of cumbersome notation such as V_{ab} and I_{cb}, we give a third rule:

Rule 3: The algebraic signs of the EMFs and the voltage (IR) drops to be used in applying Rule 2 are specified as follows:

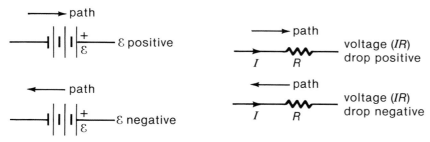

If we traverse the path *abefa* in Fig. 27–13, we find, using Rule 2 and the sign conventions in Rule 3,

$$\mathcal{E}_1 = (r_1 + R_1)I_1 - R_3 I_3$$

On the other hand, if we traverse the path in the opposite direction (that is, *afeba*), we find

$$-\mathcal{E}_1 = R_3 I_3 - (r_1 + R_1)I_1$$

which is just the negative of the first equation.

For the circuit in Fig. 27–13 there are three possible choices for paths, namely, *abcdefa*, *abefa*, and *bcdeb*. Along how many of these paths must the second rule be applied in order to solve the problem? ("Solving the problem" means finding the unknown *branch currents*, I_1, I_2, and I_3.) To attack such problems, first reduce to a minimum the number of unknown currents by repeated use of the first rule. Then, the number of remaining unknown currents is equal to the number of times the second rule must be applied. For the circuit in Fig. 27–13 the first rule gives one relationship, namely, $I_2 = I_1 + I_3$. Consider I_1 and I_3 to be the unknown currents; then, applying the second rule along *any two* of the three possible paths will suffice to solve the problem. (The third path would yield a result that is some linear combination of the first two and hence would give no new information.) It is always the case that the number of independent equations needed is equal to the number of unknowns.

Example 27–7

Determine the currents I_1, I_2, and I_3 in the circuit shown in the diagram on page 538 (which is the same as that in Example 27–6) by using Kirchhoff's rules.

Solution:

At point a Kirchhoff's first rule gives

$$I_1 = I_2 + I_3 \tag{1}$$

For application of the second rule, select the two paths P (*eacde*) and Q (*abdca*). Then, using the sign conventions in Rule 3, we find

for path P: $(1 \ \Omega + 3 \ \Omega)I_1 + (5 \ \Omega)I_3 = 6$ V $\quad$ **(2)**

for path Q: $(10 \ \Omega)I_2 - (5 \ \Omega)I_3 = 0$ $\quad$ **(3)**

Substituting for I_2 in Eq. (3) gives (with the units suppressed)

for path P: $4I_1 + 5I_3 = 6$

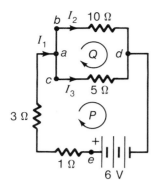

for path Q: $10I_1 - 15I_3 = 0$

Solving these equations yields

$$I_1 = \tfrac{9}{11}\,\text{A}   \quad\text{and}\quad   I_3 = \tfrac{6}{11}\,\text{A}$$

Finally, using Eq. (1), we have

$$I_2 = I_1 - I_3 = \tfrac{3}{11}\,\text{A}$$

These results are, of course, the same as those obtained in Example 27–6.

Example 27–8

Determine the currents I_1, I_2, and I_3 in the circuit of Fig. 27–13 for the EMF and resistance values shown in the diagram.

Solution:

At point b, Kirchhoff's first rule gives

$$I_2 = I_1 + I_3 \tag{1}$$

To apply the second rule, we select two paths: path P, *abefa*, and path Q, *bcdeb*. Then, using the sign conventions in Rule 3, we have

$$P:\quad (4\ \Omega + 2\ \Omega)I_1 - (10\ \Omega)I_3 = 10\ \text{V} \tag{2}$$

$$Q:\quad (6\ \Omega + 1\ \Omega)I_2 + (10\ \Omega)I_3 = 8\ \text{V} \tag{3}$$

Using Eq. (1) to substitute for I_2 in Eq. (3), we find (suppressing the units)

$$P:\quad 6\,I_1 - 10\,I_3 = 10 \tag{2'}$$

$$Q:\quad 7\,I_1 + 17\,I_3 = 8 \tag{3'}$$

Solving these two equations, we obtain

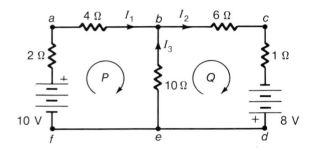

$$I_1 = \tfrac{125}{86}\,\text{A}   \quad\text{and}\quad   I_3 = -\tfrac{11}{86}\,\text{A}$$

so that

$$I_2 = I_1 + I_3 = \tfrac{57}{43}\,\text{A}$$

Notice that the result for I_3 carries a negative sign. This simply means that I_3 actually flows in the direction *opposite* to that indicated in the diagram. Thus, it is not necessary to know beforehand the directions of all the currents in a circuit. Select the directions arbitrarily and use them consistently throughout the calculation. Then, any current for which the direction was chosen incorrectly will emerge from the analysis with a negative sign.

Example 27–9

Not all two-terminal resistance networks are reducible to a single equivalent resistor by using only the combination laws discussed in Section 27–6 for parallel and series connections. Kirchhoff's rules can be used to determine the single equivalent resistor for the network illustrated in the diagram.

Solution:

First, use Kirchhoff's Rule 1 to reduce to a minimum the number of unknown currents. This is already done in the diagram; we have selected I_1 and I_2 as the unknown currents. (Because we seek the equivalent resistance of the network, the current I serves merely as a reference current.) Next, choose loops P and Q for the application of the second rule. We find

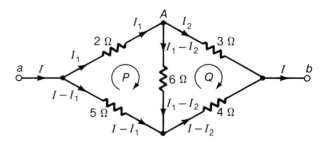

(suppressing the units)

$$P:\quad 2I_1 + 6(I_1 - I_2) - 5(I - I_1) = 0$$

$$Q:\quad 3I_2 - 4(I - I_2) - 6(I_1 - I_2) = 0$$

These equations can be rewritten as

$$P: \quad 13I_1 - 6I_2 = 5I$$
$$Q: \quad -6I_1 + 13I_2 = 4I$$

Solving for I_1 and I_2, we obtain

$$I_1 = \tfrac{89}{133}I \quad \text{and} \quad I_2 = \tfrac{82}{133}I$$

Now, we have

$$V_{ab} = V_{aA} + V_{Ab}$$
$$= 2I_1 + 3I_2 = \tfrac{424}{133}I$$

Finally, the equivalent resistance is identified as

$$R = \tfrac{424}{133}\ \Omega$$

QUESTIONS

27–1 Define and explain briefly the meaning of the terms (a) current density, (b) resistivity and conductivity, (c) mean collision time, (d) Ohm's law, (e) EMF, and (f) Joule heating.

27–2 Explain carefully the distinction between the drift velocity in a conductor and the velocity associated with random motions.

27–3 Explain why a circuit must form a closed loop if steady current is to flow. What happens if a closed-loop circuit is broken at some point? Would you expect any electrical charge to be present at the broken ends?

27–4 Under what circumstances can the terminal voltage of a battery be zero? Does this imply that its EMF is also zero?

27–5 Which of the expressions in Eqs. 27–20 and 27–23 are always true and which depend on the resistor obeying Ohm's law? Is Eq. 27–22 always true?

27–6 Suppose a particular resistive circuit element passed a current I which varied with the square of the applied voltage V; that is, $I = \gamma V^2$. Rewrite the appropriate equations to replace those of Eq. 27–23.

27–7 Suppose the current in a battery located within some cir-

cuit were to flow in a direction opposite to its EMF. What is the interpretation of Eq. 27–22 for this case?

27–8 When electric power is delivered from the generating plant to the user on transmission lines, it is always done at high voltages (and hence low currents). Why?

27–9 Define and explain briefly the meaning of the terms (a) two-terminal resistor network, (b) Kirchhoff's rules, and (c) branch currents.

27–10 Example 27–9 illustrated a case of a two-terminal network of pure resistances that cannot be reduced to a single equivalent resistance using the combination laws contained in Eqs. 27–27 and 27–30. Construct some other examples.

27–11 Consider the possibility of letting the two interconnected electrical loops shown in Fig. 27–13 represent two interconnected hydraulic lines. Discuss the appropriateness of the analogy if the EMFs represented pressure increases produced by pumps and the currents stood for water flow rate. What would be the hydraulic equivalent to resistance? To the potentials at points such as b or e? "Translate" Kirchhoff's rules as applied to the hydraulic analog. If Fig. 27–13 were an elevation view of the hydraulic system with leg abc at a higher elevation than leg fed, how would you include the gravitational effect?

PROBLEMS

Section 27–1

27–1 A copper bus bar has a cross section of 5 cm × 15 cm and carries a current with a density of 2000 A/cm². (a) What is the total current in the bus bar? (b) What amount of charge passes a given point (intersecting plane) in the bar per hour?

27–2 A small sphere that carries a charge of 8 nC is whirled in a circle at the end of an insulating string. The rotation

frequency is 100π rad/s. What average current does this rotating charge represent?

27–3 A coaxial conductor with a length of 20 m consists of an
• inner cylinder with a radius of 3.0 mm and a concentric outer cylindric tube with an inside radius of 9.0 mm. A uniformly distributed leakage current of 10 μA flows between the two conductors. Determine the leakage current density (in A/m²) through a cylindric surface (concentric with the conductors) that has a radius of 6.0 mm.

Section 27–2

27–4 The density of copper is 8.96 g/cm^3 and the atomic mass is 63.55 u. Verify the free-electron number density for copper quoted in Table 27–1.

27–5 Use the data in Table 27–1 to determine $\bar{\alpha}$ for copper in the temperature range from 273 K to 373 K. (Neglect any possible variation of n with temperature.) Compare your result with the value listed in Table 27–2.

27–6 An aluminum wire with a diameter of 0.1 mm has a uniform electric field of 0.2 V/m imposed along its entire length. The temperature of the wire is 50° C. (a) Use the information in Table 27–2 and determine the resistivity. (b) What is the current density in the wire? (c) What is the total current in the wire? (d) What is the drift speed of the conduction electrons? (e) What potential difference must exist between the ends of a 2-m length of the wire to produce the stated electric field strength?

Section 27–3

27–7 Determine the resistance at 50° C of a 1.5-m length of platinum wire that has a diameter of 0.10 mm.

27–8 The diagram shows the annulus of silicon we considered in Example 27–3 ($r_1 = 0.8$ cm, $r_2 = 3.0$ cm, and $L = 5.0$ cm). In this case the silicon is placed between two thick copper plates with a potential difference of 2.0 V between them. What is the current in the circuit? (The battery has negligible internal resistance.) Compare your result with that obtained in Example 27–3. (• Can you account qualitatively for the difference in the currents?)

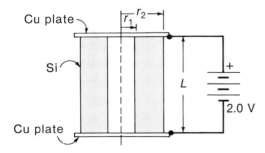

27–9 An amber rod has a diameter of 1 cm and a length of 10 cm. A conductor is attached to one end of the rod and the other end is grounded. (a) Determine the axial resistance of the rod. (This resistance will be realized only in a perfectly dry environment. Any surface moisture will lower the value considerably.) (b) If the conductor is at a potential of 10^4 V with respect to ground, what current flows in the rod? (c) At the rate determined in (b), how long would it take for a charge of 50 nC to leak from the conductor to ground?

Section 27–5

27–10 Two batteries are connected in series with a resistor R, as shown in the diagram: $\mathcal{E}_1 = 3.7$ V, $r_1 = 0.25$ Ω; $\mathcal{E}_2 = 6.1$ V, $r_2 = 0.50$ Ω. (a) Determine the current I if $R = 25$ Ω. (b) What are the values of the EMF $\mathcal{E}'$ and the internal resistance r' of the equivalent single battery that would deliver to any resistance R the same current as would the actual two batteries?

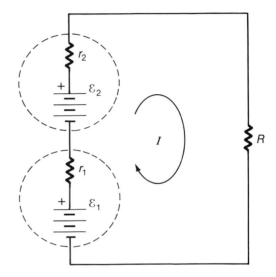

27–11 A rotating mechanical DC generator with an internal (armature) resistance of 0.65 Ω produces an EMF of 250 V. A load resistance of 10 Ω is connected to the generator terminals. (a) What current flows through the load resistance? (b) What power is delivered to the load? (c) What amount of power is lost in the internal resistance? (d) If the frictional losses (including the *windage loss* due to the rotation of the armature) amount to 600 W, what input mechanical power is required to produce the stated conditions? (e) With what efficiency is the input mechanical power delivered to the load resistor? (f) The generator has a magnetic field that is produced by an external source that requires 300 W of power. If this power is also counted as a loss, what is the overall efficiency?

Section 27–6

27–12 Three resistors (10 Ω, 20 Ω, and 30 Ω) are connected in parallel. The total current through this network is 5 A. (a) What is the voltage drop across the network? (b) What is the current through each resistor?

27–13 What additional resistance must be placed in parallel with a parallel combination of 75-Ω and 50-Ω resistors to yield a resistance of 25 Ω for the entire network?

27–14 Determine the equivalent resistance between the terminals a and b for the network illustrated.

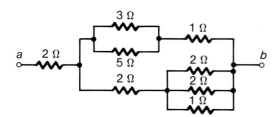

27–15 Three 100-Ω resistors are connected as shown in the diagram. The maximum power that can be dissipated in any one of the resistors is 25 W. (a) What is the maximum voltage that can be applied to the terminals a and b? (b) For the voltage determined in (a), what is the power dissipation in each resistor? What is the total power dissipation?

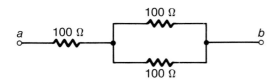

27–16 A light-emitting diode to be used in the display of a calculator must have the current through it limited to 100 mA. The power supply is regulated to 9 V. A resistor is to be placed in series with the diode to limit the current. (a) Assuming that the resistance of the diode is negligible, what should be the value of the limiting resistor? (b) Resistors are available with maximum power ratings of $\frac{1}{4}$ W, $\frac{1}{2}$ W, and 1 W. Which rating should the designer specify?

Section 27–7

27–17 Determine the current in each of the branches of the circuit shown in the diagram at the top of the next column.

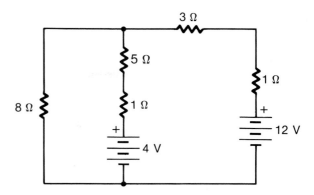

27–18 The two branch currents in the circuit shown in the diagram are $I_1 = \frac{1}{3}$ A and $I_2 = \frac{1}{2}$ A. Determine the EMFs $\mathcal{E}_1$ and $\mathcal{E}_2$.

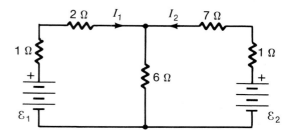

27–19 What is the EMF $\mathcal{E}_1$ of the battery in the circuit shown in the diagram?

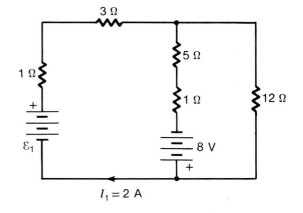

Additional Problems

27–20 In the simple Bohr model of the hydrogen atom, a single electron is imagined to move around the nuclear proton in a circular orbit with a radius of 0.53×10^{-10} m. What average current does this motion represent? [*Hint:* Calculate the electron velocity by setting the Coulomb force equal to the centripetal acceleration times the electron mass.]

27–21 An isolated parallel-plate capacitor has a plate area of 50 cm^2 and a plate separation of 5 mm. The gap is filled with a solid slab of glass that has a dielectric constant $\kappa = 5.2$ and a resistivity $\rho = 2.0 \times 10^{13}$ $\Omega \cdot$m. The capacitor is charged to a surface charge density of 15 μC/cm^2. What is the initial current that flows between the plates?

27-22 For a particular application, it is desirable to construct a resistor that is as independent of temperature as possible. The manganin wire-wound resistor described in Example 27-2(b) is placed in series with one made from constantan (60 percent Cu and 40 percent Ni) wire of 0.1 mm diameter. Use the data in Table 27-2 to calculate the length of constantan wire necessary to make the total resistance of the combination the same at 0° C and at 100° C. What is the value of the total resistance?

27-23 What must be the value of R_3 to make the equivalent resistance between the terminals a and b equal to R_1?

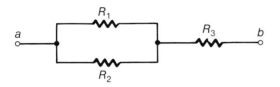

27-24 Determine the potential difference, $V_{ab} = \Phi_a - \Phi_b$, for the circuit shown in the diagram.

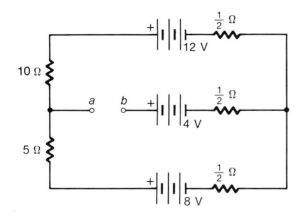

27-25 Show that the resistance between the terminals a and b
•• (the *input* resistance) for the infinite array of identical resistors shown in the diagram is $R = (1 + \sqrt{3})r$. [*Hint:* The resistor chain is infinite, so removing the first three resistors and moving the terminals to the dashed line does not alter the input resistance.]

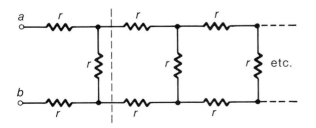

27-26 The network shown in the diagram consists of eight 1-Ω resistors. Determine the equivalent resistance between the terminals a and b. [*Hint:* If the symmetry of a network is such that two points are perceived to be at the same potential, these points may be connected by a resistanceless wire without altering the current distribution anywhere in the network.]

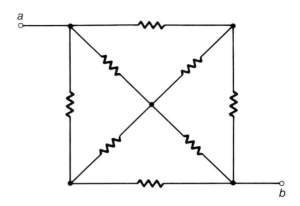

27-27 Twelve 1-Ω resistors are connected to form a cube, as
•• shown in the diagram. Show that the equivalent resistance between the terminals a and b is $\frac{5}{6}$ Ω. [*Hint:* Refer to the hint given in Problem 27-26.]

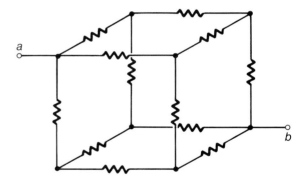

27-28 Determine the current that flows through each of the batteries in the circuit shown in the diagram.

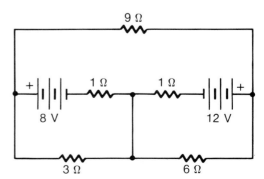

27–29 Determine the value of R in the circuit shown in the diagram such that maximum power is delivered to R. [*Hint:* Refer to Example 27–5.]

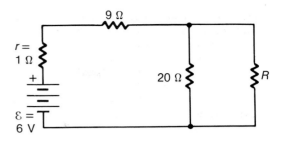

CHAPTER 28

THE MAGNETIC FIELD

In the preceding chapters electric fields were discussed for the simplifying time-independent condition. We now inquire into the behavior of magnetic fields by again considering time-independent cases. For magnetic fields this condition is achieved in situations involving *steady currents*. In this chapter, we begin the discussion of the connection between electric and magnetic effects—the subject of *electromagnetism*.

28–1 EARLY HISTORY

In the fifth century B.C., the Greeks were aware that certain natural rocks had the curious property of attracting each other and also attracting bits of iron. Because these rocks were plentiful in the district called Magnesia, in the western part of what is now Turkey, they became known as *magnets* (and the active mineral in the rocks is now called *magnetite*, Fe_3O_4).

As early as 1100 A.D. the Chinese constructed a primitive but functional magnetic compass. By the late twelfth century European navigators were using a compass consisting of a long piece of magnetite floated on a splinter of wood in a basin of water. (Navigators called these rocks *lodestones*—"leading stones.")

The earliest scientific inquiry into the nature of magnetism was carried out in the thirteenth century by a French soldier-scientist, Pierre de Maricourt, commonly called Peregrinus. He fashioned a spherical lodestone and sketched on its surface a set of lines indicating the directions taken up by an iron needle placed at various points. The lines encircled the lodestone as do the north-south meridian lines on an Earth globe, with intersections at two points. These points, which Peregrinus called the *poles* of the lodestone, are the points with the strongest power to attract bits of iron. He demonstrated that each of the fragments of a divided magnet has two poles, and that like magnetic poles repel whereas unlike magnetic poles attract one another. In this regard, magnetic poles are similar to electric charges, but, in sharp contrast with electric charges, magnetic poles always occur in closely associated pairs—isolated (single) magnetic poles *do not* exist.*

More than 300 years elapsed before the next significant contribution to the understanding of magnetism was made by William Gilbert, who was mentioned in Section 28–1 in connection with the discussion of electric charge. In his famous book *De Magnete* ("On the Loadstone and Magnetic Bodies, and On the Great Magnet, the

---

*There have been theoretical conjectures that isolated magnetic poles (magnetic monopoles) should exist, but all attempts to observe these monopoles have been unsuccessful.

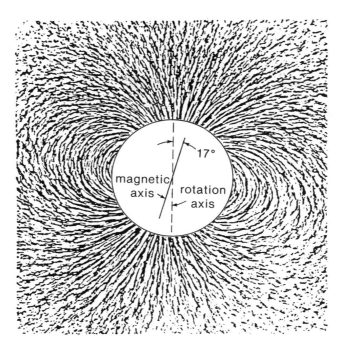

Fig. 28-1. The Earth's magnetic field closely resembles that of a magnetized sphere. The Earth's magnetic axis is inclined to the rotation axis at an angle of approximately 17°. Thus, the geomagnetic poles do not coincide with the geographic poles.

Earth,'' published in 1600), Gilbert showed that the Earth has two magnetic poles and acts as a giant magnet. Figure 28–1 shows a simulation of the Earth's magnetism (its *magnetic field*) produced by a magnetized sphere. Iron filings sprinkled on a flat piece of cardboard are induced to become tiny compass magnets that align themselves with the magnetic field. Notice the striking similarity of this field to that of an electric dipole (compare Fig. 24–6). Notice also that a compass magnet that is free to rotate about a horizontal axis will take up an equilibrium position inclined to the local vertical. Measurements of this *dip angle* around the Earth are in agreement with those expected from the dipole model shown in Fig. 28–1.

The end of a compass magnet that points toward the geomagnetic north pole is designated the N pole of the magnet; the other end is the S pole. Because opposite magnetic poles attract, the Earth's geomagnetic north pole is actually an S pole. The Earth's magnetic axis is inclined with respect to its rotation axis; thus, the geographic and geomagnetic poles do not coincide (Fig. 28–1). At present, the geomagnetic north pole lies in the vicinity of Bathurst Island, Northwest Territories, Canada (approximately 73°N and 100°W), about 1900 km from the geographic north pole.* Because the two poles do not coincide, a compass will generally point slightly east or west of true north. The difference between the directions of geographic north and geomagnetic north is called the *declination* of the particular location. Figure 28–2 illustrates an easterly declination θ_E and a westerly declination θ_W.

The *direction* of magnetic field lines is taken to be the direction in which the N pole of a compass magnet points. Thus, in the region around a bar magnet the field lines run from the N pole to the S pole, as shown in Fig. 28–3.

By using long slender magnets, the N and S poles can be effectively isolated from

Fig. 28-2. Because the Earth's magnetic north pole does not coincide with the geographic north pole, compass needles at the positions shown here exhibit an easterly declination θ_E and a westerly declination θ_W. The declination at a particular point varies slowly with time (by as much as 0.1° per year).

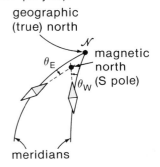

---

*The exact locations of the geomagnetic poles are difficult to determine. Moreover, the positions of the poles wander slowly with time. The first determination of the position of the geomagnetic north pole was made in 1835 by Gauss, who used dip-angle and declination measurements from 84 stations around the world to place the pole at 78°N and 69°W.

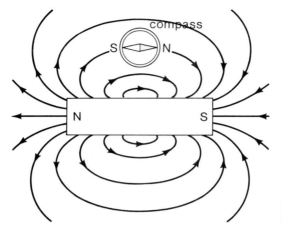

Fig. 28–3. Magnetic field lines for a simple bar magnet.

each other. Then, measurements can be made of the forces between individual like or unlike poles. Such an experiment was first carried out by the Rev. John Michell (1724– 1793), an English geologist. Michell used a torsion balance of his own invention and discovered that the magnetic force between ''isolated'' magnetic poles follows an inverse-square law with distance, just as in the case of electric charge.

A crucial discovery linking electricity and magnetism was made in 1819 by Hans Christian Oersted (1777–1851), a Danish physicist. Oersted found that a compass magnet is sensitive to the effect of current flowing in a nearby wire. That is, a current-carrying wire is surrounded by a magnetic field. A current of a few amperes produces, within a few centimeters of the wire, a magnetic field with a strength comparable to that of the Earth's field. The iron-filing technique can be used to visualize the field in the vicinity of a current-carrying wire. Figure 28–4 shows that the field lines are circles concentric with

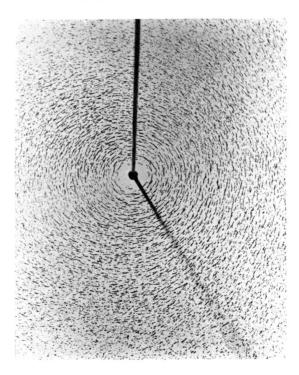

Fig. 28–4. The magnetic field lines in the vicinity of a current-carrying wire are circles concentric with the wire axis. (From *PSSC Physics,* 2nd ed., 1965: D. C. Heath & Co., and Educational Development Center, Newton, Mass.)

the wire axis. Figure 28–5 shows the direction of these field lines as indicated by a compass magnet. Also illustrated is the right-hand rule for determining this direction:

Grasp the wire with the right hand, with the extended thumb in the direction of the current in the wire; then, the fingers encircle the wire in the same direction as the field lines.

Soon after the connection between electric current and magnetism had been established by Oersted's discovery, other physicists rapidly expanded the investigation and added new details. By 1825 the mathematical theory of electromagnetism had been developed to such a degree by André-Marie Ampère (1775–1836) that his monograph on the subject is one of the most important in the history of physical science. Later, the experiments of Michael Faraday (1791–1867) and Joseph Henry (1797–1878), particularly relating to time-dependent effects, provided the foundation from which James Clerk Maxwell (1831–1879) was finally able to integrate the various discoveries into a universal theory of electrodynamics. We pursue these developments in Chapter 33.

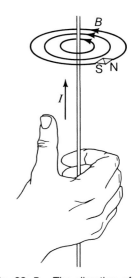

Fig. 28–5. The direction of the field lines encircling a current-carrying wire is determined by a right-hand rule.

28–2 THE MAGNETIC INDUCTION VECTOR

The electric field strength vector **E** has been defined in terms of the force exerted on a positive test charge q_0 placed in the field (see Eq. 24–1). We now give a parallel definition of the corresponding vector for the magnetic field, namely, the *magnetic induction vector* **B**.

We select a region of space in which there is known to be a static magnetic field generated by some unspecified means. Then, we place a test charge q_0 *at rest* at various points in the region and discover that the particle experiences no force at any point.* Unlike gravitational and electric fields, a static magnetic field exerts no force on a stationary charged particle. Next, we move the test charge with an arbitrary velocity **v** through various points in the region, and observe its behavior. We discover the following:

a. The particle experiences a force $\mathbf{F}_M$ that is always perpendicular to **v** regardless of the instantaneous direction of the velocity.

b. The magnitude of the force is proportional to each of the factors q_0, v, and $\sin \theta$, where θ is the angle between **v** and the direction of the field line at the position of the particle.

That is, the behavior of the force on the test charge is described by an expression of the form

$$\mathbf{F}_M = q_0 \mathbf{v} \times \mathbf{B} \qquad (28–1)$$

where **B** is a vector whose direction is the same as that of the magnetic field line that passes through the point **r**. Because the magnitude and direction of **B** are functions of the position **r**, we write it as **B(r)**. We view Eq. 28–1 as the definition of **B(r)**, the *magnetic induction* at the point **r**. The relative orientation of **v**, **B**, and $\mathbf{F}_M$ for a positive test charge is illustrated in Fig. 28–6.

In the SI system, the unit of magnetic induction is the *tesla* (T), where

$$1 \text{ T} = 1 \text{ N·s/C·m} = 1 \text{ N/A·m}$$

In Section 30–2, we introduce the concept of magnetic flux and learn that **B** is also a measure of *magnetic flux density* (indeed, this is an alternate name for **B**). The SI unit for

---

*• What additional measurements would be necessary in order to conclude that the region is free of gravitational and electric fields?

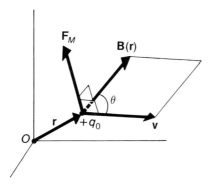

Fig. 28–6. A positive test charge q_0 with velocity **v** experiences a force $\mathbf{F}_M$ due to the presence of the magnetic field **B**. Recall that, by the definition of the vector cross product, $\mathbf{F}_M$ must be perpendicular to both **B** and **v**; its sense is determined by the right-hand rule.

magnetic flux is the *weber* (Wb), and the unit for flux density (field lines per unit area) is again the tesla (T)

$$1 \ \text{Wb/m}^2 = 1 \ \text{T}$$

A unit for magnetic induction used in an earlier version of the metric system and still used extensively is the *gauss* (G):

$$1 \ \text{T} = 10^4 \ \text{G}$$

Geophysicists and planetary physicists frequently specify magnetic fields in *gammas* (γ), where

$$1 \ \gamma = 10^{-5} \ \text{G} = 10^{-9} \ \text{T}$$

The magnitude of the Earth's magnetic field varies from about 30,000 γ at the Equator to about 60,000 γ at the Poles. The time variation of the Earth's field at a particular location can be as large as 150 γ per year. The magnitude of the Sun's magnetic field at the Earth's orbit is about 3 γ.

The Lorentz Force.   The force on a charged particle q in an electric field is $\mathbf{F} = q\mathbf{E}$ (Eq. 24–4). In the event that both electric and magnetic fields are present, the total force on the particle is the vector sum of the forces due to the electric and magnetic effects. This net electromagnetic force, which is called the *Lorentz force,* is

$$\mathbf{F} = q(\mathbf{E} + \mathbf{v} \times \mathbf{B}) \qquad (28\text{–}2)$$

The rate at which the combined electric and magnetic fields (each assumed to be time-independent) do work on a moving charged particle is $P = dW/dt = \mathbf{F}\boldsymbol{\cdot}\mathbf{v}$, or

$$P = q(\mathbf{E} + \mathbf{v} \times \mathbf{B})\boldsymbol{\cdot}\mathbf{v}$$

Now, $\mathbf{v} \times \mathbf{B}$ is a vector that is perpendicular to both **v** and **B**; consequently, $(\mathbf{v} \times \mathbf{B})\boldsymbol{\cdot}\mathbf{v}$ is identically zero. That is, a time-independent magnetic field (of any shape) *can do no work* on a moving charged particle. The work done on the particle in a combined field is due to the electric force alone.

Example 28–1

A small pellet with a mass of 0.5 mg is fired horizontally from a gun that imparts to it a velocity of 100 m/s. If the pellet carries a charge of +5 μC, what magnetic field **B** is required to cancel the Earth's gravitational force on the pellet?

Solution:

Select a coordinate frame as shown: The pellet velocity is in the direction of the positive y-axis and downward is in the direction of the negative z-axis.

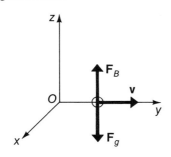

We require a magnetic field $\mathbf{B} = B_x\hat{\mathbf{i}} + B_y\hat{\mathbf{j}} + B_z\hat{\mathbf{k}}$ such that

$$\mathbf{F}_B = Q\mathbf{v} \times \mathbf{B} = -mg\hat{\mathbf{k}}$$

or

$$Q\begin{vmatrix} \hat{\mathbf{i}} & \hat{\mathbf{j}} & \hat{\mathbf{k}} \\ 0 & v & 0 \\ B_x & B_y & B_z \end{vmatrix} = -mg\hat{\mathbf{k}}$$

Evidently $B_y = B_z = 0$ and $B_x = -mg/Qv$ or

$$B_x = -\frac{(0.5 \times 10^{-6} \text{ kg})(9.80 \text{ m/s}^2)}{(5 \times 10^{-6} \text{ C})(100 \text{ m/s})} = -9.80 \times 10^{-3} \text{ T}$$

Thus

$$\mathbf{B} = -9.80 \times 10^{-3}\,\hat{\mathbf{i}} \text{ T}$$

Example 28–2

The Earth's magnetic field at Washington, D.C., has a magnitude of 56,900 γ with a dip angle of 71° and a declination of 6°W.

(a) Describe this field in local rectangular coordinates that correspond to geographic directions.

(b) Determine the force on an electron that moves horizontally and due south in this field with a velocity of 10^7 m/s.

(c) What is the acceleration of the electron in units of g?

(d) What is the kinetic energy of the electron and how does it change?

Solution:

(a) The diagram shows the field vector $\mathbf{B}$ in local rectangular coordinates with geographic north along the x-axis, west along the y-axis, and upward along the z-axis. Then,

$$B_x = (56,900 \text{ } \gamma) \cos (-71°) \cos 6° = 18,400 \text{ } \gamma$$

$$B_y = (56,900 \text{ } \gamma) \cos (-71°) \sin 6° = 1940 \text{ } \gamma$$

$$B_z = (56,900 \text{ } \gamma) \sin (-71°) = -53,800 \text{ } \gamma$$

(b) The charge of the electron is $q = -e = -1.60 \times 10^{-19}$ C and its velocity is $\mathbf{v} = -(10^7 \hat{\mathbf{i}})$ m/s. Thus, the force on the electron is $\mathbf{F} = -e\mathbf{v} \times \mathbf{B}$, or

$$\mathbf{F} = (-1.60 \times 10^{-19} \text{ C})(10^7 \text{ m/s}) \times \begin{vmatrix} \hat{\mathbf{i}} & \hat{\mathbf{j}} & \hat{\mathbf{k}} \\ -1 & 0 & 0 \\ 18.4 & 1.94 & -53.8 \end{vmatrix} \times (10^{-6} \text{ T})$$

where the components of $\mathbf{B}$ have been expressed in tesla. Then, we have

$$\mathbf{F} = (0\hat{\mathbf{i}} + 86.1\hat{\mathbf{j}} + 3.10\hat{\mathbf{k}}) \times 10^{-18} \text{ N}$$

The force is directed due west and slightly upward. (• Visualize this direction in the diagram.)

(c) The mass of the electron is $m_e = 9.11 \times 10^{-31}$ kg, so its acceleration is (in units of $g = 9.80$ m/s^2)

$$\mathbf{a} = \frac{\mathbf{F}}{m_e} = (0\hat{\mathbf{i}} + 96.4\hat{\mathbf{j}} + 3.47\hat{\mathbf{k}}) \times 10^{11} \text{ } g$$

(d) The kinetic energy is

$$K = \tfrac{1}{2}m_e v^2 = \frac{(9.11 \times 10^{-31} \text{ kg})(10^7 \text{ m/s})^2}{2(1.60 \times 10^{-19} \text{ J/eV})} = 285 \text{ eV}$$

Because no electric field is present and the magnetic field can do no work on the electron, the electron moves with this constant value of the kinetic energy.

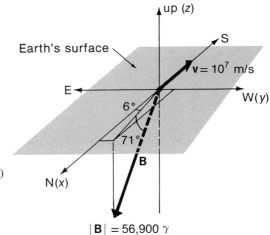

Example 28–3

An electron is traveling through the Sun's chromosphere with a speed of $\frac{1}{10}$ the speed of light and is in a locally uniform magnetic field of 0.30 T produced by a nearby sunspot. The velocity vector of the electron makes an angle of 30° with respect to the direction of the magnetic field, as shown in diagram (a). Describe the motion of the electron.

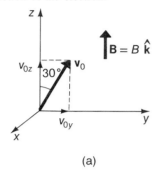

(a)

Solution:

When the instantaneous velocity of the electron is **v**, the acceleration is

$$\mathbf{a} = \frac{\mathbf{F}}{m_e} = -\frac{e}{m_e}\mathbf{v} \times \mathbf{B} = -\frac{e}{m_e}(v_x\hat{\mathbf{i}} + v_y\hat{\mathbf{j}} + v_z\hat{\mathbf{k}}) \times (B\hat{\mathbf{k}})$$

$$= -\frac{e}{m_e}B(v_y\hat{\mathbf{i}} - v_x\hat{\mathbf{j}})$$

The acceleration has no component in the z-direction, so the electron moves with a constant velocity component v_{0z} in this direction.

The component of the electron velocity that is parallel to the x-y plane is $\mathbf{v}_{xy} = v_x\hat{\mathbf{i}} + v_y\hat{\mathbf{j}}$, which is seen to be perpendicular to the acceleration **a**. (• Use the dot product to prove this statement.) Moreover, the field does no work on the electron, so its kinetic energy remains constant; hence, the value of v^2 at any instant is $v_x^2 + v_y^2 + v_z^2 = v_0^2 = $ constant. Now, $v_z = v_{0z}$; thus, at all times, we have

$$v_{xy}^2 = v_x^2 + v_y^2 = v^2 - v_z^2 = v_0^2 - v_{0z}^2 = \text{constant}$$

Thus, we see that $\mathbf{v}_{xy}$ is perpendicular to **a** and has a constant magnitude. This means that the electron moves in a path whose projection onto the x-y plane is a circle. The motion also progresses with constant velocity in the z-direction, so the path is a uniform circular helix with radius R, as shown in diagram (b).

The centripetal acceleration experienced by the electron is constant and is easily evaluated at the instant when $v_x = 0$, as in diagram (a); we find

$$|\mathbf{a}_c| = \frac{v_{0y}^2}{R} = \frac{e}{m_e}Bv_{0y}$$

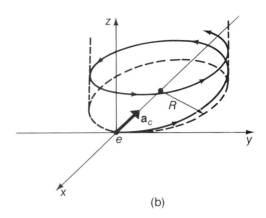

(b)

Solving for the helix radius R, we find

$$R = \frac{m_e v_{0y}}{eB}$$

$$= \frac{(9.11 \times 10^{-31} \text{ kg})(\frac{1}{10} \times 3.0 \times 10^8 \text{ m/s}) \sin 30°}{(1.60 \times 10^{-19} \text{ C})(0.30 \text{ T})}$$

$$= 2.85 \times 10^{-4} \text{ m}$$

It is instructive to calculate the period T of the motion (the time required for one turn of the helix):

$$T = \frac{2\pi R}{v_{0y}} = \frac{2\pi m_e}{eB}$$

which is seen to be independent of the velocity v_0.* The value of T in this case is

$$T = \frac{2\pi(9.11 \times 10^{-31} \text{ kg})}{(1.60 \times 10^{-19} \text{ C})(0.30 \text{ T})} = 1.19 \times 10^{-10} \text{ s}$$

The time required for the electron to travel a distance $s = 1$ m in the z-direction is

$$t = \frac{s}{v_0 \cos 30°}$$

$$= \frac{1 \text{ m}}{(\frac{1}{10} \times 3.0 \times 10^8 \text{ m/s}) \cos 30°} = 3.85 \times 10^{-8} \text{ s}$$

Therefore, for each meter of travel in the z-direction, the electron executes $(3.85 \times 10^{-8} \text{ s})/(1.19 \times 10^{-10} \text{ s}) = 323$ helical turns.

If the field is truly uniform, the electron will spiral about a particular field line. Because the helix radius here is so small, this is still the case even if the field lines are curved. Then, the electron will follow a particular field line as it curves through space. This effect is clearly evident in the photograph, which shows the giant looping arches of a solar prominence. Each arch represents a bundle of field lines about which electrons (and other charged particles) spiral.

---

*The corresponding angular frequency, $\omega_c = 2\pi/T = eB/m$, is called the *cyclotron frequency* because this is the frequency with which a particle revolves while being accelerated in a cyclotron.

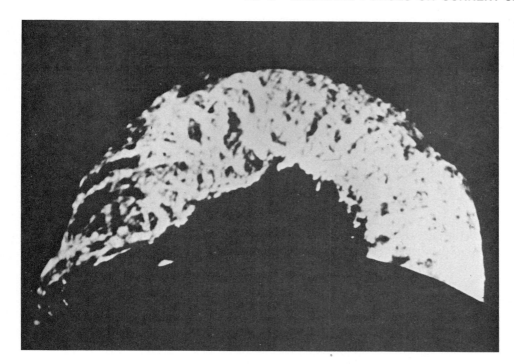

The giant solar prominence of June 4, 1946. (Courtesy of the Harvard College Observatory.)

28–3   MAGNETIC FORCES ON CURRENT-CARRYING WIRES

We mentioned in Section 28–1 Oersted's discovery of the magnetic field of a current-carrying wire. We now investigate the effect of an *external* magnetic field on such a wire. Consider a differential element $d\ell$ of a conducting wire that carries a steady current I (Fig. 28–7). When this wire is placed in a magnetic field, each of the moving charge carriers experiences a force due to the presence of the magnetic field. The sum of all the forces associated with the random thermal motion of the carriers is zero. However, the average drift velocities of the electrons and any positive charge carriers that might be present are directed along the wire. Thus, the forces associated with the motions of the charge carriers are directed perpendicular to the wire ($\mathbf{F} \perp \mathbf{v}_d$). The cumulative effect of

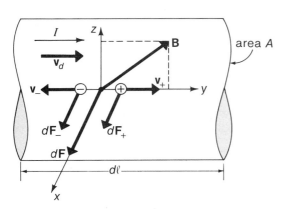

Fig. 28–7. The forces, $d\mathbf{F}_+$ and $d\mathbf{F}_-$, on the two types of charge carriers are in the same direction, namely, perpendicular to the wire. Thus, the force $d\mathbf{F}$ on the wire element is also perpendicular to the wire. For the drift velocities of the positive and negative charge carriers we use the simplified notation $\mathbf{v}_+$ and $\mathbf{v}_-$ instead of $\mathbf{v}_{d+}$ and $\mathbf{v}_{d-}$.

these individual forces is a net sideways force $d\mathbf{F}$ on the wire element. Because the positive and negative charge carriers drift in opposite directions, the magnetic force on both types of carriers is in the *same* direction, as indicated in Fig. 28–7.

In most situations involving solid conductors, the charge carriers are electrons (but positive charge carriers might be present along with the electrons).* In either case, we can identify an equivalent amount of net positive charge dq that moves with an average drift velocity $\mathbf{v}_d$ within the element of wire. Then, the force on the element $d\ell$ of the wire is

$$dF = dq\, \mathbf{v}_d \times \mathbf{B}$$

Using $I = dq/dt$, we can write

$$d\mathbf{F} = I\, dt\, \mathbf{v}_d \times \mathbf{B} \tag{28–3}$$

The quantity $\mathbf{v}_d\, dt$ represents the average distance of drift $d\ell$ experienced by the electrons during the interval dt. Then the combination, $I\,\mathbf{v}_d\, dt = I\, d\ell$, is called a current element, where $d\ell$ is a length element in the direction of the current.† Then Eq. 28–3 becomes

$$d\mathbf{F} = I\, d\ell \times \mathbf{B} \tag{28–4}$$

This result is known as *Ampère's law of magnetic force.*

If the conductor is not a thin wire (essentially a one-dimensional conductor) and has a current density $\mathbf{J}(\mathbf{r})$, the force on a volume element dV is

$$d\mathbf{F} = \mathbf{J} \times \mathbf{B}\, dV \tag{28–5}$$

Of course, isolated current elements do not exist. A real current element is always associated with a closed-circuit loop (and a source of EMF). If a wire loop has sufficient rigidity to resist any deformation by the magnetic forces, or has assumed some equilibrium shape, the total magnetic force on the loop is the integrated form of Ampère's law,

$$\mathbf{F} = I \oint d\ell \times \mathbf{B} \tag{28–6}$$

where the line integral is to be taken completely around the current loop. When the magnetic field is *uniform over the entire loop,* the integral vanishes identically (see Problem 28–9). Even though there is no net force on a current loop, the field can still exert a torque on the loop; we discuss this effect in Section 28–4.

An important case in which there *is* a net force on a current-carrying loop consists of a loop that is partly within and partly outside a uniform magnetic field. The next example shows how this situation is analyzed.

---

*When a current flows through a conductor that is in a magnetic field, the magnetic force should deflect the charge carriers and cause them to collect on one side of the conductor. Reasoning in this manner, the American physicist E. H. Hall (1855–1938) designed a series of experiments that, in 1879, demonstrated that the majority of the charge carriers in most metals bear a negative charge. (The negatively charged particles in matter—electrons—were discovered by J. J. Thomson somewhat later, in 1894.) This piling up of charge on one side of the conductor is called the *Hall effect.*

†It is important to distinguish between $d\mathbf{r}$, which is the differential element of a path or displacement, and $d\ell$, which is the differential element of a wire or circuit that carries a current. The direction of $I\, d\ell$ is always the same as the direction of current flow; the direction of $d\mathbf{r}$ is arbitrary and must be specified in each situation.

Example 28–4

A rectangular current loop of eight turns is suspended from one pan of a beam balance into a uniform magnetic field that is perpendicular to the loop, as shown in the diagram. The beam is balanced by placing weights in the left-hand pan. It is found that an additional 6.85 g must be placed in the left-hand pan when a current of 0.200 A flows through the loop in the direction indicated. Determine the magnitude B of the field.

Solution:

We apply Eq. 28–6 to the three straight-line segments of the loop that lie within the field. We see that the forces cancel for the two vertical segments, because the currents in these segments are in opposite directions and the lengths of the segments within the field are equal. There remains only the downward force on the horizontal segment of length ℓ. (• Why is the force downward?) The magnitude of this force is $F = I\ell B$ on each of the N turns in the loop. This force is balanced by the additional weight Δmg in the left-hand pan. Thus,

$$\Delta mg = IN\ell B$$

so that

$$B = \frac{\Delta mg}{IN\ell} = \frac{(6.85 \times 10^{-3} \text{ kg})(9.80 \text{ m/s}^2)}{(0.200 \text{ A})(8)(0.075 \text{ m})}$$

$$= 0.559 \text{ T or } 5590 \text{ G}$$

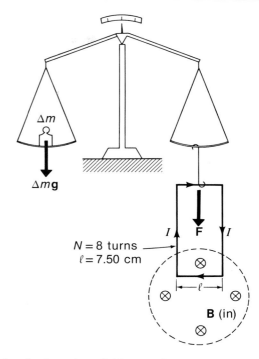

A refined version of this procedure was once a standard technique for measuring magnetic fields. Electronic methods are much simpler and more accurate and are now routinely used for such measurements.

28–4   TORQUE ON A CURRENT LOOP

Consider a rectangular loop with dimensions $w \times \ell$ whose plane makes an angle θ with the x-y plane, as indicated in Fig. 28–8. A steady current I flows in the loop in the sense shown. The entire loop is within a uniform magnetic field $\mathbf{B}$ in the z-direction. We can use Eq. 28–4 to determine the forces that act on various parts of the loop.

The force $\mathbf{F}_{ab}$ that acts on the side $\overline{ab}$ of the loop is in the x-direction and is given by

$$\mathbf{F}_{ab} = I\ell B\hat{\mathbf{i}}$$

The force F_{cd} that acts on the side $\overline{cd}$ has the same magnitude but is in the opposite direction:

$$\mathbf{F}_{cd} = -\mathbf{F}_{ab} = -I\ell B\hat{\mathbf{i}}$$

The forces on the sides $\overline{bc}$ and $\overline{da}$ are

$$\mathbf{F}_{bc} = -\mathbf{F}_{da} = IwB \cos \theta \hat{\mathbf{j}}$$

Thus, the net force that acts on the loop is zero,

$$\mathbf{F} = \mathbf{F}_{ab} + \mathbf{F}_{bc} + \mathbf{F}_{cd} + \mathbf{F}_{da} = 0$$

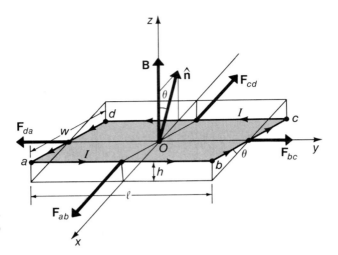

Fig. 28–8. Geometry for calculating the torque on a current loop in a magnetic field.

Moreover, $\mathbf{F}$ vanishes for any orientation of the loop.

Notice, however, that the loop experiences a net *torque* that tends to produce a rotation of the loop about the *y*-axis. Calculating this torque with respect to the origin O at the center of the loop, we find

$$\boldsymbol{\tau}_0 = 2F_{ab}h\hat{\mathbf{j}}$$

where $h = \frac{1}{2}w \sin \theta$ is the height of the side $\overline{ab}$ above the *x-y* plane. Thus, we can write

$$\boldsymbol{\tau}_0 = I\ell Bw \sin \theta \hat{\mathbf{j}} \tag{28–7}$$

Thus, the torque has the maximum magnitude, $\boldsymbol{\tau}_0 = I\ell Bw$, when $\theta = \pi/2$—that is, when the loop is in the *y-z* plane. When the loop is exactly perpendicular to $\mathbf{B}$, the torque is zero. If the direction of the current is reversed (• Can you see that this is equivalent to increasing θ by π?), the sign of the torque is also reversed.

We now introduce a useful vector quantity associated with current-carrying loops. For a planar loop, such as that in Fig. 28–8, the *magnetic moment* $\mathbf{m}$ is defined to be a vector with a magnitude equal to the product of the loop area A and the current I, with a direction along the normal $\hat{\mathbf{n}}$ to the loop:

$$\mathbf{m} = IA\hat{\mathbf{n}} \tag{28–8}$$

(The direction of the positive sense of the normal $\hat{\mathbf{n}}$ is determined by the right-hand convention applied to the direction of the current.) For the loop shown in Fig. 28–8 we have

$$\mathbf{m} = I\ell w\hat{\mathbf{n}}$$

where

$$\hat{\mathbf{n}} = -\sin \theta \hat{\mathbf{i}} + \cos \theta \hat{\mathbf{k}} \tag{28–9}$$

Using these equations to express the magnetic moment, we can write the torque as

$$\boldsymbol{\tau}_0 = \mathbf{m} \times \mathbf{B} \tag{28–10}$$

(• Verify explicitly that Eq. 28–10 is equivalent to Eq. 28–7.)

Equation 28–10 is a general result for a planar loop with *any* shape in a (locally) uniform magnetic field. Equation 28–10 is similar to the expression (Eq. 25–30) for the torque that acts on an electric dipole, namely, $\boldsymbol{\tau}_0 = \mathbf{p} \times \mathbf{E}$. This suggests that a potential energy can be associated with the relative orientation of $\mathbf{m}$ and $\mathbf{B}$ corresponding to that found for the electric dipole, namely, $U = -\mathbf{p} \cdot \mathbf{E} + U_0$ (Eq. 25–31). Indeed, the relationship for the magnetic case is (see Problem 28–10)

$$U_M = -\mathbf{m} \cdot \mathbf{B} + U_0 \qquad (28\text{–}11)$$

Example 28–5

Derive an expression for the torque on a circular loop of radius a carrying a current I. The loop is within a uniform magnetic field $\mathbf{B}$.

Solution:

A convenient orientation of the coordinate frame used for analysis can simplify the solution considerably. While maintaining complete generality we may take the loop to be in the x-y plane and the magnetic field to have only y- and z-components; i.e., $\mathbf{B} = B_y\hat{\mathbf{j}} + B_z\hat{\mathbf{k}}$. (• Prove these assertions.)

Consider a differential segment $d\ell$ of the loop, as shown in the diagram. The location of the segment is given by

$$\mathbf{r} = a(\cos \phi \hat{\mathbf{i}} + \sin \phi \hat{\mathbf{j}})$$

Also, $\qquad d\ell = a(-\sin \phi \hat{\mathbf{i}} + \cos \phi \hat{\mathbf{j}})d\phi$

The torque $\boldsymbol{\tau}_0$ about the origin O at the center of the loop is

$$\boldsymbol{\tau}_0 = \oint \mathbf{r} \times d\mathbf{F} = \oint \mathbf{r} \times (Id\ell \times \mathbf{B})$$

After substituting the expressions for $\mathbf{r}$, $d\ell$ and $\mathbf{B}$ and making some straightforward vector calculations, we find

$$\boldsymbol{\tau}_0 = a^2I \int_0^{2\pi} B_y \sin \phi(-\sin \phi \hat{\mathbf{i}} + \cos \phi \hat{\mathbf{j}})d\phi$$

The integral that involves the product $\sin \phi \cos \phi$ vanishes,

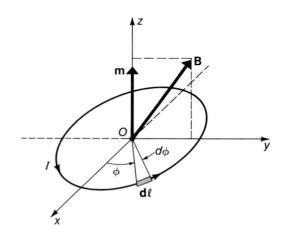

leaving

$$\boldsymbol{\tau}_0 = -a^2IB_y\hat{\mathbf{i}} \int_0^{2\pi} \sin^2 \phi \, d\phi = -\pi a^2IB_y\hat{\mathbf{i}}$$

The magnetic moment of the loop is $\mathbf{m} = \pi a^2I\hat{\mathbf{k}}$, so we again have

$$\boldsymbol{\tau}_0 = \mathbf{m} \times \mathbf{B}$$

This reinforces the assertion that Eq. 28–10 is completely general for planar loops.

The Galvanometer. A device commonly used as a meter readout in earlier electrical measuring instruments is the galvanometer. Although much present instrumentation uses electronic circuits coupled to digital meter readouts, the functioning of a galvanometer movement is a good example of magnetic forces exerted on coils (Fig. 28–9). The galvanometer coil usually consists of a few hundred turns of fine wire; because $\boldsymbol{\tau}_0$ acts equally on each turn, the meter's sensitivity is directly proportional to the number of turns. The coil is wound on an iron core that guides and concentrates the field lines. The coil assembly with the attached pointer is pivoted about an axis that is perpendicular to the plane of the page. With no current through the coil, the spring is adjusted so that the pointer lines up with the zero point of the scale.

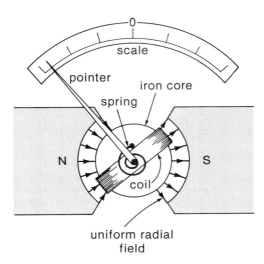

Fig. 28–9. Schematic diagram of a galvanometer movement. The concave pole faces provide a uniformly radial magnetic field in which the coil moves.

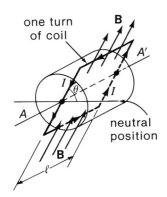

Fig. 28–10. When a steady current I flows through the coil of a galvanometer, a torque causes the coil to rotate about the axis AA'.

When a steady current I flows through the coil, a torque acts on each turn of wire, thereby producing a rotation of the coil assembly about the axis AA', as shown in Fig. 28–10. Depending on the direction of the current, the pointer will be deflected to the right or to the left of the zero point. The magnetic torque is opposed by the restoring torque due to the spring. A coiled spring can be made to produce a restoring torque that is proportional to the angular deflection. When such a spring is used in conjunction with a coil in a uniformly radial magnetic field (see Fig. 28–9), the deflection of the pointer will be a linear function of the current in the coil. (• Can you see why a uniform radial field is required for a linear response?)

Let the coil consist of N turns of wire around a core that has a length ℓ and a radius R. The field at the coil is B and the torsional constant of the spring is Γ (see Eq. 15–24). Then, the equilibrium deflection angle θ for a current I is obtained from

$$\tau_{AA'} = 2NI\ell RB = \Gamma\theta \qquad (28\text{–}12)$$

DC Motors. The fact that a current-carrying coil experiences a torque in a magnetic field is basic to the design of electric motors. The torque aids the rotation until the plane of the coil is perpendicular to the field direction. Then, $\sin\theta = 0$ and the torque vanishes. If provision is made for reversing the direction of the current in the coil whenever $\theta = 90°$ or $270°$, the rotation can be made continuous and the device becomes a motor. Figure 28–11 shows how this reversal can be accomplished in a simple way by delivering the current to the coil through a *split ring commutator*. The connection from the source of EMF to the ring is made through a pair of wire or graphite brushes. Because the split ring produces a reversal of the current in the coil every half cycle, the rotation is continuous. This is the principle of the *DC motor*.*

There exist many designs for practical electric motors. Usually, a number of multiturn coils are wound on the iron core; this assembly is called the *armature*. Also, an electromagnet is substituted for the permanent magnet shown in Fig. 28–11. One such design is illustrated in Fig. 28–12. The current I is delivered to the armature through the brushes, B_1 and B_2. The coil is wound in such a way that as the brushes slip to the next

Fig. 28–11. Schematic diagram of a DC electric motor.

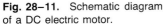

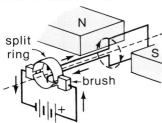

---

*The electric motor was invented in 1831 by the American physicist Joseph Henry (1797–1878).

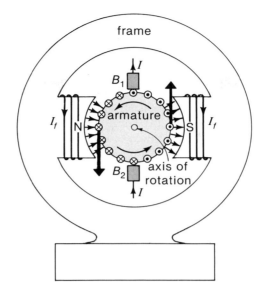

Fig. 28–12. A practical DC motor. The magnetic field is produced by the *field current* I_f in the windings that surround the pole pieces. Notice that the symbol ⊙ (suggesting the tip of an arrow) is used to indicate that the direction of the current is *out* of the plane of the page; the symbol ⊗ (suggesting the feather end of an arrow) is used to indicate that the direction of the current is *into* the plane of the page. We often use this method to indicate direction.

contact, the current always flows *into* the plane of the page in the coil sections on the N-pole side of the armature and *out of* the plane of the page in the coil sections on the S-pole side. This type of winding maintains a nearly constant torque on the armature and smooths the rotation.

28–5 ADDITIONAL APPLICATIONS OF THE LORENTZ FORCE

The Velocity Selector. Suppose that a narrow beam of identical charged particles (for example, a beam of electrons) traveling in vacuum enters a region in which there exists a uniform electric field **E** at right angles to a uniform magnetic field **B**. The particles in the beam have a spectrum of velocities, but they all enter the field region perpendicular to both field vectors, as indicated in Fig. 28–13. The electric field produces an upward force on the electrons in the beam, whereas the magnetic field produces a downward force. There is a unique velocity v_0 for which the electric and magnetic forces exactly cancel. The value of v_0 is obtained by setting the Lorentz force (Eq. 28–2) equal to zero. Because $\mathbf{v}_0 \perp \mathbf{B}$, we find

$$v_0 = E/B \qquad\qquad (28\text{–}13)$$

Electrons that have velocities less than v_0 are deflected upward and strike the exit wall at points such as a and b. Electrons that have velocities greater than v_0 are deflected downward and strike the wall at points such as c. Electrons that have the velocity v_0 (and

Fig. 28–13. Crossed electric and magnetic fields act as a *velocity selector*. Only electrons with $v_0 = E/B$ pass through the field region undeflected.

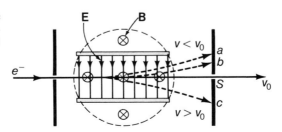

only these electrons) are undeviated and pass through the exit slit S. The arrangement of crossed electric and magnetic fields is called a *velocity selector*.

The Mass Spectrograph. In 1913, J. J. Thomson demonstrated that charged particles with different masses can be separated by using combined electric and magnetic fields. Thomson's crude device was improved upon by F. W. Aston and by A. J. Dempster. By 1919 atomic masses were being measured with a precision of 1 part in 10^3; the best modern instruments are capable of precisions of about 1 part in 10^7.

Figure 28–14 shows a schematic diagram of an early *mass spectrograph*. Low-energy positively charged ions are produced in the source chamber a and are accelerated through a potential difference V. The ions enter a region with a uniform magnetic field **B** (which, in Fig. 28–14, is directed outward from the plane of the page) and are deflected in circular arcs. The ions are recorded on a photographic plate so that the distance x from the entrance slit b (the diameter of the orbit) can be measured.

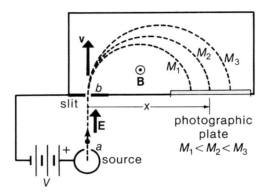

Fig. 28–14. Basic features of an early Dempster mass spectrograph. Positive ions are produced in the source a. The entire system is maintained under vacuum to prevent scattering of the ions by gas molecules.

Consider a singly ionized ion ($q = +e$) with a mass M that enters the magnetic field with a velocity v obtained from

$$\tfrac{1}{2}Mv^2 = eV$$

Within the magnetic field, the force on the ion is always perpendicular to its velocity vector, which has a constant magnitude. Therefore, the ion follows a circular path with radius $R = \tfrac{1}{2}x$. The centripetal acceleration is provided by the magnetic field. Thus,

$$\frac{v^2}{R} = \frac{v^2}{\tfrac{1}{2}x} = a_c = \frac{F}{M} = \frac{evB}{M}$$

Expressing v in terms of V and solving for M gives

$$M = \frac{eB^2}{8V}x^2 \tag{28–14}$$

Thus, the mass of the ion is obtained from measurements of B, V, and x. The relative masses of two ions accelerated at the same time involve only measurements of x and so can be obtained with high precision.

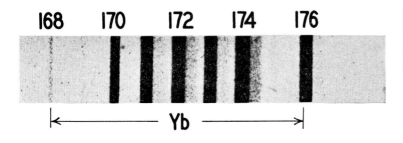

Fig. 28–15. Photographic record of ytterbium (Yb) isotopes obtained by Dempster in a mass spectrograph. (From *Introduction to Atomic Physics* by Henry Semat. Farrar & Rinehart, Inc., New York, 1939. By permission of the author, and courtesy of Prof. A. J. Dempster.)

Figure 28–15 shows the photographic record obtained by Dempster for the element ytterbium (Yb). Not all Yb atoms have the same mass (because the nuclei contain different numbers of neutrons). Thus, the mass spectrograph places each of these *isotopes* at different distances x along the plate. Dempster's results show Yb isotopes with mass numbers between 168 and 176. Notice that mass numbers 169 and 175 are absent; these isotopes are radioactive and do not occur in Nature.

Magnetic Confinement. If a charged particle is injected into a uniform magnetic field at some angle with respect to the field lines, the particle will spiral around a field line and will drift with a constant velocity along the field line (see Example 28–3). However, it is possible to produce a *nonuniform* field in which a particle is forced to move back and forth along a field line. The particle is therefore confined (or trapped) within the field.

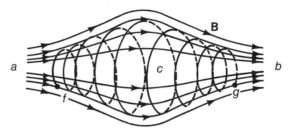

Fig. 28–16. A charged particle that is magnetically confined by the nonuniform field spirals back and forth between the points f and g.

Figure 28–16 shows a field that constitutes what is called a *magnetic bottle*. The field is strong at the ends, a and b, and is weak in the central region c. Consider a positive ion at point f. If the ion has a velocity component directed toward the right (an *axial* velocity component), it will follow the spiral path shown in the diagram. In the region between a and c, the radial velocity component of the particle interacts with the magnetic field **B** (the lines of which are *diverging* in this region) to produce an acceleration of the particle toward the right. (• Verify this by examining $\mathbf{v} \times \mathbf{B}$.) When the particle passes into the region between c and b, however, the field lines are *converging*. Then, the particle experiences an acceleration toward the left. This acceleration slows and eventually reverses the particle's axial drift toward the right. Point g represents the extreme of the rightward motion. The particle then drifts to the left until it reaches the leftward extreme at point f. Thus, the particle spirals back and forth between the reflection (or *mirror*) points, f and g. (The sense of the rotational motion is always the same.)

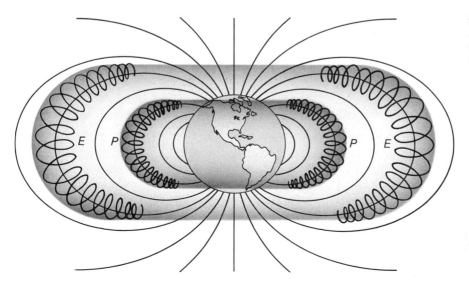

Fig. 28–17. The radiation belts consist of protons (P) and electrons (E) that are trapped in the Earth's nonuniform magnetic field.

Magnetic confinement devices are used in the study of high-temperature ionized gases (plasmas) to prevent the ions from striking and losing energy to the walls of the system. Unfortunately, particles tend to ''leak'' from these bottles for a variety of reasons, the major one having to do with ion-ion collisions at high plasma densities. These techniques may, however, eventually be perfected for use in nuclear fusion reactors that require plasmas at extremely high temperatures (10^6 K or so).

The Earth's Radiation Belts. The Earth's dipole magnetic field (see Fig. 28–1) resembles a magnetic bottle field (Fig. 28–16) that is wrapped around the globe. In 1958, during the Explorer I satellite mission, James A. Van Allen discovered that the Earth's field contains trapped charged particles. These particles—electrons and protons—are confined in broad, somewhat overlapping shells or radiation belts, as indicated in Fig. 28–17. The inner region (labeled P) primarily contains protons and is centered about 3000 km above the Earth's surface. The outer region (labeled E) primarily contains electrons and is centered about 15,000 km above the Earth's surface. The particle densities in the radiation belts range from about 10^4 to about 10^5 per cm^3.

QUESTIONS

28–1 Define and explain briefly the meaning of the terms (a) Earth's north magnetic pole, (b) Lorentz force, (c) Hall effect, (d) Ampère's law of magnetic force, and (e) magnetic moment.

28–2 A charged particle is located in a uniform magnetic field **B**. What is the force on the particle if it is (a) at rest with respect to the field, (b) moving with velocity **v** parallel to the field, or (c) moving with velocity **v** perpendicular to the field? Is it ever possible for the force on the moving charge to be in the direction of the field?

28–3 A charged particle enters a complicated (static) magnetic field with a certain kinetic energy. Explain why it will still have the same kinetic energy on leaving the region of the field if no other forces were acting on it while it was within the field.

28–4 Must magnetic lines of force always close on themselves?

28–5 Sketch the magnetic field lines for the two identical bar magnets with the orientations (a) and (b) shown in the diagrams.

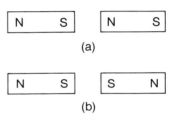

28–6 Two long parallel wires carry equal currents in opposite directions. Sketch the magnetic field lines in a plane perpendicular to the wires.

28–7 Explain how the force acting on the individual charge carriers responsible for the current in a wire placed in a magnetic field results in a macroscopic force on the wire itself.

28–8 How much work is required to rotate the rectangular current-carrying loop shown in Fig. 28–8 through 180° about the y-axis? Does this work depend on the initial angle θ?

PROBLEMS

Section 28–2

28–1 A proton moves in a uniform magnetic field **B** with a speed of 10^7 m/s and experiences an acceleration of 2×10^{13} m/s² in the $+x$-direction when its velocity is in the $+z$-direction. Determine the magnitude and direction of the field.

28–2 In a certain Cartesian reference frame the velocity of an electron is $\mathbf{v} = (2.5\hat{\mathbf{k}}) \times 10^7$ m/s and it experiences a force $\mathbf{F} = (4.0\hat{\mathbf{i}} + 3.0\hat{\mathbf{j}}) \times 10^{-13}$ N due to the presence of a uniform magnetic field. When the velocity is redirected and becomes $\mathbf{v} = (1.5\hat{\mathbf{i}} - 2.0\hat{\mathbf{j}}) \times 10^7$ m/s, the magnetic force on the electron is zero. Determine the components of the magnetic induction vector **B**. Why are both of the observations necessary to determine the field components?

28–3 A beam of protons (all with velocity **v**) emerges from a particle accelerator and is deflected in a circular arc with a radius of 0.45 m by a transverse uniform magnetic field of 0.80 T. (a) Determine the speed v of the protons in the beam. (b) What time is required for the deflection of a particular proton through an angle of 90°? (c) What is the energy of the particles in the beam?

28–4 A cosmic-ray proton in interstellar space has an energy of 10 MeV and executes a circular orbit with a radius equal to that of Mercury's orbit around the Sun (5.8 × 10^{10} m). What is the galactic magnetic field in that region of space?

28–5 The accelerating voltage that is applied to an electron gun is 15 kV, and the distance from the gun to a viewing screen is 35 cm. What is the deflection caused by the vertical component of the Earth's magnetic field (40,000 γ), assuming that any change in the horizontal component of the beam velocity is negligible?

28–6 A metal ball with a mass of 100 g is swung with an angular speed of 40π rad/s in a horizontal circle at the end of a taut string in the presence of a uniform vertical magnetic field of 100 G. (a) What amount of positive charge Q must be placed on the ball to double the tension in the string? (b) What is the direction of **B** relative to $\boldsymbol{\omega}$? (c) Would it be feasible to place the charge Q on the ball?

Section 28–3

28–7 A wire carries a steady current of 2.4 A. A straight section of the wire, with a length of 0.75 m along the x-axis, lies within a uniform magnetic field, $\mathbf{B} = (1.6\hat{\mathbf{k}})$ T. If the current flows in the +x-direction, what is the magnetic force on the section of wire?

28–8 A wire carries a steady current of 3.0 A. A portion of the wire is located within a uniform magnetic field, $\mathbf{B} = (0.65\hat{\mathbf{i}})$ T. This portion of the wire is composed of two straight-line segments, $\overline{AB}$ and $\overline{BC}$, that lie in the y-z plane. The three points are $y_A = 2.0$ m and $z_A = 1.0$ m; $y_B = 6.0$ m and $z_B = 1.0$ m; $y_C = 6.0$ m and $z_C = 7.0$ m. (a) What is the magnetic force on each segment of the wire? (b) What is the net magnetic force on the wire? (c) Consider the situation in which the wire segment in the field extends in a straight line from point A to point C. Demonstrate that the net magnetic force on the wire is the same as that found in (b).

28–9 Prove that the line integral in Eq. 28–6 vanishes identically if the loop lies entirely within a uniform magnetic field. [*Hint:* The constant vector **B** can be removed from the integral.]

Section 28–4

28–10 Begin with Eq. 28–10 for the torque on a current-carrying loop in a uniform magnetic field and derive Eq. 28–11 for the potential energy U_M relative to the condition $\theta = \pi/2$.

28–11 A coil of 10 turns of fine wire wound on a rectangular frame 8 cm × 10 cm carries a current of 10 A and is located within a uniform magnetic field of 300 G. (a) How many different geometric arrangements of coil and field produce maximum torque on the coil? Sketch one such arrangement and determine the torque τ_0. (b) How many different geometric arrangements of coil and field produce a condition of stable equilibrium? Unstable equilibrium? (c) How much work must be done by an external agency to rotate the coil from stable to unstable equilibrium?

28–12 A square loop of wire, 10 cm × 10 cm, has a mass of 50 g and pivots about an axis AA' that corresponds to a horizontal side of the square, as shown in the diagram. The loop is in equilibrium in a uniform magnetic field of 500 G directed vertically downward when $\theta = 20°$. (a) What current flows in the loop and in what direction? (b) Determine the magnitude and direction of the force exerted on the axis by the bearings.

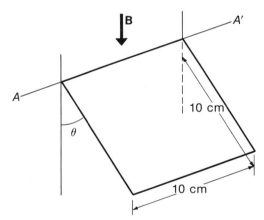

28–13 The coil of a galvanometer movement has an area of 1.0 cm^2 and consists of 100 turns of fine wire. The radial magnetic field at the position of the coil is 0.15 T, and the torsional constant of the spring is $\Gamma = 1.5 \times 10^{-7}$ N·m/deg. What is the angular deflection of the coil for a current of 1 *mA*?

Section 28–5

28–14 A crossed-field velocity selector has a magnetic field of 10^{-2} T. What electric field strength is required if 10-keV electrons are to pass through undeflected?

28–15 A Dempster mass spectrograph (Fig. 28–14) is used to determine the mass of deuterium (^{2}H) relative to the known mass of ordinary hydrogen (^{1}H; M = 1.00783 u). The line due to singly ionized hydrogen molecules (^{1}H$_2^+$) is observed on the photographic plate at x = 16.20 cm. A line corresponding to ionized atomic deuterium (^{2}H$^+$) is observed at a *smaller* value of x with a line separation of Δx = 0.06 mm. Determine the atomic mass of deuterium and compare with the value given in Fig. 19–4.

28–16 One type of mass spectrometer (developed in 1951) is based on the fact that the angular frequency of an ion in a uniform magnetic field about an axis parallel to the field direction is independent of the ion velocity. (This is the cyclotron frequency; see Example 28–3.) Low-velocity ions (K $\cong$ 10 eV) are emitted from a source in short duration bursts and are injected into a uniform magnetic field of 450 G. A measurement is made of the time required for a burst of ions to traverse 10 helical turns. How accurately must the transit time be measured in order to determine the mass of the ions to within 10^{-3} u?

28–17 A 5.0-keV proton is trapped in a magnetic bottle. At a certain instant the proton is moving in a circular path in a plane that is perpendicular to the axis of the magnetic bottle, as indicated in the diagram. The magnetic field has a magnitude of 600 G and the field line at the position of the proton makes an angle θ = 15° with respect to the bottle axis as shown. (a) What is the velocity of the proton? (b) What is the radius of the path? (c) What is the radial acceleration a_r of the proton? (d) Show that the ratio of the axial acceleration a_z to the radial acceleration is $a_z/a_r = -\tan \theta$. (e) What is the axial acceleration?

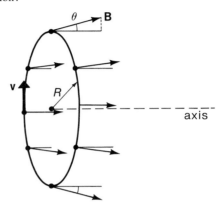

Additional Problems

28–18 One way to determine the ratio e/m for electrons (a
• method developed by H. Busch in 1922 and often used in laboratory demonstrations) involves producing monoenergetic electrons at a point source located in a uniform magnetic field. The field is adjusted until a sharp image of the source is produced at some distance L for those electrons that leave the source in directions making small angles with the field, as shown in the diagram. (a) If the electrons in the source are accelerated through a voltage V, and if the lowest focusing field is B_0, show that $e/m = 8\pi^2V/L^2B_0^2$. (b) Typical values for a laboratory exercise are L = 15.0 cm, V = 1000 V, and B_0 = 45 G. Determine the value of e/m from these data and compare with the actual value.

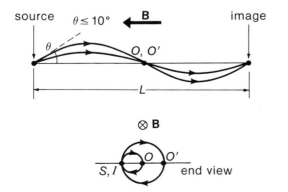

28–19 An artificial satellite circles the Earth in the equatorial
• plane at an altitude of 100 km, traveling from east to west. Along the satellite orbit the Earth's magnetic field is approximately 25,000 γ and is perpendicular to the equatorial plane. If the satellite has acquired a positive charge of 0.1 C, what force does the Earth's magnetic field exert on the satellite? Compare your result with the gravitational force on the satellite if its mass is 10^3 kg.

28–20 A small circular coil with a mass of 20 g consists of 19
•• turns of fine wire and carries a current of 1.0 A. The coil is located in a uniform magnetic field of 0.10 T with the

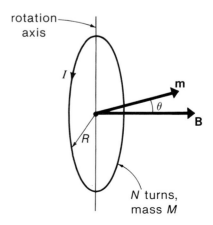

coil axis originally parallel to the field direction. (a) What is the period of small-angle oscillations of the coil axis around its equilibrium position, assuming no friction and no mechanical restoring force? (b) If the coil is rotated through an angle of 10° from its equilibrium position and then released, what will be its angular speed when it passes through the equilibrium position?

28–21 A DC motor operates from a 220-V DC power line. The motor's electromagnet is excited by a field coil whose resistance is 200 Ω. The field coil is connected in parallel with the motor armature, which has a resistance of 1.5 Ω (see the diagram). The armature current is 9.4 A. When the motor is delivering its rated output, there are frictional and windage losses amounting to 120 W. (a) What is the power input to the armature? (b) What is the total power loss in the armature, including Joule heating, friction, and windage? (c) What is the mechanical output power? (Express your result in watts and in horsepower.) (d) What is the total input current I? (e) What is the total input power? (f) What is the overall efficiency of the motor?

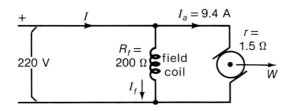

28–22 The diagram shows a modification of the Dempster mass spectrograph (Fig. 28–14) that was introduced by K. T. Bainbridge. In this device a velocity selector is placed after the entrance slit b and utilizes the same field **B** as in the main region of the spectrograph. Show that the ion mass is $M = (eB^2/2E)x$, where E is the electric field strength in the velocity selector.

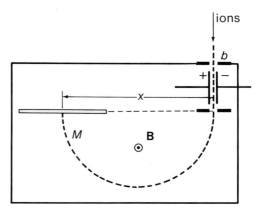

28–23 In a Dempster mass spectrograph (Fig. 28–14), the accelerating voltage is 20,000 V, and the ytterbium group of lines is centered at $\bar{x} = 25.0$ cm. (a) What magnetic field is required? (b) What is the range Δx for the isotopic lines from mass number 168 to 176?

CHAPTER 29

SOURCES OF MAGNETIC FIELDS

In the preceding chapter the discussion centered on the interaction of steady magnetic fields with individual moving charged particles and with steady currents produced by the collective motion of charge carriers in conducting media. In this chapter we examine in detail the way in which currents give rise to magnetic fields, again restricting considerations to steady currents (DC). The effects of time-dependent fields are discussed in the following chapters.

29–1 THE BIOT-SAVART LAW

It is again convenient to use the concept of a differential vector current element $Id\ell$ and to associate with it a differential magnetic induction $d\mathbf{B}$ that it causes at every point in space. The connection between these quantities is known as the *Biot-Savart law*.* In differential form the statement of the law is

$$d\mathbf{B} = K_M \frac{Id\ell \times \hat{\mathbf{u}}_r}{r^2} \tag{29–1}$$

In this expression, $d\mathbf{B}$ is the contribution to the field made by the current element $Id\ell$ at a point defined by the vector $\mathbf{r}$ from the current element, as shown in Fig. 29–1. The constant K_M is the magnetic counterpart of the constant K that appears in Coulomb's law (Eq. 23–1) and has the exact value

$$K_M = 10^{-7} \text{ T·m/A} = 10^{-7} \text{ N/A}^2$$

(Refer to Section 29–4 for additional remarks concerning K_M and the definition of the ampere.) The right-hand rule for vector cross products specifies the direction of $d\mathbf{B}$.

In the discussion of electrostatics (Section 23–3), we found it convenient to introduce the permittivity of free space, $\epsilon_0 = 1/4\pi K$, and to use this quantity in Coulomb's law. We now introduce a quantity called the *permeability of free space*,

$$\mu_0 = 4\pi K_M = 4\pi \times 10^{-7} \text{ N/A}^2$$
$$= 4\pi \times 10^{-7} \text{ T·m/A}$$

which we use in Eq. 29–1 to write the Biot-Savart law in the way it is usually stated, namely,

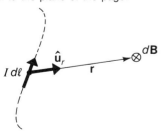

Fig. 29–1. Geometry for the Biot-Savart law. The unit vector is $\hat{\mathbf{u}}_r = \mathbf{r}/|\mathbf{r}|$. The direction of $d\mathbf{B}$ is *into* the plane of the page.

---

*Jean-Baptiste Biot (1774–1862) and Félix Savart (1791–1841) expressed the results of their investigations in terms of the forces experienced by current-carrying wires (1820). The differential formulation of the law is due to Pierre Simon Laplace (1749–1827).

$$dB = \frac{\mu_0 I}{4\pi} \frac{d\ell \times \hat{u}_r}{r^2} \qquad (29\text{--}2)$$

A steady current implies the existence of a closed current loop. To determine the field **B** at some point P in space requires the addition (by linear superposition) of all the differential field components $d\mathbf{B}$ (Eq. 29–2) produced by elements around the entire loop. That is,

$$\mathbf{B} = \frac{\mu_0 I}{4\pi} \oint_{\substack{\text{current} \\ \text{loop}}} \frac{d\ell \times \hat{u}_r}{r^2} \qquad (29\text{--}3)$$

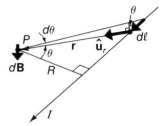

Fig. 29–2. Geometry for calculating the magnetic field produced by a steady current in a long straight wire.

Equation 29–3 can be used to calculate the magnetic field produced by a steady current in a long straight wire, as shown in Fig. 29–2. Suppose that the perpendicular distance R from the point of interest P to the wire is small compared with the length of the straight section of the wire that we consider. Thus, the wire is, in effect, infinitely long and we can ignore any contribution to the field at P from the part of the loop that connects the two ends by a return path at a great distance from P.

From the geometry in Fig. 29–2, we have

$$|d\ell \times \hat{u}_r| = d\ell \sin\left(\frac{\pi}{2} - \theta\right) = d\ell \cos\theta$$

Also, $d\ell \cos\theta = r\, d\theta$ and $r\cos\theta = R$

Using these relationships in Eq. 29–3, we find for the magnitude of **B**

$$B = \frac{\mu_0 I}{4\pi R} \int_{-\pi/2}^{\pi/2} \cos\theta\, d\theta = \frac{\mu_0 I}{2\pi R} \qquad (29\text{--}4)$$

The direction of **B** is specified by the unit azimuthal vector $\hat{u}_\phi$, as shown in Fig. 29–3. Then, we can write

$$\mathbf{B} = \frac{\mu_0 I}{2\pi R} \hat{u}_\phi \qquad \text{straight wire} \qquad (29\text{--}4a)$$

The direction of the field at the point P is perpendicular to the plane that contains the wire and the point. Moreover, **B** has a constant magnitude for a given value of R. Thus, the field lines are circles concentric with the wire. This is the same conclusion that was reached in Section 28–1 (see Fig. 28–4). The direction of the field is specified by the right-hand rule that was given in connection with the discussion of Fig. 28–5. These properties of the field are summarized in Fig. 29–3.

According to Eq. 29–4, the magnitude of the magnetic induction decreases inversely with the perpendicular distance R from the wire. At a distance of 10 cm from a wire in which the current is 30 A, we have

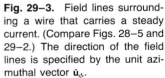

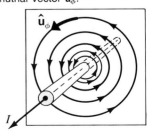

Fig. 29–3. Field lines surrounding a wire that carries a steady current. (Compare Figs. 28–5 and 29–2.) The direction of the field lines is specified by the unit azimuthal vector $\hat{u}_\phi$.

$$B = \frac{(4\pi \times 10^{-7}\ \text{T·m/A})(30\ \text{A})}{2\pi(0.10\ \text{m})}$$

$$= 6.0 \times 10^{-5}\ \text{T} = 0.60\ \text{G}$$

which is approximately the magnitude of the Earth's magnetic field near the surface.

Example 29–1

Determine the magnetic induction along the axis of a circular loop that carries a current I.

Solution:

At the point P a distance z along the loop axis, the field $d\mathbf{B}$ produced by the current element $I d\ell$ can be resolved into an axial component, $dB_\parallel = dB \sin \psi$, and a radial component, $dB_\perp = dB \cos \psi$. For each such current element there is a matching element directly opposite. The field components $dB_\perp$ exactly cancel, whereas the components $dB_\parallel$ add together. Therefore, using Eq. 29–3, we find

$$\mathbf{B} = B_\parallel \hat{\mathbf{k}} = \frac{\mu_0 I}{4\pi} \hat{\mathbf{k}} \oint \frac{a\, d\theta \sin \psi}{r^2}$$

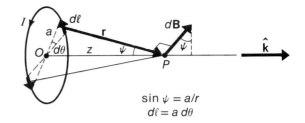

$$\sin \psi = a/r$$
$$d\ell = a\, d\theta$$

With a, ψ, and r constant, and with $\oint d\theta = 2\pi$ and $\sin \psi = a/r = a/(z^2 + a^2)^{1/2}$, we have

$$\mathbf{B} = \frac{\mu_0 I a^2}{2(z^2 + a^2)^{3/2}} \hat{\mathbf{k}}$$

The Dipole Field. In Example 29–1, notice that when $z \gg a$, the result can be expressed as

$$\mathbf{B} = \frac{\mu_0}{2\pi z^3} \mathbf{m}$$

where $\mathbf{m} = \pi a^2 I \hat{\mathbf{k}}$ is the magnetic moment of the loop (see Eq. 28–8). In Section 24–2 we found that the electric field at large distances from an electric dipole is (Eq. 24–6)

$$\mathbf{E} = \frac{1}{4\pi\epsilon_0} \left(-\frac{\mathbf{p}}{r^3} + \frac{3(\mathbf{p} \cdot \mathbf{r})\mathbf{r}}{r^5} \right) \tag{29–5}$$

Taking $\mathbf{r}$ along the z-axis (the direction of $\mathbf{p}$), we have $\mathbf{r} = z\hat{\mathbf{k}}$ and $\mathbf{p} = p\hat{\mathbf{k}}$. Using these expressions in the equation for $\mathbf{E}$, we find

$$\mathbf{E} = \frac{1}{2\pi\epsilon_0 z^3} \mathbf{p}$$

Fig. 29–4. Near field for a magnetic dipole. (• Can you define an N pole and an S pole for the dipole?)

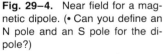

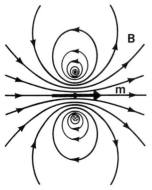

(• Verify this last result.) Thus, except for the different roles played by ϵ_0 and μ_0, the magnetic field along the axis of a current loop has the same behavior as the electric field along the axis of an electric dipole. Indeed, the distant field of a current loop has exactly the form of a dipole field, namely,

$$\mathbf{B} = \frac{\mu_0}{4\pi} \left(-\frac{\mathbf{m}}{r^3} + \frac{3(\mathbf{m} \cdot \mathbf{r})\mathbf{r}}{r^5} \right) \qquad \text{magnetic dipole} \tag{29–6}$$

That is, a current loop is a *magnetic dipole*.

Equations 29–5 and 29–6 are the *distant-field* expressions for an electric dipole and a magnetic dipole, respectively. In each case the *near field* (where r is no longer much larger than the size of the dipole) is more complicated. Compare the field lines for the finite electric dipole (Fig. 24–6) with those for the point dipole (Fig. 25–5), for which the distant-field expression is valid for all r. Figure 29–4 shows the near field of a magnetic dipole in a plane that contains the magnetic moment vector **m**. Notice that the field lines, although not circular, close on themselves just as in the case of a long straight wire (see Fig. 29–3). Do the field lines always close in this way? We answer this important question in Section 29–3.

Finally, we remark again that the Earth's magnetic field closely resembles the field of a magnetic dipole, at least for distances out to several Earth radii (see Fig. 28–1). The value of the Earth's magnetic moment is

$$m_E = 8.1 \times 10^{22} \text{ A·m}^2$$

(For distances greater than a few Earth radii, the magnetic effect of the stream of charged particles emitted by the Sun—the *solar wind*—severely distorts the dipole field of the Earth.)

Example 29–2

A *solenoid* is a helical coil that is constructed by tightly winding a single strand of fine wire around a straight, nonmagnetic cylindric shell. Usually the diameter of the wire is so small compared with the diameter of the helix that a closely spaced winding produces a helix pitch that is negligible. Determine the magnetic induction B along the axis of a solenoid with a length ℓ and a radius R that has n turns per meter of wire carrying a current I. Refer to diagram (a).

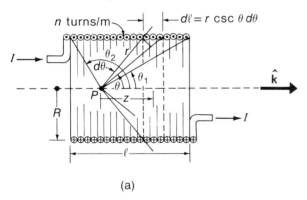

(a)

Solution:

Consider the magnetic field at a point P on the axis of the solenoid, as indicated in diagram (a). We use the result of Example 29–1 for the contribution $d\mathbf{B}$ of the turns that lie in the differential ring intercepted by two circular cones with half angles θ and $\theta + d\theta$, as shown. The axial length of this ring is $d\ell = r \csc \theta \, d\theta$, so the number of turns in the ring is $nd\ell = nr \csc \theta \, d\theta$. Thus, at point P, using $(z^2 + a^2)^{3/2} = r^3$, we have

$$d\mathbf{B} = \frac{\mu_0 nr \csc \theta \, IR^2 \, d\theta}{2r^3} \hat{\mathbf{k}}$$

Using $r = R \csc \theta$, the expression for $d\mathbf{B}$ reduces to

$$d\mathbf{B} = \tfrac{1}{2}\mu_0 nI \sin \theta \, d\theta \hat{\mathbf{k}}$$

The field **B** is obtained by integrating between $\theta = \theta_1$ and $\theta = \theta_2$, as shown in the diagram. The result is

$$\mathbf{B} = \tfrac{1}{2}\mu_0 nI(\cos \theta_1 - \cos \theta_2)\hat{\mathbf{k}}$$

$$\mathbf{B} = \mu_0 nI\hat{\mathbf{k}}$$

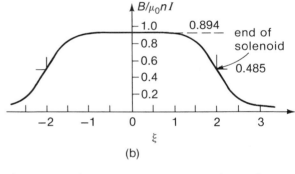

(b)

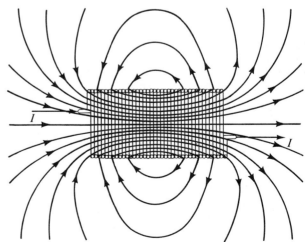

(c)

This expression for the field on the solenoid axis is valid for all axial points P, whether inside or outside the solenoid. Diagram (b) shows the axial field B in units of $\mu_0 nI$ as a function of ξ, the position of P measured from the center of the solenoid in units of R, for the case $\ell = 4R$. Diagram (c) shows the magnetic field for this case. For this geometry the field along the axis, $|\xi| < 1$, is nearly constant with the approximate value $0.86 \, \mu_0 nI$. In the

event that the solenoid is very long ($\xi \gg R$), we have $\cos \theta_1 \to 1$ and $\cos \theta_2 \to -1$. Then, the axial field is

$$\mathbf{B} = \mu_0 nI \hat{\mathbf{k}}$$

In Example 29–5 we find that for an infinite solenoid the field is constant everywhere within the solenoid.

Example 29–3

Determine the magnetic induction produced by two long, straight, parallel wires that are a distance $2a$ apart, each carrying a current I in the same direction.

Solution:

The total field is the vectorial sum of the fields produced by the individual wires (principle of superposition). Diagram (a) illustrates the calculation of $\mathbf{B} = \mathbf{B}_1 + \mathbf{B}_2$ along the x- and y- axes. Along the x-axis the individual fields are entirely in the $\pm y$-direction. Consequently, the resultant field has only a y- component, $B_y(x, 0)$. Using Eq. 29–4 for each wire, we have, in the region $|x| > a$,

$$B_y(x, 0) = \frac{\mu_0 I}{2\pi} \left(\frac{1}{x - a} + \frac{1}{x + a} \right)$$

$$= \frac{\mu_0 I}{\pi} \frac{x}{x^2 - a^2}, \qquad |x| > a$$

For $|x| < a$, we have

$$B_y(x, 0) = \frac{\mu_0 I}{2\pi} \left(\frac{1}{a + x} - \frac{1}{a - x} \right)$$

$$= -\frac{\mu_0 I}{\pi} \frac{x}{a^2 - x^2}, \qquad |x| < a$$

These expressions for $B_y(x, 0)$ are shown in diagram (b). Notice that midway between the wires, we have $B_y(0, 0) = 0$. Also, at large distances ($|x| \gg a$) the field asymptotically approaches the field produced by a single wire along the z-axis that carries a current $2I$.

Along the y-axis the y-components of the two individual fields cancel, whereas the x-components add as shown in diagram (a). Thus, for $y > 0$, we have

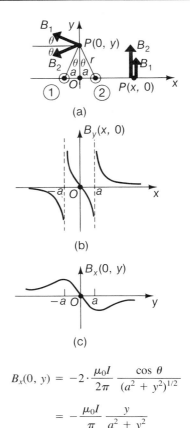

(a)

(b)

(c)

$$B_x(0, y) = -2 \cdot \frac{\mu_0 I}{2\pi} \frac{\cos \theta}{(a^2 + y^2)^{1/2}}$$

$$= -\frac{\mu_0 I}{\pi} \frac{y}{a^2 + y^2}$$

This result is shown in diagram (c). (• Can you verify that this expression is also valid for $y < 0$?) Again, notice that at large distances ($|y| \gg a$) the field asymptotically approaches the field produced by a single wire along the z-axis that carries a current $2I$.

29–2 AMPÈRE'S CIRCUITAL LAW

In the discussion of the electric field produced by a static charge distribution (Section 25–2), we learned that the line integral of $\mathbf{E} \cdot d\mathbf{r}$ around any closed path is zero (Eq. 25–14). We now consider the static magnetic case and inquire about the line integral of $\mathbf{B} \cdot d\mathbf{r}$ around a closed path.

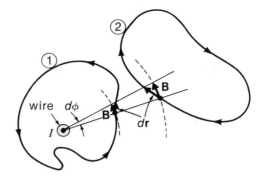

Fig. 29–5. Geometry for deriving Ampère's circuital law for a long straight wire that carries a steady current I. The line integral of $\mathbf{B} \cdot d\mathbf{r}$ does not depend on the shape of the integration path, even a path such as ① that loops back on itself. (• Can you see why?)

Figure 29–5 illustrates the case of a long straight wire that carries a steady current I directed out of the plane of the page. We wish to evaluate $\oint \mathbf{B} \cdot d\mathbf{r}$ for two closed paths—a path labeled ① that surrounds the wire and a path labeled ② that does not.* We can view $\mathbf{B} \cdot d\mathbf{r}$ as the product of B and the projection of $d\mathbf{r}$ onto $\mathbf{B}$. Because the field $\mathbf{B}$ around the straight wire has only an azimuthal component, the projection of $d\mathbf{r}$ onto $\mathbf{B}$ is $r\, d\phi$ for either path (see Fig. 29–5). Therefore, we have

$$\oint \mathbf{B} \cdot d\mathbf{r} = \oint B(r) r\, d\phi$$

Using Eq. 29–4 to write $B(r) = \mu_0 I / 2\pi r$, the line integral becomes

$$\oint \mathbf{B} \cdot d\mathbf{r} = \frac{\mu_0 I}{2\pi} \oint d\phi$$

For path ① the value of the integral is 2π, whereas for path ② the value is zero. Thus, we can express the result as

$$\oint \mathbf{B} \cdot d\mathbf{r} = \mu_0 I_p \qquad \textbf{(29–7)}$$

where I_p is the current enclosed or linked by the integration path.

The result we have obtained in Eq. 29–7 is based on the special case of a steady current in a long straight wire. It was Ampère's brilliant guess that the equation is generally true for steady currents. Indeed, this result can be derived for any field produced by steady currents by using the general form of the Biot-Savart law, and experimental tests have shown the equation to be correct.

Suppose that we construct an open surface S with the perimeter coinciding with the integration path in Eq. 29–7. The total current I_p that is enclosed by the path must all flow through the surface S, as indicated in Fig. 29–6. The current I_p can be expressed as the integral over S of the current density $\mathbf{J}$ (see Eq. 27–3); thus,

$$I_p = \int_S \mathbf{J} \cdot \hat{\mathbf{n}}\, dA$$

Then, we have

$$\oint \mathbf{B} \cdot d\mathbf{r} = \mu_0 \int_S \mathbf{J} \cdot \hat{\mathbf{n}}\, dA \qquad \textbf{(29–8)}$$

Fig. 29–6. Two arbitrary open surfaces, S and S', each with a perimeter corresponding to the integration path in Eq. 29–7 or 29–8. When a steady current flows and no charge accumulates (or disappears) between the two surfaces, it follows that the integrals of $\mathbf{J} \cdot \hat{\mathbf{n}}\, dA$ over both of the surfaces are the same. (See, however, the discussion in Section 30–4.)

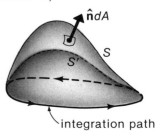

*Recall that $d\mathbf{r}$ represents an arbitrary displacement, whereas $d\boldsymbol{\ell}$, which appears only in combination with I, is always directed along a current filament.

The positive direction of $\hat{\mathbf{n}} \, dA$ is given by a right-hand rule: If the fingers of the right hand are curled in the general direction of the closed integration path, the extended thumb points in the positive sense of n. Equation 29–8 is the most general statement of *Ampère's circuital law*.

As a practical matter, Ampère's circuital law is most useful for determining unknown magnetic fields for situations in which the geometry possesses a high degree of symmetry. When such symmetry is present, it is often possible to deduce the shape of the field lines, thereby simplifying the evaluation of the line integral.* (The integral is easy to evaluate when **B** lies along or perpendicular to the path of integration.) The examples at the end of this section illustrate a few applications of this type.

Example 29–4

A long straight wire with a radius R carries a steady current I that is distributed uniformly over the cross section of the wire. Determine the magnetic induction everywhere.

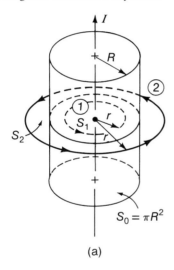

(a)

Solution:

The current density in the wire is $J = I/A = I/\pi R^2$. To calculate the magnetic induction within the wire, we use a circular Ampèrian path (labeled ① in the diagram) that is centered on the wire axis; this path lies in a plane that is perpendicular to the wire axis. Because of the cylindric symmetry, we expect that **B** is entirely azimuthal (no radial component) and is constant in magnitude along the path ①. The magnitude and radial dependence of **B** can be determined by using Eq. 29–8. The simplest surface for which ① is the perimeter is the circular disc S_1. Then,

$$\oint_① \mathbf{B} \cdot d\mathbf{r} = \mu_0 \int_{S_1} \mathbf{J} \cdot \hat{\mathbf{n}} \, dA$$

$$2\pi r B = \mu_0 \cdot \pi r^2 \cdot \frac{I}{\pi R^2}$$

from which we have

$$\mathbf{B} = \frac{\mu_0 I}{2\pi R^2} r \hat{\mathbf{u}}_\phi, \qquad r \leqslant R$$

To calculate the field outside the wire, we use the circular path ② and split the integral over the circular disc S_2 into two parts—one integral over $S_2 - S_0$ and the other over $S_0 = \pi R^2$, the cross-section of the wire. We find

$$\oint_② \mathbf{B} \cdot d\mathbf{r} = \mu_0 \int_{S_2 - S_0} \mathbf{J} \cdot \hat{\mathbf{n}} \, dA + \mu_0 \int_{S_0} \mathbf{J} \cdot \hat{\mathbf{n}} \, dA$$

$$2\pi r B = 0 + \mu_0 I$$

so that

$$\mathbf{B} = \frac{\mu_0 I}{2\pi r} \hat{\mathbf{u}}_\phi \qquad r \geqslant R$$

(• Why is the surface integral over $S_2 - S_0$ identically zero?) Notice that we could have obtained the same results in this case by using Eq. 29–7. The field outside the wire is, of course, the same as was found from the Biot-Savart law (Eq. 29–4a). The variation of B with r is shown in the diagram below.

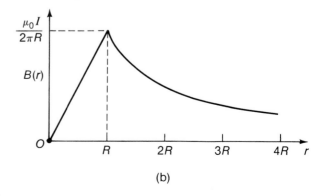

(b)

---

*This is analogous to the situation for Gauss's law; see the similar remark in Section 24–3.

Example 29–5

Determine the magnetic induction produced by an infinitely long solenoid that consists of n turns per meter of wire that carries a current I.

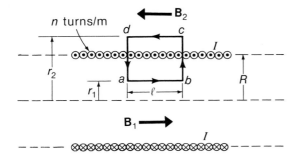

Solution:

The diagram shows a portion of the solenoid cut by a plane that contains the axis of the coil. Consider the closed rectangular Ampèrian path $abcda$ that lies in this plane.

To evaluate the line integral of $\mathbf{B} \cdot d\mathbf{r}$ we need to know the interior field $\mathbf{B}_1$ and the exterior field $\mathbf{B}_2$. What do we expect for the fields in this case? First, because the solenoid has an infinite length, the fields must be axial everywhere.* The interior field $\mathbf{B}_1$ is directed toward the right; the exterior field $\mathbf{B}_2$ (if it exists) is directed toward the left. (Refer to the diagrams in Examples 29–1 and 29–2.)

Using Eq. 29–7, we have for the path $abcda$

$$B_1 \ell + B_2 \ell = \mu_0 n \ell I$$

Now, consider r_1 fixed but allow $r_2 \to \infty$. Then, the equation above requires that B_2 be constant. (• Explain why B_2 is required to be constant.) The field cannot have a constant nonzero value extending to infinity, so we conclude that the constant value of B_2 must be zero. Therefore, we have

$$B_1 = \mu_0 n I$$

and the interior field is constant, independent of r_1. (Compare Example 29–2.)

Example 29–6

Determine the magnetic induction inside a toroidal coil that has a mean radius R. The coil has a total of N turns of wire that carries a current I.

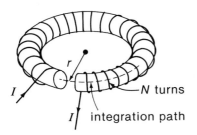

Solution:

Choose as an integration path a circle with radius r that lies within the toroid, as indicated in the diagram. The symmetry

demands that the field inside the toroid be everywhere azimuthal, that is, along the integration path. Then, the integral of $\mathbf{B} \cdot d\mathbf{r}$ around the circular path yields $2\pi r B$. The total current linked by the integration path is $I_p = NI$. Then, using Eq. 29–7, we find

$$B = B_\phi = \frac{\mu_0 NI}{2\pi r}$$

If the mean radius R of the toroid is much larger than the cross-sectional dimensions of the ring, we have $r \cong R$. Then, with $n = N/2\pi R$ equal to the number of turns per unit length, we have

$$B = B_\phi = \mu_0 n I$$

(• Can you explain why this result is the same as that for the infinite solenoid obtained in Example 29–5? What is the nature of the field outside the toroid? It is not zero. Refer to the footnote in Example 29–5.)

---

*Because there is a net current I that flows axially along the solenoid, there is an azimuthal field, $B_\phi = \mu_0 I / 2\pi r$, outside the solenoid. For reasonable values of R and n, we have $B_\phi \ll B_1$. (• Can you verify this?) In any event, $\mathbf{B}_\phi$ is perpendicular to $d\mathbf{r}$ along the integration path, so B_ϕ does not contribute to the integral of $\mathbf{B} \cdot d\mathbf{r}$. (• Can you show that $B_\phi = 0$ inside the solenoid?)

29–3 GAUSS'S LAW FOR MAGNETIC FIELDS

The quantization of electric charge is a firmly established fundamental property of matter. Electric charge is always observed to occur in units of $\pm e$. Is there a corresponding elementary unit of "magnetic charge," a magnetic *monopole*? Although there have been theoretical speculations concerning magnetic monopoles, no convincing experimental evidence of their existence has ever been obtained. Therefore, the laws of electromagnetism are formulated on the basis that *there exists no fundamental isolated magnetic charge*.

▶ If there are no magnetic monopoles, the lowest order of elementary magnetic induction is the *dipole*. On a macroscopic scale, magnetic dipole fields are set up by current loops of the type we have been discussing. On a microscopic (atomic) scale, a quantum treatment is necessary; nevertheless, a convenient model of atomic magnetism attributes the field to the effective current produced by electrons circulating around the nucleus. In addition to this orbital effect, elementary particles possess an *intrinsic* angular momentum and an associated magnetic moment. For example, the intrinsic angular momentum (or *spin*) of the electron is $s = 5.28 \times 10^{-35}$ J·s and the intrinsic magnetic dipole moment of the electron is $m_B = 9.27 \times 10^{-24}$ A·m^2, a quantity called the *Bohr magneton*. ◀

What is the effect of the absence of magnetic monopoles on the shapes of magnetic field lines? Recall that electric field lines due to static charges always originate on positive charges and terminate on negative charges. Because there are no magnetic charges to originate or terminate magnetic field lines, these lines necessarily close upon themselves. It is for this reason that in the various cases we have studied, the magnetic field lines have always been shown as nonintersecting closed loops. Although the *distant* electric and magnetic dipole fields obey the same mathematical description (see Eqs. 29–5 and 29–6), the *near* fields are quite different. Figure 29–7 shows the pattern of field lines for an electric dipole and for a magnetic dipole. Notice how the fields begin to resemble one another at large distances.

Fig. 29–7. (a) The electric field intensity due to an electric dipole. (b) The magnetic induction due to a magnetic dipole (a current loop). A Gaussian surface S is shown in each case.

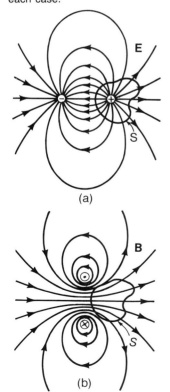

(a)

(b)

Gauss's Law. Figure 29–7 also shows a Gaussian surface S in each of the dipole fields. For the case of the electric dipole field (Fig. 29–7a), the surface encloses the positive charge $+Q$ of the dipole. According to Eq. 24–12 we have

$$\oint_S \mathbf{E} \cdot \hat{\mathbf{n}} \, dA = \frac{1}{\epsilon_0} Q$$

The fact that this integral is *positive* means that there is a net *outward* flux of field lines piercing the surface, as we expect from the geometry (Fig. 29–7a).

For the case of the magnetic dipole field (Fig. 29–7b), every field line that *enters* the volume enclosed by the surface S also *leaves* the volume. Consequently, the net *magnetic flux,* φ_M, through the surface must be zero.* Because magnetic monopoles do not exist, this is a general result: The net magnetic flux through *any* Gaussian surface in *any* magnetic field must be zero. That is, Gauss's law for magnetic fields is

$$\oint_S \mathbf{B} \cdot \hat{\mathbf{n}} \, dA = 0 \qquad \text{always} \qquad \textbf{(29–9)}$$

where S is a *closed* surface. The value of φ_M is always zero for a closed surface, but the flux can be nonzero for an open surface. (Be certain that you understand the distinction.)

---

*In analogy with the electric case (see Eq. 24–9), the integral of $\mathbf{B} \cdot \hat{\mathbf{n}} \, dA$ is called the *magnetic flux* φ_M. Note carefully the distinction between $\oint \mathbf{B} \cdot d\mathbf{r}$ (a *line integral* evaluated around a complete closed path line) and $\oint_S \mathbf{B} \cdot \hat{\mathbf{n}} \, dA$ (a *surface integral* taken over a complete volume-enclosing surface).

29-4 FORCES BETWEEN CURRENT-CARRYING WIRES

Consider two infinitely long, straight, parallel wires separated by a distance a, carrying currents I_1 and I_2 in the same direction (Fig. 29–8). What force per unit length acts between the wires? Let $\mathbf{F}_{12}$ be the force that acts on a short length ℓ of wire 1 due to the field produced by the current in wire 2. This force can be calculated by noting that the magnetic field at the position of wire 1 produced by the entire (infinite) length of wire 2 is (see Eq. 29–4)

$$B_2 = \frac{\mu_0 I_2}{2\pi a}$$

The direction of B_2 is shown in Fig. 29–8. Now, according to Eq. 28–4, the force on a current element $I d\ell$ due to a field $\mathbf{B}$ is $d\mathbf{F} = I d\ell \times \mathbf{B}$. Thus, for a length ℓ of wire 1, we have $F_{12} = I_1 \ell B_2$, or

$$F_{12}/\ell = I_1 B_2 = \frac{\mu_0 I_1 I_2}{2\pi a} \qquad \textbf{(29–10)}$$

Also, it is important to note that $\mathbf{F}_{12}$ is directed *toward* wire 2. That is, when the currents in the two wires are in the same direction, the mutual force is *attractive*; when the currents are in opposite directions, the mutual force is *repulsive*. From the symmetry of Eq. 29–10, the force $\mathbf{F}_{21}$ must be equal to $-\mathbf{F}_{12}$. This result is also required from Newton's third law.

Fig. 29–8. $\mathbf{F}_{12}$ is the force on a length ℓ of wire 1 due to the field $\mathbf{B}_2$ produced by the current in wire 2.

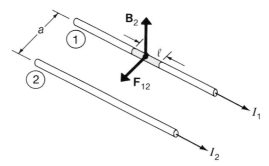

Definition of the Ampere. The definition of the ampere in use at present is as follows:

An ampere is that steady current which, when flowing in each of two long straight parallel wires separated by a distance of 1 m, causes each wire to exert on the other wire a force of 2×10^{-7} N for each meter of length of the wire.

With this definition and the expression for F_{12}/ℓ in Eq. 29–10, we see that $\mu_0/4\pi = K_M = 10^{-7}$ N/A^2, exactly, as was indicated in Section 29–1.

In Section 23–3 we gave a tentative definition of the *coulomb* in terms of the electron charge. Now that the ampere has been defined, we can give a proper definition of the coulomb:

A coulomb is the amount of charge that passes a particular point in a wire during a time of 1 s when a steady current of 1 A flows in the wire.

That is, 1 C = (1 A) × (1 s). This definition of the coulomb allows the precise evaluation of the constant K that appears in Coulomb's law. The result, as we stated in Section 23–3, is $K = 8.98755 \times 10^9$ N·m^2/C^2.

We now obtain an interesting result by calculating the ratio of the electric force constant K to the magnetic force constant K_M, namely,

$$\frac{K}{K_M} = \frac{1}{\epsilon_0 \mu_0} = \frac{8.98755 \times 10^9 \text{ N·m}^2/\text{C}^2}{10^{-7} \text{ N}/(\text{C}/\text{s})^2} = (2.99792 \times 10^8 \text{ m/s})^2 = c^2 \quad \textbf{(29–11)}$$

The quantity on the right-hand side of this expression is c^2, the square of the speed of light! This striking equality can hardly be an accident; there must be a connection between electricity and magnetism on the one hand and light on the other. The importance of the fact that the experimental value of K/K_M is close to the measured value of c^2 was first recognized by Gustav Kirchhoff. The great Scottish mathematical physicist James Clerk Maxwell (1831–1879) incorporated this idea into his theory of electromagnetism (1864) and asserted the equivalence of electromagnetic radiation and light waves. We discuss this aspect of Maxwell's theory in Chapter 33.

QUESTIONS

29–1 Define and explain briefly the meaning of the terms (a) Biot-Savart law, (b) permeability of free space, (c) solenoid, and (d) Ampère's circuital law.

29–2 An infinitely long conductor with a square cross section carries a uniform current density. Could use of Ampère's circuital law alone determine the magnetic field close to or inside the conductor? Explain why the situation might be different for the very distant field.

29–3 Assume an infinite array of parallel wires, each carrying a current I in the same direction (see diagram). If there are N wires per unit width, determine the magnetic field above and below the plane of the wires.

29–4 In Example 29–5 it was concluded that the external axial field had to vanish. Explain why B_2 (the external field) may not extend with a constant nonzero value to very large radii, i.e., to $r_2 \rightarrow \infty$.

29–5 Consider a wire coil in the form of a very long helix with a loose winding such that the spacing between adjacent turns is approximately three times the wire diameter.

Make a sketch of the magnetic field in the vicinity of a few of the turns. Discuss the behavior of the field as the spacing between turns is progressively reduced to the point where adjacent turns almost touch, thus approximating an idealized solenoid. What happens to the internal and external fields?

29–6 For an electrostatic field we found that $\oint \mathbf{E} \cdot d\mathbf{r} = 0$; hence, it was possible to define a scalar potential $\Phi(x, y, z)$ from which the electric field could be derived as the negative gradient of Φ. Discuss the possibility of deriving the magnetic field $\mathbf{B}$ from some corresponding scalar potential Φ_M.

29–7 Define and explain briefly the meaning of the terms (a) Bohr magneton, (b) magnetic flux, and (c) the practical ampere and coulomb.

29–8 When a bar magnet is broken in half, neither half becomes a magnetic monopole. Explain.

29–9 A small circular loop of wire is placed inside a larger circular loop of wire such that the two loops are coplanar but not concentric. If current circulates in both loops in the same direction, describe the motion of the smaller loop relative to the larger loop. Is there a position of stable equilibrium?

PROBLEMS

Section 29–1

29–1 Determine the magnetic induction at a point P that is a distance x from the corner of an infinitely long wire that is bent at a right angle, as shown in the diagram on page 575. The wire carries a steady current I.

29–2 A thin, flat, infinitely long ribbon of width w carries a uniform current I. Determine the magnetic induction at a point P that is in the plane of the ribbon and is a distance x from one edge, as shown in the diagram on page 575. Test your result in the limit $w \rightarrow 0$.

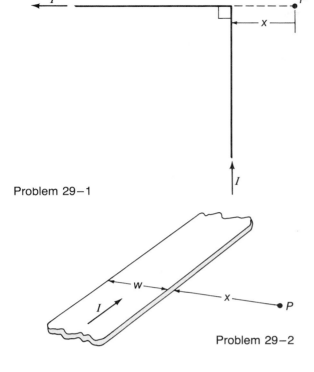

Problem 29−1

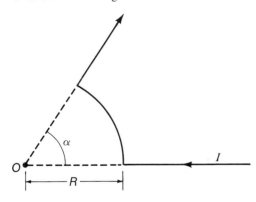

Problem 29−2

29−3 Two long straight wires are connected by a circular section that has a radius R, as shown in the diagram. All three segments lie in the same plane and carry a steady current I. Determine the magnetic induction at the center O of the circular segment.

29−4 Assume that the Earth's magnetic field is that of a dipole with moment $m_E = 8.1 \times 10^{22}$ A·m². Calculate the magnetic induction at a point on the Earth's surface at the magnetic equator.

Section 29−2

29−5 A short solenoid, with a length of 10 cm and a radius of 5 cm, consists of 200 turns of fine wire carrying a cur-

rent of 15 A. What is the magnetic induction at the center of the solenoid? For the same number of turns per unit length, what value of B would result for $\ell \to \infty$?

29−6 Use the results of Example 29−2 to determine as a function of length the magnetic induction B in the center of a solenoid with finite length and fixed radius. Make a graph of $B/\mu_0 n I$ as a function of $\eta = \ell/R$ for $0.5 \leqslant \eta \leqslant 20$.

29−7 Two infinite parallel conducting sheets carry equal surface currents of density $\mathbf{j}$ (A/m) in collinear but opposite directions, as shown in the diagram. (a) Show that the magnetic field between the sheets is uniform and that the field vanishes everywhere else. (b) Show that the magnetic induction between the sheets is $B = \mu_0 j$.

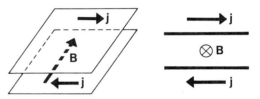

29−8 The long coaxial cable shown in the diagram carries a
••   uniform current I in the center conductor, and this current is returned in the outer cylindric shell. Determine the magnetic induction at all points inside (and outside) the cable. Sketch a graph of your results.

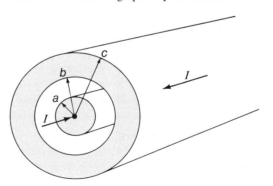

29−9 Consider a toroidal coil of the type discussed in Example
•   29−6 that consists of 500 turns with a mean radius $R = 12$ cm and a square cross section, 4 cm × 4 cm. Determine the maximum and minimum values of the magnetic induction within the coil when the current is 25 A.

Section 29−3

29−10 Calculate the magnetic flux, $\varphi_M = \int_S \mathbf{B} \cdot \hat{\mathbf{n}} \, dA$, through the flat rectangular surface S shown in the diagram on page 576, whose plane contains an infinitely long wire that carries a current I. (Notice that S is an *open* surface.)

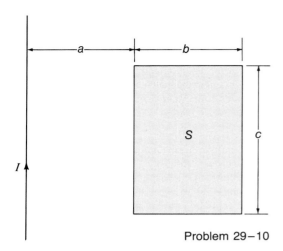

Problem 29–10

29–11 A surface in the shape of a right prism has its edge OO'
• along an infinitely long wire carrying a current I (see diagram). Calculate the magnetic flux φ_M through the shaded face $abcd$. [*Hint:* Use the geometry (top view) shown, writing the vector differential area as $\ell r \sec \theta \, d\theta \hat{n}$ and letting θ range from 0 to $\cos^{-1}(r_a/r_b)$.]

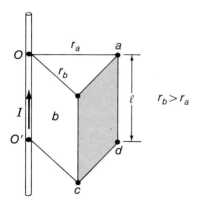

top view

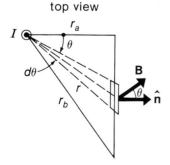

29–12 Determine by explicit calculation the magnetic flux,
•• $\varphi_M = \oint_S \mathbf{B} \cdot \hat{n} \, dA$, through the *closed* surface S shown in the diagram. Use the results obtained in Example 29–3. Assume $R \gg a$ and $\epsilon \ll a$. [*Hint:* Note carefully the behavior of the magnetic induction $\mathbf{B}$ when $R \gg a$ and when $\epsilon \ll a$.]

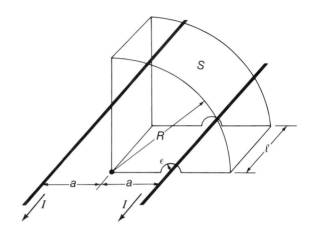

Section 29–4

29–13 Calculate the force on the rectangular loop shown in the diagram, which carries a current $I_2 = 10$ A, exerted by the field due to a long coplanar wire that carries a current $I_1 = 30$ A. (Take $a = 20$ cm, $b = 40$ cm, and $c = 50$ cm.)

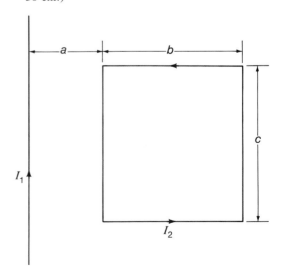

29–14 The rectangular wire frame shown in the diagram requires an additional force of $\Delta F = 10^{-3}$ N to maintain the frame in equilibrium when a current I flows. If a calibrated current of 20 A flows in the infinitely long parallel wire, what is the current I?

Problem 29-10

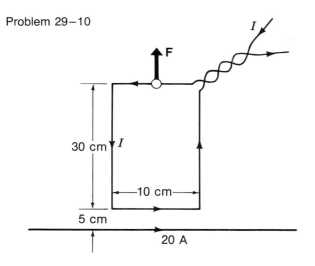

29-15 A 10-cm length of wire with a mass of 20 g is attached
• frictionlessly to the vertical segments of a wire in which
flows a current I. The surrounding uniform horizontal
field has $B = 10^4$ G and the direction shown in the diagram. What must be the current I to maintain the 10-cm
wire in an equilibrium position?

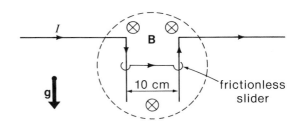

Additional Problems

29-16 Determine the magnetic induction at the center O of the
current loop shown in the diagram.

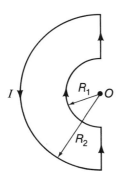

29-17 In an historic experiment conducted in 1878, the Ameri-
• can physicist Henry A. Rowland demonstrated that a
magnetic field is produced by a rotating charged disc.
Consider a uniformly charged nonconducting disc with
total charge Q and radius R that rotates with an angular
speed ω. Derive an expression for the magnetic induction at the center of the disc. Determine the value of B
for $Q = 0.1 \ \mu$C, $R = 10$ cm, and $\omega = 120\pi$ s^{-1}.
Compare this field with that of the Earth. (• Does this
comparison suggest that Rowland's experiment was easy
to perform?)

29-18 Refer to Example 29-2. Compute values of $B/\mu_0 nI$ for
$\xi = 0, \pm\frac{1}{4}, \pm\frac{1}{2}, \pm1, \pm2, \pm4$ for the case $\ell/R = 1$.
Make a graph of the results and compare with the graph
in Example 29-2, which is for the case $\ell/R = 4$.

29-19 An infinitely long cylindric conductor contains a cylin-
•• dric cavity whose center line is a distance $\mathbf{a}$ from the
center line of the conductor, as shown in the diagram.
The conductor carries a uniform current density $\mathbf{J} = J\hat{\mathbf{k}}$,
where $\hat{\mathbf{k}}$ is along the center line of the conductor. Show
that the magnetic field in the cavity is uniform and is
described by $\mathbf{B} = \frac{1}{2}\mu_0 J\hat{\mathbf{k}} \times \mathbf{a}$. [*Hint:* Calculate the field
at an arbitrary point P in the cavity by superimposing the
interior field of a *solid* cylindric conductor with a radius
R_1 and a current density $\mathbf{J}$ and the interior field of a *solid*
cylindric conductor with a radius $R_2 < R_1$ and a current
density $-\mathbf{J}$. Note also that the appropriate azimuthal unit
vectors can be expressed as $\hat{\mathbf{u}}_{\phi 1} = \hat{\mathbf{k}} \times \mathbf{r}_1$ and $\hat{\mathbf{u}}_{\phi 2} = \hat{\mathbf{k}} \times \mathbf{r}_2$.]

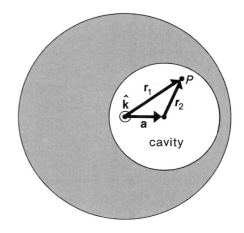

29–20 An infinitely long straight wire that carries a current $I_1 = 50$ A is partially surrounded by the loop shown in the diagram. The loop has a length $\ell = 35$ cm and a radius $R = 15$ cm, and it carries a current $I_2 = 20$ A. The axis of the loop coincides with the infinite wire. Calculate the force that acts on the loop.

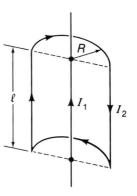

CHAPTER 30

ELECTROMAGNETIC INDUCTION

In the preceding chapters the discussions were restricted to steady-state conditions; that is, we considered only electric fields produced by static charges and magnetic fields produced by steady currents. We now remove this restriction and allow the possibility of time-varying electric and magnetic fields. Consideration of this general situation will lead us to *Maxwell's equations,* which together with the Lorentz force equation yield a complete and elegant formulation of all of the laws of electromagnetism.

30−1 THE DISCOVERY OF ELECTROMAGNETIC INDUCTION

An electric field imposed on a conductor can cause a current to flow, and this current can, in turn, produce a magnetic field. Does there exist a reciprocal effect whereby a magnetic field can produce an electric field? During the 1820's Michael Faraday* searched unsuccessfully for such an effect in steady-state situations. Finally, in 1831, he discovered that an electric field actually could be produced by a *time-varying* magnetic field. This is the phenomenon that is called *electromagnetic induction.*†

Figure 30−1 illustrates one of Faraday's early experiments. The apparatus consists of a bar magnet and a wire loop L that is part of a circuit containing a galvanometer G to indicate any current flow. If the bar magnet (whose field is permanent and constant) is stationary with respect to the loop L, the galvanometer indicates no current in the circuit. However, if the magnet is moved along the axis of the loop, as indicated in Fig. 30−1a, a current does flow in the circuit while the magnet is moving. Figure 30−1b shows that the current flows in one direction as the magnet approaches the loop and in the opposite direction after the magnet has passed through the loop, still moving along the axis.

Faraday also discovered that a current can be induced in a circuit by a changing current in a nearby circuit, without any motion of either circuit. When the circuit ① shown in Fig. 30−2 is completed by closing the switch P, a surge of current is observed in

Fig. 30−1. (a) When the magnet is in motion, a current is induced in the loop. (b) The current flows in one direction as the magnet approaches the loop and flows in the opposite direction after the magnet has passed through the loop.

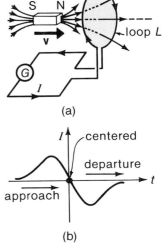

(a)

(b)

---

*Michael Faraday (1791–1867), English chemist and physicist, may have been the most gifted experimental scientist of all time. His many accomplishments included the discovery of benzene, the formulation of the laws of electrolysis, the discovery of electromagnetic induction, and the construction of the first electric generator.

†Electromagnetic induction seems actually to have been discovered first by the American physicist Joseph Henry in 1830; however, his results were not published until after Faraday's announcement in 1831, so Faraday is accorded the credit for the discovery.

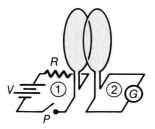

Fig. 30-2. When the switch P is closed and a current builds up in circuit ① , a current is induced to flow in circuit ② .

circuit ② .* When a steady current is eventually reached in the primary circuit, the current in the secondary circuit becomes zero. If the switch P is now opened, a new current surge is observed in the secondary circuit (but in the opposite direction). Again, an EMF is induced in the loop of the secondary circuit only while the magnetic field through the loop is *changing*. This effect is called *mutual induction*. At about the same time as Faraday's work, Henry discovered that the changing current in a loop induces an EMF in the circuit itself; this effect is called *self-induction*.

30-2 FARADAY'S LAW OF INDUCTION

Perhaps Faraday's greatest achievement in electrical science was the reduction of all the evidence from a series of experiments concerning electromagnetic induction to a simple quantitative law. *Faraday's law of induction* relates the instantaneous induced EMF, $\mathcal{E}$, in a closed circuit to the time rate of change of the magnetic flux that is enclosed by or links the circuit. This nonlocalized EMF produces the observed current in the circuit. The magnitude of the EMF is equal to the line integral around the circuit of the induced electric field intensity **E** at every point in the circuit. That is,

$$\mathcal{E} = \oint \mathbf{E} \cdot d\boldsymbol{\ell} \tag{30-1}$$

where $d\boldsymbol{\ell}$ is the differential element of path length at a point along the circuit where the induced electric field is $\mathbf{E} = \mathbf{E}(\mathbf{r})$. (Later we examine the field that is induced even when there is no circuit or conductor present.)

According to Eq. 25-14, the integral that appears in Eq. 30-1 vanishes for the electrostatic case. However, this result is no longer true when a changing magnetic field is present. Nor can we associate equipotential surfaces with such an induced electric field.

Fig. 30-3. An EMF $\mathcal{E}$ is induced in the loop when the field **B** that links the loop changes with time.

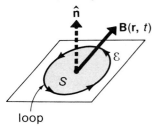

Consider the simple case of a stationary closed planar loop, as shown in Fig. 30-3. We select the direction indicated for the evaluation of the line integral around the loop. This direction, using the right-hand rule, defines the positive sense for the unit normal vector $\hat{\mathbf{n}}$ that is perpendicular to the plane surface S whose boundary is the loop. Faraday's law of induction states that

$$\mathcal{E} = \oint \mathbf{E} \cdot d\boldsymbol{\ell} = -\frac{d}{dt} \int_S \mathbf{B} \cdot \hat{\mathbf{n}} \, dA \tag{30-2}$$

The magnetic flux φ_M that links (or pierces) the open surface S is (see Section 29-3)

$$\varphi_M = \int_S \mathbf{B} \cdot \hat{\mathbf{n}} \, dA \tag{30-3}$$

Thus, the law of induction can be expressed in compact form as

$$\mathcal{E} = -\frac{d\varphi_M}{dt} \tag{30-4}$$

Magnetic flux is measured in *webers* (Wb) and EMF is measured in *volts* (V).

---

*It is customary to refer to the initiating circuit (here, circuit ①) as the *primary* circuit and the responding or coupled circuit (here, circuit ②) as the *secondary* circuit.

The negative sign that appears in Eqs. 30–2 and 30–4 expresses *Lenz's law,* discovered in 1834 by Heinrich Lenz (1804–1865) and independently by Faraday. This law states that

The direction of an induced EMF is such that it opposes the change that produced it.

For example, if the field **B** in Fig. 30–3 decreases with time, $d\varphi_M/dt$ is negative; then $\mathcal{E}$ is in the (counterclockwise) direction shown. This EMF causes a current to flow in the loop in the same (counterclockwise) direction. Comparing this situation with Fig. 29–4, we see that the magnetic field generated by the induced current (considered separately from the externally imposed field **B**) is directed upward at all points inside the loop; by the superposition principle, the induced field *adds* to the external field and tends to *oppose* the decrease. If, on the other hand, **B** is increasing with time, then the induced EMF, the induced current, and the induced field are all reversed in direction; the induced field is *subtracted* from the external field, and the increase in **B** is again opposed by the induction effect. Lenz's law is sometimes referred to as "the contrariness of Nature."

According to Eq. 29–9, we have $\oint_S \mathbf{B} \cdot \hat{\mathbf{n}}\, dA = 0$ for *any* closed surface. Consider the closed surface shown in Fig. 30–4, which consists of two parts, S_1 and S_2, whose common boundary is a closed-circuit loop. The unit normal vectors, $\hat{\mathbf{n}}_1'$ and $\hat{\mathbf{n}}_2'$, are directed outward from the surface and are therefore in the positive sense, as shown. We can write

$$\oint_{S_1+S_2} \mathbf{B} \cdot \hat{\mathbf{n}}\, dA = \int_{S_1} \mathbf{B} \cdot \hat{\mathbf{n}}_1'\, dA + \int_{S_2} \mathbf{B} \cdot \hat{\mathbf{n}}_2'\, dA = 0$$

Now, consider the calculation of the flux φ_M that pierces the surface S_1 for the case in which the line integral around the boundary loop is taken in the sense shown in Fig. 30–4. Then, $\hat{\mathbf{n}}_1$ is positive, and because $\hat{\mathbf{n}}_1' = -\hat{\mathbf{n}}_1$, we have

$$\varphi_M = \int_{S_1} \mathbf{B} \cdot \hat{\mathbf{n}}_1\, dA = -\int_{S_1} \mathbf{B} \cdot \hat{\mathbf{n}}_1'\, dA$$

Comparing this with the preceding equation, we find

$$\varphi_M = \int_{S_1} \mathbf{B} \cdot \hat{\mathbf{n}}_1\, dA = \int_{S_2} \mathbf{B} \cdot \hat{\mathbf{n}}_2\, dA$$

This means that the flux φ_M that links a particular closed circuit can be calculated by evaluating the integral of $\mathbf{B} \cdot \hat{\mathbf{n}}\, dA$ over *any open surface* whose boundary is the circuit loop. Therefore, in solving a particular problem, you can choose the simplest convenient surface.

It is often the case (especially in practical applications) that a circuit loop consists of a number N of wire turns. If the coils are tightly wound (Fig. 30–5a), they are equivalent to separate loops, each bounding a surface (Fig. 30–5b). Then the magnetic flux φ_M that links the entire coil is just N times the flux φ_M' that links a single turn. For the coil in Fig. 30–5 we have $\varphi_M = 2BA$, and, in general, the flux is

$$\varphi_M = N\varphi_M' \qquad \text{flux linkage} \qquad \text{(30–5)}$$

The magnetic flux that links a circuit loop can vary with time for several reasons. The variation may result from a changing magnetic field, from relative motion between the circuit and other circuits or magnets that produce the linking flux, or from changes in the

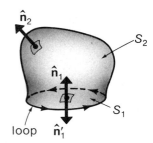

Fig. 30–4. Geometry for evaluating the flux that links a circuit loop.

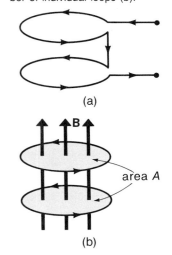

Fig. 30–5. For calculating magnetic flux linkage, a tightly wound coil of *N* turns (a) may be considered to consist of the same number of individual loops (b).

shape of the circuit itself. Moreover, any change in the loop current itelf produces a time variation in the field that links the loop (self-induction). The magnetic field **B** and the magnetic flux φ_M that appear in the different expressions of Faraday's law must include *all* sources of magnetic fields, and the time derivative that appears in Eq. 30–4 must be the *total* time derivative. These points are amplified in later discussions.

In this section we consider only the EMF induced by *external* magnetic fields generated by circuits *at rest* with respect to a rigid circuit loop selected for study. In the next section we discuss cases in which relative motion is involved. Also, for the present, self-induction effects are omitted. Because the superposition principle is valid, we may consider separately the self-induced EMF and the externally induced EMF and obtain the total induced EMF by an algebraic sum of the two when this is required. (See Chapter 31.)

There is one additional restriction. We consider only cases in which the time variation of the field is *not too fast*. That is, the magnetic field (and the current) must not change appreciably during a time interval equal to the size of the circuit divided by the speed of light. Then, we can safely neglect effects associated with the finite propagation time of the field. (Changes in the field propagate with the speed of light).

Example 30–1

Calculate the EMF in a rigid planar loop with area A induced by a uniform time-varying external magnetic field directed at an angle θ with respect to the loop normal $\hat{n}$, as shown in the diagram. The time dependence of B is given by $B_0 \sin \omega t$, where ω and B_0 are constants. Ignore self-induction effects. Assume $B_0 = 200$ G, $A = 100$ cm^2, $\theta = 30°$, and $\omega = 200\pi$ s^{-1}.

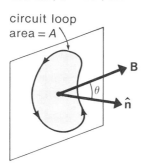

circuit loop
area $= A$

B

θ

$\hat{n}$

Solution:

The flux linkage is

$$\varphi_M = \mathbf{B} \cdot \hat{n} A = B_0 A \cos \theta \sin \omega t$$

Using Eq. 30–4, we have

$$\mathcal{E} = -\frac{d\varphi_M}{dt} = -B_0 A \,\omega \cos \theta \cos \omega t$$

$$= -(200 \times 10^{-4} \text{ T})(10^{-2} \text{ m}^2)(200\pi \text{ s}^{-1}) \cdot$$

$$\cos 30° \cos (200\pi \text{ s}^{-1})t$$

$$= -(0.109 \text{ V}) \cos (200\pi \text{ s}^{-1})t$$

The negative sign means that the sense of $\mathcal{E}$ is *opposite* to the direction of the arrows indicated for the path in the diagram, when the cosine term > 0. (• What is the situation when the cosine term < 0?)

Example 30–2

A circular loop with a radius R consists of N tight turns of wire and is linked by an external magnetic field directed perpendicular to the plane of the loop. The magnitude of the field in the plane of the loop is $B = B_0(1 - r/2R) \cdot \cos \omega t$, where R is the radius of the loop and where r is measured from the center of the loop, as shown in the diagram. Determine the EMF induced by B alone.

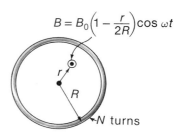

$$B = B_0\left(1 - \frac{r}{2R}\right)\cos \omega t$$

r

R

N turns

Solution:

The flux linkage is

$$\varphi_M = N\varphi'_M = NB_0 \cos \omega t \int_0^R \left(1 - \frac{r}{2R}\right) 2\pi r \, dr$$

$$= \frac{2\pi}{3} R^2 NB_0 \cos \omega t$$

Example 30–3

A long straight wire that carries a current, $I = I_0 \sin(\omega t + \delta)$, lies in the plane of a rectangular loop of N turns of wire, as shown in the diagram. The quantities I_0, ω, and δ are all constants. Determine the EMF induced in the loop by the magnetic field due to the current in the straight wire. Assume $I_0 = 50$ A, $\omega = 200\pi$ s^{-1}, $N = 100$, $a = b = 5$ cm, and $\ell = 20$ cm.

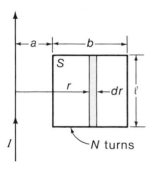

Solution:

The loop is the boundary of the plane surface S. The magnetic field produced by the current in the straight wire is perpendicular to S at all points on the surface. The magnitude of the field is (see Example 29–4)

$$B = \frac{\mu_0 I}{2\pi r}$$

Thus, the flux linkage is

$$\varphi_M = N\varphi'_M = \frac{\mu_0 NI\ell}{2\pi} \int_a^{a+b} \frac{dr}{r}$$

Then, using Eq. 30–4, we have

$$\mathcal{E} = -\frac{d\varphi_M}{dt} = \frac{2\pi}{3} \omega R^2 NB_0 \sin \omega t$$

$$\varphi_M = \frac{\mu_0 NI_0 \ell}{2\pi} \ln\left(\frac{a+b}{a}\right) \sin(\omega t + \delta)$$

Finally, the induced EMF is

$$\mathcal{E} = -\frac{d\varphi_M}{dt} = -\frac{\mu_0 NI_0 \ell \omega}{2\pi} \ln\left(\frac{a+b}{a}\right) \cos(\omega t + \delta)$$

$$= -(2 \times 10^{-7} \text{ T·m/A})(100)(50 \text{ A})(0.20 \text{ m})(200\pi \text{ s}^{-1}) \cdot$$

$$\ln\left(\frac{5+5}{5}\right) \cos(\omega t + \delta)$$

$$= -(87.1 \text{ mV}) \cos(\omega t + \delta)$$

Some discussion is necessary when ω is too large. The term $\sin(\omega t + \delta)$ in the expression for the current in the straight wire does not change appreciably when ωt changes by 0.1 rad or less. Thus, the current does not change appreciably during a time interval $\tau < 0.1/(200\pi$ s$^{-1}) = 1.6 \times 10^{-4}$ s. We define a critical length, $c\tau = (3 \times 10^8$ m/s$)(1.6 \times 10^{-4}$ s$) = 4.8 \times 10^4$ m, equal to the distance to which field changes could be propagated during an interval of 1.6×10^{-4} s. This length is so much larger than any dimension of the loop or its distance from the wire that, although we consider the straight wire to be infinitely long, we can also safely ignore the field-propagation effects in the vicinity of the loop. Moreover, the phase angle δ can be considered to be constant along the wire in the vicinity of the loop.

If the frequency ω were much larger, say, $200\pi \times 10^5$ s^{-1}, the corresponding critical length would be only 48 cm. In this situation, propagation effects would be important and the above expression for $\mathcal{E}$ would require modification. As a "rule of thumb" we can consider field-propagation effects for circuits of laboratory size to be negligible for frequencies, $\nu = \omega/2\pi$, that are less than about 10^6 Hz.

30–3 MOTIONAL EMF

To this point we have considered only cases involving stationary circuits. The time dependence present was due entirely to changing values of the magnetic fields. We wish next to consider cases involving moving circuits.

Consider the case of a conductor or circuit loop that is seen by an observer O to be in motion in the presence of an externally produced magnetic field. We restrict all speeds to be small compared with the speed of light in order to avoid relativistic effects. We begin by examining the behavior of a single charge in the presence of a magnetic field. Suppose that a charge q is seen by an observer O to be moving with a velocity $\mathbf{u}$ in some inertial frame. If there is present a magnetic field $\mathbf{B}$ (but no electric field), the Lorentz force on the charge is (see Eq. 28–2)

$$\mathbf{F} = q\mathbf{u} \times \mathbf{B}$$

Suppose that a second observer O' is in an inertial frame that has a velocity $\mathbf{u}$ with respect to the frame of observer O. The charge q is seen by O' to be at rest and therefore subject to no magnetic force due to any static magnetic field that might be present. Nevertheless, O' must agree with O that the very same value of force $\mathbf{F}$ acts on the charge. (Recall that force is a free vector and is therefore the same in all inertial frames; see Section 5–4.) Because O' sees q at rest, he concludes that an electric field $\mathbf{E}'$ must be present in order to produce the force $\mathbf{F}$; namely,

$$\mathbf{F} = q\mathbf{E}' = q\mathbf{u} \times \mathbf{B}$$

Thus, we have

$$\mathbf{E}' = \mathbf{u} \times \mathbf{B} \qquad (30\text{–}6)$$

Equation 30–6 is generally valid for all velocities u such that $u \ll c$. We conclude that if an inertial observer O determines that a magnetic field is present, another inertial observer O' in motion with respect to O will detect the additional presence of an electric field.

Let us continue examining the situation as seen by a moving observer. Suppose that a differential element $d\boldsymbol{\ell}$ of a conductor is moving with a velocity $\mathbf{u}$ in the presence of a magnetic field $\mathbf{B}$ in an inertial reference frame, as indicated in Fig. 30–6. To an observer moving with the conductor, the *only* forces that act on the charges in the conductor arise from the electric field $\mathbf{E}'$. This field produces an EMF that appears between the ends of the conductor element $d\boldsymbol{\ell}$ given by

$$d\mathcal{E} = \mathbf{E}' \cdot d\boldsymbol{\ell} = (\mathbf{u} \times \mathbf{B}) \cdot d\boldsymbol{\ell}$$

If the element $d\boldsymbol{\ell}$ is part of a closed conducting loop, the total induced EMF is

$$\mathcal{E} = \oint (\mathbf{u} \times \mathbf{B}) \cdot d\boldsymbol{\ell} \qquad (30\text{–}7)$$

Fig. 30–6. The differential element $d\boldsymbol{\ell}$ of a conductor is in motion with a velocity $\mathbf{u}$ in the presence of a magnetic field $\mathbf{B}$.

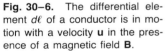
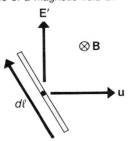

The quantity $\mathcal{E}$ is referred to as a *motional EMF*. This is an important result.

Let us apply Eq. 30–7 to the situation illustrated in Fig. 30–7. A rigid conducting loop lies partially within a uniform magnetic field $\mathbf{B}$ and is pulled with a constant velocity $\mathbf{u}$ by an external force $\mathbf{F}_e$. The integral in Eq. 30–7 is to be evaluated around the closed path *abcdefa*. The segments *bc*, *cd*, and *de* lie outside the field and therefore give no contribution. There is likewise no contribution from the segments *ab* and *ef*, because $\mathbf{E}' = \mathbf{u} \times \mathbf{B}$ is perpendicular to $d\boldsymbol{\ell}$ along these segments. The remaining segment *fa* gives

$$\mathcal{E} = B\ell u \qquad (30\text{–}8)$$

with the sense shown by the arrows along the loop in Fig. 30–7.

Equation 30–8 represents the conclusion reached by the observer O' who moves with the loop. How is the situation analyzed by the stationary observer O who sees the

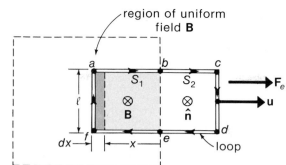

Fig. 30–7. A conducting loop moves through a magnetic field **B**. An EMF is induced in the loop with the sense indicated by the arrows. The unit normal n̂ for the surface $S = S_1 + S_2$ and the magnetic field **B** are both directed into the plane of the page.

loop in motion with a velocity **u**? Observer O notes that the magnetic flux linking the loop is decreasing. Using Lenz's law, observer O determines that the EMF induced in the loop has the same sense as that found by O'. (• Can you give the Lenz's-law argument that yields the direction of $\mathcal{E}$?)

We apply Faraday's law (Eq. 30–2) to this situation for the case in which **B** is uniform and time-independent, and in which the surface is planar; we then have

$$\mathcal{E} = -\frac{d}{dt}\int_S \mathbf{B}\cdot\hat{\mathbf{n}}\, dA = -\mathbf{B}\cdot\hat{\mathbf{n}}\,\frac{d}{dt}\int_S dA \qquad (30\text{–}9)$$

The surface S consists of two parts, $S = S_1 + S_2$, and $\mathbf{B} = 0$ in S_2 (see Fig. 30–7). Thus, the integral is simply the area of S_1, namely, ℓx. Then,

$$\mathcal{E} = -\mathbf{B}\cdot\hat{\mathbf{n}}\,\frac{d}{dt}\,(\ell x) = -B\ell\,\frac{dx}{dt}$$

Now, $dx/dt = -u$, so we have

$$\mathcal{E} = B\ell u$$

which agrees precisely with the result obtained by O'. (The fact that the observers agree on the value of the EMF is the reason we wrote $\mathcal{E}$ instead of $\mathcal{E}'$ in Eq. 30–7, even though this result was obtained by observer O'.)

Next, we inquire about the energy balance as seen by the two observers. As long as some portion of the loop is within the magnetic field and moves with the constant velocity **u**, the induced EMF remains constant. This situation is equivalent to a DC circuit in which there is a battery with an EMF $\mathcal{E}$. If the resistance of the entire loop is R, the constant current I that flows in the loop is

$$I = \frac{\mathcal{E}}{R} = \frac{B\ell u}{R}$$

The induced EMF therefore provides a power input to the loop given by

$$P = I\mathcal{E} = \frac{(B\ell u)^2}{R}$$

Both observers agree that this is the power input to the loop. According to observer O', this power is the result of the presence of the field **B** and the velocity **u** and is derived from the agency responsible for establishing the field **E**, namely, the external force $\mathbf{F}_e$.

Observer O immediately attributes the power input to $\mathbf{F}_e$, thereby agreeing with O' as to the source of the power. Because there is no acceleration of the loop, $\mathbf{F}_e$ must be balanced by the net magnetic force on the loop; that is, according to O,

$$F_e = -F_L = -\oint I \, d\boldsymbol{\ell} \times \mathbf{B} = IB\ell \frac{\mathbf{u}}{u}$$

(• Can you show this?) The power delivered by the external force is

$$P = \mathbf{F}_e \cdot \mathbf{u} = IB\ell u = I\mathcal{E} = \frac{(B\ell u)^2}{R}$$

which is the same as the result obtained in Eq. 30–10. The power P supplied to the loop is dissipated in Joule heating.

▶ The current in the loop remains constant as long as some portion of the loop is within the field and the loop moves with constant velocity. Then the magnetic field that is produced by the loop current is constant and does not give rise to any self-induced EMF. This is why we were able to ignore the field of the loop and why we obtained the simple energy relationship.

If, on the other hand, the loop is set into motion while entirely within the field, we find $\mathcal{E} = 0$ and $I = 0$. (• Can you see why?) Then, as the leading edge emerges from the field, a current begins to flow and builds up to its final value as the self-induced field of the loop is created by virtue of work performed by the force $\mathbf{F}_e$. In the preceding discussion we considered the steady-state condition that ensues, with I constant and with the full self-induced field established. Finally, when the trailing edge of the loop leaves the field, the current decreases to zero as the self-induced field decreases to zero. The energy that was stored in the self-induced field is eventually dissipated in Joule heating. As always, we find that total energy is conserved. ◀

Example 30–4

A rectangular loop of wire rotates about an axis that is perpendicular to a uniform time-independent magnetic field **B**, as shown in the diagram. Determine the induced EMF $\mathcal{E}$ at the slip-ring terminals.

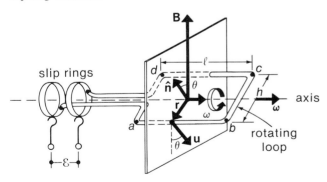

Solution:

First, obtain $\mathcal{E}$ as a motional EMF by using Eq. 30–7. Inspection of the diagram reveals that for the segments bc and da we have $\mathbf{u} \times \mathbf{B}$ perpendicular to $d\boldsymbol{\ell}$, so that these segments give zero contribution to the line integral. (We make no significant error by neglecting the small gap in the segment da. • Can you see why?) For the segments ab and cd we have $\mathbf{u} = \boldsymbol{\omega} \times \mathbf{r}$ and

$\mathbf{u} \times \mathbf{B} = \frac{1}{2}h\omega B \sin \theta$. Then, performing the line integral around the loop, we find

$$\mathcal{E} = h\ell\omega B \sin \theta$$

The loop area is $A = h\ell$; then, with $\theta = \omega t$, we have

$$\mathcal{E} = A\omega B \sin \omega t$$

The arrangement shown in the diagram is a simple alternating-current (AC) generator in which the terminal voltage varies sinusoidally with time. Notice that in the case illustrated the circuit is not completed and no current flows. Consequently, there is no self-induced field and no corresponding contribution to the EMF. In this case we call $\mathcal{E}$ the *open-circuit* EMF. (If a load is connected to the terminals, time-varying currents will flow and the *back EMF* due to the self-induced field must be considered.)

We can also use Faraday's law to obtain the open-circuit EMF. The flux that links the loop is

$$\varphi_M = AB \cos \theta = AB \cos \omega t$$

Then, using Eq. 30–4, we have

$$\mathcal{E} = -\frac{d\varphi_M}{dt} = A\omega B \sin \omega t$$

Thus, the two methods give identical results. (• Can you verify that the two methods also give the same sense of the EMF?)

Example 30—5

A conducting disc with a radius R rotates about its own axis, which is parallel to a uniform magnetic field **B**, as shown in diagram (a). Two brushes, one contacting the axle and one contacting the rim of the disc, are connected to terminals across which the induced EMF $\mathcal{E}$ is measured. This arrangement is called a *Faraday disc dynamo* or a *homopolar generator*. Determine the open-circuit EMF $\mathcal{E}$.

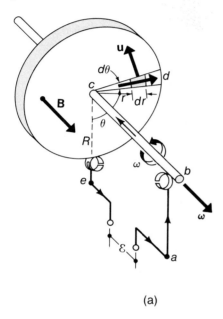

(a)

Solution:

Again, first obtain $\mathcal{E}$ as a motional EMF. Consider a thin wedge-shaped section of the disc that subtends an angle $d\theta$ at the center of the disc. Select a length dr at a radius r as the differential conductor element within this wedge. (This element is shaded in diagram (a).) In the laboratory reference frame this element has a velocity **u**, with $u = r\omega$. We have*

$$\mathbf{E}' = \mathbf{u} \times \mathbf{B} = r\omega B \hat{\mathbf{u}}_r$$

where $\hat{\mathbf{u}}_r$ is a unit vector in the radial direction; then,

$$d\mathcal{E} = \mathbf{E}' \cdot d\mathbf{r} = r\omega B \, dr$$

Integrating $d\mathcal{E}$ along the path $abcde$, only the segment cd gives a nonzero contribution. Thus,

$$\mathcal{E} = \int_0^R r\omega B \, dr = \tfrac{1}{2}\omega B R^2$$

Notice that *any* radial path along the disc would have given the same result. Even though an EMF is generated in all such wedge-shaped segments, they are all effectively in a parallel connection and so give the same total EMF.

The expression for $\mathcal{E}$ can also be obtained by using Faraday's law. Replace the integration path $abcde$ by the two paths, $abce$ and $cdec$, in which the segments ce and ec are traversed in opposite directions so their contributions to the integral cancel. No flux links the area defined by the path $abce$†; however, the flux through the area defined by the path $cdec$ is

$$\varphi_M = -\pi R^2 B \cdot \frac{\theta}{2\pi} = -\tfrac{1}{2}R^2 B\omega t$$

The negative sign arises because $\hat{\mathbf{n}}$, the unit normal vector for the surface, is directed opposite to **B**. Applying Eq. 30–4, there results

$$\mathcal{E} = -\frac{d\varphi_M}{dt} = \tfrac{1}{2}\omega B R^2$$

in agreement with the previous result.

Notice that we have calculated here the equilibrium open-circuit terminal voltage. As the disc begins to rotate, a charge separation occurs, with excess positive charge collecting around the rim of the disc and negative charge collecting at the center. The resulting electrostatic field just cancels the induced field, and at equilibrium there is no current flow. If a load is connected across the terminals of the dynamo, the current density **J** in the disc will have a complicated distribution, as indicated in diagram (b).

(b)

---

*We continue to use **u** for the velocity (because we reserve **v** for use later as another velocity); do not confuse **u** with the symbol $\hat{\mathbf{u}}$ for a unit vector.

†Notice that when calculating the flux linkage, we can tolerate the open segment in the defining path because there is zero flux through the area.

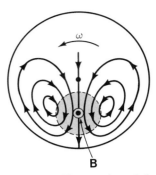

Fig. 30–8. The rotation of the conducting disc through the region of magnetic field **B** produces an eddy current in the disc.

Eddy Currents. Because the disc in Example 30–5 lies entirely within the magnetic field, an equilibrium condition is established in which no current flows. However, if the field were confined to a region of the disc, as indicated by the dashed circle in Fig. 30–8, a distribution of current—called an *eddy current*—would be set up in the disc. Figure 30–8 shows that **J** in the region of the field is almost unidirectional (nearly radial) and is directed so that it produces a torque that opposes the rotational motion of the disc. (Recall that the force per unit volume is equal to **J** × **B**—Eq. 28–5.)

It is clear from energy considerations that in all cases involving rotation the eddy currents produce torques that *oppose* the rotation. Through Joule heating effects, eddy currents dissipate mechanical energy, the ultimate source of which is the external agency that produces the rotation. If there is no coupling to such an external agency, the energy is provided from the kinetic energy of the rotating object. Thus, the result of eddy currents is always to produce a braking action.

30–4   THE MAXWELL DISPLACEMENT CURRENT

Consider the process of charging a capacitor by connecting it to a source of EMF (for example, a battery), as shown in Fig. 30–9a. When the switch P is closed, positive charge flows onto the top plate of the capacitor. The rate at which the charge flows is maximum at the instant the switch is closed and gradually decreases to zero as the capacitor charge approaches its full value, $Q_m = CV.$*

The upper plate of the capacitor may be completely surrounded by a closed surface S, as indicated in Fig. 30–9b. Conservation of charge requires that the rate at which charge accumulates on the upper plate, inside the surface S, be equal to the current that flows into the surface S. That is,

$$I(t) = \frac{dQ(t)}{dt} \tag{30–11}$$

Fig. 30–9. (a) When the switch P is closed, what current flows between the plates of the capacitor C? (b) The current $I(t)$ flows into the surface S and the displacement current flows out of the surface.

We now apply Gauss's law to the surface S, using the form expressed in Eq. 26–30 in terms of the electric displacement $\mathbf{D} = \epsilon_0\mathbf{E} + \mathbf{P}$. Thus,

$$\oint_S \mathbf{D}(t)\cdot\hat{\mathbf{n}}\, dA = Q(t)$$

This equation is valid even if the region between the plates is filled with a dielectric. Taking the time derivation of this equation and combining with Eq. 30–11, we obtain

$$I(t) = \frac{d}{dt}\oint_S \mathbf{D}(t)\cdot\hat{\mathbf{n}}\, dA \tag{30–12a}$$

In the event that the region between the plates is a vacuum, we have $\mathbf{D} = \epsilon_0\mathbf{E}$, so we can write

$$I(t) = \frac{d}{dt}\oint_S \epsilon_0\mathbf{E}(t)\cdot\hat{\mathbf{n}}\, dA \qquad \text{(vacuum)} \tag{30–12b}$$

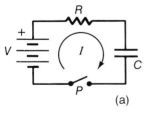

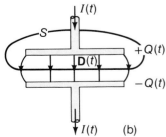

It is useful to consider the right-hand side of Eq. 30–12a or 30–12b to be a type of "current" that flows across the gap between the capacitor plates even though no actual transport of charge is involved. This permits I to be viewed as a current that flows

---

*The exact mathematical nature of $Q(t)$ is not important for the present discussion. Such matters are discussed in Chapter 32.

continuously around the entire circuit of Fig. 30–9a. The new current term is known as *Maxwell's displacement current.*

The immediate consequence of this view is to modify Ampère's circuital law. Consider the magnetic field produced by a current $I(t)$ that flows in a circuit that contains a capacitor, as indicated in Fig. 30–10. Ampère's circuital law connects the line integral of **B** around a closed path with the total current that flows through *any open surface* whose boundary is the integration path. Thus, for the integration path shown in Fig. 30–10, either of the surfaces, S_1 or S_2, can be used in Ampère's law. Whereas ordinary current flows through surface S_1, no such current flows through surface S_2. This clearly indicates the need for some modification to the original formulation of Ampère's circuital law (Eq. 29–8). Taking guidance from Eqs. 30–12, which introduce the idea of displacement current, Maxwell modified Ampère's circuital law to read

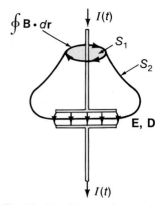

Fig. 30–10. For the integration path shown, either surface S_1 or surface S_2 may be used in Ampère's circuital law.

$$\oint \mathbf{B} \cdot d\mathbf{r} = \mu_0 \int_S \mathbf{J} \cdot \hat{\mathbf{n}}\, dA + \mu_0 \frac{d}{dt} \int_S \epsilon_0 \mathbf{E} \cdot \hat{\mathbf{n}}\, dA$$

or

$$\oint \mathbf{B} \cdot d\mathbf{r} = \mu_0 \int_S \left(\mathbf{J} + \epsilon_0 \frac{d\mathbf{E}}{dt} \right) \cdot \hat{\mathbf{n}}\, dA \qquad \text{in vacuum} \qquad \textbf{(30–13)}$$

where the open surface S can be any surface whose boundary is the integration path around which $\mathbf{B} \cdot d\mathbf{r}$ is calculated, such as S_1 or S_2 in Fig. 30–10.

If the surface S_1 in Fig. 30–10 is chosen to include some of the fringing field of the capacitor, Eq. 30–13 remains correct if the contribution of the displacement current is included in the surface integral. (• Can you see how this follows by considering the closed surface S in Fig. 30–9b to consist of the two open surfaces, $S_1 + S_2$, in Fig. 30–10? Remember the sign convention for the unit normal vectors.)

In such an elegant theory as Maxwell's theory of electromagnetism, in which electricity and magnetism are treated on an equal basis, one somehow expects a symmetry in the behavior of these field quantities. Faraday's law (Eq. 30–2) expresses the fact that a time-varying magnetic field is accompanied by an electric field. By symmetry, then, a time-varying electric field should be accompanied by a magnetic field; this is exactly the effect of the displacement-current term in Maxwell's version of Ampère's law (Eq. 30–13).

The ultimate justification for Maxwell's modification of Ampère's law by the addition of the displacement-current term is its experimental verification. It is difficult to observe directly the magnetic effects of the displacement current. However, by using very high frequencies to enhance the effect, the magnetic field in a capacitor has actually been detected. Even more important is the verification afforded by the observation of many predicted indirect effects, such as electromagnetic radiation (see Chapters 33 and 34).

30–5   MAXWELL'S FIELD EQUATIONS

In this section we summarize the equations that describe electric and magnetic fields in vacuum (or, for all practical purposes, in air). (The general relationships for fields in material media are given in Section 31–6.) The charges that are the sources of these fields are restricted to move uniformly in some inertial reference frame. We imagine that this frame is populated with as many observers as are required to make simultaneous measurements of the electric and magnetic field vectors at all points of interest by using stationary and moving test charges. That is, $E(\mathbf{r}, t)$ and $B(\mathbf{r}, t)$ are considered to be experimentally determined quantities in the inertial reference frame.

Gauss's Laws. Gauss's law for electric fields (Eq. 24–12) is a completely general law:

$$\oint_S E \cdot \hat{\mathbf{n}} \, dA = \frac{Q}{\epsilon_0} = \frac{1}{\epsilon_0} \int_V \rho \, dv \qquad (30\text{–}14)$$

where the integration is carried out at a particular instant and where Q is the instantaneous value of the net charge contained in the volume V that is bounded by the closed surface S. (As usual, the positive direction of $\hat{\mathbf{n}} \, dA$ is *outward*.)

Because magnetic monopoles do not exist, the magnetic quantities that correspond to Q and ρ are zero. Thus, Gauss's law for magnetic fields is (Eq. 29–9)

$$\oint_S \mathbf{B} \cdot \hat{\mathbf{n}} \, dA = 0 \qquad (30\text{–}15)$$

where S is any closed surface.

Ampère's Law. The integral form of Ampère's law, as modified by Maxwell to include displacement currents, is (Eq. 30–13)

$$\oint \mathbf{B} \cdot d\mathbf{r} = \mu_0 \int_S \left(\mathbf{J} + \epsilon_0 \frac{d\mathbf{E}}{dt} \right) \cdot \hat{\mathbf{n}} \, dA \qquad (30\text{–}16)$$

where the line integral can be taken around any closed path, and where S is any fixed surface that is bounded by the integration path. The positive direction of $\hat{\mathbf{n}} \, dA$ is determined by the usual right-hand rule.

Faraday's Law. Faraday's law can be used to determine the induced EMF in a circuit whether that EMF is due to a time variation of the magnetic field or the motion of a circuit through the field. We now assert that the law is completely general and does not require that the path of the line integral correspond to a physical wire or circuit. That is, we write Eq. 30–2 as

$$\oint \mathbf{E} \cdot d\mathbf{r} = -\frac{d}{dt} \int_S \mathbf{B} \cdot \hat{\mathbf{n}} \, dA \qquad (30\text{–}17)$$

where the line integral can be taken around any closed path, and where S is any stationary surface that is bounded by the integration path. Notice that we write $d\mathbf{r}$ (indicating a *general* displacement) instead of $d\ell$ (indicating a segment of a physical circuit).

The equations 30–14 through 30–17 constitute the celebrated set of Maxwell's equations that summarize the general laws of electromagnetism in vacuum (or air). Modifications are necessary when dielectric and magnetic materials are present; these are given in the next chapter.

QUESTIONS

30–1 Define and explain briefly the meaning of the terms (a) Faraday's law of induction and induced EMF, (b) Lenz's law, (c) flux linkage, (d) motional EMF, (e) homopolar generator, and (f) eddy currents.

30–2 Explain the statements made in the caption of Fig. 30–1. Would different results be observed if the magnet were stationary and the loop moved over the magnet?

30–3 Under what (extreme) circumstance could the total EMF induced in a coil be considered to be localized at one point in the loop?

30–4 The loop of wire discussed in Example 30–4 has its angular speed doubled. What changes does this produce in the open-circuit EMF?

30–5 A small conducting plate is sometimes suspended from one arm of a chemical balance so that it hangs freely between the poles of a small stationary permanent magnet, with the plane of the plate perpendicular to the magnetic field. As the balance swings, the plate moves between the poles. Explain the possible purpose of the plate and magnet.

30–6 Why should a larger and larger force be required to pull a conducting sheet out from between the poles of a magnet as you attempt to do it faster and faster?

30–7 Define and explain briefly the meaning of the terms (a) Maxwell displacement current and (b) Maxwell's field equations.

30–8 A circular disc parallel-plate capacitor is being charged at a uniform rate by a constant current. Describe the magnetic field in the presence of the capacitor fringing electric field.

30–9 A positively charged particle moves past a conducting coil traveling in a plane that contains the coil (see diagram). Describe the sense and magnitude of the current in the coil as the particle passes. Will the magnetic field produced by the induced current in the coil result in a force on the moving charge?

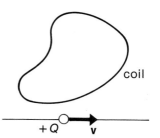

PROBLEMS

Section 30–2*

30–1 A rectangular single-turn loop with dimensions $a = 10$ cm and $b = 15$ cm lies in the x-y plane, as indicated in the diagram at right. Within the loop there is a time-dependent magnetic field

$$\mathbf{B} = B_0 (x \sin \omega t \hat{\mathbf{j}} + y \cos \omega t \hat{\mathbf{k}})$$

with $B_0 = 120$ G/m and $\omega = 10^3$ s^{-1}. Determine the EMF induced in the loop as a result of the changing magnetic field.

30–2 A toroid with a rectangular cross section ($a = 2$ cm × $b = 3$ cm, $R = 6$ cm) consists of 500 turns of wire that carries a current $I = I_0 \sin \omega t$, with $I_0 = 50$ A and a frequency $\nu = \omega/2\pi = 60$ Hz. A loop that consists of 20 turns of wire links the toroid, as shown in the diagram on page 592. Determine the EMF induced in the loop by the changing current I.

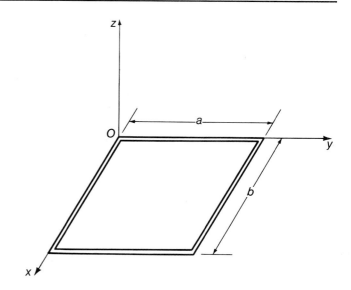

---

*In these problems consider the effect *only* of the given current; that is, neglect the effect of any self-induced EMF. If desired, we could include these effects by using superposition.

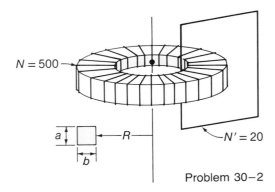

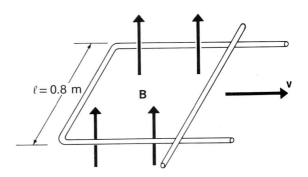

Problem 30–2

30–3 A long solenoid has n turns per meter and carries a current $I = I_0 (1 - e^{-\alpha t})$, with $I_0 = 30$ A and $\alpha = 1.6$ s^{-1}. Inside the solenoid and coaxial with it is a loop that has a radius $R = 6$ cm and that consists of a total of N turns of fine wire. What EMF is induced in the loop by the changing current? (Take $n = 400$ turns/m and $N = 250$ turns.)

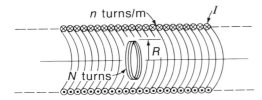

30–4 A small irregular loop enclosing area A consists of N turns of fine wire. This loop lies at the center and in the plane of a large single-turn loop with radius R that carries a current $I = I_0 (1 + \sin \omega t)$. Obtain the expression for the EMF induced in the small loop by the changing current in the large loop.

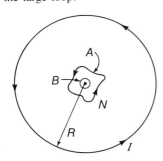

Section 30–3

30–5 A conducting U-shaped frame has a conducting cross-piece that moves with a velocity $v = 15$ m/s, as shown in the diagram. The frame is in a uniform time-independent magnetic field **B** whose direction is perpendicular to the plane of the frame and whose magnitude is 800 G. Determine the induced EMF in the loop (a) by using the idea of motional EMF and (b) by using Faraday's law.

30–6 An automobile travels due north (magnetic north) at a speed of 120 km/h in a locality where the Earth's magnetic field has a magnitude of 56,000 γ and a dip angle of 70°. Calculate the magnitude of the EMF induced between the ends of a 1.3-m axle. What is the polarity of the western end of the axle?

30–7 A conducting rod with a length of 50 cm is rotated about one end with a constant angular velocity **ω** ($\omega = 360$ s^{-1}) in the presence of a uniform magnetic field **B** ($B = 200$ G) that is parallel to **ω**. What potential difference is induced between the ends of the rod? What is the polarity of the rotating end?

30–8 A disc with radius R and resistivity ρ is located in a
•   time-varying magnetic field $B(t) = B_0 t/\tau$, that is directed parallel to the disc axis. Calculate the density **J**(r) of the eddy current induced in the disc. Assume $R = 20$ cm, $\rho = 6 \times 10^{-8}$ Ω·m, $B_0 = 2.2$ T, and $\tau = 18$ s; assume also that Ohm's law, **J** = **E**$/\rho$, is valid. (• Why is it not necessary to be concerned with self-induced fields in this situation?)

Section 30–4

30–9 A parallel-plate capacitor consists of two circular plates 20 cm in diameter spaced 5 mm apart. The capacitor is charged with a constant current of $I_0 = 2A$. Find the magnitude of the displacement current density

$$J_D = \epsilon_0 \frac{dE}{dt}$$

between the capacitor plates. Assume a uniform electric field and neglect any fringing effects.

30–10 The electric field strength in a region of space is given by **E** $= E_0 (1 - r/a)\hat{\mathbf{k}} \sin \omega t$ for $r < a$ and **E** $= 0$ for $r > a$. Determine the magnetic induction **B** everywhere within the region.

30–11 The electric field strength in a region of space is given by
•• $E = E_0 (t/\tau)\hat{\mathbf{k}}$. Determine the total displacement current through a ring with radius R whose axis coincides with the polar axis, as shown in the diagram. (a) First, use the bounding surface S, which is a flat disc whose perimeter

is the ring. (b) Next, make an explicit calculation using the bounding surface S', which is a portion of a sphere with radius $2R$. [*Hint:* Make use of the geometry shown in the diagram.]

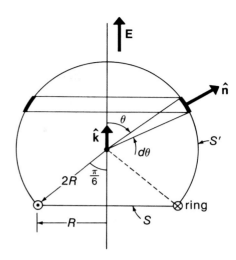

Additional Problems*

30–12 A time-dependent magnetic dipole, $\mathbf{m} = m_0\hat{\mathbf{k}} \sin \omega t$, is
 •• located on the axis of a single-turn circular loop of radius a. The distance from the dipole to the center of the loop is b, as indicated in the diagram. Determine the EMF induced in the loop by the dipole variations. [*Hint:* To determine the flux linkage with the loop, use the bounding surface S, which is a segment of a sphere with center at the position of the dipole and with radius $R = (a^2 + b^2)^{1/2}$. The relevant spherical component of the dipole field is (compare Problem 24–17 for the electric case)

$$B_r = \frac{\mu_0 m \cos \theta}{2\pi r^3}$$

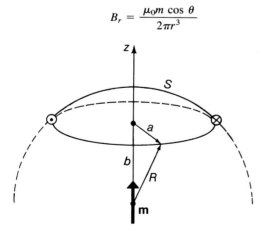

30–13 A square coil (20 cm × 20 cm) that consists of 100 turns of wire rotates about a vertical axis at 1500 rpm, as indicated in the diagram. The horizontal component of the Earth's magnetic field at the location of the loop is

20,000 γ. Calculate the maximum EMF induced in the coil by the Earth's field.

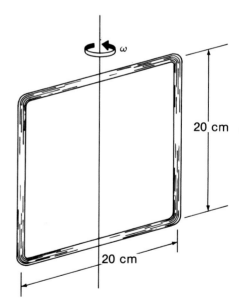

30–14 A long solenoid, whose axis coincides with the x-axis, consists of 200 turns per meter of wire that carries a steady current of 15 A. A coil is formed by wrapping 30 turns of thin wire around a frame that has a radius of 8 cm. The coil is placed inside the solenoid and mounted on an axis that is a diameter of the coil and coincides with the y-axis. The coil is then rotated with an angular speed of 4π radians per second. (The plane of the coil is

---

*In these problems consider the effects *only* of the given currents; that is, neglect the effects of any self-induced EMFs.

in the y-z plane at $t = 0$.) Determine the EMF developed in the coil.

30–15 In Problem 30–14, replace the field of the solenoid by a time-varying field, $B = (4 \times 10^{-3} \text{ T}) \sin \omega t$, where $\omega = 4\pi \text{ s}^{-1}$. Determine the EMF developed in the coil.

30–16 A helicopter is hovering with its 4.6-m rotor blades revolving at a rate of 180 rpm in a horizontal plane. At the location of the helicopter the magnitude of the Earth's magnetic field is 50,000 γ and the dip angle is 65°. What EMF is developed between the rotor hub and the tip of a blade?

30–17 A 1-m conducting rod moves with its length parallel to the x-axis and with a velocity $\mathbf{v} = v_0\hat{\mathbf{j}}$, where $v_0 = 100$ m/s. There is present a magnetic field described by $\mathbf{B} = B_0 (3\hat{\mathbf{i}} + 4\hat{\mathbf{j}} + 5\hat{\mathbf{k}})$, with $B_0 = 45$ G. What is the EMF induced between the ends of the rod?

30–18 In 1832 Faraday proposed that electric current could be
•• generated by the flowing water in the Thames using the apparatus illustrated in the diagram. Two conducting plates with lengths a and widths b are placed facing one another on opposite sides of the river, at a distance w apart. The flow velocity in the river is $\mathbf{v}$ and the vertical component of the Earth's magnetic field is $\mathbf{B}$. (a) Show that the current in the load resistor R is

$$I = \frac{abvB}{\rho + abR/w}$$

where ρ is the resistivity of the river water. (b) Calculate the short-circuit current (that is, take $R = 0$) if $a = 100$ m, $b = 5$ m, $v = 3$ m/s, $B = 50,000 \gamma$, and $\rho = 100 \text{ } \Omega\cdot\text{m}$.

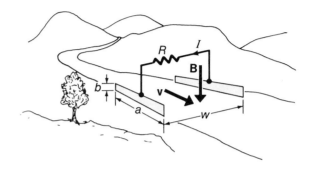

CHAPTER 31

INDUCTANCE AND MAGNETIC MATERIALS

In our previous discussions of magnetic fields, a vacuum (or air) environment was assumed. We now enlarge the scope to include the effects that magnetic materials have on the fields produced by electric currents and by changing electric fields. We also give a brief discussion of these macroscopic magnetic effects based on the behavior of the atoms that constitute the magnetic materials.

In Chapter 26 the idea of capacitance was used to facilitate the study of dielectric materials. We follow a parallel development here and first introduce the concept of *inductance*. Then, simple *inductors* are used to discuss measurements of the properties of magnetic materials.

31–1 MUTUAL INDUCTANCE

In Chapter 30 we considered the EMF induced in a circuit (the secondary) by a changing current in a nearby circuit (the primary). It was pointed out that Faraday's law of induction involves the time rate of change of the *entire* magnetic flux that links a circuit, including the flux due to any self-induced fields. However, to avoid discussing the self-induced EMFs at that point, we either omitted reference to the currents in the secondary circuits (remarking that they could be included by using superposition) or treated cases in which the secondary current was constant in time and therefore did not contribute any induced EMF of its own. In this section we proceed in the same way; in Section 31–2 we include self-induced EMFs.

Let us begin by considering a case in which no magnetic materials are present. Two coils are shown in Fig. 31–1. A (time-dependent) current I_1 flows in the primary coil ①, which consists of N_1 turns of wire and produces a magnetic induction $\mathbf{B}_1$. The flux linkage with the secondary coil ②, which consists of N_2 turns of wire, is expressed as* $N_2\varphi_{21}$, where $\varphi_{21} = \int_{S_2}\mathbf{B}_1\cdot\hat{\mathbf{n}}_2\,dA$. Using Faraday's law, the resulting induced EMF in the secondary coil is

$$\mathcal{E}_{21} = -\frac{d}{dt}(N_2\varphi_{21}) \qquad (31-1)$$

Fig. 31–1. The changing current in the primary coil ① induces an EMF in the secondary coil ②.

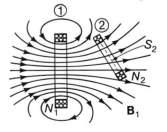

*In Eq. 30–5 we wrote this term as $N\varphi_M'$.

We restrict attention to coils and loops in *fixed* geometry. Moreover, as in Chapter 30, only slowly varying currents are allowed. Then, because $\mathbf{B}_1$ depends linearly on I_1, the EMF $\mathcal{E}_{21}$ that is induced in coil ② is directly proportional to dI_1/dt; thus,

$$\mathcal{E}_{21} = -M_{21}\frac{dI_1}{dt} \qquad \text{fixed geometry} \qquad \text{(31-2)}$$

where M_{21} is a strictly geometric factor that is called the *mutual inductance* of the pair of coils. Comparing Eqs. 31-1 and 31-2, we see that

$$M_{21} = \frac{N_2\varphi_{21}}{I_1} \qquad \text{(31-3)}$$

That is, we can interpret M_{21} as the flux linkage with coil ② per unit amount of current in coil ①.

Suppose that we reverse the situation shown in Fig. 31-1, holding the geometry fixed. Then, a time-varying current I_2 flowing in coil ② (now the primary) induces an EMF $\mathcal{E}_{12}$ in coil ① (now the secondary). Thus,

$$\mathcal{E}_{12} = -M_{12}\frac{dI_2}{dt}$$

By using techniques more advanced than those at our disposal, it can be shown that

$$M_{12} = M_{21} = M$$

where we can now drop the subscripts and write M for the mutual inductance. For simple geometries, the equality of M_{12} and M_{21} can be demonstrated by explicit calculation—see Problems 31-2 and 31-5.

Inductance is measured in units of volts per ampere per second, abbreviated to the *henry* (H), in honor of Joseph Henry. Thus,

$$1\ \text{H} = 1\ \text{V}/(\text{A}/\text{s}) = 1\ \text{V·s}/\text{A} = 1\ \text{Wb}/\text{A}$$

The henry is a rather large unit; in practical cases, the millihenry (mH) or the microhenry (μH) is a more appropriate unit.

Example 31-1

Calculate the mutual inductance M between a toroid with $N_1 = 200$ turns and a linking coil with $N_2 = 10$ turns. The toroid has an inner radius $R = 6$ cm and a rectangular cross section as shown in the diagram, with $a = 3$ cm and $b = 4$ cm.

Solution:

Using Ampère's circuital law (Eq. 29-7) for a circuit at a radius r, we obtain

$$B_1 = \frac{\mu_0 N_1 I_1}{2\pi r}$$

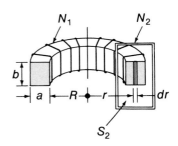

The flux φ_{21} that links the coil is

$$\varphi_{21} = \int_{S_2} \mathbf{B}_1 \cdot \hat{\mathbf{n}}_2\ dA$$

Because $\mathbf{B}_1$ is entirely azimuthal and exists only with the toroid, we have, with $dA_2 = b\,dr$,

$$\varphi_{21} = \frac{\mu_0 N_1 I_1 b}{2\pi} \int_R^{R+a} \frac{dr}{r} = \frac{\mu_0 N_1 b}{2\pi} I_1 \ln\left(\frac{R+a}{R}\right)$$

Therefore, we identify

$$M = M_{21} = \frac{N_2 \varphi_{21}}{I_1} = \frac{\mu_0 N_1 N_2 b}{2\pi} \ln\left(\frac{R+a}{R}\right)$$

Substituting the values of the various quantities, we find

$$M = (2 \times 10^{-7} \text{ H/m})(200)(10)(0.04 \text{ m}) \ln\left(\frac{0.09 \text{ m}}{0.06 \text{ m}}\right)$$
$$= 6.49 \times 10^{-6} \text{ H} = 6.49 \; \mu\text{H}$$

Notice that we have expressed μ_0 as $4\pi \times 10^{-7}$ H/m. (• Can you verify these units?) Notice also that the simplicity of the field of a toroid makes the calculation of M_{21} much easier than to do the reverse calculation of M_{12}.

31–2 SELF-INDUCTANCE

When a current I flows in an isolated coil, loop, or circuit with fixed geometry, a magnetic field is produced that links the circuit itself. If the current varies with time, so does the flux linkage. Then, according to Faraday's law, there results a self-induced EMF in the circuit. This EMF is sometimes referred to as a *back EMF* because its direction is such that it opposes any change in the driving EMF (Lenz's law).

For a coil with N turns and a linking flux, $\varphi_M = N\varphi'_M$, the self-induced EMF is (see Eq. 30–4)

$$\mathscr{E} = -\frac{d}{dt}(N\varphi'_M) \qquad \qquad \textbf{(31–4)}$$

If there are no magnetic materials present, the magnetic field and, hence, the flux linkage depend linearly on the current I in the coil, and the self-induced EMF is directly proportional to dI/dt; thus,

$$\mathscr{E} = -L\frac{dI}{dt} \qquad \qquad \textbf{(31–5)}$$

where L is a strictly geometric factor that is called the *self-inductance* (or, simply, the *inductance*) of the coil. Comparing Eqs. 31–4 and 31–5, we see that

$$L = \frac{N\varphi'_M}{I} \qquad \qquad \textbf{(31–6)}$$

An *inductor* is a compact circuit element, such as a tightly wound coil of wire, that produces electromagnetic induction effects. When electric circuits are drawn, an inductor is indicated by the symbol

Example 31–2

Determine the self-inductance of a long solenoid with a length ℓ and a radius R that consists of N turns of wire, as shown in the diagram. (Neglect end effects.)

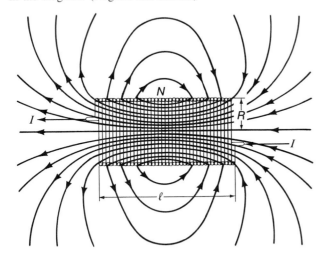

Solution:

The magnetic field inside a long solenoid is uniform and axial: $B = \mu_0(N/\ell)I$ (see Example 29–5). Therefore, the flux linking the N turns of the solenoid is

$$N\varphi_M' = N \cdot \pi R^2 \cdot \mu_0 \frac{N}{\ell} I$$

Referring to Eq. 31–6, we identify the inductance as

$$L = \frac{\mu_0 \pi R^2 N^2}{\ell}$$

▶ *End Effects.* To neglect end effects is to ignore the failure of some of the flux lines near the middle of the solenoid to link with the turns near the ends (refer to the diagram). Thus, the inductance of a solenoid with a finite length is less than the value calculated using the expression above. Recall the parallel situation with regard to the fringing field of a capacitor (Eq. 26–5a). In the same way, we write

$$L = \frac{\mu_0 \pi R^2 N^2}{\ell} f$$

where $f \leqslant 1$. For a solenoid with $\ell/R = 1.0$, we find $f = 0.53$, and when $\ell/R = 5$, we find $f = 0.85$; of course, in the limit, $\ell/R \to \infty$, and we have $f = 1$.

If the coil has $R = 4$ cm and $\ell = 20$ cm and consists of 250 turns, we find

$$L = \frac{(4\pi \times 10^{-7} \text{ H/m}) \pi (0.04 \text{ m})^2 (250)^2 (0.85)}{0.20 \text{ m}}$$

$$= 1.68 \times 10^{-3} \text{ H} = 1.68 \text{ mH} \qquad \blacktriangleleft$$

Example 31–3

Determine the self-inductance of the toroid shown in the diagram.

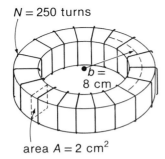

N = 250 turns

b = 8 cm

area A = 2 cm²

Solution:

According to the result obtained in Example 29–6, the magnetic induction inside an air-core toroid with mean radius b is

$$B = \mu_0 nI = \frac{\mu_0 NI}{2\pi b}$$

Then, the flux linking the N turns is

$$N\varphi_M' = N \cdot A \cdot \frac{\mu_0 NI}{2\pi b}$$

Referring to Eq. 31–6, we identify the inductance as

$$L = \frac{\mu_0 N^2 A}{2\pi b}$$

Substituting the given values, we obtain

$$L = \frac{(4\pi \times 10^{-7} \text{ H/m})(250)^2 (2 \times 10^{-4} \text{ m}^2)}{2\pi (0.08 \text{ m})}$$

$$= 3.125 \times 10^{-5} \text{ H} = 31.25 \text{ } \mu\text{H}$$

31–3   MAGNETIC FIELD ENERGY

Consider an inductor L that is connected to an external source of EMF. The resulting current $i(t)$ increases in some unspecified manner from zero to a final value I. While $i(t)$ is changing, there is a self-induced EMF (a *back* EMF), $\mathcal{E} = -L \, di/dt$, which causes energy to be absorbed by the inductor from the source of the external EMF at a rate

$$P = \mathcal{E}i = Li \, \frac{di}{dt}$$

The total amount of electrical work done by the source is

$$W = \int P \, dt = L \int_0^I i \, di$$

so that

$$W = \tfrac{1}{2}LI^2 \tag{31–7}$$

Notice that the work done does not depend on the way in which $i(t)$ varies to achieve its final value I. Notice also that when a steady-state condition is reached, we have $di/dt = 0$, and power is no longer expended by the source.

What happens to the energy that is absorbed by the inductor? It proves convenient to associate this energy with the magnetic field produced by the current flow in the inductor.* (Compare the discussion of the case of electric field energy in Section 26–3.) Therefore, the electric energy W delivered to the inductor can be equated to the energy U_M stored in the magnetic field; that is,

$$U_M = \tfrac{1}{2}LI^2 \tag{31–8}$$

Energy Density.   Inside a long solenoid the magnetic field is uniform and has the magnitude (see Example 31–2) $B = \mu_0 NI/\ell$, neglecting end effects. Thus, the energy in the field is

$$U_M = \tfrac{1}{2}LI^2 = \frac{1}{2}\left(\frac{\mu_0 \pi R^2 N^2}{\ell}\right)\left(\frac{B\ell}{\mu_0 N}\right)^2$$

$$= \frac{B^2}{2\mu_0}(\pi R^2 \ell) = \frac{B^2 V}{2\mu_0}$$

where $\pi R^2 \ell$ is the volume V of the solenoid. We define the magnetic field *energy density* to be $u_M = U_M/V$; thus,

$$u_M = \frac{B^2}{2\mu_0} = \tfrac{1}{2}\frac{\mathbf{B} \cdot \mathbf{B}}{\mu_0} \qquad \text{vacuum (or air) conditions} \tag{31–9}$$

Although this expression was obtained for the special case of a solenoid, it contains none of the parameters of the solenoid and is, in fact, quite generally true for magnetic fields in vacuum.

---

*We again restrict the current variations to be slow. Rapid variations lead to the radiation of energy away from the circuit.

Example 31–4

Determine the self-inductance of the coaxial inductor shown in the diagram. Typically, we have $c - b \ll b$, so ignore the contribution to the field in the region $b < r < c$.

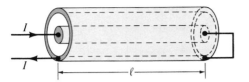

Solution:

Assume that the inductor is sufficiently long ($\ell \gg b$) or the gap is sufficiently small ($b - a \ll b$) that end effects may be neglected. Then, the magnetic induction is entirely azimuthal, and there is no field for $r > c$. Ampère's circuital law can be used to find the magnetic induction in the regions ① $r < a$ and ② $a < r < b$. Thus,

$$\mathbf{B}_1 = \frac{\mu_0 I r}{2\pi a^2} \hat{\mathbf{u}}_\phi, \qquad r < a$$

$$\mathbf{B}_2 = \frac{\mu_0 I}{2\pi r} \hat{\mathbf{u}}_\phi, \qquad a < r < b$$

Next, Eq. 31–9 is used to calculate the field energy in these regions; the results are

$$U_1 = \frac{1}{2\mu_0} \int_0^a \left(\frac{\mu_0 I r}{2\pi a^2} \right)^2 2\pi r\ell \; dr = \frac{\mu_0 I^2 \ell}{16\pi}$$

$$U_2 = \frac{1}{2\mu_0} \int_a^b \left(\frac{\mu_0 I}{2\pi r} \right)^2 2\pi r\ell \; dr = \frac{\mu_0 I^2 \ell}{4\pi} \ln \frac{b}{a}$$

The total field energy (excluding the region $b < r < c$) is $U_M = U_1 + U_2$. Then we use Eq. 31–8 to identify the self-inductance per unit length; thus,

$$\frac{L}{\ell} = \frac{\mu_0}{8\pi} \left(1 + 4 \ln \frac{b}{a} \right)$$

Typical values for a coaxial cable are $a = 0.1$ mm, $b = 3$ mm, and $c = 3.2$ mm. We find

$$\frac{L}{\ell} = \frac{\mu_0}{8\pi}(1 + 13.605) = \frac{\mu_0}{8\pi}(14.605)$$

In a calculation that includes the field contribution in the region $b < r < c$, the result is only slightly larger, namely,

$$\frac{L}{\ell} = \frac{\mu_0}{8\pi}(14.694)$$

Indeed, the major contribution to the inductance comes from the field energy in the region between the two conductors. Finally,

$$\frac{L}{\ell} = \frac{1}{8\pi}(4\pi \times 10^{-7} \text{ H/m})(14.694) = 0.735 \; \mu\text{H/m}$$

The Energy Stored by Coupled Inductors. In Section 26–4 concerning capacitor networks, we were able to give a simple treatment of parallel and series connections because fringing-field effects in most practical capacitors are negligibly small, and consequently there is very little coupling between capacitors. In the case of practical inductors, however, there may be a substantial interaction of the fields of a pair of nearby inductors. Any coupling that is present is associated with the mutual inductance of the pair.

Consider two inductors, L_1 and L_2, that have a mutual inductance M. If currents I_1 and I_2 flow in the two inductors, the energy stored in the combined field is

$$U_M = \tfrac{1}{2}L_1 I_1^2 + \tfrac{1}{2}L_2 I_2^2 \pm M I_1 I_2 \qquad \textbf{(31–10)}$$

where the $\pm$ sign arises because the mutual inductance term will add to or subtract from the self-inductance terms, depending on the relative direction of winding of the two coils.

Inductors in Series. The existence of mutual inductance between two nearby inductors complicates the addition law for combinations of inductors. For example, if two coupled inductors are connected in series so that $I_1 = I_2 = I$, Eq. 31–10 becomes

$$U_M = \tfrac{1}{2}(L_1 + L_2 \pm 2M)I^2 \qquad \textbf{(31–11)}$$

In order to have the same stored energy, a single equivalent inductor must have an inductance equal to

$$L = L_1 + L_2 \pm 2M \qquad \text{series connection} \qquad (31\text{–}12)$$

31–4   MAGNETIC MATERIALS

We now remove the restriction to vacuum conditions and inquire about magnetic fields within matter. A practical and conceptually convenient device for studying the magnetic properties of materials is the *Rowland ring*. The material to be investigated is formed into a toroid on which there is wound a tightly spaced coil of N turns, as indicated in Fig. 31–2. A secondary coil that consists of N' turns is looped around the toroid and is used to measure the EMF induced when the current in the primary coil is varied. The induction effects for a specimen material are compared with those for a toroid with an air (vacuum) core for otherwise identical conditions.

With a vacuum core, we find that the induction effects in the secondary coil are attributable to a toroidal magnetic field,

$$B_0 = \mu_0 \frac{NI}{2\pi R}$$

where $R \gg a$. With a core fabricated from the specimen material, the induction effects indicate a different magnetic field,

$$B = \kappa_M \mu_0 \frac{NI}{2\pi R}$$

The quantity

$$\kappa_M = B/B_0 \qquad (31\text{–}13)$$

is called the *relative permeability* (that is, the permeability of the material relative to vacuum). We also use a quantity μ, called the *permeability*, where

$$\mu = \mu_0 \kappa_M \qquad (31\text{–}14)$$

Finally, because κ_M is nearly unity for most materials, it is useful to introduce the magnetic susceptibility χ_M, where

$$\chi_M = \kappa_M - 1 \qquad (31\text{–}15)$$

Of course, we also have

$$\mu = (1 + \chi_M)\mu_0 \qquad (31\text{–}16)$$

Fig. 31–2. A *Rowland ring* consists of a toroidal core of the specimen material tightly wrapped with a coil of N turns. We consider a toroid with $R \gg a$.

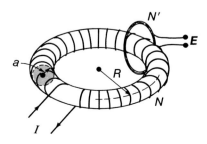

Notice the similarity of the new magnetic parameters to those introduced in the discussion of dielectric materials (Section 26-5). The dielectric constant κ is always greater than one. However, the magnetic counterpart κ_M can be greater or less than one, and in the case of ferromagnetic materials, κ_M is much greater than one. We distinguish three major categories of behavior for magnetic materials, on the simple empirical basis of the relative value of κ_M, and we discuss these in turn.

At the atomic level, there are two fundamental sources for the magnetic properties of matter, although in most cases they are interconnected in complicated ways. One source is the circulation of electrons about their parent nuclei. This orbital motion is equivalent to an atomic current and gives rise to an atomic magnetic moment. These moments are governed by quantum rules and can take on only certain discrete values. Atomic magnetic moments are *quantized* in units of the *Bohr magneton, $m_B = 9.27 \times 10^{-24}$ A·m^2.*

The second fundamental source is the magnetic moment associated with a property called the spin angular momentum of the individual electrons.* This magnetic moment is an intrinsic property of the electron, in the same way that charge and mass are intrinsic properties. The electron magnetic moment has a magnitude m_B.

Diamagnetism. Faraday discovered that a freely suspended needlelike piece of bismuth tends to align itself *perpendicular* to a uniform magnetic field instead of taking up a parallel alignment, as would a slender iron needle. Materials that exhibit this type of behavior are said to be *diamagnetic*. When an external magnetic field is imposed on a diamagnetic material, an atomic EMF is induced. The accompanying current circulates in such a way that the resulting magnetic field opposes the external field (Lenz's law). That is, the *induced magnetization* of the material is directed opposite to the imposed field. Thus, the magnetic field within a diamagnetic material is smaller than the imposed field would be in a vacuum. Consequently, diamagnetic materials have $\kappa_M < 1$. All matter is basically diamagnetic, but the effect is weak and is usually overwhelmed by either of the other effects when one is present.

Paramagnetism. An atomic system (atom or molecule) that has a permanent magnetic moment will tend to align this moment with an external magnetic field to the extent permitted by thermal agitation. This behavior is similar to that of polar molecules whose permanent electric dipole moments tend to align with an external electric field. In the event that the permanent magnetic moment of a system is either very weak or nonexistent, the diamagnetic behavior dominates. For example, the noble gases, with their completely symmetric closed electron shells, have zero permanent magnetic moments; these elements exhibit pure diamagnetism.

If the permanent magnetic moment of an atomic system is sufficiently large to overcome the inherent diamagnetism, the net alignment of the moments with the external field *increases* the magnetic induction within the sample. That is, the magnetization of the material is directed parallel to the imposed field. Such materials are said to be *paramagnetic* and have $\kappa_M > 1$. Liquid oxygen is an example of a strong paramagnetic substance; its large permanent magnetic moment is due to the combined effects of electron spin and orbital motion.

Because paramagnetism depends on the alignment of permanent magnetic moments with an applied field, the magnitude of the effect is limited by the extent to which thermal agitation permits complete alignment to take place. Thus, we expect that κ_M for paramagnetic materials will decrease with increasing temperature. (• Should diamagnetism also be temperature-dependent?)

---

*There is in addition a fundamental magnetic moment associated with the nuclear spin. But the nuclear moments are much smaller than those of the electrons, and can be neglected when describing the bulk magnetic properties of matter under ordinary conditions.

Ferromagnetism.   Certain materials exhibit a spontaneous ordering of their permanent magnetic moments within microscopic regions (called *domains*) even if no external magnetic field is applied. Because this phenomenon was first recognized for iron, it is referred to as *ferromagnetism*. This is an entirely quantum effect involving the electron spins and is restricted to a few elements. Iron, cobalt, and nickel make up one group of ferromagnetic elements; the other group consists of the rare-earth elements gadolinium and dysprosium. Ferromagnetism is also exhibited by various compounds and alloys of nonferromagnetic elements (for example, $CrBr_3$, Cu_2MnAl, and EuO).

▶ Because the ferromagnetic elements have an unusual ordering of electrons in their orbital and spin states, the angular momenta of these atoms have abnormally large electron spin components. These spin components produce a strong interaction between neighboring atoms that leads to maximum stability when the atomic magnetic moments are aligned. This spontaneous tendency to align produces microscopic regions (domains) in which the moments are all aligned. In these domains, which have volumes of about 10^{-12} to 10^{-8} m^3 and contain 10^{17} to 10^{21} atoms, the local magnetization has its maximum value. ◀

When the net magnetic moments of the individual microscopic domains are oriented at random with respect to one another, the magnetic induction of the sample as a whole is zero—the sample is unmagnetized. However, even a relatively weak external magnetic field can induce changes in the size and orientation of the domains so that a very large magnetic induction results. The value of the relative permeability κ_M can reach 10^5 or even higher. (See Fig. 31–8.) It is this property of ferromagnetic substances that makes these materials valuable as permanent magnets and as cores for electromagnets. As in the case of paramagnetism, the effects of thermal agitation limit the achievable magnetization of a ferromagnetic sample. The ferromagnetism is maximum at 0 K and decreases with increasing temperature until it becomes zero at a characteristic critical temperature called the *Curie temperature** of the material. For higher temperatures the sample is paramagnetic.

The transition from ferromagnetism to paramagnetism is an example of a phase change, not unlike the situation in which the long-range atomic order of a crystal is suddenly destroyed when melting occurs. Table 31–1 lists the Curie temperatures for the ferromagnetic elements. Notice that the two rare-earth elements are not ferromagnetic at room temperature.

Magnetization.   When a sample of bulk matter is placed in a magnetic field, there develops, due to the presence of the field, an effective magnetic moment **m** per atom (or molecule). For diamagnetic materials **m** is the induced moment, and for paramagnetic and

TABLE 31–1.   Curie Temperatures

ELEMENT	T_C (K)
Cobalt, Co	1395
Iron, Fe	1043
Nickel, Ni	631
Gadolinium, Gd	289
Dysprosium, Dy	85

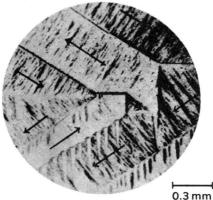

The magnetic domain structure of Permalloy (22 percent iron, 78 percent nickel) as revealed by a microscope. The arrows indicate the magnetic axes of the individual domains. (After *Physics of Magnetism* by S. Chikazumi. John Wiley & Sons, Inc., New York, 1964.)

0.3 mm

*Named for Pierre Curie (1859–1906), who discovered the effect in 1895. The relation between temperature and magnetic moment for ferromagnetic materials is called *Curie's law*.

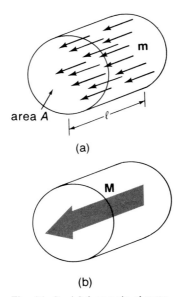

area A

(a)

M

(b)

Fig. 31–3. (a) A sample of material contains atoms (or molecules) that have effective magnetic moments **m** in the presence of a magnetic field. (b) The net effect of these moments is represented by the magnetization **M**.

Fig. 31–4. The Ampèrian currents associated with the individual magnetic moments **m** (Fig. 31–3a) cancel throughout the interior of the sample, leaving an effective surface magnetization current I_M.

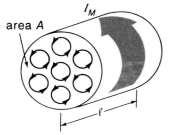

area A

I_M

ferromagnetic materials **m** includes an appropriate average over the distribution of directions of the permanent moments. Figure 31–3a shows a number of individual moments **m** in a sample of material with a volume $A\ell$. If there are $\mathcal{N}$ atoms (or molecules) per unit volume, the cumulative effect of the individual moments can be represented by the macroscopic magnetization vector **M** (Fig. 31–3b), where

$$\mathbf{M} = \mathcal{N}\mathbf{m} \qquad (31-17)$$

The quantity **M** is referred to as the *magnetization* or the *magnetic moment density.*

Magnetic moments have dimensions of *current* $\times$ *area,* so the dimensionality of **M** is *current per unit length.* Because the magnetic moment of a coil depends on the number of loops that the current makes, it is customary to express magnetization in units of ampere-turns per meter, which we abbreviate as A·turns/m.

According to Ampère's view, magnetism and magnetic effects are due to elementary current loops. Thus, the magnetic moments shown in Fig. 31–3a are considered to result from the average currents of the circulating electrons in the constituent atoms (Fig. 31–4). It is evident in Fig. 31–4 that adjacent Ampèrian current segments are oppositely directed and therefore cancel throughout the interior of the sample. On the surface, however, there is no cancellation, so there is a net *effective magnetization current* I_M that circulates around the sample (Fig. 31–4). Notice that I_M is distributed continuously along the length ℓ of the sample. The total magnetic moment of the sample is $I_M A$. The sample volume is $A\ell$, so the magnetic moment per unit volume is

$$M = \frac{I_M A}{A\ell} = \frac{I_M}{\ell} = j_M \qquad (31-18)$$

where j_M is the surface magnetization current per unit length (A/m). For the special geometry of Fig. 31–3, M is numerically equal to j_M; in the general case, it is the tangential component **M** at the surface that is equal to j_M.

Now let the sample we have been considering be a segment of a toroid with radii R and a ($R \gg a$) that is the core of a Rowland ring (Fig. 31–5). The toroid is wrapped with N turns of wire carrying a current I. If the ring core were vacuum, the current I would produce a uniform magnetic induction, $B_0 = \mu_0 NI/2\pi R$, within the core region; the magnetization would be zero. With the material core in place the induction is B, and application of Ampère's circuital law (Eq. 29–8) along the center line with radius R yields

$$2\pi RB = \mu_0(NI + 2\pi Rj_M)$$

or

$$B - \mu_0 j_M = \frac{\mu_0 NI}{2\pi R}$$

from which

$$B - \mu_0 M = B_0 \qquad (31-19)$$

Using $B = \kappa_M B_0$ and $\chi_M = \kappa_M - 1$, Eq. 31–19 can be rewritten in vector form as

$$\mu_0\mathbf{M} = \mathbf{B} - \mathbf{B}_0 \qquad (31-20a)$$

$$= (\kappa_M - 1)\mathbf{B}_0 \qquad (31-20b)$$

$$= \chi_M\mathbf{B}_0 \qquad (31-20c)$$

That is, the magnetization of the material core is proportional to the difference between the values of the induction with and without the core.

These expressions are valid for all homogeneous, isotropic magnetic media and are analogous to the expressions for uniformly polarized dielectrics (Eqs. 26–24). However, remember that in the magnetic case the susceptibility χ_M is negative for diamagnetic substances and is positive for paramagnetic and ferromagnetic substances. Values of χ_M are given in Table 31–2 for some typical diamagnetic and paramagnetic substances. The more complicated case of ferromagnetic materials is discussed in Enrichment Topic B.

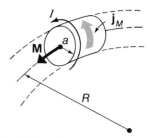

Fig. 31–5. A segment of a Rowland ring. The ring coil of N turns carries a current I. The surface magnetization current density $\mathbf{j}_M$ is associated with the magnetization $\mathbf{M}$.

TABLE 31–2. Magnetic Susceptibilities of Various Substances[a]

DIAMAGNETIC MATERIALS	χ_M
Bismuth	-1.7×10^{-4}
Mercury	-2.8×10^{-5}
Silver	-2.4×10^{-5}
Lead	-1.6×10^{-5}
Copper	-9.7×10^{-6}
Water	-9.1×10^{-6}
Hydrogen[b]	-2.2×10^{-9}
Nitrogen[b]	-6.7×10^{-9}

PARAMAGNETIC MATERIALS	χ_M
Oxygen[b]	$+1.9 \times 10^{-6}$
Magnesium	$+1.2 \times 10^{-5}$
Aluminum	$+2.0 \times 10^{-5}$
Tungsten	$+7.8 \times 10^{-5}$
Chromium	$+3.1 \times 10^{-4}$
Cerium	$+1.5 \times 10^{-3}$
Oxygen[c]	$+3.5 \times 10^{-3}$

[a] At 293 K.
[b] At 1 atm.
[c] Liquid, at 90 K.

Example 31–5

A solid toroid of bismuth has a core radius $a = 1$ cm and a toroid radius $R = 10$ cm. A winding of $N = 800$ turns around the toroid carries a steady current of 10 A. Determine the following quantities:

(a) the magnetic induction B_0;
(b) the magnetization M;
(c) the total magnetic induction B; and
(d) the total magnetization current I_M.
(e) The mass number of bismuth is $A = 209$ and its density is $\rho = 9.8$ g/cm^3. What is the effective atomic magnetic moment of bismuth expressed in Bohr magnetons?

Solution:

(a) We have $R \gg a$, so the field within the equivalent air-core toroid is essentially uniform and can be obtained from Ampère's circuital law; thus,

$$B_0 = \frac{\mu_0 N I}{2 \pi R} = \frac{(2 \times 10^{-7} \text{ T·m/A})(800)(10 \text{ A})}{0.10 \text{ m}}$$

$$= 0.016 \text{ T} = 160 \text{ G}$$

(b) Bismuth is diamagnetic, so the magnetization vector is directed opposite to the field $\mathbf{B}_0$. Using Eq. 31–20c and the value of χ_M from Table 31–2, the magnitude of the magnetization is

$$M = \frac{|\chi_M| B_0}{\mu_0} = \frac{(1.7 \times 10^{-4})(0.016 \text{ T})}{4\pi \times 10^{-7} \text{ T·m/A}}$$

$$= 2.16 \text{ A·turns/m}$$

(c) Using Eqs. 31–20a and 31–20c, we have

$$B = (\chi_M + 1) B_0 = (1 - 0.00017) B_0$$

$$= 0.99983 \, B_0$$

The susceptibility is so small that B is hardly distinguishable from B_0.

(d) The total magnetization current is

$$I_M = 2\pi R j_M = 2\pi RM = 2\pi(0.10 \text{ m})(2.16 \text{ A}\cdot\text{turns/m})$$

$$= 1.36 \text{ A}\cdot\text{turns}$$

which is to be compared with the driving ampere-turns, $NI = 8000$ A·turns.

(e) The effective atomic magnetic moment of bismuth is

$$\frac{m_{\text{eff}}}{m_B} = \frac{M}{\mathcal{N} m_B}$$

$$= \frac{(2.16 \text{ A}\cdot\text{turns/m})(209 \times 10^{-3} \text{ kg/mol})}{(6.02 \times 10^{23} \text{ mol}^{-1})(9.8 \times 10^3 \text{ kg/m}^3)(9.27 \times 10^{-24} \text{ A}\cdot\text{m}^2)}$$

$$= 8.3 \times 10^{-6}$$

where we have used $\mathcal{N} = N_0 \rho / A$

Example 31–6

A solid toroid of iron has a core radius $a = 1$ cm and a toroid radius $R = 10$ cm. A winding of $N = 800$ turns around the toroid carries a steady current of 10 A. (Compare Example 31–5, where the toroid consists of bismuth.) Assume that the maximum magnetization of iron arises from the spin alignment of two electrons in each atom with $m_0 = 2m_B$. The induction in the iron core is measured to be 1.85 T. Determine the following quantities:

(a) the magnetization M;
(b) the saturation value M_m, with all domains aligned;
(c) the magnetization current I_M; and
(d) the values of κ_M, μ, and χ_M.

Solution:

(a) The value of the induction B_0 for a vacuum core was obtained in Example 31–5: $B_0 = 0.016$ T. Then, using Eq. 31–20a, we have

$$M = \frac{B - B_0}{\mu_0} = \frac{1.85 \text{ T} - 0.016 \text{ T}}{4\pi \times 10^{-7} \text{ T}\cdot\text{m/A}}$$

$$= 1.46 \times 10^6 \text{ A}\cdot\text{turns/m}$$

(b) The maximum magnetization is

$$M_m = m_0 \mathcal{N} = 2m_B \mathcal{N}$$

Now, $\mathcal{N} = N_0 \rho / A$, where the density of iron is $\rho = 7860$ kg/m^3 and the atomic mass is $A = 55.9$ g/mol. Then,

$$M_m = \frac{2m_B N_0 \rho}{A}$$

$$= \frac{2(9.27 \times 10^{-24} \text{ A}\cdot\text{m}^2)(6.02 \times 10^{23} \text{ mol}^{-1})(7860 \text{ kg/m}^3)}{0.0559 \text{ kg/mol}}$$

$$M_m = 1.57 \times 10^6 \text{ A}\cdot\text{turns/m}$$

Thus, $\quad \dfrac{M}{M_m} = \dfrac{1.46 \times 10^6 \text{ A}\cdot\text{turns/m}}{1.57 \times 10^6 \text{ A}\cdot\text{turns/m}} = 0.93$

so that the magnetization is 93 percent of the full saturation value.

(c) The magnetization current is (see Eq. 31–18 with $\ell = 2\pi R$)

$$I_M = 2\pi RM$$

$$= 2\pi(0.10 \text{ m})(1.46 \times 10^6 \text{ A}\cdot\text{turns/m})$$

$$= 9.17 \times 10^5 \text{ A}\cdot\text{turns}$$

The direction of this magnetization current is the same as that of the coil current, the magnitude of which is $NI = 8 \times 10^3$ A·turns.

(d) From Eq. 31–13, we have

$$\kappa_M = \frac{B}{B_0} = \frac{1.85 \text{ T}}{0.016 \text{ T}} = 116$$

From Eq. 31–14, we have

$$\mu = \kappa_M \mu_0 = (116)(4\pi \times 10^{-7} \text{ T}\cdot\text{m/A})$$

$$= 1.46 \times 10^{-4} \text{ T}\cdot\text{m/A}$$

Finally, from Eq. 31–15, we have

$$\chi_M = \kappa_M - 1 = 115$$

Notice that we have $B/B_0 = 116$ in this case of a ferromagnetic material, whereas for bismuth, one of the strongest diamagnetic substances, we found $B/B_0 = 0.99983$ (Example 31–5).

31–5 THE MAGNETIC INTENSITY VECTOR

In the preceding section a modification of Ampère's circuital law was introduced in order to relate the magnetic induction **B** in a magnetic material to the actual current in an encircling coil and the state of magnetization in the sample. We then wrote (see Eq. 31–20a)

$$\mu_0 \mathbf{M} = \mathbf{B} - \mathbf{B}_0$$

where $\mathbf{B}_0$ is the induction directly produced by the (real) coil current.* This relationship is valid for all homogeneous, isotropic magnetic media. Ampère's circuital law can now be expressed in a more general form as

$$\oint \left(\frac{\mathbf{B}_0}{\mu_0} \right) \cdot d\mathbf{r} = \oint \left(\frac{\mathbf{B}}{\mu_0} - \mathbf{M} \right) \cdot d\mathbf{r} = NI \qquad \textbf{(31–21)}$$

where NI is the total (real) current through any open surface whose boundary corresponds to the path of the line integration. It is convenient to introduce a new vector quantity, called the *magnetic intensity* $\mathbf{H}$, defined as

$$\mathbf{H} = \frac{\mathbf{B}_0}{\mu_0} = \frac{\mathbf{B}}{\mu_0} - \mathbf{M} \qquad \textbf{(31–22)}$$

so that Ampère's circuital law becomes

$$\oint \mathbf{H} \cdot d\mathbf{r} = NI \qquad \text{real currents} \qquad \textbf{(31–23)}$$

We can also write (using Eqs. 31–13 and 31–14)

$$\mathbf{B} = \mu_0 \kappa_M \mathbf{H} = \mu \mathbf{H} \qquad \textbf{(31–24)}$$

The field intensity $\mathbf{H}$ has the same dimensions as $\mathbf{M}$ and is therefore measured in units of A·turns/m (or, simply, A/m). The advantage of $\mathbf{H}$ over $\mathbf{B}$ is that it can be calculated directly, for simple geometries, from the known external current I by use of Eq. 31–23.

31–6   MAXWELL'S EQUATIONS IN MATTER

In Section 30–5 Maxwell's field equations were given for the case of a vacuum environment. Table 31–3 on page 608 gives the corresponding equations for the case of a material medium.

For a linear homogeneous isotropic medium, we have the following auxiliary equations:

$$\left. \begin{array}{l} \mathbf{D} = \kappa \epsilon_0 \mathbf{E} = \epsilon \mathbf{E} \\[4pt] \mathbf{B} = \kappa_M \mu_0 \mathbf{H} = \mu \mathbf{H} \\[4pt] \mathbf{J} = \sigma \mathbf{E} \end{array} \right\} \qquad \textbf{(31–25)}$$

For these materials, the electromagnetic energy density can be expressed as

$$u = \tfrac{1}{2}(\mathbf{E} \cdot \mathbf{D} + \mathbf{B} \cdot \mathbf{H}) \qquad \textbf{(31–26)}$$

For the case of nonlinear magnetic materials, see Enrichment Topic B at the back of the book.

---

*We distinguish between the effective surface magnetization currents and the externally applied coil currents by calling the latter ''real'' currents. However, the physical sources responsible for the magnetization currents are also real and produce observable effects.

TABLE 31–3. Maxwell's Equations in Matter

LAW	EQUATION*
Faraday's law	$\oint \mathbf{E} \cdot d\mathbf{r} = -\dfrac{d}{dt} \int_S \mathbf{B} \cdot \hat{n}\, dA$
Gauss's law (electric)	$\oint_S \mathbf{D} \cdot \hat{n}\, dA = \int \rho\, dV$
Ampère's law	$\oint \mathbf{H} \cdot d\mathbf{r} = \int_S \left(\mathbf{J} + \dfrac{d\mathbf{D}}{dt} \right) \cdot \hat{n}\, dA$
Gauss's law (magnetic)	$\oint_S \mathbf{B} \cdot \hat{n}\, dA = 0$

*For stationary surfaces S.

QUESTIONS

31–1 Define and explain briefly the meaning of the terms (a) mutual inductance, (b) magnetic flux linkage, (c) self-inductance, and (d) Rowland ring.

31–2 A closed loop of wire is formed into a symmetric flat figure-8. What is the induced EMF in the loop if it is placed inside a time-varying uniform magnetic field of a long solenoid? What is the mutual inductance between the two circuits?

31–3 Two identically shaped coils of wire are placed one on top of the other, exactly overlapping. What are the relative values of L_1, L_2, and M? Is it possible to connect the two coils so that there is zero total inductance?

31–4 The smaller solenoid of Fig. 31–2 is replaced by an iron rod. What will be the nature of the force between the rod and the solenoid carrying the current I_2?

31–5 Discuss and contrast the principal characteristics of diamagnetism, paramagnetism, and ferromagnetism.

31–6 If an electrified rod is brought near an uncharged bit of paper, the paper is attracted to the rod. If a bar magnet is brought near an unmagnetized piece of iron, it too is attracted to the bar magnet. In what way are these two effects the same? In what way different? In each case, just before the attracted bit of matter strikes the rod or bar magnet, some kinetic energy has been acquired by it. What is the source of this energy?

31–7 What would you expect to happen to the magnetic field of a magnetized piece of iron if it were vigorously hammered?

31–8 Discuss and contrast the four electromagnetic vectors $\mathbf{E}$, $\mathbf{D}$, $\mathbf{B}$, and $\mathbf{H}$. What simple experiments would enable you to distinguish between $\mathbf{E}$ and $\mathbf{D}$ on the one hand and $\mathbf{B}$ and $\mathbf{H}$ on the other hand?

31–9 Gauss's law for the magnetic induction $\mathbf{B}$ is

$$\oint_S \mathbf{B} \cdot \hat{n}\, dA = 0$$

Is the corresponding expression for the magnetic intensity, $\oint_S \mathbf{H} \cdot \hat{n}\, dA$, also equal to zero? If not, give an example for which the flux of $\mathbf{H}$ through a closed surface is not zero.

PROBLEMS

Section 31–1

31–1 Two coils in fixed positions have a mutual inductance of 100 μH. What peak voltage is induced in one coil if a current, $I = I_0 \sin \omega t$, flows in the other coil? (Assume $I_0 = 10$ A and $\omega = 1000$ rad/s.)

31–2 A long solenoid consists of N_1 turns with a radius R_1. A second long solenoid (N_2 turns and radius R_2), with the same length ℓ, lies entirely within the first solenoid with the axes coinciding. Determine the mutual inductance of the pair of solenoids by first assuming that a current I flows in solenoid 1 and then assuming that the same current flows in solenoid 2. Do you find $M_{12} = M_{21}$?

31–3   Derive an expression for the mutual inductance of a long straight wire and rectangular coil of N turns with dimensions as shown in the diagram. The wire is in the plane of the coil.

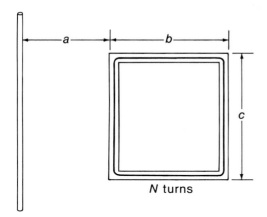

N turns

31–4   Two single-turn circular loops with radii R and $r \ll R$ have the same center and lie in the same plane. Show that the mutual inductance of the pair is $M = \mu_0 \pi r^2 / 2R$. What is the value of M for radii $r = 2$ cm and $R = 20$ cm?

31–5   A long straight wire coincides with the axis of a toroid of N turns with an inner radius R and a rectangular cross section $a \times b$, as shown in the diagram. (a) Determine the mutual inductance by considering the flux linkage produced by a current I in the long wire. (b) Determine the mutual inductance by considering the flux linkage produced by a current I in the toroidal coil, thereby demonstrating that $M = M_{12} = M_{21} = (\mu_0 Nb / 2\pi)$ ln $[(R + a)/a]$.

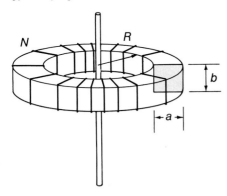

Section 31–2

31–6   What amount of self-flux links an inductor coil of 10 mH when a current of 2 A flows through the coil?

31–7   A current $I = I_0 \sin \omega t$, with $I_0 = 5$ A and $\omega / 2\pi = 60$ Hz, flows through an inductor whose inductance is 10 mH. What is the back EMF as a function of time?

31–8   An air-core Rowland ring has an inner radius $R = 10$ cm and consists of 500 turns of fine wire. The ring has a square cross section, 4 cm × 4 cm, as shown in the diagram. Determine the self-inductance of the ring.

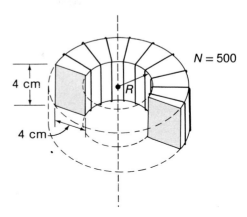

31–9   A *twin-lead* cable consists of two parallel wires with radii a separated by a distance $2b$. If the two wires carry equal currents in opposite directions, show that the self-inductance per length ℓ of the twin-lead is

$$\frac{L}{\ell} = \frac{\mu_0}{\pi} \ln \left[(2b - a)/a \right]$$

(Neglect the fields within the wires.)

Section 31–3

31–10   It proves convenient to define a *coefficient of coupling* ξ for a pair of coupled inductors such that $\xi = M / \sqrt{L_1 L_2}$, with $\xi \leq 1$. Two coupled inductors, $L_1 = 50$ mH and $L_2 = 20$ mH, with $\xi = \frac{1}{3}$, are connected in series with all induced EMFs adding. (a) What is the effective inductance of the combination? (b) What amount of energy is stored in the field when a current of 2 A flows through the combination?

31–11   An air-core solenoid with a radius of 4 cm and a length of 65 cm is tightly and uniformly wound with No. 18 wire, which has a diameter of 1.2 mm, including insulation. What amount of energy is stored in the field when the steady current equals the maximum rated value of 6 A?

31–12   An external source of EMF maintains a constant current I in an inductor as it is moved from a field-free region to a point at which a steady flux φ_M links the inductor. Show that an amount of work $I\varphi_M$ was done by the external source during the change in position. Calculate the work done if $\varphi_M = 1.6 \times 10^{-3}$ Wb and $I = 10$ A.

Section 31–4

31–13 A sample of liquid oxygen at 90 K is placed in a region in which there is a uniform magnetic field of 500 G. Determine (a) the magnetization M and (b) the field B within the sample.

31–14 A very long cylindric bar of copper is to be wrapped with wire to form a solenoid. The wire can carry a maximum of 1 A. The required field within the copper is 20 G. What should be the wrapping density, N/ℓ? (• What would be the induction without the copper core?)

31–15 A solid toroid of iron has a central radius $R = 8$ cm and a core radius $r = 8$ mm. To produce a magnetic flux of 1.5×10^{-4} Wb in the core, what current must flow through the 100-turn coil of the toroid if the relative permeability is $\kappa_M = 2400$?

Section 31–5

31–16 A long thin needle of iron ($\kappa_M = 150$) is placed in a uniform magnetic field, $B_0 = 120$ G (in air) and is oriented parallel to $\mathbf{B}_0$. Neglect end effects and determine the three magnetic vectors, $\mathbf{B}$, $\mathbf{H}$, and $\mathbf{M}$, inside the needle.

31–17 A disc of iron ($\kappa_M = 150$) with a diameter that is large compared with its thickness is placed in a uniform magnetic field, $B_0 = 120$ G (in air), and is oriented with the plane of the disc perpendicular to $\mathbf{B}_0$. Neglect end effects and determine the three magnetic vectors, $\mathbf{B}$, $\mathbf{H}$, and $\mathbf{M}$, inside the disc.

31–18 Modify the discussion leading to Eq. 31–9 to describe the situation in which the interior of the long solenoid is completely filled with a magnetic material whose relative permeability is κ_M. Show that $u_M = \frac{1}{2}\mathbf{B}\cdot\mathbf{H}$.

Additional Problems

31–19 Two long solenoids consist of interlaced windings, of N_1 turns and N_2 turns, on the same nonmagnetic cylinder. Each solenoid has a length ℓ and a radius R. (a) Calculate L_1, L_2, and M. Neglect end effects. (b) What is the value of the coefficient of coupling, $\xi = M/\sqrt{L_1 L_2}$? (See Problem 31–10.) Is there any way to change the value of ξ for solenoids wound in this way?

31–20 Two small circular current loops with radii a and b are
• separated by a distance $\mathbf{r}$, as shown in the diagram. Consider the field of each loop to be a dipole field and derive an expression for the mutual inductance.

31–21 A thin spherical shell with radius R and a uniformly distributed surface charge Q spins with an angular velocity $\boldsymbol{\omega}$ about an axis that coincides with a diameter of the sphere. Show that the magnetic moment is $\mathbf{m} = \frac{1}{3}QR^2\boldsymbol{\omega}$. Calculate ω if $Q = e$, $R = 0.53$ Å (the radius of the hydrogen atom), and m is equal to one Bohr magneton. Compare your result with the angular speed of the electron in a hydrogen atom, $\omega_H = 4.13 \times 10^{16}$ rad/s.

31–22 A saturated magnetic induction of 10,000 G is developed in a homogeneous sample of iron with $\kappa_M = 2000$. (a) What is the energy density of the field within the sample? (b) What electric field strength E would be required to develop the same energy density in a homogeneous dielectric with $\kappa = 4$?

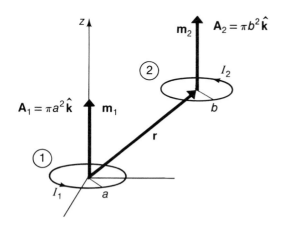

CHAPTER 32

CIRCUITS WITH TIME-VARYING CURRENTS

In Chapter 27 we were concerned with circuits in which the currents did not change with time. We now consider the general case and let the currents vary with time.

It is often possible to associate a characteristic time τ with a time-varying current. For example, if we have a sinusoidally varying current of the form $I_0 \sin \omega t$, the characteristic time is just the period $2\pi/\omega$. To avoid the complications that result when substantial amounts of electromagnetic radiation are emitted by a system, we require again (see Example 30–3) that all lengths ℓ associated with the system or circuit be small compared with a characteristic length $c\tau$. This length is the distance to which electromagnetic effects could propagate (at the speed of light c) during a time equal to the characteristic time τ of the circuit. Then, with $\ell \ll c\tau$, the amount of radiation emitted is generally quite small. When this condition is not met, phase differences throughout the circuit will result, and substantial amounts of electromagnetic energy may be radiated away. Except in some simple cases (see Chapter 33), treatment of this radiation then requires techniques beyond the scope of this text. Generally, the radiation effects are not important for laboratory-scale systems driven by oscillating currents with frequencies below about 10^6 Hz, for then the characteristic length is 300 m or more. However, for current pulses involving characteristic times of the order of 0.01 μs, we have $c\tau \approx 3$ m. Then, small circuit components and careful layout are required; also, the use of shielded cable may be necessary to suppress radiation effects.

The required condition $\ell \ll c\tau$ means that the currents in the various branches of a circuit can be assumed to obey Kirchhoff's rules at any instant of time. (• Can you see why this is so?)

When analyzing circuits we assume that all resistance, capacitance, and inductance occur only as *lumped circuit elements*. Reference to the word "lumped" means that we consider that all substantial resistance in a circuit either occurs in localized resistor elements or is associated with other well-defined, isolated sections of the circuit. Similarly, we assume that all substantial capacitance and inductance occur in specific capacitors and inductors, respectively.

▶ In a real circuit the resistors have resistance, of course, but they may also have some inductance and capacitance (especially if they are wire-wound resistors). Similarly, an inductor coil will have a small resistance and a capacitor will have a small conductivity. If the nature of the calculation requires taking into account these secondary effects, we replace each real circuit element with its approximate equivalent in terms of R, L, and C. For example, a "leaky" capacitor can be considered to be a capacitor and a (large) resistor in a parallel combination (Fig. 32–1a). An inductor coil with some resistance can be considered to be an inductor and a (small) resistor in a series combination (Fig. 32–1b).

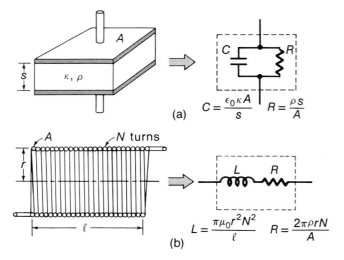

(a) $C = \dfrac{\epsilon_0 \kappa A}{s}$ $R = \dfrac{\rho s}{A}$

(b) $L = \dfrac{\pi \mu_0 r^2 N^2}{\ell}$ $R = \dfrac{2\pi \rho r N}{A}$

Fig. 32-1. (a) An ideal capacitor has only capacitance, but a real capacitor can be represented as a capacitance C and a resistance R in a parallel combination. (The resistivity ρ of a dielectric material is usually very large.) (b) The resistance R of a real inductor coil appears in series with the inductance L. (The resistivity ρ of an inductor wire is usually very small.) ◄

32-1   RC CIRCUITS

Charging a Capacitor. Consider first the process of charging a capacitor using the circuit shown in Fig. 32–2. We begin with zero charge on the capacitor. At $t = 0$ the switch S is closed to position a and a time-dependent current $I(t)$ begins to flow. The charge Q on the capacitor increases in accordance with $I = dQ/dt$. We apply Kirchhoff's second rule (see Section 27–7), proceeding clockwise around the circuit, to obtain

$$V_R + V_C = \mathcal{E}$$

or

$$IR + \frac{Q}{C} = \mathcal{E}$$

This equation can be written as

$$R\,\frac{dQ}{dt} + \frac{Q}{C} = \mathcal{E} \tag{32-1}$$

Rearranging, we obtain

$$\frac{dQ}{Q - \mathcal{E}C} = -\frac{1}{RC}\,dt$$

Integration gives

$$\ln(Q - \mathcal{E}C) = -t/RC + \ln c_1$$

or

$$Q(t) = \mathcal{E}C + c_1 e^{-t/RC} \tag{32-2}$$

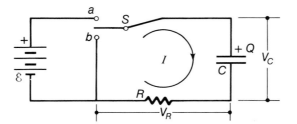

Fig. 32-2. Circuit for charging (and discharging) a capacitor C.

where c_1 is the arbitrary constant of integration that is determined by the initial condition, namely, $Q = 0$ at $t = 0$. We find $c_1 = -\mathcal{E}C$, so that

$$Q(t) = \mathcal{E}C(1 - e^{-t/RC}) \qquad (32\text{–}3)$$

By differentiating $Q(t)$ we obtain the expression for the time dependence of the current. Thus,

$$I(t) = \frac{dQ}{dt} = \frac{\mathcal{E}}{R}e^{-t/RC} \qquad (32\text{–}4)$$

The expressions for $Q(t)$ and $I(t)$ are illustrated graphically in Fig. 32–3.

The product RC has the dimensions of time and is referred to as the *time constant τ* of the circuit. For the typical values, $R = 100\ \Omega$ and $C = 10^{-9}$ F, we find $\tau = RC = 10^{-7}$ s. The characteristic length associated with this circuit is $c\tau = 30$ m. Thus, a circuit with ordinary size (say, 1 m or less) will satisfy the condition that permits the neglect of radiation and phase-difference effects.

Figure 32–3a or Eq. 32–4 shows that the initial rate of charging of the capacitor (i.e., at $t = 0$) is $dQ/dt = \mathcal{E}/R$ and that the rate decreases with time as the capacitor approaches the fully charged condition, $Q_\infty = \mathcal{E}C$. This charge is reached strictly only after an infinite time. However, for all practical purposes, the capacitor may be considered to be fully charged after a time of approximately 5τ, for then $Q/\mathcal{E}C = 1 - e^{-5} = 0.993$. Similarly, after a time of 5τ, the charging current is essentially zero.

Discharging a Capacitor. After the capacitor in Fig. 32–2 has been fully charged, the switch S is moved to position b, and the capacitor begins to discharge through the resistor R. The differential equation for the discharging condition can be obtained by simply setting $\mathcal{E} = 0$ in Eq. 32–1. The direction for positive current flow and the charge polarity are as shown in Fig. 32–2. Then, for Kirchhoff's rule around a clockwise path, we have

$$IR + \frac{Q}{C} = 0$$

or

$$I = \frac{dQ}{dt} = -\frac{Q}{RC} \qquad (32\text{–}5)$$

As expected, $I < 0$, which means that the current actually flows in the counterclockwise direction. The solution to this equation is obtained in a manner similar to that used in deriving Eq. 32–3. For $Q = Q_0$ at $t = 0$, we find

$$Q(t) = Q_0 e^{-t/RC} \qquad (32\text{–}6)$$

and the discharge current is

$$I(t) = \frac{dQ}{dt} = -\frac{Q_0}{RC}e^{-t/RC} \qquad (32\text{–}7)$$

The expressions for $Q(t)$ and $I(t)$ are illustrated graphically in Fig. 32–4.

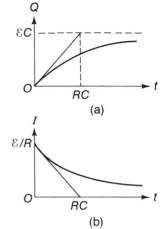

Fig. 32–3. (a) Time dependence of the charge Q on a capacitor C that is being charged through a resistance R. If the initial rate of charging were maintained, the capacitor would be fully charged after a time $\tau = RC$. (b) The current I corresponding to the rate of change of the charge in (a). (• How is the time RC related to the initial rate of charging?)

Fig. 32–4. The discharge of a capacitor through a resistance R. (a) The charge on the capacitor as a function of time. (b) The discharge current as a function of time.

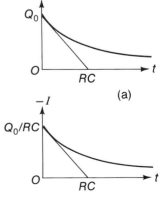

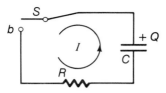

Fig. 32–5. The positive direction for current flow for a discharging capacitor can be chosen opposite to that selected in Fig. 32–2.

Sign Conventions. It is useful to summarize the important rules concerning sign conventions for Q and I, and for V_C treated as a voltage drop. When we apply Kirchhoff's second rule and proceed along a path that crosses a capacitor from the positively charged plate to the negatively charged plate, we have $V_C = +Q/C$; for a path in the opposite direction, $V_C = -Q/C$. If the direction selected for positive current corresponds to that for increasing Q, we have $I = +dQ/dt$; for a direction that corresponds to decreasing Q, $I = -dQ/dt$.

In the discussions of the charging and the discharging of a capacitor, we had, for a clockwise path in each case, $V_C = +Q/C$ and $I = +dQ/dt$. For the discharging capacitor we could have chosen the positive direction for current flow to be counterclockwise, leaving the charge polarity the same (Fig. 32–5). Then, for a counterclockwise path, we would have $I = -dQ/dt$ (decreasing Q) and $V_C = -Q/C$ (negative to positive plate); also, $V_R = +IR$. Hence,

$$IR - \frac{Q}{C} = 0$$

so that

$$R\frac{dQ}{dt} + \frac{Q}{C} = 0$$

as before. Thus, $Q(t)$ is the same as in Eq. 32–6, but $I(t)$ is now a positive quantity, as required by the choice in Fig. 32–5.

Energy Balance. The Kirchhoff equation for a charging capacitor is $Q/C + IR = \mathcal{E}$. If we multiply this equation by $dQ/dt = I$, we obtain

$$\frac{Q}{C}\frac{dQ}{dt} + I^2R = \mathcal{E}I$$

or

$$d\left(\frac{1}{2}\frac{Q^2}{C}\right) + I^2R\ dt = \mathcal{E}I\ dt \qquad (32\text{–}8)$$

The interpretation of this equation is that during a time interval dt an amount of energy $\mathcal{E}I\ dt$ supplied by the battery is expended partly in Joule heating, $I^2R\ dt$, and partly in increasing the stored energy in the electric field of the capacitor, $d(\frac{1}{2}Q^2/C)$. Thus, Eq. 32–8 expresses energy conservation for the system in which chemical energy is converted partly into heat and partly into field energy. (If electromagnetic radiation were emitted, Eq. 32–8 would no longer be adequate.)

The Kirchhoff equation for a discharging capacitor, $Q/C + IR = 0$, becomes

$$-d\left(\frac{1}{2}\frac{Q^2}{C}\right) = I^2R\ dt \qquad (32\text{–}9)$$

which states that the energy removed from the electric field of the capacitor appears in the form of Joule heat in the resistor.

Example 32–1

In the circuit of Fig. 32–2, $C = 0.1 \ \mu F$, $\mathcal{E} = 3$ V, and $R = 1000 \ \Omega$. (The internal resistance of the battery is small compared with R.) The capacitor is initially uncharged, and then the switch is closed to position a.

(a) Determine the initial current.

(b) Determine the charge Q_∞ on the capacitor when it is fully charged.

(c) Find the time required for the charge on the capacitor to reach 99 percent of Q_∞.

(d) Find the time required for the field energy in the capacitor to reach 99 percent of its final value.

Solution:

(a) The initial current is

$$I_0 = \frac{\mathcal{E}}{R} = \frac{3.0 \ \text{V}}{10^3 \ \Omega} = 3 \ \text{mA}$$

(b) The final charge is

$$Q_\infty = \mathcal{E}C = (3.0 \ \text{V})(0.1 \times 10^{-6} \ \text{F}) = 3 \times 10^{-7} \ \text{C}$$

(c) For $Q = 0.99 \ Q_\infty$, we have, using Eq. 32–3,

$$\frac{Q}{Q_\infty} = \frac{Q}{\mathcal{E}C} = 1 - e^{-t/RC} = 0.99$$

Thus,

$$e^{-t/RC} = 0.01$$

so that

$$t = -RC \ \ln 0.01 = -(10^3 \ \Omega)(10^{-7} \ \text{F})(\ln 0.01)$$

$$= 4.61 \times 10^{-4} \ \text{s}$$

(d) The stored energy is $U = \frac{1}{2}Q^2/C$. Thus

$$\frac{U}{U_\infty} = \frac{Q^2}{Q_\infty^2} = 0.99$$

so that

$$\frac{Q}{Q_\infty} = 0.995 = 1 - e^{-t/RC}$$

from which

$$e^{-t/RC} = 0.005$$

Finally,

$$t = -RC \ \ln 0.005 = -(10^3 \ \Omega)(10^{-7} \ \text{F})(\ln 0.005)$$

$$= 5.30 \times 10^{-4} \ \text{s}$$

Example 32–2

A $25\text{-}\mu F$ capacitor is left isolated after having been charged. Three hours later it is discovered that the capacitor has lost one half of its charge. What is the leakage resistance in the dielectric material of the capacitor?

Solution:

The equivalent circuit of the leaky capacitor is shown in Fig. 32–1a, and we see that the time constant is $\tau = RC$, where

R is the dielectric resistance. Thus,

$$\frac{Q}{Q_0} = e^{-t/RC}$$

from which

$$\ln \tfrac{1}{2} = -t/RC$$

Then,

$$R = \frac{-3 \times 3600 \ \text{s}}{(25 \times 10^{-6} \ \text{F})(\ln \tfrac{1}{2})}$$

$$= 6.23 \times 10^8 \ \Omega = 623 \ \text{M}\Omega$$

Example 32–3

A capacitor C_1 that carries an initial charge Q_0 is connected to an initially uncharged capacitor C_2 by closing the switch S, as indicated in the diagram. Compare the energy stored in the capacitors after equilibrium is reached with the initial stored energy.

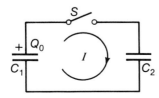

Solution:

After the switch is closed, a current $I(t)$ flows, delivering charge from C_1 to C_2, until, at equilibrium, there is a charge Q_1 on C_1 and a charge Q_2 on C_2, where $Q_1 + Q_2 = Q_0$. In this condition, the two capacitor voltages are equal; that is, $V_1 = Q_1/C_1 = V_2 = Q_2/C_2$. These two conditions lead to

$$Q_1 = \frac{C_1}{C_1 + C_2}Q_0 \quad \text{and} \quad Q_2 = \frac{C_2}{C_1 + C_2}Q_0$$

The energy originally stored in C_1 was

$$U_0 = \frac{1}{2}\frac{Q_0^2}{C_1}$$

The energy stored in the final condition is

$$U_f = U_1 + U_2 = \frac{1}{2}\frac{Q_1^2}{C_1} + \frac{1}{2}\frac{Q_2^2}{C_2}$$

$$= \frac{1}{2}\frac{Q_0^2}{C_1 + C_2}$$

Thus, $U_f < U_0$, in apparent violation of the principle of energy conservation.

In any real circuit, there would be a certain amount of resistance R in the connecting wires. Then, using Kirchhoff's second rule, we would have

$$IR - \frac{Q_1}{C_1} + \frac{Q_2}{C_2} = 0$$

Multiplying by $I = -dQ_1/dt = dQ_2/dt$ gives

$$I^2R + \frac{Q_1}{C_1}\frac{dQ_1}{dt} + \frac{Q_2}{C_2}\frac{dQ_2}{dt} = 0$$

With $U = \frac{1}{2}Q^2/C$, we can write

$$I^2R + \frac{dU_1}{dt} + \frac{dU_2}{dt} = 0$$

Integrating, we find

$$\int_0^\infty I^2R\ dt = -\int_{U_0}^{U_f} d(U_1 + U_2) = U_0 - U_f$$

Thus, the difference in stored energy between the initial and final conditions, $U_0 - U_f > 0$, is accounted for by the Joule heating in the resistance R.

Suppose that the resistance R is made very small. Then, the characteristic time τ of the circuit also becomes very small. That is, $\tau = RC \rightarrow 0$, where $C = C_1C_2/(C_1 + C_2)$ is the effective capacitance of the series connection. Thus, when $R \rightarrow 0$, less energy is dissipated as heat, but more energy is lost as radiation, so that we still have $U_f < U_0$.

▶ *The RC Differentiating Circuit.* An interesting application of the simple *RC* circuit is an often-used arrangement called a *differentiating circuit,* shown in Fig. 32–6. A time-dependent input voltage $\mathcal{E}(t)$ is applied to the *RC* circuit as indicated, and the output voltage $V(t)$ developed across the resistor is detected by a device such as an oscilloscope.* Using Kirchhoff's rule, we find

$$\frac{Q}{C} + IR = \mathcal{E}(t)$$

or $Q + IRC = C\mathcal{E}(t)$ **(32–10)**

In an appropriately designed circuit, RC is made sufficiently small that the term IRC in Eq. 32–10 can safely be neglected. Then $Q \cong C\mathcal{E}(t)$, and we have

$$V(t) = IR = R\frac{dQ}{dt} \cong RC\frac{d\mathcal{E}(t)}{dt}$$ **(32–11)**

Fig. 32–6. A differentiating circuit. The output voltage $V(t)$ is the time derivative of the input voltage $\mathcal{E}(t)$. It is assumed that no current flows at the right-hand pair of terminals; they are connections for the measuring device.

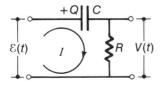

In the approximation that RC is small, the output voltage $V(t)$ is proportional to the derivative of the input voltage $\mathcal{E}(t)$. Notice that we have here an example of a frequently encountered design dilemma, namely, that the approximation cannot be improved by making RC arbitrarily small because this would decrease the magnitude of the output voltage below a useful limit.

Figure 32–7 shows the response of a differentiating circuit to a square-wave input. If the input were a perfect square wave and if the circuit were ideal ($RC \rightarrow 0$), the output voltage $V(t)$ would consist of sharp spikes. The departure of the conditions from ideal results in voltage spikes with "tails." (For a discussion of an *integrating circuit,* see Problem 32–25.)   ◀

Fig. 32–7. The response $V(t)$ of a differentiating circuit is a series of spikes when the input voltage $\mathcal{E}(t)$ is a square wave.

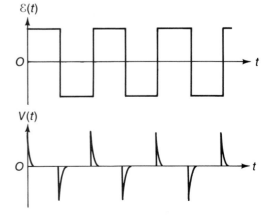

---

*An oscilloscope draws negligible current from a circuit (as do certain other kinds of voltage-sensing devices) and therefore can be ignored when Kirchhoff's rules are applied to the circuit.

32-2 *RL* CIRCUITS

Energizing an Inductor. Consider the process of applying current to an inductor using the circuit shown in Fig. 32-8. At $t = 0$ the switch S is closed to position a and a time-dependent (increasing) current $I(t)$ begins to flow. Because of this changing current in the inductor, there is produced a back EMF, $L\,dI/dt$, that opposes the driving EMF $\mathcal{E}$.

Fig. 32-8. Circuit for energizing (and de-energizing) an inductor L.

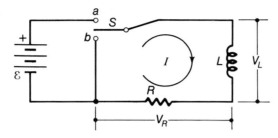

In applying Kirchhoff's rule, this back EMF appears as a voltage drop V_L. Thus, for a clockwise path in Fig. 32-8, we have

$$V_L + V_R = \mathcal{E}$$

or

$$L\frac{dI}{dt} + IR = \mathcal{E} \tag{32-12}$$

This equation has the same form as Eq. 32-1, so the solution is

$$I(t) = \frac{\mathcal{E}}{R} + c_1 e^{-(R/L)t} \tag{32-13}$$

where c_1 is the arbitrary constant of integration that is determined by the initial condition, namely, $I = 0$ at $t = 0$. We find $c_1 = -\mathcal{E}/R$, so that

$$I(t) = \frac{\mathcal{E}}{R}\left(1 - e^{-(R/L)t}\right) \tag{32-14}$$

This expression for $I(t)$ is illustrated graphically in Fig. 32-9a. (• Do you understand the physical reason why the current is zero at $t = 0$, in contrast to the case illustrated in Fig. 32-4b? Consider the back EMF produced in the inductor.)

The quotient L/R is the time constant of the circuit. When $t \gg L/R$, the current approaches the asymptotic value, $I_\infty = \mathcal{E}/R$. For all practical purposes, we may consider this current to have been reached (and the inductor fully energized) after a time of approximately 5τ, for then $I = 0.993\,\mathcal{E}/R$.

De-energizing an Inductor. After the inductor in Fig. 32-8 has been fully energized, the switch S is moved to position b.* The current decreases from its initial value I_0 in accordance with Eq. 32-12 in which we now set $\mathcal{E} = 0$. Then, $c_1 = I_0$, so we have

$$I(t) = I_0 e^{-(R/L)t} \tag{32-15}$$

The time dependence of the de-energizing current is shown in Fig. 32-9b.

Fig. 32-9. (a) Time dependence of the current in an *RL* circuit as the inductor is energized. (b) Time dependence of the current in the same *RL* circuit as the inductor is de-energized. (• Can you see the relationship between the time constant L/R and the slope of the curve in each case? Compare Figs. 32-3 and 32-4.)

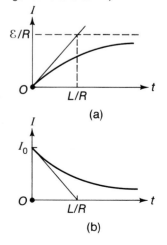

---

*The switch S is required to be of the "make before break" type; that is, contact with terminal b is made before contact with terminal a is broken. (• Can you see why this is necessary?)

Energy Balance. Consider first the process of energizing the inductor. Multiplying Eq. 32–12 by I, we obtain

$$LI \frac{dI}{dt} + I^2R = \mathcal{E}I$$

or

$$d(\tfrac{1}{2}LI^2) + I^2R \, dt = \mathcal{E}I \, dt \qquad \qquad \textbf{(32–16)}$$

The interpretation of this equation is that during a time interval dt an amount of energy $\mathcal{E}I \, dt$ supplied by the battery is expended partly in Joule heating, $I^2R \, dt$, and partly in increasing the stored energy in the magnetic field of the inductor, $d(\tfrac{1}{2}LI^2)$.

The differential equation that describes the de-energizing process is just Eq. 32–12 with $\mathcal{E} = 0$. Multiplying this equation by I leads to

$$-d(\tfrac{1}{2}LI^2) = I^2R \, dt \qquad \qquad \textbf{(32–17)}$$

which states that the energy removed from the magnetic field of the inductor appears in the form of Joule heat in the resistor.

Example 32–4

The inductor L in Fig. 32–8 is the toroid shown in the diagram, whose core consists of hot-rolled sheet iron. The resistance in the circuit is $R = 20 \, \Omega$ and the EMF of the battery is $\mathcal{E} = 6$ V. Assume a linear magnetization curve for the iron, with $B = \kappa_M \mu_0 H$ and $\kappa_M = 300$ (a constant). Neglect the battery internal resistance and the coil resistance of the inductor.

(a) What amount of energy is stored in the magnetic field of the inductor 1.0 ms after the switch is closed?

(b) What amount of energy has been supplied by the battery at $t = 1.0$ ms?

(c) What amount of energy has been expended as Joule heat in the resistor at $t = 1.0$ ms?

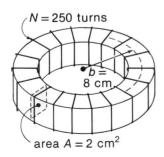

N = 250 turns

b = 8 cm

area $A = 2$ cm^2

Solution:

The self-inductance L of the iron-core toroid is*

$$L = \frac{\mu_0 \kappa_M N^2 A}{2\pi b} = 9.38 \times 10^{-3} \text{ H}$$

The magnetic induction in the iron is*

$$B = \frac{\mu_0 \kappa_M N I}{2\pi b} = \frac{LI}{NA}$$

Using $I = I_\infty = \mathcal{E}/R = (6 \text{ V})/(20 \, \Omega) = 0.3$ A, we have

$$B = \frac{LI}{NA} = \frac{(9.38 \times 10^{-3} \text{ H})(0.3 \text{ A})}{(250)(2 \times 10^{-4} \text{ m}^2)} = 5.63 \times 10^{-2} \text{ T}$$

$$= 563 \text{ G}$$

For such a low value of the maximum induction, the assumption that $B \propto H$ is quite good. Thus, κ_M is indeed constant.

(a) The time constant for the circuit is $\tau = L/R = 0.47$ ms. Then, the current at $t = 1.0$ ms is

$$I = \frac{\mathcal{E}}{R} \left(1 - e^{-t/\tau} \right) = \frac{6 \text{ V}}{20 \, \Omega} \left(1 - e^{-(1.0/0.47)} \right) = 0.264 \text{ A}$$

Hence, the amount of energy U_M stored in the magnetic field is

$$U_M = \tfrac{1}{2}LI^2 = \tfrac{1}{2}(9.38 \times 10^{-3} \text{ H})(0.264 \text{ A})^2 = 3.3 \times 10^{-4} \text{ J}$$

---

*See Example 31–3 and supply the factor κ_M to account for the presence of the iron core.

(b) The amount of energy U supplied by the battery from $t = 0$ to time t is found by using Eq. 32–14 for $I(t)$; thus,

$$U = \mathcal{E} \int_0^t I(t)\, dt = \frac{\mathcal{E}^2}{R} \int_0^t \left(1 - e^{-(R/L)t}\right) dt$$

$$= \frac{\mathcal{E}^2}{R}\left[t - \tau\left(1 - e^{-t/\tau}\right)\right]$$

When $t = 1.0$ ms, we have

$$U = \frac{(6\ \text{V})^2}{20\ \Omega}\left[(10^{-3}\ \text{s}) - (0.47 \times 10^{-3}\ \text{s})(1 - e^{-(1.0/0.47)})\right]$$

$$U = 1.05 \times 10^{-3}\ \text{J}$$

(c) The amount of energy U_R expended as Joule heat in the resistor is

$$U_R = U - U_M = (1.05 \times 10^{-3}\ \text{J}) - (3.3 \times 10^{-4}\ \text{J})$$

$$= 7.2 \times 10^{-4}\ \text{J}$$

(• Can you verify this result by integrating $RI^2(t)\,dt$?)

Example 32–5

Obtain an expression for the current $I(t)$ through the inductor L in the circuit shown in the diagram when the switch S is closed.

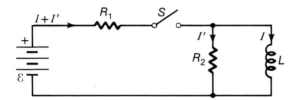

Solution:

Let the branch current in the inductor L be I, and let the branch current in the resistor R_2 be I'. Then, the battery current is $I + I'$ (Kirchhoff's first rule). Applying the second rule to the two loops, we find

$$(I + I')R_1 + L\frac{dI}{dt} = \mathcal{E}$$

$$(I + I')R_1 + I'R_2 = \mathcal{E}$$

By eliminating I' between these two equations there results

$$I\frac{R_1 R_2}{R_1 + R_2} + L\frac{dI}{dt} = \frac{R_2}{R_1 + R_2}\mathcal{E}$$

or, with obvious abbreviations,

$$IR' + L\frac{dI}{dt} = \mathcal{E}'$$

This equation has the form of Eq. 32–15, so the solution is (Eq. 32–17)

$$I(t) = \frac{\mathcal{E}'}{R'}(1 - e^{-(R'/L)t}) = \frac{\mathcal{E}}{R_1}(1 - e^{-(R'/L)t})$$

32–3   LC CIRCUITS

A useful starting point for considering more complicated circuits containing all three electrical elements—resistance, inductance, and capacitance—is the simple LC circuit shown in Fig. 32–10.

Application of Kirchhoff's second rule to the circuit yields

$$L\frac{dI}{dt} + \frac{Q}{C} = 0 \qquad (32\text{–}18)$$

with $I = +dQ/dt$. (• Verify the algebraic signs appearing in Eq. 32–18.) Eliminating the current explicitly from Eq. 32–18, we obtain

$$\frac{d^2Q}{dt^2} = -\frac{Q}{LC} \qquad (32\text{–}19)$$

We have met this common differential equation before, in Chapter 15 (Oscillatory Motion). We found that the solution to the analogous equation (Eq. 15–4)

Fig. 32–10. A simple LC circuit. The capacitor charge is $Q(t)$ and the inductor current is $I(t)$. Note that this is an idealized resistance-less, nondissipative circuit.

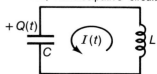

$$\frac{d^2x}{dt^2} = -\frac{\kappa}{m}x \qquad \text{(32-20)}$$

for a mass m oscillating on the end of a spring with spring constant κ was (Eq. 15–11)

$$x(t) = A \sin(\omega_0 t + \phi_0) \qquad \text{(32-21)}$$

with $\omega_0 = \sqrt{\kappa/m}$. In analogy with this result found for the mechanical system, we may write the solution to Eq. 32–19 as

$$Q(t) = Q_m \sin(\omega_0 t + \phi_0) \qquad \text{(32-22a)}$$

with $\omega_0 = \sqrt{1/LC}$. Thus the capacitor charge $Q(t)$ is a simple harmonic function of time, oscillating alternately between values $+Q_m$ and $-Q_m$ at an angular frequency ω_0. The quantity $f_0 = \omega_0/2\pi = \dfrac{1}{2\pi}\sqrt{\dfrac{1}{LC}}$ is called the *natural frequency of oscillation* of the circuit. The circuit current is

$$I(t) = +\frac{dQ}{dt} = \omega_0 Q_m \cos(\omega_0 t + \phi_0) \qquad \text{(32-22b)}$$

As in the case of the oscillatory mechanical motion, the *charge amplitude* Q_m and the *phase angle* ϕ_0 must be determined by the initial conditions. For example, if at time $t = 0$ the charge on the capacitor is zero and the current is in the direction shown in Fig. 32–10, i.e., $I_0 > 0$, then $\phi_0 = 0$. (• How does this condition on the direction of the current at $t = 0$ exclude the solution $\phi_0 = \pi$?) Further, if the amplitude of the capacitor charge is Q_m, then we have

$$Q(t) = Q_m \sin \omega_0 t \qquad \text{(32-23)}$$

The current as a function of time is

$$I = +\frac{dQ}{dt} = \omega_0 Q_m \cos \omega_0 t \qquad \text{(32-24)}$$

Fig. 32–11 shows the capacitor charge Q and the circuit current I as a function of $\omega_0 t$. Note the striking similarity in the behavior of Q and I for the LC circuit shown in Fig. 32–11 with the behavior of x and v for the simple harmonic oscillator (SHO) shown in Fig. 15–5. The close analogy between these two oscillatory systems may be further demonstrated if we consider the energy balance in the electrical circuit. Multiply Eq. 32–18 by $I = dQ/dt$ to obtain

$$\frac{Q}{C}\frac{dQ}{dt} + LI\frac{dI}{dt} = 0$$

or

$$\frac{d}{dt}\left(\frac{1}{2}\frac{Q^2}{C}\right) + \frac{d}{dt}\left(\frac{1}{2}LI^2\right) = 0$$

If we integrate this equation with respect to time, we have

$$\frac{1}{2}\frac{Q^2}{C} + \tfrac{1}{2}LI^2 = E = \text{constant} \qquad \text{(32-25)}$$

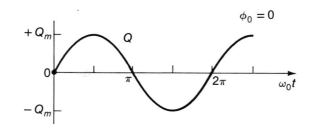

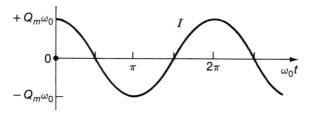

Fig. 32–11. The variation of the capacitor charge Q and the circuit current I as a function of $\omega_0 t$. The simple LC circuit has a natural angular frequency ω_0. The curves are drawn for the special case $\phi_0 = 0$.

where the constant of integration, E, is seen to represent the sum of the energy stored in the electric field of the capacitor and the energy stored in the magnetic field of the inductor. Thus, while Q and I vary with time, surging back and forth in the circuit, the total energy remains a contant. When Q has its maximum value $\pm Q_m$, I is zero, and when I has its maximum value $\pm I_m$, Q is zero; thus, Eq. 32–25 yields

$$\frac{1}{2}\frac{Q_m^2}{C} = \tfrac{1}{2}LI_m^2 = E \qquad\qquad (32\text{--}26)$$

These are just the analogous results found for the SHO, Eq. 15–5. We see that a complete analogy between the electrical and mechanical systems may be constructed, in which

$$
\begin{aligned}
L &\rightleftarrows m \\
C &\rightleftarrows 1/\kappa \\
Q &\rightleftarrows x \\
I &\rightleftarrows v \\
\frac{1}{2}\frac{Q^2}{C} &\rightleftarrows \tfrac{1}{2}\kappa x^2 \\
\tfrac{1}{2}LI^2 &\rightleftarrows \tfrac{1}{2}mv^2
\end{aligned}
\qquad\qquad (32\text{--}27)
$$

Note further that the absence of energy-dissipating friction in the mechanical system has its counterpart in the electrical system by the absence of energy-dissipating resistance.

Example 32–6

At time $t = 0$ the LC circuit shown in Fig. 32–10 has equal amounts of energy stored in the field of the capacitor and in the field of the inductor, each amounting to 500 μJ. The circuit current is in the sense shown; i.e., $I_0 > 0$. If the capacitance is 20 μF and the current amplitude is $\tfrac{1}{5}$ A,

(a) What is the natural frequency of oscillation f_0?
(b) What is the value of the inductance L?
(c) Determine the equation for $Q(t)$ and $I(t)$.

Solution:

(a) The maximum energy stored in the capacitor is the total field energy $2(500 \ \mu J) = 10^{-3}$ J. Thus, using Eq. 32–26

$$Q_m^2 = 2CE = 2(20 \times 10^{-6} \ F)(10^{-3} \ J)$$

or $\qquad Q_m = 2 \times 10^{-4}$ C

From Eq. 32–22b we see that the current amplitude is $I_m = \omega_0 Q_m$

or $\qquad \omega_0 = I_m/Q_m = 0.2 \ A/2 \times 10^{-4} \ C = 1000 \ rad/s$

and $\qquad\qquad f_0 = \omega_0/2\pi = 159$ Hz

(b) Using the defining equation for the natural angular frequency, the value of the inductance is

$$L = 1/\omega_0^2 C = \frac{1}{(1000 \ s^{-1})^2 (20 \times 10^{-6} \ F)}$$

or $\qquad\qquad\qquad L = 50$ mH

(c) The general solution for the charge and current in the LC circuit at time $t = 0$ is found to be, using Eqs. 32–22

$$Q_0 = Q_m \sin \phi_0$$

and $\qquad\qquad I_0 = \omega_0 Q_m \cos \phi_0$

Dividing the first of these equations by the second gives

$$\tan \phi_0 = \frac{\omega_0 Q_0}{I_0}$$

We also have from the statement of the problem that $\frac{1}{2}LI_0^2 = \frac{1}{2} \frac{Q_0^2}{C}$, or $Q_0^2/I_0^2 = LC = 1/\omega_0^2$, and hence $\omega_0 Q_0/I_0 = 1$.

Therefore, $\tan \phi_0 = 1$ and with $I_0 > 0$ we have $\phi_0 = \pi/4$. Thus

$$Q(t) = 2 \times 10^{-4} \sin (1000t + \pi/4)C$$

$$I(t) = 0.20 \cos (1000t + \pi/4)A$$

32–4 *LRC* CIRCUITS

Because all electrical circuits possess resistance, a more realistic study of an LC circuit should include the effects introduced by the presence of an added series resistor. When this inserted series resistance is small, the principal effect is to slowly diminish the amplitude of the oscillating charge and current due to the energy loss in the Joule heating of the resistor.

Suppose that a capacitor C, an inductor L, and a resistor R are connected in series, as shown in Fig. 32–12. Applying Kirchhoff's second rule gives

$$L \frac{dI}{dt} + RI + \frac{Q}{C} = 0$$

Now, $I = +dQ/dt$, so the Kirchhoff equation becomes

$$L \frac{d^2Q}{dt^2} + R \frac{dQ}{dt} + \frac{Q}{C} = 0 \qquad\qquad \textbf{(32–28)}$$

Equation 32–28 is an important example of a linear differential equation. Although the procedure for solving such equations is beyond the scope of this text, a critical examination of the solutions for Eq. 32–28 is of great interest.

The form of the solution to Eq. 32–28 depends on the relative values of two parameters, $\beta = R/2L$ and $\omega_0^2 = 1/LC$. The parameter β is a measure of the rate at which energy in the circuit is dissipated in Joule heating, while ω_0, as in the case of the simple LC circuit, determines the rate of transfer of the field energies between the capacitor and the inductor. We discuss in turn the three cases, $\beta^2 > \omega_0^2$, $\beta^2 = \omega_0^2$, and $\beta^2 < \omega_0^2$. Finally, we compare the results and show graphically the forms of $Q(t)$ and $I(t)$ for the different cases.

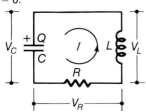

Fig. 32–12. An *LRC* circuit. The capacitor carries an initial charge Q_0, and the switch is closed at $t = 0$.

Overdamping ($\beta^2 > \omega_0^2$). The general solution to Eq. 32–28 when $\beta^2 > \omega_0^2$ is found to be*

$$Q(t) = e^{-\beta t}(c_1 e^{\gamma t} + c_2 e^{-\gamma t}) \qquad \text{overdamped} \qquad \textbf{(32–29)}$$

where

$$\beta = \frac{R}{2L}, \qquad \omega_0^2 = \frac{1}{LC}, \qquad \gamma = \sqrt{\beta^2 - \omega_0^2} \qquad \textbf{(32–30)}$$

and c_1 and c_2 are arbitrary constants.

To be specific, consider the common situation of a capacitor carrying an initial charge Q_0 suddenly connected to the inductor and resistor. In this case, at $t = 0$ we have $Q = Q_0$ and also $I = 0$. Applying these conditions to Eq. 32–29 and its time derivative, we find

$$c_1 + c_2 = Q_0$$

and

$$-\beta(c_1 + c_2) + \gamma(c_1 - c_2) = 0$$

Solving these equations for c_1 and c_2, and inserting the expressions into the general solution, we obtain

$$Q(t) = \frac{Q_0}{2\gamma} e^{-\beta t}[(\beta + \gamma)e^{\gamma t} - (\beta - \gamma)e^{-\gamma t}] \qquad \textbf{(32–31)}$$

The current in the circuit is

$$I(t) = \frac{dQ}{dt} = -\frac{Q_0 \omega_0^2}{2\gamma} e^{-\beta t}(e^{\gamma t} - e^{-\gamma t}) \qquad \textbf{(32–32)}$$

Critical Damping ($\beta^2 = \omega_0^2$). In this case, we have $\gamma = 0$, and the general solution takes the form

$$Q(t) = (c_1 + c_2 t)e^{-\beta t} \qquad \text{critically damped} \qquad \textbf{(32–33)}$$

Imposing the same initial conditions as above, we find

$$c_1 = Q_0 \qquad \text{and} \qquad -\beta c_1 + c_2 = 0$$

Then, we have

$$Q(t) = Q_0(1 + \beta t)e^{-\beta t} \qquad \textbf{(32–34)}$$

The current in the circuit is

$$I(t) = -Q_0 \beta^2 t e^{-\beta t} \qquad \textbf{(32–35)}$$

Underdamping ($\beta^2 < \omega_0^2$). In this case, $\gamma = \sqrt{\beta^2 - \omega_0^2}$ is an imaginary quantity. The situation is most easily handled by introducing a *real* quantity, ω^0, defined by

$$\omega^0 = \sqrt{\omega_0^2 - \beta^2} \qquad \textbf{(32–36)}$$

---

*Equation 32–28 is a differential equation of second order (that is, the highest-order derivative present is the second derivative), so the solution contains the two arbitrary constants c_1 and c_2, which can be determined by imposing two initial conditions.

Then, we explicitly exhibit the imaginary character of the exponents by writing the solution (Eq. 32–29) as

$$Q(t) = e^{-\beta t}(c_1 e^{i\omega^0 t} + c_2 e^{-i\omega^0 t}) \qquad \text{underdamped} \qquad (32\text{–}37)$$

where the constants, c_1 and c_2, are complex numbers. By using the Euler relation (Eq. B–6 in Appendix B), Eq. 32–37 can be expressed as

$$Q(t) = Ae^{-\beta t} \sin(\omega^0 t + \phi_0)$$

where A and ϕ_0 are real quantities that can be evaluated by imposing the initial conditions. The condition $Q = Q_0$ at $t = 0$ gives

$$A = \frac{Q_0}{\sin \phi_0}$$

and the condition $I = dQ/dt = 0$ at $t = 0$ gives

$$\phi_0 = \tan^{-1}(\omega^0/\beta)$$

We define the phase angle ϕ_0 according to the right-triangle relationship shown in Fig. 32–13. Then, we see that $\sin \phi_0 = \omega^0/\omega_0$; thus,

$$A = \frac{Q_0 \omega_0}{\omega^0}$$

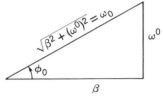

Fig. 32–13. The phase angle ϕ_0 is defined by the right-triangle relationship illustrated here.

Finally, we have

$$Q(t) = \frac{Q_0 \omega_0}{\omega^0} e^{-\beta t} \sin(\omega^0 t + \phi_0) \qquad (32\text{–}38)$$

The current in the circuit is

$$I(t) = -Q_0 \frac{\omega_0^2}{\omega^0} e^{-\beta t} \sin \omega^0 t \qquad (32\text{–}39)$$

Comparison of the Solutions. Figure 32–14a shows the time dependence of the charge $Q(t)$ for the three cases we have just analyzed. The factor $e^{-\beta t}$ causes $Q(t)$ to decrease in all of the cases, but only for underdamping (case ③) is the general decrease accompanied by oscillation. The case of critical damping is of particular interest because it represents the most rapid approach to zero charge. That is, when the circuit is critically damped, the charge reaches (and remains smaller than) any prescribed value in a time shorter than that required for overdamped or underdamped circuits with the same value of ω_0. When a circuit (or a mechanical system) is required to recover in the least time from some disturbance away from equilibrium, the conditions are adjusted to produce critical damping.

Figure 32–14b shows the time dependence of the capacitor discharge current, $-I(t)$, for the three cases. The peak currents have been arbitrarily made equal for ease of comparison. (The degree of distortion can be judged from the fact that the integral of $I(t)$ over the range $0 \leq t \leq \infty$ is just Q_0. Thus, the area under each of the $I(t)$ curves must be the same.)

Again, we should note the close analogy between the behavior of the electrical *LRC* circuit and the mechanical system consisting of the damped harmonic oscillator discussed in Section 15–5. The damping of the mechanical system was the result of the energy

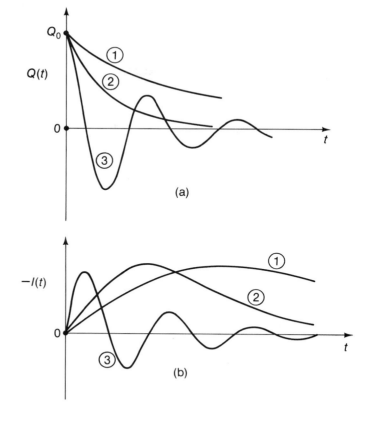

Fig. 32–14. (a) Time dependence of the charge $Q(t)$ for the three cases: ① overdamping, ② critical damping, and ③ underdamping. (b) Time dependence of the current $I(t)$ for the same three cases.

dissipation in the velocity-dependent friction. The frictional force $bv = b\dfrac{dx}{dt}$ is the analog of the voltage drop in the resistor $RI = R\dfrac{dQ}{dt}$.

Comparing quantities in Eq. 32–28 with those in Eq. 15–28 term by term again verifies the analogs stated in Eq. 32–27. We should note in addition that each term in Eq. 15–28 is a force, while each term in Eq. 32–28 is a voltage drop. We may therefore extend the table of analog quantities (Eq. 32–27) and include

$$\left.\begin{array}{c} b \rightleftarrows R \\ \text{Force} \rightleftarrows \text{Voltage drop and EMF} \end{array}\right\}\qquad \textbf{(32–40)}$$

32–5 SINUSOIDALLY DRIVEN *LRC* CIRCUITS

The circuit of Fig. 32–10 is driven only by the initial charge Q_0 of the capacitor. Now we consider an *LRC* circuit that is driven by an external source of repetitive voltage $\mathcal{E}(t)$. Any repetitive voltage, of whatever form, can be decomposed by Fourier analysis into a sum of sinusoidal terms (see Section 16–8). Consequently, if we obtain the response of the circuit to a general sinusoidal driving voltage, $\mathcal{E}(t) = \mathcal{E}_0 \cos \omega t$, the response to any repetitive driving voltage can be generated by the appropriate superposition of responses with different values of $\mathcal{E}_0$ and ω.

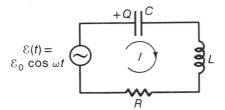

Fig. 32–15. An *LRC* circuit is driven by a sinusoidal voltage.

Consider the circuit shown in Fig. 32–15. The relevant Kirchhoff equation is

$$L\frac{dI}{dt} + RI + \frac{Q}{C} = \mathcal{E}_0 \cos \omega t \qquad (32\text{–}41)$$

The general solution to this equation consists of the sum of two terms: a *transient* solution, which is simply the nondriven response we have just discussed, and a *particular* solution constructed specifically to reproduce the driving term on the right-hand side of the equation. In the analysis here, we restrict attention to the particular or steady-state solution.

In Section 15–3 we found it useful to relate linear simple harmonic motion to uniform circular motion. Projecting the position, velocity, and acceleration of the circular motion onto its diameter faithfully reproduced the corresponding displacement, velocity, and acceleration of the SHM. An analytic variant of this technique may be used to determine the solution to Eq. 32–41. This elegant procedure is widely used in engineering practice and involves the use of the special properties of *complex numbers*, as explained in Appendix B.

As indicated in Fig. B–2, complex numbers may be represented as vectors in the complex plane. The operation of taking the real part of a complex number corresponds to projecting the vector onto the real axis of the *Argand diagram* (Fig. 32–14), and taking the complex part corresponds to projecting the vector onto the imaginary axis.* Consider the complex quantity $e^{i\omega t}$. It corresponds to a complex vector with a constant magnitude of 1 that rotates in the counterclockwise direction with constant angular speed ω. Such a vector is called a *phasor*. Its projection onto the real axis is **Re** $e^{i\omega t} = \cos \omega t$. Thus we note that the real part of the phasor corresponds to a sinusoidal or harmonic time variation. It is therefore apparent that an analytic formulation of the relationship between a uniform circular rotation and its projected linear harmonic variation should be possible using the phasor $e^{i\omega t}$.

The procedure is to treat the quantities $Q(t)$, $I(t)$, and $\mathcal{E}(t)$ as complex quantities, always understanding that the true physical values are obtained by taking the real part. Thus, for example†

$$Q(\text{physical}) \equiv \mathbf{Re}\,\hat{Q}(\text{complex})$$

For the driving voltage in Eq. 32–41 we would write

$$\hat{\mathcal{E}} = \mathcal{E}_0 e^{i\omega t}$$

since $$\mathbf{Re}\,\hat{\mathcal{E}} = \mathbf{Re}(\mathcal{E}_0 e^{i\omega t}) = \mathcal{E}_0 \cos \omega t$$

---

*The projection operators **Re** and **Im** are defined by Eq. B–28 of Appendix B; **Re** corresponds to projection onto the real axis and **Im** corresponds to projection onto the imaginary axis. Thus **Re** $\rho e^{i\theta} = \rho \cos \theta$ and **Im** $\rho e^{i\theta} = \rho \sin \theta$, if ρ and θ are real quantities. If you are not certain that you know the algebra of complex numbers, read Appendix B before continuing to study this section.

†A caret (^) over a symbol indicates a complex representation of the quantity.

is just the originally specified voltage. The relationship between $\mathcal{E}$(physical) and $\hat{\mathcal{E}}$(complex) is shown in Fig. 32–16.

The time differentiation and projection operations may be performed in either order; thus,

$$I = \mathbf{Re}\,\hat{I} = \mathbf{Re}\,\frac{d\hat{Q}}{dt} = \frac{d}{dt}\,\mathbf{Re}\,\hat{Q} = \frac{dQ}{dt}$$

Therefore, we can express Eq. 32–41 as

$$L\frac{d\hat{I}}{dt} + R\hat{I} + \frac{\hat{Q}}{C} = \hat{\mathcal{E}}$$

In practice we almost always measure the current in a circuit, not the charge on a capacitor. Accordingly, we convert Eq. 32–41 into an equation involving only the current by differentiating each term with respect to time. The result is

$$L\frac{d^2\hat{I}}{dt^2} + R\frac{d\hat{I}}{dt} + \frac{\hat{I}}{C} = i\omega\mathcal{E}_0 e^{i\omega t} \qquad \textbf{(32–42)}$$

We expect the response, that is, $\hat{I}(t)$, to have the same sinusoidal time variation as the driving voltage. Therefore, we assume a steady-state solution of the form, $\hat{I}(t) = \hat{I}_0 e^{i\omega t}$. Any possible phase difference between the current and the voltage is incorporated in the complex constant $\hat{I}_0$. Then, we obtain

$$\left(-\omega^2 L\hat{I}_0 + i\omega R\hat{I}_0 + \frac{\hat{I}_0}{C}\right)e^{i\omega t} = i\omega\mathcal{E}_0 e^{i\omega t} \qquad \textbf{(32–43)}$$

Cancelling the phasors $e^{i\omega t}$ and solving for $\hat{I}_0$, we find

$$\hat{I}_0 = \frac{\mathcal{E}_0}{R + i(\omega L - 1/\omega C)} = \frac{\mathcal{E}_0}{R + i(X_L - X_C)} \qquad \textbf{(32–44)}$$

where

$$X_L = \omega L \qquad \textbf{(32–45a)}$$

is the *inductive reactance,* and where

$$X_C = \frac{1}{\omega C} \qquad \textbf{(32–45b)}$$

is the *capacitive reactance.*

The fact that cancelling the factor $e^{i\omega t}$ in Eq. 32–43 leaves the expression for $\hat{I}_0$ (Eq. 32–44) with no time dependence justifies the assumption that

$$\hat{I}(t) = \hat{I}_0 e^{i\omega t}$$

Equation 32–44 can be written in a form that resembles Ohm's law if we introduce a quantity called the *complex impedance* $\hat{Z}$, which in the present case is

$$\hat{Z} = Ze^{i\delta} = R + iX \qquad \textbf{(32–46)}$$

where X is the *total reactance,*

$$X = X_L - X_C = \omega L - \frac{1}{\omega C} \qquad \textbf{(32–47a)}$$

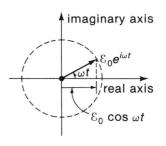

Fig. 32–16. The complex representation of the driving voltage, $\hat{\mathcal{E}} = \mathcal{E}_0 e^{i\omega t}$. The physical voltage, $\mathcal{E}_0 \cos \omega t$, is given by the projection of $\mathcal{E}_0 e^{i\omega t}$ onto the real axis.

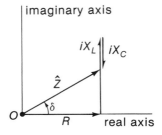

$$\hat{Z} = Ze^{i\delta}$$
$$\hat{Z} = |\hat{Z}| = \sqrt{R^2 + X^2}$$
$$\delta = \tan^{-1}\frac{X_L - X_C}{R}$$
$$X_L = \omega L; \; X_C = \frac{1}{\omega C}$$
$$X = X_L - X_C$$

Fig. 32–17. The complex imped-ance, $\hat{Z} = R + i(X_L - X_C)$, for a series LRC circuit driven at an angular frequency ω.

and where

$$\delta = \tan^{-1}(X/R) \tag{32–47b}$$

The magnitude of $\hat{Z}$ is (see Eq. B–19 in Appendix B)

$$|\hat{Z}| = Z = \sqrt{R^2 + X^2} \tag{32–48}$$

The inductive reactance X_L, the capacitive reactance X_C, and the impedance Z are expressed in the same unit, the ohm, as the resistance R. Note also that for constant values of R, L, and C, the reactances X_L and X_C, as well as the impedance Z, are functions of the external source frequency ω.

Figure 32–17 shows $\hat{Z}$ and its components in an Argand diagram. Using the complex impedance in Eq. 32–44 we have the Ohm's-law form

$$\hat{I}_0 = \frac{\mathcal{E}_0}{\hat{Z}} = \frac{\mathcal{E}_0}{Z}e^{-i\delta} = I_0 e^{-i\delta} \tag{32–49}$$

where

$$I_0 = \frac{\mathcal{E}_0}{Z} = \frac{\mathcal{E}_0}{[R^2 + (X_L - X_C)^2]^{1/2}} \tag{32–50}$$

Then, we also have

$$\hat{I}(t) = \hat{I}_0 e^{i\omega t} = I_0 e^{i(\omega t - \delta)} \tag{32–51}$$

Finally, the physical solution is

$$I(t) = \mathbf{Re}\,\hat{I}(t) = I_0 \cos(\omega t - \delta)$$

or

$$I(t) = \frac{\mathcal{E}_0}{(R^2 + X^2)^{1/2}} \cos(\omega t - \delta) \tag{32–52}$$

In this equation, δ is the phase difference between the driving voltage, $\mathcal{E}_0 \cos \omega t$, and the response current $I(t)$. This phase difference is introduced by the presence of the capacitive and inductive reactances; in a circuit that contains only purely resistive elements, the current is in phase with the driving voltage ($\delta = 0$). The more common situation is that X_C and X_L are both nonzero; δ can still be zero, however, if both reactances have the same magnitude (Eq. 32–47a). This is discussed further in the section on resonance.

Let us rewrite Eq. 32–43 by dividing each term by $i\omega e^{i\omega t}$; the result is

$$i\omega L\hat{I} + R\hat{I} + \frac{\hat{I}}{i\omega C} = \hat{\mathcal{E}}$$

We can use Eqs. B–9 in Appendix B to write $i = e^{i(\pi/2)}$ and $i^{-1} = e^{-i(\pi/2)}$. Then, we have

$$X_L \hat{I}e^{i(\pi/2)} + RI + X_C \hat{I}e^{-i(\pi/2)} = \hat{\mathcal{E}} \tag{32–53}$$

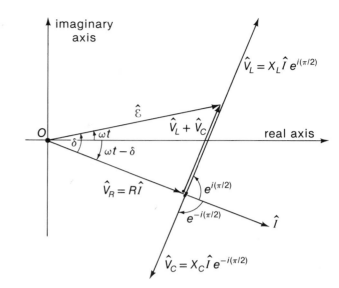

Fig. 32–18. The Argand diagram (complex representation) of Eqs. 32–44 and 32–46 for a series *LRC* circuit driven at an angular frequency ω. Notice that multiplying a complex quantity by $e^{i\theta}$ rotates that quantity counterclockwise through an angle θ in the Argand diagram. Thus, for example, $\hat{V}_L = X_L\hat{I}e^{i(\pi/2)}$ is directed perpendicular to $\hat{I}$, as shown.

Now, define the complex voltage drops to be

$$\hat{V}_L = X_L\hat{I}e^{i(\pi/2)}, \qquad \hat{V}_R = R\hat{I}, \qquad \hat{V}_C = X_C\hat{I}e^{-i(\pi/2)} \qquad \textbf{(32–54)}$$

Hence, Eq. 32–53 becomes

$$\hat{V}_L + \hat{V}_R + \hat{V}_C = \hat{\mathcal{E}} \qquad \textbf{(32–55)}$$

Figure 32–18 shows Eqs. 32–53 and 32–55 represented in an Argand diagram. Here, we see graphically the various phase relationships among the applied EMF and the current and the three voltage drops.

Power.   The instantaneous power dissipation in a circuit is*

$$P(t) = \mathcal{E}(t)I(t) = [\textbf{Re } \hat{\mathcal{E}}(t)][\textbf{Re } \hat{I}(t)] \qquad \textbf{(32–56)}$$

Using $\hat{\mathcal{E}}(t) = \mathcal{E}_0 e^{i\omega t}$ and $\hat{I}(t) = I_0 e^{i(\omega t - \delta)}$, we have

$$P(t) = \mathcal{E}_0 I_0 \cos \omega t \cos (\omega t - \delta)$$
$$= \mathcal{E}_0 I_0(\cos^2 \omega t \cos \delta + \cos \omega t \sin \omega t \sin \delta)$$

With the identity, $2 \cos \theta \sin \theta = \sin 2\theta$, the expression for $P(t)$ can be written as

$$P(t) = (\mathcal{E}_0 I_0 \cos \delta) \cos^2 \omega t + (\tfrac{1}{2}\mathcal{E}_0 I_0 \sin \delta) \sin 2\omega t \qquad \textbf{(32–57)}$$

The first term in this equation is always positive and represents the power dissipation in the resistive part of the circuit. The second term is oscillatory (with frequency 2ω) and represents the surging of energy back and forth between the electromagnetic fields associated with the capacitive and inductive parts of the circuit, and also the energy absorbed from and returned to the source.

The average power dissipation can be obtained by averaging Eq. 32–57 for $P(t)$ over one complete period, $2\pi/\omega$. The average of $\sin 2\omega t$ is zero; the average of $\cos^2 \omega t$ is $\tfrac{1}{2}$.

---

*The right-hand side of this equation is *not* the same as $\textbf{Re}[\hat{\mathcal{E}}(t)\hat{I}(t)]$. (• Can you demonstrate the correctness of this statement?)

Therefore,

$$\overline{P} = \tfrac{1}{2}\mathscr{E}_0 I_0 \cos \delta \tag{32–58}$$

The root-mean-square (rms) value calculated over a complete period for a sinusoidally varying quantity, such as $\xi(t) = \xi_0 \cos \omega t$ or $\xi_0 \sin \omega t$, is

$$\xi_{\text{rms}} = \left[\frac{\omega}{2\pi} \int_0^{\omega/2\pi} \xi_0^2 \cos^2 \omega t \; dt \right]^{1/2} = \frac{\xi_0}{\sqrt{2}} \tag{32–59}$$

In engineering practice it is customary to quote rms values for currents and voltages in AC circuits. (Most AC meters read rms values of current and voltage.) Thus, we can express the power dissipation as

$$\overline{P} = \mathscr{E}_{\text{rms}} I_{\text{rms}} \cos \delta \tag{32–60}$$

where $\mathscr{E}_{\text{rms}} = \dfrac{\mathscr{E}_0}{\sqrt{2}}$ and $I_{\text{rms}} = \dfrac{I_0}{\sqrt{2}}$ \hfill (32–61)

The peak voltage $\mathscr{E}_0$ for most household circuits is maintained at approximately 170 V, so that $\mathscr{E}_{\text{rms}} = (170 \text{ V})/\sqrt{2} = 120$ V. For this reason, we refer to "120-V" power lines.*

The term $\cos \delta$ in the expressions for $\overline{P}$ is referred to as the *power factor* for the circuit. For example, consider the case of a factory that is equipped with large pieces of rotating machinery. Such machines represent very inductive loads; then, with X_L large compared with X_C and R, δ approaches $\pi/2$ and $\cos \delta$ becomes small. Then, the current I_0 that is necessary to operate the machines is much greater than would be the case for a power factor near unity. That is, the required current is large even though the average power remains modest. Because of the large value of I_0, the resistive power loss in the transmission lines is increased. Under such conditions, the factory must correct the poor power factor or pay a surcharge to the power company to compensate for the line losses. One way to correct the power factor in this case would be to connect a low-loss capacitive load across the line.

Resonance. The maximum value of the current amplitude I_0 in a series *LRC* circuit occurs when $X = X_L - X_C = 0$. This is easy to see in the expression for $\hat{I}_0$ (Eq. 32–44) or that for I (Eq. 32–52). This condition is called *resonance,* and it occurs when the external source frequency is equal to the *natural resonance frequency,* designated by ω_0. At resonance, we have $X_L = \omega_0 L = 1/\omega_0 C = X_C$. That is,

$$\omega_0 = \frac{1}{\sqrt{LC}} \tag{32–62}$$

Resonance also yields $\delta = \tan^{-1}(X/R) = 0$ and $\cos \delta = 1$. Thus, the power factor is unity at resonance, so this is also the condition of maximum absorbed power.

Although the resonance frequency is determined only by the values of L and C, the behavior of the circuit at other frequencies depends on the relative values of L, C, and R.

---

*The line voltage is sometimes decreased by as much as 5 percent in peak-load situations to reduce power consumption (thereby producing a *brownout*). If the rms voltage is reduced below about 108 V, induction motors (of the type used in refrigerators and air conditioners) may be seriously damaged.

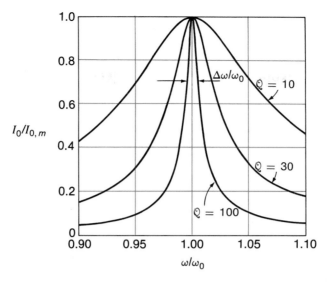

Fig. 32–19. The current amplitude for a series *LRC* circuit in the vicinity of the resonance frequency ω_0 for three different values of the Q of the circuit. The maximum current amplitude is $I_{0,m} = \mathcal{E}_0/R$. The full width at the one-half power points occurs for $I_0 = I_{0,m}/\sqrt{2}$ and is equal to $\Delta\omega/\omega_0 = 1/Q$.

We define a quantity called the *quality factor* (or, simply, the Q) of the circuit according to

$$Q = \frac{\omega_0 L}{R} = \frac{1}{R}\sqrt{\frac{L}{C}} \tag{32-63}$$

Then, we have (see Problem 32–15)

$$I_0(\omega) = \frac{\mathcal{E}_0/R}{\left[1 + \frac{\omega^2}{\omega_0^2}Q^2\left(1 - \frac{\omega_0^2}{\omega^2}\right)^2\right]^{1/2}} \tag{32-64}$$

It is apparent that the denominator of the expression in Eq. 32–64 has its minimum value (namely, unity) when $\omega = \omega_0$. At this frequency, the current amplitude has its maximum value, $I_{0,m} = \mathcal{E}_0/R$. The resonance behavior of $I_0(\omega)$ is illustrated in Fig. 32–19 for three different values of Q. Although the current amplitude achieves its maximum value at $\omega = \omega_0$, the charge, $Q_0(\omega) = I_0(\omega)/\omega$, achieves its maximum value at a slightly lower frequency (see Problem 32–17).

Figure 32–20 shows the behavior of the phase angle δ (see Problem 32–15) in the

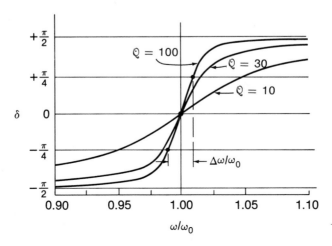

Fig. 32–20. The phase angle δ in the vicinity of the resonance frequency ω_0 for the same three values of the Q of the circuit as used in Fig. 32–19. Notice that the one-half power points correspond to $\delta = \pm\pi/4$.

vicinity of the resonance frequency ω_0. Notice that below resonance ($\omega < \omega_0$) the capacitive reactance dominates, which produces $-\pi/2 < \delta < 0$, whereas above resonance ($\omega > \omega_0$) the inductive reactance dominates and $0 < \delta < \pi/2$.

In Fig. 32–17 we see that $\cos \delta = R/Z$. Therefore, the average power delivered to the resistor is (Eq. 32–58)

$$\overline{P}(\omega) = \tfrac{1}{2}\mathcal{E}_0 I_0 \cos \delta = \tfrac{1}{2}\mathcal{E}_0 \left(\frac{\mathcal{E}_0}{Z}\right)\left(\frac{R}{Z}\right)$$

$$= \frac{\mathcal{E}_0^2 R}{2Z^2} = \frac{\tfrac{1}{2}\mathcal{E}_0^2 R}{R^2 + (\omega L - 1/\omega C)^2} \qquad \textbf{(32–65)}$$

At resonance we have $Z = R$, so that

$$\overline{P}_m = \frac{\mathcal{E}_0^2}{2R} \qquad \textbf{(32–66)}$$

The two frequencies, ω_1 and ω_2, at which the power absorbed from the source is one half the maximum power, $\overline{P} = \tfrac{1}{2}\overline{P}_m$, are called the *one-half power points* and are such that (see Problem 32–16)

$$\omega_2 - \omega_1 = \Delta\omega = \frac{\omega_0}{\mathcal{Q}} \qquad \textbf{(32–67)}$$

Thus, the larger the value of $\mathcal{Q}$, the narrower is the resonance curve; this feature is evident in Fig. 32–19.

Example 32–7

The *LRC* circuit shown in the diagram has $L = 10$ mH, $C = 1$ μF, $R = 3.3$ Ω, and $\mathcal{E}_0 = 1$ V.

(a) Determine the resonance frequency ω_0, the $\mathcal{Q}$ of the circuit, and the frequency interval between the one-half power points.

(b) Calculate the current amplitude at resonance, and determine the average power dissipated per period at resonance.

(c) Calculate the current amplitude and the average power dissipated per period at a frequency that is 5 percent lower than the resonance frequency.

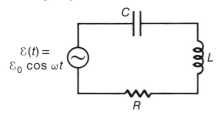

Solution:

(a) The resonance frequency is

$$\omega_0 = \frac{1}{\sqrt{LC}} = \frac{1}{\sqrt{(10^{-2}\text{ H})(10^{-6}\text{ F})}} = 10^4 \text{ rad/s}$$

The $\mathcal{Q}$ of the circuit is

$$\mathcal{Q} = \frac{\omega_0 L}{R} = \frac{(10^4\text{ s}^{-1})(10^{-2}\text{ H})}{3.3\ \Omega} = 30$$

The half-power interval is

$$\Delta\omega = \frac{\omega_0}{\mathcal{Q}} = \frac{10^4\text{ s}^{-1}}{30} = 333 \text{ s}^{-1}$$

This situation is represented by the middle curve in Fig. 32–19.

(b) At resonance, we have

$$I_{0,m} = \frac{\mathcal{E}_0}{R} = \frac{1\text{ V}}{3.3\ \Omega} = 0.30 \text{ A}$$

$$\overline{P}_m = \frac{\mathcal{E}_0^2}{2R} = \frac{(1\text{ V})^2}{2(3.3\ \Omega)} = 0.15 \text{ W}$$

(c) When $\omega = 0.95\ \omega_0 = 9500\text{ s}^{-1}$, we have

$$X = X_L - X_C = \omega L - \frac{1}{\omega C}$$

$$= (9500\text{ s}^{-1})(10^{-2}\text{ H}) - \frac{1}{(9500\text{ s}^{-1})(10^{-6}\text{ F})}$$

$$= -10.3\ \Omega$$

(The fact that X is negative indicates that the load is dominantly capacitive.) Then, the impedance is

$$Z = \sqrt{R^2 + X^2} = \sqrt{(3.3\ \Omega)^2 + (-10.3\ \Omega)^2}$$

$$= 10.8\ \Omega$$

Thus, the current amplitude is

$$I_0 = \frac{\mathcal{E}_0}{Z} = \frac{1\ \text{V}}{10.8\ \Omega} = 0.093\ \text{A}$$

which is approximately one third of the maximum current $I_{0,m}$.

The phase angle is

$$\delta = \tan^{-1}\frac{X}{R} = \tan^{-1}\frac{(-10.3\ \Omega)}{(3.3\ \Omega)}$$

$$= -1.26\ \text{rad} \qquad (-72.2°)$$

Then, the average power is

$$\bar{P} = \tfrac{1}{2}\mathcal{E}_0 I_0 \cos\delta = \tfrac{1}{2}(1\ \text{V})(0.093\ \text{A}) \cos 72.2°$$

$$= 0.014\ \text{W}$$

which is about 10 percent of the maximum power $\bar{P}_m$.

Example 32–8

An electric generator operates at a frequency $\nu = 60$ Hz and develops a peak EMF of $120\sqrt{2}$ V. The generator has an internal resistance $r = 5\ \Omega$ and an effective series inductive reactance $X_L = 15\ \Omega$.

(a) What should be the load impedance for maximum power transfer to the load, which consists of resistive and capacitive elements in series, as shown in the diagram?

(b) Under the conditions determined in (a), what is the average power delivered to the load?

(c) What would be the power transfer if all capacitance were removed from the load?

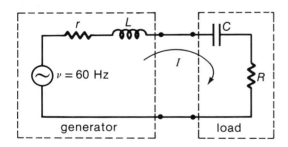

Solution:

(a) For maximum power delivered to the entire circuit, we require a unity power factor, which corresponds to resonance for both the power and the current; this occurs for $X_C = X_L$. We also require $R = r$ for maximum power transfer to the load, as we found earlier for DC circuits (see Example 27–5). Thus,

$$X_C = X_L = 15\ \Omega$$

so that

$$C = \frac{1}{\omega X_C} = \frac{1}{2\pi(60\ \text{Hz})(15\ \Omega)} = 177\ \mu\text{F}$$

Also, $\qquad\qquad R = r = 5\ \Omega$

In this condition, we have *impedance matching* between the generator and the load. The required capacitance (177 μF) is actually quite large. (Calculate the plate area required for a spacing of 1 mm and $\kappa = 3$.) This is a common problem that occurs in power transmission at the relatively low frequency of 60 Hz.

(b) The average power delivered to the entire circuit at resonance is

$$\bar{P}_m = \frac{\mathcal{E}_0^2}{2(r+R)} = \frac{(120\sqrt{2}\ \text{V})^2}{2(10\ \Omega)} = 1.44\ \text{kW}$$

This is the total power dissipated in the resistive circuit elements. Because $R = r$, one half of the power is delivered to the load; that is,

$$\bar{P}_m(\text{load}) = 0.72\ \text{kW}$$

(c) Removing the capacitance of the load but leaving the resistance, we have

$$X = X_L = 15\ \Omega, \qquad R + r = 10\ \Omega$$

Then, using Eq. 32–65, we have

$$\bar{P} = \frac{\mathcal{E}_0^2(R+r)}{2Z^2} = \frac{\mathcal{E}_0^2(R+r)}{2[(R+r)^2 + X_L^2]}$$

$$= \frac{(120\sqrt{2}\ \text{V})^2(10\ \Omega)}{2[(10\ \Omega)^2 + (15\ \Omega)^2]} = 0.44\ \text{kW}$$

and one half of this power is delivered to the load; thus,

$$\bar{P}(\text{load}) = 0.22\ \text{kW}$$

That is, the addition of the 177-μF capacitor to the purely resistive load increases the power delivered to the load by a factor of more than 3.

32–6   TRANSFORMERS

In Chapter 31 we discussed the coupling of circuits by magnetic induction. An important application of magnetically coupled circuits is in the common device called a *transformer*. The purpose of a transformer is to increase or decrease the input (AC line) voltage to the value required by the circuit receiving power. If the output voltage of the transformer is greater than the input voltage, the device is called a *step-up* transformer. Conversely, if the voltage is lowered, the device is a *step-down* transformer.

Figure 32–21 is a schematic diagram of a step-up transformer. Two coils of wire are wound on an iron core. The *primary* coil (which is connected to an AC generator or power line) consists of N_1 turns; the *secondary* coil (which can be connected to the load R) consists of N_2 turns ($N_2 > N_1$).

We begin with the switch S open, so that no current flows in the secondary circuit. Moreover, we assume ideal conditions in which the primary circuit is without resistance or capacitance, so that the reactance is due entirely to the inductance of the primary coil. According to Eq. 32–47b, we have $\delta = \tan^{-1}(X/R)$; with $R = 0$, we see that $\delta = \pi/2$. This means that the voltage $V_1(t)$ across the primary coil and the current $I_1(t)$ in the primary are 90° out of phase. Then, the power factor is $\cos \delta = \cos(\pi/2) = 0$; that is, there is no power expenditure when the secondary is an open circuit.

The (small) magnetizing current that flows in the primary coil induces a substantial magnetic flux φ_M in the iron core. We assume that there is no flux leakage from the core, so that the entire flux φ_M links the secondary coil. According to Faraday's law (Section 30–2), the EMF per turn is the same for both the primary and secondary windings. Thus, the voltages across the primary and secondary coils are related by

$$\frac{V_1(t)}{N_1} = \frac{V_2(t)}{N_2} \qquad \text{or} \qquad \frac{V_{1,\text{rms}}}{N_1} = \frac{V_{2,\text{rms}}}{N_2} \qquad \textbf{(32–68)}$$

Now, suppose that the switch S is closed, thereby connecting the resistive load R to the secondary coil. A current $I_2(t)$ flows in the secondary circuit and power is delivered to the load (see Eq. 32–60):

$$\overline{P} = V_{2,\text{rms}} I_{2,\text{rms}} \cos \delta$$

Because the flux in the iron core is large, the inductance of each coil is also large (see Eq. 31–6). Then, in general, we have $X_L = \omega L_2 \gg R$. This means that the phase angle in the secondary circuit is essentially zero and that the power factor, $\cos \delta$, is nearly unity.

The time-varying secondary current, $I_2(t)$, induces a back EMF that opposes the primary voltage, $V_1(t)$. However, because conditions are ideal, with no resistance or capacitance in the primary circuit, the voltage $V_1(t)$ must always equal the applied EMF, $\mathcal{E}(t)$. Thus, a new current, $I_1'(t)$, must flow in the primary circuit with the proper magnitude and phase to cancel the back EMF generated in the primary coil by the secondary current, $I_2(t)$. The current $I_1(t)$ in the primary circuit now consists of the sum of the magnetizing current (small in magnitude and in phase with V_1) and the induced current, $I_1'(t)$ (larger in magnitude and in phase with V_1). That is, $I_1(t) \cong I_1'(t)$, and the power factor is essentially unity.

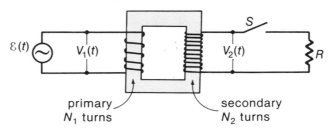

Fig. 32–21. Schematic of an iron-core transformer. The voltage $V_1(t)$ across the primary is just equal to the output voltage $\mathcal{E}(t)$ of the generator.

A detailed mathematical analysis of the relatively complicated behavior discussed above is given in Problem 32–32.

Energy conservation requires the generator to deliver power equal to that dissipated in the load R. Thus, we have, with $\delta = 0$ in both circuits,

$$\overline{P}_1 = V_{1,\text{rms}}I_{1,\text{rms}} = \mathcal{E}_{\text{rms}}I_{1,\text{rms}}$$

$$= V_{2,\text{rms}}I_{2,\text{rms}} = \overline{P}_2 \qquad (32\text{-}69)$$

Using Eq. 32–68, we can write

$$V_{2,\text{rms}} = \frac{N_2}{N_1}\mathcal{E}_{\text{rms}} \qquad (32\text{-}70a)$$

$$I_{2,\text{rms}} = \frac{N_1}{N_2}I_{1,\text{rms}} \qquad (32\text{-}70b)$$

QUESTIONS

32–1 Define and explain briefly the meaning of the terms (a) RC time constant, (b) differentiating circuit, (c) R/L time constant, and (d) natural frequency of oscillation of a simple LC circuit.

32–2 Mention was made of the fact that real inductors and capacitors have some resistance associated with them. Explain why you would expect that real circuits, even simple ones consisting of just hook-up wire, would inevitably also have some inductance and capacitance associated with them.

32–3 Refer to the circuit for charging a capacitor, shown in Fig. 32–2. Draw the graph of $V_C(t)$ and compare it with Fig. 32–3a; draw the graph of $V_R(t)$ and compare it with Fig. 32–3b. What do you conclude about the distribution of potential in the circuit as a function of time?

32–4 Define and explain briefly the meaning of the terms overdamped, critically damped, and underdamped LRC circuits.

32–5 Explain how it is possible in a simple LC circuit to have current through the inductor yet at the same time no charge on the capacitor, and to have charge on the capacitor yet at the same time no current through the inductor.

32–6 Define and explain briefly the meaning of the terms (a) phasor, (b) inductive and capacitive reactance, (c) impedance, (d) power factor, (e) series resonance, and (f) impedance matching.

32–7 What are the limiting values of X_L and X_C for $\omega \to \infty$ and $\omega \to 0$?

32–8 In a single-loop DC circuit of series resistors and a bat-

tery, the voltage drop in any position of the circuit is always less than the battery EMF. Explain how it is possible for a driven series LRC circuit to have voltage drops across some elements that are greater than $\mathcal{E}_0$ of the source voltage.

32–9 Why would a power company be justified in leveling a surcharge on top of the charge for the actual power consumed in a factory if the power factor were low, say $\cos \delta \approx 0.5$?

32–10 Describe in detail the analog relationships between electrical circuit elements—current, charge, and voltage—on the one hand and mechanical system elements—velocity, displacement, and force—on the other hand.

32–11 Describe the motion of the dial pointer of a galvanometer-type ammeter if inserted into the circuit shown in Fig. 32–2. It is observed that the dial pointer requires a short time to reach the deflection corresponding to the value of the current given by Eq. 32–4 at $t = 0$. What mechanical properties would control the initial motion of the dial pointer?

32–12 Discuss some uses of step-up and step-down transformers.

32–13 Can a transformer operate with DC currents? What would most likely happen if a transformer designed for a 100-V AC primary were connected to a 100-V DC power source?

32–14 Why is electric power now almost universally supplied as alternating current (AC) rather than direct current (DC)?

PROBLEMS

Section 32–1

32–1 A 6.0-V battery with negligible internal resistance is used to charge a 0.5-μF capacitor that is in series with a 100-Ω resistor. (a) How much energy is eventually stored in the capacitor? (b) How much energy was expended in heating the resistor? (c) How much energy was supplied by the battery? (d) By what factor(s) would the above results be altered if R were halved to 50 Ω?

32–2 The parallel-plate capacitor shown in Fig. 32–1a has a glass dielectric completely filling the space between the plates. If the resistivity of the glass is $\rho = 10^{13}$ Ω·m and the dielectric constant is $\kappa = 3.40$, for how long will the isolated capacitor continue to hold more than half of its original charge?

32–3 A capacitor C is charged to a value $Q_0 = \mathcal{E}C$ with the polarity indicated in the diagram. Obtain an expression for the charge $Q(t)$ as a function of time t after the switch S is closed. Determine the instant at which $Q(t) = 0$. [*Hint:* Be careful with sign conventions.]

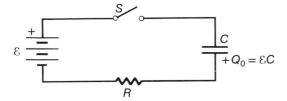

32–4 A capacitor C is charged in the circuit shown in Fig. 32–2. (a) After how many time constants will the energy stored in the capacitor equal 80 percent of the final value? (b) At the instant found in (a), what is the current expressed as a fraction of the initial current?

Section 32–2

32–5 When an inductor L is energized in the circuit shown in Fig. 32–8, the current reaches 90 percent of its asymptotic value 3.0 ms after the switch is closed. If $R = 10$ Ω, what is the inductance L?

32–6 In the energizing circuit of Fig. 32–8, we have $L = 350$ mH, $R = 5$ Ω, and $\mathcal{E} = 6.0$ V. Determine the following quantities 0.1 s after the switch is closed: (a) the rate at which energy is being stored in the magnetic field of the inductor; (b) the instantaneous power delivered to the resistor; and (c) the instantaneous rate at which energy is being expended by the battery. (d) Comment on the energy balance in the system at the instant in question.

32–7 To avoid any possible damage to the current-carrying coil of an electromagnet if the circuit is broken, the usual practice is to place a non-Ohmic device called a *diode* (often in the form of a neon bulb) in parallel with the coil, as indicated in the diagram. If the diode is connected as shown, it draws very little current when the switch S is closed because the back resistance R (in the direction opposite the "arrowhead") is very large. Now, if the circuit is broken (or the switch is opened), the magnet current is diminished slowly as it passes through the forward resistance r of the diode. (• What would happen if the diode were not present?) Suppose that $\mathcal{E} = 12$ V, $R_0 = 2$ Ω, and $L = 5.0$ H. What should the value of r be if the voltage across the magnet is not to exceed 1500 V? At what rate must the diode be able to dissipate heat?

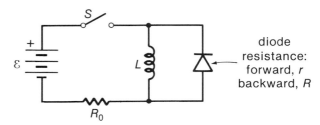

Section 32–3

32–8 The AM radiofrequency broadcast band ranges from 550 kHz to 1550 kHz. Given a 50-μH inductor, what range of capacitance should a variable capacitor span in order to "tune" the LC–natural frequency of a radio receiver to cover the AM broadcast band?

32–9 An inductor having $L = 12$ mH and a capacitor form an LC circuit with a natural frequency of 10^4 Hz. (a) What value of capacitance is required for the capacitor? (b) If the amplitude of the capacitor voltage is 10 V, what is the energy in the circuit?

32–10 An LC circuit consists of an inductor with $L = 25$ mH and a capacitor with $C = 0.1$ μF. At $t = 0$ the charge on the capacitor is 10 nC and the circuit current is zero. (a) What is the natural frequency of oscillation? (b) What is the total energy stored in the circuit? (c) What is the amplitude of the circuit current? (d) Write the equation for $Q(t)$ and $I(t)$.

Section 32–4

32–11 An LRC circuit has the values $L = 5$ mH, $C = 0.5$ μF, and $R = 10$ Ω. At $t = 0$ the capacitor carries a charge Q_0, and the circuit current is $I_0 = 0$. (a) After what elapsed time will the amplitude of the oscillating circuit current be reduced to $\frac{1}{10}$ the initial amplitude? (b) Approximately how many current oscillations have occurred during this time interval?

32–12 An LC circuit with $L = 5$ mH and $C = 0.5$ μF is criti-
• cally damped by inserting a series resistance R. At $t = 0$ the capacitor carries a charge $Q_0 = 2$ μC and the circuit current is $I_0 = 0$. (a) What is the frequency $f_0 = \omega_0/2\pi$? (b) What is the value of R? (c) After what elapsed time is the circuit current at its maximum? (d) What is the maximum value of the circuit current?

32–13 The LC circuit of Problem 32–11 has a resistance $R = 100$ Ω inserted in series with the inductor and capacitor.
• At $t = 0$ the capacitor carries a charge $Q_0 = 2$ μC and the circuit current is $I_0 = 0$. (a) What is the angular frequency ω^0? (b) After what elapsed time is the circuit current at its maximum? (c) What is the maximum value of the circuit current? (d) Write the expression for the circuit current $I(t)$.

Section 32–5

32–14 Consider a driven series LRC circuit. Use the fact that $d\hat{Q}/dt = \hat{I}(t) = I_0 e^{i(\omega t - \delta)}$ to show that

$$\hat{Q}(t) = \frac{\mathcal{E}_0}{i\omega Z} e^{i(\omega t - \delta)} = \frac{\mathcal{E}_0}{\omega Z} e^{i(\omega t - \delta - \pi/2)}$$

Then, show that the physical charge is

$$Q(t) = \frac{\mathcal{E}_0}{\omega Z} \sin(\omega t - \delta)$$

32–15 (a) Show that for an LRC circuit

$$\tan \delta = \frac{X_L - X_C}{R} = \frac{\omega}{\omega_0} \mathcal{Q} \left(1 - \frac{\omega_0^2}{\omega^2}\right)$$

(b) Use the result of part (a) and Eq. 32–50 to verify Eq. 32–64.

32–16 Use Eq. 32–65 to show that the frequencies ω_1 and ω_2
• for which $\overline{P}(\omega) = \frac{1}{2}\overline{P}_m$ are given by

$$\omega_{1,2} = \pm \frac{R}{2L} + \sqrt{\frac{R^2}{4L^2} + \frac{1}{LC}}$$

and, hence, that

$$\Delta\omega = \omega_2 - \omega_1 = \frac{R}{L} = \frac{\omega_0}{\mathcal{Q}}$$

32–17 Express the complex charge $\hat{Q}(t)$ and the physical charge $Q(t)$ as in Problem 32–14. Show that the charge amplitude $Q_0 = \mathcal{E}_0/\omega Z$ is maximum at the frequency $\omega_Q = \sqrt{\omega_0^2 - R^2/2L^2}$. Verify that $\omega_Q \cong \omega_0$ when the $\mathcal{Q}$ of the circuit is large. [*Hint:* Set $\dfrac{d}{d\omega}(Q_0^2)$ equal to zero.]

32–18 A 30-Ω resistor, an inductor with a reactance of 40 Ω at
• 60 Hz, and a capacitor with a reactance of 50 Ω at 60 Hz are connected in series to a 120-V, 60-Hz power line (peak voltage $\mathcal{E}_0 = 120\sqrt{2}$ V). (a) What is the impedance of the circuit? (b) What is the peak current in the circuit? (c) What is the phase difference between the current and the applied voltage? (d) What average power per cycle is absorbed by the circuit?

32–19 A piece of rotating machinery is connected to a 240-V,
•• 60-Hz power line (peak voltage $\mathcal{E}_0 = 240\sqrt{2}$ V). The load impedance of the unit is $(5 + i8)$ Ω. What value of capacitance must be placed in parallel with the machine to produce unity power factor? What amount of power is absorbed by the machine? If the machine delivers shaft power with an efficiency of 82 percent, what is the output mechanical power (in horsepower)? [*Hint:* Unity power factor results when the parallel combination behaves like a pure resistance.]

Section 32–6

32–20 A transformer has $N_1 = 350$ turns and $N_2 = 2000$ turns. If the input voltage is $\mathcal{E}(t) = (170\ \mathrm{V})\cos \omega t$, what rms voltage is developed across the secondary coil?

32–21 A small plant draws 200 kW of power, for a substantially resistive load, from a power substation 5 km away. The resistance of the aluminum wire used as the transmission line connecting the plant with the substation is 0.42 Ω/km. (a) Determine the power loss in the transmission line if the power is delivered at 220 V; if delivered at 13,500 V. (b) Express these results in percent of the power used by the plant. Would you recommend the use of a transformer to provide high-voltage power transmission to match low-voltage power use? [Assume ordinary *single-phase* AC use. In practice, a load this large would be serviced from a *three-phase* circuit, leading to somewhat different numerical results.]

32–22 The generator in the diagram has an internal resistance
• $r = 0.5$ Ω and produces a voltage $\mathcal{E}(t) = (170\ \mathrm{V})\cos \omega t$, with $\omega/2\pi = 60$ Hz. (a) If a load resistor, $R = 20$ Ω, is

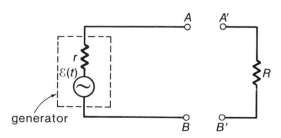

generator

connected to the generator terminals, what average power will be delivered to the load? (b) In Example 32–8 we found that maximum power is delivered to a load when the load impedance (here, the resistance R) is equal to the generator impedance (here, the resistance r).

Additional Problems

32–23   Verify by direct substitution that Eq. 32–3 for $Q(t)$ satisfies the differential equation, Eq. 32–1.

32–24   Use the fact that $e^{i(A+B)} = e^{iA}e^{iB}$ to show that

$$\cos (A + B) = \cos A \cos B - \sin A \sin B$$

and

$$\sin (A + B) = \sin A \cos B + \cos A \sin B$$

32–25   The diagram shows an *integrating circuit* in which the time constant, $\tau = RC$, is large compared with the time interval during which $\mathcal{E}(t)$ exhibits any substantial change. (a) Show that

$$V(t) = \frac{1}{C} \int I \, dt \cong \frac{1}{RC} \int \mathcal{E}(t) \, dt$$

(b) Sketch $V(t)$ for the case in which $\mathcal{E}(t)$ is a square-wave voltage (see Fig. 32–7).

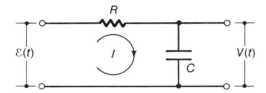

32–26   A resistor draws a peak current of 1.0 A when connected to a 120-V, 60-Hz power line. (The peak AC voltage is $\mathcal{E}_0 = 120\sqrt{2}$ V.) (a) What capacitance C must be placed in series with the resistor to reduce the peak current to 0.5 A? (b) Under the conditions determined in (a), what is the phase difference between the current and the applied voltage?

32–27   A small AC motor has an inductance of 40 mH and an effective series resistance of 25 Ω. The motor is connected to a 120-V, 60-Hz power line (peak voltage $\mathcal{E}_0 = 120\sqrt{2}$ V). (a) What is the inductive reactance X_L of the motor? (b) Determine the peak current I_0 and the phase angle δ_L between the current and the applied voltage.

32–28   The parallel-plate capacitor shown in the diagram has a dielectric with constant κ and a capacitance κC. At $t = 0$, the dielectric is suddenly removed. (a) How much work was required to remove the dielectric? (b) Obtain expressions for the charge $Q(t)$ on the capacitor and for the current $I(t)$ in the circuit. Make sketches of $Q(t)$ and

$I(t)$. (c) How much energy was dissipated as heat in the resistor? (d) How much work was done on the battery? (e) Account for the total energy balance.

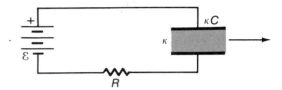

32–29   A peak current of 12 A flows through an inductor coil
•   when it is connected to a 120-V, 60-Hz power line (peak voltage $\mathcal{E}_0 = 120\sqrt{2}$ V). If the coil is placed in series with a 10-Ω resistor and again connected to the power line, the peak current is reduced to 8 A. Calculate the resistance and the inductance of the coil.

32–30   A 30-Ω resistor, a 100-mH inductor, and a 30-μF capac-
••   itor are connected in series to a 120-V, 60-Hz power line (peak voltage $\mathcal{E}_0 = 120\sqrt{2}$ V). Consider an instant when the line voltage is zero. (a) What amount of energy is stored in the capacitor? (b) What amount of energy is stored in the inductor? (c) What is the power delivered to the resistor? [*Hint:* $(\mathbf{Re}\ \hat{Q})^2$ is *not* equal to $\mathbf{Re}\ \hat{Q}^2$; similarly for $\hat{I}$.]

32–31   The impedance of a series circuit is $\hat{Z}_1 = (25\ \Omega)e^{-i\pi/4}$ at
•• $\omega_1 = 1000\ \text{s}^{-1}$ and is $\hat{Z}_2 = (35.4\ \Omega)e^{i\pi/3}$ at $\omega_2 = 2000\ \text{s}^{-1}$. Determine L, R, and C for the circuit. What is the resonance frequency ω_0? What is the $\mathcal{Q}$ of the circuit?

32–32   The diagram shows the electrical equivalent circuit of a
••   transformer being driven by a sinusoidal voltage with frequency ω. Analysis of this *coupled circuit* gives solutions for the primary current $\hat{I}_1$ and the secondary current $\hat{I}_2$ that are

$$\hat{I}_1 = \frac{(R_2 + i\omega L_2)\hat{E}}{R_1 R_2 + \omega^2(M^2 - L_1 L_2) + i\omega(R_1 L_2 + R_2 L_1)}$$

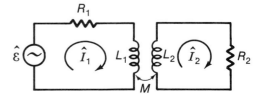

and

$$\hat{I}_2 = \frac{-i\omega M \hat{E}}{R_1 R_2 + \omega^2 (M^2 - L_1 L_2) + i\omega(R_1 L_2 + R_2 L_1)}$$

When there is negligible flux leakage in the transformer core, $M^2 = L_1 L_2$; thus the term in the denominators in ω^2 vanishes. With negligible flux leakage we also have $L_1/L_2 = (N_1/N_2)^2$ and $M/L_2 = N_1/N_2$. (a) Prove that if $\omega L_1 \gg R_1$ and $\omega L_2 \gg R_2$, the currents $\hat{I}_1$ and $\hat{I}_2$ are in phase with $\hat{E}$. (b) Prove Eq. 32–70.

32–33 A driving force $F(t) = F_0 \cos \omega t$ is applied to the me-
• chanical system shown in the diagram. Show that the solution for the displacement $x(t)$ is

$$x(t) = A \sin(\omega t - \delta)$$

with

$$A = F_0/\sqrt{(\kappa - m\omega^2)^2 + b^2\omega^2}$$

and

$$\tan \delta = \frac{1}{b}(m\omega - \kappa/\omega)$$

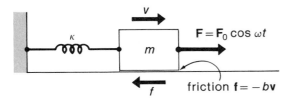

32–34 The horizontal spring-block system of Problem 32–33
•• has $m = 0.50$ kg, $\kappa = 250$ N/m, and $b = 5$ N·s/m. A sinusoidal driving force $F(t) = F_0 \cos \omega t$, with $F_0 =$

7.5 N, is applied to the block. (a) Find the frequency ω_Q for which the amplitude is a maximum (see Problem 32–17). This condition is referred to as amplitude resonance. (b) Calculate the amplitude at amplitude resonance. (c) Find the frequency ω_0 for which the velocity of the block is in resonance. (d) Calculate the amplitude at the frequency ω_0.

32–35 The drawing shows a simple rotational mechanical system that has a rotational inertia $\mathcal{I}$ about the system axis, a frictional resistance coefficient r, and a torque constant Γ ($1/\Gamma$ is the *rotational compliance*). The driving torque $\tau(t)$ produces a time-dependent angular displacement from equilibrium $\theta(t)$. The equation of motion is

$$\mathcal{I}\frac{d^2\theta}{dt^2} + r\frac{d\theta}{dt} + \Gamma\theta = \tau(t)$$

and the angular speed is

$$\eta = \frac{d\theta}{dt}$$

Construct a set of analog equivalences to those of Eqs. 32–27 and 32–40 for this mechanical system.

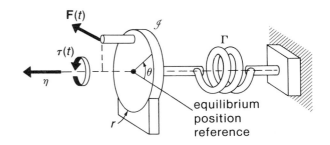

CHAPTER 33

ELECTROMAGNETIC WAVES

In previous chapters the discussions of electric and magnetic fields were limited primarily to fields produced by static charges or steady currents. When field variations were considered, we required the changes to be slow in order to avoid the complications associated with radiation effects.

In this chapter we enlarge the discussion to include situations in which a system produces time-varying electric and magnetic fields that propagate through space. This combined field constitutes *electromagnetic radiation* and carries energy away from the source. We examine first the simple case of a monochromatic (single-frequency) plane electromagnetic wave that travels through a medium in which there are no charges or currents. We then turn our attention to simple radiating systems.

(Before proceeding, you may find it useful to review the main features of mechanical waves in Chapter 16 and of sound in Chapter 17.)

33–1 ELECTROMAGNETIC PLANE WAVES

Basic Considerations. Many of the important characteristics of electromagnetic waves are revealed by a study of the simplest type of propagating electromagnetic disturbance, namely, a *plane wave*. We assume that the observer of the wave is far from the source. Under this condition, the field vectors, $\mathbf{E}(\mathbf{r}, t)$ and $\mathbf{H}(\mathbf{r}, t)$, that describe the electromagnetic field are, at any instant, uniform over a plane that is perpendicular to the line connecting the source and the observer (Fig. 33–1). The medium through which the wave propagates has a uniform dielectric constant κ and a uniform permeability κ_M. The source, the medium, and the observer are all at rest in the same inertial reference frame. We consider a *monochromatic* wave; that is, the field variations are sinusoidal and are described by a single frequency ω.

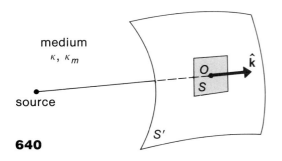

Fig. 33–1. An electromagnetic field disturbance radiates in all directions from a source. The surface S is a small portion of the spherical surface S' whose center is the source. If O is sufficiently far from the source, S can be considered to be planar, and an observer will find that the field vectors, $\mathbf{E}$ and $\mathbf{H}$, are essentially uniform over S. Thus, the electromagnetic wave in the vicinity of the observer appears to be a *plane wave*. The direction of propagation of the wave is parallel to the line connecting the source and O; we take this line to be the z-axis ($\hat{\mathbf{k}}$) of the coordinate frame used to describe the wave.

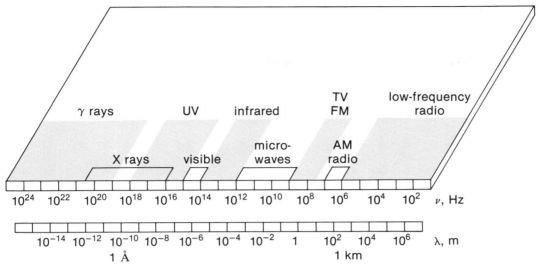

Fig. 33–2. Frequencies and wavelengths for some of the bands in the electromagnetic spectrum. For most of the bands the end-points are not well defined.

Although all electromagnetic waves have the same basic character regardless of frequency, it is convenient to assign different names to the waves in different frequency regimes. Some typical frequencies and wavelengths are indicated in Fig. 33–2. We give special attention to the visible part of the electromagnetic spectrum when the subject of light is discussed in Chapter 34.

To describe the wave we use a coordinate frame in which the z-axis is in the direction of propagation of the plane wave, and we select the x-axis to coincide with the direction of the electric vector **E**. An electromagnetic wave is said to be *polarized* in the direction of its *electric* vector. Thus, the wave we are considering here is polarized in the x-direction. We have

$$\mathbf{E}(\mathbf{r},\, t) = \hat{\mathbf{i}} E_0 \cos(kz - \omega t) \qquad (33\text{–}1)$$

where $k = 2\pi/\lambda$ is the magnitude of the *propagation vector*,* $\mathbf{k} = k\hat{\mathbf{k}}$, whose direction is the direction of propagation of the wave. The quantity k is called the *wave number* and has dimensions $[k] = L^{-1}$ (see Section 16–4). The propagation velocity (or phase velocity) of the wave is the product of the frequency, ν, and the wavelength, λ, or

$$v = \nu\lambda = (\omega/2\pi)\lambda = \omega/k \qquad (33\text{–}2)$$

We assert that a plane electromagnetic wave whose electric field is described by Eq. 33–1 has a magnetic field strength described by†

$$\mathbf{H}(\mathbf{r},\, t) = \hat{\mathbf{j}} H_0 \cos(kz - \omega t) \qquad (33\text{–}3)$$

Thus, **H** is perpendicular to both **E** and **k** and is in phase with **E**. We proceed to show that the expressions given here for **E** and **H** satisfy Maxwell's equations, thereby justifying the

---

*Do not be confused by this expression! It means simply that the vector **k** has magnitude k and is in the z-direction ($\hat{\mathbf{k}}$).

†For the applications we study, it turns out to be more convenient to write the electromagnetic wave relationships in terms of **H** instead of **B**.

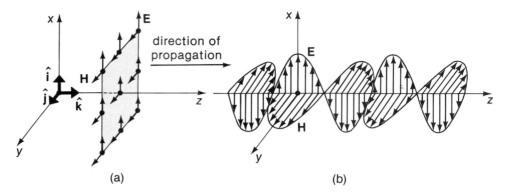

Fig. 33–3. (a) A wavefront of **E** and **H** at a particular instant. Both **E** and **H** are transverse to the direction of propagation ($\hat{k}$) of the wave. (b) The variation of **E** and **H** along the z-axis at a particular instant. The wave propagates in the $+z$-direction.

ad hoc manner in which they have been introduced. In this process we discover that the propagation velocity v for electromagnetic waves in vacuum is the same as the velocity of light c.

Figure 33–3a shows a portion of the wavefront of **E** and **H**. Over the wavefront, **E** is everywhere in the x-direction and **H** is everywhere in the y-direction. Figure 33–3b shows, at a particular instant, the variation of **E** and **H** along any line parallel to the direction of propagation (here, the z-direction). Notice that **E** has a changing orientation in the $\pm x$-direction and that **H** has a similar variation in the $\pm y$-direction.

Faraday's Law. We will verify that Eqs. 33–1 and 33–3 together satisfy each of Maxwell's equations in vacuum (see Section 30–6). Let us begin with Faraday's law. Consider two fixed rectangular plane surfaces with sides a and b, parallel to the x-y plane, and separated by a distance Δz, as shown in Fig. 33–4. As the plane wave that we have been considering propagates along the z-direction, the electric and magnetic field strengths at some instant take on the values indicated in the diagram. Faraday's law must be valid, so we equate the line integral of $\mathbf{E} \cdot d\mathbf{r}$ around the indicated path to the rate of change of magnetic flux through the area S that is bounded by the integration path. Thus,[*]

$$\oint \mathbf{E} \cdot d\mathbf{r} = -\frac{\partial \varphi_M}{\partial t} = -\mu_0 \kappa_M \frac{\partial \varphi_H}{\partial t} \tag{33–4}$$

where $\varphi_H = a \, \Delta z H$ is the flux of **H** through the area S (compare Eq. 30–2), and where we have used $\mathbf{B} = \mu_0 \kappa_M \mathbf{H}$ (Eq. 31–24). Traversing the path in the counterclockwise sense, as shown in Fig. 33–4, yields

$$\left(E + \frac{\Delta E}{2} \right) a + 0 - \left(E - \frac{\Delta E}{2} \right) a + 0 = -\mu_0 \kappa_M \, a \, \Delta z \frac{\partial H}{\partial t}$$

Fig. 33–4. Geometry for applying Faraday's law to the case of a plane electromagnetic wave.

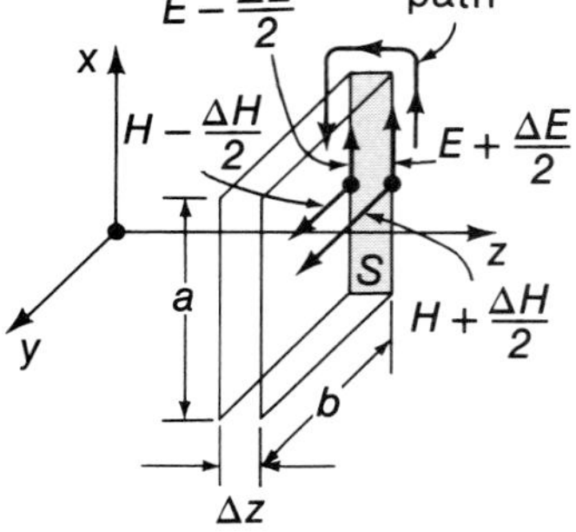

from which

$$\frac{\Delta E}{\Delta z} = -\mu_0 \kappa_M \frac{\partial H}{\partial t}$$

In the limit, $\Delta z \to 0$, we have

$$\frac{\partial E}{\partial z} = -\mu_0 \kappa_M \frac{\partial H}{\partial t} \tag{33–5}$$

This means that Faraday's law requires the *spatial* rate of change of E to be proportional to the negative of the *time* rate of change of H. Substituting into this equation the expressions

---

[*] Because the field and, hence, φ_M and φ_H are functions of both position and time, the time derivative that appears in Eq. 30–2 must be written here as a *partial* derivative. Moreover, the space and time derivates of E and H are also partial derivatives, as in Eq. 33–5.

for $E(\mathbf{r}, t)$ and $H(\mathbf{r}, t)$ (Eqs. 33–1 and 33–3) gives

$$-kE_0 \sin (kz - \omega t) = -\mu_0 \kappa_M \omega H_0 \sin (kz - \omega t) \qquad \textbf{(33–6)}$$

We originally assumed, by writing Eqs. 33–1 and 33–3 with the same factor, $kz - \omega t$, that $\mathbf{E}$ and $\mathbf{H}$ were *in phase*. Suppose that a phase difference δ is introduced by writing $\mathbf{H} = H_0 \hat{\mathbf{j}} \cos (kz - \omega t + \delta)$. Then, we obtain an equation corresponding to Eq. 33–6 that contains $\sin (kz - \omega t + \delta)$ on the right-hand side and $\sin (kz - \omega t)$ on the left-hand side. But we require this equation to be valid for any time t. For this to be true, δ must be zero and the electric and magnetic field strengths are indeed in phase. Figure 33–3b has been drawn in accordance with this requirement.

From Eq. 33–6 we see that

$$\frac{E_0}{H_0} = \frac{\mu_0 \kappa_M \, \omega}{k} = \mu_0 \kappa_M v \qquad \textbf{(33–7)}$$

This condition *must* be met, so that the expressions for $\mathbf{E}(\mathbf{r}, t)$ and $\mathbf{H}(\mathbf{r}, t)$ will be consistent with Faraday's law.

Ampère's Law. Figure 33–5 shows the same situation we have just discussed, except that a different integration path and surface S are now indicated. Using $\mathbf{D} = \epsilon_0 \kappa \mathbf{E}$ (Eq. 26–31), Ampère's law (see Table 31–3) becomes

$$\oint \mathbf{H} \cdot d\mathbf{r} = \epsilon_0 \kappa \frac{\partial \varphi}{\partial t} \qquad \textbf{(33–8)}$$

The integration path shown in Fig. 33–5 is traversed in the clockwise sense, and φ is the electric flux through the area S. Thus,

$$\left(H + \frac{\Delta H}{2}\right)b + 0 - \left(H - \frac{\Delta H}{2}\right)b + 0 = -\epsilon_0 \kappa b \, \Delta z \frac{\partial E}{\partial t}$$

The negative sign on the right-hand side of this equation results from the fact that the normal to the surface S is in the $-x$-direction ($\hat{\mathbf{n}} = -\hat{\mathbf{i}}$). Simplifying, we have

$$\frac{\Delta H}{\Delta z} = -\epsilon_0 \kappa \frac{\partial E}{\partial t}$$

In the limit, $\Delta z \to 0$, there results

$$\frac{\partial H}{\partial z} = -\epsilon_0 \kappa \frac{\partial E}{\partial t} \qquad \textbf{(33–9)}$$

This equation is analogous to Eq. 33–5. (• Express it in words.) Substituting into this equation the expressions for $E(\mathbf{r}, t)$ and $H(\mathbf{r}, t)$ gives

$$-kH_0 \sin (kz - \omega t) = -\epsilon_0 \kappa \omega E_0 \sin (kz - \omega t) \qquad \textbf{(33–10)}$$

(If the phase difference were introduced between E and H, we would again find $\delta = 0$, indicating that the field vectors are in phase.)

From Eq. 33–10 we see that

$$\frac{E_0}{H_0} = \frac{k}{\epsilon_0 \kappa \omega} = \frac{1}{\epsilon_0 \kappa v} \qquad \textbf{(33–11)}$$

This condition must also be met, so that the expressions for $\mathbf{E}(\mathbf{r}, t)$ and $\mathbf{H}(\mathbf{r}, t)$ are consistent with Ampère's law.

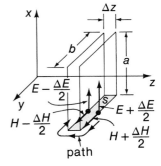

Fig. 33–5. Geometry for applying Ampère's circuital law to the case of a plane electromagnetic wave.

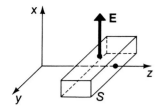

Fig. 33–6. Geometry for applying Gauss's law to the case of a plane electromagnetic wave.

The Gauss Laws. Because there is no free charge in the medium, Gauss's law for the electric field (Eq. 24–12) becomes

$$\oint_S \mathbf{E} \cdot \hat{\mathbf{n}} \, dA = 0 \tag{33-12}$$

Consider the Gaussian surface S shown in Fig. 33–6. The electric field vector $\mathbf{E}$ has only an x-component, so there is flux only through the sides of S that are parallel to the y-z plane. However, $\mathbf{E}$ has the same instantaneous value everywhere on any plane that is parallel to the x-y plane. Consequently, just as much electric flux emerges from the upper surface of S as enters the lower surface. Thus, the net flux that penetrates S is zero, so the expression for $\mathbf{E}(r, t)$ is consistent with Gauss's law for the electric field.

The same argument can be made for the magnetic field. (• Can you see how to do this?) Thus, the expression for $\mathbf{H}(r, t)$ is consistent with Gauss's law for the magnetic field.

The Propagation Velocity. Comparison of Eqs. 33–7 and 33–11 shows that Maxwell's equations require that

$$\frac{E_0}{H_0} = \mu_0 \kappa_M v = \frac{1}{\epsilon_0 \kappa v}$$

Solving for v gives

$$v = \frac{1}{\sqrt{\epsilon_0 \mu_0 \kappa \kappa_M}} \tag{33-13}$$

In vacuum, we have $\kappa = 1$ and $\kappa_M = 1$. Therefore, substituting the values of ϵ_0 and μ_0, we find (compare Eq. 29–11)

$$v = \frac{1}{\sqrt{(8.85419 \times 10^{-12} \text{ F/m})(4\pi \times 10^{-7} \text{ T·m/A})}}$$

$$= 2.99792 \times 10^8 \text{ m/s}$$

That is,

$$v = \frac{1}{\sqrt{\epsilon_0 \mu_0}} = c \quad \text{in vacuum} \tag{33-14}$$

Thus, the propagation velocity of an electromagnetic wave in vacuum is just c, the velocity of light in vacuum, a result we anticipated in the remarks following Eq. 29–11. No restriction has been placed on the value of ω, so the result applies to *all* forms of electromagnetic radiation. Experiments have verified that, in vacuum, radio waves, microwaves, infrared radiation, visible light, and X rays all travel with the velocity c.

From Eqs. 33–11 and 33–13 we have

$$\frac{E_0}{H_0} = \sqrt{\frac{\mu_0 \kappa_M}{\epsilon_0 \kappa}} \tag{33-15}$$

Notice that E_0 has dimensions of V/m, whereas H_0 has dimensions of A/m. Thus, E_0/H_0 has dimensions of V/A or *ohms*. For vacuum conditions, we have $\kappa = 1$ and $\kappa_M = 1$;

then,

$$\frac{E_0}{H_0} = \sqrt{\frac{\mu_0}{\epsilon_0}} \equiv Z_0 \qquad \text{vacuum} \qquad \text{(33–16)}$$

Substituting the values for μ_0 and ϵ_0, we find $E_0/H_0 = Z_0 = 377 \ \Omega$. The quantity Z_0 is called the *impedance of free space* and is an important concept in radio-wave propagation and antenna theory. The impedance Z of a material medium is defined as the ratio E_0/H_0 in Eq. 33–15.

For electromagnetic waves propagating through a material medium characterized by κ and κ_M, the velocity is

$$v = \frac{c}{\sqrt{\kappa \kappa_M}} = \frac{c}{n} \qquad \text{(33–17)}$$

The quantity $n = \sqrt{\kappa \kappa_M}$ is referred to as the *index of refraction* of the medium. In Chapter 34 we see how n plays a prominent role in the description of the propagation of light through transparent media. Both κ and κ_M (and hence n) have some dependence on frequency.

Example 33–1

A plane sinusoidal electromagnetic wave from a distant radio station is incident normally on a dielectric wall that has $\kappa = 1.25$ and $\kappa_M = 1.00$ for the particular frequency of the wave. The amplitude in air of the electric field of the incident wave (the *signal strength*) is $E_{0i} = 15 \ \mu\text{V/m}$. The amplitude of the transmitted electric field E_{0t} (that is, the field that emerges from the other side of the wall) is given by* $E_{0t}/E_{0i} = 2/(n + 1)$.

(a) Determine the amplitude H_{0i} of the associated incident magnetic field.

(b) What is the amplitude E_{0t} of the transmitted wave?

(c) What is the electromagnetic impedance Z_w of the wall?

(d) Determine the amplitude H_{0t} of the transmitted magnetic field.

Solution:

(a) The dielectric properties of air are essentially the same as those of vacuum, so we take the impedance of air to be $Z_0 = 377 \ \Omega$ and use Eq. 33–16:

$$H_{0i} = E_{0i}/Z_0 = (15 \times 10^{-6} \ \text{V/m})/(377 \ \Omega)$$

$$= 3.98 \times 10^{-8} \ \text{A/m}$$

(b) We have $n = \sqrt{\kappa \kappa_M} = \sqrt{\kappa} = \sqrt{1.25} = 1.12$; then,

$$E_{0t} = 2 E_{0i}/(n + 1) = 2(15 \times 10^{-6} \ \text{V/m})/2.12$$

$$= 14.2 \ \mu\text{V/m}$$

(c) The wall impedance is found from Eq. 33–15:

$$Z_w = E_{0t}/H_{0t} = Z_0/\sqrt{\kappa} = (377 \ \Omega)/1.12 = 337 \ \Omega$$

(d) Thus, the magnetic field amplitude is

$$H_{0t} = E_{0t}/Z_w = (14.2 \ \mu\text{V/m})/(337 \ \Omega) = 4.21 \times 10^{-8} \ \text{A/m}$$

---

*This relationship for the ratio E_{0t}/E_{0i} is developed in Chapter 34. For the purposes here, the derivation is not important.

33–2 ENERGY AND MOMENTUM IN ELECTROMAGNETIC WAVES

Energy Density. According to Eq. 31–26, the energy density for combined electric and magnetic fields in a linear medium is

$$u = \tfrac{1}{2}\mathbf{E} \cdot \mathbf{D} + \tfrac{1}{2}\mathbf{B} \cdot \mathbf{H} = \tfrac{1}{2}\epsilon_0 \kappa E^2 + \tfrac{1}{2}\mu_0 \kappa_M H^2$$

Squaring and rearranging Eq. 33–15, we can write

$$\tfrac{1}{2}\mu_0 \kappa_M H^2 = \tfrac{1}{2}\epsilon_0 \kappa E^2$$

We see that the energy density in an electromagnetic wave consists of equal contributions from the electric and magnetic fields. For a plane wave, the combined energy density is itself a propagating wave,*

$$u(z, t) = \epsilon_0 \kappa E^2 = \epsilon_0 \kappa E_0^2 \cos^2 (kz - \omega t) \qquad \textbf{(33–18)}$$

The time average if $u(z, t)$ over one complete period, $T = 2\pi/\omega$, at any point z is

$$\bar{u} = \frac{1}{T} \cdot \epsilon_0 \kappa E_0^2 \int_t^{t+T} \cos^2 (kz - \omega t)\, dt \qquad z \text{ constant}$$

so that

$$\bar{u} = \tfrac{1}{2}\epsilon_0 \kappa E_0^2 \qquad \textbf{(33–19)}$$

This expression gives the average energy density in a monochromatic plane wave, in units of W/m^3.

The Poynting Vector. The fact that the energy density in an electromagnetic wave is itself a propagating wave (Eq. 33–18) means that the field energy of a wave is carried along as the wave propagates through the medium. We define a vector **N** in the direction of propagation, which is the *energy flux* of a wave, the amount of energy that flows past a fixed point per unit time and per unit area (perpendicular to the flow direction). Accordingly, the units of **N** are W/m^2.

Consider the geometry shown in Fig. 33–7. During a time interval Δt a wavefront advances a distance $\Delta z = v\, \Delta t$. Thus, the field energy within the volume $A\, \Delta z$ passes through the fixed area A (shown shaded) during the interval Δt. Therefore, we can write

$$NA = \frac{uA\, \Delta z}{\Delta t}$$

from which we have

$$N = uv = \epsilon_0 \kappa v E_0^2 \cos^2 (kz - \omega t)$$

Using Eq. 33–11, this expression becomes

$$N = E_0 H_0 \cos^2 (kz - \omega t) = EH \qquad \textbf{(33–20)}$$

We can represent the vector character of **N** by writing

$$\mathbf{N} = \mathbf{E} \times \mathbf{H} \qquad \textbf{(33–21)}$$

Fig. 33–7. Geometry for discussing the flow of energy through the fixed area A. In vacuum, $v = c$. (• What is the direction of **N** when **E** is in the $-x$-direction and **H** is in the $-y$-direction?)

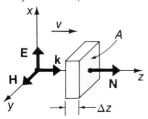

_____

*Recall that any function of the form $f(kz - \omega t)$ represents a propagating wave; here, the function f is the square of the cosine function.

which shows explicitly that the direction of **N** is the direction of propagation of the wave. (Check the direction in Fig. 33–3.) The vector **N** is called the *Poynting vector.** In vacuum we have $v = c$, so that

$$\mathbf{N} = uc\hat{\mathbf{k}} \qquad (33\text{–}21a)$$

Although the Poynting vector was introduced in the context of energy flow in an electromagnetic plane wave, this vector has a more general and far greater significance. Maxwell's equations may be used to obtain an energy conservation equation for a closed region of space with a volume V bounded by a surface S, namely,

$$\oint_S \mathbf{N} \cdot \hat{\mathbf{n}}\, dA + \frac{\partial}{\partial t} \int_V (\tfrac{1}{2}\mathbf{E} \cdot \mathbf{D} + \tfrac{1}{2}\mathbf{B} \cdot \mathbf{H})\, dv + \int_V \mathbf{E} \cdot \mathbf{J}\, dv = 0 \qquad (33\text{–}22)$$

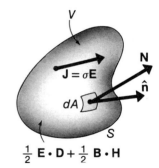

Fig. 33–8. Energy changes within the volume V are described by the integral of the Poynting vector **N** over the bounding surface S.

In this equation, the surface integral of the Poynting vector **N** represents the outward flow of electromagnetic energy through the surface S (Fig. 33–8). The second term in the equation gives the rate of increase of electromagnetic energy within the volume V. (When there is an *outward* flow of energy the second term is *negative*.) The last term represents the Joule heating within V. In this way we can view *all* transfers of electromagnetic energy to take place via fields instead of current-carrying conductors. Wires and conductors are necessary, not for the purpose of transmitting electromagnetic energy, but simply to give to the fields the shapes that are required. Then, the Poynting vector describes the energy transfer across the bounding surface of any volume in which electromagnetic energy is either stored or dissipated. Moreover, as shown in Example 33–4, the energy equation (Eq. 33–22) may be applied in cases that involve steady currents (and, therefore, no electromagnetic waves at all).

▶ *Interactions with Matter.* We have been considering the propagation of waves in material media whose electromagnetic properties are characterized by κ and κ_M. The resulting equations based on these quantities have an obvious simplicity. However, the application of these equations to the interactions of electromagnetic waes with the atoms and molecules in a material medium is actually far from simple. In fact, many of the various types of interactions that can take place—particularly for short-wavelength radiations—require quantum theory for a correct description.

Even for those cases in which the interaction of a wave with

the medium can be characterized adequately by bulk properties that are treated according to the laws of classical physics, the analysis is still complicated. One of the primary reasons is that the bulk electric and magnetic properties of real substances cause the phase velocity of the wave and the absorption of the wave energy to vary with frequency. Only when the wavelength of the radiation becomes relatively long (radio waves) does the index of refraction n approach the value $\sqrt{\kappa \kappa_M}$ for nonconducting media. Then, we find $\kappa_M \cong 1$ and $n \cong \sqrt{\kappa}$, where κ is the DC value (that is, the value for $\nu \to 0$) listed in Table 26–1.   ◀

Electromagnetic Momentum and Radiation Pressure.   It is natural to expect a wave that transports energy to carry momentum as well. Indeed, a treatment of electromagnetic wave phenomena more detailed than we give here demonstrates that there exists a simple relationship connecting the Poynting vector **N** for a propagating wave in vacuum and the linear momentum density **G** of the wave, namely,

$$\mathbf{G} = \frac{1}{c^2}\mathbf{N} \qquad (33\text{–}23)$$

where **G** is momentum per unit volume and is measured in units of J·s/m^4. Because $v = c$ for a wave in vacuum, we have $N = uc$; thus, the momentum density for a wave that

---

*After the English physicist John Henry Poynting (1852–1914).

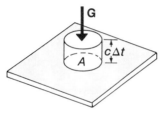

Fig. 33–9. By transferring momentum to a surface, an electromagnetic wave exerts a *radiation pressure*.

propagates in the z-direction can also be expressed as

$$\mathbf{G} = \frac{1}{c} u \hat{\mathbf{k}} \qquad (33\text{–}24)$$

If a wave traveling in a vacuum is incident on the surface of a substance that totally absorbs the wave, a pressure will be exerted on the surface. As indicated in Fig. 33–9, the total wave momentum contained within a volume $Ac\,\Delta t$ will be absorbed by the surface (which is perpendicular to the direction of $\mathbf{G}$) during a time interval Δt. Thus, a force F is exerted by the wave on an area A of the surface, where

$$\text{Impulse} = (\text{Momentum density}) \cdot (\text{Volume})$$

$$F\,\Delta t = \frac{1}{c} u \times Ac\,\Delta t$$

The force per unit area is the *radiation pressure p_{rad}*,

$$p_{\text{rad}} = \frac{F}{A} = u \qquad (33\text{–}25)$$

As usual, pressure is measured in N/m^2 or Pa. The result expressed by Eq. 33–25 is independent of the nature of the absorption process.

The first spacecraft deliberately equipped for solar sailing were the twin U.S. Mariner 3 and 4 (shown) probes, launched in 1964. Vanes affixed to each craft's four solar panels were designed to let light-pressure help with attitude-control. (Courtesy of National Aeronautics and Space Administration.)

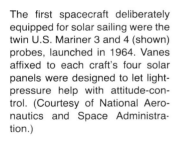

Example 33–2

A pulsed source of microwaves produces bursts of 20-Ghz radiation, with each burst having a duration of 1 ns. A parabolic reflector ($R = 6$ cm) is used to produce a parallel beam of radiation, as indicated in the diagram. The average output power of the source during each pulse is 25 kW.

(a) What is the wavelength of the radiation?

(b) What is the total energy U in each pulse?

(c) Determine the average energy density $\bar{u}$ and the average magnitude of the Poynting vector, $\bar{N}$, during each pulse.

(d) Determine the average momentum density $\bar{G}$ and the average radiation pressure $\bar{p}$ on a totally absorptive surface that is perpendicular to the direction of propagation of the pulse.

(e) Calculate the values of E_0 and H_0 in the wave.

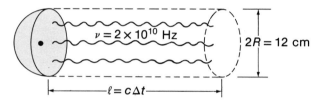

$$\nu = 2 \times 10^{10} \text{ Hz}$$

$$2R = 12 \text{ cm}$$

$$\ell = c\Delta t$$

Solution:

(a) The wavelength of the radiation is

$$\lambda = \frac{c}{\nu} = \frac{3.00 \times 10^8 \text{ m/s}}{2 \times 10^{10} \text{ s}^{-1}} = 0.015 \text{ m}$$

which is in the microwave part of the electromagnetic spectrum (see Fig. 33–2).

(b) The total energy in the pulse is

$$U = \bar{P}\,\Delta t = (2.5 \times 10^4 \text{ J/s})(10^{-9} \text{ s})$$

$$= 2.5 \times 10^{-5} \text{ J}$$

(c) The length of the pulse is

$$\ell = c\,\Delta t = (3.00 \times 10^8 \text{ m/s})(10^{-9} \text{ s})$$

$$= 0.30 \text{ m}$$

so that the average energy density is

$$\bar{u} = \frac{U}{\pi R^2 \ell} = \frac{2.5 \times 10^{-5} \text{ J}}{\pi (0.06 \text{ m})^2 (0.30 \text{ m})}$$

$$= 7.37 \times 10^{-3} \text{ J/m}^3$$

Then, we have

$$\bar{N} = \bar{u}c = (7.37 \times 10^{-3} \text{ J/m}^3)(3.00 \times 10^8 \text{ m/s})$$

$$= 2.21 \times 10^6 \text{ W/m}^2$$

(d) The average momentum density is

$$\bar{G} = \frac{1}{c}\bar{u} = \frac{7.37 \times 10^{-3} \text{ J/m}}{3.00 \times 10^8 \text{ m/s}}$$

$$= 2.46 \times 10^{-11} \text{ J·s/m}^4$$

and the average radiation pressure is

$$\bar{p}_{\text{rad}} = \bar{u} = 7.37 \times 10^{-3} \text{ Pa}$$

(e) The value of E_0 is obtained from Eq. 33–19 with $\kappa = 1$:

$$E_0 = \sqrt{\frac{2\bar{u}}{\epsilon_0}} = \sqrt{\frac{2(7.37 \times 10^{-3} \text{ J/m}^3)}{8.85 \times 10^{-12} \text{ C/V·m}}}$$

$$= 4.08 \times 10^4 \text{ V/m}$$

Then, using Eq. 33–16, we have

$$H_0 = \frac{E_0}{Z_0} = \frac{4.08 \times 10^4 \text{ V/m}}{377 \ \Omega}$$

$$= 1.08 \times 10^2 \text{ A/m}$$

Example 33–3

A dish antenna with a diameter of 20 m is used to receive (at normal incidence) a radio signal from a distant source, as shown in the diagram. The signal is a continuous sinusoidal wave with amplitude $E_0 = 0.2 \ \mu\text{V/m}$. (This is a relatively low signal strength; most radio receivers require signal strengths of a few $\mu\text{V/m}$ for acceptable signal-to-noise ratios.) Assume that the antenna absorbs all of the radiation that is incident on its geometric cross section.

(a) Determine the value of H_0 for the incident wave.

(b) What is the flux of incident energy (that is, what is the average magnitude of the Poynting vector)?

(c) At what average power level is energy delivered to the antenna?

(d) What is the force exerted on the antenna by the incident radiation?

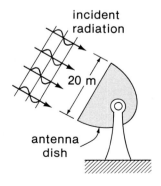

incident radiation

20 m

antenna dish

Solution:

(a) Both κ and κ_M are essentially unity for the medium (air) through which the wave propagates. Then, using Eq. 33–16, we have

$$H_0 = \frac{E_0}{Z_0} = \frac{2 \times 10^{-7} \text{ V/m}}{377 \ \Omega} = 5.31 \times 10^{-10} \text{ A/m}$$

(b) The average magnitude of the Poynting vector can be obtained from $N = EH$ (Eq. 33–20); thus,

$$\bar{N} = \left(\frac{E_0}{\sqrt{2}}\right)\left(\frac{H_0}{\sqrt{2}}\right) = \tfrac{1}{2}E_0H_0$$

$$= \tfrac{1}{2}(2 \times 10^{-7} \text{ V/m})(5.31 \times 10^{-10} \text{ A/m})$$

$$= 5.31 \times 10^{-17} \text{ W/m}^2$$

(c) The average power delivered by the wave to the antenna is

$$\bar{P} = \pi R^2 \, \bar{N} = \pi (10 \text{ m})^2 (5.31 \times 10^{-17} \text{ W/m}^2)$$

$$= 1.67 \times 10^{-14} \text{ W}$$

▶ A major source of noise in receiver circuits is the effect of thermal agitation in resistors. The average noise power generated in this way may be shown to be $\bar{P}_N = 4kT \, \Delta\nu$, where k is Boltzmann's constant, T is the absolute temperature, and $\Delta\nu$ is the *band width* of the receiver, that is, the frequency interval over which the receiver circuit is tuned to respond. For a wide-band receiver operating at ultra-high frequencies (UHF), we might have $\Delta\nu = 10^6$ Hz. Then, with $T = 300$ K, we find $\bar{P}_N = 4(1.37 \times 10^{-23} \text{ J/K})(300 \text{ K})(10^6 \text{ s}^{-1}) = 1.64 \times 10^{-14}$ W. This noise power is about equal to the power delivered by the wave to the antenna in this example. Then the signal-to-noise ratio is $\bar{P}/\bar{P}_N \cong 1$, and special techniques are necessary to extract meaningful signals from the background noise generated within the receiving equipment itself. ◀

(d) the average radiation pressure exerted on the antenna by the wave is (using Eqs. 33–21a and 33–25) $\bar{p}_{\text{rad}} = \bar{u} = \bar{N}/c$. Then, the average force due to the incident radiation is

$$\bar{F} = \pi R^2 \times \bar{p}_{\text{rad}} = \frac{\pi R^2 \, \bar{N}}{c}$$

$$= \frac{\pi (10 \text{ m})^2 (5.31 \times 10^{-17} \text{ W/m}^2)}{3.00 \times 10^8 \text{ m/s}}$$

$$= 5.56 \times 10^{-23} \text{ N}$$

which is a very small force indeed! ($\bar{F}$ corresponds to the weight of about 3400 hydrogen atoms.)

Example 33–4

Consider a long straight wire with a radius a that carries a uniform steady current I. Show that the integral of the Poynting vector over the surface of the wire gives a rate of input of field energy through the surface just equal to the Joule heating within the wire.

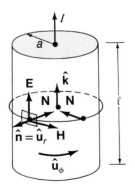

Solution:

Using Ampère's law, we find the magnetic field strength at the surface of the wire to be

$$\mathbf{H} = \frac{I}{2\pi a}\,\hat{\mathbf{u}}_\phi$$

as shown in the diagram. Whatever the resistance characteristics of the wire, homogeneity alone insures that the electric field, $\mathbf{E} = E\hat{\mathbf{k}}$, is constant everywhere inside and on the surface of the wire. Therefore, the Poynting vector on the surface of the wire is

$$\mathbf{N} = \mathbf{E} \times \mathbf{H} = -\frac{EI}{2\pi a}\,\hat{\mathbf{u}}_r$$

where we have used $\hat{\mathbf{k}} \times \hat{\mathbf{u}}_\phi = -\hat{\mathbf{u}}_r$. We integrate $\mathbf{N} \cdot \hat{\mathbf{u}}_r \, dA$ over the surface of a section of the wire with length ℓ, obtaining

$$\oint_S \mathbf{N} \cdot \hat{\mathbf{u}}_r \, dA = -(2\pi a \ell)\left(\frac{EI}{2\pi a}\right) = -EI\ell$$

The negative sign means that the energy flow is *inward*. The decrease in electric potential over the length ℓ of wire is $V = E\ell$. Thus, we see that $EI\ell = VI$ is just the Joule heating in the wire. If the material of the wire obeys Ohm's law, we have $E = J/\sigma = I/\pi a^2 \sigma$, where σ is the conductivity; then,

$$EI\ell = \left(\frac{\ell}{\pi a^2 \sigma}\right)I^2 = I^2 R$$

where R is the resistance of the length ℓ of the wire.

Notice that the result obtained here agrees with the energy-conservation equation (Eq. 33–22) for the steady-state conditions of this problem.

33-3 RADIATION FROM AN ACCELERATING CHARGE (Optional)

Electromagnetic fields and electromagnetic waves originate from distributions of electric charge. In fact, we now demonstrate that electromagnetic radiation is produced by charges that undergo *acceleration*. By restricting attention to radiations with relatively long wavelengths (for example, radio waves), a satisfactory description can be made in terms of a classical model that involves alternating currents in conductors, called *antennas*. The electrons in these antennas are accelerated back and forth, thereby generating electromagnetic waves.

Before considering the effect of accelerating a charge, let us inquire about the nature of the electric field associated with a charged particle moving at a constant velocity **v** in some inertial frame. In a second inertial frame moving with the charge, the charge is at rest and its field is the uniform radial electrostatic field we have studied since Chapter 24. If the relative velocity between the two frames (i.e., the velocity of the particle) is well below the speed of light, a simple Galilean transformation requires the electric field lines of the moving charge also to be a uniform radial field centered on the charge. (• Would you also expect to associate a magnetic field with the moving charge?)

Figure 33–10a shows the electric field lines associated with a small positive charge q moving at a constant velocity **v** in some inertial frame. The field energy and momentum also move along with the charge. If the velocity of the charged particle is suddenly altered, the fields surrounding the charge must also respond. However, the changes in the field propagate at the velocity of light; thus, field strengths and energy and momentum densities change in an outward-spreading wave traveling at the speed of light, c. Figure 33–10b shows the behavior of the electric field of a positively charged particle originally at rest, receiving an acceleration of short duration τ, and thereafter proceeding at the acquired velocity **v**. The reference frame is the one in which the charge was originally at rest. The figure shows a time t after acceleration when the field disturbance has reached a spreading wavefront at a distance $r = ct$ from the moving charge. The field lines at distances greater than r diverge from the original stationary position of the charge, while those at distances less than r are centered on the moving charge. Consequently, a "kink" must appear in the field lines (which are continuous). We shall discover in the following treatment that this kink represents the radiating field.

Figure 33–11 shows a particular electric field line, with segments corresponding to $t < 0$, $0 < t < \tau$, and $t > \tau$. We consider the case in which $\tau \ll t$, so that $\overline{PP'} \ll \overline{PQ} \cong vt$, where $v = a\tau$. From the diagram we see that

$$\overline{AD} = \overline{QG} = \overline{PQ} \sin \theta = a\tau t \sin \theta$$

The kink **E′** propagates outward with the velocity c, so we also have

$$\overline{DB} = c\tau$$

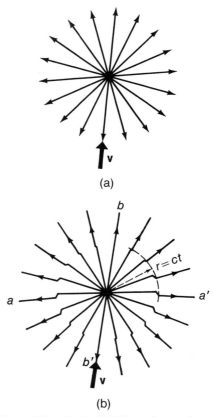

(a)

(b)

Fig. 33–10. (a) The electric field lines of a small positive charge moving at a constant velocity **v**. (b) The electric field lines for a small positive charge at a time t after having received an acceleration of short duration at time $t = 0$.

Fig. 33–11. Geometry of an electric field line for a charge that undergoes a uniform acceleration **a** for a time interval $\tau \ll t$.

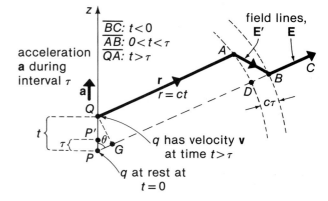

The field line $\mathbf{E}'$ has a tangential component E_t' and a radial component E_r'. The ratio of these components is $\overline{AD}/\overline{DB}$ (see Fig. 33–11); thus,

$$\frac{E_t'}{E_r'} = \frac{\overline{AD}}{\overline{DB}} = \frac{at}{c} \sin \theta$$

The radial component of the field is

$$E_r' = \frac{q}{4\pi\epsilon_0 r^2}$$

It follows that

$$E_t' = \frac{qat \sin \theta}{4\pi\epsilon_0 c r^2}$$

Using $r = ct$, we obtain the desired expression for the tangential component of the field, namely,

$$E_t' = \frac{qa \sin \theta}{4\pi\epsilon_0 c^2 r} \qquad (33\text{–}26)$$

The outward-propagating disturbance that results from the

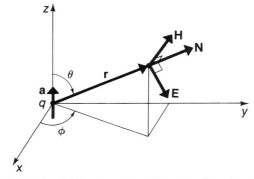

Fig. 33–12. The field vectors, **E** and **H**, and the Poynting vector **N** for the outward propagating disturbance due to an accelerating charge.

acceleration is just the tangential component E_t' of the kink. Notice that $E_t' \propto 1/r$, *not* $1/r^2$.* In Fig. 33–12 the situation in Fig. 33–11 is redrawn to show only E_t' (now designated simply **E**) and the accompanying vectors, **H** and **N**. The electric vector **E** is in the direction $\hat{\mathbf{u}}_\theta$, and **N** is directed outward ($\hat{\mathbf{u}}_r$). Thus, **H** is in the direction $\hat{\mathbf{u}}_\phi$, as indicated in Fig. 33–12. The magnitude of **H** is obtained from $|\mathbf{H}|/|\mathbf{E}| = \sqrt{\epsilon_0/\mu_0}$ (see Eq. 33–16); thus,

$$\mathbf{E} = \frac{qa}{4\pi\epsilon_0 c^2} \frac{\sin \theta}{r} \hat{\mathbf{u}}_\theta \qquad (33\text{–}27a)$$

$$\mathbf{H} = \frac{qa}{4\pi c} \frac{\sin \theta}{r} \hat{\mathbf{u}}_\phi \qquad (33\text{–}27b)$$

$$\mathbf{N} = \mathbf{E} \times \mathbf{H} = \frac{q^2 a^2}{16\pi^2\epsilon_0 c^3} \frac{\sin^2 \theta}{r^2} \hat{\mathbf{u}}_r \qquad (33\text{–}27c)$$

The method by which we have obtained these expressions is not rigorous; however, a precise calculation based directly on Maxwell's equations yields exactly the same results.

A polar diagram of $|\mathbf{N}|$ as a function of θ is called the *radiation pattern* for the particular system. Figure 33–13 shows the pattern for a uniformly accelerated charge, $|\mathbf{N}| \propto \sin^2 \theta$.

Notice that $\mathbf{N} \propto \hat{\mathbf{u}}_r/r^2$. This dependence insures that the total energy flux through a spherical surface of radius r is the same for all distances r. (• Can you verify this intuitively on the basis of energy conservation and analytically by integration over θ?)

Fig. 33–13. Radiation pattern for a charge undergoing uniform acceleration: $|\mathbf{N}| \propto \sin^2 \theta$. Notice that no radiation is emitted along the line of acceleration.

Example 33–5

A charge q undergoes a uniform acceleration $\mathbf{a}$. What is the total power radiated?

Solution:

The polar axis of a spherical coordinate frame is chosen to coincide with the direction of the acceleration vector $\mathbf{a}$, as shown in the diagram on the opposite page. The Poynting vector **N** is given by Eq. 33–27c. Thus, we have

$$P = \oint_S \mathbf{N} \cdot \hat{\mathbf{u}}_r \, dA = \frac{q^2 a^2}{16\pi^2\epsilon_0 c^3} \int_0^{2\pi} d\phi \int_0^{\pi} \sin^2 \theta \cdot \sin \theta \, d\theta$$

$$= \frac{q^2 a^2}{8\pi\epsilon_0 c^3} \int_0^{\pi} \sin^3 \theta \, d\theta$$

---

*Near an accelerating charge, a portion of the field disturbance does have a dependence on $1/r^2$. At large distances, as we consider here, this component is negligible compared with $E_t' \propto 1/r$.

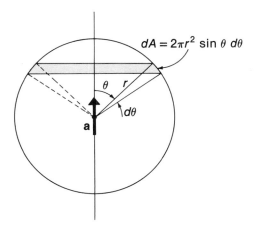

$$dA = 2\pi r^2 \sin \theta \, d\theta$$

The integral is 4/3 (see the table inside the rear cover). Thus,

$$P = \frac{q^2 a^2}{6\pi\epsilon_0 c^3}$$

This is the *Larmor formula,* first derived by Sir J. J. Larmor in 1897.

Much of the preceding discussion of radiation from accelerated charges may be applied to the study of radiation from antennas. The particular case of an antenna with small dimensions compared with the wavelength of the radiation is treated in Enrichment Topic C at the back of the book.

QUESTIONS

33–1   Define and explain briefly the meaning of the terms (a) electromagnetic plane wave, (b) radiation energy density, (c) the Poynting vector, and (d) radiation pressure.

33–2   Give some typical values for the wavelengths of X rays, visible light, infrared radiation, ultraviolet radiation, and AM radio waves.

33–3   The electric field of a particular electromagnetic wave in vacuum is

$$\mathbf{E}(\mathbf{r}, t) = \hat{\jmath} E_0 \sin (kx - \omega t)$$

What is the direction of propagation of the wave? What is the polarization of the wave? Write the expression for the associated magnetic field strength vector $\mathbf{H}(\mathbf{r}, t)$.

33–4   A low-frequency electromagnetic wave enters a nonmagnetic material with a dielectric constant $\kappa = 4$. What is the index of refraction of the material? What is the wave velocity in the material? Compare the frequency and wavelength of the wave with the frequency and wavelength it would have in a vacuum.

33–5   Describe how your everyday world would appear if your eyes were sensitive to infrared radiation rather than to visible light.

33–6   An electromagnetic wave in air and a sound wave in air each have a wavelength of 10 cm. Compare the physical properties of these two waves, such as frequency, momentum, and energy density. In what different ways would you react physiologically to these waves? Why are these two waves of the same wavelength so different in their properties?

33–7   The Sun is continuously radiating huge amounts of electromagnetic energy. Since such radiation carries momentum, why doesn't the momentum of the Sun change with time?

33–8   It was stated in the text that "Wires and conductors are necessary, not for the purpose of transmitting electromagnetic energy, but simply to give to the fields the shapes that are required." Explain the meaning of this statement in terms of the Poynting vector and Eq. 33–22.

33–9   Under what circumstances will a charged particle radiate electromagnetic energy? When radiation occurs, what happens to the energy and momentum of the particle?

PROBLEMS

Section 33–1

33–1 A monochromatic plane electromagnetic wave in vacuum has a propagation vector $\mathbf{k} = k\hat{\mathbf{k}}$. At time t, the electric vector is $\mathbf{E}(\mathbf{r}, t) = (200 \text{ V/m})\hat{\mathbf{i}} \cos (kz - \omega t)$. Determine the magnetic vector $\mathbf{H}(\mathbf{r}, t)$.

33–2 The magnetic vector for a monochromatic plane electromagnetic wave in vacuum has only an x-component, $H_x = H_0 \cos (ky - \omega t)$, with $H_0 = 1.6 \times 10^{-6}$ T. What is the direction of the propagation vector $\mathbf{k}$? Determine the electric vector $\mathbf{E}(\mathbf{r}, t)$.

33–3 What is the propagation velocity of an electromagnetic wave in a medium for which $\kappa = 3.1$ and $\kappa_M = 200$?

33–4 • Start with Eqs. 33–5 and 33–9 and show that E and H satisfy the wave equations

$$\frac{\partial^2 E}{\partial z^2} = \epsilon_0\mu_0\kappa\kappa_M \frac{\partial^2 E}{\partial t^2}$$

$$\frac{\partial^2 H}{\partial z^2} = \epsilon_0\mu_0\kappa\kappa_M \frac{\partial^2 H}{\partial t^2}$$

[*Hint:* Consider differentiating Eqs. 33–5 and 33–9 with respect to z and t.]

Section 33–2

33–5 A monochromatic plane electromagnetic wave in vacuum has $E_0 = 15.0$ V/m. What is the energy flux of the wave?

33–6 Monochromatic electromagnetic radiation is emitted isotropically at a rate of 40 W from a source whose size is small compared with the wavelength of the radiation ($\lambda = 1.0$ mm). At a distance $r = 2.0$ m, the waves in a small area are essentially plane waves. Evaluate the wave amplitudes, E_0 and H_0, at this distance.

33–7 A radio wave propagates through a dielectric medium with dielectric constant κ and $\kappa_M = 1$. The field vectors, $\mathbf{E}$ and $\mathbf{H}$, for the plane wave are given by Eqs. 33–1 and 33–3, in which $E_0 = 1$ V/m, $H_0 = 3.2 \times 10^{-3}$ A/m, and $\nu = \omega/2\pi = 10^6$ Hz. (a) What is the dielectric constant of the medium? (b) What is the phase velocity of the wave? (c) What is the wavelength of the wave? (d) What is the average energy density in the wave? (e) What is the average energy transport (in W/m²)?

33–8 •• The circular parallel-plate capacitor shown in the diagram is slowly charged by a current $I(t)$. (a) Use Ampère's circuital law and the idea of displacement current to show that $\mathbf{E} = E\hat{\mathbf{k}}$ and $\mathbf{H} = U\hat{\mathbf{u}}_\phi$ at some point $r = a$ are related by $H(a) = \frac{1}{2}\epsilon_0 a \, dE/dt$. (Assume vacuum conditions. Assume also that $I(t)$ varies sufficiently slowly that $\mathbf{E}$ is essentially uniform in the region between the plates.) (b) Evaluate the surface integral of the Poynting vector $\mathbf{N}$ over the outer cylindric surface of the capacitor gap at $r = a$. Show that there is an *inward* flux,

$$-\oint_S \mathbf{N} \cdot d\mathbf{A} = \frac{d}{dt} \int_V \tfrac{1}{2}\epsilon_0 E^2 \, dv$$

Interpret this result. (c) If the capacitor is charged by a battery through a series resistor R, giving a time constant $\tau = RC$, the field E, instead of being constant, will have a radial variation. In such a case, it can be shown that $E = E(r) \cong E_0 \cos (r/\sqrt{2}c\tau)$, for $r \ll c\tau$. Assume $R = 1$ MΩ, $a = 5$ cm, and $b = 0.5$ cm, and estimate the ratio $[E(a) - E_0]/E_0$. Are radiation effects negligible? Is the situation changed if $R = 10 \, \Omega$?

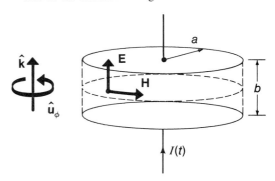

Section 33–3

33–9 A small ball with a mass of 100 g carrying a charge of 50 nC is dropped from a height of 10 m. (a) Find the total radiated power. (b) Find the energy lost during the 10-m fall. (c) Compare the energy radiated to the kinetic energy at the end of the fall. (• Are radiation effects under such simple circumstances important?)

33–10 An electron is released from rest in an electric field of strength 10^3 V/m. Calculate the rate at which the electron radiates energy as it accelerates.

Additional Problems

33–11 A sinusoidal plane wave with wavelength $\lambda = 10$m travels in vacuum parallel to the y-axis. The wave is polarized in the x-direction, and the average energy transport is 0.20 W/m². (a) Determine E_0 and H_0. (b) Write the equations for **E** and **H**.

33–12 A sinusoidal plane-polarized radio wave has a frequency $\nu = 500$ kHz and transports energy through air at a rate of 1 W/m². (a) Determine E_0 and H_0. (b) What is the energy density u in the wave? (c) What amount of field energy is contained within a cube with sides equal to the wavelength λ of the wave?

33–13 A 150-μs burst of 1-MW laser radiation is directed horizontally at a 10-mg bob that hangs at the end of a thin fiber, whose length is 50 cm. Assume that the radiation is completely absorbed by the bob (and that there is negligible vaporization of the bob!). What is the maximum angle of deflection of the fiber?

33–14 Extend the discussion in Example 33–4 to the interior of
  • the conductor by considering what happens on a cylindric surface with $r < a$.

33–15 A long solenoid consists of many tightly wound turns of
  •• wire. A slowly increasing current $I(t)$ flows through the wire, as indicated in the diagram. Assume vacuum conditions. Assume also that the variation of $I(t)$ is sufficiently slow that **H** may be considered to be uniform inside the solenoid. (a) Use Faraday's law of induction to show that $\mathbf{E} = -E\hat{\mathbf{u}}_\phi$ and $\mathbf{H} = H\hat{\mathbf{k}}$ at some circumferential point, $r = a$, are related by $E(a) = \frac{1}{2}\mu_0 a \, dH/dt$. (b) Evaluate the integral of $\mathbf{N} \cdot \hat{\mathbf{n}} \, dA$ over the outer cylindric surface of the solenoid at $r = a$ and show that the inward flux is

$$-\oint_S \mathbf{N} \cdot \hat{\mathbf{n}} \, dA = \frac{d}{dt} \int_V \tfrac{1}{2}\mu_0 H^2 \, dv$$

Interpret this result.

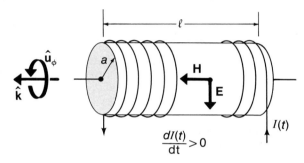

CHAPTER 34

LIGHT

In this and the next three chapters we single out for study that segment of the electromagnetic spectrum to which the human eye is sensitive. This portion of the spectrum is termed *visible light*. We begin by studying the nonquantum properties of light, emphasizing its wavelike character. Then, we proceed to the topic of *physical optics,* involving interference and diffraction wave effects. Finally, we describe *geometric optics,* in which the behavior of light is governed by the laws of reflection and refraction of light rays.

34–1 BRIEF HISTORY

Light plays such an important role in human existence that optical phenomena have been studied since the earliest times. The ancient Greeks knew that when a light ray is reflected from a mirror, the angle of incidence is equal to the angle of reflection. They were also aware of the bending (or refraction) of light as it passes from one transparent medium to another. The refraction of light was studied extensively in the eleventh century by the Arabian mathematician Alhazen. In about 1280 spectacle lenses were invented and were adapted to correct faulty vision. In the seventeenth century, Galileo (1564–1642) devised a practical telescope and van Leeuwenhoek (1632–1723) developed the first microscope. These inventions paved the way for the detailed investigation of large-scale and small-scale aspects of the Universe.

A casual observation reveals that an object produces a sharp shadow when placed in the light from a pointlike or nearly pointlike source, such as the Sun. That is, light does not appear to bend around (or be *diffracted* by) such an object, in contrast to the obvious diffraction of water waves when an obstacle is encountered. Thus, light seems to exhibit rectilinear propagation, and this apparent behavior led to the early notion that light consists of corpuscular matter. Although Newton did not commit himself to either a wave or a corpuscular description of light, he developed a theory of refraction that successfully accounted for the bending of light rays at the interface between two media in terms of an interface force on light corpuscles. However, this theory predicted that the velocity of the corpuscles would be greater in a medium such as glass than in air; later measurements showed that this prediction is contrary to the actual case (see Problem 34–25).

A competing theory that treated light as a wave phenomenon was devised in 1678 (published 1690) by the Dutch physicist Christiaan Huygens (1629–1695), a contemporary of Newton. This theory did not gain much support because it failed to account for polarization. (Huygens had assumed light to be in the form of *longitudinal* waves, similar to sound waves.) The wave nature of light was finally convincingly demonstrated in 1801 by the interference experiments of Thomas Young (1773–1829). Additional evidence in support of the wave theory was supplied soon thereafter, particularly in a series of papers by Augustin Fresnel (1788–1827). By 1826 the corpuscular theory was dead. The theoretical analysis of Maxwell (1864) established a firm foundation for the wave theory of light based on the principles of electrodynamics.

The wave theory was unchallenged until the beginning of the twentieth century, when it was found that a radical new idea was required to explain various puzzling phenomena involving the interaction of light with atomic systems. To account for the experimental results, Max Planck (1858–1947) and Albert Einstein (1879–1955) found it necessary to suppose that light consists ultimately of elementary *quanta* (or *photons*). We can think of a photon as a bundle of electomagnetic radiation that oscillates with a definite frequency and travels through space with the speed of light. Individual quanta carry energy and momentum, so light has some particle-like properties. But when the number of identical photons is very large, they exhibit the properties of a continuous wave with the same definite frequency and propagation speed as the quanta. Deviations from the predictions of the classical wave theory occur only when the photons interact individually with atoms. In 1913 the Danish physicist Niels Bohr (1885–1962) devised an atomic model for the emission and absorption of light quanta. This led eventually, in the 1920's, to the merger of the quantum theory of light with the emerging general quantum theory of matter.

34–2 PHYSICAL PROPERTIES OF LIGHT

Many of the physical properties of light are shared by the other types of electromagnetic waves. In the following, however, our emphasis is placed on the study of light.

Spectral Characteristics. The portion of the electromagnetic spectrum termed *visible light* corresponds approximately to the wavelength region from 4000 Å (violet) to 7000 Å (red).* The various colors of light in this range are identified in Table 34–1. "White" light is a mixture of light of all these wavelengths.

The Sun emits electromagnetic radiation with a much wider range of wavelengths than can be detected by the eye. However, the broad maximum in the solar intensity spectrum is at a wavelength that closely matches the central region of visible light (see Fig. 34–1). Presumably, genetic adaptation during the course of evolutionary history has been responsible for adjusting the sensitivity of the eye to match the solar emission maximum.

There are sources of light, such as the Sun, fire, and incandescent lamps, that emit radiation with a continuous spectrum of wavelengths. There are also sources, such as discharge tubes and arc lamps, that emit light with discrete wavelengths. For example, a hydrogen discharge tube produces intense lines† at wavelengths $\lambda = 656.3$ nm, 486.1 nm, 434.0 nm, 410.1 nm, and so forth. Also, in the following we refer frequently to the yellow light (the intense D lines) from luminous sodium vapor: $\lambda_{D_1} = 589.6$ nm and $\lambda_{D_2} = 589.0$ nm. (These lines are so close in wavelength that they appear as a single line in many instruments.)

The physical processes that are responsible for continuous spectra and for line spectra

TABLE 34–1.
Approximate Wavelengths for the Various Colors in the Visible Spectrum

COLOR	WAVELENGTH RANGE (Å)
Violet	4000–4500
Blue	4500–5000
Green	5000–5500
Yellow	5500–6000
Orange	6000–6500
Red	6500–7000

---

*Wavelengths of radiation in or near the visible region are expressed variously in angstroms (10^{-10} m), micrometers (μm), and nanometers (nm). Thus, 4000 Å = 400 nm = 0.400 μm. You should be familiar with all of these units.

† The term "line" is used to denote light with a discrete wavelength because when a spectrograph is used to observe the light, each wavelength corresponds to a line image on the recording film.

Fig. 34–1. Comparison of the solar radiation intensity spectrum with the sensitivity of the human eye. Notice that the solar spectrum at sea level contains numerous dips due to the absorption of radiation by various molecular species in the atmosphere.

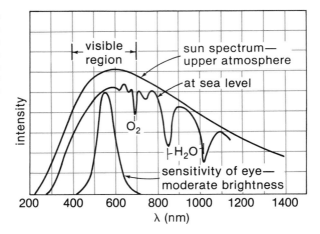

are quite different. Line spectra arise from a type of resonance phenomenon in atoms and molecules, whereas the origin of continuous spectra is more akin to noise generation.

States of Polarization. The state of polarization of light is described in the same manner as for all electromagnetic radiation, which is in terms of the direction along which the electric field vector oscillates (see Section 33–1). An electromagnetic wave is said to be *linearly polarized* if its electric vector $\mathbf{E}(\mathbf{r}, t)$ always oscillates parallel to a fixed direction in space.* For example, a wave whose electric vector is expressed as

$$\mathbf{E}(\mathbf{r}, t) = E_0 \cos (kz - \omega t)\hat{\mathbf{i}} \qquad \textbf{(34–1a)}$$

is a plane wave that is linearly polarized in the *x*-direction. Figure 34–2a shows a portion of this wave at the instant $t = 0$ (compare Fig. 33–3). The electric vector vibrates parallel to the *x*-axis with values between $+E_0$ and $-E_0$. It is customary to indicate this vibrational character in diagrams by means of a short line segment with double arrowheads, as in Fig. 34–2b. The direction of the line segment corresponds to the direction of linear polarization.

The wave described by Eq. 34–1a is polarized in the *x*-direction and propagates in the *z*-direction. Because the direction of propagation corresponds to the direction of $\mathbf{E} \times \mathbf{H}$, it follows that the magnetic field vector is in the *y*-direction:

$$\mathbf{H}(\mathbf{r}, t) = H_0 \cos (kz - \omega t)\hat{\mathbf{j}} \qquad \textbf{(34–1b)}$$

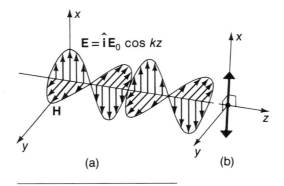

Fig. 34–2. (a) A linearly polarized electromagnetic wave. (b) The direction of linear polarization is indicated by a line segment with double arrowheads.

---

*We use here the term *linearly polarized* wave, but elsewhere you may find these waves called *plane-polarized* waves; the terms are synonymous.

Because the magnetic vector of a plane wave can always be deduced from a knowledge of the electric vector $\mathbf{E}(\mathbf{r}, t)$ and the propagation vector $\mathbf{k}$, we usually do not specifically exhibit $\mathbf{H}(\mathbf{r}, t)$. In any event, the interaction of light with matter is governed almost exclusively by the electric field of the wave.

If an electric dipole antenna is driven by a sinusoidal current, the resulting radiation at a large distance in a certain direction from the antenna consists of approximately plane waves with a definite polarization (see Section 33–3). In the case of light, however, the radiating sources are atoms, and each atomic emission event produces a single wave train with finite length. Thus, an ordinary light beam consists of a very large number of uncorrelated wave trains, each arising from an individual atom. Each of these wave trains has a specific linear polarization, but the polarization directions are randomly distributed in a plane that is perpendicular to the direction of propagation of the beam. Light that consists of wave trains with random polarizations is said to be *unpolarized* light.

The Speed of Light.   The speed of light was of interest to and was speculated on by the ancient Greeks. It was clear that the speed of light must be much greater than the speed of sound (perhaps infinite?) because there was a time delay between seeing a lightning flash and hearing the accompanying thunder. However, no direct measurements were attempted for many centuries.

The first recorded attempt to measure the speed of light was made by Galileo in about 1600. His method, which involved signalling with lanterns between himself and a companion on two separate hilltops, failed to find a finite value. The speed of light is far too great to be measured in this crude way. (See Problem 34–1.)

The earliest reliable determination of the speed of light was made in 1676 by the Danish astronomer Olaus Roemer (1644–1710) in connection with his study of Jupiter's moons. His data yielded a value of 2.3×10^8 m/s for the speed of light. Although the currently accepted value is considerably higher, his accomplishment is remarkable when allowance is made for the relatively poor knowledge of astronomical distances available to him.

A second astronomical determination of the speed of light was reported in 1729 by the English astronomer James Bradley (1693–1762). He found that if a telescope is pointed in a certain direction to observe a particular star on one night, six months later the telescope must be pointed in a slightly different direction to observe the same star. This effect is called the *aberration of light* and depends on the relative values of the speed of light and the Earth's orbital speed. Bradley's results determined the speed of light to be 3.06×10^8 m/s.

The first entirely terrestrial measurement of the speed of light was made by the French physicist Armand Fizeau (1819–1896) in 1849. The apparatus that was used is shown schematically in Fig. 34–3. Light from a source S passes through a lens L_1 and is reflected to the left by a mirror M_1. The lens L_1 focuses the light at the position of a rotating toothed wheel. Light that passes through a gap in the wheel travels a distance ℓ to a mirror M_2, where it is reflected and returns along the same path to the toothed wheel.

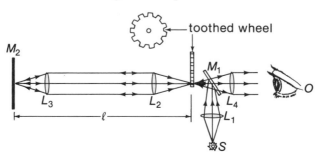

Fig. 34–3. Fizeau's toothed-wheel apparatus for measuring the speed of light. The flight path for the light was $\ell = 8.63$ km.

If it is not blocked by a tooth of the wheel, some of the light is transmitted through the mirror M_1 and reaches the observer O. (M_1 is "half-silvered" and so partially reflects and partially transmits incident light.) The rotation speed of the wheel is adjusted until the returning light is entirely blocked by a tooth of the wheel and the observer sees no light. That is, the light travels the distance 2ℓ during the time required for the wheel to rotate from a gap to the next toothed section. In this way Fizeau obtained the value $c = 3.13 \times 10^8$ m/s (see Problem 34–2).

An improvement in Fizeau's method was made by J. B. L. Foucault (1819–1868), who substituted a rotating many-sided mirror for the toothed wheel. Foucault's value for the speed of light, obtained in 1862, was $c = 2.98 \times 10^8$ m/s, a value within 0.6 percent of the best modern determination. Also, by inserting a water-filled tube into the light path, Foucault demonstrated that the speed of light in water is less than that in air.

The rotating-mirror technique was carried to its height by Michelson,* who mounted an eight-sided mirror at the observatory on Mount Wilson in southern California. The return mirror was located on Mount San Antonio, 35 km away. The measurements made during the period 1923–1927 produced the value $c = 2.99798 \times 10^8$ m/s, with an uncertainty of 4 in the last decimal.

The most precise modern methods for determining the speed of light involve the simultaneous measurement of wavelength λ and frequency ν for a radiation by direct comparison with the atomic standards for length and time (see Section 1–1); then, the value of c is obtained from $c = \lambda\nu$. These techniques are capable of precisions of a few parts per *billion*. The currently accepted value is

$$c = 2.99792458 \times 10^8 \text{ m/s}$$

with an uncertainty of only 1 in the last decimal. In this book we usually use approximate values,

$$c \cong 2.998 \times 10^8 \text{ m/s} \qquad \text{or} \qquad c \cong 3.00 \times 10^8 \text{ m/s}$$

34–3 WAVEFRONTS AND RAYS

Consider a monochromatic plane wave that propagates in a direction defined by the wave vector $\mathbf{k}$ with a time-dependent phase, $\phi(t) = \mathbf{k}\cdot\mathbf{r} - \omega t + \phi_0$ (see Section 16–4). At a particular time t, a plane of constant phase (a *wavefront*) is defined by all points $P(\mathbf{r})$ for which $\phi(t)$ has a particular fixed value. Figure 34–4a illustrates the case of a wave propagating in the z-direction so that $\mathbf{k} = k\hat{\mathbf{k}}$, where $k = \omega/c$. Values of $\mathbf{r}$ for which the phase is constant at a particular time t must have a constant value for $\mathbf{k}\cdot\mathbf{r} = kr\cos\theta = kz$. That is, the wavefronts are planes that are parallel to the x-y plane. In diagrams, a series of such planes are usually shown for wavefronts that differ in phase by 2π and are therefore spaced apart by one wavelength.

Rays are lines that are drawn perpendicular to the wavefronts and in the direction of the wave propagation (Fig. 34–4b). It is often convenient to describe optical effects in terms of rays instead of wavefronts. For this purpose we sometimes isolate a single ray, the idealization of a narrow beam of light.

Figure 34–5 shows the case of spherical waves expanding from a point source. At any point $P(\mathbf{r})$, the vectors $\mathbf{k}$ and $\mathbf{r}$ are parallel. Therefore, $\phi(t) = kr - \omega t + \phi_0$, and for

Fig. 34–4. (a) A *wavefront* of a plane wave consists of all points $P(\mathbf{r})$ for which the phase, $\phi(t) = \mathbf{k}\cdot\mathbf{r} - \omega t + \phi_0$, is constant. For the case illustrated, $\mathbf{k} = k\hat{\mathbf{k}}$, and the wavefronts, such as $ABCD$, are parallel to the x-y plane. (b) A *ray* is a line drawn perpendicular to the wavefronts and in the direction of the wave propagation vector $\hat{\mathbf{k}}$.

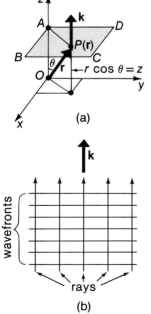

*Albert A. Michelson (1852–1931), one of the most renowned American scientists of his era and the first American to be awarded a Nobel Prize in science (1907). Michelson is also known for the famous ether-drift experiments performed with E. W. Morley.

a constant phase we must have r = constant. Thus, the wavefronts are concentric spheres and the rays are radial lines.

The Huygens-Kirchhoff Principle. About 1680 Christiaan Huygens devised a clever scheme for determining the way in which a wavefront advances. Although intended to represent the propagation of mechanical waves in material media, Huygens' idea is also valid for light waves. According to *Huygens' construction,* the way in which a wavefront of arbitrary shape will move forward can be determined by considering every point on a given wavefront at any instant t to be the source of a secondary, outward-propagating spherical wave (or *wavelet*). By drawing spheres with radii $c \Delta t$ about every point on the original wavefront and then constructing the *forward-going* envelope (which is tangent to the forward edge of each wavelet), the new position of the wavefront at time $t + \Delta t$ is determined (Fig. 34-6).

Huygens' construction is an incomplete concept because it does not address the question of why there are no backward-going wavelets. Kirchhoff eliminated this defect and converted Huygens' idea into a rigorous principle in 1882. The Huygens-Kirchhoff principle and its variations are used regularly in discussions of diffraction.

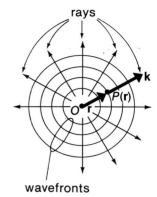

Fig. 34-5. The wavefronts expanding from a point source are concentric spheres, and the rays are radial lines.

Fig. 34-6. Huygens' construction for determining the advance of a wavefront of arbitrary shape. Every point on the wavefront *AB* at time *t* is considered to be the source of a spherical wavelet. By drawing spheres with radii $c \Delta t$ and then constructing the *forward-going* envelope, the wavefront *A'B'* at time $t + \Delta t$ is determined.

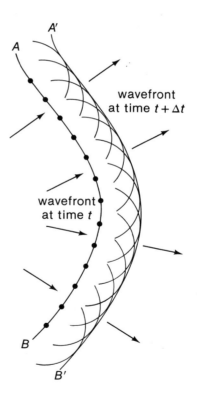

34-4 THE LAWS OF REFLECTION AND REFRACTION

A nonluminous object can be observed by virtue of the redirected (and perhaps altered) light that originated elsewhere. This process involves a variety of interactions between light and matter. Simple *diffuse reflection* occurs when light is incident on a rough irregular surface and is scattered in many directions, with some of the light scattered in the direction of the observer. For a white surface, diffuse reflection occurs equally for all wavelengths, so the reflected light has the same color or colors as the incident light.

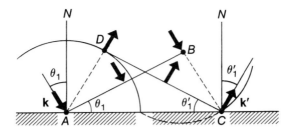

Fig. 34–7. Detail of the way in which the wavefront $\overline{AB}$ is transformed into the wavefront $\overline{DC}$ by reflection. The triangles $\triangle ABC$ and $\triangle ADC$ are congruent, so that $\theta_1 = \theta'_1$.

If a surface appears to have a particular color when irradiated with white light, selective absorption or scattering of some of the component wavelengths has taken place. More complicated effects such as fluorescence and luminescence may also occur, in which the atoms of the irradiated material are induced to emit radiation with a characteristic color.

Specular Reflection. The simplest form of reflection is *specular reflection,* which occurs when light is incident on a polished surface such as a mirror or a metal plate. A light ray that undergoes specular reflection will leave the surface in a well-defined direction.

Figure 34–7 shows an incident wavefront $\overline{AB}$ being reflected by the mirror surface $\overline{AC}$. Notice that the *angle of incidence* θ_1 may be defined either as the angle that an incident wavefront makes with the mirror or as the angle that an incident ray makes with the normal to the mirror. The incident wavefront $\overline{AB}$ emerges a time Δt later as the reflected wavefront $\overline{DC}$.

The interval Δt is the time required for a Huygens wavelet emitted at point B on the incident wavefront to reach point C on the mirror. The segment $\overline{BC}$ corresponds to an incident ray. During this same interval Δt, the wavelet emitted at point A travels to point D. Thus, $\overline{DC}$ corresponds to the reflected wavefront.

The ray segments $\overline{BC}$ and $\overline{AD}$ each have length $v\,\Delta t$, where v is the wave velocity in the medium above the surface of the mirror. The angles $\angle ADC$ and $\angle ABC$ are both right angles. Therefore, the triangles $\triangle ABC$ and $\triangle ADC$, which have the common side $\overline{AC}$, are congruent. Thus, we conclude that the angle of incidence θ_1 is equal to the angle of reflection θ'_1; that is,

$$\theta_1 = \theta'_1 \qquad \text{specular reflection} \qquad (34-2)$$

This is the *law of reflection.*

The wavelength λ of the radiation is unaltered by the reflection process. (• Can you see why this statement is true?)

Refraction. When a light ray is incident on the interface between two transparent media, as in Fig. 34–7, it is observed that part of the light is reflected in the usual way, with $\theta_1 = \theta'_1$, whereas some of the light is transmitted into the second medium with $\theta_2 \neq \theta_1$. If medium 1 is air and medium 2 is a substance such as water or glass, it is observed that $\theta_2 < \theta_1$. This bending of light at an interface is called *refraction,* and θ_2 is the *angle of refraction.* Thus, when a light ray enters water from air, the ray is refracted *toward* the normal to the surface.

The analysis of refraction using Huygens' construction proceeds in a manner parallel to that for reflection. The Huygens wavelet emitted at point B (Fig. 34–8) arrives at point C on the interface after time $\Delta t = \overline{BC}/v_1$, where v_1 is the wave velocity in medium 1. During this same interval Δt, the wavelet emitted at point A travels to point D. The new (refracted) wavefront $\overline{DC}$ makes an angle θ_2 with the interface. In Fig. 34–8 we see that

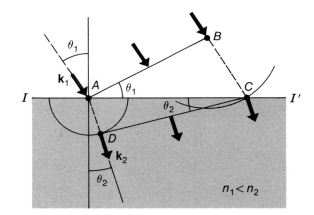

Fig. 34-8. Detail of the way in which the wavefront $\overline{AB}$ is transformed into the wavefront $\overline{DC}$ by refraction at the interface II'. The angles θ_1 and θ_2 are related by Snell's law. (For simplicity, the reflected wavefronts have been omitted; they are as shown in Fig. 34-7.)

$$\sin \theta_1 = \frac{\overline{BC}}{\overline{AC}}$$

$$\sin \theta_2 = \frac{\overline{AD}}{\overline{AC}}$$

Dividing one of these equations by the other yields

$$\overline{AD} \sin \theta_1 = \overline{BC} \sin \theta_2 \qquad (34-3)$$

Now, $\overline{BC} = v_1 \Delta t$ and $\overline{AD} = v_2 \Delta t$, where v_2 is the wave velocity in medium 2. Using these expressions in Eq. 34-3 and multiplying both sides by c, we have

$$\frac{c}{v_1} \sin \theta_1 = \frac{c}{v_2} \sin \theta_2$$

The ratio of the velocity of light in vacuum to the velocity in a medium is usually designated by the letter n. Thus, the *law of refraction* can be expressed as

$$n_1 \sin \theta_1 = n_2 \sin \theta_2 \qquad (34-4)$$

This result is referred to as *Snell's law*.* Because the ratio

$$n = \frac{c}{v} \qquad (34-5)$$

appears in the law of refraction, the quantity n is called the *index of refraction* or the *refractive index* of the medium. It is the same quantity defined in Eq. 33-17. Note that for light, $n > 1$ for all materials.

The value of the refractive index for a material depends somewhat on the wavelength of the incident light. (This effect gives rise to the phenomenon of *dispersion*—see Section 34-6.) When an index is quoted without specification of the wavelength, the value is, by

---

*After the Dutch mathematician Willebrord Snell (1591–1626), who discovered the result experimentally in about 1621.

TABLE 34–2. Refractive Indexes for Various Materials (for sodium light, $\lambda = 5893$ Å)

GASES (at 0° C and 1 atm)

Gas	n
Hydrogen (H_2)	1.000132
Air	1.000293
Ammonia (NH_3)	1.000376
Methane (CH_4)	1.000444
Carbon dioxide (CO_2)	1.000451
Sulfur dioxide (SO_2)	1.000686
Chlorine (Cl_2)	1.000773

LIQUIDS (at 20° C)

Liquid	n
Methyl alcohol (CH_3OH)	1.329
Water (H_2O)	1.333
Ethyl alcohol (C_2H_5OH)	1.362
Benzene (C_6H_6)	1.501
Carbon disulfide (CS_2)	1.625
Methylene iodide (CH_2I_2)	1.756

SOLIDS (at 20° C)

Solid	n
Ice (H_2O)	1.310
Fluorite (CaF_2)	1.434
Fused quartz (SiO_2)	1.458
Plastic (lucite)	1.491
Glass (ordinary crown)	1.517
Rock salt (NaCl)	1.544
Plastic (polystyrene)	1.595
Glass (dense flint)	1.655
Diamond (C)	2.417

custom, the value appropriate for yellow light (in particular, the yellow light emitted by sodium with $\lambda = 5893$ Å). Table 34–2 lists values of n for materials of various types.

A light wave (in fact, any type of wave) undergoes a change in wavelength upon refraction, whereas no such change occurs upon reflection. To insure a time-independent match of the boundary conditions of the electromagnetic field components, the frequency of a wave must be the same on either side of a boundary. However, the wavelength will be different on the two sides if the wave velocity is different (see Section 16–6). Thus,

$$\nu_1 = \frac{v_1}{\lambda_1} = \nu_2 = \frac{v_2}{\lambda_2}$$

so that

$$\lambda_2 = \frac{v_2}{v_1}\lambda_1 = \frac{n_1}{n_2}\lambda_1 \qquad \textbf{(34–6)}$$

Example 34–1

A narrow beam of sodium yellow light is incident from air on a smooth surface of water at an angle $\theta_1 = 35°$. Determine the angle of refraction θ_2 and the wavelength of the light in water. Refer to Fig. 34–8.

Solution:

For all practical purposes, the index of refraction of air is 1 (see Table 34–2), so $n_1 = 1$. Also from Table 34–2, we have n_2 (water) = 1.333. Then, using Snell's law (Eq. 34–4), we have

$$1 \cdot \sin \theta_1 = 1.333 \sin \theta_2$$

so that

$$\sin \theta_2 = \frac{1}{1.333} \sin 35°$$

from which

$$\theta_2 = 25.5°$$

The average wavelength of sodium yellow light in vacuum (or air) is $\lambda_1 = 5893$ Å. Therefore, using Eq. 34–6, we find the wavelength in water to be

$$\lambda_2 = \frac{n_1}{n_2}\lambda_1 = \frac{1}{1.333}(5893 \text{ Å}) = 4421 \text{ Å}$$

Example 34–2

A narrow beam of light passes through a plate of glass with a thickness $s = 4$ mm and a refractive index $n = 1.52$. The beam enters from air at an angle $\theta_1 = 30°$. Calculate the deviation d of the ray, as indicated in the diagram.

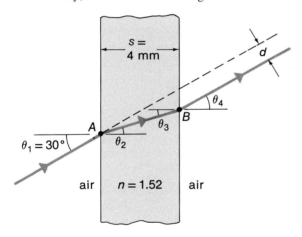

Solution:

Because the surfaces of the glass plate are parallel, the interior angles, θ_2 and θ_3, are equal. Then, applying Snell's law at A

and B, we obtain

$$\sin \theta_1 = n \sin \theta_2 = n \sin \theta_3 = \sin \theta_4$$

Thus, $\theta_4 = \theta_1$, and the emergent ray is parallel to the incident ray. There is a deviation d of the ray but no angular displacement. (A glass plate therefore has no lenslike properties.)

The deviation d is

$$d = \overline{AB} \sin (\theta_1 - \theta_2)$$

We also have

$$s = \overline{AB} \cos \theta_2$$

Dividing one of these equations by the other gives

$$d = s\frac{\sin (\theta_1 - \theta_2)}{\cos \theta_2}$$

From Snell's law we have

$$\sin \theta_2 = \frac{1}{n} \sin \theta_1 = \frac{1}{1.52} \sin 30°$$

from which $\theta_2 = 19.2°$ and $\theta_1 - \theta_2 = 10.8°$. Thus,

$$d = (4 \text{ mm})\frac{\sin 10.8°}{\cos 19.2°} = 0.794 \text{ mm}$$

Total Internal Reflection. Figure 34–8 shows that when the refractive index of the incident medium is less than that of the second medium (that is, $n_1 < n_2$), the ray is refracted *toward* the normal. In this circumstance, the ray will penetrate into the second medium for any incident angle θ_1. However, the situation is different when $n_2 < n_1$, the case illustrated in Fig. 34–9. Ray a is incident normally on the interface and passes through undeflected, whereas ray b is deflected in accordance with Snell's law, giving $\theta_2 > \theta_1$. As θ_1 is increased, there is eventually reached a critical angle θ_c, corresponding

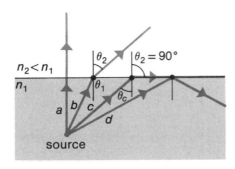

Fig. 34–9. When $n_2 < n_1$, total internal reflection occurs for all incident angles greater than the critical angle θ_c. (The reflected parts of rays *a*, *b*, and *c* are not shown.)

to ray *c* in Fig. 34–9, for which $\theta_2 = 90°$. Then, $\sin \theta_2 = (n_1/n_2) \sin \theta_c = 1$, whence

$$\theta_c = \sin^{-1} (n_2/n_1) \qquad (34-7)$$

If θ_1 is made greater than θ_c, Snell's law would require that $\sin \theta_2 > 1$, an unphysical situation. Thus, when $\theta_1 > \theta_c$, no refraction occurs and the ray is completely reflected at the interface (ray *d* in Fig. 34–9). This phenomenon is called *total internal reflection*.

▶ *Fiber Optics.* When light is projected into the end of a glass or plastic rod with a small diameter as in Fig. 34–10, the angles with which the various rays strike the surface are greater than the critical angle. Such rays are "piped" through the rod by successive internal reflections, even along a curved path, and eventually emerge from the opposite end. Bundles of tiny rods (called

A 3-foot coherent fiber optic bundle with an 8- × 10-mm format consisting of 27,000 60-micron multifibers. (Courtesy of Galileo Electro-Optics Corp., Sturbridge, Mass.)

light pipes or *optical fibers*) are used in a variety of applications that do not permit normal straight-line propagation. (For example, optical fibers allow a physician to see inside a patient's stomach.) Also, special fibers with low-loss characteristics are beginning to be used in the communications industry. A bundle of fibers can carry hundreds of times more telephone calls using light signals than can a bundle of copper wires of the same size using ordinary electrical signals. ◀

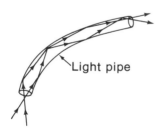

Fig. 34–10. Light is transmitted through a curved plastic rod (a *light pipe* or *optical fiber*) by successive internal reflections.

Example 34–3

As shown in the diagram, a light ray is incident normally on one face of a 30°–60°–90° block of dense flint glass (a *prism*) that is immersed in water.

(a) Determine the exit angle θ_4 of the ray.

(b) A substance is dissolved in the water to increase the index of refraction. At what value of n_2 will total internal reflection cease at point *P*?

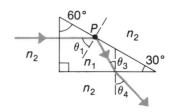

Solution:

(a) From the geometry of the prism we see that $\theta_1 = 60°$ and $\theta_3 = 30°$. The refractive indexes are (see Table 34-2) $n_1 = 1.655$ and $n_2 = 1.333$. Then, using Snell's law, we find

$$\sin \theta_4 = \frac{n_1}{n_2} \sin \theta_3 = \frac{1.655}{1.333} \sin 30° = 0.621$$

so that $\theta_4 = 38.4°$

(b) Total internal reflection occurs when $(n_1/n_2) \sin \theta_1 > 1$. (For the initial situation here, $(n_1/n_2) \sin \theta_1 = 1.075$.) Total internal reflection will cease when $(n_1/n_2) \sin \theta_1 = 1$. Thus,

$$n_2 = n_1 \sin \theta_1 = (1.655) \sin 60° = 1.433$$

Fermat's Principle. A powerful principle that specifies the paths followed by light rays was formulated in 1657 by the French mathematician Pierre de Fermat (1601–1665). *Fermat's principle* states that a light ray that connects two points within a medium follows a path for which the transit time of the light is minimum.

Figure 34-11 shows various conceivable paths that light might follow from point P to point Q. The time required for light to travel between the points along the particular path C is

$$t = \int_P^Q \frac{ds}{v}$$

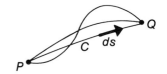

Fig. 34-11. According to Fermat's principle, light travels from P to Q along the path for which the transit time is least.

If the medium through which the light passes has a variable index of refraction, we can use $v = c/n$ to express the integral as

$$t = \frac{1}{c} \int_P^Q n \, ds \qquad (34\text{-}8)$$

According to Fermat's *principle of least time,* the integral, $\int_P^Q n \, ds$, called the *optical path length,* must be minimum.

Figure 34-12 shows the application of Fermat's principle to the refraction of light at a plane surface. Consider the passage of light from point A in a uniform medium with index n_1 to point B in a second uniform medium with index n_2, through an arbitrary interface point P. From A to P the least-time path is clearly a straight line whose optical path length is $n_1 \ell_1$. Similarly, the least-time path from P to B is a straight line with optical path length $n_2 \ell_2$. It remains to determine the location of point P, specified by the coordinate x. The total optical path length is

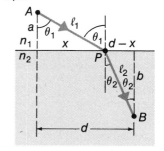

Fig. 34-12. Geometry for deriving Snell's law from Fermat's principle.

$$L = \int_A^B n \, ds = n_1 \ell_1 + n_2 \ell_2$$

$$= n_1 \sqrt{a^2 + x^2} + n_2 \sqrt{b^2 + (d - x)^2}$$

The minimum value of L is obtained by setting dL/dx equal to zero. Thus,

$$\frac{n_1 x}{\sqrt{a^2 + x^2}} = \frac{n_2(d - x)}{\sqrt{b^2 + (d - x)^2}}$$

or

$$n_1 \sin \theta_1 = n_2 \sin \theta_2$$

which we recognize to be Snell's law.

The law of reflection (Eq. 34-2) is also easy to obtain from Fermat's principle (see Problem 34-9).

34–5 REFLECTION AND TRANSMISSION (Optional)

In Section 16–6 we discussed the reflection and transmission of waves on a string when they encounter a discontinuity in the mass density of the string. The behavior of the reflected and transmitted waves depends strongly on whether the incident wave approached the heavier portion of the string from the lighter portion or the other way around.

In the present section we examine the similar phenomenon of reflection and transmission of light waves that are incident on a boundary between two media with different indexes of refraction. The situation for light is more complicated because of the possibility of two states of polarization of the waves when they are in oblique incidence on the interface between the two media. You should again note, however, the importance of whether the wave leaves the lower-index medium and enters the higher-index medium or vice versa.

When light is incident on the interface between two media, the fraction of the wave intensity that is reflected and the fraction that is transmitted depend on the refractive indexes of the media. Expressions for these fractions (called the *reflectance R* and the *transmittance T*) can be derived by examining the behavior of the field vectors at the interface.

Normal Incidence. Consider light in the form of a plane electromagnetic wave that is incident normally on the flat interface between two media, represented by the x-y plane in Fig. 34–13. The propagation vector $\mathbf{k}_1$ of the incident wave has the direction of $\mathbf{E}_i \times \mathbf{H}_i$ (see Fig. 33–7). Moreover, the propagation vector for the reflected wave is $-\mathbf{k}_1$, corresponding to the direction of $\mathbf{E}_r \times \mathbf{H}_r$. Because an electromagnetic wave always propagates in the direction of $\mathbf{E} \times \mathbf{H}$, it is clear (see Fig. 34–13) that there must be a change of orientation of the electric and magnetic vectors upon reflection. Figure 34–13 shows a *change of phase* for the magnetic vector but not for the electric vector; as we shall see, this is the case if $n_1 > n_2$.

For the orientation of the electric and magnetic vectors

Fig. 34–13. The electric and magnetic vectors of a plane electromagnetic wave at the boundary between two dielectric media for a particular instant. (The vector sets have been displaced from $z = 0$ for clarity.) The diagram represents the case $n_1 > n_2$, and the subscripts represent the incident wave (*i*), the reflected wave (*r*), and the transmitted wave (*t*). The magnetic vector undergoes a phase change upon reflection, but the electric vector does not.

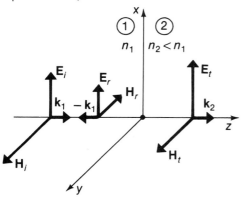

shown in Fig. 34–13, the physical conditions at the interface between the media require that

$$\left.\begin{array}{c} \mathbf{E}_i + \mathbf{E}_r = \mathbf{E}_t \\ \mathbf{H}_i + \mathbf{H}_r = \mathbf{H}_t \end{array}\right\} \qquad \textbf{(34–9)}$$

at any instant of time. In terms of the amplitudes, these conditions become

$$\left.\begin{array}{c} E_i^0 + E_r^0 = E_t^0 \\ H_i^0 - H_r^0 = H_t^0 \end{array}\right\} \qquad \textbf{(34–10)}$$

It is readily shown (see Problem 34–13) that these conditions lead to the amplitude ratios

$$\boxed{\begin{array}{c} E_r^0 / E_i^0 = \dfrac{n_1 - n_2}{n_1 + n_2} \\[2mm] E_t^0 / E_i^0 = \dfrac{2n_1}{n_1 + n_2} \end{array}} \qquad \textbf{(34–11)}$$

(Compare these expressions with those obtained for the case of waves on a string, Eqs. 16–16.)

The quantity E_t^0 / E_i^0 is called the *amplitude transmission coefficient*. Equation 34–11 shows that $E_t^0 / E_i^0 > 0$ for all values of the refractive indexes. Thus, at the interface, the vectors $\mathbf{E}_i$ and $\mathbf{E}_t$ always have the same direction. That is, the incident and transmitted waves are in phase for all n_1 and n_2. On the other hand, Eq. 34–11 also shows that the *amplitude reflection coefficient* E_r^0 / E_i^0 is positive only when $n_1 > n_2$ (the case illustrated in Fig. 34–13). In this case, $\mathbf{E}_r$ and $\mathbf{E}_i$ have the same direction at the interface and are therefore *in phase*. When $n_1 < n_2$ (for example, when light is incident on glass or plastic from air or water), we have $E_r^0 / E_i^0 < 0$. Then, $\mathbf{E}_r$ and $\mathbf{E}_i$ have opposite directions at the interface, and the incident and reflected waves *differ in phase* by π. In summary,

$$\begin{array}{l} n_1 > n_2: \left\{\begin{array}{l} \mathbf{E}_r \text{ and } \mathbf{E}_i \text{ in phase, } E_r^0/E_i^0 > 0 \\[1mm] \mathbf{E}_t \text{ and } \mathbf{E}_i \text{ in phase, } E_t^0/E_i^0 > 0 \end{array}\right. \\[5mm] n_1 < n_2: \left\{\begin{array}{l} \mathbf{E}_r \text{ and } \mathbf{E}_i \text{ differ in phase by } \pi, \\ \quad E_r^0/E_i^0 < 0 \\[1mm] \mathbf{E}_t \text{ and } \mathbf{E}_i \text{ in phase, } E_t^0/E_i^0 > 0 \end{array}\right. \end{array} \qquad \textbf{(34–12)}$$

The *reflectance R* is defined to be the fraction of the incident energy flux that is reflected at the boundary. This energy flux is represented by the Poynting vector, $\mathbf{N} = \mathbf{E} \times \mathbf{H}$ (Eq. 33–21), the time average of which is proportional to $E^0 H^0$. The calculation of $R = \overline{N_i}$ for normal incidence yields

$$\boxed{R = \left(\frac{n_1 - n_2}{n_1 + n_2}\right)^2} \qquad \textbf{(34–13)}$$

The *transmittance* T is defined in a similar way as $T = \overline{N_t}/\overline{N_i}$ and is

$$T = \frac{n_2}{n_1}\left(\frac{2n_1}{n_1 + n_2}\right)^2 \qquad \text{(34-14)}$$

Because energy is conserved, we have

$$R + T = 1 \qquad \text{(34-15)}$$

(• Can you verify this conservation condition by using Eqs. 34-13 and 34-14?)

Notice that R and T remain the same if n_1 and n_2 are interchanged. That is, the reflectance and transmittance values for normal incidence at a particular interface do not depend on the direction in which the light is incident on the interface.

Example 34-4

Ordinary crown glass has a refractive index of 1.517 (Table 34-2). When light is incident normally from air on a plate of crown glass, what fraction of the light is reflected and what fraction is transmitted?

Solution:

We have $n_1 = 1.000$ and $n_2 = 1.517$. Then, using Eqs. 34-13 and 34-14, we find

$$R = \left(\frac{n_1 - n_2}{n_1 + n_2}\right)^2 = \left(\frac{1.000 - 1.517}{1.000 + 1.517}\right)^2 = 0.0422$$

$$T = \frac{n_2}{n_1}\left(\frac{2n_1}{n_1 + n_2}\right)^2 = \frac{1.517}{1.000}\left(\frac{2 \times 1.000}{1.000 + 1.517}\right)^2 = 0.9578$$

from which we see that $R + T = 1$, as required. The small value of R agrees with the observation that ordinary glass readily transmits light in normal incidence.

Oblique Incidence. When light is incident normally on a surface, the electric and magnetic vectors of the wave are always tangential to the surface. However, when light is incident at an oblique angle, the situation is more complicated. In this case it is convenient to define a *plane of incidence*, which is the plane that contains the propagation vectors, $\mathbf{k}_1$, $\mathbf{k}_1'$, and $\mathbf{k}_2$. The electric vector $\mathbf{E}_i$ of the incident wave can have any polarization around the direction defined by $\mathbf{k}_1$, but $\mathbf{E}_i$ can always be resolved into components, $\mathbf{E}_{i\perp}$ and $\mathbf{E}_{i\parallel}$, that are perpendicular and parallel, respectively, to the plane of incidence.

The perpendicular and parallel vector components can be treated separately. The resulting expressions for the amplitude reflection and transmission coefficients were first obtained by Fresnel in 1823. We show here, in Fig. 34-14, only the graphic results for a case of particular interest, namely, light incident from air ($n_1 = 1$) on glass ($n_2 = 1.5$). Figure 34-14a shows that the amplitude reflection coefficient for the perpendicular component is negative for all θ_1. This means that there is a phase change of π upon reflection of the perpendicular component for $n_2 > n_1$, in agreement with the conclusion reached for normal incidence (see Eq. 34-14a). There is no phase change upon transmission.

Figure 34-14b shows the peculiar behavior of the amplitude reflection coefficient for the parallel component. For incident angles θ_1 near 0°, the reflected wave exhibits a phase change, as does the perpendicular component. However, at an incident angle of about 56° in this case, the reflection coefficient E_r^0/E_i^0 changes sign. The angle, $\theta_B = \tan^{-1}(n_2/n_1)$, at which this occurs is called *Brewster's angle,** and it corresponds to *zero* amplitude reflection coefficient for the parallel component of the incident beam.

If light containing both perpendicular and parallel components is incident at Brewster's angle, only the perpendicular component is reflected. Thus, the reflected beam is completely polarized in the direction perpendicular to the plane of incidence. For this reason Brewster's angle is sometimes called the *polarizing angle*. The complete transmission of the parallel component at Brewster's angle occurs both for $n_2 > n_1$ and $n_2 < n_1$. In each case the angle between the transmitted ray and the direction of specular reflection is 90°. We return to this important phenomenon in Section 34-7, where methods for producing linearly polarized light are discussed.

---

*The relationship connecting the refractive indexes and θ_B was discovered experimentally in 1811 by the Scottish physicist, Sir David Brewster (1781-1868).

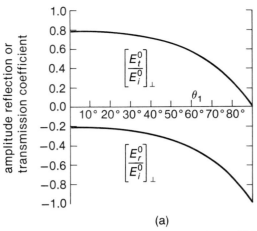 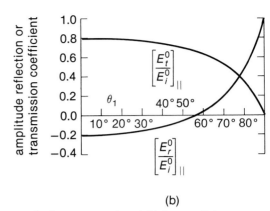

(a) (b)

Fig. 34–14. Amplitude reflection and transmission coefficients for (a) the perpendicular component and (b) the parallel component of a wave incident from air ($n_1 = 1$) on glass ($n_2 = 1.5$). The reflection coefficient $(E_r^0/E_i^0)_{\parallel}$ becomes zero at Brewster's angle.

34–6 DISPERSION

The index of refraction of most substances varies somewhat with the wavelength of the incident radiation. Figure 34–15 shows this variation in the visible region for three important types of glass and for fused quartz. When white light, which consists of light with all colors (wavelengths), enters a substance, the variation of the refractive index with wavelength causes the various colors to be spatially separated or *dispersed*.

Prisms. A simple device that exhibits dispersion effectively is a triangular piece of glass, called a *prism*. Figure 34–16 illustrates the passage of a narrow monochromatic light beam or ray through a prism with a refractive angle α. The angle δ measures the

Fig. 34–15. The variation with wavelength (measured in air) of the index of refraction for three important types of glass and for fused quartz. Wavelengths near 600 nm = 6000 Å correspond to yellow light.

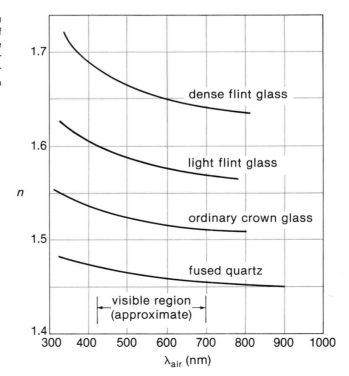

deviation of the ray. As the incident angle θ_1 is varied, δ reaches a minimum value when $\overline{AP} = \overline{AQ}$. Under this condition of minimum deviation, we have (see Problem 34–15)

$$n \sin \frac{\alpha}{2} = \sin \frac{\alpha + \delta_{min}}{2} \qquad (34\text{–}16)$$

The measurement of δ_{min} for incident light with different wavelengths is one method of obtaining curves of n versus λ such as those in Fig. 34–15.

When white light is incident on a prism, the deviation angle δ is different for each color. Then, the white light is dispersed into a rainbow spectrum of colors, as indicated in Fig. 34–17. Because $n(\text{red}) < n(\text{violet})$ for most substances (see Fig. 34–15), red light is deviated less than violet light in passing through the prism.

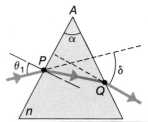

Fig. 34–16. A light ray is deviated by an amount δ upon passing through a prism. If the prism has the shape of an isosceles triangle, at minimum deviation the ray *PQ* is parallel to the base.

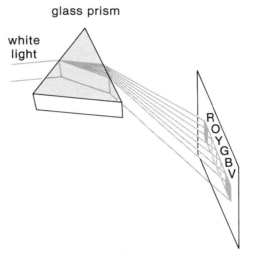

Fig. 34–17. Dispersion of white light by a prism. The order of colors in the spectrum is red, orange, yellow, green, blue, and violet.

34–7 POLARIZERS

Ordinary sources of light emit unpolarized light. Such light consists of individual wave trains that have randomly oriented planes of electric vector vibrations or states of polarization. Figure 34–18 shows some possible planes of the electric vectors for such wave trains traveling in a direction perpendicular to the page. The average behavior of E_x and E_y are indistinguishable for any selection of x- and y-axes. Thus, we have $\overline{E_x^2} = \overline{E_y^2}$, regardless of the directions in which the x- and y-axes are chosen. The total energy flux in the beam of unpolarized light may be considered to be proportional to the sum $\overline{E^2} = \overline{E_x^2} + \overline{E_y^2}$.

Many interesting phenomena based on the transverse character of light waves require polarized beams. In this section we examine methods that may be used to produce polarized beams from unpolarized light. Devices used to achieve this effect are known as *polarizers*.

Polarized light may be produced from unpolarized light by three phenomena of major importance: (a) reflection, (b) scattering, and (c) selective absorption.* In this section we give a brief discussion of these effects.

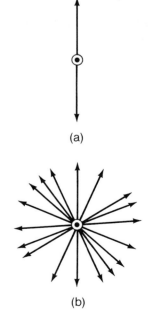

Fig. 34–18. (a) A linearly polarized wave train moving normal to the page. (b) The random superposition of many wave trains present in unpolarized light.

(a)

(b)

---

*Another phenomenon, of historical importance, for producing polarized light is *birefringence*. Certain crystals such as calcite and tourmaline transmit light with two different speeds, depending on the state of polarization of the light and its direction of propagation relative to the crystal axes. This effect may be used to split an incident beam of unpolarized light into two emerging beams of different polarization.

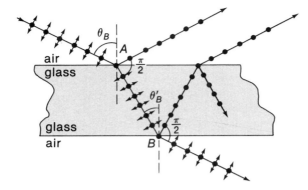

Fig. 34–19. An unpolarized wave is incident on a glass plate at Brewster's angle θ_B. The reflected wave is linearly polarized in the direction perpendicular to the plane of incidence. Notice that the double arrowheads indicate the parallel component and that the dots indicate the perpendicular component.

Reflection. In Section 34–5 we learned that when a light wave is incident on the interface between two media at an angle equal to Brewster's angle θ_B, the amplitude reflection coefficient for the component with polarization parallel to the plane of incidence is zero. The relevant condition is

$$\tan \theta_B = \frac{n_2}{n_1} \qquad \text{Brewster's angle} \qquad (34\text{–}17)$$

where the light is incident on medium 2 from medium 1. When light is incident on glass ($n_2 = 1.50$) from air ($n_1 = 1$), Brewster's angle is 56.3°. When light is incident on an air-glass interface from the glass side, we find $\theta_B' = 33.7°$.

The existence of Brewster's angle provides a practical means for producing polarized light. Figure 34–19 shows the geometry of the incident, reflected, and transmitted waves when the angle of incidence is equal to Brewster's angle. Because none of the parallel component is reflected at point A, the reflected wave is completely polarized in the perpendicular direction and the transmitted wave is partially polarized in the parallel direction.* If the glass plate has parallel surfaces, an incident angle equal to Brewster's angle θ_B at the air-glass interface guarantees that the transmitted wave is incident on the glass-air interface at Brewster's angle θ_B' appropriate for that interface (see Problem 34–27). Consequently, only the perpendicular component is reflected at B, and the transmitted wave becomes further depleted in the perpendicular component. By using a stack of glass plates, as in Fig. 34–20, this effect is enhanced so that the transmitted wave is almost completely polarized in the parallel direction. The reflected wave is completely

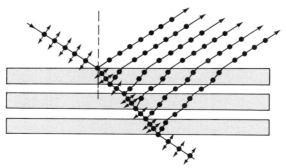

Fig. 34–20. An unpolarized incident wave can be separated into two polarized waves by a stack of glass plates. The reflected wave is completely polarized in the perpendicular direction. Each reflection depletes the perpendicular component of the transmitted wave so that the emergent wave is almost completely polarized in the parallel direction.

---

*This effect for the reflected beam was discovered in 1808 by the French engineer and physicist Étienne Louis Malus (1775–1812). The efficiency of polarization by this method is not high because only about 15 percent of the perpendicular component in the incident beam is actually reflected. (See Fig. 34–14.)

polarized in the perpendicular direction. Even if the incident angle is not equal to Brewster's angle, each beam will have a substantial partial polarization if the incident angle is not close to 0° or 90°.

Example 34–5

Sunlight is reflected from a pool of water standing in a roadway. An observer finds that the reflected sunlight is completely polarized. What is the angle of the Sun above the horizon?

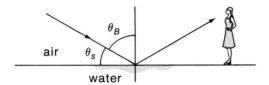

Solution:

The refractive index for water is $n_2 = 1.33$, so Brewster's angle is

$$\theta_B = \tan^{-1} \frac{n_2}{n_1} = \tan^{-1} 1.33 = 53.1°$$

Then, the angle of the Sun above the horizon is

$$\theta_S = 90° - 53.1° = 36.9°$$

The observer would need some type of polarization-sensitive device to determine that the reflected light is polarized, because human eyes (except for those of some exceptional persons) do not distinguish between polarized and unpolarized light. It has been demonstrated, however, that some types of ants and beetles, and particularly the honeybee, can sense polarization.

Scattering. If a narrow beam of natural light is passed through air, water, or some other convenient transparent medium that contains a suspension of ultramicroscopic particles, the scattered light is found to be partially polarized, with the degree of polarization depending on the angle of scattering. Figure 34–21 shows an unpolarized beam that is directed along the z-axis. The atomic electrons in the scattering center (particle) located at O are set into oscillation in response to the electric field of the incident wave. Because the incident beam is unpolarized, the electrons oscillate with a uniform distribution in the x-y plane. Now, according to the result obtained in Section 33–4, an oscillating dipole does not radiate along its direction of motion. Thus, the radiation emitted along the y-axis in Fig. 34–21 is due entirely to electron oscillations parallel to the x-axis and is therefore

Fig. 34–21. Unpolarized incident light that is scattered through 90° is linearly polarized in the direction perpendicular to the plane of incidence.

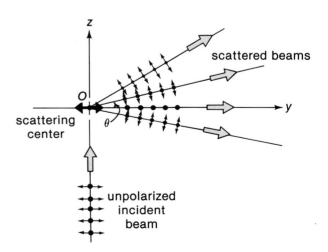

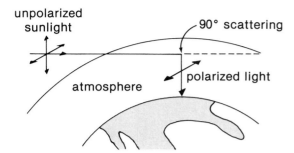

Fig. 34–22. Unpolarized sunlight becomes almost completely polarized upon scattering through an angle of 90° by air molecules.

linearly polarized in the *x*-direction. In fact, all of the radiation emitted in the *x*-*y* plane is polarized perpendicular to the plane of incidence associated with the particular direction of radiation. In other directions, the scattered light is partially polarized in accordance with the $\sin^2 \theta$ distribution of an oscillating electric dipole.

Polarization by scattering occurs also for sunlight scattered by air molecules. The maximum effect is observed on a clear day, when there is little water vapor or dust in the air, and when the Sun is near the horizon (Fig. 34–22). Then the light reaching an observer on the ground from directly overhead can be polarized to the extent of 70 or 80 percent.

Selective Absorption—Dichroism. In 1815 Biot discovered that certain mineral crystals, such as *tourmaline,* exhibit selective absorption of light depending on its state of polarization. These crystals (called *dichroic crystals*) are anisotropic; they strongly absorb light that is polarized in a direction parallel to a particular plane in the crystal, but freely transmit light that is polarized in the perpendicular direction. Thus, when an incident wave of unpolarized light is passed through a dichroic crystal, one of the polarization components is absorbed and the emergent wave is highly polarized in the perpendicular direction (Fig. 34–23).

The most common type of dichroic polarizer in use is Polaroid H-sheet, invented by E. H. Land* in 1938. A plastic sheet of long-chain molecules of polyvinyl alcohol is heated and then stretched to many times its original length. During this process, the molecules become aligned along the direction of stretching. The sheet is then cemented to a rigid piece of plastic to stabilize its size and is next dyed with iodine. The iodine atoms attach themselves to the polyvinyl alcohol molecules and in effect become long parallel

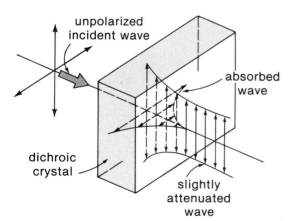

Fig. 34–23. Polarization by selective absorption of one polarization component in a dichroic crystal.

---

*Edwin H. Land (1909–   ), who formed the Polaroid Corporation to manufacture and distribute polarizing materials, later developed the popular "picture-in-a-minute" system of photography.

conducting chains. When unpolarized light is incident on a piece of H-sheet, the polariza-tion component that is parallel to the chains of iodine atoms induces currents in the conducting chains and is therefore strongly absorbed. Consequently, the transmitted beam is polarized in the direction perpendicular to the direction of the molecules. The direction of polarization of the transmitted beam is called the *polarization axis* or *transmission axis* of the material.

34–8   THE LAW OF MALUS

Suppose that a beam of unpolarized light with intensity I_0 is passed through a piece of Polaroid H-sheet (called a *polarizer*). The emerging beam is then linearly polarized in the direction of the polarization axis of the H-sheet and carries an energy flux $I_1 = \frac{1}{2}I_0$, as indicated in Fig. 34–24. The beam next passes through a second piece of H-sheet (called an *analyzer*), whose axis is at an angle θ with respect to the polarization direction of the entering beam. What is the intensity of the light transmitted through the analyzer? The answer to this question is readily deduced. The inset in Fig. 34–24 shows the electric vector $\mathbf{E}'$ of one of the wave trains in the beam entering the analyzer. The component of $\mathbf{E}'$ that is perpendicular to the transmission axis, namely, $E'_x = E' \sin\theta$, is extinguished, whereas the parallel component, $E'_y = E' \cos\theta$, is transmitted with only a slight attenua-tion. The associated energy flux is proportional to E'^2_y, so the total transmitted flux is given by the average over all such wave trains. The result is

$$I_2 = I_1 \cos^2\theta \qquad\qquad (34\text{–}18)$$

This is the *law of Malus,* discovered in 1809.

(• Can you see how Polaroid sunglasses reduce glare? Remember, reflected sunlight is partially polarized. What should be the direction of the transmission axis in the sun-glasses?)

▶ Ideally, the transmittance T_d for the *desired* component should be unity and the transmittance T_u for the *undesired* com-ponent should be zero. For H-sheet, neither of these ideal values is realized. In fact, both T_d and T_u vary somewhat with wave-length. Averaged over the visible spectrum, the transmittance values for one type of H-sheet (not including reflective losses at the surfaces) are $\overline{T_d} \cong 0.83$ and $\overline{T_u} \cong 0.001$. By adding iodine and thereby reducing $\overline{T_d}$ to about 0.5, the value of $\overline{T_u}$ can be reduced to about 10^{-6}. Thus, the transmitted beam is reduced to about half the ideal intensity, but the polarization is essentially complete.   ◀

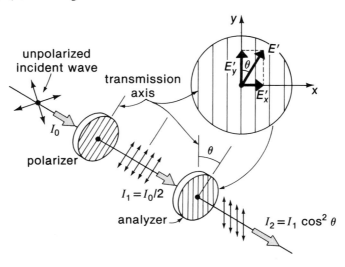

Fig. 34–24. Illustrating the law of Malus.

QUESTIONS

34–1   Define and explain briefly the meaning of the terms (a) linear polarization, (b) Huygens' construction, (c) Snell's law, (d) total internal reflection, (e) fiber optics, and (f) Fermat's principle.

34–2   In what ways are all electromagnetic radiations the same? In what ways are they different?

34–3   How far apart would Galileo and his assistant have had to be in order to detect a time lag equal to the reaction time of 0.2 s and thus at least produce an indication of the speed of light?

34–4   How would the lunar landscape appear to an astronaut if there were only specular reflection at the surfaces illuminated by sunlight? Explain the importance of diffuse reflection and the scattering of sunlight by the atmosphere in our ability to "see" the world around us.

34–5   Explain why a straight rod, when partially submerged in water at an angle, appears to be bent at the surface of the water.

34–6   Name some practical applications of fiber optics.

34–7   Define and explain briefly the meaning of the terms (a) reflectance and transmittance, (b) Brewster's angle, (c) dispersion, (d) dichroism, (e) law of Malus, and (f) birefringence.

34–8   Explain why white light incident from air onto an irregular glass object is reflected as white light but transmitted through the object with a separation of colors.

34–9   Explain how a Polaroid sheet is able to produce linearly polarized light.

34–10 In "line-of-sight" reception of a radio-frequency transmission from a vertical antenna, how should the receiving antenna be oriented for maximum reception of the signal?

34–11 An unpolarized radio wave is normally incident on a grid of parallel wires held in a conducting frame, as in the diagram. (The spacing between the wires in the grid is small compared with the wavelength of the wave.) What will be the polarization of the transmitted beam? [*Hint:* Consider any possible currents induced in the grid.]

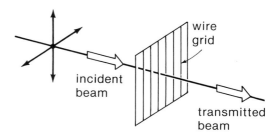

34–12 A light beam consists of equal intensities of linearly polarized and unpolarized light. The beam is passed through a Polaroid sheet at normal incidence. Sketch the measured transmitted intensity as the sheet is rotated through 360°.

34–13 How would the morning overhead sky appear if viewed through a Polaroid sheet that is slowly rotated in a plane perpendicular to the viewing direction?

PROBLEMS

Section 34–2

34–1   In Galileo's attempt to determine the speed of light, he and his companion were located on hilltops 3 km apart. What time interval between uncovering his lantern and seeing the light from his companion's lantern would Galileo have had to measure in order to obtain a value for the speed of light?

34–2   In his measurement of the speed of light, Fizeau used a wheel with 720 teeth and a wheel-to-mirror distance of 8633 m. The slowest angular speed at which the light was blocked by a tooth and did not reach the observer was found to be 756 rpm. What value did Fizeau obtain for c?

34–3   In 1932 Michelson and his collaborators determined the speed of light by using a 32-sided rotating mirror to flash bursts of light down a path with a total length of 15 km. At what speed must the mirror rotate if the flash produced by light shining on one mirror face is reflected from the distant stationary mirror and returns to strike the adjacent face when it has moved to the exact position of the first face?

34–4   Write an expression for $\mathbf{E}$ that describes a wave linearly polarized in a direction that makes an angle of 120° with respect to the x-axis.

Section 34–4

34–5 Light is incident normally on a 1-cm layer of water that lies on top of a flat Lucite plate whose thickness is 0.5 cm. How much longer is required for light to pass through this double layer than is required to traverse the same distance in air?

34–6 A light ray is incident from air on the surface of a clear soft drink ($n_2 = 1.330$) that stands in a glass ($n_3 = 1.520$), as shown in the diagram. The angle of incidence is $\theta_1 = 50°$. Determine the angles θ_2 and θ_3.

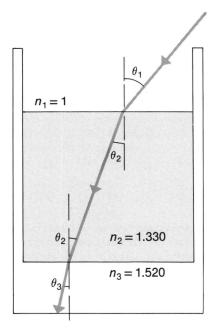

34–7 A cylindric tank with an open top has a diameter of 3 m and is completely filled with water. When the setting Sun reaches an angle of 28° above the horizon, sunlight ceases to illuminate the bottom of the tank. How deep is the tank?

34–8 Light is incident from below on the interface between glass with index $n_1 = 1.520$ and a drop of liquid with

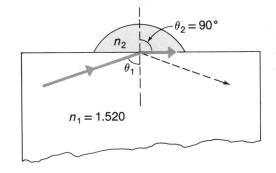

index n_2, as shown in the diagram. When the angle of incidence θ_1 is increased to 74.0°, total internal reflection occurs (that is, $\theta_2 = 90°$). Determine the refractive index of the liquid.

34–9 Use Fermat's principle to derive the law of reflection, • $\theta_1 = \theta_1'$. Refer to the diagram.

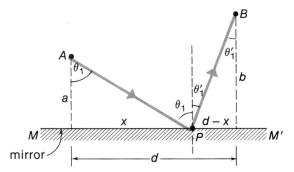

Section 34–5

34–10 When light is incident normally from air on a certain type of clear plastic, 94 percent of the light energy is transmitted. What is the refractive index of the plastic?

34–11 A beam of light is incident normally on a polished face of a diamond. (a) What is the ratio of the amplitude of the reflected electric field to that of the incident electric field? (b) What fraction of the incident (time-averaged) energy flux is reflected?

34–12 Show that when θ_1 is equal to Brewster's angle θ_B, there is an angle of 90° between the direction of the transmitted ray and the direction of the reflected ray. (For $\theta_1 = \theta_B$, reflection occurs only for the perpendicular component of the incident beam.)

34–13 Using the boundary conditions for normal incidence, Eqs. 34–10, derive the amplitude transmission and reflection coefficients given by Eqs. 34–11. [*Hint:* Make use of Eq. 33–15 relating H^0 to E^0 and Eqs. 33–13 and 33–14. Show that the second equation of Eqs. 34–9 becomes $n_1E_i^0 - n_1E_r^0 = n_2E_t^0$, etc.]

Section 34–6

34–14 A prism made from fused quartz ($n = 1.4585$) has an apex angle $\alpha = 30°$. (a) What is the incident angle θ_1 for minimum deviation of sodium yellow light ($\lambda = 5893$ Å)? (b) Determine the minimum deviation δ.

34–15 Derive Eq. 34–16 which relates the minimum deviation •• δ_{min} to the refractive index n and the apex angle α of the prism. [*Hint:* Refer to the diagram. It is required that $d\delta/d\theta_1 = 0$. Use the relationship connecting δ with θ_1

and θ_2 to show that this condition implies $d\theta/d\theta_1 = -1$. Then, write $d\theta_2/d\theta_1 = (d\theta_2/d\gamma)\cdot(d\gamma/d\beta)\cdot(d\beta/d\theta_1) = -1$. Use Snell's law at points A and B to obtain the required relationships.]

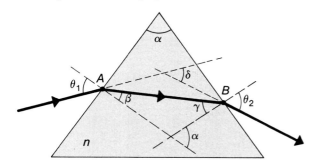

34–16 A prism made from fused quartz has an apex angle of
• 30°. A beam of white light is incident on the prism with $\theta_1 = 22.178°$, which yields minimum deviation for yellow light ($\lambda = 5893$ Å). The refractive indexes for quartz for other wavelengths are $n(7000\text{ Å}) = 1.4553$ and $n(4000\text{ Å}) = 1.4702$. Determine the deviation δ for $\lambda = 4000$ Å and $\lambda = 7000$ Å. What is $\Delta\delta$ between the violet and red ends of the spectrum? [*Hint:* Refer to the diagram for Problem 34–15. Find β, γ, θ_2, in that order, for each wavelength. Calculate to five significant figures.]

Section 34–7

34–17 An unpolarized beam of light is reflected from a glass surface. It is found that the reflected beam is linearly polarized when the light is incident from air at an angle of 58.6°. What is the refractive index of the glass?

Additional Problems

34–23 When the Sun is directly overhead, a narrow shaft of light enters a Mithran temple through a small hole in the ceiling and forms a spot on the floor 10.0 m below. (a) At what speed (in cm/min) does the spot move across the (flat) floor? (b) If a mirror is placed on the floor to intercept the light, at what speed will the reflected spot move across the ceiling?

34–24 A beam of sodium yellow light ($\lambda = 5893$ Å) is incident normally on an optically flat fluorite plate that has a thickness of 20.00 μm. (a) What is the optical path length through the fluorite plate? (b) What is the wavelength of the light in the plate? (c) What phase change does the light wave undergo in passing through the plate compared with a wave that travels the same distance in air?

34–25 According to Newton's corpuscular theory of light, a
• "light particle" that passes into a medium from vacuum

34–18 A certain substance has a critical angle for internal reflection equal to θ_c and a Brewster's angle equal to θ_B. Show that $\sin \theta_c \tan \theta_B = 1$.

34–19 A glass plate ($n_g = 1.52$) is submerged in water ($n_w = 1.33$). What is the polarizing angle for light incident on the glass plate?

Section 34–8

34–20 A beam of unpolarized light is incident normally on two pieces of H-sheet that act as a polarizer-analyzer combination. What fraction of the incident energy flux is transmitted through the pair for angles 0, $\pi/6$, $\pi/4$, $\pi/3$, and $\pi/2$ between the two transmission axes? Assume that the transmittance for the desired component is $\overline{T_d} = 0.83$.

34–21 What angle between the transmission axes of a polarizer-analyzer pair of H-sheet should be chosen so that the emergent energy flux is 0.25 of the incident flux? Assume that the transmittance for the desired component is $\overline{T_d} = 0.83$.

34–22 Two pieces of H-sheet are positioned with their transmission axes at right angles, so no light is transmitted through the pair. A third piece of H-sheet is placed between the original two pieces. Derive an expression for the intensity of light transmitted through the three-piece combination as a function of the angle θ between the transmission axes of the polarizer and the middle sheet. The incident light is unpolarized and has intensity I_0. Assume that the H-sheet is ideal and transmits 100 percent of the desired component of the incident beam.

(air) experiences an attractive force at the interface due to the medium. The work W done on the light particle as it passes through the interface results in a negative potential energy, $U = -W$, relative to vacuum. Refer to the

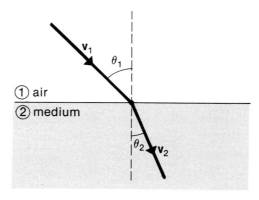

diagram and show that (a) $v_2 = \sqrt{v_1^2 - 2U/m}$ and (b) $n = (\sin \theta_1)/(\sin \theta_2) = v_2/v_1 = \sqrt{1 - 2U/mv_1^2}$ where m is the mass of a light corpuscle. Although the result for n gives $n > 1$ in agreement with experiment, the theory attributes this to $v_2 > v_1$ or $v_2 > c$, instead of $v_2 < c$, as given by the wave theory. [*Hint:* The attractive force that acts on the corpuscle can have no component tangent to the interface. (• Can you see why this is so?)]

34–26 A light beam that is polarized parallel to the plane of incidence is incident at Brewster's angle on a glass plate with parallel surfaces, as indicated in the diagram. There is no reflection at the air-glass interface. Show that there is also no reflection at the glass-air interface, so that there is 100 percent transmission through the plate. (A glass plate oriented at Brewster's angle with respect to an incident beam constitutes a *Brewster window* for the parallel component of the beam.)

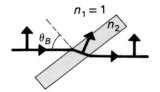

34–27 Two pieces of H-sheet are positioned with their transmission axes perpendicular. A third piece of H-sheet is placed between the original two pieces. Derive an expression for the intensity of light transmitted through the three-piece combination as a function of the angle θ between the transmission axis of the first sheet and the middle sheet. The incident light is unpolarized and has intensity I_0. Assume that the H-sheet is ideal and transmits 100 percent of the desired component of the incident beam.

CHAPTER 35

INTERFERENCE

Maxwell's equations, which describe the behavior of all electromagnetic phenomena, are *linear* equations. This property permits the wide (and exact) application of the superposition principle in the analysis of electromagnetic effects. Thus, we may view a complex situation involving the simultaneous presence at a point of several different electromagnetic wavefronts in terms of the vector sum of the separate field components. The overlapping waves that are combined might be the different Huygens' wavelets imagined to constitute a single propagating wavefront, or they might be waves that originate in separate and independent sources. At any particular instant the vector addition of the field components will give rise to constructive and/or destructive interference effects.*

Phenomena that involve the superposition of waves are usually classified either as *diffraction* effects or as *interference* effects, although there is no completely satisfactory way to make the distinction. If a wavefront initiated by a single source is allowed to propagate unimpeded through a homogeneous medium, the superposition of the Huygens' secondary wavelets results in straight-line propagation of the wave. However, if such a wave is partially blocked by an obstacle or is passed through an aperture, some of the Huygens' secondary sources are eliminated. This results in the spreading of the wave around the obstacle or from the aperture and can also produce constructive and/or destructive interference effects within the transmitted beam. We choose to refer to such phenomena as *diffraction* effects.

When the superposition involves two or more isolated portions of a wavefront, or waves from independent sources, we refer to the resulting variation in the combined field amplitude as an *interference* effect. It should be emphasized, however, that diffraction and interference effects both result from the same basic process of superposition and are therefore not completely distinct effects. In this chapter we examine "interference" effects; Chapter 36 is concerned with "diffraction" effects.

35–1 CONDITIONS FOR INTERFERENCE

In order for interference effects to be observed, it is necessary that several conditions be met. The most important of these is that the actual or effective sources must have the same frequency and a constant phase difference; this is the condition of *coherence*. It is relatively easy to establish coherence for electronically generated electromagnetic waves.

---

*Some aspects of interference phenomena have already been treated in connection with sound (Section 17–4).

For example, special electronic circuits are used to "frequency lock" and "phase lock" the oscillators that drive a pair of radio antennas. Light, on the other hand, consists of radiation from individual atoms. Each atomic emission event produces a train of wave oscillations that has a finite length. In ordinary sources,* these emissions occur randomly from atom to atom. Consequently, the resulting wave trains have random phases. If two such wave trains overlap, interference appropriate for the particular phase difference of the two waves will occur. However, light from a real source, averaged over the very large number of wave trains present, each with a random phase, produces no observable interference effect. Thus, *interference effects involving light ultimately require that the overlapping waves be generated by the same source*. This requirement for a single source can be met by using slits, prisms, and lenses to split the individual wave trains from a primary source into two or more secondary wave trains that are coherent (see Section 35-2).

If the pattern formed by constructive interference (intensity maxima) and destructive interference (intensity minima) is to be easily observed, additional (approximate) conditions should be satisfied. For the maximum effect produced by destructive interference, the vector sum of the two overlapping field components should be zero. The sum of two vectors can be zero only if the vectors are antiparallel and have the same magnitude. Hence, the interfering waves should have wavefronts that are parallel, or nearly so, and the field vectors should have the same polarization.

Finally, two light waves will interfere only if the wave trains overlap in space. Figure 35-1 is a schematic representation of a wave train for which the electric vector is polarized in a plane that is perpendicular to the x-axis.† Within the nominal length L, the wave train is essentially sinusoidal. In order for an oscillation to be a pure monochromatic wave, the frequency must be the same wherever and whenever it is measured; such a wave must have an infinite length. Any wave with finite length, such as the wave train in Fig. 35-1, consists of the superposition of waves with slightly different frequencies distributed in such a way that there is constructive interference in the region of length L and destructive interference everywhere else. Such waves are said to be *quasi-monochromatic* ("almost monochromatic"). The wave trains from ordinary light sources have lengths that are typically a fraction of a meter to several meters.

The physical effects that are produced by incident light depend on the energy flux delivered to a surface. The Poynting vector $\mathbf{N}$ (W/m²) was introduced in Section 33-2 as a measure of energy flux. In the following discussions, we often refer to the *time-averaged* energy flux, a quantity called the *wave intensity* (symbol, I). For the wave train in Fig. 35-1, with $E = E_0 \cos (kx - \omega t)$, the wave intensity at a particular plane, $x = x_0$, is

$$I = \overline{N} = \overline{EH}$$

Fig. 35-1. Schematic representation of a wave train (a plane wave) with finite length. This wave train consists of only a few oscillations, but an actual wave train would typically have a few million oscillations within its length L.

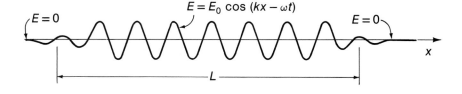

----

*An exception is the class of devices that exhibit *laser* action.

†Because the electrons in matter acquire only small velocities through interactions with light waves, magnetic forces are almost always negligible in comparison with electric forces. Therefore, in describing interference and related effects, we concentrate on the behavior of the electric vectors.

Now, $H = H_0 \cos (kx - \omega t)$ and $H_0 = \sqrt{\epsilon_0 / \mu_0} E_0 = E_0 / Z_0$ (Eq. 33–16), where $Z_0 = 377 \; \Omega$, for a wave propagating in vacuum (or air). Then,

$$I = \overline{E^2} / Z_0 \qquad (35\text{–}1)$$

so that

$$I = (E_0^2 / Z_0) \; \overline{\cos^2 (kx - \omega t)}$$

or

$$I = \tfrac{1}{2} E_0^2 / Z_0 \qquad \text{wave intensity} \qquad (35\text{–}1a)$$

where the units of I are W/m^2 and those of E_0 are V/m.

Example 35–1

Two coherent equal-amplitude electromagnetic plane waves are propagating in the x-direction with parallel polarizations. The waves arrive at a plane $x = x_0$ with their wavefronts essentially parallel. The phase difference of the two waves at that plane is δ. What is the wave intensity on the plane $x = x_0$? (For convenience, consider the waves to be infinite in extent, i.e., purely monochromatic.)

$E_1 = E_0 \cos (kx_0 - \omega t)$

$E_2 = E_0 \cos (kx_0 - \omega t - \delta)$

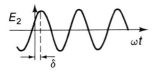

Solution:

The diagram shows the time variation of the electric vectors, E_1 and E_2, for the two waves on the plane $x = x_0$. The

superposition of the two waves gives

$$E = E_1 + E_2 = E_0[\cos (kx_0 - \omega t) + \cos (kx_0 - \omega t - \delta)]$$

from which

$$E = 2E_0 \cos \left(kx_0 - \omega t - \frac{\delta}{2} \right) \cos \frac{\delta}{2}$$

The (time-averaged) wave intensity is given by Eq. 35–1 as $I = \overline{E^2} / Z_0$. Now, in this case,

$$\overline{E^2} = 4E_0^2 \cos^2 \frac{\delta}{2} \; \overline{\cos^2 \left(kx_0 - \omega t - \frac{\delta}{2} \right)}$$

$$= 2E_0^2 \cos^2 \frac{\delta}{2}$$

It is customary to use I_0 to represent the intensity of one wave alone. Thus, $I_0 = \tfrac{1}{2} (E_0^2 / Z_0)$, so that

$$I = 4I_0 \cos^2 \frac{\delta}{2}$$

Note the important point that the intensity of two equal-amplitude superimposed waves that are in phase ($\delta = 0$) is *four times* the intensity of either wave alone, not simply twice the individual intensity.

35–2 YOUNG'S DOUBLE-SLIT EXPERIMENT

In an important experiment performed by Thomas Young in 1801, interference effects produced by light were clearly demonstrated. This was the first of several crucial experiments that eventually established the wave theory of light. From this experiment, Young was also able to determine approximately the wavelength of light.

The (nearly) monochromatic light from a source such as glowing sodium vapor is collimated (that is, formed into a narrow beam) by a narrow slit S in a barrier, as shown in

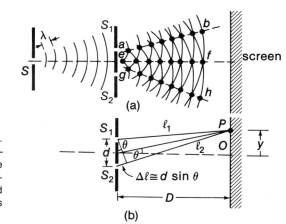

(a)

(b)

Fig. 35-2. (a) Interference of the coherent waves from the slits, S_1 and S_2. Wavefronts corresponding to amplitude maxima of a particular sign (and therefore spaced apart by one wavelength λ) are shown for a particular instant. To the right of the slits, the dots correspond to instantaneous intensity maxima produced by constructive interference. (Compare Fig. 17-8 for the case of sound waves.) The lines $\overline{ab}$, $\overline{ef}$, and $\overline{gh}$ are lines of antinodes. (b) Geometry of Young's double-slit experiment.

Fig. 35-2a; this slit acts as the primary light source.* Diverging waves radiate from this slit and impinge on the pair of narrow, parallel slits, S_1 and S_2, that are separated by a distance d and are equidistant from S. The result is that S_1 and S_2 become sources of coherent secondary waves of equal amplitude, which radiate outward and interfere in the region to the right of the slits, as indicated in Fig. 35-2a. The light that passes through the slits falls on a screen at a distance D from the slits (Fig. 35-2b). At a point $P(y)$ on the screen, the path length ℓ_2 from S_2 is longer than the path length ℓ_1 from S_1 by an amount, $\Delta\ell \cong d \sin\theta$, when θ is small. (• Can you see how to obtain this expression for $\Delta\ell$?) Because the waves are in phase when they leave the slits S_1 and S_2, the wave from S_2 arrives at the point $P(y)$ with a phase lag δ relative to the wave from S_1. To convert $\Delta\ell$ into a phase difference, first divide by λ to measure the path-length difference in wavelengths and then multiply by 2π to convert to radians. Thus,

$$\delta = \frac{2\pi d}{\lambda}\sin\theta \qquad (35-2)$$

When θ is a small angle, the wavefronts of the interfering waves are nearly parallel. Then, we have the conditions of Example 35-1, so the intensity at the point $P(y) = P(\theta)$ is

$$I(\theta) = 4I_0 \cos^2\left(\frac{\pi d}{\lambda}\sin\theta\right) \qquad (35-3)$$

An intensity maximum occurs for $\theta = 0$, corresponding to the intersection of the central line of antinodes $\overline{ef}$ with the screen (Fig. 35-2a). The next intensity maxima, on either side of the central maximum, have $\sin\theta = \pm\lambda/d$ and correspond to lines of antinodes $\overline{ab}$ and $\overline{gh}$. (• What is the value of the argument of the cosine function in Eq. 35-3 at this angle θ?) In fact, constructive interference and, hence, intensity maxima occur on the screen when

$$\sin\theta = \pm\frac{\lambda}{d}m, \qquad m = 0, 1, 2, 3, \ldots \qquad \text{maxima} \qquad (35-4a)$$

---

*Young actually used sunlight passed through pinholes, but the essential features of the two arrangements are the same.

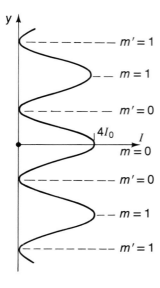

Fig. 35–3. Intensity pattern along the screen in Young's double-slit experiment. The intensity maximum (bright fringe) labeled $m = 0$ is the central or zero-order bright fringe. The average intensity across the screen is $2I_0$.

Destructive interference and, hence, intensity minima (*zero* intensity) occur when

$$\sin \theta = \pm \frac{\lambda}{d}(m' + \tfrac{1}{2}), \qquad m' = 0, 1, 2, 3, \dots \qquad \text{minima} \qquad \textbf{(35–4b)}$$

The result of this interference is to produce on the screen a series of alternating bright and dark bands or *interference fringes*. The central bright fringe ($m = 0$) is called the *zero-order fringe;* the next pair of bright fringes ($m = \pm 1$) are the *first-order fringes;* and so forth. (See Fig. 35–3.)

The intensity at each maximum is $4I_0$. Because the intensity curve has a cosine-squared shape (Eq. 35–3), the average intensity over the screen is $2I_0$, which is the same as the intensity would be if the sources were incoherent.

As a practical matter, with simple arrangements involving ordinary sources of (nearly) monochromatic light and ordinary slits, only the first five to 10 bright fringes on either side of the central maximum will be clearly visible. With intense, highly monochromatic laser sources and very narrow slits, many more fringes are visible. If sunlight is used, as in Young's original experiment, all fringes with $m \geqslant 1$ exhibit color structure and only the first few fringes are discernible. (• Can you give the reason for this limitation?)

The analysis we have made is based on the implicit assumption that the slits are sufficiently narrow that they represent line sources of light. Actually, the slits have finite size, and this leads to diffraction effects that are superimposed upon the interference effects. The required modifications of the results are discussed in Section 36–2.

Example 35–2

In a double-slit experiment, the slits are separated by an amount $d = 0.4$ mm and the distance to the screen is $D = 2.0$ m. Determine the spacing between adjacent bright fringes if yellow sodium light ($\lambda = 5.89 \times 10^{-7}$ m) is used.

Solution:

The spacing is

$$\Delta y = D(\tan \theta_{m+1} - \tan \theta_m)$$

Now, θ is a small angle, so $\tan \theta \cong \sin \theta$; then,

$$\Delta y = D(\sin \theta_{m+1} - \sin \theta_m)$$

Using Eq. 35-4a, we obtain

$$\Delta y = D \cdot \frac{\lambda}{d} [(m + 1) - m]$$

$$\Delta y = D \cdot \frac{\lambda}{d} = \frac{(2.0 \text{ m})(5.89 \times 10^{-7} \text{ m})}{0.4 \times 10^{-3} \text{ m}} = 2.94 \text{ mm}$$

The spacing between the dark fringes is the same.

35-3   INTERFERENCE PRODUCED BY THIN FILMS

It is a common experience to see colored bands and rings in soap bubbles and in thin films of oil floating on water. These effects are the result of the interference between rays of light that are reflected into the viewer's eye from the two surfaces of a thin film.

Consider the reflection of light from a thin wedge-shaped film, as shown in Fig. 35-4. Each wave train of light from a source S is partially reflected and partially transmitted at points such as a, b, and c. All portions of each original wave train that eventually reach the observer's eye are coherent and will interfere.

Fig. 35-4. Interference of light reflected from the surfaces of a thin, transparent film in air.

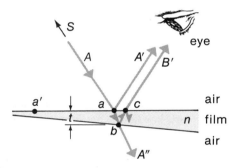

Figure 35-5 shows the geometry of the reflection in more detail. The ray AB' clearly follows a longer optical path than does the ray AA'. For normal incidence, $\theta_1 = 0$, the path difference is $2nt$. For the general case of oblique incidence, we have* $\Delta \ell = n(\overline{ab} + \overline{bc}) - \overline{ad} = 2nt \cos \theta$. This path difference produces a phase difference, $\delta = (2\pi/\lambda_a) \Delta \ell = (2\pi/\lambda_a) \cdot 2nt \cos \theta$, where λ_a is the wavelength of the light in air. An additional phase difference is introduced because the reflection at a involves a phase change of π, whereas no phase change occurs for the reflection at b or for the transmissions at a and c. (Refer to Section 34-5 for a discussion of the behavior of the phase upon reflection. When the medium from which the ray is incident has the smaller index of refraction and the incident angle is near $0°$, there is a phase change of π. When the incident medium has the greater index of refraction, there is no phase change.) Thus, the phase difference between rays AA' and AB' is

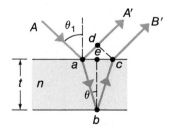

Fig. 35-5. Details of the reflection of light from a thin film near the point of incidence a.

$$\delta = \frac{2\pi}{\lambda_a} \Delta \ell + \pi = \frac{2\pi}{\lambda_a} (2nt \cos \theta) + \pi \qquad \textbf{(35-5)}$$

For near-normal incidence, the amplitudes of the two rays differ by only a few percent. When the eye is focused on the plane of the film, interference fringes are observed.

---

*We can see this as follows: First, $\overline{ab} = \overline{bc}$, so that $\Delta \ell = 2n \overline{ab} - \overline{ad}$. Now, $\overline{ad} = 2\overline{ae} \cos (\pi/2 - \theta_1) = 2\overline{ae} \sin \theta_1 = 2n\overline{ae} \sin \theta$, where we have used Snell's law at point a. Then, $\overline{ad} = 2n \overline{ab} \sin^2 \theta$, and $\Delta \ell = 2n \overline{ab} - 2n \overline{ab} \sin^2 \theta = 2n \overline{ab}(1 - \sin^2 \theta) = 2n \overline{ab} \cos^2 \theta = 2nt \cos \theta$, where $t = \overline{ab} \cos \theta$. At normal incidence, $\Delta \ell = 2nt$, as required.

Each fringe of a particular color (or wavelength λ_a) corresponds to a band along which $nt \cos \theta$ is constant and, hence, the phase difference δ has a constant value. As the point of focus is moved along the film surface from a to a' (Fig. 35–4), both the thickness t and the value of $\cos \theta$ decrease. (We assume that n is constant.) Consequently, the phase difference δ will remain constant only for light with decreasing wavelength. That is, the interference fringe varies in color from the red end of the spectrum at a toward the blue end at a'. For the next band, the additional path length difference is one wavelength λ_a. These variations produce the colored bands and rings that are observed in thin films.

The process just described will produce strong (high-visibility) fringe patterns with extended light sources, such as the Sun or a flame, only for films whose thickness is not much more than a few wavelengths of visible light. Fringes are observable under these conditions because only light within a narrow range of incident angles can enter the pupil of the eye. For example, if an eye with a pupil diameter of 4 mm is located 1 m from a film, the angular diameter of the pupil is only 0.2°. Then, the value of $\cos \theta$ across any pupil diameter is essentially constant. On the other hand, if the film is thick, even small variations in $\cos \theta$ produce large changes in the phase difference δ. Then, the fringes become too closely spaced to be observed with the unaided eye. Consequently, interference fringes are seen clearly only for very thin films. (The aid of a telescope in near-normal geometry is required for the observation of fringes for very large path-length differences.)

Antireflective Coating. The interference effects that are produced by thin films can be put to practical use. The reflectance of light incident normally on glass is about 0.04 (see Example 34–4). By applying a suitable thin coating to the surface, the reflectance can be reduced to about 0.01. The essential details are shown in Fig. 35–6. If the coating material is chosen such that

$$n_c \cong \sqrt{n_g} \qquad (35-6)$$

the relevant reflection and transmission coefficients for normal incidence combine to give nearly equal amplitudes for the first reflected ray AA' and the sum of the subsequent reflected waves such as AB' (see Problem 35–21). In this case we have $n_a < n_c < n_g$, so that both rays AA' and AB' undergo a phase change of π at the points a and b, respectively. (• Can you verify this statement?) Thus, the phase difference between rays AA' and AB' is due entirely to the path length of AB' within the coating. For normal incidence, this path length difference is $\Delta\ell = 2n_c t$. If the thickness of the coating is such that $\Delta\ell = \frac{1}{2}\lambda_a$, the combining out-of-phase amplitudes (which are almost the same in magnitude) result in nearly complete destructive interference. However, the cancellation can be complete only for one wavelength. Antireflective coatings are usually applied with thicknesses to cancel yellow-green light with $\lambda_a = 550$ nm, in the middle of the visible spectrum. Then the reduction of the reflectance for red and blue light is substantial but not complete. It is for

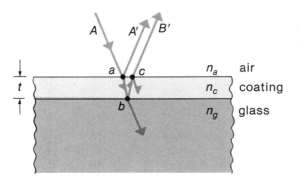

Fig. 35–6. An antireflective coating with refractive index $n_c \cong \sqrt{n_g}$ is applied to a glass surface.

this reason that surfaces with antireflective coatings (for example, camera lenses) appear slightly purple—a combination of reflected red and blue light.

An effective coating material must satisfy Eq. 35–6 and must also be mechanically and chemically stable, moisture-resistant, and so forth. The most commonly used antireflective material is magnesium fluoride, MgF_2, which has $n_c = 1.38$. When applied to dense flint glass ($n_g = 1.656$ for $\lambda_a = 550$ nm—see Fig. 34–15), this substance is quite effective because $\sqrt{n_g} = 1.29$, which is close to n_c. A coating thickness of

$$t = \frac{\lambda_a}{4n_c} = \frac{550 \times 10^{-9} \text{ m}}{4 \times 1.38} = 100 \text{ nm}$$

which is called a *quarter-wavelength coating*, reduces the overall reflectance from about 4 percent to about 1 percent.

It is possible to construct a highly *reflective* surface by using a sack of quarter-wavelength films of alternating high refractive index and low refractive index. (• Can you see why this arrangement produces a large reflectance?) Such a film stack has the advantage, not possessed by a metallic surface, that little light is lost by absorption. It is possible to construct a stack that has $R = 0.995$ and $T = 0.003$.

Newton's Rings. It is possible to observe interference effects due to an air gap between two glass surfaces. This air gap may be a wedge-shaped space between two flat plates (see Example 35–3), or it may involve a curved surface. An example of the latter is the space between a flat plate and the spherical surface of a plano-convex lens (Fig. 35–7). Newton (as well as his contemporaries, Robert Boyle and Robert Hooke) noted that colored rings (called *Newton's rings*) are observed when looking straight down on such a lens that is illuminated by a nearby source. The interference effects are pronounced if monochromatic light is used and if the source and observer are close to the normal at the point of contact of the lens, as in Fig. 35–7. (Newton's rings are also observable in transmitted light.)

If the radius R of the spherical lens surface is very large, the thickness of the air gap near the point of contact is of the order of the wavelength of light, and the analysis for the case of thin films applies. Consider a light ray at near-normal incidence. This ray will be partially reflected at points a and b (see Fig. 35–7), producing two interfering rays with a path-length difference (in air) of $2t$. The optical path-length difference is thus $4\pi t/\lambda_a$. The reflection at b involves a phase change of π, whereas no phase change results from the reflection at a. Then, interference maxima occur at radii r for which $4\pi t/\lambda_a = (2m + 1)\pi$,

Fig. 35–7. The geometry of Newton's rings. The lens has one flat surface and is called a *plano-convex* lens. (Photo courtesy of Bausch & Lomb.)

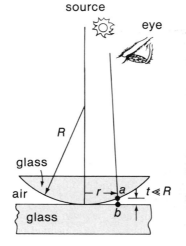

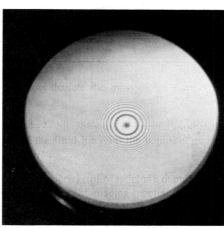

with $m = 0, 1, 2, \ldots$ .Now, we also have $R^2 = (R - t)^2 + r^2$; for $t \ll R$, this expression gives $t \cong r^2/2R$. Thus, interference maxima produce bright rings with radii

$$r_m = \sqrt{(m + \tfrac{1}{2})R\lambda_a}, \qquad m = 0, 1, 2, \ldots \qquad \textbf{(35-7)}$$

At the point of contact $(r = 0)$, the medium is effectively continuous, so there is no reflection. Consequently, the center of the ring system is dark, as shown in Fig. 35-7.

Example 35-3

Two flat glass plates are separated at one edge by a thin spacer with a thickness h. For the case of near-normal incidence, locate the interference maxima with respect to the contact edge.

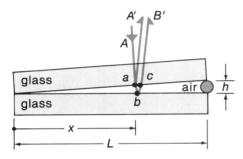

Solution:

The thickness of the glass plates is large compared with the wavelength of the incident light, so the observable interference effects are due entirely to the wedge-shaped air gap. The ray AB' undergoes a phase change of π upon reflection at point b. The

thickness of the air gap at this point, $t = (x/L)h$, introduces an additional phase difference, $\delta = 4\pi xh/L\lambda_a$. Hence, interference maxima occur at positions for which

$$\frac{4\pi xh}{L\lambda_a} + \pi = 2m\pi, \qquad m = 1, 2, 3, \ldots$$

At the contact edge, the medium is essentially continuous, so there is no reflection and this edge is dark.

The spacing Δx between successive bright fringes is given by

$$\frac{4\pi(x + \Delta x)h}{L\lambda_a} - \frac{4\pi xh}{L\lambda_a} = 2(m + 1)\pi - 2m\pi$$

from which

$$\Delta x = \frac{L}{2h}\lambda_a$$

For $L = 20$ cm, $h = 0.1$ mm, and $\lambda_a = 589$ nm (sodium yellow light), we find

$$\Delta x = \frac{20 \text{ cm}}{2(0.01 \text{ cm})}(5.89 \times 10^{-7} \text{ m}) = 5.89 \times 10^{-4} \text{ m}$$

$$= 0.589 \text{ mm}$$

35-4   THE MICHELSON INTERFEROMETER

In 1881, Albert A. Michelson used the idea of light interference to devise an instrument capable of making distance measurements with extremely high precision (within a fraction of a wavelength of light). A diagram of the essential features of a *Michelson interferometer* is shown in Fig. 35-8. Light from a monochromatic source is incident on a half-silvered glass plate P_1 (the ''beam splitter'') where the original beam is partially reflected and partially transmitted. The reflected portion, labeled ① in the diagram, travels a distance L_1 to mirror M_1, which reflects the beam back toward P_1. There it is partially transmitted and arrives finally at the eye of the observer. The transmitted portion of the original beam, labeled ②, travels a distance L_2 to mirror M_2, which reflects the beam back toward P_1. It is partially reflected and arrives finally at the eye of the observer. The plate P_2 (the ''compensator plate'') is cut from the same piece of glass as P_1 and is introduced into beam ② to equalize the optical path lengths of the two beams. The two

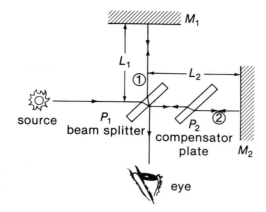

Fig. 35–8. Schematic diagram of the arrangement of plates and mirrors in a Michelson interferometer.

wave trains, having traveled these two different paths, are coherent when they arrive at the eye and therefore produce observable interference effects. The mirrors, M_1 and M_2, are nearly but *not exactly* perpendicular; the small angular difference results in a slight inclination, at the eye, of one image of the extended source with respect to the other. The observer then sees a series of parallel interference fringes. (The effect is similar to that produced by the air gap of varying thicknesses discussed in Example 35–3.)

In a practical Michelson interferometer, the mirror M_1 can be translated along the direction perpendicular to its surface by means of a precision screw. As the length L_1 is changed, the pattern of interference fringes is observed to shift. By using a telescope with a cross-hair eyepiece, it is easy to see the fringes shift as L_1 is changed. A shift of the pattern by an amount corresponding to the distance between two successive bright fringes is observed for a change in L_1 equal to $\frac{1}{2}\lambda_a$. The distance that M_1 is moved (in units of the wavelength of the source light) is determined by counting the number of fringes that pass the cross-hair as the movement is made. A careful observer using a well-constructed instrument can detect a shift as small as 1/40 of a fringe spacing. This means that distances can be measured to about 75 Å.

It was pointed out in Section 35–1 that a wave train of light has a length that is typically of the order of a meter (see Fig. 35–1). Thus, when a particular wave train enters a Michelson interferometer and is divided by the "beam splitter" into two portions, these portions will not eventually overlap and interfere if the path-length difference, $L_1 - L_2$, is greater than the length of the wave train. This path-length difference in which some observable degree of interference can occur is called the *coherence length* of the radiation.

35–5 HOLOGRAPHY (Optional)

In ordinary photography, a piece of film is used to record the intensity distribution of light from a source object. In a special type of photography, called holography,* the wavefronts themselves (both the phase and the intensity) are recorded by means of interference effects. A holograph therefore contains much more information than does a photograph. When a holographic image is reproduced, it presents a remarkable three-dimensional appearance. The image stands in space and may be viewed from different angles and distances. The effect is the same as viewing the object itself with all of its three-dimensional aspects.

Figure 35–9a illustrates the process of recording the holographic information on film. The incident light is a highly monochromatic plane wave such as may be produced by a *laser*. One portion of the incident wavetrain is directed at the film and constitutes the *reference plane wave*. The other portion illuminates the objects studied. At each point on these objects, such as point

---

*From the Greek *holographos*, "written in full."

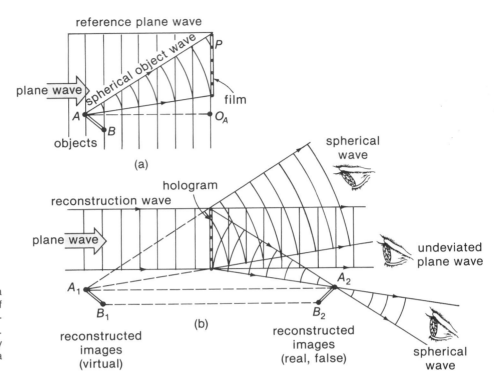

(a)

(b)

Fig. 35–9. (a) Preparation of a hologram by the interference of the reference wave and the reflected object wave at the film. (b) Reconstruction of the image by irradiating the hologram with a plane wave of laser light.

A shown in Fig. 35–9a, a reflected *spherical object wave* is initiated, and a portion of this wave reaches the film. Laser radiation is highly monochromatic and has a coherence length of several meters. Consequently, if the scene that is viewed has dimensions of a few meters or less, the reference wave and the object waves will overlap and interfere at the film. For example, point *P* on the film is one point on a circular arc centered on O_A along which the intensity is maximum for object point *A*. The intensity distribution that is produced by object point *A* is determined both by the brightness of the reflected wave from *A* and by the phase relationship between the object wave and the reference wave at the film. The total pattern recorded on the film is the linear superposition of the contributions from all object points such as *A* and *B*. The transparent positive print of the developed film is called a *holograph* or *hologram* (see the photograph at right).

The reconstruction or viewing process is illustrated in Fig. 35–9b. A plane wave of monochromatic laser light with the *same wavelength* as the light used to expose the film is incident normally on the hologram. The transmitted radiation can be analyzed in terms of three components. In the particular version of holography depicted in Fig. 35–9b, these components are (1) a reduced-intensity undeviated continuation of the incident plane wave (which is slightly diffracted); (2) a diverging spherical wave (formed by diffraction) that appears to originate in a virtual image that corresponds in location and brightness to the object point (for example, point *A*); and (3) a spherical wave (also formed by diffraction) that converges to form a real image of the

object (and thereafter diverges); this image is reversed and is referred to as a *false image*.

In the process just described, we considered the use of a thin flat film to obtain an essentially two-dimensional pattern on the film. A hologram prepared in this way can be viewed only with monochromatic light. If the image is to be the same size as the object, the reference and reconstruction waves must both be

Hologram of a barracuda. (Courtesy of Metrologic Instruments, Inc.)

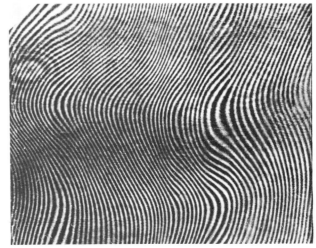

plane waves* with the same wavelength. Thus, if the hologram is prepared using red laser light, the full-size image will appear in red light.† However, if a film with a thick emulsion is used, the standing-wave pattern produced throughout the emulsion volume by interference can be recorded. Full-color holography can then be achieved by exposing the film using three lasers that emit red, green, and blue light (the colors used for color television), and then viewing the hologram with white light.‡

To prepare a hologram using a continuously operating laser,

it is necessary that the objects remain absolutely stationary. A movement of a fraction of a wavelength will produce a blurred holographic image. However, three-dimensional motion-picture holography may be feasible by using *pulsed* lasers.§ When recording, a laser burst with a duration of about 5 ns for each motion-picture frame should arrest sufficiently any ordinary stage motion. On reconstruction, each frame could be viewed using a longer burst duration in order to reduce flicker. (See Problem 35–15.)

QUESTIONS

35–1   Define and explain briefly the meaning of the terms (a) coherence, (b) quasi-monochromatic waves, (c) interference fringes, (d) antireflective coating, (e) the Michelson interferometer, and (f) holography.

35–2   Monochromatic, but incoherent, light strikes a narrow slit in a barrier. Why can the light emerging from the slit be considered coherent?

35–3   In what way are interference effects produced by sound waves, water waves, and light similar and in what ways are they different?

35–4   When two sound waves of slightly different frequencies are superimposed, beat frequencies are produced. Why would it be difficult to observe beat effects with light?

35–5   Young performed his famous experiment on interference using two pinholes and sunlight. Describe and sketch the appearance of the fringe intensity pattern you would expect to see. Estimate how many orders of interference patterns you would expect to resolve.

35–6   If Young's experiment with two slits employed "white" light, and if one slit had a blue filter placed in front of it and the other a red filter placed in front of it, describe the intensity pattern you would expect to see on the screen.

35–7   Suppose Young's double-slit experiment were performed with each slit covered by a Polaroid sheet, with the transmission axes of the sheets oriented perpendicular to each other. Describe the intensity pattern you would expect to see on the screen.

35–8   Suppose Young's double-slit experiment were performed with linearly polarized monochromatic incident light. If the slit spacing and wavelength were such that the low-order minima were at angles greater than, say, 15°, how close to zero intensity would be observed at the minima for (a) polarization in the direction parallel to the slits and (b) polarization perpendicular to the slits?

35–9   What effect on the screen intensity pattern results if Young's double-slit arrangement has the slit separation doubled? If light of twice the wavelength is used? If both the slit separation and wavelength are doubled?

35–10   Design a suitable Young's double-slit arrangement using 10-cm radio waves.

35–11   Describe how the Newton's ring pattern would change if the lens radius were doubled; if the wavelength of the light were doubled; if both the lens radius and wavelength were doubled.

PROBLEMS

Section 35–1

35–1   Reconsider Example 35–1 for the case of two waves with unequal amplitudes, so that

$$E = E_1 \cos (kx_0 - \omega t) + E_2 \cos (kx_0 - \omega t - \delta)$$

Show that the time-averaged value of E^2 is

$$\overline{E^2} = \tfrac{1}{2}E_1^2 + \tfrac{1}{2}E_2^2 + E_1E_2 \cos \delta$$

Show that this expression reduces to that obtained in Example 35–1 for the case $E_1 = E_2 = E_0$.

---

* In methods that use spherical waves, magnified (or diminished) images are possible, and the perspective of the images may also be altered.

† If the hologram is irradiated with white light, each wavelength produces an image of a different size and no apparent reconstruction results.

‡ This method is a variation of the color photography process invented in 1891 by the French physicist Gabriel Lippmann (1845–1921), for which he received the 1908 Nobel Prize in physics.

§ Such motion pictures were first made in 1976 by NIFKI, the Soviet Cine and Photo Research Institute.

Section 35–2

35–2 Young's experiment is performed by passing monochromatic light through two narrow parallel slits that are separated by 0.30 mm. The bright fringes on a screen 1.5 m away are observed to have a spacing of 2.99 mm. What is the wavelength of the incident light?

35–3 Two narrow parallel slits are separated by 0.85 mm and are illuminated by light with $\lambda = 6000$ Å. What is the phase difference between the two interfering waves on a screen (2.8 m away) at a point 2.50 mm from the central bright fringe? What is the ratio of the intensity at this point to the intensity at the center of a bright fringe?

35–4 Young's experiment is performed using white light. The two narrow slits are separated by 0.40 mm. The screen is at a distance of 3.0 m. (a) Locate the first-order maximum for yellow-green light ($\lambda = 550$ nm). (b) What is the separation between the first-order maxima for red light ($\lambda = 700$ nm) and violet light ($\lambda = 400$ nm)? (c) At what point nearest the central maximum will a maximum for yellow light ($\lambda = 600$ nm) coincide with a maximum for violet light ($\lambda = 400$ nm)? Identify the order of each maximum.

35–5 Consider the interference arrangement shown in the diagram, called Fresnel's double mirror. It consists of two plane mirrors with a very small angle α between them. (a) Show that the distance between successive bright fringes on the screen is given by $\Delta y = \lambda D/2a\alpha$. (b) Locate the first-order bright fringe if $a = 25$ cm, $D = 1.25$ m, $\lambda = 5 \times 10^{-7}$ m, and $\alpha = 5'$ (5 arc minutes).

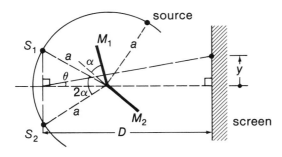

Section 35–3

35–6 A soap film ($n = 1.45$) is contained within a rectangular wire frame. The frame is held vertically so that the film drains downward due to gravity and becomes thicker at the bottom than at the top, where the thickness is essentially zero. The film is viewed in white light with near-normal incidence, and the first violet ($\lambda = 420$ nm) interference band is observed 3 cm from the top edge of the film. (a) Locate the first red ($\lambda = 680$ nm) interference band. (b) Determine the film thickness at the posi-

tions of the violet and red bands. (c) What is the wedge angle of the film?

35–7 A soap film such as that in Problem 35–6 is viewed in yellow light ($\lambda = 589$ nm) with near-normal incidence. Interference fringes with a uniform spacing of 4.5 mm are observed. What is the variation in thickness of the film per centimeter of vertical distance?

35–8 A coating of magnesium fluoride is to be used on a glass plate to reduce the reflection of yellow light ($\lambda = 580$ nm) that is incident at an angle of 30°. What thickness should the coating be?

35–9 A thin film of glass ($n = 1.52$) with a thickness of 0.42 μm is viewed in white light at near-normal incidence. Visible light with what wavelength is most strongly reflected by the film?

35–10 The air in the gap between the glass plates of Example 35–3 is replaced by methyl alcohol ($n = 1.329$). When sodium yellow light ($\lambda = 589$ nm) is normally incident, fringes with a spacing of 0.45 mm are observed. What is the wedge angle between the plates?

35–11 A plano-convex lens with a radius of curvature $R = 3.0$ m is in contact with a flat plate of glass. A light source and the observer's eye are located close to the normal, as shown in Fig. 35–7. The radius of the 50th of the bright Newton's rings is found to be 9.8 mm. What is the wavelength of the light produced by the source?

Section 35–4

35–12 A Michelson interferometer (Fig. 35–8) is illuminated with green light ($\lambda = 5300$ Å). How many fringes will be seen to pass the cross-hair in the viewing telescope as the mirror M_1 is moved from one end of a 1.0-cm gauge block to the other end? A new light source with a different color is substituted, and the observer counts 36,200 fringes as the mirror is returned to its original position. What is the wavelength of the light from the new source?

35–13 An evacuated glass tube with parallel glass end plates and an inside length of 10.0 cm is placed into one arm of a Michelson interferometer that uses sodium vapor light (Fig. 35–8). A gas is slowly admitted into the tube and the shift of the fringes is observed. When a gas pressure of 1.0 atm is reached, 99.5 fringes have passed the cross-hair of the viewing telescope. What is the refractive index of the gas? Give a possible identification of the gas. (Use Table 34–2; the temperature is 0° C.)

35–14 The interference pattern observed in a Michelson interferometer that uses sodium yellow light ($\lambda = 589$ nm) has a fringe spacing of 0.60 mm. What amount of deviation from an angle of 90° between the mirrors, M_1 and M_2, could produce this fringe spacing?

Section 35–5

35–15 A holographic motion picture is to be made of a scene in which all motion takes place with a speed less than 10 m/s. The illuminating laser produces light with $\lambda = 5000$ Å. The amount of motion during the record-ing of any frame of the film cannot exceed $\lambda/10$. (a) What is the maximum allowable duration of the laser pulse? (b) What is the length of the corresponding wave train? Is this adequate for recording a scene with typical dimensions?

Additional Problems

35–16 A *Rayleigh refractometer,* shown schematically in the diagram, is a device used in precision measurements of refractive indexes of gases. Two identical glass cells, A and B, are placed behind a pair of narrow slits that are illuminated by a plane wave of blue-green light ($\lambda = 550$ nm). Each cell has an inside length $\ell = 2.500$ cm and $\ell \ll D$. When the cells are evacuated, the optical path lengths are identical and the central bright fringe is located at P. Gas is slowly admitted into cell A and the central fringe is observed to move upward. At a pressure of 1.00 atm (and a temperature of $0°$ C) the central fringe is at Q, a distance equal to 20.5 fringe spacings from P. Determine the refractive index of the gas. Give a possible identification of the gas. (Refer to Table 34–2.)

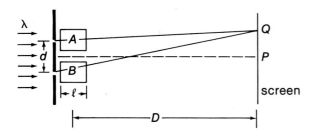

35–17 Young's double-slit experiment is performed with sodium yellow light ($\lambda = 5890$ Å) and with a slits-to-screen distance of 2.0 m. The tenth interference minimum (dark fringe) is observed to be 7.26 mm from the central maximum. Determine the spacing of the slits.

35–18 The diagram shows a modification of Young's experiment using the so-called *Billet split lens.* It consists of a lens that is cut in half, with the halves slightly displaced. This arrangement is used with light of wavelength 475 nm. The focal length of the thin lens is $f = +25$ cm, and the lens halves are separated by an opaque spacer

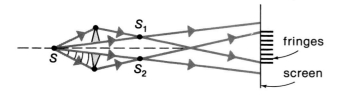

with a thickness of 0.30 mm. The primary source S is 50 cm behind the lens, and the screen is 1.5 m in front of the lens. (a) Explain how in effect two coherent sources, S_1 and S_2, are formed. (b) Locate the secondary sources, S_1 and S_2, and determine their separation. (c) What is the spacing of the fringes on the screen?

35–19 A 0.10-μm layer of carbon disulfide ($n = 1.625$) covers a glass plate ($n = 1.520$). What is the longest wavelength of light for which complete destructive interference will result when a normally incident beam is reflected?

35–20 A thin layer of methylene iodide ($n = 1.756$) is sandwiched between two flat parallel plates of glass. What must be the thickness of the liquid layer if normally incident light with $\lambda = 600$ nm is to be strongly reflected?

35–21 Consider light at normal incidence on a glass plate
 • coated with an antireflective material (Fig. 35–6). Show that equating the amplitude reflection coefficients at the air-coating interface and the coating-glass interface lead to Eq. 35–6. (This result is only approximate because it ignores the contribution from multiple reflections and assumes that the reflection coefficients are very much smaller than the transmission coefficients.)

35–22 Two glass plates with $L = 10$ cm are in contact along one edge and are separated by a 0.10-mm thick spacer at the opposite edge, as in Example 35–3. When monochromatic light is incident normally, fringes with a spacing of 0.34 mm are observed. What is the wavelength of the light?

35–23 Monochromatic light with continuously variable wavelength is incident normally on a thin film of oil ($n = 1.30$) that covers a glass plate. As the wavelength of the light is varied from 450 nm to 750 nm, complete destructive interference is observed for $\lambda = 500$ nm and for $\lambda = 700$ nm. Determine the thickness of the oil film. (Note that $n_{\text{oil}} < n_{\text{glass}}$.)

35–24 Sodium yellow light ($\lambda = 589$ nm) is used to produce Newton's rings in the arrangement shown in Fig. 35–7. The radius of the 30th bright ring is found to be 8.0 mm. What is the thickness of the air film at the position of this ring? What is the radius of curvature of the lens?

CHAPTER 36

DIFFRACTION

Diffraction effects occur when incident waves interact with obstacles or apertures with finite size. Then, the waves bend around the obstacles or spread out from the apertures in the familiar manner of water waves. Although many effects involving light can be analyzed in terms of rectilinear propagation, careful investigation reveals that light waves also diffract into regions that would not be illuminated if only straight-line propagation were possible.*

It is convenient to separate the phenomena associated with the diffraction of light into two categories. *Fresnel diffraction* refers to cases in which the viewing screen or the observer's eye is relatively close to the diffracting obstacle or aperture. In this case, the light is converging when it reaches the eye or screen (see Fig. 36–1a).

Considerably simpler to analyze than Fresnel diffraction is the case in which both the source and the eye or screen are at very large distances from the diffracting obstacle or aperture. Then, the incident radiation is in the form of plane waves, as is the diffracted radiation that is directed toward an observing point. This situation is referred to as *Fraunhofer diffraction*.† It is usually arranged by using lenses to produce and to focus the parallel rays, thereby removing the source and the observation point to optical infinity (Fig. 36–1b). The discussion in this chapter is concerned exclusively with Fraunhofer diffraction because of the simpler analysis possible for this case and its wide range of practical applications.

36–1 FRAUNHOFER DIFFRACTION

Consider the Fraunhofer diffraction produced by a single slit with a finite width, as shown in Fig. 36–1b. Light from a primary point source S of monochromatic light strikes a converging lens‡ L_1, which directs a beam of plane waves toward the slit A. The light that passes through the slit is focused by a second converging lens L_2 onto a screen. The details of this process are shown in Fig. 36–2.

An important feature of ideal focusing lenses, such as those shown in Fig. 36–1, is the constancy of the optical path length from the source S along any ray to the wavefront

---

*The first detailed observations of diffraction were made by an Italian physicist, Francesco Grimaldi (1618–1663), who noted the broadening of a beam of light upon passing through a narrow slit. Even earlier, Leonardo da Vinci (1452–1519) had recorded the observation of diffraction effects in his famous *Notebooks*.

†After the German optician and physicist Joseph von Fraunhofer (1787–1826).

‡The properties of lenses will be treated in Chapter 37. For now, it is necessary to know only that it is possible to select a lens that will convert diverging spherical waves to plane waves, and vice versa.

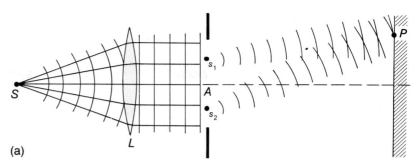

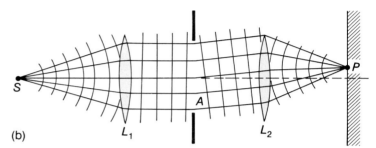

Fig. 36–1. (a) Fresnel diffraction. Light diverging from the source S is rendered parallel by lens L so that the radiation incident on the aperture A consists of plane waves. Radiation from the Huygens' secondary sources, such as s_1 and s_2, that are distributed over A combine at the observation point P. (b) Fraunhofer diffraction. As in (a), the radiation incident on the aperture A consists of plane waves. The diffracted beam of parallel rays is brought to a focus at P by lens L_2. Because plane waves are incident on A, the phase of the radiation is the same over the full extent of the aperture plane.

of the plane wave arriving at the aperture. This results from the application of Fermat's principle, discussed in Section 34–4, to the image-forming properties of lenses. Details of this behavior will be examined in Section 37–5. The existence of equal optical path lengths to the aperture wavefront insures the in-phase relationship of all electromagnetic vibrations on the wavefront.

A plane wavefront that arrives at the slit from lens L_1 may be replaced by an array of secondary (Huygens') line sources in the plane of the slit, each radiating into the region to the right of the slit. One such line source, a strip with a width dx parallel to the slit sides, is shown in Fig. 36–2. A narrow bundle of rays pq from this secondary source is focused by lens L_2 onto the viewing screen at point P. Additional rays from other similar secondary sources are also focused at P, and the light intensity there is obtained by superimposing all such rays.

All of the secondary Huygens' sources in the plane of the slit are coherent and radiate in phase. Choose the central ray rsu as the reference ray for determining the phase differences of the other rays as they arrive at P. The optical path length $\overline{pqP}$ is the same as $\overline{suP}$, again a result of the operation of Fermat's principle. Thus, the rays from the source at dx

Fig. 36–2. The geometry for describing the Fraunhofer diffraction produced by a single slit.

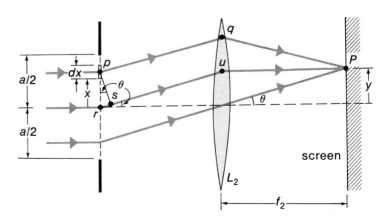

arrive at P with a phase lead, compared with the reference ray, corresponding to the optical path difference $\overline{rs} \cong x \sin \theta$.

The slit is illuminated uniformly by the light from lens L_1. Consequently, the electric field amplitude dE_0 at point P due to the secondary line source dx may be expressed as

$$dE_0 = \xi \, dx$$

where ξ is a constant with appropriate dimensions. The reference ray is considered to have zero phase angle. If the angle θ is very small, all contributions to the electric field intensity on the screen may be considered to be parallel and to lie in the plane of the screen. Thus, the time-dependent electric field component at P due to the segment dx is

$$dE = \xi \, dx \cos \left(\omega t + \frac{2\pi}{\lambda} x \sin \theta \right)$$

In terms of a convenient new parameter, the phase angle α,*

$$\alpha = \frac{\pi a}{\lambda} \sin \theta \qquad\qquad (36\text{–}1)$$

the superposition of all contributions due to the secondary sources from $x = -a/2$ to $x = a/2$ is

$$E = \xi \int_{-a/2}^{a/2} \cos \left(\omega t + \frac{2\alpha}{a} x \right) dx$$

$$= \frac{\xi a}{2\alpha} [\sin (\omega t + \alpha) - \sin (\omega t - \alpha)]$$

$$= \xi a \left(\frac{\sin \alpha}{\alpha} \right) \cos \omega t$$

where we have used Eqs. A–20 and A–21 in Appendix A. The intensity I is equal to E^2/Z_0 (Eq. 35–1), so we can write

$$I = I_0 \left(\frac{\sin \alpha}{\alpha} \right)^2 \qquad \text{single slit} \qquad\qquad (36\text{–}2)$$

where $\frac{1}{2}\xi^2 a^2/Z_0$ has been set equal to I_0. (Compare Eq. 35–1a.)

Figure 36–3 shows the intensity I that is observed on the viewing screen. The central position ($\alpha = 0$) corresponds to a maximum because

$$\lim_{\alpha \to 0} \frac{\sin \alpha}{\alpha} = 1$$

In addition, there are secondary maxima that are given closely (but not exactly) by the condition $\sin \alpha = \pm 1$; that is,

$$\alpha = \frac{\pi a}{\lambda} \sin \theta \cong \pm\tfrac{1}{2}(2m + 1)\pi, \qquad m = 1, 2, 3, \ldots \qquad \text{approximate maxima}$$

$$(36\text{–}3a)$$

---

*Be careful to distinguish between the phase angle α and the geometric angle θ. (• Verify that α is the phase difference between the reference ray rsu and the ray that leaves one edge of the slit.)

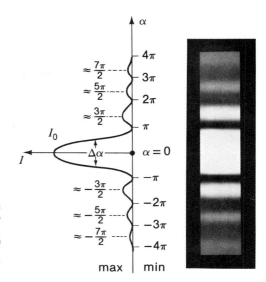

Fig. 36-3. Intensity pattern on a viewing screen due to the Fraunhofer diffraction produced by a single slit. The parameter α is defined by Eq. 36-1. The intensity minima occur at points for which $(\sin \alpha)/\alpha = 0$, but there is no minimum for $\alpha = 0$. Thus, the angular width (from minimum to minimum) of the central peak is twice that of the secondary peaks. (Photo from *Atlas of Optical Phenomena* by M. Cagnet, M. Francon, and J. C. Thrierr. 1962: Springer-Verlag.)

(• Can you see why this relationship is only approximate?) The intensity minima occur when $\sin \alpha = 0$ (excluding $\alpha = 0$); that is,

$$\alpha = \frac{\pi a}{\lambda} \sin \theta = \pm m\pi, \qquad m = 1, 2, 3, \ldots \qquad \text{minima} \qquad \textbf{(36-3b)}$$

It is customary to define the "half-width" $\Delta\alpha$ of the central maximum to be twice the value of α for which $I = \frac{1}{2}I_0$. Thus (see Problem 36-3),

$$\Delta\alpha = 2.783 \text{ rad} \qquad \textbf{(36-4)}$$

If the infinite slit we have been considering becomes a rectangular aperture, the diffraction pattern will contain contributions from the width a and the length b. In the direction perpendicular to the length of the rectangle, the intensity maxima are given by Eq. 36-3a. To find the maxima in the direction perpendicular to the width, the width a in Eq. 36-3a must be replaced by the length $b > a$, which means that the spacing between maxima in this direction (determined by the values of θ) will be *less* than in the other direction. This effect is shown clearly in the photograph on page 698.

The condition for the location of the intensity minima in a Fraunhofer pattern (Eq. 36-3b) can be obtained without reference to the expression for the intensity given in Eq. 36-2. In Fig. 36-4 the slit aperture is divided into a large number of adjoining Huygens'

Fig. 36-4. Geometry for locating the intensity minima in a Fraunhofer diffraction pattern.

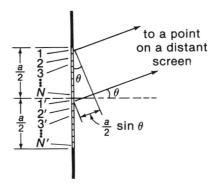

TABLE 36-1. Positions of Diffraction Maxima and Minima

SINGLE SLIT		
	α	I/I_0
1st max.	0	1
1st min.	$\pi = 3.14$	0
2nd max.	$1.43\pi = 4.49$	0.0472
2nd min.	$2\pi = 6.28$	0
3rd max.	$2.46\pi = 7.72$	0.0169
3rd min.	$3\pi = 9.42$	0
4th max.	$3.47\pi = 10.90$	0.0083
CIRCULAR APERTURE		
	α	I/I_0
1st max.	0	1
1st min.	$1.22\pi = 3.83$	0
2nd max.	$1.64\pi = 5.14$	0.0175
2nd min.	$2.23\pi = 7.02$	0
3rd max.	$2.68\pi = 8.42$	0.0042
3rd min.	$3.24\pi = 10.17$	0
4th max.	$3.67\pi = 11.62$	0.0016

equivalent slits. There are $2N$ such slits, labeled 1 through N in the upper half of the aperture and $1'$ through N' ($= N$) in the lower half. The contributions from 1 and $1'$ will cancel at a point on a distant screen if

$$\frac{a}{2} \sin \theta = \frac{\lambda}{2}$$

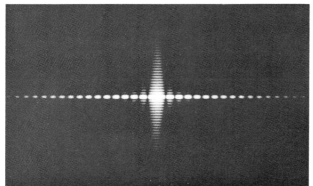

Diffraction pattern from a rectangular slit. (From *Atlas of Optical Phenomena* by M. Cagnet, M. Francon, and J. C. Thrierr. 1962: Springer-Verlag.)

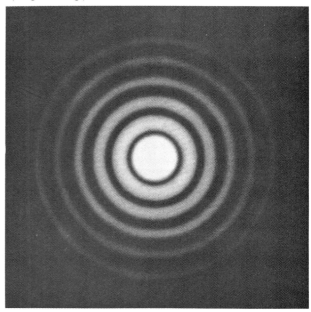

Diffraction pattern from a circular aperture. (From *Atlas of Optical Phenomena* by M. Cagnet, M. Francon, and J. C. Thrierr. 1962: Springer-Verlag.)

This same condition applies for the cancellation of the contributions from the pairs $2-2'$, $3-3', \ldots, N-N'$, and hence from the entire aperture. The condition we have obtained is just Eq. 36–3b with $m = +1$. (• Can you deduce the condition for $m = \pm 2, \pm 3, \ldots$?)

The diffraction pattern produced by a circular aperture was first calculated in precise form in 1835 by Sir George Airy (1801–1892), the English Astronomer Royal. The result is similar to that for the infinite slit, but the expression (Eq. 36–2) contains a complicated function instead of the simple sine function. The secondary intensity maxima occur for approximately the same values of θ, but the magnitudes for the case of the circular aperture are smaller than those for the case of the infinite slit (see Table 36–1).*

Example 36–1

A slit with a width of 0.030 mm is illuminated with plane waves of monochromatic light ($\lambda = 589$ nm) in the geometry shown in Fig. 36–2. The distance from lens L_2 to the screen (the *focal length* of the lens) is $f_2 = 60$ cm.

(a) Determine the locations on the screen of the first minima.

(b) Determine the approximate locations on the screen of the first secondary maxima.

(c) Find the approximate ratio of the intensity of the first secondary maxima to the intensity of the central maximum.

(d) What is the width of the central maximum at half maximum intensity?

Solution:

(a) From Eq. 36–3b for $m = 1$, we find for the first minima,

$$\frac{\pi a}{\lambda} \sin \theta' = \pm \pi$$

or

$$\theta' = \pm \sin^{-1} \left(\frac{5.89 \times 10^{-7} \text{ m}}{3.0 \times 10^{-5} \text{ m}} \right) = \pm 0.0196 \text{ rad}$$

The position on the screen, measured from the central maximum, is obtained by referring to the ray that passes undeviated through the center of lens L_2 (see Fig. 36–2):

$$y' = \pm f_2 \tan \theta' \cong \pm (60 \text{ cm})(0.0196)$$

$$= \pm 1.18 \text{ cm}$$

(b) The approximate locations of the first secondary maxima can be found by using Eq. 36–3a with $m = 1$; thus,

$$\theta'' \cong \pm \sin^{-1} \frac{3\lambda}{2a} = \pm \sin^{-1} \frac{3(5.89 \times 10^{-7} \text{ m})}{2(3.0 \times 10^{-5} \text{ m})}$$

$$= \pm 0.0295 \text{ rad}$$

from which

$$y'' = \pm f_2 \tan \theta'' \cong \pm 1.77 \text{ cm}$$

(c) The approximate intensity ratio is found by using Eq. 36–2,

$$\frac{I''}{I_0} = \left(\frac{\sin \alpha''}{\alpha''} \right)^2$$

together with Eq. 36–3a,

$$\alpha'' = \frac{\pi a}{\lambda} \sin \theta'' \cong \frac{3}{2} \pi$$

Thus,

$$\frac{I''}{I_0} \cong \left(\frac{2}{3\pi} \sin (3\pi/2) \right)^2 = \frac{4}{9\pi^2} = 0.045$$

Compare Table 36–1.

(d) All relevant values of θ are small ($\theta \ll 1$), so we can write

$$\alpha = \frac{\pi a}{\lambda} \sin \theta \cong \frac{\pi a}{\lambda} \theta$$

and

$$\Delta \alpha = \frac{\pi a}{\lambda} \Delta \theta$$

Also,

$$\Delta y \cong f_2 \Delta \theta = \frac{f_2 \lambda}{\pi a} \Delta \alpha$$

Then, using $\Delta \alpha = 2.783$ (Eq. 36–4), we find

$$\Delta y \cong \frac{(60 \text{ cm})(5.89 \times 10^{-7} \text{ m})}{\pi (3.0 \times 10^{-5} \text{ m})} \cdot (2.783) = 1.04 \text{ cm}$$

Note carefully that although the various values of θ in this example are all very small (this is the usual situation), the corresponding values of the phase angle α are *not small*—indeed, we often have $\alpha \approx 1$. Therefore, caution must be exercised when making small-angle approximations.

---

*Solutions for problems involving axial symmetry (as for circular apertures) are most readily expressed in terms of a special function, called the *Bessel function*, instead of sines and cosines. The first zero of this function corresponds to $\alpha = (\pi a/\lambda) \sin \theta = 1.220\pi$.

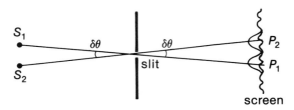

Fig. 36–5. Two incoherent sources, S_1 and S_2, produce separate and independent single-slit Fraunhofer patterns on the screen.

Angular Resolution. Suppose that a narrow slit is illuminated with light from two incoherent sources, S_1 and S_2, that have an angular separation $\delta\theta$, as indicated in Fig. 36–5. The source-to-slit and slit-to-screen distances are sufficiently great that the conditions for Fraunhofer diffraction apply. Separate and independent contributions to the intensity, in the form of single-slit Fraunhofer patterns, P_1 and P_2, are observed on the screen. If $\delta\theta$ is made smaller and smaller, the degree of overlap of the patterns will eventually reach a point at which it will no longer be possible to determine that two sources are present. That is, it will no longer be possible to *resolve* the two sources.

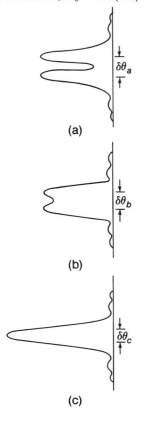

Fig. 36–6. Intensity distribution patterns on the screen of Fig. 36–5 for three different values for the angular separation of the two equal-intensity sources: (a) clearly resolved, $\delta\theta_a = 1.25(\lambda/a)$; (b) just resolved, $\delta\theta_b = \lambda/a$; and (c) unresolved, $\delta\theta_c = 0.75(\lambda/a)$.

Figure 36–6 shows the intensity patterns for three different values of the angular separation of the two sources. What is the smallest angular separation $\delta\theta_{min}$ for which two incoherent sources can just be resolved? The *Rayleigh criterion* (an arbitrary rule) states that two point sources of monochromatic radiation with the same wavelength are just resolved if the position of the first minimum of one diffraction image coincides with the position of the central maximum of the other diffraction image. This is the situation illustrated in Fig. 36–6b, which shows a slight dip between the two maxima. If $\delta\theta$ is reduced by a small amount (Fig. 36–6c), the dip disappears and it is no longer obvious that two maxima are present. Referring to Fig. 36–3, we can see that the Rayleigh criterion for a single slit means $\delta\alpha_{min} \cong \pi$, or

$$\delta\theta_{min} \cong \frac{\lambda}{\pi a} \delta\alpha_{min} \cong \frac{\lambda}{a} \qquad \text{single slit} \qquad (36–5a)$$

The value of $\delta\theta_{min}$ is called the *angular resolution* of the slit. In Fig. 36–6, the intensity patterns correspond to $\delta\theta_a$, equal to 1.25, 1.00, and 0.75 times $\delta\theta_{min} = \lambda/a$. Thus, the Rayleigh criterion is seen to be a reasonable rule for specifying the angular separation of two just-resolved sources.

The position of the first minimum of a diffraction pattern relative to the position of the central maximum is slightly different for a circular aperture than for a single slit (see Table 36–1). When the Rayleigh criterion is applied, the angular resolution of a circular aperture with diameter d is given by

$$\delta\theta_{min} = \frac{1.22\,\lambda}{d} \qquad \text{circular aperture} \qquad (36–5b)$$

Diffraction effects can place a limit on the usefulness of certain high-resolution telescopes (see Problem 36–4).

Notice from Eqs. 36–5 that the angular resolution improves (that is, $\delta\theta_{min}$ becomes smaller) as the wavelength of the radiation is made smaller in comparison with the size of the aperture. For a particular aperture, two sources can be resolved better in blue light than in red light.

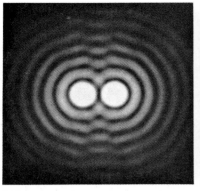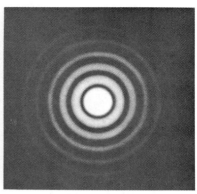

Diffraction patterns corresponding to the distributions in Fig. 36–6. (From *Atlas of Optical Phenomena* by M. Cagnet, M. Francon, and J. C. Thrierr. 1962: Springer-Verlag.)

Example 36–2

A spy satellite circles the Earth at an altitude of 150 km and carries out its surveillance using a special high-resolution telescopic camera with a lens diameter of 35 cm. If the angular resolution of this camera is limited by diffraction, estimate the separation of two small objects on the Earth's surface that are just resolved in yellow-green light ($\lambda = 5500$ Å).

Solution:

The minimum angular separation is (Eq. 36–5b)

$$\delta\theta_{min} \cong \frac{1.22\,\lambda}{d} = \frac{1.22(5.50 \times 10^{-7}\ \text{m})}{0.35\ \text{m}}$$

$$\delta\theta_{min} = 1.92 \times 10^{-6}\ \text{rad}$$

Hence, the minimum linear separation between the objects is

$$\delta x_{min} = h\delta\theta_{min} \cong (150 \times 10^3\ \text{m})(1.92 \times 10^{-6}\ \text{rad})$$

$$\cong 29\ \text{cm}$$

Factors other than diffraction usually place the limits on the definition that can be achieved by cameras. Nevertheless, the angular resolution of these special surveillance cameras is exceptionally good.

Example 36–3

At what maximum distance can the headlights of an approaching automobile be resolved by the eye, if the limitation is set by diffraction effects? Assume that the separation of the headlights is $\ell = 1.42$ m and that the diameter of the pupil of the human eye (in low-light conditions) is 4.0 mm. The mean wavelength of the light is 600 nm.

Solution:

Using Eq. 36–5b, we find

$$\delta\theta_{min} = \frac{1.22\,\lambda}{d} = \frac{1.22(6.0 \times 10^{-7}\ \text{m})}{4.0 \times 10^{-3}\ \text{m}}$$

$$= 1.83 \times 10^{-4}\ \text{rad}$$

Hence, the maximum distance at which the headlights can be resolved is

$$y_m = \frac{\ell}{\delta\theta_{min}} = \frac{1.42\ \text{m}}{1.83 \times 10^{-4}} = 7.76 \times 10^3\ \text{m}$$

$$= 7.76\ \text{km}$$

Again, factors other than diffraction will set a smaller limit.

36–2 THE DIFFRACTION GRATING

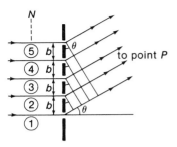

Fig. 36–7. Fraunhofer diffraction by a diffraction grating consisting of N equally spaced slits.

The wavelength of light can be determined with high precision by using a device called a *diffraction grating*. Such gratings are usually prepared by ruling a large number of narrow, closely spaced parallel grooves on a polished glass plate. Diffraction occurs for both reflected and transmitted light.*

Consider the Fraunhofer diffraction of light with wavelength λ by a transmission grating that consists of N equally spaced narrow slits (Fig. 36–7).† The *grating spacing* (the distance between successive slits) is b, and the direction to the point P of interest on the distant viewing screen is specified by the angle θ. For convenience, the ray that passes through the first slit, labeled ① in Fig. 36–7, is taken to be the reference ray with zero phase angle. The ray labeled ② has a path length that is shorter by an amount $b \sin \theta$ and therefore has a phase lead, compared with the reference ray, of $(2\pi b / \lambda) \sin \theta$. The ray labeled ③ has a phase lead of twice this amount, and so forth. If the contribution to the electric field E at point P due to the first slit is $E_0 \cos \omega t$, the superposition of the contributions from all N slits gives

$$E = E_0 \sum_{n=0}^{N-1} \cos \left(\omega t + \frac{2\pi}{\lambda} nb \sin \theta \right) \tag{36–6}$$

▶ The sum in Eq. 36–6 can be evaluated in a simple way by making use of the properties of complex numbers. In complex number notation, the expression for E becomes

$$E = E_0 \, \mathbf{Re} \sum_{n=0}^{N-1} e^{i(\omega t + 2n\beta)}$$

$$= E_0 \, \mathbf{Re} \, e^{i\omega t} \sum_{n=0}^{N-1} e^{i2n\beta} \tag{36–7}$$

where

$$\beta = (\pi b / \lambda) \sin \theta \tag{36–8}$$

The summation in Eq. 36–7 is a simple geometric progression, the sum of which is (see Eq. A–3 in Appendix A)

$$a + ar + ar^2 + ar^3 + \ldots + ar^{N-1} = \frac{a(r^N - 1)}{r - 1}$$

where, in the present case, $a = 1$ and $r = e^{i2\beta}$. Thus,

$$E = E_0 \, \mathbf{Re} \, e^{i\omega t} \cdot \frac{e^{i2N\beta} - 1}{e^{i2\beta} - 1} \tag{36–9}$$

This equation can be put in the form (see Problem 36–10),

$$E = E_0 \, \mathbf{Re} \, e^{i[\omega t + (N-1)\beta]} \cdot \frac{\sin N\beta}{\sin \beta} \tag{36–10}$$

◀

The closed form of Eq. 36–6 is

$$E = E_0 \frac{\sin N\beta}{\sin \beta} \cos [\omega t + (N-1)\beta] \tag{36–11}$$

where $\beta = (\pi b / \lambda) \sin \theta$. The intensity at P is proportional to the time average of E^2; thus,

$$I = \frac{I_0}{N^2} \left(\frac{\sin N\beta}{\sin \beta} \right)^2 \tag{36–12}$$

where I_0 is the intensity for $\beta = 0$ (that is, $\theta = 0$).

---

*The gratings are referred to as *reflection gratings* and *transmission gratings,* respectively. In each case, the set of undisturbed strips of surface between the grooves constitutes the array of line sources. Reflection gratings are often prepared by ruling lines on metallic mirrors. The first gratings were made by Fraunhofer (and consisted of parallel fine wires), but the first really successful ruled gratings were prepared by Henry Rowland in the 1870's. Rowland's ruling engine could score almost 6000 grooves per centimeter.

†For the moment we consider only the diffraction effects produced by the array of N slits, ignoring the single-slit diffraction due to each individual slit. The complete diffraction pattern, formed by the superposition of all effects, is discussed later in this section.

The *principal maxima* of the intensity I occur when $\sin \beta = 0$; thus,

$$\beta = \frac{\pi b}{\lambda} \sin \theta = \pm m\pi, \qquad m = 0, 1, 2, \ldots \qquad \text{principal maxima} \qquad \textbf{(36-13)}$$

To calculate the value of I at the principal maxima, we use the fact that (see Problem 36-18)

$$\lim_{\beta \to \pm m} \left(\frac{\sin N\beta}{\sin \beta} \right)^2 = N^2 \qquad \textbf{(36-14)}$$

Then, from Eq. 36-12, we have

$$I = I_0 \qquad \text{principal maxima} \qquad \textbf{(36-15)}$$

In Fig. 36-7 we can see how these maxima arise. For the constructive superposition of all of the rays, the difference in path length between successive rays must be an integer number of wavelengths; that is, the condition for the principal maxima is $b \sin \theta = \pm m\lambda$, in agreement with Eq. 36-13.

The value $m = 0$ corresponds to the central maximum. For $m = 1$, the two resulting maxima are referred to as the *first-order* maxima; for $m = 2$, the maxima are *second order;* and so forth. Between each pair of principal maxima are $N - 2$ additional maxima that correspond to $\sin N\beta \approx 1$; because $\sin \beta$ is also approximately equal to unity, the intensity of these secondary maxima, which does have a regular variation, is nevertheless of the order of I_0/N^2 (see Eq. 36-12). Between each pair of secondary maxima is a minimum for which $\sin N\beta = 0$ (but $\beta \neq \pm m\pi$). Figure 36-8 shows the principal and secondary maxima for the case $N = 9$.

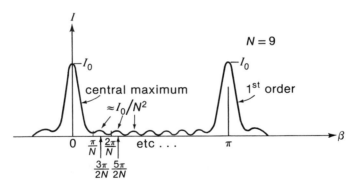

Fig. 36-8. Intensity pattern for a grating with $N = 9$.

As N is increased, the intensity at a principal maximum increases as N^2. This effect is due to two factors. First, the total power that reaches the viewing screen is proportional to N, the number of slits that are open to the incident plane wave. Second, the width of each principal maximum, which can be defined as the distance between the minima on either side, is proportional to $1/N$ (see Fig. 36-8). Thus, the power density (the intensity) at a principal maximum increases as N^2. Because the intensity of the secondary maxima decreases as $1/N^2$, increasing the number of slits in the grating both narrows the principal maxima and heightens the contrast with the secondary maxima. This is a point of the greatest importance for applications of diffraction gratings.

Resolving Power.   The quality of a diffraction grating can be judged by its resolving power, which is a measure of the ability of the grating to produce a principal maximum for radiation with wavelength λ that is separated from one for radiation with a nearby wavelength, $\lambda' = \lambda + \Delta\lambda$. The resolving power of a grating is defined to be $R_g = \lambda/\Delta\lambda$, where $\Delta\lambda$ is the smallest difference between resolvable wavelengths. The angular separation $\Delta\beta$ between a principal maximum and the adjacent minimum is $\Delta\beta = \pi/N$ (see Eq. 36–12 or Fig. 36–8). Then, using the definition of β (Eq. 36–8) we have

$$\Delta\beta = \frac{\pi b}{\lambda} \cos\theta \, \Delta\theta = \frac{\pi}{N} \tag{36-16}$$

from which the angular change $\Delta\theta$ required to descend from a principal maximum to the adjacent zero of intensity may be calculated. Suppose that there is a nearby principal maximum due to the presence in the source of radiation with wavelength $\lambda' = \lambda + \Delta\lambda$. According to the Rayleigh criterion, the two maxima are just resolved (in the same order m) when the separation is $\Delta\beta = \pi/N$. Differentiating Eq. 36–13 gives

$$\Delta\lambda = \frac{b}{m} \cos\theta \, \Delta\theta \tag{36-17}$$

Combining Eqs. 36–16 and 36–17, there results

$$R_g = \frac{\lambda}{\Delta\lambda} = mN \tag{36-18}$$

That is, the *resolving power* of a grating in any order m is proportional to the total number of slits in the grating. Practical considerations usually limit the size of simple gratings of the type discussed here to about 15 cm in length with a maximum of about 400,000 ruled grooves. Special gratings of different design may have first-order resolving powers as large as 10^6.

The Effect of Finite Slit Width. To this point we have considered the slits in the diffraction grating to have negligible width. When the finite width of the slits is included in the calculation, the intensity given by Eq. 36–12 is modulated (multiplied) by the intensity function for diffraction by a single slit (Eq. 36–2). The complete expression for the intensity is just the product of these equations, namely,

$$I = \frac{I_0}{N^2} \left(\frac{\sin N\beta}{\sin\beta}\right)^2 \left(\frac{\sin\alpha}{\alpha}\right)^2 \quad \text{grating } N \text{ rulings} \tag{36-19}$$

where $\alpha = (\pi a/\lambda) \sin\theta$ and where a is the slit width.

For the case of a double slit ($N = 2$), Eq. 36–19 can be reduced to (see Problem 36–19)

$$I = I_0 \cos^2\beta \left(\frac{\sin\alpha}{\alpha}\right)^2 \quad \text{double slit} \tag{36-19a}$$

Figure 36–9 shows the individual terms, $\cos^2\beta$ and $(1/\alpha^2)\sin^2\alpha$, together with their product, for the case $b = 4a$.

If the slit spacing is large compared with the slit width ($b \gg a$), the width of the central maximum of the diffraction term (the term $(1/\alpha^2)\sin^2\alpha$ in Eq. 36–19a) is large compared with the distance between successive zeroes of the interference term ($\cos^2\beta$). Then, for small angles θ, the diffraction term is essentially constant, so that the variation of the intensity is due almost entirely to the interference term. This is the case that was discussed in Section 35–2 (see Fig. 35–3).

Fig. 36-9. Intensity variation in the Fraunhofer pattern due to a double slit with $b = 4a$. (a) The diffraction term, $(1/\alpha^2) \sin^2 \alpha$. (b) The interference term, $\cos^2 \beta$. (c) The product of (a) and (b) produces the observed intensity pattern. Compare with the single-slit pattern in Fig. 36-3.

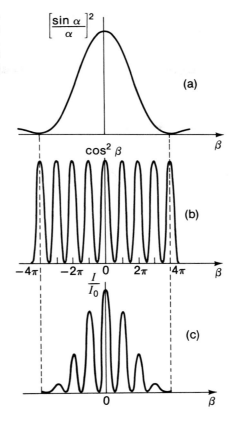

Double slit with $b = 10a$. (Photo from *Atlas of Optical Phenomena* by M. Cagnet, M. Francon, and J. C. Thrierr. 1962: Springer-Verlag.)

When $N > 2$, the individual secondary maxima, corresponding to those in Fig. 36-9c, become sharper and better defined. Figure 36-10 shows the intensity pattern for the case $N = 9$ and $b/a = 2.25$. Notice the important point that whereas the resolving power is greater in the higher orders ($R \propto m$, Eq. 36-18), the intensity at the principal maxima in these orders may be substantially diminished.

Fig. 36-10. Intensity pattern for a grating with $N = 9$ and a ratio of slit spacing to slit width of $b/a = 2.25$. The dashed curve represents the modulation of the grating pattern by the diffraction due to the finite width of the slits.

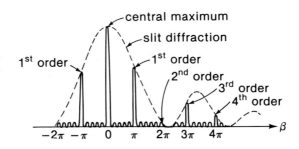

Example 36–4

A 2-cm diffraction grating contains a total of 10,000 equally spaced grooves, each of width $a = 0.8\ \mu$m. The grating is used in the Fraunhofer geometry with light that has a wavelength $\lambda = 550$ nm and with a focusing lens L_2 that has a focal length $f_2 = 50$ cm.

(a) What is the position on the viewing screen of the first-order principal maximum?

(b) What is the full width of the principal maxima?

(c) Determine the intensity at the first-order principal maximum relative to that at the central maximum.

(d) Determine the resolving power of the grating in first order.

(e) Principal maxima of how many orders are visible on the viewing screen?

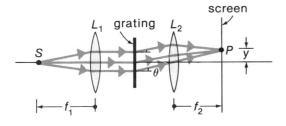

Solution:

(a) The grating spacing b is

$$b = \frac{2 \times 10^{-2}\ \text{m}}{10{,}000} = 2 \times 10^{-6}\ \text{m}$$

The first-order principal maximum corresponds to (Eq. 36–13)

$$\frac{b}{\lambda} \sin \theta_1 = 1$$

so that

$$\theta_1 = \sin^{-1} (\lambda/b) = \sin^{-1} \left(\frac{5.50 \times 10^{-7}\ \text{m}}{2 \times 10^{-6}\ \text{m}} \right)$$

$$= 16.0°$$

Referring to the diagram, we see that

$$y_1 = f_2 \tan \theta_1 = (50\ \text{cm}) \tan 16.0°$$

$$= 14.3\ \text{cm}$$

measured from the position of the central maximum. Conversely, the wavelength of the incident light can be determined by measuring the position on the screen of any principal maximum. However, wavelength determinations are usually made by measuring the separation between a maximum of the unknown radiation and a maximum (perhaps in a different order) due to radiation with a precisely known wavelength.

(b) The "full width" of a principal maximum means the distance between the zero intensity points on either side of the maximum. Using Eq. 36–16, we have for the first-order maximum

$$2\ \Delta\theta = \frac{2\lambda}{Nb \cos \theta_1} = \frac{2(5.50 \times 10^{-7}\ \text{m})}{10^4 (2 \times 10^{-6}\ \text{m}) \cos 16.0°}$$

$$= 5.72 \times 10^{-5}\ \text{rad}$$

Thus,

$$\Delta y = f_2 \frac{d(\tan \theta)}{d\theta} \times (2\ \Delta\theta)$$

$$= \frac{f_2 \times 2\ \Delta\theta}{\cos^2 \theta_1} = \frac{(50\ \text{cm})(5.72 \times 10^{-5})}{\cos^2 16.0°}$$

$$= 3.10 \times 10^{-3}\ \text{cm}$$

(c) The slit width is $a = 0.8\ \mu$m, so the value of α at the first-order principal maximum is

$$\alpha = \frac{\pi a}{\lambda} \sin \theta_1 = \frac{\pi (8 \times 10^{-7}\ \text{m}) \sin 16.0°}{5.5 \times 10^{-7}\ \text{m}}$$

$$= 1.26\ \text{rad}$$

Thus, the intensity ratio is

$$\frac{I}{I_0} = \left(\frac{\sin \alpha}{\alpha} \right)^2 = \left(\frac{\sin 1.26}{1.26} \right)^2 = 0.571$$

(d) According to Eq. 36–18,

$$R_g = mN = 1 \times 10{,}000 = 10{,}000$$

Hence, the principal maxima due to radiation with wavelength 550.055 nm or 549.945 nm could just be resolved from those of the given radiation ($\lambda = 550$ nm).

(e) According to Eq. 36–13,

$$m = \frac{b}{\lambda} \sin \theta$$

The largest conceivable value of θ is 90°, which corresponds to the ends of an infinite viewing screen. Then, $\sin \theta = 1$, so that the order m cannot exceed the value

$$m = \frac{b}{\lambda} = \frac{2 \times 10^{-6}\ \text{m}}{5.5 \times 10^{-7}\ \text{m}} = 3.6$$

Thus, principal maxima of only the first three orders appear on the viewing screen.

36-3   X-RAY DIFFRACTION (Optional)

In 1895, Roentgen* discovered that when cathode rays (later called electrons) are rapidly stopped in a piece of metal, a penetrating radiation is produced. It was soon established that this radiation, called *X rays* by Roentgen, is a type of electromagnetic radiation with very short wavelength. X rays are usually produced by accelerating electrons through a potential difference of 10 to 300 kV, then allowing them to strike a metallic target, such as tungsten or molybdenum. The large deceleration of the electrons in the target results in the production of electromagnetic radiation in the manner discussed in Section 33-3. The X rays so produced have a continuous distribution of wavelengths with values in the fractional angstrom range.†

In order to observe the effects of diffraction, while avoiding the difficulties associated with using very small angles of incidence (*grazing* incidence), the grating spacing b must be of the order of the wavelength of the radiation analyzed. The regular array of atoms in a crystal forms a natural diffraction grating with a spacing that is typically a few angstroms. The scattering of X rays from the atoms in a crystalline lattice gives rise to diffraction effects very similar to those observed with visible light incident on ordinary gratings.

The suggestion that crystals could be used to diffract X rays was first made by von Laue.‡ The experiment, carried out in 1912, confirmed that a collimated beam of X rays incident on a single crystal are strongly scattered only in certain very sharply defined directions. A photographic film exposed to the scattered radiation reveals a pattern of spots that correspond to the directions of preferential scattering (Fig. 36-11). These patterns of

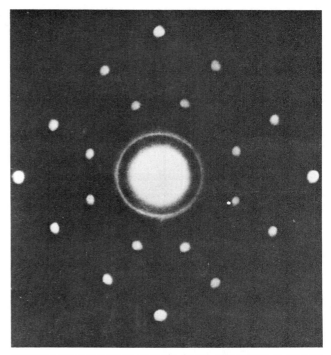

Neutron Laue photograph of NaCl. (From *Introduction to Modern Physics* by Elmer E. Anderson. 1982: Saunders College Publishing. Courtesy of the author and Dr. E. O. Wollan.)

spots, called *Laue patterns,* are related in complicated ways to the geometric properties of the target crystals.

Shortly after the experiment suggested by von Laue was successfully performed, W. L. Bragg§ used a different experimental arrangement to study X-ray diffraction. A single crystal cleaved along one of its lattice planes was used to *reflect* the X-ray beam (instead of *transmit* the beam, as in the von Laue experiment). This type of experiment is much easier to analyze in terms of the crystal geometry than is the transmission experiment, so we now consider this case.

The experimental arrangement used by Bragg is shown schematically in Fig. 36-12. Notice that the incident angle θ and the diffraction angle ϕ are measured from the crystal surface. For simplicity, consider the scattering of X rays from the atoms arranged in the symmetry planes of a simple cubic crystal, as indi-

Fig. 36-11. A monochromatic beam of X rays incident on a single crystal is diffracted strongly in certain directions determined by the geometry of the crystalline lattice.

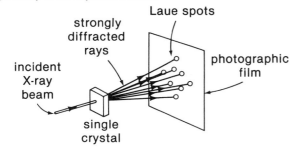

---

*Wilhelm Konrad Roentgen (1845–1923), German physicist, who received the Nobel Prize in physics in 1901 for his pioneering studies of X rays. Roentgen made the first X-ray photograph (of his own hand) in 1895.

†Such radiation is referred to as *bremsstrahlung* (German, "braking radiation"). The interaction of the energetic electrons with individual target atoms may also produce X rays with well-defined characteristic wavelengths.

‡Max von Laue (1879–1960), German physicist and winner of the 1914 Nobel Prize in physics for his work in X-ray analysis.

§William L. Bragg (1890–1971), British physicist, who shared the 1915 Nobel Prize in physics with his father, William H. Bragg (1862–1942), for their research on X rays and crystal structure.

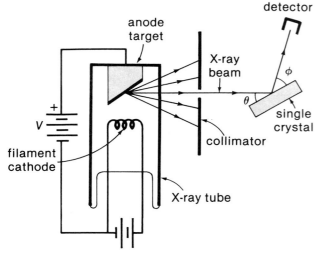

Fig. 36–12. The experimental arrangement used by Bragg to study X-ray diffraction by reflection from a single crystal.

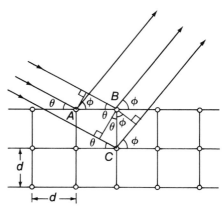

Fig. 36–13. The Bragg scattering of X rays from the atomic planes of a cubic crystal.

cated in Fig. 36–13. For constructive interference of the rays scattered from adjacent atoms in the horizontal plane, such as atoms A and B, the condition is

$$d \cos \theta - d \cos \phi = n\lambda, \qquad n = 0, 1, 2, \ldots$$

For adjacent atoms in the vertical plane, such as atoms B and C, the condition is

$$d \sin \theta + d \sin \phi = m\lambda, \qquad m = 1, 2, 3, \ldots$$

The condition for simultaneous constructive interference from *all* atomic centers is obtained by eliminating the angle ϕ between

the two equations above. This gives

$$\lambda_{nm} = \frac{2d}{n^2 + m^2}(n \cos \theta + m \sin \theta) \qquad \begin{cases} n = 0, 1, 2, \ldots \\ m = 1, 2, 3, \ldots \end{cases}$$

$$(36-20)$$

This relationship is known as the *Bragg equation*.

When $n = 0$, Eq. 36–20 becomes

$$\lambda_{0m} = \frac{2d}{m} \sin \theta, \qquad m = 1, 2, 3, \ldots \quad (36-21)$$

The condition, $n = 0$, makes ϕ equal to θ and yields the equivalent of ''specular reflection'' from each horizontal plane. The value of m is referred to as the *order* of Bragg diffraction.

Example 36–5

The experimental arrangement shown in Fig. 36–12 is used as a *monochromator* to select a particular wavelength from the continuous distribution of wavelengths incident on the crystal. The X-ray source and the detector are placed so that $\theta = \phi = 10°$. The crystal is rock salt, which has a cubic structure with $d = 2.82$ Å. What is the wavelength of the strongly reflected radiation in first order?

Solution:

Specular reflection, $\phi = \theta$, requires $n = 0$. Then, using Eq. 36–21 with $m = 1$, we find

$$\lambda_{01} = 2d \sin \theta = 2(2.82 \text{ Å}) \sin 10°$$

$$= 0.979 \text{ Å}$$

Example 36–6

When monochromatic X rays are incident on a sample that consists of a microcrystalline powder (a randomly oriented collection of microscopically small crystals), each diffraction spot for a single crystal of the same material becomes a diffraction *ring*, as indicated in the diagram. (• Can you see why rings are formed? Each single-crystal diffraction spot corresponds to a di-

rection that makes the same angle with the direction of the incident beam, as shown in the diagram.) X rays with $\lambda = 0.75$ Å are incident on a sample that is 15 cm from the plane of the photographic film. The crystal structure of the sample is cubic with an atomic spacing $d = 2.75$ Å. Determine the radii of the rings that correspond to first- and second-order specular reflection.

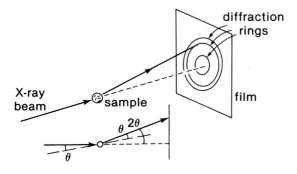

diffraction rings

X-ray beam → sample

film

θ 2θ

θ

Solution:

Specular reflection requires $n = 0$, so we use Eq. 36−21 with $m = 1$ and $m = 2$. Thus,

$$\lambda_{01} = 2d \sin \theta_1 \quad \text{and} \quad \lambda_{02} = d \sin \theta_2$$

But $\lambda_{01} = \lambda_{02} = 0.75$ Å, so that

$$\theta_1 = \sin^{-1} \left(\frac{0.75 \text{ Å}}{2(2.75 \text{ Å})} \right) = 7.84°$$

$$\theta_2 = \sin^{-1} \left(\frac{0.75 \text{ Å}}{2.75 \text{ Å}} \right) = 15.8°$$

Now, the total scattering angle in each case is 2θ (see the diagram). Therefore, the ring radii are

$$r_1 = (15 \text{ cm}) \tan 2\theta_1 = (15 \text{ cm}) \tan 15.68°$$
$$= 4.21 \text{ cm}$$

$$r_2 = (15 \text{ cm}) \tan 2\theta_2 = (15 \text{ cm}) \tan 31.6°$$
$$= 9.23 \text{ cm}$$

QUESTIONS

36−1 Define and explain the meaning of the terms (a) Fresnel and Fraunhofer diffraction, (b) angular resolution and the Rayleigh criterion, (c) diffraction grating, (d) X-ray diffraction, and (e) the Bragg equation.

36−2 Describe the effect on the Fraunhofer diffraction pattern produced by a single slit if (a) the slit width is doubled, (b) the wavelength is doubled, and (c) both the slit width and wavelength are doubled.

36−3 What would be the effect on the diffraction effects produced by a single slit if "white" light instead of monochromatic light were used?

36−4 A large steel-framed structure is directly in line between you and a radio transmitter. Would the line-of-sight reception of signals from the transmitter be stronger for AM or FM broadcasting?

36−5 Contrast diffraction effects produced by sound waves with those produced by light. Sound of frequency 1000 Hz passes through a slit aperture and produces a first-order minimum at an angle $\theta = 30°$. How wide is the slit?

36−6 List and discuss factors other than lens resolution that might limit the image definition of a spy or surveillance camera such as the one described in Example 36−2.

36−7 Why do radio telescopes have collecting "dishes" tens to hundreds of meters in diameter?

36−8 Identify and carefully distinguish the angles θ, β, and α used to describe the diffraction effects of a grating.

36−9 Discuss how the resolution of a diffraction grating depends on the total number and on the spacing of the rulings.

PROBLEMS

Section 36−1

36−1 A plane wave with wavelength $\lambda = 500$ nm illuminates a narrow slit, and the diffracted light is focused on a screen by a lens with focal length $f_2 = 1.5$ m, as in Fig. 36−2. The first secondary maximum is located 4 mm from the central maximum. (a) What is the width of the slit? (b) Locate the approximate positions of the maxima with $m = 2$ and $m = 3$.

36−2 A plane wave with wavelength $\lambda = 650$ nm is incident on a narrow slit, and the diffracted light has its first minimum at $\theta = \pi/2$ and thus illuminates the entire space beyond the slit. (a) Determine the width of the slit. (b) At what angle θ is the intensity equal to one half the maximum intensity?

36−3 Prove the result stated in Eq. 36−4 by solving the equation $\sin \alpha = \alpha/\sqrt{2}$. Use an iterative procedure on an electronic calculator to obtain $\alpha = 1.39156$ and $\Delta\alpha = 2\alpha = 2.783$ rad.

36-4   What is the smallest angular separation between two stars that can be resolved by the Hale telescope at Mount Palomar (diameter $= 5.08$ m), if the limitation is set by diffraction effects? Assume that the observation is made in yellow-green light ($\lambda = 550$ nm), and give your answer in seconds of arc.

Section 36-2

36-5   A 5-cm diffraction grating has a total of 15,000 grooved lines, each with a width of 1.0 μm. The grating is illuminated with light that has a wavelength of 480 nm. (a) Determine the angles θ_m that correspond to the principal maxima of the first three orders. (b) How many complete orders are present in the entire space beyond the grating? (c) How many complete orders are present in the range $-\pi \leqslant \alpha \leqslant \pi$?

36-6   Young's interference experiment is performed (using Fraunhofer geometry) with two slits of width 0.20 mm, whose centers are separated by 0.80 mm. How many maxima (including the central maximum) lie in the region $|\alpha| \leqslant \pi$? Explain why these are the lines of high visibility.

36-7   A certain grating produces a first-order maximum with
 •   an intensity equal to one half that of the central maximum. Determine the intensities of the maxima of orders 2, 3, 4, and 5.

36-8   A 3-cm diffraction grating with 250 grooves per millimeter is used in third order. Monochromatic light produces a strong maximum at $\theta = 18.0°$. (a) What is the wavelength of the light? (b) A second source of mono-

chromatic light is now added. What is the difference in wavelength between the two radiations if they are just resolved in third order?

36-9   A diffraction grating with 100 grooved lines per millimeter is used to observe the yellow doublet emitted by sodium vapor ($\lambda_1 = 5895.93$ Å and $\lambda_2 = 5889.96$ Å). What is the minimum length of the grating if these spectral lines are to be resolved in first order?

36-10 Supply the steps necessary to convert Eq. 36-9 into Eq. 36-10. [*Hint:* Use Eq. B-12 in Appendix B.]

Section 36-3

36-11 The highest-quality diffraction gratings have about 25,000 grooves/cm. Suppose that such a grating were used in the standard transmission mode (Fig. 36-7) to analyze X rays with a wavelength of 0.50 Å. At what angle θ would the first-order maximum appear? Comment on your result. (Neglect the practical question of whether a suitably thin grating could be constructed to transmit the X rays.)

36-12 X rays with $\lambda = 1.35$ Å are diffracted by a crystal with cubic lattice structure and exhibit a strong second-order maximum ($n = 0$, $m = 2$) at $\theta = 35.0°$. (a) What is the crystal lattice spacing d? (b) Determine the angles θ_m for all of the Bragg orders that are possible (for $n = 0$).

36-13 An X-ray diffraction experiment is performed in the geometry shown in Example 36-6 with the sample-to-film distance equal to 50 cm. The lattice spacing of the sample crystalline material is 2.20 Å. The third-order ring ($n = 0$, $m = 3$) has a radius of 8.0 cm. What is the wavelength of the X radiation?

Additional Problems

36-14 Plane waves of light with $\lambda = 600$ nm illuminate a single slit. The diffraction pattern is projected onto a screen by means of a lens with focal length $f_2 = 1.50$ m, as in Fig. 36-3. The separation on the screen between adjacent minima is 4.0 mm. Determine the width of the slit.

36-15 Calculate the angular width $\Delta\theta$ at half maximum intensity for the central maximum of a Fraunhofer single-slit pattern for the cases $a = \lambda$, $a = 2\lambda$, and $a = 5\lambda$.

36-16 What is the smallest diameter of a telescope lens that would permit resolving two pointlike objects separated by 1 km on the central visible surface of the Moon, assuming that the limitation is set by diffraction? The objects are viewed in yellow-green light ($\lambda = 550$ nm). (The Earth-Moon distance is 3.84×10^5 km.)

36-17 According to Eq. 36-3a, the first secondary maximum
 ••   ($m = 1$) occurs for $\alpha = 3\pi/2$. Use Eq. 36-2 and show that this maximum actually occurs for $\alpha = 0.954 \times 3\pi/2$. (Compare Table 36-1.) [*Hint:* Set $dI/d\alpha = 0$. Make a sketch of the resulting expression to determine a reasonable starting point for a successive approximation calculation.]

36-18 Prove Eq. 36-14 (a) by using l'Hôpital's rule of calcu-
 •   lus and (b) by making a series expansion of $\sin N\beta$.

36-19 Show that Eq. 36-19 for the case of Young's double-slit experiment ($N = 2$) can be expressed as (Eq. 36-19a)

$$I = I_0 \cos^2 \beta \left(\frac{\sin \alpha}{\alpha} \right)^2$$

36–20 Consider a single slit of width a to be equivalent to a diffraction grating with N grooves and a grating spacing $b = a/N$. Show that the expression for the intensity due to a diffraction grating (Eq. 36–12) reduces to that for a single slit (Eq. 36–2) in the limit $N \to \infty$.

36–21 Consider a diffraction grating with $N = 9$ (see Fig. 36–
  • 10). What value of b/a will cause the fourth-order secondary maximum to occur at the position of the first-

order secondary maximum of the diffraction term? (This change will enhance the intensity of the $m = 4$ maximum compared with that shown in Fig. 36–10.)

36–22 Monochromatic X rays with $\lambda = 1.50$ Å are incident on
  • a crystal of rock salt ($d = 2.82$ Å). What are the angles θ and ϕ that correspond to the strong reflection in the mode with $n = 1$ and $m = 2$?

CHAPTER 37

GEOMETRIC OPTICS

A variety of optical devices have been developed to aid the eye and extend its capabilities. These devices have the common feature that they employ curved surfaces to form images by reflection or refraction. The study of such image formation is called *geometric optics* or *ray optics* and is the subject of this chapter.

In all cases of interest, diffraction effects are usually small, owing to the smallness of the wavelength of light compared with the size of everyday objects. (In some situations, particularly those involving the fine details of images, diffraction sets the ultimate limitation on the quality of the images.) In examining the image-forming properties of optical systems, therefore, we shift attention from wavefronts to *rays*, as defined in Section 34–3.

37–1 PLANE MIRRORS

We introduce the idea of image formation by examining the (specular) reflection by a plane mirror of light from a point source. Figure 37–1a shows two of the rays that emanate from the point source O, the *object*. These rays are reflected by the mirror MM'. The normal ray OA is reflected back toward O. The oblique ray OB is reflected with equal angles of incidence and reflection θ. The backward extension of the reflected ray BC intersects the extension of ray OA at the point I. Notice that the intersection point I is independent of the angle θ. Thus, all of the rays from O that are reflected by the mirror appear to have originated at the point I, the *image point* of O. Because the triangle $\triangle OBI$ is isosceles, the image point I is the same distance behind the mirror that the object point O is in front of the mirror; that is,

$$s' = -s \qquad (37-1)$$

where we use the convention that distances are reckoned from the mirror surface and are positive if they are measured to a point behind the mirror.

Figure 37–1b shows the observer's eye intercepting several reflected rays from O. The observer perceives the source of the rays to be I, the image of the object O. Because the rays do not actually originate at the point I but only appear to do so, the image is said to be *virtual*. A virtual image can be seen by the eye (Figure 37–1b) but it cannot be projected onto a screen or piece of film. If the eye is moved to a different position, the image point I remains fixed. (• Can you see why this is the case?)

An extended object may be considered to be a collection of point sources of light. A mirror will form an image of each of these sources, thereby producing the image of the entire object. For example, consider the four point sources, O, A, B, and C, located at the

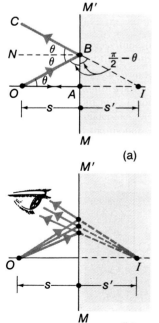

Fig. 37–1. (a) The reflection of two rays, OA and OB, from an object O define the image point I. (b) An observer sees the reflected rays from O diverging from I, the (virtual) image of O.

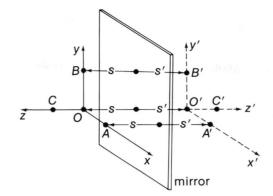

Fig. 37-2. The image of the right-handed coordinate frame *OABC* is the reversed or left-handed frame *O'A'B'C'*. The mirror is parallel to the *x-y* plane.

origin *O* and along the three axes of the right-handed coordinate frame shown in Fig. 37–2. The images of these sources are located at *O'*, *A'*, *B'*, and *C'*, respectively, and define a left-handed coordinate frame.* Thus, the mirror forms a reversed image of the object, in agreement with everyday experience.

Example 37-1

What is the minimum height of a mirror in which a 5'10" (1.77-m) student can observe her full image?

Solution:

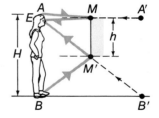

The top of the mirror at *M* needs to be no higher than the mid-height between *E* and *A*. The bottom of the mirror at *M'* needs to be no lower than the mid-height between *E* and *B*. Therefore, the height *h* of the mirror needs to be no more than half the height *H* of the student, or $h = \frac{1}{2}H = 2'11''$ (0.885 m).

37-2   SPHERICAL MIRRORS

Spherical reflecting surfaces—both concave and convex—possess interesting focusing properties. If the rays from a point source of light are brought together at a common point by reflection (or refraction) at a certain surface, that point is the *image* of the object point. Unlike a plane mirror, which forms an exact image of an object, a spherical mirror suffers from an intrinsic imperfection called *spherical aberration*. Figure 37–3a shows that a wide bundle of rays from a point source are not all reflected through the same point by a spherical mirror, so no well-defined image point results. However, if the bundle of rays emitted by a source is confined to a small region around the axis, the blurring of the image can be made suitably small (Fig. 37–3b). The rays that are incident along or near the axis of an optical system are called *paraxial rays*.† We consider here

---

*A right-handed coordinate frame is one for which the unit vectors obey the relationship $\hat{\imath} \times \hat{\jmath} = \hat{k}$; a left-handed frame is one for which $\hat{\imath} \times \hat{\jmath} = -\hat{k}$.

†The first systematic and reasonably complete description of image formation by paraxial rays was given by Gauss (1778–1855).

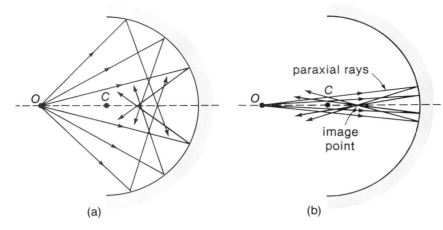

(a) (b)

Fig. 37–3. (a) Spherical aberration. The rays in a wide bundle from a point source are not all reflected through the same point. The image of the object O is therefore blurred. (b) A narrow bundle of rays near the axis does produce a sharp image. For a point source, the *axis* of the mirror is the line defined by the center C of the mirror and the object point O.

only such rays, although in diagrams we often exaggerate the beam size in the interests of clarity. (See, however, Problem 37–34.)

Concave Spherical Mirrors. Figure 37–4 shows a concave spherical mirror with radius of curvature R and center C. The line joining the point source at O and the center of the mirror at C determines the mirror axis, which intersects the mirror at the *vertex V*. The location of the source O is specified by its distance s from the vertex.

The ray OP is reflected from the mirror surface at P exactly as it would be from a flat mirror oriented tangent to the spherical surface at P. Thus, the incident and reflected rays at P make equal angles θ with the normal N to the spherical surface. All other rays from O for which the angle α is small ($\alpha \ll 1$) also intersect the axis (approximately) at I. Thus, I is the image point of O. The distance to I from the vertex V is labeled s'.

Notice that the angle β is an exterior angle of the triangle ΔOPC and that the angle γ is an exterior angle of the triangle ΔCPI. Thus, $\beta = \alpha + \theta$ and $\gamma = \beta + \theta$; hence,

$$2\beta = \alpha + \gamma \tag{37–2}$$

Let the arc length $\overparen{PV}$ be ℓ; then, because $\overline{CV} = R$, we have $\beta = \ell/R$. If, in addition, $\alpha \ll 1$ and $\gamma \ll 1$, we have $\alpha \cong \ell/s$ and $\gamma \cong \ell/s'$. When these expressions for α and γ are substituted into Eq. 37–2, the result is

$$\frac{1}{s} + \frac{1}{s'} = \frac{2}{R} \tag{37–3}$$

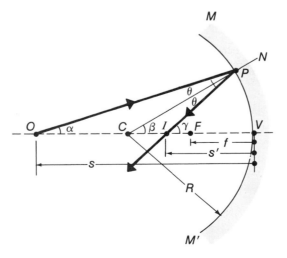

Fig. 37–4. Geometry for locating the focal point F of a concave spherical mirror. The ray OP is a paraxial ray ($\alpha \ll 1$); the angle α has been exaggerated for clarity.

The point at which an optical system causes incident parallel rays to converge is called the *focal point* of the system. When the source O in Fig. 37−4 is moved far away from the mirror ($s \to \infty$), the incident rays become parallel and the image distance s' approaches the limiting value $R/2$. Thus, the focal point F of the spherical mirror is located a distance $f = R/2$ from the vertex V. Equation 37−3 may now be expressed as

$$\frac{1}{s} + \frac{1}{s'} = \frac{1}{f} \qquad \text{concave mirror} \qquad \textbf{(37−4)}$$

with

$$f = \tfrac{1}{2}R \qquad \text{concave mirror} \qquad \textbf{(37−5)}$$

It is important to note that the object and image points can be interchanged (that is, $s \to s'$ and $s' \to s$) without affecting Eq. 37−4. This is a property of any focusing system and is referred to as the *principle of reversibility*. Thus, the directions of the rays in Fig. 37−4 can be reversed and the diagram will still represent a realizable situation.

When the object is placed between the focal point F and the vertex V, negative values of s' result. The meaning of $s' < 0$ is illustrated in Fig. 37−5, where the ray geometries for $s' > 0$ and $s' < 0$ are compared. Figure 37−5a shows that $s' > 0$ corresponds to the case in which I is located in front of the mirror and the rays converge to form a real image. Figure 37−5b shows that $s' < 0$ corresponds to the case in which I is located behind the mirror and the rays diverge from a virtual image. Finally, notice that for a plane mirror, we have $R = \infty$ and, hence, $f = \infty$. Then, for any object position, we find $s' = -s$. This is just the sign convention that was introduced in writing Eq. 37−1.

Convex Spherical Mirrors. Consider the convex spherical mirror with radius of curvature R shown in Fig. 37−6. The ray OP is reflected from the mirror surface at P with equal angles θ of incidence and reflection. Regardless of the position of the object O, the reflected rays appear to diverge from an image located behind the mirror. Thus, for any s, we have $s' < 0$, and the image formed by a convex spherical mirror is always virtual.

We now extend the sign convention we have been using. The points C, F, and I are all located *behind* the mirror,* so the distances to these points as measured from the vertex V are taken to be *negative*. Thus, the focal length is $f = \overline{VF} < 0$, and the image distance is $s' = \overline{VI} < 0$. We also make a distinction between the *radius of curvature R* of a surface, which is always *positive*, and the distance $r = \overline{VC}$, which locates the center of curvature and is called the *surface radius*. For the case illustrated in Fig. 37−6, we have $\overline{VC} = r = -R$. (For the case of a concave mirror, we would have $r = +R$.) Following an analysis similar to that for the concave mirror we find the result

$$\frac{1}{s} + \frac{1}{s'} = \frac{2}{r} = -\frac{2}{R} \qquad \textbf{(37−6)}$$

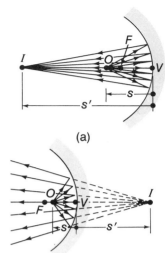

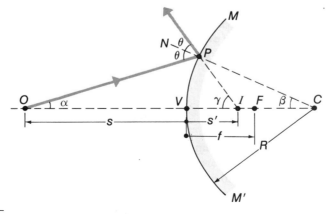

Fig. 37−5. (a) A concave spherical mirror produces a real image when the object O lies outside the focal point F (that is, when $s > f$ and $s' > 0$). (b) A virtual image is formed when O lies inside F (that is, when $s < f$ and $s' < 0$).

Fig. 37−6. Geometry for locating the focal point F of a convex spherical mirror with radius of curvature R.

---

*"In front of" and "behind" are judged by looking along the direction of the incident light.

or
$$\frac{1}{s} + \frac{1}{s'} = \frac{1}{f} \qquad \text{convex mirror} \tag{37-7}$$

with
$$f = -\tfrac{1}{2}R \qquad \text{convex mirror} \tag{37-8}$$

Equations 37–4 and 37–7 are identical, so we can write for *all* spherical mirrors the so-called *Descartes formula*,

$$\frac{1}{s} + \frac{1}{s'} = \frac{1}{f} \tag{37-9}$$

with
$$f = \tfrac{1}{2}r = \begin{cases} +\tfrac{1}{2}R & \text{concave mirror} \\ -\tfrac{1}{2}R & \text{convex mirror} \end{cases} \tag{37-10}$$

The distances s, s', and f are positive when measured in front of the mirror and negative when measured behind the mirror. Images formed in front of the mirror ($s' > 0$) are real, and images formed behind the mirror ($s' < 0$) are virtual. A convex mirror can form only virtual images of real objects.

The sign conventions for mirrors are summarized in Fig. 37–7.

Fig. 37–7. Sign conventions for mirrors. The object distance s for a real object is always positive.

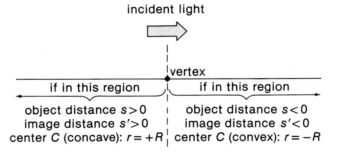

incident light

vertex

if in this region	if in this region
object distance $s>0$	object distance $s<0$
image distance $s'>0$	image distance $s'<0$
center C (concave): $r=+R$	center C (convex): $r=-R$

Example 37–3

(a) A point source of light is located 20 cm in front of a concave spherical mirror with $R = 30$ cm. Determine the position and character of the image point.

(b) If the concave mirror is replaced by a convex mirror with $R = 30$ cm, determine the position and character of the new image point.

Solution:

(a) For the concave mirror, we have $f = \tfrac{1}{2}R = \tfrac{1}{2}(30 \text{ cm}) = 15$ cm and $s = +20$ cm. Then, Eq. 37–9 gives

$$\frac{1}{s'} = \frac{1}{f} - \frac{1}{s} = \frac{1}{15 \text{ cm}} - \frac{1}{20 \text{ cm}} = \frac{1}{60 \text{ cm}}$$

so that $s' = 60$ cm. The image point is located in front of the mirror and is *real*.

(b) For the convex mirror, we have $f = -\tfrac{1}{2}R = -\tfrac{1}{2}(30 \text{ cm}) = -15$ cm and $s = +20$ cm. Then, Eq. 48–9 gives

$$\frac{1}{s'} = \frac{1}{f} - \frac{1}{s} = -\frac{1}{15 \text{ cm}} - \frac{1}{20 \text{ cm}} = -\frac{7}{60 \text{ cm}}$$

so that $s' = -60/7$ cm. The image point is located behind the mirror and is *virtual*.

Example 37–4

The inner surface of a hollow sphere with radius R is coated with a reflecting substance. A point source of light is located at a point O_1, which is $3R/7$ from the center C of the sphere. Deter-

mine the position and character of the image formed by rays that lie near the axis $\overline{O_1 C}$ and are reflected first from the surface nearest O_1.

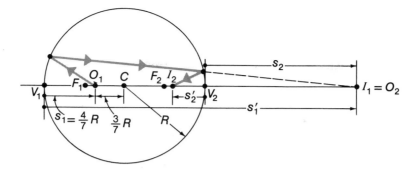

Solution:

The source point O_1 is a distance $s_1 = R - 3R/7 = 4R/7$ from the vertex V_1. If the right-hand hemisphere were missing, an image would be formed at I_1, a distance s_1' from V_1. The points O_1 and I_1 are to the right of the vertex V_1, but they are in *front* of the mirror according to our convention. Thus, s_1, s_1', r, and f are all positive. Using Eq. 37–9, we find

$$\frac{1}{s_1'} = \frac{1}{f} - \frac{1}{s_1} = \frac{2}{R} - \frac{7}{4R} = \frac{1}{4R}$$

so that

$$s_1' = 4R$$

Because the right-hand hemisphere is present, the image at I_1 is not actually formed. However, the rays inside the sphere that converge toward I_1 define a *virtual object* O_2 for the second reflection. The corresponding object distance is *negative*: $s_2 = -(4R - 2R) = -2R$. For the right-hand mirror surface, $f = R/2$. Using Eq. 37–9 again, we obtain

$$\frac{1}{s_2'} = \frac{1}{f} - \frac{1}{s_2} = \frac{2}{R} - \frac{1}{-2R} = \frac{5}{2R}$$

so that

$$s_2' = \frac{2}{5}R$$

Thus, the final image at I_2 is *real* and lies in front of the right-hand mirror surface.

(• Where would the image be formed if the source point were moved to F_1?)

Objects with Finite Size. The image formed by a mirror of an object with extent can be determined by considering the special geometry of certain easily constructed rays called *principal rays*. Figure 37–8 shows these rays emanating from the tip X of an object in the form of a luminous arrow. Ray a leaves X parallel to the axis and upon reflection passes through the focal point F. Ray v is directed toward the vertex V and is reflected at an angle α with respect to the axis equal to the angle of incidence. Ray b passes through the focal point F and upon reflection emerges parallel to the axis. Finally, ray c passes through the mirror center C and is reflected back along its original path. Any two of these four principal rays can be used to locate graphically the image point X' of the arrow tip.

Fig. 37–8. Any two of the four principal rays (a, v, b, and c) may be used to locate the image of an off-axis object point.

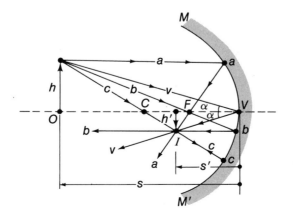

When solving any type of image problem, you should always make a sketch similar to Fig. 37–8. (This recommendation applies also for problems involving *lenses*.)

The height of the object arrow measured from the axis is h and the height of the image arrow is h'. Then, using the triangles formed by the incident and reflected portions of the vertex ray v, we have

$$\tan \alpha = \frac{h}{s} = -\frac{h'}{s'}$$

The negative sign is included because we choose arbitrarily to measure distances above the axis as *positive* and those below the axis as *negative*. In Fig. 37–8, we have $h > 0$ and $h' < 0$. When $h'/h < 0$, the image is *inverted* with respect to the object. We define the *transverse* or *lateral magnification* m to be

$$m = \frac{h'}{h} = -\frac{s'}{s} \qquad \text{transverse magnification} \qquad \text{(37–11)}$$

When $m < 0$, the image is inverted.

▶ *Axial Magnification.* For a small object we can also define an *axial* or *longitudinal magnification* μ. Consider an object that has an axial length Δs and is a mean distance s from the vertex; the image has a mean position s' and a corresponding axial length $\Delta s'$. The general relationship is (Eq. 37–9)

$$\frac{1}{s} + \frac{1}{s'} = \frac{1}{f}$$

Taking differentials, with f fixed, we find

$$-\frac{ds}{s^2} - \frac{ds'}{s'^2} = 0$$

We approximate ds by Δs and ds' by $\Delta s'$; then, the axial magni-

fication is defined to be

$$\mu = \frac{\Delta s'}{\Delta s} = -\left(\frac{s'}{s}\right)^2 = -m^2 \qquad \text{axial magnification} \qquad \text{(37–12)}$$

Notice that m may be either positive or negative, depending on the relative sign of s and s'; however, $\mu < 0$ always. (That is, the image is always reversed front-to-back with respect to the object.) Equations 37–11 and 37–12 are completely general and are valid for all mirror and object geometries. Because μ and m are different, the image is inescapably distorted. ◀

Figure 37–8 shows that when an object is located outside the center C of a concave mirror (so that $s > 2f$), the image formed is real, inverted, and diminished in size (i.e., $-1 < m < 0$). (• Can you apply the reversibility principle to Fig. 37–8 to deduce how the magnification depends on whether the object is outside the center C or is between C and F?) When the object is moved inside the focal point (so that $s < f$), the image becomes virtual, erect (that is, oriented the same as the object), and magnified, as shown in Fig. 37–9a. A concave mirror is therefore a *magnifying mirror* for objects located

Fig. 37–9. (a) When an object is located inside the focal point of a concave mirror, the image is virtual, erect, and magnified. (b) A convex mirror always forms a virtual, erect, and diminished image, regardless of the location of the object.

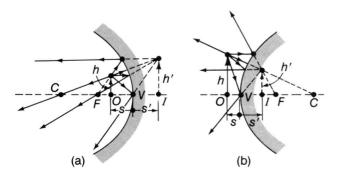

(a)   (b)

inside the focal point. (Your dentist uses such a mirror to examine the inner surfaces of your teeth.)

Figure 37–9b illustrates the case for a convex mirror. Regardless of the position of the object, the image is always virtual, erect, and diminished.

Example 37–5

A small luminous object is located in front of a concave mirror whose radius of curvature is 30 cm. The image projected by the mirror onto a screen is three times as large as the object. Determine the location of the object and the screen.

Solution:

In view of the discussion concerning Fig. 37–8, a real image with $|m| > 1$ requires $s' > s > 0$ and an object position between C and F. It follows that $m = -3$, and using Eq. 37–11,

we have $s' = 3s$. With $R = 30$ cm, Eq. 37–10 gives

$$f = \tfrac{1}{2}R = 15 \text{ cm}$$

Finally, using Eq. 37–9, we have

$$\frac{1}{s} + \frac{1}{3s} = \frac{1}{15 \text{ cm}}$$

from which $s = 20$ cm

and $s' = 3s = 60$ cm

Because $m < 0$, the image is inverted.

37–3   REFRACTION AT SPHERICAL SURFACES

One of the most useful optical elements is the *lens*. A lens is usually a thin, circular piece of glass (or other transparent substance) with spherical surfaces. To understand the image-forming properties of lenses, it is necessary first to examine, in this section, the nature of the refraction process at a spherical interface. In Section 37–4 we discuss the properties of different types of lenses.

Figure 37–10 shows a convex spherical interface with radius of curvature R that separates two media, one with index of refraction n and one with index n'. A point source of light is located at O. The line that connects O and the center C of the surface defines the axis of the system and the vertex V. It is again convenient to measure distances from the vertex, so the object distance is $\overline{OV} = s$ and the image distance is $s' = \overline{VI}$.

For the case $n' > n$ (Fig. 37–10a) an incident ray OP may be sufficiently refracted toward the normal at P that the ray will intersect the axis at the image point I. Thus, a *real* image is formed *behind* the refracting surface. On the other hand, if $n' < n$ (Fig. 37–10b), an incident ray will be refracted away from the normal and will always appear to diverge from the image point I. Thus, a *virtual* image is formed in *front* of the refracting

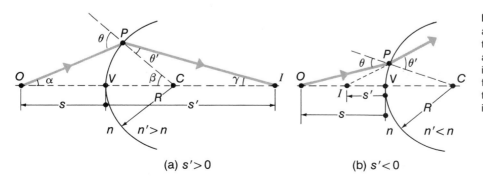

Fig. 37–10. Image formation by a convex spherical interface between two media. (a) $n' > n$, and an object position such that a real image is formed behind the interface: $s' > 0$. (b) $n' < n$, and a virtual image is formed in front of the interface: $s' < 0$.

(a) $s' > 0$ (b) $s' < 0$

Fig. 37–11. Sign conventions for spherical interface between refracting media. The object distance s for a real object is always positive.

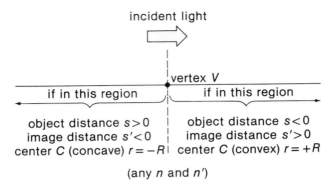

surface. The situation is therefore the opposite of that for image formation by mirrors, in which real images appear in front of mirrors and virtual images appear behind. For refracting systems, the "front" surface is the surface on which the light is first incident.

We can preserve the sign convention adopted for mirrors by making the interpretation that the image distance s' is positive for real images and negative for virtual images (Fig. 37–11). Moreover, the surface radius is $r = +R$ when the center C is located behind the vertex, as in Fig. 37–10. (If the surface were concave,* we would have $r = -R$.)

With these conventions, we can derive the equivalent of the Descartes formula (Eq. 37–9) for a refracting surface. Referring to Fig. 37–10a, we see that $\theta = \alpha + \beta$ and $\beta = \theta' + \gamma$. For paraxial rays, we can write $\sin \theta \cong \theta$ and $\sin \theta' \cong \theta'$; then, Snell's law becomes

$$n\theta \cong n'\theta' \tag{37–13}$$

Using the preceding expressions for θ and θ', Eq. 37–13 becomes

$$n\alpha + n'\gamma = (n' - n)\beta \tag{37–14}$$

We also have $\overparen{PV} = \ell$, $\alpha \cong \ell/s$, $\gamma \cong \ell/s'$, and $\beta = \ell/r = \ell/R$ because $r = +R$ and all the angles are positive. Then, Eq. 37–14 can be expressed as

$$\frac{n}{s} + \frac{n'}{s'} = \frac{n' - n}{R} \qquad \text{convex surface} \tag{37–15a}$$

For a concave surface, the derivation proceeds in a similar manner, except that $r = -R$. The resulting equation is (see Problem 37–10)

$$\frac{n}{s} + \frac{n'}{s'} = -\frac{n' - n}{R} \qquad \text{concave surface} \tag{37–15b}$$

The equation for both the convex and concave cases can be combined in the form

$$\frac{n}{s} + \frac{n'}{s'} = \frac{n' - n}{r} \tag{37–16}$$

with

$$r = +R \qquad \text{convex surface}$$
$$r = -R \qquad \text{concave surface} \tag{37–17}$$

---

*"Concave" and "convex" refer to the shape of the surface when looking along the direction of the incident light.

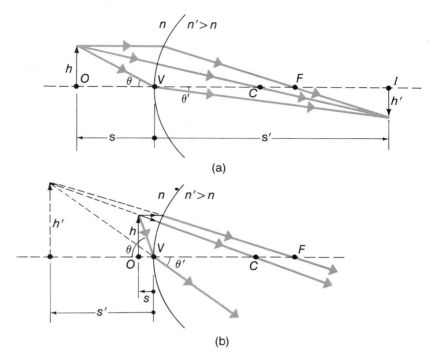

(a)

(b)

Fig. 37-12. Image formation by a convex refracting surface. The three principal rays are shown. (Compare Figs. 37-9 and 37-10.) (a) The image is real, inverted, and diminished. (b) The object is moved close to the surface, and the image is virtual, erect, and magnified. Notice that the location of the focal point F is determined by the incident ray that is parallel to the axis OVC.

For either a convex or a concave surface, $s' > 0$ corresponds to a real image located behind the surface and $s' < 0$ corresponds to a virtual image located in front of the surface.

Objects with Finite Size. As is the case for mirrors, the image formed by a refracting surface of an object with extent can be determined by considering the geometry of the principal rays. Figure 37-12 shows the three principal rays emanating from the tip of an object arrow. (• Why is there no fourth principal ray as in the case of a mirror— Figs. 37-8 and 37-9?) In Fig. 37-12a a real image is formed behind the lens. When the object is moved close to the same refracting surface, as in Fig. 37-12b, a virtual image is formed in front of the surface.

Because the rays are paraxial in Fig. 37-12, we see that $h \cong s\theta$ and $h' \cong s'\theta'$. Then, using Eq. 37-13, the lateral magnification m can be expressed as

$$m = \frac{h'}{h} = \frac{s'\theta'}{s\theta} = -\frac{ns'}{n's} \qquad \textbf{(37-18)}$$

where the negative sign again means that the image is inverted when $s > 0$ and $s' > 0$.

▶ The axial magnification μ is obtained by taking differentials of Eq. 37-16 with n and n' fixed; we obtain

$$-\frac{n\,ds}{s^2} - \frac{n'\,ds'}{s'^2} = 0$$

Thus, we find

$$\mu = \frac{\Delta s'}{\Delta s} = \frac{ds'}{ds} = -\frac{n'}{n}m^2 \qquad \textbf{(37-19)} ◀$$

Example 37-6

A diver wears thin plastic goggles that, when underwater, assume a spherical shape with radius of curvature R = 25 cm due to the water pressure. A minnow with a length h = 5 cm swims directly in front of the diver's goggles at a distance s = 75 cm. What are the apparent location and size of the minnow as perceived by the diver? (Neglect any optical effect due to the composition of the goggles.)

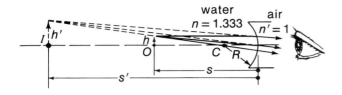

Solution:

Using Eq. 37–16, we have, with $r = -R$,

$$\frac{1}{s'} = \frac{1}{n'}\left(\frac{n' - n}{r} - \frac{n}{s}\right) = \frac{1 - 1.333}{-25 \text{ cm}} - \frac{1.333}{75 \text{ cm}}$$

$$\frac{1}{s'} = -\frac{0.111}{25 \text{ cm}} = -\frac{1}{225 \text{ cm}}$$

or

$$s' = -225 \text{ cm}$$

The lateral magnification, from Eq. 37–18, is

$$m = -\frac{ns'}{n's} = \frac{(1.333)(-225 \text{ cm})}{(1)(75 \text{ cm})} = +4$$

Therefore, the apparent length of the minnow is

$$h' = mh = (4)(5 \text{ cm}) = 20 \text{ cm}$$

Example 37–7

A point source of light (in air) is placed at O_1, a distance $2R$ from the center of a glass sphere with radius of curvature R and refractive index $n = 1.50$. Find the image point of the source formed by paraxial rays.

Solution:

The rays from the source at O_1 first strike the convex face of the sphere. Thus, in using Eq. 37–16, we have $r = +R$ for the surface radius. Also, $n_1 = 1$, $n_1' = 1.50$, and $s_1 = R$; we have

$$\frac{1}{s_1'} = \frac{1}{1.50}\left(\frac{1.50 - 1}{R} - \frac{1}{R}\right) = -\frac{1}{3R}$$

so that

$$s_1' = -3R$$

Hence, the image formed by the first refraction is virtual and is located at I_1, a distance $3R$ in front of the vertex V_1.

The refracted ray within the sphere appears to have originated at the point I_1; thus, this point serves as the *virtual object** for the second refraction. The ray strikes the concave surface of the interface with vertex V_2. Thus, in using Eq. 37–16, we have $r = -R$ for the surface radius. Also, $n_2 = 1.50$, $n_2' = 1$, and $s_2 = 3R + 2R = 5R$. Equation 37–16 now gives

$$\frac{1}{s_2'} = \frac{1 - 1.50}{-R} - \frac{1.50}{5R} = \frac{1}{5R}$$

so that

$$s_2' = 5R$$

Thus, the final image is located a distance $5R$ behind the rear surface of the sphere, and the image is real.

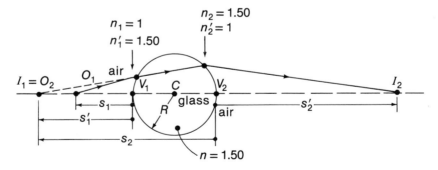

37–4 THIN LENSES

The laws that govern refraction at spherical interface surfaces can be used to examine the behavior of lenses of various types. In order to simplify the discussion, we restrict attention to thin lenses and paraxial rays. The study of these simplified cases exhibits all of the basic features of image formation by lenses. However, modern lens design is not

---

*Rays actually intersect at the position of a *real* object or a *real* image; otherwise, we have a *virtual* object or a *virtual* image.

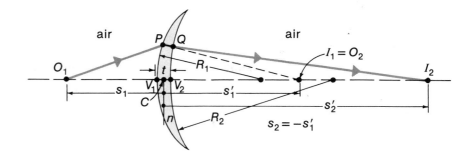

Fig. 37-13. Geometry for determining the image point formed by a thin lens of a point source located at O_1 on the lens axis. O_2 is a *virtual object* for the refraction at Q, and $s_2 < 0$.

limited by these restrictions because the time-consuming and tedious calculations that are necessary to analyze real lenses without approximations are now made with fast electronic computers. Not only is the design of lenses now carried out by computers, but so is the control of the manufacturing process. One result of this modern approach to lens production is that camera lenses today have much better optical quality and yet are less expensive than they were 20 or 30 years ago.

A *thin lens* is one whose thickness along any line parallel to the lens axis is small compared with the radii of curvature of the lens surfaces. For example, the lens shown in Fig. 37-13 is *thin* because $t \ll R_1$ and $t \ll R_2$. Thus, it is permissible to consider the vertexes V_1 and V_2 to coincide with the center of the lens at C. Lenses are commonly used in air, so we take $n_1 = 1$ and set $n_2 = n$, the refractive index of the lens glass.

Figure 37-13 shows the ray O_1P refracted by the interface surface at P. The object distance is $s_1 = \overline{O_1V_1} \cong \overline{O_1C}$, and the surface radius is $r_1 = +R_1$. The image that would be formed by this surface alone, if the medium n were extended, is real and its position is given by Eq. 37-16; thus,

$$\frac{1}{s_1} = \frac{n-1}{R_1} - \frac{n}{s_1'} \qquad \textbf{(37-20)}$$

A second refraction of the ray occurs at Q. The (virtual) object O_2 for this refraction is the image I_1 that results from the first refraction. The interface surface is again convex, so the surface radius is $r_2 = +R_2$. Equation 37-16 gives

$$\frac{n}{s_2} = \frac{1-n}{R_2} - \frac{1}{s_2'} \qquad \textbf{(37-21)}$$

Now, $|s_1'| = \overline{V_1I_1} \cong \overline{CI_1}$ and $|s_2| = \overline{V_2O_2} \cong \overline{CI_1}$. According to the sign convention, $s_1' > 0$ and $s_2 < 0$; therefore, $s_2 = -s_1'$. It is important to realize that this condition is always true. For example, if the image point I_1 is in front of V_1, we would have $s_1' < 0$; but, then, O_2 would be in front of V_2, so that $s_2 > 0$. Substituting $s_2 = -s_1'$ into Eq. 37-21, we have

$$-\frac{n}{s_1'} = \frac{1-n}{R_2} - \frac{1}{s_2'}$$

and substituting this result into Eq. 37-20 gives

$$\frac{1}{s_1} + \frac{1}{s_2'} = (n-1)\left(\frac{1}{R_1} - \frac{1}{R_2}\right)$$

This equation can be rewritten to refer to the lens as a whole by dropping the subscripts on s and s'. Moreover, the equation applies for either convex or concave surfaces if we

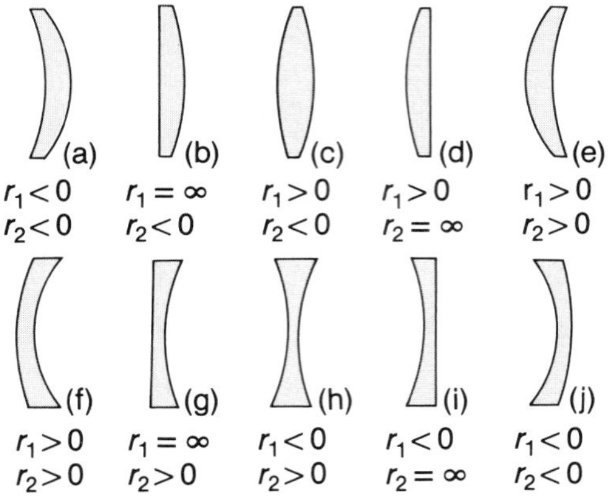

Fig. 37–14. Various lens surface combinations. Lenses (a) through (e) are *converging* (or *positive*) *lenses* with $|r_1| > |r_2|$ for (a) and $|r_2| > |r_1|$ for (e). Lenses (f) through (j) are *diverging* (or *negative*) *lenses* with $|r_1| > |r_2|$ for (f) and $|r_2| > |r_1|$ for (j). In general, converging lenses are thicker at the center than at the edges, whereas diverging lenses are thinner at the center than at the edges.

(a) $r_1 < 0$, $r_2 < 0$
(b) $r_1 = \infty$, $r_2 < 0$
(c) $r_1 > 0$, $r_2 < 0$
(d) $r_1 > 0$, $r_2 = \infty$
(e) $r_1 > 0$, $r_2 > 0$

(f) $r_1 > 0$, $r_2 > 0$
(g) $r_1 = \infty$, $r_2 > 0$
(h) $r_1 < 0$, $r_2 > 0$
(i) $r_1 < 0$, $r_2 = \infty$
(j) $r_1 < 0$, $r_2 < 0$

replace R_1 and R_2 by r_1 and r_2. Thus,

$$\frac{1}{s} + \frac{1}{s'} = (n - 1)\left(\frac{1}{r_1} - \frac{1}{r_2}\right)$$

(37–22)

where s and s' are measured from the center C of the lens and, along with r_1 and r_2, follow the sign convention previously established. If either refracting surface of the lens is concave, the surface radius r for that surface is a negative quantity.

Several different lens surface combinations are illustrated in Fig. 37–14. The lenses (a) through (e) are called *converging lenses* because they make a beam of parallel rays converge toward a focal point F' located behind the lens (Fig. 37–15a). The lenses (f) through (j) are *diverging lenses*; they cause a beam of parallel rays to diverge from a focal point F' located in front of the lens (Fig. 37–15b). Notice in both cases that the point toward which the incident rays converge or away from which they diverge is labeled F'. This point is located behind a converging lens (Fig. 37–15a) and in front of a diverging lens (Fig. 37–15b). If the rays were incident from the right, the focal points would be the points labeled F. For a particular thin lens, the focal points, F and F', are equidistant from the lens.

As is the case for spherical mirrors, it is useful to define the focal length f of a lens as the distance s' when $s \to \infty$ (or vice versa). Thus, from Eq. 37–22, we have

$$\frac{1}{f} = (n - 1)\left(\frac{1}{r_1} - \frac{1}{r_2}\right)$$

(37–23)

which is known as the *lens maker's formula*. Notice that for all of the lenses (a) through (e) in Fig. 37–14, Eq. 37–23 shows that $f > 0$; for this reason, such lenses are referred to

Fig. 37–15. (a) A parallel beam that is incident on a converging lens is brought to a focus at F'. (b) A parallel beam that is incident on a diverging lens appears to diverge from a focal point F'.

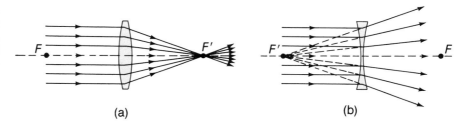

(a)   (b)

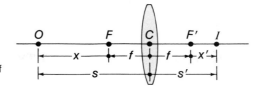

Fig. 37-16. Geometry for the Newtonian form of the lens equation, $xx' = f^2$.

as *positive lenses*. Similarly, all of the lenses (f) through (j) have $f < 0$ and are called *negative lenses*.

The lens equation (Eq. 37–22) now becomes

$$\frac{1}{s} + \frac{1}{s'} = \frac{1}{f}$$

(37–24)

This result is known as the *Gaussian form* of the lens equation. Notice that Eq. 37–24 is identical to the mirror equation (Eq. 37–9).

In the Gaussian form of the lens equation, the distances s and s' are measured from the center C of the lens. It is sometimes convenient to use a lens equation in which the object and image distances are measured from the focal points. We define (see Fig. 37–16)

$$x = s - f \quad \text{and} \quad x' = s' - f$$

(37–25)

Then, in terms of x and x', Eq. 37–24 becomes

$$xx' = f^2$$

(37–26)

This result is known as the *Newtonian form* of the lens equation. The previous sign conventions apply, with the proper signs for x and x' given by Eqs. 37–25.

Objects with Finite Size.   The image-forming properties of a thin lens for an object with extent are determined by considering the behavior of the principal rays, just as for the case of a single interface (Fig. 37–12). Figure 37–17 shows how the three principal rays from the tip of an object are used to determine the position of the image. Notice that the ray c passes essentially *undeviated* through the center C of the lens; this is the case only if the lens is *thin* (see Example 34–2).

In the ray construction shown in Fig. 37–17, notice that the straight central ray c forms two similar triangles with the lens axis and the heights h and h'. Thus, the lateral magnification can be expressed as

$$m = \frac{h'}{h} = -\frac{s'}{s}$$

(37–27)

where the negative sign again indicates an inverted image when $s > 0$ and $s' > 0$.

The location and the character of an image formed by a single converging lens is determined by the position of the object with respect to the focal point. Figure 37–18 illustrates the image formation by such a lens. If the object is located outside the focal

Fig. 37-17. The three principal rays, *a, b,* and *c,* are used to determine the position of the image *I*.

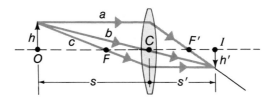

Fig. 37–18. Image formation by a single converging lens. (a) When $s > f$, we have $s' > f$; the image is real and inverted. (b) When $s < f$, we have $s' < 0$ and $|s'| > f$; the image is virtual and erect.

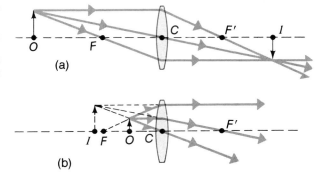

(a)

(b)

point F (that is, farther from the lens than F), the image is real and inverted (Fig. 37–18a). (• How does the magnification change with object position?) However, if the object is located inside the focal point, as in Fig. 37–18b, the image is virtual, erect, and magnified. Thus, a simple converging lens acts as a *magnifying glass* when a small object is placed close to and inside the focal point. Notice that a single converging lens does not form images between its focal points.

Figure 37–19 shows that a single diverging lens produces an erect virtual image inside the focal point F' regardless of the position of the object.

Fig. 37–19. Image formation by a single diverging lens. The image is virtual, erect, and diminished for both $s > |f|$ and $s < |f|$.

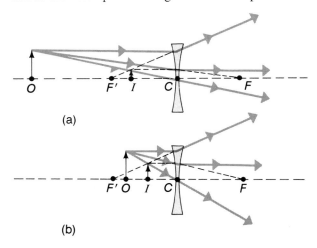

(a)

(b)

Example 37–8

The thin lens shown in the diagram is made of glass ($n = 1.520$) and has radii of curvature $R_1 = 10$ cm and $R_2 = 4$ cm.
(a) Determine the focal length f of the lens.
(b) A luminous object with a height $h = 2$ cm is located

30 cm in front of the lens. Determine the position, the character, and the size of the image.

Solution:

(a) Using the lens maker's formula (Eq. 37–23), we find

$$\frac{1}{f} = (n - 1)\left(\frac{1}{r_1} - \frac{1}{r_2}\right)$$

$$= (1.520 - 1)\left(\frac{1}{10 \text{ cm}} - \frac{1}{(-4 \text{ cm})}\right)$$

from which $\qquad f = +5.49$ cm

(b) Using Eq. 48–24 with $s = 30$ cm, we obtain

$$\frac{1}{s'} = \frac{1}{f} - \frac{1}{s} = \frac{1}{5.49 \text{ cm}} - \frac{1}{30 \text{ cm}}$$

from which $s' = +6.72$ cm

Alternatively, we can determine the value of s' by using Eq. 37–26. Thus,

$$x' = \frac{f^2}{x} = \frac{f^2}{s - f} = \frac{(5.49 \text{ cm})^2}{(30 \text{ cm}) - (5.49 \text{ cm})}$$

$$= +1.23 \text{ cm}$$

Then, $s' = x' + f = (1.23 \text{ cm}) + (5.49 \text{ cm})$

$$= +6.72 \text{ cm}$$

as before.

The magnification is

$$m = -\frac{s'}{s} = -\frac{6.72 \text{ cm}}{30 \text{ cm}} = -0.224$$

so that the image is real and inverted with a height

$$h' = mh = (-0.224)(2 \text{ cm}) = -0.448 \text{ cm}$$

where the negative sign means that h' is measured from the axis in the direction opposite that for h.

Example 37–9

Rework Example 37–8 in the event that the second surface of the lens is convex; that is, $r_2 = 4$ cm instead of -4 cm.

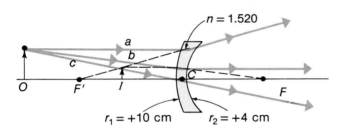

$r_1 = +10$ cm $r_2 = +4$ cm

(a) The focal length is now

$$\frac{1}{f} = (1.520 - 1)\left(\frac{1}{10 \text{ cm}} - \frac{1}{4 \text{ cm}}\right)$$

from which $f = -12.82$ cm

Thus, the focal point F' is located 12.82 cm in front of the lens, as indicated in the diagram.

(b) Using Eq. 37–24 with $s = 30$ cm, we obtain

$$\frac{1}{s'} = \frac{1}{(-12.82 \text{ cm})} - \frac{1}{30 \text{ cm}}$$

from which $s' = -8.98$ cm

(• What do you find for s' by using Eq. 37–26?)

The magnification is

$$m = -\frac{s'}{s} = -\frac{(-8.98 \text{ cm})}{30 \text{ cm}} = +0.299$$

so that the image is virtual and erect with a height,

$$h' = mh = (0.299)(2 \text{ cm}) = 0.598 \text{ cm}$$

Notice that, in each example, ray a leaves the object parallel to the axis and emerges from the lens directed toward or away from F'. Similarly, ray b leaves the object directed toward F and emerges from the lens parallel to the axis.

37–5   FERMAT'S PRINCIPLE AND GEOMETRIC ABERRATIONS (Optional)

Spherical mirrors and lens elements with spherical surfaces are subject to an intrinsic defect called spherical aberration, which is the inability to bring all rays from an object point to the same focus. Only paraxial rays, which make relatively small angles of divergence with the axis, form a reasonably sharp image. As more divergent rays are permitted to participate, the aberration becomes increasingly worse.

What type of reflecting and refracting surfaces would be completely free from spherical aberration? That is, what type of surfaces are required such that *all* of the rays emanating from an object O are brought precisely to focus at an image point I? According to Fermat's principle (Section 34–4), the optical path

length connecting points O and I, namely, $L = \int_O^I n \, ds$, must be a minimum in comparison with all neighboring paths of the same general character (that is, reflected or refracted from the same surface). Now, if *all* rays from O are to reach I, it follows that the optical path length must be the same for each ray. Thus, successful optical design consists of contouring surfaces to shapes that lead to equal optical path lengths. High-quality camera lenses achieve the same result by making use of many separate lens elements.

As an illustration of Fermat's principle for refraction at a single air-glass interface, refer to Fig. 37–20. It proves convenient to specify the optical path length of the ray in terms of the

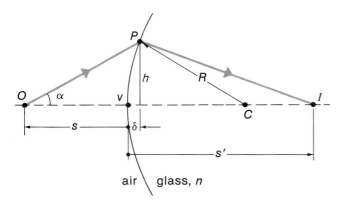

Fig. 37–20. Fermat's principle can be used to demonstrate that only paraxial rays from the object point O are brought to a sharp focus at I.

small parameter δ defined in Fig. 37–20. The axial ray OVI that connects the object point O with the image point I has the optical path length

$$L_O = \oint_O^I n\,ds = s + ns'$$

The optical path length of the ray OPI is

$$L_P = \sqrt{(s + \delta)^2 + h^2} + n\sqrt{(s' - \delta)^2 + h^2} \quad \textbf{(37–28)}$$

We also have $h^2 + (R - \delta)^2 = R^2$, or $h^2 = 2R\delta - \delta^2$. Using this expression for h^2 in Eq. 37–28 gives

$$L_P = s[1 + 2\delta(R + s)/s^2]^{1/2} + ns'[1 + 2\delta(R - s')/s'^2]^{1/2}$$

This expression for L_P is exact; however, by expanding in powers of δ and retaining only the linear term, we can show that L_P is approximately equal to L_O. Thus, to first order in δ, we have

$$L_P \cong s + ns' + \delta R\left(\frac{1}{s} + \frac{n}{s'}\right) + \delta(1 - n) \quad \textbf{(37–29)}$$

Now, Eq. 37–16 for the present case, with $r = +R$, is

$$\frac{1}{s} + \frac{n}{s'} = \frac{n - 1}{R}$$

Consequently, the last two terms in Eq. 37–29 sum to zero. There remains

$$L_P \cong s + ns'$$

Thus, $L_P \cong L_O$, as we set out to show. To the extent that spherical aberration can be ignored (or corrected), it follows that the wavefronts associated with *all* rays that leave the object at a given instant arrive at the image *in phase*. (This is an important point in the study of diffraction; see Chapter 36.)

This result means that only paraxial rays (for which δ is small) are focused at point I. Rays that make larger angles with the axis suffer spherical aberration and contribute to blurring of the image. Spherical aberration is only one of the various types of aberrations that affect refracting devices. (Another is *chromatic aberration*; see Problem 37–35.)

37–6   THICK LENSES AND LENS COMBINATIONS (Optional)

Principal Planes and Principal Points. The discussion in Section 37–4 was concerned with *thin* lenses, idealized lenses whose thickness along the optic axis is negligibly small. We now turn attention to cases in which the lens thickness is sufficiently large that it must be explicitly included in any calculation. The treatment of thick lenses is also applicable to systems of separated thin lenses, which are found in many optical instruments. Although the situation becomes more complicated, it is still possible to make a simple analysis of thick-lens systems by introducing some new geometric features of the lenses.

Figure 37–21 shows a thick lens that brings an incident parallel ray to the focal point F' in the region called the *image space* of the lens. The actual ray is refracted at each surface of the lens and follows the path $OABF'$. We can simplify the description of this ray within the lens by representing the transit in terms of a single refraction (at point B'). To do this, we introduce the *principal plane $H'B'$* and the *principal point H'* of the image

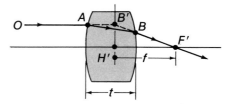

Fig. 37–21. The intersection of the extended rays, OAB' and $B'BF'$, locates the principal plane $B'H'$ in the image space of a thick lens.

space. As indicated in Fig. 37–21, the principal plane in the image space is located by the intersection of the incident parallel ray with the emergent ray which passes through the focal point F', when both rays are extended within the lens (dashed lines in Fig. 37–21). The principal plane and principal point in the object space (not shown in Fig. 37–21; see Fig. 37–22) are located

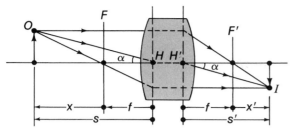

Fig. 37-22. Geometric construction to locate the image of an object produced by a thick lens. The focal planes are indicated by F and F'.

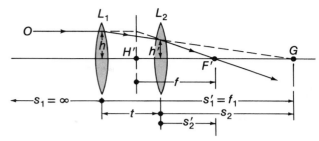

Fig. 37-23. Geometry for the calculation of the focal length f of a two-lens system.

by a similar construction with a ray incident from the right.

Figure 37-22 shows the geometry of a thick lens and the construction that locates the image I of the object O. For a particular lens, when the locations of the principal planes are specified and the focal length f is given, the positions of the focal points, F and F', are then determined (see Fig. 37-22). These parameters are intrinsic properties of a particular lens and determine the image-forming properties of the lens.

We now consider the formation of an image by a thick lens. It is convenient to measure all distances from the focal planes or the principal planes. Then, both the Gaussian and Newtonian forms of the lens equation apply (see Eqs. 37-24 and 37-26):

$$\frac{1}{s} + \frac{1}{s'} = \frac{1}{f} \qquad \text{Gaussian formula} \qquad \textbf{(37-30a)}$$

$$xx' = f^2 \qquad \text{Newtonian formula} \qquad \textbf{(37-30b)}$$

The principal rays that are used to locate the image of the object O are shown in Fig. 37-22. Notice that the "central" ray $OHH'I$ makes equal angles α with the lens axis; that is, OH and $H'I$ are parallel but offset rays. The hypothetical rays within the lens are again shown dashed. Notice also that in the limit of zero lens thickness, the principal planes coincide with the central plane of the resulting thin lens.

A Simple Two-Element Lens. Consider a simple lens system that consists of two separated thin lenses, as shown in Fig. 37-23. The lenses are separated by a distance t and have focal lengths f_1 and f_2. What is the focal length f for the lens combination? The treatment of the properties of a thick lens does not explicitly involve the fact that the lens has only two air-glass interfaces. Consequently, we may discuss in the same way any

system of lenses arranged in fixed geometry. Thus, the principal and focal planes are located just as for a thick lens. The parallel ray OF' originates in an object at an infinite distance so that $s_1 = \infty$. The lens L_1, if alone, would form an image at G; thus, G becomes the virtual object for lens L_2. Then, as indicated in Fig. 37-23, we have* $s_2 = -(f_1 - t)$. Now, the lens equation for the second lens is $1/s_2 + 1/s_2' = 1/f_2$, which becomes

$$\frac{1}{s_2'} = \frac{1}{f_2} + \frac{1}{f_1 - t} \qquad \textbf{(37-31)}$$

Also, from the similar triangles, we have

$$\frac{h}{s_1'} = \frac{h'}{-s_2} \qquad \textbf{(37-32a)}$$

and

$$\frac{h}{f} = \frac{h'}{s_2'} \qquad \textbf{(37-32b)}$$

where f is measured from the principal point H'. Dividing Eq. 37-32b by Eq. 37-32a, we find

$$\frac{1}{f} = -\left(\frac{1}{s_1'}\right)\left(\frac{s_2}{s_2'}\right) \qquad \textbf{(37-33)}$$

Then, using Eq. 37-31 together with $s_1' = f_1$ and $s_2 = -(f_1 - t)$, Eq. 37-33 gives

$$\frac{1}{f} = \frac{1}{f_1} + \frac{1}{f_2} - \frac{t}{f_1 f_2} \qquad \textbf{(37-34)}$$

Because the location of the focal point F' is determined by s_2' (which is known from Eq. 37-31), the position of the principal point H' is specified by the value of f calculated from Eq. 37-34.

Example 37-10

The lens shown in the diagram on page 730 has a refractive index $n = 1.520$ and a thickness $t = 3.5$ cm, measured between the vertexes V_1 and V_2. The surface radii are $r_1 = +12.0$ cm and $r_2 = -16.0$ cm. Determine the focal length f and the locations of the focal point F' and the principal point H'.

---

*According to the convention on signs (see Section 37-4), s_2 is negative because G is to the right of lens L_2.

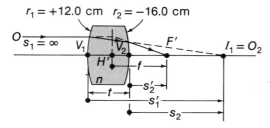

$r_1 = +12.0$ cm $r_2 = -16.0$ cm

Solution:

Consider an object O at infinity so that $s_1 = \infty$. Equation 37–16 for a spherical interface states that

$$\frac{n}{s} + \frac{n'}{s'} = \frac{n' - n}{r} \qquad (1)$$

Applying this equation to the first interface of the thick lens yields

$$\frac{1}{\infty} + \frac{1.520}{s_1'} = \frac{1.520 - 1}{12.0 \text{ cm}}$$

from which $\qquad s_1' = +35.1$ cm

The real image I_1 that would be formed by the first interface alone is the virtual object O_2 for the second interface. Thus,

$$s_2 = -(35.1 \text{ cm} - 3.5 \text{ cm}) = -31.6 \text{ cm}$$

Using Eq. (1) for this interface gives

$$\frac{1.520}{-31.6 \text{ cm}} + \frac{1}{s_2'} = \frac{1 - 1.520}{-16.0 \text{ cm}}$$

from which $\qquad s_2' = +12.4$ cm

That is, the focal point F' is 12.4 cm to the right of vertex V_2.

The focal length f can be obtained from Eq. 37–33, which is valid for thick lenses; thus,

$$\frac{1}{f} = -\left(\frac{1}{35.1 \text{ cm}}\right)\left(\frac{-31.6 \text{ cm}}{12.4 \text{ cm}}\right)$$

from which $\qquad f = 13.8$ cm

These results for s_2' and f determine the position of the principal point H', which is 13.8 cm − 12.4 cm = 1.4 cm to the left of vertex V_2.

The focal point F and the principal point H in the object space can be located by a similar calculation for a parallel ray incident from the right. The results are that F is located 12.7 cm to the left of vertex V_1 and that the principal point H is 13.8 cm − 12.7 cm = 1.1 cm to the right of vertex V_1. (See Problem 37–20.)

Example 37–11

A telephoto lens for a camera should have a long focal length in order to produce a large image on the film surface. (Recall that the lateral magnification m is proportional to the focal length, $m = -f/x$; see Problem 37–16.) A second requirement is that the image plane F' should be close to the lens system so that the overall length of the telephoto lens plus camera body will not be inconveniently large. These seemingly contradictory design aims can be achieved by a lens combination whose principal point H' (from which the focal length is measured) lies on the *object* side of the lens system. Determine the focal length and the locations of the focal points and principal points for the two-element telephoto lens shown in the diagram.

$f_1 = +10$ cm $f_2 = -3$ cm

F   H   H'   F'

L₁   L₂

$t =$
8cm

Solution:

According to Eq. 37–34, the focal length f of the lens combination is given by

$$\frac{1}{f} = \frac{1}{10 \text{ cm}} + \frac{1}{-3 \text{ cm}} - \frac{8 \text{ cm}}{(10 \text{ cm})(-3 \text{ cm})}$$

from which $\qquad f = +30$ cm

Consider an object at an infinite distance to the left of lens L_1 so that $s_1 = \infty$; then, $s_1' = f_1 = 10$ cm. This image position is the position of the virtual object for lens L_2; hence, $s_2 = -(10 \text{ cm} - 8 \text{ cm}) = -2$ cm, measured from L_2. The lens equation then gives

$$\frac{1}{s_2'} = \frac{1}{-3 \text{ cm}} - \frac{1}{-2 \text{ cm}}$$

from which $\qquad s_2' = 6$ cm

Thus, the focal point F' is 6 cm to the right of lens L_2. Because the distance from H' to F' is $f = +30$ cm, the principal point H' is therefore 30 cm − (8 cm + 6 cm) = 16 cm to the *left* of lens L_1, as indicated in the diagram.

A similar calculation using a parallel ray incident from the right shows that the focal point F is 110 cm to the left of L_1 and that the principal point H is 80 cm to the left of L_1.

Notice that the physical length of the telephoto system, from the first lens L_1 to the focal plane, is 14 cm. If a single lens with the same focal length were used instead, the lens-to-focal-plane distance would be more than twice as large (30 cm).

37-7   SIMPLE MAGNIFIERS

For an object to be seen in greatest detail, the largest possible image must be formed on the retina of the eye. This is easily accomplished by placing the object close to the eye. However, a normal human eye cannot focus properly on an object that is closer than about 25 cm. An object at this distance is said to be at the *near point* of the eye. (• Try to measure the minimum distance at which your eye can focus on a small object, such as the print on this page.) Figure 37–24a shows an object being examined by an unaided eye; with the object at the near point, the subtended angle is θ. Figures 37–24b and 37–24c illustrate two ways to view the same object with a simple magnifier (or *magnifying glass*), which consists of a single converging lens. The object in Fig. 37–24b is positioned just inside the focal point of the lens to produce a virtual image at the near point of the eye. This method of viewing can produce eyestrain because the eye muscles must work to provide the necessary focusing effect. A more satisfactory arrangement (Fig. 37–24c) is to position the object at the focal point so that the virtual image appears at infinity; then the retinal image is formed with the eye muscles completely relaxed.

By convention, the magnifying power of a device that produces a virtual image is defined in terms of the *angular magnification* γ. We define the angular magnification of an optical device to be the ratio of the tangents of the angles θ_{lens} and θ_{eye}, where θ_{lens} is the angle subtended by the image when viewed through the lens system, and θ_{eye} is the angle subtended by the object itself when placed at the near point of the eye. Thus,*

$$\gamma = \frac{\tan \theta_{lens}}{\tan \theta_{eye}} \qquad (37-35)$$

Referring to Figs. 37–24b and 37–24c, we see that there are two possibilities for γ depending on the use made of the lens; that is,

$$\gamma = \frac{\tan \theta_b}{\tan \theta} \qquad \text{and} \qquad \gamma_\infty = \frac{\tan \theta_c}{\tan \theta}$$

where θ is the subtended angle for the unaided eye (Fig. 37–24a); that is,

$$\tan \theta = \frac{h}{25 \text{ cm}} \qquad (37-36)$$

where we follow the usual practice and express the near-point distance in cm. We assume that the position of the eye coincides with that of the lens. Then, for the case illustrated in Fig. 37–24b, we have $s_b' = -25$ cm; hence,

$$\frac{1}{s_b} = \frac{1}{f} + \frac{1}{25 \text{ cm}}$$

Fig. 37–24. (a) An object at the near point is viewed by an unaided eye. (b) A simple magnifier produces a virtual image at the near point. (c) The object is positioned to produce a virtual image at infinity, which can be viewed by a relaxed eye.

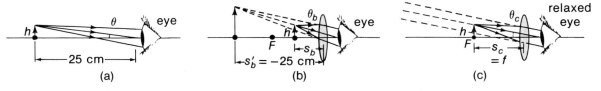

<hr>

* It is also common practice to define the angular magnification as the ratio of the angles themselves; that is, $\gamma = \theta_{lens}/\theta_{eye}$. For small angles, $\tan \theta \cong \theta$, and the definitions are essentially equivalent.

Thus,
$$\tan \theta_b = \frac{h}{s_b} = h\left(\frac{1}{f} + \frac{1}{25 \text{ cm}}\right) \tag{37-37}$$

For the case illustrated in Fig. 37–24c, we have
$$\tan \theta_c = \frac{h}{f} \tag{37-38}$$

It follows that the angular magnifications (Eqs. 37–35) for these two cases can be expressed as

$$\gamma = \frac{25 \text{ cm}}{f} + 1 \qquad \text{image at near point} \tag{37-39a}$$

$$\gamma_\infty = \frac{25 \text{ cm}}{f} \qquad \text{image at infinity} \tag{37-39b}$$

The distance from the object to the lens can be varied between s_b and s_c and the eye can still produce a retinal image in sharp focus. Typical values of γ are 2 or 3 for simple lenses, because aberrations usually preclude the use of lenses with shorter focal lengths and larger magnifications. For these low values of γ, the change in magnification as the object distance is changed between s_b and s_c is readily observable. Specially shaped single lenses can have values of γ as large as 10 or so.

37–8 MICROSCOPES

To obtain magnifications greater than those possible with simple magnifiers, it is necessary to use some form of *compound microscope*. The simplest such instrument consists of two lens elements, an *objective* and an *ocular* (or *eyepiece*), as shown in Fig. 37–25. We treat the objective and ocular as simple lenses, although in a practical microscope they would be multielement lenses (as shown in Fig. 37–26). The photograph on page 733 shows a cutaway view of a modern compound microscope.

The angular magnification Γ of a compound microscope is defined to be the ratio $\Gamma = \tan \theta_e / \tan \theta$, where θ_e is the angle subtended by the final image (see Fig. 37–25) and θ is the angle that the object height h would subtend at the eye if placed at the near point (see Fig. 37–24a). Now,

$$\tan \theta = \frac{h}{25 \text{ cm}}$$

and
$$\tan \theta_e = \frac{h'}{f_e}$$

where f_e is the focal length of the eyepiece (Fig. 37–25). Thus,

$$\Gamma = \frac{\tan \theta_e}{\tan \theta} = \frac{h'}{h}\frac{25 \text{ cm}}{f_e}$$

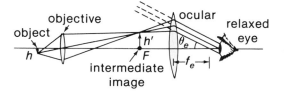

Fig. 37–25. A simple two-lens microscope. The intermediate real image, formed at the focal point F of the ocular, serves as the object for the ocular, which in turn forms a virtual image at infinity.

The ratio h'/h is just the lateral magnification m of the objective (Eq. 37–27), and $(25 \text{ cm})/f_e$ is the angular magnification γ_∞ of the ocular (Eq. 37–39b); therefore, we have for the overall angular magnification of the microscope

$$\Gamma = m\gamma_\infty \qquad \text{microscope} \qquad \text{(37–40)}$$

for viewing by a relaxed eye (final image at infinity).* For a final image at the near point, the angular magnification is

$$\Gamma = m\gamma \qquad \text{(37–40a)}$$

It is customary to refer to $|m|$ as the *magnifying power* of the objective and to γ_∞ as the magnifying power of the ocular. The values of $|m|$ and γ_∞ are both expressed as $\times 5$, $\times 10$, $\times 40$, and so forth.

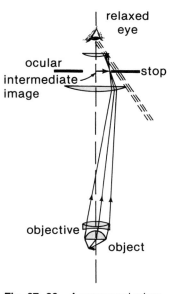

Fig. 37–26. A compound microscope with multielement lenses. The "stop" is an aperture that limits the extreme off-axis rays. The lenses are positioned to produce a virtual image at infinity for viewing by a relaxed eye.

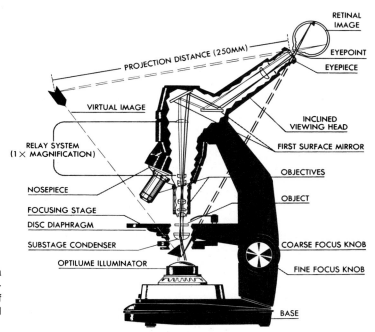

Sectional view of a modern compound microscope. (Courtesy of Bausch & Lomb Optical Co.)

Linear Resolution. Diffraction effects, primarily in the objective system, determine the ultimate resolution of a microscope and set the limit on the maximum usable magnification. For a discussion of microscope objectives, it is more convenient to cast in a different form the expression for the resolution that was developed in Section 36–1. The points O and O' shown in Fig. 37–27 are two object points that are just resolved according to the Rayleigh criterion (Section 36–1). The separation distance ξ is then the *linear resolution* of the lens.

Fig. 37–27. Geometry for discussing the resolving power of a microscope objective.

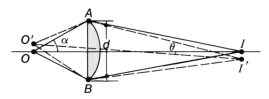

---

*For the situation shown in Fig. 37–25, we have $h'/h < 0$; thus, $\Gamma < 0$, and the final image is inverted.

We omit here the derivation and simply quote the result for a circular lens, namely

$$\xi = \frac{1.22 \lambda}{2 \sin \alpha}$$ (37–41)

Note carefully the distinction between the angle θ (the angle to which Eq. 36–5b applies) and the angle α (the half angle of the lens subtended at the object).

Example 37–12

A compound microscope consists of a thin objective lens with focal length $f_1 = 1$ cm and diameter $d_1 = 1.5$ cm, together with a thin ocular lens with focal length $f_2 = 5$ cm. The lenses are mounted at opposite ends of a tube with length $\ell = 15$ cm. (Refer to Fig. 37–25.)

(a) At what distance from the objective should a specimen be placed to produce a final image at the near point of the eye?

(b) For an image at the near point, what is the overall angular magnification of the microscope?

(c) What is the linear resolution of the objective?

Solution:

(a) With the ocular considered to coincide with the eye, and with the final image at the near point, we have $s_2' = -25$ cm (see Fig. 37–24b); hence,

$$\frac{1}{s_2} = \frac{1}{f_2} - \frac{1}{s_2'} = \frac{1}{5 \text{ cm}} + \frac{1}{25 \text{ cm}}$$

which yields $\qquad s_2 = 4.17$ cm

Then the distance from the objective to the intermediate image (which serves as the object for the ocular lens) is

$$s_1' = \ell - s_2 = 15 \text{ cm} - 4.17 \text{ cm} = 10.83 \text{ cm}$$

Finally, $\qquad \dfrac{1}{s_1} = \dfrac{1}{f_1} - \dfrac{1}{s_1'} = \dfrac{1}{1 \text{ cm}} - \dfrac{1}{10.83 \text{ cm}}$

from which $\qquad s_1 = 1.10$ cm

This is the required specimen-to-objective distance.

(b) Using Eq. 37–40a, we have (see Problem 37–24)

$$\Gamma = m\gamma = -\frac{s_1'}{s_1} \left(\frac{25 \text{ cm}}{f_2} + 1 \right)$$

$$= -\frac{10.83 \text{ cm}}{1.10 \text{ cm}} \left(\frac{25 \text{ cm}}{5 \text{ cm}} + 1 \right) = -59.1$$

The minus sign indicates that the final image is inverted.

(c) We have

$$\alpha = \tan^{-1} \frac{d_1}{2s_1} = \tan^{-1} \frac{1.5 \text{ cm}}{2(1.10 \text{ cm})} = 34.3°$$

and $\qquad \sin \alpha = \sin 34.3° = 0.564$

Thus, the linear resolution of the objective for light with $\lambda = 550$ nm is (Eq. 37–41)

$$\xi = \frac{1.22(550 \text{ nm})}{2(0.564)} = 595 \text{ nm}$$

37–9 TELESCOPES

Telescopes perform two important functions. One is to produce an enlarged image of a distant object, and the other is to gather more light from the object than is possible with the unaided eye. A large useful magnification requires an objective with a large diameter in order to reduce diffraction effects and improve the resolution of the instrument. A large light-gathering power also requires a large-diameter objective lens (or mirror in the case of reflecting telescopes).

Figure 37–28 shows two types of simple telescopes that use lenses (*refracting telescopes*). Diagram (a) illustrates an *astronomical telescope,* which produces an inverted final image, and diagram (b) shows a *Galilean telescope,* which produces an erect final image. Both telescopes are shown giving images at infinity for a distant object (parallel

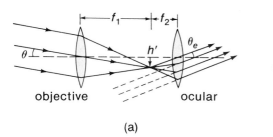

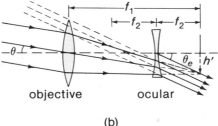

objective   ocular

(a)

objective   ocular

(b)

Fig. 37–28. (a) An astronomical telescope produces an inverted image of a distant object. (b) A Galilean telescope produces an erect image of a distant object.

incident rays); this is the condition appropriate for viewing by a relaxed eye. The erect final image that is produced by a Galilean telescope has an obvious advantage for terrestrial observations. For viewing astronomical objects, an inverted image presents no disadvantage.

The angular magnification Γ of a telescope is defined in the same way as for a microscope, namely, $\Gamma = \tan \theta_e / \tan \theta$, where θ is the angle that the (distant) object subtends at the unaided eye and θ_e is the angle subtended by the final image. In Fig. 37–28 we see that

$$\tan \theta = h'/f_1 \qquad \text{and} \qquad \tan \theta_e = -h'/f_2$$

Hence,*

$$\Gamma = \frac{\tan \theta_e}{\tan \theta} = -\frac{f_1}{f_2} \qquad \text{telescope} \qquad (37\text{–}42)$$

Newton recognized that a mirror does not suffer from chromatic aberration (the focusing of different wavelengths at different focal planes) and therefore has an advantage over a lens as the objective of a telescope. Moreover, the reflectance of a mirror is generally superior to the transmittance of a lens, yielding a brighter final image. Also, spherical aberration can be considerably reduced (indeed, eliminated entirely for rays parallel to the optic axis) by using a paraboloidal mirror surface. Newton constructed the first *reflecting telescope* in 1667. Several methods have been devised for viewing an object with a reflecting telescope; three such arrangements are shown in Fig. 37–29.

Fig. 37–29. Different methods for viewing an object with a reflecting telescope. (a) Observing at the primary focus. (This method is limited to very large telescopes, such as the Mt. Palomar instrument. See the photograph on page 736.) (b) The Newtonian method of observation. (c) The Cassegrainian method. All reflecting telescopes have long focal lengths; these diagrams are not to scale. Also, the mirror area blocked by the observing apparatus is generally small, to maximize the light-gathering area.

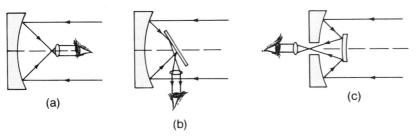

(a)

(b)

(c)

Example 37–13

The Hale telescope (Fig. 37–30) has a mirror diameter of 5.08 m and primary focal length of 17 m. (By using secondary mirrors, the focal length can be increased to 150 m.) For viewing at the primary focus (Fig. 37–29a), an observation cage with a diameter of 1.8 m is used to hold the observer. (See the

---

*Notice that the ocular of the Galilean telescope is a negative lens; thus $f_2 < 0$, which gives a positive value for Γ, indicating an erect final image.

photograph.) Suppose that the ocular that is used has a focal length of 4 cm.

(a) What is the ratio of the light-gathering power of the telescope to that of the unaided eye? (Assume a pupil diameter of 2 mm.)

(b) What is the angular magnification for the two focal lengths given above?

(c) What is the angular resolution of the Hale telescope, if diffraction-limited?

Solution:

(a) Allowing for the light blocked by the observer's cage, the ratio of the light-gathering power of the telescope to that of the unaided eye is equal to the ratio of the collecting areas:

$$\frac{L_t}{L_e} = \frac{\pi[(5.08 \text{ m})^2 - (1.8 \text{ m})^2]}{\pi(2 \times 10^{-3} \text{ m})^2} = 5.6 \times 10^6$$

(b) Equation 37–42 applies for both mirror and lens objectives. (• Can you see why?) For the two mirror focal lengths given, we have

$$\Gamma_{\text{primary}} = -\frac{f_1}{f_2} = -\frac{17 \text{ m}}{0.04 \text{ m}} = -425$$

$$\Gamma_{\text{secondary}} = -\frac{150 \text{ m}}{0.04 \text{ m}} = -3750$$

(c) Using Eq. 36–5b with $\lambda = 550$ nm, we have for the Hale telescope,

$$\delta\theta_{\text{min}} = \frac{1.22 \lambda}{d} = \frac{1.22(550 \times 10^{-9} \text{ m})}{5.08 \text{ m}}$$

$$= 1.32 \times 10^{-7} \text{ rad} \quad \text{or} \quad 0.0272 \text{ arc sec}$$

The actual resolution of the Hale telescope is limited by factors other than diffraction (for example, atmospheric turbulence), so the achieved value is about 5×10^{-6} rad or 1 arc sec. The main advantage of a large-diameter telescope is not that the diffraction effects are small but that the light-gathering power is large.

An astronomer rides in the observer's cage at the primary focus of the Hale telescope at the Palomar Observatory. (Courtesy of Palomar Observatory, California Institute of Technology.)

QUESTIONS

37-1 Define and explain briefly the meaning of the terms (a) paraxial rays, (b) principle of reversibility, (c) Descartes' formula for spherical mirrors, (d) transverse and axial magnification, and (e) Gaussian and Newtonian lens equations.

37-2 When you look into a plane mirror, your image is inverted from left to right. If several plane mirrors were available, how would you arrange them to observe an image of yourself as others see you?

37-3 Do the image-forming properties of a spherical mirror change if the mirror is immersed in water? Explain.

37-4 How would you determine experimentally whether an image is real or virtual?

37-5 In Example 37-4 the image I_1 was described as forming a *virtual object* O_2 for the second reflection. Why is this a reasonable description?

37-6 A real object is used with a convex mirror. Prove that the image is always virtual, always erect, always reduced, and always located inside the focal point, no matter where the object is located.

37-7 A converging lens may be used to start a fire by concentrating the sun's rays on a clear day. Explain how such a ''burning glass'' works.

37-8 Describe some simple, rapid ways to determine experimentally the focal length of a lens.

37-9 Sketch the graph of the behavior of the image distance s' as a function of the object distance s for a convex spherical refracting surface. Refer to Eq. 37-16. (a) Assume $n' > n$. (b) Assume $n' < n$.

37-10 An air bubble in water can serve as a lens. Explain. Would the lens so formed be diverging or converging?

37-11 Do the image-forming properties of a simple lens with one element depend on the wavelength of the light? What about a simple spherical mirror?

37-12 When point sources are used to demonstrate Fraunhofer diffraction, a converging lens is often employed. Explain with the aid of Fermat's principle why phase changes for different rays do not occur. Would the situation be different if large lens apertures with nonparaxial rays were involved? Explain.

37-13 Define and explain briefly the meaning of the terms (a) principal point, (b) telephoto lens, (c) near point of eye, (d) angular magnification, (e) compound microscope, and (f) astronomical telescope.

37-14 Explain why the angular magnification of a simple magnifier depends on how the eye is focused.

37-15 Why would it be advantageous to use blue light with a compound microscope?

37-16 List the advantages and disadvantages of a refracting telescope compared with a reflecting telescope.

PROBLEMS

Section 37-1

37-1 A plane mirror is displaced a distance L perpendicular to its surface. Show that the image point for any fixed object point moves a distance $2L$.

37-2 A plane mirror is rotated through an angle α about an axis that lies in the plane of the mirror. Show that the reflected ray for any particular incident ray that is perpendicular to the axis rotates through an angle 2α.

37-3 Two plane mirrors intersect at an angle of 90°. The line of intersection is the x-axis of a coordinate frame, and the y- and z-axes lie in the planes of the two mirrors. An object is at the point (3, 4, 5) cm. Determine the positions of the image points.

Section 37-2

37-4 A concave spherical mirror has a radius of curvature $R = 50$ cm. A small luminous object is placed on the axis at various distances s from the mirror vertex. Determine the position, the size, and the character of the image for s equal to (a) 100 cm, (b) 35 cm, (c) 15 cm, (d) 2 cm, and (e) $s \rightarrow 0$. Comment on the latter result.

37-5 An object with a height of 3 cm is located on and is perpendicular to the axis of a mirror. A real image, 2 cm in height, is formed 100 cm from the object. What is the focal length of the mirror? Is the mirror concave or convex?

37-6 A concave mirror has a radius of curvature $R = 36$ cm. Find the two positions of the object for which the lateral size of the image is twice that of the object. What is the character of each of these images?

37-7 An object with a height of 2 cm is placed 37.5 cm in front of a concave spherical mirror that has a focal length of 25 cm. A plane mirror, whose surface is perpendicular to the axis, faces the spherical mirror at a distance of 50 cm from the vertex. Determine the position, the size, and the character of the image that is formed by one reflection in each mirror.

Section 37–3

37–8   A transparent sphere of unknown composition is observed to form an image of the Sun on the surface of the sphere opposite the Sun. What is the refractive index of the sphere material?

37–9   A hemispherical piece of glass with radius $R = 10$ cm and index $n = 1.520$ has its plane face in contact with oil of the same index. Where in the oil should a luminous object be placed if the image is to be twice the size of the object? Is the image real or virtual?

37–10   The interface between two media with indexes n and n' is in the shape of a concave spherical surface with radius R, as shown in the diagram. Show that, for paraxial rays,

$$\frac{n}{s} + \frac{n'}{s'} = \frac{n' - n}{r}$$

where $r = -R$.

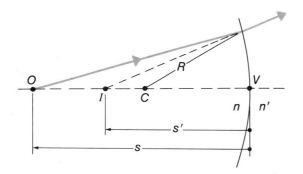

37–11   A plastic sphere with radius $R = 6$ cm and index $n = 1.360$ is situated in air. A point source of light is placed 20 cm from the center of the sphere. Determine the position of the image point.

37–12   A glass sphere with radius $R = 10$ cm and index $n = 1.520$ is coated with a reflecting layer over one hemisphere. An object with a height $h = 1.0$ cm is placed 15 cm in front of the clear surface of the sphere. Determine the position, the size, and the character of the final image. Sketch the principal rays.

Section 37–4

37–13   What is the focal length of a thin lens of the type shown in Fig. 37–14j if $r_1 = -10$ cm, $r_2 = -30$ cm, and $n = 1.520$?

37–14   An object with a height of 0.50 cm is placed 100 cm in front of a converging lens with a focal length $f = +30$ cm. Determine the position, the size, and the character of the image. Repeat the calculation for s reduced to 20 cm. Sketch the principal rays for each case.

37–15   A thin lens with refractive index n is immersed in a liquid with index n'. Show that the focal length f of the lens is given by

$$\frac{1}{f} = \left(\frac{n}{n'} - 1\right)\left(\frac{1}{r_1} - \frac{1}{r_2}\right)$$

37–16   Use the Newtonian form of the lens equation (Eq. 37–26) to show that the lateral magnification can be expressed as $m = -f/x = -x'/f$. [*Hint:* Construct the principal rays.]

37–17   An object has a height of 2 cm. Its image is formed by a converging lens on a screen that is 20 cm from the lens. When the lens is moved 10 cm closer to the screen (with the object fixed), another image is formed on the screen. Determine the focal length of the lens, the distance from the screen to the object, and the sizes of the two images. [*Hint:* Use the principle of reversibility.]

Section 37–5

37–18   Rays that are parallel to the axis of a paraboloidal mirror are brought to a *precise* focus at its focus F. Show that this follows from Fermat's principle and the definition of a parabola. [*Hint:* A parabola is a curve every point of which is equidistant from a fixed point F and a straight line $\overline{LL'}$; i.e., $r_1 = r_2$.]

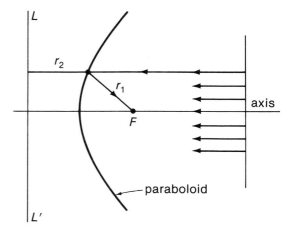

Section 37–6

37–19   A symmetric thick lens has a focal length of 5 cm and principal planes that are separated by 2 cm. Locate the image of an object that is placed 4 cm from the center of the lens. Is the image real or virtual? Erect or inverted?

37–20 Verify the position of the focal point F and the principal point H in the object space of the lens of Example 37–10.

37–21 Two thin lenses, one with $f_1 = 50$ cm and the other with $f_2 = 30$ cm, are placed in contact with their optic axes coincident. Determine the focal length of the combination. Locate both focal points and both principal points.

Section 37–7

37–22 A magnifying glass with a focal length of 10 cm is used by a person with normal vision. (a) Determine the angular magnifications γ and γ_∞. (b) For what range of object distance, s_b to s_c, will the image be in sharp focus? (Refer to Fig. 37–24.)

Section 37–8

37–23 The microscope described in Example 37–12 is refocused to permit viewing with a relaxed eye. (a) What is the new distance from the specimen to the objective? (b) What is the new magnification?

37–24 A compound microscope (Fig. 37–25) consists of an objective with a focal length of 6 mm and an eyepiece with a focal length of 20 mm. The tube length (the distance between the two lenses) is 15 cm. (a) What is

the magnification of the microscope when adjusted for viewing by a relaxed eye? (b) For the condition in (a), what is the specimen-to-objective distance? (c) If the diameter of the objective is 3 mm, what is the linear resolution ξ for $\lambda = 550$ nm?

37–25 A compound microscope is focused so that the final image appears at the near point of the viewer's eye (25 cm). Show that the angular magnification is

$$\Gamma = -\frac{s_1'}{s_1}\left(\frac{25 \text{ cm}}{f} + 1\right)$$

Section 37–9

37–26 The distance between the objective and the ocular of a simple astronomical telescope (Fig. 37–28a) is 40 cm, and the ocular has a magnification ×10. What is the angular magnification of the telescope?

37–27 A small Galilean telescope (Fig. 37–28b) has an objective with focal length $f_1 = 15$ cm and an ocular with focal length $f_2 = -3$ cm. A distant object is viewed with a relaxed eye. (a) What is the angular magnification of the telescope? (b) What is the distance between the lenses?

37–28 A reflecting telescope (Fig. 37–29a) consists of a spherical mirror with a radius of curvature of 1.5 m and a diameter of 10 cm, together with an ocular with magnification ×15. The telescope is used to view a light source with $\lambda = 550$ nm. (a) Determine the angular magnification of the telescope. (b) What is the angular resolution (expressed in seconds of arc)?

Additional Problems

37–29 The Moon's diameter subtends an angle of 9.04×10^{-3} rad at the position of the Earth. The mirror of a telescope is a concave spherical surface with $R = 80$ m. What is the diameter of the Moon's image formed on the focal plane?

37–30 A small object is located 3 m below the surface of a pool of water. How far below the surface will the object appear to be when viewed from directly above? (a) Use Snell's law, assuming the angles to be small. (b) Show that the same result is obtained by using Eq. 37–16.

37–31 A glass rod has a diameter of 10 cm, a length of 30 cm, and a refractive index $n = 1.520$. One end of the rod is flat and the other end has a hemispherical convex surface (with $R = 5$ cm). An object with a height of 1 mm is placed on the rod axis, 40 cm in front of the vertex of the hemispherical surface. Determine the po-

sition, the size, and the character of the image. Sketch the principal rays. (As usual, the image formed by the first refraction serves as the object for the second refraction.)

37–32 A parallel beam of light enters a glass hemisphere perpendicular to the flat face, as shown in the diagram. The radius is $R = 6$ cm and the index of refraction is $n = 1.560$. Determine the point at which the beam is focused. (Assume paraxial rays.)

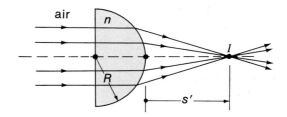

37–33 A thin lens of the type shown in Fig. 37–14d is made from glass with $n = 1.560$. When an object is placed 36 cm in front of the lens, a real image is formed 48 cm behind the lens. (a) What is the focal length of the lens? (b) What is the radius r_1 of the first surface?

37–34 *Geometric construction for refraction at a spherical sur-*
•• *face*. The path of a ray upon crossing a spherical inter-face between two media can be obtained by the following geometric construction. The spherical surface has a radius R. Construct two auxiliary concentric spherical surfaces with radii $(n'/n)R$ and $(n/n')R$, as shown in the diagram. The incident ray intersects the actual surface at P. Extend the incident ray until it intersects the outer sphere at A. Connect A with the common center C, thereby generating the intersection point B on the inner sphere. Then, the ray within the medium n' follows the path PB and intersects the optic axis at F. Show that this construction is consistent with Snell's law at point P. [*Hint:* Show that the triangles $\triangle CPA$ and $\triangle CPB$ are similar, and use the law of sines.] Use this procedure to construct the paths of three different parallel incident rays and demonstrate the occurrence of spherical aberration. (Choose $n = 1$ and $n' = 1.5$.)

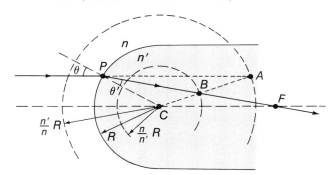

37–35 Because the index of refraction for all substances depends on the wavelength of the incident light, the focal length of a lens is different for different colors of light. This effect, called *chromatic aberration*, causes a blur-ring of the colors in an image. What is the fractional difference in focal length $\Delta f/f$ between yellow and red light for a lens made from dense flint glass? Between yellow and violet light? Use $\lambda(\text{red}) = 750$ nm, $\lambda(\text{yellow}) = 600$ nm, and $\lambda(\text{violet}) = 450$ nm; refer to Fig. 34–15.

37–36 The dimensions of a thick plano-convex lens are shown
•   in the diagram. (a) Locate the principal points and the focal points. (b) An object is placed 10 cm to the left of the vertex of the convex surface. Locate the image.

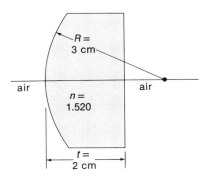

37–37 A magnifying glass is used by a person with normal vision. When the glass is held close to the eye (the situation assumed in obtaining Eqs. 37–39), the angular magnification is $\gamma = 4$. The glass is next moved a distance of 1.5 cm from the eye. If the image is located at the near point of the eye, what is the new value of γ?

37–38 A telescope is focused to give a final image at the near
•• point of a normal eye. Show that the angular magnification is

$$\Gamma = -f_1 \left(\frac{1}{f_2} + \frac{1}{25 \text{ cm}} \right)$$

37–39 Use the result of Problem 37–38 to calculate the angular
•   magnification of the Galilean telescope described in Problem 37–27 when the object viewed is at a distance of 15 m and the final image is at the near point of the eye.

GENERAL PLANETARY ORBITS

We wish to discuss the general elliptic orbit solutions to the problem involving two gravitating masses. Although the analysis is complicated, the treatment is necessary for understanding the orbital behavior of space probes and satellites as well as a more detailed analysis of planetary and lunar orbits.

In a majority of the cases of interest, one of the gravitating masses is much larger than the other. To a good approximation, we may take the center of mass of such a system to be fixed at the position of the larger mass, M, and consider the smaller mass, m, to be moving about the mass M with coordinates r and ϕ in the orbital plane. As an example, we select the planet-Sun system for study.

The Effective Potential Energy Function and the Characteristics of Planetary Orbits. The total energy of the planet-Sun system in our approximation is

$$E = K + U = \tfrac{1}{2}mv^2 + U(r)$$

$$= \tfrac{1}{2}m(v_r^2 + v_\phi^2) + U(r)$$

where the kinetic energy has been expressed in terms of the radial velocity v_r and the azimuthal velocity v_ϕ of the planet. Using $v_\phi = r\omega$ and $L = mr^2\omega$, which is a constant of the motion, we can write

$$E = \tfrac{1}{2}mv_r^2 + \frac{L^2}{2mr^2} - G\frac{mM}{r^2} \qquad \textbf{(EA-1)}$$

The term $L^2/2mr^2$ represents the kinetic energy of the system due to its angular motion.† It is often convenient to combine this term with $U(r)$ to form an *effective potential energy function V(r)*:

$$V(r) = \frac{L^2}{2mr^2} - \frac{k}{r} \qquad \textbf{(EA-2)}$$

where $k = GmM$. Thus,

---

*Supplementary topics in this book appear both in optional sections within the main body of the text and in these Enrichment Topics. The objective in both cases is to provide options for course requirements or choices for subject emphasis. The optional sections are generally shorter or less demanding.

†$L^2/2mr^2$ is sometimes called the *centrifugal energy*.

Fig. EA–1. The effective potential energy function $V(r)$ for the case of planetary motion. The ordinary potential energy function is $U(r) = -GmM/r = -k/r$.

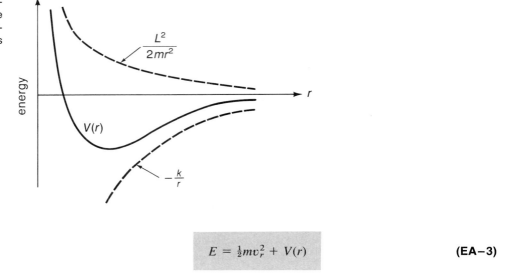

$$E = \tfrac{1}{2}mv_r^2 + V(r) \qquad \text{(EA–3)}$$

The effective potential energy function and its contributing parts are shown in Fig. EA–1. Notice how the terms combine to produce a minimum. $V(r)$ approaches zero as $r \to \infty$.

Figure EA–2 shows more details of the motion in terms of $V(r)$. First, notice that the radial velocity v_r must be a real quantity; that is, $v_r^2 \geq 0$. Equation EA–3 shows that the total energy E has its minimum value when $v_r \equiv 0$ and $dV/d_r = 0$. Differentiating Eq. EA–2 (recall that $L =$ constant) and setting the result equal to zero gives (after some algebra)

$$E_{min} = -\frac{mk^2}{2L^2} \qquad \text{(EA–4)}$$

Fig. EA–2. The effective potential energy function for the case of planetary motion, showing radius values for two energies, corresponding to elliptic and circular orbits.

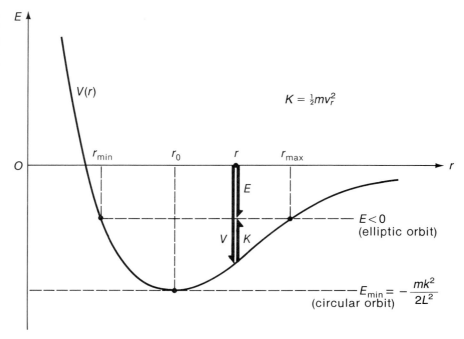

as indicated in Fig. EA–2. The fact that the radial velocity v_r is identically zero means that the value of r never changes; the radius remains fixed at $r = r_0 = L^2/mk$. Thus, the motion at minimum total energy follows a *circular path*.

Next, increase the total energy above the minimum value but subject to the restriction $E_{min} < E < 0$. Then, v_r can be different from zero, and the radius changes with time between the values r_{min} and r_{max} (see Fig. EA–2). At an intermediate value of r, notice how K, V, and E are represented in the diagram. As r approaches either r_{min} or r_{max}, the kinetic energy K approaches zero; that is, v_r decreases and finally equals zero at $r = r_{min}$ and at $r = r_{max}$. These positions are called the *turning points* of the motion, the points at which v_r changes sign. This type of motion takes place along an *elliptic path*. (It is often said that the particle moves in the *potential energy well* $V(r)$ between the limits r_{min} and r_{max}.) The student will find it valuable to review Section 8–3 at this point.

The values of r_{min} and r_{max} can be obtained from Eq. EA–1 by setting $v_r = 0$. Multiplying the resulting expression by r^2 and dividing by E, we obtain

$$r^2 + \frac{k}{E}r - \frac{L^2}{2mE} = 0 \qquad \textbf{(EA–5)}$$

This is a quadratic equation for r, the solutions of which are

$$r = -\frac{k}{2E}\left[1 \pm \left(1 + \frac{2EL^2}{mk^2}\right)^{1/2}\right] \qquad \textbf{(EA–6)}$$

These values of r, which we designate r_{min} and r_{max}, are both real (and positive) because E is negative and lies between $-mk^2/2L^2$ and zero. For planetary motion, the point r_{min} at which the planet is nearest to the Sun is called the *perihelion* and the point r_{max} of greatest separation is called the *aphelion*. (The corresponding terms for motion around the Earth are *perigee* and *apogee*.)

The *eccentricity e* of an elliptic orbit is defined as

$$e = \frac{r_{max} - r_{min}}{r_{max} + r_{min}} = \frac{r_{max} - r_{min}}{2a} \qquad \textbf{(EA–7)}$$

where a is the semimajor axis of the orbit (see Fig. EA–3). The eccentricity uniquely specifies the *shape* of the orbit; when this is combined with the value of a (which gives the *size* of the orbit), and with the orientation of the axes in space, the entire orbit is known (see Example EA–1).

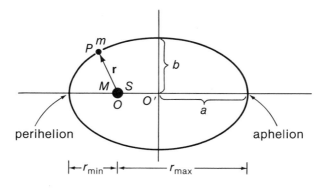

Fig. EA–3. Geometry of an elliptic planetary orbit. The center of the Sun S is located essentially at one focus of the ellipse (because $M \gg m$). The distance $\overline{SO'}$ as drawn is much exaggerated. The perihelion is located at r_{min} and the aphelion at r_{max}. The semimajor axis is a and the semiminor is b, with $b = a\sqrt{1 - e^2}$ for an eccentricity e.

If the energy of an orbiting particle is made equal to zero or becomes positive, its motion is no longer bounded. In these cases, there is a minimum allowed value for r, but no maximum value. Thus, if the body moves toward O, the separation will decrease until the minimum value is reached. Then, v_r changes sign and the body thereafter recedes from O, never to return. (Compare the discussion of Fig. 8–2.) The motion for $E > 0$ is hyperbolic, and for $E = 0$ it is parabolic.

Kepler's Laws. In Section 13–1, we summarized the three laws of planetary motion that were discovered by Johannes Kepler in the early seventeenth century. Now that we have studied some of the details of planetary motion, we find that we must modify the laws as originally stated by Kepler.

Kepler believed that the planets move in elliptic orbits with the center of the Sun at one focus. We now see that the statement of this first law of Kepler is indeed correct:

I. *A planet moves about the Sun in an elliptic path with the focus at the position of the Sun.*

From the discussions beginning in Section 10–3, we know that angular momentum is a conserved quantity for motion due to any central force—in particular, the gravitational force. Consider a particle that moves along the path PP' in Fig. EA–4. During the time Δt, the position line $r(t)$ sweeps out the triangular shaded region, the area of which is (approximately)

$$\Delta A = \tfrac{1}{2} r^2 \, \Delta \phi$$

Dividing by Δt and proceeding to the limit, $\Delta t \to 0$, we have

$$\operatorname*{Lim}_{\Delta t \to 0} \frac{\Delta A}{\Delta t} = \tfrac{1}{2} r^2 \operatorname*{Lim}_{\Delta t \to 0} \frac{\Delta \phi}{\Delta t}$$

or

$$\frac{dA}{dt} = \tfrac{1}{2} r^2 \omega = \frac{L}{2m} = \text{constant} \qquad \text{(EA–8)}$$

where we have used $L = mr^2 \omega = \text{constant}$. Thus, we see that Kepler's second law is valid if stated in the following way:

II. *The position vector for a planet, measured from the Sun, sweeps out equal areas in equal time intervals; that is, $dA/dt = \text{constant}$.*

Notice that Kepler's second law depends only on the fact that the force of interaction

Fig. EA–4. Illustrating Kepler's second law.

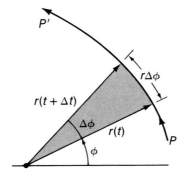

between the bodies is a *central force* so that angular momentum is conserved. In fact, the second law is valid for *any* central force.

Next, we write Eq. EA–8 in the form,

$$dt = \frac{2m}{L} dA$$

If we integrate this expression over one complete period T, we find $T = (2m/L)A$. For the case of an elliptic orbit, we have $A = \pi ab$ and hence $T = 2\pi mab/L$. Using the geometry of Fig. EA–3 we have

$$T^2 = 4\pi^2 m^2 a^4 (1 - e^2)/L^2 \qquad \text{(EA–9)}$$

In Problem EA–8 the result

$$a(1 - e^2) = L^2/m^2 GM$$

is derived; thus, substituting into Eq. EA–9 yields

$$T^2 = \frac{4\pi^2}{GM} a^3 \qquad \text{(EA–10)}$$

The result expressed by Eq. EA–10, which states that T^2/a^3 is a constant for all planets, is Kepler's third law:

III. *The ratio of the square of the period* (T^2) *to the cube of the semimajor axis* (a^3) *is the same for all planets.*

For a circular orbit, the eccentricity is $e = 0$ and both the semiminor axis b and the semimajor axis a are equal to the orbit radius r. Note that with $a = r$, Eq. EA–10 is the same as Eq. 13–16 found earlier for circular orbits.

Planetary Data.   Some of the important information concerning planetary motion and the properties of the planets is summarized in Table EA–1. Notice that the orbital

TABLE EA–1.   Planetary Data*

PLANET	DIAMETER (EARTH = 1)	MASS (EARTH = 1)	SEMIMAJOR AXIS a (A.U.)	SIDEREAL PERIOD (YEARS)	ECCENTRICITY OF ORBIT
Mercury	0.3824	0.0553	0.3871	0.24084	0.2056
Venus	0.9489	0.8150	0.7233	0.61515	0.0068
Earth	1	1	1	1.00004	0.0167
Mars	0.5326	0.1074	1.5237	1.8808	0.0934
Jupiter	11.194	317.89	5.2028	11.862	0.0483
Saturn	9.41	95.17	9.5388	29.456	0.0560
Uranus	4.4	14.56	19.1914	84.07	0.0461
Neptune	3.8	17.24	30.0611	164.81	0.0100
Pluto†	0.4	0.02	39.5294	248.53	0.2484

*The diameters and masses are those recommended by the International Astronomical Union in 1976 (except for more recent values for the diameter of Uranus and the mass of Pluto). The remaining data are 1974 values. (From J. M. Pasachoff, *Contemporary Astronomy*, 2nd ed., Saunders College Publishing, Philadelphia, 1981.)

†The eccentricity of Pluto's orbit is so large that once each revolution it actually dips inside Neptune's orbit. This happened in 1979. Pluto will remain closer to the Sun than Neptune until 1999, when once again it will become the "ninth" planet.

semimajor axes are given in *astronomical units* (A.U.), equal to the length of the Earth's semimajor axis:

$$1 \text{ A.U.} = 1.496 \times 10^{11} \text{ m}$$

Also, all diameters and masses are expressed in terms of Earth units:

$$1 \text{ Earth diameter*} = 1.274 \times 10^7 \text{ m}$$

$$1 \text{ Earth mass} = 5.974 \times 10^{24} \text{ kg}$$

$$\text{Mass of Sun} = 1.989 \times 10^{30} \text{ kg}$$

The periods given are *sidereal periods,* that is, periods for motion with respect to the fixed stars. The units are *solar years* (approximately $365\frac{1}{4}$ mean solar days). The sidereal year is 20 min 24 s longer than the solar year because of the slow precession of the Earth's rotation axis with respect to the stars.

Example EA–1

Use the information in Table EA–1 to calculate the total energy E and the angular momentum L for Mars, and also the speed of the planet at perihelion and at aphelion.

Solution:

From the definition (Eq. EA–7) of the eccentricity e and using $a = \frac{1}{2}(r_{max} + r_{min})$ (see Fig. EA–3) we find

$$r_{max} = a(1 + e) \quad \text{and} \quad r_{min} = a(1 - e)$$

(• Can you obtain these expressions?) The values of e and a are given in Table EA–1; then,

$$r_{max} = a(1 + e)$$

$$= (1.5237 \text{ A.U.})(1.496 \times 10^{11} \text{ m/A.U.})(1 + 0.0934)$$

$$= 2.492 \times 10^{11} \text{ m}$$

Solving for E in the first of these equations gives

$$E = -\frac{GMm}{r_{max} + r_{min}}$$

$$= -\frac{(6.673 \times 10^{-11} \text{ N·m}^2/\text{kg}^2)(1.989 \times 10^{30} \text{ kg})(0.1074 \times 5.974 \times 10^{24} \text{ kg})}{(2.492 + 2.067) \times 10^{11} \text{ m}}$$

$$= -1.868 \times 10^{32} \text{ J}$$

Then, the second equation gives

$$L = \sqrt{2m|E|r_{max}r_{min}}$$

$$= \sqrt{2(6.416 \times 10^{23} \text{ kg})(1.868 \times 10^{32} \text{ J})(2.492 \times 10^{11} \text{ m})(2.067 \times 10^{11} \text{ m})}$$

$$= 3.514 \times 10^{39} \text{ kg·m}^2/\text{s}$$

$$r_{min} = a(1 - e)$$

$$= (1.5237 \text{ A.U.})(1.496 \times 10^{11} \text{ m/A.U.})(1 - 0.0934)$$

$$= 2.067 \times 10^{11} \text{ m}$$

Equation EA–5 can be written in standard quadratic form as

$$Ar^2 + Br + C = 0$$

where $A = E$, $B = k = GMm$, and $C = -L^2/2m$, where M is the mass of the Sun and m is the mass of Mars. According to Eq. A–7 in Appendix A, the sum of the roots of a quadratic equation is equal to $-B/A$ and the product is equal to C/A. Therefore,

$$r_{max} + r_{min} = -\frac{k}{E} = -\frac{GMm}{E}$$

and

$$r_{max}r_{min} = -\frac{L^2}{2mE}$$

---

*This is an average value. "The radius" of the Earth usually means the equatorial radius, 6378 km. The polar radius is 6356 km.

At perihelion and at aphelion the planetary velocity is entirely azimuthal. Thus,

$$L = mv_\phi(p)r_{min} = mv_\phi(a)r_{max}$$

At perihelion

$$v_\phi(p) = \frac{L}{mr_{min}} = \frac{3.514 \times 10^{39} \text{ kg·m}^2/\text{s}}{(6.416 \times 10^{23} \text{ kg})(2.067 \times 10^{11} \text{ m})}$$

$$v_\phi(p) = 26.5 \text{ km/s}$$

At aphelion

$$v_\phi(a) = \frac{L}{mr_{max}} = \frac{3.514 \times 10^{39} \text{ kg·m}^2/\text{s}}{(6.416 \times 10^{23} \text{ kg})(2.492 \times 10^{11} \text{ m})}$$

$$= 22.0 \text{ km/s}$$

PROBLEMS

EA–1   The Earth satellite Cosmos #358 was placed in a near-circular orbit ($e = 0.002$) with a semimajor axis of 13,823 km. Calculate its period based on both a circular orbit and the actual elliptic orbit, and compare the two results.

EA–2   The Earth satellite ATS #2 was placed in an eccentric orbit with $e = 0.455$ and a semimajor axis of 24,123 km. (a) Calculate the period. (b) Determine r_{max} and r_{min}, and the corresponding altitudes above sea level.

EA–3   The planet Mercury has an orbital period of 88.0 days
●●   and an orbital velocity of 59.0 km/s at its perihelion. (a) Use this information to determine the eccentricity of the orbit and the lengths of the semimajor and semiminor axes in A.U. Compare with the data in Table EA–1. (b) What is the orbital velocity at the aphelion?

EA–4   A satellite is placed into orbit around the Moon. The
●●   period of the motion is 124.2 min. The apolune (the lunar equivalent of *apogee*) is 310 km above the Moon's surface, and the perilune is 20 km above the surface. The radius of the (spherical) Moon is 1738 km. (a) What is the eccentricity of the satellite's orbit? (b) Calculate the mass and the density of the Moon.

EA–5   On the basis of the results of Example EA–1 show that for any solar body of mass m in orbit around the Sun, the speed at perihelion $v_\phi(p)$ and the speed at aphelion $v_\phi(a)$ are given by

$$v_\phi(p)r_{min} = v_\phi(a)r_{max} = \sqrt{\frac{2GMr_{min}r_{max}}{r_{min} + r_{max}}}$$

where M is the mass of the Sun and $m/M \ll 1$.

EA–6   The period of Halley's comet is not exactly the same
●   from one orbit to the next, due primarily to the perturbations caused by the massive planets Jupiter and Saturn. The variation is between 74 and 79 y; the average period is 76.5 y. The eccentricity of the orbit is 0.967. (a) Find the perihelion and aphelion distances in A.U. (b) Calculate the azimuthal velocities at perihelion and at aphelion. Compare these with the Earth's average orbital velocity. [*Hint:* Make use of the results of Problem EA–5.]

EA–7   An artificial satellite with a mass of 220 kg is in an
●●   elliptic orbit around the Earth. The apogee distance (from the Earth's center) is 14,000 km; at perigee the satellite descends to 200 km above the Earth's surface. Determine the following: (a) the period of the orbital motion; (b) the eccentricity of the orbit; (c) the total energy of the satellite; (d) the angular momentum of the satellite; and (e) the speed of the satellite at perigee and at apogee.

EA–8   Use the geometry of Fig. EA–3 and Eq. EA–7 to show
●   that

$$r_{min} = a(1 - e) \quad \text{and} \quad r_{max} = a(1 + e)$$

Then use the results of Example EA–1 for ($r_{max} + r_{min}$) and $r_{max}r_{min}$ to show that

$$a(1 - e^2) = \frac{L^2}{m^2GM}$$

PRACTICAL MAGNETIC MATERIALS

Ferromagnetic materials, because they can achieve very high states of magnetization, find many important applications in magnetic devices of various sorts. Usually, these devices are fabricated from iron or from special iron alloys. These materials have complicated nonlinear magnetization properties that depend strongly on the fabrication process and on the magnetic history of a sample.

B-H Curves. A convenient way to present the magnetic characteristics of a ferromagnetic material is in terms of a *B-H curve*, a plot of the magnetic induction B versus the magnetic intensity H. Figure EB–1 shows such curves for several ferromagnetic materials. Notice that a log-log graph is used to cover the wide range of values. Also shown for convenience are dashed lines along which the relative permeability κ_M is constant. The significant feature of each curve is that it rises rapidly within a small range of H values and then reaches a plateau, called *saturation*.

Fig. EB–1. Magnetization curves for various types and alloys of iron. Note the logarithmic scales. (a) Supermalloy. (b) Mumetal. (c) 45 Permalloy. (d) USS motor steel, 26 gauge. (e) Pure iron sheet. (f) Hot-rolled iron sheet. The intensity H depends on the real current applied, and the induction B is the resulting field in the sample. Thus, "H is what you pay for, and B is what you get."

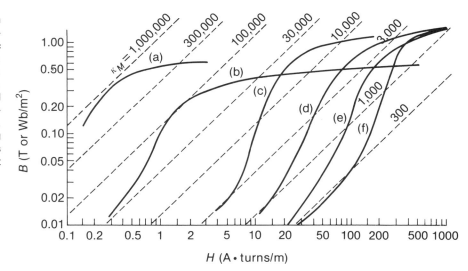

Example EB–1

A long solenoid with a radius $R = 4$ cm carries a current of 1 A in a coil that has a wrapping density of 200 turns per meter. Along the axis of the solenoid there is a cylinder with a radius $a = 2$ cm that is made from hot-rolled iron sheet. Assume that the *B-H* curve of Fig. EB–1f applies.

(a) Determine the field vectors, **H**, **B** and **M**, in the three regions indicated in the diagram.

(b) Compare the magnetic flux φ_M in the iron core with that in region ②.

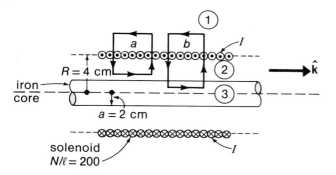

Solution:

(a) We assume that the solenoid is infinitely long, so all of the field quantities are zero in region ① outside the solenoid. The magnetic intensity **H** in regions ② and ③ can be determined by applying Eq. 31–23 around the paths *a* and *b*, respectively. The results for the two paths are the same because the solenoid current *I* is the only real current enclosed. Thus,

$$\mathbf{H}_2 = \mathbf{H}_3 = \frac{NI}{\ell}\hat{\mathbf{k}} = (200 \text{ A·turns/m})\hat{\mathbf{k}}$$

With $H_3 = 200$ A·turns/m, Fig. EB–1f gives $B_3 = 0.20$ T, so that $\mathbf{B}_3 = (2000 \text{ G})\hat{\mathbf{k}}$. On the other hand,

$$\mathbf{B}_2 = \mu_0\mathbf{H}_2 = (4\pi \times 10^{-7} \text{ T·m/A})(200 \text{ A·turns/m})\hat{\mathbf{k}}$$

$$\mathbf{B}_2 = (2.51 \times 10^{-4} \text{ T})\hat{\mathbf{k}} = (2.51 \text{ G})\hat{\mathbf{k}}$$

Thus, the induction in the iron core is much greater than the induction in the air region; that is, $B_3 \gg B_2$.

The value of the induction in region ② is just the value we would have in region ③ if the iron core were not present. Therefore, in the core, we have

$$\kappa_M = \frac{B_3}{B_2} = \frac{0.20 \text{ T}}{2.51 \times 10^{-4} \text{ T}} \cong 800$$

which is in agreement with Fig. EB–1f.

The magnetization **M** is given by Eq. 31–22, namely,

$$\mathbf{M} = \frac{\mathbf{B}}{\mu_0} - \mathbf{H}$$

In region ② we have $\mathbf{B}_2 = \mu_0\mathbf{H}_2$, so that $\mathbf{M}_2 = 0$. In region ③ we have

$$M_3 = \frac{0.20 \text{ T}}{4\pi \times 10^{-7} \text{ T·m/A}} - (200 \text{ A·turns/m})$$

so that $\mathbf{M}_3 = (1.590 \times 10^5 \text{ A·turns/m})\hat{\mathbf{k}}$

(b) The magnetic flux in the iron core is

$$\varphi_3 = \pi a^2 B_3 = \pi(0.02 \text{ m})^2(0.20 \text{ T})$$

$$= 2.51 \times 10^{-4} \text{ Wb}$$

whereas the flux in the air region ② is

$$\varphi_2 = \pi(R^2 - a^2)B_2$$

$$= \pi[(0.04 \text{ m})^2 - (0.02 \text{ m})^2](2.51 \times 10^{-4} \text{ T})$$

$$= 9.46 \times 10^{-7} \text{ Wb}$$

Thus, the air region carries only 0.38 percent of the total magnetic flux. This illustrates the general feature that the magnetic flux in a system tends to be concentrated in any ferromagnetic material that is present.

Hysteresis. When a ferromagnetic material is magnetized, some of the domains are driven beyond the point at which the changes are reversible. Then, when the imposed field intensity *H* is decreased to zero, the material retains some degree of magnetization (that is, $B \neq 0$ even though $H = 0$). If the coil current that produces the imposed field is slowly cycled between maximum values in either direction, the corresponding values of *H* and *B* trace out a closed curve called a *hysteresis loop** (Fig. EB–2). Starting from an initial

---

*The term *hysteresis* is used generally to describe a system whose state depends on its previous history. Hysteresis also occurs in dielectric and elastic materials.

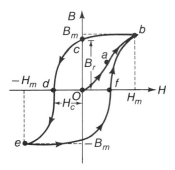

Fig. EB–2. The hysteresis loop for a typical ferromagnetic material. The various quantities indicated are defined in the text. The permeability μ of the material is not constant and is equal to the slope of the B-H curve at any point.

unmagnetized state O, the magnetization of the sample follows the path $\overline{Oab}$, arriving at a maximum (saturated) induction B_m when the imposed field intensity is H_m. When H is reduced to zero, the corresponding value of B (point c) is called the *remanence* B_r. The value of the reverse field intensity that is necessary to reduce the induction to zero at point d is called the *coercive force* H_c (although H_c is not actually a ''force''). The part of the curve between points c and d is referred to as the *demagnetization curve*. If the coil current is varied slowly, the curve efb is a rotated image of $bcde$.

In order for a material to be suitable for fabricating a good permanent magnet, the remanence must be large; that is, B_r must be a large fraction of the saturated induction B_m. The material should also have a large coercive force H_c so that relatively weak stray fields will not substantially affect the state of magnetization. A variety of special alloys are used in preparing permanent magnets.

Electromagnets. To perform various tests and experiments it is often necessary to have a strong (and variable) magnetic field within a certain working volume in air. This can be accomplished by bending a bar or rod of ferromagnetic material into some shape that brings the ends close together, forming a small air gap, as shown in Fig. EB–3a. A winding of N turns is wrapped around the core material and carries a current I. Most of the magnetic flux generated by the magnetizing current is guided by the core to the air gap. The stray or *leakage* flux, indicated by the field lines a, b, c, and d, usually represents a negligible part of the core flux φ_M. If the air gap is relatively small (that is, $\ell_g < \sqrt{A}$), the amount of fringing field around the gap is also small and may be neglected (see Fig. EB–3b). Since the lines of the magnetic induction **B** are continuous we have $B_g \cong B_F$.

Ampère's circuital law (Eq. 31–23) can be applied along the centerline of the core with the result

$$NI = \ell_F H_F + \ell_g H_g$$

Now, we have $H_F = B_F/\mu_F$ and $H_g = B_g/\mu_0$; also, $\varphi_M = B_F A \cong B_g A$. Then, the expression for Ampère's law becomes

$$NI = \frac{\varphi_M}{A}\left(\frac{\ell_F}{\mu_F} + \frac{\ell_g}{\mu_0}\right)$$

$$= \frac{\varphi_M}{A\mu_0}\left(\frac{\ell_F}{\kappa_M} + \ell_g\right)$$

(EB–1)

This equation can be used to predict the current required to produce a given flux within a magnet of specified geometry.

Fig. EB–3. (a) An electromagnet with a small air gap ℓ_g. (b) Detail of the magnetic induction in the vicinity of the air gap. The normal component of **B** is continuous, so $B_g \cong B_F$.

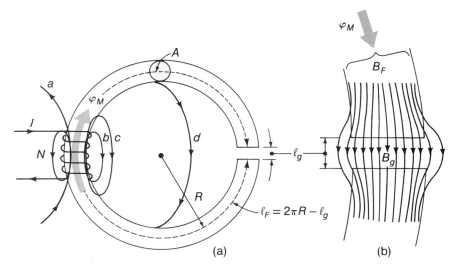

PROBLEMS

EB–1   A C-shaped electromagnet of the type shown in Fig. EB–3 has a core made of hot rolled iron. The core cross section is 4.0 cm² and the path lengths are $\ell_F = 30$ cm for the core and $\ell_g = 2$ mm for the air gap. The exciting coil has $NI = 200$ A·turns. Find the magnetic induction **B** in the air gap. [*Hint:* The curve shown in Fig. EB–1f must be used in a trial-and-error procedure.]

EB–2   A C-shaped yoke fabricated from pure iron has a 500-turn magnetizing coil, as shown in the diagram. This electromagnet is used to convert a rectangular rod of Alnico-5 alloy into a permanent magnet. During assembly of the yoke and rod, two 0.1-mm air gaps were inadvertently left at the ends of the rod. For the dimensions shown in the diagram and for a desired magnetic induction of 10,000 G in the Alnico rod (which occurs with $\kappa_M = 15$), determine the required current I in the magnetizing coil. Assume that the median path (shown dashed in the diagram) can be used for the application of Ampère's law with the mean values of the fields. Assume also that fringing-field effects can be ignored. Is it worthwhile to attempt to reduce the air gaps or to improve the magnetic characteristics or geometry of the yoke?

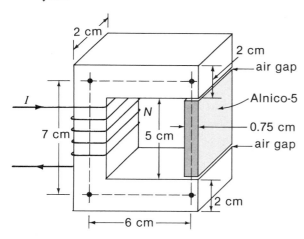

EB–3   A common geometry for an electromagnet is shown in the diagram. The yoke is fabricated from pure iron. If a magnetic induction, $B = 0.5$ T, is required in the air gap, determine the current I that must flow in each of

the identical magnetizing coils. Use Fig. EB–1. Neglect fringing-field effects and flux leakage. [*Hint:* At points a and b the main (gap) flux divides equally. Thus, Ampère's law may be applied either to path *abcda* or path *abefa*.]

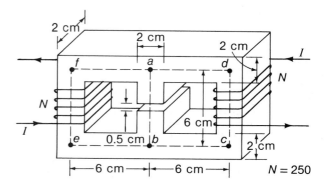

EB–4   An iron-core Rowland ring has a central radius R and a cross-sectional area A, and is wrapped uniformly with N turns of wire (see Fig. 31–3). The input of electric energy to the coil is $dW = i(t)e(t)\,dt$, where $i(t)$ and $e(t)$ are the instantaneous values of the coil current and the back EMF, respectively. (a) Use Faraday's law to show that $dW = NAi(t)\,dB$. (b) Use Ampère's law to show that $dW = (2\pi RA)H\,dB$. (c) Show that the input electric energy W per cubic meter is equal to the area in a *B-H* diagram between the *B*-axis and the magnetization curve, $W/V = \int_0^B H\,dB$, where $V = 2\pi RA$. (d) Assume that the iron core is a linear magnetic material (that is, $B = \mu_0\kappa_M H$, with κ_M not dependent on H). Does the result of (c) agree with that of Problem 31–16, namely, that $\mu_M = \frac{1}{2}\mathbf{B}\cdot\mathbf{H}$? (e) Show that the energy loss per cubic meter in carrying the iron sample through one complete hysteresis loop is equal to the area of the loop in a *B-H* diagram.

EB–5   A long needlelike cavity (filled with air) is present in a uniformly magnetized piece of hot-rolled sheet iron. The needle axis is parallel to the magnetization **M** in the sample; $M = 1.5 \times 10^4$ A·turns/m. (a) Determine H and B in the sample. (b) Determine H and B in the cavity. (Neglect end effects.) [*Hint:* A trial-and-error method will be necessary. Use Fig. EB–1.]

ELECTRIC DIPOLE ANTENNA

A practical device for producing electromagnetic radiation consists of an oscillator and a dipole antenna, as illustrated in Fig. EC–1. To simplify the analysis we assume that the length 2ℓ of the antenna is short compared with the wavelength λ of the emitted radiation,* that is, $2\ell \ll \lambda$. An antenna of this sort is called a *Hertzian dipole*.† For such a dipole, we can consider the phase of the current in each antenna element to be the same at any instant along the entire length ℓ.

When the current I is directed as indicated in Fig. EC–1, positive charge $+q$ accumulates at the upper tip of the antenna while negative charge $-q$ accumulates at the lower tip. We can deduce the resulting radiation pattern by using the current I to determine $\mathbf{H}$ or by using the charge q to determine $\mathbf{E}$. For distances $r \gg \ell$, the distant-field analysis is valid and we again have plane waves with $|\mathbf{H}|/|\mathbf{E}| = \sqrt{\epsilon_0/\mu_0}$, for vacuum (see Eq. 33–16).

Suppose that the driving oscillator produces an antenna current, $I(t) = I_0 \sin \omega t$. Because $2\ell \ll \lambda$, the current at any instant has essentially the same magnitude at every point on the antenna. The current causes charges $\pm q_0 \cos \omega t$ to accumulate at the tips

Fig. EC–1. Geometry of a Hertzian dipole antenna and the field vectors at a particular instant. Notice that the oscillator is inductively coupled to the antenna; this is common practice.

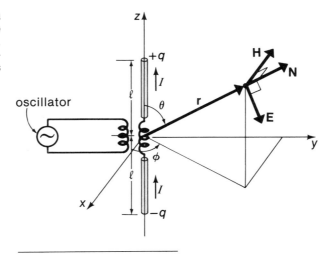

---

*This requirement is called the *dipole approximation* and is used in the analysis of ideal antennas. However, in practical situations the requirement can be relaxed to $2\ell \lesssim \lambda/4$ without seriously degrading the results.

†After the German physicist Heinrich Hertz (1857–1894), who extensively studied the propagation of electromagnetic radiation in the 1880's.

of the antenna, where $q_0 = I_0/\omega$. Thus, we have in effect an oscillating electric dipole that consists of a pair of time-varying charges separated by a fixed distance 2ℓ. This dipole produces the same distant field as does a dipole that consists of a pair of fixed charges with magnitude $q_0 = I_0/\omega$ separated by a time-varying distance, $z = 2\ell \cos \omega t$. In this equivalent situation, the acceleration of each charge has the magnitude $a_z = d^2z/dt^2 = 2\ell\omega^2 \cos \omega t$.

Look again at Eqs. 33–27; we expect the product qa that appears in these equations now to be replaced by $q_0 \times 2\ell\omega^2 \cos \omega t$. To obtain $\mathbf{E}(\mathbf{r}, t)$ and $\mathbf{H}(\mathbf{r}, t)$, we must modify this expression for qa to take account of the fact that there is a finite propagation time between the source and the field point $\mathbf{r}$. This is the same effect we discussed in Section 17–3 for spherical sound waves. Thus, the term $\cos \omega t$ must be replaced by $\cos (kr - \omega t)$ where $k = 2\pi/\lambda$. (Compare Eq. 17–12a.) Altogether, then, we have for a Hertzian dipole antenna,

$$\mathbf{E}(\mathbf{r}, t) = \frac{I_0\ell\omega}{2\pi\epsilon_0 c^2} \frac{\cos (kr - \omega t)}{r} \sin \theta \ \hat{\mathbf{u}}_\theta \qquad \textbf{(EC–1)}$$

$$\mathbf{H}(\mathbf{r}, t) = \frac{I_0\ell\omega}{2\pi c} \frac{\cos (kr - \omega t)}{r} \sin \theta \ \hat{\mathbf{u}}_\phi \qquad \textbf{(EC–2)}$$

These expressions are valid for $2\ell \ll \lambda$ and $r \gg \lambda$. As before, the field vectors are mutually perpendicular, and the Poynting vector, $\mathbf{N} = \mathbf{E} \times \mathbf{H}$, is directed radially outward. Thus, the radiation intensity in a direction specified by the angle θ is proportional to $\sin^2 \theta$ (a *dipole radiation pattern*).

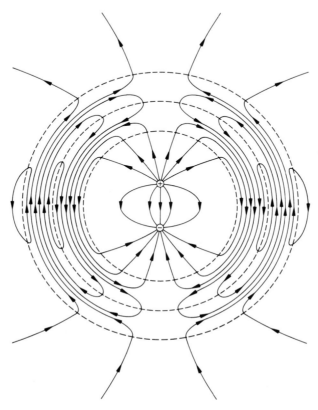

Fig. EC–2. The electric field for a sinusoidally driven Hertzian dipole after oscillating through one complete cycle and returning to the static condition. The near field is a static dipole field, but at large distance the field pattern approaches that for a plane wave and propagates outward with the speed c. (From *Principles of Physics* by J. P. Hurley and C. Garrod. Boston: Houghton Mifflin Co., 1978.)

Fig. EC–3. The electric field for a sinusoidally driven Hertzian dipole.

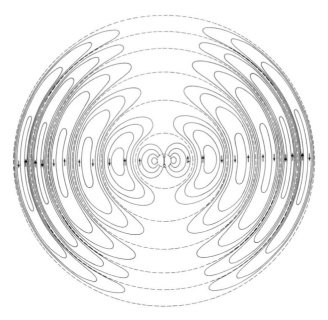

A precise calculation using Maxwell's equations produces expressions for the field vectors in the distant field that are the same as Eqs. EC–1 and EC–2. In addition, however, such a treatment reveals the presence of complicated near fields in which power surges back and forth between the field and the antenna.

Consider a Hertzian dipole that undergoes sinusoidal oscillations. Figure EC–2 shows the electric field lines for the radiation from such a dipole that has completed one cycle of oscillation and thereafter is static. The complete electric field is obtained by rotating this field pattern about a vertical axis through the dipole. Notice that the field lines near the dipole are the same as those for a static dipole (see Fig. 24–6). For the largest distances shown, a segment of the field begins to resemble a plane wave. The disturbance propagates outward with the speed c.

When the sinusoidal current oscillations are allowed to continue indefinitely, the radiation pattern for the electric field becomes that illustrated in Fig. EC–3.

Example EC–1

A Hertzian dipole has $2\ell = 5$ cm and is driven at a frequency $\nu = 150$ MHz. The peak current in the antenna is $I_0 = 30$ A. Determine the average radiated power and the effective resistance (the *radiation resistance*) of the antenna.

Solution:

We could obtain the desired result by determining the Poynting vector from Eqs. EC–1 and EC–2 for **E** and **H**, then integrating $\mathbf{N} \cdot d\mathbf{A}$ over a convenient surface. Instead, it is easier to take the result of Example 33–5 for an accelerating charge, namely, $P = q^2 a^2 / 6\pi\epsilon_0 c^3$, and make the substitution of $2I_0\ell\omega$ for the peak value of qa. Then, the peak power is

$$P_0 = \frac{2I_0^2\ell^2\omega^2}{3\pi\epsilon_0 c^3}$$

Now, we use $\omega = 2\pi\nu = 2\pi c/\lambda$ to obtain the average power, $\bar{P} = \frac{1}{2}P_0$; thus,

$$\bar{P} = \frac{4\pi}{3}\frac{I_0^2}{\epsilon_0 c}\left(\frac{\ell}{\lambda}\right)^2$$

The wavelength of the radiation is $\lambda = c/\nu = 2.00$ m. Then, we find

$$\bar{P} = \frac{4\pi(30 \text{ A})^2}{3(8.85 \times 10^{-12} \text{ F/m})(3.00 \times 10^8 \text{ m/s})} \cdot \left(\frac{0.025 \text{ m}}{2 \text{ m}}\right)^2$$

$$= 221.9 \text{ W}$$

The radiation resistance R_{rad} of the antenna is defined by the relationship

$$\bar{P} = \frac{1}{2}I_0^2 R_{rad}$$

Using the preceding expression for $\bar{P}$, we have

$$R_{rad} = \frac{8\pi}{3\epsilon_0 c} \left(\frac{\ell}{\lambda}\right)^2$$

For vacuum, we have $c = 1/\sqrt{\epsilon_0\mu_0}$. Thus,

$$R_{rad} = \frac{8\pi}{3} \sqrt{\frac{\mu_0}{\epsilon_0}} \left(\frac{\ell}{\lambda}\right)^2 = \frac{8\pi}{3} Z_0 (\ell/\lambda)^2$$

For the present case, in which $\lambda = 80\,\ell$, we find

$$R_{rad} = 0.493\ \Omega$$

The Ohmic resistance of the antenna might be several ohms. Then, more power is expended in heat than is radiated into space. Consequently, a Hertzian dipole is not an efficient radiator. For efficient radiation, the antenna length must be comparable with the wavelength. But then the dipole approximation ($2\ell \ll \lambda$) is no longer valid and a more complicated analysis is required.

Example EC–2

Two identical Hertzian dipole antennas, separated by a distance d, are mounted parallel, as shown in diagram (a), where the axes of the dipoles are perpendicular to the plane of the page. The current in each antenna has the same magnitude, but the current in antenna B has a phase lag α compared with the current in antenna A. Determine the field strength E for the combination as a function of the azimuthal angle ϕ in the equatorial plane, $\theta = \pi/2$ (the plane of the page in the diagram).

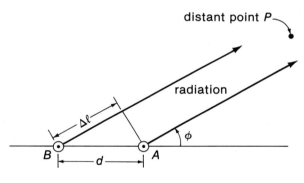

distant point P

radiation

B d A

ϕ

$\Delta\ell$

view in equatorial plane

(a)

Taking the ratio of the peak values of E and either E_A or E_B, we find

$$\left.\frac{E}{E_A}\right|_{peak} = 2\cos\left(\frac{\pi d}{\lambda}\cos\phi + \tfrac{1}{2}\alpha\right)$$

For the case, $d = \lambda/4$ and $\alpha = \pi/2$, this ratio function becomes a *cardioid*, as shown in diagram (b). Other values of d and α lead to the patterns shown in diagram (c).

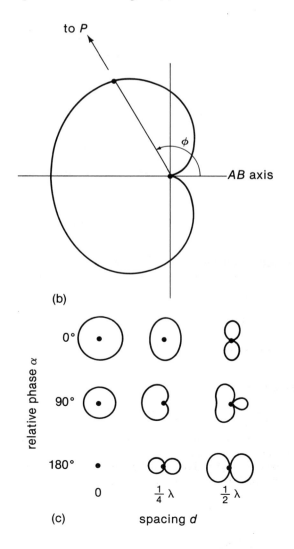

to P

ϕ

AB axis

(b)

(c)

spacing d

relative phase α

$0°$ $90°$ $180°$

0 $\frac{1}{4}\lambda$ $\frac{1}{2}\lambda$

Solution:

To arrive at a distant point P, the radiation emitted by antenna B must travel a distance different from that traveled by the radiation emitted by antenna A. This difference in path length, $\Delta\ell = d\cos\phi$, introduces a phase difference, $\delta = (2\pi d/\lambda)\cos\phi$, between the waves superimposed at P. In addition, the wave from antenna B has a phase lag α. Thus, the electric field strengths of the two waves are

$$E_A = E_0 \cos(kr - \omega t)$$

$$E_B = E_0 \cos\left(kr + \frac{2\pi d}{\lambda}\cos\phi - \omega t + \alpha\right)$$

Adding, we find

$$E = E_A + E_B$$

$$= 2E_0 \cos\left(\frac{\pi d}{\lambda}\cos\phi + \tfrac{1}{2}\alpha\right)\cos\left(kr - \omega t + \frac{\pi d}{\lambda}\cos\phi + \tfrac{1}{2}\alpha\right)$$

PROBLEMS

EC–1 A horizontal Hertzian dipole operating in the 20-m radio band has an antenna length $2\ell = 50$ cm and carries a peak current of $I_0 = 100$ A. (a) Calculate the peak electric and magnetic fields E_0 and H_0 at a location that is 5 km away and in a direction having $\theta = 30°$. (b) What is the average power received at that point?

EC–2 An airborne radio transmitter utilizes a Hertzian dipole antenna with length $2\ell = \lambda/20$. A receiver located at ground level 10 km directly below the transmitter requires a peak signal strength of 20 μV/m for the desired signal-to-noise ratio. (a) What peak current must flow in the antenna? (b) What is the required average power input to the antenna? (c) What is the radiation resistance of the antenna?

EC–3 A Hertzian dipole can be used as a receiving antenna as well as a transmitting antenna. For an antenna with a length 2ℓ, an EMF of $2\ell E_0 \cos \omega t$ is developed when the antenna intercepts a sinusoidal plane wave with peak amplitude E_0 and with **E** parallel to the antenna. This antenna geometry and the equivalent circuit are shown in the diagram. Maximum power is absorbed from the wave when the load resistance R_L is equal to the radiation resistance R_{rad} of the antenna. (If the antenna has any reactance, this must be balanced by adding the complex-conjugate reactance to the load; that is, the antenna must be *tuned*.) Then, one half of the power in the wave is absorbed by the load and the other half is dissipated in the radiation resistance by being reradiated into space. (The incident wave induces currents in the antenna, causing it to act as a transmitter just as if the currents were induced by an attached generator.) Thus, the average power delivered to the load is

$$\bar{P} = \frac{(2\ell E_0)^2}{2} \times \frac{1}{4R_{rad}}$$

where $R_{rad} = (8\pi/3)Z_0(\ell/\lambda)^2$, as we found in Example EC–1 for a Hertzian dipole ($2\ell \ll \lambda$). Show that, independent of ℓ, the antenna presents an effective area A to the incident wave given by $A = (3/8\pi)\lambda^2$. [*Hint:* $\bar{P} = \bar{N}A$.]

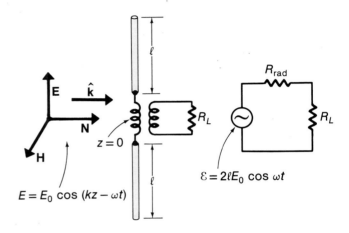

EC–4 An omnidirectional antenna aboard a satellite radiates an average power of 10 W at a frequency of 30 MHz. The satellite signal is received at a distance of 100 km by an antenna that has an effective cross-sectional area of $(3/8\pi)\lambda^2$. (Refer to Problem EC–3.) (a) Determine the average rate at which energy is intercepted by the antenna. (b) What is the peak value E_0 of the electric field of the intercepted wave?

BASIC ALGEBRA AND TRIGONOMETRY

A–1 SERIES

The *arithmetic progression* or *series* S_n, to n terms, can be summed, with the result

$$S_n^A = a + (a + d) + (a + 2d) + (a + 3d) + \cdots$$
$$+ [a + (n - 1)d]$$
$$= na + \tfrac{1}{2}n(n - 1)d$$
$$= \tfrac{1}{2}n[(\text{first term}) + (\text{last term})] \qquad \textbf{(A–1)}$$

A *polynomial* is a series whose n terms are in ascending positive-integer powers of a quantity, say r, which we can write in the form

$$S_n = a_0 + a_1 r + a_2 r^2 + a_3 r^3 + \cdots + a_{n-1} r^{n-1} \qquad \textbf{(A–2)}$$

where the *coefficients,* $a_0, a_1, a_2, \ldots, a_{n-1}$, are *real* numbers. When the number of terms is allowed to increase without limit ($n \to \infty$: "when n becomes indefinitely large"), the resulting *infinite series* is referred to as a *power series*.

The simplest polynomial, to n terms, occurs when all the coefficients, $a_0, a_1, a_2, \ldots, a_{n-1}$, are equal to a number, say a. This defines the *geometric progression* or *series* S_n^G which can also be summed, with the result

$$S_n^G = a + ar + ar^2 + ar^3 + \cdots + ar^{n-1}$$
$$= \frac{a(1 - r^n)}{1 - r} \qquad \textbf{(A–3)}$$

If $r^2 < 1$, the powers of r decrease in numerical value as n increases, and in the limit $n \to \infty$, r^n vanishes. Then,

$$S_n^G (n \to \infty) = \frac{a}{1 - r} \qquad \textbf{(A–4)}$$

A–2 SIGMA NOTATION

It is customary to represent the sum of a series of quantities by means of the so-called *sigma notation*. Then, Eq. A–2 is expressed as

$$S_n = \sum_{i=0}^{n-1} a_i r^i$$
$$= a_0 + a_1 r + a_2 r^2 + a_3 r^3 + \cdots + a_{n-1} r^{n-1} \qquad \textbf{(A–5)}$$

That is, the Greek capital sigma (Σ) means "sum all of the quantities of the form that follows over the range of the index specified below and above this symbol." In this case, the *index* is the letter i (but, of course, the index can be any letter or symbol because it does not appear in the result). Sometimes we economize by writing

$$\sum_i p_i \qquad \text{or} \qquad \sum p_i$$

where it is understood that we sum over the entire allowed range of i.

A–3 QUADRATIC EQUATIONS

A *quadratic equation* has the form

$$Ax^2 + Bx + C = 0$$

which we can express in terms of its *roots,* x_1 and x_2, as

$$(x - x_1)(x - x_2) = 0$$

Then,

$$x_{1,2} = \frac{-B \pm \sqrt{B^2 - 4AC}}{2A} \qquad \textbf{(A–6)}$$

If $B^2 - 4AC > 0$, the roots are real and unequal; if $B^2 - 4AC = 0$, the roots are real and equal; and if $B^2 - 4AC < 0$, the roots are complex.

Comparing the two forms of the quadratic equation, we see that the sum of the roots and the product of the roots can be expressed as

$$x_1 + x_2 = -\frac{B}{A} \quad \text{and} \quad x_1 x_2 = \frac{C}{A} \quad \textbf{(A–7)}$$

A–4   LOGARITHMS

The logarithm to the base m of a number a is the power to which m must be raised to produce a. That is,

$$\text{if} \quad m^y = a, \quad \text{then} \quad y = \log_m a$$

For example, $10^3 = 1000$, so we say that the logarithm to the base 10 of 1000 is 3.

For the real numbers a, b, and n, logarithms obey the following rules:

$$\left.\begin{aligned} \log_m ab &= \log_m a + \log_m b \\[4pt] \log_m \frac{a}{b} &= \log_m a - \log_m b \\[4pt] \log_m (a^n) &= n \log_m a \\[4pt] \log_m 1 &= 0 \end{aligned}\right\} \quad \textbf{(A–8)}$$

Common logarithms have base 10, and we write $\log_{10} a$ or, simply, $\log a$. *Natural* or *Naperian logarithms* have base e ($e = 2.71828\ldots$), and we write $\ln a$. The above rules apply for logarithms with any base. The conversion between common and natural logarithms can be made by using

$$\ln a = \log_{10} a / \log_{10} e = 2.3026 \log_{10} a$$

We also note that

$$\log_{10} 10 = 1 \quad \text{and} \quad \ln e = 1$$

A useful infinite power series expansion for $\ln (1 + x)$ with $x^2 \leq 1$ is

$$\ln (1 + x) = x - \frac{x^2}{2} + \frac{x^3}{3} - \frac{x^4}{4} + \cdots$$
$$+ (-1)^{n-1} \frac{x^n}{n} + \cdots \quad \textbf{(A–9)}$$

where we show the alternation of sign by the device of using $(-1)^{n-1}$.

A–5   EXPONENTIAL FUNCTIONS

For the real numbers a, n, and m, exponentials obey the following rules:

$$(a^n)(a^m) = a^{n+m}$$
$$(a^n)^m = a^{nm}$$
$$a^0 = 1$$

We also have

$$\ln e^x = x \ln e = x \quad \text{and} \quad e^{\ln x} = x \quad \textbf{(A–10)}$$

A useful infinite power series expansion for e^x with $x^2 < \infty$ is

$$e^x = 1 + \frac{x}{1!} + \frac{x^2}{2!} + \frac{x^3}{3!} + \cdots + \frac{x^n}{n!} + \cdots \quad \textbf{(A–11)}$$

Sometimes the notation $\exp x$ is used for e^x. An exclamation mark denotes the factorial, where $n! = 1 \cdot 2 \cdot 3 \cdots (n - 1) \cdot n$; by definition, $0! = 1$.

A–6   BINOMIAL THEOREM

The general rule for calculating the nth power of the sum of two algebraic quantities, a and b, is given by the *binomial theorem*. If we let $b/a = \xi$ (with $\xi^2 \leq 1$), we can express the binomial theorem in the form

$$(a + b)^n = a^n \left(1 + \frac{b}{a}\right)^n = a^n (1 + \xi)^n$$

$$= a^n \left[1 + n\xi + \frac{n(n - 1)}{2!} \xi^2 + \frac{n(n - 1)(n - 2)}{3!} \xi^3 + \cdots\right.$$

$$\left. + \frac{n(n - 1)(n - 2) \cdots (n - r + 1)}{r!} \xi^r + \cdots\right] \quad \textbf{(A–12)}$$

When n is a positive integer, the series terminates and the rth term can be written as $\dfrac{n!}{(n - r)! r!} \xi^r$; then the final term in the series is ξ^n. When $\xi = 1$, Eq. A–12 is valid only for $n > 0$; otherwise, the relationship is valid for any n. For example, if $n = \frac{1}{2}$ and $\xi^2 \leq 1$, we have

$$(1 \pm \xi)^{1/2} = 1 \pm \frac{1}{2}\xi + \frac{\frac{1}{2}(\frac{1}{2} - 1)}{2} \xi^2$$

$$\pm \frac{\frac{1}{2}(\frac{1}{2} - 1)(\frac{1}{2} - 2)}{3 \cdot 2} \xi^3 + \cdots$$

$$= 1 \pm \frac{1}{2}\xi - \frac{1}{8}\xi^2 \pm \frac{1}{16}\xi^3 - \cdots \quad \textbf{(A–13)}$$

The case $n = -\frac{1}{2}$ and $\xi^2 < 1$ is often encountered:

$$(1 \pm \xi)^{-1/2} = 1 \mp \frac{1}{2}\xi + \frac{3}{8}\xi^2 \mp \frac{5}{16}\xi^3 + \cdots \quad \textbf{(A–14)}$$

Other useful expansions are

$$(1 \pm \xi)^{-1} = 1 \mp \xi + \xi^2 \mp \xi^3 + \cdots \quad \textbf{(A–15)}$$

$$(1 \pm \xi)^{-3/2} = 1 \mp \frac{3}{2}\xi + \frac{15}{8}\xi^2 \mp \frac{35}{16}\xi^3 + \cdots \quad \textbf{(A–16)}$$

$$(1 \pm \xi)^{-2} = 1 \mp 2\xi + 3\xi^2 \mp 4\xi^3 + \cdots \quad \textbf{(A–17)}$$

In these equations, one takes either the upper or the lower sign throughout.

A–7   BASIC TRIGONOMETRIC DEFINITIONS

In order to write down the basic trigonometric relationships, consider the right triangle shown in Fig. A–1. Then,

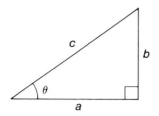

Fig. A–1. A right triangle.

$$\sin \theta = \frac{b}{c}, \quad \cos \theta = \frac{a}{c}, \quad \tan \theta = \frac{\sin \theta}{\cos \theta} = \frac{b}{a}$$

$$\csc \theta = \frac{1}{\sin \theta} = \frac{c}{b}, \quad \sec \theta = \frac{1}{\cos \theta} = \frac{c}{a}, \quad \text{(A–18)}$$

$$\text{ctn } \theta = \frac{\cos \theta}{\sin \theta} = \frac{a}{b}$$

Using the Pythagorean theorem, $a^2 + b^2 = c^2$, we find

$$\cos^2 \theta + \sin^2 \theta = 1$$
$$1 + \tan^2 \theta = \sec^2 \theta \qquad \text{(A–19)}$$
$$1 + \text{ctn}^2 \theta = \csc^2 \theta$$

A–8   TRIGONOMETRIC IDENTITIES

The basic trigonometric functions for the sum and difference of angles are

$$\sin (x + y) = \sin x \cos y + \cos x \sin y \qquad \text{(A–20)}$$

$$\sin (x - y) = \sin x \cos y - \cos x \sin y \qquad \text{(A–21)}$$

$$\cos (x + y) = \cos x \cos y - \sin x \sin y \qquad \text{(A–22)}$$

$$\cos (x - y) = \cos x \cos y + \sin x \sin y \qquad \text{(A–23)}$$

$$\tan (x + y) = \frac{\tan x + \tan y}{1 - \tan x \tan y} \qquad \text{(A–24)}$$

$$\tan (x - y) = \frac{\tan x - \tan y}{1 + \tan x \tan y} \qquad \text{(A–25)}$$

The addition rules are

$$\sin x + \sin y = 2 \sin \tfrac{1}{2} (x + y) \cdot \cos \tfrac{1}{2} (x - y) \qquad \text{(A–26)}$$

$$\sin x - \sin y = 2 \cos \tfrac{1}{2} (x + y) \cdot \sin \tfrac{1}{2} (x - y) \qquad \text{(A–27)}$$

$$\cos x + \cos y = 2 \cos \tfrac{1}{2} (x + y) \cdot \cos \tfrac{1}{2} (x - y) \qquad \text{(A–28)}$$

$$\cos x - \cos y = -2 \sin \tfrac{1}{2} (x + y) \cdot \sin \tfrac{1}{2} (x - y) \qquad \text{(A–29)}$$

$$\tan x + \tan y = \frac{\sin (x + y)}{\cos x \cos y} \qquad \text{(A–30)}$$

$$\tan x - \tan y = \frac{\sin (x - y)}{\cos x \cos y} \qquad \text{(A–31)}$$

The half-angle formulas are

$$\sin^2 \tfrac{1}{2}x = \tfrac{1}{2}(1 - \cos x) \qquad \text{(A–32)}$$

$$\cos^2 \tfrac{1}{2}x = \tfrac{1}{2}(1 + \cos x) \qquad \text{(A–33)}$$

The double-angle formulas are

$$\sin 2x = 2 \sin x \cos x \qquad \text{(A–34)}$$

$$\cos 2x = 1 - 2 \sin^2 x \qquad \text{(A–35)}$$

A–9   TRIANGLE RULES

Consider the general triangle shown in Fig. A–2. We have

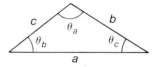

Fig. A–2

Law of sines: $\dfrac{a}{\sin \theta_a} = \dfrac{b}{\sin \theta_b} = \dfrac{c}{\sin \theta_c}$ **(A–36)**

Law of cosines: $a^2 = b^2 + c^2 - 2bc \cos \theta_a$ **(A–37)**

Area of triangle: $A = \tfrac{1}{2}bc \sin \theta_a$ **(A–38)**

A–10   RADIAN MEASURE

The radian measure of an angle θ, such as that between the lines $A'O$ and $B'O$ in Fig. A–3, is defined as the ratio of the length of the intercepted arc $\widehat{AB}$ on any circle centered on O to the radius r of that circle. That is,

$$\theta = \frac{\widehat{AB}}{r} \qquad (\theta \text{ expressed in radians})$$

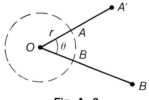

Fig. A–3

When θ is increased to $360°$, the arc length $\widehat{AB}$ equals the circumference of the circle. Then,

$$360° = 2\pi \text{ rad}$$

from which $1° = \dfrac{2\pi}{360} = 0.01745329 \ldots \text{ rad}$

and $1 \text{ rad} = \dfrac{360°}{2\pi} = 57.2957795 \ldots °$ **(A–39)**

A–11 TRIGONOMETRIC SERIES

Some useful trigonometric series are

$$\sin \theta = \theta - \frac{\theta^3}{3!} + \frac{\theta^5}{5!} - \frac{\theta^7}{7!} + \cdots \quad \theta^2 < \infty \qquad \textbf{(A–40)}$$

$$\cos \theta = 1 - \frac{\theta^2}{2!} + \frac{\theta^4}{4!} - \frac{\theta^6}{6!} + \cdots \quad \theta^2 < \infty \qquad \textbf{(A–41)}$$

$$\tan \theta = \theta + \frac{\theta^3}{3} + \tfrac{2}{15}\theta^5 + \tfrac{17}{315}\theta^7 + \cdots \quad \theta^2 < \frac{\pi^2}{4} \qquad \textbf{(A–42)}$$

In these equations the angle θ is expressed in *radians*.

From these expressions we see that, when $\theta \ll 1$,

$$\sin \theta \cong \theta \qquad \cos \theta \cong 1 \qquad \tan \theta \cong \theta \qquad \textbf{(A–43)}$$

Also,

$$\lim_{\theta \to 0} \frac{\sin \theta}{\theta} = 1 \qquad \textbf{(A–44)}$$

Other useful expansions are

$$\sin^{-1} x = x + \frac{1}{2}\frac{x^3}{3} + \frac{1.3}{2.4}\frac{x^5}{5} \qquad \textbf{(A–45)}$$
$$+ \frac{1 \cdot 3 \cdot 5}{2 \cdot 4 \cdot 6}\frac{x^7}{7} + \cdots$$

$$\cos^{-1} x = \frac{\pi}{2} - \sin^{-1} x \qquad \textbf{(A–46)}$$

$$\tan^{-1} x = x - \frac{x^3}{3} + \frac{x^5}{5} - \frac{x^7}{7} + \cdots \qquad \textbf{(A–47)}$$

COMPLEX NUMBERS

Basic Definitions. A *complex number* is an expression of the form $a + ib$, where a and b are ordinary real numbers and i is the imaginary unit defined by $i^2 = -1$. The *real part* of the complex number $a + ib$ is a, and the *imaginary part* is b.

Graphic Representations of Complex Numbers. A complex number may be plotted as a point in a *complex plane* defined by a horizontal (real) x-axis and a vertical (imaginary) y-axis. A general complex number may then be represented as $z = x + iy$. (In these discussions we reserve the letter z to represent a complex quantity.) Figure B–1, called an *Argand diagram,* shows several points plotted in the complex plane.

Another convenient way to represent complex numbers is in polar form, namely,

$$z = \rho(\cos\theta + i\sin\theta) \tag{B-1}$$

where

$$\left.\begin{array}{l} x = \rho\cos\theta \\ y = \rho\sin\theta \end{array}\right\} \tag{B-2}$$

Also,

$$\rho = |z| = \sqrt{x^2 + y^2} \tag{B-3}$$

Fig. B–1. An Argand diagram representation of several complex numbers $z = x + iy$.

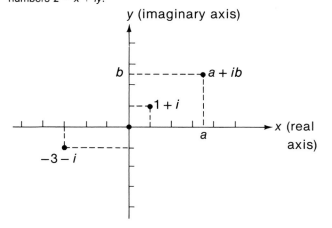

and*

$$\tan\theta = \frac{y}{x} \tag{B-4}$$

The various polar quantities are illustrated in Fig. B–2.

The quantity ρ is the *absolute value* (or *amplitude*) of z, $\rho = |z|$. The angle θ, which is measured counterclockwise from the (real) x-axis, is defined in the domain,† $0 \leqslant \theta \leqslant 2\pi$, and is referred to as the *argument* (or *phase*) of z.

An alternative form for writing a complex number (due to Euler) is

$$z = \rho e^{i\theta} \tag{B-5}$$

This form is obtained by noting that we can *formally* express the exponential function of a complex number, $\zeta = i\theta$, in terms of the power series

$$e^\zeta = 1 + \zeta + \frac{\zeta^2}{2!} + \frac{\zeta^3}{3!} + \cdots + \frac{\zeta^n}{n!} + \cdots$$

Fig. B–2. The polar representation of a complex number $z = x + iy$.

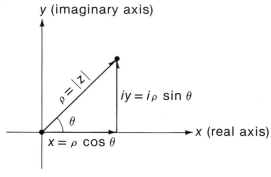

---

*The possible sign ambiguity that results when calculating $\theta = \tan^{-1}(y/x)$ is resolved by noting the quadrant in which $x + iy$ lies. Recall the similar problem encountered in the discussion of vectors in two dimensions (Section 5–4).

†The domain of θ is sometimes defined to be $-\pi \leqslant \theta \leqslant \pi$.

$$e^{\zeta} = 1 + i\theta + \frac{i^2\theta^2}{2!} + \frac{i^3\theta^3}{3!} + \frac{i^4\theta^4}{4!} + \cdots$$

Now, we have $i^2 = 1$, $i^3 = -i$, $i^4 = 1$, and so forth. Thus,

$$e^{i\theta} = \left(1 - \frac{\theta^2}{2!} + \frac{\theta^4}{4!} - \cdots\right) + i\left(\theta - \frac{\theta^3}{3!} + \frac{\theta^5}{5!} - \cdots\right)$$

We also have

$$\cos\theta = 1 - \frac{\theta^2}{2!} + \frac{\theta^4}{4!} - \cdots$$

$$\sin\theta = \theta - \frac{\theta^3}{3!} + \frac{\theta^5}{5!} - \cdots$$

Hence, $$e^{i\theta} = \cos\theta + i\sin\theta \qquad \textbf{(B–6)}$$

so that $$z = \rho e^{i\theta} = \rho(\cos\theta + i\sin\theta) \qquad \textbf{(B–7)}$$

To be more general, we can allow ρ and θ to become complex. For example, suppose that $\theta = \alpha + i\beta$, where α and β are real numbers. Then, with $\rho = 1$, we have

$$e^{i\theta} = e^{i(\alpha + i\beta)} = e^{i\alpha}e^{-\beta}$$
$$= e^{-\beta}(\cos\alpha + i\sin\alpha) \qquad \textbf{(B–8)}$$

Notice that

$$|e^{i\theta}|^2 = |\cos\theta + i\sin\theta|^2$$
$$= \cos^2\theta + \sin^2\theta = 1$$

Thus, the quantity $e^{i\theta}$, with unit magnitude, is a convenient way to specify the phase of a complex number. Accordingly, $e^{i\theta}$ is sometimes called a *phasor* (or *rotator*) and is particularly useful when $\theta = \omega t$, for then $e^{i\theta} = e^{i\omega t}$ represents a rotation at a rate ω about the origin, as indicated in Fig. B–3.

Some useful results are

$$\left.\begin{array}{l} e^{i0} = \cos 0 + i\sin 0 = 1 \\[6pt] e^{i(\pi/2)} = \cos\dfrac{\pi}{2} + i\sin\dfrac{\pi}{2} = i \\[6pt] e^{i\pi} = \cos\pi + i\sin\pi = -1 \\[6pt] e^{i(3\pi/2)} = \cos\dfrac{3\pi}{2} + i\sin\dfrac{3\pi}{2} = -i \end{array}\right\} \quad \textbf{(B–9)}$$

$$\left.\begin{array}{l} e^{i\theta'}\ (\theta' > 2\pi) = e^{i(\theta' - 2\pi)} \\[6pt] e^{-i\theta} = e^{i(2\pi - \theta)} \end{array}\right\} \quad \textbf{(B–10)}$$

Thus,

$$e^{-i(\pi/2)} = e^{i(3\pi/2)} = -i$$

The Euler Formulas. Using Eq. B–6, namely,

$$e^{i\theta} = \cos\theta + i\sin\theta$$

we can write

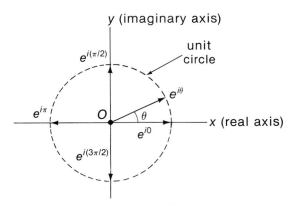

Fig. B–3. Use of the phasor $e^{i\theta}$ to represent rotation.

$$e^{-i\theta} = \cos(-\theta) + i\sin(-\theta)$$
$$= \cos\theta - i\sin\theta$$

Adding these two expressions and solving for $\cos\theta$, we have

$$\cos\theta = \frac{e^{i\theta} + e^{-i\theta}}{2} \qquad \textbf{(B–11)}$$

Similarly, by subtracting we find

$$\sin\theta = \frac{e^{i\theta} - e^{i\theta}}{2i} \qquad \textbf{(B–12)}$$

Equations B–11 and B–12 are known as the *Euler formulas*.

Equality. Two complex quantities are *equal* if and only if the real parts and the imaginary parts are separately equal. Thus, if we have

$$a + ib = c + id$$

then it is required that $a = c$ and $b = d$. A complex quantity is zero if and only if both the real part and the imaginary part are separately equal to zero.

Addition and Subtraction. Complex quantities are added and subtracted in accordance with the rule

$$(a + ib) \pm (c + id) = (a \pm c) + i(b \pm d) \quad \textbf{(B–13)}$$

Complex Conjugation. If a complex quantity is written exclusively in terms of real quantities and the imaginary unit i, the *complex conjugate* is formed by everywhere replacing i by $-i$. The conjugate is indicated by a superscript asterisk. If $z = x + iy$, the conjugate is

$$z^* = (x + iy)^* = x - iy \qquad \textbf{(B–14)}$$

Also, $$z^* = (\rho e^{i\theta})^* = \rho(\cos\theta + i\sin\theta)^*$$

$$= \rho(\cos\theta - i\sin\theta) = \rho e^{-i\theta} \qquad \textbf{(B–15)}$$

Multiplication. The multiplication of complex quantities follows the rules of ordinary algebra. Thus,

$$(a + ib)(c + id) = ac + iad + ibc + i^2bd$$

$$= (ac - bd) + i(ad + bc) \quad \textbf{(B--16)}$$

In Euler form we have

$$\left(\rho_1 e^{i\theta_1}\right)\left(\rho_2 e^{i\theta_2}\right) = \rho_1\rho_2 e^{i(\theta_1 + \theta_2)} \quad \textbf{(B--17)}$$

If a complex quantity, $z = e^{i\theta}$, is multiplied by $e^{i\phi}$ (ϕ real), the magnitude is unchanged but the phasor is rotated counterclockwise by an angle ϕ in the Argand diagram. Thus,

$$z = \rho e^{i\theta} e^{i\phi} = \rho e^{i(\theta + \phi)} \quad \textbf{(B--18)}$$

Absolute Values. The square of the absolute value of a complex quantity (that is, $|z|^2$, which is a *real* quantity) is obtained by multiplying the quantity by its complex conjugate. Thus,

$$zz^* = (a + ib)(a - ib) = a^2 - iab + iba - i^2b^2$$

$$= a^2 + b^2 = |z|^2 \quad \textbf{(B--19)}$$

In Euler form, we have

$$zz^* = (\rho e^{i\theta})(\rho e^{-i\theta}) = \rho^2 = |z|^2 \quad \textbf{(B--20)}$$

Division and Rationalization of a Complex Denominator. The division of complex quantities is most easily given in Euler form, namely,

$$\frac{z_1}{z_2} = \frac{\rho_1 e^{i\theta_1}}{\rho_2 e^{i\theta_2}} = (\rho_1/\rho_2)e^{i(\theta_1 - \theta_2)} \quad \textbf{(B--21)}$$

Setting $z_1 = 1$, we have for the reciprocal of a complex quantity,

$$z^{-1} = \frac{1}{\rho e^{i\theta}} = \rho^{-1} e^{-i\theta} \quad \textbf{(B--22)}$$

(• Can you use this result to show that $i^{-1} = -i$?)

If a complex quantity is expressed in Cartesian form, the reciprocal is obtained by a process called *rationalization of the denominator*. This consists of multiplying both numerator and denominator by the complex conjugate of the denominator. Thus,

$$\frac{1}{a + ib} = \frac{1}{a + ib} \cdot \frac{a - ib}{a - ib}$$

$$= \frac{a}{a^2 + b^2} - i\frac{b}{a^2 + b^2} \quad \textbf{(B--23)}$$

The division of one complex quantity by another expressed in Cartesian form is also accomplished by using the rationalization procedure. Thus,

$$\frac{c + id}{a + ib} = \frac{ac + bd}{a^2 + b^2} + i\frac{ad - bc}{a^2 + b^2} \quad \textbf{(B--24)}$$

De Moivre's Theorem. Using the Euler form of a complex quantity, the nth power is expressed as

$$z^n = (\rho e^{i\theta})^n = \rho^n e^{in\theta}$$

$$= \rho^n(\cos n\theta + i \sin n\theta) \quad \textbf{(B--25)}$$

when n is a positive integer. If $\rho = 1$, we have de Moivre's theorem, namely,

$$(e^{i\theta})^n = (\cos\theta + i\sin\theta)^n$$

$$= \cos n\theta + i\sin n\theta \quad \textbf{(B--26)}$$

Any multiple of 2π can be added to the angle θ without affecting the value of $e^{i\theta}$; that is,

$$e^{i\theta} = e^{i(\theta + 2\pi k)}$$

where k is any integer. Then, the general expression for $z^{1/n}$, the nth root of z, is

$$z^{1/n} = \rho^{1/n} e^{i(\theta + 2\pi k)/n}$$

$$= \rho^{1/n}\left(\cos\frac{\theta + 2\pi k}{n} + i\sin\frac{\theta + 2\pi k}{n}\right) \quad \textbf{(B--27)}$$

The n distinct roots of z are obtained by giving k the values $0, 1, 2, \ldots, (n - 1)$, successively. Setting $z = 1$, we see that the n roots of unity are spaced uniformly around the unit circle in an Argand diagram.

The Projection Operators, Re and Im. The operators **Re** and **Im** project the real and imaginary parts, respectively, of the complex quantity on which they act. Thus,

$$\left.\begin{array}{l} \textbf{Re}\,(a + ib) = a \\ \textbf{Im}\,(a + ib) = b \end{array}\right\} \quad \textbf{(B--28)}$$

Notice that

$$z_1 z_2^* + z_1^* z_2 = \rho_1\rho_2(e^{i(\theta_1 - \theta_2)} + e^{-i(\theta_1 - \theta_2)})$$

$$= 2\rho_1\rho_2 \cos(\theta_1 - \theta_2)$$

$$= 2\textbf{Re}(\rho_1\rho_2 e^{i(\theta_1 - \theta_2)})$$

$$= 2\textbf{Re}(z_1 z_2^*) \quad \textbf{(B--29)}$$

Then, we can write

$$|z_1 + z_2|^2 = (z_1 + z_2)(z_1^* + z_2^*)$$

$$= z_1 z_1^* + z_1 z_2^* + z_1^* z_2 + z_2 z_2^*$$

$$= |z_1|^2 + |z_2|^2 + 2\textbf{Re}(z_1 z_2^*) \quad \textbf{(B--30)}$$

TABLES OF CONVERSION FACTORS

ANGLE

	RADIAN	rev	°	'	"
1 RADIAN =	1	0.1592	57.30	3438	2.063×10^5
1 revolution =	6.283	1	360	2.160×10^4	1.296×10^6
1 degree =	1.745×10^{-2}	2.778×10^{-3}	1	60	3600
1 minute =	2.909×10^{-4}	4.630×10^{-5}	1.667×10^{-2}	1	60
1 second =	4.848×10^{-6}	7.716×10^{-7}	2.778×10^{-4}	1.667×10^{-2}	1

SOLID ANGLE

1 sphere = 4π steradians = 12.57 sr

LENGTH

	METER	km	cm	in	ft	mi
1 METER =	1	10^{-3}	100	39.37	3.281	6.214×10^{-4}
1 kilometer =	1000	1	10^5	3.937×10^4	3281	0.6214
1 centimeter =	10^{-2}	10^{-5}	1	0.3937	3.281×10^{-2}	6.214×10^{-6}
1 inch =	2.54×10^{-2}	2.540×10^{-5}	2.540	1	8.333×10^{-2}	1.578×10^{-5}
1 foot =	0.3048	3.048×10^{-4}	30.48	12	1	1.894×10^{-4}
1 statute mile =	1609	1.609	1.609×10^5	6.336×10^4	5280	1

1 angstrom = 10^{-10} m
1 nautical mile = 1.1508 mi = 6076.1 ft = 1852 m
1 astronomical unit = 1.495×10^8 km
1 light year = 9.461×10^{12} km

1 parsec = 3.084×10^{13} km
1 fathom = 6 ft
1 yard = 3 ft
1 inch = 1000 mils = 6 picas = 72 points

AREA

	METER^2	cm^2	ft^2	in^2
1 SQUARE METER =	1	10^4	10.76	1550
1 square centimeter =	10^{-4}	1	1.076×10^{-3}	0.1550
1 square foot =	9.290×10^{-2}	929.0	1	144
1 square inch =	6.452×10^{-4}	6.452	6.944×10^{-3}	1

1 hectare $= 10^4 \text{m}^2 = 2.471$ acres
1 acre $= 43,560 \text{ ft}^2$
1 square mile $= 2.788 \times 10^7 \text{ ft}^2 = 640$ acres
1 circular mil $= 7.854 \times 10^{-7} \text{ in}^2 = 5.067 \times 10^{-10} \text{ m}^2$

VOLUME

	METER^3	cm^3	ft^3	in^3
1 CUBIC METER =	1	10^6	35.31	6.102×10^4
1 cubic centimeter =	10^{-6}	1	3.531×10^{-5}	6.102×10^{-2}
1 cubic foot =	2.832×10^{-2}	2.832×10^4	1	1728
1 cubic inch =	1.639×10^{-5}	16.39	5.787×10^{-4}	1

1 liter $= 1000 \text{ cm}^3 = 10^{-3} \text{ m}^3$
1 U.S. fluid gallon $= 4$ quarts $= 8$ pints $= 128$ fluid ounces $= 231 \text{ in}^3$
                    $= 3.785$ liters
1 British Imperial gallon $= 277.4 \text{ in}^3$
1 oil barrel $= 42$ U.S. gallons

TIME

	SECOND	min	h	d	y
1 SECOND =	1	1.667×10^{-2}	2.778×10^{-4}	1.157×10^{-5}	3.169×10^{-8}
1 minute =	60	1	1.667×10^{-2}	6.944×10^{-4}	1.901×10^{-6}
1 hour =	3600	60	1	4.167×10^{-2}	1.141×10^{-4}
1 day =	8.640×10^4	1440	24	1	2.738×10^{-3}
1 year =	3.156×10^7	5.260×10^5	8.766×10^3	365.2	1

SPEED

	METER/SECOND	km/h	ft/s	mi/h
1 METER/SECOND =	1	3.6	3.281	2.237
1 kilometer/hour =	0.2778	1	0.9113	0.6214
1 foot/second =	0.3048	1.097	1	0.6818
1 mile/hour =	0.4470	1.609	1.467	1

1 knot = 1 nautical mile/hour = 1.151 mi/h = 1.852 km/h
1 mi/min = 88.00 ft/s = 96.54 km/h

MASS

	KILOGRAM	g	slug	lb (mass)*
1 KILOGRAM =	1	10^3	6.852×10^{-2}	2.205
1 gram =	10^{-3}	1	6.852×10^{-5}	2.205×10^{-3}
1 slug =	14.59	1.459×10^4	1	32.17
1 pound (mass)* =	0.4536	453.6	3.108×10^{-2}	1

*Quantities in shaded area refer to the mass equivalents of pound weights; for example, 1 pound = 0.4536 kg means that a mass of 0.4536 kg weighs 1 pound under standard conditions of gravity (g = 9.8067 m/s^2).

1 tonne (metric ton) = 1000 kg = 2205 lb (mass)
1 atomic mass unit = 1.6606×10^{-27} kg
1 ton (mass) = 2000 lb (mass) = 907.2 kg
1 long ton (mass) = 2205 lb (mass)
1 avoirdupois pound = 16 avoirdupois ounces = 7000 grains = 0.4536 kg
1 troy pound = 12 troy ounces = 5760 grains = 0.8229 avoirdupois pound

FORCE

	NEWTON	dyne	lb
1 NEWTON =	1	10^5	0.2248
1 dyne =	10^{-5}	1	2.248×10^{-6}
1 pound =	4.448	4.448×10^5	1

Note: A 1-kg mass is accelerated 1 m/s^2 by an applied force of 1 N.

A 1-g mass is accelerated 1 cm/s^2 by an applied force of 1 dyne.

A 1-slug mass is accelerated 1 ft/s^2 by an applied force of 1 lb.

PRESSURE

	PASCAL	atm	cm Hg	lb/in² (psi)
1 PASCAL =	1	9.869×10^{-6}	7.501×10^{-4}	1.450×10^{-4}
1 atmosphere =	1.013×10^{5}	1	76	14.70
1 cm mercury* =	1333	1.316×10^{-2}	1	0.1934
1 pound per in² =	6.895×10^{3}	6.805×10^{-2}	5.171	1

*At 0° C and under standard conditions of gravity ($g = 9.8067$ m/s²).

1 torr = 1 mm Hg* = 133.3 Pa
1 bar = 10^5 Pa
1 atm = 29.92 in Hg* = 1.013 bar

ENERGY

	JOULE	Cal	kW·h	ft·lb	BTU
1 JOULE =	1	2.389×10^{-4}	2.778×10^{-7}	0.7376	9.481×10^{-4}
1 Calorie* =	4186	1	1.163×10^{-3}	3087	3.968
1 kilowatt-hour =	3.6×10^{6}	860.0	1	2.655×10^{6}	3413
1 foot-pound =	1.356	3.239×10^{-4}	3.766×10^{-7}	1	1.285×10^{-3}
1 British thermal unit =	1055	0.2520	2.930×10^{-4}	777.9	1

*International Table Calorie

1 erg = 10^{-7} J
1 eV = 1.602×10^{-19} J
1 thermochemical Calorie = 4184 J
1 horsepower-hour = 2.685×10^{6} J

POWER

	WATT	kW	Cal/s	ft·lb/s	BTU/h	hp
1 WATT =	1	10^{-3}	2.389×10^{-4}	0.7376	3.413	1.341×10^{-3}
1 kilowatt =	10^{3}	1	0.2389	737.6	3413	1.341
1 Cal/s =	4186	4.186	1	3087	1.429×10^{4}	5.613
1 ft·lb/s =	1.356	1.356×10^{-3}	3.239×10^{-4}	1	4.628	1.818×10^{-3}
1 BTU/h =	0.2930	2.930×10^{-4}	7.000×10^{-5}	0.2161	1	3.929×10^{-4}
1 horsepower =	745.7	0.7457	0.1782	550	2545	1

1 ton (refrigeration) = 12000 BTU/h

ANSWERS TO SELECTED PROBLEMS

CHAPTER 1

2. 88.5 km/h   **5.** 7913 mi   **7.** 10.4 s   **11.** 128.2 y
14. 1.58 kg   **16.** 198.6 kg   **18.** (a) $\rho_E = 5.49 \times 10^3$ kg/m^3;   (b) $\rho_S = 1.41 \times 10^3$ kg/m^3

CHAPTER 2

2. $\frac{9}{2}$ m; 4 m; 0   **4.** (a) 0.0411 m/s;   (b) 2.5 m/s,
3.33 m/s   **6.** (a) 6.52 m/s;   (b) 3.83×10^{-3} s, yes
9. 1.125 m/s   **11.** 2 m/s^2 **13.** $v_m = 15$ cm/s,
$a_m = 45$ cm/s^2 **15.** (a) 2.41 m/s^2;   (b) 14.4 s;
(c) 60.3 m/s   **18.** (a) 10.2 s;   (b) 130 m
19. (a) 33.3 s;   (b) 926 m; (c) 55.6 m/s
22. (a) 145.5 m;   (b) 5.45 s   **24.** −16.9 m/s
25. 2.14 m   **27.** $s(t) = \frac{1}{2}t^3 + t^2 + \frac{1}{2}t + 1$
32. $v_s = 0.707\, v_g$ **33.** (a) 17.9 m;   (b) −12 m/s;
(c) 0.612 g   **35.** 0.203 m/s^2 **39.** (a) 1.51 s;
(b) 6.95 m;   (c) the first ball is falling   **41.** $v_0 = 20.1$ m,
$h = 20.7$ m   **43.** 14 km

CHAPTER 3

2. (a) 2.88 units, 33.7° west of north;   (b) 5.37 units, 63.4°
east of north   **6. c** $= -2\hat{i} - 9\hat{j}$ **8.** $C = 7.27\hat{i} + 4.77\hat{j}$
10. (a) $C = 5\hat{i} - \hat{j}$;   (b) $D = +17\hat{j}$ **12.** 74.7°
14. $A_u = 3.13$ units   **20.** $\dfrac{ar^2}{x^2 y^2}\hat{i}$
23. r $= (1.64\hat{i} + 1.15\hat{j})$ m   **27.** 41.41°   **30.** The
three vectors form a right triangle.   **32.** (a) $x = 2$ m;
(b) 2.41 m

CHAPTER 4

3. $\Delta r = -6\hat{i} + 6\hat{j}$ **5.** (a) $\mathbf{v} = 3\alpha\hat{i} + 2\beta\hat{j}$;   (b) 11.3 m
8. v $= (6t + 2)\hat{i} + (4t + 4)\hat{j}$, **a** $= 6\hat{i} + 4\hat{j}$; $|\mathbf{a}| = 7.21$ m/s^2;
$\theta = 33.69°$ **11. v** $= R(\omega - \omega \cos \omega t)\,\hat{i} + R\omega \sin \omega t\,\hat{j}$,
$\mathbf{a} = R\omega^2 \sin \omega t\,\hat{i} + R\omega^2 \cos \omega t\,\hat{j}$; $t = \dfrac{2n\pi}{\omega}$,
$n = 0, 1, 2, \ldots$; $r = 2\pi Rn\,\hat{i}$, $n = 0, 1, 2, \ldots$;

$\mathbf{a} = R\omega^2\,\hat{j}$; no.   **13.** (a) 7.10 m;   (b) 38.6 m;
(c) $v = 20.96$ m/s, $t = 2.22$ s   **14.** 44.7 m/s
18. $\alpha = 53.6°$, $d = 23.6$ m   **19.** $\alpha = 56.6°$ or 27.7°
22. 35.4 rpm; 18.5 m/s   **25.** (a) $\phi = 3.14$ rad;
(b) $\omega = 3.14$ s^{-1}; (c) $v = 0.314$ m/s;   (d) $a_\phi = 0.314$ m/s^2,
$a_\rho = 0.987$ m/s^2;   (e) $|\mathbf{a}| = 1.036$ m/s^2 **28.** 0.85 m/s
30. 23.6 m/s, 26.5 m/s   **32.** 567 m and −87.3 m
35. 6.67 m/s   **37.** 7.20 m   **39.** $v_0 = \sqrt{Rg}$
41. 1.75×10^{-5} m   **45.** (a) 3 km/h;   (b) 7 km/h
46. (a) 0.782 s;   (b) −7.67 m/s;   (c) −5.67 m/s;
(d) 1.44 m   **47.** $t = 60$ s; 180 m and 240 m

CHAPTER 5

3. $\sqrt{5}$ N   **5.** 10 m/s   **8.** 188 m   **10.** 10^6 N; 10^4 N;
$(101 - n) \times 10^4$ N   **13.** (a) 19.6 N;   (b) 25.6 N;
(c) 15.6 N   **16.** 0.544 m   **19.** 1.66×10^6 N
22. $T = (8t^2 - 40t + 50)$ N   **24.** (a) 6.25 s;   (b) 15.6 m;
(c) $\mathbf{v} = (1.6\hat{i} + 10\hat{j})$ m/s   **26.** $F_g = 1.46 \times 10^4$ N,
$f_g = 659$ N, $v = 7350$ m/s   **29.** 83.4   **31.** 9.77 N
33. (a) 405 m;   (b) 7.63 s

CHAPTER 6

2. (a) 1.96 m/s^2;   (b) 23.5 N;   (c) 47.0 N
4. (a) 3.84 m/s^2;   (b) 68.2 N;   (c) 54.6 N;   (d) 61.4 N
7. 862 N   **9.** 1.25 m/s^2 **11.** (a) 4.88 m/s^2;
(b) 24.6 N   **14.** $F_{max} = 29.7$ N, $F_{min} = 1.18$ N
17. $v = 18(9 + 2t)^{-1}$ m/s   **19.** 400 N   **22.** (a) $a_1 = 0$,
$a_2 = 0$, $a_p = 0$;   (b) $a_1 = 7.70$, $a_2 = 0$, $a_p = 3.85$ m/s^2;
(c) $a_1 = 25.2$ m/s^2, $a_2 = 4.20$ m/s^2, $a_p = 14.7$ m/s^2
23. 0.600   **26.** 1.61 m   **28.** $a = 2.88$ m/s^2,
$T = 1.59$ N   **30.** (a) 0.445;   (b) 1.38 kg   **33.** 1.98 m

CHAPTER 7

5. $W_{g1} = 49$ J, $W_{g2} = -39.2$ J; $W_1 = 5.45$ J,
$W_2 = 4.36$ J   **7.** 1.88×10^9 J   **8.** 1,960 J   **12.** 2.67 J
13. 3.22 m; 1.61 m   **18.** $W_g = 24.5$ J, $K = 24.5$ J,
$v = 14.0$ m/s   **21.** 7.92 m/s^2 **22.** 12.1 m; 7.00 m

23. 2.52 m   **25.** 33.1 m/s or 74.0 mi/h   **27.** 23.7 W; 33.1 W   **34.** $\frac{5}{2}R$ **37.** $v = \sqrt{6\,g\ell}$; 4mg

CHAPTER 8

3. (a) $U_1 = mg(h + \frac{3}{4}\ell)$, $U_2 = mg\left(h + \dfrac{\ell}{4}\right)$

(b) $y_1 = h + \frac{3}{4}\ell$, $y_2 = h + \dfrac{\ell}{4}$; same point $\dfrac{\ell}{4}$ away

6. 9.89 cm   **9.** (a) -160 J;   (b) 73.5 J;   (c) 28.8 N; (d) 0.679   **10.** 2.17 m/s   **12.** 0.377   **15.** 735 J **16.** $x = 0$, $\frac{4}{3}$ m; $x = 0$ stable, $x = \frac{4}{3}$ m unstable; $\frac{16}{27}$ J **20.** $y = \frac{1}{3}R$ **21.** 2.60 m/s   **24.** $D = 0.398$ N·m^2 **27.** 10.1 m/s   **29.** 0.272 J

CHAPTER 9

5. 12.0 N   **8.** (a) 2.28 kg·m/s;   (b) 456 N **10.** 382 m/s   **12.** 5 m/s   **14.** (a) 0.0556 m; (b) 2.09 m/s   **16.** (a) 1.15 m/s;   (b) 0.346 m/s **18.** (a) 7.97×10^{-3};   (b) 6.70×10^{-23} **20.** 454 km **24.** (a) 40 m from the target;   (b) 5.33 s after first **26.** 100 m north of a point directly under the balloon and 1.69 km below the balloon.   **27.** (a) 1.61 m/s; (b) 2 m/s;   (c) $v_1 = 1.48$ m/s, $v_2 = 2.88$ m/s **28.** 523 kg/s; 1570 kg/s; 1.57×10^6 N and 4.70×10^6 N   **31.** (a) 4.42×10^5 kg;   (b) 1.92×10^4 kg **33.** 35.4 N   **36.** 331 m/s   **39.** 8.52 m **41.** (a) 1 m/s;   (b) 3 J;   (c) $\frac{2}{3}$ for A, $\frac{1}{3}$ for B;

(d) 24 N/m   **44.** $v_0 \geqslant \sqrt{\dfrac{2E}{m}\left(1 + \dfrac{M}{m}\right)}$

47. (a) 19.9 m;   (b) 3.67 s

CHAPTER 10

3. (a) $\mathbf{v} = 15\hat{\mathbf{j}} - 20\hat{\mathbf{i}}$;   (b) Equation of plane, $z = 5$ m **6.** (a) $\mathbf{r}(2) = (6\hat{\mathbf{i}} - 3\hat{\mathbf{j}})$ m;   (b) $\boldsymbol{\ell} = (-36\hat{\mathbf{k}})$ kg·m^2/s **9.** (a) 720 kg·m^2/s;   (b) 360 N·m   **11.** $a = g \sin \alpha$ **12.** (a) 7.63 rad/s;   (b) 100.8°;   (c) $\Delta E = 2.21$ J **15.** (b) 20 rad/s^2;   (c) 8 rad/s   **17.** ω decreases by 26 parts in 10^6 **20.** (a) 2 rad/s;   (b) 10 J;   (c) 210.7°

CHAPTER 11

3. $\frac{3}{2}MR^2$ **6.** (a) 3.0 N·m;   (b) 5.56 rad/s^2; (c) 16.67 rad/s   **8.** 7.84 m/s   **11.** (a) 35 rad/s^2; (b) 70 rad/s; (c) 0.70 N   **14.** $g(M_1 - M_2)/(M_1 + M_2 + m/2)$ **17.** (a) 1.12 m/s;   (b) 0.063 N;   (c) 24.3 cm **19.** $a = 7.84$ m/s^2; $\alpha_1 = \alpha_2 = 49.0$ rad/s^2; 15.7 N **20.** 0.667 m   **23.** (a) 3.26×10^7 J;   (b) 24.3 min **25.** (a) 0.50 rad/s;   (b) 5.00×10^4 N;   (c) 0.90 **27.** (b) 144 J;   (c) 2.45 m   **29.** $a_M = 0.889$ m/s^2; $a_m = 1.78$ m/s^2 **31.** Angle with vertical is 219.6°; $X = 10.0$ m, $Y = -4.75$ m measured from table edge.

CHAPTER 12

4. (a) 424 N;   (b) $F_h = 424$ N, $F_v = 490$ N **5.** $F_{v1} = F_{v2} = 122.5$ N; $F_{h1} = 76.6$ N, $F_{h2} = -76.6$ N **7.** 849 N   **9.** 0.70   **11.** (a) 73.5 N;   (b) $F_v = 0$, $F_h = -T$ **13.** 0.789   **16.** (a) $F_v = 490$ N, $F_h = 849$ N; (b) $T = 849$ N   **18.** 2/3

CHAPTER 13

3. $\dfrac{GMm}{R^2} \cdot \dfrac{\sin \alpha}{\alpha}$ **5.** 2.25×10^{-7} **6.** (a) 0.165g;

(b) 27.9g   **10.** 43.6 km/s   **12.** (a) 7.91 km/s; (b) 13.7 km/s   **16.** 4.22×10^7 m   **18.** 316 M_E **20.** 1.58×10^{10} J   **23.** 7.78 m/s^2 **25.** (a) $+1.21 \times 10^{-3}$;   (b) $+3.67 \times 10^{-4}$;   (c) -1.94×10^{-4} **28.** -6.36×10^{35} J

CHAPTER 14

2. 5.29×10^{18} kg   **4.** 5.09×10^4 N   **7.** 710 N **9.** 77.4 cm^2 **11.** (a) 1.02×10^4 kg;   (b) 19.9×10^6 Pa   **13.** 10.5 m   **15.** 1.08 cm   **17.** 1.12×10^3 Pa   **20.** 611 g   **22.** 3.21 N   **23.** (a) 6.67 m/s; (b) 0.280 m/s   **25.** 14.8 m/s   **27.** 2.23 m **30.** 3.91 cm   **33.** 7.90 W/m; 21%   **36.** 1.24 atm **40.** (a) 5.48 m;   (b) 1.75 atm   **42.** 1.75 cm Hg **43.** 0.424 cm   **44.** 18.7 cm^3; 19.8 g   **46.** (a) 4.36 km; (b) 1.86×10^3/m^3;   (c) 1.30×10^3 kg **49.** (a) $F = 2\pi d^2 R^4(p - p_0)/(4R^4 - d^4)$ (b) 12.7 atm **51.** 6.52×10^{-4} Pa·s

CHAPTER 15

4. (a) $v_m = 1.88$ m/s, $a_m = 35.5$ m/s^2;   (b) $v = 1.13$ m/s, $a = -28.4$ m/s^2;   (c) 34.1 ms **5.** (b) 41.8 cm;   (c) 3.74 m/s;   (d) 29.6 cm; (e) 8.94 rad/s, 0.702 s;   (f) 2.28 rad; (g) $x = 0.418 \sin (8.94t + 2.28)$ m   **7.** (a) 2.29 s;

(b) 0.109 m/s;   (c) $y = 0.040 \sin \left(2.74t + \dfrac{3\pi}{2}\right)$ m

8. (a) 75 cm;   (b) $y = 0.75 \sin (2t + \pi)$ m **11.** (b) 2.41 rad;   (c) 0.003 rad;   (d) $t_b = 0.252$ s, $t_c = 0.737$ s   **13.** (a) 0.202 rad; (b) $\theta = 0.202 \sin (2.65t)$ rad   **15.** 0.689 s **18.** (a) 13.4 N·s/m;   (b) 81.9 ms   **23.** 6.62 cm **25.** 35.7 cm   **26.** (a) 15.8 rad/s;   (b) 5.23 cm;

(c) 1.31 cm, $\phi = \dfrac{3\pi}{2}$ **27.** (a) 0.397 s;

(b) $y = 0.0363 \sin (15.8t + 2.72)$ m;   (c) 0.132 J **29.** 9.75×10^{-4} N **32.** $\kappa_a = \kappa_1 + \kappa_2$; $\kappa_b = \kappa_1\kappa_2/(\kappa_1 + \kappa_2)$; $\kappa_c = \kappa_1 + \kappa_2$

35. (a) $x = 0.15 \sin \left(10\pi t + \dfrac{7\pi}{4}\right)$;   (b) 740 N

37. $L/\sqrt{12}$

CHAPTER 16

2. 397 m; 28%   **4.** (a) 20 m/s;   (b) 8 cm
7. $v = \sqrt{xg}$ **10.** 2.79×10^3 m/s
15. (a) $y(0,t) = 2A \cos\left(\omega t - \dfrac{5\pi}{6}\right)$;

(b) 89.4 m/s, 89.4 cm;   (c) $A = 4.77$ cm;

(d) $y^+ (x, t) = 0.0477 \cos\left(7.03x - 200\pi t + \dfrac{5\pi}{6}\right)$ m,

$y^-(x, t) = 0.0477 \cos\left(7.03x + 200\pi t - \dfrac{5\pi}{6}\right)$

17. (a) 35.6 cm;   (b) 90 N;   (c) 84.3 Hz;   (d) 2.83 cm
18. (a) 8.28 cm, same;   (b) 1.72 cm, inverted;
(c) 2.94 W   **22.** (a) 4 times;   (b) 11.1 cm
25. (a) 17.5 Hz;   (b) 13.6 cm;   (d) $T_p = 2.01$ s
29. (b) 1.67 m;   (c) 6.12 g;   (d) 0.60 N;   (e) 1.2%
30. 631 N   **35.** Shortening string by 3.81 cm; 3.9%

CHAPTER 17

3. 2.33 m   **5.** (a) 212 Hz;   (b) 0.69 ℓ **8.** 339.7 m/s;
13.7° C   **11.** 13.3 kW   **14.** 3.1%   **16.** (a) The
hyperbola: $60x^2 - 4y^2 = 15\lambda^2$ **18.** 67.0 km/h
20. 309 km/h   **23.** 50%   **25.** 1.09×10^{-11} m;
(b) 1.09×10^{-5} m   **28.** (a) 71.2 Hz;
(b) $\kappa_a = 2.80 \times 10^3$ N/m;   (c) 110.2 Hz
29. $120 \times 72 \times 48$ cm   **31.** 33.4° C   **33.** 257.5 Hz
35. (a) 93.0 dB;   (b) 1.30 Pa;   (c) 28.3 m

CHAPTER 18

3. −259.22° C   **5.** 29.9 cm   **6.** 7.39 s
8. −25.2° C   **10.** 764.4 torr   **11.** 44.4 Cal
13. (a) 8.38 Cal;   (b) 69.3° C   **15.** 87.0° C
17. 100.4 m³/h   **20.** 5.39×10^{-6} **24.** 0.0507 kg/h
27. 67.0 m   **30.** (b) 3.72×10^{-4} deg⁻¹
31. 21.6 cm   **34.** 5.1° C   **38.** 1.24 kW/m
40. (a) 11.9 Cal/h;   (b) 66.2° C   **42.** (a) 8.90° C;
(b) 4.44×10^3 Cal;   (c) 6.48×10^3 Cal;
(d) 2.04×10^3 Cal   **43.** 3.29 cm   **44.** 39.5° C

CHAPTER 19

4. 131.28 g   **6.** 16.2 Å   **9.** 4.002×10^5 Pa, 43.4 psi;
4.48×10^5 Pa, 50.4 psi   **10.** (a) 2.24 m;   (b) 9.02 atm
12. (a) $p(He) = 0.315$ atm, $p(O_2) = 0.685$ atm;
(b) 94.6%   **14.** $r(N) = 1.812$ Å, $r(O) = 1.727$ Å
16. 2.30 atm   **19.** 53.8%   **21.** 156.8° C
23. 203.6 kg   **25.** (a) $m(Ar) = 1.556$ g,
$m(CO_2) = 4.163$ g;   (b) 2.86 kg/m³   **28.** (a) 100.75° C;
(b) −2.72° C   **32.** 0.362 ℓ **34.** 46.0%
36. 21.6 g/m³

CHAPTER 20

4. (a) $v_{rms}(H_2) = 1838$ m/s, $v_{rms}(He) = 1305$ m/s,
$v_{rms}(CO_2) = 393.5$ m/s;   (b) $v_{rms}(O_2) = 539.3$ m/s;
$v_{rms}(Ar) = 482.7$ m/s, $v_{rms}(NH_3) = 739.3$ m/s
9. $c_v = 0.175$ Cal/kg·deg
11. $T(diamond) = 465$ K, $T(Al) = 99.5$ K, $T(Cd) = 42$ K,
$T(Pb) = 22$ K

CHAPTER 21

4. 2019 J   **7.** (a) 4.99×10^4 J   **8.** (a) 99.97 R;
(b) 250.43 R;   (c) 150.46 R;   (d) 150.00 R
10. 4145 J = 1 Cal   **15.** (a) 40.7° C;   (b) 3 Cal;
(c) 2 Cal;   (d) 409.5;   (e) 2.15   **18.** (a) 1.83×10^7 J;
(b) 1.65×10^6 J; 27.4 min   **21.** (a) 1316 MW;
(b) 3290 MW, 1974 MW;   (c) 55.1%, actual 30.4%;
(d) 2300 barrels;   (e) 0.13 deg   **22.** 9.39 ℓ/h; 10.1 km/ℓ
(23.8 mi/gal)   **25.** 59.2 W   **27.** (a) 5.56 Cal;
(b) 19.44 Cal;   (c) 0.230 m³;   (d) 78.5 deg
29. (b) 2.74×10^4 J;   (c) 0;   (d) 2.74×10^4 J;
(e) 98.2 m³   **33.** Neon compartment 0.769 m long
34. (a) 0.499 mol;   (b) $T = 507°$ C, $p = 79.8$ atm;
(c) 135° C;   (d) 3830 J;   (e) 76.2%

CHAPTER 22

3. 30.6 J   **4.** 6.6%   **8.** −38.0 J/deg
11. (a) 1.10×10^4 J;   (b) 7.36×10^3 J;
(c) 3.68×10^3 J   **13.** +2.00 Cal/deg

CHAPTER 23

3. (a) and (b) 3.232 m beyond the −3 μC charge
7. 0.204 μC   **10.** $\omega^2 = q^2/\pi\epsilon_0 m\ell^3$ **11.** $r = \ell/\sqrt{2}$

CHAPTER 24

5. (a) 1079 N/C, 1.08 N/C;   (b) 2.70×10^4 N/C, 270 N/C;
(c) 25, 250   **7.** 7.08×10^{-12} C
15. $a = 5.28 \times 10^{17}$ m/s²; 1.71 mm   **18.** 91.4% in LiF and
4.9% in HI   **19.** 32.8%

CHAPTER 25

3. (a) 0.389 J;   (b) 2.03×10^5 V;   (c) 0   **4.** (a) Between
the charges 35.7 cm from the 5 μC charge   **8.** (a) 6 m;
(b) −2 μC   **13.** 167 μC   **18.** (a) 47.2×10^{-12} C;
(b) 1.51×10^{-16} C   **20.** 5.12×10^4 V
21. (a) 1.80×10^5 V;   (b) 1.27×10^5 V
24. (a) 33.3 cm;   (b) 37.1 μC

CHAPTER 26

2. 1240 V **4.** 32.4 pF **7.** (a) 131 pC; (b) 746 V
10. (a) 5 μC on larger sphere; (b) 8.99×10^4 V
11. (a) 39.8 J/m^3; (b) 3.98×10^{-2} N
13. (a) 4.5×10^{-2} J; (b) 0.135 J **17.** C = $\epsilon_0 A/(s - d)$
20. (a) 1.95×10^5 V/m; (b) 2.70 μC/m^2;
(c) $\sigma = 4.42$ μC/m^2, $\sigma_P = 2.70$ μC/m^2 **21.** (a) 45.0 C;
(b) 41.4 C **28.** (a) 709 μF; (b) 709 C **30.** 2 μF

CHAPTER 27

3. 13.3 μA/m^2 **6.** (a) 3.21×10^{-8} Ωm;
(b) 6.23×10^6 A/m^2; (c) 48.9 mA;
(d) $v_d = 2.15 \times 10^{-4}$ m/s; (e) 0.4 V **8.** 164 μA
11. (a) 23.5 A; (b) 5.51 kW; (c) 358 W; (d) 6.47 kW;
(e) 85.2%; (f) 81.4% **13.** 150 Ω **15.** (a) 75 V;
(b) $P_1 = 25$ W, $P_2 = P_3 = 6.25$ W; 37.5 W
17. $I_1 = \frac{11}{13}$ A, $I_2 = \frac{6}{13}$ A, $I_3 = \frac{17}{13}$ A **19.** 32/3 V
21. 8.15×10^{-7} A **23.** $R_3 = R_1^2/(R_1 + R_2)$
26. 2/3 Ω **29.** 20/3 Ω

CHAPTER 28

2. $\mathbf{B} = (-0.075\hat{\mathbf{i}} + 0.100\hat{\mathbf{j}})$ T **4.** 7.88×10^{-12} T
6. (a) 1260 C; (b) $\mathbf{B} \parallel \boldsymbol{\omega}$ **7.** $-2.88\hat{\mathbf{j}}$ N **13.** 10°
15. 2.01417 u **17.** (a) 9.78×10^5 m/s; (b) 17.6 cm;
(c) 5.48×10^{12} m/s^2; (e) 1.47×10^{12} m/s^2
19. 0.0204 N **20.** (a) 0.257 s; (b) 4.26 rad/s
21. (a) 2.068 kW; (b) 253 W; (c) 1.815 kW;
(d) 10.5 A; (e) 2.310 kW; (f) 78.6%

CHAPTER 29

2. $B = \dfrac{\mu_0 I}{2\pi w} \ln\left(\dfrac{x + w}{x}\right)$ **4.** 31,200 γ **5.** 267 G,
378 G **9.** 0.0250 T and 0.0179 T
10. $B = \dfrac{\mu_0 I c}{2\pi} \ln\left(\dfrac{b + a}{a}\right)$ **15.** 1.96 A
17. 7.54×10^{-11} T **20.** 9.33×10^{-4} N

CHAPTER 30

2. 0.306 cos ωt V **3.** 68.2 $e^{-1.6t}$ mV **7.** 0.90 V
8. 2.10×10^5 A/m^2 **13.** 12.6 mV **15.** 30.3 cos $8\pi t$ mV
17. 2.25 V

CHAPTER 31

3. $M = \dfrac{\mu_0 N c}{2\pi} \ln\left(\dfrac{a + b}{a}\right)$ **4.** 3.95×10^{-9} H
8. 0.673 mH **11.** 0.0513 J **15.** 1.24 A
17. $B = 0.012$ T, $H = 63.7$ A/m, $M = 9.49 \times 10^3$ A/m

20. $M = \dfrac{\pi\mu_0 a^2 b^2}{4r^3}(3\cos^2\theta - 1)$ **22.** (a) 199 J/m^3;
(b) 3.35×10^6 V/m

CHAPTER 32

4. (a) 2.25; (b) 10.6% **6.** (a) 1.31 W; (b) 4.16 W;
(c) 5.48 W **9.** (a) 2.11×10^{-8} F; (b) 1.06×10^{-6} J
12. (a) 3180 Hz; (b) 200 Ω; (c) 50 μs; (d) 14.7 mA
18. (a) $(30 - 10i)$ Ω; (b) 5.37 A; (c) $-18.4°$;
(d) 432 W **19.** (a) 238 μF; 3.24 kW; 3.56 hp
22. (a) 688 W; (b) $N_1/N_2 = 0.158$ **26.** (a) 9.00 μF;
(b) $-60°$ **29.** 7.5 Ω, 31.8 mH **30.** (a) 0.126 J;
(b) 0.154 J; (c) 92.2 W **31.** 26.3 mH, 22.7 μF,
17.7 Ω **34.** (a) 21.2 rad/s; (b) 6.88 cm;
(c) 22.4 rad/s; (d) 6.70 cm

CHAPTER 33

1. $\mathbf{H} = 0.531 \cos(kz - \omega t)\hat{\mathbf{j}}$ (A/m)
3. 1.20×10^7 m/s **5.** 0.299 W/m^2 **7.** (a) 1.453;
(b) 2.49×10^8 m/s; (c) 249 m; (d) 6.43×10^{-12} J/m^3;
(e) 1.60 mW/m^2 **10.** 1.76×10^{-25} W
12. (a) $E_0 = 27.5$ V/m, $H_0 = 72.8$ mA/m;
(b) 3.33×10^{-9} J/m^3; (c) 0.720 J **13.** 1.29°

CHAPTER 34

2. 3.13×10^8 m/s **3.** 3.75 rpm **6.** $\theta_2 = 35.2°$,
$\theta_3 = 30.3°$ **8.** 1.461 **11.** (a) -0.415; (b) 17.2%
14. (a) 22.18°; (b) 14.36° **16.** $\Delta\delta = 0.477°$
19. 48.8° **21.** 31.6° **24.** (a) 28.68 μm; (b) 4109 Å;
(c) $2\pi \times 14.73 = 2\pi \times 0.73 = 4.59$ rad **27.** $\dfrac{I_0}{8} \sin^2(2\theta)$

CHAPTER 35

3. 7.95 rad; 0.453 **7.** 226 nm **9.** 511 nm (green
light) **11.** 646 nm ($m = 49$) (orange light)
13. n = 1.000293, air **15.** (a) 5 ns; (b) 1.5 m
17. 1.54 mm **20.** 85.4 nm **22.** 680 nm (red light)
24. 8.54 μm ($m = 29$); $R = 3.75$ m

CHAPTER 36

2. (a) 650 nm; (b) 26.3° **4.** 0.0272″ of arc
7. $I_2 = 0.0158$ I_0, $I_3 = 0.0424$ I_0, $I_4 = 0.0139$ I_0,
$I_5 = 0.0081$ I_0 **9.** 9.88 mm **11.** 25.8″ of arc
13. 0.116 Å **15.** 52.5°, 25.6°, 10.2°
16. 25.8 cm **22.** $\theta = 9.92°$, $\phi = 63.1°$

CHAPTER 37

5. 120 cm, concave **7.** Real image midway between mirrors, inverted and 4 cm in height **9.** For an erect image 4.62 cm below lens, for an inverted image 33.9 cm below lens. Both images are virtual. **12.** Final image is virtual, inverted and 0.39 cm in height, located 19.7 cm inside sphere measured from clear surface vertex. **14.** (a) $s' = +42.9$ cm, $m = -0.429$, real; (b) $s' = -60$ cm, $m = +3$, virtual. **17.** $f = 6.67$ cm, $\ell = 30$ cm, $m_1 = -2$ and $m_2 = -1/2$ **19.** Image is virtual and erect, located 6.5 cm from center of lens on the same side as the object. **22.** (a) $\gamma = 3.5$, $\gamma_\infty = 2.5$; (b) From $s_b = 7.14$ cm to $s_c = 10$ cm **24.** (a) $\Gamma = -258$; (b) 0.629 cm; (c) 1.45 μm **27.** (a) $\Gamma = +5$; (b) 12 cm **30.** 2.25 m **32.** 10.7 cm beyond curved face vertex **33.** (a) 20.6 cm; (b) $r_1 = +11.5$ cm **36.** (a) Referring to diagram shown: H' at 1.31 cm left of flat face vertex and F' at 4.45 cm to the right. H at the curved face vertex and F at 5.76 cm in front of this vertex. (b) $x' = 7.82$ cm beyond F'. **37.** 4.08

INDEX

SI UNITS*

QUANTITY	NAME	SYMBOL (Alternate Expression)
SI Base Units		
Length	meter	m
Mass	kilogram	kg
Time	second	s
Electric current	ampere	A
Temperature	kelvin	K
Amount of a substance	mole	mol
SI Derived Units with Special Names		
Frequency	hertz	Hz (s^{-1})
Force	newton	N $(kg \cdot m/s^2)$
Pressure	pascal	Pa (N/m^2)
Energy, work	joule	J $(N \cdot m)$
Power	watt	W (J/s)
Electric charge	coulomb	C $(A \cdot s)$
Electric potential	volt	V (W/A)
Capacitance	farad	F (C/V)
Electric resistance	ohm	Ω (V/A)
Conductance	siemens	S (A/V)
Magnetic flux	weber	Wb $(V \cdot s)$
Magnetic field strength (flux density)	tesla	T (Wb/m^2)
Inductance	henry	H (Wb/A)
SI Derived Units Expressed by Means of Special Names		
Viscosity	pascal·second	$Pa \cdot s$
Surface tension	newton per meter	N/m
Specific heat capacity	joule per kilogram·meter	$J/kg \cdot m$
Entropy	joule per kelvin	J/K
Thermal conductivity	watt per meter·kelvin	$W/m \cdot K$
Electric field strength	volt per meter	V/m
Units to be Used with SI System Temporarily		
(Length)	ångstrom	$Å$ $(10^{-10}$ m$)$
(Pressure)	atmosphere	atm $(1.013 \times 10^5$ Pa$)$
(Radioactive disintegration rate)	curie	Ci $(3.7 \times 10^{10}$ s^{-1}, exactly$)$
Units Used with SI System Whose Values in SI Units Are Obtained from Experiment		
(Energy)	electronvolt	eV $(1.602 \times 10^{-19}$ J$)$
(Mass)	unified atomic mass unit	u $(1.661 \times 10^{-27}$ kg$)$
(Length)	astronomical unit	A.U. $(1.496 \times 10^{11}$ m$)$

---

*As recommended by the General Conference of Weights and Measures on the International System of Units (1974).